Physiology and Pathology of Chloride Transporters and Channels in the Nervous System

From Molecules to Diseases

Cover: Confocal image of hippocampal CA1 pyramidal cells in a transgenic mouse expressing the chloride indicator Clomeleon. The micrograph was obtained with a Leica SP2 confocal microscope at the imaging facility of the Max-Planck-Institute for Medical Research, Heidelberg, Germany. Image kindly supplied by Drs. Thomas Kuner, Ken Berglund and George Augustine, modified from the cover of *Brain Cell Biology*, Vol. 35, No. 4–6, December 2006, reproduced with permission of Springer Science and Business Media.

Physiology and Pathology of Chloride Transporters and Channels in the Nervous System

From Molecules to Diseases

F. Javier Alvarez-Leefmans
and
Eric Delpire

ELSEVIER

AMSTERDAM • BOSTON • HEIDELBERG • LONDON • NEW YORK • OXFORD
PARIS • SAN DIEGO • SAN FRANCISCO • SINGAPORE • SYDNEY • TOKYO
Academic Press is an imprint of Elsevier

Academic Press is an imprint of Elsevier
32 Jamestown Road, London NW1 7BY, UK
30 Corporate Drive, Suite 400, Burlington, MA 01803, USA
525 B Street, Suite 1900, San Diego, CA 92101-4495, USA

First edition 2009

British Library Cataloguing-in-Publication Data
A catalogue record for this book is available from the British Library

Library of Congress Cataloging-in-Publication Data
A catalog record for this book is available from the Library of Congress

ISBN: 978-0-12-374373-2

For information on all Academic Press publications
visit our website at www.elsevierdirect.com

Typeset by Macmillan Publishing Solutions
www.macmillansolutions.com

Printed and bound in China

09 10 11 12 13 10 9 8 7 6 5 4 3 2 1

Contents

Preface

This book is about how chloride ions are regulated and how they cross the plasma membrane of neurons, glial and epithelial cells. It spans from the molecular structure and function of carriers and channels involved in chloride transport to their role in various neurological diseases. The importance of chloride ions in cell physiology has not been fully recognized until recently. This is in spite of the fact that chloride, in addition to bicarbonate, is the most abundant free anion in animal cells, and performs or determines fundamental biological functions in all tissues. For many years it was thought that chloride was distributed in thermodynamic equilibrium across the plasma membrane of most cells. It took several decades to eradicate this erroneous view that had become a textbook dogma. This is probably one of the reasons why it has taken such a long time to begin to recognize the importance of chloride ions in neuronal function and dysfunction, compared with the weight given to the cations sodium, calcium and potassium. Research carried out during the last couple of decades has led to a dramatic change in this simplistic view. We now know that most animal cells, neurons included, exhibit a non-equilibrium distribution of chloride across their plasma membranes. Thus, far from being passively distributed across the plasma membrane in accordance with the membrane potential, chloride is actively transported and tightly regulated in virtually all animal cells. The level of intracellular chloride results from a delicate functional balance between chloride channels and carriers present in the plasma membrane. Alterations in this balance underlie various nervous system dysfunctions.

Over the last 10 to 15 years, with the spectacular growth of molecular biology and the advent of new optical and electrophysiological methods, an enormous amount of exciting new information has become available on the molecular structure and function of chloride channels and carriers and their involvement in various diseases. In nerve cells, chloride channels and carriers play key functional roles in GABA- and glycine-mediated synaptic signaling in the de- and hyperpolarizing directions, including pre- and postsynaptic inhibition. They also play a central role in neuronal growth and development, extracellular potassium scavenging, regulation of intracellular pH, sensory-transduction including nociception, neurotransmitter uptake, and cell volume control. Disruption of chloride homeostasis in neurons, glial and epithelial cells underlies diverse pathological conditions such as epilepsy, deafness, imbalance, brain edema, and neurogenic inflammation. Further, primary brain tumor cells (e.g. glioma cells) migrate within the brain utilizing chloride channels and transporters expressed in their plasma membrane. Accordingly, drugs that target chloride channels (e.g. chlorotoxin) are being studied as possible therapeutic agents in primary brain tumor growth and invasion. Further, drugs that block cation-coupled-chloride cotransporters, such as 5-sulfamoyl benzoic acid derivatives (e.g. bumetanide), that have been used for many years as diuretics, are now being investigated as potential neuroprotective compounds in brain ischemic damage, as therapeutic agents in neonatal epileptic seizures, and as possible analgesic and anti-inflammatory agents in acute tissue injury.

This book brings together most of the primary research information about the molecular physiology and pathology of chloride carriers and channels scattered throughout the scientific literature. The forerunner of the present book, "Chloride Channels and Carriers in Nerve and Muscle, and Glial Cells" (Alvarez-Leefmans, F.J. and Russell, J.M. Eds., Plenum, New York, 1990), appeared prior to the molecular biology era. Since its publication a virtual explosion in our understanding of these transport protein molecules has occurred. However, until now no book or monograph has effectively integrated the emerging concepts in a way that is useful to both general and specialized neuroscience communities. To address this shortcoming the present book has been assembled with contributions by world leaders in the field. Our goal has been to create a book that makes available to both specialists and non-specialists detailed information and fundamental concepts being developed in this exponentially expanding field. The contributors present their topics of study not just by reviewing their own findings but also by putting them in perspective so that others wishing to enter this exciting field of neuroscience may quickly see the obvious areas where more work is needed.

The book is divided into five parts and thirty chapters. Many chapters discuss their topics from a historical perspective. This interesting exercise shows that many current concepts were proposed or discovered long ago. These basic concepts were the result of considerable solid work and insights into cellular physiology without which current molecular data would be meaningless. Extensive introductory sections cover basic thermodynamic and kinetics aspects of chloride transport, as well as current methods for studying chloride regulation, from fluorescent dyes in single cells to knock-out models.

All chapters have been carefully edited and cross-referenced to avoid unnecessary repetition and to provide the reader with easy referencing to sections of the book where related information can be found. We have strived to attain continuity between the chapters and have emphasized the use of a unified nomenclature. Nevertheless, as editors, not censors, we have respected the views expressed by the authors although they do not always reflect our own.

Needless to say, this volume is the result of a collective effort. The editors are particularly grateful to the contributing authors, and to Dr. Johannes Menzel, Ms. Clare Caruana, Ms. Deena Burgess and Ms. Kim Lander at Elsevier for their support throughout the editing and production process. We would also like to express our thanks to our respective institutions, Wright State University Boonshoft School of Medicine, and Vanderbilt University School of Medicine, for their continual support of the work dedicated to putting together the present volume.

Francisco Javier Alvarez-Leefmans
Eric Delpire

List of Contributors

Norma C. Adragna Department of Pharmacology and Toxicology and Cell Biophysics Group, Boonshoft School of Medicine, Wright State University, Dayton, OH, USA

Norio Akaike Research Division for Life Sciences, Kumamoto Health Science University, Kumamoto, Japan

Francisco Javier Alvarez-Leefmans Department of Pharmacology & Toxicology, Boonshoft School of Medicine, Wright State University, Dayton, OH, USA

George J. Augustine Department of Neurobiology, Duke University Medical Center, Durham, NC, USA; Laboratory of Synaptic Circuitry, Duke-NUS Graduate Medical School, Singapore; A*STAR/Duke-NUS Neuroscience Research Partnership, Singapore, and Department of Physiology, National University of Singapore, Singapore.

Ken Berglund Department of Neurobiology, Duke University Medical Center, Durham, NC, USA

Emmanuel J. Botzolakis The Medical Scientist Training Program, Vanderbilt University School of Medicine, Nashville, TN, USA

Fiona C. Britton Department of Physiology & Cell Biology, University of Nevada School of Medicine, Reno, NV, USA

Peter D. Brown Faculty of Life Sciences, University of Manchester, Manchester, UK

Min-Hwang Chang Physiology & Biomedical Engineering, Mayo Clinic College of Medicine, Rochester, MN, USA

Sarah L. Davies Faculty of Life Sciences, University of Manchester, Manchester, UK

Yves De Koninck Department of Psychiatry, Division of Cellular Neurobiology, Centre de recherche université Laval Robert-Giffard, Québec, QC, Canada

Eric Delpire Departments of Anesthesiology and Molecular Physiology & Biophysics, Vanderbilt University School of Medicine, Nashville, TN, USA

Mauricio Di Fulvio Department of Pharmacology and Toxicology, Boonshoft School of Medicine, Wright State University, Dayton, OH, USA

Amal K. Dutta Department of Cell Physiology, National Institute for Physiological Sciences, Okazaki, Japan

J. Clive Ellory Department of Physiology, Anatomy & Genetics, University of Oxford, Oxford, UK

Stephan Frings Department of Molecular Physiology, University of Heidelberg, Heidelberg, Germany

Kenneth Gagnon Departments of Anesthesiology and Radiation Oncology, Vanderbilt University School of Medicine, Nashville, TN, USA

Gerardo Gamba Molecular Physiology Unit, Instituto Nacional de Ciencias Médicas y Nutrición Salvador Zubirán and Instituto de Investigaciones Biomédicas, Universidad Nacional Autónoma de México, Mexico City, Mexico

Nicole Garbarini Department of Anesthesiology, Vanderbilt University School of Medicine, Nashville, TN, USA

John S. Gibson Department of Veterinary Medicine, University of Cambridge, Cambridge, UK

Steffen Hamann Nordic Centre for Water Imbalance Related Disorders, Institute of Cellular and Molecular Medicine, The Panum Institute, University of Copenhagen, Denmark

H. Criss Hartzell Department of Cell Biology and Center for Neurodegenerative Disease, Emory University School of Medicine, Atlanta, GA,USA

Hana Inoue Department of Cell Physiology, National Institute for Physiological Sciences, Okazaki, Japan

Thomas J. Jentsch Leibniz-Institut für Molekulare Pharmakologie (FMP) and Max-Delbrück-Centrum für Molekulare Medizin (MDC), Berlin, Germany

Kristopher T. Kahle Department of Neurosurgery, Massachusetts General Hospital and Harvard Medical School, Boston, MA, USA

James L. Kenyon Departments of Physiology & Cell Biology and Pharmacology, University of Nevada School of Medicine, Reno, NV, USA

Douglas B. Kintner Department of Neurological Surgery, University of Wisconsin School of Medicine and Public Health, Madison, WI, USA

Thomas Kuner Institute of Anatomy and Cell Biology, University of Heidelberg, Heidelberg, Germany

Peter K. Lauf Cell Biophysics Group, Boonshoft School of Medicine, Wright State University, Dayton, OH, USA

Normand Leblanc Department of Pharmacology, University of Nevada School of Medicine, Reno, NV, USA

Nanna MacAulay Nordic Centre for Water Imbalance Related Disorders, Institute of Cellular and Molecular Medicine, The Panum Institute, University of Copenhagen, Denmark

Robert L. Macdonald Departments of Neurology, Pharmacology, and Molecular Physiology & Biophysics, Vanderbilt University School of Medicine, Nashville, TN, USA

Daniel C. Marcus Cellular Biophysics Laboratory, Department of Anatomy & Physiology, Kansas State University, Manhattan, KS, USA

Ian D. Millar Faculty of Life Sciences, University of Manchester, Manchester, UK

David Mount Renal Division, Brigham and Women's Hospital, Harvard Institutes of Medicine, Boston MA, USA

Gaia Novarino Leibniz-Institut für Molekulare Pharmakologie (FMP) and Max-Delbrück-Centrum für Molekulare Medizin (MDC), Berlin, Germany

Martha O'Donnell Department of Physiology and Membrane Biology, School of Medicine, University of California, Davis, CA, USA

Yasunobu Okada Department of Cell Physiology, National Institute for Physiological Sciences, Okazaki, Japan

John A. Payne Department of Physiology and Membrane Biology, School of Medicine, University of California, Davis, CA, USA

Brooks B. Pond Department of Pharmaceutical Sciences, East Tennessee State University College of Pharmacy, Johnson City, TN, USA

Michael F. Romero Physiology & Biomedical Engineering, and Nephrology & Hypertension, Mayo Clinic College of Medicine, Rochester, MN, USA

John M. Russell Department of Biology, Life Sciences Complex, Syracuse University, Syracuse, NY, USA

Ravshan Z. Sabirov Laboratory of Molecular Physiology, Institute of Physiology and Biophysics, Tashkent, Uzbekistan; and Department of Cell Physiology, National Institute for Physiological Sciences, Okazaki, Japan.

Kaori Sato Department of Cell Physiology, National Institute for Physiological Sciences, Okazaki, Japan

Harald Sontheimer Department of Neurobiology and Center for Glial Biology in Medicine, University of Alabama, Birmingham, AL, USA

Kevin J. Staley Division of Pediatric Neurology, Massachusetts General Hospital, and Harvard Medical School Boston, MA, USA

Tobias Stauber Leibniz-Institut für Molekulare Pharmakologie (FMP) and Max-Delbrück-Centrum für Molekulare Medizin (MDC), Berlin, Germany

Dandan Sun Department of Neurological Surgery, University of Wisconsin School of Medicine and Public Health, Madison, WI, USA

Makoto Suzuki Edogawabashi Suzuki Clinic, Tokyo, Japan

Abduqodir H. Toychiev Department of Cell Physiology, National Institute for Physiological Sciences, Okazaki, Japan

Noga Vardi Department of Neuroscience, School of Medicine, University of Pennsylvania, Philadelphia, PA, USA

Alan S. Verkman Departments of Medicine and Physiology, Cardiovascular Research Institute, University of California, San Francisco, CA, USA

Philine Wangemann Cellular Physiology Laboratory, Department of Anatomy & Physiology, Kansas State University, Manhattan, KS, USA

Thomas Zeuthen Nordic Centre for Water Imbalance Related Disorders, Institute of Cellular and Molecular Medicine, The Panum Institute, University of Copenhagen, Denmark

Ling-Li Zhang Department of Neuroscience, School of Medicine, University of Pennsylvania, Philadelphia, PA, USA

PART I

OVERVIEW OF CHLORIDE TRANSPORTERS AND CHANNELS

CHAPTER

1

Chloride Channels: An Historical Perspective

H. Criss Hartzell

OUTLINE

I. INTRODUCTION

One reason I became a scientist was that other disciplines, such as history, frightened me because their realities seemed variable, dictated by perception rather than by hard data. Although I have learned that science, too, is encumbered by perceptual distortions, it was nevertheless unsettling for me to accept an invitation to write a chapter with "history" in the title. I simply do not trust my historical perceptions. So, at the outset, I apologize deeply to those I have misrepresented or forgotten. This chapter is not meant to be a comprehensive history of Cl^- channels, but rather is meant to present my personal view of how Cl^- channels have finally begun to attract the attention they deserve. For more comprehensive, scholarly, and balanced approaches, the reader should consult a number of reviews on Cl^- channels (Nilius et al., 1997; Hume et al., 2000; Nilius and Droogmans, 2001; Welsch et al., 2001; Eggermont et al., 2001; Jentsch et al., 2002; Nilius and Droogmans, 2003; Faundez and Hartzell, 2004; Hartzell et al., 2005; Sile et al., 2006; Okada et al., 2006; Puljak and Kilic, 2006; Gadsby et al., 2006; Zifarelli and Pusch, 2007; Hartzell et al., 2008; Jentsch, 2008) as well as Chapters 12–15 in this book.

II. CHLORIDE "PASSIVITY"

Most ion channel specialists will agree, I think, that our understanding of how anion channels work still lags significantly behind our understanding of cation channels. In Bertil Hille's first edition of the "bible" on ion channels published in 1984 (Hille, 1984), fewer than three pages were devoted to Cl^- channels. Why was this? Certainly, there is a natural tendency to embrace positivity over negativity, but the reason is certainly deeper. As a graduate student in the late 1960s, I *remember* learning that Cl^- channels did not exist. Actually, it is highly unlikely that such a statement was ever made by my professors, but it is how my naïve mind heard it. Probably I was told that Cl^- moved through membranes by a purely electro-diffusive mechanism and that Cl^- was distributed passively across the plasma membrane. To me, the statement that Cl^- was both "passive" and "negative" had the same effect as the kiss of death. If Cl^- was passively distributed, it seemed useless: it could not provide energy for transport of other substances and its passive distribution precluded a signaling function. But clearly, I was way off-base in thinking that Cl^- channels did not exist just because Cl^- seemed to be distributed passively. It was well known that muscle, neurons, and other cell types exhibited Cl^- conductances. And Hodgkin and Horowicz (Hodgkin and Horowicz, 1959) referred to the Cl^- permeation pathway they studied in skeletal muscle as "the Cl^- channel", which we now know is ClC-1 (see Chapter 12, this volume).

This view of Cl^- "passivity" came about because, at the time, ideas about Cl^- were dominated by work on skeletal muscle and erythrocytes. In resting skeletal muscle, Cl^- permeability is extremely high: P_{Cl} is more than twice that of P_K. Because of this high permeability, Cl^- distributes passively according to the membrane potential (Hodgkin and Horowicz, 1959; Hutter and Noble, 1960; Adrian, 1960, 1961) and any active transport of Cl^- would need to work very hard to counteract this high permeability. In erythrocytes the intracellular Cl^- concentration is also close to electrochemical equilibrium due to a large Cl^- conductance and to Cl^-/HCO_3^- exchange. The prevalent perception, at least in textbooks (e.g. Ruch and Patton, 1965; Hille, 1984), that Cl^- was distributed passively in most cells was also bolstered by earlier experiments supporting this idea in squid axon (Steinbach, 1941). It is important to note, however, that the apparent passive distribution of Cl^- did not rule out active transport of Cl^-, it only meant that the passive leak was much greater than any active transport that might occur. It now is clear that although Cl^- may appear to be in electrochemical equilibrium, Cl^- is actively transported in muscle as well as other cells. For example, blocking muscle Cl^- channels with 9AC reveals the presence of active Cl^- transport (Aickin, 1990; Alvarez-Leefmans, 2001; Gosmanov et al., 2003).

Nevertheless, at the time, with skeletal muscle, erythrocytes, and squid axon all weighing in on the side of Cl^- being in electrochemical equilibrium, other data to the contrary seemed not to receive much attention. Results from a number of tissues did not fit into the equilibrium model. Contradicting the data of Steinbach, Richard D. Keynes (Keynes, 1963) showed clearly that intracellular Cl^- concentration in squid axon was 2–3 times higher than equilibrium and that active Cl^- uptake must occur to explain this. But, to the public, perhaps a two-fold gradient seemed unimportant compared to the 10- to 1000-fold gradients that cations normally exhibited. Strong evidence also existed that Cl^- was actively transported in gastric and intestinal epithelia (Hogben, 1959), but the idea that the stomach secreted 1-N HCl may have seemed bizarre enough in itself to bar Cl^- from entering the mainstream by this route.

The discovery that GABA hyperpolarized neurons by opening ligand-gated Cl^- channels (Kuffler and Edwards, 1958; Boistel and Fatt, 1958) showed that E_{Cl} was not equal to resting E_m in neurons. However, the measured reversal potentials of GABA-produced i.p.s.p.'s were so close to the resting potential that is was concluded that "normally Cl^- ions are in electrochemical equilibrium across the membrane" (Coombs et al., 1955). Later, it was recognized that GABA produced depolarizing responses in some neurons, like dorsal root ganglion (De Groat et al., 1972), and it became clear that intracellular Cl^- concentration could change, and therefore Cl^- must be actively transported, at least under certain conditions, as discussed throughout this book and specifically in Chapters 2, 5, 17 and 22.

III. ACTIVE CHLORIDE TRANSPORT

Along with the realization that the electrochemical equilibrium of Cl^- ($E_{Cl} = E_m$) in skeletal muscle and erythrocytes was due to their high Cl^- permeability came the discovery that in cells with low resting Cl^- permeability, active Cl^- transport mechanisms were present and often generated electrochemical Cl^- gradients ($E_{Cl} > E_m$ or $E_{Cl} < E_m$), as illustrated in Fig. 1.1. Some cells, notably epithelial cells, immature neurons, and mature sensory neurons, express transporters such as the Na^+-K^+-$2Cl^-$ cotransporter (NKCC), the Cl^-/HCO_3^- exchangers, or the Na^+-Cl^- cotransporter (NCC) that typically accumulate Cl^- at concentrations above electrochemical equilibrium inside the cell. Indeed, the accumulation of intracellular Cl^- above electrochemical equilibrium is the basis for Cl^--driven

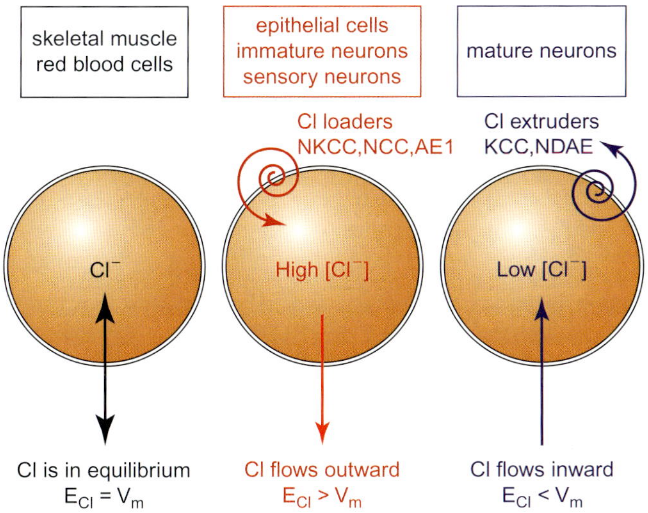

FIGURE 1.1 Control of cytosolic Cl^- in different cell types. In skeletal muscle and erythrocytes, Cl^- appears to be in electrochemical equilibrium because there is a large resting Cl^- conductance that masks active Cl^- transport. In epithelial cells, immature neurons, and adult sensory neurons, Cl^- loaders establish a higher than electrochemical equilibrium intracellular Cl^- concentration. In most mature central neurons, Cl^- extruders establish a lower than electrochemical equilibrium intracellular Cl^- concentration.

fluid secretion by epithelial cells. In secretory epithelial cells, basolateral NKCC accumulates Cl^- in the cell above electrochemical equilibrium. Then, opening of Cl^- channels in the apical membrane results in efflux of Cl^- down its electrochemical gradient into the lumen, followed by paracellular fluxes of Na^+ and water. Other cell types, such as many mature neurons, express transporters that tend to extrude Cl^-, like the K^+-Cl^- cotransporters (KCCs) discussed in Chapter 17 and the Na^+-dependent anion exchangers (NDAE) discussed in Chapter 4. The resulting lower than electrochemical equilibrium intracellular Cl^- concentration is the basis for the typical hyperpolarizing action of GABA in mature neurons, for example. Indeed, in immature neurons that express Cl^- loaders, and mature primary sensory neurons, GABA produces depolarizing postsynaptic potentials, rather than hyperpolarizing ones. The depolarizing responses may be important in stabilizing synapses during development (Ben-Ari et al., 2007) in CNS neurons, and in presynaptic inhibition in the central terminals of mature primary sensory neurons, as discussed in Chapter 22 in this volume.

The recognition that Cl^- concentration differs in different cell types depending on the complement of expressed Cl^- transporters has placed Cl^- in what may be a special position among biological ions. Although the concentrations of Na^+, K^+, Ca^{2+} and Mg^{2+} can change with cellular activity, the concentrations of these cations under "resting" conditions are very similar in different cell types. In contrast, the intracellular concentrations of Cl^- in different cell types can vary over nearly an order of magnitude (see Fig. 13.3 in Chapter 13). Also, although the concentration of cations, most notably Ca^{2+}, can change transiently, these changes are generally fast and local. It seems, however, that changes in $[Cl^-]_i$ occur over a slower time scale. Incidentally, this may be another reason why Cl^- has been neglected: if one is looking for changes on the time scale of seconds, things that take hours to happen will be missed. It has already been mentioned that Cl^- concentration in neurons varies during development (see also Chapters 7 and 19 in this volume), but it also may change dramatically in response to synaptic activity (Kuner and Augustine, 2000; Isomura et al., 2003; Berglund et al., 2006 and 2008), as discussed in Chapter 7 in this volume.

IV. TECHNICAL HURDLES TO STUDYING Cl⁻ CHANNELS

There are also technical reasons why the study and understanding of Cl^- channels and transporters have lagged behind that of cation channels. One reason is that the pharmacology of Cl^- channels was, and still is, distressingly inadequate. Unlike many cation channels that have very specific drugs that can be used to block them (like TTX for voltage-gated Na^+ channels, charybdotoxin for large-conductance K^+ channels and conotoxins for voltage-gated Ca^{2+} channels), Cl^- channel blockers are notorious for their low affinity and low specificity. Until recently, the Cl^- channel pharmacopeia has consisted of a selection of relatively low-affinity dirty molecules, including the stilbene disulfonate derivatives such as the amino reactive agents DIDS and SITS, the diphenylamine-2-carboxylate (DPC) derivatives such as 5-nitro-2-(3-phenylpropylamino) benzoic acid (NPPB) and anthracene-9-carboxylate (9-AC); fenamic acids like niflumic acid (NFA) and flufenamic acid FFA; and indanyloxyacetic acid 94 (IAA-94). But these drugs have neither high affinity nor high selectivity. Very recently, there has been progress in identifying extremely high affinity (nanomolar) drugs for some Cl^- channels, notably CFTR and calcium-activated chloride channels (CaCCs) (Yang et al., 2003; Muanprasat et al., 2004; Verkman et al., 2006; Muanprasat et al., 2007; De La Fuente et al., 2008). Also, peptide toxins for CFTR and ClC-2 have recently been found (Thompson et al., 2005; Fuller et al., 2007). Experience with these toxins is presently limited, but it is likely that these toxins will be extremely useful in studying these channels and opens the door to searching for other native toxins directed at Cl^- channels (see Chapter 26 in this volume).

The lack of good pharmacological agents, coupled with limitations of techniques available to measure Cl^- concentration and flux, conspired to suppress the acknowledgement of Cl^- as an important biological ion. Cl^- fluxes were laborious and difficult to measure because the principal radioisotope of chlorine (^{36}Cl) has a low specific activity and is expensive. Cl^--sensitive electrodes were not developed until the 1970s (Alvarez-Leefmans et al., 1990). Although they are a very accurate method to measure $[Cl^-]_i$, they are invasive and their use is restricted to large and robust cells (Alvarez-Leefmans et al., 1990; Coles, 1995).

Progress in studying Cl^- fluxes has been aided significantly by development of small fluorescent molecules and yellow fluorescent protein mutants that change their fluorescence in response to different anion concentrations (Biwersi and Verkman, 1991; Biwersi et al., 1992, 1994; Mansoura et al., 1999; Kuner and Augustine, 2000; Galietta et al., 2001) as discussed in Chapters 6 and 7 in this volume. Some of the small molecule Cl^- sensors are membrane permeant and can be loaded into cells for non-invasive measurement of Cl^- concentrations and fluxes in living cells (e.g. Chapter 22). The Cl^--sensitive yellow fluorescent protein sensors are genetically encoded and can be transfected into cells. Transgenic animals have been created that express the Cl^--sensitive indicator Clomeleon (Berglund et al., 2008), as discussed in detail in Chapter 7.

The availability of these fluorescent probes for Cl^- has now made it possible to develop high-throughput screens for Cl^- channel drugs. These screens have already been productive in finding higher affinity Cl^- channel agonists and antagonists, especially for CFTR and CaCCs (Yang et al., 2003; Muanprasat et al., 2004; Verkman et al., 2006; Muanprasat et al., 2007; De La Fuente et al., 2008).

V. THE CHLORIDE AWAKENING

Even into the 1990s, Cl^- channels continued to receive less press than cation channels. In the second edition of his book, Bertil Hille (Hille, 1992) relegates Cl^- channels to the back of the bus by saying that they have "uncertain physiological significance in many cell types". However, by the 3rd edition (Hille, 2001), the tide was beginning to turn: the book has a section entitled "Cl^- Channels Have Multiple Functions". What happened between 1992 and 2001? Cl^- channel respectability received a boost from three major arenas: the biophysical characterization of single Cl^- channels, relationship of Cl^- channels to human disease, and the cloning of two important sets of Cl^- channels, CFTR and the ClC family (see Chapters 12 and 13, this volume).

A. Single Channel Properties of ClC-0

The rigorous biophysical characterization by Chris Miller in the mid-1980s of the gating of *Torpedo* electroplax Cl^- channels (now known as ClC-0) incorporated into lipid bilayers, raised interest because these channels had the peculiar property, unlike cation channels, of being double-barreled (Miller and White, 1984). This finding raised considerable interest because it suggested that Cl^- channels were genuinely different, and weirder, than cation channels. Also, the demonstration that Cl^- transport actually could be visualized at the single channel level forced people like me, having misinterpreted my professors, to realize that Cl^- channels actually existed!

B. Myotonia Congenita

But, more than the single channel measurements, the Cl^- channel field received a strong boost from the understanding that Cl^- channel dysfunction could be associated with disease. Studies in the 1960s and 1970s by Shirley Bryant established that inherited myotonia was related to a defect in the Cl^- conductance of skeletal muscle (Lipicky and Bryant, 1966; Bryant, 1969; Lipicky et al., 1971). His first studies had shown that the muscle fibers of the famous "fainting" myotonic goats of Tennessee (amusing movies can be found on www.youtube.com) had an increased membrane resistance and that normal muscle fibers bathed in low Cl^- medium behaved like myotonic fibers; that is, they were hyperexcitable (had a reduced firing threshold) and tended to fire action potentials spontaneously. Later studies showed directly that Cl^- conductance was reduced in muscle fibers from both myotonic goats and humans. Thomas Jentsch's expression cloning in 1990 of the double-barreled *Torpedo* ClC-0 channel (Jentsch et al., 1990, 2002) then paved the way for showing that myotonia congenita is caused by mutations in the ClC-1 channel which is the major Cl^- conductance in skeletal muscle (Steinmeyer et al., 1991; Koch et al., 1992) as discussed in Chapter 12 in this book. Some of the myotonia-causing ClC-1 mutations alter voltage sensitivity (Pusch et al., 1995), whereas others alter relative cation/anion permeability (Fahlke et al., 1997).

The mechanism of myotonia illustrates well the subtlety of Cl^- channel function. In muscle, the Na^+-K^+ ATPase sets up the K^+ gradients, and inwardly rectifying K^+ channels establish the resting membrane potential ($E_m \sim E_K$). Because active Cl^- transport is relatively small and the resting Cl^- conductance is high, Cl^- then distributes passively across the plasma membrane and $E_{Cl} = E_m \sim E_K$. But this passive distribution of Cl^- does not mean that it is useless, as I naïvely thought long ago. Because the Cl^- conductance is several times larger than the K^+ conductance, the Cl^- battery is actually a very strong buffer of membrane potential when E_K changes transiently, as it does during muscle electrical activity when K^+ accumulates extracellularly in the lumen of the transverse tubules. In myotonic muscle, this accumulated K^+ depolarizes the muscle fiber in accordance with the Nernst equation, whereas in normal muscle the Cl^- battery keeps the membrane near E_{Cl} (Adrian and Bryant, 1974). It turns out that being passive and negative is not such a bad thing, after all!

C. Cystic Fibrosis

At roughly the same time, developments in the field of cystic fibrosis (CF) revealed that this disease was also a Cl^- channelopathy (Quinton, 1999). The first indication that cystic fibrosis was related to an electrolyte disturbance was the discovery in 1953 that the sweat of CF patients was unusually salty (Di Sant'Agnese et al., 1953). It took about five years for this finding to become accepted as a fundamental part of the disease, but it took even longer (30 years) before it was proposed that the disease was caused by a defect in Cl^- transport (Quinton, 1983) when it was found that the sweat duct and airways exhibited very low Cl^- permeabilities (Quinton, 1983; Knowles et al., 1983a, b). In 1989, the gene (*CFTR*) responsible for CF was positionally cloned (Riordan et al., 1989; Kerem et al., 1989; Rommens et al., 1989), but it was two years later before a definitive demonstration that the *CFTR* gene product functioned as a Cl^- channel when expressed heterologously (Anderson et al., 1991). These authors showed that mutation of K95 and K335 altered the relative anionic selectivity of the channel. Even though this was a more rigorous demonstration that CFTR was a channel than was previously required for acceptance of any cloned cation channel (e.g. the nicotinic acetylcholine receptor (Noda et al., 1982), the voltage-gated Na^+ channel (Noda et al., 1986), the Shaker K^+ channel (Papazian et al., 1987), some investigators were still not convinced (e.g. Hipper et al., 1995). Skepticism remained partly because CFTR was a member of the ABC transporter family and did not "look" like a "respectable" ion channel. It was not until it was incorporated into artificial lipid bilayers that it was finally generally conceded that it was a Cl^- channel (Bear et al., 1992). But, even though it is clear that CFTR is a Cl^- channel, its evolutionary roots suggest that it evolved from a transporter (Gadsby et al., 2006; Jordan et al., 2008) (see below).

We now know that CFTR is the intestinal channel that is responsible for secretory diarrheas, like that of cholera. Cholera toxin and other enterotoxins activate the cAMP cascade that results in phosphorylation of CFTR and activation of Cl^- (and HCO_3^-) secretion. Given the long history of cholera (Field, 2003), one might expect that the importance of transepithelial Cl^- fluxes would have been recognized early. Although electrolyte replacement therapy was used as early as 1832 to treat cholera (Field, 2003), the rationale for its use was mechanistically primitive (Awad et al., 2008). It was not until the 1970s that it was shown that cholera toxin increased Cl^- flux across the intestinal mucosa (Field et al., 1972), but it remained unclear for some time whether the increased Cl^- flux was mediated by increased secretion or decreased absorption and whether the fluxes were active or passive (Naftalin and Simmons, 1979).

VI. Cl^- CHANNEL GENES

Although hundreds of mammalian Na^+, Ca^+ and K^+ channel subtypes have been cloned, Cl^- channel genes are limited to nine ClCs, one CFTR, four bestrophins and 25 ligand-gated anion channels. The recently described TMEM16A/anoctamin family may add as many as 10 genes to this list (Yang et al., 2008; Caputo et al., 2008; Hartzell, 2008; Schroeder et al., 2008). In addition, it appears that some members of the SLC26 family of anion transporters may be Cl^- channels. Specifically, SLC26A7 may be a pH-sensitive Cl^- channel (Kim et al., 2005) and SLC26A9 may be an HCO_3^- sensitive Cl^- channel (Dorwart et al., 2007; Loriol et al., 2008). But, still, the number of Cl^- channel genes falls short of the number of cation channel genes that have been found. Is this because there are fewer kinds of Cl^- channels? It seems more likely that there are still Cl^- channel genes waiting to be identified. For example, although bestrophins and CLCAs have been proposed as Ca^{2+}-activated Cl^- channels, neither of these seem to fit the bill precisely (Hartzell et al., 2005, 2008). Recent reports suggest that the TMEM16 family is likely to contain the molecular correlates of Ca^{2+}-activated Cl^- channels (Yang et al., 2008; Caputo et al., 2008; Hartzell, 2008; Schroeder et al., 2008). Single channel data suggests that there are at least four different classes of Ca^{2+}-activated Cl^- channels (Hartzell et al., 2005), and as of this writing, it remains unclear whether the TMEM16 family explains all of these and, if so, which TMEM16 family members or splice variants correspond to which channel phenotypes (for further discussion see Chapter 13 in this volume). The volume regulated anion channel (VRAC) also has no clear molecular counterpart. VRACs have different properties in different cell types (Nilius et al., 1997), so there may also be multiple families of these as well. Further, there is the 40-pS outwardly rectifying Cl^- channel (ORCC) of CFTR fame (Egan et al., 1992) and the large-conductance Cl^- channel (Franciolini and Nonner, 1994; Suzuki, 2006) that so far have no proven molecular identity. There are also cAMP-activated Cl^- currents that have been described in choroid plexus of mice lacking functional CFTR expression, suggesting that there are additional kinds of cAMP-activated Cl^- channels (Kibble et al., 1997). Table 1.1 lists the kinds of Cl^- channels that have been identified biophysically and their molecular counterparts.

VII. STRUCTURE AND FUNCTION

Considerable excitement was generated in 2002 when the crystal structure of a ClC homolog from the bacterium *Salmonella typhimurium* was solved (Dutzler et al., 2002; Dutzler, 2006). But, quite surprisingly, this structure has turned out not to be a Cl^- channel, but rather a Cl^-/H^+ exchanger! Although at the time this protein was crystallized, it was thought to be a Cl^- channel, it was recently discovered that it is a Cl^-/H^+ exchanger with a stoichiometry of two Cl^- ions exchanged for one H^+ (Accardi et al., 2004). This means that, to date, there are no *bona fide* Cl^- channels whose crystal structure has been solved, although the 3D structure of the cytosolic C-terminal fragments of some vertebrate ClCs are known (Meyer and Dutzler, 2006). Disappointingly, even 20 years after cloning CFTR, we have little 3D structural information about the pore of this channel. Parts of CFTR have been crystallized and an homology model has been developed based on ABC transporter structure (Mendoza and Thomas, 2007). Just by comparison, of the 2047 "ion channel" structure hits in RSCB Protein Data Bank (http://www.rcsb.org) (December 26, 2008), only 20 are "chloride channels".

The Cl^- channel community has still not fully recovered from the shock elicited by the discovery that the canonical member of the ClC family was not a Cl^- channel but rather a Cl^-/H^+ antiporter (Accardi and Miller, 2004) and that ClC-4 and ClC-5 are also Cl^-/H^+ antiporters (Scheel et al., 2005; Picollo and Pusch, 2005). Although it was first believed that ClCs with intracellular functions (ClC-3 to ClC-7) are transporters and the others (ClC-0, ClC-1, ClC-2, ClC-Ka and ClC-Kb) are Cl^- channels, recent findings have blurred this distinction. ClC-1 also exhibits a small proton flux (Picollo and Pusch, 2005) and the asymmetric gating of ClC-0 is dependent on a real, but immeasurably small, proton flux through the channel (Lisal and Maduke, 2008). These findings are very intriguing, because traditionally it was thought that transporters and ion channels operated by completely distinct mechanisms: channels behave as an aqueous tunnel open to both sides of the membrane at the same time, whereas transporters function by an "alternating access" mechanism where the transported ions bind to a site on one side of the membrane and are then shuttled by a conformational

TABLE 1.1 Correspondence between Cl^- Channels and Genes

Cl^- channel type	Known genes	Potential genes	Rejected
Ligand-gated	$GABA_A$ Glycine 5 HT Glutamate		
Ca-activated	TMEM16A [ANO1] TMEM16B [ANO2]	CLCA, bestrophins	
PKA-activated	CFTR	unknown	
Volume-regulated		CLC-2, ClC-3, bestrophins	pICln, mdr
Voltage-gated	ClC-1, ClC-2		
Acid-activated		ClC-2, ClC-3, SLC26A7	
Large-conductance		Tweety	
Intracellular	VDAC, ClC-3,4,5,6,7	CLICs	
Ca- and cGMP-activated		Bestrophin-3	
ORCC		Unknown	

change to the other side. But it seems that ClC-0, and probably ClC-1, have properties that combine those of channels and transporters. These findings have led to the proposal that ClC chloride channels are "broken transporters" and probably evolved from transporters (Chen and Hwang, 2008; Lisal and Maduke, 2008). Similarly, CFTR is a member of the large ABC superfamily of transporters (Gadsby et al., 2006) and thus is likely to have evolved from an ancestral transporter (Jordan et al., 2008). If ClC Cl^- channels and CFTR both evolved from a transporter, it suggests that these Cl^- channels may be relatively recent additions to the genome. Consistent with this expectation, it is interesting that although ClC Cl^-/H^+ antiporters and numerous cation channels have been found in prokaryotes, no bona fide Cl^- channels have been unambiguously identified outside the eukaryotic kingdom.

One might expect that, if Cl^- channels evolved from transporters, their mechanisms might be more complicated than those of typical channels. This is certainly true of the gating of CFTR and ClCs: CFTR gating is dependent on both phosphorylation and ATP binding and hydrolysis at two nucleotide binding domains (Gadsby et al., 2006) and gating of at least some ClC channels depends in a complex way on permeating Cl^- and H^+. But, the permeation and selectivity mechanisms of anion channels are relatively primitive. Most anion channels are rather non-selective for anions and it appears that the mechanisms of ion discrimination are rather simple. The permeation path of CFTR, for example, appears to be simply a polarizable tunnel that can stabilize a partially dehydrated ion as it passes through (Dawson et al., 1999; Smith et al., 1999). This contrasts with the bacterial K^+ channel that selects for K^+ by means of a well-defined structure consisting of a tetrahedral array of oxygen ligands (Zhou et al., 2001). In the K^+ channel, the channel is structurally specialized to recognize the ion, but in CFTR and probably other anion channels as well, selectivity is based simply on the physics of hydration and the dielectric properties of the pore of the channel. In this regard, then, anion channels are more simple, and perhaps more primitive, than K^+ or Ca^{2+} channels for their respective cations. Certainly, this primitive mechanism may have occurred because the predominant anions in biological systems are Cl^-, HCO_3^- and large organic anions. Because these ions differ greatly in their size, selective pressure to evolve a complicated and specific selectivity filter may have been absent.

The functions that Cl^- channels do, compared to cation channels, can also be viewed as relatively primitive. Cation channels that readily spring to mind are involved in highly evolved functions like action potential conduction and synaptic transmission. In contrast, Cl^- channels have more "basic" cellular functions like cell-volume regulation, secretion and proliferation. This difference may explain Hille's statement in 1992 (Hille, 1992) that Cl^- channels had "uncertain physiological significance in many cell types". Coming from the cation-centric view of the world, researchers were primed to think of ion channels as performing specialized functions, like synaptic transmission, while other, slower and more ubiquitous, functions seemed "uncertain".

VIII. DISEASES CAUSED BY DISORDERS OF Cl^- CHANNELS AND TRANSPORTERS

Certainly, one of the major driving forces for interest in Cl^- channels was the realization that a variety of human diseases are caused by disorders in Cl^- channel and transporter function (Table 1.2). On the heels of the discovery that cystic fibrosis and myotonia congenita are caused by defects in Cl^- channels, there is now a growing list of human diseases that are caused by defects in Cl^- channels and transporters, as further discussed in Chapters 12, 21, 24, 26 and 27 in this volume.

IX. INTRACELLULAR Cl^- CHANNELS

The first Cl^- channel ever isolated was the mitochondrial voltage-dependent anion channel, VDAC, which was purified biochemically and incorporated into artificial lipid bilayers in 1976 (Schein et al., 1976; Colombini, 1980, 1983). Mitochondria were an obvious choice for purification of an intracellular channel because mitochondria were easily isolated in large quantity and purity and because there was a considerable amount known about the permeability of the mitochondrial membrane from studies on the electron transport chain. By the mid-1980s it had become clear that other intracellular membranes also had channels. Lysosomes and endosomes were known to have an acidic lumenal pH that was generated by proton ATPases (Al Awqati, 1986). It was also recognized that Cl^- channels often provided a counter-ion shunt to dissipate the transmembrane potential generated by the proton pump (reviewed in Faundez and Hartzell, 2004). The cloning of the ClC channel family provided a large boost to the field of intracellular channels as it was realized that several of the ClC channel subtypes, most notably ClC-3 through ClC-7, were localized in the organelles in the endosomal–lysosomal pathway (Jentsch, 2007 and

TABLE 1.2 Disorders of Chloride Transport

Human disease	Protein	Gene	Defective function
Hyperekplexia	GLRA1	GLRA1	Synaptic inhibition
Juvenile myoclonus epilepsy	GABAα1	GABRA1	Synaptic inhibition
Epilepsy	GABA-γ2	GABRG2	Synaptic inhibition
Myotonia cogenita	ClC-1	CLCN1	Membrane potential
Dent's disease	ClC-5	CLCN5	Endosomal acidification
Neuronal ceroid lipofuscinosis	CLC-7	CLCN7	Lysosomal dysfunction
Osteopetrosis	CLC-7, Ostm1	CLCN7 OSTM1	Acid secretion by osteoclasts
Bartter syndrome type III	ClC-Kb	CLCNKB	Renal salt balance
Bartter syndrome type IV with deafness	Barttin	BSND	Renal salt loss, endolymph secretion
Vitelliform macular dystrophy	Bestrophin-1	VMD2	RPE Cl^- transport?
Spherocytosis	AE1	SLC4A1	Membrane stability
Distal renal tubule acidosis	AE1	SLC4A1	Renal pH balance
Anderman syndrome	KCC3	SLC12A6	Neural development
Deafness	KCC4	SLC12A7	Potassium recycling
Bartter syndrome type I	NKCC	SLC12A1	Renal salt balance
Congenital chloride diarrhea	SLC26A3	SLC26A3	Intestinal fluid secretion

Chapter 12, this volume). These ClC family members play a role in the acidification of endosomes and lysosomes. There is a gradient of pH from clathrin-coated vesicles (pH 6.8) to early endosomes (pH 6.2), late endosomes (pH 5.5) and lysosomes (pH 5) (Faundez and Hartzell, 2004), and different ClCs appear to be expressed at different points along the endosomal–lysosomal pathway. For example, ClC-5 is localized in apical endosomes, whereas ClC-7 is localized in lysosomes (Jentsch, 2007). Dent's disease, a renal disorder characterized by loss of protein, phosphate, and Ca^{2+}, is caused by a defect in ClC-5 and can be explained by defective endocytosis, which may be related to defective acidification of the endosome (Piwon et al., 2000; Jentsch, 2007). Although it has been proposed that ClC-7 plays a key role in regulating acidification of lysosomes (Graves et al., 2008), the neurodegeneration that is caused by defects in ClC-7 cannot be explained by defective lysosomal acidification because lysosomal pH is normal in ClC-7 knockout mice (Kaspar et al., 2005). This observation suggests that intra-lumenal Cl^- concentration itself may be important in lysosomal function (Faundez and Hartzell, 2004; Jentsch, 2007). Intra-lumenal Cl^- concentration has been shown to change along the endosomal–lysosomal pathway (Sonawane et al., 2002; Faundez and Hartzell, 2004). If it turns out to be true that intra-lumenal Cl^- is important in the function of various intracellular organelles, this will open a new vista in Cl^- channel research.

There are other channels that have been proposed to be intracellular ion channels. In particular, the reader should consult reviews on the CLICs (Ashley, 2003). And, recently, a new Cl^- channel, GPHR, has been identified to play a role in acidification of the Golgi apparatus (Caldwell and Howell, 2008; Maeda et al., 2008).

X. CHLORIDE MAY REGULATE PROTEIN FUNCTION

If intra-lumenal Cl^- concentration is important in endosomal or lysosomal function, this implies that Cl^- may regulate the activity of proteins. It is well established that various proteins have Cl^--ion binding sites and that Cl^- binding alters the conformation or activity of these proteins. Proteins that have Cl^--binding sites or whose function is altered by $[Cl^-]$ include hemoglobin (Prange et al., 2001), glycogen synthase phosphatase (Meijer et al., 1992), glucose-6-phosphatase (Pederson et al., 1998), α-amylase (Feller et al., 1996), ANP (atrionaturietic peptide) receptor (Misono, 2000; van den Akker et al., 2000), various ion channels and transporters (Pazoles et al., 1980; Moriyama and Nelson, 1987; Yuan et al., 2000; Sangan et al., 2002), angiotensin-converting enzyme I (Liu et al., 2001), AML1 (acute myeloid leukemia-1) transcription factors (Sayer et al., 2003), G-proteins (Nakajima et al., 1992), cathepsin-C (Cigic and Pain, 1999) and certain kinases (Treharne et al., 1994) including WNK kinases (Kahle et al., 2008; Ponce-Coria et al., 2008).

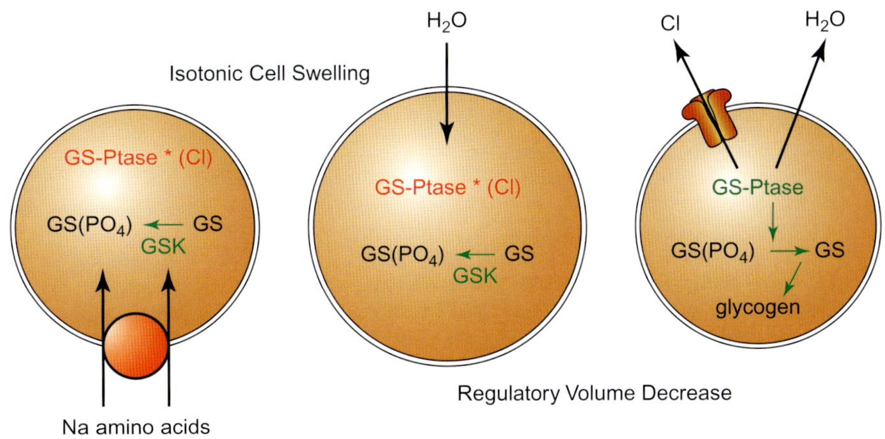

FIGURE 1.2 Cl^- as a second messenger: amino acids stimulate glycogen synthesis. In the absence of extracellular amino acids, glycogen synthase (GS) is phosphorylated by glycogen synthase kinase (GSK) and the GS phosphatase (GS-Ptase) is inhibited by high extracellular Cl^-. Na^+-dependent uptake of amino acids causes isotonic cell swelling. Cell swelling stimulates activation of VRAC and regulatory volume decrease, resulting in a loss of cytosolic Cl^-. The loss of cytosolic Cl^- stimulates GS-Ptase, which dephosphorylates GS, and glycogen synthesis is stimulated.

One example where Cl^- plays a role as a "second messenger" is in glycogen metabolism (Meijer et al., 1992; Pederson et al., 1998), as illustrated in Fig. 1.2. Liver glycogen synthesis is stimulated by extracellular amino acids that are transported into the cell by Na^+-dependent amino acid transporters. This accumulation of intracellular Na^+ and amino acids results in osmotic water influx and cell swelling. Cell swelling then stimulates the activation of volume-regulated anion channels (VRACs), efflux of Cl^- and regulatory volume decrease. The loss of intracellular Cl^- then results in a stimulation of glycogen synthase. Glycogen synthase is stimulated because its activity is regulated by glycogen synthase phosphatase, which is directly regulated by Cl^- ions.

Another example, although less well understood, is regulation of the balance between influx and efflux of Cl^- by intracellular Cl^- through WNK kinases. It has been known for some time that NKCC is activated by low intracellular Cl^- by a Cl^--sensitive protein kinase (Lytle and Forbush, 1996). More recently, the WNK kinases have been shown to be the likely candidates for the Cl^--sensitive kinases (Kahle et al., 2008; Ponce-Coria et al., 2008), as discussed in detail in Chapter 18 in this volume.

XI. OTHER FUNCTIONS OF Cl^- CHANNELS

As noted above, Cl^- channels generally exhibit rather low selectivity for anions (e.g. Hartzell et al., 2005). For this reason they should more precisely be called "anion channels", but "Cl^- channel" often seems an appropriate default because Cl^- is the predominant anion in biological systems. But the name shouldn't blind us to the fact that there are other, perhaps more important but lower abundance anions present. Certainly, HCO_3^- is a very important biological anion. The importance of CFTR-dependent HCO_3^- secretion in the pathogenesis of cystic fibrosis has received considerable attention, although questions remain how much HCO_3^- transport occurs via CFTR itself and HCO_3^- transporters such as SLC26A3 that are regulated by CFTR (Ko et al., 2002; Shcheynikov et al., 2006). HCO_3^- is significantly permeant through overexpressed bestrophin channels (e.g. Qu and Hartzell, 2008, but the physiological relevance of this observation remains to be demonstrated. Recently, it has been suggested that ClC-3 may be a pathway by which reactive oxygen species, such as superoxide, exit endosomes (Moreland et al., 2006; Hawkins et al., 2007; Miller et al., 2007; Lassegue, 2007; Mumbengegwi et al., 2008). If certain anion channels are also involved in reactive oxygen signaling, this will certainly elevate Cl^- channels to a new level of respectability.

XII. CONCLUDING REMARKS

Cl^- channels have come a very long way in the 40 years I have been studying ion channels. From having uncertain physiological significance in the 1970s to now having 50 or more genes and one crystal structure, the physiological functions are becoming well defined. But, it seems that the future and challenge of Cl^- channels lies in understanding how they are mechanistically linked to their functions. For example, we understand little how cell volume is actually regulated and how Cl^- channels may be linked to cell proliferation. And, although we have gained tremendous insights about the role of Cl^- channels in human disease, therapeutic interventions aimed at Cl^- channels remain primitive. The next few decades are likely to see new and exciting developments in the Cl^- channel arena. The chapters that follow in this book provide the foundation for these developments.

References

Accardi, A., Kolmakova-Partensky, L., Williams, C., and Miller, C. (2004). Ionic currents mediated by a prokaryotic homologue of CLC Cl^- channels. *J. Gen. Physiol.* **123**, 109–119.

Accardi, A. and Miller, C. (2004). Secondary active transport mediated by a prokaryotic homologue of ClC Cl^- channels. *Nature* **427**, 803–807.

Adrian, R.H. (1960). Potassium chloride movement and the membrane potential of frog muscle. *J. Physiol,* **151**, 154–185.

Adrian, R.H. (1961). Internal chloride concentration and chloride efflux of frog muscle. *J. Physiol.* **156**, 623–632.

Adrian, R.H. and Bryant, S.H. (1974). On the repetitive discharge in myotonic muscle fibres. *J. Physiol.* **240**, 505–515.

Aickin, C. (1990). Choride transport across the sarcolemma of vertebrate smooth and skeletal muscle. In *Chloride Channels and Carriers in Nerve, Muscle and Glial Cells* (Alvarez-Leefmans, F.J. and Russell, J.M., eds), pp. 209–249. Plenum Press, New York.

Alvarez-Leefmans, F.J. (2001). Intracellular chloride regulation. In *Cell Physiology Sourcebook: A Molecular Approach* (Sperelakis, N., ed.), pp. 301–318. Academic Press, San Diego.

Alvarez-Leefmans, F.J., Russell, J.M., and Giraldez, F. (1990). Methods for measuring chloride transport across nerve, muscle and glial cells. In *Chloride Channels and Carriers in Nerve, Muscle, and Glial Cells* (Alvarez-Leefmans, F.J. and Russell, J.M., eds), pp. 3–66. Plenum Press, New York.

Al Awqati, Q. (1986). Proton-translocating ATPases. *Ann. Rev. Cell Biol.* **2**, 179–199.

Anderson, M.P., Gregory, R.J., Thompson, S., Souza, D.W., Paul, S., Mulligan, R.C., Smith, A.E., and Welsh, M.J. (1991). Demonstration that CFTR is a chloride channel by alteration of its anion selectivity. *Science* **253**, 202–205.

Ashley, R.H. (2003). Challenging accepted ion channel biology: p64 and the CLIC family of putative intracellular anion channel proteins. *Mol. Membr. Biol.* **20**, 1–11.

Awad, S., Allison, S.P., and Lobo, D.N. (2008). The history of 0.9% saline. *Clin. Nutr.* **27**, 179–188.

Bear, C.E., Li, C.H., Kartner, N., Bridges, R.J., Jensen, T.J., Ramjeesingh, M., and Riordan, J.R. (1992). Purification and functional reconstitution of the cystic fibrosis transmembrane conductance regulator (CFTR). *Cell* **68**, 809–818.

Ben-Ari, Y., Gaiarsa, J.L., Tyzio, R., and Khazipov, R. (2007). GABA: a pioneer transmitter that excites immature neurons and generates primitive oscillations. *Physiol. Rev.* **87**, 1215–1284.

Berglund, K., Schleich, W., Krieger, P., Loo, L.S., Wang, D., Cant, N.B., Feng, G., Augustine, G.J., and Kuner, T. (2006). Imaging synaptic inhibition in transgenic mice expressing the chloride indicator, Clomeleon. *Brain Cell Biol.* **35**, 207–228.

Berglund, K., Schleich, W., Wang, H., Feng, G., Hall, W.C., Kuner, T., and Augustine, G.J. (2008). Imaging synaptic inhibition throughout the brain via genetically targeted Clomeleon. *Brain Cell Biol.* **36**, 101–118.

Biwersi, J., Farah, N., Wang, Y.X., Ketcham, R., and Verkman, A.S. (1992). Synthesis of cell-impermeable Cl-sensitive fluorescent indicators with improved sensitivity and optical properties. *Am. J. Physiol.* **262**, C242–C250.

Biwersi, J., Tulk, B., and Verkman, A.S. (1994). Long-wavelength chloride-sensitive fluorescent indicators. *Anal. Biochem.* **219**, 139–143.

Biwersi, J. and Verkman, A.S. (1991). Cell-permeable fluorescent indicator for cytosolic chloride. *Biochem.* **30**, 7879–7883.

Boistel, J. and Fatt, P. (1958). Membrane permeability change during inhibitory transmitter action in crustacean muscle. *J. Physiol.* **144**, 176–191.

Bryant, S.H. (1969). Cable properties of external intercostal muscle fibres from myotonic and nonmyotonic goats. *J. Physiol.* **204**, 539–550.

Caldwell, J.H. and Howell, K.E. (2008). Pores galore for the Golgi. *Nat. Cell Biol.* **10**, 1125–1126.

Caputo, A., Caci, E., Ferrera, L., Pedemonte, N., Barsanti, C., Sondo, E., Pfeffer, U., Ravazzolo, R., Zegarra-Moran, O., and Galietta, L.J. (2008). TMEM16A, a membrane protein associated with calcium-dependent chloride channel activity. *Science* **322**, 590–594.

Chen, T.Y. and Hwang, T.C. (2008). CLC-0 and CFTR: chloride channels evolved from transporters. *Physiol. Rev.* **88**, 351–387.

Cigic, B. and Pain, R.H. (1999). Location of the binding site for chloride ion activation of cathepsin C. *Eur. J. Biochem.* **264**, 944–951.

Coles, J.A. (1995). Measurement of cytosolic chloride activity by ion-selective microelectrodes. *Meth. Neurosci.* **27**, 340–360.

Colombini, M. (1980). Structure and mode of action of a voltage dependent anion-selective channel (VDAC) located in the outer mitochondrial membrane. *Ann. NY Acad. Sci.* **341**, 552–563.

Colombini, M. (1983). Purification of VDAC (voltage-dependent anion-selective channel) from rat liver mitochondria. *J. Membr. Biol.* **74**, 115–121.

Coombs, J.S., Eccles, J.C., and Fatt, P. (1955). The specific ionic conductances and the ionic movements across the motoneuronal membrane that produce the inhibitory post-synaptic potential. *J. Physiol.* **130**, 326–373.

Dawson, D.C., Smith, S.S., and Mansoura, M.K. (1999). CFTR: mechanism of anion conduction. *Physiol. Rev.* **79**, S47–S75.

De Groat, W.C., Lalley, P.M., and Saum, W.R. (1972). Depolarization of dorsal root ganglia in the cat by GABA and related amino acids: antagonism by picrotoxin and bicuculline. *Brain Res.* **44**, 273–277.

De La Fuente, R., Namkung, W., Mills, A., and Verkman, A.S. (2008). Small-molecule screen identifies inhibitors of a human intestinal calcium-activated chloride channel. *Molec. Pharmacol.* **73**, 758–768.

Di Sant'Agnese, P.A., Darling, R.C., Perera, G.A., and Shea, E. (1953). Abnormal electrolyte composition of sweat in cystic fibrosis of the pancreas; clinical significance and relationship to the disease. *Pediatrics* **12**, 549–563.

Dorwart, M.R., Shcheynikov, N., Wang, Y., Stippec, S., and Muallem, S. (2007). SLC26A9 is a Cl^- channel regulated by the WNK kinases. *J. Physiol.* **584**, 333–345.

Dutzler, R. (2006). The ClC family of chloride channels and transporters. *Curr. Opin. Struct. Biol.* **16**, 439–446.

Dutzler, R., Campbell, E.B., Cadene, M., Chait, B.T., and MacKinnon, R. (2002). X-ray structure of a ClC chloride channel at 3.0 A reveals the molecular basis of anion selectivity. *Nature* **415**, 287–294.

Egan, M., Flotte, T., Afione, S., Solow, R., Zeitlin, P.L., Carter, B.J., and Guggino, W.B. (1992). Defective regulation of outwardly rectifying Cl^- channels by protein kinase A corrected by insertion of CFTR. *Nature* **358**, 581–584.

Eggermont, J., Trouet, D., Carton, I., and Nilius, B. (2001). Cellular function and control of volume-regulated anion channels. *Cell Biochem. Biophys.* **35**, 263–274.

Fahlke, C., Beck, C.L., and George, A.L., Jr. (1997). A mutation in autosomal dominant myotonia congenita affects pore properties of the muscle chloride channel. *Proc. Natl. Acad. Sci. USA* **94**, 2729–2734.

Faundez, V. and Hartzell, H.C. (2004). Intracellular chloride channels: determinants of function in the endosomal pathway. *Sci STKE* **2004**, re8.

Feller, G.O., Bussy, C., Houser, C., and Gerday, C. (1996). Structural and functional aspects of chloride binding to Altermonas haloplanctis alpha-amylase. *J. Biol. Chem.* **271**, 23836–23841.

Field, M. (2003). Intestinal ion transport and the pathophysiology of diarrhea. *J. Clin. Invest.* **111**, 931–943.

Field, M., Fromm, D., al-Awqati, Q., and Greenough, W.B., III (1972). Effect of cholera enterotoxin on ion transport across isolated ileal mucosa. *J. Clin. Invest.* **51**, 796–804.

Franciolini, F. and Nonner, W. (1994). A multi-ion permeation mechanism in neuronal background chloride channels. *J. Gen. Physiol.* **104**, 725–746.

Fuller, M.D., Thompson, C.H., Zhang, Z.R., Freeman, C.S., Schay, E., Szakacs, G., Bakos, E., Sarkadi, B., McMaster, D., French, R.J., Pohl, J., Kubanek, J., and McCarty, N.A. (2007). State-dependent inhibition of cystic fibrosis transmembrane conductance regulator chloride channels by a novel peptide toxin. *J Biol. Chem.* **282**, 37545–37555.

Gadsby, D.C., Vergani, P., and Csanady, L. (2006). The ABC protein turned chloride channel whose failure causes cystic fibrosis. *Nature* **440**, 477–483.

Galietta, L.J., Haggie, P.M., and Verkman, A.S. (2001). Green fluorescent protein-based halide indicators with improved chloride and iodide affinities. *FEBS Lett.* **499**, 220–224.

Gosmanov, A.R., Lindinger, M.I., and Thomason, D.B. (2003). Riding the tides: K^+ concentration and volume regulation by muscle Na^+-K^+-$2Cl^-$ cotransport activity. *News Physiol. Sci.* **18**, 196–200.

Graves, A.R., Curran, P.K., Smith, C.L., and Mindell, J.A. (2008). The Cl^-/H^+ antiporter ClC-7 is the primary chloride permeation pathway in lysosomes. *Nature* **453**, 788–792.

Hartzell, C., Putzier, I., and Arreola, J. (2005). Calcium-activated chloride channels. *Annu. Rev. Physiol.* **67**, 719–758.

Hartzell, H.C. (2008). Physiology. CaCl-ing channels get the last laugh. *Science* **322**, 534–535.

Hartzell, H.C., Qu, Z., Yu, K., Xiao, Q., and Chien, L.T. (2008). Molecular physiology of bestrophins: multifunctional membrane proteins linked to Best disease and other retinopathies. *Physiol. Rev.* **88**, 639–672.

Hawkins, B.J., Madesh, M., Kirkpatrick, C.J., and Fisher, A.B. (2007). Superoxide flux in endothelial cells via the chloride channel-3 mediates intracellular signaling. *Mol. Biol. Cell* **18**, 2002–2012.

Hille, B. (1984). *Ionic Channels of Excitable Membranes*. Sinaur Associates Inc, Sunderland, MA.

Hille, B. (1992). *Ionic Channels of Excitable Membranes*. Sinaur Associates Inc, Sunderland, MA.

Hille, B. (2001). *Ion Channels of Excitable Membranes*. Sinaur Associates Inc, Sunderland, MA.

Hipper, A., Mall, M., Greger, R., and Kunzelmann, K. (1995). Mutations in the putative pore-forming domain of CFTR do not change anion selectivity of the cAMP activated Cl^- conductance. *FEBS Lett.* **374**, 312–316.

Hodgkin, A.L. and Horowicz, P. (1959). The influence of potassium and chloride ions on the membrane potential of single muscle fibres. *J. Physiol.* **148**, 127–160.

Hogben, C.A. (1959). Physiochemical aspects of hydrochloric acid formation. *Am. J. Dig. Dis.* **4**, 184–193.

Hume, J.R., Duan, D., Collier, M.L., Yamazaki, J., and Horowitz, B. (2000). Anion transport in heart. *Physiol. Rev.* **80**, 31–81.

Hutter, O.F. and Noble, D. (1960). The chloride conductance of frog skeletal muscle. *J. Physiol.* **151**, 89–102.

Isomura, Y., Sugimoto, M., Fujiwara-Tsukamoto, Y., Yamamoto-Muraki, S., Yamada, J., and Fukuda, A. (2003). Synaptically activated Cl^- accumulation responsible for depolarizing GABAergic responses in mature hippocampal neurons. *J. Neurophysiol.* **90**, 2752–2756.

Jentsch, T.J. (2007). Chloride and the endosomal-lysosomal pathway: emerging roles of CLC chloride transporters. *J. Physiol* **578**, 633–640.

Jentsch, T.J. (2008). CLC chloride channels and transporters: from genes to protein structure, pathology and physiology. *Crit Rev. Biochem. Mol. Biol.* **43**, 3–36.

Jentsch, T.J., Stein, V., Weinreich, F., and Zdebik, A.A. (2002). Molecular structure and physiological function of chloride channels. *Physiol. Rev.* **82**, 503–568.

Jentsch, T.J., Steinmeyer, K., and Schwarz, G. (1990). Primary structure of Torpedo marmorata chloride channel isolated by expression cloning in Xenopus oocytes. *Nature* **348**, 510–514.

Jordan, I.K., Kota, K.C., Cui, G., Thompson, C.H., and McCarty, N.A. (2008). Evolutionary and functional divergence between the cystic fibrosis transmembrane conductance regulator and related ATP-binding cassette transporters. *Proc. Natl. Acad. Sci. USA* **105**, 18865–18870.

Kahle, K.T., Ring, A.M., and Lifton, R.P. (2008). Molecular physiology of the WNK kinases. *Annu. Rev. Physiol.* **70**, 329–355.

Kaspar, D., Planells-Cases, R., Fuhrmann, J.C., Scheel, O., Zeitz, O., Ruether, K., Poet, M., Steinfield, R., Schweizer, M., Kornak, U., and Jentsch, T.J. (2005). Loss of the chloride channel ClC-7 leads to lysosomal storage disease and neurodegeneration. *EMBO J.* **24**, 1079–1091.

Kerem, B., Rommens, J.M., Buchanan, J.A., Markiewicz, D., Cox, T.K., Chakravarti, A., Buchwald, M., and Tsui, L.C. (1989). Identification of the cystic fibrosis gene: genetic analysis. *Science* **245**, 1073–1080.

Keynes, R.D. (1963). Chloride in the squid giant axon. *J. Physiol.* **169**, 690–705.

Kibble, J.D., Garner, C., Colledge, W.H., Brown, S., Kajita, H., Evans, M., and Brown, P.D. (1997). Whole cell Cl^- conductances in mouse choroid plexus epithelial cells do not require CFTR expression. *Am. J. Physiol.* **272**, C1899–C1907.

Kim, K.H., Shcheynikov, N., Wang, Y., and Muallem, S. (2005). SLC26A7 is a Cl^- channel regulated by intracellular pH. *J. Biol. Chem.* **280**, 6463–6470.

Knowles, M., Gatzy, J., and Boucher, R. (1983a). Relative ion permeability of normal and cystic fibrosis nasal epithelium. *J. Clin. Invest.* **71**, 1410–1417.

Knowles, M.R., Stutts, M.J., Spock, A., Fischer, N., Gatzy, J.T., and Boucher, R.C. (1983b). Abnormal ion permeation through cystic fibrosis respiratory epithelium. *Science* **221**, 1067–1070.

Ko, S.B., Shcheynikov, N., Choi, J.Y., Luo, X., Ishibashi, K., Thomas, P.J., Kim, J.Y., Kim, K.H., Lee, M.G., Naruse, S., and Muallem, S. (2002). A molecular mechanism for aberrant CFTR-dependent HCO_3^- transport in cystic fibrosis. *EMBO J.* **21**, 5662–5672.

Koch, M.C., Steinmeyer, K., Lorenz, C., Ricker, K., Wolf, F., Otto, M., Zoll, B., Lehmann-Horn, F., Grzeschik, K.H., and Jentsch, T.J. (1992). The skeletal muscle chloride channel in dominant and recessive human myotonia. *Science* **257**, 797–800.

Kuffler, S.W. and Edwards, C. (1958). Mechanism of gamma aminobutyric acid (GABA) action and its relation to synaptic inhibition. *J. Neurophysiol.* **21**, 589–610.

Kuner, T. and Augustine, G.J. (2000). A genetically encoded ratiometric indicator for chloride: capturing chloride transients in cultured hippocampal neurons. *Neuron* **27**, 447–459.

Lassegue, B. (2007). How does the chloride/proton antiporter ClC-3 control NADPH oxidase? *Circ. Res.* **101**, 648–650.

Lipicky, R.J. and Bryant, S.H. (1966). Sodium, potassium, and chloride fluxes in intercostal muscle from normal goats and goats with hereditary myotonia. *J. Gen. Physiol.* **50**, 89–111.

Lipicky, R.J., Bryant, S.H., and Salmon, J.H. (1971). Cable parameters, sodium, potassium, chloride, and water content, and potassium efflux in isolated external intercostal muscle of normal volunteers and patients with myotonia congenita. *J. Clin. Invest.* **50**, 2091–2103.

Lisal, J. and Maduke, M. (2008). The ClC-0 chloride channel is a "broken" Cl^-/H^+ antiporter. *Nat. Struct. Mol. Biol.* **15**, 805–810.

Liu, X., Farnandez, M., Wouters, M.A., Heyberger, S., and Husain, A. (2001). Arg^{1098} is critical for the chloride dependence of human angiotensin I-converting enzyme C-domain catalytic activity. *J. Biol. Chem.* **276**, 33518–33525.

Loriol, C., Dulong, S., Avella, M., Gabillat, N., Boulukos, K., Borgese, F., and Ehrenfeld, J. (2008). Characterization of SLC26A9, facilitation of Cl^- transport by bicarbonate. *Cell Physiol. Biochem.* **22**, 15–30.

Lytle, C. and Forbush, B. (1996). Regulatory phosphorylation of the secretory Na-K-Cl cotransporter: modulation by cytoplasmic Cl. *Am. J. Physiol.* **270**, C437–C448.

Maeda, Y., Ide, T., Koike, M., Uchiyama, Y., and Kinoshita, T. (2008). GPHR is a novel anion channel critical for acidification and functions of the Golgi apparatus. *Nat. Cell Biol.* **10**, 1135–1145.

Mansoura, M.K., Biwersi, J., Ashlock, M.A., and Verkman, A.S. (1999). Fluorescent chloride indicators to assess the efficacy of CFTR cDNA delivery. *Hum. Gene Ther.* **10**, 861–875.

Meijer, A.J., Baquet, A., Gustafson, L., van Woerkom, G.M., and Hue, L. (1992). Mechanism of activation of liver glycogen synthase by swelling. *J. Biol. Chem.* **267**, 5823–5828.

Mendoza, J.L. and Thomas, P.J. (2007). Building an understanding of cystic fibrosis on the foundation of ABC transporter structures. *J. Bioenerg. Biomembr.* **39**, 499–505.

Meyer, S. and Dutzler, R. (2006). Crystal structure of the cytoplasmic domain of the chloride channel ClC-0. *Structure* **14**, 299–307.

Miller, C. and White, M.M. (1984). Dimeric structure of single chloride channels from Torpedo electroplax. *Proc. Natl. Acad. Sci.* **81**, 2772–2775.

Miller, F.J., Jr., Filali, M., Huss, G.J., Stanic, B., Chamseddine, A., Barna, T.J., and Lamb, F.S. (2007). Cytokine activation of nuclear factor kappa B in vascular smooth muscle cells requires signaling endosomes containing Nox1 and ClC-3. *Circ. Res.* **101**, 663–671.

Misono, K.S. (2000). Atrial natriuretic factor binding to its receptor is dependent on chloride concentration: a possible feedback-control mechanism in renal salt regulation. *Circ. Res.* **86**, 1135–1139.

Moreland, J.G., Davis, A.P., Bailey, G., Nauseef, W.M., and Lamb, F. (2006). Anion channels, including ClC-3, are required for normal neutrophil oxidative function, phagocytosis, and transendothelial migration. *J. Biol. Chem.* **281**, 12277–12288.

Moriyama, Y. and Nelson, N. (1987). The purified ATPase from chromaffin granule membranes is an anion-dependent proton pump. *J. Biol. Chem.* **262**, 9175–9180.

Muanprasat, C., Kaewmokul, S., and Chatsudthipong, V. (2007). Identification of new small molecule inhibitors of cystic fibrosis transmembrane conductance regulator protein: in vitro and in vivo studies. *Biol. Pharm. Bull.* **30**, 502–507.

Muanprasat, C., Sonawane, N.D., Salinas, D., Taddei, A., Galietta, L.J., and Verkman, A.S. (2004). Discovery of glycine hydrazide pore-occluding CFTR inhibitors: mechanism, structure-activity analysis, and in vivo efficacy. *J. Gen. Physiol.* **124**, 125–137.

Mumbengegwi, D.R., Li, Q., Li, C., Bear, C.E., and Engelhardt, J. F. (2008). Evidence for a superoxide permeability pathway in endosomal membranes. *Molec. Cell. Biol.* **28**, 3700–3712.

Naftalin, R.J. and Simmons, N.L. (1979). The effects of theophylline and choleragen on sodium and chloride ion movements within isolated rabbit ileum. *J. Physiol.* **290**, 331–350.

Nakajima, T., Sugimoto, T., and Kurachi, Y. (1992). Effects of anions on the G protein-mediated activation of the muscarinic K^+ channel in the cardiac atrial cell membrane. Intracellular chloride inhibition of the GTPase activity of GK. *J. Gen. Physiol.* **99**, 665–682.

Nilius, B. and Droogmans, G. (2001). Ion channels and their functional role in vascular endothelium. *Physiol. Rev.* **81**, 1415–1459.

Nilius, B. and Droogmans, G. (2003). Amazing chloride channels: an overview. *Acta Physiol. Scand.* **177**, 119–147.

Nilius, B., Eggermont, J., Voets, T., Buyse, G., Manolopoulos, V., and Droogmans, G. (1997). Properties of volume-regulated anion channels in mammalian cells. *Prog. Biophys. Mol. Biol.* **68**, 69–119.

Noda, M., Ikeda, T., Suzuki, H., Takeshima, H., Takahashi, T., Kuno, M., and Numa, S. (1986). Expression of functional sodium channels from cloned cDNA. *Nature* **322**, 826–828.

Noda, M., Takahashi, H., Tanabe, T., Toyosato, M., Furutani, Y., Hirose, T., Asai, M., Inayama, S., Miyata, T., and Numa, S. (1982). Primary structure of alpha-subunit precursor of Torpedo californica acetylcholine receptor deduced from cDNA sequence. *Nature* **299**, 793–797.

Okada, Y., Shimizu, T., Maeno, E., Tanabe, S., Wang, X., and Takahashi, N. (2006). Volume-sensitive chloride channels involved in apoptotic volume decrease and cell death. *J. Membr. Biol.* **209**, 21–29.

Papazian, D.M., Schwarz, T.L., Tempel, B.L., Jan, Y.N., and Jan, L. Y. (1987). Cloning of genomic and complementary DNA from Shaker, a putative potassium channel gene from Drosophila. *Science* **237**, 749–753.

Pazoles, C.J., Creutz, C.E., Ramu, A., and Pollard, H.B. (1980). Permeant anion activation of MgATPase activity in chromaffin granules. Evidence for direct coupling of proton and anion transport. *J. Biol. Chem.* **255**, 7863–7869.

Pederson, B.A., Nordlie, M.A., Foster, J.D., and Nordlie, R.C. (1998). Effects of ionic strength and chloride ion on activities of the glucose-6-phosphatase system: regulation of the biosynthetic activity of glucose-6-phosphatase by chloride ion inhibition/deinhibition. *Arch. Biochem. Biophys.* **353**, 141–151.

Picollo, A. and Pusch, M. (2005). Chloride/proton antiporter activity of mammalian CLC proteins ClC-4 and ClC-5. *Nature* **436**, 420–423.

Piwon, N., Gunther, W., Schwake, M., Bosl, M.R., and Jentsch, T.J. (2000). ClC-5 Cl^--channel disruption impairs endocytosis in a mouse model for Dent's disease. *Nature* **408**, 369–373.

Ponce-Coria, J., San-Cristobal, P., Kahle, K.T., Vazquez, N., Pacheco-Alvarez, D., de Los Heros, P., Juarez, P., Munoz, E., Michel, G., Bobadilla, N.A., Gimenez, I., Lifton, R.P., Hebert, S.C., and Gamba, G. (2008). Regulation of NKCC2 by a chloride-sensing mechanism involving the WNK3 and SPAK kinases. *Proc. Natl. Acad. Sci. USA* **105**, 8458–8463.

Prange, H.D., Shoemaker, J.L., Westen, E.A., Horstkotte, D.G., and Pinshow, B. (2001). Physiological consequences of oxygen-dependent chloride binding to hemoglobin. *J. Appl. Physiol.* **91**, 33–38.

Puljak, L. and Kilic, G. (2006). Emerging roles of chloride channels in human diseases. *Biochim. Biophys. Acta* **1762**, 404–413.

Pusch, M., Steinmeyer, K., Koch, M.C., and Jentsch, T.J. (1995). Mutations in dominant human myotonia congenita drastically alter the voltage dependence of the CIC-1 chloride channel. *Neuron* **15**, 1455–1463.

Qu, Z. and Hartzell, H.C. (2008). Bestrophin Cl^- channels are highly permeable to HCO_3^-. *Am. J. Physiol. Cell Physiol.* **294**, C1371–C1377.

Quinton, P.M. (1983). Chloride impermeability in cystic fibrosis. *Nature* **301**, 421–422.

Quinton, P.M. (1999). Physiological basis of cystic fibrosis: a historical perspective. *Physiol. Rev.* **79**, 3–22.

Riordan, J.R., Rommens, J.M., Kerem, B., Alon, N., Rozmahel, R., Grzelczak, Z., Zielenski, J., Lok, S., Plavsic, N., Chou, J.L. et al. (1989). Identification of the cystic fibrosis gene: cloning and characterization of complementary DNA. *Science* **245**, 1066–1073.

Rommens, J.M., Iannuzzi, M.C., Kerem, B., Drumm, M.L., Melmer, G., Dean, M., Rozmahel, R., Cole, J.L., Kennedy, D., Hidaka, N. et al. (1989). Identification of the cystic fibrosis gene: chromosome walking and jumping. *Science* **245**, 1059–1065.

Ruch, T.C. and Patton, H.D. (1965). *Physiology and Biophysics*. W.B. Saunders and Co, Philadelphia.

Sangan, P., Rajendran, V.M., Geibel, J.P., and Binder, H.J. (2002). Cloning and expression of a chloride-dependent Na^+-H^+ exchanger. *J. Biol. Chem.* **277**, 9668–9675.

Sayer, J.A., Carr, G., Pearce, S.H., Goodship, T.H., and Simmons, N. L. (2003). Disordered calcium crystal handling in antisense CLC-5-treated collecting duct cells. *Biochem. Biophys. Res. Commun.* **300**, 305–310.

Scheel, O., Zdebik, A.A., Lourdel, S., and Jentsch, T.J. (2005). Voltage-dependent electrogenic chloride/proton exchange by endosomal CLC proteins. *Nature* **436**, 424–427.

Schein, S.J., Colombini, M., and Finkelstein, A. (1976). Reconstitution in planar lipid bilayers of a voltage-dependent anion-selective channel obtained from paramecium mitochondria. *J. Membr. Biol.* **30**, 99–120.

Schroeder, B.C., Cheng, T., Jan, Y.N., and Jan, L.Y. (2008). Expression cloning of TMEM16A as a calcium-activated chloride channel subunit. *Cell* **134**, 1019–1029.

Shcheynikov, N., Ko, S.B., Zeng, W., Choi, J.Y., Dorwart, M.R., Thomas, P.J., and Muallem, S. (2006). Regulatory interaction between CFTR and the SLC26 transporters. *Novartis. Found. Symp.* **273**, 177–186.

Sile, S., Vanoye, C.G., and George, A.L. (2006). Molecular physiology of renal ClC chloride channels/transporters. *Curr. Opin. Nephrol. Hypertens.* **15**, 511–516.

Smith, S.S., Steinle, E.D., Meyerhoff, M.E., and Dawson, D.C. (1999). Cystic fibrosis transmembrane conductance regulator. Physical basis for lyotropic anion selectivity patterns. *J. Gen. Physiol.* **114**, 799–818.

Sonawane, N.D., Thiagarajah, J.R., and Verkman, A.S. (2002). Chloride concentration in endosomes measured using a ratioable fluorescent Cl^- indicator: evidence for chloride accumulation during acidification. *J. Biol. Chem.* **277**, 5506–5513.

Steinbach, H.B. (1941). Chloride in the giant axons of squid. *J. Cell. Comp. Physiol.* **17**, 57–64.

Steinmeyer, K., Ortland, C., and Jentsch, T.J. (1991). Primary structure and functional expression of a developmentally regulated skeletal muscle chloride channel. *Nature* **354**, 301–304.

Suzuki, M. (2006). The Drosophila tweety family: molecular candidates for large-conductance Ca^{2+}-activated Cl- channels. *Exp. Physiol.* **91**, 141–147.

Thompson, C.H., Fields, D.M., Olivetti, P.R., Fuller, M.D., Zhang, Z.R., Kubanek, J., and McCarty, N.A. (2005). Inhibition of ClC-2 chloride channels by a peptide component or components of scorpion venom. *J. Membr. Biol.* **208**, 65–76.

Treharne, K.J., Marshall, L.J., and Mehta, A. (1994). A novel chloride-dependent GTP-utilizing protein kinase in plasma membranes from human respiratory epithelium. *Am. J. Physiol.* **267**, L592–L601.

van den Akker, F., Zhang, X., Miyagi, M., Huo, X., Misono, K.S., and Yee, V.C. (2000). Structure of the dimerized hormone-binding domain of a guanylyl-cyclase-coupled receptor. *Nature* **406**, 101–104.

Verkman, A.S., Lukacs, G.L., and Galietta, L.J. (2006). CFTR chloride channel drug discovery-inhibitors as antidiarrheals and activators for therapy of cystic fibrosis. *Curr. Pharm. Des.* **12**, 2235–2247.

Welsch, M.J., Ramsey, B.W., Accurso, F., and Cutting, G.R. (2001). Cystic fibrosis. In *The Metabolic and Molecular Bases of Inherited Disease* (Scriver, C.R., Beaudet, A.L., Sly, W.S., and Valle, D., eds), Vol. 3, p. 5121. McGraw-Hill, New York.

Yang, H., Shelat, A.A., Guy, R.K., Gopinath, V.S., Ma, T., Du, K., Lukacs, G.L., Taddei, A., Folli, C., Pedemonte, N., Galietta, L.J., and Verkman, A.S. (2003). Nanomolar affinity small molecule correctors of defective Delta F508-CFTR chloride channel gating. *J. Biol. Chem.* **278**, 35079–35085.

Yang, Y.D., Cho, H., Koo, J.Y., Tak, M.H., Cho, Y., Shim, W.S., Park, S.P., Lee, J., Lee, B., Kim, B.M., Raouf, R., Shin, Y.K., and Oh, U. (2008). TMEM16A confers receptor-activated calcium-dependent chloride conductance. *Nature* **455**, 1210–1215.

Yuan, A., Dourado, M., Butler, A., Walton, N., Wei, A., and Salkoff, L. (2000). SLO-2, a K^+ channel with an unusual Cl^- dependence. *Nature Neurosci.* **3**, 771–779.

Zhou, Y., Morais-Cabral, J.H., Kaufman, A., and MacKinnon, R. (2001). Chemistry of ion coordination and hydration revealed by a K^+ channel-Fab complex at 2.0 A resolution. *Nature* **414**, 43–48.

Zifarelli, G. and Pusch, M. (2007). CLC chloride channels and transporters: a biophysical and physiological perspective. *Rev. Physiol Biochem. Pharmacol.* **158**, 23–76.

CHAPTER

2

Sodium-Coupled Chloride Cotransporters: Discovery and Newly Emerging Concepts

John M. Russell

OUTLINE

I. INTRODUCTION

Sodium-coupled chloride cotransporters (NCCCs) are secondary active transporters. Secondary active transport is a remarkable and widespread cellular mechanism used to move solutes across biological membranes against their concentration and/or electrochemical gradients using the potential energy stored in the gradient of another solute, usually that of sodium. The concept that the potential energy stored in the electrochemical transmembrane gradient of sodium (generated by the sodium pump, a primary active transporter) might be captured by mechanisms residing in the plasma membrane to power the uphill transport of another solute has been well accepted for some time now. It was not always so because, as with most new ideas, this one met with considerable early resistance. Many argued that such a mechanism somehow broke the Second Law of Thermodynamics. Furthermore, when such mechanisms were extended to include uphill movements of the chloride anion, a deep-seated bias was revealed among cell biologists. Namely, that Cl^- could not be distributed out of electrochemical equilibrium. This bias resulted from the fact that the early dominant models of cellular ionic distributions were the red blood cell and frog skeletal muscle. Both of these cells have a $[Cl^-]_i$ that is at, or very near, electrochemical equilibrium, a pattern most investigators assumed was representative of all animal cells. However, we now know that these two cell types are actually among the relatively few that exhibit such an apparent passive distribution of Cl^- across the plasmalemma.

The nature of the preceding objections reveals an important truth about how scientific progress is made and measured. New ideas have to force their way into our thinking by proof of their utility, feasibility and ultimately objective experimental proof. They almost always have to compete against older ideas, ideas that have apparently served scientists well for some time. That is a fundamental part of how we make scientific progress. Although it may be a bit grandiose to consider the idea of Na^+-coupled cotransport as a paradigm shift in the sense Thomas Kuhn used the term (Kuhn, 1962), it was certainly an important shift in the way cell physiologists think about solute transport mechanisms.

Several families of Na^+-coupled transporters which also transport Cl^- are in various states of characterization at this time. These include the Na^+-coupled amino acid transporters (e.g. alanine), the Na^+-coupled neurotransmitter transporters (e.g. GABA), and the Na^+-coupled chloride cotransporters. I will limit discussion in this chapter to the latter group, which belongs to the SLC12 gene family. Specifically, I will focus on the Na^+-K^+-Cl^- cotransporters (NKCC1, gene SLC12A2; NKCC2, gene SLC12A1) and the Na^+-Cl^- cotransporter (NCC; gene SLC12A3). Under normal physiological conditions, these cotransporters promote active Cl^- influx into the cell, that is, they move Cl^- into the cell against its electrochemical gradient (Chapter 5). All three share the structural feature of having 12 membrane-spanning regions and share a high degree of sequence identity (see Fig. 16.1, Chapter 16). However, they differ significantly in their tissue/cell type location, in their pharmacology and in their particular functions.

The history of the discovery, physiological and molecular characterization of the Na^+-coupled Cl^- cotransporters has been well detailed in recent years (e.g. Russell, 2000; Hebert et al., 2003; Gamba, 2005). Briefly, functional evidence of the existence of these Na^+-coupled Cl^- cotransporters first appeared in the late 1960s and early 1970s. By the early 1980s, the physiological properties of these cotransporters had been broadly outlined. Each was shown to be electroneutral; transporting equal numbers of positive and negative charges during each turnover of the transport cycle. Each has an absolute requirement for Na^+ and Cl^- in the extracellular fluid (Geck et al., 1980; Renfro, 1977; Stokes et al., 1984). It was also known by that time that NKCC function could be completely inhibited by sulfamoylbenzoic acid diuretics (bumetanide being the "gold standard") whereas NCC function could be inhibited by thiazide diuretics. As another example of how our thinking is influenced by the current models, an early report of what turned out to be activity of the NKCC was hypothesized to be a form of the sodium pump termed sodium pump 2 (Hoffman and Kregenow, 1966) which was the only well-characterized ion transporter at that time. During this early period there was speculation that the NKCC, the NCC and K^+-Cl^- cotransport might simply represent different modes of a single functional entity. When it became clear that they were not the same entity operating in different modes, and well before the proteins responsible for the ion movements were characterized at the molecular level, it was attractive to imagine that the Na^+-K^+-Cl^- cotransporter was comprised of subunits of the Na^+-Cl^- and the K^+-Cl^- cotransporters. Unfortunately, this idea, which fits Occam's razor so well, does not fit the actual facts in the matter.

Returning to the idea of Cl^- being distributed out of electrochemical equilibrium, we know that the net electrochemical driving force on Na^+ is directed inwardly across the cell membrane. So, it follows that both the NKCCs and the NCC will move Cl^- into the cell interior resulting in a higher-than-equilibrium intracellular concentration of Cl^-. We have known for nearly 20 years that these cotransporters (especially the NKCC1; see below) are responsible for many cells having higher than passive $[Cl^-]_i$ (reviewed in Russell, 2000). Thus, it is disappointing that many textbooks of physiology and cell biology still contain tables purporting to show the "normal" cell ionic composition that list intracellular Cl^- at or near equilibrium concentrations (Alberts et al., 2003; Silverthorne, 2004; but see Boron and Boulpaep, 2003).

II. LOCALIZATION AND REGULATION

A. The Na^+-Cl^- Cotransporter

The best known location of this Na^+-coupled cotransporter is the apical membrane of the distal convoluted tubule of the renal nephron (Gamba, 2005). However, there are recent reports that it may be found in other epithelia or epithelia-like tissues. Osteoblasts have recently been reported to express NCC protein (Dvorak et al., 2007). In addition, there are reports that NCC message and protein have been identified in the endolymphatic sac of the inner ear (Akiyama et al., 2008) and in human lens epithelial cells (Lauf, 2008). Nevertheless, at present the best physiological, biochemical and genetic evidence limits it to the distal convoluted tubule.

The NCC and the two NKCC isoforms share the motif of 12 membrane-spanning segments (as judged by Kyte-Doolittle hydropathy analysis; see Chapters 11 and 16). As with the NKCCs, the amino and carboxyl terminals of the NCC are believed to be intracellular, as evidenced by the need to permeabilize cells for antibodies to access N- and C-terminal epitopes. All three proteins are glycosylated on an extracelluar loop between transmembrane segments 7 and 8. Recently, it has been shown that the elimination of the two glycosylation sites on the NCC resulted in a reduction of membrane expression of the NCC, an increase in its affinity for the inhibitor metolazone and a small increase in the affinity for extracellular Cl^- (Hoover et al., 2003).

B. The Na^+-K^+-$2Cl^-$ Cotransporters

Two isoforms of the NKCC have been identified. One, the NKCC1 (SLC12A2), is widely distributed, being found in the plasmalemma of most animal cell types (e.g. Randall et al., 1997; Russell, 2000; Gamba, 2005; Pedersen et al., 2008). The other, NKCC2 (SLC12A1), has thus far been found only in the thick ascending limb of the mammalian nephron (see Chapter 16). As mentioned above, both isoforms are believed to have 12 transmembrane-spanning segments. Both have prominent intracellular carboxyl and amino tails within each of which are several protein kinase-consensus binding sites. More about these potential phosphorylation sites below. As mentioned previously both NKCC isoforms have a prominent extracellular loop between transmembrane segments 7 and 8 which contain two putative sites for N-glycosylation. Recently, it was reported that mutation of the two glycosylation sites on the NKCC2 protein resulted in a significant reduction of ion transport activity and a reduction in the surface expression of the cotransporter (Paredes et al., 2006). The NKCC1 is the larger protein of the two, being ~100 amino acid residues or ~10 kDa larger. The two NKCC isoforms share ~60% amino acid sequence identity with each other and ~50% identity with the NCC.

NKCC2 has three alternatively spliced variants that have rather specific regional localizations within the kidney nephron (Payne and Forbush, 1994). Early physiological results (e.g. Rocha and Kokko, 1973; Burg, 1982; Reeves et al., 1988) strongly suggested that different regions of the thick ascending limb of the nephron had different transport properties with respect to the diuretic-sensitive uptake of Na^+ and Cl^- in this region. In 2002, two groups independently showed that these functional differences are likely to be the result of the different kinetic properties of the alternatively spliced variants of the NKCC2 (Giménez et al., 2002; Plata et al., 2002). Of the three renal-specific NKCC2 variants, isoform B has the higher affinity for Na^+, Cl^- and bumetanide, and is located in the distal region of the thick ascending limb. Since fluid in that region of the tubule has the lowest concentration of these ions as a result of their being removed by the other NKCC variants in the more proximal regions, locating the variant with the highest ion affinity in the more distal region makes good engineering sense as it permits the mechanism to continue to "dilute" the tubular fluid by removal of additional Na^+ and Cl^-.

Much of what we know of the NKCCs and NCC comes from work on epithelial tissues. Epithelial tissues provide rich sources of the transporters and the function of the NCC and the NKCC (all isoforms) to permit epithelia to absorb or secrete NaCl-rich fluids is clearly quite important. However, this emphasis undoubtedly limits our appreciation for other functions of the NKCC. After all, the NKCC1 is found in nearly every tissue examined, including many that are not epithelial in nature, a fact that strongly suggests it provides a function (or functions) not limited to trans-epithelial transport.

III. FUNCTIONS OF THE Na^+-COUPLED COTRANSPORTERS

At the most fundamental level, both the NKCCs and the NCC act to move Na^+ and Cl^- across the plasmalemma into cells. As noted already, they act to transport Cl^- up its electrochemical gradient by harnessing the potential energy of the Na^+ chemical gradient. In epithelial tissues, both transporters are members of multicomponent solute transport processes that play critical roles in the function of the particular epithelial tissue. The critical nature of these epithelial Na^+-coupled Cl^- cotransporters to the organism is underlined by the fact that several diseases are associated with their genetically linked malfunction. Thus, Bartter's syndrome is a renal disease characterized by mutations of the NKCC2 isoform (among several ion transporter-encoding genes) whereas Gitelman's syndrome is characterized by mutations of the NCC gene that result in a loss of function of that cotransporter. In addition, it has become increasingly obvious that one NKCC isoform (NKCC1) is critically important for neuronal development and function, as discussed below and in several chapters of the present volume.

A. The NCC Functions to Move Na^+ across Epithelia

Although the NCC was one of the first of this family to undergo functional characterization (Renfro, 1975; Stokes et al., 1984), detailed functional studies on this transporter have proven difficult owing to the often complicated anatomy of its locations. In fact, much of the original basic functional characterizations were made using the teleost urinary bladder (Renfro, 1977; Stokes et al., 1984) as a surrogate for the distal mammalian nephron. Thus, its best-understood function is in the distal tubule of the renal nephron where it participates in the fine control of total body Na^+ content. It is this function that serves as the target of the thiazide diuretics. Further, Gitelman's syndrome is caused by a relatively rare loss-of-function mutation of the NCC. Clinically, this syndrome presents as a metabolic acidosis with low blood potassium (hypokalemia) and low blood chloride (hypochloremia). It is also characterized by low urinary calcium and low blood magnesium and a high sodium-content diuresis. For a detailed description of NCC function, see Chapters 16 and 18 in this volume.

B. The NKCC has Multiple Functions

Both the NKCC1 and NKCC2 act in concert with other transport mechanisms to effect trans-epithelial water and salt transport. In addition, there have been three other functional roles suggested for the NKCC1. They are: (1) to maintain $[Cl^-]_i$ at levels above electrochemical equilibrium; (2) to increase cell volume in response to cell shrinkage; (3) a role in the cell division. Each will be discussed below.

C. The NKCC Functions to Maintain $[Cl^-]_i$ above Electrochemical Equilibrium

1. *Epithelial Tissue*

The ability of the NCC and the NKCCs to increase $[Cl^-]_i$ is a key factor in the their roles for trans-epithelial salt and water transport; the Na^+-coupled cotransporters act to move Cl^- across one membrane into the polarized epithelial cell thereby increasing $[Cl^-]_i$. The increased $[Cl^-]_i$ provides the electrochemical driving force for the movement of Cl^- across the membrane on the other side of the epithelial cell. This movement of Cl^- through specialized channels then sets up an electrical potential that supports the movement of a counter-cation. The overall process thereby results in the trans-epithelial movement of these ions coupled often to the osmotic movement of water.

2. *Neuronal Function, Intracellular [Cl⁻] and the NKCC*

The role of the NKCC1 in nervous tissue is currently a very active and exciting field of research. What follows is a brief overview of the background and current status. This volume contains several other chapters that address various aspects of these issues. The squid giant axon, the prototypical neuronal model, was the first animal cell for which $[Cl^-]_i$ was shown to be higher than expected from electrochemical equilibrium considerations. In 1963, Richard Keynes confirmed that axoplasmic $[Cl^-]$ is 130–150 mM. Given the extracellular $[Cl^-]$ of ~480 mM (squids live in the ocean), and a resting membrane potential of ~− 60 to −70 mV, $[Cl^-]_i$ would have to be about 35 mM in order to be in electrochemical equilibrium (Keynes, 1963). It was clear that $[Cl^-]_i$ in the squid axon was nearly four times higher than can be explained by its thermodynamically passive distribution.

In the squid giant axon a series of reports have demonstrated that Cl^- influx is bumetanide sensitive, ATP dependent and dependent upon extracellular Na^+ and K^+ (Russell, 1979, 1983; Altamirano and Russell, 1987; Breitwieser et al., 1996). That is, it is effected by an NKCC mechanism. In addition, squid axon was the first preparation in which the inhibitory effect of intracellular Cl^- on NKCC function was demonstrated (Russell, 1979; Breitwieser et al., 1990). Figure 2.1 shows that raising the intracellular $[Cl^-]$ not only inhibits bumetanide-sensitive ^{36}Cl influx, but also inhibits its efflux. The fact that intracellular Cl^- inhibited both unidirectional fluxes showed that its effect was not merely kinetic in nature, e.g. end-product inhibition, but was an effect on the transport mechanism itself. We now know that high intracellular $[Cl^-]$ reduces phosphorylation of the NKCC protein at key sites (see below for further discussion). It is very interesting to note that the NCC has recently also been shown to undergo phosphorylation and transport activation upon reducing the $[Cl^-]_i$ (Pacheco-Alvarez et al., 2006), yet another operational similarity between these two cotransporter mechanisms.

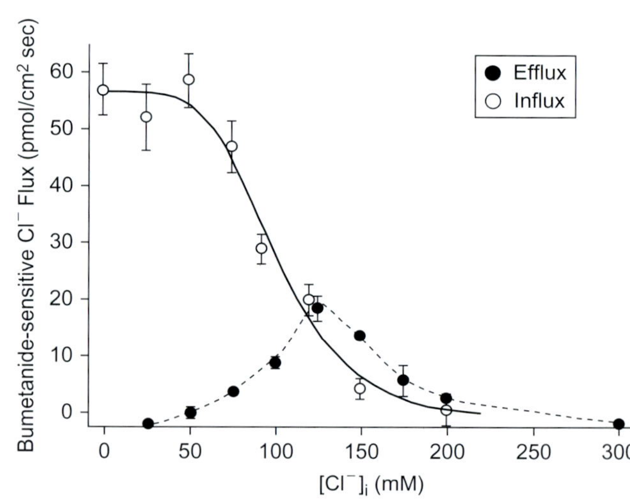

FIGURE 2.1 Bumetanide-sensitive unidirectional ^{36}Cl influx and efflux are inversely dependent upon the $[Cl^-]_i$ using the internally dialyzed squid giant axon. These results were the first to demonstrate that inhibition of the NKCC by increased $[Cl^-]_i$ could not be simple kinetic "end-product inhibition". (From Russell, 2000 with permission.)

Considerable direct and indirect evidence has since been reported to satisfy us that an NKCC mechanism, much like that studied in the squid axon, is also responsible for the above-equilibrium $[Cl^-]_i$ values reported for many different neurons. In amphibian and mammalian nervous tissue, the NKCC1 isoform has been identified as the likely mediator of the elevated $[Cl^-]_i$ values (Alvarez-Leefmans et al., 1988, 2001; Plotkin et al., 1997; Sung et al., 2000; Kanaka et al., 2001; Rocha-González et al., 2008).

More recently, the fact that the $[Cl^-]_i$ in certain CNS neurons is higher than the electrochemical equilibrium point has been shown to have important functional and developmental consequences. Intracellular Cl^- concentration is an important determinant of the transmembrane electrical potential across the neuronal (as well as other cell types) plasmalemma. Because many neurons have imbedded in their plasma membranes ligand-gated Cl^- channels that respond to neurotransmitters (gamma-amino butyric acid and/or glycine), the ratio $[Cl^-]_o/[Cl^-]_i$ will determine whether the synaptic response to the neurotransmitter will be depolarizing or hyperpolarizing. When $[Cl^-]_i$ is higher than equilibrium, the response will be depolarizing and often excitatory. Conversely, when $[Cl^-]_i$ is at or below equilibrium, the response will be either no change in transmembrane voltage or a hyperpolarization, i.e. an inhibitory post-synaptic potential leading to synaptic inhibition.

As will be discussed in much more detail in Chapter 14 of this volume, we now know that the neurotransmitter gamma-amino butyric acid (GABA) opens $GABA_A$-mediated Cl^- channels. During embryonic development of central neurons, the mechanisms that determine their $[Cl^-]_i$ undergo radical change. Early in embryonic and postnatal development, $[Cl^-]_i$ is above electrochemical equilibrium (e.g. Berglund et al., 2006). During this period GABA elicits excitatory post-synaptic potentials. Later in development, and throughout adulthood, these same cells respond to GABA with inhibitory post-synaptic potentials. This remarkable change in the steady-state $[Cl^-]_i$ is now known to be the result of a reversal in the relative expression levels and function of NKCC1 and KCC2 proteins (Clayton et al., 1998; Rivera et al., 1999; Zhu et al., 2005). This fascinating developmental pattern is covered in detail in Chapters 17 and 19 of this volume and will not be further discussed here.

Primary afferent neurons are notable exceptions to this developmental metamorphosis that occurs in the brain and spinal cord. Alvarez-Leefmans' group has been studying these neurons for some time and has clearly demonstrated that they always maintain their $[Cl^-]_i$ at above-equilibrium levels, meaning that GABA will always elicit a depolarizing response (Alvarez-Leefmans et al., 1988; Rocha-González et al., 2008). As these neurons are involved with gating the flow of nociceptive and other sensory information, it is not surprising that the NKCC1 has been implicated in pain perception. For instance, blockers of the NKCC such as bumetanide and furosemide have been reported to increase the threshold for perception of pain in rats by mechanisms that are both central and peripheral (Granados-Soto et al., 2005). Further, there is evidence to support the view that hypersensitivity to pain might result, among other factors, from increased $[Cl^-]_i$ in spinal dorsal horn nociceptive neurons; peripheral inflammation is correlated with a reduction in the expression of spinal cord K^+-Cl^- cotransporters without any effect on the expression levels of NKCC1 protein (Zhang et al., 2008), a combination expected to yield increased $[Cl^-]_i$. This topic will be covered in detail in Chapters 22 and 23 of this volume.

D. The NKCC and Cell Volume Regulation

The NKCC has long been implicated in cell volume regulation in a wide variety of animal cells (e.g. Russell, 2000; Friedrich et al., 2006). This property, although widely accepted, is still incompletely understood. It has been repeatedly demonstrated that cell shrinkage stimulates NKCC ion transport activity, and is associated with increased levels of phosphorylation of the NKCC protein (Haas et al., 1995; Lytle, 1997). In this sense, it appears that cell volume activation of the NKCC is like other modes of NKCC stimulation. For example, lowering $[Cl^-]_i$ results in increased phosphorylation of the cotransporter (see below). This presents a bit of a conundrum because hypertonic cell shrinkage would be expected to *increase* $[Cl^-]_i$ as the result of a reduced intracellular water content, and increased $[Cl^-]_i$ is well known to inhibit NKCC ion transport presumably by reducing the phosphorylation of the NKCC protein (e.g. Haas et al., 1995). Yet, the effect of cell shrinkage is to (in most cases) activate the cotransporter and increase NKCC protein phosphorylation (e.g. Lytle, 1997). Part of the answer to this apparent conundrum is seen in Fig. 2.2, which shows that operationally cell shrinkage reduces the inhibitory effect of the increased $[Cl^-]_i$. Thus, as the cell is shrunken, it takes a higher $[Cl^-]_i$ to cause the same degree of inhibition. Given the current working model for the regulation of the NKCC activity (see below for further discussion), this might be the result of reducing the phosphatase activation by increased $[Cl^-]_i$ or of offsetting stimulation of the SPAK (also known as PASK or STK39) kinase by the cell shrinkage. It is important

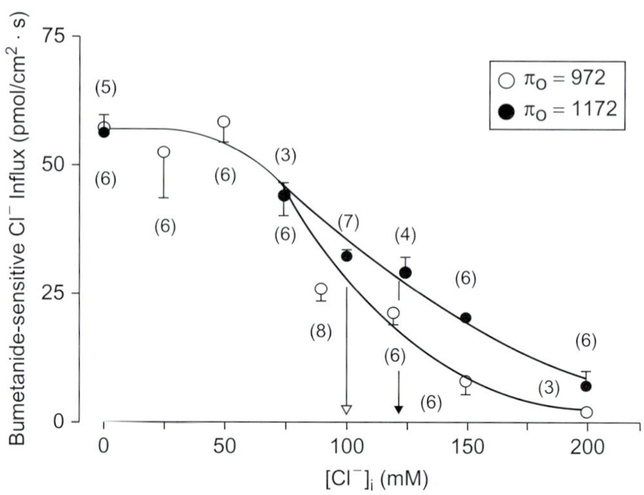

FIGURE 2.2 The ability of increased $[Cl^-]_i$ to inhibit NKCC-mediated ^{36}Cl influx is reduced by cell shrinkage. The $[Cl^-]_i$ of squid giant axons was set at various levels by means of intracellular dialysis. Increasing $[Cl^-]_i$ led to reduced influx in axons exposed to either normotonic fluid (for squid = 972 mOsm/kg) or hypertonic fluid (1172 mOsm/kg). However, the inhibition by any given level of $[Cl^-]_i$ was less when cells were exposed to hypertonic fluid. (From Breitwieser et al., 1990, with permission.)

to emphasize that the $[Cl^-]_i$ set-point is not the result of a thermodynamic equilibrium of the cotransporter, i.e. where the inwardly directed thermodynamic driving force is zero (Russell, 2000; Rocha-Gonzalez et al., 2008), but of a kinetic "brake" as will be discussed in more detail in Chapter 5. The important point in this discussion is that this effect of increased $[Cl^-]_i$ can be overcome by reducing cell volume. However, the effects of reducing $[Cl^-]_i$ and cell shrinkage are not additive as shown in squid axon (see Fig. 2.2). That is, the highest flux rate observed under conditions of 0 mM $[Cl^-]_i$ are not further increased by cell shrinkage presumably because at 0 mM $[Cl^-]_i$ the NKCC protein is maximally phosphorylated, i.e. the transport mechanism is saturated.

Another unsolved issue related to the cell-volume regulating function is the nature of the signal that leads to activation of the NKCC, i.e. what is the nature of the volume sensor? Some evidence exists that implicates cytoskeletal elements (e.g. Jessen and Hoffman, 1992; Mathews et al., 1998). Further, evidence has been presented showing that the serine–threonine protein kinase oxidative stress-responsive 1 (OSR1) phosphorylates and activates the NKCC (and other members of the SLC12A family) in response to cell shrinkage. This kinase is in turn activated by one of the "with-no-lysine" (WNK) protein kinases (Anselmo et al., 2006). These observations still leave unanswered the question of exactly where and how does a cell monitor its own volume.

E. The NKCC and Cell Division

An exciting new area of interest for ion transport aficionados has been the strong possibility that at least some ion transport mechanisms, including the NKCC1, play important roles in the normal (and pathological) conduct of the cell cycle and cell division.

Cell division requires an increase of cell volume even prior to cytokinesis. There are reports of a 40–50% increase in cell volume during late G_1 phase suggesting that mammalian cells, like yeast, may have a size checkpoint prior to entering the S-phase (e.g. Dolznig et al., 2004). Given the ability of the NKCC to effect cell volume increase (discussed above), it is logical to postulate that the NKCC may play a role in cell volume increases that occur during cell division (e.g. Lang, 2007) either alone or in parallel with other solute transporters. Interestingly, a recent report suggested that following a cell volume increase during the G_1/S transition, there is a volume decrease leading to condensation of the cytoplasm and nucleus that occurs during the M-phase of the cell cycle resulting from Cl^- loss through Cl^- channels in a regulatory-volume-decrease-like mechanism (Habela and Sontheimer, 2004). Such a mechanism implies that $[Cl^-]_i$ must be higher than predicted from thermodynamic equilibrium considerations which is exactly the state that can be achieved through the activation of the NKCC. Further circumstantial evidence linking the NKCC1 to cell division is the simple fact of its wide distribution across nearly all cell types. This suggests it may serve a fundamental function. Cell division is certainly one of the most fundamental functions carried out by cells and considerable evidence has accrued over the past 10–15 years that inorganic ions and their transmembrane movements play important permissive roles in the normal progression of the cell cycle and subsequent cell division (e.g. Lang, 2007).

The experimental evidence for such a role for the NKCC1 is of three types. First is the often-repeated observation that cells treated with mitogens exhibit increased ion transporter activity of the NKCC (discussed in Russell, 2000; Yang et al., 2001) among several ion transporters and ion channels. The second type of evidence is that, at least in some cells, treatment with bumetanide or furosemide slows down cell division (e.g. discussed in Russell, 2000; Lee et al., 2003; Iwamoto et al., 2004; Shiozaki et al., 2006). More recently, evidence has been presented to indicate this effect of NKCC inhibitors is brought about by prolonging the proliferating cells' time in the G_1 phase of the cell cycle (e.g. Panet et al., 2000; Iwamoto et al., 2004). The third kind of evidence has come from observations of the connection between carcinogenic-like cell proliferation behavior and the activity of the

NKCC. For example, Panet et al. (2000) reported that overexpression of NKCC1 protein into mouse fibroblasts resulted in a phenotype with certain characteristics of cancer cells, including the ability to proliferate under conditions of very low serum concentrations, loss of contact inhibition, and the development of the ability to form cell colonies growing in soft agar. This latter "clonogenic" activity was prevented by treatment with bumetanide.

Thus, a body of evidence exists to indicate a potential role for the NKCC1 in cell division. In sum, this evidence is supportive, but not yet definitive. The mechanism of the effect is still unknown. However, it seems likely that the NKCC1 might be part of a mechanism with built-in redundancy since not all cells tested showed effects on their cell cycles by bumetanide treatment. Whether the NKCC role is as conceptually simple as participating to achieve a necessary cell size increase prior to progressing into the S-phase or is more complex, as suggested by reports of effects of furosemide on the phosphorylation of MAPK kinase (Panet et al., 2006), only further studies will tell. Finally, as will be discussed in a later chapter, it is possible that some virally mediated effects on host cells may be dependent upon inactivating the cell cycle (see Chapter 27).

IV. NKCC REGULATION

This topic will be covered in much more detail in Chapter 18 of this volume. What follows is a historical overview with strong emphasis on the NKCC1 isoform, since study of this isoform has led our understanding of the regulation of several other members of the SLC12 gene family.

As noted above, among the earliest demonstrations of "regulation" of the NKCC were the effects of lowering the $[Cl^-]_i$ of the cell or shrinking the cell. Both treatments increase ion transport activity by the NKCC1. The first step towards understanding the mechanisms of these effects was the observation that although the ion movements fit the criteria for an Na^+-coupled cotransporter, ATP was an absolute requirement for transport activity (Russell, 1979, 1983). The internally dialyzed squid giant axon allowed us to remove intracellular ATP while maintaining intracellular ion concentrations steady. In the absence of ATP, bumetanide-sensitive ^{36}Cl influx was non-existent. Lytle and Forbush (1992) demonstrated that the biochemical basis of this functional observation was the phosphorylation of the NKCC1 protein at threonine and serine sites. The first demonstration that NKCC protein phosphorylation was Cl^- dependent was by Haas et al. (1995). As many as 20 consensus phosphorylation sites were identified along the N- and C-termini of the NKCC1 protein for protein kinase A, protein kinase C and casein kinase II (Delpire et al., 1994; Xu et al., 1994). Although some evidence was presented for stimulation via PKA and PKC (e.g. Russell, 2000) and even for myosin light-chain kinase (Klein and O'Neill, 1995) mechanisms, each tended to be tissue specific. Lytle (1998), following a suggestion by Jennings and Al-Rohil (1990), was the first to explicitly postulate that there must be a final common volume-sensitive protein kinase that actually phosphorylated and activated the NKCC protein. This notion was later expanded to include the ideas that (1) the "common kinase" must also be sensitive to changes in $[Cl^-]_i$ and (2) that it reciprocally controlled NKCC and K^+-Cl^- cotransport functions. However, this common pathway or kinase remained elusive for several years.

This is where things stood when in the early part of the present decade, reports from two labs (Piechotta et al., 2002, 2003; Dowd and Forbush, 2003; Gagnon et al., 2005) revealed a new set of players in, as it turns out, the regulation of most of the major members of the SLC12 gene family including NKCC1, NKCC2, NCC and certain of the K^+-Cl^- cotransporters. The evidence to date supports the following general model which is summarized in Fig. 2.3. A protein kinase known as SPAK and a related protein kinase known as OSR1 (oxidative stress response protein 1) were first identified by Delpire's group (Piechotta et al., 2002, 2003) as interacting with the amino terminals of NKCC and of the K^+-Cl^- cotransporter via a specialized binding motif. Dowd and Forbush (2003) reported that transfecting cells with an inactive SPAK (they called PASK) resulted in the loss of intracellular Cl^- sensitivity of the NKCC1 and that the NKCC1 protein from cells transfected with the defective SPAK (PASK) gene exhibited much less phosphorylation. Interestingly, heterologous expression of SPAK (PASK) in *Xenopus* oocytes that co-expressed NKCC1 did not result in enhanced activity by the NKCC1 (Gagnon et al., 2006). This suggested there might be another player in the regulatory mix. These authors then showed that co-expressing SPAK along with WNK4 caused a large increase of NKCC1-mediated ^{86}Rb uptake and, perhaps most interestingly, this flux was no longer volume sensitive. Co-expression of WNK4 alone with NKCC1 had no such effect. Significantly, co-expressing these two kinases along with the K^+-Cl^- cotransporter, isoform 2, resulted in the opposite functional effect, a result in good agreement with the reciprocal effects of cell shrinkage on the native functions of these two members of the SLC12A family of cation-chloride cotransporters.

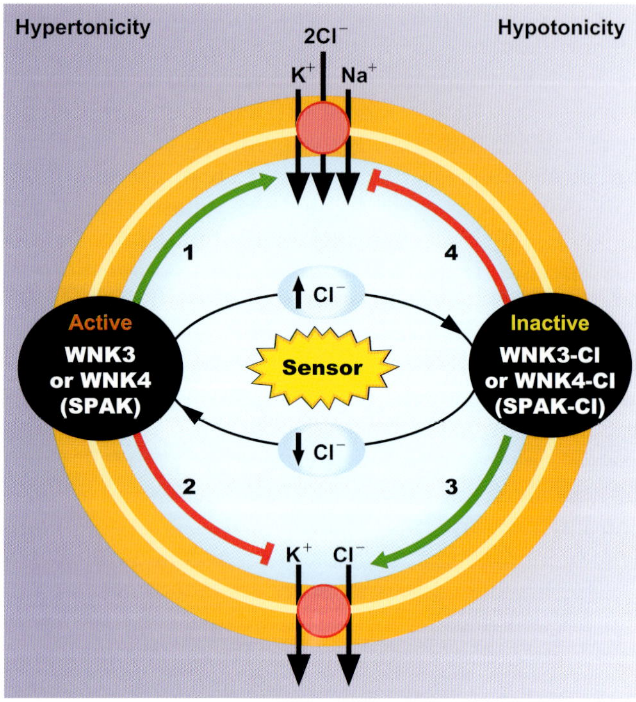

FIGURE 2.3 Intracellular Cl^- modulates Na^+-K^+-$2Cl^-$ and K^+-Cl^- cotransporter activities through serine–threonine kinases SPAK and WNK. A decrease in $[Cl^-]_i$ activates the kinases leading to NKCC activation (green arrow 1) and KCC inhibition (red inhibitory line 2). Upon increase in $[Cl^-]_i$, kinases are deactivated, leading to cotransporters' dephosphorylation by phosphatases (not shown). This results in inhibition of NKCC (red inhibitory line 4), and activation of KCC (green arrow 3). NKCC is activated by hypertonicity through SPAK and WNK kinases, whereas KCC is activated by hypotonicity through dephosphorylation by phosphatases (not shown). (Modified from Kahle et al., 2006.)

Figure 2.3 summarizes the current working model of the regulation of the Na^+-dependent and -independent chloride ion transporters by changes in cell volume and $[Cl^-]_i$. The final common kinase is believed to be SPAK and the WNK kinases either directly or indirectly "sense" cell volume or $[Cl^-]_i$ and then act to phosphorylate either the SPAK (PASK) or a phosphatase. Thus, the binding sites on the cotransporters and the protein kinases provide the appropriate scaffolding to effect the appropriate physiological effect on the various members of the SLC12A family of ion transporters.

References

Akiyama, K., Miyashita, T., Mori, T., Inamoto, R., and Mori, N. (2008). Expression of thiazide-sensitive Na^+-Cl^- cotransporter in the rat endolymphatic sac. *Biochim. Biophys. Res. Comm.* **371**, 649–653.

Alberts, B., Bray, D., Hopkin, K., Johnson, A., Lewis, J., Raff, M., Roberts, K., and Walter, P. (2003). Chap. 12, Table 12-1. Membrane transport. In *Essential Cell Biology*, 2nd edition, p. 390. Garland Science, NY.

Altamirano, A.A. and Russell, J.M. (1987). Coupled Na/K/Cl efflux: "reveres" unidirectional fluxes in squid giant axons. *J. Gen. Physiol.* **89**, 669–686.

Alvarez-Leefmans, F.J., Gamiño, S.M., Giraldez, F., and Noguerón, I. (1988). Intracellular chloride regulation in amphibian dorsal root ganglion neurons studied with ion-selective microelectrodes. *J. Physiol. (Lond.)* **406**, 225–246.

Alvarez-Leefmans, F.J., León-Olea, M., Mendoza-Sotelo, J., Alvarez, F.J., Antón, B., and Garduño, R. (2001). Immunolocalization of the Na^+-K^+-$2Cl^-$ cotransporter in peripheral nervous tissue of vertebrates. *Neuroscience* **104**, 569–582.

Anselmo, A.N., Earnest, S., Chen, W., Juang, Y.C., Kim, S.C., Zhao, Y., and Cobb, M.H. (2006). WNK1 and OSR1 regulate the Na^+, K^+, $2Cl^-$ cotransporter in HeLa cells. *Proc. Natl. Acad. Sci. USA* **103**, 10883–10888.

Berglund, K., Schleich, W., Krieger, P., Loo, L.S., Wang, D., Cant, N.B., Feng, G., Augustine, G.J., and Kuner, T. (2006). Imaging synaptic inhibition in transgenic mice expressing the chloride indicator Clomeleon. *Brain Cell Biol.* **35**, 207–228.

Boron, W.F. and Boulpaep, E.L. (2003). Chap 3, Table 3-2. Physiology of membranes. In *Medical Physiology*, p. 52. Saunders, Philadelphia.

Breitwieser, G.E., Altamirano, A.A., and Russell, J.M. (1990). Osmotic stimulation of Na^+-K^+-Cl^- cotransport in squid giant axons is $[Cl^-]_i$-dependent. *Am. J. Physiol.* **258**, C749–C753 (Cell. Physiol. **27**).

Breitweiser, G.E., Altamirano, A.A., and Russell, J.M. (1996). Elevated $[Na^+]_i$ and $[Cl^-]_i$ inhibit Na^+, K^+,Cl^- cotransport by different mechanisms in squid giant axons. *J. Gen. Physiol.* **107**, 261–270.

Burg, M.B. (1982). Thick ascending limb of Henle's loop. *Kidney Int.* **22**, 454–464.

Clayton, G.H., Owens, G.C., Wolff, J.S., and Smith, R.L. (1998). Ontogeny of cation-Cl- cotransporter expression in rat neocortex. *Brain Res. Dev. Brain Res.* **8**, 281–292.

Delpire, E., Rauchman, M.I., Beir, D.R., Hebert, S.C., and Gullans, S.R. (1994). Molecular cloning and chromosome localization of a putative basolateral Na^+-K^+-$2Cl^-$ cotransporter from mouse inner medullary collecting duct (mIMCD-3) cells. *J. Biol. Chem.* **269**, 25677–25683.

Dolznig, H., Grebien, F., Sauer, T., Beugl, H., and Müllner, E.W. (2004). Evidence for a size-sensing mechanism in animal cells. *Nat. Cell Biol.* **6**, 899–905.

Dowd, B.F. and Forbush, B. (2003). PASK (proline-alanine-rich STE20-related kinase), a regulatory kinase of the Na-K-Cl cotransporter (NKCC1). *J. Biol. Chem.* **278**, 27347–27353.

Dvorak, M.M., De Joussineau, C., Carter, D.H., Pisitkun, T., Knepper, M.A., Gamba, G., Kemp, P.J., and Riccardi, D. (2007). Thiazide diuretics directly induce osteoblast differentiation and mineralized nodule formation by interacting with a sodium chloride co-transporter in bone. *J. Am. Soc. Neprol.* **18**, 2509–2516.

Friedrich, B., Matskevich, I., and Lang, F. (2006). Cell volume regulatory mechanisms. *Contrib. Nephrol.* **152**, 1–8.

Gagnon, K.B.E., England, R., and Delpire, E. (2006). Volume sensitivity of cation-Cl^- cotransporters is modulated by the interaction of two kinases: Ste20-related proline-alanine rich kinase and WNK4. *Am. J. Physiol. Cell Physiol.* **290**, C134–C142.

Gamba, G. (2005). Molecular physiology and pathophysiology of electroneutral cation-chloride cotransporters. *Physiol. Rev.* **85**, 423–493.

Geck, P., Pietrzyk, C., Burckhardt, B.-C., Pfeiffer, B., and Heinz, E. (1980). Electrically silent cotransport of Na^+, K^+ and Cl^- in Ehrlich cells. *Biochim. Biophys. Acta* **600**, 432–447.

Giménez, I., Isenring, P., and Forbush, B. (2002). Spatially distributed alternative splice variants of the renal Na-K-Cl cotransporter exhibit dramatically different affinities for the transported ions. *J. Biol. Chem.* **277**, 8767–8770.

Granados-Soto, V., Arguelles, C.F., and Alvarez-Leefmans, F.J. (2005). Peripheral and central antinocioceptive action of Na^+-K^+-$2Cl^-$ cotransporter blockers on formalin-induced nociception in rats. *Pain* **114**, 231–238.

Haas, M., McBrayer, D., and Lytle, C. (1995). $[Cl^-]_i$-dependent phosphorylation of the Na-K-Cl cotransport protein of dog tracheal epithelial cells. *J. Biol. Chem.* **270**, 28955–28961.

Habela, C.W. and Sontheimer, H. (2004). Cytoplasmic volume condensation is an integral part of mitosis. *Nat. Cell Biol.* **6**, 899–905.

Hebert, S.C., Mount, D.B., and Gamba, G. (2003). Molecular physiology of cation-coupled Cl^- cotransport: the SLC12 family. *Pflugers Arch.* **447**, 580–593.

Hoffman, J.F., and Kregenow, F.M. (1966). The characterization of new energy dependent cation transport processes in red blood cells. *Ann. N. Y. Acad. Sci.* **137**, 566–576.

Hoover, R.S., Poch, E., Monroy, A., Vasquez, N., Nishio, T., Gamba, G., and Hebert, S.C. (2003). N-Glycosylation at two sites critically alters thiazide binding and activity of the rat thiazide-sensitive Na^+:Cl^- cotransporter. *J. Am. Soc. Nephrol.* **14**, 271–282.

Iwamoto, L.M., Fujiwara, N., Nakamura, K.T., and Wada, R.K. (2004). Na-K-2Cl cotransporter inhibition impairs human lung cellular proliferation. *Am. J. Physiol. Lung Cell Mol. Physiol.* **287**, L510–L514.

Jennings, M.L. and Al-Rohil, N. (1990). Kinetics of activation and inactivation of swelling-stimulated K^+/Cl^- transport. *J. Gen. Physiol.* **95**, 1021–1040.

Jessen, F. and Hoffman, E.K. (1992). Activation of the $Na^+/K^+/Cl^-$ cotransport system by reorganization of the actin filaments in Ehrlich ascites tumor cells. *Biochim. Biophys. Acta* **1110**, 199–201.

Kahle, K.T., Rinehart, J., Ring, A., Giménez, I., Gamba, G., Hebert, S.C., and Lifton, R.P. (2006). WNK protein kinases modulate cellular Cl^- flux by altering the phosphorylation state of the Na-K-Cl and K-Cl cotransporters. *Physiology* **21**, 326–335.

Kanaka, C., Ohno, K., Okabe, A., Kuriyama, K., Itoh, T., Fukuda, A., and Sato, K. (2001). The differential expression patterns of messenger RNAs encoding K-Cl cotransporters (KCC1,2) and Na,K,2Cl cotransporter (NKCC1) in the rat nervous system. *Neuroscience* **104**, 933–946.

Keynes, R.D. (1963). Chloride in the squid giant axon. *J. Physiol. (Lond.)* **169**, 690–705.

Klein, J.D. and O'Neill, W.C. (1995). Volume-sensitive myosin phosphorylation in vascular endothelial cells: correlation with Na-K-2Cl cotransport. *Am. J. Physiol.* **269**, C1524–C1531.

Kuhn, T.J. (1962). *The Structure of Scientific Revolutions*. University of Chicago Press.

Lang, F. (2007). Mechanisms and significance of cell volume regulation. *J. Am. Coll. Nutr.* **26**, 613S–623S.

Lauf, P.K., Chimote, A.A., and Adragna, N.C. (2008). Lithium fluxes indicate presence of Na-Cl cotransport (NCC) in human lens epithelial cells. *Cell Physiol. Biochem.* **21**, 335–346.

Lee, H.A., Jeong, H., Kim, E.Y., Nam, M.Y., Yoo, Y.J., Seo, J.T., Shin, D.M., Ohk, S.H., and Lee, S.I. (2003). Bumetanide, the specific inhibitor of Na^+-K^+-$2Cl^-$ cotransport, inhibits 1alpha,25-dihydroxyvitamin D3-induced osteoclastogenesis in a mouse co-culture system. *Exp. Physiol.* **88**, 569–574.

Lytle, C. and Forbush, B., III (1992). The Na-K-Cl cotransport protein of shark rectal gland. II. Regulation by direct phosphorylation. *J. Biol. Chem.* **267**, 25438–25443.

Lytle, C. (1997). Activation of the avian erythrocyte Na-K-Cl cotransport protein by cell shrinkage, cAMP, fluoride, and calyculin-A involves phosphorylation at common sites. *J. Biol Chem.* **272**, 15069–15077.

Lytle, C. (1998). A volume-sensitive protein kinase regulates the Na-K-2Cl cotransporter in duck red blood cells. *Am. J. Physiol.* **274**, C1002–C1010.

Matthews, J.B., Smith, J.A., Mun, E.C., and Sicklick, J.K. (1998). Osmotic regulation of the Na^+-K^+-Cl^- cotransport: role of Cl^- and actin. *Am. J. Physiol. Cell Phuysiol.* **274**, C697–C706.

Pacheco-Alvarez, D., San Cristóbal, P., Meade, P., Moreno, E., Vazquez, N., Muñoz, E., Díaz, A., Juárez, M.E., Giménez, I., and Gamba, G. (2006). The Na^+:Cl^- cotransporter is activated and phosphorylated at the amino-terminal domain upon intracellular chloride depletion. *J. Biol. Chem.* **281**, 28755–28763.

Panet, R., Eliash, M., and Atlan, H. (2006). $Na^+/K^+/Cl^-$ cotransporter activates MAP-kinase cascade downstream to protein kinase C, and upstream to MEK. *J. Cell Physiol.* **206**, 578–585.

Panet, R., Markus, M., and Atlan, H. (2000). Overexpression of the $Na^+/K^+/Cl^-$ cotransporter gene induces cell proliferation and phenotypic transformation in mouse fibroblasts. *J. Cell Physiol.* **182**, 109–118.

Paredes, A., Plata, C., Rivera, M., Moreno, E., Vazquez, N., Muñoz-Clares, R., Hebert, S.C., and Gamba, G. (2005). Activity of the renal Na^+-K^+-$2Cl^-$ cotransporter is reduced by mutagenesis of N-glycosylation sites: role for protein surface charge in Cl^- transport. *Am. J. Physiol. Renal Physiol.* **290**, 1094–1102.

Payne, J.A. and Forbush, B., III (1994). Alternatively spliced isoforms of the putative renal Na-K-Cl cotransporter are differentially distributed within the rabbit kidney. *Proc. Natl. Acad. Sci. USA* **91**, 4544–4548.

Pedersen, M., Carmosino, M., and Forbush, B. (2008). Intramolecular and intermolecular fluorescence resonance energy transfer in fluorescent protein-tagged Na-K-Cl cotransporter (NKCC1). Sensitivity to regulatory conformational change and cell volume. *J. Biol. Chem.* **283**, 2663–2674.

Piechotta, K., Lu, J., and Delpire, E. (2002). Cation chloride cotransporters interact with the stress-related kinases Ste20-related proline-alanine-rich kinase (SPAK) and oxidative stress response 1 (OSR1). *J. Biol. Chem.* **277**, 50812–50819.

Piechotta, K., Garbarini, N., England, R., and Delpire, E. (2003). Characterization of the interaction of the stress kinase SPAK with the Na^+-K^+-$2Cl^-$ cotransporter in the nervous system: evidence for a scaffolding role of the kinase. *J. Biol. Chem.* **278**, 52848–52856.

Plata, C., Meade, P., Vazquez, N., Hebert, S.C., and Gamba, G. (2002). Functional properties of the apical Na^+-K^+-$2Cl^-$ cotransporter isoforms. *J. Biol. Chem.* **277**, 11004–11012.

Plotkin, M.D., Snyder, E.Y., Hebert, S.C., and Delpire, E. (1997). Expression of the Na-K-2Cl cotranporter is developmentally regulated in postnatal rat brains: a possible mechanism underlying GABA's excitatory role in immature brain. *J. Neurobiol.* **33**, 781–795.

Randall, J., Thorne, J., and Delpire, E. (1997). Partial cloning and characterization of Slc12a2: the gene encoding the secretory Na^+-K^+-$2Cl^-$ cotransporter. *Am. J. Physiol. Cell Physiol.* **273**, C1267–C1277.

Reeves, W.B., Moloney, D.A., and Andreoli, T.E. (1988). Diluting power of thick limbs of Henle. III. Modulation of *in vitro* diluting power. *Am. J. Physiol. Renal Fluid Electrolyte Physiol.* **255**, F1145–F1154.

Renfro, J.L. (1977). Interdependence of active Na^+ and Cl^- transport by the isolated urinary bladder of the teleost, *Pseudopleuronectes americanus*. *J. Exp. Zool.* **199**, 383–390.

Rivera, C., Voipio, J., Payne, J.A., Ruusuvuori, E., Lahtinen, H., Lamsa, K., Pirvola, U., Saarma, M., and Kaila, K. (1999). The K^+/Cl^- co-transporter KCC2 renders GABA hyperpolarizing during neuronal maturation. *Nature* **397**, 251–255.

Rocha, A.S. and Kokko, J.P. (1973). Sodium chloride and water transport in the medullary thick ascending limb of Henle.

Evidence for active chloride transport. *J. Clin. Invest.* **52**, 612–623.

Rocha-González, H.I., Mao, S., and Alvarez-Leefmans, F.J. (2008). Na^+, K^+, $2Cl^-$ cotransport and intracellular chloride regulation in rat primary sensory neurons: thermodynamic and kinetic aspects. *J. Neurophysiol.* **100**, 169–184.

Russell, J.M. (1979). Chloride and sodium influx: a coupled uptake mechanism in the squid giant axon. *J. Gen. Physiol.* **73**, 801–818.

Russell, J.M. (1983). Cation-coupled chloride influx in squid axon: role of potassium and stoichiometry of the process. *J. Gen Physiol.* **81**, 909–925.

Russell, J.M. (2000). Sodium-potassium-chloride cotransport. *Physiol. Rev.* **80**, 211–276.

Shiozaki, A., Miyazaki, H., Niisato, N., Nakahari, T., Iwasaki, Y., Itoi, H., Ueda, Y., Yamgishi, H., and Marunaka, Y. (2006). Furosemide, a blocker of $Na^+/K^+/2Cl^-$ cotransporter, diminishes proliferation of poorly differentiated human gastric cancer cells by affecting G_0/G_1 state. *J. Physiol. Sci.* **56**, 401–406.

Silverthorne, D. (2004). Chap 8, Table 8-2. Neurons: Cellular and network properties. In *Human Physiology: An Integrated Approach*, 3rd edition, p. 248. Pearson/Benjamin Cummins, San Francisco.

Stokes, J.B., Lee, I., and D'Amico, M. (1984). Sodium chloride absorption by the urinary bladder of the winter flounder. A thiazide-sensitive, electrically neutral transport system. *J. Clin. Invest.* **74**, 7–16.

Sung, K.W., Kirby, M., McDonald, M.P., Lovinger, D.M., and Delpire, E. (2000). Abnormal $GABA_A$ receptor-mediated currents in dorsal root ganglion neurons isolated from Na-K-2Cl cotransporter null mice. *J. Neurosci.* **20**, 7531–7538.

Xu, J.C., Lytle, C., Zhu, T., Payne, J.A., Benz, E., and Forbush, B., III (1994). Molecular cloning and functional expression of the bumetanide-sensitive Na-K-Cl cotransporter. *Proc. Natl. Acad. Sci. USA* **91**, 2201–2205.

Yang, H., Wang, Z., Miyamoto, Y., and Reinach, P.S. (2001). Cell signaling pathways mediating epidermal growthfactor stimulation of Na:K:2Cl cotransport activity in rat corneal epithelial cells. *J. Membr. Biol.* **183**, 93–101.

Zhang, W., Liu, L.Y., and Xu, T.L. (2008). Reduced potassium-chloride co-transporter expression in spinal cord dorsal horn neurons contributes to inflammatory pain hypersensitivity in rats. *Neuroscience* **152**, 502–510.

Zhu, L., Lovinger, D., and Delpire, E. (2005). Cortical neurons lacking KCC2 expression show impaired regulation of intracellular chloride. *J. Neurophysiol.* **93**, 1557–1568.

CHAPTER

3

Pathophysiology of the K^+-Cl^- Cotransporters: Paths to Discovery and Overview

John S. Gibson, J. Clive Ellory, Norma C. Adragna and Peter K. Lauf

OUTLINE

I. THE EMERGENCE OF K^+-Cl^- COTRANSPORT AS A FUNCTIONAL TRANSPORT ENTITY

Secondary active transport, where the flux of one of the transported species moving downhill provides the driving force for movement of another ion or non-electrolyte, is a common membrane transport mechanism. Ionic species providing the driving force include Na^+, H^+ and K^+. Coupled substrates include different sugars, amino acids and other organic molecules translocated by discrete families of transporters. In the present context, the family of cation-chloride cotransporters (CCCs) represents a series of membrane proteins, the SLC12 superfamily, capable of the electroneutral coupled transport of Cl^- with Na^+ (NCC), K^+ (KCCs) or both Na^+ and K^+ (NKCCs) (Gamba, 2005). Originally, these transporters were recognized and studied in RBCs and epithelia, but more recent work has broadened their known distribution to include neuronal tissue, endothelial and vascular smooth muscle cells, and other cell types, including certain neoplasias. Molecular biology has also revealed a variety of CCC isoforms, with differing tissue distribution and physiological functions.

Recent knockout mouse studies have confirmed some important functions and revealed some surprises. We continue to discover new physiological roles for these transporters and how they contribute to disease states when their normal function is perturbed. The present chapter summarizes the paths to KCC discovery and highlights some features of the KCCs.

Early experiments on human RBCs by Wieth suggested that there was a component of the passive cation (Na^+ and K^+) fluxes which was Cl^- dependent (Funder and Wieth, 1967). RBCs are unusual in having a high (dominant) Cl^- permeability, making it difficult to measure Cl^- transport directly (with tracer experiments requiring a millisecond time course at 37°C). However, in Funder and Wieth's study, substituting Cl^- with another anion such as NO_3^- or SCN^- inhibited the Na^+ and K^+ fluxes, thus establishing the interaction between Cl^- and these monovalent cations. Before the establishment of volume- and thiol-stimulated Cl^--dependent K^+ fluxes (Kregenow, 1971, 1977; Ellory and Dunham, 1980; Lauf and Theg, 1980; Lauf, 1984), later termed K^+-Cl^- cotransport, there was evidence for cation "leaks" in RBCs, as they were called at the time, which did not behave in a classical passive diffusional manner. In this context, Hoffman proposed the pump 2 concept to explain non-linear behavior of ouabain-insensitive cation transport in human RBCs (Hoffman, 1966). Beauge and Adragna (1971, 1974) and Sachs (1971) found ouabain-insensitive Na^+-dependent inward and outward Rb^+/K^+ fluxes in human RBCs. Subsequently, Wiley and Cooper (1974) confirmed that passive Na^+ and K^+ fluxes in human RBCs were at least partially interdependent and not simple electrodiffusive leaks and, in addition, showed that they were coupled. They also introduced furosemide as an inhibitor of these fluxes. Despite these early evidences for non-electrogenic mechanisms, the idea that the membrane potential played a key role in passive movements of K^+, Na^+ and Cl^- persisted for some time (Schmidt and McManus, 1977) until, in 1980, elegant studies on Ehrlich ascites tumor cells by the group of Heinz showed beyond doubt that the fluxes of these ions through the NKCC were electroneutral (Geck and Heinz, 1980). Around the same time, studies with nucleated RBCs (Kregenow, 1971, 1980) also produced evidence for Cl^-dependent K^+ transport while the kinetic characteristics of volume-dependent Cl^--dependent cation fluxes in human RBCs were also studied by Dunham and Ellory and others (Dunham and Ellory, 1980; Chipperfield, 1980). Resolution of KCC properties was achieved using LK sheep RBCs, where NKCC is known to be absent. Activation of K^+ fluxes, but not those of Na^+, follows cell swelling,

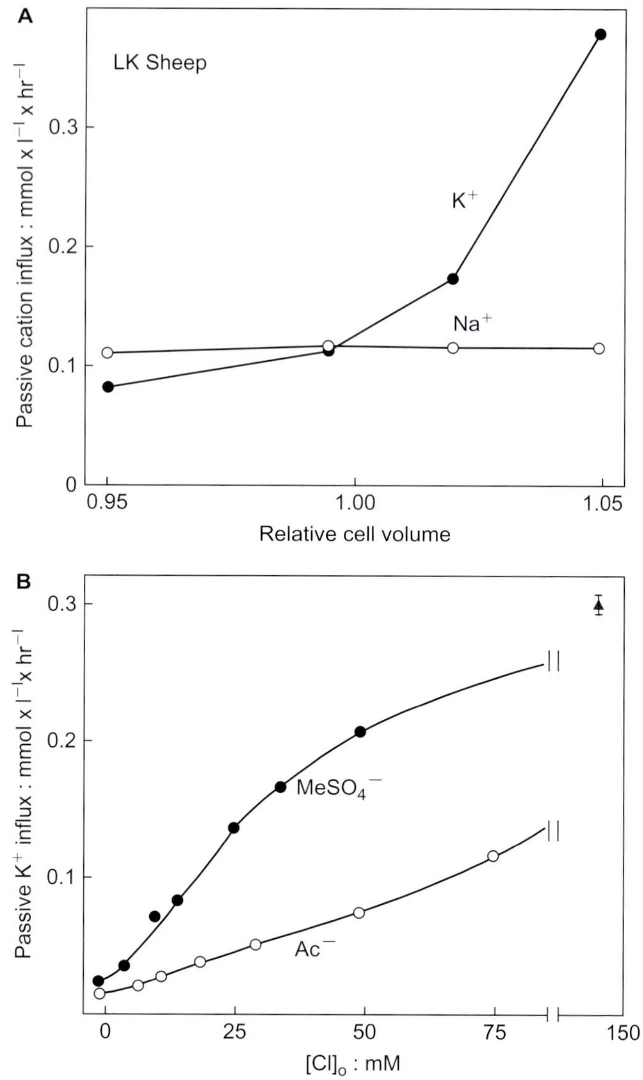

FIGURE 3.1 Swelling-activated Cl^--dependent K^+ transport in LK sheep RBCs. (A) Cell swelling in hypotonic solutions activates K^+ influx but has no effect on Na^+ influx. (B) Equimolar Cl^- replacement by methyl sulphonate ($MeSO_4^-$) or acetate (Ac^-) inhibits swelling activated K^+ influx. (Redrawn from Ellory and Dunham, 1980.)

an effect that was Cl^- dependent (Ellory and Dunham, 1980; Dunham and Ellory, 1981 – see Fig. 3.1). Likewise, a thiol reaction with *N*-ethylmaleimide (NEM) causes activation of K^+ fluxes, and not those of Na^+ (Lauf and Theg, 1980; see Fig. 3.2), probably by inactivating inhibitory kinases (see below). Studies were extended to RBCs from other species (dog, rabbit, mouse, goat, horse: Adragna et al., 2004a) and to human pathophysiology, importantly including sickle cell disease (Brugnara, 2004).

Most attempts to identify membrane transporters relied heavily on the availability of more-or-less specific pharmacological inhibitors. In this context,

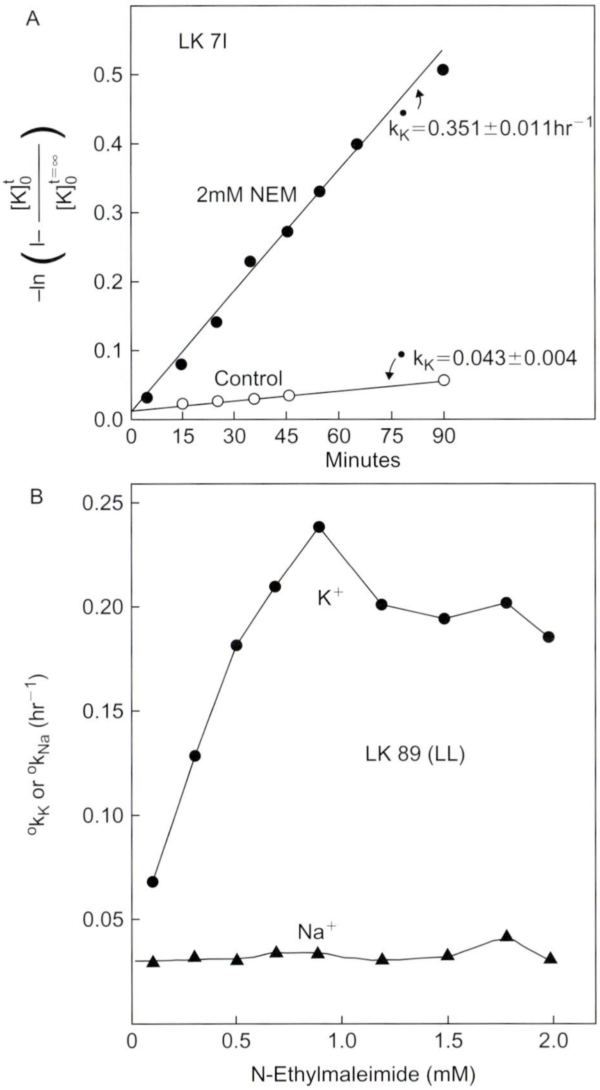

FIGURE 3.2 *N*-ethylmaleimide (NEM) selectively activates K^+ efflux in LK sheep RBCs. (A) First order rate constants of K^+ efflux before and after treatment with NEM (2 mM) for 15 min at 37°C. (B) First order rate constants of K^+ or Na^+ efflux as function of increasing NEM concentrations. (Redrawn from Lauf and Theg, 1980.)

ouabain as a defining Na^+/K^+ pump inhibitor is a good example. Although some success was achieved with loop diuretics as NKCC and KCC inhibitors, these molecules (furosemide, bumetanide) were not discriminatory, inhibiting both CCCs and also the AE1 (Band 3) anion exchanger (Brazy and Gunn, 1976). Thiazides proved to be specific in epithelial studies on NCC (reviewed in Gamba, 2005). For KCC, however, such specific inhibitors await identification. Dihydroindenyloxyalkanoic acid (DIOA) was thought to represent the best KCC inhibitor identified to date (Garay et al., 1988). More recently, however, it has become apparent that DIOA, at least in non-RBCs, is more promiscuous, for example inhibiting NKCC in HEK 293 cells (Gillen and Forbush, 1999) and also other transport processes together with various secondary effects (Anfinogenova et al., 2001; Lauf et al., 2008). Other bumetanide analogs have proved equally non-specific (Cabantchik and Greger, 1992). In addition, while di-isocyanostilbene-disulfonate (DIDS) is able to inhibit KCC in sheep RBCs in the micromolar range (Delpire and Lauf, 1992), it is, of course, more effective and better known as an AE1 inhibitor. Abolishing protein phosphatase activity provides a powerful indirect means to inhibit KCCs with, for example, calyculin A, preventing their activation in tissues and in heterologous systems (reviewed in Gamba, 2005). Given the multiple cellular roles of KCCs and their participation in a number of pathologies, the discovery of effective inhibitors has been eagerly awaited.

Since its definition in RBCs, KCC as a transport entity has now been recognized across many (if not all) tissues, including neurons, vascular smooth muscle, endothelium, epithelia (such as kidney, small intestine and placenta), heart and skeletal muscle (Mercado et al., 2000) and, recently, in human lens epithelial cells (Lauf et al., 2006; Misri et al., 2006). In many of these tissues, KCC is assumed to function as a volume regulatory device, mediating coupled K^+ and Cl^- efflux and hence bringing about regulatory volume decrease (RVD) (Hallows and Knauf, 1994), as first shown in sheep reticulocytes (Lauf and Bauer, 1987). It is often difficult to show participation of KCC in RVD in non-erythroid cells, however, due to the activity of K^+ and Cl^- channels which are more effective at carrying out this function. In epithelial cells, KCC may also contribute to RVD but the cellular asymmetry results in trans-epithelial salt transport (Greger and Schlatter, 1983; Ellory et al., 1984; Eveloff and Warnock, 1987; Reuss and Cotton, 1994). Latterly, other roles for KCC have also emerged, encompassing such areas as regulation of ion homeostasis, especially that of intracellular $[Cl^-]$ and extracellular $[K^+]$ in neurons (Payne, 1997; Reid et al., 2001), as discussed in detail in Chapter 17. More recently, an association with growth of cancer cells has been recognized (Shen et al., 2000) and may indicate an important role for KCC in cell proliferation, migration and control of cell cycle (Chapters 26 and 27). In a reciprocal way, intracellular $[Cl^-]$ may control CCC activity (reviewed in Kahle et al., 2006; see also Chapters 2, 16 and 17). There are now a number of diseases whose pathology involves altered activity of KCCs (see later). Notwithstanding, in some ways, in the tissue in which its transport function was first acknowledged, the RBC, the pathophysiology of KCC continues to be enigmatic.

II. MOLECULAR IDENTIFICATION OF K^+-Cl^- COTRANSPORTERS

KCC belongs to the electroneutral cation-chloride-coupled cotransporter (CCC) superfamily (SLC12; from the acronym SLC for *so*lute *c*arriers) of which there are nine members (*SLC12A1-9*: Mount et al., 1998; Mercado et al., 2000; Gamba, 2005; see Chapter 16). Three represent Na^+-coupled Cl^- cotransporters: the thiazide-sensitive Na^+-Cl^- cotransporter (NCC: *SLC12A3*) which in mammals was first reported in renal distal tubules but has now also been found in the intestine (Bazzini et al., 2005), osteoblasts (Dvorak et al., 2007), and human lens epithelial cells (Lauf et al., 2008). Of the two bumetanide-sensitive Na^+-K^+-Cl^- cotransporters, the NKCC1 (*SLC12A2*) is widespread while NKCC2 (*SLC12A1*) appears to be renal specific, found in the apical membrane of cells from the thick ascending limb of the loop of Henle. Two members are orphan, *SLC12A8-9*, in that their transport function remains unknown as they fail to augment Na^+, K^+ or Cl^- transport upon heterologous expression. The family is completed by the four KCC isoforms, KCC1-4 (*SLC12A4-7*).

KCC1 (Gillen et al., 1996) became the first member of the KCC family to be cloned. So far, four KCC genes have been found, KCC1-4 (*SLCA4-7*), with around 70% identity to each other (see Fig. 16.3 in Chapter 16). KCC1 shares most homology to KCC3; KCC2 with KCC4. (Note: there is some confusion in the early literature over the nomenclature. Three groups independently cloned KCC3 in 1999: KCC3 was termed KCC4 by Mount et al., 1999; latterly the terminology of Hiki and Race – Race et al., 1999; Hiki et al., 1999 – has been adopted in deference to their earlier publication.) All four members share the same topology with 12 putative membrane spanning domains, a large extracellular loop between fifth and sixth transmembrane domains (TMs), a short hydrophilic cytoplasmic amino-terminus, together with a long carboxy-terminus. This structure is remarkably similar to that of the Na^+-coupled Cl^- cotransporters, despite a relatively low level of sequence homology (<50%) (see Chapter 16). Between KCCs, the TMs show the greatest identity. In addition, alterations in the amino acid sequence of TMs between isoforms are linked with differences in ion affinity – notably those of TM2, 4 and 7 (e.g. Mercado et al., 2000, 2005). The predicted protein for KCC1 is 1085 amino acids long, of about 120 kDa, increasing to 150 kDa when glycosylated. All KCC proteins have a variety of potential phosphorylation sites, a significant feature since regulation by phosphorylation is anticipated (see below), notwithstanding the fact that evidence for direct phosphorylation of the KCC transporter (aside from KCC2 – Wake et al., 2007) remains lacking (see section III in Chapter 17). Splice variants of KCC isoforms are multiple, and are discussed in detail in Chapter 17.

III. TISSUE DISTRIBUTION OF K^+-Cl^- COTRANSPORTER ISOFORMS

The first cloned member of the KCC family, KCC1, was found in all tissues tested (Gillen et al., 1996). These include brain, colon, heart, kidney, liver, lung, spleen, placenta, muscle and pancreas (Gillen et al., 1996), as well as RBCs (Pellegrino et al., 1998). It is generally considered to represent an ubiquitous house-keeping protein, involved in cell volume regulation. KCC2, recognized as a neuronal-specific isoform (Payne et al., 1996; Payne, 1997) is discussed in great detail in Chapter 17.

Like KCC1, both KCC3 and KCC4 are found in numerous tissues (Hiki et al., 1999; Mount et al., 1999; Race et al., 1999). For KCC3, KCC3a is abundant in brain, muscle, lung and heart; KCC3b is more abundant in the kidney than in any other tissue (Pearson et al., 2001). Nevertheless, the kidney also expresses KCC3a, albeit at lower levels, as well as other KCC isoforms. KCC3 is extensively expressed in the brain including hypothalamus, cerebellum, brain stem, cerebral cortex, white matter and choroid plexus (Pearson et al., 2001). Recently, both KCC3 and KCC4 have been shown in lens epithelial cells (Lauf et al., 2008; Misri et al., 2006). KCC4 is also expressed in several tissues. High levels are found in the heart and kidney, while levels in the brain are very low (Mount et al., 1999). Notwithstanding limited expression in the brain, KCC4 is found in certain areas of the CNS such as ventricular zones, the nucleus of the trigeminal nerve, choroid plexus and peripheral ganglia (Li et al., 2002; Karadsheh et al., 2004). In the kidney, it is localized to the basolateral membrane of proximal tubules and also in type-A intercalated cells of the collecting duct (Boettger et al., 2002).

Following cloning of the KCC isoforms, the identity of the RBC isoform(s) was eagerly awaited. While the initial isoform cloned from the human erythroleukemia cell line K562 and also from the mouse erythroleukemia cell line MEL proved to be KCC1 (Pellegrino et al., 1998), it is unlikely to be the sole, or even predominant, RBC isoform. Thus KCC1 was not detected in human or mouse reticulocytes (Pellegrino et al., 1998). Subsequently, proteins of both KCC3 and KCC1 have been reported as expressed in sheep RBCs (Lauf et al., 2001), K562 cells and bone marrow (Mercado et al., 2005). RBCs from patients with complete or partial agenesis of the corpus callosum and peripheral neuropathy (ACCPN syndrome), a disease linked to KCC3 mutation, have subtle alterations in their KCC activity (see also Chapter 17). In RBCs, all three non-neuronal KCCs (KCC1, 3 and 4) appear to be present, with a number of splice variants evident (Crable et al., 2005). Comparison of the affinities and ionic selectivities of expressed KCCs with those observed in RBCs also suggest that KCC1 is not the main functional isoform (Hebert et al., 2004; Mercado et al., 2005). Finally, in KCC1 knockout mice, RBC KCC activity was unaffected, while it was decreased in KCC3 knockouts, and almost completely abolished in mice lacking both isoforms (Rust et al., 2007). It is probably true to say that the dominant isoform(s), at least of mature RBCs, remains to be ascribed unambiguously while very little is known about regulation of expression in this tissue.

IV. TRANSPORT MODES, SELECTIVITY AND KINETICS

Implicit in the original definition as a Cl^--dependent K^+ flux and from membership of the CCC superfamily, the KCC transporters are generally considered to mediate an obligatory coupled movement of K^+ with Cl^- (Fig. 3.1B). Certainly, the necessity for the presence of both ions has been demonstrated in many instances, although certain substitutions support transport. Whether their co-dependence necessarily indicates coupled movement is less easy to demonstrate unequivocally but available evidence supports obligate linkage of K^+ and Cl^- movements. Thus uphill K^+ movement can be driven by a Cl^- gradient (Brugnara et al., 1989); K^+ transport is unaffected by membrane potential indicative of an electroneutral process both in human and sheep RBCs (Kaji, 1993; Lauf and Adragna, 1996); and a 1:1 stoichiometry of K^+ and Cl^- movement was demonstrated in NEM-treated RBCs (Jennings and Adame, 2001). That endogenous *Xenopus* oocyte KCC activity was shown to have Hill co-efficients close to unity for both K^+ and Cl^- is again supportive of 1:1 coupled electroneutral transport (Mercado et al., 2001).

The K^+-Cl^- cotransporter is capable of operating bidirectionally and thereby mediates either influx or efflux. Net flow will be dependent on the chemical gradients for K^+ and Cl^- and also their affinity for the transporter, which may differ on either face of the membrane. In human RBCs, unidirectional isotopic flux measurements can also include an element of exchange fluxes (K^+/ K^+) as well as net transport (Kaji, 1989), but in low K^+ (LK) sheep RBCs transport through the loaded carrier is rate limiting (Delpire and Lauf, 1991a). The ability to mediate influx as well as efflux may be important in extracellular K^+ homeostasis in the CNS, where accumulation of extracellular K^+ following periods of high neuronal activity may be buffered by KCl uptake (see Chapter 17 as well as Payne, 1997; Reid et al., 2001). Under physiological conditions, the transporter usually mediates KCl efflux, thereby lowering intracellular $[Cl^-]$ below equilibrium so that anion channel opening results in Cl^- influx and membrane hyperpolarization.

As is generally the case for coupled transporters, K^+ and Cl^- can be substituted by similar ions. Thus Rb^+ is a good replacement for K^+ (Dunham and Ellory, 1981; Lauf, 1983), whereas Na^+ and choline$^+$ are largely ineffective. Its anion preference is $Cl^- > Br^- > SCN^- > I^- > NO_3^- > MeSO_4^-$ (Ellory et al., 1982). Ionic selectivity is also subtly different for the various KCC isoforms when expressed in heterologous systems. For instance, in *Xenopus* oocytes, only KCC3 prefers Br^- over Cl^- (Mercado et al., 2000, 2002; Song et al., 2002), which has been taken as supportive evidence for a major role of this isoform in RBCs. The difference may in part be due to secondary effects of ion substitutions, however, which should always be considered. For instance, fewer permeable ions will result in RBC shrinkage, depolarization and alkalinization, while, in sheep RBCs, NO_3^- is reported to reduce ATP levels (Flatman et al., 1996). Probably the best non-transported Cl^- substitutes for RBCs are the organic anions methylsulfate and methylsulfonate (Payne et al., 1990).

Determination of the affinity constants for KCCs has also been extensively studied in RBCs, in which ion substitutions are relatively simple; and also to a lesser extent for various isoforms expressed in heterologous systems such as *Xenopus* oocytes. In RBCs, there is considerable variability across different species, with apparent affinities at the extracellular site of around 10–100 mM for K^+ (typically around 30 mM) and 15–90 mM for Cl^- (Lauf et al., 1992). Affinities for the intracellular binding site may differ, along with order of ion binding, and also maximal transport rates for influx and efflux (Kaji, 1989) and the kinetics of operation (Delpire and Lauf, 1991a, b) seems to be different from the gliding symmetry model proposed for NKCC (Lytle et al., 1998). An additional intriguing variable which might also influence affinity in native cells is the possible existence of heteroligomers (Casula et al., 2001; Simard et al., 2007). Transport affinities for non-erythroid cells have also been studied. For instance, the apparent affinities for K^+ and Cl^- are shown in Fig. 3.3 for the swelling-activated K^+-Cl^- cotransport in C6 glioma cell line

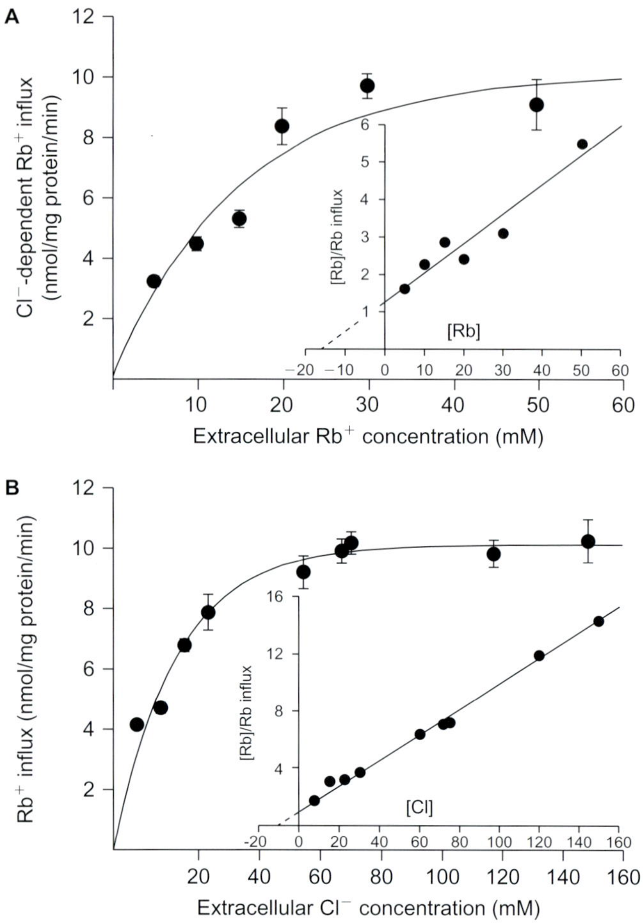

FIGURE 3.3 Kinetics of K$^+$-Cl$^-$ cotransport in glial cells. Rb$^+$ influx as a function of (A) external Rb$^+$ and (B) Cl$^-$ concentrations. Insets show Hanes-Woolf plots, giving affinities (K_ms) of 15 mM for Rb$^+$ and 13 mM for Cl$^-$. Ouabain and bumetanide were present throughout. RT – PCR experiments showed the presence of mRNA for KCC1, 3 and 4. (From Gagnon et al., 2007, reproduced with permission.)

(Gagnon et al., 2007). For heterologously expressed KCCs, the extracellular affinities for KCC1 are about 20 mM for both K$^+$ and Cl$^-$ (Gillen and Forbush, 1999; Mercado et al., 2000; Bergeron et al., 2003). For KCC2, K$^+$ affinity is high ($<$10 mM), but reported values for Cl$^-$ affinity range from $>$100 mM in HEK-293 cells (Payne, 1997) to $<$10 mM (Song et al., 2002) in *Xenopus* – a difference which may be due to the formation of heterodimers between expressed KCC2 and the endogenous KCCs in HEK cells or in oocytes, or may be due to different cytoplasmic environments and interacting proteins in amphibians and mammals. KCC3 has a K$^+$ affinity close to 10 mM but again a range of Cl$^-$ affinities, with a mean around 30 mM (Race et al., 1999; Mercado et al., 2002; Bergeron et al., 2003). Finally, KCC4 has a K$^+$ affinity probably at around 10 mM with that for Cl$^-$ similar or slightly higher (Mercado et al., 2000). A direct comparison of different isoforms gave an overall affinity profile (for both K$^+$ and Cl$^-$) of KCC2 = KCC4 = KCC3 > KCC1 (Gamba, 2005).

V. PHYSIOLOGICAL PROPERTIES, MODULATION AND REGULATION OF K$^+$-Cl$^-$ COTRANSPORTERS

A. Volume Sensitivity

Cl$^-$-dependent K$^+$ flux consistent with KCC activity has been described as volume sensitive in RBCs from LK sheep and humans. In these cells, a small volume increase stimulates KCC activity several-fold (Dunham and Ellory, 1981), as shown in Fig. 3.1 A.Volume sensitivity is also found in the basolaterally located KCC of salt absorbing epithelia (Greger and Schlatter, 1983) such as the kidney, and in a variety of other tissues including vascular smooth muscle cells (Saitta et al., 1990; Adragna et al., 2000), Ehrlich tumour cells (Thornhill and Laris, 1984; Hoffmann et al., 1988), cardiac cells (Taouil and Hannaert, 1999), thymocytes (Soler et al., 1993), lens (Lauf et al., 2006), cornea (Capo-Aponte et al., 2007) and glial cells (Gagnon et al., 2007). Volume changes can occur under anisotonic conditions, for example in the renal medulla or epithelia exposed to anisotonic bathing solution. Isosmotic swelling is probably more common in salt transporting epithelia. It can also occur in other cell types, e.g. in neurons in response to exposure to NH_3/NH_4^+ resulting in net accumulation of intracellular NH_4^+ (Alvarez-Leefmans et al., 2006), and also in RBCs due to Na$^+$ and Cl$^-$ entry following β-adrenergic stimulation of Na$^+$/H$^+$ exchange coupled with Cl$^-$/OH$^-$ exchange (Bourne and Cossins, 1982; Borgese et al., 1987). In some systems, isosmotic swelling preferentially activates KCC, whereas anisotonic swelling results in activation of volume-sensitive K$^+$ and Cl$^-$ channels (see Chapter 15). All K-Cl cotransporters are stimulated by anisosmotic swelling in a variety of tissues. Whether or not they participate in regulatory volume decrease under physiological conditions is debatable (see Chapter 17). In some cases, KCCs are regarded as one of the main pathways mediating RVD, especially in epithelial cells where rates of salt influx and efflux at the two faces are coupled, but can become unbalanced (see Reuss and Cotton, 1994). Volume-sensitive KCC in red cells is enriched in younger cells, with activity becoming quiescent as cells mature (Hall and Ellory, 1986b; Brugnara and Tosteson, 1987; Canessa et al., 1987; Lauf and Bauer, 1987). It has therefore been associated with

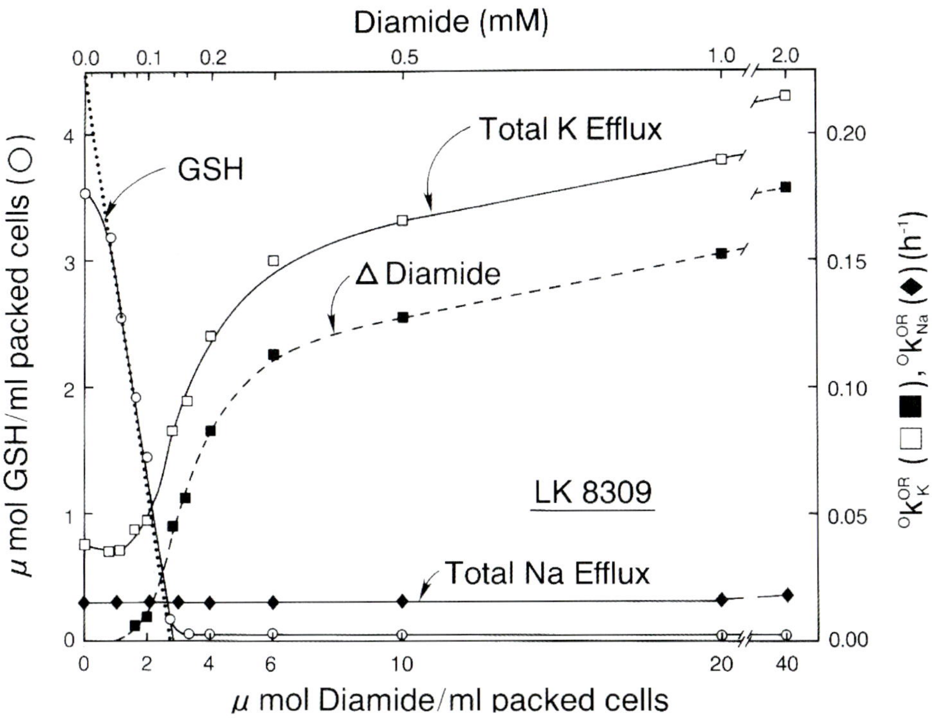

FIGURE 3.4 Evidence for modulation of K$^+$-Cl$^-$ cotransport by redox potential. Diamide reversibly oxidizes glutathione (GSH) to its disulfide. Lowering the free GSH concentrations is associated with a four-fold stimulation of K$^+$ efflux without affecting Na$^+$ efflux, both expressed as rate constants k_K or k_{Na}. This K$^+$-efflux stimulation is Cl$^-$-dependent. (From Lauf, 1988, reproduced with permission.)

the reduction in red cell volume which is a feature of red cell development. In contrast to the three other isoforms, the neuron-specific KCC2 is constitutively active under isotonic conditions at normal volume, which does not preclude further activation by cell swelling (Strange et al., 2000; Song et al., 2002).

B. pH

In RBCs, pH represents a second modulatory influence which may be as important, if not more so, than volume (Brugnara et al., 1986; Ellory et al., 1989). There is also some interaction between the effects of volume and pH change, usually synergism (Speake et al., 1997). KCC activity shows a bell-shaped relationship against pH, typically with an optimum activity near neutrality, which decreases at higher or lower pH values. Acidification will cause a secondary increase in cell volume but it would appear that the main signal is intracellular pH (Zade-Oppen and Lauf, 1990; Ellory et al., 1991). Although the Jacob-Stewart cycle of RBCs couples extracellular and intracellular pH (Hladky and Rink, 1977), this coupling can be broken using impermeable anions such as gluconate, which result in Cl$^-$ efflux and OH$^-$ influx and hence intracellular alkalinization. Thus incubation at an extracellular pH of 6.1 in gluconate solution, gives an intracellular pH close to the normal 7.1 (Ellory et al., 1991). H$^+$ sensitivity could reside in the regulatory apparatus controlling KCC activity (notably the kinases and phosphatases – see below) or within KCC itself (since pH dependence is not completely lost following NEM treatment – Zade-Oppen and Lauf, 1990; Lauf and Adragna, 1996, 1998). Effects of pH have also been observed in expressed KCCs, while lack of acid inhibition of the KCC4 isoform in oocytes (Bergeron et al., 2003) may indicate that KCC1 and KCC3 are responsible for the pH sensitivity of KCC in RBCs.

C. Redox Potential

Thiol reactions, in particular with NEM, were identified early on as a stimulus for KCC in RBCs (Lauf and Theg, 1980). This effect is also seen in other tissues like vascular smooth muscle (Adragna et al., 2004a) and also in heterologously expressed KCCs (Gillen et al., 1996; Lauf et al., 2001; Gamba, 2005). Subsequently, a number of oxidants (notably H_2O_2, NO, nitrite and peroxynitrite) have also been shown to stimulate KCC activity (Bize and Dunham, 1995; Adragna and Lauf, 1998). Their target, however, remains unclear. Possibilities include GSH (Fig. 3.4), NADH and NADPH, as well as thiol residues on the KCC transporter or some regulatory protein.

D. Other KCC Modulators

A number of other important effectors have been identified in RBCs. For instance, Mg^{2+} depletion stimulates KCC activity and alters the response to several modulators presumably via effects on kinases (Delpire and Lauf, 1991c). High hydrostatic pressure also increases activity probably also by interrupting the signaling pathways (Hall and Ellory, 1986a). O_2 tension is interesting as it has significant effects on KCC activity, as it does on many aspects of RBC physiology (Barvitenko et al., 2005). RBCs, of course, experience large fluctuations in O_2 tension as they circulate between lung and metabolically active peripheral tissues. In the case of RBC KCC, low O_2 tension inhibits and high O_2 stimulates (Borgese et al., 1991). These higher levels of O_2 are in fact often required to activate the transporter, and to enable it to respond to other stimuli such as volume change. Interestingly, the other major RBC CCC, NKCC, is reciprocally controlled by O_2 such that it is stimulated by deoxygenation (see Gibson et al., 2000 for a review). High temperature will also stimulate KCC, to a greater extent to that expected from a simple increase in maximal transport velocity (Khan and Ellory, 2000), probably through effects on regulatory kinases and phosphatases. However, the temperature effect may be irreversible (Orlov et al., 1993). High urea concentrations, at levels found in the medulla of the concentrating kidney, also increase activity (Kaji and Gasson, 1995). Finally, there are indications that physiological levels of bicarbonate may inhibit KCC activity (Godart et al., 1997). This intriguing finding may be of high relevance to non-erythroid KCCs, like KCC2, which participates in the regulation of $[Cl^-]_i$ in neurons that express $GABA_A$-receptor anion channels that are permeable to Cl^- and HCO_3^- (Fatima-Shad and Barry, 1993).

E. Growth Factors

KCC activity may be modulated by growth factors. Examples include insulin-like growth factor 1 (IGF-1) in cervical cells (Shen et al., 2004; Hsu et al., 2007b), vascular endothelial growth factor (VEGF) in endothelial cells (Hiki et al., 1999) and platelet-derived growth factor (PDGF) in vascular smooth muscle cells (Zhang et al., 2003; Adragna et al., 2006).

F. Regulation

Like several other members of the SCL12 family, protein phosphorylation occupies an important position in the regulation of KCC activity. The transporter sequences contain several consensus sites for phosphorylation, although these may vary between KCC isoforms and even between splice variants of single isoforms. *In situ* numerous kinases and phosphatases appear to be involved. Notwithstanding, the target(s) of phosphorylation remain unclear. In contrast to NKCC, the presence of phospho-residues on the KCC transporter *per se* has only been reported recently, and that solely for KCC2 isoform (Wake et al., 2007). In addition, it should also be noted that under certain circumstances KCC activity can be modulated in the absence of altered phosphorylation (Sachs and Martin, 1993).

Early evidence suggesting a role for phosphorylation originally came from experiments involving ATP or Mg^{2+} depletion, both of which are associated with stimulation of KCC activity (Lauf, 1985; Delpire and Lauf, 1991c). The main impetus, however, was generated from observations that specific protein phosphatase inhibitors (including calyculin A and okadaic acid) inhibit KCC activity (Jennings and Al-Rohil, 1990; Jennings and Schulz, 1991; Kaji and Tsukitani, 1991; O'Neill, 1991). The delay between swelling and full KCC activation led these workers to hypothesize that swelling activation of KCC resulted from inhibition of a volume-sensitive kinase (Jennings and Al-Rohil, 1990). Subsequently, a large body of pharmacological evidence implicating a role for protein phosphorylation has accumulated. Inhibitors of serine–threonine kinases (e.g. staurosporine) stimulate (Bize and Dunham, 1994) while those acting on tyrosine kinases inhibit (Cossins et al., 1994; Flatman et al., 1996), probably acting upstream of the serine–threonine target. A stimulatory tyrosine kinase could account for the increase in KCC activity following addition of Mg^{2+}-ATP to either RBC "ghosts" or intact RBC cells (Delpire and Lauf, 1991c; Sachs and Martin, 1993). In addition, ML-7, an inhibitor of myosin-light-chain kinase (MLCK), stimulates KCC (Kelley and Dunham, 1996), suggesting a role for this kinase. The serine–threonine phosphatases PP1 and PP2a may both be involved in transporter activation (Bize et al., 1999, 2000). Latterly, roles for PKC and PKG have also been suggested (Ferrell et al., 2000; Adragna et al., 2002), especially in VSM cells. PKG is activated via NO (Adragna and Lauf, 1998). Elaborate hypotheses have been proposed to account for these findings, involving a hierarchical scheme of often conjugate pairs of enzymes acting at both serine–threonine and tyrosine residues (Dunham et al., 1993; Flatman et al., 1996; Adragna et al., 2004a; Adragna and Lauf, 2007).

Molecular evidence of KCC regulation is rapidly emerging although it is still scarce. Knockout mice lacking the Src-family tyrosine kinases Hck and Fgr display an abnormally high RBC KCC activity (De Franceschi et al., 1997), possibly through loss of inhibition of PP1a (De Franceschi et al., 2006). More recently, a central role for the "with no lysine kinases" (WNKs) has been proposed for regulation of CCCs (Kahle et al., 2005, 2006). These may act via stress-related serine–threonine kinases (the ste20-related proline–alanine-rich kinase SPAK) and the oxidative stress-responsive kinase 1 (OSR1) (reviewed in Flatman, 2007 and Delpire and Gagnon, 2008). In particular, WNK3 emerges as a key regulator of kinase/phosphatase signaling pathways which can bypass tonicity requirements for KCC (de los Heros et al., 2006).

In the context of regulation, while it appears certain that protein phosphorylation is involved in modulating the activity of KCC, how these enzymes are coupled into the physiological stimuli to which KCC responds, and particularly cell volume, remains unclear. Four important areas of study are represented by macromolecular crowding, fluctuations of free [Mg^{2+}], variation in intracellular [Cl^-] and modification of the cytoskeleton. Evidence for all four exists. Macromolecular crowding, first postulated by Parker and co-workers, is particularly significant in RBC because of their high hemoglobin content (Colclasure and Parker, 1992; Minton, 1994). However, the concept is applicable to the cytoplasm of all cells (Zhou et al., 2008). The crowded environment of proteins and other molecules within the RBC is envisaged as altering enzyme activity, like kinases and phosphatases. Small changes in RBC water content will modulate this interaction.

Intracellular Mg^{2+}, likewise, may change with cell volume, or pH, and is also altered by changes in O_2 tension (Flatman, 1980). It may thereby alter KCC activity directly or via Mg^{2+}-ATP and phosphorylation. Notwithstanding, changes in Mg^{2+} do not appear to be involved in the O_2 response of RBC KCC, as Mg^{2+}-clamped red cells still respond to oxygenation/deoxygenation (Muzyamba et al., 2006). The involvement of WNK3 kinase has identified a Cl^- dependence in regulation by phosphorylation (Kahle et al., 2005). Finally, a role for the cytoskeleton has also been proposed, since maneuvers which disrupt its integrity such as cytochalasin B, high hydrostatic pressure or heat treatment alter KCC activity in various tissues (Hall and Ellory, 1986a; Garay et al., 1988; Orlov et al., 1993).

VI. PATHOLOGY AND K^+-Cl^- COTRANSPORTERS

Given their widespread expression, it is not surprising that mutations, altered expression or abnormal regulation of KCCs have been associated with a variety of diseases involving a number of different organ systems. At the time of writing this chapter, the first member of the KCC family that was cloned, KCC1, lacks an association with disease, but the others, i.e. KCC2, 3 and 4, have all been linked with naturally occurring diseases or pathologies in *null* mice. Selected diseases are reviewed below.

A. Sickle Cell Disease

Although sickle cell disease (SCD) is one of the commonest inherited disorders affecting millions worldwide, its pathogenesis together with the extent to which altered KCC activity contributes remains unclear. The genetic cause has been understood for some time and it has been termed "the first molecular disease" (Pauling et al., 1949; Bunn and Forget, 1986). In most cases, a single base mutation results in a single amino acid substitution in the beta chain of hemoglobin (β6: glutamic acid to valine). The resulting loss of a negative charge facilitates the formation of hemoglobin polymers upon deoxygenation, distorting the RBC into bizarre shapes including the eponymous sickle. Notwithstanding its simple molecular definition, the ensuing pathology is complex, extensive and less well understood (Stuart and Nagel, 2004). RBCs have a shorter lifespan which contributes to chronic anemia. In addition, ischemic disorders are also characteristic, and may affect multiple organ systems including brain, retina, kidney, bone, lung and penis. In part, microvascular occlusion results from the increased rigidity of RBCs containing HbS polymers. Other abnormalities, however, include increased RBC adhesiveness and involvement of other cell types such as white cells and endothelium. A particular feature of sickle cells is their increased propensity for dehydration (Ellory et al., 1998), important because it raises the intracellular concentration ([HbS]), thus reducing the lag time for polymerization upon deoxygenation (Eaton and Hofrichter, 1987). Increased permeability of the RBC membrane causes solute loss, accompanied by osmotically obliged water. The Gardos channel (a Ca^{2+}-activated K^+ channel) and P_{sickle} (the deoxygenation-induced cation pathway) are involved (Lew and Bookchin, 2005). In addition, the increased activity and abnormal properties of the RBC KCC contribute to this feature (Brugnara, 2004). The underlying cause(s) is not clear. An altered interaction between HbS and an intracellular target has been suggested (Olivieri et al., 1992). Immunoblotting has revealed an increase in KCC protein in the RBC membrane (Su et al., 1999), while various maneuvers including deoxygenation result in a different pattern of phosphoresidues compared to normal red cells (Merciris et al., 2001). Differences in the expression of KCC isoforms or splice variants are also present (Crable et al., 2005), together with the possibility of altered interactions through heterooligomers. In support of a role for KCC, SAD mice lacking both KCC1 and KCC3 show normalized cell volumes, although the proportion of dense cells is unchanged (Rust et al., 2007). Finally, regulation of KCC is abnormal (Joiner, 1993), notably with high activity even in deoxygenated red cells when KCC is usually inactive (Gibson et al., 1998; Gibson, 2001). In this context, it is also worth noting that altered KCC activity is also seen in other hemoglobinopathies (Olivieri et al., 1992).

B. Nervous System

Adermann's disease is a rare neurological condition involving hereditary motor and sensory neuropathy, with mental retardation and dysmorphic features. There is also complete or partial agenesis of the corpus callosum (HMSN/ACC or ACCPN). It therefore involves both developmental problems with neurons and also neurodegeneration, particularly in the white matter. The disease is transmitted as an autosomal recessive. Prevalence is particularly high in Quebec, Canada. It appears to be a disease associated with inactivating mutations in KCC3 (Howard et al., 2002), while missense mutations have also been reported (Uyanik et al., 2006). KCC3 has been proposed to play a role in myelination (Pearson et al., 2001). Two KCC3 knockout mouse models reproduce some, but not all, of these features (Howard et al., 2002; Boettger et al., 2003; Byun and Delpire, 2007). Interestingly, agenesis of the corpus callosum is not observed, but this is a notably variable feature in humans. Like KCC4 knockouts (Boettger et al., 2002), the mice also have inner ear deficiencies, in this case becoming deaf during their first year of life (Boettger et al., 2003).

In neurons, NKCC1 and KCC2 functionally interact to determine intracellular [Cl^-]. During development, it is believed that increased KCC2 activity reduces intracellular [Cl^-] so that activation of GABA/glycine-gated channels causes the usual inhibitory hyperpolarization (Rivera et al., 1999; Hubner et al., 2001; Reid et al., 2001; Ueno et al., 2002; Woo et al., 2002). The consequences of KCC2 gene inactivation in mice is thoroughly discussed in Chapters 17 and 24.

Considering the importance of KCC2 and KCC3 in the CNS, an association between KCC2 or KCC3 and naturally occurring CNS disease in humans (such as idiopathic epilepsy) represents an obvious possibility. Some epilepsy and other neurological syndromes are indeed linked to the same chromosome location as that encoding for KCC3 (15q14), although this does not necessarily indicate the involvement of KCC3 mutations (Mercado et al., 2004). However, an association between rare variants of KCC3 and bipolar disorder has been reported (Meyer et al., 2005; Moser et al., 2008).

Finally, KCC4, despite its limited brain expression, appears involved in CNS development. Thus, KCC4 *null* mice, which also show a renal phenotype (see below), also exhibit hearing loss, in this case early onset, becoming deaf by the second week of life (cf. those lacking KCC3). These animals have complete loss of the outer hair cells of basal turns of the cochlea and the organ of Corti (Boettger et al., 2002). KCC4 loss presumably causes death of these cells by osmotic perturbation or membrane depolarization. Thus both CCCs, KCC3 and KCC4 (Boettger et al., 2002, 2003), as well as NKCC1 (Delpire et al., 1999), are important in cochlear development and function.

C. Renal Function and Hypertension

Bartter's and Gittelman's syndromes are well recognized diseases which in some cases are associated with mutations in the Na^+-coupled CCCs (Zelikovic, 2001; Kamel and Halperin, 2002). The kidney is also rich in KCCs, particularly KCC3b but also others including KCC4. KCC4 has been localized to the basolateral membrane of proximal tubule and alpha-intercalated cells of the collecting duct (Boettger et al., 2002) with expected participation in transepithelial transport (Mount and Gamba, 2001). Thus KCC3 knockouts have impaired transport in their proximal tubules, while KCC4 *null* mice develop a distal tubular metabolic acidosis (Boettger et al., 2002). These findings indicate the importance of KCC in renal functions such as acid–base balance. The kidney, of course, is also crucial in maintenance of normal blood pressure. Renal KCC mutations may act via this organ to alter blood pressure. In addition, however, KCC may also have its effects elsewhere. Thus NO stimulates KCC in LK sheep RBCs and in VSM cells with concomitant relaxation of porcine coronary arteries (Adragna and Lauf, 1998; Adragna et al., 2000, 2004b). A number of NO donors also increased expression of both KCC1 and particularly KCC3 (Di Fulvio et al., 2002). It follows that KCC mutations, via reduced vasorelaxation and increased resistance to blood flow, might contribute to some cases of hypertension. In support of this hypothesis, mouse KCC3 knockouts, created as a model for ACCPN, are severely hypertensive (Boettger et al., 2003; Adragna et al., 2004b). On the other hand, non-vascular neurogenic mechanisms have also been proposed to account for the high blood pressure in these animals (Rust et al., 2006). Further work is required in this area.

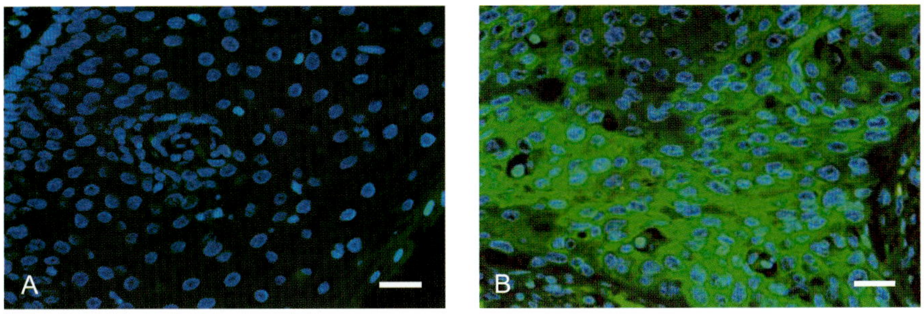

FIGURE 3.5 Immunocytochemistry of K^+-Cl^- cotransporters (KCCs) in cervical tissue. Photomicrographs show immunostaining with anti-KCC antibodies in normal (A) and cancerous (B) cervical tissue. Nuclei were stained blue with Hoeschst 33258; the primary antibody was produced to KCC1 (amino acids 1–14) and reacted with a green Alexa-labeled secondary antibody (Molecular Probes). Scale bar 20μm (Modified from Shen et al., 2003)

D. Cancer

Altered KCC activity has also been associated with neoplastic behaviour (Shen et al., 2000). Thus certain cervical cancers show up-regulation of KCC1 (Fig. 3.5), KCC3 and KCC4, while expression of loss-of-function KCC mutants inhibit cell growth of cervical cancer cells (Shen et al., 2003, 2004). Equilibration of cervical SiHa cells with $^{36}Cl^-$ shows that inhibition of KCC following infection with dominant-negative KCC results in elevated levels of intracellular [$^{36}Cl^-$], consistent with reduction in KCl efflux via KCC (Fig. 3.6). KCC activity also appears to affect cell cycle gene products including retinoblastoma and cdc2 kinase. More recently, KCC3 overexpression has been found to promote epithelial–mesenchymal transition, a critical cellular event in malignancy (Hsu et al., 2007a). Epithelial markers such as E-cadherin and β-catenin, which localize to cell–cell junctions, are decreased, whereas mesenchymal markers (vimentin) are up-regulated. Insulin-like growth factor-1 (IGF-1) may also promote growth and spread of neoplasms via KCC activation (Shen et al., 2004), as also proposed in breast cancer (Hsu et al., 2007b). The involvement of KCC in other neoplastic conditions awaits elucidation.

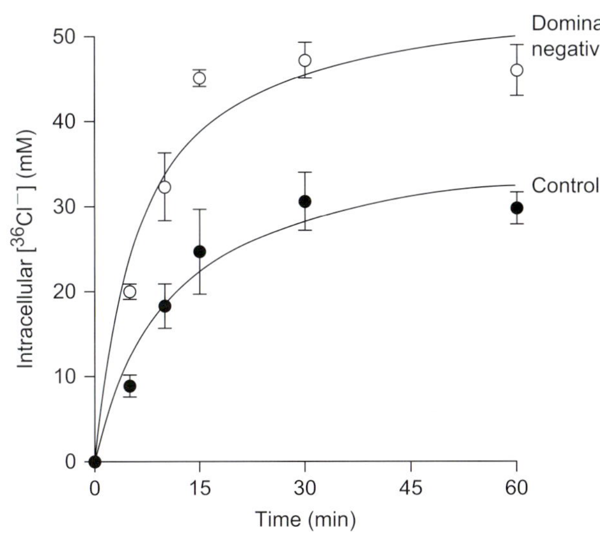

FIGURE 3.6 Effect of K^+-Cl^- cotransport on intracellular chloride levels in cervical cancer cell lines (SiHa cells). Radioisotope equilibration ([$^{36}Cl^-$]) against time in wild-type SiHa cells is compared to SiHa cells transfected with dominant-negative KCC, and shows that suppression of KCC function to inhibit KCl efflux results in elevation of intracellular [$^{36}Cl^-$] (Data redrawn from Shen et al., 2003)

VII. FUTURE PERSPECTIVES

The story of KCC has progressed considerably from the initial observations in RBCs through its molecular identification, to the subsequent demonstration of its participation in a number of diseases. KCCs are widely distributed in numerous tissues, with some isoforms relatively promiscuous while others, like KCC2b and KCC3b, so far show marked tissue specificity. Considerable biological importance can be anticipated from the widespread tissue distribution of the KCCs, the presence of multiple isoforms and sub-isoforms together with the relationship with Na^+-linked CCCs and transport proteins. Cellular roles include volume regulation, transepithelial transport and intracellular ion homeostasis, such that KCC is involved in numerous functions from that of the kidney to CNS development. Its pathophysiological significance is beginning to become apparent. Further advances will include the elucidation of significant roles for KCC in a variety of other tissues and organs (the eye, the ear, selective regions of the nervous, cardiovascular, respiratory, endocrine, digestive and reproductive systems), and the final identification of the elusive phosphorylation sites and signaling pathways responsible for regulating these proteins. The development of a high throughput screening assay to assist the identification of selective inhibitors will lead to pharmacological agents that could play a role in controlling the renal aspects of hypertension, and the proliferation of epithelial cancers among others.

Acknowledgements

We thank the Wellcome Trust, the British Heart Foundation, Action Medical Research, the National Institutes of Health and the American Heart Association for supporting the work of our laboratories.

References

Adragna, N.C., Di Fulvio, M., and Lauf, P.K. (2004a). Regulation of K-Cl cotransport: from function to genes. *J. Membr. Biol* **201**, 109–137.

Adragna, N.C., Chen, Y., Delpire, E., Lauf, P.K., and Morris, M. (2004b). Hypertension in K-Cl cotransport-3 knockout mice. In Cell Volume and Signaling, *Adv. Exper. Med. Biol.*, **559**, 379–385 (Lauf, P.K. and Adragna, N.C., eds).

Adragna, N.C., Ferrell, C.M., Zhang, J., Di Fulvio, M., Temprana, C.F., Sharma, A., Fyffe, R.E., Cool, D.R., and Lauf, P.K. (2006). Signal transduction mechanisms of K^+-Cl^- cotransport regulation and its relationship to disease. *Acta Physiol. (Oxford)* **187**, 125–139.

Adragna, N.C. and Lauf, P.K. (1998). Role of nitric oxide derivative in K-Cl activation of low-potassium sheep red blood cells. *J. Memb. Biol.* **166**, 157–167.

Adragna, N.C. and Lauf, P.K. (2007). K-Cl cotransport function and its potential contribution to cardiovascular disease. *Pathophysiology* **14**, 135–146.

Adragna, N.C., White, R.E., Orlov, S.N., and Lauf, P.K. (2000). K-Cl cotransport in vascular smooth muscle and erythrocytes: possible implications in vasodilation. *Am. J. Cell Physiol.* **278**, C381–C390.

Adragna, N.C., Zhang, J., Di Fulvio, M., Lincoln, T.M., and Lauf, P.K. (2002). KCl cotransport regulation and protein kinase G in cultured vascular smooth muscle cells. *J. Membr. Biol.* **187**, 157–165.

Alvarez-Leefmans, F.J., Herrera-Perez, J.J., Marquez, M.S., and Blanco, V.M. (2006). Simultaneous measurement of water volume and pH in single cells using BCECF and fluorescence imaging microscopy. *Biophys. J.* **90**, 608–618.

Anfinogenova, Y.J., Rodriguez, X., Grygorczyk, R., Adragna, N.C., Lauf, P.K., and and Orlov, S.N. (2001). Swelling-induced K^+ fluxes in vascular smooth muscle cells are mediated by charybdotoxin-sensitive K^+ channels. *Cell. Physiol. Biochem.* **11**, 285–352.

Barvitenko, N.N., Adragna, N.C., and Weber, R.E. (2005). Erythtocyte signal transduction pathways, their oxygenation dependence and functional significance. *Cell. Physiol. Biochem.* **15**, 1–18.

Bazzini, C., Vezzoli, V., Sironi, C., Dossena, S., Ravasio, A., De Biasi, S., Garavaglia, M., Rodighiero, S., Meyer, G., Furst, J., Ritter, M., and Paulmichl, M. (2005). Thiazide-sensitive NaCl cotransporter in the intestine: possible role of hydrochlorothiazide in intestinal Ca^{2+} uptake. *J. Biol. Chem.* **280**, 19902–19910.

Beauge, L.A. and Adragna, N.C. (1971). The kinetics of ouabain inhibition and the partition of Rb influx in human red blood cells. *J. Gen. Physiol.* **57**, 576–592.

Beauge, L.A. and Adragna, N.C. (1974). pH dependence of ruibidium influx in human red blood cells. *Biochim. Biophys. Acta* **352**, 441–447.

Bergeron, M.J., Gagnon, E., Wallendorff, B., Lapointe, J.Y., and Isenring, P. (2003). Ammonium transport and pH regulation by K^+-Cl^- cotransporters. *Am. J. Renal Physiol.* **285**, F68–F78.

Bize, I. and Dunham, P.B. (1994). Staurosporine, a protein kinase inhibitor, activates K-Cl cotransport in LK sheep erythrocytes. *Am. J. Physiol.* **266**, C759–C770.

Bize, I and Dunham, P.B. (1995). H_2O_2 activates red blood cell K-Cl cotransport via stimulation of a phosphatase. *Am. J. Physiol.* **269**, C849–C855.

Bize, I., Guvenc, B., Buchbinder, G., and Brugnara, C. (2000). Stimulation of human erythrocyte K-Cl cotransport and protein phosphatase type 2 A by n-ethylmaleimide: role of intracellular Mg^{2+}. *J. Membr. Biol.* **177**, 159–168.

Bize, I., Guvenc, B., Robb, A., Buchbinder, G., and Brugnara, C. (1999). Serine/threonine protein phosphatases and regulation of K-Cl cotransport in human erythrocytes. *Am. J. Physiol.* **277**, C926–C936.

Boettger, T., Hubner, C.A., Maier, H., Rust, M.B., Beck, F.X., and Jentsch, T.J. (2002). Deafness and renal tubular acidosis in mice lacking the K-Cl cotransporter Kcc4. *Nature* **416**, 874–878.

Boettger, T., Rust, M.B., Maier, H., Seidenbecher, T., Schweizer, M., Keating, D.J., Faulhaber, J., Ehmke, H., Pfeffer, C., Scheel, O., Lemcke, B., Horst, J., Leuwer, R., Pape, H.C., Volkl, H., Hubner, C.A., and Jentsch, T.J. (2003). Loss of K-Cl co-transporter KCC3 causes deafness, neurodegeneration and reduced seizure threshold. *EMBO J* **22**, 5422–5434.

Borgese, F., Garcia-Romeu, F., and Motais, R. (1987). Control of cell volume and ion transport by beta-adrenergic catecholamines in erythrocytes of rainbow trout Salmo gairdneri. *J. Physiol.* **382**, 123–144.

Borgese, F., Motais, R., and Garcia-Romeu, F. (1991). Regulation of Cl-dependent K transport by oxy-deoxyhemoglobin transitions in trout red cells. *Biochim. Biophys. Acta* **1066**, 252–256.

Bourne, P.K. and Cossins, A.R. (1982). On the instability of K^+ influx in erythrocytes of the rainbow trout *Salmo gairdneri*, and the role of catecholamine hormones in maintaining *in vivo* influx activity. *J. Exper. Biol.* **101**, 93–104.

Brazy, P.C. and Gunn, R.B. (1976). Furosemide inhibition of chloride transport in human red blood cells. *J. Gen. Physiol.* **68**, 583–599.

Brugnara, C. (2004). Sickle cell disease. In *Red Cell Membrane Transport in Health and Disease* (Bernhardt, I. and Ellory, J.C., eds), pp. 549–567. Springer Verlag, Berlin.

Brugnara, C., Bunn, H.F., and Tosteson, D.C. (1986). Regulation of erythrocyte cation and water content in sickle cell anemia. *Science* **232**, 388–390.

Brugnara, C. and Tosteson, D.C. (1987). Cell volume, K transport, and cell density in human erythrocytes. *Am. J. Physiol.* **252**, C269–C276.

Brugnara, C., Van Ha, T., and Tosteson, D.C. (1989). Role of chloride in potassium transport through the K-Cl cotransport system in human red blood cells. *Am. J. Physiol.* **256**, C944–C1003.

Bunn, H.F. and Forget, B.G. (1986). *Hemoglobin: Molecular, Genetic and Clinical Aspects*. Saunders, Philadelphia.

Byun, N. and Delpire, E. (2007). Axonal and periaxonal swelling precede peripheral neurodegeneration in KCC3 knockout mice. *Neurobiol. Disord.* **28**, 39–51.

Cabantchik, Z.I. and Greger, R. (1992). Chemical probes for anion transporters of mammalian cell membranes. *Am. J. Physiol.* **262**, C803–C827.

Canessa, M., Fabry, M.E., Blumenfeld, N., and Nagel, R.L. (1987). Volume-stimulated, Cl-dependent K efflux is highly expressed in young human red cells containing normal hemoglobin or HbS. *J. Membr. Biol.* **97**, 97–105.

Capo-Aponte, J.E., Wang, Z., Bildin, V.N., Iserovich, P., Pan, Z., Zhang, F.Q., Pokorny, K.S., and Reinach, P.S. (2007). Functional and molecular characterization of multiple K-Cl cotransporter isoforms in corneal epithelial cells. *Exper. Eye Res.* **84**, 1090–1103.

Casula, S., Shmukler, B.E., Wilhelm, S., Stuart-Tilley, A.K., Su, W., Chernova, M.N., Brugnara, C., and Alper, S.L. (2001). A dominant negative mutant of the KCC1 K-Cl cotransporter. *J. Biol. Chem.* **276**, 41870–41878.

Chipperfield, A.R. (1980). An effect of chloride on (Na + K) co-transport in human red blood cells. *Nature* **286**, 281–282.

Colclasure, G.C. and Parker, J.C. (1992). Cytosolic protein concentration is the primary volume signal for swelling-induced [K-Cl] cotransport in dog red cells. *J. Gen. Physiol.* **100**, 1–10.

Cossins, A.R., Weaver, Y.R., Lykkeboe, G., and Nielsen, O.B. (1994). Role of protein phosphorylation in control of K flux pathways of trout red blood cells. *Am. J. Physiol.* **267**, C1641–C1650.

Crable, S.C., Hammond, S.M., Papes, R., Rettig, R.K., Zhou, G.-P., Gallagher, P.G., Joiner, C.H., and Anderson, K.P. (2005). Multiple isoforms of the KCl cotransporter are expressed in sickle and normal erythroid cells. *Exper. Hematol.* **33**, 624–631.

De Franceschi, L., Fumagalli, L., Olivieri, O., Corrocher, R., Lowell, C.A., and Berton, G. (1997). Deficiency of Src family kinases Fgr and Hck results in activation of erythrocyte K/Cl cotransport. *J. Clin. Invest.* **99**, 220–227.

De Franceschi, L.D., Villa-Moruzzi, E., Biondani, A., Siciliano, A., Brugnara, C., Alper, S.L., Lowell, C.A., and Berton, G. (2006). Regulation of K-Cl cotransport by protein phosphatase 1a in mouse erythrocytes. *Pflugers Arch.* **451**, 760–768.

de los Heros, P., Kahle, K.T., Rinehart, J., Bobadilla, N.A., Vazquez, N., San Cristobal, P., Mount, D.B., Lifton, R.P., Hebert, S.C., and Gamba, G. (2006). WNK3 bypasses the tonicity requirement for K-Cl cotransporter activation via a phosphatase-dependent pathway. *Proc. Nat. Acad. Sci. USA* **103**, 1976–1981.

Delpire, E. and Gagnon, K.B. (2008). d OSR1: STE20 kinases involved in the regulation of ion homoeostasis and volume control in mammalian cells. *Biochem. J.* **409**, 321–331.

Delpire, E. and Lauf, P.K. (1991a). Kinetics of Cldependent K fluxes in hyposmotically low K sheep erythrocytes. *J. Gen. Physiol.* **97**, 173–193.

Delpire, E. and Lauf, P.K. (1991b). Transeffects of cellular K and Cl on ouabainresistant Rb(K) influx in low K sheep red blood cells: further evidence for asymmetry of KCl cotransport. *Pflügers Arch.* **419**, 540–542.

Delpire, E. and Lauf, P.K. (1991c). Magnesium and ATP dependence of K-Cl co-transport in low K^+-sheep red blood cells. *J. Physiol.* **441**, 219–231.

Delpire, E. and Lauf, P.K. (1992). Kinetics of DIDS inhibition of swelling-activated K-Cl cotransport in low K sheep erythrocytes. *J. Membr. Biol.* **126**, 89–96.

Delpire, E., Lu, J., England, R., Dull, C., and Thorne, T. (1999). Deafness and imbalance associated with inactivation of the secretory Na-K-2Cl co-transporter. *Nat. Gen.* **22**, 192–195.

Di Fulvio, M., Lauf, P.K., and Adragna, N.C. (2002). Protein kinase G (PKG) signaling pathway regulates the potassium-chloride cotransporter-3 (KCC3a and KCC3b) mRNA expression in primary cultures of rat vascular smooth muscle cells (VSMCs). *FASEB J.* **16**, A421.

Dunham, P.B. and Ellory, J.C. (1980). Chloride-activated potassium transport in human erythrocytes. *Proc. Nat. Acad. Sci. USA* **77**, 1711–1715.

Dunham, P.B. and Ellory, J.C. (1981). Passive potassium transport in low potassium sheep red cells: dependence upon cell volume and chloride. *J. Physiol.* **318**, 511–530.

Dunham, P.B., Klimczak, J., and Logue, P.J. (1993). Swelling activation of K-Cl cotransport in LK sheep erythrocytes: a three-state process. *J. Gen. Physiol.* **101**, 733–766.

Dvoark, M.M., De Joussineau, C., Carter, D.H., Pisitkun, T., Knepper, M.A., Gamba, G., Kemp, P.J., and Riccardi, D. (2007). Thiazide diuretics directly induce osteoblast differentiation and mineralized nodule formation by interacting with a sodium chloride cotransporter in bone. *J. Am. Soc. Nephrol.* **18**, 2509–2516.

Eaton, W.A. and Hofrichter, J. (1987). Hemoglobin S gelation and sickle cell disease. *Blood* **70**, 1245–1266.

Ellory, J.C. and Dunham, P.B. (1980). Volume-dependent passive potassium transport in LK sheep red cells. In *Membrane Transport in Erythrocytes. Alfred Benzon Symposium 14* (Lassen, U.V., Ussing, J.O., and Wieth, J.O., eds), pp. 409–423. Munksgaard, Copenhagen.

Ellory, J.C., Dunham, P.B., Logue, P.J., and Stewart, G.W. (1982). Anion-dependent cation transport in erythrocytes. *Philos. Trans. R. Soc. Lond. B Biol. Sci.* **299**, 483–495.

Ellory, J.C., Gibson, J.S., and Lau, K.R. (1984). Chloride absorption by marine teleost intestine: the role of the basolateral cell membrane. *J. Physiol.* **354**, 35P.

Ellory, J.C., Gibson, J.S., and Stewart, G.W. (1998). Pathophysiology of abnormal cell volume in human red cells. *Contrib. Nephrol.* **123**, 220–239.

Ellory, J.C., Hall, A.C., and Ody, S.A. (1989). Is acid a more potent activator of KCl co-transport than hypotonicity in human red cells? *J. Physiol.* **420**, 149P.

Ellory, J.C., Hall, A.C., Ody, S.A., deFigueiredos, C.E., Chalder, S., and Stuart, J. (1991). KCl cotransport in HbAA and HbSS red cells: activation by intracellular acidity and disappearance during maturation. In *Red Blood Cell Aging* (Mangani, M. and DeFlora, A., eds), pp. 47–57. Plenum Press, New York.

Eveloff, J. and Warnock, D.G. (1987). K-Cl transport systems in rabbit renal basolateral membrane vesicles. *Am. J. Ren. Physiol.* **252**, F883–F889.

Fatima-Shad, K. and Barry, P.H. (1993). Anion permeation in GABA- and glycine-gated channels of mammalian cultured hippocampal neurons. *Proc. Biol. Sci.* **253**, 69–75.

Ferrell, C.M., Lauf, P.K., Wilson, B.A., and Adragna, N.C. (2000). Lithium and protein kinase C modulators regulate swelling-activated K-Cl cotransport and reveal a complete phosphotidylinositol cycle in low K sheep erythrocytes. *J. Membr. Biol.* **177**, 81–93.

Flatman, P.W. (1980). The effect of buffer composition and deoxygenation on the concentration of ionized magnesium inside human red blood cells. *J. Physiol.* **300**, 19–30.

Flatman, P.W. (2007). Cotransporters, WNKs and hypertension: important leads from the study of monogenetic disorders of blood pressure regulation. *Clin. Sci.* **112**, 203–216.

Flatman, P.W., Adragna, N.C., and Lauf, P.K. (1996). Role of protein kinases in regulating sheep erythrocyte K-Cl cotransport. *Am. J. Physiol.* **271**, C255–C263.

Funder, J. and Wieth, J.O. (1967). Effects of some monovalent anions on Na and K, and on glucose metabolism of ouabain-treated human red cells. *Acta Physiol. Scand.* **71**, 168–185.

Gagnon, K.B., England, R., and Delpire, E. (2006). Volume sensitivity of cation-Cl- cotransporters is modulated by the interaction of two kinases: Ste20-related proline-alanine-rich kinase and WNK4. *Am. J. Physiol. Cell Physiol.* **290**, C134–C142.

Gagnon, K.B.E., Adragna, N.C., Fyffe, R.E.W., and Lauf, P.K. (2007). Characterisation of glial cell K-Cl cotransport. *Cell. Physiol. Biochem.* **20**, 121–130.

Gamba, G. (2005). Molecular physiology and pathophysiology of electroneutral cation-chloride cotransporters. *Physiol. Rev.* **85**, 423–493.

Garay, R.P., Nazaret, C., Hannaert, P.A., and Cragoe, E.J. (1988). Demonstration of a $[K^+,Cl^-]$-cotransport system in human red cells by its sensitivity to [(dihydroindenyl)oxy]alkanoic acids: regulation of cell swelling and distinction from the bumetanide-sensitive $[Na^+K^+,Cl^-]$-cotransport system. *Mol. Pharmacol.* **33**, 696–701.

Geck, P. and Heinz, E. (1980). Coupling of ion flows in cell suspension systems. *Ann. NY Acad. Sci.* **341**, 57–66.

Gibson, J.S. (2001). Oxygen-sensitive cation transport in sickle cells. *Blood Cells, Mol. Dis.* **27**, 112–120.

Gibson, J.S., Cossins, A.R., and Ellory, J.C. (2000). Oxygen-sensitive membrane transporters in vertebrate red cells. *J. Exper. Biol.* **203**, 1395–1407.

Gibson, J.S., Speake, P.F., and Ellory, J.C. (1998). Differential oxygen sensitivity of the KCl cotransporter in normal and sickle human red blood cells. *J. Physiol.* **511**, 225–234.

Gillen, C.M., Brill, S., Payne, J.A., and Forbusch, I.B. (1996). Molecular cloning and functional expression of the KCl cotransporter from rabbit, rat and human. *J. Biol. Chem.* **217**, 16237–16244.

Gillen, C.M. and Forbush, B., III (1999). Functional interaction of the K-Cl cotransporter (KCC1) with the Na-K-Cl cotransporter in HEK-293 cells. *Am. J. Physiol.* **276**, C328–C336.

Godart, H., Dormandy, A., and Ellory, J.C. (1997). Do HbSS erythrocytes lose KCl in physiological conditions? *Brit. J. Haematol.* **98**, 25–31.

Greger, R. and Schlatter, E. (1983). Properties of the basolateral membrane of the cortical thick ascending limb of Henle's loop of rabbit kidney. A model for secondary active chloride transport. *Pflugers Arch.* **396**, 325–334.

Hall, A.C. and Ellory, J.C. (1986a). Effect of high hydrostatic pressure on "passive" monovalent cation transport in human red cells. *J. Membr. Biol.* **94**, 1–17.

Hall, A.C. and Ellory, J.C. (1986b). Evidence for the presence of volume-sensitive KCl transport in "young" human red cells. *Biochim. Biophys. Acta* **858**, 317–320.

Hallows, K.R. and Knauf, P.A. (1994). Principles of cell volume regulation. In *Cellular and Molecular Physiology of Cell Volume Regulation* (Strange, K. ed.), pp. 3–29. CRC Press, Florida.

Hebert, S.C., Mount, D.B., and Gamba, G. (2004). Molecular physiology of cation-coupled Cl$^-$ cotransport: the SLC12 family. *Pflugers Arch.* **447**, 580–593.

Hiki, K., D'Andrea, R.J., Furez, J., Crawford, J., Woollatt, E., Sutherland, G.R., Vadas, M.A., and Gamble, J.R. (1999). Cloning, characterization and chromosomal location of a novel human K$^+$-Cl$^-$ cotransporter. *J.Biol. Chem.* **274**, 10661–10667.

Hladky, S.B. and Rink, T.J. (1977). pH equilibrium across the red cell membrane. In *Membrane Transport in Red Cells* (Ellory, J.C. and Lew, V.L., eds), pp. 115–135. Academic Press, London.

Hoffman, J.F. (1966). The red cell membrane and the transport of sodium and potassium. *Am. J. Med.* **41**, 666–680.

Hoffmann, E.K., Lambert, I.H., and Simonsen, L.O. (1988). Mechanisms of volume regulation in Ehrlich ascites tumour cells. *Renal Physiol. Biochem.* **11**, 221–247.

Howard, H.C., Mount, D.B., Rochefort, D., Byun, N., Dupre, N., Lu, J., Fan, X., Song, L., Riviere, J.B., Prevost, C., Horst, J., Simonati, A., Lemcke, B., Welch, R., England, R., Zhang, F.Q., Mercado, A., Siesser, W.B., George, A.L., McDonald, M.P., Bouchard, J.P., Mathieu, J., Delpire, E., and Rouleau, G.A. (2002). The K-Cl cotransporter KCC3 is mutant in a severe peripheral neuropathy associated with agenesis of the corpus callosum. *Nat. Gen.* **32**, 384–393.

Hsu, Y.-M., Chen, Y.-F., Chou, C.-Y., Tang, M.-J., Chen, J.H., Wilkins, R.J., Ellory, J.C., and Shen, M.-R. (2007a). KCl cotransporter-3 down-regulates E-cadherin/beta-catenin complex to promote epithelial–mesenchymal transition. *Cancer Res.* **67**, 11064–11073.

Hsu, Y.-M., Chou, C.-Y., Chen, H.H., Lee, W.Y., Chen, Y.F., Lin, P.W., Alper, S.L., Ellory, J.C., and Shen, M.-R. (2007b). IGF-1 upregulates electroneutral K-Cl cotransporter KCC3 and KCC4 which are differentially required for breast cancer cell proliferation and invasiveness. *J. Cell Physiol.* **210**, 626–636.

Hubner, C.A., Stein, V., Hermans-Borgmeyer, I., Meyer, T., Ballanyi, K., and Jentsch, T.J. (2001). Disruption of KCC2 reveals an essential role of K-Cl cotransport already in early synaptic inhibition. *Neuron* **30**, 515–524.

Jennings, M.L. and Adame, M.F. (2001). Direct estimate of 1:1 stoichiometry of K(+)-Cl(−) cotransport in rabbit erythrocytes. *Am. J. Physiol.* **281**, C825–C832.

Jennings, M.L. and Al-Rohil, N. (1990). Kinetics of activation and inactivation of swelling-stimulated K/Cl transport: The volume-sensitive parameter is the rate constant for inactivation. *J. Gen. Physiol.* **95**, 1021–1040.

Jennings, M.L. and Schulz, R.K. (1991). Okadaic acid inhibition of KCL cotransport: evidence that protein dephosphorylation is necessary for activation of transport by either swelling or N-ethylmaleimide. *J. Gen. Physiol.* **97**, 799–817.

Joiner, C.H. (1993). Cation transport and volume regulation in sickle red blood cells. *Am. J. Physiol.* **264**, C251–C270.

Kahle, K.T., Rinehart, J., de los Heros, P., Louvi, A., Meade, P., Vazquez, N., Hebert, S.C., Gamba, G., Gimenez, I., and Lifton, R.P. (2005). WNK3 modulates transport of Cl$^-$ in and out of cells: implications for control of cell volume and neuronal excitability. *Proc. Nat. Acad. Sci. USA* **102**, 16783–16788.

Kahle, K.T., Rinehart, J., Ring, A., Gimenez, I., Gamba, G., Hebert, S. C., and Lifton, R.P. (2006). WNK protein kinases modulate cellular Cl$^-$ flux by altering the phosphorylation state of the Na-K-Cl and K-Cl cotransporters. *Physiology* **21**, 326–335.

Kaji, D.M. (1989). Kinetics of volume-sensitive K transport in human erythrocytes: evidence for asymmetry. *Am. J. Physiol.* **256**, C1214–C1223.

Kaji, D.M. (1993). Effect of membrane potential on K-Cl transport in human erythrocytes. *Am. J. Physiol.* **264**, C376–C382.

Kaji, D.M. and Gasson, C. (1995). Urea activation of K-Cl cotransport in human erythrocytes. *Am. J. Physiol.* **268**, C1018–C1025.

Kaji, D.M. and Tsukitani, Y. (1991). Role of protein phosphorylation in activation of KCl cotransport in human erythrocytes. *Am. J. Physiol.* **260**, C178–C182.

Kamel, K.S. and Halperin, M.L. (2002). Bartter's, Gitelman's and Gordon's syndromes. From physiology to molecular biology and back, yet still some unanswered questions. *Nephron* **92** (Supplement 1), 18–27.

Karadsheh, M.F., Byun, N., Mount, D.B., and Delpire, E. (2004). Localization of the KCC4 K-Cl cotransporter in the nervous system. *Neuroscience* **123**, 381–392.

Kelley, S.J. and Dunham, P.B. (1996). Mechanism of swelling activation of K-Cl cotransport in inside-out vesicles of LK sheep erythrocyte membranes. *Am. J. Physiol.* **270**, C1122–C1130.

Khan, A. and Ellory, J.C. (2000). Elevated temperatures enhance KCC1 activity in sickle cells. *Bioelectrochemistry* **52**, 127–131.

Kregenow, F.M. (1971). The response of duck erythrocytes to ninhemolytic hypotonic media. Evidence for a volume-controlling mechanism. *J. Gen. Physiol.* **58**, 372–395.

Kregenow, F.M. (1977). Transport in avian red cells. In *Membrane Transport in Red Cells* (Lew, V.L. and Ellory, J.C., ed.), pp. 383–426. Academic, London.

Kregenow, F.M. (1980). Osmoregulatory salt transport mechanisms: control of cell volume in anisotonic media. *Annu. Rev. Physiol.* **43**, 493–505.

Lauf, P.K. (1983). Thiol-dependent passive K/Cl transport in sheep red cells I. Dependence on chloride and external ions. *J. Membr. Biol.* **73**, 237–246.

Lauf, P.K. (1984). Thiol dependent, passive K/Cl transport in sheep erythrocytes. IV. Furosemide inhibition and the role of external Rb$^+$, Na$^+$ and Cl$^-$. *J. Membr. Biol.* **77**, 57–62.

Lauf, P.K. (1985). Passive K$^+$-Cl$^-$ fluxes in low-K$^+$ sheep erythrocytes: modulation by A23187 and bivalent cations. *Am. J. Physiol.* **249**, C271–C278.

Lauf, P.K. (1988). Thioldependent K:Cl transport in sheep red cells: VIII. Activation through metabolically and chemically reversible oxidation by diamide. *J. Memb. Biol.* **101**, 179–188.

Lauf, P.K. and Adragna, N.C. (1996). A thermodynamic study of electroneutral K-Cl cotransport in pH- and volume-clamped low K sheep erythrocytes with normal and low internal magnesium. *J. Gen. Physiol.* **108**, 341–350.

Lauf, PK. and Adragna, N.C. (1998). Functional evidence for a pH sensor of erythrocyte K-Cl cotransport through inhibition of internal protons and diethylcarbonate. *Cell. Physiol. Biochem.* **8**, 46–60.

Lauf, P.K. and Bauer, J. (1987). Direct evidence for chloride-dependent volume reduction in macrocytic sheep reticulocyte. *Biochem. Biophys. Res. Comm.* **144**, 849–855.

Lauf, P.K., Bauer, J., Adragna, N.C., Fujise, H., Martin, A., Zade-Oppen, M., Ryu, K.H., and Delpire, E. (1992). Erythrocyte K-Cl cotransport: properties and regulation. *Am. J. Physiol.* **263**, C917–C932.

Lauf, P.K., Misri, S., Chimote, A.A., and Adragna, N.C. (2008). Apparent intermediate K conductance channel hyposmotic activation in human lens epithelial cells. *Am. J. Physiol.* **294**, C820–C832.

Lauf, P.K. and Theg, B.E. (1980). A chloride dependent K^+ flux induced by *N*-ethylmaleimide in genetically low K^+ sheep and goat erythrocytes. *Biochem. Biophys. Res. Comm.* **92**, 1422–1428.

Lauf, P.K., Warwar, R., Brown, T.L., and Adragna, N.C. (2006). Regulation of potassium transport in human lens epithelial cells. *Exper. Eye Res.* **82**, 55–64.

Lauf, P.K., Zhang, J., Delpire, E., Fyffe, R.E.W., Mount, D.B., and Adragna, N.C. (2001). Erythrocyte K-Cl cotransport: immunocytochemical and functional evidence for more than one KCC isoform in HK and LK sheep red blood cells. *Comp. Biochem. Physiol.* **130**, 499–509.

Lew, V.L. and Bookchin, R.M. (2005). Ion transport pathology in the mechanism of sickle cell dehydration. *Physiol. Rev.* **85**, 179–200.

Li, H., Tornberg, J., Kaila, K., Airaksinen, M.S., and Rivera, C. (2002). Patterns of cation-chloride cotransporter expression during embryonic rodent CNS development. *European J. Neurosci.* **16**, 2358–2370.

Lytle, C., McManus, T.J., and Haas, M. (1998). A model of Na-K-2Cl cotransport based on ordered ion binding and glide symmetry. *Am. J. Physiol.* **274**, C299–C309.

Mercado, A., de los Heros, P., Vazquez, N., Meade, P., Mount, D.B., and Gamba, G. (2001). Functional and molecular characterization of the K-Cl cotransporter of *Xenopus leavis* oocytes. *Am. J. Cell Physiol.* **281**, C670–C680.

Mercado, A., Mount, D.B., Cortes, R., Vazquez, N., and Gamba, G. (2002). Functional characterization of two alternative isoforms of the KCC-3 K-Cl cotransporter. *FASEB J.* **16**, A58.

Mercado, A., Mount, D.B., and Gamba, G. (2004). Electroneutral cation-chloride cotransporters in the central nervous system. *Neurochem. Res.* **29**, 17–25.

Mercado, A., Song, L., Vazquez, N., Mount, D.B., and Gamba, G. (2000). Functional comparison of the K^+-Cl^- cotransporters KCC1 and KCC4. *J. Biol. Chem.* **275**, 30326–30334.

Mercado, A., Vazquez, N., Song, L., Cortes, R., Enck, A.H., Welch, R., Delpire, E., Gamba, G., and Mount, D.B. (2005). NH_2-terminal heterogeneity in the KCC3 K^+-Cl^- cotransporter. *Am. J. Cell Physiol.* **289**, F1246–F1261.

Merciris, P., Hardy-Dessources, M.D., and Giraud, F. (2001). Deoxygenation of sickle cells stimulates Syk tyrosine kinase and inhibits a membrane tyrosine phosphatase. *Blood* **98**, 3121–3127.

Meyer, J., Johannssen, K., Freitag, C.M., Schraut, K., Teuber, I., Hahner, A., Mainhardt, C., Mossner, R., Volz, H.-P., Wienker, T.F., McKeane, D., Stephan, D.A., Rouleau, G., Reif, A., and Lesch, K.-P. (2005). Rare variants of the gene encoding the potassium chloride co-transporter 3 are associated with bipolar disorder. *Int. J. Neuropsychopharmacol.* **8**, 495–504.

Minton, A.P. (1994). Influence of macromolecular crowding on intracellular association reactions: possible role in volume regulation. In *Cellular and Molecular Physiology of Cell Volume Regulation* (Strange, K. ed.), pp. 181–190. CRC, Boca Raton.

Misri, S., Chimote, A.A., Adragna, N.C., and Lauf, P.K. (2006). KCC isoforms in a human lens epithelial cell line (B3) and lens tissue extracts. *Exper. Eye Res.* **83**, 1287–1294.

Moser, D., Ekawardhani, S., Kumsta, R., Palmason, H., Bock, C., Athanassiadou, Z., Lesch, K.-P., and Meyer, J. (2008). Functional analysis of a potassium-chloride co-transporter 3 (SLC12A6) promoter polymorphism leading to an additional DNA methylation site. *Neuropsychopharmacology* (in press).

Mount, D.B., Delpire, E., Gamba, G., Hall, A.E., Poch, E., Hoover, R. S.J., and Hebert, S.C. (1998). The electroneutral cation-chloride cotransporters. *J. Exper. Biol.* **201**, 2091–2102.

Mount, D.B. and Gamba, G. (2001). Renal potassium-chloride cotransporters. *Curr. Opin. Nephrol. Hypertension* **10**, 685–691.

Mount, D.B., Mercado, A., Song, L., Xu, J., George, A.L., Delpire, E., and Gamba, G. (1999). Cloning and characterization of KCC3 and KCC4, new members of the cation-chloride cotransporter gene family. *J. Biol. Chem.* **274**, 16355–16362.

Muzyamba, M.C., Campbell, E.H., and Gibson, J.S. (2006). Effect of intracellular magnesium and oxygen tension on K + -Cl− cotransport in normal and sickle human red cells. *Cell. Physiol. Biochem.* **17**, 121–128.

O'Neill, W.C. (1991). Swelling-activated K-Cl cotransport: metabolic dependence and inhibition by vanadate and fluoride. *Am. J. Physiol.* **260**, C308–C315.

Olivieri, O., Vitoux, D., Galacteros, F., Bachir, D., Blouquit, Y., Beuzard, Y., and Brugnara, C. (1992). Hemoglobin variants and activity of (K^+Cl^-) cotransport system in human erythrocytes. *Blood* **79**, 793–797.

Orlov, S.N., Kolosova, I.A., Cragoe, E.J., Gurlo, T.G., Mongin, A. A., Aksentsev, S.L., and Konev, S.V. (1993). Kinetics and peculiarities of thermal inactivation of volume-induced Na^+/H^+ exchange, Na^+,K^+,$2Cl^-$ cotransport and K^+,Cl^- cotransport in rat erythrocytes. *Biochim. Biophys. Acta* **1151**, 186–192.

Pauling, L., Itano, H.A., Singer, S.J., and Wells, I.C. (1949). Sickle cell anaemia, a molecular disease. *Science* **110**, 1141–1152.

Payne, J.A. (1997). Functional characterization of the neuronal-specific K-Cl cotransporter: implications for $[K^+]_o$ regulation. *Am. J. Physiol.* **273**, C1516–C1525.

Payne, J.A., Lytle, C., and McManus, T.J. (1990). Foreign anion substitution for chloride in human red blood cells: effect on ionic and osmotic equilibria. *Am. J. Physiol.* **259**, C819–C827.

Payne, J.A., Stevenson, T.J., and Donaldson, L.F. (1996). Molecular characterization of a putative K-Cl cotransporter in rat brain. *J. Boil. Chem.* **271**, 16245–16252.

Pearson, M.M., Lu, J., Mount, D.B., and Delpire, E. (2001). Localisation of the K^+-Cl^- cotransporter, KCC3, in the central and peripheral nervous systems: expression in the choroid plexus, large neurons and white matter tracts. *Neuroscience* **103**, 481–491.

Pellegrino, C.M., Rybicki, A.C., Musto, S., Nagel, R.L., and Schwartz, R.S. (1998). Molecular identification of erythroid K:Cl cotransporter in human and mouse erythroleukemic cells. *Blood Cells, Mol. Dis.* **24**, 31–40.

Race, J.E., Makhlouf, F.N., Logue, P.J., Wilson, F.H., Dunham, P.B., and Holtzman, E.J. (1999). Molecular cloning and functional characterization of KCC3, a new K-Cl cotransporter. *Am. J. Physiol.* **277**, C1210–C1219.

Reid, K.H., Li, G.-Y., Payne, R.S., Schurr, A., and Cooper, N.G.F. (2001). The mRNA level of the potassium-chloride cotransporter KCC2 covaries with seizure susceptibility in inferior colliculus of the post-ischemic audiogenic seizure-prone rat. *Neurosci. Lett.* **308**, 29–32.

Reuss, L. and Cotton, C.U. (1994). Volume regulation in epithelia: transcellular transport and cross-talk. In *Cellular and Molecular Physiology of Cell Volume Regulation* (Strange, K. ed.), pp. 31–47. CRC Press, Florida.

Rivera, C., Voipio, J., Payne, J.A., Ruusuvuori, E., Lahtinen, H., Lamsa, K., Pirvola, U., Saarma, M., and Kaila, K. (1999). The K^+/Cl^- cotransporter KCC2 renders GABA hyperpolarizing during neuronal maturation. *Nature* **397**, 251–255.

Rust, M.B., Alper, S.L., Rudhard, Y., Shmukler, B.E., Vicente, R., Brugnara, C., Trudel, M., Jentsch, T.J., and Hubner, C.A. (2007). Disruption of erythroid K-Cl cotransporters alters erythrocyte volume and partially rescues erythrocyte dehydration in SAD mice. *J. Clin. Invest.* **117**, 1708–1717.

Rust, M.B., Faulhaber, J., Budack, M., Pfeffer, C., Maritzen, T., Didie, M., Beck, F.-X., Boettger, T., Schubert, R., Ehmke, H., Jentsch, T.J., and Hubner, C.A. (2006). Neurogenic mechanisms contribute to hypertension in mice with disruption of the K-Cl cotransporter KCC3. *Circ. Res.* **98**, 549–556.

Sachs, J.R. (1971). Ouabain-insensitive sodium movements in the human red blood cells. *J. Gen. Physiol.* **57**, 259–282.

Sachs, J.R. and Martin, D.W. (1993). The role of ATP in swelling-stimulated K-Cl cotransport in human red cell ghosts. *J. Gen. Physiol.* **102**, 551–573.

Saitta, M., Cavalier, S., Garay, R., Cragoe, E.J., and Hannaert, P. (1990). Evidence for a DIOA-sensitive [K^+-Cl^-]-cotransport system in cultured vascular smooth muscle cells. *Am. J. Hypertension* **3**, 939–942.

Schmidt, W.F. and McManus, T.J. (1977). Ouabain-insensitive salt and water movements in duck red cells. Norepinephrine stimulation of sodium plus potassium cotransport. *J. Gen. Physiol.* **70**, 81–97.

Shen, M.-R., Chou, C.-Y., and Ellory, J.C. (2000). Volume-sensitive KCl cotransport associated with human cervical carcinogenesis. *Pflugers Arch.* **440**, 751–760.

Shen, M.-R., Chou, C.-Y., Hsu, K.F., Chiu, W.T., Tang, M.-J., Alper, S.L., and Ellory, J.C. (2003). KCl cotransport is an important modulator of human cervical growth and invasion. *J. Biol. Chem.* **278**, 39941–39950.

Shen, M.-R., Lin, A.C., Hsu, Y.-M., Chang, T.J., Tang, M.-J., Alper, S.L., Ellory, J.C., and Chou, C.-Y. (2004). Insulin-like growth factor 1 stimulates KCl cotransport, which is necessary for invasion and proliferation of cervical cancer and ovarian cancer cells. *J. Biol. Chem.* **279**, 40017–40025.

Simard, C.F., Bergeron, M.J., Frenette-Cotton, R., Carpentier, G.A., Pelchat, M.-E., Caron, L., and Isenring, P. (2007). Homooligomeric and heterooligomeric associations between K^+-Cl^- cotransporter isoforms and between K^+-Cl^- and Na^+-K^+-Cl^- cotransporters. *J. Biol. Chem.* **282**, 18083–18093.

Soler, R.R., Hannaert, P., Cragoe, E.J., and Garay, R.P. (1993). Volume-dependent K^+ and Cl^- fluxes in rat thymocytes. *J. Physiol.* **465**, 387–401.

Song, L., Mercado, A., Vazquez, N., Xie, Q., Desai, R., George, A.L., Gamba, G., and Mount, D.B. (2002). Molecular, functional and genomic characterization of human KCC2, the neuronal K-Cl cotransporter. *Mol. Brain Res.* **103**, 91–105.

Speake, P.F., Roberts, C.A., and Gibson, J.S. (1997). Effect of respiratory blood parameters on equine red blood cell K-Cl cotransporter. *Am. J. Physiol.* **273**, C1811–C1818.

Strange, K., Singer, T.D., Morrison, R., and Delpire, E. (2000). Dependence of KCC2 K-Cl cotransporter activity on a conserved carboxy terminus tyrosine residue. *Am. J. Physiol. Cell Physiol.* **279** (3), C860–C867.

Stuart, M.J. and Nagel, R.L. (2004). Sickle-cell disease. *The Lancet* **364**, 1343–1360.

Su, W., Shmukler, B.E., Chernova, M.A., Stuart-Tilley, A.K., De Franceschi, L., Brugnara, C., and Alper, S.L. (1999). Mouse K-Cl cotransporter KCC1: cloning, mapping, pathological expression, and functional regulation. *Am. J. Physiol.* **277**, C899–C912.

Taouil, K. and Hannaert, P. (1999). Evidence for the involvement of K+ channels and K(+)-Cl- cotransport in the regulatory volume decrease of newborn rat cardiomyocytes. *Pflugers Arch.* **439**, 56–66.

Thornhill, W.B. and Laris, P.C. (1984). KCl loss and cell shrinkage in the Ehrlich ascites tumour cell induced by hypotonic media, 2-deoxyglucose and propanolol. *Biochim. Biophys. Acta* **773**, 207–218.

Ueno, T., Okabe, A., Akaike, N., Fukuda, A., and Nabekura, J. (2002). Diversity of neuron-specific K^+-Cl^- cotransporter expression and inhibitory postsynaptic potential depression in rat motorneurons. *J. Biol. Chem.* **277**, 4945–4950.

Uyanik, G., Elcioglu, N., Penzien, J., Gross, C., Yilmaz, Y., Olmez, A., Demir, E., Wahl, D., Scheglmann, K., Winner, B., Bogdahn, U., Topaloglu, H., Hehr, U., and Winkler, J. (2006). Novel truncating and missense mutations of the KCC3 gene associated with Andermann syndrome. *Neurology* **66**, 1044–1048.

Wake, H., Watanabe, M., Moorhouse, A.J., Kanematsu, T., Horibe, S., Matsukawa, N., Asai, K., Ojika, K., Hirata, M., and Nabekura, J. (2007). Early changes in KCC2 phosphorylation in response to neuronal stress result in functional downregulation. *J. Neurosci.* **27**, 1642–1650.

Wiley, J.S. and Cooper, R.A. (1974). A furosemide-sensitive cotransport of sodium plus potassium in the human red cell. *J. Clin. Invest.* **53**, 745–755.

Woo, N.S., Lu, J., England, R., McClelland, R., Dufour, S., Mount, D.B., Deutch, A.Y., Lovinger, D.M., and Delpire, E. (2002). Hyperexcitability and epilepsy associated with disruption of the mouse neuronal-specific K-Cl cotransporter gene. *Hippocampus* **12**, 258–268.

ZadeOppen, A.M.M. and Lauf, P.K. (1990). Thioldependent passive K:Cl transport in sheep red blood cells: IX. Modulation by pH in the presence and absence of DIDS and the effect of NEM. *J. Membr. Biol.* **118**, 143–151.

Zelikovic, I. (2001). Molecular pathophysiology of tubular transport disorders. *Pediat. Nephrol.* **16**, 919–935.

Zhang, J., Lauf, P.K., and Adragna, N.C. (2003). Platelet-derived growth factor regulates K-Cl cotransport in vascular smooth muscle cells. *Am. J. Physiol.* **284**, C674–C680.

Zhou, H.X., Rivas, G., and Minton, A.P. (2008). Macromolecular crowding and confinement: biochemical, biophysical, and potential physiological consequences. *Annu. Rev. Biophys.* **37**, 375–397.

CHAPTER

4

From Cloning to Structure, Function, and Regulation of Chloride-dependent and Independent Bicarbonate Transporters

Michael F. Romero, Min-Hwang Chang and David B. Mount

OUTLINE

I. INTRODUCTION

Regulation of acid–base homeostasis as well as other ionic concentrations, such as Na^+, is critical for animal life. For example, humans' normal blood pH is closely regulated within 7.35–7.45. Since vertebrates, mammals in particular, generate significant amounts of acid via metabolism, these organisms must excrete acid (H^+) or increase systemic HCO_3^- concentration to buffer this metabolic acid. Bicarbonate, along with CO_2, is the major pH buffering system of biological fluids. Bicarbonate coexists with CO_2 gas in solution. The reaction $CO_2 + H_2O \leftrightarrow H_2CO_3 \leftrightarrow H^+ + HCO_3^-$ is readily reversible. In biological systems, the interconversion of CO_2 to HCO_3^- is catalyzed by the family of carbonic anhydrase enzymes (Nakhoul et al., 1996; Sly and Hu, 1995). In some epithelia, bicarbonate in ultrafiltrate luminal fluid with Na^+-H^+ exchanger secretes H^+ (Alpern and Rector, 1996) and forms CO_2 and H_2O. Carbon dioxide is vented by the lung. The lungs and the kidneys play complementary roles in acid–base homeostasis (Romero et al., 1997).

In kidney renal proximal tubules most of the filtered HCO_3^- is reabsorbed. The mechanisms by which HCO_3^- moves from the proximal tubule cell back into the blood were unclear before 1980. The first electrogenic Na^+-HCO_3^- cotransporter (NBC1) was discovered by Boron and Boulpaep (Boron and Boulpaep, 1983) in the renal proximal tubule of the salamander. This new basolateral Na^+-HCO_3^- cotransporter is called "electrogenic" because it transports multiple HCO_3^- with each Na^+, and it is Cl^- independent. It is inhibited by stilbene sulfonate derivatives such as DIDS. A similar cotransporter activity was later reported in several mammals: bovine corneal endothelia cells (Jentsch et al., 1984), rat proximal convoluted tubules (Alpern, 1985) and rabbit proximal straight tubules.

However, the molecular characterization of this cotransporter protein was not revealed until 1996 (Romero et al., 1996, 1997) when salamander (*Ambystoma tigrinum*) kidney was used in expression cloning in *Xenopus* oocytes. At the molecular level this electrogenic NBC1 sequence is related to the electroneutral band-3-like proteins – the anion exchangers AE1, AE2, and AE3 (Romero et al., 1998, 1997). The molecular homology reveals a topologically related transporter family now known as the bicarbonate transporter superfamily SLC4. Other members in the family including: electroneutral Na^+-HCO_3^- cotransporters (NBC3), a second electrogenic Na^+-HCO_3^- cotransporter (NBC4), and Na^+-driven Cl^-/HCO_3^- exchangers (NDAE1 and NDCBE1).

II. HISTORY OF ANION EXCHANGERS

The history of anion exchangers is concurrent with analytical tools developed for studying membrane proteins, particularly those of the red blood cell (Yawata, 2003). Why the red blood cell? It had been known since the early 1900s that red cell shape could be changed from the typical "biconcave disc" to other shapes. These set of hereditary pathophysiologies is known as hereditary spherocytosis. For about 60 years, the scientific debate over the nature of these cell deformations focused on ATP usage and cationic permeabilities. Major contributions were made by the laboratories of Hoffman, Parker, Passow, Skou and Tosteson. Many of these studies began to make use of the isolate red cell membrane (red cell ghost), developed by Dodge (Dodge et al., 1963), which allowed control over both the extracellular as well as the intracellular solution composition. The biochemical tools of the time were somewhat limited in that membrane proteins were often not solubilized. It was Maddy who initially proposed the possibility of using a charged detergent, sodium dodecyl sulfate (SDS), to solubilize membrane proteins. This method was of course then used by Laemmli (Laemmli, 1970) and Fairbanks (Fairbanks et al., 1971) to isolate membrane proteins from the red blood cell. From these classic and initial SDS-PAGE experiments, these investigators began to characterize the membrane components of the red cell ghosts. These major proteins were counted from the top (highest molecular weight) of the SDS gel:

band 1	α-spectrin
band 2	β-spectrin
band 2.1	ankyrin
band 3	anion exchanger 1 (AE1, SLC4A1)
band 4.1	protein 4.1
band 4.2	protein 4.2
band 5	actin
band 6	glyceraldehyde-3-phosphate dehydrogenase (G-3-PD)
band 7	stomatin or aquaporin (CHIP28)

As indicated above, "band 3" turned out to be a Cl^-/HCO_3^- exchanger or more generally and anion exchanger, i.e. anion exchanger 1.

These methods of isolating red cell ghosts and identifying band 3 as a membrane transport protein continues to be experimentally exploited. Early experiments to characterize the function of this anion exchange protein and its mechanism of membrane transport were performed by the groups of Michael Tanner (Bristol) (Tanner and Boxer, 1972; Tanner and Gray, 1971), Philip Low (Low, 1978; Snow et al., 1978), Michael Jennings (Jennings, 1976, 1978; Ku et al., 1979), Phil Knauf (Knauf and Rothstein, 1971; Rothstein et al., 1976), Reinhart Reithmeier (Boodhoo and Reithmeier, 1984; Reithmeier, 1983) and others (Cabantchik and Rothstein, 1974; Lepke and Passow, 1972; Schwoch and Passow, 1973).

III. MODES OF MEMBRANE ANION TRANSPORT – THEORETICAL

Membranes present a barrier to the movement of ions and other solutes. The cell membrane is a mosaic of phospholipids, cholesterol and proteins. The phospholipids make the bulk of the membrane and form a bilayer. This lipid bilayer has the hydrophilic (water-liking) phosphate and sugar head-groups facing both the outside world and the inside world. The middle of the lipid bilayer contains the C_{14}–C_{22} acyl chains and is hydrophobic (water repelling). This bilayer functions like an electrical capacitor, separating charge and thus creating a membrane potential (electrical gradient). However, unlike a capacitor, an intact bilayer without proteins will not pass ions or solutes (uncharged molecules), thus creating an additional solute concentration gradient (chemical gradient).

"Nature" has devised four major ways to move ions and solutes across this otherwise non-permeable membrane. One mode is primary active transport, which consumes ATP to move solutes against their gradients, and will not be considered further. The three other modes of transmembrane movement make use of the "potential energy" (technically the Gibbs free energy, ΔG) stored in the form of the electrochemical gradient or electrochemical potential difference. As the names imply, the electrochemical potential difference ($\Delta\mu_{ion}$)[a] has an electrical part ($zF\Delta\Psi$)[b] and a chemical part ($RT\ln\{[\text{ion}]_{inside}/[\text{ion}]_{outside}\}$)[c]. This equation can be used to predict whether ions and solutes will move into or out of a cell (or vesicle). The "rules" governing which constituents are in the $\Delta\mu$ equation are determined by which protein is performing the transport. If there is no protein, there is no transmembrane movement of ions.

The first transmembrane route is "diffusion". Diffusion implies that the particle is moving freely in an aqueous solution down its electrochemical gradient. For a membrane this "diffusive pathway" is basically a channel, which has defined opening and closing characteristics. When the channel is open, both the concentration gradient and the electrical gradient of the permeable ion determine whether the ion will move into or out of the cell. For example, the movement of Cl^- through a Cl^- channel in a non-excitable cell will be determined by

$$\Delta\mu_{Cl} = RT\ln\left(\frac{[Cl^-]_{inside}}{[Cl^-]_{outside}}\right) + (-1)FV_m \qquad \text{(Eq. 1)}$$

The second transmembrane route for ions and solutes is exchange. As the name implies, "exchange" means that at least one ion from the outside moves to the inside *and* there is an obligate movement of an ion from the inside to the outside. The band 3, Cl^-/HCO_3^- exchanger is an example of a 1 Cl^- for 1 HCO_3^- exchange. This type of transport mode is also subject to the driving force of the electrochemical potential. However, since this exchange activity is found in a single protein, the movements of the two ions are coupled. In the "$\Delta\mu_{ion}$" form:

$$\Delta\mu_{Cl/HCO_3} = \Delta\mu_{Cl} - \Delta\mu_{HCO_3} \qquad \text{(Eq. 2)}$$

Or,

$$\Delta\mu_{Cl/HCO_3} = RT\ln\left(\frac{[Cl^-]_i}{[Cl^-]_o}\right) + (-1)FV_m - \left\{RT\ln\left(\frac{[HCO_3^-]_i}{[HCO_3^-]_o}\right) + (-1)FV_m\right\}$$

Or,

$$\Delta\mu_{Cl/HCO_3} = RT\ln\left(\frac{[Cl^-]_i[HCO_3^-]_o}{[Cl^-]_o[HCO_3^-]_i}\right) - FV_m + FV_m$$

Hence,

$$\Delta\mu_{Cl/HCO_3} = RT\ln\left(\frac{[Cl^-]_i[HCO_3^-]_o}{[Cl^-]_o[HCO_3^-]_i}\right) \qquad \text{(Eq. 3)}$$

In other words for this example, the charges moving in both directions sum to zero. Therefore, this electroneutral Cl^-/HCO_3^- exchanger is predicted to be *unaffected* by membrane potential. However, if there is a net charge movement by the exchange process, then the exchanger will also be effected by membrane potential (see section VI.C.4; further discussion in Chapter 5, section II.E).

The third transmembrane route for ions and solutes is cotransport. Cotransport is also known as "symport", i.e. transport in the same direction (two or more ions into the cell; or two or more ions out of the cell). An example is K^+-Cl^- cotransport (see Chapters 5 and 17). Again, the ion movements are coupled in the same protein.

The driving force for the cotransport is given by the following equation:

$$\Delta\mu_{K\text{-}Cl} = \Delta\mu_{K} + \Delta\mu_{Cl} \qquad \text{(Eq. 4)}$$

If follows that,

$$\Delta\mu_{K\text{-}Cl} = RT\ln\left(\frac{[K^+]_i}{[K^+]_o}\right) + (+1)FV_m + RT\ln\left(\frac{[Cl^-]_i}{[Cl^-]_o}\right) + (-1)FV_m$$

[a] $\Delta\mu_{ion} = RT\ln\{[\text{ion}]_{inside}/[\text{ion}]_{outside}\} + zF\Delta\Psi$.
[b] "z" is the net charge on a molecule, F is the Faraday constant and $\Delta\Psi = \Psi(\text{inside}) - \Psi(\text{outside})$, which is also equal to the membrane potential (V_m).
[c] "R" is the gas constant, "T" is temperature in kelvin degrees and "ln" is the natural log.

Hence,

$$\Delta\mu_{\text{K-Cl}} = RT\ln\left(\frac{[\text{K}^+]_i[\text{Cl}^-]_i}{[\text{K}^+]_o[\text{Cl}^-]_o}\right) + FV_m - FV$$

Or,

$$\Delta\mu_{\text{K-Cl}} = RT\ln\left(\frac{[\text{K}^+]_i[\text{Cl}^-]_i}{[\text{K}^+]_o[\text{Cl}^-]_o}\right) \quad \text{(Eq. 5)}$$

For this example as well, the net charge moving in the same direction is zero, and this K^+-Cl^- cotransport would be *unaffected* by membrane potential.

These are the "simple" transmembrane transport modes. Of course, it is also possible to have complex transporters. One "complex" transporter which will be discussed below is the Na^+-driven Cl^-/HCO_3^- exchanger (also known generally as Na^+-driven anion exchanger). Even though several genes encoding this activity have been cloned and the proteins functionally expressed, there is still discussion over two major transport models. Model 1 predicts 1 Na^+ and 1 HCO_3^- being exchanged for 1 Cl^- and 1 H^+. Model 2 predicts 1 Na^+ and 2 HCO_3^- being exchanged for 1 Cl^-. In both cases, there is no net charge movement, so the transporters should not be affected by membrane potential. Nevertheless, the $\Delta\mu_{\text{Model 1}}$ (Eq. 8) and $\Delta\mu_{\text{Model 2}}$ (Eq. 10) are quite different.

$$\Delta\mu_{\text{Model 1}} = \Delta\mu_{\text{Na}} + \Delta\mu_{\text{HCO}_3} - (\Delta\mu_{\text{Cl}} + \Delta\mu_{\text{H}}) \quad \text{(Eq. 6)}$$

Or,

$$\Delta\mu_{\text{Model 1}} = RT\ln\left(\frac{[\text{Na}^+]_i[\text{HCO}_3^-]_i[\text{Cl}^-]_o[\text{H}^+]_o}{[\text{Na}^+]_o[\text{HCO}_3^-]_o[\text{Cl}^-]_i[\text{H}^+]_i}\right) \quad \text{(Eq. 7)}$$

Or,[d]

$$\Delta\mu_{\text{Model 1}} = 2.3\,(\text{pH}_i - \text{pH}_o)\left(RT\log_{10}\left(\frac{[\text{Na}^+]_i[\text{HCO}_3^-]_i[\text{Cl}^-]_o}{[\text{Na}^+]_o[\text{HCO}_3^-]_o[\text{Cl}^-]_i}\right)\right) \quad \text{(Eq. 8)}$$

$$\Delta\mu_{\text{Model 2}} = \Delta\mu_{\text{Na}} + 2\Delta\mu_{\text{HCO}_3} - \Delta\mu_{\text{Cl}} \quad \text{(Eq. 9)}$$

[d]Remember that $RT\ln(X) = 2.3\{RT\log_{10}(X)\}$.

Or,

$$\Delta\mu_{\text{Model 2}} = RT\ln\left(\frac{[\text{Na}^+]_i[\text{HCO}_3^-]_i^2[\text{Cl}^-]_o}{[\text{Na}^+]_o[\text{HCO}_3^-]_o^2[\text{Cl}^-]_i}\right) \quad \text{(Eq. 10)}$$

Model 1 is transformed to use pH (Eq. 8) rather than $[H^+]$ (Eq. 7). It should become clear that dependence on $[HCO_3^-]$ and pH (Model 1) is different than just $[HCO_3^-]$. In particular, this "simple" change implies that Model 2 should not be affected by pH directly and should in theory be *pH-insensitive*!

While this is a somewhat simple model comparison, it is an important comparison. One discovery (discussed at the end of this chapter) makes this discussion particularly important. Accardi and Miller found that the activity of *E. coli* ecClC, crystallized by R. MacKinnon's group (Dutzler et al., 2002), is that of a $2Cl^-$ for H^+ exchanger (Accardi and Miller, 2004) rather than a "typical" ClC-channel. Recently, the groups of Jentsch (Scheel et al., 2005) and Pusch (Picollo and Pusch, 2005) have provided compelling evidence that mammalian ClC-4 and ClC-5 are transporters behaving like ecClC as a $2Cl^-/H^+$ exchanger (see Chapter 12). For direct comparison, let's examine the two models of movement. The Cl^- channel will behave as diffusive movement (Eq. 1), whose activity (or macroscopic current, "*I*") will be dictated by the opening and closing of the channel (open probability, P_o), number of channels (n) and the single channel current (i). The model for $2Cl^-/H^+$ exchange is quite different, as discussed above for exchangers.

$$\Delta\mu_{\text{2Cl/H}} = 2\Delta\mu_{\text{Cl}} - \Delta\mu_{\text{H}} \quad \text{(Eq. 11)}$$

Or,

$$\Delta\mu_{\text{2Cl/H}} = RT\ln\left(\frac{[\text{Cl}^-]_i^2[\text{H}^+]_o}{[\text{Cl}^-]_o^2[\text{H}^+]_i}\right) + 2(-1)FV_m - (+1)FV_m$$

Or,

$$\Delta\mu_{\text{2Cl/H}} = 2.3RT\log_{10}\left(\frac{[\text{Cl}^-]_i^2[\text{H}^+]_o}{[\text{Cl}^-]_o^2[\text{H}^+]_i}\right) - 3FV_m \quad \text{(Eq. 12)}$$

Or,

$$\Delta\mu_{\text{2Cl/H}} = 2.3\,(\text{pH}_i - \text{pH}_o)\,RT\log_{10}\left(\frac{[\text{Cl}^-]_i^2}{[\text{Cl}^-]_o^2}\right) - 3FV_m \quad \text{(Eq. 13)}$$

Eq. 13 is quite different to Eq. 1 due to both the dependence on the membrane pH gradient ($pH_i - pHo$) and also the very steep dependence on membrane potential ($3FV_m$). These characteristics turn out to be well suited for endosomes (ClC-4 and ClC-5) as well as *E. coli* ecClC.

IV. THE FOCUS OF THE REVIEW

Epithelial cells and cells of the nervous system are special in that the plasma membrane surrounding the cell is not uniform. This "non-uniformity" exists at several levels. First, the lipids and phospholipids of the inner bilayer leaflet are typically different to those in the outer leaflet. Second, membranes facing the blood, typically called basolateral, are of different protein and lipid composition than the inner organ, typically called apical/luminal membrane. Epithelia separate two or more body compartments. On the other hand, cells in the nervous system either connect or separate two or more regions of the brain or body.

Several organ systems have evolved to be specialized in HCO_3^- transport – notably the kidney and the pancreas. The kidney has the job of absorbing 178–179 L of the 180 L of filtrate from glomeruli. With a blood [HCO_3^-] of 25 mM, this volume represents ~4500 mmoles or >350 g HCO_3^- absorbed each day (see proximal tubule model in Fig. 4.5).[e] Likewise, the pancreatic ducts move CO_2 and HCO_3^- from the blood into pancreatic secretions. In humans, pancreatic fluid, at pH 8.1, is virtually isotonic $NaHCO_3$, i.e. 150 mM. These epithelia organs are the "workhorses" for several other organs and specialized cells have developed HCO_3^- transport systems. Bicarbonate secretion also occurs in salivary glands, stomach, intestine, prostate, uterus, and eye. Many of the molecular entities participating in these varied HCO_3^- transport functions have been identified. To limit the scope of this review, we will focus on epithelial transporters in vertebrates – especially human. We review functional characterization of different Na^+ dependent HCO_3^- transporters. Na^+-independent HCO_3^- transporters will not be discussed in detail. Detailed reviews of both the SLC4 (Romero et al., 2004) and the SLC26 (Mount and Romero, 2004) gene families, i.e. the known HCO_3^- transporter families, were published several years ago. Since the focus of this book is Cl^- transport, we will discuss the Cl^- transporters of the SLC4 and SLC26 gene/protein families. Many proteins in these two families, though not all, transport Cl^-. Unique for these transporters is that Cl^- transport, when present, is *thermodynamically coupled to exchange of* HCO_3^-.

[e] 350 g of HCO_3^- is ~1 lb of $NaHCO_3$ absorbed daily.

V. SLC4 ANION EXCHANGERS AND COTRANSPORTERS

A. General Properties of SLC4 Anion Exchangers

1. Structural Features

Band 3 (AE1) accounts for ~25% of red blood cells (RBC), making it the most abundant RBC membrane protein. In 1985, Kopito and Lodish cloned murine AE1 (Kopito and Lodish, 1985), allowing the homology cloning of AE2 and AE3. Human AE1 is 911 amino acids and is the founding member of the Slc4 family. Since that time, several Na^+-coupled HCO_3^- transporters (electrogenic and electroneutral) as well as a Na^+-driven Cl^-/HCO_3^- exchanger have been (i) cloned, (ii) found to be in the Slc4 protein family and (iii) functionally characterized.

2. Topology

This SLC4 gene and protein family contains 10 mammalian members (Fig. 4.1). All but one member (SLC4A11) are HCO_3^- transporters. These SLC4 transporters couple HCO_3^- movements in exchange for Cl^-, cotransport with Na^+ or both (see Fig. 4.1). SLC4A11

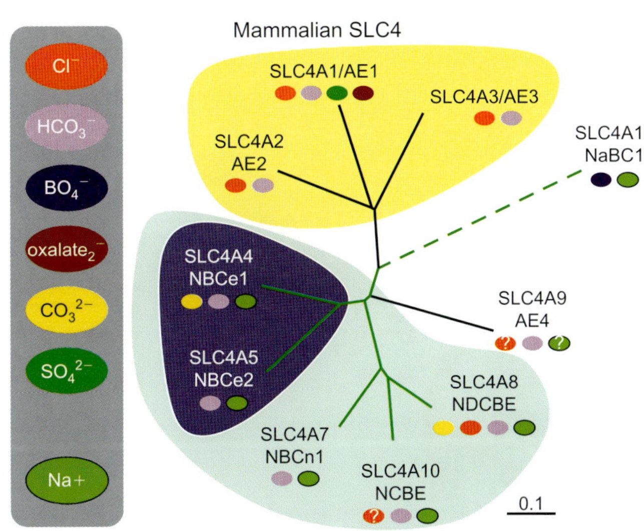

FIGURE 4.1 The SLC4 protein family of bicarbonate transporters. Dendrogram and transported species.

has been characterized as an electrogenic Na^+ borate cotransporter (Park et al., 2004) which is unaffected by the presence or absence of HCO_3^- or Cl^-.

Zhang and Low purified and crystallized the cytoplasmic N-terminus of AE1 (Zhang et al., 2000). Zhu and Casey have modeled the AE1, cytoplasmic N-terminus (404 amino acids long) as an anchorage site for several proteins, including components of the cytoskeleton, some glycolytic enzymes, and hemoglobin (Zhu et al., 2003). In that same year, Chang and Low structurally identified an ankrin binding loop in this domain (Chang and Low, 2003). Recently, Chang and coworkers (Chang et al., 2008) demonstrated that Zhang's structure was also a good model for SLC4A4, an electrogenic Na^+-HCO_3^- cotransport (NBCe1 in Fig. 4.2, see also section V.D). This most recent structural model (Chang et al., 2008) combined with functional analysis seems to indicate that the Zhang and Low model for the AE1 N-terminus may accurately describe the structures of most Slc4 N-termini associated with HCO_3^- transport (Chang et al., 2008).

3. *Pharmacology of SLC4 Transporters*

AE1 and most of the SLC4 protein family are functionally inhibited by stilbene derivatives such as DIDS and SITS (Barzilay et al., 1979; Cabantchik and Greger, 1992; Cabantchik and Rothstein, 1972). Both DIDS and SITS have an isothiocyano group which can covalently react with free amines: SITS has one group and DIDS has two groups. SITS interacts with AE1 in two steps: (1) a rapid, reversible electrostatic ionic interaction, and (2) a slower irreversible covalent reaction with an AE1 lysine. In human AE1, the target of the covalent reaction is the lysine at position 539 (K539), near the extracellular end of TM #5 (Bartel et al., 1989b; Landolt-Marticorena et al., 1995; Okubo et al., 1994; Schopfer and Salhany, 1995; Wood et al., 1992). Based on sequence homology of the AEs, Kopito (Kopito et al., 1989) identified a putative "DIDS-reaction motif" adjacent to TM #5 (KLXK, where X is I or Y).

While the stilbenes have been used extensively to characterize anion transport and HCO_3^- transport in particular, they covalently interact with tertiary amines, e.g. in lysine (K). Other compounds have also been used as anion transport inhibitors: diphenylamine carboxylic acid (DPC), niflumic acid (NF) and tenidap. Even amiloride analogs such as benzamil (Ducoudret et al., 2001) inhibits one of the Na^+-coupled HCO_3^- transports (NBCe1) in the Slc4 family. Unlike the stilbenes, none of these other compounds covalently interact with anion transporters and thus their inhibition is completely reversible after removal. DPC and NF seem to inhibit most anion transport systems. Tenidap, on the other hand, works very well for reversible inhibition of NBCe1 and seems to be effective on other Na^+-coupled HCO_3^- transport systems (Lu and Boron, 2007; Romero et al., 1997).

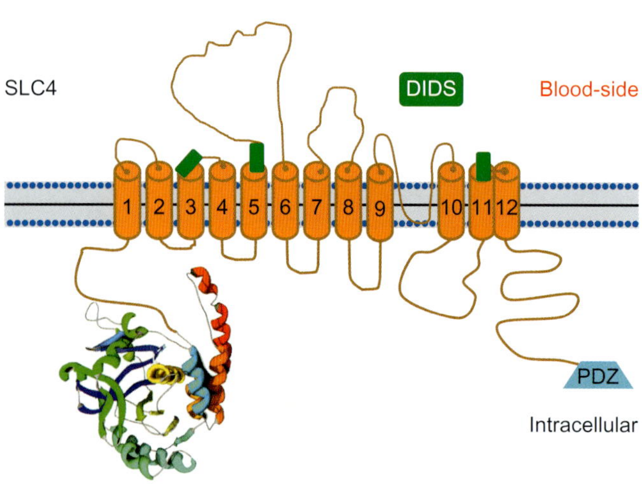

FIGURE 4.2 General topology for SLC4 protein family. Hydropathy model with N-terminus from AE1 crystal structure.

4. *Signature Sequences*

With the human genome complete and all Slc4 proteins identified, two general sequence features come to light: (i) "signature" sequences of the SLC4 family and (ii) additional potential DIDS-reaction-site motifs (see Table 4.1 and Romero and Boron, 1999; Romero et al., 1997). There are several areas of amino acid similarity or identity: ETARWIKFEE, AITFGGLLG, VREQRVTG and FLYMGV. These sequences are also interesting in that they appear invariant among vertebrates. Components of "ETARWIKFEE" in particular have been implicated in pH sensitivity of AE2 (Stewart et al., 2004) and are critical for the N-terminal globular domain (Chang et al., 2008).

A "DIDS interaction motif" – KL(X)K (X = I,V,Y) – was noted with the cloning of AE3 (Kopito et al., 1989). As discussed below, DIDS is an isothiocynostilbene derivative which aided biochemical analysis of AE1 (Bartel et al., 1989a; Okubo et al., 1994). The cognate sequence in NBCe1 (SLC4A4) is KMIK (558–561) (Romero and Boron, 1999; Romero et al., 1997). There is a more C-terminal site (identified originally in NBCe1 (Romero et al., 1997)) in SLC4A4-10 which has the motif sequence K-(Z)-FK. (R-(Z)-FK in AE1-3), where Z = C, K, Q, T. It is unclear if this site can be modified in any of the SLC4 proteins. A third C-terminal site is k-(Y)(X)-K (Romero et al., 1997), where Y = M, L and X = I, V, Y. Lu and Boron have demonstrated that this is the reaction site in NBCe1 (Lu and Boron, 2007). In SLC4A3, SLC4A9 and SLC4A11, R is

TABLE 4.1 The three cognate sequences of the 'DIDS-binding motifs' in human SLC4 members. The consensus 'DIDS-binding motifs' (see Fig 4.2) are indicated at the top of each column. As indicated in the text while there are three consensus motifs, only one site in each SLC4 protein seems to actually interact with the stilbenes such as DIDS. Which site binds needs to be determined clone by clone and has only been definitively shown for AE1 and NBCe1.

Clone	KMIK	K-(Z)-FK	k-(Y)(X)-K
SLC4A1 (AE1)	^{539}KLIK	^{589}RKFK	^{695}KMVK
SLC4A2 (AE2)	^{841}KLVK	^{918}RKFK	^{1024}MLQK
SLC4A3 (AE3)	^{842}KLYK	^{911}RKFR	^{1017}RLLK
SLC4A4 (NBCe1-A)	^{558}KMIK	^{667}KKFK	^{768}KLKK
SLC4A5 (NBCe2-D)	^{655}KMIG	^{762}KKFK	^{863}KLKK
SLC4A7 (NBCn1)	^{742}KLFD	^{843}KQFK	^{944}KLKK
SLC4A8 (NDCBE)	^{612}KLIH	^{713}KTFK	^{814}KLKK
SLC4A9 (AE4)	^{511}KMLN	^{608}KCVK	^{709}RLQK
SLC4A10 (NCBE)	^{613}KLFE	^{714}KQFK	^{815}KLKK
SLC4A11 (NaBC1)	^{509}GTVK	^{592}YQFK	^{689}RLVK

substituted for K (R still has a potentially reactive amino-group). In SLC4A2 (AE2), M replaces K and is thus unlikely to react with DIDS (see Table 4.1). The NBCe1 sequence data originally suggested a more generalized DIDS-binding motif. However, sequence comparison of the entire SLC4 family, DIDS-inhibition data and mutagenisis (e.g. as in Lu and Boron, 2007) will be necessary for identification of which, if any, of these residues are modified to result in transport inhibition.

B. The Na^+ Independent Cl^-/HCO_3^- (Anion) Exchangers

There are three identified anion exchangers – AE1, AE2 and AE3 (SLC4A1, SLC4A2 and SLC4A3, respectively, see Fig. 4.1). As previously noted, ions and solutes do not typically "diffuse" across the plasma membrane of cells. A "pathway", in our case a transporter, is needed for these ions and solutes to cross the hydrophobic membrane barrier. AE1-3 mediate the electroneutral exchange of one monovalent anion for another across the plasma membrane. As discussed in section III electroneutral transport means that the transporter does *not* make direct use of the energy derived from the potential difference (V_m) across the membrane in which the transporter is present.

AE1-3 prefer Cl^- and HCO_3^- as substrates, although the AEs can also transport OH^-, and AE1 can cotransport SO_4^{2-} plus H^+ in exchange for Cl^-, although at a very low rate (Jennings, 1976). Physiologically, the important transport activity is Cl^- exchange for HCO_3^-, with only the transmembrane chemical gradients of the two ions determining the net transport direction. In most cells, there is an inward chemical gradient for Cl^- which dominates, which in turn drives exchange for intracellular HCO_3^-. Disulfonic stilbenes (SITS, DIDS) inhibit all of the transport activity of all three AEs (see section V.A.3). The AE literature has been reviewed in detail by Alper (Alper et al., 2002) and the SLC4 family has also been recently reviewed (Pushkin and Kurtz, 2006; Romero et al., 2004).

1. *AE1 (SLC4A1, Band 3)*

The Cl/HCO_3^- exchanger of erythrocytes (Band 3, AE1, SLC4A1) was one of the first transporters physiologically identified and characterized. It was one of the first transporters of any type to be cloned (Kopito and Lodish, 1985). AE1 is the founding Slc4 family member, and accordingly is also one of the most intensively investigated Slc4 proteins.

Band 3 (AE1) accounts for ~25% of red blood cells (RBC), making it the most abundant RBC membrane protein. In 1985, Kopito and Lodish cloned murine AE1 (Kopito and Lodish, 1985), allowing the homology cloning of AE2 and AE3. Human AE1 is 911 amino acids. Zhang and Low purified and crystallized the cytoplasmic N-terminus of AE1 (Zhang et al., 2000). Zhu and Casey have modeled the AE1, cytoplasmic N-terminus (404 amino acids long) as an anchorage site for several proteins, including components of the cytoskeleton, some glycolytic enzymes, and hemoglobin (Zhu et al., 2003). As mentioned above, it has been demonstrated that Zhang's structure is a also a good model for NBCe1. These recent structure-function analyses indicate that the Zhang and Low model for the AE1 N-terminus accurately describe the structures of most SLC4 N-termini endowed with HCO_3^- transport capacity (Chang et al., 2008).

The transmembrane domain of human AE1 is ~475 amino acids (Zhu et al., 2003), and encodes the anion-exchange function of the protein (Groves et al., 1996; Groves and Tanner, 1995). The C-terminus of human AE1 (40 aa) is cytoplasmic and can bind carbonic anhydrase II (CA-II) (Vince and Reithmeier, 1998).

In erythrocytes, AE1 is the key transporter to move CO_2 from the systemic tissues to the lungs. Cellular metabolism generates CO_2, which moves from mitochondria to the blood plasma and then this same CO_2 enters the erythrocyte or RBC (Fig. 4.3). CA-II then converts CO_2 and H_2O into HCO_3^- and H^+. AE1 disposes of the newly generated HCO_3^- by exchanging it for Cl^- . In this process, de-oxygenated hemoglobin (that has higher H^+ affinity than oxygenated hemoglobin) then binds and buffers the resulting H^+. These simultaneous processes permit the RBC to take up additional CO_2 (Jacobs–Stewart cycle). This cycle reverses in the lungs, i.e. CO_2 is released into the blood for exchange into alveoli and ultimate elimination via ventilation. At the cell membrane level, CA-II interacts with the AE1 C-terminus. RBCs also contain high levels of the aquaporin 1 water channel (Preston and Agre, 1991), which is permeable to CO_2 (Cooper and Boron, 1998; Nakhoul et al., 1998). This RBC complex (AE1, CA-II, AQP1, and hemoglobin) has been called the HCO_3^- metabolon (Sterling et al., 2001) and does seem to physiologically expedite CO_2 transport. This concept is presently still controversial (Fang et al., 2002).

Aside from RBCs, collecting duct α intercalated cells of the kidney have the highest AE1 expression. These polarized epithelial cells sort AE1 to the basolateral membrane (van Adelsberg et al., 1993) where it facilitates the final absorption of HCO_3^- from the forming urine to the blood. This "kidney" isoform of AE1 results from an alternate promoter, from the erythroid mRNA (eAE1), producing an N-terminal variant (kAE1) missing the first 65 amino acids (Brosius et al., 1989; Sahr et al., 1994). Low amounts of AE1 mRNA are found in the colon (Papageorgiou et al., 2001) and heart (Sabolic et al., 1997).

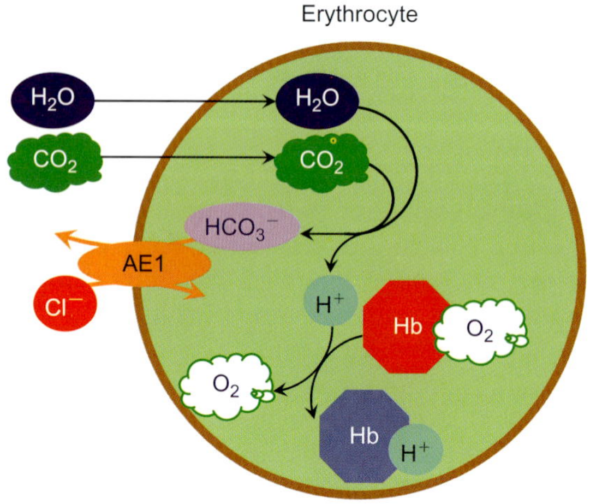

FIGURE 4.3 Model of anion transport in erythrocytes.

Since AE1 is a dominant membrane protein of the red cell and it is anchored to its cytoskeleton, many AE1 mutations in humans have been identified by misshapen RBCs. These naturally occurring mutations can alter RBC function as well as HCO_3^- absorption by the renal collecting duct intercalated cells (Alper, 2002; Shayakul and Alper, 2000). For example, one deletion Δ400–408 (cytoplasmic N-terminus and beginning of transmembrane span 1, see Fig. 4.2) results in a non-functional transporter. Yet, hetero-dimerization of the mutant and wild-type AE1 has function. This mutant AE1 protein increases the RBC membrane rigidity and may help protect against cerebral malaria. These RBCs have also altered shape (Southeast Asian ovalocytosis, SAO), but exhibit no clinical pathology. Clinically, the altered HCO_3^- absorption is manifest as a distal renal tubular acidosis.

Many mutations throughout AE1 protein result in a form of autosomal-dominant hereditary spherocytosis (HS), i.e. increased RBC fragility and hemolytic anemia. Renal pathophysiology is not concurrent with HS. There are also several AE1 mutations such as membrane-spanning domain, R589 mutations (Bruce et al., 1997; Jarolim et al., 1998; Karet et al., 1998), S613F (Bruce et al., 1997), ΔV850 (Bruce et al., 2000), and an 11-amino-acid cytoplasmic C-terminal, deletion (Jarolim et al., 1998) which result in autosomal-dominant distal renal tubular acidosis (dRTA) and no associated RBC pathology. When functionally tested in *Xenopus* oocytes, the AE1-dominant mutations show only moderately decreased function. The recessive dRTA-mutation in *Xenopus* oocytes has altered plasma membrane trafficking suggesting that this is the reason for the renal pathophysiology in the intercalated cells. Interestingly, co-expression of glycophorin A, an AE1-binding membrane protein, enables the surface delivery of these recessive AE1 mutations (Tanphaichitr et al., 1998; Young et al., 2000). Apparently, glycophorin A facilitates RBC trafficking of AE1 as well (Groves and Tanner, 1992), explaining why RBC pathophysiology is not present in recessive dRTA.

2. *AE2 (SLC4A2)*

Demuth and coworkers reported a fragment of "non-erythroid" band 3 from human kidney and lymphoma

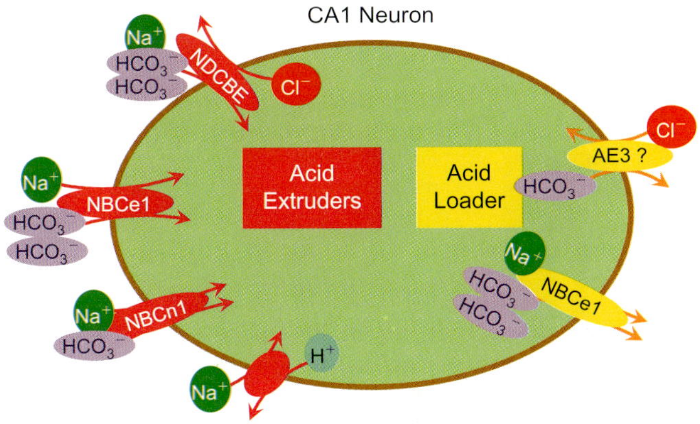

FIGURE 4.4 Model of CA1 neuron anion transport. Proteins which transport HCO_3^- or H^+ either bring net acid into the cell ("Acid Loader") or remove net acid from the cell ("Acid Extruder"). Each cell type has its own complement of these acid–base transporters.

cells (Demuth et al., 1986). This cDNA was later called AE2 when Alper and his group cloned a full-length mouse ortholog from kidney (Alper et al., 1988).

The three AEs (AE1, AE2, AE3 – see below) are most similar in their membrane-spanning domains. The cytoplasmic N-termini show the largest differences. Human and mouse AE2 have three alternative promoters (a, b, and c) resulting in five isoforms: a, b1, b2, c1, and c2 (the presence of c1 and c2 in humans has not been not confirmed). Yet, the physiological significance of these AE2 isoforms has not been determined. AE2 isoforms all have cytoplasmic N-termini that are significantly longer than that of AE1 (or AE3). For example, the elongated AE2a-N-terminus accounts for the increased length (1240 residues) compared to human erythrocyte AE1 (911 residues).

AE2 also shows the largest tissue distribution of the AEs (SLC4A1-A3). AE2 is found at basolateral membranes of most epithelial cells. That said, mRNA and protein are especially high in gastric parietal cells (Stuart-Tilley et al., 1994), choroid-plexus epithelial cells (Alper et al., 1994), colon surface enterocytes (Alper et al., 1999), and renal collecting ducts (Alper et al., 1997; Stuart-Tilley et al., 1998). In gastric parietal cells, AE2 presumably facilitates HCl secretion by exporting some HCO_3^- into the blood thereby balancing the H^+ pumped into the lumen of the gastric gland. AE2 may similarly aid H^+ secretion in the renal thick ascending limb (TAL). That is, by having AE2 in parallel with a Na^+/H^+ exchanger, AE2 contributes to net NaCl absorption. For epithelial cells, e.g. the choroid plexus, this mechanism can be part of the transepithelial Na^+ and Cl^- movement.

In addition to intracellular pH (pH_i) homeostasis, AE2 is involved in cell-volume homeostasis. This cell-volume regulation (volume-regulatory increases, VRI) is due to the uptake of Cl^- followed by osmotic water movement. When expressed in *Xenopus* oocytes, the ^{36}Cl efflux activity can be increased raising either intra-/extra-cellular pH (Stewart et al., 2002); hypertonicity (Humphreys et al., 1995); or NH_4^+ (ammonium, which paradoxically decreases the oocyte pH_i) (Humphreys et al., 1997). These pH_i and VRI responses seem controlled by the WRETARWIKFEE motif (Romero et al., 1997) (which is highly conserved in the SLC4 protein family) in the cytoplasmic N-terminus (Chernova et al., 2003b; Stewart et al., 2002).

3. *AE3 (SLC4A3)*

In 1989 Kopito and coworkers cloned a neuronal homolog to AE1, which has since been named AE3 (Kopito et al., 1989). There are two AE3 isoforms: brain form (bAE3) (Kopito et al., 1989) and cardiac form (cAE3) (Linn et al., 1992). This nomenclature is historical rather than descriptive, i.e. both isoforms may be in brain, heart and other tissues. AE3 is similar to AE1 and AE2 in the membrane-spanning domains, but differs most significantly at the N-terminus. The cytoplasmic N-terminus bAE3, like that of AE2a and AE2b, is substantially longer (1232 residues) than AE1. AE3 is the dominant AE paralog in excitable tissues: brain (Kopito et al., 1989; Kudrycki et al., 1990), heart (Linn et al., 1992; Yannoukakos et al., 1994), retina (Kobayashi et al., 1994) and smooth muscle (Brosius et al., 1997). AE3 isoform mRNAs are, nevertheless, expressed in both renal and gut epithelial cells.

AE3 is further activated by increases in pH_i as is AE2. Thus, it is likely that AE3 also contributes to pH_i regulation by extruding HCO_3^- in response to an intracellular alkali load. The A867D-AE3 polymorphism (in 5/6 extracellular loop, Fig. 4.2) is implicated in susceptibility to idiopathic generalized epilepsy (Sander et al., 2002) (Fig. 4.4). Interestingly, while the AE3 knockout mouse appears superficially normal, this mouse has a decreased seizure threshold after exposure to bicuculline, pentylenetetrazole, or pilocarpine (Hentschke et al., 2006). This pharmacologic change has been attributed to loss of AE3 in the pyramidal cell layer of the hippocampal CA3 region (Hentschke et al., 2006). The mechanisms underlying this increased in seizure susceptibility of $AE3^{-}/^{-}$ mice are not understood. The seizure threshold is altered, presumably due to loss of CA3 neuron Cl^-/HCO_3^- exchange activity, which may result in increased intracellular [HCO_3^-] in CA3 pyramidal neurons. Hentschke and co-workers hypothesized that HCO_3^- could account for this phenotype because $GABA_A$ receptors are also HCO_3^- permeable (Hentschke et al.,

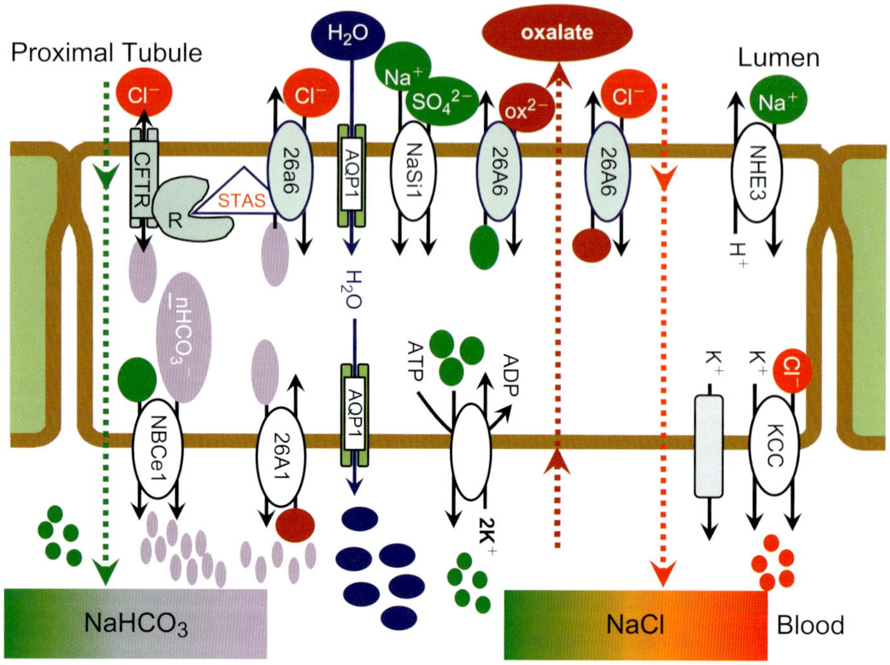

FIGURE 4.5 Renal proximal tubule anion transport. Model epithelial cell of the proximal tubule (PT). Only SLC4 or SLC26 anion transporters which appear to be dominant for isotonic NaCl (green→red) or $NaHCO_3$ (green→light purple) absorption are illustrated. Ions are color coded and then represented as appropriately colored dots.

2006). However, $[Cl^-]_i$ decreased would result in enhanced GABAergic inhibition (Hentschke et al., 2006). Clearly, more research is needed to understand the mechanisms by which AE3 is involved in neuronal excitability. Interestingly, *ae3*$^{-/-}$ mice exhibit a lower respiratory rate (Meier et al., 2007), apparently due to AE3 activity in chemosensory neurons.

C. Na^+ Dependent Anion Transporters – Molecular Entities

To date, all of the molecular entities functionally demonstrated as Na^+ dependent HCO_3^- transporters are members of the SLC4 family (for SLC family descriptions, see http://www.bioparadigms.org/slc/menu.asp). Thus, these transporters have additional similarity other than the obvious transport of Na^+ and HCO_3^-. The amino acids sequence identities among them are higher than ~30%, except BTR1 (Romero et al., 2004). According to the hydropathy analysis, all SLC4 family members have 10–14 transmembrane (TM) segments in between N- and C-terminal hydrophilic domains, both of which are intracellular. They all have a consensus N-glycosylation site on the third extracellular loop. Despite the similarities among them, Na^+ dependent HCO_3^- transporters differ from each other in several important ways: ion specificity, transport direction, and cellular location.

Na^+ dependent HCO_3^- transporters can be placed in two major groups: electrogenic and electroneutral. There are two known electrogenic Na^+-HCO_3^- cotransporters: NBCe1 (SLC4A4) and NBCe2 (SLC4A5). The electroneutral Na^+-HCO_3^- cotransporters can be subdivided into Cl^- independent (at least NBCn1/SLC4A7) and Cl^- dependent (NDAE1, NDCBE), i.e. Na^+-driven Cl^-/HCO_3^- exchangers. First, activity of clones (ion specificity and typical transport direction) will be discussed in the order indicated above. Next, function and clone assignment in various epithelial tissues will be discussed.

D. The Electrogenic Na^+-HCO_3^- Cotransporters

There are two identified electrogenic Na^+-HCO_3^- cotransporters in mammals: NBCe1 (SLC4A4) and NBCe2 (NBC4/SLC4A5). There is a third NBCe in the invertebrate nervous system, i.e. isolated from the squid giant fiber lobe (Piermarini et al., 2007). These NBCe proteins cotransport Na^+ and 2–3HCO_3^- across epithelial membranes or neuronal membranes (below "2NBCe" designates 1Na^+ + 2HCO_3^- (Eqs 14 and 15); and "3NBCe" designates 1Na^+ + 3HCO_3^- (Eqs 16 and 17). This stoichiometry indicates (based on thermodynamics, see section III) that both membrane potential (V_m) and the ion gradients (Na^+ and HCO_3^-) determine the electrochemical gradient and thus dictate the direction of ion transport (see below).

1. *Transport Models*

$$2{:}1 \quad \Delta\mu_{2NBCe} = 2\Delta\mu_{HCO_3} + \Delta\mu_{Na} \qquad \text{(Eq. 14)}$$

Or,

$$\Delta\mu_{2NBCe} = RT\ln\left(\frac{[HCO_3^-]_i^2[Na^+]_i}{[HCO_3^-]_o^2[Na^+]_i}\right) + 2(-1)\,FV_m + (+1)FV_m$$

Or,

$$\Delta\mu_{2NBCe} = 2.3RT\log_{10}\left(\frac{[HCO_3^-]_i^2[Na^+]_o}{[HCO_3^-]_o^2[Na^+]_i}\right) - FV_m \qquad \text{(Eq. 15)}$$

$$3{:}1 \quad \Delta\mu_{3NBCe} = 3\Delta\mu_{HCO_3} + \Delta\mu_{Na} \qquad \text{(Eq. 16)}$$

Or,

$$\Delta\mu_{3NBCe} = RT\ln\left(\frac{[HCO_3^-]_i^3[Na^+]_i}{[HCO_3^-]_o^3[Na^+]_i}\right) + 3(-1)FV_m + (+1)FV_m$$

Or,

$$\Delta\mu_{3NBCe} = 2.3RT\log_{10}\left(\frac{[HCO_3^-]_i^3[Na^+]_o}{[HCO_3^-]_o^3[Na^+]_i}\right) - 2FV_m \qquad \text{(Eq. 17)}$$

Based on these thermodynamic considerations, it is widely held that a 3:1 stoichiometry is needed for HCO_3^- absorption by the renal proximal tubule (movement from cell *out to* the blood). Likewise, these equations are also used to indicate that a 2:1 stoichiometry would mean Na^+ and HCO_3^- cotransport from the blood *into* the cell.

2. *NBCe1 (SLC4A4)*

Boron and Boulpaep, using the perfused renal, proximal tubule of the salamander *Ambystoma tigrinum*, were the first to report an activity which they call "electrogenic Na^+/ HCO_3^- cotransport" (Boron and Boulpaep, 1983). This activity was subsequently reported in a variety of mammalian tissues, notably the renal proximal tubule (Baum, 1989; Geibel et al., 1989; Preisig and Alpern, 1989) and corneal endothelial cells (Jentsch et al., 1984). About 15 years later, Romero and colleagues (Boron, Boulpaep and Hediger) using an expression-cloning approach, cloned the first cDNA encoding an Na^+-coupled HCO_3^- transporter, *NBC* for Na-Bicarbonate Cotransporter (Romero et al., 1997). Since that time, other NBCs have been cloned and characterized. In the more systematic nomenclature, this initial clone is called "aNBCe1-A" – "a" for *Ambystoma*, "e" for electrogenic, "1" to distinguish the two NBCe genes, and "A" for the first gene isoform. Alternatively, aNBCe1-A may also be called "Slc4a4-A".

After *Ambystoma* NBCe1-A was cloned, orthologs of NBCe1-A were cloned from humans (Burnham et al., 1997) and rats (Romero et al., 1998). The NBCe1 amino acid sequence is 34% identical to members of the AE branch (Fig. 4.1). Human SLC4A4[f] gene is located at 4q21, has 26 exons and spans 450 kb (Abuladze et al., 2000). Three NBCe1 isoforms have been identified (A–C). NBCe1-A/SLC4A4-A (kNBCe1) is the "kidney" isform and arises from an alternated promoter and coding sequence within intron 3 (Abuladze et al., 2000). NBCe1-B (SLC4A4-B, pNBC) is found in the gut (small intestine and pancreas) (Abuladze et al., 1998), heart (Choi et al., 1999), eye (Bok et al., 2001; Li et al., 2005; Romero, 2001; Romero et al., 2009; Sciortino, 2001; Sun and Bonanno, 2003; Sun et al., 2000; Vorum et al., 2003), and lung and epidydimus (Jensen et al., 1999). NBCe1-B begins with 85 novel, amino acids than the first 41 amino acids of NBCe1-A. NBCe1-B is the most widespread NBCe1 isoform; reported activity was found in the pancreas (Abuladze et al., 1998), eye (Romero, 2001), and heart (Choi et al., 1999). NBCe1-C (SLC4A4-C, rb2NBC1) was cloned from rat brain (Bevensee et al., 2000) is identical to NBCe1-B at the N-terminus but varies at the C-terminus (61 unique, C-terminal amino acids due to a 97-base pair (bp) deletion before the 3′ end of the A/B open reading frame. NBCe1-B and NBCe1-C isoforms have similar physiology to NBCe1-A: voltage dependent (Sciortino and Romero, 1999), require and transport Na^+ (Sciortino and Romero, 1999), require and transport HCO_3^- (Grichtchenko et al., 2000; Sciortino and Romero, 1999) and are blocked by stilbene compounds (Abuladze et al., 1998; Bevensee et al., 2000; Burnham et al., 1997; Choi et al., 1999; Grichtchenko et al., 2000; Müller-Berger et al., 1998; Romero et al., 1998, 1997; Sciortino and Romero, 1999). NBCe1-B seems to have a lower bicarbonate transport capacity (Tatishchev et al., 2003) which is reversed by co-expression of IRBIT, an inositol 1,4,5-*tris*-phosphate receptor-binding protein (Shirakabe et al., 2006). The N-terminal interaction with IRBIT results in NBCe1-A-like activity. The novel C-terminus of NBCe1-C seems to reduce transport due

[f]All capital letters are used to designate a human clone (e.g. SLC4A4) whereas Slc4a4 would designate another species NBCe1 ortholog.

to decrease plasma membrane expression (McAlear et al., 2006).

Mutations of human NBCe1 (SLC4A4; OMIM #603345) cause a severe proximal renal tubular acidosis combined with bilateral glaucoma (Demirci et al., 2006; Dinour et al., 2004; Igarashi et al., 1999, 2001, 2000, 2002; Inatomi et al., 2004) and dental malformation (Dinour et al., 2004). When *nbce1* is knocked out in mice, several acidosis and dental malformations are maintained, although the mice are hypotensive and die early of gut malabsorption (Gawenis et al., 2007). This particular study is yet another illustration that the "mouse disease" does not always recapitulate the human pathophysiology.

The ocular pathophysiology resulting from recessive NBCe1 mutations is more complicated to understand than the resulting metabolic acidosis. Igarashi and colleagues have hypothesized that inactivation or decreased activity of NBCe1 could increase the HCO_3^- concentration in the corneal stroma thus increasing pH (Igarashi et al., 2002). The elevated pH and CO_3^{2-} concentration would facilitate Ca^{2+} and Mg^{2+} deposition and precipitation in the Bowman membrane, leading to band keratopathy. NBCe1 inactivation could also disrupt homeostasis of the lens by affecting this active transport, resulting in cataract formation. These same investigators speculated that NBCel inactivation could alter the contractile properties of the human trabecular meshwork cells leading to increased resistance to aqueous humor outflow (Igarashi et al., 2002). It is noteworthy that fluid transport through the human trabecular meshwork differs from that found in rodents. The patient with the homozygous Q29X nonsense mutation has pRTA and glaucoma without band keratopathy or cataracts (Igarashi et al., 2001). This indicates that the NBCe1-A isoform is causative, and that the corneal phenotype (human corneal endothelia) is likely dictated by NBCe1-B.

3. *NBCe2 (SLC4A5)*

A second electrogenic Na^+-HCO_3^- cotransporter (NBC4/NBCe2/SLC4A5) was cloned from human heart (Pushkin et al., 2000a). NBCe2 and NBCe1 are ~53% identical, while NBCe2 is ~29% identical to the AE family (Fig. 4.1). There are six NBCe2 isoforms (SLC4A5A-F) reported (Pushkin et al., 2000a, 2000b; Virkki et al., 2002; Xu et al., 2003). Sassani and co-workers (Sassani et al., 2002) and Virkki and colleagues (Virkki et al., 2002) have characterized NBC4c as electrogenic, Na^+ and HCO_3^- dependent and DIDS inhibitable. Whereas, Xu et al. have reported that NBC4e is an electroneutral Na^+-HCO_3^- cotransporter (Xu et al., 2003). The NBCe2 protein nomenclature is confusing because several isoforms seem to be from incompletely spliced pre-mRNA (Virkki et al., 2002). Like other Slc4 proteins, NBCe2 is >1000 amino acids and likely has 12 membrane-spanning domains (Fig. 4.2). NBCe2 is located at the 2p13 human gene locus. SLC4A5-C isoform display similar electrogenicity and stoichiometry as NBCe1 (Sassani et al., 2002; Virkki et al., 2002) with the exception that SLC4A5-E isoform is characterized as electroneutral (Xu et al., 2003) (meaning *no* voltage dependence). NBCe2 mRNA is most abundant in liver, testes, and spleen, while lesser amounts are in heart, lung, and kidney (Pushkin et al., 2000a). There is also detected NBCe2-C mRNA in brain, pancreas, muscle, and peripheral blood leukocytes (Sassani et al., 2002). NBCe2-C (NBC4c) protein has been localized to basolateral membrane in hepatocytes, the apical membrane in intrahepatic bile ducts and the apical membrane of renal-pelvis uroepithelial cells (Abuladze et al., 2004).

E. The Electroneutral Na^+-coupled HCO_3^- Cotransporters

The second major branch of Na^+-HCO_3^- cotransporters in the SLC4 family are electroneutral proteins (Fig. 4.1). The electroneutral Na^+-HCO_3^- cotransporters can be subdivided into Cl^- independent (e.g. NBCn1/SLC4A7) and Cl^- dependent (e.g. NDAE1, NDCBE/SLC4A8), also known as Na^+-driven Cl^-/HCO_3^- exchangers. The Cl^- independent NBCn's are characterized by a $1Na^+$:$1HCO_3^-$ transport stoichiometry. The Cl^- dependent electroneutral Na^+-HCO_3^- cotransporters move no net charge but appear to cotransport 1 Na^+ and 2 HCO_3^- in exchange for 1 Cl^- (see Eqs 8 and 9). Some SLC4 family members have a controversial function (see section V.G). Of interest is that these Slc4 proteins with controversial function all group with the Na^+ dependent part of the Slc4 tree (Fig. 4.1).

1. *NBCn1 (SLC4A7)*

The SLCA4A7 gene is located at 3p22 human gene locus and corresponding cDNA were originally isolated from cardiac, skeletal, and smooth muscles (Choi et al., 2000). They are ~55% identical to the NBCe's and 33–43% identical to the AEs. So far, there have been 4 NBCn1 isoforms (SLC4A7A-D) reported. Choi and co-workers cloned out three rat isoforms from smooth muscle and demonstrated that SLC4A7 functions as an electroneutral NBC, NBCn1 (Choi et al., 2000). The first reported SLC4A7 (NBCn1-D) was called NBC3 because the experiments did not test electrogenicity

(Pushkin et al., 1999). Choi et al. pointed out that the NBCn1-B isoform has a putative PDZ-binding motif in the C-terminius (Choi et al., 2000). Unlike most of other Slc4 proteins, the human NBCn1-D isoform is not inhibited by stilbenes but rather sensitive to ethyliso-propyl amiloride (EIPA) (Pushkin et al., 1999). Rat NBCn1-B is slightly DIDS sensitive and EIPA insensitive (Choi et al., 2000; Park et al., 2002). Northern blot analysis detected human NBCn1 mRNA only in heart and skeletal muscle (Pushkin et al., 1999), while rat NBCn1 mRNA was found in spleen and testis with lesser amounts in heart, brain, lung, liver and kidney but not in skeletal muscle (Choi et al., 2000).

F. The Na^+-coupled Cl^-/HCO_3^- Exchangers

The Na^+-coupled Cl^-/HCO_3^- exchangers are the third branch in the SLC4 gene family. Their transport activity is equivalent to the electroneutral exchange of $NaHCO_3$ for HCl or more likely $Na(HCO_3)_2^-$ exchange for Cl^- (Grichtchenko et al., 2001) (see Eqs 8 and 9).

1. *NDAE1 (Drosophila SLC4)*

The first cDNA encoding an Na^+-driven Cl^-/HCO_3^- exchanger was cloned from *Drosophila melanogaster* and was functionally characterized as a Na^+-driven anion exchanger (NDAE1) (Romero et al., 2000). The NDAE1 gene is found on *Drosophila* chromosome 2 in the 54A region. NDAE1 encodes a 1030 amino acid, a protein of ~115 kDa, comprising multiple putative membrane-spanning domains, and has both sequence homology and predicted membrane topology similar to both the AEs and NBCs (Fig. 4.2). Functionally, when NDAE1 is expressed in *Xenopus* oocytes, this membrane protein mediates the transport of Cl^-, Na^+, H^+, and HCO_3^-. Interestingly, OH^- seems to substitute for at the HCO_3^- site(s) in NDAE1. NDAE1 mediated transport is blocked by DIDS and has a leak current (~one charge per 1000 turnovers; Romero et al., 2000). NDAE1 is expressed throughout *Drosophila* development in the central and peripheral nervous systems, sensilla, and the alimentary tract (Malpighian tubules, gut, and salivary glands) which initially suggested that a "mammalian NDAE1" would be responsible for the Na^+ dependent Cl^-/HCO_3^- exchange activity characterized in mammalian neurons, kidney, and fibroblasts (Sciortino et al., 2001). Subsequently, NDCBE was identified as a mammalian Na^+-driven Cl^-/HCO_3^- exchanger (see below). NDAE1 is one of two SLC4 genes in *Diptera* (flies and mosquitos) (Romero et al., 2000), this gene does not have a direct ortholog in mammals.

2. *NDCBE (SLC4A8)*

The Na^+-Driven Chloride/Bicarbonate Exchanger (NDCBE) was the first key transporter shown to regulate intracellular pH in squid axons (Boron and De Weer, 1976; Boron and Russell, 1983; Russell and Boron, 1976), snail neurons (Thomas, 1976a, 1976b, 1977), and barnacle muscle (Boron, 1977). Human NDCBE (SLC4A8) maps to chromosome 12q13 and encodes 1044 amino acids (Grichtchenko et al., 2001). NDCBE cloned from human brain has higher identity to NBC transporters (Fig. 4.1; ~50% to NBCe1-A from salamander, rat and humans; ~73% NBCn1; 71% to mouse NCBE, and 47% to *Drosophila* NDAE1) than Na^+ independent Cl^-/HCO_3^- exchangers (e.g. 34% to AE2). Human NDCBE mRNA showed a strong ~12 kb band in all major regions of human brain and in testis with lesser signal in kidney and ovary (Grichtchenko et al., 2001). NDCBE is an electroneutral transporter and normally mediates the uptake of 1 Na^+ and 2 HCO_3^- in exchange for 1 Cl^- (Eqs 9 and 10, Fig. 4.4). Unlike NDAE1, NDCBE displays an absolute dependence on Na^+, HCO_3^- and Cl^-, and is inhibited by DIDS (Grichtchenko et al., 2001). Wang and colleagues subsequently cloned mouse NDCBE (kNBC3, 1089 amino acids) (Wang et al., 2001) but Cl^- dependence was not tested. Recently, an Na^+-driven Cl^-/HCO_3^- exchanger from squid giant fiber lobe was also cloned and characterized to be NDCBE ortholog using *Xenopus* oocytes expression system (Virkki et al., 2003).

3. *NCBE (SLC4A10) – NBCn2?*

In 2000, a novel SLC4 clone, Na^+-Cl^-/Bicarbonate Exchanger (designated NCBE), was reported from the mouse insulin-secreting cell line MIN6 cDNA library (Wang et al., 2000). The NCBE protein consists of 1088 amino acids. NCBE mRNA is expressed at high levels in the brain and low levels in the pituitary, testis, kidney, and ileum. Functional analyses of the NCBE protein expressed in *Xenopus laevis* oocytes and HEK293 cells demonstrate that it transports extracellular Na^+ and HCO_3^- into cells in exchange for intracellular Cl^- and H^+. Recently, human ortholog consisting of two splice variants, NCBE-A and NCBE-B, have been identified (Choi et al., 2002). These preliminary studies indicated that human NCBE-B, in which 30 amino acids are inserted in the cytoplasmic C-terminus, as expressed in *Xenopus* oocytes, is sensitive to DIDS and has absolute requirements for Na^+ and HCO_3^- but appears not to require Cl^- (Choi et al., 2002).

The function of SLC4A10 (NCBE) has been revisited. Parker and co-workers used *Xenopus* oocytes to compare NCBE function to other SLC4 proteins

(Parker et al., 2008). The major difference in these experiments was that the investigators also monitored surface Cl^- activity in addition to intracellular pH. This technique allowed them to detect the initial surface transients as part of their transport assay. The experimental design consisted of clones that are both Cl^- coupled and not Cl^- coupled. Their data indicate that activity of NCBE does not elicit a surface Cl^- response. However, $^{36}Cl^-$ uptake was higher in these NCBE oocytes. At the present time, it seems most likely that NCBE's function is predominantly that of an electroneutral NBC (perhaps NBCn2). That said, the transporter and channel literature are starting to more frequently contain reports of "auxiliary" functions (Picollo and Pusch, 2005; Scheel et al., 2005).

G. Other SLC4 HCO_3^- Transporters with Controversial Function

1. *SLC4A9 (AE4 or NBCn3?)*

A novel NBC-like cDNA was cloned from α-intercalated cells in rabbit kidney by Tsuganezawa (Tsuganezawa et al., 2001). The cDNA is more closely related to the Na^+-coupled HCO_3^- transporters than to the Cl^-/HCO_3^- exchanger. However, when expressed in Cos-7 cells, AE4 appears to behave as a Na^+-dependent Cl^-/HCO_3^- exchanger. Following the cloning cDNA of AE4 in rabbit kidney, a human version of AE4 was cloned by Lipovich et al. (Lipovich et al., 2001) and by Parker et al. (Parker et al., 2001). The AE4 gene maps to chromosome 5q23-31 and encodes a 104 kDa protein expressed mainly in the kidney. Human AE4 shares 84% identity with reported rabbit AE4. Functional studies in oocytes indicated that rabbit AE4 mediate a modest level of sodium independent and DIDS insensitive $^{36}Cl^-$ uptake (Tsuganezawa et al., 2001). Ko et al. (Ko et al., 2002a) cloned and localized rat AE4 in the renal cortical collecting duct. Expression rat AE4 in HEK-293 and LLC-PK1 cells showed that AE4 is targeted to the plasma membrane. Measurement of intracellular pH revealed that AE4 indeed functions as a Cl^-/HCO_3^- exchanger. However, surprisingly, AE4 activity was inhibited by DIDS.

In the small intestine, AE4 is immunolocalized to the apical membranes of surface cells in both mouse and rabbit stomach and duodenum. Functional studies in oocytes indicated that AE4 functions as a Cl^--HCO_3^- exchanger. These data indicate that AE4 is an apical Cl^--HCO_3^- exchanger in gastric mucous cells and duodenal villus cells. Based on its function and location we propose that AE4 may play an important role in mucosal protection (Xu et al., 2003).

Preliminary experiments from Boron's group (Parker et al., 2002), however, indicate that Na^+ rather than Cl^- is coupled to HCO_3^- transport. Perhaps this means that AE4 is actually "NBCn3". Once again, the apparently "simple" issue of Cl^- transport (coupling) or not is still not defined. While this distinction has impact for the kidney α-intercalated cells, Slc4a9's (AE4 or NBCn3) presence or role in the CNS is not defined. As implied for SLC4A10, other cellular factors may be able to account for the apparently divergent functions. Nevertheless, to date no such factor has been identified, keeping this hypothesis as speculation.

H. SLC4 HCO_3^- Transporters and CNS Function

HCO_3^- transporters have two major functions: transport of the HCO_3^- anion and H^+ buffering by HCO_3^-. Figure 4.4 illustrates which SLC4 transporters are involved in H^+ buffering (AE3, NBCe1 and NDCBE) in hippocampal pyramidal neurons. Recently, NBCn1 was also found to be present in hippocampal neurons (Cooper et al., 2005). Figure 4.6 develops these complementary functions more fully for the case of CNS synapses. One of the major controls in neuronal cells is the ability to change membrane potential (V_m). Why is this V_m control a key factor? As discussed above, the direction of Na^+ and HCO_3^- transport by NBCe1 is V_m dependent. That is, if V_m is more negative than -60 mV, then Na^+ and HCO_3^- (via NBCe1) will move *out of* the cell. On the other hand, if V_m is more positive than -40 mV, then Na^+ and HCO_3^- (via NBCe1) will move *into* the cell. Why do we care? This directionality of transport will either move additional HCO_3^- buffer into the cell or additional HCO_3^- buffer into the extracellular space (Grichtchenko and Chesler, 1994a, b; Huang et al., 1995). Both of these processes can also change pH, which in turn can alter $GABA_A$-receptor anion channel activity (Chen and Chesler, 1990, 1992; Kaila et al., 1992, 1989b, 1993; Mason et al., 1990). To further complicate this cellular coupling, $GABA_A$-anion channels are permeable to HCO_3^- (Chen and Chesler, 1990; Kaila et al., 1989a, b, 1993) as discussed in Chapter 5.

Cellular buffering has been shown to increase metabolic substrate uptake (Becker et al., 2004; Becker and Deitmer, 2004; Deitmer, 2000), e.g. lactate. As indicated above, V_m changes alter the direction of NBCe1 transport, which in turn alter the amount of monocarboxylates (MC^-) which can enter the cell presumably by MCT1 (Becker et al., 2004; Becker and Deitmer, 2004) (Fig. 4.6). The Na^+-HCO_3^- *influx* also increases the cellular buffering which prevents intracellular pH changes with MC^- influx (Becker and Deitmer, 2004).

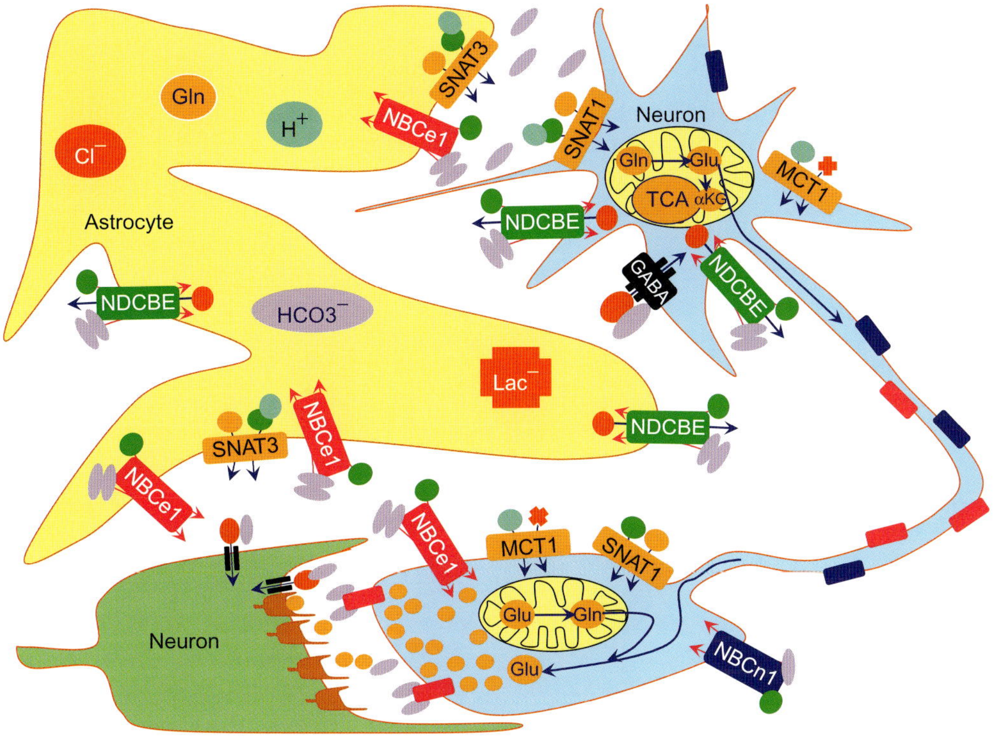

FIGURE 4.6 Slc4 transporters and CNS synapses. Synaptic transmission is complex. Slc4 HCO_3^- transporters appear to fulfill two major roles at these CNS cell–cell interphases: H^+ buffering and anion transport. A simplified schematic of two HCO_3^- transporters known to be in neurons and astrocytes (NBCe1 – purple and NDCBE – dark green), illustrates one model of transmission. Buffering can facilitate transport of monocarboxylate substrates (e.g. lactate, Lac^-) needed for the TCA (Krebs) cycle. The buffering of the extracellular space at the nerve terminal can also affect the activation of $GABA_A$-receptor channels and subsequent uptake of both Cl^- and HCO_3^-. The activity of these $GABA_A$-anion channels in turn can alter the membrane potential of the neurons. As discussed in the NBCe sections, the direction of NBCe1 mediated Na^+/HCO_3^- cotransport is voltage dependent. Consequently, at more hyperpolarized potentials NBCe1 will bring Na^+ and HCO_3^- *into the cell*, and at relatively depolarized potentials (more positive than −40 mV) NBCe1 remove Na^+ and HCO_3^- *from the cell.*

The neuronal cellular physiology is further complicated by the presence of carbonic anhydrases, in particular carbonic anhydase II (CA-II). As has been reported with the AEs (Sterling et al., 2001), coexpression of CA-II with NBCe1 causes an increase in NBCe1 activity (Becker and Deitmer, 2007). CA-II also seems to have a non-enzymatic role in "H^+ shuttling" which can increase the activity of MCT1 (Becker and Deitmer, 2008). Expression of the Na^+/H^+/glutamine (Gln) transporter SNAT3 in *Xenopus* oocytes elicits a Na^+ or H^+ conductance (Schneider et al., 2007). When these cells are coinjected with CA-II and exposed to Gln in a CO_2/HCO_3^- buffer, the cation conductance is suppressed (Weise et al., 2007). Furthermore, coexpression of NBCe1 and SNAT3 increased H^+-coupled Gln transport by SNAT3 in a CO_2/HCO_3^- buffer (Wendel et al., 2008). Again, this system is made yet more complex by the neuron's ability to change V_m, which not only would alter the magnitude of HCO_3^-, H^+ and Gln flux, but also the direction of the flux (Fig. 4.6). This "metabolon" consisting of transporters and enzymes can thus meet the dynamic changes of the CNS.

I. Outlook and Pharmaceutical Considerations – SLC4 Transporters

As discussed in this chapter, SLC4 HCO_3^- transporters are crucial to many key physiological processes. It is attractive to speculate that AE1 and AE2 as pharmacologic targets would aid in the control of distal renal tubular acidosis (metabolic). That said, the HCO_3^- buffer titrations that these proteins participate in result in a somewhat minor acidosis. As noted, NBCe1 (SLC4A4) defects result in a severe *proximal* renal tubular acidosis (blood pH less than 7.1 and blood [HCO_3^-] of 3–12 mM). Since the proximal tubule handles the bulk of salt and water absorption, it would seem that only minor adjustments in NBCe1 activity should have a more pronounced effect on systemic acid–base balance than alteration of AE1 or AE2 activity. Perhaps more important though is that controlling NBCe1 function might be significant in controlling some and perhaps most forms of glaucoma (a neuropathy associated with increased ocular pressure which can lead to irreversible blindness).

NBCe2 (SLC4A5) seems to be the dominant HCO_3^- transporter at the basolateral membrane of hepatocytes as well as the apical membranes of intrahepatic bile ducts and uroepithelial of the renal pelvis. Activators of NBCe2 might prove useful augmenters of hepatobiliary function, particularly in the solubilization of fat and drug detoxification. NBCe2 is found in the testis and NBCe1 is found in the epidydimus and prostate. Thus, it is also possible that local targeting of these NBCe proteins could alter the critical, nurturing environment for sperm and result in a quite effective male contraceptive.

In the CNS, there appear to be three major Slc4 HCO_3^- transporters (Fig. 4.4): AE3 (Slc4a3), NBCe1 (SLC4A4) and NDCBE (Slc4a8). As indicated in Fig. 4.4 (CA1 neuron), these HCO_3^- transporters can be grouped into "acid loaders" (*export* HCO_3^- ions) or "acid extruders" (*import* HCO_3^- ions). Additionally, lack of "Cl^- loading" due to AE3 activity is associated with idiopathic epilepsy (Sander et al., 2002) and altered seizure threshold (Hentschke et al., 2006). Presumably, if AE3 and NDCBE are found in the same cells, their functions could oppose one another in terms of acid–base balance as well as intra- and extracellular Cl^- balance. It is attractive to speculate that NDCBE antagonism could counteract some of these seizure phenotypes.

Clearly, physiology and pathophysiology are usually not controlled by one protein. Future experiments to better understand transport mechanisms, transport regulation and protein interactions will need to precede development of any therapeutics. It will also be increasingly important to have detailed clinical assessments of patients identified with mutations and perhaps even polymorphisms in Slc4 genes and proteins.

VI. SLC26 ANION EXCHANGERS/ TRANSPORTERS/CHANNELS

A. Overview of SLC26 Proteins

The SLC26 anion transporters are a relatively young gene family; the first four members were identified in the mid-late 1990s, SLC26A5 in 2000, and the rest of the gene family emerged in 2002. These proteins are not homologous to the SLC4 family, with which they share anion specificity and pharmacology. A partial list of physiological processes in which SLC26 transporters play critical roles includes skeletal development (Hastbacka et al., 1994), synthesis of thyroid hormone (Everett et al., 1997), transepithelial Na^+/Cl^- transport (Hoglund et al., 1996; Knauf et al., 2001; Xie et al., 2002) (Fig. 4.5), bicarbonate excretion by the distal nephron (Royaux et al., 2001), intestinal oxalate secretion (Jiang et al., 2006), bronchial mucus production (Nakagami et al., 2008; Nakao et al., 2008), and bicarbonate secretion by the exocrine pancreas (Ko et al., 2002c). Several paralogs are expressed within the brain, yet nothing is known about the role of these transporters in neurological function. Emphasis here is on the physiological role of paralogs that function as Cl^- transporters, in addition to the evolving understanding of the functional characteristics and regulation of these important transporters. We also discuss the SLC26A5 "anion sensor" prestin, given its expression in brain and novel functional attributes.

SLC26 exchangers transport a number of monovalent and divalent anions, including sulfate (SO_4^{2-}), chloride (Cl^-), iodide (I^-), formate, oxalate, hydroxyl ion (OH^-), and bicarbonate (HCO_3^-) (Bissig et al., 1994; Jiang et al., 2002; Karniski et al., 1998; Moseley et al., 1999; Satoh et al., 1998; Scott and Karniski, 2000; Soleimani et al., 2001; Xie et al., 2002) (Fig. 4.7). Individual paralogs differ significantly in anion specificity, such that SLC26A6 is capable of transporting all of the substrates above (Jiang et al., 2002; Xie et al., 2002). SLC26A4 in turn transports monovalent anions such as Cl^-, I^-, and formate, but not divalent anions such as SO_4^{2-} and oxalate (Scott and Karniski, 2000; Scott et al., 1999). Several paralogs function in Cl^-/OH^- and Cl^-/HCO_3^- exchange, with increasingly important roles in transepithelial Na^+, Cl^- and HCO_3^- cotransport. Where direct comparisons have been made, the SLC26 proteins appear to be more

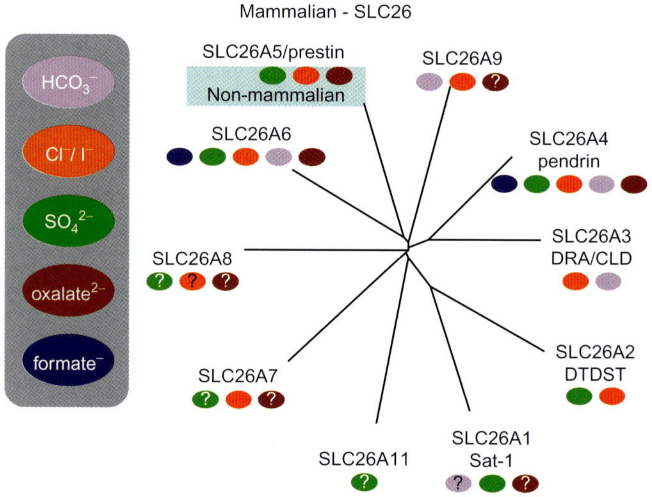

FIGURE 4.7 The SLC26 protein family of anion transporters and channels. Dendrogram illustrating the "relatedness" of human SLC26 proteins and their transported ionic species.

potent exchangers of Cl^- with HCO_3^- than with OH^- (Ko et al., 2002b; Wang et al., 2002; Xie et al., 2002).

A major, unexpected divergence from the SLC4 exchangers is that SLC26-dependent Cl^-/HCO_3^- exchange is electrogenic (Ko et al., 2002b; Xie et al., 2002), with differing stoichiometries for individual paralogs. SLC26A3, SLC26A4, and SLC26A6 appear to exchange 2 HCO_3^- for 1 Cl^- anion, yet SLC26A9 exchanges 1 HCO_3^- for 2 Cl^- (Romero et al., 2006; Shcheynikov et al., 2006b). Furthermore, some SLC26 paralogs appear to have predominant (SLC26A9) (Chang et al., 2009; Dorwart et al., 2007; Romero et al., 2006; Xu et al., 2008) or exclusive (SLC26A7) (Kim et al., 2005) channel-like characteristics. This latter characteristic is of course highly reminiscent of other transporter gene families (e.g. SLC6, the neurotransmitter transporter family; Carvelli et al., 2004; Chen et al., 2004). As noted previously (section III), the CLC "channel" family (Jentsch, 2008; Jentsch et al., 2005, 2002), wherein individual family members function as Cl^- channels and others as electrogenic Cl^-/H^+ exchangers (Accardi et al., 2004, 2005; Accardi and Miller, 2004; Jentsch, 2008; Picollo and Pusch, 2005; Scheel et al., 2005), as discussed in Chapter 12 in this volume.

B. Structural Features of SLC26 "Transporters"

One feature of the SLC26 proteins is the relatively low conservation between orthologs in mice and humans; the percent amino acid identity ranges from a low of 76% (SLC26A8) to a high of 90% (SLC26A9), versus the reported median of 86% for mouse and human orthologous genes (Makalowski and Boguski, 1998). The HUGO nomenclature is used herein; "SLC26A-" denotes a human gene/protein, "Slc26a-" denotes a rodent ortholog, and where appropriate the genes *per se* are written in italics. Although clearly homologous, the SLC26 proteins in *Drosophila* and *C. elegans* are only 25–40% identical to the mammalian proteins, such that one cannot in most instances discern orthologs for the individual mammalian exchangers; perhaps the one exception is SLC26A5/prestin, which appears to have identifiable orthologs in these genomes. Regardless, several orthologs do begin to emerge in genomes from fish and amphibians, e.g. in *Danio rerio* (zebrafish) or *Xenopus laevis*. All ten of the mammalian SLC26 proteins (see Fig. 4.7) predict a hydrophobic core, for which the details of membrane topology are as yet unknown. However, given that the amino- and carboxy-terminal domains of SLC26A3, SLC26A5, and SLC26A6 are intracellular (Dorwart et al., 2008; Lohi et al., 2003; Zheng et al., 2001), it is expected that these proteins span the plasma membrane from 10 to 14 times (Moseley et al., 1999; Saier et al., 1999) (see Fig. 4.8). Higher-order structure is somewhat controversial, with one paper suggesting an oligomeric structure for SLC26A5 (Zheng et al., 2006) and another indicating that this and other paralogs function as homomeric dimers (Detro-Dassen et al., 2008). Notably, when coexpressed, heterodimeric D154N-D342Q prestin exhibits voltage-dependent capacitances that differ from the respective homomeric mutants, suggesting functional interactions between the two homomeric subunits of SLC26 proteins (Detro-Dassen et al., 2008).

Much of the homology between SLC26 exchangers is found within the hydrophobic core of transmembrane (TM) domains. One region of homology within TM1-3 encompasses the 22-amino-acid "sulfate transport" consensus signature (Prosite, PS01130), which was initially defined by the comparison of the first mammalian family members with homologs in lower organisms (Fig. 4.7). Although not all members of the mammalian family conform to the exact consensus sequence, this region contains several invariant residues which are presumably critical for anion transport. There is a second cluster of invariant residues at the C-terminal end of the hydrophobic core of the proteins, in a conserved segment defined by Saier et al. (Saier et al., 1999) (see Fig. 4.7). This region includes the triplet -NQE-, residues 417–419 of Slc26a2, which is conservatively variable only in Slc26a8 (-NQD-). Three highly conserved or invariant residues in this section, E419, N425, and L483 in Slc26a2, have been shown to have functional significance in SHST1, an SLC26 homolog from the plant *S. hamata* (Khurana et al.,

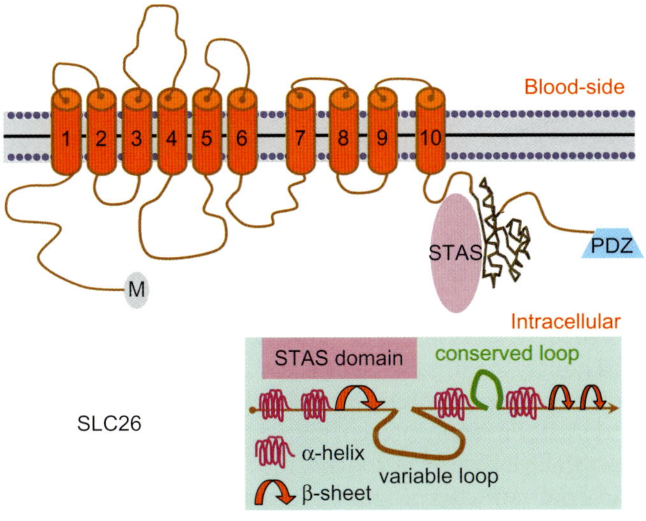

FIGURE 4.8 General topology of SLC26 proteins. Hydropathy model with C-terminal STAS domain extrapolated from SPOIIA.

2000). Moreover, two of these invariant residues are mutated (N425D and L483P) in patients with a severe defect in human SLC26A2, causing achondrogenesis type 1B and/or atelosteogenesis type 2; the SLC26A2 N425D mutant has been shown to be non-functional in *Xenopus laevis* oocytes (Karniski, 2001).

Structure-function analysis of transport function has largely been confined to the plant paralog SHT1. The five cysteines of SHT1 are not required for activity of this transporter (Howitt, 2005), suggesting that substituted cysteine mutagenesis will be a productive avenue for studying the function of this and other SLC26 transporters. Detailed mutagenesis of the first three TM domains of SHT1 has shown that conserved prolines in this region are critically important for activity, suggesting a role for proline-induced kinks in the proper conformation of these TM domains (Shelden et al., 2001). Systematic mutagenesis of charged residues within or near transmembrane domains suggests critical charge interactions between residues in TM3, TM6, and TM8, in addition to within TM1-3 (Shelden et al., 2001). Subsequent mutation of polar residues within TM1-2 suggest that interactions between these residues are critical for function (Leves et al., 2008). Notably, mutagenesis of this region in SLC26A5 reveals marked size-dependent functional effects of the substituted residues, suggesting a tight packing of the helices in this region (Rajagopalan et al., 2006). Therefore, TM1-3 is particularly important for the function of all the SLC26 transporters. Notably, however, the determinants of anion selectivity and the variable stoichiometry of the mammalian SLC26 transporters are as yet unknown.

Many of the SLC26 proteins end with a class I PDZ interaction motif (Songyang et al., 1997) (Fig. 4.8); the exceptions include human SLC26A1 (mouse and rat Slc26a1 end with S-A-L, a typical class I motif), SLC26A2, SLC26A4, SLC26A5, and SLC26A11. These C-terminal motifs are dispensable for transport function in *Xenopus* oocytes (Alper et al., 2001; Lohi et al., 2002b), but appear to tether SLC26 exchangers to regulatory proteins and other transporters, including NHE3 and CFTR (Ahn et al., 2001; Ko et al., 2002b; Lohi et al., 2003). Indeed, the reciprocal-regulation and co-association of SLC26A3 and CFTR is attenuated by mutation of their C-terminal PDZ interaction motifs (see also below) (Ko et al., 2004).

The C-terminal cytoplasmic domain of all ten SLC26 proteins includes the STAS (Sulfate Transporter and Anti-Sigma) domain (Fig. 4.8), uncovered due to the homology between the SLC26 gene family and bacterial anti-sigma factor antagonists. The physiological and/or mechanistic roles of the STAS domain in the SLC26 exchangers are only just beginning to emerge; however, the existence of disease-associated mutations in this domain underscore its potential importance (Aravind and Koonin, 2000). Structural features have been predicted from the NMR analysis of the anti-sigma factor SPOIIAA, and include a characteristic α-helical handle. There is also a highly conserved loop interspersed between a β-pleated sheet and an α-helical region, just upstream of the α-helical handle. This loop and β-pleated sheet have been proposed to play a role in nucleotide binding and hydrolysis, by extension from the known biochemistry of the anti-sigma factor antagonists (Aravind and Koonin, 2000) (Fig. 4.8, inset). The loop is highly conserved in the ten SLC26 proteins and contains two invariant residues, D660 and L667 in the Slc26a2 protein. The STAS domain also contains a highly variable loop just proximal to the β-pleated sheet and putative nucleotide binding loop. This variable loop is the site of significant insertions in the SLC26 proteins, of as much as 150 amino acids in the case of SLC26A8. No such insertion is present in bacterial SLC26 homologs, and this loop is also much shorter in SLC26A11 and in the *Drosophila* and *C. elegans* paralogs; how this variable loop contributes to paralog-specific function and/or regulation is not known.

Deletion of the STAS domain (including the full C-terminus) leads to loss of function in SLC26A3, which otherwise tolerates removal of up to ~40 C-terminal amino acids (Chernova et al., 2003a). Random mutagenesis of a plant SLC26 sulfate transporter revealed two general classes of mutant effects within the STAS domain (Shibagaki and Grossman, 2006). Mutation within the β-pleated sheet region tends to reduce intracellular accumulation of the protein, suggesting a role in biosynthesis or stability of the protein. In contrast, mutation of the N-terminal end of the α-helical regions was associated with impaired function despite expression at the plasma membrane (Shibagaki and Grossman, 2006). A more recent study of disease-associated mutations in the STAS domain of SLC26A3 indicates a predominant effect on biosynthesis, affecting different steps in the folding and/or trafficking pathway of the protein (Shibagaki and Grossman, 2006). Collectively, these studies indicate that the presence, folding, and structure of the STAS domain are important for transport function of the SLC26 transporters.

The STAS domain has also emerged as a critical domain for protein–protein interaction (Ko et al., 2004). Provocative studies by Muallem and coworkers have thus revealed a startling intermolecular interaction between the STAS domain of SLC26A3 and the R domain of CFTR. As discussed in the section on SLC26A6, there is a reciprocal regulatory interaction between these SLC26 exchangers and CFTR. Full functional and biochemical interaction between SLC26A3 and CFTR requires the C-terminal PDZ interaction motifs (Ko et al., 2004). However, mutants lacking these motifs weakly coassociate when expressed at high levels; this interaction occurs via binding between the R and STAS domains. Notably, phosphorylation of the R domain in CFTR enhances interaction with the STAS domains of SLC26A3 (Ko et al., 2004). Finally, SLC26A3 and SLC26A6 activate the chloride channel activity of CFTR; this activation requires the R domain and can even occur after coexpression with the isolated STAS domains (Ko et al., 2004).

C. Transport Properties, Expression and Function of SLC26 "Transporters"

1. *SLC26A3 (DRA)*

SLC26A3 was initially identified as a potential tumor suppressor that was down-regulated in adenoma (hence the alias DRA) and abundant in normal colonic mucosa (Schweinfest et al., 1993). Down-regulation of SLC26A3 in neoplastic cells is attributed more to its status as a marker of differentiated colonic epithelium than to a role in cellular proliferation (Byeon et al., 1996; Silberg et al., 1995). Although the risk of colonic malignancy in kindreds with mutations in *SLC26A3* is not dramatically increased (Hemminki et al., 1998), inducible overexpression of the SLC26A3 protein leads to growth suppression of cultured cell lines (Chapman et al., 2002). Targeted deletion of murine Slc26a3 also results in abnormal growth of intestinal epithelial cells, with a marked expansion of the colonic crypt proliferative zone (Schweinfest et al., 2006). Therefore, the SLC26A3 protein may indeed play a role in the development and/or proliferation of epithelial cells.

The SLC26A3 protein is most similar to SLC26A4, with 44% identity and 60% similarity. Reports that SLC26A3 transports both SO_4^{2-} and oxalate when expressed in *Xenopus* oocytes (Silberg et al., 1995) are thus somewhat surprising, given that SLC26A4 transports only monovalent anions (Scott and Karniski, 2000). Notably, despite evidence that SO_4^{2-} and oxalate transport by SLC26A3 is DIDS sensitive and mutually *cis* inhibitory, the absolute values in pmol/oocyte/hr in these two studies (Silberg et al., 1995) are as much as several hundred-fold lower than those reported for other members of the family (Karniski, 2001; Scott and Karniski, 2000; Scott et al., 1999; Xie et al., 2002). Furthermore, the reported chloride uptakes in SLC26A3-expressing oocytes were only ~two-fold higher than background, considerably less than that achieved by this (D.B. Mount, unpublished data) and other groups (Chernova et al., 2003a).

The evidence that SLC26A3 functions as a Cl^-/base exchanger is a good deal more convincing, coming as it does from several transport physiology laboratories (Alper et al., 2001; Ko et al., 2002b; Melvin et al., 1999). Indirect evidence that Cl^-/HCO_3^- mediated by murine Slc26a3 is electroneutral (Melvin et al., 1999) appears to have been misleading, since direct measurement unveils significant changes in membrane voltage in response to HCO_3^- and Cl^- addition or removal (Ko et al., 2002b). Changes in membrane voltage under equivalent conditions are opposite to those seen in SLC26A6 oocytes (Ko et al., 2002b; Xie et al., 2002), and the stoichiometries of these two exchangers are opposite (Shcheynikov et al., 2006b). The relevance of this physiology to HCO_3^- transport is discussed in the SLC26A6 section below.

SLC26A3 transcript is predominantly expressed in the digestive system, where it is particularly abundant in duodenum and colon (Jacob et al., 2002). SLC26A3 is expressed at the apical membrane of both surface and crypt cells in the colon (Haila et al., 2001; Moseley et al., 1999; Rajendran et al., 2000). Cl^--base exchange via SLC26A3 functions in transepithelial salt transport by the colon, in collaboration with Na^+-H^+ exchange mediated by NHE3 (Schultheis et al., 1998). Targeted deletion of the murine Slc26a3 gene thus results in reduction in Cl^-/base exchange in colonic apical membrane vesicles, in addition to a high-chloride diarrhea (Schweinfest et al., 2006). Flux studies reveal that Cl^- absorption is essentially abolished in the small intestine of Slc26a3 knockout mice, versus a 20% reduction in mice lacking Slc26a6 (Walker et al., 2008). Recessive loss-of-function mutations in the human *SLC26A3* gene in turn cause severe congenital chloride diarrhea, CLD (OMIM: 21470 and 126650) (Hoglund et al., 1996). The majority of cases come from three geographic areas, Finland, Poland, and Arabic countries; distinct founder effect mutations have been characterized from all three regions (Makela et al., 2002). Loss of transport function has been reported for several mutations by Moseley et al. (Moseley et al., 1999), with the caveats discussed above. More recently, Chernova and coworkers and Ko and coworkers have demonstrated loss of function for selected mutants (Chernova et al., 2003a; Ko et al., 2002b); disease-associated mutations in the STAS domain of SLC26A3 appear to cause defects in biosynthesis and/or protein stability (Dorwart et al., 2008). Unlike SLC26A2 (Karniski, 2001) and SLC26A4 (Scott et al., 2000), a detailed comparison of the multiple SLC26A3 mutations reported in CLD (Makela et al., 2002) has not been published. However, it seems unlikely that an equivalent phenotypic series will be uncovered, since even affected siblings with identical mutations can have divergent clinical phenotypes (Hoglund et al., 2001).

Down-regulation of SLC26A3 appears to play an important role in acquired diarrhea. Infection of human intestinal cells and mice with enteropathogenic *E. coli* thus results in marked redistribution of the transporter protein away from the apical membrane, causing a reduction in apical Cl^-/base exchange (Gill et al., 2007). Similar down-regulation of SLC26A3 occurs in an animal model of inflammatory bowel disease and in human ulcerative colitis (Yang et al., 1998).

Finally, SLC26A3 is expressed in the pancreas (see below for SLC26A6), eccrine sweat glands (Haila et al., 2000), seminal vesicles (Haila et al., 2000), kidney (Wedenoja et al., 2008), and spermatocytes (Chen et al., 2008). In the kidney, the protein appears to localize to the apical membrane of non-A/non-B intercalated cells (Wedenoja et al., 2008). Patients with CLD can develop nephrocalcinosis and mild chronic kidney disease; this is due to the lifelong diarrhea, volume depletion, and hypochloremic alkalosis associated with this disorder, rather than to loss of the renal function of SLC26A3. Superficially reminiscent of the association of CFTR mutations with congenital absence of the vas deferens and male infertility (Cuppens et al., 1998), male patients with CLD have reduced fertility (Hoglund et al., 2006). A recent report implicates SLC26A3 in the HCO_3^- influx that initiates the capacitation of sperm (Fig. 4.9) (Chen et al., 2008). Slc26a8 is also highly expressed in sperm, with male infertility in the relevant murine knockout mice (Toure et al., 2007). Notably, however, transport function has yet to be convincingly demonstrated for Slc26a8, and this murine phenotype appears to involve a defect in flagellar differentiation (Toure et al., 2007), rather than sperm capacitation (Figs 4.9 and 4.10).

2. *SLC26A4 (Pendrin)*

SLC26A4 (Pendrin) was cloned by positional characterization of the gene for Pendred syndrome (OMIM #274600) (Coyle et al., 1996; Sheffield et al., 1996), the

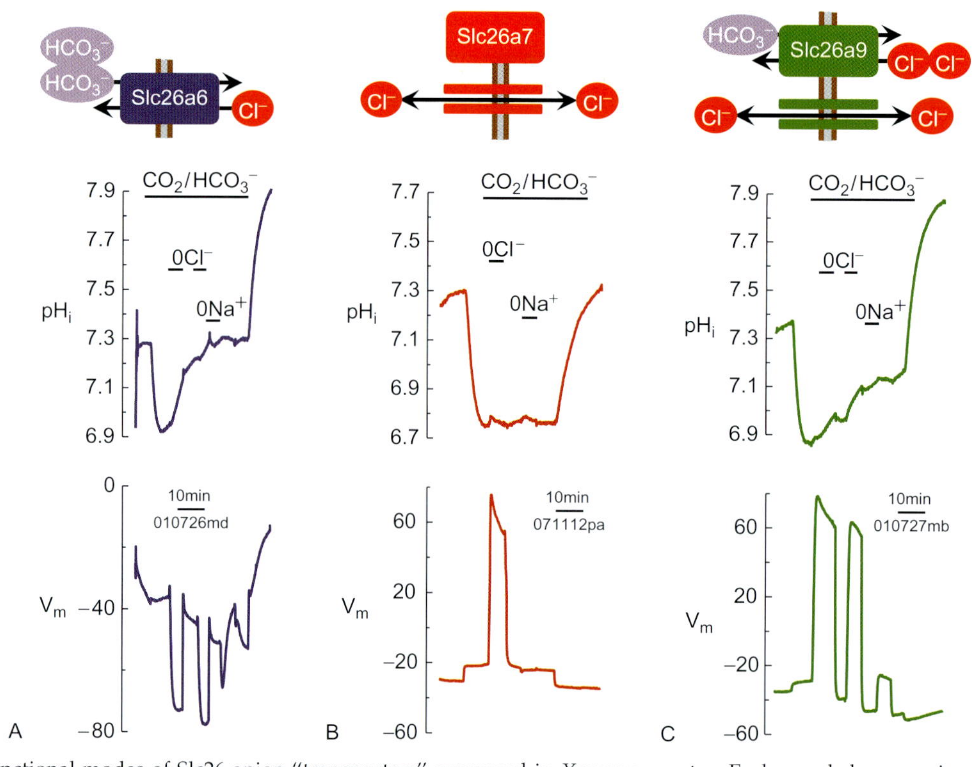

FIGURE 4.9 Functional modes of Slc26 anion "transporters" expressed in *Xenopus* oocytes. Each panel shows an ion-selective microelectrode experiment from a single *Xenopus* oocyte. Oocytes were exposed to 5% CO_2/33 mM HCO_3^- (pH 7.5) for the indicated period of time. Gluconate was the Cl^- replacement ("0Cl^-"); and choline was the Na^+ replacement ("0Na^+"). For each panel, the top diagram illustrates the transport mode. **A**. The Slc26a6 oocyte displays electrogenic $Cl^-/nHCO_3^-$ exchange activity (Xie et al., 2002) since Cl^- removal causes an increase of intracellular pH *and* hyperpolarization. **B**. The Slc26a7 oocyte does not change intracellular pH with removal of Cl^- or removal of Na^+, but does show large depolarizations with Cl^- removal ("channel-like" activity) (Kim et al., 2005). **C**. The Slc26a9 oocyte, like the Slc26a6 oocyte, shows an intracellular pH increase with Cl^- removal. However, Slc26a9 (like Slc26a7) depolarizes when Cl^- is removed. This Slc26a9 protein has recently been shown to have open and close states measure by patch clamp (Chang et al., 2009).

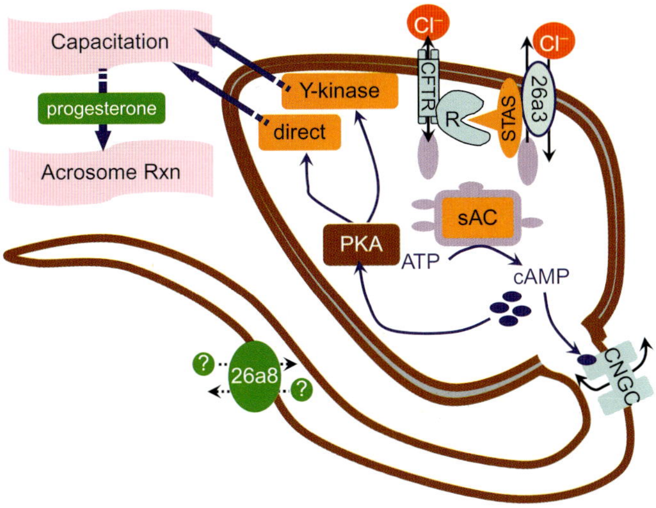

FIGURE 4.10 Anion transport and sperm capacitation. Activation of Slc26a3 (DRA/CLD) by the CFTR Cl^- channel, via a protein interaction, results in sperm capacitation. Additionally, Slc26a8 plays a role in flagellar motility, but the mode of action is currently unknown.

most common hereditary cause of syndromic deafness (see Chapter 21). Pendred syndrome is an autosomal recessive syndrome comprising sensorineural deafness and an enlarged thyroid (goiter), sometimes accompanied by hypothyroidism (Reardon et al., 1999). Biochemical studies of an involved thyroid suggested a defect in organification of iodide in the thyroid, without impairment in iodide uptake or TSH signaling (Sheffield et al., 1996); comparable insights into the associated deafness were not available prior to the cloning of SLC26A4, although patients with Pendred syndrome were known to exhibit a dilated vestibular aqueduct and/or Mondini dysplasia, a complex abnormality in cochlear structure. The SLC26A4 gene first emerged in chromosome 7-specific sequence data from the Human Genome Project, and was found to reside only 49 kb from the previously identified SLC26A3 gene (Everett et al., 1997). In a subsequent paper, four recurrent mutations in European families with Pendred syndrome accounted for 74% of the cases, suggesting multiple founder effects (Coyle et al., 1998). The chromosomal localization of Pendred syndrome was known to be very close to that of DFNB4, an autosomal recessive form of non-syndromic deafness (Coyle et al., 1996); involvement of SLC26A4 in DFNB4 was confirmed soon after the initial cloning (Li et al., 1998). Many if not all patients with DFNB4 due to mutations in SLC26A4 have dilated vestibular aqueducts, as seen in Pendred syndrome (Reardon et al., 2000).

Mutations in the SLC26A4 protein that are clinically associated with Pendred syndrome cause complete loss of transport function when studied in *Xenopus* oocytes, whereas those exclusively associated with DFNB4 have residual transport activity (Scott et al., 2000). Although this suggests that the phenotype is largely determined by the severity of the defect in SLC26A4, other genetic and/or environmental (e.g. iodide intake) factors may also play a role (Masmoudi et al., 2000; Taylor et al., 2002). There is also considerable variation in vestibular phenotype in both humans and mice, suggesting other modifying influences on the inner ear physiology (Everett et al., 2001). The majority of disease-associated mutations in SLC26A4 appear to cause a processing defect, with failure to reach the plasma membrane (Rotman-Pikielny et al., 2002; Taylor et al., 2002); however, differences in the cellular localization of specific disease-associated mutants suggest a heterogeneity of responsible mechanisms (Yoon et al., 2008). Notably, several nonconservative coding SNPs in "normal-hearing" individuals occur in highly conserved residues; some but not all of these mutations reduce function, emphasizing the difficulty in predicting the functional effect of coding SNPs in this and other membrane transporters (Pera et al., 2008).

Human SLC26A4 expressed in both *Xenopus* oocytes and Sf9 cells mediates Cl^- and iodide (I^-) transport, exhibiting mutual *cis*-inhibition consistent with Cl^-/I^- exchange (Scott et al., 1999). SLC26A4 appears to be selective for monovalent anions, also transporting formate but neither SO_4^{2-} (Scott et al., 1999) nor oxalate (Scott and Karniski, 2000). Intracellular pH measurements in response to manipulation of bath HCO_3^- and/or Cl^- have demonstrated that SLC26A4 also functions as a Cl^-/base exchanger (Ko et al., 2002b; Soleimani et al., 2001). However, unlike its close relative SLC26A3, SLC26A4 appears to mediate *electroneutral* Cl^-/base exchange, i.e. with a stoichiometry of 1:1 (Shcheynikov et al., 2008); Cl^-/I^- exchange is also electroneutral, and the exchanger appears to prefer I^- over Cl^- (Shcheynikov et al., 2008).

The functional characteristics of SLC26A4 correlate with its physiological roles in thyroid and kidney, two tissues in which this gene is expressed at significant levels. SLC26A4 protein is detected at the apical membrane of thyroid follicles, facing the colloid lumen (Pera et al., 2008). Iodide is thus taken up at the basolateral membrane by the Na^+-linked I^- transporter (NIS, SLC5A5) and secreted through SLC26A4 into the follicular lumen, wherein organification occurs. The expression of SLC26A4 in the mammary gland (Rillema and Hill, 2003), endometrium, and placenta (Bidart et al., 2000) reflects a similar role in iodide transport by these tissues. It seems likely that SLC26A4 is not the sole efflux pathway for I^- in the thyroid, given that targeted deletion of the murine

Slc26a4 gene does not result in thyroid abnormalities (Everett et al., 2001). Moreover, the modest I^- transport exhibited by the mutant SLC26A4 proteins associated with DFNB4 is evidently sufficient to maintain normal thyroid histology (Scott et al., 2000). Notably, within the parotid gland, targeted deletion of murine Slc26a4 completely abolishes I^- secretion; genetic loss of Slc26a6 has no effect (Pera et al., 2008).

Considerable excitement was initially generated by the observation that SLC26A4 mediates Cl^-/formate exchange, given possible involvement in transepithelial Na^+-Cl^- transport by the proximal tubule (see section VI.C.4). However, expression of SLC26A4 in the proximal tubule is minimal or absent in several species, except perhaps for rat (Soleimani et al., 2001), and murine Slc26a4 is quite clearly not involved in formate-stimulated Na^+-Cl^- transport in this nephron segment (Karniski et al., 2002). There is, however, robust expression of SLC26A4 in distal type B intercalated cells (Royaux et al., 2001), which are known to excrete HCO_3^- via a Cl^-/HCO_3^- exchanger with functional similarities to SLC26A4 (Emmons, 1999). Targeted deletion of murine Slc26a4 reduces distal HCO_3^- secretion in alkali-loaded knockout mice (Royaux et al., 2001). Along the distal nephron, SLC26A4 protein is detectable in both sub-apical vesicles and at the apical membrane, within both type B and non-A/non-B intercalated cells (Kim et al., 2002; Wall et al., 2003); systemic acidosis induces an intracellular shift in type B cells, whereas alkalosis increases the proportion of apical staining for SLC26A4 (Wagner et al., 2002).

Slc26a4 has been shown to be exquisitely sensitive to dietary intake and tubular delivery of Cl^-, suggesting an important role for dysregulation of this transporter in salt-sensitive hypertension. Quentin and coworkers thus demonstrated that increased Cl^- intake decreased Slc26a4 expression in intercalated cells (Quentin et al., 2004). In addition, mice lacking Slc26a4 become hypotensive when treated with dietary Cl^- restriction (Verlander et al., 2003). Under these same conditions, plasma pH and $[HCO_3^-]$ are increased while plasma $[Cl^-]$ is unchanged compared with the same parameters in wild-type animals. Slc26a4-deficient mice do, however, excrete more Cl^- and their urinary pH is acidic compared with wild-type animals. Nevertheless, these differences between wild-type and knockout mice are not present in animals on high-salt diets (Verlander et al., 2003). Slc26a4 knockout mice are also resistant to mineralocorticoid-induced hypertension (Verlander et al., 2003). Finally, dietary Cl^- restriction with provision of Na^+-HCO_3^- results in Cl^- wasting in Slc26a4 knockout mice and increased apical expression of Slc26a4 protein in the type B intercalated cells of normal littermate controls (Verlander et al., 2006).

As expected from the phenotype of Pendred syndrome, targeted deletion of murine Slc26a4 results in profound deafness and variable deficits in vestibular function (Everett et al., 2001). Slc26a4 transcript is expressed throughout the endolymphatic duct and sac, suggesting a role in endolymphatic fluid reabsorption (Everett et al., 1999). Structural abnormalities in the inner ear in Pendred syndrome/DFNB4 had been attributed in part to arrested development, and the availability of a murine model affords an important opportunity to follow changes in inner ear structure during pre- and post-natal development. The inner ear in Slc26a4-null mice is normal until embryonic day 15, at which point they begin to develop a dilated endolymphatic duct and sac. There is almost a complete absence of normal otoconia with frequent "giant" otoconia; the latter are considered a marker for biochemically abnormal endolymph in which the kinetics of otoconia dissolution and aggregation are altered (Everett et al., 2001). In this regard, ultrastructural characterization of mitochondria-rich endolymphatic cells indicates a subtype analogous to type B intercalated cells of the distal nephron (Peters et al., 2002), suggesting that defects in SLC26A4 might be associated with marked changes in the pH of endolymphatic fluid. Slc26a4-deficient mice do in fact demonstrate an acidic endolymph (Nakaya et al., 2007; Wangemann et al., 2007). This acidic milieu leads to reduced absorption of Ca^{2+} via the acid-sensitive TRPV5 and TRPV6 channels (Nakaya et al., 2007; Wangemann et al., 2007); cellular calcium overload may lead to the observed cellular degeneration in the cochlea of Slc26a4-knockout mice. Associated oxidative stress, possibly generated by extracellular acidity, also leads to reduced expression of the Kcnij10 K^+ channel in *stria vascularis*. Decreased K^+ channel function in turn leads to a reduction in the endocochlear potential and deafness (Singh and Wangemann, 2008), as discussed in detail in Chapter 21.

Finally, two provocative, landmark studies recently identified a critical role for SLC26A4 in bronchial mucus production (Nakagami et al., 2008; Nakao et al., 2008). SLC26A4 expression in human primary tracheal epithelial cells is thus strongly induced by IL-13 (Nakao et al., 2008), a Th2-type cytokine with a key role in asthma; similar pulmonary induction of Slc26a4 is detectable in murine models of asthma and chronic obstructive pulmonary disease (COPD) (Kuperman et al., 2005; Nakao et al., 2008). Slc26a4 knockout mice are less prone to allergen-induced airway hyperreactivity and inflammation than control mice (Nakagami et al., 2008). Overexpression of SLC26A4 in turn induces expression of MUC5AC, a major component of mucus in asthma and COPD, cultured airway epithelial cells and stimulates mucus production in murine lungs (Nakao et al., 2008). Furthermore, SLC26A4 expression is increased almost five-fold during infection with rhinovirus, the common cold. Collectively, these studies identify SLC26A4 as a major pharmacological target in COPD, asthma, and respiratory infections, major causes of morbidity and mortality.

3. *SLC26A5 (Prestin)*

The outer hair cells (OHCs) in the mammalian cochlea (inner ear) function as critical amplifiers of the auditory signal, as discussed in detail in Chapter 21. Striking changes occur in the shape of OHCs in response to changes in membrane voltage; depolarization shortens the cell whereas hyperpolarization lengthens the cell (Brownell et al., 1985). This somatic "electromotility" is thought to play a critical role in active amplification, enhancing detection and discrimination of acoustic frequencies beyond 70 kHz. Several lines of evidence supported the existence of a "molecular motor" in the lateral plasma membrane of OHCs, generating changes in the area of the plasma membrane as a function of membrane potential (Ashmore, 2008). However, the identity of the protein(s) responsible was elusive for the better part of a decade.

In a landmark study published in 2000, Dallos and coworkers identified SLC26A5 (or "prestin", as in "presto" for fast) as a candidate for the molecular motor of OHCs (Zheng et al., 2000). This approach utilized a subtractive cDNA cloning strategy for messages enriched in the outer hair cell of the gerbil ear (Zheng et al., 2000). Slc26a5 transcripts represented ~10% of clones differentially expressed in inner vs. outer hair cell. Although Slc26a5 is highly expressed in the ear, transcripts are detected in brain, heart, spleen, and testis (Zheng et al., 2003); the role of Slc26a5 in these tissues is as yet unknown. The protein is expressed at the basal and lateral membrane of OHCs but not at the apical membrane (Yu et al., 2006), as expected for the molecular motor responsible for electromotility (Ashmore, 2008).

Sequence and hydropathy analysis reveals that Slc26a5 is similar to other SLC26 proteins. However, differences within the conserved hydrophobic regions discussed above indicated that Slc26a5 might have transport properties that are unique within the SLC26 family (Zheng et al., 2000). Heterologous expression of the transporter in HEK293 cells revealed that the transporter has electrophysiological and electromotile behavior consistent with that of the molecular structure of OHCs. These investigators thus detected a "gating current" and non-linear capacitance (NLC) change in cells transfected with Slc26a5 but not SLC26A4. Cells also demonstrated a voltage-dependent electromotile response; NLC and this electromotile response were both blocked by 10 mM Na^+-salicylate, which blocks these responses in OHCs. In another landmark study, Oliver and coworkers demonstrated that this non-linear capacitance change is mediated by intracellular halides, especially Cl^- and HCO_3^- (Oliver et al., 2001). In the absence of Cl^- and HCO_3^- these charge movements were not present, nor were they seen in cells expressing human SLC26A6. These capacitance changes were frequency dependent and were also found in inner hair cells, with both systems blocked by salicylate. These investigators hypothesized that prestin is the motor protein of inner hair cells and that intracellular Cl^- and HCO_3^- acts as the voltage "sensor" for the protein (Oliver et al., 2001). Voltage-dependent movement of anions during a truncated transport cycle would thus switch prestin from a contracted to an expanded state, generating electromotility. Notably, reversible effects of *in vivo* changes in intracellular Cl^- on electromotility and cochlear amplification have also been obtained in OHCs (Santos-Sacchi et al., 2006).

Several lines of evidence suggest that SLC26A5 functions as a "partial" or "modified" anion exchanger (Schaechinger and Oliver, 2007). First, there is the evident homology to other SLC26 transporters, particularly Slc26a6. Second, millimolar intracellular Cl^- or HCO_3^- modulates the voltage dependency of NLC in Slc26a5-expressing cells (Oliver et al., 2001); more recent data suggest that intracellular SO_4^{2-} can also modulate NLC in OHCs (Rybalchenko and Santos-Sacchi, 2008). Third, many of the properties of Slc26a5 can be accounted for by modeling it as a Cl^-/SO_4^{2-} exchanger (Muallem and Ashmore, 2006). Fourth, and perhaps most compelling, is the recent report that zebrafish (zSLC26A5) and chicken SLC26A5 function as salicylate-sensitive Cl^-/SO_4^{2-} and a Cl^-/ox^{2-} exchanger (Schaechinger and Oliver, 2007, transport mode illustrated for proximal tubule in Fig. 4.5). This has led to the hypothesis that the anion and voltage sensitivity of Slc26a5 is a consequence of an incomplete anion exchange cycle; given the effect of intracellular Cl^-, HCO_3^-, and SO_4^{2-} (Rybalchenko and Santos-Sacchi, 2008; Schaechinger and Oliver, 2007), it seems plausible that mammalian SLC26A5 has lost the ability to bind or translocate SO_4^{2-} across the membrane (Schaechinger and Oliver, 2007). Notably, zSLC26A5 has a weak NLC activity but no electromotile activity (Albert et al., 2007), suggesting that comparative mutational analysis of this and related orthologs will begin to yield the structural determinants of anion transport, non-linear capacitance, and electromotility.

Genetic evidence clearly implicates SLC26A5 in mammalian hearing. Mutations in human SLC26A5 have thus been linked to deafness in two separate kindreds with non-syndromic hearing loss (Liu et al., 2003). Targeted deletion of murine Slc26a5 results in impaired electromotility and a deafness phenotype (Liberman et al., 2002). However, these mice also demonstrate shortening of OHCs and considerable loss of hair cells in high-frequency regions, i.e. the evidence for a specific role for altered electromotility was not definitive. Furthermore, subsequent analysis revealed dramatic changes in the intrinsic stiffness of OHCs, suggesting that loss of Slc26a5 alters the mechanical properties of these cells (Mellado Lagarde et al., 2008). These issues were elegantly addressed by the generation and characterization of knock-in mice for a mutation in Slc26a5 that impairs NLC and the electromotile response, but not expression at the cell membrane. The morphology and stiffness of OHCs is not affected in these knock-in mice, whereas cochlear function is profoundly affected (Dallos et al., 2008). Therefore, the electromotile activity of Slc26a5 is required for cochlear amplification.

4. *SLC26A6 (PAT-1, CFEX)*

SLC26A6 was the first member of this family to be identified exclusively through database mining. Everett and Green had initially noted the existence of multiple SLC26 ESTs that were evidently derived from novel members of the family (Everett and Green, 1999); partial characterization of many of these genes was subsequently reported by Lohi et al., in addition to the cloning of a full-length human SLC26A6 cDNA (Lohi et al., 2000). Murine Slc26a6 was then identified as a primary candidate for the apical Cl^-/formate exchanger of the renal proximal tubule (Knauf et al., 2001). SLC26A6 is one of the more widely expressed members of the family, with particularly abundant transcript in kidney, pancreas, intestine, heart, muscle, and placenta (Everett and Green, 1999; Knauf et al., 2001; Ko et al., 2002b; Lohi et al., 2000; Xie et al., 2002). Significant heterogeneity appears to be generated by alternative splicing of the SLC26A6 gene, with two major alternatively spliced isoforms differing in the presence or absence of an N-terminal extension of 23 amino acids in both mice and humans (Everett and Green, 1999; Lohi et al., 2000; Xie et al., 2002). Three other functional splice variants were also described in human SLC26A6 (Lohi et al., 2002b). Of considerable concern, however, was that a splice variant of SLC26A6 that lacks TM5-6 was supposedly fully functional (Lohi et al., 2003), a bizarre, provocative result that was not replicated by Alper and coworkers using this same construct (Chernova et al., 2005). Immunolocalization has mostly been reported for epithelial tissues, where SLC26A6 protein is apically expressed (Knauf et al., 2001; Lohi et al., 2000; Petrovic et al., 2002; Wang et al., 2002). SLC26A6 is also heavily expressed in cardiac myocytes, wherein it appears to be the dominant Cl^--base exchanger (Alvarez et al., 2004).

SLC26A6 is the most convincingly versatile anion exchanger of the family, functioning in Cl^-/oxalate, SO_4^{2-}/oxalate, SO_4^{2-}/Cl^-, Cl^-/formate, Cl^-/HCO_3^-, and Cl^-/OH^- exchange when expressed in *Xenopus* oocytes (Jiang et al., 2002; Ko et al., 2002b; Wang et al., 2002; Xie et al., 2002) (Fig. 4.5). SLC26A6 was a candidate for the long-elusive apical Cl^- entry site involved in proximal tubular Na^+-Cl^- absorption (Fig. 4.5), since it is expressed at the apical membrane of proximal tubule cells (Knauf et al., 2001) and appears to mediate the multiple modes of anion exchange that have been implicated in this process (Jiang et al., 2002; Xie et al., 2002); the relevant historical background has been reviewed (Kurtz et al., 1994). In addition to the exchange of Cl^- with various anions, SLC26A6 may also function in formate/OH^- exchange (Saleh et al., 1996), which is required for apical formate recycling (see Fig. 4.5). Notably, the analysis of Slc26a6 knockout mice indicated no effect on baseline fluid absorption (J_v) across perfused proximal tubules (Wang et al., 2005). However, the usual stimulation of J_v by perfusion with luminal oxalate was completely abolished in Slc26a6 knockout mice (Wang et al., 2005), along with loss of Cl^-/oxalate exchange in isolated brush border vesicles (Jiang et al., 2006); Slc26a6 is thus the dominant Cl^-/oxalate exchanger in apical membranes of the proximal tubule. Slc26a6 also contributes considerably to apical Cl^-/base (Wang et al., 2005) and Cl^-/formate (Jiang et al., 2006; Wang et al., 2005) exchange in the proximal tubule; however, other as-yet-unidentified anion exchangers appear to also mediate these activities. There is considerable functional and pharmacological heterogeneity of anion exchange pathways in proximal brush border vesicles (Karniski and Aronson, 1987), consistent perhaps with this underlying molecular heterogeneity.

A major surprise was that Cl^--base exchange mediated by SLC26 family members is electrogenic, with opposite apparent stoichiometries for SLC26A6 and SLC26A3/A4 (Ko et al., 2002b; Xie et al., 2002). Thus whereas SLC26A6-expressing cells hyperpolarize upon removing extracellular Cl^- in the presence of HCO_3^- (Eqs 18 and 19) (Xie et al., 2002), SLC26A3-expressing cells exhibit a pronounced *depolarization* (Eqs 20 and 21) (Ko et al., 2002b); a modest hyperpolarization is also seen during Cl^-/OH^- exchange mediated by Slc26a6 (Xie et al., 2002). It appears that functional CFTR is required for the activation of SLC26 Cl^-/base exchange by cyclic-AMP (Ko et al., 2002b); this observation may be of profound significance for cystic fibrosis, given the role for abnormal bicarbonate transport in this disease (Choi et al., 2001). Epithelial HCO_3^- transport has both an electrogenic component, generally ascribed to the HCO_3^- permeability of CFTR itself, and an electroneutral component attributed to Cl^-/HCO_3^- exchange (see Fig. 4.5). CFTR-regulated Cl^-/HCO_3^- exchange appears to be electroneutral (Lee et al., 1999); however, cyclic-AMP has no effect on Cl^-/HCO_3^- exchange mediated by the electroneutral SLC4 anion exchangers AE1-3 (Eqs 2 and 3) (Ko et al., 2002b). CFTR does, however, markedly activate Cl^-/OH^- exchange mediated by SLC26A3, SLC26A4, and SLC26A6 (Ko et al., 2002b). Like SLC26A6 (Lohi et al., 2000), SLC26A3 protein is detected at the apical membrane of pancreatic ductal cells (Greeley et al., 2001). In cell membranes expressing an equal proportion of SLC26A6 and SLC26A3, one would expect net Cl^-/HCO_3^- exchange to be electroneutral; variation in the activity and/or ratio of these and other SLC26 exchangers would lead to net electrogenic HCO_3^- transport (Ko et al., 2002b). The apparent stoichiometry of these paralogs (Shcheynikov et al., 2006b) suggests that SLC26A6 is best suited to HCO_3^- excretion in the early portion of the pancreatic duct, with SLC26A3 (HCO_3^-:Cl^- of 1:2) (Eqs 20 and 21) better suited for generation of the low Cl^- and HCO_3^- concentration of the final secreted fluid (Ko et al., 2002b) (Fig. 4.5).

$$\begin{aligned}\Delta\mu_{Slc26a6} &= \Delta\mu_{Cl/2HCO_3} = \Delta\mu_{Cl} - 2\Delta\mu_{HCO_3}\\ &= \Delta\mu_{Cl} - \Delta\mu_{oxalate}\\ &= \Delta\mu_{Cl} - \Delta\mu_{sulfate}\\ &= \Delta\mu_{Cl} - \Delta\mu_{formate^{g}}\end{aligned} \qquad \text{(Eq. 18)}$$

Or,

$$\begin{aligned}\Delta\mu_{Cl/2HCO_3} &= RT\ln\left(\frac{[Cl^-]_i}{[Cl^-]_o}\right) + (-1)FV_m\\ &\quad - 2RT\ln\left(\frac{[HCO_3^-]_i}{[HCO_3^-]_o}\right) + (-1)FV_m\end{aligned}$$

Or,

$$\begin{aligned}\Delta\mu_{Cl/2HCO_3} &= RT\ln\left(\frac{[Cl^-]_i[HCO_3^-]_o^2}{[Cl^-]_o[HCO_3^-]_i^2}\right)\\ &\quad - FV_m + 2FV_m\end{aligned}$$

Or,

$$\Delta\mu_{Cl/2HCO_3} = RT\ln\left(\frac{[Cl^-]_i[HCO_3^-]_o^2}{[Cl^-]_o[HCO_3^-]_i^2}\right) + FV_m \qquad \text{(Eq. 19)}$$

$$\begin{aligned}\Delta\mu_{Slc26a3} &= \Delta\mu_{Slc26a9} = \Delta\mu_{2Cl/HCO_3}\\ &= 2\Delta\mu_{Cl} - \Delta\mu_{HCO_3}\end{aligned} \qquad \text{(Eq. 20)}$$

$$\begin{aligned}\Delta\mu_{2Cl/HCO_3} &= 2RT\ln\left(\frac{[Cl^-]_i}{[Cl^-]_o}\right) + (-1)FV_m\\ &\quad - RT\ln\left(\frac{[HCO_3^-]_i}{[HCO_3^-]_o}\right) + (-1)FV_m\end{aligned}$$

$$\begin{aligned}\Delta\mu_{2Cl/HCO_3} &= RT\ln\left(\frac{[Cl^-]_i^2[HCO_3^-]_o}{[Cl^-]_o^2[HCO_3^-]_i}\right)\\ &\quad - 2FV_m + FV_m\end{aligned}$$

$$\Delta\mu_{2Cl/HCO_3} = RT\ln\left(\frac{[Cl^-]_i^2[HCO_3^-]_o}{[Cl^-]_o^2[HCO_3^-]_i}\right) - FV_m \qquad \text{(Eq. 21)}$$

Analysis of knockout mice has revealed a slightly more complicated picture. Targeted deletion of Slc26a6 results in an unexpected *increase* in Cl^-/HCO_3^- exchange in pancreatic ducts (Ishiguro et al., 2007; Wang et al., 2006). There are considerable differences between these two studies, involving separate knockout strains. Ishiguro et al. report that basal and stimulated Cl^-/HCO_3^- exchange is increased in their Slc26a6 knockout mice (Ishiguro et al., 2007). Notably, whereas the *influx* mode of Cl^-/HCO_3^- exchange was dramatically increased, the *efflux* appeared to be decreased; this was accompanied by a marked increase in Slc26a3 expression, suggesting that this is the dominant exchanger in these Slc26a6-deficient cells. In contrast, Wang and coworkers (Wang et al., 2006) report a considerable increase in basal Cl^-/HCO_3^- exchange and fluid secretion, but a marked attenuation of stimulated secretion. No change in the expression of Slc26a3 was detected in pancreatic ducts versus whole pancreas as shown by Ishiguro and colleagues (Ishiguro et al., 2007). Knockdown and pharmacological inhibition of CFTR indicates that CFTR activity is paradoxically enhanced in Slc26a6 knockout mice (Wang et al., 2006). These combined results were explained by a removal of tonic inhibition of CFTR in the resting duct and a reduced activation of the absent Slc26a6 by CFTR in stimulated ducts (Wang et al., 2006). The complexity of these findings underscores the importance of the reciprocal interactions between CFTR and SLC26A6, mediated by binding between the R domain of CFTR

[g] The Cl-formate exchange of Slc26a6 is *not* obviously electrogenic, even in heterologous expression systems.

and the STAS domain of SLC26A6 (Ko et al., 2004). The pancreatic phenotype of Slc26a3-deficient and Slc26a3/Slc26a6 double knockout mice will be of particular interest, and should begin to clarify the relative role of these two paralogs in pancreatic HCO_3^- secretion.

The apical expression of SLC26A3 and SLC26A6 within the gastrointestinal tract overlaps significantly, with robust expression of both in the duodenum (Jacob et al., 2002; Wang et al., 2002) and variable coexpression elsewhere. Jacob et al. have suggested that the relative surfeit of SLC26A3 transcript in duodenum compared to NHE2 and NHE3 reflects the role of this segment in secretion of HCO_3^- (Jacob et al., 2002). This intestinal mode of HCO_3^- secretion involves many of the same molecular players as those used for proximal tubule HCO_3^- absorption (Kurita et al., 2008) (Fig. 4.5). SLC26A6 colocalizes with H^+/K^+-ATPase in gastric parietal cells, with dramatic attenuation of expression in *H^+/K^+-ATPase*-null mice (Wang et al., 2002). Analysis of Slc26a6 knockout mice reveals important contributions of this exchanger to duodenal HCO_3^- excretion (Simpson et al., 2007) and jejunal Na^+ and Cl^- absorption (Seidler et al., 2008).

Consistent with its evident role in HCO_3^- homeostasis, SLC26A6 has been shown to interact with carbonic anhydrase II (CA-II) (Alvarez et al., 2005); this is reminiscent of the putative and disputed (see section V) interactions between SLC4 transporters and carbonic anhydrases. Inhibition of carbonic anhydrase with acetazolamide inhibits Cl^-/HCO_3^- exchange in HEK293 cells expressing SLC26A6; notably, although this is consistent with an effect mediated by carbonic anhydrase inhibition, direct inhibition of the transporter is difficult to exclude. Regardless, Slc26a6 appears to interact with CAII via a motif with sequence similarity to CA-II binding sites in SLC4 transporters (Alvarez et al., 2005). No such site is found in SLC26A3, which does not bind CA-II but appears to require cytosolic carbonic anhydrase activity for full Cl^-/HCO_3^- exchange activity (Sterling et al., 2002). Phosphorylation of an adjacent serine in SLC26A6 by the protein kinase C site appears to disrupt the interaction with CA-II and reduce Cl^-/HCO_3^- exchange (Alvarez et al., 2005); notably, however, when expressed in *Xenopus* oocytes, the alanine mutant of this serine retains sensitivity to protein kinase C (Hassan et al., 2007).

Finally, Slc26a6 has been shown to transport oxalate (Fig. 4.5, Eq. 18) (Jiang et al., 2002; Xie et al., 2002), and the overall transport characteristics are very similar to those reported for the oxalate exchanger of ileal brush border (Knickelbein et al., 1986). This suggested a dominant role for SLC26A6 and other coexpressed SLC26 exchangers in the absorption of dietary oxalate by the small intestine, a major determinant of urinary excretion and thus the risk for renal stones (Holmes et al., 2001). Unexpectedly, targeted deletion of murine Slc26a6 results in marked hyperoxaluria and kidney stones; this phenotype appears to be due to a loss of oxalate secretion in the small intestine, tilting the balance in favor of oxalate absorption (Freel et al., 2006; Jiang et al., 2006). A key issue is the identity of the absorptive mechanism for oxalate in the small intestine; regardless, these results indicate a critical role for SLC26 transporters in oxalate homeostasis and kidney stones.

5. *SLC26A7-SLC26A11*

Several other novel SLC26 paralogs have more recently been described (see Fig. 4.7). Although Lohi et al. included a gene now denoted SLC26A10 on chromosome 12 in their initial report (Lohi et al., 2000), our own sequencing of EST cDNAs in both mouse and human suggests that SLC26A10 is an expressed pseudogene (DBM, unpublished data). There are also several GenBank cDNA accessions for this gene, none of which contains an uninterrupted open reading frame. Many of the other SLC26 paralogs have tissue-specific or highly restricted expression patterns; SLC26A7 in kidney (Lohi et al., 2002a; Vincourt et al., 2003), SLC26A8 in testis and spermatocytes (Lohi et al., 2002a; Toure et al., 2001), and SLC26A9 in lung and stomach (Lohi et al., 2002a). SLC26A11 is more widely expressed (Vincourt et al., 2003).

Modest anion transport has been reported for all four of the novel paralogs. SLC26A7, SLC26A8, and SLC26A9 reportedly transport Cl^-, SO_4^{2-}, and oxalate (Lohi et al., 2002a). Of some concern, SO_4^{2-} transport mediated by SLC26A9 is not *cis* inhibited by oxalate (Lohi et al., 2002a), which is at least superficially inconsistent with the ability of this exchanger to mediate oxalate transport. Furthermore, we have not been able to reproducibly demonstrate SO_4^{2-} transport mediated by SLC26A7-9 or SLC26A11. SLC26A7 and SLC26A9 do, however, function as robust chloride transporters.

SLC26A7 has been reported to function in Cl^-/HCO_3^- exchange (Petrovic et al., 2003); however, this has not been replicated by our group (Chang et al., 2009; Romero et al., 2006) or that of Muallem et al. (Kim et al., 2005). Rather, SLC26A7 appears to function as a pH-regulated chloride channel with minimal conductivity for OH^- or HCO_3^- and no capacity for Cl^--base exchange (Kim et al., 2005). Notably, SLC26A1 and SLC26A7 are unique among the SLC26 family in trafficking to the basolateral membrane of epithelia, gastric parietal cells (Kosiek et al., 2007; Petrovic et al., 2003) and renal alpha intercalated cells (Petrovic et al.,

2004) in the case of SLC26A7. Surprisingly, murine Slc26a7 protein has also been localized at the apical membrane of the proximal tubule, suggesting a role in proximal tubular Na^+-Cl^- transport (Dudas et al., 2006). The role of SLC26A7 in these cells awaits the characterization of Slc26a7-deficient knockout mice.

SLC26A9 is highly expressed in stomach and lung, but detectable by RT-PCR in several other tissues. Functional characterization in *Xenopus* oocytes indicates that murine Slc26a9 transports Cl^- but not formate, SO_4^{2-}, or oxalate (Romero et al., 2006). These uptake data for mouse Slc26a9 differ from that published for human SLC26A9 (Lohi et al., 2002a). Lohi et al. thus reported that SLC26A9-injected cells mediate uptake of SO_4^{2-}, Cl^-, ox^{2-} and formate (Lohi et al., 2002a). Perhaps some differences are due to species variation, although mouse and human Slc26a9 are ~90% identical. Notably, Lohi's experiments lack a positive control (e.g. Slc26a6) for the substrates tested. In addition, the reported *cis*-inhibition data by Lohi disagree with uptake measurements in the same study, i.e. oxalate was a substrate but did not *cis* inhibit $^{35}SO_4^{2-}$ uptake. Regardless, four groups have reported that SLC26A9 exhibits Cl^- channel-like characteristics (Chang et al., 2009; Dorwart et al., 2007; Loriol et al., 2008; Romero et al., 2006; Xu et al., 2008). Recent, single channel measurements show that Slc26a9 displays discrete open and close states (Chang et al., 2009). We have found that Slc26a9 also mediates Cl^-/HCO_3^- exchange, with an evident stoichiometry opposite to that of Slc26a6 (Chang et al., 2009; Romero et al., 2006). Notably, we have also utilized Na^+ electrodes and uptakes to demonstrate that Slc26a9 has a cation dependence (Chang et al., 2009). Slc26a9 thus has three discrete physiological modes: nCl^-/HCO_3^- exchanger, Cl^- channel, and Na^+-anion cotransporter (Chang et al., 2009). How these modalities relate to the physiological role of Slc26a9 is not as yet known. Moreover, it is not clear if Slc26a9 form different species (even mammalian species), will have identical physiology. That said, targeted deletion of murine Slc26a9 causes achlorhydria, evidently due to effects on the viability or development of secretory canaliculi and/or via an effect on chloride secretion (Xu et al., 2008). Clearly, these Slc26a9$^{-/-}$ animals will be of great interest.

D. Outlook and Pharmaceutical Considerations – SLC26 "Transporters"

Slc26a1 (Sat-1) was cloned by functional expression as an Na^+ independent SO_4^{2-} transporter (Bissig et al., 1994), and the resulting Slc26 gene family was initially called the "sulfate permease" family. Over the last 15 years, nine other Slc26 genes have been identified to complete the mammalian gene family (Fig. 4.7). Thus far, SLC26A2 (DTDST) (Hastbacka et al., 1992, 1994; Superti-Furga et al., 1996), SLC26A3 (DRA/CLD) (Silberg et al., 1995) and SLC26A4 (pendrin) (Everett et al., 1997) are genetic disorders with Mendelian inheritance (Everett and Green, 1999). Additionally, SLC26A5 is associated with another category of human deafness (Liu et al., 2003).

Many SLC26 proteins have overlapping tissue expression and some similar transport functions. For example, at least four family members, Slc26a2, Slc26a3, Slc26a6 and Slc26a9 function, in part, as Cl^-/base exchangers and are expressed at the apical membrane of pancreatic ductal cells (Chang et al., 2009; Greeley et al., 2001; Haila et al., 2001; Ko et al., 2002b; Lohi et al., 2000; Xie et al., 2002). HCO_3^- exchange by both SLC26A3 and SLC26A6 is apparently regulated by CFTR (Ko et al., 2002b), and thus implicating them in CFTR-dependent HCO_3^- secretion and pancreatitis (Choi et al., 2001). Thus, CFTR-independent activation of SLC26 Cl^-/HCO_3^- exchange is likely a worthy goal in cystic fibrosis (Ko et al., 2004). Alternatively, specific activation of the Slc26a9 channel activity could also prove a useful therapeutic goal (Chang et al., 2009). Slc26a9 activity can apparently be altered by WNK kinases (Dorwart et al., 2007). While other Slc26 transporter regulation by WNKs have not been reported, it is attractive to speculate that some of the WNK-associated hypertension phenotypes (Choate et al., 2003; Kahle et al., 2008, 2003; Rinehart et al., 2005; Wilson et al., 2001) might alter Slc26 protein function in the kidney. Accordingly, blood pressure may be regulated by Slc26 transport antagonists. Additionally, the DIDS-sensitive nature of intestinal anion exchangers, potentially SLC26 transporters, could provide insight to the intestinal transport of drugs such as salicylate (Ogihara et al., 1999) and ciprofloxacin (Cavet et al., 1997).

More recently, knockout mice have revealed several other possible disease associations. Deletion of *Slc26a6* in mice results in calcium-oxalate kidney stones apparently due to decreased intestinal oxalate secretion (Freel et al., 2006; Jiang et al., 2006). In humans, one SLC26A6 polymorphism (206M) is associated with calcium-oxalate kidney stones (Corbetta et al., 2009). The Slc26a6$^{-/-}$ mice also have reduced small intestine NaCl absorption (Seidler et al., 2008). Recently, Singh and co-workers have found that fructose absorption by the small intestine is functionally linked to NaCl absorption (Singh et al., 2008). Notably, these investigators found that the Slc26a6$^{-/-}$ mice did *not* have altered intestinal NaCl absorption after fructose feeding (Singh et al., 2008). Given the widespread use of high-fructose corn syrup as a food sweetener, Slc26a6 polymorphisms may be associated with certain aspects of "Metabolic Syndrome". These findings, however, present a physiological quandary. Therapeutic intervention of Slc26a6 may reduce the risk of metabolic syndrome and hypertension, but increase the risk of kidney stone formation. Clearly, therapeutics which are tissue specific would be needed.

Other Slc26 knockout mice have physiological phenotypes which may also facilitate therapeutic development. Deletion of *Slc26a8 (tat-1)* results in male infertility due to lack of sperm flagellar movement (Toure et al., 2007). This result could imply that a testis directed drug could be a reversible, male contraceptive. The *Slc26a9*$^{-/-}$ mouse shows decreased parietal cell (stomach) H^+ secretion (Xu et al., 2008). Perhaps targeting Slc26a9 function in the stomach could be another form of antacid.

Even though the function of several Slc26 transporter/channels has been studied (some in great detail), there are some gaps and discrepancies remaining in the published work. For example, what are the stoichiometries of all the Slc26 proteins which can function as Cl^-/HCO_3^- exchangers? Are these modes static? For SLC26A6, how and why does stoichiometry (dictating thermodynamics of ion and solute transport, see Eqs 18 and 19) and electrophysiology differ in its multiple exchange modes? For Slc26a9, are the distinct transport modes ($n Cl^-/HCO_3^-$ exchange, Cl^- channel and Na^+ anion cotransporter) cell type specific? Are these modes controlled by interaction proteins such as CFTR or PDZ scaffolds? What are the regulatory (Dorwart et al., 2008; Ko et al., 2004; Shcheynikov et al., 2006a) and/or mechanistic implications (Sindic et al., 2006) of the cytoplasmic STAS domain? How does the primary structure of the SLC26 proteins confer anion specificity and/or versatility? Ultimately, one hopes that the study of bacterial and/or *Drosophila* paralogs of the Slc26 transporter/channels will include the determination of crystal structure, to help solve these important questions and identify critical residues for ion binding sites.

Acknowledgements

We thank the members of the Romero and Mount laboratories for excellent technical support as well as help with this work: N. Angle, G. Babcock, H. Holmes, C. Plata, E. Scileppi, A. Sindic, L. Song, Q. Xie, R. Welch and M. Sanders. This work was supported by NIH grants DK056218, DK060845, EY017732 (MFR); DK038226, DK070756, DK57708 (DBM). Other funding includes an Advanced Career Development Award from the Veterans Administration (DBM); postdoctoral fellowship from AHA (M-HC); and the Cystic Fibrosis Foundation (ROMERO-06G0).

References

Abuladze, N. et al. (1998). Molecular Cloning, Chromosomal Localization, Tissue Distribution, and Functional Expression of the Human Pancreatic Sodium Bicarbonate Cotransporter. *J. Biol. Chem.* **273**, 17689–17695.

Abuladze, N., Pushkin, A., Tatishchev, S., Newman, D., Sassani, P., and Kurtz, I. (2004). Expression and localization of rat NBC4c in liver and renal uroepithelium. *Am. J. Physiol. Cell Physiol.* **287**, C781–C789.

Abuladze, N. et al. (2000). Structural organization of the human NBC1 gene: kNBC1 is transcribed from an alternative promoter in intron 3. *Gene* **251**, 109–122.

Accardi, A., Kolmakova-Partensky, L., Williams, C., and Miller, C. (2004). Ionic currents mediated by a prokaryotic homologue of CLC Cl- channels. *J. Gen. Physiol.* **123**, 109–119.

Accardi, A. and Miller, C. (2004). Secondary active transport mediated by a prokaryotic homologue of ClC Cl- channels. *Nature* **427**, 803–807.

Accardi, A., Walden, M., Nguitragool, W., Jayaram, H., Williams, C., and Miller, C. (2005). Separate ion pathways in a Cl^-/H^+ exchanger. *J. Gen. Physiol.* **126**, 563–570.

Ahn, W. et al. (2001). Regulatory interaction between the cystic fibrosis transmembrane conductance regulator and HCO_3^- salvage mechanisms in model systems and the mouse pancreatic duct. *J. Biol. Chem.* **276**, 17236–17243.

Albert, J.T. et al. (2007). Voltage-sensitive prestin orthologue expressed in zebrafish hair cells. *J. Physiol.* **580**, 451–461.

Alper, S.L. (2002). Genetic diseases of acid-base transporters. *Annu. Rev. Physiol.* **64**, 899–923.

Alper, S.L., Chernova, M.N., and Stewart, A.K. (2001). Regulation of $Na^{(+)}$-Independent $Cl^{(-)}/HCO_{(3)}^{(-)}$ Exchangers by pH.. *Jop* **2**, 171–175.

Alper, S.L., Darman, R.B., Chernova, M.N., and Dahl, N.K. (2002). The AE gene family of Cl/HCO_3^- exchangers. *J. Nephrol.* **15** (Suppl 5), S41–S53.

Alper, S.L., Kopito, R.R., Libresco, S.M., and Lodish, H.F. (1988). Cloning and characterization of a murine band 3-related cDNA from kidney and from a lymphoid cell line. *J. Biol. Chem.* **263**, 17092–17099.

Alper, S.L., Rossmann, H., Wilhelm, S., Stuart-Tilley, A.K., Shmukler, B.E., and Seidler, U. (1999). Expression of AE2 anion exchanger in mouse intestine. *Am. J. Physiol.* **277**, G321–G332.

Alper, S.L., Stuart-Tilley, A., Simmons, C.F., Brown, D., and Drenckhahn, D. (1994). The fodrin-ankyrin cytoskeleton of choroid plexus preferentially colocalizes with apical $Na^+ K^+$-ATPase rather than with basolateral anion exchanger AE2. *J. Clin. Invest.* **93**, 1430–1438.

Alper, S.L., Stuart-Tilley, A.K., Biemesderfer, D., Shmukler, B.E., and Brown, D. (1997). Immunolocalization of AE2 anion exchanger in rat kidney. *Am. J. Physiol.* **273**, F601–F614.

Alpern, R.J. (1985). Mechanism of basolateral membrane $H^+/OH^-/HCO_3^-$ transport in the rat proximal convoluted tubule. A sodium-coupled electrogenic process. *J. Gen. Physiol.* **86**, 613–636.

Alpern, R.J. and Rector, F.C. Jr., (1996). Renal Acidification Mechanisms. In *The kidney* (Brenner, B.M. and Rector, F.C., eds), Vol. 1. W.B. Saunders Co, Philadelphia, PA.

Alvarez, B.V., Kieller, D.M., Quon, A.L., Markovich, D., and Casey, J.R. (2004). Slc26a6: a cardiac chloride-hydroxyl exchanger and predominant chloride-bicarbonate exchanger of the mouse heart. *J. Physiol.* **561**, 721–734.

Alvarez, B.V., Vilas, G.L., and Casey, J.R. (2005). Metabolon disruption: a mechanism that regulates bicarbonate transport. *Embo J.* **24**, 2499–2511.

Aravind, L. and Koonin, E.V. (2000). The STAS domain - a link between anion transporters and antisigma-factor antagonists. *Curr. Biol.* **10**, R53–R55.

Ashmore, J. (2008). Cochlear outer hair cell motility. *Physiol. Rev.* **88**, 173–210.

Bartel, D., Hans, H., and Passow, H. (1989a). Identification by site-directed mutagenesis of Lys-558 as the covalent attachment site of H2DIDS in the mouse erythroid band 3 protein. *Biochim. Biophys. Acta.* **985**, 355–358.

Bartel, D., Lepke, S., Layh-Schmitt, G., Legrum, B., and Passow, H. (1989b). Anion transport in oocytes of Xenopus laevis induced by expression of mouse erythroid band 3 protein--encoding cRNA and of a cRNA derivative obtained by site-directed mutagenesis at the stilbene disulfonate binding site. *Embo J.* **8**, 3601–3609.

Barzilay, M., Ship, S., and Cabantchik, Z.I. (1979). Anion transport in red blood cells. I. Chemical properties of anion recognition sites as revealed by structure-activity relationships of aromatic sulfonic acids. *Membr. Biochem.* **2**, 227–254.

Baum, M. (1989). Axial heterogeneity of rabbit proximal tubule luminal H^+ and basolateral HCO_3^- transport. *Am. J. Physiol.* **256**, F335–F341.

Becker, H.M., Broer, S., and Deitmer, J.W. (2004). Facilitated Lactate Transport by MCT1 when Coexpressed with the Sodium Bicarbonate Cotransporter (NBC) in Xenopus Oocytes. *Biophys. J.* **86**, 235–247.

Becker, H.M. and Deitmer, J.W. (2004). Voltage dependence of h+ buffering mediated by sodium bicarbonate cotransport expressed in Xenopus oocytes. *J. Biol. Chem.* **279**, 28057–28062.

Becker, H.M. and Deitmer, J.W. (2007). Carbonic anhydrase II increases the activity of the human electrogenic Na^+/HCO_3^- cotransporter. *J. Biol. Chem.* **282**, 13508–13512.

Becker, H.M. and Deitmer, J.W. (2008). Nonenzymatic proton handling by carbonic anhydrase II during H^+-lactate cotransport via monocarboxylate transporter 1. *J. Biol. Chem.* **283**, 21655–21667.

Bevensee, M.O., Schmitt, B.M., Choi, I., Romero, M.F., and Boron, W.F. (2000). An electrogenic Na/HCO_3^- cotransporter (NBC) with a novel C terminus, cloned from rat brain. *Am. J. Physiol. Cell Physiol.* **278**, C1200–C1211.

Bidart, J.M. et al. (2000). Expression of Na^+/I^- symporter and Pendred syndrome genes in trophoblast cells. *J. Clin. Endocrinol Metab.* **85**, 4367–4372.

Bissig, M., Hagenbuch, B., Stieger, B., Koller, T., and Meier, P.J. (1994). Functional expression cloning of the canalicular sulfate transport system of rat hepatocytes. *J. Biol. Chem.* **269**, 3017–3021.

Bok, D. et al. (2001). Immunolocalization of electrogenic sodium-bicarbonate cotransporters pNBC1 and kNBC1 in the rat eye. *Am. J. Physiol. Renal Physiol.* **281**, F920–F935.

Boodhoo, A. and Reithmeier, R.A. (1984). Characterization of matrix-bound Band 3, the anion transport protein from human erythrocyte membranes. *J. Biol. Chem.* **259**, 785–790.

Boron, W.F. (1977). Intracellular pH transients in giant barnacle muscle fibers. *Am. J. Physiol.* **233**, C61–C73.

Boron, W.F. and Boulpaep, E.L. (1983). Intracellular pH regulation in the renal proximal tubule of the salamander. Basolateral HCO_3^- transport. *J. Gen. Physiol.* **81**, 53–94.

Boron, W.F. and De Weer, P. (1976). Active proton transport stimulated by CO_2/HCO_3^-, blocked by cyanide. *Nature* **259**, 240–241.

Boron, W.F. and Russell, J.M. (1983). Stoichiometry and ion dependencies of the intracellular-pH-regulating mechanism in squid giant axons. *J. Gen. Physiol.* **81**, 373–399.

Brosius, F.C., III. et al. (1997). AE anion exchanger mRNA and protein expression in vascular smooth muscle cells, aorta, and renal microvessels. *Am. J. Physiol.* **273**, F1039–F1047.

Brosius, F.C.d., Alper, S.L., Garcia, A.M., and Lodish, H.F. (1989). The major kidney band 3 gene transcript predicts an amino-terminal truncated band 3 polypeptide. *J. Biol. Chem.* **264**, 7784–7787.

Brownell, W.E., Bader, C.R., Bertrand, D., and de Ribaupierre, Y. (1985). Evoked mechanical responses of isolated cochlear outer hair cells. *Science* **227**, 194–196.

Bruce, L.J. et al. (1997). Familial distal renal tubular acidosis is associated with mutations in the red cell anion exchanger (Band 3, AE1) gene. *J. Clin. Invest.* **100**, 1693–1707.

Bruce, L.J. et al. (2000). Band 3 mutations, renal tubular acidosis and South-East Asian ovalocytosis in Malaysia and Papua New Guinea: loss of up to 95% band 3 transport in red cells. *Biochem. J.* **350** (Pt 1), 41–51.

Burnham, C.E., Amlal, H., Wang, Z., Shull, G.E., and Soleimani, M. (1997). Cloning and functional expression of a human kidney $Na^+:HCO_3^-$ cotransporter. *J. Biol. Chem.* **272**, 19111–19114.

Byeon, M.K. et al. (1996). The down-regulated in adenoma (DRA) gene encodes an intestine-specific membrane glycoprotein. *Oncogene* **12**, 387–396.

Cabantchik, Z.I. and Greger, R. (1992). Chemical probes for anion transporters of mammalian cell membranes. *Am. J. Physiol.* **262**, C803–C827.

Cabantchik, Z.I. and Rothstein, A. (1972). The nature of the membrane sites controlling anion permeability of human red blood cells as determined by studies with disulfonic stilbene derivatives. *J. Membr. Biol.* **10**, 311–330.

Cabantchik, Z.I. and Rothstein, A. (1974). Membrane proteins related to anion permeability of human red blood cells. I. Localization of disulfonic stilbene binding sites in proteins involved in permeation. *J. Membr. Biol.* **15**, 207–226.

Carvelli, L., McDonald, P.W., Blakely, R.D., and Defelice, L.J. (2004). Dopamine transporters depolarize neurons by a channel mechanism. *Proc. Natl. Acad. Sci. U. S. A.* **101**, 16046–16051.

Cavet, M.E., West, M., and Simmons, N.L. (1997). Fluoroquinolone (ciprofloxacin) secretion by human intestinal epithelial (Caco-2) cells. *Br. J. Pharmacol.* **121**, 1567–1578.

Chang, M.-H., DiPiero, J.M., Sönnichsen, F.D., and Romero, M.F. (2008). Entry to "HCO_3 tunnel" revealed by human mutation and structural model. *J. Biol. Chem.* **283**, 18402–18410.

Chang, M.-H. et al. (2009). Slc26A9 - anion exchanger, channel and Na^+ transporter. *J. Membr. Biol.* **128**, 125–140.

Chang, S.H. and Low, P.S. (2003). Identification of a critical ankyrin-binding loop on the cytoplasmic domain of erythrocyte membrane band 3 by crystal structure analysis and site-directed mutagenesis. *J. Biol. Chem.* **278**, 6879–6884.

Chapman, J.M., Knoepp, S.M., Byeon, M.K., Henderson, K.W., and Schweinfest, C.W. (2002). The Colon Anion Transporter, Down-Regulated in Adenoma, Induces Growth Suppression That Is Abrogated by E1A. *Cancer Res.* **62**, 5083–5088.

Chen, J.C. and Chesler, M. (1990). A bicarbonate-dependent increase in extracellular pH mediated by GABAA receptors in turtle cerebellum. *Neurosci Lett.* **116**, 130–135.

Chen, J.C. and Chesler, M. (1992). Modulation of extracellular pH by glutamate and GABA in rat hippocampal slices. *J. Neurophysiol.* **67**, 29–36.

Chen, N.H., Reith, M.E., and Quick, M.W. (2004). Synaptic uptake and beyond: the sodium- and chloride-dependent neurotransmitter transporter family SLC6. *Pflugers Arch.* **447**, 519–531.

Chen, W.Y. et al. (2008). Cl^- Is Required for HCO_3^- Entry Necessary for Sperm Capacitation in Guinea Pig: Involvement of a Cl^-/HCO_3^- Exchanger (SLC26A3) and CFTR. *Biol. Reprod.*

Chernova, M.N. et al. (2005). Functional comparison of mouse slc26a6 anion exchanger with human SLC26A6 polypeptide variants: Differences in anion selectivity, regulation, and electrogenicity. *J. Biol. Chem.* **280**, 8564–8580.

Chernova, M.N. et al. (2003a). Acute regulation of the SLC26A3 congenital chloride diarrhoea anion exchanger (DRA) expressed in Xenopus oocytes. *J. Physiol.* **549**, 3–19.

Chernova, M.N., Stewart, A.K., Jiang, L., Friedman, D.J., Kunes, Y. Z., and Alper, S.L. (2003b). Structure-Function Relationships of AE2 Regulation by the Ca^{2+}-sensitive Stimulators, NH_4^+ and Hypertonicity. *Am. J. Physiol. Cell Physiol.* **284**, C1235–C1246.

Choate, K.A., Kahle, K.T., Wilson, F.H., Nelson-Williams, C., and Lifton, R.P. (2003). WNK1, a kinase mutated in inherited hypertension with hyperkalemia, localizes to diverse Cl^--transporting epithelia. *Proc. Natl. Acad. Sci. U. S. A.* **100**, 663–668.

Choi, I., Aalkjaer, C., Boulpaep, E.L., and Boron, W.F. (2000). An electroneutral sodium/bicarbonate cotransporter NBCn1 and associated sodium channel. *Nature* **405**, 571–575.

Choi, I., Rojas, J.D., Kobayashi, C., and Boron, W.F. (2002). Functional characterization of "NCBE", a Na/HCO_3 cotransporter. *FASEB J.* **16**, A796.

Choi, I., Romero, M.F., Khandoudi, N., Bril, A., and Boron, W.F. (1999). Cloning and characterization of an electrogenic Na/HCO_3 cotransporter from human heart. *Am. J. Physiol.* **274**, C576–C584.

Choi, J.Y., Muallem, D., Kiselyov, K., Lee, M.G., Thomas, P.J., and Muallem, S. (2001). Aberrant CFTR-dependent HCO_3^- transport in mutations associated with cystic fibrosis. *Nature* **410**, 94–97.

Cooper, D.S. et al. (2005). Molecular and functional characterization of the electroneutral Na/HCO_3 cotransporter NBCn1 in rat hippocampal neurons. *J. Biol. Chem.* **280**, 17823–17830.

Cooper, G.J. and Boron, W.F. (1998). Effect of PCMBS on CO_2 permeability of xenopus oocytes expressing aquaporin 1 or its C189S mutant. *Am. J. Physiol.* **275**, C1481–C1486.

Corbetta, S. et al. (2009). Analysis of the 206M polymorphic variant of the SLC26A6 gene encoding a Cl^- oxalate transporter in patients with primary hyperparathyroidism. *Eur. J. Endocrinol.* **160**, 283–288.

Coyle, B. et al. (1996). Pendred syndrome (goitre and sensorineural hearing loss) maps to chromosome 7 in the region containing the nonsyndromic deafness gene DFNB4. *Nat. Genet.* **12**, 421–423.

Coyle, B. et al. (1998). Molecular analysis of the PDS gene in Pendred syndrome. *Hum. Mol. Genet.* **7**, 1105–1112.

Cuppens, H. et al. (1998). Polyvariant mutant cystic fibrosis transmembrane conductance regulator genes. The polymorphic (Tg)m locus explains the partial penetrance of the T5 polymorphism as a disease mutation. *J. Clin. Invest.* **101**, 487–496.

Dallos, P. et al. (2008). Prestin-based outer hair cell motility is necessary for mammalian cochlear amplification. *Neuron* **58**, 333–339.

Deitmer, J.W. (2000). Glial strategy for metabolic shuttling and neuronal function. *Bioessays* **22**, 747–752.

Demirci, F.Y., Chang, M.-H., Mah, T.S., Romero, M.F., and Gorin, M.B. (2006). Proximal renal tubular acidosis and ocular pathology: L522P, a novel missense mutation in the gene (SLC4A4) for Sodium Bicarbonate Cotransporter protein (NBCe1). *Mol. Vision* **12**, 324–330.

Demuth, D.R. et al. (1986). Cloning and structural characterization of a human non-erythroid band 3-like protein. *Embo J.* **5**, 1205–1214.

Detro-Dassen, S. et al. (2008). Conserved Dimeric Subunit Stoichiometry of SLC26 Multifunctional Anion Exchangers. *J. Biol. Chem.* **283**, 4177–4188.

Dinour, D. et al. (2004). A novel missense mutation in the sodium bicarbonate cotransporter (NBCe1/SLC4A4) causes proximal tubular acidosis and glaucoma through ion transport defects. *J. Biol. Chem.* **279**, 52238–52246.

Dodge, J.T., Mitchell, C., and Hanahan, D.J. (1963). The preparation and chemical characteristics of hemoglobin-free ghosts of human erythrocytes. *Arch. Biochem. Biophys.* **100**, 119–130.

Dorwart, M.R., Shcheynikov, N., Baker, J.M., Forman-Kay, J.D., Muallem, S., and Thomas, P.J. (2008). Congenital chloride-losing diarrhea causing mutations in the stas domain result in misfolding and mistrafficking of SLC26A3. *J. Biol. Chem.* **283**, 8711–8722.

Dorwart, M.R., Shcheynikov, N., Wang, Y., Stippec, S., and Muallem, S. (2007). SLC26A9 is a Cl channel regulated by the WNK kinases. *J. Physiol.* **584**, 333–345.

Ducoudret, O., Diakov, A., Mueller-Berger, S., Romero, M.F., Boron, W.F., and Frömter, E. (2001). The renal $Na-HCO_3^-$ cotransporter expressed in Xenopus laevis oocytes: Inhibition by tenidap and benzamil and effect of temperature on transport rate and stoichiometry. *Pflügers Arch.* **442**, 709–717.

Dudas, P.L., Mentone, S., Greineder, C.F., Biemesderfer, D., and Aronson, P.S. (2006). Immunolocalization of anion transporter Slc26a7 in mouse kidney. *Am. J. Physiol. Renal Physiol.* **290**, F937–F945.

Dutzler, R., Campbell, E.B., Cadene, M., Chait, B.T., and MacKinnon, R. (2002). X-ray structure of a ClC chloride channel at 3.0 A reveals the molecular basis of anion selectivity. *Nature* **415**, 287–294.

Emmons, C. (1999). Transport characteristics of the apical anion exchanger of rabbit cortical collecting duct beta-cells. *Am. J. Physiol.* **276**, F635–F643.

Everett, L.A. et al. (2001). Targeted disruption of mouse Pds provides insight about the inner-ear defects encountered in Pendred syndrome. *Hum. Mol. Genet.* **10**, 153–161.

Everett, L.A. et al. (1997). Pendred syndrome is caused by mutations in a putative sulphate transporter gene (PDS). *Nat. Genet.* **17**, 411–422.

Everett, L.A. and Green, E.D. (1999). A family of mammalian anion transporters and their involvement in human genetic diseases. *Hum. Mol. Genet.* **8**, 1883–1891.

Everett, L.A., Morsli, H., Wu, D.K., and Green, E.D. (1999). Expression pattern of the mouse ortholog of the Pendred's syndrome gene (Pds) suggests a key role for pendrin in the inner ear. *Proc. Natl. Acad. Sci. U. S. A.* **96**, 9727–9732.

Fairbanks, G., Steck, T.L., and Wallach, D.F. (1971). Electrophoretic analysis of the major polypeptides of the human erythrocyte membrane. *Biochemistry* **10**, 2606–2617.

Fang, X., Yang, B., Matthay, M.A., and Verkman, A.S. (2002). Evidence against aquaporin-1-dependent CO_2 permeability in lung and kidney. *J. Physiol.* **542**, 63–69.

Freel, R.W., Hatch, M., Green, M., and Soleimani, M. (2006). Ileal oxalate absorption and urinary oxalate excretion are enhanced in Slc26a6 null mice. *Am. J. Physiol. Gastrointest Liver Physiol.* **290**, G719–G728.

Gawenis, L.R. et al. (2007). Colonic anion secretory defects and metabolic acidosis in mice lacking the NBC1 Na^+/HCO_3^- cotransporter. *J. Biol. Chem.* **282**, 9042–9052.

Geibel, J., Giebisch, G., and Boron, W.F. (1989). Basolateral sodium-coupled acid-base transport mechanisms of the rabbit proximal tubule. *Am. J. Physiol.* **257**, F790–F797.

Gill, R.K. et al. (2007). Mechanism underlying inhibition of intestinal apical Cl/OH exchange following infection with enteropathogenic E. coli. *J. Clin. Invest.* **117**, 428–437.

Greeley, T., Shumaker, H., Wang, Z., Schweinfest, C.W., and Soleimani, M. (2001). Downregulated in adenoma and putative anion transporter are regulated by CFTR in cultured pancreatic duct cells. *Am. J. Physiol. Gastrointest Liver Physiol.* **281**, G1301–G1308.

Grichtchenko, I.I. and Chesler, M. (1994a). Depolarization-induced acid secretion in gliotic hippocampal slices. *Neuroscience* **62**, 1057–1070.

Grichtchenko, I.I. and Chesler, M. (1994b). Depolarization-induced alkalinization of astrocytes in gliotic hippocampal slices. *Neuroscience* **62**, 1071–1078.

Grichtchenko, I., II., Choi, I., Zhong, X., Bray-Ward, P., Russell, J. M., and Boron, W.F. (2001). Cloning, characterization and chromosomal mapping of a human electroneutral Na^+-driven Cl^- HCO_3 exchanger. *J. Biol. Chem.* **276**, 8358–8363.

Grichtchenko, I., Romero, M.F., and Boron, W.F. (2000). Extracellular HCO_3 dependence of electrogenic Na/HCO_3 cotransporters (NBC) cloned from salamander and rat kidney. *J. Gen. Physiol.* **115**, 533–546.

Groves, J.D., Falson, P., le Maire, M., and Tanner, M.J. (1996). Functional cell surface expression of the anion transport domain of human red cell band 3 (AE1) in the yeast Saccharomyces cerevisiae. *Proc. Natl. Acad. Sci. U. S. A.* **93**, 12245–12250.

Groves, J.D. and Tanner, M.J. (1992). Glycophorin A facilitates the expression of human band 3-mediated anion transport in Xenopus oocytes. *J. Biol. Chem.* **267**, 22163–22170.

Groves, J.D. and Tanner, M.J. (1995). Co-expressed complementary fragments of the human red cell anion exchanger (band 3, AE1) generate stilbene disulfonate-sensitive anion transport. *J. Biol. Chem.* **270**, 9097–9105.

Haila, S., Hastbacka, J., Bohling, T., Karjalainen-Lindsberg, M.L., Kere, J., and Saarialho-Kere, U. (2001). Slc26a2 (diastrophic dysplasia sulfate transporter) is expressed in developing and mature cartilage but also in other tissues and cell types. *J. Histochem. Cytochem.* **49**, 973–982.

Haila, S. et al. (2000). The congenital chloride diarrhea gene is expressed in seminal vesicle, sweat gland, inflammatory colon epithelium, and in some dysplastic colon cells. *Histochem. Cell Biol.* **113**, 279–286.

Hassan, H.A., Mentone, S., Karniski, L.P., Rajendran, V.M., and Aronson, P.S. (2007). Regulation of anion exchanger Slc26a6 by protein kinase C. *Am. J. Physiol. Cell Physiol.* **292**, C1485–C1492.

Hastbacka, J., de la Chapelle, A., Kaitila, I., Sistonen, P., Weaver, A., and Lander, E. (1992). Linkage disequilibrium mapping in isolated founder populations: diastrophic dysplasia in Finland. *Nat. Genet.* **2**, 204–211.

Hastbacka, J. et al. (1994). The diastrophic dysplasia gene encodes a novel sulfate transporter: positional cloning by fine-structure linkage disequilibrium mapping. *Cell* **78**, 1073–1087.

Hemminki, A. et al. (1998). Intestinal cancer in patients with a germline mutation in the down-regulated in adenoma (DRA) gene. *Oncogene* **16**, 681–684.

Hentschke, M. et al. (2006). Mice with a targeted disruption of the Cl^-/HCO_3^- exchanger AE3 display a reduced seizure threshold. *Mol. Cell. Biol.* **26**, 182–191.

Hoglund, P. et al. (1996). Mutations of the Down-regulated in adenoma (DRA) gene cause congenital chloride diarrhoea. *Nat. Genet.* **14**, 316–319.

Hoglund, P., Hihnala, S., Kujala, M., Tiitinen, A., Dunkel, L., and Holmberg, C. (2006). Disruption of the SLC26A3-mediated anion transport is associated with male subfertility. *Fertil. Steril.* **85**, 232–235.

Hoglund, P., Holmberg, C., Sherman, P., and Kere, J. (2001). Distinct outcomes of chloride diarrhoea in two siblings with identical genetic background of the disease: implications for early diagnosis and treatment. *Gut* **48**, 724–727.

Holmes, R.P., Goodman, H.O., and Assimos, D.G. (2001). Contribution of dietary oxalate to urinary oxalate excretion. *Kidney Int.* **59**, 270–276.

Howitt, S.M. (2005). The role of cysteine residues in the sulphate transporter, SHST1: construction of a functional cysteine-less transporter. *Biochim. Biophys. Acta.* **1669**, 95–100.

Huang, W., Smith, S.E., and Chesler, M. (1995). Addition of carbonic anhydrase augments extracellular pH buffering in rat cerebral cortex. *J. Neurophysiol.* **74**, 1806–1809.

Humphreys, B.D., Chernova, M.N., Jiang, L., Zhang, Y., and Alper, S.L. (1997). NH_4Cl activates AE2 anion exchanger in Xenopus oocytes at acidic pHi. *Am. J. Physiol.* **272**, C1232–C1240.

Humphreys, B.D., Jiang, L., Chernova, M.N., and Alper, S.L. (1995). Hypertonic activation of AE2 anion exchanger in Xenopus oocytes via NHE- mediated intracellular alkalinization. *Am. J. Physiol.* **268**, C201–C209.

Igarashi, T. et al. (1999). Mutations in SLC4A4 cause permanent isolated proximal renal tubular acidosis with ocular abnormalities. *Nat. Genet.* **23**, 264–266.

Igarashi, T. et al. (2001). Novel nonsense mutation in the Na^+/HCO_3^- cotransporter gene (SLC4A4) in a patient with permanent isolated proximal renal tubular acidosis and bilateral glaucoma. *J. Am. Soc. Nephrol.* **12**, 713–718.

Igarashi, T., Inatomi, J., Sekine, T., Takeshima, Y., Yoshikawa, N., and Endou, H. (2000). A Nonsense Mutation in the Na^+/HCO_3^- Cotransporter Gene (SLC4A4) in a Patient with Permanent Isolated Proximal Renal Tubular Acidosis and Bilateral Glaucoma. *J. Am. Soc. Nephrol.* **11**, A0573.

Igarashi, T., Sekine, T., Inatomi, J., and Seki, G. (2002). Unraveling the molecular pathogenesis of isolated proximal renal tubular acidosis. *J Am Soc Nephrol.* **13**, 2171–2177.

Inatomi, J. et al. (2004). Mutational and functional analysis of SLC4A4 in a patient with proximal renal tubular acidosis. *Pflugers Arch.* **448**, 438–444.

Ishiguro, H. et al. (2007). Effect of Slc26a6 deletion on apical Cl^-/HCO_3^- exchanger activity and cAMP-stimulated bicarbonate secretion in pancreatic duct. *Am J Physiol Gastrointest Liver Physiol.* **292**, G447–G455.

Jacob, P. et al. (2002). Down-regulated in adenoma mediates apical Cl^-/HCO_3^- exchange in rabbit, rat, and human duodenum. *Gastroenterology* **122**, 709–724.

Jarolim, P. et al. (1998). Autosomal dominant distal renal tubular acidosis is associated in three families with heterozygosity for the R589H mutation in the AE1 (band 3) Cl^-/HCO_3^- exchanger. *J. Biol. Chem.* **273**, 6380–6388.

Jennings, M.L. (1976). Proton fluxes associated with erythrocyte membrane anion exchange. *J. Membr. Biol.* **28**, 187–205.

Jennings, M.L. (1978). Characteristics of CO_2-independent pH equilibration in human red blood cells. *J. Membr. Biol.* **40**, 365–391.

Jensen, L.J. et al. (1999). Localization of sodium bicarbonate cotransporter (NBC) protein and messenger ribonucleic acid in rat epididymis. *Biol. Reprod.* **60**, 573–579.

Jentsch, T.J. (2008). CLC chloride channels and transporters: from genes to protein structure, pathology and physiology. *Crit. Rev. Biochem. Mol. Biol.* **43**, 3–36.

Jentsch, T.J., Keller, S.K., Koch, M., and Wiederholt, M. (1984). Evidence for coupled transport of bicarbonate and sodium in cultured bovine corneal endothelial cells. *J. Membr. Biol.* **81**, 189–204.

Jentsch, T.J., Neagoe, I., and Scheel, O. (2005). CLC chloride channels and transporters. *Curr. Opin. Neurobiol.* **15**, 319–325.

Jentsch, T.J., Stein, V., Weinreich, F., and Zdebik, A.A. (2002). Molecular structure and physiological function of chloride channels. *Physiol. Rev.* **82**, 503–568.

Jiang, Z. et al. (2006). Calcium oxalate urolithiasis in mice lacking anion transporter Slc26a6. *Nat. Genet.* **38**, 474–478.

Jiang, Z., Grichtchenko, II., Boron, W.F., and Aronson, P.S. (2002). Specificity of anion exchange mediated by mouse Slc26a6. *J. Biol. Chem.* **277**, 33963–33967.

Kahle, K.T., Ring, A.M., and Lifton, R.P. (2008). Molecular physiology of the WNK kinases. *Annu. Rev. Physiol.* **70**, 329–355.

Kahle, K.T. et al. (2003). WNK4 regulates the balance between renal NaCl reabsorption and K^+ secretion. *Nat. Genet.* **35**, 372–376.

Kaila, K., Paalasmaa, P., Taira, T., and Voipio, J. (1992). pH transients due to monosynaptic activation of GABAA receptors in rat hippocampal slices. *Neuroreport* **3**, 105–108.

Kaila, K., Pasternack, M., Saarikoski, J., and Voipio, J. (1989a). Influence of GABA-gated bicarbonate conductance on potential, current and intracellular chloride in crayfish muscle fibres. *J. Physiol. (Lond)* **416**, 161–181.

Kaila, K., Pasternack, M., Saarikoski, J., and Voipio, J. (1989b). Influence of HCO_3^- on the postsynaptic actions of GABA at the crayfish neuromuscular synapse. *Acta. Physiol. Scand. Suppl.* **582**, 18.

Kaila, K., Voipio, J., Paalasmaa, P., Pasternack, M., and Deisz, R.A. (1993). The role of bicarbonate in GABAA receptor-mediated IPSPs of rat neocortical neurones. *J. Physiol. (Lond)* **464**, 273–289.

Karet, F.E. et al. (1998). Mutations in the chloride-bicarbonate exchanger gene AE1 cause autosomal dominant but not autosomal recessive distal renal tubular acidosis. *Proc. Natl. Acad. Sci. U. S. A.* **95**, 6337–6342.

Karniski, L.P. (2001). Mutations in the diastrophic dysplasia sulfate transporter (DTDST) gene: correlation between sulfate transport activity and chondrodysplasia phenotype. *Hum. Mol. Genet.* **10**, 1485–1490.

Karniski, L.P. and Aronson, P.S. (1987). Anion exchange pathways for Cl^- transport in rabbit renal microvillus membranes. *Am. J. Physiol.* **253**, F513–F521.

Karniski, L.P., Lotscher, M., Fucentese, M., Hilfiker, H., Biber, J., and Murer, H. (1998). Immunolocalization of sat-1 sulfate/oxalate/bicarbonate anion exchanger in the rat kidney. *Am. J. Physiol.* **275**, F79–F787.

Karniski, L.P., Wang, T., Everett, L.A., Green, E.D., Giebisch, G., and Aronson, P.S. (2002). Formate-stimulated NaCl absorption in the proximal tubule is independent of the pendrin protein. *Am. J. Physiol.* **283**, F952–F956.

Khurana, O.K., Coupland, L.A., Shelden, M.C., and Howitt, S.M. (2000). Homologous mutations in two diverse sulphate transporters have similar effects. *FEBS Lett.* **477**, 118–122.

Kim, K.H., Shcheynikov, N., Wang, Y., and Muallem, S. (2005). SLC26A7 Is a Cl^- Channel Regulated by Intracellular pH. *J. Biol. Chem.* **280**, 6463–6470.

Kim, Y.H. et al. (2002). Immunocytochemical localization of pendrin in intercalated cell subtypes in rat and mouse kidney. *Am. J. Physiol. Renal Physiol.* **283**, F744–F754.

Knauf, F., Yang, C.L., Thomson, R.B., Mentone, S.A., Giebisch, G., and Aronson, P.S. (2001). Identification of a chloride-formate exchanger expressed on the brush border membrane of renal proximal tubule cells. *Proc. Natl. Acad. Sci. U. S. A.* **98**, 9425–9430.

Knauf, P.A. and Rothstein, A. (1971). Chemical modification of membranes. I. Effects of sulfhydryl and amino reactive reagents on anion and cation permeability of the human red blood cell. *J. Gen. Physiol.* **58**, 190–210.

Knickelbein, R.G., Aronson, P.S., and Dobbins, J.W. (1986). Oxalate transport by anion exchange across rabbit ileal brush border. *J. Clin. Invest.* **77**, 170–175.

Ko, S.B. et al. (2002a). AE4 is a DIDS-sensitive Cl^-/HCO_3^- exchanger in the basolateral membrane of the renal CCD and the SMG duct. *Am. J. Physiol. Cell Physiol.* **283**, C1206–C1218.

Ko, S.B. et al. (2002b). A molecular mechanism for aberrant CFTR-dependent HCO_3^- transport in cystic fibrosis. *Embo J.* **21**, 5662–5672.

Ko, S.B. et al. (2004). Gating of CFTR by the STAS domain of SLC26 transporters. *Nat. Cell. Biol.* **6**, 343–350.

Ko, S.B.H., Choi, J.Y., Lee, M.G., Thomas, P.J., and Muallem, S. (2002c). A Molecular Mechanism for Aberrant CFTR-Dependent HCO_3^- transport. *FASEB J.* **16**, a819.

Kobayashi, S., Morgans, C.W., Casey, J.R., and Kopito, R.R. (1994). AE3 anion exchanger isoforms in the vertebrate retina: developmental regulation and differential expression in neurons and glia. *J. Neurosci.* **14**, 6266–6279.

Kopito, R.R., Lee, B.S., Simmons, D.M., Lindsey, A.E., Morgans, C. W., and Schneider, K. (1989). Regulation of intracellular pH by a neuronal homolog of the erythrocyte anion exchanger. *Cell* **59**, 927–937.

Kopito, R.R. and Lodish, H.F. (1985). Primary structure and transmembrane orientation of the murine anion exchange protein. *Nature* **316**, 234–238.

Kosiek, O. et al. (2007). SLC26A7 Can function as a chloride-loading mechanism in parietal cells. *Pflugers Arch.*.

Ku, C.P., Jennings, M.L., and Passow, H. (1979). A comparison of the inhibitory potency of reversibly acting inhibitors of anion transport on chloride and sulfate movements across the human red cell membrane. *Biochim. Biophys. Acta.* **553**, 132–141.

Kudrycki, K.E., Newman, P.R., and Shull, G.E. (1990). cDNA cloning and tissue distribution of mRNAs for two proteins that are related to the band 3 Cl^-/HCO_3^- exchanger. *J. Biol. Chem.* **265**, 462–471.

Kuperman, D.A. et al. (2005). Dissecting asthma using focused transgenic modeling and functional genomics. *J. Allergy Clin. Immunol.* **116**, 305–311.

Kurita, Y. et al. (2008). Identification of intestinal bicarbonate transporters involved in formation of carbonate precipitates to stimulate water absorption in marine teleost fish. *Am. J. Physiol. – Comp. & Reg. Physiol.* **284**, R1402–R1412.

Kurtz, I., Nagami, G., Yanagawa, N., Li, L., Emmons, C., and Lee, I. (1994). Mechanism of apical and basolateral Na(+)-independent Cl-/base exchange in the rabbit superficial proximal straight tubule. *J. Clin. Invest.* **94**, 173–183.

Laemmli, U.K. (1970). Cleavage of structural proteins during the assembly of the head of bacteriophage T4. *Nature* **227**, 680–685.

Landolt-Marticorena, C., Casey, J.R., and Reithmeier, R.A. (1995). Transmembrane helix-helix interactions and accessibility of H2DIDS on labelled band 3, the erythrocyte anion exchange protein. *Mol. Membr. Biol.* **12**, 173–182.

Lee, M.G. et al. (1999). Regulation of Cl^-/HCO_3^- exchange by cystic fibrosis transmembrane conductance regulator expressed in NIH 3T3 and HEK 293 cells. *J. Biol. Chem.* **274**, 3414–3421.

Lepke, S. and Passow, H. (1972). The effect of pH at hemolysis on the reconstitution of low cation permeability in human erythrocyte ghosts. *Biochim. Biophys. Acta.* **255**, 696–702.

Leves, F.P., Tierney, M.L., and Howitt, S.M. (2008). Polar residues in a conserved motif spanning helices 1 and 2 are functionally important in the SulP transporter family. *Int. J. Biochem. Cell Biol.* **40**, 2596–2605.

Li, J., Sun, X.C., and Bonanno, J.A. (2005). Role of NBC1 in apical and basolateral HCO_3^- permeabilities and transendothelial HCO_3^- fluxes in bovine corneal endothelium. *Am. J. Physiol. Cell Physiol.* **288**, C739–C746.

Li, X.C. et al. (1998). A mutation in PDS causes non-syndromic recessive deafness. *Nat. Genet.* **18**, 215–217.

Liberman, M.C., Gao, J., He, D.Z., Wu, X., Jia, S., and Zuo, J. (2002). Prestin is required for electromotility of the outer hair cell and for the cochlear amplifier. *Nature* **419**, 300–304.

Linn, S.C., Kudrycki, K.E., and Shull, G.E. (1992). The predicted translation product of a cardiac AE3 mRNA contains an N terminus distinct from that of the brain AE3 Cl^-/HCO_3^- exchanger. Cloning of a cardiac AE3 cDNA, organization of the AE3 gene, and identification of an alternative transcription initiation site. *J. Biol. Chem.* **267**, 7927–7935.

Lipovich, L., Lynch, E.D., Lee, M.K., and King, M.C. (2001). A novel sodium bicarbonate cotransporter-like gene in an ancient duplicated region: SLC4A9 at 5q31. *Genome Biol.* **2**.

Liu, X.Z. et al. (2003). Prestin, a cochlear motor protein, is defective in non-syndromic hearing loss. *Hum. Mol. Genet.* **12**, 1155–1162.

Lohi, H., Kujala, M., Kerkela, E., Saarialho-Kere, U., Kestila, M., and Kere, J. (2000). Mapping of five new putative anion transporter genes in human and characterization of SLC26A6, a candidate gene for pancreatic anion exchanger. *Genomics* **70**, 102–112.

Lohi, H. et al. (2002a). Functional characterization of three novel tissue-specific anion exchangers SLC26A7, -A8, and -A9. *J. Biol. Chem.* **277**, 14246–14254.

Lohi, H. et al. (2002b). Isoforms of the anion exchanger SLC26A6 (PAT1) mediate chloride and sulfate transport and have functional PDZ interaction domains. *Am. J. Physiol. Cell Physiol.*.

Lohi, H. et al. (2003). Isoforms of SLC26A6 mediate anion transport and have functional PDZ interaction domains. *Am. J. Physiol. Cell Physiol.* **284**, C769–C779.

Loriol, C. et al. (2008). Characterization of SLC26A9, facilitation of $Cl^{(-)}$ transport by bicarbonate. *Cell Physiol. Biochem.* **22**, 15–30.

Low, P.S. (1978). Specific cation modulation of anion transport across the human erythrocyte membrane. *Biochim. Biophys. Acta.* **514**, 264–273.

Lu, J. and Boron, W.F. (2007). Reversible and irreversible interactions of DIDS with the human electrogenic Na/HCO_3 cotransporter NBCe1-A: role of lysines in the KKMIK motif of TM5. *Am. J. Physiol. Cell Physiol.* **292**, C1787–C1798.

Makalowski, W. and Boguski, M.S. (1998). Evolutionary parameters of the transcribed mammalian genome: an analysis of 2,820 orthologous rodent and human sequences. *Proc. Natl. Acad. Sci. U. S. A.* **95**, 9407–9412.

Makela, S., Kere, J., Holmberg, C., and Hoglund, P. (2002). SLC26A3 mutations in congenital chloride diarrhea. *Hum. Mutat.* **20**, 425–438.

Masmoudi, S. et al. (2000). Pendred syndrome: phenotypic variability in two families carrying the same PDS missense mutation [In Process Citation]. *Am. J. Med. Genet.* **90**, 38–44.

Mason, M.J., Mattsson, K., Pasternack, M., Voipio, J., and Kaila, K. (1990). Postsynaptic fall in intracellular pH and increase in surface pH caused by efflux of formate and acetate anions through GABA-gated channels in crayfish muscle fibres. *Neuroscience* **34**, 359–368.

McAlear, S.D., Liu, X., Williams, J.B., McNicholas-Bevensee, C.M., and Bevensee, M.O. (2006). Electrogenic Na/HCO_3 cotransporter (NBCe1) variants expressed in xenopus oocytes: functional comparison and roles of the amino and carboxy termini. *J. Gen. Physiol.* **127**, 639–658.

Meier, S., Hubner, C.A., Groeben, H., Peters, J., Bingmann, D., and Wiemann, M. (2007). Expression of anion exchanger 3 influences respiratory rate in awake and isoflurane anesthetized mice. *J. Physiol. Pharmacol.* **58** (Suppl. 5), 371–378.

Mellado Lagarde, M.M., Drexl, M., Lukashkin, A.N., Zuo, J., and Russell, I.J. (2008). Prestin's role in cochlear frequency tuning and transmission of mechanical responses to neural excitation. *Curr. Biol.* **18**, 200–202.

Melvin, J.E., Park, K., Richardson, L., Schultheis, P.J., and Shull, G. E. (1999). Mouse down-regulated in adenoma (DRA) is an intestinal Cl^-/HCO_3^- exchanger and is up-regulated in colon of mice lacking the NHE3 Na^+/H^+ exchanger. *J. Biol. Chem.* **274**, 22855–22861.

Moseley, R.H. et al. (1999). Downregulated in adenoma gene encodes a chloride transporter defective in congenital chloride diarrhea. *Am. J. Physiol.* **276**, G185–G192.

Mount, D.B. and Romero, M.F. (2004). The SLC26 gene family of multifunctional anion exchangers. *Pflügers Arch.* **447**, 710–721.

Muallem, D. and Ashmore, J. (2006). An anion antiporter model of prestin, the outer hair cell motor protein. *Biophys. J.* **90**, 4035–4045.

Müller-Berger, S., Heyer, M., Romero, M.F., Boron, W.F., and Frömter, E. (1998). Stoichiometry of rat Na^+-HCO_3^- cotransporter (rkNBC) overexpressed in *Xenopus laevis* oocytes measured using giant patches. *J. Am. Soc. Nephrol.* **9**, 9A.

Nakagami, Y. et al. (2008). The epithelial anion transporter pendrin is induced by allergy and rhinovirus infection, regulates airway surface liquid, and increases airway reactivity and inflammation in an asthma model. *J. Immunol.* **181**, 2203–2210.

Nakao, I. et al. (2008). Identification of pendrin as a common mediator for mucus production in bronchial asthma and chronic obstructive pulmonary disease. *J. Immunol.* **180**, 6262–6269.

Nakaya, K. et al. (2007). Lack of pendrin HCO_3^- transport elevates vestibular endolymphatic $[Ca^{2+}]$ by inhibition of acid-sensitive TRPV5 and TRPV6 channels. *Am. J. Physiol. Renal Physiol.* **292**, F1314–F1321.

Nakhoul, N.L., Davis, B.A., Romero, M.F., and Boron, W.F. (1998). Effect of expressing the water channel aquaporin-1 on the CO_2 permeability of Xenopus oocytes [see comments]. *Am. J. Physiol.* **274**, C543–C548.

Nakhoul, N.L. et al. (1996). Processing and functional expression of carbonic anhydrase isoforms in *Xenopus laevis* oocytes. *FASEB J.* **10**, A88.

Ogihara, T., Tamai, I., and Tsuji, A. (1999). Structural characterization of substrates for the anion exchange transporter in caco-2 cells. *J. Pharm. Sci.* **88**, 1217–1221.

Okubo, K., Kang, D., Hamasaki, N., and Jennings, M.L. (1994). Red blood cell band 3. Lysine 539 and lysine 851 react with the same H2DIDS (4,4′-diisothiocyanodihydrostilbene-2,2′-disulfonic acid) molecule. *J. Biol. Chem.* **269**, 1918–1926.

Oliver, D. et al. (2001). Intracellular anions as the voltage sensor of prestin, the outer hair cell motor protein. *Science* **292**, 2340–2343.

Papageorgiou, P., Shmukler, B.E., Stuart-Tilley, A.K., Jiang, L., and Alper, S.L. (2001). AE anion exchangers in atrial tumor cells. *Am. J. Physiol. Heart. Circ. Physiol.* **280**, H937–H945.

Park, M. et al. (2002). The cystic fibrosis transmembrane conductance regulator interacts with and regulates the activity of the HCO_3^- salvage transporter human Na^+ -HCO_3^- cotransport isoform 3. *J. Biol. Chem.* **277**, 50503–50509.

Park, M., Li, Q., Shcheynikov, N., Zeng, W., and Muallem, S. (2004). NaBC1 Is a Ubiquitous Electrogenic $Na^{(+)}$-Coupled Borate Transporter Essential for Cellular Boron Homeostasis and Cell Growth and Proliferation. *Mol. Cell.* **16**, 331–341.

Parker, M.D., Boron, W.F., and Tanner, M.J.A. (2002). Characterization of human 'AE4' as an electroneutral, sodium-dependent bicarbonate transporter. *FASEB J.* **16**, A796.

Parker, M.D., Musa-Aziz, R., Rojas, J.D., Choi, I., Daly, C.M., and Boron, W.F. (2008). Characterization of human SLC4A10 as an electroneutral Na/HCO_3 cotransporter (NBCn2) with Cl^- self-exchange activity. *J. Biol. Chem.* **283**, 12777–12788.

Parker, M.D., Ourmozdi, E.P., and Tanner, M.J. (2001). Human BTR1, a new bicarbonate transporter superfamily member and human AE4 from kidney. *Biochem. Biophys. Res. Commun.* **282**, 1103–1109.

Pera, A. et al. (2008). Functional assessment of allelic variants in the SLC26A4 gene involved in Pendred syndrome and nonsyndromic EVA. *Proc. Natl. Acad. Sci. U. S. A.*.

Peters, T.A., Tonnaer, E.L., Kuijpers, W., Cremers, C.W., and Curfs, J.H. (2002). Differences in endolymphatic sac mitochondria-rich cells indicate specific functions. *Laryngoscope* **112**, 534–541.

Petrovic, S. et al. (2004). SLC26A7: a basolateral Cl^-/HCO_3^- exchanger specific to intercalated cells of the outer medullary collecting duct. *Am. J. Physiol. Renal Physiol.* **286**, F161–F169.

Petrovic, S. et al. (2002). Colocalization of the apical Cl^-/HCO_3^- exchanger PAT1 and gastric H-K-ATPase in stomach parietal cells. *Am. J. Physiol. Gastrointest Liver Physiol.* **283**, G1207–G1216.

Petrovic, S., Wang, Z., Ma, L., and Soleimani, M. (2003). Regulation of the apical Cl^-/HCO_3^- exchanger pendrin in rat cortical collecting duct in metabolic acidosis. *Am. J. Physiol. Renal Physiol.* **284**, F103–F112.

Picollo, A. and Pusch, M. (2005). Chloride/proton antiporter activity of mammalian CLC proteins ClC-4 and ClC-5. *Nature.* **436**, 420–423.

Piermarini, P.M., Choi, I., and Boron, W.F. (2007). Cloning and characterization of an electrogenic Na/HCO_3^- cotransporter from the squid giant fiber lobe. *Am. J. Physiol. Cell Physiol.* **292**, C2032–C2045.

Preisig, P.A. and Alpern, R.J. (1989). Basolateral membrane H-OH-HCO_3 transport in the proximal tubule. *Am. J. Physiol.* **256**, F751–F765.

Preston, G.M. and Agre, P. (1991). Isolation of the cDNA for erythrocyte integral membrane protein of 28 kilodaltons: member of an ancient channel family. *Proc. Natl. Acad. Sci. U. S. A.* **88**, 11110–11114.

Pushkin, A., Abuladze, N., Lee, I., Newman, D., Hwang, J., and Kurtz, I. (1999). Cloning, Tissue Distribution, Genomic Organization, and Functional Characterization of NBC3, a New Member of the Sodium Bicarbonate Cotransporter Family. *J. Biol. Chem.* **274**, 16569–16575.

Pushkin, A., Abuladze, N., Newman, D., Lee, I., Xu, G., and Kurtz, I. (2000a). Cloning, characterization and chromosomal assignment of NBC4, a new member of the sodium bicarbonate cotransporter family. *Biochim. Biophys. Acta.* **1493**, 215–218.

Pushkin, A., Abuladze, N., Newman, D., Lee, I., Xu, G., and Kurtz, I. (2000b). Two C-terminal variants of NBC4, a new member of the sodium bicarbonate cotransporter family: cloning, characterization, and localization. *IUBMB Life.* **50**, 13–19.

Pushkin, A. and Kurtz, I. (2006). SLC4 base (HCO_3^-, CO_3^{2-}) transporters: classification, function, structure, genetic diseases, and knockout models. *Am. J. Physiol. Renal Physiol.* **290**, F580–F599.

Quentin, F. et al. (2004). The Cl^-/HCO_3^- exchanger pendrin in the rat kidney is regulated in response to chronic alterations in chloride balance. *Am. J. Physiol. Renal Physiol.* **287**, F1179–F1188.

Rajagopalan, L. et al. (2006). Essential helix interactions in the anion transporter domain of prestin revealed by evolutionary trace analysis. *J. Neurosci.* **26**, 12727–12734.

Rajendran, V.M. et al. (2000). Regulation of DRA and AE1 in rat colon by dietary Na depletion. *Am. J. Physiol. Gastrointest Liver Physiol.* **279**, G931–G942.

Reardon, W., CF, O.M., Trembath, R., Jan, H., and Phelps, P.D. (2000). Enlarged vestibular aqueduct: a radiological marker of pendred syndrome, and mutation of the PDS gene. *Qjm.* **93**, 99–104.

Reardon, W. et al. (1999). Prevalence, age of onset, and natural history of thyroid disease in Pendred syndrome. *J. Med. Genet.* **36**, 595–598.

Reithmeier, R.A. (1983). Inhibition of anion transport in human red blood cells by 5,5′- dithiobis(2-nitrobenzoic acid). *Biochim. Biophys. Acta.* **732**, 122–125.

Rillema, J.A. and Hill, M.A. (2003). Prolactin regulation of the pendrin-iodide transporter in the mammary gland. *Am. J. Physiol. Endocrinol. Metab.* **284**, E25–E28.

Rinehart, J. et al. (2005). WNK3 kinase is a positive regulator of NKCC2 and NCC, renal cation-Cl^- cotransporters required for normal blood pressure homeostasis. *Proc. Natl. Acad. Sci. U. S. A.* **102**, 16777–16782.

Romero, M.F. (2001). The Electrogenic Na^+/HCO_3^- Cotransporter, NBC. *J. Pancreas* **2**, 182–191.

Romero, M.F. and Boron, W.F. (1999). Electrogenic Na/HCO_3 cotransporters: Expression cloning and physiology. *Ann. Rev. Physiol.* **61**, 699–723.

Romero, M.F. et al. (2006). "Physiology of electrogenic SLC26 paralogs" In - Epithelial Anion Transport in Health and Disease: the role of the SLC26 transporters family. *Novartis Found. Symp.* **273**, 126–147.

Romero, M.F., Fong, P., Berger, U.V., Hediger, M.A., and Boron, W. F. (1998). Cloning and functional expression of rNBC, an electrogenic Na^+-HCO_3^- cotransporter from rat kidney. *Am. J. Physiol.* **274**, F425–F432.

Romero, M.F., Fulton, C.M., and Boron, W.F. (2004). The SLC4 gene family of HCO_3^- transporters. *Pflügers Arch.* **447**, 495–509.

Romero, M.F., Hediger, M.A., Boulpaep, E.L., and Boron, W.F. (1996). Expression cloning of the renal electrogenic Na/HCO_3 cotransporter (NBC-1) from *Ambystoma tigrinum*. *FASEB J.* **10**, A89.

Romero, M.F., Hediger, M.A., Boulpaep, E.L., and Boron, W.F. (1997). Expression cloning and characterization of a renal electrogenic Na^+/HCO_3^- cotransporter. *Nature* **387**, 409–413.

Romero, M.F., Henry, D., Nelson, S., Harte, P.J., Dillon, A.K., and Sciortino, C.M. (2000). Cloning and characterization of a Na^+ driven anion exchanger (NDAE1): a new bicarbonate transporter. *J Biol. Chem.* **275**, 24552–24559.

Romero, M.F. et al. (2009). Slc4 and Slc26 bicarbonate transporter function and localization in the eye. *IVOS*.

Rothstein, A., Cabantchik, Z.I., and Knauf, P. (1976). Mechanism of anion transport in red blood cells: role of membrane proteins. *Fed. Proc.* **35**, 3–10.

Rotman-Pikielny, P. et al. (2002). Retention of pendrin in the endoplasmic reticulum is a major mechanism for Pendred syndrome. *Hum. Mol. Genet.* **11**, 2625–2633.

Royaux, I.E. et al. (2001). Pendrin, encoded by the Pendred syndrome gene, resides in the apical region of renal intercalated cells and mediates bicarbonate secretion. *Proc. Natl. Acad. Sci. U. S. A.* **98**, 4221–4226.

Russell, J.M. and Boron, W.F. (1976). Role of choloride transport in regulation of intracellular pH. *Nature* **264**, 73–74.

Rybalchenko, V. and Santos-Sacchi, J. (2008). Anion control of voltage sensing by the motor protein prestin in outer hair cells. *Biophys. J.* **95**, 4439–4447.

Sabolic, I., Brown, D., Gluck, S.L., and Alper, S.L. (1997). Regulation of AE1 anion exchanger and $H^{(+)}$-ATPase in rat cortex by acute metabolic acidosis and alkalosis. *Kidney Int.* **51**, 125–137.

Sahr, K.E., Taylor, W.M., Daniels, B.P., Rubin, H.L., and Jarolim, P. (1994). The structure and organization of the human erythroid anion exchanger (AE1) gene. *Genomics* **24**, 491–501.

Saier, M.H., Jr. et al. (1999). Phylogenetic characterization of novel transport protein families revealed by genome analyses. *Biochim. Biophys. Acta.* **1422**, 1–56.

Saleh, A.M., Rudnick, H., and Aronson, P.S. (1996). Mechanism of $H^{(+)}$-coupled formate transport in rabbit renal microvillus membranes. *Am. J. Physiol.* **271**, F401–F407.

Sander, T. et al. (2002). Association of the 867Asp variant of the human anion exchanger 3 gene with common subtypes of idiopathic generalized epilepsy. *Epilepsy Res.* **51**, 249–255.

Santos-Sacchi, J., Song, L., Zheng, J., and Nuttall, A.L. (2006). Control of mammalian cochlear amplification by chloride anions. *J. Neurosci.* **26**, 3992–3998.

Sassani, P. et al. (2002). Functional characterization of NBC4: a new electrogenic sodium- bicarbonate cotransporter. *Am. J. Physiol. Cell Physiol.* **282**, C408–C416.

Satoh, H., Susaki, M., Shukunami, C., Iyama, K., Negoro, T., and Hiraki, Y. (1998). Functional analysis of diastrophic dysplasia sulfate transporter. Its involvement in growth regulation of chondrocytes mediated by sulfated proteoglycans. *J. Biol. Chem.* **273**, 12307–12315.

Schaechinger, T.J. and Oliver, D. (2007). Nonmammalian orthologs of prestin (SLC26A5) are electrogenic divalent/chloride anion exchangers. *Proc. Natl. Acad. Sci. U. S. A.* **104**, 7693–7698.

Scheel, O., Zdebik, A.A., Lourdel, S., and Jentsch, T.J. (2005). Voltage-dependent electrogenic chloride/proton exchange by endosomal CLC proteins. *Nature* **436**, 424–427.

Schneider, H.P., Broer, S., Broer, A., and Deitmer, J.W. (2007). Heterologous expression of the glutamine transporter SNAT3 in Xenopus oocytes is associated with four modes of uncoupled transport. *J. Biol. Chem.* **282**, 3788–3798.

Schopfer, L.M. and Salhany, J.M. (1995). Characterization of the stilbenedisulfonate binding site on band 3. *Biochemistry.* **34**, 8320–8329.

Schultheis, P.J. et al. (1998). Renal and intestinal absorptive defects in mice lacking the NHE3 Na^+/H^+ exchanger. *Nat. Genet.* **19**, 282–285.

Schweinfest, C.W., Henderson, K.W., Suster, S., Kondoh, N., and Papas, T.S. (1993). Identification of a colon mucosa gene that is down-regulated in colon adenomas and adenocarcinomas. *Proc. Natl. Acad. Sci. U. S. A.* **90**, 4166–4170.

Schweinfest, C.W. et al. (2006). slc26a3 (dra)-deficient mice display chloride-losing diarrhea, enhanced colonic proliferation, and distinct up-regulation of ion transporters in the colon. *J. Biol. Chem.* **281**, 37962–37971.

Schwoch, G. and Passow, H. (1973). Preparation and properties of human erythrocyte ghosts. *Mol. Cell. Biochem.* **2**, 197–218.

Sciortino, C.M. (2001). Characterization and localization of the sodium mediated bicarbonate transporters NBC and NDAE1. In *Physiology & Biophysics*. Case Western Reserve University, Cleveland, OH, p. 252.

Sciortino, C.M., Fletcher, B.R., Shrode, L.D., Harte, P.J., and Romero, M.F. (2001). Localization of Endogenous and Recombinant Na^+-driven Anion Exchanger Protein, NDAE1, from Drosophila melanogaster. *Am. J. Physiol. Cell Physiol.* **281**, C449–C463.

Sciortino, C.M. and Romero, M.F. (1999). Cation and voltage dependence of rat kidney, electrogenic Na^+/HCO_3^- cotransporter, rkNBC, expressed *in* oocy*tes*. *Am. J. Physiol.* **277**, F611–F623.

Scott, D.A. and Karniski, L.P. (2000). Human pendrin expressed in Xenopus laevis oocytes mediates chloride/formate exchange. *Am. J. Physiol. Cell Physiol.* **278**, C207–C211.

Scott, D.A. et al. (2000). Functional differences of the PDS gene product are associated with phenotypic variation in patients with Pendred syndrome and non-syndromic hearing loss (DFNB4). *Hum. Mol. Genet.* **9**, 1709–1715.

Scott, D.A., Wang, R., Kreman, T.M., Sheffield, V.C., and Karniski, L. P. (1999). The Pendred syndrome gene encodes a chloride-iodide transport protein. *Nat. Genet.* **21**, 440–443.

Seidler, U. et al. (2008). Sodium and chloride absorptive defects in the small intestine in Slc26a6 null mice. *Pflugers Arch.* **455**, 757–766.

Shayakul, C. and Alper, S.L. (2000). Inherited renal tubular acidosis. *Curr. Opin. Nephrol. Hypertens.* **9**, 541–546.

Shcheynikov, N. et al. (2006a). Regulatory interaction between CFTR and the SLC26 transporters. *Novartis Found. Symp.* **273**, 177–186 discussion 186-92, 261-4.

Shcheynikov, N. et al. (2006b). Coupling Modes and Stoichiometry of Cl^-/HCO_3^- Exchange by slc26a3 and slc26a6. *J. Gen. Physiol.* **127**, 511–524.

Shcheynikov, N. et al. (2008). The Slc26a4 transporter functions as an electroneutral $Cl^-/I^-/HCO_3^-$ exchanger: role of Slc26a4 and Slc26a6 in I- and HCO_3^- secretion and in regulation of CFTR in the parotid duct. *J. Physiol.* **586**, 3813–3824.

Sheffield, V.C. et al. (1996). Pendred syndrome maps to chromosome 7q21-34 and is caused by an intrinsic defect in thyroid iodine organification. *Nat. Genet.* **12**, 424–426.

Shelden, M.C., Loughlin, P., Tierney, M.L., and Howitt, S.M. (2001). Proline residues in two tightly coupled helices of the sulphate transporter, SHST1, are important for sulphate transport. *Biochem. J.* **356**, 589–594.

Shibagaki, N. and Grossman, A.R. (2006). The role of the STAS domain in the function and biogenesis of a sulfate transporter as probed by random mutagenesis. *J. Biol. Chem.* **281**, 22964–22973.

Shirakabe, K. et al. (2006). IRBIT, an inositol 1,4,5-trisphosphate receptor-binding protein, specifically binds to and activates pancreas-type Na^+/HCO_3^- cotransporter 1 (pNBC1). *Proc. Natl. Acad. Sci. U. S. A.* **103**, 9542–9547.

Silberg, D.G., Wang, W., Moseley, R.H., and Traber, P.G. (1995). The Down regulated in Adenoma (dra) gene encodes an intestine-specific membrane sulfate transport protein. *J. Biol. Chem.* **270**, 11897–11902.

Simpson, J.E. et al. (2007). PAT-1 (Slc26a6) is the predominant apical membrane Cl^-/HCO_3^- exchanger in the upper villous epithelium of the murine duodenum. *Am. J. Physiol. Gastrointest Liver Physiol.* **292**, G1079–G1088.

Sindic, A., Plata, C., Sussman, C.R., Chang, M.-H., and Romero, M.F. (2006). Localization of Slc26a9 and role of the STAS domain. *FASEB J.* **20**, a839.

Singh, A.K. et al. (2008). Fructose-induced hypertension: Essential role of chloride and fructose absorbing transporters PAT1 and Glut5. *Kidney Int.* **74**, 438–447.

Singh, R. and Wangemann, P. (2008). Free radical stress-mediated loss of Kcnj10 protein expression in stria vascularis contributes to deafness in Pendred syndrome mouse model. *Am. J. Physiol. Renal Physiol.* **294**, F139–F148.

Sly, W.S. and Hu, P.Y. (1995). Human carbonic anhydrases and carbonic anhydrase deficiencies. *Annu. Rev. Biochem.* **64**, 375–401.

Snow, J.W., Brandts, J.F., and Low, P.S. (1978). The effects of anion transport inhibitors on structural transitions in erythrocyte membranes. *Biochim. Biophys. Acta.* **512**, 579–591.

Soleimani, M. et al. (2001). Pendrin: an apical $Cl^-/OH^-/HCO_3^-$ exchanger in the kidney cortex. *Am. J. Physiol. Renal Physiol.* **280**, F356–F364.

Songyang, Z. et al. (1997). Recognition of unique carboxyl-terminal motifs by distinct PDZ domains. *Science* **275**, 73–77.

Sterling, D., Brown, N.J., Supuran, C.T., and Casey, J.R. (2002). The functional and physical relationship between the DRA bicarbonate transporter and carbonic anhydrase II. *Am. J. Physiol. Cell Physiol.* **283**, C1522–C1529.

Sterling, D., Reithmeier, R.A., and Casey, J.R. (2001). A transport metabolon. Functional interaction of carbonic anhydrase II and chloride/bicarbonate exchangers. *J. Biol. Chem.* **276**, 47886–47894.

Stewart, A.K., Chernova, M.N., Shmukler, B.E., Wilhelm, S., and Alper, S.L. (2002). Regulation of AE2-mediated Cl(-) Transport by Intracellular or by Extracellular pH Requires Highly Conserved Amino Acid Residues of the AE2 NH(2)-terminal Cytoplasmic Domain. *J. Gen. Physiol.* **120**, 707–722.

Stewart, A.K., Kerr, N., Chernova, M.N., Alper, S.L., and Vaughan-Jones, R.D. (2004). Acute pH-dependent regulation of AE2-mediated anion exchange involves discrete local surfaces of the NH2-terminal cytoplasmic domain. *J. Biol. Chem.* **279**, 52664–52676.

Stuart-Tilley, A., Sardet, C., Pouyssegur, J., Schwartz, M.A., Brown, D., and Alper, S.L. (1994). Immunolocalization of anion exchanger AE2 and cation exchanger NHE-1 in distinct adjacent cells of gastric mucosa. *Am. J. Physiol.* **266**, C559–C568.

Stuart-Tilley, A.K., Shmukler, B.E., Brown, D., and Alper, S.L. (1998). Immunolocalization and tissue-specific splicing of AE2 anion exchanger in mouse kidney. *J. Am. Soc. Nephrol.* **9**, 946–959.

Sun, X.C. and Bonanno, J.A. (2003). Identification and cloning of the Na/HCO_3^- cotransporter (NBC) in human corneal endothelium. *Exp. Eye. Res.* **77**, 287–295.

Sun, X.C., Bonanno, J.A., Jelamskii, S., and Xie, Q. (2000). Expression and localization of $Na^{(+)}$-$HCO_{(3)}^{(-)}$ cotransporter in bovine corneal endothelium. *Am. J. Physiol. Cell Physiol.* **279**, C1648–C1655.

Superti-Furga, A. et al. (1996). Achondrogenesis type IB is caused by mutations in the diastrophic dysplasia sulphate transporter gene. *Nat. Genet.* **12**, 100–102.

Tanner, M.J. and Boxer, D.H. (1972). Separation and some properties of the major proteins of the human erythrocyte membrane. *Biochem. J.* **129**, 333–347.

Tanner, M.J. and Gray, W.R. (1971). The isolation and functional identification of a protein from the human erythrocyte 'ghost'. *Biochem. J.* **125**, 1109–1117.

Tanphaichitr, V.S. et al. (1998). Novel AE1 mutations in recessive distal renal tubular acidosis. Loss-of-function is rescued by glycophorin A. *J. Clin. Invest.* **102**, 2173–2179.

Tatishchev, S. et al. (2003). Identification of Membrane Topography of the Electrogenic Sodium Bicarbonate Cotransporter pNBC1 by in Vitro Transcription/Translation. *Biochemistry.* **42**, 755–765.

Taylor, J.P., Metcalfe, R.A., Watson, P.F., Weetman, A.P., and Trembath, R.C. (2002). Mutations of the PDS gene, encoding pendrin, are associated with protein mislocalization and loss of iodide efflux: implications for thyroid dysfunction in Pendred syndrome. *J. Clin. Endocrinol Metab.* **87**, 1778–1784.

Thomas, R.C. (1976a). The effect of carbon dioxide on the intracellular pH and buffering power of snail neurones. *J. Physiol. (Lond)* **255**, 715–735.

Thomas, R.C. (1976b). Ionic mechanism of the H^+ pump in a snail neurone. *Nature* **262**, 54–55.

Thomas, R.C. (1977). The role of bicarbonate, chloride and sodium ions in the regulation of intracellular pH in snail neurones. *J. Physiol. (Lond)* **273**, 317–338.

Toure, A. et al. (2007). The Testis Anion Transporter 1 (Slc26a8) is required for sperm terminal differentiation and male fertility in the mouse. *Hum. Mol. Genet.*.

Toure, A., Morin, L., Pineau, C., Becq, F., Dorseuil, O., and Gacon, G. (2001). Tat1, a novel sulfate transporter specifically expressed in human male germ cells and potentially linked to rhogtpase signaling. *J. Biol. Chem.* **276**, 20309–20315.

Tsuganezawa, H. et al. (2001). A new member of the HCO_3^- transporter superfamily is an apical anion exchanger of β -intercalated cells in the kidney. *J. Biol. Chem.* **276**, 8180–8189.

van Adelsberg, J.S., Edwards, J.C., and al-Awqati, Q. (1993). The apical Cl/HCO_3 exchanger of beta intercalated cells. *J. Biol. Chem.* **268**, 11283–11289.

Verlander, J.W. et al. (2003). Deoxycorticosterone upregulates PDS (Slc26a4) in mouse kidney: role of pendrin in mineralocorticoid-induced hypertension. *Hypertension* **42**, 356–362.

Verlander, J.W. et al. (2006). Dietary Cl(-) restriction upregulates pendrin expression within the apical plasma membrane of type B intercalated cells. *Am. J. Physiol. Renal Physiol.* **291**, F833–F839.

Vince, J.W. and Reithmeier, R.A.F. (1998). Carbonic anhydrase II binds to the carboxyl terminus of human band 3, the erythrocyte Cl^-/HCO_3^- exchanger. *J. Biol. Chem.* **273**, 28430–28437.

Vincourt, J.B., Jullien, D., Amalric, F., and Girard, J.P. (2003). Molecular and functional characterization of SLC26A11, a sodium-independent sulfate transporter from high endothelial venules. *Faseb J.* **17**, 890–892.

Virkki, L.V., Choi, I., Davis, B.A., and Boron, W.F. (2003). Cloning of a Na^+-driven Cl/HCO_3 exchanger from squid giant fiber lobe. *Am. J. Physiol. Cell Physiol.* **285**, C771–C780.

Virkki, L.V., Wilson, D.A., Vaughan-Jones, R.D., and Boron, W.F. (2002). Functional characterization of human NBC4 as an electrogenic Na/HCO_3 cotransporter (NBCe2). *Am. J. Physiol. Cell Physiol.* **282**, C1278–C1289.

Vorum, H., Aalkjaer, C., Hager, H., Nielsen, S., and Maunsbach, A.B. (2003). Electrogenic Na^+/HCO_3^- cotransporter rkNBC1 immunolocalized in rat eye. *Ann. N. Y. Acad. Sci.* **986**, 646–648.

Wagner, C.A. et al. (2002). Regulation of the expression of the Cl^-/ anion exchanger pendrin in mouse kidney by acid-base status. *Kidney Int.* **62**, 2109–2117.

Walker, N.M. et al. (2008). Down-regulated in adenoma Cl/HCO_3 exchanger couples with Na/H exchanger 3 for NaCl absorption in murine small intestine. *Gastroenterology* **135**, 1645–1653 e3.

Wall, S.M. et al. (2003). Localization of pendrin in mouse kidney. *Am. J. Physiol. Renal Physiol.* **284**, F229–F241.

Wang, C.Z., Yano, H., Nagashima, K., and Seino, S. (2000). The Na^+-driven Cl^-/HCO_3^- Exchanger. Cloning, tissue distribution, and functional characterization. *J. Biol. Chem.* **275**, 35486–35490.

Wang, Y. et al. (2006). Slc26a6 regulates CFTR activity in vivo to determine pancreatic duct HCO_3^- secretion: relevance to cystic fibrosis. *Embo J.* **25**, 5049–5057.

Wang, Z., Conforti, L., Petrovic, S., Amlal, H., Burnham, C.E., and Soleimani, M. (2001). Mouse Na^+ : HCO_3^- cotransporter isoform NBC-3 (kNBC-3): Cloning, expression, and renal distribution. *Kidney Int.* **59**, 1405–1414.

Wang, Z., Petrovic, S., Mann, E., and Soleimani, M. (2002). Identification of an apical Cl^-/HCO_3^- exchanger in the small intestine. *Am. J. Physiol. Gastrointest Liver Physiol.* **282**, G573–G579.

Wang, Z. et al. (2005). Renal and intestinal transport defects in Slc26a6-null mice. *Am. J. Physiol. Cell Physiol.* **288**, C957–C965.

Wangemann, P. et al. (2007). Loss of cochlear HCO_3^- secretion causes deafness via endolymphatic acidification and inhibition of Ca^{2+} reabsorption in a Pendred syndrome mouse model. *Am. J. Physiol. Renal Physiol.* **292**, F1345–F1353.

Wedenoja, S. et al. (2008). The impact of sodium chloride and volume depletion in the chronic kidney disease of congenital chloride diarrhea. *Kidney Int.* **74**, 1085–1093.

Weise, A., Becker, H.M., and Deitmer, J.W. (2007). Enzymatic suppression of the membrane conductance associated with the glutamine transporter SNAT3 expressed in Xenopus oocytes by carbonic anhydrase II. *J. Gen. Physiol.* **130**, 203–215.

Wendel, C., Becker, H.M., and Deitmer, J.W. (2008). The sodium-bicarbonate cotransporter NBCe1 supports glutamine efflux via SNAT3 (SLC38A3) co-expressed in Xenopus oocytes. *Pflugers Arch.* **455**, 885–893.

Wilson, F.H. et al. (2001). Human hypertension caused by mutations in WNK kinases. *Science* **293**, 1107–1112.

Wood, P.G., Muller, H., Sovak, M., and Passow, H. (1992). Role of Lys 558 and Lys 869 in substrate and inhibitor binding to the murine band 3 protein: a study of the effects of site-directed mutagenesis of the band 3 protein expressed in the oocytes of Xenopus laevis. *J. Membr. Biol.* **127**, 139–148.

Xie, Q., Welch, R., Mercado, A., Romero, M.F., and Mount, D.B. (2002). Molecular and functional characterization of the Slc26A6 anion exchanger, functional comparison to Slc26a1. *Am. J. Physiol. Renal Physiol.* **283**, F826–F838.

Xu, J. et al. (2008). Deletion of the chloride transporter Slc26a9 causes loss of tubulovesicles in parietal cells and impairs acid secretion in the stomach. *Proc. Natl. Acad. Sci. U. S. A.* **105**, 17955–17960.

Xu, J. et al. (2003). Expression of the Na^{+}-HCO_3^{-} cotransporter NBC4 in rat kidney and characterization of a novel NBC4 variant. *Am. J. Physiol. Renal. Physiol* **284**, F41–F50.

Yang, H. et al. (1998). Intestinal inflammation reduces expression of DRA, a transporter responsible for congenital chloride diarrhea. *Am. J. Physiol.* **275**, G1445–G1453.

Yannoukakos, D., Stuart-Tilley, A., Fernandez, H.A., Fey, P., Duyk, G., and Alper, S.L. (1994). Molecular cloning, expression, and chromosomal localization of two isoforms of the AE3 anion exchanger from human heart. *Circ. Res.* **75**, 603–614.

Yawata, Y. (2003). Introduction: History of Red Cell Membrane Research. In *Cell Membrane: The Red Blood Cell as a Model* (Yawata, Y. ed.), WILEY-VCH Verlag GmbH & Co, Weinheim, pp. 1–25.

Yoon, J.S. et al. (2008). Heterogeneity in the processing defect of SLC26A4 mutants. *J. Med. Genet.* **45**, 411–419.

Young, M.T., Beckmann, R., Toye, A.M., and Tanner, M.J. (2000). Red-cell glycophorin A-band 3 interactions associated with the movement of band 3 to the cell surface. *Biochem. J.* **350** (Pt 1), 53–60.

Yu, N., Zhu, M.L., and Zhao, H.B. (2006). Prestin is expressed on the whole outer hair cell basolateral surface. *Brain Res.* **1095**, 51–88.

Zhang, D., Kiyatkin, A., Bolin, J.T., and Low, P.S. (2000). Crystallographic structure and functional interpretation of the cytoplasmic domain of erythrocyte membrane band 3. *Blood* **96**, 2925–2933.

Zheng, J. et al. (2006). Analysis of the oligomeric structure of the motor protein prestin. *J. Biol. Chem.* **281**, 19916–19924.

Zheng, J., Long, K.B., Matsuda, K.B., Madison, L.D., Ryan, A.D., and Dallos, P.D. (2003). Genomic characterization and expression of mouse prestin, the motor protein of outer hair cells. *Mamm. Genome.* **14**, 87–96.

Zheng, J., Long, K.B., Shen, W., Madison, L.D., and Dallos, P. (2001). Prestin topology: localization of protein epitopes in relation to the plasma membrane. *Neuroreport* **12**, 1929–1935.

Zheng, J., Shen, W., He, D.Z., Long, K.B., Madison, L.D., and Dallos, P. (2000). Prestin is the motor protein of cochlear outer hair cells. *Nature* **405**, 149–155.

Zhu, Q., Lee, D.W., and Casey, J.R. (2003). Novel topology in C-terminal region of the human plasma membrane anion exchanger, AE1. *J. Biol. Chem.* **278**, 3112–3120.

CHAPTER

5

Thermodynamics and Kinetics of Chloride Transport in Neurons: An Outline

Francisco Javier Alvarez-Leefmans and Eric Delpire

OUTLINE

I. INTRODUCTION

A. Biological Significance of the Non-passive Distribution of Cl^- across Neuronal Membranes

It is well established that Cl^- is not distributed in thermodynamic equilibrium across the plasma membrane of most, if not all, animal cells, including neurons. That is, Cl^- does not distribute itself passively across the plasma membrane in accordance with the membrane potential (E_m). On the contrary, it is actively transported and tightly regulated in virtually all animal cells (Alvarez-Leefmans, 2001). Some cells actively extrude Cl^-, others actively accumulate it, but few cells ignore it. The level of intracellular Cl^- results from a delicate functional balance between Cl^- channels and carriers present in the plasma membrane. Further, the intracellular Cl^- concentration, $[Cl^-]_i$, may also be affected by Cl^- channels and carriers expressed in the membrane of intracellular organelles (Chapter 12).

By virtue of being distributed out of electrochemical equilibrium, Cl^- serves as a key player in a variety of cellular functions such as intracellular pH regulation (Chapter 4); cell volume regulation (Chapter 15); and transepithelial salt transport (Chapters 18). In addition to these functions, in the nervous system Cl^- gradients are fundamental in synaptic signaling mediated by ligand-gated anion channels (in both the de- and hyper-polarizing directions); neuronal growth, migration and targeting; and K^+ buffering of extracellular fluids (e.g. choroid plexus, and possibly Schwann's cells). Abnormalities in cell Cl^- homeostasis underlie many pathological conditions that are discussed throughout this book, such as epilepsy (Chapter 24), deafness and motor disturbances of vestibular origin (Chapter 21), pain and neurogenic inflammation (Chapters 22 and 23). The present chapter is divided in three main sections. The aim of the first two sections is to discuss some of the fundamental principles of thermodynamics of Cl^- transport that determine the intracellular concentration of this anion in nerve cells and its impact on synaptic signaling. In the third section, we discuss some kinetic properties of cation-coupled carrier-mediated Cl^- transport.

The notion that Cl^- is non-passively distributed and actively transported across cell membranes, including of course neurons, may sound obvious nowadays. However, the path to the acceptance of this idea has been long and tortuous, as eloquently discussed by Hartzell and Russell in Chapters 1 and 2 of this volume. It is an example of how a wrong idea, sprouting from a generalization based on a few observations, can remain in scientific thought for a long time, in spite of abundant evidence against it. It has taken more than 50 years to eradicate the "passive distribution dogma", and until recently most cell physiology and neuroscience textbooks continued to include this error.

B. Origin and Collapse of the "Passive Cl^- Distribution Dogma" in Neurons

The origin of the "passive Cl^- distribution dogma" in neurons can be traced back to the work of H. Burr Steinbach (1905–1981), who in 1941 reported that in freshly dissected squid axons the average $[Cl^-]_i$ was 36 mmol/Kg axoplasm. (Steinbach, 1941). Taking the water content of squid axoplasm as 880 g/Kg (Keynes, 1963), this corresponded to 41 mM, a value close to what would be expected if Cl^- was in electrochemical equilibrium across the axolemma. In the late 1950s and early 1960s several physiologists from the British School rigorously demonstrated that the resting Cl^- conductance of frog skeletal muscle membrane was twice as large as that for K^+ (Hodgkin and Horowicz, 1959; Hutter and Noble, 1960). Around the same time, Richard H. Adrian (1927–1995) showed that the $[Cl^-]_i$ in frog skeletal muscle was consistent with a distribution of Cl^- determined by a purely passive mechanism (Adrian, 1960, 1961). Thus, E_{Cl} , the Nernst equilibrium potential for Cl^- as defined by Eq. 1, was found to be equal to E_m. It is true than in skeletal muscle E_{Cl} closely follows E_m (Vaughan-Jones, 1982), as illustrated in Fig. 5.1. However, the fact that E_m = E_{Cl} does not preclude the presence of carrier-mediated Cl^- transport in skeletal muscle or in any cell type. In fact, there is compelling

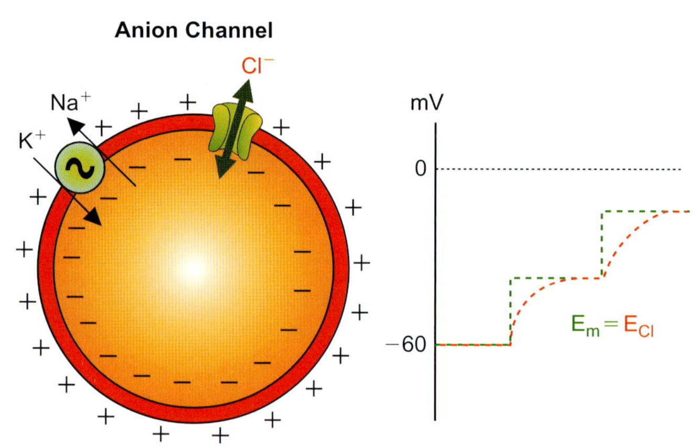

FIGURE 5.1 Passive Cl^- distribution across the cell membrane of a cell having only anion channels and a Na^+/K^+ ATPase (left). The resting membrane potential (E_m) is −60 mV. Right: E_m is changed in steps of ~20 mV, from the initial resting membrane potential. Cl^- redistributes passively, following E_m. At each steady state E_m = E_{Cl} (see text for details).

evidence for an inward transport of Cl^- mediated by NKCC in skeletal muscle fibers (Aickin, 1990; Aickin et al., 1989; Fu et al., 1999; Gosmanov et al., 2003; Gosmanov and Thomason, 2003; Harris and Betz, 1987; Jurkat-Rott et al., 2006; Wong et al., 1999). However, as shown in Fig. 5.2, the relatively high channel-mediated permeability to Cl^- masks any effects that the operation of this transport system may have on $[Cl^-]_i$. Accordingly, reducing the Cl^- conductance unmasks a powerful active Cl^- uptake known to be mediated mostly by Na^+, K^+, Cl^- cotransport (Aickin, 1990).

Returning to the story in squid axons, Steinbach had reported in the same 1941 paper that during the first 30 minutes after isolating the axon $[Cl^-]_i$ rose from 36 to 75 mmol/Kg axoplasm and remained steady at this high level for "long periods of time…". This doubling of $[Cl^-]_i$ observed by Steinbach prompted Keynes to reanalyze the issue. In his landmark paper of 1963 Keynes (Fig. 5.3) wrote: "In December 1960 a number of analyses was made of the chloride content of extruded axoplasm from *Loligo forbesi*, and it soon became clear that Steinbach's low values for $[Cl]_i$ could not be confirmed, even in the freshest material the average value obtained being over 100 m-mole/kg. Since Steinbach's figures were consistent with a simple passive distribution of chloride, the new ones could not be, for the corresponding value for E_{Cl}, was less than −40 mV. Hence it had to be supposed either that the internal chloride was partly bound, or that chloride was being transported inwards against the electrochemical gradient by some form of active transport mechanism. As will be seen, measurements of the chloride activity in extruded axoplasm ruled out the first explanation, and tracer experiments provided support for the second hypothesis, without, however, settling finally the question of the exact nature of the transport process." The question of the "exact nature of the transport process" was elucidated nearly 20 years later by John Russell who carried out a series of careful studies in which squid axons were internally dialyzed (Russell, 1979, 1983), an approach that has the unique advantage of controlling the intracellular ionic composition while measuring unidirectional fluxes of Na^+, K^+ and Cl^-. The active Cl^- uptake mechanism turned out to be an electroneutral Na^+-K^+-Cl^- cotransporter with a stoichiometry of $2Na^+$:$1K^+$:$3Cl^-$. About the same time and independently, Geck, Heinz and co-workers made the first description in a mammalian cell (Ehrlich cells) of an "electrically silent" Na^+, K^+ and Cl^- cotransport system with a stoichiometry of $1Na^+$:$1K^+$:$2Cl^-$ (Geck and Heinz, 1986; Geck et al., 1980).

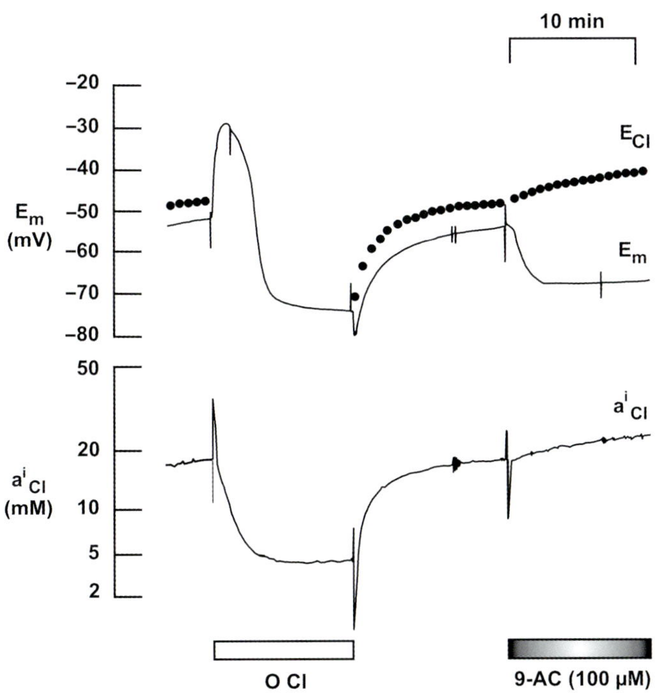

FIGURE 5.2 Active accumulation of Cl^- in rat skeletal muscle unmasked by reducing the resting Cl^- conductance. Membrane potential (E_m) and the intracellular Cl^- activity (a^i_{Cl}) were measured simultaneously in a rat lumbrical muscle, during the removal and re-addition of external Cl^-, and during inhibition of the Cl^- conductance by application of 9-anthracene carboxylic acid (9-AC). The Cl^- equilibrium potential (E_{Cl}), calculated from the recorded a^i_{Cl} and a^o_{Cl}, using Eq. 1 was plotted as filled circles above the recording of E_m. Note that E_{Cl} was positive with respect to E_m, indicating that a^i_{Cl} is maintained above electrochemical equilibrium. Upon reduction of the Cl^- conductance, the difference $E_m - E_{Cl}$ increased significantly. The recording was made with a double-barreled microelectrode, one barrel sensing E_m and the other the a^i_{Cl}. The preparation was maintained in the nominal absence of CO_2 in solutions equilibrated with 100% O_2. The mechanism responsible for the uphill accumulation of Cl^- is a Na^+, K^+, Cl^- cotransporter (modified from Aickin, 1990).

In vertebrate neurons, the first description of an electrically silent Na^+-K^+-Cl^- cotransporter (NKCC) was made in frog dorsal root ganglion neurons in the mid 1980s (Alvarez-Leefmans et al., 1986, 1988), as further discussed below and in Chapter 22. The simultaneous measurements of E_m and intracellular Cl^- activity (a^i_{Cl}) in single neurons were done by means of double-barreled microelectrodes, one barrel sensing E_m and the other a^i_{Cl}. Thus, it was possible to measure E_m directly and calculate E_{Cl} from direct measurement of a^i_{Cl} in real time. These experiments demonstrated that "uphill" accumulation of Cl^- occurred without changes in E_m (Fig. 5.4).

The current idea is that the outward directed electrochemical gradient for Cl^- in neurons, whether generated by NKCCs or other anion active transport mechanisms, seems to be crucial in the development and maturation of neurons (Chapter 19). In adult neurons, outward Cl^- gradients appear to be pivotal for olfactory sensory transduction (Chapter 20), and for

690

J. Physiol. (1963), **169**, *pp.* 690–705
With 1 *text-figure*
Printed in Great Britain

CHLORIDE IN THE SQUID GIANT AXON

BY R. D. KEYNES

From the Laboratory of the Marine Biological Association, Plymouth, and the A.R.C. Institute of Animal Physiology, Babraham, Cambridge

(*Received* 16 *May* 1963)

The classical experiments of Boyle & Conway (1941) first established that the distribution in frog muscle fibres of chloride as well as potassium ions depends partly on a double Donnan system, the majority of the internal anions and of the external cations being able to penetrate the membrane much less readily than Cl^- and K^+ ions respectively. Subsequent work on the chloride permeability of frog muscle has fully confirmed that the membrane is permeable to Cl^- ions, and in a resting fibre equilibrated with normal Ringer's solution the chloride conductance of the membrane is about twice as large as the potassium conductance (Hodgkin & Horowicz, 1959; Hutter & Noble, 1960). Moreover, Adrian (1960, 1961) has shown that in this tissue the size of the intracellular chloride concentration $[Cl]_i$ is consistent with a distribution of Cl^- ions determined by a purely passive mechanism. Thus the Nernst equilibrium potential for chloride, calculated as $E_{Cl} = RT/F \ln [Cl]_i/[Cl]_o$, is equal to the resting membrane potential E_m.

It has usually been assumed that in the squid giant axon chloride is passively distributed in a similar way. The evidence supporting this assumption is Steinbach's (1941) report that in a freshly dissected axon $[Cl]_i$ was about 36 m-mole/kg axoplasm; taking the water content of squid axoplasm as 880 g/kg (see p. 693) this corresponds to 41 m-mole Cl/kg H_2O. In squid plasma $[Cl]_o$ is 560 m-mole/kg H_2O (see Hodgkin, 1951), whence E_{Cl} would be -66 mV. This is close to the value of E_m recorded *in vivo* by Hodgkin & Keynes (see Hodgkin, 1958) and Moore & Cole (1960). Steinbach (1941) also found that during the first 30 min after isolating the axon $[Cl]_i$ rose to 75 m-mole/kg axoplasm, and thereafter remained steady for some time. Such a doubling of $[Cl]_i$ would correspond to a reduction of E_{Cl} to just under -50 mV, a figure not wholly incompatible with the lower resting potential normally observed in isolated axons. Against the view that chloride is passively distributed in squid axons should be set the arguments presented in a brief note by Mauro (1954) concerning the potential recorded internally with a silver/silver chloride electrode, and

FIGURE 5.3 Richard D. Keynes (1919–). Eminent British physiologist from Cambridge University who revolutionized the way of thinking about chloride transport across cell membranes. The right panel reproduces the first page of his epoch-making paper "Chloride in the squid giant axon", published in 1963 in the *Journal of Physiology*. (Photograph of R.D. Keynes reproduced with permission from Oxford University Press. First page of the 1963 paper reproduced with permission from *The Journal of Physiology*.)

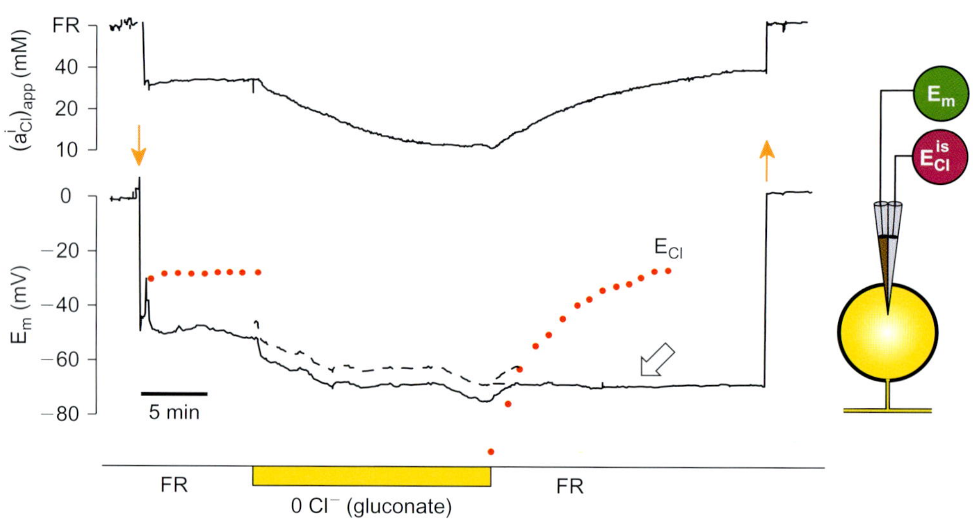

FIGURE 5.4 Electrically silent, transmembrane movements of Cl^- mediated through Na^+, K^+, Cl^- cotransport in a frog dorsal root ganglion (DRG) neuron. The intracellular Cl^- activity, a^i_{Cl}, was measured with a double-barrel Cl^--selective microelectrode in the neuronal cell body (scheme on the right). After obtaining steady-state readings, the effects of removing and re-adding external Cl^- on a^i_{Cl} and E_m were recorded. The upper trace shows the differential signal ($E_m - E^{is}_{Cl}$), where E^{is}_{Cl} is the potential recorded through the ion-sensitive barrel. The differential signal indicates $(a^i_{Cl})_{app}$, the apparent intracellular Cl^- activity. The lower trace is the membrane potential (E_m). The dashed line above a segment of the E_m trace is the E_m value after subtraction of liquid junction potentials. At the periods indicated on the bottom marks the bathing solution was changed from standard frog Ringer (FR) to a Cl^--free solution in which Cl^- was replaced with gluconate. The red filled circles superimposed on the E_m trace are the estimated E_{Cl} calculated from measured a^o_{Cl} and a^i_{Cl}. Large light-gray arrow indicates that the "uphill" accumulation of Cl^- following Cl^- depletion occurred without changes in E_m. Small orange arrows indicate electrode penetration and withdrawal from the cell. (Modified from Alvarez-Leefmans et al., 1988.)

presynaptic inhibition and nociception in primary sensory neurons (Chapter 22). Further, reversal of the Cl^- electrochemical gradient from inward to outward in certain neurons underlies pathological conditions such as neuropathic pain (Chapter 23) and some types of epilepsy (Chapter 24).

Back in 1982 Clive Ellory and Richard Keynes organized a meeting at the Royal Society in London. This effort culminated in the publication of a complete volume of the "Philosophical Transactions of the Royal Society". One can still profit from reading the papers collected in that volume that had the propitious title: "The binding and transport of anions in living tissues." It was again Keynes who, in the Introductory Remarks of the meeting, started by remembering a previous Discussion Meeting he had organized in 1970, concerned with the active transport of salts and water in living tissues (Keynes, 1982). "...the focus of our deliberations – said Keynes – was principally on the transport of the dominant biological cations: sodium in vertebrates, and potassium in insects. The anions chloride and bicarbonate were not totally neglected... ." "...But on looking through the published proceedings, one gets the impression that all the time it was sodium that was in the limelight with the thick black arrows, while chloride was relegated to the indignity of being passively transported, and was shuffled off with dashed lines in an obscure corner of the diagram. One of the reasons for this emphasis was, of course, the accident of the way in which the subject had happened to develop historically." "...the discovery of Na, K-ATPase by Skou (Skou, 1960), not unassisted by the availability of admirably specific blocking agents for sodium transport (Glynn, 1957; Schatzmann, 1953), quickly brought the sodium pump to the fore, and it has monopolized the centre of the stage ever since." "...Why is it that every cell, at least in the animal kingdom, has a sodium pump, and only a few can afford the luxury of a chloride pump?"

C. An Overview of the Ionic Mechanism of GABA- and Glycine-mediated Hyperpolarizing Postsynaptic Inhibition

The examples considered so far are for cases in which $[Cl^-]_i$ is maintained at values *higher* than predicted for electrochemical equilibrium generating an outward Cl^- gradient. However, in many instances $[Cl^-]_i$ is actively maintained at values *lower* than electrochemical equilibrium, generating an inward Cl^- gradient, as is the case for most mature vertebrate neurons. In fact, this is the basis for GABA- and glycine-mediated *hyperpolarizing inhibition*; these neurotransmitters activate receptors that are coupled to anion channels. When these channels are opened by the inhibitory neurotransmitter there is a Cl^- influx that hyperpolarizes the membrane. The hyperpolarization raises the threshold thereby decreasing the neuronal excitability.

Hyperpolarizing inhibition was discovered in 1951 by John C. Eccles (1903–1997) and his collaborators Lawrence G. Brock (1923–1996) and John S. Coombs (1917–1993), at the time they were working at University of Otago, New Zealand. Coombs was an electronic engineer recruited from the physics department, and Brock a young Otago medical graduate who had completed a first degree in chemistry. A turning point in the history of neuroscience research occurred when Brock successfully used glass microelectrodes to record intracellular voltage changes in the cat spinal motoneurons while Coombs designed and operated the electronic stimulating and recording apparatus (Todman, 2008). Their groundbreaking discoveries were published in a series of papers between 1952 and 1953 (Brock et al., 1952a, b, c, 1953). They reported that "inhibitory synaptic action evoked a hyperpolarization of the neuronal membrane", ever since named "the inhibitory postsynaptic potential" (IPSP).

In 1951, Eccles was appointed chair of Physiology at The John Curtin School of Medical Research in Canberra, Australia, but remained "loaned" for some time at Otago (Fenner and Curtis, 2001). In 1952, he and Coombs moved to Canberra where he was joined by the brilliant biophysicist Paul Fatt, who had earned his PhD at University College London under Bernard Katz (1911–2003). By the time Fatt went to Australia, he and Katz had already pioneered the use of intracellular microelectrodes to study the membrane ion permeability at excitatory synapses of frog neuromuscular junction (Fatt and Katz, 1951, 1952b) and at inhibitory synapses on crustacean neuromuscular junction (Fatt and Katz, 1952a, 1953). The new collaboration with Eccles propelled the second breakthrough in the study of hyperpolarizing inhibition; they impaled motoneurons with double-barreled microelectrodes, one barrel measured E_m (resting membrane potential, action and synaptic potentials) and the other barrel was used to pass current across the motoneuron membrane to set E_m above or below the resting level. Additionally, ions could be ejected through one of the barrels to alter intracellular anion and cation levels.

The collaboration between Fatt and Eccles produced many epoch-making papers, among which was one particularly relevant to the present discussion, "The specific ion conductances and the ionic movements across the motoneuronal membrane that produced the inhibitory postsynaptic potential" (Coombs et al., 1955). This was the first study addressing the ionic mechanisms underlying the generation of IPSPs.

They studied the IPSPs evoked in spinal motoneurons by stimulation of Group Ia fibers from an antagonistic muscle nerve, or by activation of Renshaw cells. They tested the effect of changing the steady level of E_m on the IPSP and found that at "normal" resting potential, activation of inhibitory pathways produced a brief hyperpolarization of about 5 mV. On depolarization of the membrane, the amplitude of the IPSP increased, the extent of this increase being approximately proportional to the E_m displacement. On hyperpolarization the IPSP decreased to the point of vanishing at a displacement of -7 mV from the resting E_m and then reversed in polarity on further displacement of E_m. A plot of the peak amplitude of the IPSP against E_m yielded a curve that crosses through zero at a membrane potential which they named "the reversal potential" for the IPSP (E_{IPSP}), and which was approximately at -80 mV. Further, they made the crucial observation that when motoneurons were impaled with microelectrodes filled with concentrated Cl^- solutions (e.g. KCl), the IPSPs were initially recorded as hyperpolarizing, but gradually became smaller, reversed their polarity and became depolarizing. They concluded that Cl^- diffusion from the microelectrodes increased $[Cl^-]_i$ thereby shifting E_{Cl} toward depolarizing values, and that therefore chloride ions were carrying the current for the generation of the IPSP. However, the "passive distribution dogma" for Cl^- once more got in the way; they erroneously concluded that both K^+ and Cl^- were carrying the current for the generation of the IPSP. This conclusion was based on the observation that E_{IPSP} was -80 mV whereas the after-hyperpolarization which followed action potentials, assumed to be generated by a K^+ current, reversed at -90 mV, which they estimated to be the K^+ equilibrium potential (E_K). Thus, they assumed that Cl^- was "passively distributed" and since E_{IPSP} lay between E_{Cl} (assumed to be equal to E_m) and E_K, both K^+ and Cl^- were carrying the current for the generation of the IPSP.

In subsequent years, Eccles and his collaborators extended their studies to the IPSP evoked in hippocampal pyramidal cells (Allen et al., 1977). Unfortunately, in these studies they perpetuated the same error born in the mid 1950s, namely that Cl^- dependent IPSPs resulted from inhibitory transmitter opening of channels that were permeable to both Cl^- and K^+ and that, as in motoneurons, Cl^- was passively distributed across the membrane. According to this hypothesis, the normal hyperpolarizing nature of the IPSP was determined by E_K. Despite strenuous experimental attempts to demonstrate an involvement of K^+ in the generation of Cl^- dependent IPSPs, no convincing evidence was or has ever been produced (Allen et al., 1977; Eccles et al., 1964a, b; Eccles, 1964a).

Fatt departed from the Eccles' views and in 1958, in collaboration with Boistel, published another landmark paper addressing the ionic mechanisms underlying inhibitory transmitter action in crustacean (crayfish *Astacus fluviatilis*) neuromuscular junction (Boistel and Fatt, 1958). They first showed that the inhibitory transmitter changes the crustacean muscle membrane permeability specifically to Cl^- ions ("and probably to other small anions"). They also showed that GABA and the inhibitory transmitter produced the same changes in Cl^- permeability of the postsynaptic membrane. In subsequent studies from other groups it was confirmed that GABA was the inhibitory transmitter and that it acted by increasing the membrane permeability to anions (Earl and Large, 1974; Takeuchi and Takeuchi, 1967). In their paper, Boistel and Fatt discussed and compared their observations in crayfish muscle with those made in motoneurons in his previous collaboration with Coombs and Eccles (Coombs et al., 1955). They stated that if "the conductance increase during inhibitory action involves mainly Cl, as has been found for crayfish muscle, the existence of a Cl pump extruding Cl from the cell must be postulated". In a review published in 1974 he concluded: "It seems reasonable to propose on the basis of available experimental information that the inhibitory synaptic activity of motoneurons consists of an increase in membrane permeability which is exclusive for anions and the occurrence of the reversal potential for the response at a level of membrane potential more negative than the resting potential may be taken as indicating the existence of a metabolically-driven Cl^- pump, moving Cl^- outward across the membrane" (Fatt, 1974).

In the early 1970s, Russell and Brown, working on neurons of the sea hare *Aplysia californica*, first demonstrated the existence of an active Cl^- extrusion mechanism. They directly measured intracellular Cl^- activity (a^iCl) and E_m in single neurons using ion-sensitive microelectrodes and found that E_{Cl} was consistently more negative than E_m by about 6 mV (Russell and Brown, 1972). Around the same time, Lux, Llinás and associates published a series of papers which challenged the model proposed by Eccles to explain the ionic mechanisms underlying hyperpolarizing IPSPs (Llinas and Baker, 1972; Llinas et al., 1974; Lux, 1971; Lux et al., 1970). The basic conclusions derived from their work on mammalian motoneurons were: (1) that the hyperpolarizing IPSP was generated by the influx of Cl^-, (2) that the IPSP was normally hyperpolarizing because E_{Cl} was more negative than E_m, (3) that the driving force for the IPSP, that is $E_{Cl} - E_m$, was generated and maintained by an outwardly directed active Cl^- transport mechanisms referred to as an outward "Cl^- pump", and (4) that the "Cl^- pump" was inhibited by NH_4^+. Many other groups confirmed and extended their observations and conclusions (reviewed in Alvarez-Leefmans, 1990).

D. The Membrane Permeability to Cl^- can Increase without Producing Changes in E_m: Shunting Inhibition

As mentioned above, when anion channels are opened by GABA or glycyne, if E_{Cl} is more negative than E_m, the inhibitory neurotransmitter produces a Cl^- influx that hyperpolarizes the membrane. The hyperpolarization raises the threshold thereby decreasing the neuronal excitability. Thus, *hyperpolarizing* inhibition mediated by Cl^- can be defined as that resulting from a hyperpolarizing influx of Cl^- that decreases neuronal excitability.

There are cases, however, in which the GABA-gated Cl^- conductance increases without producing any changes in E_m. The neurotransmitter-evoked increase in membrane conductance decreases the postsynaptic cell input resistance and transiently "clamps" E_m to the value of E_{IPSP}. Although generally overlooked in the literature, shunting inhibition was discovered by Fatt and Katz working on crab neuromuscular junction (Fatt and Katz, 1953). In this preparation muscle fibers are innervated by inhibitory and excitatory nerves. "...the main effect of inhibitory impulses is to *attenuate the 'end plate potentials'*, i.e. to diminish the local depolarization produced by motor impulses. Inhibitory impulses do not by themselves change the resting potential of the muscle fibre, unless this has previously been displaced from its normal level. But even though no potential change may be recorded, the inhibitory impulse...always produces a transient increase in membrane conductance (or 'ion permeability')... ." In the discussion section of this seminal paper, they stated "the mechanism of synaptic inhibition described here differs substantially from that found by Brock et al. (1952) – they were referring to (Brock et al., 1952b) – in the cat's spinal cord. A 'hyperpolarization' of the postsynaptic membrane, superficially resembling that described by Brock et al., is occasionally observed in crustacean muscle, but only when the fibres are partly depolarized, and in any case the effect is small and must be regarded as a by-product, rather than an agent, of the inhibitory process."

The type of effect they were describing for the first time (Fig. 5.5) has since then been referred to as "shunting inhibition" because the synaptic conductance short-circuits currents that are generated at adjacent excitatory synapses. If a shunting inhibitory synapse is activated, the input resistance is reduced locally and, following Ohm's law, the amplitude of subsequent excitatory postsynaptic potentials (EPSPs) is reduced. Unlike hyperpolarizing inhibition that decays more slowly and is determined by the "cable properties" of neurons, the shunting inhibitory effect lasts for only as long as the ion channels mediating the inhibitory event are open.

Shunting inhibition has recently attracted considerable attention because it is one of the mechanisms that determine precise timing, in the millisecond range, of spike generation in neurons, a property that is fundamental in shaping neuronal network oscillations (Mann and Paulsen, 2007). Further, shunting inhibition seems to be the main inhibitory mechanism in GABAergic interneurons (Banke and McBain, 2006). Further, there is evidence supporting the hypothesis that shunting inhibition mediates firing rate gain control in neurons (Prescott and De Koninck, 2003).

In conclusion, the hyperpolarization produced by the inhibitory neurotransmitter is not necessary to produce inhibition. This phenomenon has at least two possible explanations. If the inhibitory neurotransmitter (GABA or glycine) increases the permeability selectively to Cl^- without changing E_m, this implies that $E_{Cl} = E_{IPSP} = E_m$. This does not preclude Cl^- transport across the plasma membrane as discussed above, particularly if there is a constitutively active resting anion permeability that overwhelms the active Cl^- extrusion mechanism (see below, Section II.F). Alternatively, if the ligand-gated anion channels are permeated by Cl^- and HCO_3^- (Bormann et al., 1987; Fatima-Shad and Barry, 1993; Inomata et al., 1986; Kaila et al., 1993), this could also result in $E_m = E_{IPSP}$. This is because E_{Cl} is slightly more negative than E_m, but E_{HCO3} is more positive than E_m, and E_{IPSP} could just lie in between (see below, section II.H). In any event, shunting inhibition results because the action of the inhibitory neurotransmitter transiently "clamps" E_m to E_{IPSP} and short-circuits the membrane thereby decreasing the amplitude of EPSPs as well as the probability of reaching the threshold for action potential firing (i.e. a decrease in excitability).

E. Depolarizing Inhibition Mediated by Cl^-

There are several cases in which GABA or glycine can have a depolarizing action, not only in immature neurons, but also in adult neurons. One example of *depolarizing inhibition* in adult neurons is GABA-mediated presynaptic inhibition of primary afferent terminals in the spinal cord. GABA depolarizes primary sensory neurons because these cells maintain

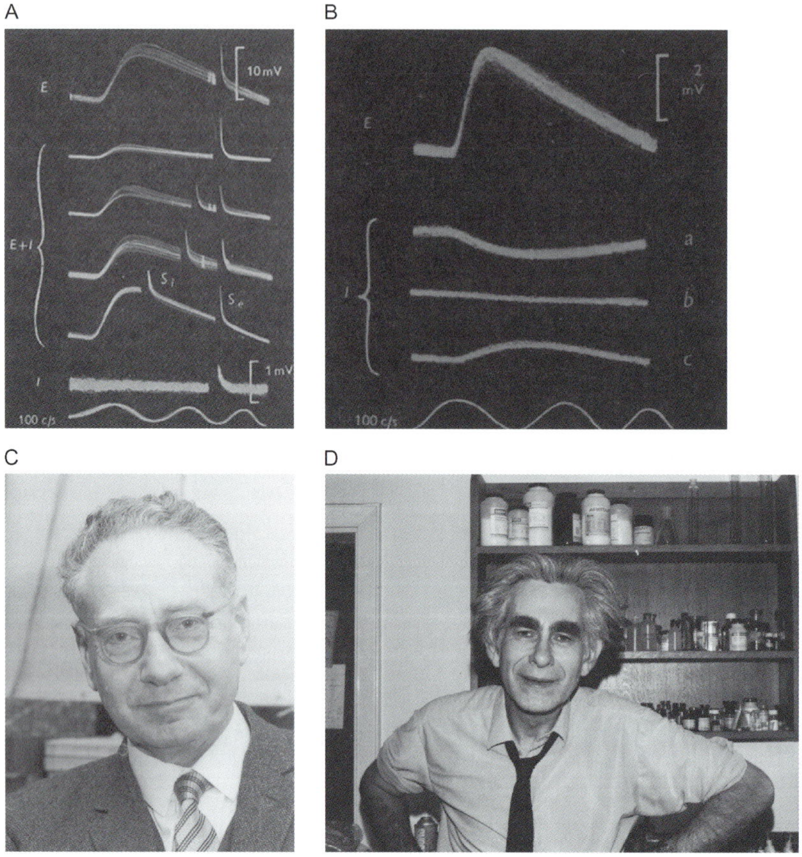

FIGURE 5.5 The discovery of "shunting inhibition" by Paul Fatt and Bernard Katz. **A.** Effect of inhibitory nerve stimulation on the end plate potential (e.p.p.) recorded with an intracellular microelectrode in a single muscle fiber of opener of claw of hermit crab. Stimulation at 33/sec. Each record is a photograph of multiple oscilloscope sweeps. Se and Si are the stimulus artifacts appearing on the falling phases of preceding responses. Top trace, labeled E shows e.p.p.s produced by stimulation of the excitatory axon. Bottom trace, labeled I (just above the sinusoidal time calibration) shows the lack of change in E_m upon stimulation of the inhibitory axon alone. Traces in the middle (E + I) show the effect of combined stimulation of both excitatory (E) and inhibitory (I) axons, producing attenuated e.p.p.s. The E stimulus was fixed (right-hand artifact Se) whereas the I stimulus (Si) preceded by varying intervals. **B.** Intracellular potentials produced by stimulation of the excitatory motor axon E. I: transmembrane potential changes due to stimulation of the inhibitory axon. The membrane potential was depolarized to −48 mV in "a", and hyperpolarized to −90 mV in "c". The resting potential in record "b" was −73 mV. Note the dependence of the I-response on the level of E_m: "its absence at the normal level" ("b"), and its reversal during hyperpolarization ("c") and depolarization ("a"). **C.** Photograph of Sir Bernard Katz (1911–2003) taken ca. 1979 at University College London. **D.** Photograph of Professor Paul Fatt taken in 1979 at his laboratory, University College London. (A and B are from Fatt and Katz, 1953; reproduced with permission from *The Journal of Physiology*. C Present from Professor Katz to one of the authors (F.J.A.-L). D Photograph taken by one of the authors (F.J.A.-L.)

$[Cl^-]_i$ above electrochemical equilibrium generating an outward gradient which, as discussed in Chapter 22, makes possible primary afferent depolarization (PAD). The latter, in turn, is thought to block the invasion of action potential into the terminals by a mixture of Na^+ channel inactivation and membrane shunting (Labrakakis et al., 2003; Lev-Tov et al., 1983, 1988; Rocha-González et al., 2008; Rudomin and Schmidt, 1999; Willis, 2006, 1999). Another example of depolarizing inhibition mediated by GABA in vertebrate neurons is that of the secretory nerve terminals from rat posterior pituitary. GABA depolarizes these terminals and inhibits neurosecretion; the depolarization inactivates Na^+ channels sufficiently to block action potential invasion into the terminals and thereby suppresses transmitter release (Zhang and Jackson, 1993; Jackson and Zhang, 1995). Depolarizing inhibition has also been described in the cell bodies of neurons in avian cochlear nucleus where the underlying mechanism seems to be accommodation (Monsivais and Rubel, 2001). In invertebrates, the examples of GABA-mediated depolarizing inhibition are innumerable (e.g. French et al., 2006).

In sum, GABA- or glycine-gated Cl^- currents can be depolarizing, hyperpolarizing or shunting. As Eccles concluded in 1964 (Eccles, 1964b) "…It is important to recognize that the IPSP may be in either the hyperpolarizing or the depolarizing direction in accord with the electrochemical gradients, or, of course, it may be virtually zero."

F. The Nature of the "Cl^- Pumps" in Neurons: Primary and Secondary Active Transport

From the 1950s to the late 1970s, cell physiology was dominated by the Na^+/K^+ ATPase, the sodium pump. Therefore, it was logical to think that active Cl^- transport had to be effected by an ATP-driven pump and neurobiologists referred to them at that time as "Cl^- pumps". The name "pump" in transport terminology is reserved to denote *primary active transport mechanisms*. "Active transport" refers to carrier-mediated transport processes that are capable of bringing about the net transfer of an uncharged solute from a region of low concentration to one of higher concentration (i.e. against a chemical gradient), or to the transfer of a charged solute against combined chemical and electrical driving forces. Thus, active transport processes are capable of counteracting or reversing the direction of diffusion and therefore perform work and consume energy. "Primary active transport" implies that the carrier mechanism responsible for the movement of a solute against a concentration difference (gradient) or a combined concentration and electrical potential difference (for the case of charged solutes, like ions) is *directly coupled to metabolic energy*, usually derived from an exergonic chemical reaction, such as ATP hydrolysis. The quintessential example of primary active transport is of course the Na^+/K^+ ATPase. Carrier-mediated Cl^- transport in animal cells is performed, as far as we know, only by *secondary active transport processes* and not by Cl^- pumps (however see Gerencser and Zhang, 2003).

Secondary active transport refers to the processes that mediate the *uphill* movements of solutes but are not directly coupled to metabolic energy; instead, the energy required is derived from the *downhill* movement of another solute. More specifically, the driving force for the uphill movement of a solute in secondary active transport is provided by coupling the uphill movement of that solute to the downhill movement of one or more solutes for which a favorable electrochemical potential energy difference exists (Geck and Heinz, 1989; Heinz, 1989). For example, the energy stored in the Na^+ gradient, which itself is set up by a primary active transporter (i.e. the Na^+/K^+ ATPase), drives the secondary active transport of another solute, say Cl^-, e.g. the Na^+-Cl^- cotransporter. The terms *countertransport, antiport or exchange* are used as synonyms when the translocation of the solute is coupled to the translocation of another solute that uses the same carrier, in the opposite direction, for example, the Na^+-independent anion exchangers (AEs in Fig. 5.7). When the translocation of the solutes occurs in the same direction, through the same carrier protein, the terms *cotransport* or *symport* are used. A typical example is the Na^+-K^+-$2Cl^-$ cotransporter, as illustrated in Fig. 5.6.

II. THERMODYNAMICS OF Cl^- TRANSPORT

For simplicity, we will focus on the definition of thermodynamic equilibrium conditions for only anion channels and three carrier-mediated anion transport mechanisms that are present in neurons: (a) Na^+-K^+-$2Cl^-$ cotransport (NKCC), (b) K^+-Cl^- cotransport (KCC) and (c) Na^+-independent anion exchange (AE). Each of these mechanisms is considered separately. Obviously, in a cell they coexist and interact with each other, and modeling can become extremely complex. With the possible exception of KCC, these carriers operate away from thermodynamic equilibrium because they are kinetically regulated. Nevertheless, estimation of the thermodynamic equilibrium brackets the upper and lower limits that $[Cl^-]_i$ can attain by virtue of the operation of each of these transport systems. Thermodynamics aspects of AE are dealt with in detail in Chapter 4, and those for KCC in Chapter 17.

A. Passive and Non-passive Cl^- Distribution across the Plasma Membrane

The Cl^- equilibrium potential, E_{Cl}, is defined by the Nernst equation:

$$E_{Cl} = \frac{RT}{zF} \ln \frac{[Cl^-]_i}{[Cl^-]_o} \qquad \text{(Eq. 1)}$$

where z (=−1) is the valence of Cl^-, R is the gas constant, T is absolute temperature and F is the Faraday's constant.

Consider the idealized cell shown in Fig. 5.1, which has only anion channels and Na^+ pumps, but no active (uphill) Cl^- transporters. In this cell, Cl^- distributes passively following E_m, the resting membrane potential. The Cl^- flux through the channel is driven by the electrochemical potential gradient for Cl^-, $\Delta\mu_{Cl}$, which is the sum of a chemical and an electrical component. This is formally defined by Eq. 2:

$$\Delta\mu_{Cl} = RT \ln \frac{[Cl^-]_i}{[Cl^-]_o} + zFE_m \qquad \text{(Eq. 2)}$$

Since the cell is negative inside with respect to the outside (in the model shown $E_m = -60\,mV$, and z, the

Cl^- valence, is −1), the electrical potential gradient will tend to drive Cl^- out of the cell. However, the chemical potential gradient is inward; the external Cl^- concentration $[Cl^-]_o$ is higher than the intracellular concentration $[Cl^-]_i$. Consequently, the chemical gradient will drive Cl^- towards the cell interior. When the two forces cancel each other (they are equal and opposite), the system reaches equilibrium. At that point $\Delta\mu_{Cl} = 0$.

When $\Delta\mu = 0$

$$E_m = \frac{RT}{F}\ln\frac{[Cl^-]_i}{[Cl^-]_o} = E_{Cl} \qquad \text{(Eq. 3)}$$

That is, when Cl^- is passively distributed, E_m will have the same value as E_{Cl}.

B. Calculation of $[Cl^-]_i$ at Equilibrium for an Idealized Cell having only Anion Channels

The predicted intracellular Cl^- concentration at electrochemical equilibrium $[Cl^-]_i^{eq}$ is obtained from Eq. 4:

$$[Cl^-]_i^{eq} = [Cl^-]_o \exp\left(\frac{E_m F}{RT}\right) \qquad \text{(Eq. 4)}$$

Thus, Eq. 4 predicts the value that $[Cl^-]_i$ will reach if Cl^- is *passively* distributed across the plasma membrane. Evidently, this value will change with E_m, as illustrated in Fig. 5.1. For example, if $[Cl^-]_o$ is 113 mM (in mammalian cerebrospinal fluid $[Cl^-]$ ranges between 113 and 120 mM; see Table 29.1 in Chapter 29) and E_m is constant at −60 mV, RT/F at 37°C will be 26.73 mV, and $[Cl^-]_i^{eq} \approx 12$ mM. Clearly, both E_m and E_{Cl} are −60 mV. If the cell is depolarized, say by 20 mV, to −40 mV, Cl^- redistributes across the membrane until reaching equilibrium, which for the example considered will occur when $[Cl^-]_i^{eq} = 25.3$ mM, and thus $E_{Cl} = -40$ mV.

C. Equal Values of E_m and E_{Cl} do not Preclude the Existence of Carrier-mediated Cl^- Transport

If Cl^- is *non-passively distributed* it means that it is *actively transported* and this implies energy consumption. Active Cl^- transport (i.e. transport against Cl^- electrochemical equilibrium or "uphill" transport) requires demonstrating that $[Cl^-]_i$ is maintained at a level different from that predicted by Eq. 4, and therefore E_{Cl} and E_m will have different values. In most cells, and certainly in neurons, $[Cl^-]_i$ is maintained at a value different from that predicted from Eq. 4. However, as mentioned above, equal values of E_m and E_{Cl} do not preclude the presence of carrier-mediated Cl^- transport. A clear example illustrating this notion is skeletal muscle cells which, as mentioned above, have a substantial channel-mediated (electrodiffusional) Cl^- permeability that masks the effects of an inwardly directed active Cl^- transport system (Harris and Betz, 1987; Aickin, 1990), as illustrated in Fig. 5.2. It is predictable that neurons in which GABA produces no change in E_m, but produces shunting inhibition, do have Cl^- transport systems that serve functions different than setting E_{Cl} at a different value than E_m.

D. Thermodynamic Equilibrium Conditions and General Functional Properties of Na^+-K^+-Cl^- cotransporters

The Na^+-K^+-Cl^- cotransporters, referred to here generically as NKCC, mediate the coupled, electrically neutral movement of Na^+, K^+ and Cl^- ions across the membrane of many animal cells, including neurons. In vertebrates, the Na^+-K^+-Cl^- cotransporters exhibit a transport stoichiometry of $1Na^+:1K^+:2Cl^-$ (see below). They belong to the SLC12 gene family of electroneutral cation-chloride-cotransporters. Two NKCC isoforms, encoded by separate genes have been identified: NKCC1 (SLC12A2) and (NKCC2) (SLC12A1), as discussed in Chapters 11 and 16. Under normal physiological conditions, NKCC works as an active ("uphill") transport system for accumulation of Cl^- into cells. NKCC maintains $[Cl^-]_i$ at levels *above the predicted electrochemical equilibrium but below the value that would attain if the cotransporter reached thermodynamic equilibrium.* Thus, kinetic constraints inactivate the cotransporter even in the presence of a substantial thermodynamic driving force, as will be discussed below. Although NKCC is a secondary active transporter and does not consume ATP (Geck et al., 1980), it is inhibited in cells whose ATP content has been reduced (Adragna et al., 1985; Altamirano et al., 1988; Flatman, 1991). This need for ATP reflects a long known role for phosphorylation in regulating transport (Lytle and Forbush, 1992), as discussed in detail in Chapter 18.

Besides its functions in cell water volume control, in neurons NKCC makes possible the depolarizing action of transmitters that open Cl^- channels, such as GABA and glycine, a function that is important in nervous system development (Chapters 7 and 19), sensory transduction (Chapters 20 and 22) and some types of epilepsy (Chapter 24). In some cell types there is substantial evidence showing that NKCC functions to offset osmotically induced cell shrinkage by mediating the net influx of ions and water (Hoffmann et al., 2009; Koivusalo et al., 2009). Whether it serves to maintain cell volume under euvolemic conditions is less clear. A recent study showed that in rat primary sensory neurons isosmotic removal of external Cl^- leads to cell shrinkage following Cl^- depletion. Recovery from shrinkage is mediated by a Na^+-dependent mechanism that is entirely blocked by 10 μM bumetanide suggesting that NKCC is mediating cell water volume recovery in isosmotic media and that it is responsible for the maintenance of cell volume constant in isosmotic media (Rocha-González et al., 2008). It will be interesting to know if other cell types behave like DRG neurons.

NKCC may also play important roles in the cell cycle; overexpression of the NKCC1 cotransporter gene has been shown to induce cell proliferation and phenotypic transformation in mouse fibroblasts (Panet et al., 2000), as discussed in Chapter 27. NKCC has also been suggested to function as a K^+ uptake mechanism thereby keeping $[K^+]_i$ at constant levels or as an extracellular K^+ buffer, keeping the $[K^+]_o$ constant and so preventing its accumulation in the extracellular space (Alvarez-Leefmans et al., 2001; Haas and McManus, 1985; Payne et al., 1995; Walz, 1992; Wong et al., 1999; Wu et al., 1998), as discussed in Chapter 22 in this volume.

1. *Basic Features of the NKCCs*

Four functional features define the NKCCs (Alvarez-Leefmans, 2001; Russell, 2000). (1) Ion translocation by the NKCC requires the simultaneous presence of all three ions (Na^+, K^+ and Cl^-) on the same side of the membrane as has been shown in squid axons and in vertebrate neurons (Achilles et al., 2007; Alvarez-Leefmans et al., 1988; Rocha-González et al., 2008) and many other cell types (Lytle, 2003; Russell, 2000). (2) The cotransport process is electroneutral, and in vertebrate cells it has a stoichiometry of $1Na^+:1K^+:2Cl^-$ in nearly all cases (see below and Fig. 5.4). (3) The transporter can work in forward and backward modes, i.e. net transport may occur into or out of the cells, the magnitude and direction of this transport being determined by the sum of the chemical potential gradients of the transported ions (Lytle et al., 1998). (4) Transport is inhibited by the so-called "loop diuretics", all of which are derivatives of 5-sulfamoyl benzoic acid. They include furosemide, bumetanide and benzmetanide. The sensitivity to loop diuretics, however, varies widely (Russell, 2000). It depends on NKCC isoform (Gamba, 2005), whether it is assessed under basal conditions or on activation of the cotransporter (Hannaert et al., 2002), or on the degree of glycosylation (Paredes et al., 2006).

2. *Electroneutral Cotransport Process and Stoichiometry of NKCC*

Geck and coworkers (1980), working on Ehrlich ascites tumor cells, provided the first comprehensive evidence for the existence of the electrically neutral cotransport of Na^+, K^+ and Cl^- ions, with a stoichiometry of 1Na:1K:2Cl. A solidly documented exception is the squid axon cotransporter which has a stoichiometry of 2Na:1K:3Cl, probably reflecting its distinct molecular structure (Russell, 1983, 2000). The 1Na:1K:2Cl stoichiometry has been found in all vertebrate cells in which it has been studied (Flatman, 2004; Gamba, 2005; Hoffmann, 2001; Lytle, 2003; O'Grady et al., 1987; Russell, 2000).

There is general agreement that the NKCC is electrically silent, which means that the overall transport process of Na^+, K^+ and Cl^- via the cotransporter is not driven by the transmembrane voltage, nor does the transport process directly generate a membrane current that may affect the transmembrane voltage. As correctly pointed out by Russell (2000), the electrical silence of the NKCC is now well accepted by workers in the field, yet direct evidence for the assertion is surprisingly sparse. Electrical silence is often inferred from the apparent stoichiometry of the cotransport process, but arriving at this conclusion involves a somewhat circular argument.

Conclusive proof of electrical silence requires direct and simultaneous measurement of transmembrane potential and cotransporter mediated ion fluxes. There are only two reports in the literature fulfilling these requirements, one for the squid axon NKCC cotransporter (Russell, 1984) and one for frog dorsal root ganglion cells (Alvarez-Leefmans et al., 1988). The latter is, so far, the only existing direct evidence for the electroneutrality of Cl^- transport through NKCC in a vertebrate neuron (Fig. 5.4, thick arrow). Since the cell membrane of DRG neurons has a relatively low resting Cl^- permeability (Alvarez-Leefmans, 1990; Staley et al., 1996), the movements of Cl^- occurred largely through a non-electrodiffusional pathway, i.e. an electroneutral carrier mechanism, which required external Na^+ and K^+ for uphill accumulation of Cl^-, and was bumetanide sensitive, i.e. an NKCC. The presence of NKCC protein in frog DRG cells was corroborated later on using immunolabeling and Western blot analyses with monoclonal antibodies (Alvarez-Leefmans et al., 2001, 1998).

The magnitude and direction of the cotransport process is determined by the sum of the *chemical* potential gradients of the transported ions. NKCC is electrically neutral and thus net cotransport neither affects nor is affected by the transmembrane potential. Consequently, the magnitude and direction of the driving force for net ion movement must be calculated from the chemical rather than the electrochemical potential gradients of all the transported ions.

The NKCC is a secondary active transport process and hence it is reversible, that is, it mediates ion fluxes into or out of the cell, the net direction of cotransport depending on the overall net free energy (ΔG) of the system. If ΔG is negative, it means that the direction of the cotransport will be inward, i.e. favoring uptake. If ΔG is zero, the system is at equilibrium. Since NKCC is electroneutral and has a 1Na:1K:2Cl stoichiometry, ΔG is given by the following equation:

$$\Delta G = \Delta\mu_{Na,K,Cl} = \Delta\mu_{Na} + \Delta\mu_{K} + 2\Delta\mu_{Cl} \quad \text{(Eq. 5)}$$

In other words, the driving force for the electroneutral $1Na^+,1K^+,2Cl^-$ cotransport ($\Delta\mu_{Na,K,Cl}$) is the sum of the *chemical* potential differences of Na^+ ($\Delta\mu_{Na}$) and K^+ ($\Delta\mu_K$) plus twice the Cl^- potential difference ($\Delta\mu_{Cl}$), since two Cl^- ions are translocated per cycle. From the definition of chemical potential for each of the ions involved, it follows:

$$\Delta G = \Delta\mu_{Na,K,Cl} = RT\ln\frac{[Na^+]_i}{[Na^+]_o} + RT\ln\frac{[K^+]_i}{[K^+]_o} + 2RT\ln\frac{[Cl^-]_i}{[Cl^-]_o} \quad \text{(Eq. 6)}$$

Hence

$$\Delta G = RT\ln\frac{[Na^+]_i\,[K^+]_i\,[Cl^-]_i^2}{[Na^+]_o[K^+]_o[Cl^-]_o^2} \quad \text{(Eq. 7)}$$

In most vertebrate cells, under physiological conditions and for a stoichiometry $1Na^+:1K^+:2Cl^-$, ΔG is always *negative*, therefore favoring *net ion uptake*. Russell has computed ΔG for a variety of vertebrate cells; it ranges from $-0.71\,KJ\,mol^{-1}$ in bovine aortic endothelial cells up to $-8.86\,KJ\,mol^{-1}$ in rabbit ventricular myocytes (Russell, 2000). ΔG has been estimated to be $-1.29\,KJ\,mol^{-1}$ and $-2.8\,KJ\,mol^{-1}$ in rat and frog DRG neurons, respectively (Rocha-González et al., 2008; Russell, 2000). As pointed out by Lytle (2003), the rare exception may be red blood cells, which have an usually high $[Cl^-]_i$ (75–100 mM). This reflects their high anion permeability, mainly mediated by the anion exchanger protein AE1, which has translocation rates and characteristics similar to channels (Ellory et al., 2009). In human, duck and ferret red blood cells under physiological conditions $\Delta\mu_{Na,K,Cl}$ is nearly 0, leaving little if any driving force to run the cotransporter (Lytle, 2003).

The NKCC cotransport process reaches thermodynamic equilibrium when $\Delta G = \Delta\mu_{Na,K,Cl} = 0$, that is, when the product of the concentrations of Na^+ and K^+ times the square of the Cl^- concentration on one side of the membrane is matched by the product of these ions in the *trans* side:

$$[Na^+]_i[K^+]_i[Cl^-]_i^2 = [Na^+]_o[K^+]_o[Cl^-]_o^2 \quad \text{(Eq. 8)}$$

Since the cotransporter is electroneutral, thermodynamic equilibrium will not be affected by E_m. Assuming for the sake of argument that Na^+-K^+-Cl^- cotransport is the only Cl^- transport system operating at steady state in a cell, that it has a $1Na^+:1K^+:2Cl^-$ stoichiometry, and that there are no kinetic constraints, the predicted $[Cl^-]_i$ when the cotransport reaches thermodynamic equilibrium is given by the following equation derived from Eq. 8:

$$[Cl^-]_i = \sqrt{\frac{[Na^+]_o[K^+]_o[Cl^-]_o^2}{[Na^+]_i[K^+]_i}} \quad \text{(Eq. 9)}$$

For example, consider the model cell shown in Fig. 5.6, in which NKCC is the sole mechanism actively transporting Cl^- and the Na^+/K^+ pump maintains $[Na^+]_i$ and $[K^+]_i$ constant. The cell has a resting E_m of -60 mV and the relevant ion concentrations (in mM) are: $[Na^+]_o = 132$; $[K^+]_o = 3$; $[Cl^-]_o = 113$; $[Na^+]_i = 10$

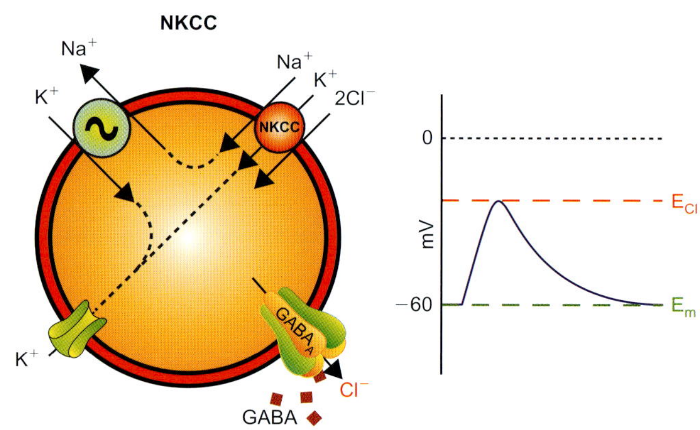

FIGURE 5.6 Model of a cell expressing Na^+, K^+, $2Cl^-$ cotransport (NKCC) and a $GABA_A$-gated anion channel. The energy stored mainly in the inward Na^+ chemical gradient maintained by the Na^+/K^+ ATPase is used by NKCC for the active Cl^- uptake process. The Na^+/K^+ ATPase expels the Na^+ entering through NKCC. The K^+, entering through NKCC diffuses out of the cell via K^+ channels. Cl^- is accumulated above electrochemical equilibrium, shifting E_{Cl} to values more positive than E_m. Opening of GABA-gated anion channels produce a membrane depolarization (blue trace) due to Cl^- efflux.

and $[K^+]_i = 120$. Under these conditions, when the cotransporter reaches thermodynamic equilibrium, without any kinetic constraints (e.g. phosphorylation state), the $[Cl^-]_i$ will be ~65 mM, and this would be the *maximal possible* $[Cl^-]_i$ under these conditions. Thus, we can conclude from the examples considered in Figs 5.1 and 5.6 that if Cl^- was in electrochemical equilibrium, $[Cl^-]_i$ calculated from Eq. 4 would be 12 mM, but if we add NKCC to the same cell (Fig. 5.6), assuming that the resting channel-mediated permeability to anion channels is negligible, the $[Cl^-]_i$ can increase up to about five times above its equilibrium level, and E_{Cl} would now be ~−12 mV. Since E_m is kept constant, at −60 mV, opening of ligand-gated Cl^- channels (GABA in the example shown in Fig. 5.6) will produce an efflux of Cl^-. Since Cl^- is negatively charged, its electrodiffusional efflux produces an inward (depolarizing) current. If an imaginary experimenter estimated in our model cell a $[Cl^-]_i$ *higher* than 65 mM under the above conditions, an estimate that could be obtained from E_{GABA} determinations (assuming that the anion channels opened by GABA are perfectly selective to Cl^-) or even by direct measurement with a fluorescent probe or an intracellular ion-sensitive microelectrode, it is very likely that his/her measurements would be in error.

In reality, $[Cl^-]_i$ in the model of Fig. 5.6 is expected to be below NKCC thermodynamic equilibrium (as defined by Eq. 9), because the cotransporter is *kinetically* regulated, as discussed in Chapters 2, 18 and 22. Under ionic conditions close to those in the model discussed, it has been demonstrated by direct measurement of $[Cl^-]_i$ with a fluorescent probe in rat dorsal root ganglion cells that if NKCC attained thermodynamic equilibrium, $[Cl^-]_i$ would be ~83 mM but the actual value measured was ~48 mM (Rocha-González et al., 2008).

3. *Extracellular K^+ Concentration and NKCC Operation*

One functional feature of NKCC is its sensitivity to $[K^+]_o$. Slight increases in $[K^+]_o$, say between 3 and 5.5 mM result in a significant increase in $(\Delta\mu_{Na, K, Cl})$, and consequently in the $[Cl^-]_i$ that can be potentially attained when NKCC reaches equilibrium, as defined by Eqs 8 and 9. Going back to the cell model shown in Fig. 5.6, the $[Cl^-]_i$ when NKCC is in equilibrium is ~65 mM when $[K^+]_o = 3$ mM. By increasing $[K^+]_o$ just by 1 mM, to 4 mM, the $[Cl^-]_i$ predicted from Eq. 9 will be 75 mM, and for a $[K^+]_o$ of 5.5 mM, $[Cl^-]_i$ will be ~88 mM! This feature, in conjunction with the K_d for K^+ of NKCCs, make these transport proteins ideally suited for buffering extracellular K^+ (Alvarez-Leefmans et al., 2001; Lytle, 2003; Walz, 1992; Wu et al., 1998).

4. *Impact of Intracellular Na^+ Concentration on NKCC Operation*

Another important physiological feature of NKCCs is the effect that changes in $[Na^+]_i$ have on the thermodynamic equilibrium set point. A decrease in $[Na^+]_i$ could occur in neurons upon activation of the electrogenic Na^+/K^+ pump (Brumback and Staley, 2008). In our cell model (Fig. 5.6), a decrease in $[Na^+]_i$ from 10 to 5 mM leads to a predicted increase in $\Delta\mu_{Na, K, Cl}$, and consequently in the $[Cl^-]_i$ from 65 mM to 91 mM, if there were no kinetic constraints.

E. Thermodynamic Equilibrium Conditions of Na^+-independent Anion Exchangers

Another variable that affects $[Cl^-]_i$, and hence the driving force for NKCC, is the intracellular pH (pH_i). This is partly due to the presence of Cl^--coupled pH regulating transport proteins, i.e. the anion exchangers. We will not dwell on the thermodynamic and molecular aspects of these proteins since they are covered in Chapter 4 of this volume. Further, we only consider here, as an example, the Na^+-independent Cl^-/HCO_3^- exchangers (AEs) belonging to the SLC4A gene family. These transport proteins were proposed a long time ago to be a possible additional mechanism for uphill accumulation of intracellular Cl^- in neurons (Alvarez-Leefmans, 1990), and there is evidence suggesting that they may play that role in the maintenance of $[Cl^-]_i$ above equilibrium levels in developing rat auditory brainstem neurons (Becker et al., 2003), and in embryonic chick motoneurons (Gonzalez-Islas et al., 2009).

Figure 5.7 illustrates a model of a cell endowed only with a Cl^-/HCO_3^- exchanger, a system that is electroneutral, with a stoichiometry of $1{:}Cl/1{:}HCO_3^-$. Therefore, the driving force for this anion transport pathway is the sum of the *chemical* potential gradients of the anions involved. Because the anions are transported in opposite directions, the combined chemical gradient ($\Delta\mu_{Cl/HCO_3}$) is given by the *difference* of the Cl^- and HCO_3^- chemical potential gradients $\Delta\mu_{Cl}$ and $\Delta\mu_{HCO_3}$, respectively:

$$\Delta\mu_{Cl/HCO_3} = \Delta\mu_{Cl} - \Delta\mu_{HCO_3} \qquad \text{(Eq. 10)}$$

Hence:

$$\Delta G = \Delta\mu_{Cl/HCO_3} = RT\ln\frac{[Cl^-]_i}{[Cl^-]_o} - RT\ln\frac{[HCO_3^-]_i}{[HCO_3^-]_o} \qquad \text{(Eq. 11)}$$

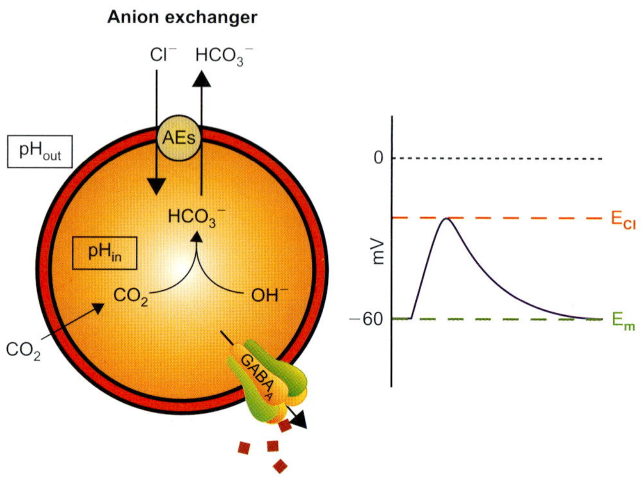

FIGURE 5.7 Model of a neuron expressing only Na^+-independent anion exchangers (AEs) and a Na^+/K^+ ATPase (not shown). The pH across the plasma membrane (pH_{in}–pH_{out}) is constant. Resting $E_m = -60$ mV. The Na^+-independent AEs mediate active accumulation of Cl^- in exchange for HCO_3^-. Intracellular $[Cl^-]$ is maintained at higher than passive levels. Consequently, E_{Cl} is more positive than E_m. GABA-gated anion channels have a depolarizing action mediated by electrodiffusional Cl^- efflux. (See text for details.)

The transport process reaches thermodynamic equilibrium when $\Delta G = \Delta\mu_{Cl/HCO_3} = 0$, and Eq. 11 becomes:

$$\frac{[Cl^-]_i}{[Cl^-]_o} = \frac{[HCO_3^-]_i}{[HCO_3^-]_o} \qquad \text{(Eq. 12)}$$

Eq. 12 relates the intracellular and extracellular Cl^- and HCO_3^- concentrations when the Cl^-/HCO_3^- exchange process has reached equilibrium, that is, when no further net movement of Cl^- and HCO_3^- can occur across the membrane. For the model shown in Fig. 5.7, $pH_o = 7.3$ and $pH_i = 7.1$ (i.e. $\Delta pH = 0.2$), and $[HCO_3^-]_o$ is taken as 22 mM, the concentration measured in mammalian cerebrospinal fluid at 37°C (see Table 29.1, Chapter 29 in this volume). In the presence of a CO_2/HCO_3^--buffered solution, the distribution of Cl^- will be determined by the distribution of HCO_3^- across the membrane. This is because CO_2 is freely permeable through the membrane, and at constant PCO_2 ($[CO_2]_i = [CO_2]_o$) the ratio of intracellular and extracellular $[HCO_3^-]$ is determined by ΔpH. Under these conditions E_H, the equilibrium potential for H^+, is equal to E_{HCO_3}, the equilibrium potential for bicarbonate (Bevensee and Boron, 1998). In the model considered here, the cell possesses only Na^+-independent anion exchangers (AEs in Fig. 5.7). Thus the $[Cl^-]_i$ when the anion exchangers reach equilibrium can be predicted by knowing $[Cl^-]_o$ and the ΔpH. Since we know ΔpH we can calculate E_H, which is given by the Nernst equation,

$$E_H = \frac{RT}{F} \ln (pH_i - pH_o) \qquad \text{(Eq. 13)}$$

At 37°C, and taking $pH_o = 7.3$ and $pH_i = 7.1$, $E_H = -12.3$ mV. Since $E_H = E_{HCO_3}$, we can calculate the $[HCO_3^-]_i$ at equilibrium using the following expression:

$$[HCO_3^-]_i = [HCO_3^-]_o \exp\left(\frac{E_{HCO_3^-}F}{RT}\right) \qquad \text{(Eq. 14)}$$

Hence,

$$[HCO_3^-]_i = 22\,\text{mM} \exp\left(\frac{-12.3\,\text{mV}F}{RT}\right) = 13.9\,\text{mM} \qquad \text{(Eq. 14.1)}$$

We can now calculate $[Cl^-]_i$ when the Cl^-/HCO_3^- reaches equilibrium, for a $[Cl^-]_o = 113$ mM from the following expression derived from Eq. 12:

$$[Cl^-]_i = \frac{[HCO_3^-]_i\,[Cl^-]_o}{[HCO_3^-]_o} = \frac{13.9 \times 113}{22} = 71.4\,\text{mM} \qquad \text{(Eq. 15)}$$

Thus, the maximal possible $[Cl^-]_i$ due to the AE under the conditions considered here would be 71.4, that is, ~6 mM more than the $[Cl^-]_i$ estimated when NKCC reaches equilibrium using Eq. 9. If we now consider a cell expressing both NKCC and AE, $[Cl^-]_i$ could be higher than predicted by the sole operation of NKCC, not only because NKCC reaches equilibrium at a slightly lower $[Cl^-]_i$ than the AE in the example considered, but, more importantly, because NKCC and AEs may be kinetically inhibited at different set points for $[Cl^-]_i$ (Alper et al., 2002; Kurschat et al., 2006; Stewart et al., 2002, 2004). Obviously, in a cell endowed with Na^+-independent AEs, activation of $GABA_A$ receptors would produce a depolarizing Cl^- efflux (Fig. 5.7).

F. Impact of Cl^- Channels on $[Cl^-]_i$

The idealized conditions considered above for NKCC and AE do not take into consideration resting Cl^- conductances. In DRG neurons the resting Cl^- conductance due to voltage-sensitive Cl^- channels (e.g. ClC-2) is negligibly small (Alvarez-Leefmans, 1990); in fact these neurons do not express ClC-2 (Staley et al., 1996). In central neurons, ClC-2 has been proposed to participate in intracellular Cl^- regulation (Staley et al., 1996). This is because ClC-2 gating is sensitive to voltage and $[Cl^-]_i$ (Haug et al., 2003; Niemeyer et al., 2004; Pusch et al., 1999); a rise in $[Cl^-]$i increases ClC-2 open probability towards positive voltages (see

Chapter 12 in this volume). Further, in central neurons, or in the intraspinal terminals of DRG neurons the resting Cl^- conductance may be continuously active due to tonic release of GABA. Extrasynaptic receptors mediate "tonic" $GABA_A$ receptor activated Cl^- conductance. Delta subunit-containing extrasynaptic $GABA_A$ channels are highly sensitive to GABA (submicromolar) and do not desensitize and thus may contribute a substantial fraction of the resting input conductance (Glykys and Mody, 2007; Kullmann et al., 2005). In neurons having $[Cl^-]_i$ above equilibrium, a tonic Cl^- conductance could actually be stimulating NKCC; this cotransporter is activated by decreasing $[Cl^-]_i$ (Rocha-González et al., 2008; Russell, 2000).

G. Thermodynamic Equilibrium Conditions of K^+-Cl^- Cotransporters

Let us now turn our attention to the KCCs. Thermodynamics aspects of Cl^- transport through this type of transporter are discussed in detail in Chapters 3 and 17. The K^+-Cl^- cotransporters (KCCs) are integral membrane proteins that mediate the obligatorily coupled, electroneutral movement of K^+ and Cl^- across the plasma membrane of many animal cells. As an electroneutral secondary active transport mechanism, the direction of the net movement of K^+ and Cl^- by the cotransporter is determined solely by the sum of the chemical potential gradients of the two ions. Under normal physiological conditions, the outwardly directed K^+ chemical potential gradient maintained by the Na^+/K^+ pump drives Cl^- uphill against its chemical potential gradient. Hence, under normal conditions the KCC is an efflux pathway for K^+ and Cl^-. However, the cotransporter is bi-directional and can mediate net ion efflux or influx, depending upon the prevailing K^+ and Cl^- chemical potential gradients. In neurons, this is crucial inasmuch as KCC works near its thermodynamic equilibrium, and therefore it has the potential to mediate Cl^- fluxes in both directions (see Fig. 17.4 in Chapter 17 in this volume). The stoichiometry of KCC has been determined in the isoform expressed in rabbit red blood cells (Jennings and Adame, 2001), and its $1K^+$:$1Cl^-$. This means that the overall transport process of K^+ and Cl^- via the cotransporter is not driven by the transmembrane voltage, nor does the transport process directly generate a membrane current that may change the transmembrane voltage (Brugnara et al., 1989; Kaji, 1993).

Of the four isoforms known, KCC2 and its splice variants are particularly relevant in neurons inasmuch as there is substantial evidence supporting the notion that KCC2 is the isosform that actively extrudes Cl^- from nerve cells in isosmotic conditions, rendering E_{Cl} more negative than E_m and thus making possible

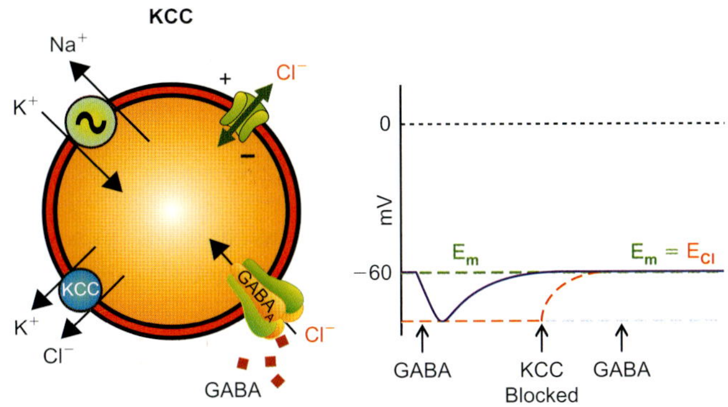

FIGURE 5.8 Model of a neuron expressing only K^+-Cl^- cotransporter (KCC), Na^+/K^+ ATPase, voltage-gated anion channels and GABA-gated anion channels. KCC actively extrudes Cl^- from the cell shifting E_{Cl} (red dashed line on the right) to a more negative value than E_m. Right: A puff of GABA produces a hyperpolarizing IPSP (blue trace). When KCC is blocked, $[Cl^-]_i$ raises until reaching passive equilibrium ($E_m = E_{Cl}$). A second GABA application under these conditions does not produce a change in E_m but an increase in membrane conductance which is the basis of "shunting inhibition".

hyperpolarizing inhibition (Chapter 17), i.e. KCC2 is the "outward Cl^- pump" of earlier literature (see above Section I.F).

Assuming that the K^+-Cl^- cotransport is the sole system transporting Cl^-, and that $[K^+]_i$ is held constant by the Na^+/K^+ pump (Fig. 5.8), the driving force for the cotransport $\Delta\mu_{K,Cl}$ is the sum of the chemical potential differences of K^+ and Cl^-, as expressed in the following equation:

$$\Delta\mu_{K,Cl} = \Delta\mu_k + \Delta\mu_{Cl} \quad \text{(Eq. 16)}$$

It follows that

$$\Delta\mu_{K,Cl} = RT\ln\frac{[K^+]_i}{[K^+]_o} + RT\ln\frac{[Cl^-]_i}{[Cl^-]_o} \quad \text{(Eq. 17)}$$

This system attains thermodynamic equilibrium when

$$\Delta\mu_{K,Cl} = 0 \quad \text{(Eq. 18)}$$

and therefore for a 1:1 stoichiometry

$$\frac{[K^+]_o}{[K^+]_i} = \frac{[Cl^-]_i}{[Cl^-]_o} \quad \text{(Eq. 19)}$$

and hence the $[Cl^-]_i$ at equilibrium will be

$$[Cl^-]_i = \frac{[Cl^-]_o[K^+]_o}{[K^+]_i} \quad \text{(Eq. 20)}$$

The $[K^+]_o$ at which the flux reverses from outward to inward or vice versa (flux reversal point or FRP) can be determined using the following equation (Lauf and Adragna, 1996):

$$\text{FRP} = [K^+]_o = \frac{[K^+]_i\,[Cl^+]_i}{[Cl^+]_o} \qquad \text{(Eq. 21)}$$

From Eq. 20 we can estimate the $[Cl^-]_i$ when KCC reaches equilibrium. In our model (Fig. 5.8), $[Cl^-]_o = 113\,mM$, $[K^+]_o = 3\,mM$ and $[K^+]_i = 120\,mM$, and hence, from Eq. 20, $[Cl^-]_i$ will be ~3 mM and E_{Cl} would be ~− 98, that is ~38 mV more negative than the resting E_m (−60 mV). Thus, the driving force for the IPSP ($E_{Cl} - E_m$) in the example considered would be more than 30 mV. Further, opening of GABA-gated anion channels would produce a membrane hyperpolarization, i.e. an IPSP due to Cl^- influx (outward current). Of course the actual $[Cl^-]_i$ predicted by Eq. 20 would be slightly higher in real neurons, because $[K^+]_i$ may be lower than 120 mM, the "textbook" value used in our calculations. Another point that is important to consider is that $[Cl^-]_i$ in a mammalian neuron coexpressing KCC and a resting Cl^- conductance will be determined by the relative contribution of each of these mechanisms. If the resting Cl^- conductance is large enough, it could overwhelm the active extrusion mechanism (i.e. KCC) and $[Cl^-]_i$ will appear in equilibrium, although never above electrochemical equilibrium, unless there is an inward active transport put into operation.

The above discussion leads us to the next consideration. What happens when KCC is blocked, or its functional expression is negligible (e.g. immature neurons)? If we block KCC in the model of Fig. 5.8, Cl^- will redistribute across the plasma membrane following the resting potential and $E_m = E_{Cl}$. Thus, activation of GABA-gated anion channels under these conditions will produce a decrease in membrane conductance without changing E_m (shunting inhibition), as illustrated in Fig. 5.8. There can be no Cl^-- mediated depolarization unless there is an active Cl^- uptake mechanism coexpressed with KCC. Inhibition of KCC *per se* cannot result in a $GABA_A$- or glycine-evoked depolarization, as often mentioned in the literature. Which are the possible explanations for the observation that block or decreased expression of KCC turns GABA depolarizing? (1) Upon KCC inhibition $[Cl^-]_i$ increases until $E_{Cl} = E_m$ (i.e. Cl^- becomes passively distributed), but the GABA-gated anion channels are permeable to HCO_3^-. Since E_{HCO3} is more positive than E_m, GABA could generate a small depolarization due to HCO_3^- efflux via GABA-gated anion channels (Kaila et al., 1993; Staley and Proctor, 1999; Staley et al., 1995). This of course will depend on the relative permeability of the GABA-gated channels to HCO_3^-, a parameter which appears to be extremely variable between neurons and most of the time is unknown. An equivalent situation could result from passive accumulation of intracellular Cl^- upon repeated exposure to GABA in adult neurons (Cordero-Erausquin et al., 2005). In this case, passive Cl^- influx via GABA-gated channels would exceed KCC transport capacity. (2) Reversing the flux through KCC (e.g. if there is an abnormal increase in extracellular $[K^+]$). (3) More likely, because KCC and NKCC (or another active Cl^- uptake mechanism) coexist in the same cell (Fig. 5.9). In this situation, ablation or decreased expression of KCC leaves NKCC unrestrained, generating an increase in $[Cl^-]_i$ sufficient to shift E_{Cl} to depolarized values with respect to E_m (Khirug et al., 2008; Munoz et al., 2007).

As mentioned above, under physiological conditions KCCs function as active Cl^- extruders but work near thermodynamic equilibrium. Only under non-physiological or pathological high levels of extracellular K^+ can KCCs reverse their normal operation and accumulate Cl^- (DeFazio et al., 2000; Kakazu et al., 2000). Under physiological conditions extracellular K^+ is tightly buffered in the brain (Somjen, 2004). Therefore, KCCs cannot

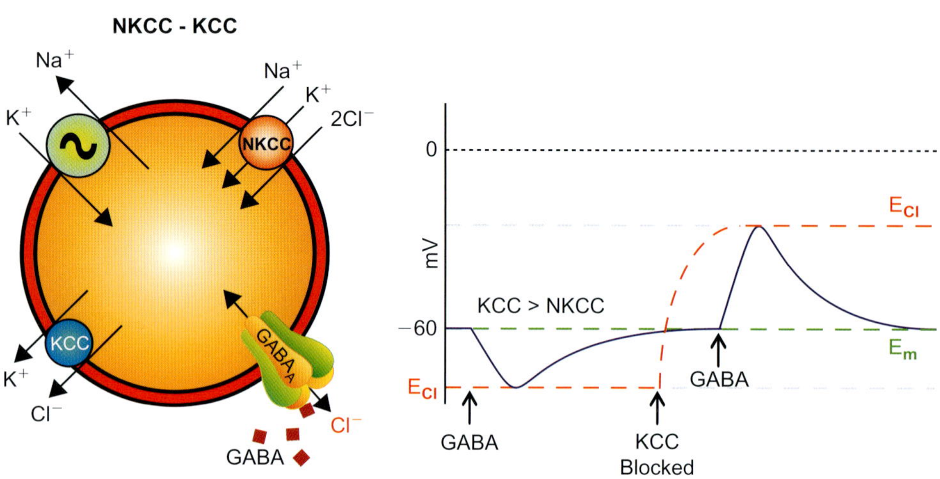

FIGURE 5.9 Model of a neuron coexpressing NKCC and KCC. In this model KCC dominates over NKCC and $[Cl^-]_i$ is maintained below electrochemical equilibrium. Therefore, E_{Cl} (red dashed line in right panel) is more negative than E_m (green dashed line in right panel). At the time indicated by the first arrow in the right panel, a puff of GABA produces a hyperpolarizing IPSP. Upon blockage of KCC (indicated by the second arrow labeled "KCC Blocked"), NKCC is left unrestrained and $[Cl^-]_i$ increases shifting E_{Cl} to values more positive than E_m. Under these conditions, a second puff of GABA (indicated by the third arrow) will produce a membrane depolarization.

be considered key players in normal active Cl^- accumulation in neurons. However, KCCs may influence the levels of intracellular Cl^- by counteracting a coexisting Cl^- uptake mechanism. In fact, the coexistence of active Cl^- extruders and loaders in a given cell is the situation expected for a tightly regulated ion like Cl^-.

Functional interaction between Cl^- extruders and loaders coexpressed in a cell ultimately determines $[Cl^-]_i$ (Gillen and Forbush, 1999; Lytle and McManus, 2002). Equally important in this context is the fact that cytoplasmic Cl^- is a negative feedback regulator of NKCC1 activity in many cell types, including neurons (Breitwieser et al., 1990, 1996; Gillen and Forbush, 1999; Haas et al., 1995; Lytle and Forbush, 1996; Rocha-González et al., 2008; Russell, 2000). The consequences of functional interaction between KCCs and NKCCs in mammalian cells were tested in an epithelial cell line (HEK293) transfected with rabbit KCC1. Overexpression of KCC1 led to a decrease in $[Cl^-]_i$ that resulted in activation of endogenous NKCC (Gillen and Forbush, 1999). A steep relation between cell Cl^- and NKCC activity was found over the physiological ranges supporting a primary role for $[Cl^-]_i$ in activation of NKCC. Interestingly, activation of $GABA_A$ receptors in immature rat cortical neurons and in olygodendrocytes increases NKCC1 activity (Schomberg et al., 2003; Wang et al., 2003). Decreases in $[Cl^-]_i$ produced by GABA have been measured in neurons that have higher than passive $[Cl^-]_i$ (Ballanyi and Grafe, 1985; Chub et al., 2006; Schomberg et al., 2003). Moreover, in peripheral neurons, the GABA-evoked depolarization activates K^+ channels resulting in K^+ efflux and a decrease in $[K^+]_i$ (Ballanyi and Grafe, 1985; Deschenes and Feltz, 1976). Thus, activation of $GABA_A$ receptors may produce net efflux of K^+ and Cl^- with consequent osmotic water efflux and cell shrinkage (Alvarez-Leefmans et al., 1998). $GABA_A$-induced shrinkage has been demonstrated in populations of embryonic brain stem neurons (Momose-Sato et al., 1998) and in single oligodendrocytes in culture (Wang et al., 2003). However, activation of $GABA_A$ receptors in immature cortical neurons in culture produces a decrease in $[Cl^-]_i$ and stimulation of NKCC1 without measurable shrinkage suggesting that shrinkage in these cells is not a necessary condition for activating NKCC1 (Schomberg et al., 2003).

H. Influence of GABA-induced Bicarbonate Permeability on E_{IPSP}: Why E_{GABA} is not Equal to E_{Cl}

Due to the technical difficulties in directly measuring $[Cl^-]_i$ and E_m with fluorescent probes or ion-sensitive microelectrodes, the practice of estimating $[Cl^-]_i$ using gramicidin perforated patches (Chapter 8) is widespread. The underlying assumption is that E_{Cl} is equal to E_{GABA}, and hence, knowing E_{GABA} it is possible to calculate $[Cl^-]_i$. Although this approach has provided valuable information, the $[Cl^-]_i$ estimated with this method is at best just a rough approximation of the actual $[Cl^-]_i$ in neurons. This is because it is in fact difficult to validate that $E_{Cl} = E_{GABA}$. As already mentioned, it has long been known that the anion channels gated by GABA are permeable primarily to Cl^- but to a variable extent they are also permeated by HCO_3^- (Bormann et al., 1987; Fatima-Shad and Barry, 1993; Inomata et al., 1986; Kaila et al., 1993; Kelly et al., 1969). Thus, the reversal potential of a $GABA_A$-mediated IPSP (E_{IPSP}), or the reversal potential of a GABA-induced current (E_{GABA}) mediated by activation of $GABA_A$ receptor-channels is more accurately described by the following expression based on the Goldman-Hodgkin-Katz equation:

$$E_{IPSP} = E_{GABA} = -\frac{RT}{zF}\ln\frac{[Cl^-]_o + \alpha[HCO_3^-]_o}{[Cl^-]_i + \alpha[HCO_3^-]_i} \quad \text{(Eq. 22)}$$

where α is the ratio of permeabilities P_{HCO_3}/P_{Cl}, and other variables have been defined previously.

Let us turn our attention again to the model illustrated in Fig. 5.8. Assume, for simplicity, that the resting anion permeability of this neuron is negligible, that the cell has only KCC, GABA-gated anion channels and, of course, a Na^+/K^+ pump that keeps the Na^+ and K^+ gradients constant. In this cell $[Cl^-]_i$ can be altered in a predictable manner by changing $[K^+]_o$ (Eq. 20). At the same time, in our imaginary experiment we also measure GABA-induced currents with a gramicidin patch pipette (not shown). Activation of $GABA_A$ receptors at different holding potentials in the model allows measurement of E_{GABA}, which is equal to E_{IPSP}. Further, imagine that by an independent method (e.g. a fluorescent dye), we are also able to directly measure $[Cl^-]_i$ in the same cell. For the purpose of this imaginary experiment, $[HCO_3^-]_o$ is taken as 22 mM, $[HCO_3^-]_i = 13.9$ mM (Eq. 14.1, above) and $\alpha = 0.1$. With the data obtained from the model, it is possible to plot $[Cl^-]_i$, E_{IPSP} (or E_{GABA}) and E_{Cl} as a function of $[K^+]_o$ using the following equations: $[Cl^-]_i$ from Eq. 20, E_{IPSP} from Eq. 22 and E_{Cl} from Eq. 1. The ion concentrations used in the calculations are the same as those we have been using above, and the results are plotted in Fig. 5.10.

The first conclusion that can be reached by examining the graphs shown in Fig. 5.10 is that at physiological $[K^+]_o$, i.e. 3 mM, $[Cl^-]_i = 2.8$, $E_{Cl} = -97$ mV and $E_{IPSP} = -87$ mV. The difference between E_{Cl} and E_{IPSP}

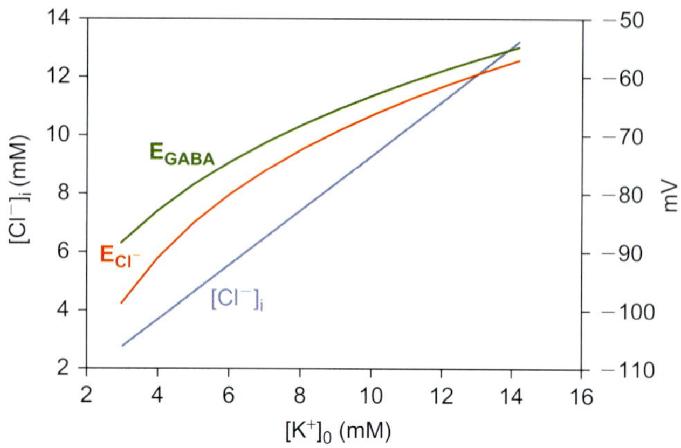

FIGURE 5.10 Effect of changing $[K^+]_o$ on $[Cl^-]_i$, E_{Cl} and E_{GABA} in the cell model shown in Fig. 5.8. $E_{IPSP} = E_{GABA}$. The GABA-induced currents were recorded with a gramicidin patch pipette (not shown). Activation of $GABA_A$ receptors at different holding potentials in the model allows measurement of E_{GABA}, the reversal of the GABA-induced membrane current. When the currents are evoked by activation of GABAergic synapsis, the reversal potential of the IPSP (E_{IPSP}) will have the same value as E_{GABA}. The intracellular Cl- concentration $[Cl^-]_i$, is measured simultaneously with an intracellular probe. The parameters of the model are as follows: $[Cl^-]_o = 113$mM; $[K^+]_i = 120$mM; $pH_o = 7.3$; $pH_i = 7.1$; $[HCO_3^-]_o = 22$mM; $[HCO_3^-]_i = 13.9$mM, α, the permeability ratio P_{HCO_3}/P_{Cl} was 0.1. Temperature 37°C. (See text for further details.)

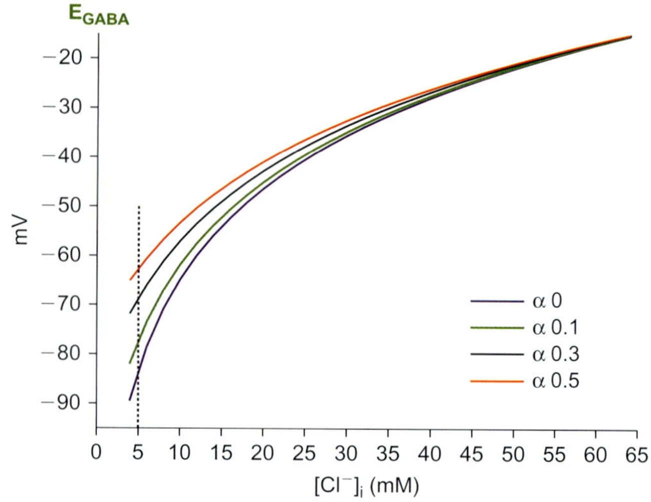

FIGURE 5.11 GABA reversal potential (E_{GABA}) plotted as a function of $[Cl^-]_i$ for different values of α, the permeability ratio P_{HCO_3}/P_{Cl}. Vertical dotted line crosses the curves at a $[Cl^-]_i$ of 5mM, the basal concentration measured in mature CNS neurons by Berglund and collaborators (2006).

(which is the same as E_{GABA} in this case) is significant for the whole range of $[Cl^-]_i$ (from ~2.8 up to ~13mM). Note also that if we assume, as is generally done, that GABA-mediated $E_{IPSP} = E_{Cl}$, there is a systematic error in the estimation of $[Cl^-]_i$ which amounts to ~ 4mM when $[K^+]_o = 3$mM, i.e. under physiological conditions when $\alpha = 0.1$. Doing this experiment using HEPES-buffered solutions equilibrated with air instead of a HCO_3^-/CO_2-containing solution, aiming at decreasing $[HCO_3]_i$ may reduce the difference between E_{IPSP} and E_{Cl} to an extent that is difficult to predict. This is because the CO_2 that is dissolved in HEPES (usually equilibrated with air) plus the CO_2 that is produced inside the cell by metabolism (there are many types of neurons that exhibit high metabolic rates and produce large quantities of CO_2 and HCO_3^- catalyzed by intracellular carbonic anhydrases) can yield intracellular HCO_3^- concentrations that are difficult to predict quantitatively. In estimating E_{IPSP} in Fig. 5.10 we used $\alpha = 0.1$, but as already mentioned, measured α in neurons ranges from 0.18 in mouse cultured spinal neurons (Bormann et al., 1987) to 0.44 in hippocampal pyramidal neurons obtained from P0 rats (Fatima-Shad and Barry, 1993). We do not know the factors that determine the permeability to HCO_3^- in GABA-gated channels in neurons and if this permeability is the same in all neurons and/or if it changes during development. We know that the composition of $GABA_A$ receptor subunits changes during development (Galanopoulou, 2008; Ma et al., 1993) but whether α changes with subunit composition is unknown (Farrant and Kaila, 2007). Bestrophin Cl^- channels can be highly permeable to HCO_3^-, with α values up to 1.1 (Qu and Hartzell, 2008).

Figure 5.11 shows a plot of E_{GABA} calculated using Eq. 22, as a function of $[Cl^-]_i$ within the physiological range (4–65mM) and for various values of α ranging from 0 to 0.5. The vertical dotted line crosses the curves at a $[Cl^-]_i = 5$mM. The corresponding reversal potentials where the dotted line crosses are -83.3mV when $\alpha = 0$, i.e. E_{Cl}, and -62.5mV when $\alpha = 0.5$. If we assume that $E_{IPSP} = E_{GABA} = E_{Cl}$ the estimated $[Cl^-]_i$ ranges from 5 to 10.9mM, respectively. The error in estimating $[Cl^-]_i$ (based on the assumption that $E_{GABA} = E_{Cl}$) decreases with $[Cl^-]_i$ and becomes negligible only at intracellular $[Cl^-]$ of about 50mM. In conclusion, estimates of $[Cl^-]_i$ from gramicidin patches are just an approximation to the actual $[Cl^-]_i$. The error of these estimates depends on α and $[Cl^-]_i$ among other factors. The error can be significant in adult central neurons that have relatively low levels of intracellular Cl^-, but negligible for neurons that have relatively high $[Cl^-]_i$ like immature central neurons or adult primary sensory neurons.

III. KINETICS OF CATION-Cl^- COTRANSPORT

Whereas thermodynamic parameters dictate the overall electrochemical force driving ion movements through transport mechanisms, kinetic parameters are intrinsic properties of these transporters. Obviously, these intrinsic properties also affect ion movements. They include the affinity of the transport mechanism for specific transported species, the affinity of the transport mechanism for ions at modulatory sites, the effect of ions at the opposite site (*trans*-side) of the membrane, the turnover rate of single transport molecules, and the overall rate of transport. All of these parameters can be obtained through detailed transport kinetic studies. Unfortunately, very little of this work has been done in vertebrate neuronal tissues due to the combination of two factors: the relatively recent discovery of chloride transporters in vertebrate neurons as compared to other tissues, and the difficulty in measuring unidirectional ion fluxes in neuronal cells. For these reasons, most of the work described here involves non-neuronal tissues.

A. Definitions and Simple Derivations

Kinetics experiments are typically performed to extract binding affinities and maximal rates of transport. This is possible because transporters are enzyme-like mechanisms, and the rate or velocity of transport can be expressed as a function of substrate concentration. Given that the Michaelis–Menten relationship, and other derivatives and plots, can be applied to membrane transporters, valuable information on the modalities of transport can be gained from transport kinetic studies.

The simplest enzymatic reaction comprises three distinct steps: (1) the binding of the substrate to the enzyme, (2) the transformation of the substrate by the enzyme and (3) the release of the product from the enzyme. It is possible to substitute the term *enzyme* with *transporter*, and repeat the three steps as follows: (1) the binding of the substrate to the transporter, (2) the translocation of the transporter to the other side of the membrane and (3) the release of the substrate. Thus, from this point on, we will leave our enzyme-transporter analogy and focus solely on transport kinetics. There are two critical assumptions made at this junction that help simplifying the derivation of transport velocities. First, the translocation step (step 2) is rate-limiting, meaning that it occurs with much lower speeds than the binding and release steps (steps 1 and 3). Second, the movement of substrate is followed in one direction (unidirectional measurements, see Chapter 9) under *zero-trans* conditions, meaning that no substrate is present on the other side at time zero). While the first condition applies to most transport systems studied, the second assumption is not always met, and this has consequences that will be discussed below. Nevertheless, based on these assumptions, rapid equilibrium kinetics do apply, and simple derivations can be made. Finally, because the translocation step is slow, steps 2 and 3 are usually combined into a single translocation-release step.

This leaves us with the simple reaction: $S + T \longleftrightarrow TS \rightarrow T + S$, where S is the substrate and T is the transporter. Note that there is a forward binding reaction (with a rate-constant: k_{12}), and a reverse de-binding reaction (with a rate-constant: k_{21}). The overall rate constant is K or the ratio of k_{12}/k_{21}. The translocation has a forward rate-constant: k_p. Rapid equilibrium kinetics are based on the existence of at least one rate limiting step which renders the velocity of transport dependent upon the concentration of the intermediate feeding the reaction (in this case TS, or transporters loaded with their substrate and ready to translocate). All other upstream (binding) reaction steps, which are orders of magnitude faster than the rate-limiting step, can therefore be lumped together as they must reach near equilibrium, as the overall reaction reaches a steady state. This is not to say that these rapid partial reactions are not important as they define some key kinetic behavior, e.g. sensitivity of ion affinities to other transported species.

The velocity of the reaction v depends on the concentration of the intermediate $[TS]$ and the rate constant kp. This translates into the following equation:

$$v = \mathrm{kp}[\mathrm{TS}] \qquad \text{(Eq. 23)}$$

Let's now define the reaction at its maximal rate, based on the total number of transporters. TS do not constitute the total number of transporters, since there are also transporters in the reaction that are free of substrate (T). Thus, the total number of transporters is given by the following expression:

$$[\mathrm{Tot}] = [\mathrm{T}] + [\mathrm{TS}] \qquad \text{(Eq. 24)}$$

Dividing each side of Eq. 23 by $[Tot]$, we obtain:

$$\frac{v}{[\mathrm{Tot}]\,\mathrm{kp}} = \frac{[\mathrm{TS}]}{[\mathrm{T}] + [\mathrm{TS}]} \qquad \text{(Eq. 25)}$$

Defining the maximal velocity (V_{max}) as if all transporters were ready to translocate, e.g. $V_{max} = [Tot]\ kp$, Eq. 25 becomes:

$$\frac{v}{[V_{max}]} = \frac{[\mathrm{TS}]}{[\mathrm{T}] + [\mathrm{TS}]} \qquad \text{(Eq. 26)}$$

Finally, as $[TS]$ is defined as the product of the concentrations of free transporters $[T]$ and substrates $[S]$ divided by the rate constant K, Eq. 26 becomes:

$$\frac{v}{[V_{max}]} = \frac{[\mathrm{T}][\mathrm{S}]/\mathrm{K}}{[\mathrm{T}] + [\mathrm{T}][\mathrm{S}]/\mathrm{K}} \qquad \text{(Eq. 27)}$$

$[T]$ can be eliminated and the equation becomes:

$$\frac{v}{[V_{max}]} = \frac{[S]/K}{1 + [S]/K} \quad \text{(Eq. 28)}$$

Dividing both the numerator and the denominator by $[S]/K$ yields the Michaelis–Menten equation:

$$v = \frac{V_{max}}{1 + K/[S]} \quad \text{(Eq. 29)}$$

When using a double-reciprocal plot of $1/v$ versus $1/[S]$ (Fig. 5.12), the intercept on the y-axis provides the V_{max} value ($1/V_{max}$), whereas the intercept on the x-axis provides the value ($-1/K$), from which the substrate affinity (K_m) can be extracted. This simple derivation, which as we will see later is only a first approximation, provides the basis for the measurement of maximal velocities of transport, and in a great number of studies the transport affinity for the substrate.

B. Measurements of Ion Affinities for Cation-Coupled Chloride-Cotransporters

By measuring ion transport flux at different ion concentrations, and by using the Michaelis–Menten equation and related plots, the affinities for Na^+, K^+ and Cl^- were measured for various cation-coupled chloride-cotransporters. The values obtained from several studies are summarized in Table 5.1. Note that the affinities reported for K^+ are higher for NKCCs versus KCCs. However, KCC2 has a significantly higher affinity for K^+ compared, and significantly higher for

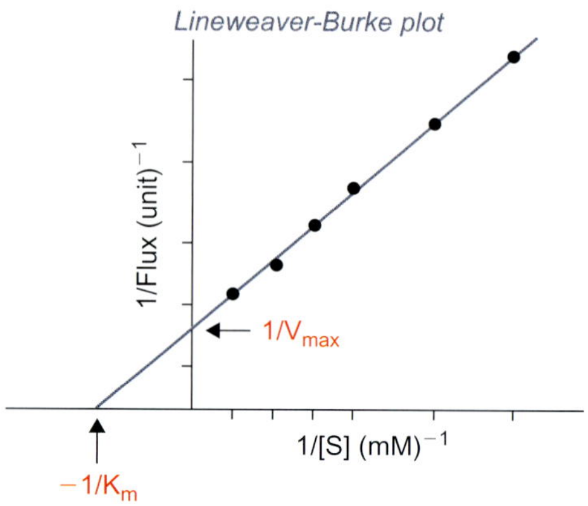

FIGURE 5.12 Lineweaver-Burke (double-reciprocal) plot for the transport of a substrate S. The transport or flux of S through a transporter is measured for varying concentrations of the substrate [S]. The double-reciprocal relationship is linear, the intercept of the y-axis provides the reciprocal of the maximal flux (V_{max}) and the intercept on the x-axis provides the negative reciprocal of the K_m (binding affinity).

TABLE 5.1 Partial list of ion affinities measured for the different cation-Cl− cotransporters

Transporter		Na^+	K^+	Cl^-	References
K^+-Cl^- cotransport	sheep RBC	N/A	54	26	Delpire and Lauf (1991a)
	human RBC	N/A	115–140		Kaji (1989)
KCC1	rat/HEK293	N/A	25		Gillen et al. (1996)
	human/oocytes	NA	25.5	17.2	Mercado et al. (2000)
KCC2	rat/HEK293	N/A	5.2		Payne et al. (1996)
	rat/HEK293	N/A	2.1	26	Delpire et al. (2009)
KCC3	human/oocytes	N/A	10.7	7.3	Mercado et al. (2005)
KCC4	human/oocytes	N/A	17.5	16.1	Mercado et al. (2000)
NKCC1	winter flounder	5–7	4.5	46	O'Grady et al. (1986)
	human RBC		5.5		Chipperfield (1981)
	rat RBC		3.5		Duhm and Göbel (1984)
NKCC2 A, B	rabbit/oocytes	<5	2–3	10–15	Gagnon et al. (2003)
NKCC2	rabbit/oocytes	50–55	8	110	Gagnon et al. (2003)
NKCC2 A, B	shark/oocytes	<5	<2	40	Gagnon et al. (2003)
NKCC2 F	shark/oocytes	75	8	60	Gagnon et al. (2003)
NCC	winter fl/oocytes	25	N/A	13.6	Gamba et al. (1993)
	rat DCT	9	N/A	12	Velazquez et al. (1984)

Additional values for K^+-Cl^- cotransporter under several modes of activation are also available in Lauf et al. (1992).

KCC2 compared to the other K^+-Cl^- cotransporters. The significance of a higher K^+ affinity for KCC2 is discussed in Chapter 17.

It is important to realize at this point that the affinities provided in these studies, although very informative, are *apparent* affinities and not *true* affinities. The true affinity, defined as K in the previous section, may be different from the affinity obtained experimentally. Indeed, there are several factors that affect ion affinity measurements. First, remember that we assume that no substrate is present at the *trans*-side. This is actually not the case in most measurements performed under physiological conditions, especially for influx measurements, where the transported ion is typically present inside the cell. Thus, our rate-limiting reaction is truly not unidirectional as transport also occurs in the reverse direction. More complex kinetics, such as steady-state kinetics, are needed to account for the effect of reverse transport, as discussed below. Second, as cation-chloride cotransporters couple the movement of multiple ions, the binding of one ion to the transporter can modify the binding of another ion, thus making the measurement of true affinities difficult.

C. Rapid Equilibrium Kinetics and K^+-Cl^- Cotransport

From the previous paragraph, it will be obvious to most readers that the kinetics of a two-substrate transporter will be somewhat more complex than that for a one-substrate transporter. First, because two ions are cotransported, so that different modalities of ion binding can be envisioned. For instance, an existing order of ion binding to the transporter could exist, i.e. Cl^- binding first, followed by K^+, or the inverse. Alternatively, either ion could bind to the transporter first, followed by the other ion. Also, the binding of one ion could affect the affinity of the cotransporter for the binding of the second ion, and vice versa, or, alternatively, the binding could be of no consequence. Those possibilities can be clearly expressed as distinct kinetic models, and the obvious question is whether these models can be identified through specific behaviors when plotting experimental data. The answer is yes, each model with its sets of conditions gives rise to equations with distinct properties. As an example, we will describe the kinetic model showing asymmetry of K^+-Cl^- cotransport in sheep red cells (Delpire and Lauf, 1991a, b). As seen in Fig. 5.13A, we determined ordered binding at the

A

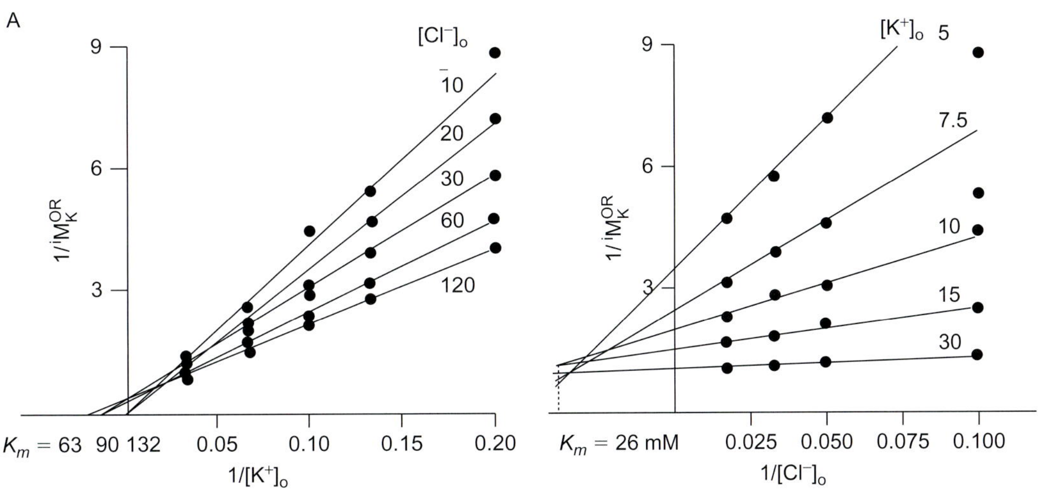

B

$$\frac{1}{v} = \frac{K_1}{V_{max}}\left\{1 + \frac{K_2}{[Cl^-]_o}\right\}\frac{1}{[K^+]_o} + \frac{1}{V_{max}}$$

$$\frac{1}{v} = \frac{K_2}{V_{max}}\,\frac{K_1}{[K^+]_o}\,\frac{1}{[Cl^-]_o} + \frac{1}{V_{max}}\left\{1 + \frac{K_1}{[K^+]_o}\right\}$$

C

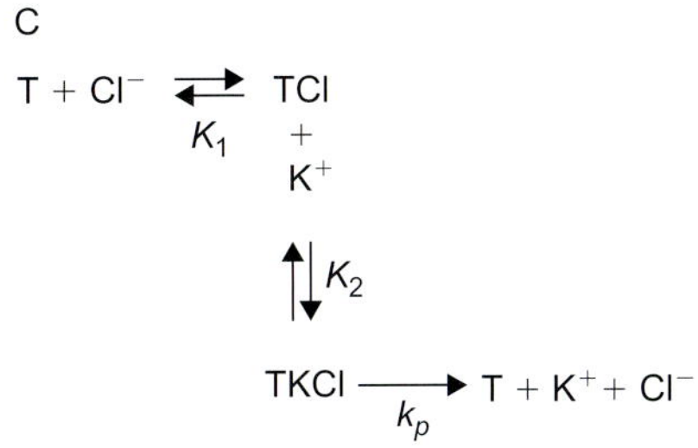

FIGURE 5.13 Ordered ion binding for swollen activated K^+-influx through K^+-Cl^- cotransporter in sheep red blood cells. **A.** Lineweaver Burke plots of ouabain-resistant K^+ influx varying external $[K^+]$ for different external $[Cl^-]$ and inversely, varying external $[Cl^-]$ for different external $[K^+]$. Note the different behaviors of the two families of lines, with a common intercept close to the *y*-axis in the *left* plot, whereas intercept in the second quadrant for the *right* plot. **B.** Equations derived for ordered binding, with Cl^- first followed by K^+. Note that when $1/[K^+]$ is *null* (*top* equation), $1/v$ equals $1/V_{max}$. Thus, the intercept is unique on the *y*-axis. In contrast, when $1/[Cl^-]$ is *null* (*bottom* equation), $1/v$ is dependent on external $[K^+]$. Thus, the intercept on the *y*-axis varies with $[K^+]$. **C.** Kinetic model of ordered binding fitting the data plotted in panel A.

extracellular side by measuring ion fluxes at different ion concentrations, with Cl^- binding first, followed by K^+. This was based on the behavior of double-reciprocal plots which showed a unique intercept of a family of lines on the *y*-axis when $[K^+]$ was varied for different $[Cl^-]$ concentrations, and a unique intercept of a family of lines in the second quadrant when $[Cl^-]$ was varied for different $[K^+]$. As the equations show (Fig. 5.13B), this behavior is typical of *ordered* binding with Cl^- ions binding first, followed by K^+. The model of ordered binding is depicted in Fig. 5.13C. Note that this behavior was not observed in a recent study examining the effect of an inhibitor on NEM-activated KCC2 expressed in HEK293 cells (Delpire et al., 2009). This suggests the possibility of different kinetic modalities for different K^+-Cl^- cotransporters, or different kinetic modalities for different modes of activation/function. This interesting aspect will need further study. When K^+ efflux measurements via swollen-activated sheep-red-cell K^+-Cl^- cotransport were performed while varying intracellular ions, the intercepts were located on the *x*-axis in both plots, indicating *random* binding at the internal aspect of the membrane (Delpire and Lauf, 1991a).

D. Steady-state Kinetics and K^+-Cl^- Cotransport

As mentioned above, rapid equilibrium kinetics assumes that the rate-limiting reaction of transport is unidirectional, i.e. the flux in the reverse direction is null or minimal. This is a rather unique assumption, as ion transport measurements are often performed under physiologic conditions, where reverse transport is not insignificant. Steady-state transport kinetics have been developed to account for a significant transport in the reverse direction. Note, however, that equations derived under both types of kinetics often provide similar results. The steady-state equations are far more complex than rapid equilibrium equations, but they can be simplified under *zero-trans* conditions. Fig. 5.14 shows the steady-state equations derived for the influx of K^+ for the *ordered* binding model (Delpire and Lauf, 1991a). It can be seen that under *zero-trans* conditions, the equation simplifies. The *ordered* binding could be verified via the measurement of *trans* effects. Indeed, when measuring K^+ efflux, one could determine that the efflux was not affected by external K^+ when Cl^- was absent, whereas the efflux was affected by external Cl^- whether or not external K^+ was present. These data could only be explained if K^+ was unable to interact with the transporter in the absence of Cl^-, i.e. Cl^- binds first.

The steady-state equations agreed with the experimental data. The reverse set of experiments verifying *random* binding inside was published in a subsequent study, as special manipulations needed to be worked out to obtain zero-trans conditions for Cl^- on the inside of the red cell. This was achieved by using a DIDS pre-treatment to permanently shut off the robust activity of the Cl^-/HCO_3^- exchanger, in order to vary the inside Cl^- concentration. The experiment was possible as DIDS effect on K^+-Cl^- cotransporter is reversible (Delpire and Lauf, 1992). The data showed *trans* inhibition of K^+ influx by both internal ions in the absence of their partner, confirming, through an independent kinetic method, the *random* binding of ions on the inside of the cells (Delpire and Lauf, 1991b).

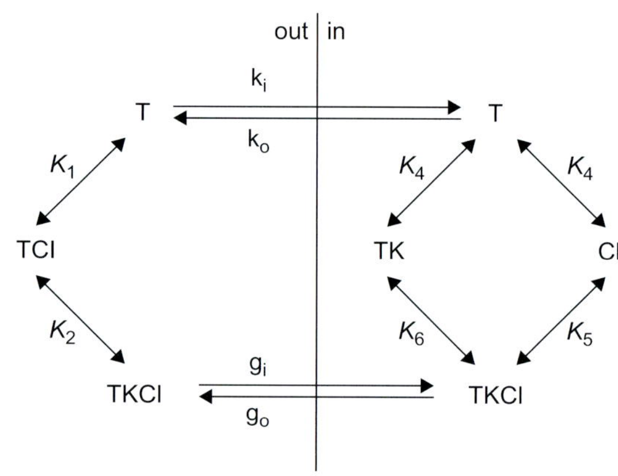

$$v_{in} = \frac{[Z_i Cl_o K_o] + Cl_i\, Cl_o\, K_i\, K_o}{Z_i Z_o R + Z_i R_i\, Cl_o K_o + Z_o R_o Cl_i\, K_i + R_e Cl_i\, Cl_o K_i\, K_o}$$

with $nR = (1 + Cl_i/K_3)/K_o + (1 + Cl_o/K_1)/k_i$

$$nR_i = (1 + Cl_i/K_3)/K_o + 1/gi$$

$$nR_o = (1 + K_6/Cl_i)/g_o = (1 + Cl_o/K_1)/k_i$$

$$nR_e = (1 + K_6/Cl_i)/g_o + 1/g_i$$

$$Z_i = K_3 K_5 k_o / g_o$$

$$Z_o = K_1 K_2 k_i / g_i$$

under *zero trans* Cl, $$v_{in} = \frac{Cl_o K_o}{Z_o R_o + R_i\, Cl_o K_o}$$

FIGURE 5.14 Steady-state model of swelling-activated K^+-Cl^- cotransport in sheep red blood cells. The model considers *ordered* binding outside, whereas *random* binding inside. Note that the free transporter and the fully loaded transporter are allowed to move through the membrane in both directions (steady-state equilibrium). The equation for the influx of K^+ is presented in its full form, and under *zero* Cl^- *trans* condition. Redrawn from Delpire and Lauf (1991a).

E. Rapid Equilibrium Kinetics and K^+-Cl^- Cotransport Inhibition

Adding the binding of an inhibitor to models of two-substrate transport mechanism increases even further the degree of complexity, both for the possible models and their associated equations. Indeed, as we have seen in the previous two sections, the transporter exists as a free molecule, as a molecule binding one ion and as a molecule binding both ions. Among many possible models, experimental data will dictate the viable kinetic model and determine with which form(s) of the transporter the inhibitor interacts. For example, a detailed analysis of a novel K^+-Cl^- cotransporter inhibitory compound, PubChem SID# 56405457, binding to KCC2 expressed in HEK293 cells, revealed interesting behaviors suggesting that the drug cannot bind to a transporter occupied by Cl^-. Furthermore, the data are only consistent with a model where the drug can bind to either a free transporter, or a transporter occupied by K^+. As mentioned before, this suggests that in this case, K^+ can also bind to a free transporter, making the ion binding outside *random*. The model of inhibitor binding to K-Cl cotransport is represented in Fig. 5.15, and the corresponding equations can be found in Delpire et al. (2009). Note that these data are in agreement with those obtained in red blood cells with furosemide, which suggest competitive inhibition with K^+ ions and non-competitive inhibition with Cl^- ions (Lauf, 1984a).

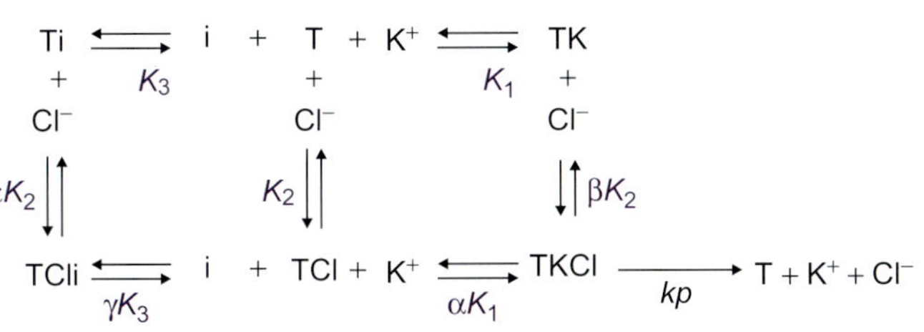

FIGURE 5.15 Kinetic model of KCC2 inhibition. The model was based on experiments measuring K^+ uptake at varying concentrations of inhibitor and external ions (either $[K^+]$ or $[Cl^-]$) in HEK-293 cells overexpressing KCC2 and stimulated by pre-treatment with N-ethylmaleimide. Note that the model allows for inhibitor binding to the free transporter or the transporter occupied by K^+, but not to any transporter occupied by Cl^-. Redrawn from Delpire et al. (2009).

F. Kinetics of Na^+-K^+-Cl^- Cotransport

The kinetics of Na^+-K^+-$2Cl^-$ cotransport is even more complex than that of K^+-Cl^- cotransport, as four ions are involved in the translocation process. Like K^+-Cl^- cotransport, Na^+-K^+-$2Cl^-$ cotransport requires the presence of all three ions to promote the translocation and transport from one side of the membrane to the other. However, in some species, the Na^+-K^+-$2Cl^-$ cotransporter, under proper ionic conditions, can also mediate partial reactions, such as K^+/K^+ exchange and Na^+/Na^+ exchange reactions. Kinetics of Na^+-K^+-$2Cl^-$ cotransport have been best studied in duck red blood cells (Lytle et al., 1998). The analysis is based mainly on the behavior of K^+/K^+ exchange in duck red cells containing high $[K^+]$, as well as Na^+/Na^+ exchange in duck red cells loaded with high $[Na^+]$ and low $[K^+]$. Based on the fact that K^+/K^+ exchange, which is measured in the absence of external Na^+, requires internal Na^+, whereas Na^+/Na^+ exchange, which is measured in the absence of internal Na^+, requires external K^+, a glide symmetry model was proposed (Fig. 5.16). The model suggests *ordered* binding with Na^+ binding first, followed by one Cl^-, followed by K^+, which is then followed by a second Cl^-. Glide symmetry implies that the *release* order is identical on the other side of the membrane, i.e. Na^+ first and so on. As rigorously discussed by Russell (Russell, 2000), the model accounts for most of the experimental data.

G. Hill Coefficients

The degree of cooperativity of substrate binding to a transporter is obtained from Hill's equation (Hill, 1910). A coefficient of unity indicates completely independent binding and, for a transporter, this is typically interpreted as indicative of a single binding site. In contrast, a coefficient greater than unity indicates that more than one ion binds to the transporter, as the binding of one ion increases the binding of the second (cooperativity). Hill's coefficients for K^+ and Cl^- have been measured for K^+-Cl^- cotransporter in sheep and human red blood cells (Dunham and Ellory, 1981; Kaji, 1989; Lauf, 1984b), and for KCC1 and KCC4

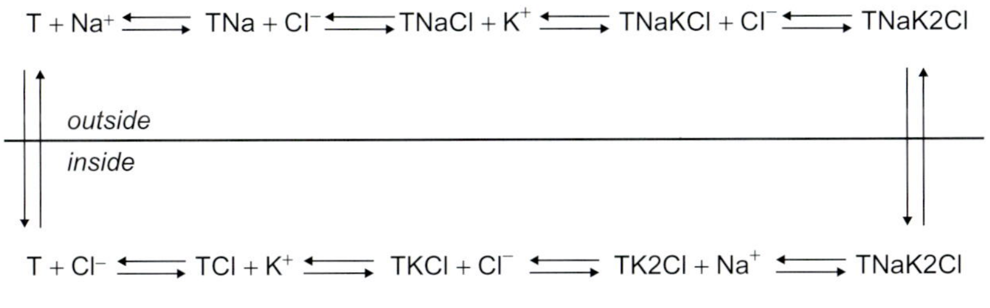

FIGURE 5.16 Kinetic model of Na^+-K^+-$2Cl^-$ cotransporter in duck red blood cells. The model describes *ordered* binding with Na^+ binding first, followed by one Cl^-, then K^+, and finally a second Cl^-. Note that the release on the other side follows the inverse order: Cl^- first, followed by K^+, the second Cl^-, and finally Na^+. This is the glide symmetry model described in Lytle et al. (1998).

expressed in *Xenopus laevis* oocytes (Mercado et al., 2000). Measured coefficients for K^+ were unanimously close to unity. For Cl^-, the Hill's coefficient was often found to have values closer to two, suggesting the binding of multiple Cl^- ions to the K^+-Cl^- cotransporter. Similarly, Hill coefficients of unity were also determined for Na^+ and Cl^- in the Na^+-Cl^- cotransporter (Gamba et al., 1993). Finally, Hill's coefficients of unity for Na^+ and K^+ but two for Cl^- were determined for the Na^+-K^+-$2Cl^-$ cotransporter (O'Grady et al., 1986). These data are in agreement with the traditionally accepted stoichiometry of the three types of cation-Cl^- cotransporters in vertebrates.

IV. CONCLUSIONS

Knowing the basic thermodynamic and kinetic properties of Cl^- transport mechanisms is fundamental to our understanding of how these proteins regulate intracellular Cl^-. This is particularly true in neurons where ligand-gated anion channels, such as those coupled to $GABA_A$ and glycine receptors, play critical roles in synaptic signaling during nervous system development, in mature neurons and under various pathologic conditions. A complete understanding of Cl^- homeostasis will only be possible when all the Cl^- carriers and channels, their mutual interactions and regulatory mechanisms, are considered acting jointly in a cell. We have seen that Cl^- homeostasis is not only determined by the thermodynamic forces that drive the transport of Cl^-, but also by the kinetic properties of each carrier. Other determining factors include pH, CO_2/HCO_3^-, K^+, Na^+, transmembrane potential and the permeability properties of anion channels.

Acknowledgements

The work of FJA-L was partially supported by the National Institute of Neurological Disorders and Stroke Grant NS29227 and by Wright State University Boonshoft School of Medicine Research Challenge and Seed Grant Programs. The work of ED was partially supported by the National Institutes of Health Grants NS36758, NS53658 and GM74771. We are grateful to Professor Jonathan Ashmore FRS for helpful discussions and for reading the manuscript.

References

Achilles, K., Okabe, A., Ikeda, M., Shimizu-Okabe, C., Yamada, J., Fukuda, A., Luhmann, H.J., and Kilb, W. (2007). Kinetic properties of Cl uptake mediated by Na^+-dependent K^+-2Cl cotransport in immature rat neocortical neurons. *J. Neurosci.* **27**, 8616–8627.

Adragna, N.C., Perkins, C.M., and Lauf, P.K. (1985). Furosemide-sensitive Na^+-K^+ cotransport and cellular metabolism in human erythrocytes. *Biochim. Biophys. Acta* **812**, 293–296.

Adrian, R.H. (1960). Potassium chloride movement and the membrane potential of frog muscle. *J. Physiol.* **151**, 154–185.

Adrian, R.H. (1961). Internal chloride concentration and chloride efflux of frog muscle. *J. Physiol.* **156**, 623–632.

Aickin, C.C. (1990). Chloride transport across the sarcolemma of vertebrate smooth and sketal. In *Chloride Channels and Carriers in Nerve, Muscle, and Glial Cells* (Alvarez-Leefmans, F.J. and Russell, J.M., eds), pp. 209–249. Plenum Press, New York.

Aickin, C.C., Betz, W.J., and Harris, G.L. (1989). Intracellular chloride and the mechanism for its accumulation in rat lumbrical muscle. *J. Physiol.* **411**, 437–455.

Allen, G.I., Eccles, J., Nicoll, R.A., Oshima, T., and Rubia, F.J. (1977). The ionic mechanisms concerned in generating the i.p.s.ps of hippocampal pyramidal cells. *Proc. R. Soc. Lond. B Biol Sci.* **198**, 363–384.

Alper, S.L., Chernova, M.N., and Stewart, A.K. (2002). How pH regulates a pH regulator: a regulatory hot spot in the N-terminal cytoplasmic domain of the AE2 anion exchanger. *Cell Biochem. Biophys.* **36**, 123–136.

Altamirano, A.A., Breitwieser, G.E., and Russell, J.M. (1988). Vanadate and fluoride effects on Na^+-K^+-Cl^- cotransport in squid giant axon. *Am. J. Physiol.* **254**, C582–C586.

Alvarez-Leefmans, F.J. (1990). Intracellular Cl^- regulation and synaptic inhibition in vertebrate and invertebrate neurons. In *Chloride Channels and Carriers in Nerve, Muscle, and Glial Cells* (Alvarez-Leefmans, F.J. and Russell, J.M., eds), pp. 109–158. Plenum Press, New York.

Alvarez-Leefmans, F.J. (2001). Intracellular chloride regulation. In *Cell Physiology Sourcebook: A Molecular Approach* (Sperelakis, N., ed.), pp. 301–318. Academic Press, San Diego.

Alvarez-Leefmans, F.J., Gamiño, S.M., and Giraldez, F. (1986). Direct demonstration that chloride ions are not passively distributed across the membrane of dorsal root ganglion cells of the frog: preliminary studies on the nature of the chloride pump. *Biophys. J.* **49**, 413a.

Alvarez-Leefmans, F.J., Gamino, S.M., Giraldez, F., and Nogueron, I. (1988). Intracellular chloride regulation in amphibian dorsal root ganglion neurones studied with ion-selective microelectrodes. *J. Physiol.* **406**, 225–246.

Alvarez-Leefmans, F.J., Leon-Olea, M., Mendoza-Sotelo, J., Alvarez, F.J., Anton, B., and Garduno, R. (2001). Immunolocalization of the Na^+-K^+-$2Cl^-$ cotransporter in peripheral nervous tissue of vertebrates. *Neuroscience* **104**, 569–582.

Alvarez-Leefmans, F.J., Nani, A., and Márquez, S. (1998). Chloride transport, osmotic balance, and presynaptic inhibition. In *Presynaptic Inhibition and Neural Control* (Rudomin, P., Romo, R., and Mendell, L.M., eds), pp. 50–79. Oxford University Press, New York.

Ballanyi, K. and Grafe, P. (1985). An intracellular analysis of gamma-aminobutyric-acid-associated ion movements in rat sympathetic neurones. *J. Physiol.* **365**, 41–58.

Banke, T.G. and McBain, C.J. (2006). GABAergic input onto CA3 hippocampal interneurons remains shunting throughout development. *J. Neurosci.* **26**, 11720–11725.

Becker, M., Nothwang, H.G., and Friauf, E. (2003). Differential expression pattern of chloride transporters NCC, NKCC2, KCC1, KCC3, KCC4, and AE3 in the developing rat auditory brainstem. *Cell Tissue Res.* **312**, 155–165.

Berglund, K., Schleich, W., Krieger, P., Loo, L.S., Wang, D., Cant, N.B., Feng, G., Augustine, G. J., and Kuner, T. (2006). Imaging synaptic inhibition in transgenic mice expressing the chloride indicator, Clomeleon. *Brain Cell Biol* **35**, 207–228.

Bevensee, M.O. and Boron, W.F. (1998). Thermodynamics and physiology of cellular pH regulation. In *pH and Brain Function* (Kaila, K. and Ransom, B.R., eds), pp. 173–194. Wiley-Liss, New York.

Boistel, J. and Fatt, P. (1958). Membrane permeability change during inhibitory transmitter action in crustacean muscle. *J. Physiol.* **144**, 176–191.

Bormann, J., Hamill, O.P., and Sakmann, B. (1987). Mechanism of anion permeation through channels gated by glycine and gamma-aminobutyric acid in mouse cultured spinal neurones. *J. Physiol.* **385**, 243–286.

Breitwieser, G.E., Altamirano, A.A., and Russell, J.M. (1990). Osmotic stimulation of Na^+-K^+-Cl^- cotransport in squid giant axon is $[Cl^-]_i$ dependent. *Am. J. Physiol.* **258**, C749–C753.

Breitwieser, G.E., Altamirano, A.A., and Russell, J.M. (1996). Elevated $[Cl^-]_i$, and $[Na^+]_i$ inhibit Na^+, K^+, Cl cotransport by different mechanisms in squid giant axons. *J. Gen. Physiol.* **107**, 261–270.

Brock, L.G., Coombs, J.S., and Eccles, J.C. (1952a). The nature of the monosynaptic excitatory and inhibitory processes in the spinal cord. *Proc. R. Soc. Lond. B Biol. Sci.* **140**, 170–176.

Brock, L.G., Coombs, J.S., and Eccles, J.C. (1952b). The recording of potentials from motoneurones with an intracellular electrode. *J. Physiol.* **117**, 431–460.

Brock, L.G., Coombs, J.S., and Eccles, J.C. (1952c). Synaptic excitation and inhibition. *J. Physiol.* **117**, 8p.

Brock, L.G., Coombs, J.S., and Eccles, J.C. (1953). Intracellular recording from antidromically activated motoneurones. *J. Physiol.* **122**, 429–461.

Brugnara, C., Van Ha, T., and Tosteson, D.C. (1989). Role of chloride in potassium transport through a K-Cl cotransport system in human red blood cells. *Am. J. Physiol.* **256**, C994–C1003.

Brumback, A.C. and Staley, K.J. (2008). Thermodynamic regulation of NKCC1-mediated Cl^- cotransport underlies plasticity of $GABA_A$ signaling in neonatal neurons. *J. Neurosci.* **28**, 1301–1312.

Chipperfield, A.R. (1981). Chloride dependence of frusemide- and phloretin-sensitive passive sodium and potassium fluxes in human red cells. *J. Physiol. (Lond.)* **312**, 435–444.

Chub, N., Mentis, G.Z., and O'Donovan, M.J. (2006). Chloride-sensitive MEQ fluorescence in chick embryo motoneurons following manipulations of chloride and during spontaneous network activity. *J. Neurophysiol.* **95**, 323–330.

Coombs, J.S., Eccles, J.C., and Fatt, P. (1955). The specific ionic conductances and the ionic movements across the motoneuronal membrane that produce the inhibitory post-synaptic potential. *J. Physiol.* **130**, 326–374.

Cordero-Erausquin, M., Coull, J.A., Boudreau, D., Rolland, M., and De Koninck, Y. (2005). Differential maturation of GABA action and anion reversal potential in spinal lamina I neurons: impact of chloride extrusion capacity. *J. Neurosci.* **25**, 9613–9623.

DeFazio, R.A., Keros, S., Quick, M.W., and Hablitz, J.J. (2000). Potassium-coupled chloride cotransport controls intracellular chloride in rat neocortical pyramidal neurons. *J. Neurosci.* **20**, 8069–8076.

Delpire, E., Days, E., Mi, D., Lewis, M., Kim, K., Lidnsley, C., and Weaver, C.D. (2009). Small molecule screen identifies inhibitors of the neuronal K-Cl cotransporter KCC2. *Proc. Natl. Acad. Sci. USA* **106**, 5383–5388.

Delpire, E. and Lauf, P.K. (1991a). Kinetics of Cl-dependent K fluxes in hyposmotically low K sheep erythrocytes. *J. Gen. Physiol.* **97**, 173–193.

Delpire, E. and Lauf, P.K. (1991b). Trans-effects of cellular K and Cl on ouabain-resistant Rb(K) influx in low K sheep red blood cells: further evidence for asymmetry of K-Cl cotransport. *Pflügers Arch.* **419**, 540–542.

Delpire, E. and Lauf, P.K. (1992). Kinetics of DIDS inhibition of swelling-activated K-Cl cotransport in low K sheep erythrocytes. *J. Membrane Biol.* **126**, 89–96.

Deschenes, M. and Feltz, P. (1976). GABA-induced rise of extracellular potassium in rat dorsal root ganglia: an electrophysiological study in vivo. *Brain Res.* **118**, 494–499.

Duhm, J. and Göbel, B.O. (1984). Na^+-K^+ transport and volume of rat erythrocytes under dietary K^+ deficiency. *Am. J. Physiol.* **246**, C20–C29.

Dunham, P.B. and Ellory, J.C. (1981). Passive potassium transport in low potassium sheep red cells: dependence upon cell volume and chloride. *J. Physiol. (Lond.)* **318**, 511–530.

Earl, J. and Large, W.A. (1974). Electrophysiological investigation of GABA-mediated inhibition at the hermit crab neuromuscular junction. *J. Physiol.* **236**, 113–127.

Eccles, J., Eccles, R.M., and Ito, M. (1964a). Effects of intracellular potassium and sodium injections on the inhibitory postsynaptic potential. *Proc. R. Soc. Lond. B Biol. Sci.* **160**, 181–196.

Eccles, J., Eccles, R.M., and Ito, M. (1964b). Effects produced on inhibitory postsynaptic potentials by the coupled injections of cations and anions into motoneurons. *Proc. R. Soc. Lond. B Biol. Sci.* **160**, 197–210.

Eccles, J.C. (1964a). Ionic mechanism of postsynaptic inhibition. *Science* **145**, 1140–1147.

Eccles, J.C. (1964b). *The Physiology of Synapses*. Springer, Berlin.

Ellory, J.C., Guizouarn, H., Borgese, F., Bruce, L.J., Wilkins, R.J., and Stewart, G.W. (2009). Review. Leaky Cl^--HCO_3^- exchangers: cation fluxes via modified AE1. *Philos. Trans. R. Soc. Lond. B Biol. Sci.* **364**, 189–194.

Farrant, M. and Kaila, K. (2007). The cellular, molecular and ionic basis of GABA(A) receptor signalling. *Prog. Brain Res.* **160**, 59–87.

Fatima-Shad, K. and Barry, P.H. (1993). Anion permeation in GABA- and glycine-gated channels of mammalian cultured hippocampal neurons. *Proc. Biol. Sci.* **253**, 69–75.

Fatt, P. (1974). Postsynaptic cell characteristics determining membrane potential changes. In *Physics and Mathematics of the Nervous System: Proceedings of a Summer School Organized by the International Centre for Theoretical Physics, Trieste, and the Institute for Information Sciences, University of Tübingen, held at Trieste, August 21–31, 1973* (Conrad, M., Güttinger, W., and Dal Cin, M., eds), pp. 150–170. Springer-Verlag, Berlin, New York.

Fatt, P. and Katz, B. (1951). An analysis of the end-plate potential recorded with an intracellular electrode. *J. Physiol.* **115**, 320–370.

Fatt, P. and Katz, B. (1952a). The action of inhibitory nerve impulses on the surface membrane of crustacean muscle fibres. *J. Physiol.* **118**, 47P–48P.

Fatt, P. and Katz, B. (1952b). Spontaneous subthreshold activity at motor nerve endings. *J. Physiol.* **117**, 109–128.

Fatt, P. and Katz, B. (1953). The effect of inhibitory nerve impulses on a crustacean muscle fibre. *J. Physiol.* **121**, 374–389.

Fenner, F. and Curtis, D. (2001). *The John Curtin School of Medical Research. The First Fifty Years 1948–1998*. Brolga Press, Gundaroo.

Flatman, P.W. (1991). The effects of metabolism on Na^+-K^+-Cl^- co-transport in ferret red cells. *J. Physiol.* **437**, 495–510.

Flatman, P.W. (2004). Regulation of Na-K-2Cl cotransport in red cells. *Adv. Exp. Med. Biol.* **559**, 77–88.

French, A.S., Panek, I., and Torkkeli, P.H. (2006). Shunting versus inactivation: simulation of GABAergic inhibition in spider mechanoreceptors suggests that either is sufficient. *Neurosci. Res.* **55**, 189–196.

Fu, L., Wong, J.A., Schneider, E.G., and Thomason, D.B. (1999). Unique 5′-end of a Na^+-K^+-$2Cl^-$ cotransporter-like mRNA expressed in rat skeletal muscle. *DNA Seq.* **10**, 127–132.

Gagnon, E., Forbush, B., Caron, L., and Isenring, P. (2003). Functional comparison of renal Na-K-Cl cotransporters between distant species. *Am. J. Physiol. Cell Physiol.* **284**, C365–C370.

Galanopoulou, A.S. (2008). $GABA_A$ receptors in normal development and seizures: friends or foes? *Curr. Neuropharmacol.* **6**, 1–20.

Gamba, G. (2005). Molecular physiology and pathophysiology of electroneutral cation-chloride cotransporters. *Physiol. Rev.* **85**, 423–493.

Gamba, G., Saltzberg, S.N., Lombardi, M., Miyanoshita, A., Lytton, J., Hediger, M.A., Brenner, B.M., and Hebert, S.C. (1993). Primary structure and functional expression of a cDNA encoding the thiazide-sensitive, electroneutral sodium-chloride cotransporter. *Proc. Natl. Acad. Sci. USA* **90**, 2749–2753.

Geck, P. and Heinz, E. (1986). The Na-K-2Cl cotransport system. *J. Membr. Biol.* **91**, 97–105.

Geck, P. and Heinz, E. (1989). Secondary active transport: introductory remarks. *Kidney Int.* **36**, 334–341.

Geck, P., Pietrzyk, C., Burckhardt, B.C., Pfeiffer, B., and Heinz, E. (1980). Electrically silent cotransport on Na^+, K^+ and Cl^- in Ehrlich cells. *Biochim. Biophys. Acta* **600**, 432–447.

Gerencser, G.A. and Zhang, J. (2003). Existence and nature of the chloride pump. *Biochim. Biophys. Acta* **1618**, 133–139.

Gillen, C.M., Brill, S., Payne, J.A., and Forbush, B.I. (1996). Molecular cloning and functional expression of the K-Cl cotransporter from rabbit, rat, and human A new member of the cation-chloride cotransporter family. *J. Biol. Chem.* **271**, 16237–16244.

Gillen, C.M. and Forbush, B., 3rd (1999). Functional interaction of the K-Cl cotransporter (KCC1) with the Na-K-Cl cotransporter in HEK-293 cells. *Am. J. Physiol.* **276**, C328–C336.

Glykys, J. and Mody, I. (2007). Activation of $GABA_A$ receptors: views from outside the synaptic cleft. *Neuron* **56**, 763–770.

Glynn, I.M. (1957). The action of cardiac glycosides on sodium and potassium movements in human red cells. *J. Physiol.* **136**, 148–173.

Gonzalez-Islas, C., Chub, N., and Wenner, P. (2009). NKCC1 and AE3 appear to accumulate chloride in embryonic motoneurons. *J. Neurophysiol.* **101**, 507–518.

Gosmanov, A.R., Schneider, E.G., and Thomason, D.B. (2003). NKCC activity restores muscle water during hyperosmotic challenge independent of insulin, ERK, and p38 MAPK. *Am. J. Physiol. Regul. Integr. Comp. Physiol.* **284**, R655–R665.

Gosmanov, A.R. and Thomason, D.B. (2003). Regulation of Na^+-K^+-$2Cl^-$ cotransporter activity in rat skeletal muscle and intestinal epithelial cells. *Tsitologiia* **45**, 812–816.

Haas, M., McBrayer, D., and Lytle, C. (1995). $[Cl^-]_i$-dependent phosphorylation of the Na-K-Cl cotransport protein of dog tracheal epithelial cells. *J. Biol. Chem.* **270**, 28955–28961.

Haas, M. and McManus, T.J. (1985). Effect of norepinephrine on swelling-induced potassium transport in duck red cells Evidence against a volume-regulatory decrease under physiological conditions. *J. Gen. Physiol.* **85**, 649–667.

Hannaert, P., Alvarez-Guerra, M., Pirot, D., Nazaret, C., and Garay, R.P. (2002). Rat NKCC2/NKCC1 cotransporter selectivity for loop diuretic drugs. *Naunyn Schmiedebergs Arch. Pharmacol.* **365**, 193–199.

Harris, G.L. and Betz, W.J. (1987). Evidence for active chloride accumulation in normal and denervated rat lumbrical muscle. *J. Gen. Physiol.* **90**, 127–144.

Haug, K., Warnstedt, M., Alekov, A.K., Sander, T., Ramirez, A., Poser, B., Maljevic, S., Hebeisen, S., Kubisch, C., Rebstock, J., Horvath, S., Hallmann, K., Dullinger, J.S., Rau, B., Haverkamp, F., Beyenburg, S., Schulz, H., Janz, D., Giese, B., Muller-Newen, G., Propping, P., Elger, C.E., Fahlke, C., Lerche, H., and Heils, A. (2003). Mutations in CLCN2 encoding a voltage-gated chloride channel are associated with idiopathic generalized epilepsies. *Nat. Genet.* **33**, 527–532.

Heinz, E. (1989). The unfinished story of secondary active transport. In *Membrane Transport: People and Ideas* (Tosteson, D.C., ed.), pp. 237–250. American Physiologycal Society, Bethesda, Maryland.

Hill, A.V. (1910). The possible effects of the aggregation of the molecules of hemoglobin on its dissociation curves. *J. Physiol. (Lond.)* **40**, 4–7.

Hodgkin, A.L. and Horowicz, P. (1959). The influence of potassium and chloride ions on the membrane potential of single muscle fibres. *J. Physiol.* **148**, 127–160.

Hoffmann, E.K. (2001). The pump and leak steady-state concept with a variety of regulated leak pathways. *J. Membr. Biol.* **184**, 321–330.

Hoffmann, E.K., Lambert, I.H., and Pedersen, S.F. (2009). Physiology of cell volume regulation in vertebrates. *Physiol. Rev.* **89**, 193–277.

Hutter, O.F. and Noble, D. (1960). The chloride conductance of frog skeletal muscle. *J. Physiol.* **151**, 89–102.

Inomata, N., Oomura, Y., Akaike, N., and Edwards, C. (1986). The anion selectivity of the gamma-aminobutyric acid controlled chloride channel in the perfused spinal ganglion cell of frog. *Neurosci. Res.* **3**, 371–383.

Jackson, M.B. and Zhang, S.J. (1995). Action potential propagation and propagation block by GABA in rat posterior pituitary nerve terminals. *J. Physiol.* **483**, 597–611.

Jennings, M.L. and Adame, M.F. (2001). Direct estimate of 1:1 stoichiometry of K^+-Cl^- cotransport in rabbit erythrocytes. *Am. J. Physiol. Cell Physiol.* **281**, C825–C832.

Jurkat-Rott, K., Fauler, M., and Lehmann-Horn, F. (2006). Ion channels and ion transporters of the transverse tubular system of skeletal muscle. *J. Muscle Res. Cell Motil.* **27**, 275–290.

Kaila, K., Voipio, J., Paalasmaa, P., Pasternack, M., and Deisz, R.A. (1993). The role of bicarbonate in GABAA receptor-mediated IPSPs of rat neocortical neurones. *J. Physiol.* **464**, 273–289.

Kaji, D. (1989). Kinetics of volume-sensitive K transport in human erythrocytes: evidence for asymmetry. *Am. J. Physiol.* **256**, C1214–C1223.

Kaji, D.M. (1993). Effect of membrane potential on K-Cl transport in human erythrocytes. *Am. J. Physiol.* **264**, C376–C382.

Kakazu, Y., Uchida, S., Nakagawa, T., Akaike, N., and Nabekura, J. (2000). Reversibility and cation selectivity of the K(+)-Cl(−) cotransport in rat central neurons. *J. Neurophysiol.* **84**, 281–288.

Kelly, J.S., Krnjevic, K., Morris, M.E., and Yim, G.K. (1969). Anionic permeability of cortical neurones. *Exp. Brain Res.* **7**, 11–31.

Keynes, R.D. (1963). Chloride in the squid giant axon. *J. Physiol.* **169**, 690–705.

Keynes, R.D. (1982). Introductory remarks. *Phil. Trans. R. Soc. Lond. B* **299**, 367–368.

Khirug, S., Yamada, J., Afzalov, R., Voipio, J., Khiroug, L., and Kaila, K. (2008). GABAergic depolarization of the axon initial segment in cortical principal neurons is caused by the Na-K-2Cl cotransporter NKCC1. *J. Neurosci.* **28**, 4635–4639.

Koivusalo, M., Kapus, A., and Grinstein, S. (2009). Sensors, transducers, and effectors that regulate cell size and shape. *J. Biol. Chem.* **284**, 6595–6599.

Kullmann, D.M., Ruiz, A., Rusakov, D.M., Scott, R., Semyanov, A., and Walker, M.C. (2005). Presynaptic, extrasynaptic and axonal $GABA_A$ receptors in the CNS: where and why? *Prog. Biophys. Mol. Biol.* **87**, 33–46.

Kurschat, C.E., Shmukler, B.E., Jiang, L., Wilhelm, S., Kim, E.H., Chernova, M.N., Kinne, R.K., Stewart, A.K., and Alper, S.L. (2006). Alkaline-shifted pHo sensitivity of AE2c1-mediated anion exchange reveals novel regulatory determinants in the AE2 N-terminal cytoplasmic domain. *J. Biol. Chem.* **281**, 1885–1896.

Labrakakis, C., Tong, C.K., Weissman, T., Torsney, C., and MacDermott, A.B. (2003). Localization and function of ATP and GABAA receptors expressed by nociceptors and other postnatal sensory neurons in rat. *J. Physiol.* **549**, 131–142.

Lauf, P.K. (1984a). Thiol-dependent passive K/Cl transport in sheep red cells. IV. Furosemide inhibition as a function of external Rb^+, Na^+, and Cl^-. *J. Membr. Biol.* **77**, 57–62.

Lauf, P.K. (1984b). Thiol-dependent passive K^+Cl^- transport in sheep red blood cells: VI. Functional heterogeneity and immunologic identity with volume-stimulated $K^+(Rb^+)$ fluxes. *J. Membr. Biol.* **82**, 167–178.

Lauf, P.K. and Adragna, N.C. (1996). A thermodynamic study of electroneutral K-Cl cotransport in pH- and volume-clamped low K sheep erythrocytes with normal and low internal magnesium. *J. Gen. Physiol.* **108**, 341–350.

Lauf, P.K., Bauer, J., Adragna, N.C., Fujise, H., Zade-Oppen, A.M.M., Ryu, K., and Delpire, E. (1992). Erythrocyte K-Cl cotransport: properties and regulation. *Am. J. Physiol.* **263**, C917–C932.

Lev-Tov, A., Fleshman, J.W., and Burke, R.E. (1983). Primary afferent depolarization and presynaptic inhibition of monosynaptic group Ia EPSPs during posttetanic potentiation. *J. Neurophysiol.* **50**, 413–427.

Lev-Tov, A., Meyers, D.E., and Burke, R.E. (1988). Modification of primary afferent depolarization in cat group Ia afferents following high frequency intra-axonal tetanization of individual afferents. *Brain Res.* **438**, 328–330.

Llinás, R. and Baker, R. (1972). A chloride-dependent inhibitory postsynaptic potential in cat trochlear motoneurons. *J. Neurophysiol.* **35**, 484–492.

Llinás, R., Baker, R., and Precht, W. (1974). Blockage of inhibition by ammonium acetate action on chloride pump in cat trochlear motoneurons. *J. Neurophysiol.* **37**, 522–532.

Lux, H.D. (1971). Ammonium and chloride extrusion: hyperpolarizing synaptic inhibition in spinal motoneurons. *Science* **173**, 555–557.

Lux, H.D., Loracher, C., and Neher, E. (1970). The action of ammonium on postsynaptic inhibition of cat spinal motoneurons. *Exp. Brain Res.* **11**, 431–447.

Lytle, C. (2003). Na^+-K^+-$2Cl^-$ cotransport. In *Red Cell Membrane Transport in Health and Disease* (Bernhardt, I. and Ellory, J.C., eds), pp. 173–195. Springer, Berlin; New York.

Lytle, C. and Forbush, B., 3rd (1992). The Na-K-Cl cotransport protein of shark rectal gland II. Regulation by direct phosphorylation. *J. Biol. Chem.* **267**, 25438–25443.

Lytle, C. and Forbush, B., 3rd (1996). Regulatory phosphorylation of the secretory Na-K-Cl cotransporter: modulation by cytoplasmic Cl. *Am. J. Physiol.* **270**, C437–C438.

Lytle, C. and McManus, T. (2002). Coordinate modulation of Na-K-2Cl cotransport and K-Cl cotransport by cell volume and chloride. *Am. J. Physiol. Cell Physiol.* **283**, C1422–C1431.

Lytle, C., McManus, T.J., and Haas, M. (1998). A model of Na-K-2Cl cotransport based on ordered ion binding and glide symmetry. *Am. J. Physiol.* **274**, C299–C309.

Ma, W., Saunders, P.A., Somogyi, R., Poulter, M.O., and Barker, J.L. (1993). Ontogeny of $GABA_A$ receptor subunit mRNAs in rat spinal cord and dorsal root ganglia. *J. Comp. Neurol.* **338**, 337–359.

Mann, E.O. and Paulsen, O. (2007). Role of GABAergic inhibition in hippocampal network oscillations. *Trends Neurosci.* **30**, 343–349.

Mercado, A., Song, L., Vazquez, N., Mount, D.B., and Gamba, G. (2000). Functional comparison of the K^+-Cl^- cotransporters KCC1 and KCC4. *J. Biol. Chem.* **275**, 30326–30334.

Mercado, A., Vazquez, N., Song, L., Cortes, R., Enck, A.H., Welch, R., Delpire, E., Gamba, G., and Mount, D.B. (2005). NH2-terminal heterogeneity in the KCC3 K^+-Cl^- cotransporter. *Am. J. Physiol. Renal Physiol.* **289**, F1246–F1261.

Momose-Sato, Y., Sato, K., Hirota, A., and Kamino, K. (1998). GABA-induced intrinsic light-scattering changes associated with voltage-sensitive dye signals in embryonic brain stem slices: coupling of depolarization and cell shrinkage. *J. Neurophysiol.* **79**, 2208–2217.

Monsivais, P. and Rubel, E.W. (2001). Accommodation enhances depolarizing inhibition in central neurons. *J. Neurosci.* **21**, 7823–7830.

Munoz, A., Mendez, P., DeFelipe, J., and Alvarez-Leefmans, F.J. (2007). Cation-chloride cotransporters and GABA-ergic innervation in the human epileptic hippocampus. *Epilepsia* **48**, 663–673.

Niemeyer, M.I., Yusef, Y.R., Cornejo, I., Flores, C.A., Sepulveda, F. V., and Cid, L.P. (2004). Functional evaluation of human ClC-2 chloride channel mutations associated with idiopathic generalized epilepsies. *Physiol. Genomics* **19**, 74–83.

O'Grady, S.M., Musch, M.W., and Field, M. (1986). Stoichiometry and ion affinities of the Na-K-Cl cotransport system in the intestine of the winter flounder (Pseudopleuronectes americanus). *J. Membr. Biol.* **91**, 33–41.

O'Grady, S.M., Palfrey, H.C., and Field, M. (1987). Characteristics and functions of Na-K-Cl cotransport in epithelial tissues. *Am. J. Physiol.* **253**, C177–C192.

Panet, R., Marcus, M., and Atlan, H. (2000). Overexpression of the Na^+-K^+-Cl^- cotransporter gene induces cell proliferation and phenotypic transformation in mouse fibroblasts. *J. Cell Physiol.* **182**, 109–118.

Paredes, A., Plata, C., Rivera, M., Moreno, E., Vazquez, N., Munoz-Clares, R., Hebert, S.C., and Gamba, G. (2006). Activity of the renal Na^+-K^+-$2Cl^-$ cotransporter is reduced by mutagenesis of N-glycosylation sites: role for protein surface charge in Cl^- transport. *Am. J. Physiol. Renal Physiol.* **290**, F1094–F1102.

Payne, J.A., Stevenson, T.J., and Donaldson, L.F. (1996). Molecular characterization of a putative K-Cl cotransporter in rat brain A neuronal-specific isoform. *J. Biol. Chem.* **271**, 16245–16252.

Payne, J.A., Xu, J.C., Haas, M., Lytle, C.Y., Ward, D., and Forbush, B., 3rd (1995). Primary structure, functional expression, and chromosomal localization of the bumetanide-sensitive Na-K-Cl cotransporter in human colon. *J. Biol. Chem.* **270**, 17977–17985.

Prescott, S.A. and De Koninck, Y. (2003). Gain control of firing rate by shunting inhibition: roles of synaptic noise and dendritic saturation. *Proc. Natl. Acad. Sci. USA* **100**, 2076–2081.

Pusch, M., Jordt, S.E., Stein, V., and Jentsch, T.J. (1999). Chloride dependence of hyperpolarization-activated chloride channel gates. *J. Physiol.* **515**, 341–353.

Qu, Z. and Hartzell, H.C. (2008). Bestrophin Cl^- channels are highly permeable to HCO3. *Am. J. Physiol. Cell Physiol.* **294**, C1371–C1377.

Rocha-González, H.I., Mao, S., and Alvarez-Leefmans, F.J. (2008). Na^+, K^+, $2Cl^-$ cotransport and intracellular chloride regulation in rat primary sensory neurons: thermodynamic and kinetic aspects. *J. Neurophysiol.* **100**, 169–184.

Rudomin, P. and Schmidt, R.F. (1999). Presynaptic inhibition in the vertebrate spinal cord revisited. *Exp. Brain Res.* **129**, 1–37.

Russell, J.M. (1979). Chloride and sodium influx: a coupled uptake mechanism in the squid giant axon. *J. Gen. Physiol.* **73**, 801–818.

Russell, J.M. (1983). Cation-coupled chloride influx in squid axon Role of potassium and stoichiometry of the transport process. *J. Gen. Physiol.* **81**, 909–925.

Russell, J.M. (1984). Chloride in the squid giant axon. *Curr. Top. Membr. Transp.* **22**, 177–193.

Russell, J.M. (2000). Sodium-potassium-chloride cotransport. *Physiol. Rev.* **80**, 211–276.

Russell, J.M. and Brown, A.M. (1972). Active transport of chloride by the giant neuron of the Aplysia abdominal ganglion. *J. Gen. Physiol.* **60**, 499–518.

Schatzmann, H.J. (1953). [Cardiac glycosides as inhibitors of active potassium and sodium transport by erythrocyte membrane.] *Helv. Physiol. Pharmacol. Acta* **11**, 346–354.

Schomberg, S.L., Bauer, J., Kintner, D.B., Su, G., Flemmer, A., Forbush, B., and Sun, D. (2003). Cross talk between the GABA(A) receptor and the Na-K-Cl cotransporter is mediated by intracellular Cl. *J. Neurophysiol.* **89**, 159–167.

Skou, J.C. (1960). Further investigations of Mg^{++} + Na^{+}-activated adenosinetriphosphatase, possibly related to the active, linked transport of Na^{+} and K^{+} across the nerve membrane. *Biochim. Biophys. Acta* **42**, 6–23.

Somjen, G.G. (2004). *Ions in the Brain: Normal Function, Seizures, and Stroke*. Oxford University Press, Oxford; New York.

Staley, K., Smith, R., Schaack, J., Wilcox, C., and Jentsch, T.J. (1996). Alteration of $GABA_A$ receptor function following gene transfer of the CLC-2 chloride channel. *Neuron* **17**, 543–551.

Staley, K.J. and Proctor, W.R. (1999). Modulation of mammalian dendritic $GABA_A$ receptor function by the kinetics of Cl^- and HCO_3^- transport. *J. Physiol.* **519** (Pt 3), 693–712.

Staley, K.J., Soldo, B.L., and Proctor, W.R. (1995). Ionic mechanisms of neuronal excitation by inhibitory $GABA_A$ receptors. *Science* **269**, 977–981.

Steinbach, H.B. (1941). Chloride in the giant axons of the squid. *J. Cell. Comp. Physiol.* **17**, 57–64.

Stewart, A.K., Chernova, M.N., Shmukler, B.E., Wilhelm, S., and Alper, S.L. (2002). Regulation of AE2-mediated Cl^- transport by intracellular or by extracellular pH requires highly conserved amino acid residues of the AE2 NH2-terminal cytoplasmic domain. *J. Gen. Physiol.* **120**, 707–722.

Stewart, A.K., Kerr, N., Chernova, M.N., Alper, S.L., and Vaughan-Jones, R.D. (2004). Acute pH-dependent regulation of AE2-mediated anion exchange involves discrete local surfaces of the NH2-terminal cytoplasmic domain. *J. Biol. Chem.* **279**, 52664–52676.

Takeuchi, A. and Takeuchi, N. (1967). Anion permeability of the inhibitory post-synaptic membrane of the crayfish neuromuscular junction. *J. Physiol.* **191**, 575–590.

Todman, D. (2008). John Eccles (1903–97) and the experiment that proved chemical synaptic transmission in the central nervous system. *J. Clin. Neurosci.* **15**, 972–977.

Vaughan-Jones, R.D. (1982). Chloride activity and its control in skeletal and cardiac muscle. *Philos. Trans. R. Soc. Lond. B Biol. Sci.* **299**, 537–548.

Velazquez, H., Good, D.W., and Wright, F.S. (1984). Mutual dependence of sodium and chloride absorption by renal distal tubule. *Am. J. Physiol.* **247**, F904–F911.

Walz, W. (1992). Role of Na/K/Cl cotransport in astrocytes. *Can. J. Physiol. Pharmacol.* **70** (Suppl), S260–S262.

Wang, H., Yan, Y., Kintner, D.B., Lytle, C., and Sun, D. (2003). GABA-mediated trophic effect on oligodendrocytes requires Na-K-2Cl cotransport activity. *J. Neurophysiol.* **90**, 1257–1265.

Willis, W.D. (2006). John Eccles' studies of spinal cord presynaptic inhibition. *Prog. Neurobiol.* **78**, 189–214.

Willis, W.D., Jr. (1999). Dorsal root potentials and dorsal root reflexes: a double-edged sword. *Exp. Brain Res.* **124**, 395–421.

Wong, J.A., Fu, L., Schneider, E.G., and Thomason, D.B. (1999). Molecular and functional evidence for Na(+)-K(+)-2Cl(−) cotransporter expression in rat skeletal muscle. *Am. J. Physiol.* **277**, R154–R161.

Wu, Q., Delpire, E., Hebert, S.C., and Strange, K. (1998). Functional demonstration of Na^+-K^+-Cl^- cotransporter activity in isolated, polarized choroid plexus cells. *Am. J. Physiol.* **275**, C1565–C1572.

Zhang, S.J. and Jackson, M.B. (1993). GABA-activated chloride channels in secretory nerve endings. *Science* **259**, 531–534.

PART II

CURRENT METHODS FOR STUDYING CHLORIDE REGULATION

CHAPTER

6

Chemical and GFP-based Fluorescent Chloride Indicators

Alan S. Verkman

OUTLINE

I. INTRODUCTION

Cl^- is a major constituent of cells and extracellular compartments. Cl^- transport across cell membranes is crucial to many key cellular processes such as cell volume and pH regulation, transepithelial fluid transport and synaptic transmission. Within cells, Cl^- transport is involved in organellar acidification. A variety of human diseases are associated with mutations in Cl^- transporting proteins, such as cystic fibrosis, which is caused by loss-of-function mutations in the cystic fibrosis transmembrane conductance regulator (CFTR) protein, a cAMP-regulated Cl^- channel.

Several approaches are available to measure cellular [Cl^-] and membrane Cl^- transport. Patch-clamp is the gold standard for elucidation of the biophysical properties of Cl^- channels (see Chapter 8), as are double-barreled Cl^- sensing microelectrodes for measurement of $[Cl^-]_i$ (Coles, 1995). Both methods require technical expertise and specialized instrumentation, and are not easily amenable for high-throughput applications. Radioactive $^{36}Cl^-$ has been used to study Cl^- transport, though its low specific activity and the requirement of separating internal and external radioactivity limit its utility (see Chapter 9). Other approaches for studying Cl^- transport for specialized applications, which are not discussed further here, include membrane potential-sensitive dyes, nuclear magnetic resonance, and X-ray probe electron microanalysis.

Here, we review available Cl^--sensitive fluorescent indicators, which have been used widely in measurements of cellular $[Cl^-]$ and Cl^- transport. Cl^- indicators provide an alternative and sometimes exclusive approach to study Cl^-, such as in measurements of endosomal $[Cl^-]$ (Sonawane et al., 2003) and mapping of $[Cl^-]$ kinetics in brain cortex (Inglefield and Schwartz-Bloom, 1997). Fluorescent Cl^- indicators are also of utility for high-throughput discovery Cl^- transport modulators for drug development and as laboratory reagents for "chemical knockout".

II. SMALL-MOLECULE CHLORIDE-SENSITIVE FLUORESCENT INDICATORS

A. Collisional Quenching Mechanism of Quinolinium Cl^- Indicators

The available chemical-type Cl^- indicators involve halide quenching of the fluorescence of heterocyclic organic compounds containing a quaternary nitrogen (reviewed in Verkman, 1990). Fluorescence quenching occurs by a collisional mechanism with a linear Stern-Volmer relation, $F_o/F = 1 + K_{hal}[hal]$, where F_o is the fluorescence in the absence of halide (hal), F is the fluorescence in the presence of halide, and K_{hal} is the Stern-Volmer quenching constant (in M^{-1}). Collisional quenching involves transient interaction between Cl^- and the indicator, without binding. For quinolinium indicators such as SPQ (Fig. 6.1), collisional quenching occurs by a charge-transfer mechanism involving transient formation of a charge-transfer complex (Jayaraman and Verkman, 2000). Advantages of a collisional quenching mechanism are the absence of Cl^- binding and hence buffering, and rapid, nanosecond response times. However, in collisional quenching the fluorescence lifetime decreases in proportion to intensity without a change in spectral shape, precluding multi-wavelength ratiometric measurement of $[Cl^-]$.

The first and most widely used chemical-type fluorescent Cl^- indicator is SPQ (Illsley and Verkman, 1987). SPQ is excited at ultraviolet wavelengths with absorbance maxima at 318 and 350 nm and molar extinction coefficients of 5430 and 3470 M^{-1}, respectively. SPQ fluorescence is greatest at 450 nm, with a quantum yield of 0.69 in the absence of halides. The Stern-Volmer constant for quenching of SPQ by Cl^- in aqueous buffers is 118 M^{-1}, producing 50% reduction in fluorescence at 8 mM Cl^-. SPQ fluorescence is not altered by cations, phosphate, nitrate and sulfate, but is quenched weakly by some monovalent anions including acetate, gluconate and bicarbonate.

Improved and specialized Cl^- indicators, such as cell permeable and ratioable indicators, were developed following structure-activity analysis of quinolinium Cl^- indicators. The effects of heterocyclic backbone structure and the nature and position of substituents on Cl^- sensitivity and spectral properties were systematically examined (Biwersi et al., 1992; Krapf et al., 1988; Verkman et al., 1989a). A positively

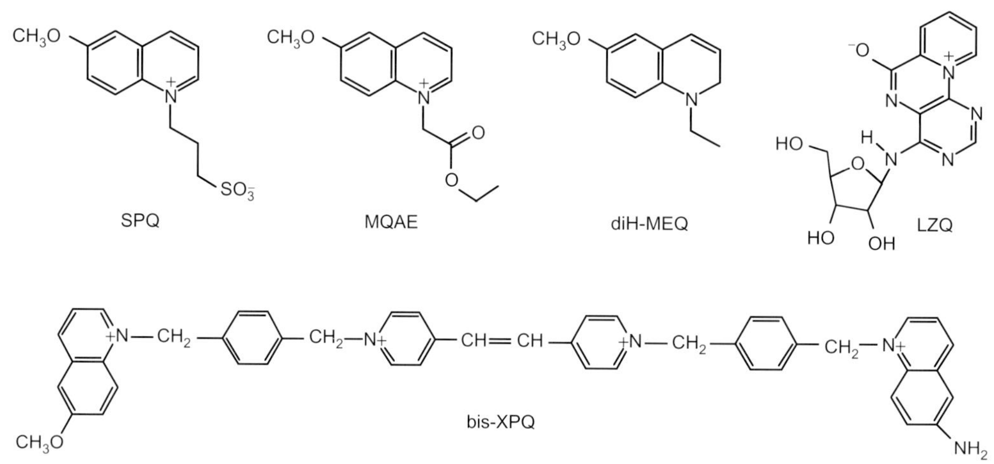

FIGURE 6.1 Structures of chemical-type fluorescent halide indicators.

charged quaternary nitrogen in the heterocyclic ring is necessary for Cl^- sensitivity. Indicators with a bicyclic quinoline backbone had better Cl^- sensitivity compared to isoquinoline and the tricyclic backbones 5,6-benzoquinoline and phenantridine. The position and the nature of the substituents on the quinoline ring altered the Cl^- sensitivity remarkably. Substitution of N-sulfopropylquinolinium with the electron-donating substituents methyl and methoxy at positions 2–6 in the quinoline ring increased Cl^- sensitivity, whereas substitution with the electron-withdrawing groups Cl^- or NO_2^- reduced sensitivity, as did substitution at positions 7 or 8. K_{Cl} was also sensitive to the charge of the N-substituent, probably because the charge density alters the Cl^- density near the positively charged nitrogen involved in the quenching process. Replacement of the 3′-sulfopropyl group of 6-methoxyquinolinium by an uncharged ethyl group increased Cl^- sensitivity 1.3-fold whereas replacement by 3′-trimethylammoniumpropyl increased Cl^- sensitivity by >three-fold. The quinolinium MQAE (Fig. 6.1) synthesized from these structure-activity studies has been widely used for cellular Cl^- measurements (e.g. Chapter 22).

B. Cell-permeable and Trappable Quinolinium Cl^- Indicators

A limitation of quinolinium Cl^- indicators is the necessity of invasive or slow diffusive cell loading (see below) because the fixed positively charged nitrogen confers high polarity and membrane impermeability. We thus designed modified quinoliniums for rapid, non-invasive loading into living cells, followed by chemical/enzymatic reaction in cytoplasm rendering them polar and membrane impermeable. A cell-permeable Cl^- indicator was synthesized by masking the positive charge on the nitrogen by reduction of the quinolinium moiety to the uncharged 1,2-dihydroquinoline (Biwersi and Verkman, 1991), giving the compound 6-methoxy-N-ethyl-1,2-dihydroquinoline (diH-MEQ) (Fig. 6.1). diH-MEQ is non-polar and enters the cell rapidly where it is oxidized to the membrane-impermeable and Cl^--sensitive compound 6-methoxy-N-ethylquinolinium MEQ, which has similar properties to SPQ.

C. Dual-wavelength Quinolinium Cl^- Indicators for Ratio Imaging

When measured at a single excitation and emission wavelength, indicator fluorescence depends on both indicator concentration and $[Cl^-]$. Ratiometric normalization for indicator concentration permits measurement of absolute ion concentration in different cells and different regions of the same cell, as done for Ca^{2+} determination using fura-2 and pH determination using BCECF (O'Connor and Silver, 2007). As mentioned above, the collisional quenching mechanism of quinolinium-based Cl^- indicators precludes their use in ratiometric applications. To allow ratio imaging of $[Cl^-]$, we synthesized a series of dual-wavelength "hybrid" Cl^- indicators (Jayaraman et al., 1999a). Cl^--sensitive and -insensitive chromophores and spacer groups were screened to yield conjugates that were fluorescent, suitable for ratio imaging, non-toxic, easy to load and not metabolized in cells. One of these compounds (bis-XPQ) is shown in Fig. 6.1. 6-methoxyquinolinium (as in SPQ) was used as the Cl^--sensitive chromophore and 6-aminoquinolinium as the Cl^--insensitive chromophore, the latter having similar excitation but red-shifted emission spectra compared to 6-methoxyquinolinium. The chromophores were linked covalently by a spacer. Upon excitation at 365 nm, fluorescence emission at 450 nm (methoxyquinolinium fluorescence) is Cl^- sensitive whereas emission at 560 nm (aminoquinolinium fluorescence) is Cl^- insensitive. Changes in $[Cl^-]_i$ can be recorded by emission ratio imaging (450/560) at 365 nm excitation wavelength. Cell permeable/trappable dual-wavelength indicators were also synthesized by the reduction–oxidation strategy used for diH-MEQ above.

D. Long-wavelength Halide Indicators

Though widely used for cellular applications, a limitation of quinolinium-type Cl^- indicators is the need for ultraviolet excitation and their relatively dim fluorescence. Ultraviolet excitation can be associated with autofluorescence background, and photobleaching and photodynamic cell injury. A series of long-wavelength polar fluorophores was screened to identify compounds with high Cl^- and/or I^- sensitivity in cells, bright fluorescence, low toxicity, uniform loading of cell cytoplasm, chemical stability in cells and minimal leakage out of cells. A class of 9-substituted acridinium compounds was identified (Biwersi et al., 1994; Legg and Hercules, 1970), which have bright fluorescence and high Cl^- sensitivity (K_{Cl} up to 390 M^{-1}). Although 9-substituted acridiniums were excellent for measurement of Cl^- transport in liposomes, extracellular fluid compartments, and endosomes in living cells, as described below, they are unstable in cytoplasm because of a hydroxylation reaction and so not suitable for measurements of $[Cl^-]_i$. Of many screened compounds, the best indicators identified for use in cytoplasm were of the luminarine class (Jayaraman et al., 1999b; Skalski et al., 1989). LZQ (Fig. 6.1) is brightly fluorescent with excitation and emission maxima at 428 and 533 nm, respectively, molar extinction 9000 $M^{-1}cm^{-1}$ and quantum yield 0.47. LZQ fluorescence is quenched by I^- by a collisional mechanism with Stern-Volmer constant 70 M^{-1}, but not by other halides (including Cl^-), NO_3^-, cations or changes in pH (range 4–9). LZQ gives robust fluorescence signals in cells with little background, and is thus is the preferred method to measure halide transport when I^- can be substituted for Cl^-.

E. Applications to Cellular Cl^- Measurements

In addition to the optical requirements mentioned above, indicators for use in living cells should ideally be non-toxic, chemically stable, trappable, uniformly distributed, and sensitive and specific for cytoplasmic Cl^- or I^-. The quinolinium and luminarine compounds are non-toxic and chemically stable in cytoplasm. In one study, large quantities of SPQ were injected intravenously into rabbits, without apparent toxicity, for loading of endosomes from kidney proximal tubule (Bae and Verkman, 1990). To follow are practical considerations on indicator loading, Cl^- sensitivity and specificity, and design and analysis of cell experiments.

SPQ and other compounds with low membrane permeability such as LZQ can be loaded into cells by slow passive diffusion (e.g. 5 mM SPQ overnight in culture medium), hypotonic shock (e.g. 50% hypotonic medium containing 5–10 mM SPQ for 3–5 min), or direct microinjection. Slow passive diffusion is based on the long half-time for indicator equilibration between the cytoplasmic and external compartments by transmembrane diffusion and endocytosis. The cell permeable/trappable indicator diH-MEQ can be loaded into cells by brief incubation (2–10 min) at low concentration (generally 25–100 μM). Several published studies have utilized diH-MEQ (Inglefield and Schwartz-Bloom, 1997; Woll et al., 1996). The choice of loading methods is generally optimized empirically for the cell type being studied, and related practical considerations such as the need for rapid loading in freshly isolated cells and tissues.

Cl^- transport measurements using fluorescent indicators are generally made on cell layers or individual cells grown or immobilized on a solid transparent support. A convenient approach is to grow cells on a round coverglass in a well of a 6- or 12-well plate. The coverglass is mounted in a closed chamber with the cells facing the solution and the cell-free surface facing the objective lens (details in Chao et al., 1989). The thin coverglass permits viewing of the cells with a short working distance, high numerical aperture objective lens having efficient light collection. The cell layer is superfused continuously by gravity or a perfusion pump. Polarized epithelial cells can be grown on transparent, low-autofluorescence permeable supports and mounted in a double perfusion chamber in which the apical and basolateral cell surfaces are perfused independently. The details for construction of a double perfusion chamber are given in Verkman et al. (1992). Cl^- measurements can also be made by cuvette fluorimetry, fluorescence activated cell sorting and *in vivo* fiberoptic methods.

Experimental protocols for Cl^- transport measurements depend on the transporter being studied. Figure 6.2A shows a commonly used strategy for measurement of CFTR-mediated Cl^- transport in cells loaded with a Cl^- indicator. Initially, the cells are bathed in a saline solution containing physiological [Cl^-]. The fluorescence of the cytoplasmic Cl^- indicator

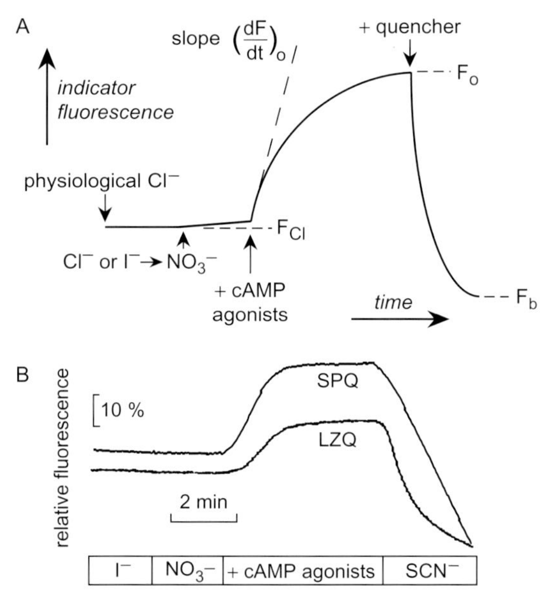

FIGURE 6.2 Measurement of cell membrane Cl^- transport using chemical-type Cl^--sensitive fluorescent indicators. **A**. Strategy for measurement of cAMP-stimulated CFTR Cl^- channel function. Time course of intracellular indicator fluorescence in response to Cl^-/NO_3^- exchange, addition of cAMP agonists and quenching of indicator fluorescence. F_o, fluorescence at zero Cl^-; F_{Cl}, fluorescence at initial intracellular Cl^-; F_b, background fluorescence; $(dF/dt)_o$ initial rate of fluorescence increase after cAMP agonist addition. **B**. Experimental data from CFTR-transfected fibroblasts. I^- and NO_3^- are transported by CFTR, but only I^- quenches indicator fluorescence. CFTR activation by cAMP agonists results in rapid I^- efflux. SCN^- quenches indicator fluorescence to give background fluorescence. Adapted from Jayaraman et al. (1999b).

is initially F_{Cl}. After replacement of the saline solution with an isosmolar solution in which Cl^- is replaced by NO_3^-, an anion that is transported by CFTR but does not quench SPQ fluorescence, Cl^- can exit from cells through a functional Cl^- transporter. If CFTR is the principal membrane Cl^- transporter but is inactive, minimal Cl^- efflux occurs and fluorescence increases little. CFTR Cl^- channel activation following addition of cAMP agonists results in Cl^- efflux and NO_3^- influx, producing an increase in fluorescence. Because external Cl^- concentration is zero, fluorescence ultimately increases to F_o, corresponding to zero intracellular Cl^-. To compute absolute Cl^- flux, a second reference fluorescence value is needed, which can be obtained by replacing the NO_3^--containing solution with SCN^-, a highly permeable ion that strongly quenches SPQ fluorescence. The "background" fluorescence signal with complete indicator quenching is denoted F_b.

For computation of net Cl^- flux (J_{Cl}, in M s^{-1}) from measured fluorescence and from the Stern-Volmer relation, $[Cl^-]_i$ is $[(F_o - F_b)/(F_{Cl} - F_b) - 1]/K_{Cl}$. It follows that,

$$J_{Cl} = -(dF/dt)_o(F_o - F_b)/[(K_{Cl}(F_{Cl} - F_b)^2] \quad \text{(Eq. 1)}$$

where $(dF/dt)_o$ is the slope of the fluorescence vs. time curve, and K_{Cl} (in M^{-1}) is the Stern-Volmer constant for quenching of intracellular indicator Cl^-. It is assumed in Eq. 1 that cell volume remains constant and that only changes in Cl^- affect indicator fluorescence. As discussed in Chao et al. (1989), the Stern-Volmer constant for SPQ in cells (generally 12–20 M^{-1}) is considerably less than that in acellular solutions (118 M^{-1}). From analysis of nanosecond fluorescence lifetimes, it was concluded that the reduced sensitivity of Cl^- indicators in cytoplasm vs. solution is due to quenching of indicator fluorescence by non-Cl^- intracellular anions including proteins and organic solutes. Thus, SPQ fluorescence is mildly sensitive to changes in cell volume (which alters concentration of non-Cl^- anions) and pH (which alters intracellular protein charge). These effects generally preclude use of SPQ and related indicators for studies of cell volume regulation where substantial changes in cell volume occur.

It is noted that J_{Cl} depends on the electrochemical gradient driving Cl^- transport. According to the Goldman equation,

$$J_{Cl} = (P_{Cl}/[1 - e^{-\psi F/RT}])([Cl]_o - [Cl]_i e^{-\psi F/RT}) \quad \text{(Eq. 2)}$$

where P_{Cl} is an apparent Cl^- permeability coefficient, $[Cl]_i$ and $[Cl]_o$ are intracellular and external Cl^- concentrations, respectively, ψ is membrane potential, and RT/F is approximately −60 mvolts at 37°C.

Measurement of K_{Cl} in cells is needed for determination of absolute $[Cl^-]$ and J_{Cl}. Determination of K_{Cl} has been accomplished by measurement of indicator fluorescence vs. Cl^- activity using the Cl^-/OH^- ionophore tributyltin together with buffers containing high K^+ and the K^+/H^+ ionophore nigericin (Chao et al., 1989). Under these conditions, extracellular and intracellular $[Cl^-]$ are nearly equalized. An alternative approach is to replace tributyltin with a combination of ionophores (e.g. valinomycin, nigericin, monensin) that equalize cytoplasmic and extracellular K^+, Na^+ and H^+, secondarily forcing equalization of Cl^-.

Figure 6.2B shows an example of Cl^- transport data obtained using intracellular SPQ and LZQ in Swiss 3T3 fibroblasts expressing wildtype human CFTR. Indicators were loaded by overnight incubation. Where indicated, extracellular I^- was replaced by NO_3^- and forskolin ("+cAMP agonists") was added. After maximum fluorescence was reached corresponding to zero intracellular Cl^-, 150 mM KSCN was added to quench indicator fluorescence.

Cl^- indicators have been used in numerous published studies of Cl^- transport in cultured and freshly isolated cells and tissues (reviewed in Mansoura et al., 1999). In the cystic fibrosis field, for example, Cl^- indicators have been used to establish CFTR function, determine effects of mutagenesis, screen activators and inhibitors, and evaluate the efficacy of gene delivery (Gill et al., 1997; Munkonge et al., 2004; Zamecnik et al., 2007).

III. Cl^- SENSING MACROMOLECULAR CONJUGATES

A. Bisacridinium Cl^- Indicators for Measurements of Organellar $[Cl^-]$

The biacridinium (BAC) chromophore was found to have excellent properties for measurements of organellar $[Cl^-]$. Figure 6.3A shows the structure of a BAC-TMR–dextran conjugate, in which the green-fluorescing Cl^--sensitive BAC chromophore was conjugated to amino dextran together with the red-fluorescing reference chromophore tetramethylrhodamine (TMR) (Sonawane et al., 2002). BAC green fluorescence is quenched by Cl^- by a collision mechanism (Stern-Volmer constant 36 M^{-1}), reducing its fluorescence emission over a wide range of $[Cl^-]$. BAC fluorescence was insensitive to pH in the range appropriate for cellular measurements, and insensitive to cations, non-halide anions (nitrate, phosphate, bicarbonate, sulfate) and albumin.

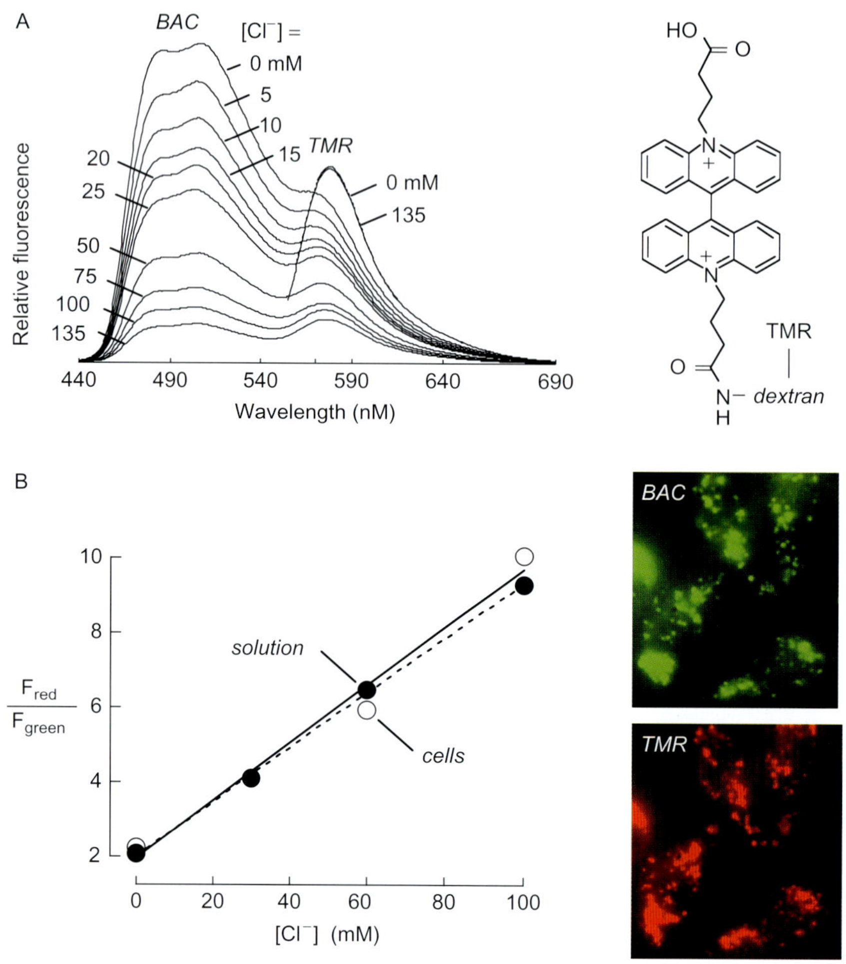

FIGURE 6.3 [Cl⁻] in endosomes measured by ratio imaging BAC-TMR–dextran. **A.** Fluorescence emission spectra of BAC-TMR–dextran (excitation 434 nm-BAC, 540 nm-TMR) at indicated [Cl⁻]. Structure shown on the right. **B.** (left) Calibration of BAC-TMR–dextran fluorescence vs. [Cl⁻] in solution (filled circles) and living cells (open circles). Indicator red-to-green fluorescence ratios shown as a function of [Cl⁻]. J774 cells were pulse-labeled with BAC-TMR–dextran for 2 min, and incubated for 1 hour with buffers containing specified [Cl⁻] and ionophores. (right) Micrographs showing TMR red fluorescence and BAC green fluorescence in BAC-TMR–dextran labeled endosomes. Adapted from Sonawane et al. (2002).

Figure 6.3B (left) shows a calibration of BAC-TMR–dextran red-to-green fluorescence ratio vs. [Cl⁻] in aqueous solution and cells. Endosomes were labeled by fluid-phase endocytosis for 2 min with BAC-TMR–dextran. Representative cell images in Fig. 6.3B (right) show endosomes as distinct bright spots on a dark background. The fluorescent spots became larger with increasing chase time, corresponding to progression from early endosomes to multivesicular bodies to lysosomes. Endosomal [Cl⁻] increased over 45 min from 17 to 53 mM in J774 cells and 28 to 73 mM in CHO cells, during which time endosomal pH decreased from 6.95 to 5.30 (J774) and 6.92 to 5.60 (CHO). The acidification and increased [Cl⁻] were blocked by bafilomycin, an inhibitor of vacuolar (V-type) H^+ ATPases. Together with ion substitution and buffer capacity measurements, it was concluded that Cl^- transport accounts quantitatively for the electrical shunt during endosomal acidification.

To study receptor-mediated endocytosis BAC–dextran was conjugated to various ligands, such as transferrin (marker of early/recycling endosomes) and alpha-2-macroglobulin (marker of late endosomes) as shown in Fig. 6.4A (top) (Sonawane and Verkman, 2003). Direct BAC conjugation to proteins was not possible because of protein quenching effects. Figure 6.4A (bottom) shows the kinetics of endosomal [Cl⁻] following pulse-labeling of a cohort of endosomes with the alpha-2-macroglobulin conjugate. Interestingly, [Cl⁻] was initially low, increased over time in parallel to endosomal acidification, and decreased following proton pump inhibition by bafilomycin. Together with additional data, the results suggested than an interior-negative Donnan potential was responsible for the

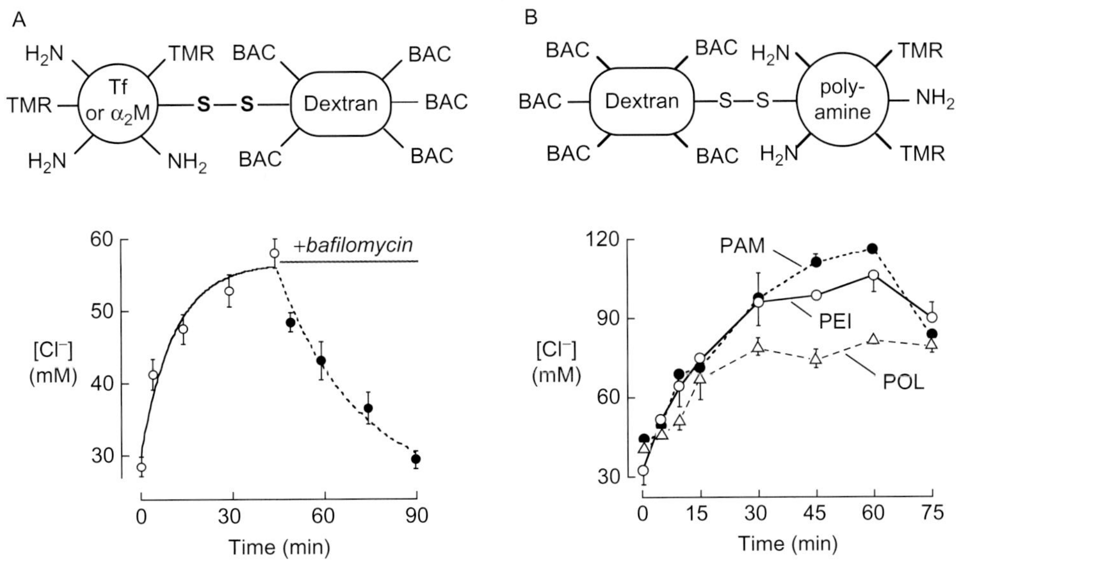

FIGURE 6.4 Ratioable Cl^--sensing protein and polyplex conjugates for [Cl^-] measurements in receptor-mediated endosomes. **A.** (top) TMR-labeled transferrin (Tf) and alpha-2-macroglobulin (α2M) were conjugated to BAC-labeled dextran by a disulfide linker to generate dual-wavelength Cl^- indicators. (bottom) Time course of endosomal [Cl^-] after labeling at 4°C with BAC–dextran–α2M-TMR and perfusion at 37°C. After the initial incubation chase period of 45 min, bafilomycin was added to the perfusate. **B.** (top) BAC–dextran was conjugated to TMR-labeled polyamines as non-viral delivery vehicles. (bottom) Time course [Cl^-] in endosomes containing indicated DNA polyplexes. Polylysine (POL), polyethyleneimine (PEI), polyamidoamine (PAM). Adapted from Sonawane and Verkman (2003) and Sonawane et al. (2003).

low [Cl^-] just after endocytosis. The reduced [Cl^-] and volume in early endosomes was proposed as an elegant mechanism by which endosomal acidification and [Cl^-] accumulation can occur without endosome lysis. Follow-up studies using similar methods provided evidence for the involvement of ClC-type Cl^- channels in the Cl^- shunt that allows endosomal acidification (Hara-Chikuma et al., 2005).

Another interesting study involved the mechanism by which non-viral gene delivery occurs, testing the "proton sponge hypothesis" whereby DNA-polyplexes with high buffer capacity are internalized by endocytosis and ultimately swell/burst because of Cl^- uptake during acidification. For these studies various polyamine conjugates of BAC–dextran were generated (Fig. 6.4B, top) and complexed with DNA (Sonawane et al., 2003). Figure 6.4B (bottom) shows large increases in [Cl^-] with the highly buffering polyamines PAM and PEI, compared to polylyine, which has little buffer capacity at endosomal pH. Together with addition data showing increased volume and lysis of PAM and PEI-containing endosomes, the results provided direct support for the proton sponge hypothesis.

B. Measurements of Extracellular Space [Cl^-]

Another application of Cl^--sensitive dextran conjugates is in the measurement of extracellular space [Cl^-]. One example is [Cl^-] in the airway surface liquid (ASL), the thin layer of fluid at the interface between airway epithelial cells and inhaled/exhaled gases. ASL composition is of importance in the pathophysiology of various airway diseases including cystic fibrosis. Figure 6.5A shows a dextran conjugate used to measure ASL [Cl^-], consisting of a blue-fluorescing quinolinium and a red-fluorescing reference dye (Jayaraman et al., 2001). Indicator red-to-blue fluorescence ratio provided a quantitative measure of [Cl^-], which was used to measured ASL [Cl^-] in tracheal epithelial cell cultures (Fig. 6.5B), as well as in airways in living mice and isolated lung preparations (Song et al., 2003).

IV. GREEN FLUORESCENT PROTEIN-BASED HALIDE INDICATORS

A. Cl^- Sensing by the GFP Mutant YFP-H148Q

The green fluorescent protein (GFP) is a genetically encoded, intrinsically fluorescent protein of ~30 kDa isolated from the jellyfish *Aequoria Victoria* (Tsien, 1998). Compared to chemical probes, genetically encoded fluorescent proteins permit stable, non-invasive staining at specific sites with little cellular toxicity. The GFP chromophore is autocatalytically

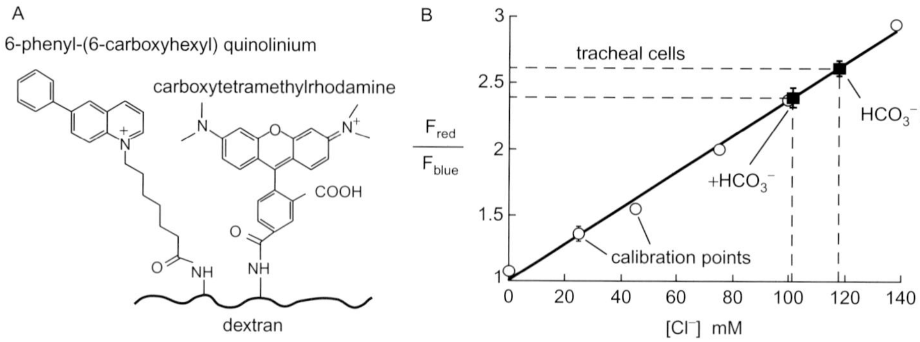

FIGURE 6.5 Airway surface liquid [Cl$^-$] measured using a ratioable dextran conjugate. **A.** Dextran conjugate containing a Cl$^-$ sensing quinolinium (blue fluorescing) and a reference chromophore (carboxytetramethylrhodamine, red fluorescing). **B.** Calibration of red-to-blue fluorescence ratio with [Cl$^-$] shown together measurements in airway surface liquid of tracheal cells. Adapted from Jayaraman et al. (2000).

generated by the post-translational cyclization and oxidation of three residues encoded within its primary sequence. Amino acids throughout GFP have been mutated to generate an array of fluorescent proteins with altered spectral and biophysical properties. Yellow fluorescent protein (YFP), the GFP variant from which available halide sensors are derived, was generated using a rational mutagenic strategy based upon crystallographic data (Wachter et al., 1998). Aromatic amino acids were introduced at Thr203 to extend the π system of the chromophore, lowering its excited state energy and consequently increasing its emission wavelength. YFP contains tyrosine at position 203 as well as mutations that increase its folding efficiency (S65G, V68L, S72A). Maximum fluorescence emission of YFP is at 528 nm, ~20 nm higher than that of GFP-S65T.

Crystal structures of several GFP variants indicate a cylindrical (~40 × 20 Å) structure composed of an 11-standed β-barrel that encloses the chromophore. Although not directly accessible to solvent, the phenolic group within the chromophore of most GFP variants is pH sensitive, with quenching of fluorescence by protonation. The phenolic portion of the chromophore is located near the His148 imidazole ring, where irregularities are found in the β-strand structure. The crystal structure of YFP-H148G (Wachter et al., 1998) showed a cavity in the vicinity of the chromophore permitting solvent access and fluorescence quenching by some halides (Wachter and Remington, 1999). The H148Q mutant of YFP demonstrated characteristics most suitable for cell-based assays of Cl$^-$ flux (Jayaraman et al., 2000).

Fluorescence titrations of purified recombinant YFP-H148Q indicated a pK_a of ~7 in the absence of Cl$^-$ that increased to ~8 at 150 mM Cl$^-$ (Fig. 6.6A) (Jayaraman et al., 2000). At pH 7.5, YFP-H148Q fluorescence decreased with increasing [Cl$^-$] and [I$^-$] with 50% quenching at ~100 mM Cl$^-$ and 21 mM I$^-$. The anion selectivity sequence for YFP-H148Q quenching ($F^- > I^- > NO_3^- > Cl^- > Br^-$) suggested strong binding of weakly hydrated chaotropic ions. YFP-H148Q was insensitive to large ions (including gluconate, sulfate, phosphate and isothionate). A static quenching mechanism involving Cl$^-$ binding to YFP-H148Q was established by biophysical studies (Jayaraman et al., 2000). The YFP-H148Q fluorescence lifetime was insensitive to Cl$^-$, whereas YFP-H148Q molar absorbance decreased with increasing Cl$^-$. Crystallographic studies of YFP-H148Q in the presence of iodide identified a discrete binding site for halides (Wachter et al., 2000). Bound iodide was shown to interact with the chromophore and the phenol group of T203Y with nearby amino acids helping to stabilize the YFP-H148Q/halide interaction. Halide binding to YFP-H148Q thus stabilizes the deprotonated form of the chromophore mimicking a lower pH at the chromophore.

Stopped-flow fluorescence analysis established the kinetics and mechanism of YFP-H148Q quenching by Cl$^-$. In the absence of Cl$^-$, YFP-H148Q fluorescence changes rapidly ($t_{1/2} < 10$ ms) in response to pH changes, whereas the fluorescence response was biexponential in the presence of 100 mM Cl$^-$, with fast ($t_{1/2} < 10$ ms) and slower ($t_{1/2}$~100 ms) components. Cl$^-$ dissociation and association with the deprotonated chromophore (pH 8) was relatively slow with $t_{1/2}$~200 ms, but faster ($t_{1/2}$~70 ms) for the protonated chromophore (pH 6.4). A kinetic model incorporating these data contains four equilibria involving YFP-H148Q protonation and Cl$^-$ binding (Fig. 6.6B). The rapid kinetics of these changes in fluorescence permits application of the YFP-H148Q to the generally much slower cell-based assays of Cl$^-$ flux.

YFP-H148Q was expressed in tissue culture cell lines to test its utility in cell-based assays of [Cl$^-$]$_i$ (Fig. 6.6C). YFP-H148Q fluorescence was observed throughout the cell cytoplasm and nucleus. *In vivo*

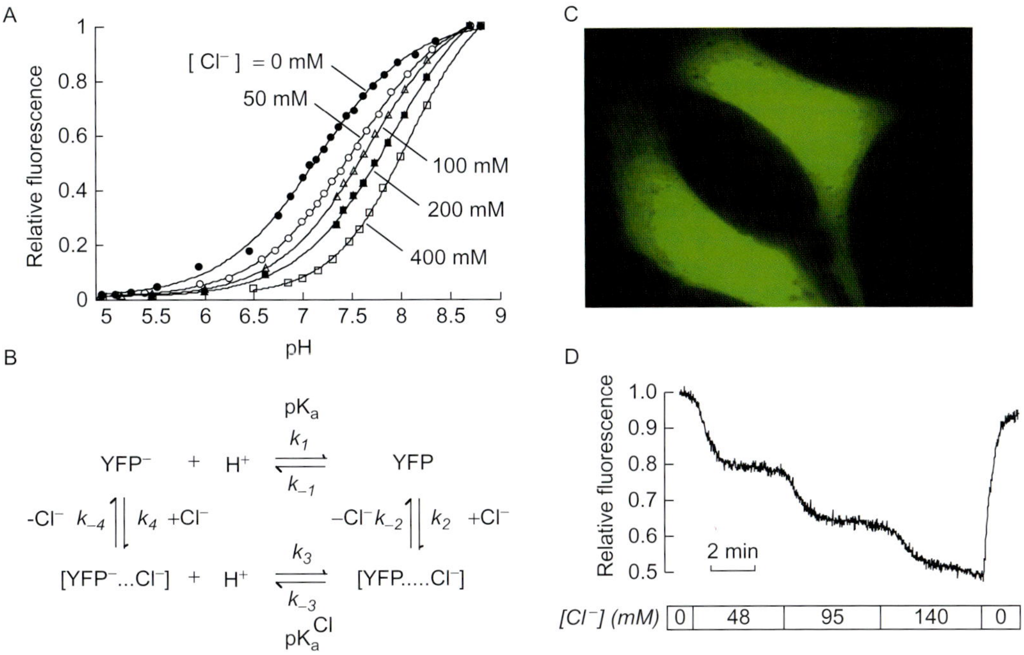

FIGURE 6.6 YFP-H148Q as a genetically encoded Cl^- sensing protein. **A.** Dependence of YFP-H148Q fluorescence on pH at indicated $[Cl^-]$. **B.** Kinetic scheme for YFP-H148Q interaction with Cl^- deduced from kinetic data. Rate constants: $k_1 = 1.4 \times 10^7\ s^{-1}$; $k_{-1} = 2.7 \times 10^7\ s^{-1}$; $k_2 = 5.6 \times 10^8\ M^{-1}s^{-1}$; $k_{-2} = 1.7 \times 10^7\ s^{-1}$; $k_3 = 3.5 \times 10^6\ s^{-1}$; $k_{-3} = 2.3 \times 10^6\ s^{-1}$; $k_4 = 3.6 \times 10^7\ M^{-1}s^{-1}$; $k_{-4} = 9.3 \times 10^6\ s^{-1}$; relative fluorescence: $[YFP^-] = 1$; $[YFP^- ... Cl^-] = 0.3$; YFP and $[YFP ... Cl^-] = 0$. **C.** Confocal fluorescence micrograph of YFP-H148Q transfected fibroblasts expressing CFTR showing uniform cytoplasmic and nuclear staining. **D.** Intracellular fluorescence Cl^- titration at pH 7.4 using high K^+ buffer containing ionophores and indicated $[Cl^-]$. Adapted from Jayaraman et al. (2000).

calibration using ionophore-treated cell cultures indicated similar pH and Cl^- sensitivities of YFP-H148Q in cells and aqueous solutions (Fig. 6.6D).

B. YFP Mutagenesis Identifies Improved YFP Halide Sensors

Although a good probe for some cell-based assays of Cl^-/halide flux, we sought to discover variants of YFP-H148Q with increased sensitivity. In particular, the anion conductance of the cystic fibrosis-causing ΔF508-CFTR channel is greatly reduced when compared to wildtype CFTR, requiring a very sensitive YFP-H148Q variant for high-throughput screening. Based upon the structural data random mutations were introduced in six hydrophobic residues lining the YFP-H148Q halide binding site in an attempt to modify the polarity and/or size of the cavity and thus halide binding affinity (Fig. 6.7) (Galietta et al., 2001). Degenerate primers were used to generate YFP-H148Q libraries containing mutations in the residue pairs V150/I152, V163/F165, and L201/T203. The mutagenesis procedure generated a diverse library as indicated by the different fluorescence of bacterial colonies (Fig. 6.7, left, middle). Colonies were grown, replicated to agar plates, and lysed *in situ*. K_d values for Cl^- and I^- binding were determined using a fluorescence plate reader.

Screening of >1000 colonies yielded YFP-H148Q mutants with significantly different Cl^- and I^- sensitivities compared to YFP-H148Q. The mutants V150T, I152L/Y, V163T/L, V150A/I152L, V163A/F165Y and V163T/F165Y had K_d values for I^- of <15 mM. We characterized the I152L mutant further because of its low K_d for I^- of 3 mM. Dissociation constants for Cl^-, I^- and NO_3^- at cytoplasmic pH, determined by fluorescence titrations using purified recombinant YFP-H148Q/I152L, were 85, 10 and 2 mM, respectively (Fig. 6.8A, left). As found for YFP-H148Q, indicator pK_a decreased with increasing Cl^- (Fig. 6.8A, right); for I152L, the pK_a was 6.95 in the absence of Cl^-, increasing to 7.70 and 7.89 in the presence of 75 and 150 mM Cl^-, respectively. Stopped-flow fluorescence analysis indicated a rapid response of YFP-H148Q/I152L fluorescence to changes in Cl^- concentration (Fig. 6.8A, inset).

YFP-H148Q/I152L was transfected into CFTR-expressing cells to test its suitability for screening applications. Replacement of 20 mM Cl^- by I^- produced a slow decline in fluorescence due to basal CFTR activity, which increased rapidly with addition of the cAMP agonist forskolin (Fig. 6.8B). The maximum

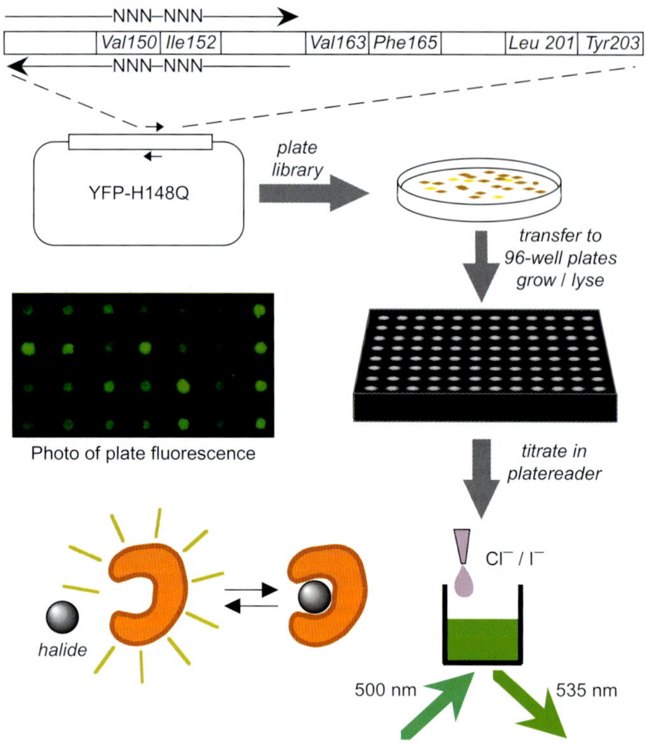

FIGURE 6.7 Strategy for generating and screening of YFP mutational libraries. Indicated pairs of amino acids were randomly mutated using appropriate primers. Transformed bacterial colonies were transferred to 96-well plates for growth, replication to agar plates, lysis, and screening. (left, middle) Photograph of a section of a replicate agar plate showing bacteria with differing amounts of fluorescence.

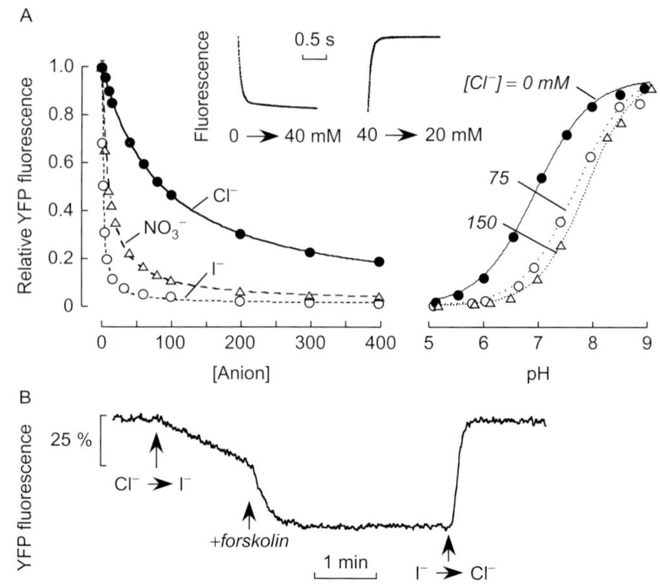

FIGURE 6.8 Characterization of purified YFP-H148Q/I152L and application to cell-based assay of CFTR halide transport. **A.** Titration of YFP-H148Q/I152L with Cl^-, I^- and NO_3^- done at cytoplasmic pH (left). Fluorescence pH titrations of YFP-H148Q/I152L done at indicated $[Cl^-]$ (right). Inset shows Cl^- association and dissociation kinetics measured by stopped-flow fluorimetry. **B.** Time course of YFP-H148Q/I152L fluorescence in CFTR-expressing cells in response to exchange of 20 mM Cl^- for I^- and addition of forskolin. Adapted from Galietta et al. (2001).

fluorescence decrease of ~50% was much greater than that of <10% in an identical experiment done using YFP-H148Q.

C. YFP Applications to Cl^- Transport Modulator Discovery by High-throughput Screening

The halide-sensing YFPs were used in several high-throughput screening projects to identify Cl^- channel activators and inhibitors. In an initial project to discover new classes of CFTR modulators with improved potency and selectivity, 60,000 diverse drug-like small molecules were screened (Ma et al., 2002b). Compounds were tested in epithelial cells co-expressing human wildtype CFTR and the YFP sensor. Primary screening consisted of short-term stimulation of cells with 10 μM test compound and 0.5 μM forskolin, followed by I^- challenge (Fig. 6.9A). Figure 6.9B shows representative fluorescence data from single wells of 96-well plates, showing saline control, dose–response of the known CFTR activator apigenin, and examples of test compounds have different activities. The screen yielded 57 strong activators (greater activity than reference compound apigenin), most of which were unrelated in chemical structure to known CFTR activators. Secondary analysis yielded 14 compounds that activated CFTR without cAMP elevation or phosphatase inhibition, suggesting direct CFTR interaction. Figure 6.9C shows two examples of CFTR activators from the screen with submicromolar potency. Similar activator screens were applied to identify potentiators (correctors of defective Cl^- channel gating) and correctors (correctors of defective protein processing) of ΔF508-CFTR, the most common CFTR mutation causing cystic fibrosis. The screens involved measurements of I^- influx in epithelial cells co-expressing human ΔF508-CFTR and YFP-H148Q/I152L. The screens yielded potentiators with nanomolar potency (Pedemonte et al., 2005b; Yang et al., 2003) and correctors with low micromolar potency (Pedemonte et al., 2005a) with promising drug-like properties for pre-clinical development.

Similar screens were successful in identifying the first potent CFTR inhibitors. Compounds were tested for their ability to inhibit I^- influx in CFTR-expressing cells following CFTR activation by cAMP agonists. Out of 200,000 compounds screened, one class of

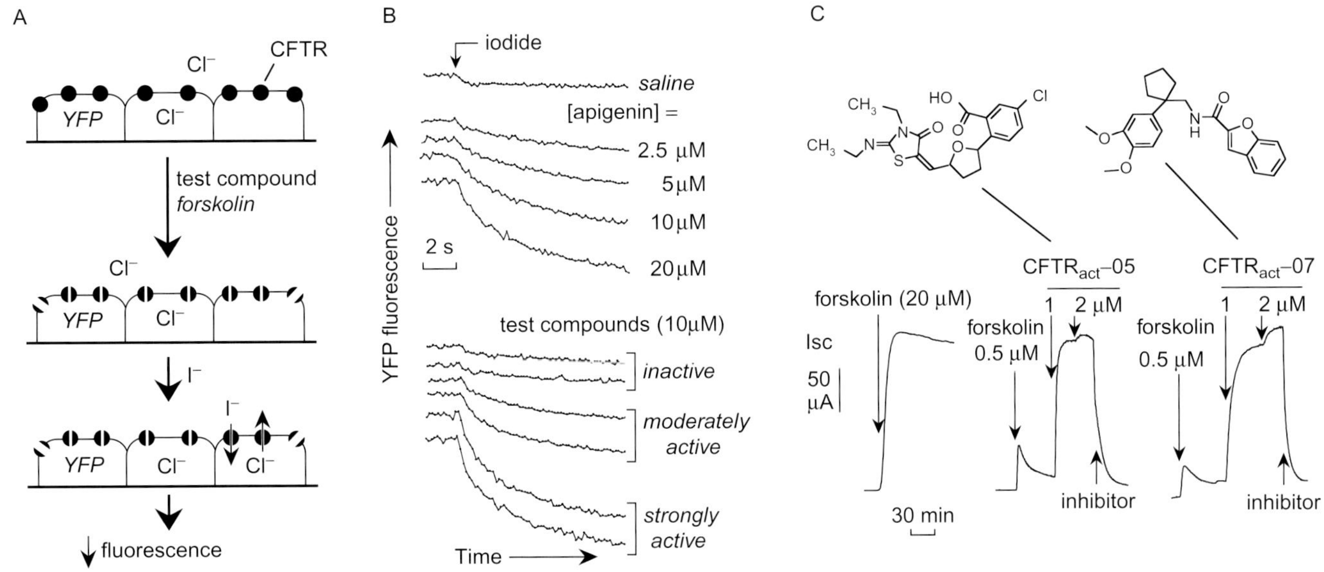

FIGURE 6.9 Identification of small-molecule CFTR activators by high-throughput screening. **A.** Screening protocol. Cells co-expressing CFTR and YFP-H148Q in 96-well microplates were exposed to a low concentration of forskolin together with test compounds (at 10 μM). I^- influx was induced by adding an I^- containing solution. **B.** Representative fluorescence data from individual wells showing saline controls (no agonist), dose–response of the agonist apigenin, and examples of test compounds. **C.** Structures of two CFTR agonists identified in a screen of 60,000 compounds (top), with short-circuit current measurement of Cl^- conductance showing submicromolar activation potency (bottom). Adapted from Ma et al. (2002b).

molecules, the thiazolidinones (such as $CFTR_{inh}$-172), inhibited CFTR with IC50~200 nM (Ma et al., 2002a). These compounds act from the cell interior and stabilize the channel closed state. A second class of molecules, the glycine hydrazides, inhibit CFTR from the cell exterior by blocking the CFTR pore (Muanprasat et al., 2004). Membrane-impermeant glycine hydrazides were synthesized by conjugation to polyethylene glycol (Sonawane et al., 2006) and various lectins (Sonawane et al., 2007), which fully inhibited CFTR when added externally. The CFTR inhibitors emerging from compound screening and follow-up chemistry have potential applications to the therapy of enterotoxin-mediated secretory diarrheas such as cholera and Travelers' diarrhea (Sonawane et al., 2007; Thiagarajah et al., 2004) and polycystic kidney disease (Yang et al., 2008). Similar YFP cell-based inhibitor screening identified the first potent inhibitors of the calcium-activated Cl^- channel (de la Fuente et al., 2008).

V. CONCLUSIONS

As reviewed in this chapter, chemical and GFP-based Cl^- sensing probes are available for a wide variety of cellular applications, including measurements of $[Cl^-]$ in cells, organelles and extracellular space compartments. Though these indicators are now widely used, there remains a need for improved indicators to overcome limitations such as intrinsic pH sensitivity and relatively dim fluorescence of some currently used indicators.

References

Bae, H.R. and Verkman, A.S. (1990). Protein kinase A regulates chloride conductance in endocytic vesicles from proximal tubule. *Nature* **348**, 637–639.

Biwersi, J. and Verkman, A.S. (1991). Cell-permeable fluorescent indicator for cytosolic chloride. *Biochemistry* **30**, 7879–7883.

Biwersi, J., Farah, N., Wang, Y.X., Ketcham, R., and Verkman, A.S. (1992). Synthesis of cell-impermeable Cl-sensitive fluorescent indicators with improved sensitivity and optical properties. *Am. J. Physiol.* **262**, C242–C250.

Biwersi, J., Tulk, B., and Verkman, A.S. (1994). Long-wavelength chloride-sensitive fluorescent indicators. *Anal. Biochem.* **219**, 139–143.

Chao, A.C., Dix, J.A., Sellers, M.C., and Verkman, A.S. (1989). Fluorescence measurement of chloride transport in monolayer cultured cells. *Mechanisms of chloride transport in fibroblasts. Biophys. J.* **56**, 1071–1081.

Coles, J.A. (1995). Measurement of cytosolic chloride activity by ion-selective microelectrodes. In *Methods in Neurosciences* (Kraicer, J. and Dixon, S.J., eds), Vol. 27, *Measurements and Manipulation of Intracellular Ions*, pp. 340–360. Academic Press, Orlando.

de la Fuente, R., Namkung, W., Mills, A., and Verkman, A.S. (2008). Small-molecule screen identifies inhibitors of a human intestinal calcium-activated chloride channel. *Mol. Pharmacol.* **73**, 758–768.

Galietta, L.J., Haggie, P.M., and Verkman, A.S. (2001). Green fluorescent protein-based halide indicators with improved chloride and iodide affinities. *FEBS Lett.* **499**, 220–224.

Gill, D.R., Southern, K.W., Mofford, K.A., Seddon, T., Huang, L., Sorgi, F., Thomson, A., MacVinish, L.J., Ratcliff, R., Bilton, D., Lane, D.J., Littlewood, J.M., Webb, A.K., Middleton, P.G., Colledge, W.H., Cuthbert, A.W., Evans, M.J., Higgins, C.F., and Hyde, S.C. (1997). A placebo-controlled study of liposome-mediated gene transfer to the nasal epithelium of patients with cystic fibrosis. *Gene Ther.* **4**, 199–209.

Hara-Chikuma, M., Yang, B., Sonawane, N.D., Sasaki, S., Uchida, S., and Verkman, A.S. (2005). ClC-3 chloride channels facilitate endosomal acidification and chloride accumulation. *J. Biol. Chem.* **280**, 1241–1247.

Illsley, N.P. and Verkman, A.S. (1987). Membrane chloride transport measured using a chloride-sensitive fluorescent probe. *Biochemistry* **26**, 1215–1219.

Inglefield, J.R. and Schwartz-Bloom, R.D. (1997). Confocal imaging of intracellular chloride in living brain slices: measurement of GABA receptor activity. *J. Neurosci. Meth.* **75**, 127–135.

Jayaraman, S., Haggie, P., Wachter, R.M., Remington, S.J., and Verkman, A.S. (2000). Mechanism and cellular application of a green fluorescent protein-based halide sensor. *J. Biol. Chem.* **275**, 6047–6050.

Jayaraman, S., Biwersi, J., and Verkman, A.S. (1999a). Synthesis and characterization of dual-wavelength chloride sensitive fluorescent indicators for ratio imaging. *Am. J. Physiol.* **276**, C747–C757.

Jayaraman, S., Song, Y., Vetrivel, L., Shankar, L., and Verkman, A.S. (2001). Non-invasive in vivo fluorescence measurement of airway surface liquid depth, salt concentration, and pH. *J. Clin. Invest* **107**, 317–324.

Jayaraman, S., Teitler, L., Skalski, B., and Verkman, A.S. (1999b). Long-wavelength iodide-sensitive fluorescent indicators for measurement of functional CFTR expression in cells. *Am. J. Physiol.* **277**, C1008–C1018.

Jayaraman, S. and Verkman, A.S. (2000). Charge transfer mechanism for quenching of quinolinium fluorescence by halides. *Biophys. Chem.* **85**, 49–57.

Krapf, R., Illsley, N.P., Tseng, H.C., and Verkman, A.S. (1988). Structure–activity relationships of chloride-sensitive fluorescent indicators for biological application. *Anal. Biochem.* **169**, 142–150.

Legg, K.D. and Hercules, D.M. (1970). Quenching of lucigenin fluorescence. *J. Phys. Chem.* **74**, 2114–2121.

Ma, T., Thiagarajah, J.R., Yang, H., Sonawane, N.D., Folli, C., Galietta, L.J., and Verkman, A.S. (2002a). Thiazolidinone CFTR inhibitor identified by high-throughput screening blocks cholera toxin-induced intestinal fluid secretion. *J. Clin. Invest* **110**, 1651–1658.

Ma, T., Vetrivel, L., Yang, H., Pedemonte, N., Zegarra-Moran, O., Galietta, L.J., and Verkman, A.S. (2002b). High-affinity activator of cystic fibrosis transmembrane conductance regulator (CFTR) chloride conductance identified by high-throughput screening. *J. Biol. Chem.* **277**, 37235–37241.

Mansoura, M.K., Biwersi, J., Ashlock, M.A., and Verkman, A.S. (1999). Fluorescent chloride indicators to assess the efficacy of CFTR delivery. *Hum. Gene Ther.* **10**, 861–875.

Muanprasat, C., Sonawane, N.D., Salinas, D., Taddei, A., Galietta, L.J., and Verkman, A.S. (2004). Discovery of glycine hydrazide pore-occluding CFTR inhibitors: mechanism, structure–activity analysis, and in vivo efficacy. *J. Gen. Physiol.* **124**, 125–137.

Munkonge, F., Alton, E.W., Andersson, C., Davidson, H., Dragomir, A., Edelman, A., Farley, R., Hjelte, L., McLachlan, G., Stern, M., and Roomans, G.M. (2004). Measurement of halide efflux from cultured and primary airway epithelial cells using fluorescence indicators. *J. Cyst. Fibros.* **3** (2), 171–176.

O'Connor, N. and Silver, R.B. (2007). Ratio imaging: practical considerations for measuring intracellular Ca^{2+} and pH in living cells. *Methods Cell Biol.* **81**, 415–433.

Pedemonte, N., Lukacs, G.L., Du, K., Caci, E., Zegarra-Moran, O., Galietta, L.J., and Verkman, A.S. (2005a). Small molecule correctors of defective ΔF508-CFTR cellular processing identified by high-throughput screening. *J. Clin. Invest* **115**, 2564–2571.

Pedemonte, N., Sonawane, N.D., Taddei, A., Hu, J., Zegarra-Moran, O., Suen, Y.F., Galietta, L.J., and Verkman, A.S. (2005b). Phenylglycine and sulfonamide correctors of defective deltaF508 and G551D cystic fibrosis transmembrane conductance regulator chloride-channel gating. *Mol. Pharmacol.* **67**, 1797–1807.

Skalski, B., Paszyc, S., Adamiak, R.W., Steer, R.P., and Verrall, R.E. (1989). Photophysical studies of luminarosine – a new, highly fluorescent ribonucleoside with pteridine-like betaine as the aglycone. *J. Chem. Soc. Perkin Trans.* **II**, 1691–1696.

Sonawane, N. and Verkman, A.S. (2003). Determinants of [Cl^-] in recycling and late endosomes and Golgi complex measured using fluorescent ligands. *J. Cell Biol.* **160**, 1129–1138.

Sonawane, N., Szoka, F.C., and Verkman, A.S. (2003). Chloride accumulation and swelling in endosomes enhances DNA transfer by polyamine-DNA polyplexes. *J. Biol. Chem.* **278**, 44826–44831.

Sonawane, N., Thiagarajah, J.R., and Verkman, A.S. (2002). Chloride concentration in endosomes measured using a ratioable fluorescent Cl^- indicator: evidence for Cl^- accumulation during acidification. *J. Biol. Chem.* **277**, 5506–5513.

Sonawane, N., Zhao, D., Zegarra-Mora, O., Galietta, L.J.V., and Verkman, A.S. (2007). Lectin conjugates as potent, nonabsorbable CFTR inhibitors for reducing intestinal fluid secretion in cholera. *Gastroenterology* **132**, 1234–1244.

Sonawane, N.D., Hu, J., Muanprasat, C., and Verkman, A.S. (2006). Luminally-active, nonabsorbable CFTR inhibitors as potential therapy to reduce intestinal fluid losses in cholera. *FASEB J.* **20**, 130–132.

Song, Y., Thiagarajah, J., and Verkman, A.S. (2003). Ionic concentrations, pH and depth of airway surface liquid in distal airways. *J. Gen. Physiol.* **122**, 511–519.

Thiagarajah, J.R., Broadbent, T., Hsieh, E., and Verkman, A.S. (2004). Prevention of toxin-induced intestinal ion and fluid secretion by a small-molecule CFTR inhibitor. *Gastroenterology* **126**, 511–519.

Tsien, R.Y. (1998). The green fluorescent protein. *Annu. Rev. Biochem.* **67**, 509–544.

Verkman, A.S. (1990). Development and biological applications of chloride-sensitive fluorescent indicators. *Am. J. Physiol.* **259**, C375–C388.

Verkman, A.S., Chao, A.C., and Hartmann, T. (1992). Hormonal regulation of Cl transportin polar airway epithelia measured by a fluorescent indicator. *Am. J. Physiol.* **262**, C23–C31.

Verkman, A.S., Sellers, M.C., Chao, A.C., Leung, T., and Ketcham, R. (1989a). Synthesis and characterization of improved chloride-sensitive fluorescent indicators for biological applications. *Anal. Biochem.* **178**, 355–361.

Wachter, R.M. and Remington, S.J. (1999). Sensitivity of the yellow variant of green fluorescent protein to halides and nitrate. *Curr. Biol.* **9**, R628–R629.

Wachter, R.M., Elsliger, M.-A., Kallio, K., Hanson, G.T., and Remington, S.J. (1998). Structural basis of spectral shifts in the yellow-emission variants of green fluorescent protein. *Structure* **6**, 1267–1277.

Wachter, R.M., Yarbough, D., Kallio, K., and Remington, S.J. (2000). Crystallographic and energetic analysis of binding of selected anions to the yellow variants of green fluorescent protein. *J. Mol. Biol.* **301**, 157–171.

Woll, E., Gschwentner, M., Furst, J., Hofer, S., Buemberger, G., Jungwirth, A., Frick, J., Deetjen, P., and Paulmichl, M. (1996). Fluorescence-optical measurements of chloride movements in cells using the membrane potential dye diH-MEQ. *Pflugers Arch.* **432**, 486–493.

Yang, H., Shelat, A.A., Guy, R.K., Gopinath, V.S., Ma, T., Du, K., Lukacs, G.L., Taddei, A., Folli, C., Pedemonte, N., Galietta, L.J.V., and Verkman, A.S. (2003). Nanomolar-affinity small-molecular activators of ΔF508-CFTR chloride channel gating. *J. Biol. Chem.* **278**, 35079–35085.

Yang, B., Sonawane, N.D., Zhao, D., Somlo, S., and Verkman, A.S. (2008). Small-molecule CFTR inhibitors slow cyst growth in polycystic kidney disease. *J. Am. Soc. Nephrol.* **19**, 1300–1310.

Zamecnik, P.C., Raychowdhury, M.K., Tabatadze, D.R., and Cantiello, H.F. (2007). Reversal of cystic fibrosis phenotype in a cultured ΔF508 cystic fibrosis transmembrane conductance regulator cell line by oligonucleotide insertion. *Proc. Natl. Acad. Sci. USA* **25**, 8150–8155.

CHAPTER

7

Clomeleon, a Genetically Encoded Chloride Indicator

Ken Berglund, Thomas Kuner and George J. Augustine

OUTLINE

I. INTRODUCTION

The regulation of intracellular chloride concentration ($[Cl^-]_i$) is tightly linked to multiple cellular mechanisms: pH regulation, potassium distribution, membrane potential, transmembrane Cl^- fluxes, cell volume regulation and oxygen metabolism. Because of this intricate association with multiple regulatory loops, it is essential to be able to measure $[Cl^-]_i$ directly. However, limitations in the properties of organic indicator dyes have made direct measurement of $[Cl^-]_i$ challenging (see Chapter 6 in this volume).

Here we summarize the properties of Clomeleon, a genetically encoded fluorescent indicator that permits non-invasive, spatiotemporally resolved optical recordings of steady-state as well as dynamic changes of $[Cl^-]_i$. We also review the application of Clomeleon to $[Cl^-]_i$ measurements in neurons, the cell type that interests us the most.

II. PROPERTIES OF CLOMELEON

A. Fluorescence Resonance Energy Transfer

Clomeleon is a fusion protein consisting of the cyan fluorescent protein (CFP) and the topaz variant of yellow fluorescent protein (YFP) (Kuner and Augustine, 2000). A linker of 24 amino acid residues containing an rTEV (recombinant tobacco etch virus protease) cleavage site connects the C-terminus of CFP to the N-terminus of YFP (Fig. 7.1a). This linker keeps the two fluorescent proteins in close proximity,

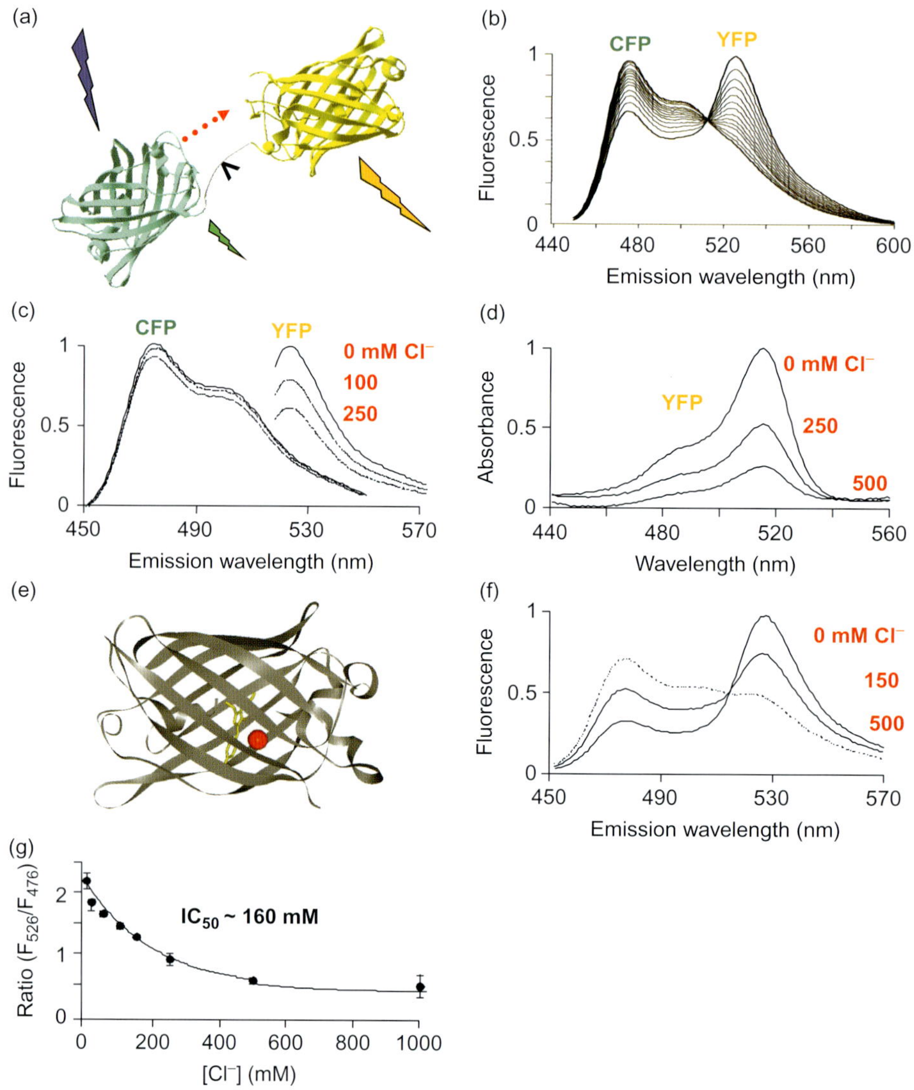

FIGURE 7.1 Design and basic properties of Clomeleon. (a) Structure of Clomeleon. CFP (cyan) and Topaz (yellow) are connected with a short linker (gray). The protease site recognized by the rTEV is marked by an arrowhead. Excitation of CFP (blue lightning bolt) causes direct emission from CFP (cyan lightning bolt) and indirect emission from Topaz (yellow lightning bolt) through FRET (dotted red arrow). (b) Emission spectrum of Clomeleon during proteolysis of the linker region. CFP was excited by 434 nm light while the recombinant Clomeleon protein was treated with the rTEV protease for variable lengths of time. Progressive loss of the YFP moiety of Clomeleon by the protease caused loss of YFP fluorescence and increasing CFP fluorescence, due to a loss of FRET. (c) Emission spectra of recombinant CFP and YFP excited at 434 nm and 510 nm, respectively. CFP and YFP spectra are normalized to peak emission measured in the absence of Cl^-. (d) Absorbance of recombinant YFP at different [Cl^-]. Absorbance is normalized to the peak absorption measured in the absence of Cl^-. (e) Crystal structure of YFP(H148Q). Shown are the backbone of the protein in ribbon representation (black), the chromophore in stick representation (yellow), and one of iodide ions (red). This figure was drawn using the program 3D Molecule Viewer (Invitrogen) based on the atomic coordinates found in the RCSB Protein Data Bank (accession code: 1F09; Wachter et al., 2000). (f) Emission spectra of recombinant Clomeleon in the presence of different [Cl^-]. In all cases, Clomeleon was excited at 434 nm and the emission spectra were normalized to the 527 nm emission peak measured in the absence of Cl^-. (g) Clomeleon fluorescence emission ratio (526 nm/476 nm) is a unique function of [Cl^-]. Points indicate mean ± SEM (n = 3, same batch of protein) and the curve is an exponential fit to the data. Panels c, f and g are reproduced from Kuner and Augustine (2000).

permitting fluorescence resonance energy transfer (FRET) between CFP and YFP. Thus, illumination of Clomeleon with blue light (~430–460 nm, blue lightning bolt in Fig. 7.1a) excites CFP, resulting in radiationless FRET (red dotted arrow in Fig. 7.1a) to YFP. In this case, CFP fluoresces weakly while YFP produces a robust emission (cyan and yellow lightning bolts in Fig. 7.1a). After cleavage of the linker peptide by rTEV protease, illumination with blue light results only in emission of cyan fluorescence but no more yellow emission, directly demonstrating that YFP must be excited via FRET from the excited CFP (Fig. 7.1b). The ratio between the YFP (530 nm) and CFP (480 nm) peak emissions decreases as the YFP acceptor is removed from the CFP donor, until the spectrum of CFP alone defines a minimal ratio (R_{min}). The maximal ratio (R_{max}) is set by the FRET efficiency, mostly limited by the minimal distance between the two fluorophores imposed by the barrel structure of GFP. The protease sensitivity was engineered into Clomeleon for calibration purposes and is not required for its function as a Cl^- indicator.

B. Cl^- Sensing

The sensor function of Clomeleon is mediated by an intrinsic sensitivity of YFP to halides (Wachter and Remington, 1999; Kuner and Augustine, 2000). Exposing YFP protein to solutions with increasing concentrations of Cl^- causes a reduction of YFP emission, while CFP emission remains largely unaffected (Fig. 7.1c). How does Cl^- reduce YFP fluorescence? Cl^- binding puts the YFP fluorophore into a non-absorbing state, as can be seen by the reduced light absorbance of YFP at increasing Cl^- concentrations (Fig. 7.1d). This is consistent with the crystal structure of YFP, which shows that one of the two Cl^- binding sites of YFP is close to the chromophore and, in fact, the chromophore contributes to coordination of halide ions (Fig. 7.1e). Binding of Cl^- alters the protonation state of the fluorophore, and thereby prevents its excitation (Wachter et al., 2000). Hence, Cl^- binding to Clomeleon removes YFP as an acceptor for CFP and therefore changes the YFP/CFP emission ratio, similar to the effect caused by rTEV cleavage of Clomeleon into its constituent fluorescent proteins, as described above. These features provide the basis for Clomeleon as a ratiometric indicator for Cl^-.

When linked with CFP, YFP can act as a FRET acceptor for CFP in the nominal absence of Cl^-, evident as a large peak of YFP fluorescence emission and a small peak of CFP fluorescence emission (Fig. 7.1f). Increasing concentrations of Cl^- progressively remove YFP as an acceptor and thereby increase CFP fluorescence. The YFP to CFP peak ratios change with the concentration of Cl^- and can be described by the Hill equation (Fig. 7.1g). Therefore, the ratio of YFP and CFP fluorescence emission can be directly translated into $[Cl^-]$ using an appropriate calibration curve, and thus Clomeleon can be used as a ratiometric indicator for $[Cl^-]$.

C. Comparison to Other Genetically Encoded Indicators

Clomeleon differs from most other genetically encoded indicators (Hasan et al., 2004; Pologruto et al., 2004; Reiff et al., 2005), because the sensor function is provided by the fluorescent protein itself and not by a distinct sensing moiety. Apart from pHluorin, a pH-sensitive variant of GFP that can be used for detection of H^+ (Miesenböck et al., 1998), all other genetically encoded indicators are based on the principle that binding of the ligand results in changes in the distance and/or orientation between CFP and YFP. For example, Ca^{2+} binding to Cameleon results in a conformational change mediated by the M13 helix, pulling CFP and YFP closer together (Miyawaki et al., 1997). Because FRET efficiency changes with the 6th power of the radius between the fluorophores, minute conformational changes caused by ligand binding can be detected. While sensing of Cl^- by Clomeleon is based on FRET, the Cl^--induced change in FRET does not require any such conformational change.

D. Ionic Selectivity

In addition to Cl^-, the halides F^-, I^- and Br^- also bind to YFP and change Clomeleon's fluorescence emission (Fig. 7.2a). The sequence of affinities is 6 ± 2, 46 ± 14, 111 ± 21 and 167 ± 13 mM (mean ± SEM) for F^-, I^-, Br^- and Cl^-, respectively (Kuner and Augustine, 2000). Among these, Cl^- is the only physiologically relevant halide because most tissues and cells do not contain F^-, I^- or Br^- in concentrations relevant for binding to Clomeleon. However, we frequently use the high-affinity binding of F^- to Clomeleon as a way to conveniently define R_{min} in cellular calibration experiments (see below).

Other intracellular anions – such as glutamate, ATP, HCO_3^-, PO_4^{2-}, SO_4^{2-} and NO_3^- – do not affect Clomeleon at their physiological concentrations (Fig. 7.2b). Also anions commonly used in electrophysiology, such as gluconate and methylsulfate, do not affect the readout of Clomeleon. Hence, under physiological conditions, Cl^- is the only anion that is sensed by Clomeleon.

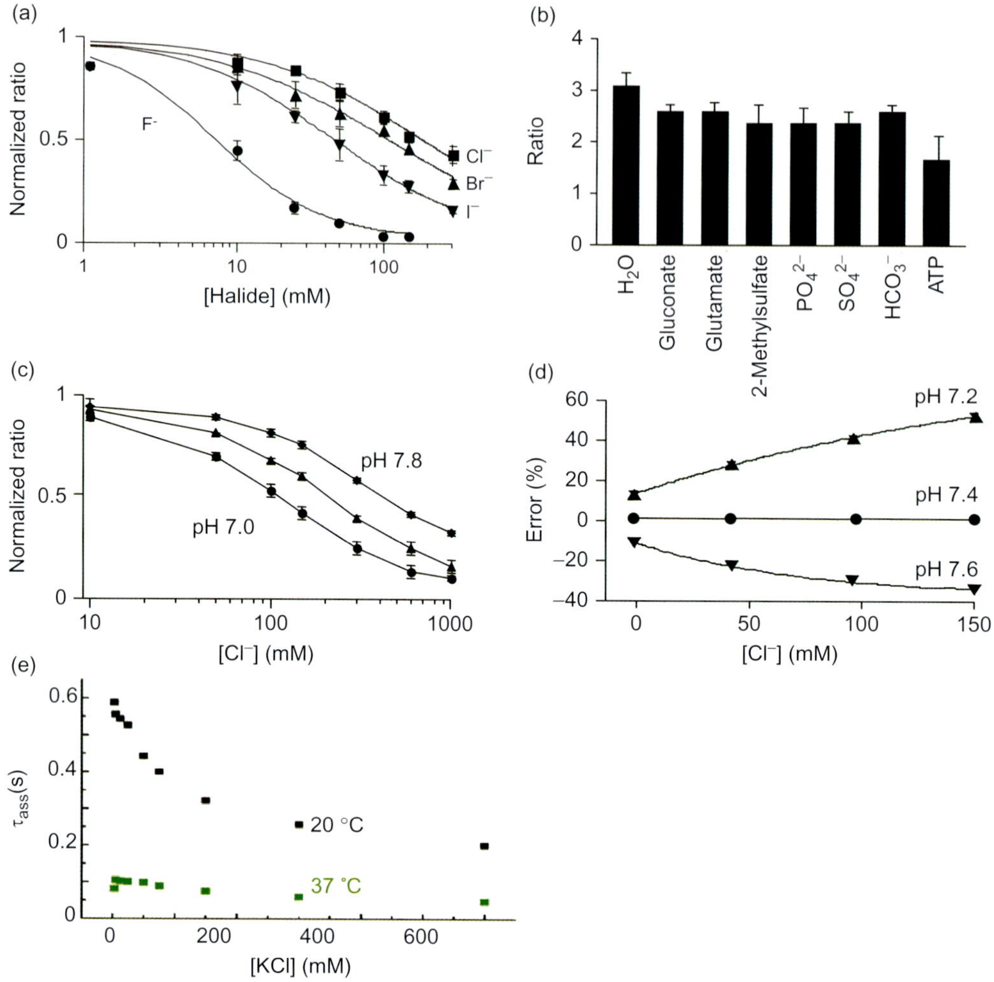

FIGURE 7.2 Ionic selectivity of Clomeleon. (a) Titration curves for recombinant Clomeleon protein in the presence of various halides. Symbols designate mean ± SEM (n = 3–5). Curves are Hill equation fits to the data. Ratios were normalized for each experiment to R_{max} obtained in K-gluconate solution (150 mM) and R_{min} obtained in KF solution (150 mM). (b) Anion selectivity of Clomeleon. Values (mean ± SEM; n = 3) indicate Clomeleon emission ratio under the indicated ionic conditions. (c) Parallel shifts of the Cl^- titration curves at various pH. Each data point represents the mean ± SEM (n = 3) and data are normalized as in (a). (d) Relative error in the readout of [Cl^-] at pH 7.2 and 7.6 for the case where Clomeleon is calibrated at pH 7.4. (e) Stopped-flow measurements of the association time constants for Clomeleon binding to Cl^-, determined at various [Cl^-] and the two indicated temperatures. Panels a, c and d are reproduced from Kuner and Augustine (2000).

E. Effects of pH

The changes in absorbance caused by Cl^- binding to YFP are mediated by changes in the protonation state of the fluorophore (Wachter and Remington, 1999). Hence, changes in pH affect the apparent affinity to Cl^-. The IC_{50} of Cl^- increases with pH (Fig. 7.2c) over a wide concentration range. For example, the IC_{50} ranges from 15 mM at pH 6, up to 900 mM at pH 8. The pK_a of the YFP variant used as a sensor in Clomeleon will influence the IC_{50} of Cl^- and define the extent of pH dependence. For example, the Topaz variant (pK_a of ~6.4) has a larger IC_{50} but is less affected by physiological changes in pH (Kuner and Augustine, 2000), while the YFP (H148Q) variant (pK_a of 7.14) has a smaller IC_{50} but is more susceptible to pH changes (Jayaraman et al., 2000).

In a cellular environment pH is usually tightly regulated, but intracellular pH can change under certain functional states (Chesler and Kaila, 1992). If such changes remain undetected, Clomeleon will report [Cl^-] erroneously. To minimize such errors, it is necessary to determine intracellular pH under the conditions in which Clomeleon is being used to measure [Cl^-]. However, for many experimental situations this may not be necessary, in particular when [Cl^-] is less than ~25 mM. In this range, the readout error introduced by inadvertent changes in pH of ±0.2 pH units will remain below ±20% (Fig. 7.2d) and may be compatible with the experimental goals. When measuring

$[Cl^-]$ higher than 25 mM, however, the measurement error will increase to 50% for pH changes of ±0.2 pH units. In such cases, pH should be measured independently in the same experimental conditions; this can conveniently be done by using conventional, organic pH indicator dyes (Pond et al., 2006).

The known relationship between the IC_{50} of Cl^- and the pH can be used to correct ratio readouts of Clomeleon for known changes in pH. For example, if an IC_{50} has been obtained in a certain cell type at a pH of 7.2, it can still be used to accurately determine $[Cl^-]$ in other cell types at different resting pH. The following relation can be used to calculate the IC_{50} at a given pH (pH′):

$$IC'_{50} = IC_{50} \times 10^{0.82 \times (pH' - 7.2)}$$

The corrected IC'_{50} can then be used to determine the actual $[Cl^-]$ (for details see Berglund et al., 2006; Pond et al., 2006). This approach could also be used to correct the $[Cl^-]$ readout dynamically in experiments where pH and $[Cl^-]$ are measured simultaneously.

In conclusion, the pH dependence of Clomeleon needs to be considered, in particular when working under conditions of high $[Cl^-]_i$. In experimental conditions when pH changes occur, corrections must be used to determine $[Cl^-]_i$ under conditions of changing pH. When large pH shifts are known to be absent and when $[Cl^-]_i$ is lower than approximately 25 mM, Clomeleon will accurately report $[Cl^-]_i$ without requiring such corrections. Thus, judiciously used, Clomeleon can provide a reliable tool to measure $[Cl^-]_i$ regardless of pH or $[Cl^-]_i$ level.

F. Temporal Resolution

The temporal resolution of Clomeleon is limited by the kinetics of binding of Cl^- to YFP. Kinetic analyses of recombinant Clomeleon, based on stopped-flow measurements, reveal binding time constants of 500–600 ms at 20°C for physiologically relevant $[Cl^-]$ (Fig. 7.2e). The slow binding of Cl^- is also evident in simultaneous imaging and electrophysiological recordings from cultured hippocampal neurons. While rapid application of GABA produces an immediate increase of the Cl^- current, changes in $[Cl^-]_i$ reported by Clomeleon are delayed by about 1 s (Fig. 5A in Kuner and Augustine, 2000). For example, the limited temporal resolution would preclude direct visualization of rapid, highly localized changes in $[Cl^-]_i$ during inhibitory synaptic transmission. A faster binding time constant of 100–200 ms can be attained by increasing the temperature to 37°C (Fig. 7.2e).

The slow binding of Cl^- appears to be caused by steric constraints of the Cl^- binding site (Wachter et al., 2000), suggesting that an improved design of the Cl^- binding site may accelerate Cl^- binding. Indeed, mutations of YFP that confer faster binding kinetics have been identified (Galietta et al., 2001). For example, the YFP variant with I152L mutation has a binding time constant of 20 ms at 37°C. The use of such a mutant should be considered for cases where high temporal resolution is required.

III. STRATEGIES FOR CLOMELEON EXPRESSION

A rich variety of molecular genetic methods can be used to express Clomeleon in any cell type. Expression plasmids provide the simplest approach to express Clomeleon in cultured cells on a bulk scale using any standard DNA transfection protocol (Kuner and Augustine, 2000) or electroporation (Teruel et al., 1999). *In vivo* electroporation (Inoue and Krumlauf, 2001; Wei et al., 2003) of Clomeleon plasmids can be used to achieve widespread expression of Clomeleon in the intact brain.

For the case of neurons, the main challenge is often to selectively express Clomeleon only in the type of neuron of interest. Due to the lack of cellular specificity of the approaches described above, other strategies of expressing high levels of Clomeleon are required. For example, single-cell electroporation of plasmids (Rathenberg et al., 2003; Kitamura et al., 2008) could allow expression of Clomeleon in morphologically identified neurons. Viral gene transfer combined with stereotaxic delivery (Wimmer et al., 2004) can also yield some degree of control over the selectivity of Clomeleon expression (Fig. 7.3a). For example, a neuron-specific expression can be achieved with recombinant adeno-associated virus (AAV) pseudotyped with the AAV-1 and AAV-2 serotypes (McCown, 2005). Good spatiotemporal control of Clomeleon expression can be attained by precise stereotaxic delivery to certain brain areas. Particularly strong expression of Clomeleon can be accomplished using an expression cassette employing the CMV-enhancer-chicken-β-actin promoter (CBA), WPRE element and bGH polyadenylation site (Klugmann et al., 2005).

Transgenic approaches provide a powerful means of cell-type specific expression of Clomeleon in mice. A large set of BAC transgenic mice (www.gensat.org) provides a rich choice of cell- or tissue-specific expression patterns (Heintz, 2004) that could be utilized for Clomeleon expression. Thus far, we have used three

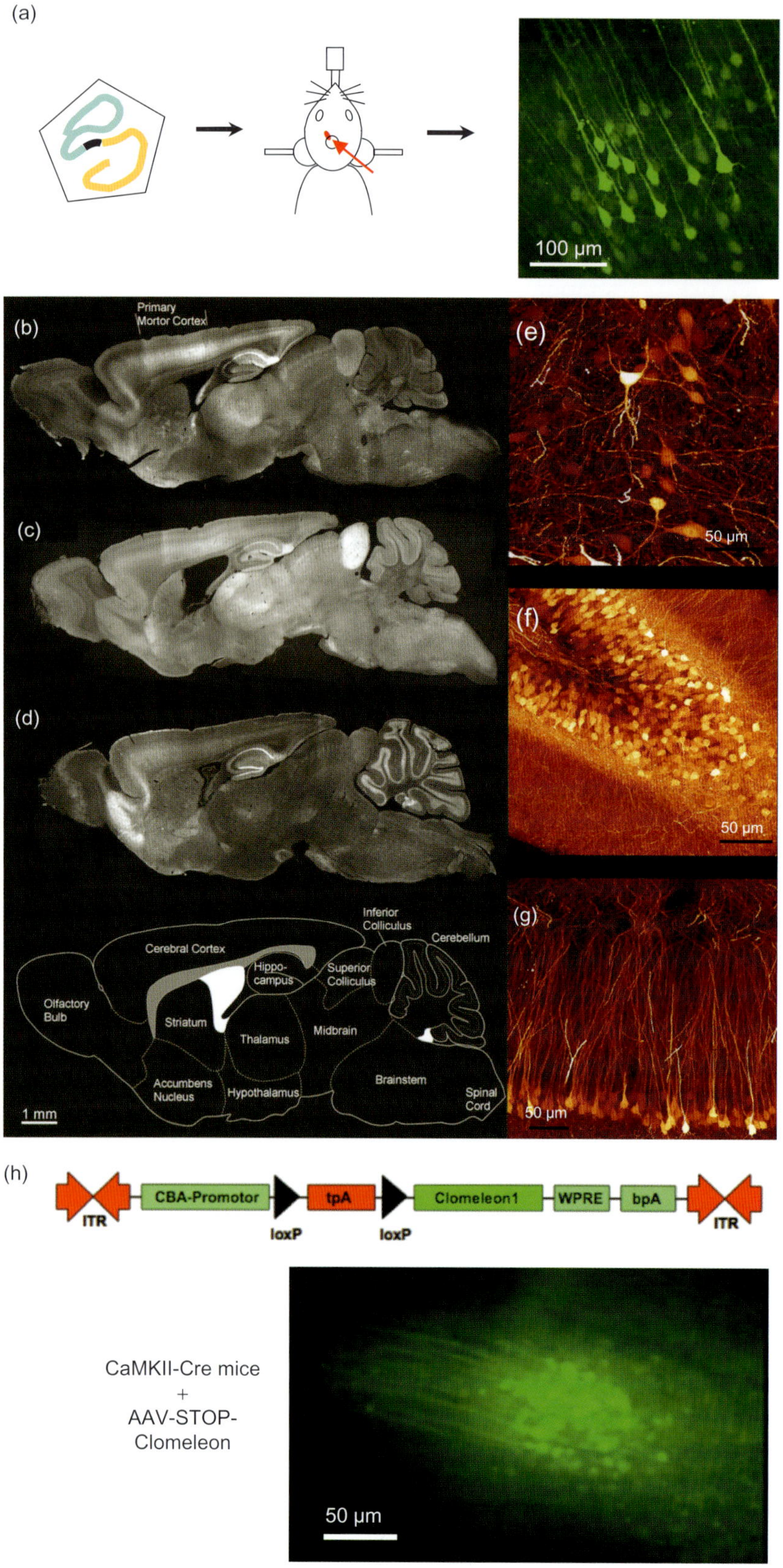

FIGURE 7.3 Methods for expressing Clomeleon in neurons. (a) Stereotaxic delivery of AAV1/2-Clomeleon. AAV1/2 containing Clomeleon transgene (left) was delivered to the neocortex of a mouse mounted on a stereotaxic device (middle). Clomeleon expression was observed in neocortical layer 5 neurons (right). Projection of an image stack obtained with confocal microscopy of aldehyde-fixed tissue. (b–d) Differential expression of Clomeleon in thy1::Clomeleon mouse lines. YFP fluorescence images of paramedial sagittal sections of fixed brains of adult mice were obtained from lines CLM11, 12 and 13. Diagram below illustrates anatomy of the mouse brain at this plane of section. (e–g) Confocal images of sections taken from CLM11, shown as maximal projections of an image stack: hilus (e), dentate gyrus (f) and CA1 neurons (g) of the hippocampus. (h) Clomeleon expression by combination of adenovirus and Cre recombinase. Stereotaxic injection of AAV1/2-STOPflox-Clomeleon (top) into a CaMKII-Cre mouse resulted in Clomeleon expression in neocortical layer 5 neurons (bottom). A wide-field epifluorescence image was obtained in live slice preparation. Panels a and h are reproduced from Berglund et al. (2008).

other strategies to achieve selective expression of Clomeleon in defined groups of neurons in mice.

The first transgenic strategy was based on using the thy1 promoter, which is known to mediate strong, neuron-specific expression of transgenes within the brain (Caroni, 1997). Pronounced positional variegation of this promoter results in striking line-to-line variations in neuronal expression, which can be used to advantage for fluorescence imaging of individual neuron types (Feng et al., 2000). We characterized seven thy1::Clomeleon mouse lines, each of them showing a unique pattern of neuronal Clomeleon expression. For example, line CLM11 shows strong Clomeleon expression in the CA1 area of the hippocampus and in neocortical areas (Fig. 7.3b), line CLM12 shows strong expression in the superior and inferior colliculi (Fig. 7.3c) and line CLM13 shows intense Clomeleon expression in the dentate gyrus, cerebellar mossy fibers and piriform cortex (Fig. 7.3d). The number of Clomeleon-expressing cells in any brain region ranges from sparse patterns with only single neurons expressing Clomeleon to regions with dense expression (Figs 7.3e–g). Several studies of neuronal $[Cl^-]_i$ have been conducted with these thy1::Clomeleon transgenic mice (Berglund et al., 2006; Duebel et al., 2006; Pond et al., 2006, see below).

The second strategy for generating a Clomeleon indicator mouse line utilized the Cre-loxP system and mice lines with cell-type specific expression of the Cre recombinase. Clomeleon expression was rendered conditional by placing it after a loxP-flanked transcriptional stop cassette. This cassette was targeted into the ROSA26 locus (Soriano, 1999; Srinivas et al., 2001) using homologous recombination (Berglund et al., 2008). This locus is active in all cells and can, in principle, yield Clomeleon expression in any cell type. The transcriptional stop cassette restricts expression of Clomeleon to only those cells expressing Cre recombinase. Hence, this universal Clomeleon indicator mice line (UniAct-Clomeleon) allows specific expression of Clomeleon in cell populations expressing Cre recombinase. The rather low levels of Clomeleon expression in neurons mediated by the ROSA26 promoter limits intensity-based ratiometric imaging, but suffices for two-photon lifetime imaging of the cyan donor fluorescence (Jose et al., 2007).

A third strategy combines cell-type specific expression of Cre recombinase with viral expression of the STOPflox-Clomeleon cassette described above (Fig. 7.3h). The advantage of this approach is that high levels of Clomeleon expression can be attained through viral expression while achieving cell-type selectivity via the expression of Cre recombinase. The resultant strong expression of Clomeleon allows intensity-based ratiometric imaging to be conveniently employed. Stereotaxic delivery of virus provides an additional means of controlling expression density, for example by injecting only small amounts of virus. This strategy is best suited in combination with BAC-mediated expression of Cre, because highly selective BAC promoters often are rather weak and may not produce high levels of Clomeleon expression.

IV. APPROACHES TO CLOMELEON IMAGING

A. Microscopy

Clomeleon imaging thus far has been successfully implemented via four different approaches: (1) wide-field epifluorescence imaging, using either two PMT detectors (Kuner and Augustine, 2000), a two-channel beam-splitter/filter-wheel and one CCD camera (Berglund et al., 2006; Pond et al., 2006), or two CCD cameras; (2) confocal fluorescence imaging using a spinning disk microscope (Berglund et al., 2006); (3) two-photon imaging (Duebel et al., 2006); and (4) two-photon lifetime imaging (Jose et al., 2007).

Each of these approaches offers distinct advantages and the method of choice to measure $[Cl^-]_i$ depends on the specific questions to be addressed and the available experimental conditions. For example, the wide-field epifluorescence approach is suitable for imaging primary neuronal cultures or acute brain slices with few neurons expressing Clomeleon at high levels. Because of the relative efficiency of wide-field fluorescence imaging, the measurements sensitivity of this approach is high, allowing imaging at a high temporal resolution of about 20 ms when using a highly efficient CCD camera. Imaging with a spinning disk confocal microscope makes it possible to define more precisely the volume from which the Clomeleon signal emerges, which is important in acute brain slices containing many neurons expressing high levels of Clomeleon. However, because of the relative inefficient photon throughput of spinning disk confocal microscopes, the signal-to-noise ratio is low even in cells with high levels of Clomeleon expression. In practice, this limits the temporal resolution to about 200 ms per frame. Two-photon imaging of Clomeleon works well in cultured neurons, acute brain slices and in whole-mounted retinas. Two-photon microscopy has also yielded preliminary images of Clomeleon fluorescence in the intact brain of mice (see below). Finally, Clomeleon lifetime imaging could be useful to image cells with low levels of Clomeleon expression.

However, the presence of multiple decay components in the excited state lifetime of CFP may require design of new versions of Clomeleon based on a donor fluorophore with a single excited state lifetime.

B. Bleaching and Photophysics of Clomeleon

When using Clomeleon, it is important to know that YFP bleaches more rapidly than CFP. Interestingly, this bleaching of YFP has been found to be reversible over time (Dickson et al., 1997; Berglund et al., 2005; Sinnecker et al., 2005). Such differential bleaching will affect the fluorescence emission ratio of Clomeleon, leading to an apparent increase in $[Cl^-]_i$ (Berglund et al., 2005). Thus, to avoid this problem, the amount of excitation of Clomeleon should be reduced as much as practicable. The simplest means of avoiding this problem is to reduce the intensity of the excitation light; we have found that it is possible to obtain good signal-to-noise ratios when using levels of excitation light that yield stable Clomeleon fluorescence with minimal bleaching. Considering this issue, the level of Clomeleon expression is possibly one of the most important considerations for the outcome of Cl^- imaging experiments. The signal-to-noise ratio will be ideal if Clomeleon expression is high, to allow the excitation light to be dim. Because the levels of Clomeleon expression (typically micromolar) are orders of magnitude less than $[Cl^-]_i$ (typically millimolar), buffering of $[Cl^-]_i$ by Clomeleon is not an issue (Kuner and Augustine, 2000). So the general goal should be to express as much Clomeleon as possible in the cell(s) of interest. In addition, to further increase the contrast between cells that express Clomeleon, ideally only few neurons in a given tissue should express the Cl^- indicator.

C. Calibration

The calibration of $[Cl^-]$ measurements in intact cells is challenging because cellular homeostasis mechanisms make it difficult to clamp $[Cl^-]_i$ at a defined concentration. Two main approaches have been used for this purpose: equilibration of $[Cl^-]_i$ via a whole-cell recording pipette or by treating the cell with Cl^--selective ionophores/antiporters that make the cell membrane permeable to Cl^-.

With the patch pipette calibration method, the cell is dialyzed with a solution containing a defined $[Cl^-]$ and the Clomeleon emission ratio is measured at this defined $[Cl^-]$. The R_{min} value can be conveniently determined with 100 mM F^- in the pipette, which abolishes YFP fluorescence completely (see Fig. 7.2a). Determining R_{max} at low $[Cl^-]_i$ is more difficult, because it is unclear to what extent the low $[Cl^-]$ in the pipette can overwhelm cellular Cl^- homeostatic mechanisms and thereby keep intracellular $[Cl^-]$ low. The main difficulty with this method is that each experiment can determine only one point on the Clomeleon calibration curve, meaning that a substantial number of experiments are needed to establish a reliable calibration curve.

In the ionophore/antiporter calibration approach, the cells are exposed to the Cl^-/OH^- antiporter tributyltin, which forms pores in the cell membrane and allows external Cl^- to equilibrate with intracellular Cl^-. Because of the OH^- permeability of tributyltin, it is also necessary to clamp intracellular pH. For this purpose, the H^+/K^+ ionophore, nigericin, must also be applied to the cells. This method (Krapf et al., 1988) has been widely used in brain slices (Inglefield and Schwartz-Bloom, 1997, 1999; Marandi et al., 2002; Berglund et al., 2006, 2008; Pond et al., 2006). Again, R_{min} values can be determined using an F^- solution, while R_{max} and values for the ratio at other levels of $[Cl^-]_i$ can be determined simply by manipulating $[Cl^-]$. One convenient feature of this approach is that several points on the calibration curve can be determined from each cell, making determination of the calibration curve much faster and easier. However, in our experience the ability of the tributyltin/nigericin treatment to control intracellular Cl^- can be variable. The precision of this approach depends critically on a number of factors, including how the ionophore solution is prepared and applied. For best results, freshly prepared nigericin should be fully dissolved in a saline where Na^+ has been replaced with K^+ and *N*-methyl-D-glucamine ($NMDG^+$) to block Na^+–H^+ exchange and thereby maintain intracellular pH (Boyarsky et al., 1988). A detailed description of the use of this method to calibrate Clomeleon signals in brain slices can be found in Berglund et al. (2006).

V. APPLICATIONS OF CLOMELEON

Clomeleon has been used in a variety of studies of neuronal $[Cl^-]_i$. In this section, we summarize all applications of Clomeleon that have been published so far.

A. Measurement of Resting $[Cl^-]_i$ in Mature Neurons

Many studies have shown that resting $[Cl^-]_i$ in mature central nervous system (CNS) neurons is generally maintained at low levels, i.e. below the predicted electrochemical equilibrium. This allows inhibitory neurotransmitters that work by increasing the membrane conductance to Cl^-, such as GABA and glycine, to produce hyperpolarizing Cl^- influx. However, a common problem with all these studies has been the difficulty in determining the absolute value of resting $[Cl^-]_i$. The use of Clomeleon circumvents this problem allowing $[Cl^-]_i$ to be measured and not just inferred from the reversal of GABA or glycine currents (see also Chapters 6 and 22 in this volume).

In thy1::Clomeleon transgenic mice, Clomeleon is expressed in a wide variety of neurons, allowing direct measurements of resting $[Cl^-]_i$ in mammalian neurons. Using this indicator, resting $[Cl^-]_i$ has been measured in pyramidal cells in the hippocampus, pyramidal cells in the amygdala, granule cells in the cerebellum and premotor neurons in the superior colliculus (Berglund et al., 2006, 2008). Resting $[Cl^-]_i$ in these mature central neurons was found to range from 5 to 10 mM, which is similar to the values estimated using indirect methods. These values are also consistent with the observed hyperpolarizing inhibition produced by neurotransmitters that open Cl^- channels in mature central neurons.

B. Shift in $[Cl^-]_i$ during Neuronal Development

Many studies have documented that, over the course of development, GABAergic transmission in CNS neurons switches from depolarizing to hyperpolarizing (Ben-Ari et al., 1989; Cherubini et al., 1991; Owens et al., 1996). This transition is presumably due to a progressive decrease in $[Cl^-]_i$ as suggested by experiments measuring the reversal potential of GABA- or glycine-induced Cl^- currents using the gramicidin-perforated patch method (see Chapter 8 in this volume). However, with the possible exception of amacrine and ganglion neurons in the developing mouse retina (Zhang et al., 2006; see also Chapter 19 in this volume), the hypothesis of the developmental shift in $[Cl^-]_i$ had not yet been validated using direct measurement of $[Cl^-]_i$ in cortical neurons.

Using Clomeleon, it has been possible to directly measure $[Cl^-]_i$ in developing CNS neurons. In initial studies, $[Cl^-]_i$ was measured in rat hippocampal neurons that were cultured and subsequently transfected with Clomeleon (Kuner and Augustine, 2000). In these cultured cells, $[Cl^-]_i$ was higher in younger neurons and then declined over the next 2 weeks. In a subsequent study, resting $[Cl^-]_i$ was measured in CA1 pyramidal neurons in slices of hippocampus of thy1::Clomeleon transgenic mice (Berglund et al., 2006). At postnatal day 5 (P5), $[Cl^-]_i$ varied widely over, in the range of 20–30 mM. At older ages the distribution of $[Cl^-]_i$ became narrower, suggesting a more homogeneous population of neurons at more mature stages. Mean $[Cl^-]_i$ was 17 mM at P10 and further decreased to 6 mM at P20. This corresponds to a drop in the Cl^- equilibrium potential of more than 30 mV between P5 and P20.

These results are entirely consistent with the previously reported developmental changes in polarity of GABA-induced currents in CNS neurons. These developmental changes are associated with differential expression profiles of Cl^- cotransporters KCC2 and NKCC1 (Rivera et al., 1999; Stein et al., 2004), which presumably are responsible for the developmental drop in $[Cl^-]_i$. In conclusion, non-invasive optical imaging with Clomeleon provides direct measurements of $[Cl^-]_i$ that support the hypothesis that postnatal changes in the Cl^- electrochemical gradient underlie the functional switch in GABAergic transmission during development.

C. Compartmentalization of $[Cl^-]_i$ in Neurons

A great advantage of Clomeleon is its spatial resolution, which provides a new dimension of measurement that can reveal $[Cl^-]_i$ gradients between cell compartments (see also Chapter 19 in this volume).

The first hint of such gradients in neurons came from the observation that $[Cl^-]_i$ is higher in the dendrites than in the soma of cultured hippocampal neurons (Kuner and Augustine, 2000). Measurements of $[Cl^-]_i$ in hippocampal slices have reiterated this conclusion: resting $[Cl^-]_i$ was 6–10 mM in apical dendrites of CA1 pyramidal cells and 8 mM in basal dendrites, whereas it was 5 mM in the cell bodies of these neurons, although the difference was not statistically different (Berglund et al., 2006). $[Cl^-]_i$ gradients across cell compartments have been reported in other neurons as well (Derdikman et al., 2003; Szabadics et al., 2006; Khirug et al., 2008), as discussed below and in Chapter 19.

Imaging resting $[Cl^-]_i$ in ON-type bipolar neurons in the retina has revealed a particularly interesting and important type of $[Cl^-]_i$ gradient. Although somewhat controversial (Satoh et al., 2001; Billups and Attwell, 2002), it has been proposed that dendritic $[Cl^-]_i$ of these cells is high enough for GABA to produce membrane depolarization (Miller and Dacheux, 1983). To address this possibility, Duebel et al. (2006) used two-photon imaging to measure resting $[Cl^-]_i$ in ON-type bipolar neurons, employing retina from thy1::Clomeleon transgenic mice that express Clomeleon in these cells (Haverkamp et al., 2005; Berglund et al., 2006).

They found that resting $[Cl^-]_i$ is relatively high in the dendrites of these cells (type 9 bipolar cells), averaging approximately 45 mM. Remarkably, resting $[Cl^-]_i$ in the somata of these cells is 20 mM lower on average, leading to steep standing gradients in resting $[Cl^-]_i$ between the soma and the dendrites (Fig. 7.4a), as discussed in Chapter 19 in this volume. Thus, the ratiometric nature of Clomeleon imaging was key in providing the first direct demonstration of high $[Cl^-]_i$ in the dendrites of these neurons. Further, the effects of GABA on $[Cl^-]_i$ in these cells were determined by applying GABA while using Clomeleon to measure the resulting changes in $[Cl^-]_i$. GABA caused a reduction in $[Cl^-]_i$ in the dendrites of these cells (Fig. 7.4b, red trace), showing that dendritic $[Cl^-]_i$ is actually higher than electrochemical equilibrium and thus GABA causes a depolarizing Cl^- efflux. Although GABA is depolarizing in the dendrites of ON-type bipolar cells, GABA released on to axon terminals at the other end of the same cell has a conventional hyperpolarizing inhibitory action. This difference in polarity of GABA actions at the two ends of the ON-type bipolar neurons arises from differences in $[Cl^-]_i$. Compared to the dendrites, $[Cl^-]_i$ at the soma is lower and thus the soma responds to puff application of GABA with an increase in $[Cl^-]_i$ (Fig. 7.4b, blue trace), consistent with hyperpolarizing action of GABA in the axon.

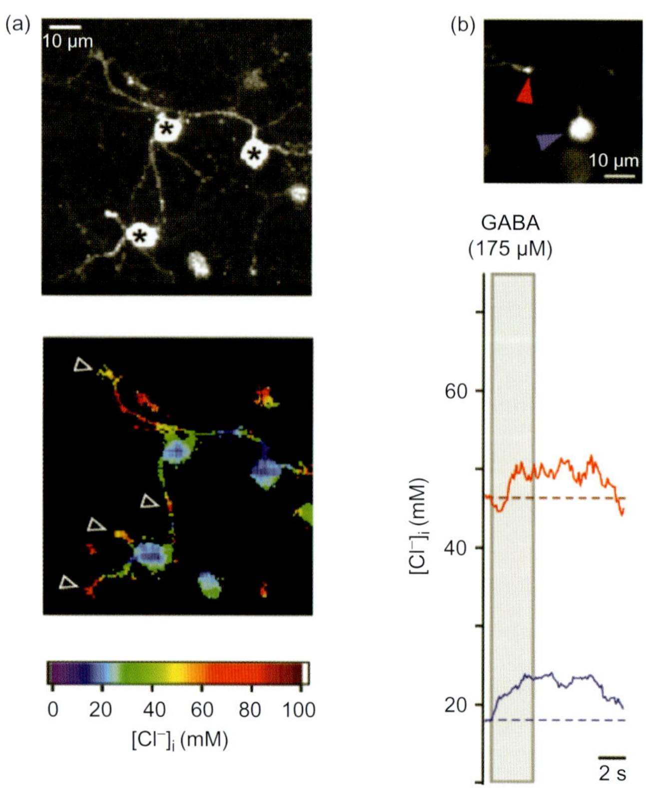

FIGURE 7.4 Effect of GABA application on $[Cl^-]_i$ in retinal bipolar cells. (a) $[Cl^-]_i$ in Clomeleon-expressing type 9 bipolar cells in the whole-mounted retina. Shown are morphology (top) of a group of type 9 bipolar cells (asterisks) in a collapsed image stack and $[Cl^-]_i$ (pseudo-color scale; bottom) in the same cells. $[Cl^-]_i$ was high in varicosities (arrowheads). (b) Traces showing dendritic (red) and somatic (blue) $[Cl^-]_i$ in a type 9 bipolar cell during GABA application (shaded time period). Images at top illustrate the location of the two measurement sites (arrowheads; $n = 7$ trials). Reproduced from Duebel et al. (2006).

D. Imaging Cl^- Fluxes Associated with Synaptic Inhibition

Major forms of synaptic inhibition are due to transmitters that activate receptors coupled to Cl^- channels, resulting in transmembrane Cl^- fluxes that should transiently change $[Cl^-]_i$. One of the most promising applications of Clomeleon in neurons is to image these changes in $[Cl^-]_i$ that are associated with synaptic inhibition. Several experiments evaluated the capacity of Clomeleon to detect synaptic inhibition by combining patch clamp recording and simultaneous Clomeleon imaging in hippocampal neurons. The first such experiments examined responses of dissociated cultured neurons to application of GABA; the resulting activation of $GABA_A$ receptors produced Cl^- currents that could be measured by patch clamp recordings, and $[Cl^-]_i$ changes that were detected by Clomeleon (Kuner and Augustine, 2000). It was observed that both the polarity and magnitude of $[Cl^-]_i$ changes were well correlated with the GABA-induced Cl^- currents, regardless of whether Cl^- was entering a neuron or was leaving it. Changes in $[Cl^-]_i$ occurred with a delay of approximately 1 s, presumably limited by the kinetics of Clomeleon, as described above. These changes in $[Cl^-]_i$ returned to the baseline approximately 10 s after the Cl^- current ceased, indicating the relatively slow rate of Cl^- removal from these cells by transporters (Staley and Proctor, 1999).

To examine the ability of Clomeleon to detect responses associated with activation of GABAergic inhibitory synapses, responses were measured in CA1 pyramidal neurons in hippocampal slices from thy1::Clomeleon transgenic mice (Berglund et al., 2006). A local inhibitory pathway (Freund and Buzsáki, 1996; Somogyi and Klausberger, 2005) was activated by an extracellular stimulating electrode that was placed at the distal part of the pyramidal cell apical dendrites while excitatory synaptic transmission was eliminated by treating the slices with a glutamate receptor antagonist, kynurenic acid. Under voltage-clamp conditions, a train of electrical stimuli evoked summating inhibitory postsynaptic currents (IPSCs). These IPSCs elevated $[Cl^-]_i$ in the soma of individual pyramidal cells by as much as 4 mM; as was the case in cultured neurons, the $[Cl^-]_i$ changes were well correlated with the total charge carried by Cl^-. As also observed in cultured neurons, the rise in $[Cl^-]_i$ lagged behind the current response by ~1 s and the change in $[Cl^-]_i$ lasted much longer than the duration of the Cl^- current.

Such results indicate that Clomeleon can report elevation of Cl^- upon activation of $GABA_A$ receptors when the inwardly directed driving force on Cl^- is kept relatively constant under voltage-clamp conditions. However, in physiological conditions the changes in membrane potential produced by inhibitory postsynaptic potentials (IPSPs) will cause the driving force for Cl^- influx to decline during the IPSP, thereby reducing net Cl^- influx. Because this effect could limit the ability to image inhibitory transmission, we next asked whether Clomeleon could image inhibitory synaptic transmission in neurons whose membrane potential was not kept constant by voltage clamping (Berglund et al., 2006). Under these conditions, a single stimulus caused an IPSP that hyperpolarized the resting membrane potential from −71 mV to −76 mV. Trains of stimuli evoked summating IPSPs. However, simultaneous Clomeleon imaging revealed barely detectable changes in $[Cl^-]_i$ in the soma of the same cell. The small magnitude of these responses is consistent with the expected reduction in Cl^- driving force in unclamped cells. When $[Cl^-]_i$ signals were collected and averaged from many somata, the signal-to-noise ratio of the measurement was greatly improved, allowing responses to brief trains of IPSPs to be detected. Maximal changes in $[Cl^-]_i$ were of the order of 2 mM, which is smaller than those in voltage-clamped cells consistent with the reduced driving force on synaptic Cl^- fluxes in unclamped cells. These results indicate that Clomeleon can report changes in $[Cl^-]_i$ during inhibitory synaptic transmission in unclamped neurons. Averaging $[Cl^-]_i$ signals from 10–20 cells over four trials doubles the signal-to-noise ratio, yielding a detection limit of a 90 μM rise in $[Cl^-]_i$. This corresponds to the response to approximately three IPSPs. The experiments above demonstrate that Clomeleon enables imaging of Cl^--mediated inhibitory transmission without any manipulation of the Cl^- driving force.

Such measurements can obviously be performed in the absence of any recording electrodes. In such conditions, interneuron stimulation produces the expected increases in $[Cl^-]_i$ in all layers of area CA1. The sensitivity of the method is sufficient to detect recruitment of presynaptic interneurons: increasing stimulus intensity, to activate larger numbers of interneurons, causes progressively larger changes in $[Cl^-]_i$ (Figs 7.5a–d). The magnitude of the spatially averaged $[Cl^-]_i$ change is proportional to the stimulus intensity (Fig. 7.5e) and is blocked by $GABA_A$-specific antagonist, SR95531, indicating that the $[Cl^-]_i$ changes resulted from progressive activation of GABAergic interneurons by stronger stimuli.

The spatiotemporal dynamics of inhibitory synaptic responses can be quantified by line scan analysis. This analysis displays the magnitude of synaptic $[Cl^-]_i$ changes as a function of both space and time (Fig. 7.6). The area of analysis shown in Fig. 7.6 covers several compartments within the hippocampal CA1 region, including the distal parts of pyramidal cell apical dendrites (in the stratum lacunosum-moleculare), the proximal

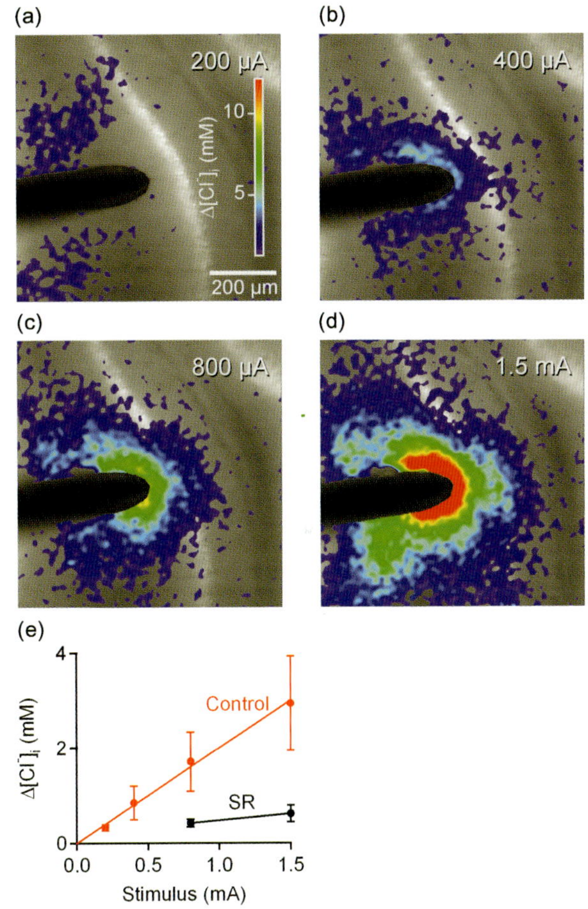

FIGURE 7.5 Clomeleon imaging of Cl^- changes produced by synaptic inhibition. (a–d) Images showing Cl^- changes (pseudocolor scale) superimposed on YFP fluorescence image (gray scale) measured in CLM1 hippocampal slice. Electrical stimulation was delivered for 1 s with current intensities indicated in the upper right corner of each image through a concentric bipolar electrode (the black rod on the left). Images were acquired 4 s after each stimulus. (e) Relationship between $[Cl^-]_i$ changes, averaged across the whole CA1 field, and stimulus intensity. Points (mean±SEM; $n = 4$) were measured before (control) and during application of the GABA receptor antagonist, SR95531 (SR; 10 μM). Reproduced from Berglund et al. (2006).

parts of apical dendrites (in the stratum radiatum), the pyramidal cell bodies (in the stratum pyramidale) and the basal dendrites (in the stratum oriens). This analysis reveals that $[Cl^-]_i$ changes are largest in the apical dendrites, the location closest to the site of stimulation (Fig. 7.6a). The time course of these changes varies across compartments, as can be seen most clearly by superimposing plots of $[Cl^-]_i$ changes over time within each compartment (Fig. 7.6b). While $[Cl^-]_i$ changes in the distal part of apical dendrites rise rapidly and reach a maximum at the end of the train of stimuli, $[Cl^-]_i$ rises more slowly in basal dendrites and in the proximal part of apical dendrites. Changes in $[Cl^-]_i$ rise more slowly in somata, probably due to slow diffusion of Cl^- from the dendrites and/or from the surface-to-volume ratio of the cell bodies. Fig. 7.6c and d show that $[Cl^-]_i$ changes associated with synaptic inhibition of CA1 pyramidal cells are larger and faster in the apical dendrites and smaller and slower in the somata and in the basal dendrites. These differences presumably reflect differences in the location of active inhibitory synapses on the pyramidal cells, in the surface-to-volume ratio of the different compartments, and/or in spatial variations in the driving force on Cl^- (Staley and Proctor, 1999).

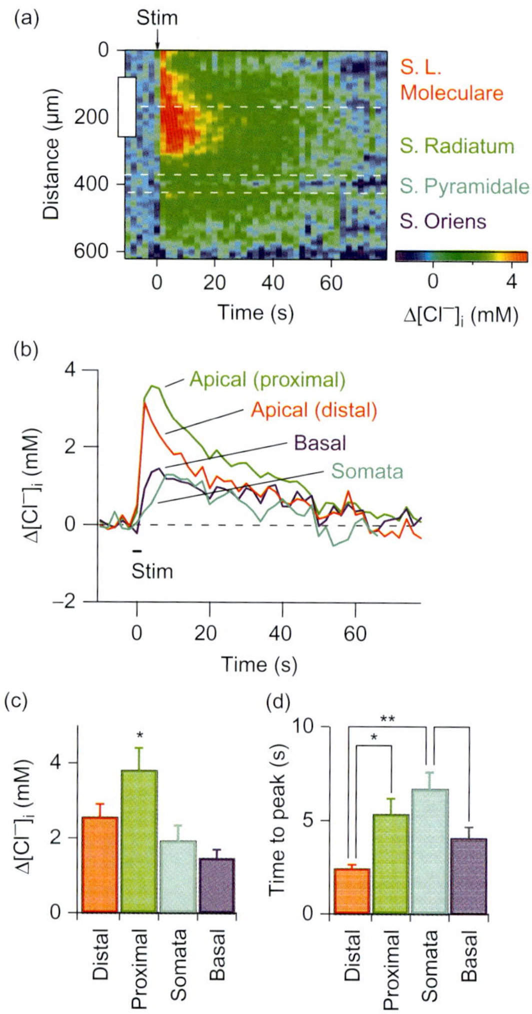

FIGURE 7.6 Spatiotemporal profiles of synaptic Cl^- transients. Responses of hippocampal CA1 pyramidal cells to stimulation of presynaptic interneurons were imaged as in Fig. 7.5. (a) A line-scan of $[Cl^-]_i$ changes measured across the layers of area CA1. Resting $[Cl^-]_i$ measured before stimulation was subtracted to show relative changes caused by inhibitory synaptic activity (stimulus applied at time indicated by the arrow). White rectangle on the ordinate denotes the size and location of the stimulating electrode. (b) Time course of changes in $[Cl^-]_i$ calculated from each layer. Colors of the plots correspond to those of the labels in (a). Duration of the 2 s stimulus train (Stim) is indicated by bar. (c) Peak magnitude of $[Cl^-]_i$ changes in the four compartments of pyramidal cells in each layer. Mean±SEM ($n = 7$); asterisk denotes a significant difference ($p < 0.05$) from other compartments. (d) Time to peak of $[Cl^-]_i$ changes in the same four compartments of pyramidal cells. Reproduced from Berglund et al. (2006).

In conclusion, the above results document that Clomeleon can detect changes in postsynaptic $[Cl^-]_i$ in hippocampal neurons that result from synaptic inhibition mediated by $GABA_A$ receptors. These results support the utility of Clomeleon as a tool for imaging Cl^--mediated synaptic inhibition in the brain.

E. Pathological Changes of $[Cl^-]_i$ Detected by Clomeleon

Abnormal regulation of neuronal $[Cl^-]_i$ contributes to the pathology of a multitude of diseases (Payne et al., 2003). For example, chronic pain is partly mediated by an increase in resting $[Cl^-]_i$ in spinal nociceptive neurons, resulting in disinhibition of these neurons (Coull et al., 2003, 2005; Torsney and MacDermott, 2005), as discussed in Chapter 23 in this volume. Hypoxia and ischemia also induce pathological alteration in neuronal $[Cl^-]_i$ and GABAergic transmission, leading to excitotoxicity in the brain (Allen and Attwell, 2004; Allen et al., 2004), as discussed in detail in Chapter 25 in this volume. Hippocampal slices from thy1::Clomeleon mice have been successfully used to track pathological increases of $[Cl^-]_i$ resulting from oxygen deprivation. The ability to measure these changes in $[Cl^-]_i$ also permitted identification of the source of the increased $[Cl^-]_i$, which is an influx of Cl^- resulting from activation of the NKCC1 transporter by phosphorylation (Pond et al., 2006).

F. Clomeleon Imaging *in Vivo*

The high expression of Clomeleon in thy1::Clomeleon transgenic mice also permits non-invasive brain imaging *in vivo* (Helmchen and Denk, 2005). Two-photon microscopy has enabled the visualization

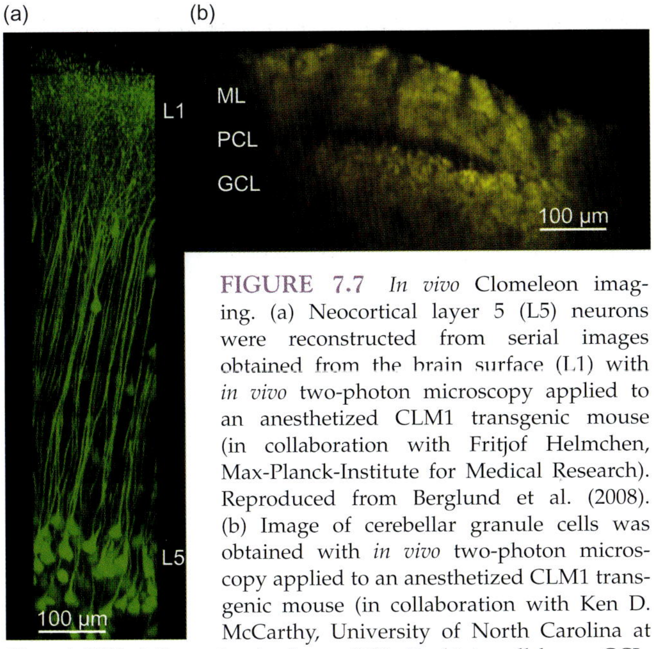

FIGURE 7.7 *In vivo* Clomeleon imaging. (a) Neocortical layer 5 (L5) neurons were reconstructed from serial images obtained from the brain surface (L1) with *in vivo* two-photon microscopy applied to an anesthetized CLM1 transgenic mouse (in collaboration with Fritjof Helmchen, Max-Planck-Institute for Medical Research). Reproduced from Berglund et al. (2008). (b) Image of cerebellar granule cells was obtained with *in vivo* two-photon microscopy applied to an anesthetized CLM1 transgenic mouse (in collaboration with Ken D. McCarthy, University of North Carolina at Chapel Hill). ML: molecular layer; PCL: Purkinje cell layer; GCL: granule cell layer. Note expression of Clomeleon in ML and GCL, where granule cells reside, while Clomeleon expression is excluded from Purkinje cells in this mouse line.

of neocortical layer 5 pyramidal neurons in living mice (Fig. 7.7a; Krieger et al., 2007; Berglund et al., 2008). This approach has also been used to image across many cell layers in the whole cerebellum of thy1::Clomeleon transgenic mice that express Clomeleon in granule cells (Fig. 7.7b). This approach opens the possibility of imaging synaptic inhibition in the brains of intact behaving mice.

VI. OUTLOOK: OPTIMIZING CLOMELEON IMAGING

The initial version of Clomeleon has provided a range of possibilities for directly measuring $[Cl^-]_i$ in neurons and other cell types. Nevertheless, several features of Clomeleon could be further improved:

1. *Cl^- sensitivity*. Clomeleon has an IC_{50} of approximately 160 mM, which is about 20 times higher than the resting $[Cl^-]_i$ in mature CNS neurons. An IC_{50} for Cl^- of about 10 mM would be ideal for the range of $[Cl^-]_i$ in mature neurons and would allow better detection of the small changes in $[Cl^-]_i$ associated with synaptic inhibition. Variants of YFP with decreased IC_{50} for Cl^- have been already developed (Galietta et al., 2001) and similar mutations improve the Cl^- sensitivity of Clomeleon (Markova et al., 2008; J. Casad and G.J. Augustine, unpublished results).
2. *pH dependence*. Creating YFP variants with reduced sensitivity to protons, but unaltered Cl^- sensitivity, would be ideal for measuring $[Cl^-]_i$ in cells with fluctuating pH. However, because Cl^- acts via the protonation of the YFP fluorophore, a dissociation of both properties may not be possible. Alternatively, an approach using a genetically encoded pH sensor simultaneously may yield an experimental situation in which both pH and Cl^- could be measured in parallel to facilitate correction of Clomeleon signals for pH changes.
3. *Dynamic range*. A larger change in the YFP/CFP emission ratio between Cl^--bound and Cl^--free forms of Clomeleon would provide better resolution of $[Cl^-]$. An optimized design of the CFP-YFP tandem and the use of CFP variants with higher quantum yields, such as Cerulean (Rizzo et al., 2004), should improve the dynamic range of Clomeleon.

References

Allen, N.J. and Attwell, D. (2004). The effect of simulated ischaemia on spontaneous GABA release in area CA1 of the juvenile rat hippocampus. *J. Physiol.* **561**, 485–498.

Allen, N.J., Rossi, D.J., and Attwell, D. (2004). Sequential release of GABA by exocytosis and reversed uptake leads to neuronal swelling in simulated ischemia of hippocampal slices. *J. Neurosci.* **24**, 3837–3849.

Ben-Ari, Y., Cherubini, E., Corradetti, R., and Gaiarsa, J.L. (1989). Giant synaptic potentials in immature rat CA3 hippocampal neurones. *J. Physiol.* **416**, 303–325.

Berglund, K., Dunbar, R.L., Lee, P., Feng, G., and Augustine, G.J. (2005). A practical guide: imaging synaptic inhibition with Clomeleon, a genetically encoded chloride indicator. In *Imaging in Neuroscience and Development: A Laboratory Manual* (Konnerth, A. and Yuste, R., eds), pp. 595–598. Cold Spring Harbor Laboratory Press, Cold Spring Harbor.

Berglund, K., Schleich, W., Krieger, P., Loo, L.S., Wang, D., Cant, N.B., Feng, G., Augustine, G.J., and Kuner, T. (2006). Imaging synaptic inhibition in transgenic mice expressing the chloride indicator, Clomeleon. *Brain Cell Biol.* **35**, 207–228.

Berglund, K., Schleich, W., Wang, H., Feng, G., Hall, W.C., Kuner, T., and Augustine, G.J. (2008). Imaging synaptic inhibition throughout the brain via genetically targeted Clomeleon. *Brain Cell Biol.* **36**, 101–118.

Billups, D. and Attwell, D. (2002). Control of intracellular chloride concentration and GABA response polarity in rat retinal ON bipolar cells. *J. Physiol.* **545**, 183–198.

Boyarsky, G., Ganz, M.B., Sterzel, R.B., and Boron, W.F. (1988). pH regulation in single glomerular mesangial cells. I. Acid extrusion in absence and presence of HCO_3^-. *Am. J. Physiol.* **255**, C844–C856.

Caroni, P. (1997). Overexpression of growth-associated proteins in the neurons of adult transgenic mice. *J. Neurosci. Methods* **71**, 3–9.

Cherubini, E., Gaiarsa, J.L., and Ben-Ari, Y. (1991). GABA: an excitatory transmitter in early postnatal life. *Trends Neurosci.* **14**, 515–519.

Chesler, M. and Kaila, K. (1992). Modulation of pH by neuronal activity. *Trends Neurosci.* **15**, 396–402.

Coull, J.A., Beggs, S., Boudreau, D., Boivin, D., Tsuda, M., Inoue, K., Gravel, C., Salter, M.W., and De Koninck, Y. (2005). BDNF from microglia causes the shift in neuronal anion gradient underlying neuropathic pain. *Nature* **438**, 1017–1021.

Coull, J.A., Boudreau, D., Bachand, K., Prescott, S.A., Nault, F., Sík, A., De Koninck, P., and De Koninck, Y. (2003). Trans-synaptic shift in anion gradient in spinal lamina I neurons as a mechanism of neuropathic pain. *Nature* **424**, 938–942.

Derdikman, D., Hildesheim, R., Ahissar, E., Arieli, A., and Grinvald, A. (2003). Imaging spatiotemporal dynamics of surround inhibition in the barrels somatosensory cortex. *J. Neurosci.* **23**, 3100–3105.

Dickson, R.M., Cubitt, A.B., Tsien, R.Y., and Moerner, W.E. (1997). On/off blinking and switching behaviour of single molecules of green fluorescent protein. *Nature* **388**, 355–358.

Duebel, J., Haverkamp, S., Schleich, W., Feng, G., Augustine, G.J., Kuner, T., and Euler, T. (2006). Two-photon imaging reveals somatodendritic chloride gradient in retinal ON-type bipolar cells expressing the biosensor Clomeleon. *Neuron* **49**, 81–94.

Feng, G., Mellor, R.H., Bernstein, M., Keller-Peck, C., Nguyen, Q.T., Wallace, M., Nerbonne, J.M., Lichtman, J.W., and Sanes, J.R. (2000). Imaging neuronal subsets in transgenic mice expressing multiple spectral variants of GFP. *Neuron* **28**, 41–51.

Freund, T.F. and Buzsáki, G. (1996). Interneurons of the hippocampus. *Hippocampus* **6**, 347–470.

Galietta, L.J., Haggie, P.M., and Verkman, A.S. (2001). Green fluorescent protein-based halide indicators with improved chloride and iodide affinities. *FEBS Lett.* **499**, 220–224.

Hasan, M.T., Friedrich, R.W., Euler, T., Larkum, M.E., Giese, G., Both, M., Duebel, J., Waters, J., Bujard, H., Griesbeck, O., Tsien, R.Y., Nagai, T., Miyawaki, A., and Denk, W. (2004). Functional fluorescent Ca^{2+} indicator proteins in transgenic mice under TET control. *PLoS Biol.* **2**, e163.

Haverkamp, S., Wassle, H., Duebel, J., Kuner, T., Augustine, G.J., Feng, G., and Euler, T. (2005). The primordial, blue-cone color system of the mouse retina. *J. Neurosci.* **25**, 5438–5445.

Heintz, N. (2004). Gene expression nervous system atlas (GENSAT). *Nat. Neurosci.* **7**, 483.

Helmchen, F. and Denk, W. (2005). Deep tissue two-photon microscopy. *Nat. Methods* **2**, 932–940.

Inglefield, J.R. and Schwartz-Bloom, R.D. (1997). Confocal imaging of intracellular chloride in living brain slices: measurement of $GABA_A$ receptor activity. *J. Neurosci. Methods* **75**, 127–135.

Inglefield, J.R. and Schwartz-Bloom, R.D. (1999). Using confocal microscopy and the fluorescent indicator, 6-methoxy-*N*-ethylquinolinium iodide, to measure changes in intracellular chloride. *Methods Enzymol.* **307**, 469–481.

Inoue, T. and Krumlauf, R. (2001). An impulse to the brain – using *in vivo* electroporation. *Nat. Neurosci.* **4** (Supplement), 1156–1158.

Jayaraman, S., Haggie, P., Wachter, R.M., Remington, S.J., and Verkman, A.S. (2000). Mechanism and cellular applications of a green fluorescent protein-based halide sensor. *J. Biol. Chem.* **275**, 6047–6050.

Jose, M., Nair, D.K., Reissner, C., Hartig, R., and Zuschratter, W. (2007). Photophysics of Clomeleon by FLIM: discriminating excited state reactions along neuronal development. *Biophys. J.* **92**, 2237–2254.

Khirug, S., Yamada, J., Afzalov, R., Voipio, J., Khiroug, L., and Kaila, K. (2008). GABAergic depolarization of the axon initial segment in cortical principal neurons is caused by the Na-K-2Cl cotransporter NKCC1. *J. Neurosci.* **28**, 4635–4639.

Kitamura, K., Judkewitz, B., Kano, M., Denk, W., and Hausser, M. (2008). Targeted patch-clamp recordings and single-cell electroporation of unlabeled neurons in vivo. *Nat. Methods* **5**, 61–67.

Klugmann, M., Symes, C.W., Leichtlein, C.B., Klaussner, B.K., Dunning, J., Fong, D., Young, D., and During, M.J. (2005). AAV-mediated hippocampal expression of short and long Homer 1 proteins differentially affect cognition and seizure activity in adult rats. *Mol. Cell. Neurosci.* **28**, 347–360.

Krapf, R., Berry, C.A., and Verkman, A.S. (1988). Estimation of intracellular chloride activity in isolated perfused rabbit proximal convoluted tubules using a fluorescent indicator. *Biophys. J.* **53**, 955–962.

Krieger, P., Kuner, T., and Sakmann, B. (2007). Synaptic connections between layer 5B pyramidal neurons in mouse somatosensory cortex are independent of apical dendrite bundling. *J. Neurosci.* **27**, 11473–11482.

Kuner, T. and Augustine, G.J. (2000). A genetically encoded ratiometric indicator for chloride: capturing chloride transients in cultured hippocampal neurons. *Neuron* **27**, 447–459.

Marandi, N., Konnerth, A., and Garaschuk, O. (2002). Two-photon chloride imaging in neurons of brain slices. *Pflügers Archiv.* **445**, 357–365.

Markova, O., Mukhtarov, M., Real, E., Jacob, Y., and Bregestovski, P. (2008). Genetically encoded chloride indicator with improved sensitivity. *J. Neurosci. Methods* **170**, 67–76.

McCown, T.J. (2005). Adeno-associated virus (AAV) vectors in the CNS. *Curr. Gene Ther.* **5**, 333–338.

Miesenböck, G., De Angelis, D.A., and Rothman, J.E. (1998). Visualizing secretion and synaptic transmission with pH-sensitive green fluorescent proteins. *Nature* **394**, 192–195.

Miller, R.F. and Dacheux, R.F. (1983). Intracellular chloride in retinal neurons: measurement and meaning. *Vision Res.* **23**, 399–411.

Miyawaki, A., Llopis, J., Heim, R., McCaffery, J.M., Adams, J.A., Ikura, M., and Tsien, R.Y. (1997). Fluorescent indicators for Ca^{2+} based on green fluorescent proteins and calmodulin. *Nature* **388**, 882–887.

Owens, D.F., Boyce, L.H., Davis, M.B., and Kriegstein, A.R. (1996). Excitatory GABA responses in embryonic and neonatal cortical slices demonstrated by gramicidin perforated-patch recordings and calcium imaging. *J. Neurosci.* **16**, 6414–6423.

Payne, J.A., Rivera, C., Voipio, J., and Kaila, K. (2003). Cation-chloride co-transporters in neuronal communication, development and trauma. *Trends Neurosci.* **26**, 199–206.

Pologruto, T.A., Yasuda, R., and Svoboda, K. (2004). Monitoring neural activity and $[Ca^{2+}]$ with genetically encoded Ca^{2+} indicators. *J. Neurosci.* **24**, 9572–9579.

Pond, B.B., Berglund, K., Kuner, T., Feng, G., Augustine, G.J., and Schwartz-Bloom, R.D. (2006). The chloride transporter Na^+-K^+-Cl^- cotransporter isoform-1 contributes to intracellular chloride increases after *in vitro* ischemia. *J. Neurosci.* **26**, 1396–1406.

Rathenberg, J., Nevian, T., and Witzemann, V. (2003). High-efficiency transfection of individual neurons using modified electrophysiology techniques. *J. Neurosci. Methods* **126**, 91–98.

Reiff, D.F., Ihring, A., Guerrero, G., Isacoff, E.Y., Joesch, M., Nakai, J., and Borst, A. (2005). *In vivo* performance of genetically encoded indicators of neural activity in flies. *J. Neurosci.* **25**, 4766–4778.

Rivera, C., Voipio, J., Payne, J.A., Ruusuvuori, E., Lahtinen, H., Lamsa, K., Pirvola, U., Saarma, M., and Kaila, K. (1999). The K^+/Cl^- co-transporter KCC2 renders GABA hyperpolarizing during neuronal maturation. *Nature* **397**, 251–255.

Rizzo, M.A., Springer, G.H., Granada, B., and Piston, D.W. (2004). An improved cyan fluorescent protein variant useful for FRET. *Nat. Biotechnol.* **22**, 445–449.

Satoh, H., Kaneda, M., and Kaneko, A. (2001). Intracellular chloride concentration is higher in rod bipolar cells than in cone bipolar cells of the mouse retina. *Neurosci. Lett.* **310**, 161–164.

Sinnecker, D., Voigt, P., Hellwig, N., and Schaefer, M. (2005). Reversible photobleaching of enhanced green fluorescent proteins. *Biochem.* **44**, 7085–7094.

Somogyi, P. and Klausberger, T. (2005). Defined types of cortical interneurone structure space and spike timing in the hippocampus. *J. Physiol.* **562**, 9–26.

Soriano, P. (1999). Generalized lacZ expression with the ROSA26 Cre reporter strain. *Nat. Genet.* **21**, 70–71.

Srinivas, S., Watanabe, T., Lin, C.S., William, C.M., Tanabe, Y., Jessell, T.M., and Costantini, F. (2001). Cre reporter strains produced by targeted insertion of EYFP and ECFP into the ROSA26 locus. *BMC Dev. Biol.* **1**, 4.

Staley, K.J. and Proctor, W.R. (1999). Modulation of mammalian dendritic $GABA_A$ receptor function by the kinetics of Cl^- and HCO_3^- transport. *J. Physiol.* **519**, 693–712.

Stein, V., Hermans-Borgmeyer, I., Jentsch, T.J., and Hübner, C.A. (2004). Expression of the KCl cotransporter KCC2 parallels neuronal maturation and the emergence of low intracellular chloride. *J. Comp. Neurol.* **468**, 57–64.

Szabadics, J., Varga, C., Molnár, G., Oláh, S., Barzó, P., and Tamás, G. (2006). Excitatory effect of GABAergic axo-axonic cells in cortical microcircuits. *Science* **311**, 233–235.

Teruel, M.N., Blanpied, T.A., Shen, K., Augustine, G.J., and Meyer, T. (1999). A versatile microporation technique for the transfection of cultured CNS neurons. *J. Neurosci. Methods* **93**, 37–48.

Torsney, C. and MacDermott, A.B. (2005). Neuroscience: a painful factor. *Nature* **438**, 923–925.

Wachter, R.M. and Remington, S.J. (1999). Sensitivity of the yellow variant of green fluorescent protein to halides and nitrate. *Curr. Biol.* **9**, R628–R629.

Wachter, R.M., Yarbrough, D., Kallio, K., and Remington, S.J. (2000). Crystallographic and energetic analysis of binding of selected anions to the yellow variants of green fluorescent protein. *J. Mol. Biol.* **301**, 157–171.

Wei, F., Xia, X.M., Tang, J., Ao, H., Ko, S., Liauw, J., Qiu, C.S., and Zhuo, M. (2003). Calmodulin regulates synaptic plasticity in the anterior cingulate cortex and behavioral responses: a microelectroporation study in adult rodents. *J. Neurosci.* **23**, 8402–8409.

Wimmer, V.C., Nevian, T., and Kuner, T. (2004). Targeted in vivo expression of proteins in the calyx of Held. *Pflügers Archiv.* **449**, 319–333.

Zhang, L.L., Pathak, H.R., Coulter, D.A., Freed, M.A., and Vardi, N. (2006). Shift of intracellular chloride concentration in ganglion and amacrine cells of developing mouse retina. *J. Neurophysiol.* **95**, 2404–2416.

CHAPTER

8

Gramicidin Perforated Patch

Norio Akaike

OUTLINE

I. INTRODUCTION

Developments and improvements of the conventional whole-cell patch technique during the last decades yielded many insights on various aspects of membrane physiology (Akaike et al., 1978; Hamill et al., 1981; Kostyuk et al., 1975; Lee et al., 1979). The patch clamp has permitted quantitative analysis of a variety of voltage-gated- and ligand-gated-ion channels, the electrogenic Na^+/K^+ pump and the Na^+/Ca^{2+} exchanger to mention a few examples. While whole cell patch technique allows one to control the composition of the intracellular milieu by equilibrating it with that of the artificial pipette solutions, the dialysis or the dilution of intracellular components disturbs the native cytoplasmic biochemical environment which contains substances that are key for regulation and function of various transporters, ion channels and metabotropic receptors (Korn and Horn, 1989; Kurachi et al., 1989; Sims et al., 1991; Ueno et al., 1992; Wendt et al., 1992; Ye and Akaike, 1993); G-protein and second messenger mediated responses (Harata and Akaike, 1993; Shirasaki et al., 1994; Uneyama et al., 1993); and intracellular Ca^{2+} buffering systems (Akaike and Uneyama, 1994; Mistry and Hablitz, 1990; Omura et al., 1993). These disadvantages of the conventional whole-cell patch recording configuration were overcome by the development of the "perforated patch" recording configuration using pore-forming antibiotics that are selectively permeable to certain ions. These antibiotics include nystatin (Akaike and Harata, 1994; Horn and Marty, 1988), amphotericin B (Rae et al., 1991) and gramicidin (Abe et al., 1994; Akaike, 1996; Ebihara et al., 1995; Kakazu et al., 1999, 2000; Nabekura et al., 2002; Rhee et al., 1994; Tajima et al., 1996; Wake et al., 2007). Nystatin forms small aqueous pores (radius 0.4 nm) on the cell membrane that are permeable to small monovalent cations with the following selectivity sequence: $Rb^+ > K^+ > Cs^+ > Na^+ > Li^+$, and to anions such as Cl^-. When inserted into patch pipettes, the permeability properties of nystatin allow electrical access to the cell while preserving relatively large intracellular molecules such as cAMP and ATP. Thus, the nystatin-perforated patch technique allows the study of functional channel responses in their native environment, without altering the cellular contents which includes key second messenger molecules (Akaike and Harata, 1994). However, since the nystatin pores are permeable to monovalent anions, the intracellular Cl^- concentration ($[Cl^-]_i$) is determined by the Cl^- concentration in the patch pipette solution, and thus the native $[Cl^-]_i$ is altered.

Cl^- is one of the major and most important constituents of living cells; it participates in a number of functions such as cell volume and pH_i regulation (Chapters 15 and 4, respectively), salt secretion and absorption (Chapters 16 and 29), G-protein-coupled signal transduction (Higashijima et al., 1987) and modulation of network excitability via GABA and glycine receptor-channel complexes (e.g. Chapters 14, 19 and 24 in this volume). Considering the functional significance of Cl^-, a method that allowed electrophysiological measurements of membrane currents carried by Cl^- while preserving native $[Cl^-]_i$ was required. These needs prompted the development of the gramicidin-perforated patch recording technique (Akaike, 1996). Gramicidin is a polypeptide antibiotic that forms small pores in the cell membrane similar to nystatin, but allows only the movement of monovalent cations excluding anions (Hladky and Haydon, 1972). The cation selectivity sequence of gramicidin in cell membranes is similar to that measured in lipid bilayers: $H^+ > NH_4^+ > Cs^+ > Rb^+ > K^+ > Na^+ > Li^+$ (Myers and Haydon, 1972; Tajima et al., 1996). Gramicidin is not permeable to Cl^- in lipid bilayers or in living cells (Myers and Haydon, 1972; Tajima et al., 1996). Thus, by adding gramicidin to the patch pipette solution, we can observe native responses from electrically and chemically excitable cells, while maintaining native $[Cl^-]_i$, independently of the Cl^- concentration in the patch pipette solution.

II. APPLICATION OF GRAMICIDIN PERFORATED PATCH TO CNS NEURONS

A. Getting the Patch

The stock solution of gramicidin is prepared by dissolving gramicidin D (Sigma), which is a mixture of gramicidin A, B and C, into methanol at 10 mg/ml. After ultrasonication for a few seconds, the stock solution is directly dissolved into the patch pipette solution, resulting in a final concentration 0.1 mg/ml. Patch pipettes are fabricated from glass tubes (Narishige, G-1.5) on a two-stage vertical pipette puller. The pipette tips are initially filled with gramicidin-free internal solution by immersion, and then the remainder of the pipette is backfilled with the same patch pipette solution containing gramicidin. The resistance of a pipette having an internal diameter of about 1 μm is 5~8 MΩ. As gramicidin diffuses to the pipette tip and enters the cell membrane, a time-dependent change in the capacitative current is observed (Fig. 8.1A). In small neurons dissociated from rat CNS, where the soma diameter is 10–15 μm, the membrane perforation by gramicidin requires more than 30 min after making a GΩ seal. However, when patch pipettes having larger internal diameter (2~3 MΩ resistance) are applied to larger cell bodies (diameter ~30 μm) the current recording is initiated some 20 min after the GΩ seal formation. Once gramicidin ionophores are formed on the patch cell membrane, the ion conducting channels can be constantly maintained for ~2 h. It should be noted that gramicidin in the patch pipette solution loses its activity within 2 h. Thus the gramicidin pipette solution should be refreshed frequently.

B. GABA Responses

It is well established that both GABA and glycine are the most important inhibitory neurotransmitters in the mammalian CNS (Lynch, 2004; Olsen and Sieghart, 2009). Application of the gramicidin-perforated patch technique to mammalian CNS neurons makes possible the recording of native GABA- or glycine-gated currents (Abe et al., 1994; Ebihara et al., 1995; Kakazu et al., 1999). Figure 8.1B shows the GABA-induced hyperpolarization observed in a neuron dissociated from rat substantia nigra *pars reticulata* (SNR) recorded under current-clamp by means of a gramicidin-perforated patch pipette. The resting membrane potential (E_m) in this cell was −58 mV (Ebihara et al., 1995). Under voltage-clamp conditions, GABA elicited an outward current at a holding potential (V_H) of −50 mV. Since Cl^- is an anion, the outward current implies Cl^- flux into the cell. To change the recording mode to conventional whole cell patch configuration the patched membrane was ruptured by increasing suction in the pipette interior. This resulted in a clear-cut change in the direction of the GABA-induced current from outward to inward

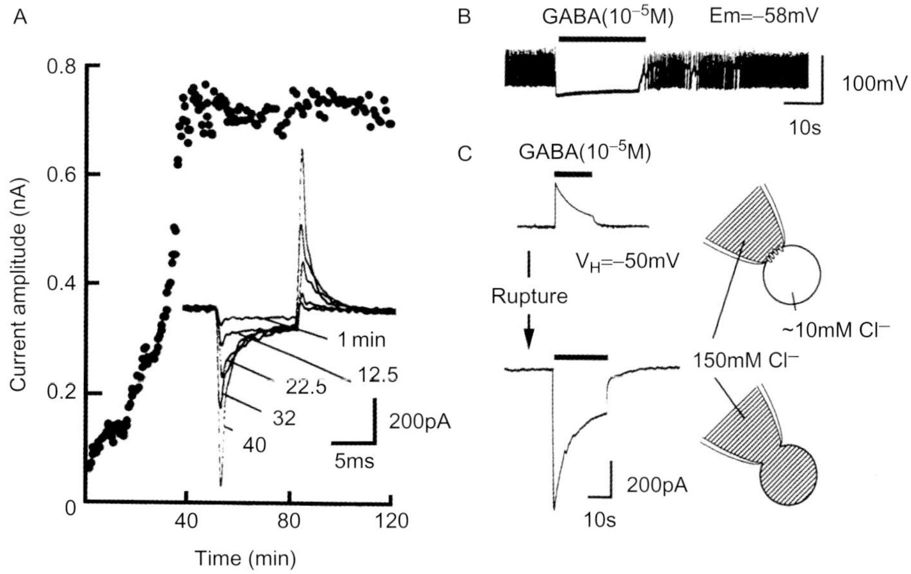

FIGURE 8.1 Gramicidin-perforated patch recording from substantia nigra *pars reticulata* (SNR) neurons. **A.** Relationship between the time after establishment of a GΩ seal and the peak capacitative current amplitude. Transient capacitative currents were elicited by 10 ms hyperpolarizing voltage steps from −60 to −70 mV every 30 s. The inset shows the capacitative currents 1–40 min after forming the GΩ seal. The holding potential (V_H) was −60 mV. **B.** Recording of membrane potentials using gramicidin-perforated patch technique under current-clamp conditions. GABA (10^{-5} M) was exogenously applied during the period indicated by the horizontal bar. Resting membrane potential (E_m) was −58 mV. **C.** GABA-induced outward (upper panel) and inward (lower panel) currents at a V_H of −50 mV before and after the rupture of the patch membrane, respectively. The upper panel shows GABA-induced outward current recorded by the gramicidin-perforated patch technique. Under these conditions, the intracellular Cl^- concentration ($[Cl^-]_i$) was not disturbed by the 150 mM Cl^- concentration in the patch pipette solution. The lower panel illustrates GABA-induced inward current in the same neuron after rupture of the patch membrane by increasing the negative pressure of the patch pipette interior. The patch pipette Cl^- rapidly diffused into the cell interior within 5 min of the rupture and the $[Cl^-]_i$ reached 150 mM. (Reproduced from Ebihara et al., 1995, with permission.)

(Fig. 8.1C). These results clearly indicate that Cl^- from the patch pipette solution (150 mM Cl^-) diffuses into the neuronal cell body of SNR following rupture of the membrane patch. The resulting increase in $[Cl^-]_i$ shifts E_{Cl} toward more depolarized values than V_H causing Cl^- efflux through $GABA_A$ receptor-Cl^- channel complexes. Several observations have demonstrated that Cl^- cannot permeate the gramicidin-formed pores, strongly suggesting that native $[Cl^-]_i$ is unaltered under gramicidin patch conditions (Akaike and Harata, 1994; Hladky and Haydon, 1972). Further validation of the method has been provided by using independent measurements of $[Cl^-]_i$ by means of fluorescent indicators. These reports show that $[Cl^-]_i$ inferred from the reversal of GABA responses measured through gramicidin patches is close to that measured with fluorescent dyes (see Chapter 7 in this volume and Achilles et al., 2007).

C. Developmental Change in Glycine and GABA Responses

The lateral superior olive (LSO) neurons in rat brainstem, one of the central auditory nuclei, receives auditory information from both the contralateral side as inhibitory glycinergic inputs and the ipsilateral side as excitatory glutamatergic inputs. This arrangement is important for sound localization. In immature LSO neurons (P0–P2), glycinergic synaptic inputs or exogeneous application of glycine induces membrane depolarization in these neurons (Kakazu et al., 1999; Kandler and Friauf, 1995). The depolarizing response turns into a hyperpolarizing one with maturation (Fig. 8.2A). Using gramicidin perforated patch recording it was possible to infer the native $[Cl^-]_i$ of LSO neurons at each developmental stage, from the reversal potential (E_{Gly}) of the glycine responses measured from current/voltage (I/V) curves before and during glycine application under voltage-clamp conditions (Ehrlich et al., 1999; Kakazu et al., 1999). In >50% of LSO neurons at P0–P2 and P6–P8, E_{Gly} was more depolarized than the resting membrane potential (E_m). At P13–P15 E_{Gly} became more hyperpolarized than E_m in 90% of LSO neurons. The differences in the mean values of E_{Gly} observed between P0–P2 and P13–P15 and between P6–P8 and P13–P15 were statistically significant ($p < 0.05$) (Fig. 8.2B). The glycine-induced responses in rat LSO neurons were likely to result from Cl^- gradients favoring the flux of Cl^- through

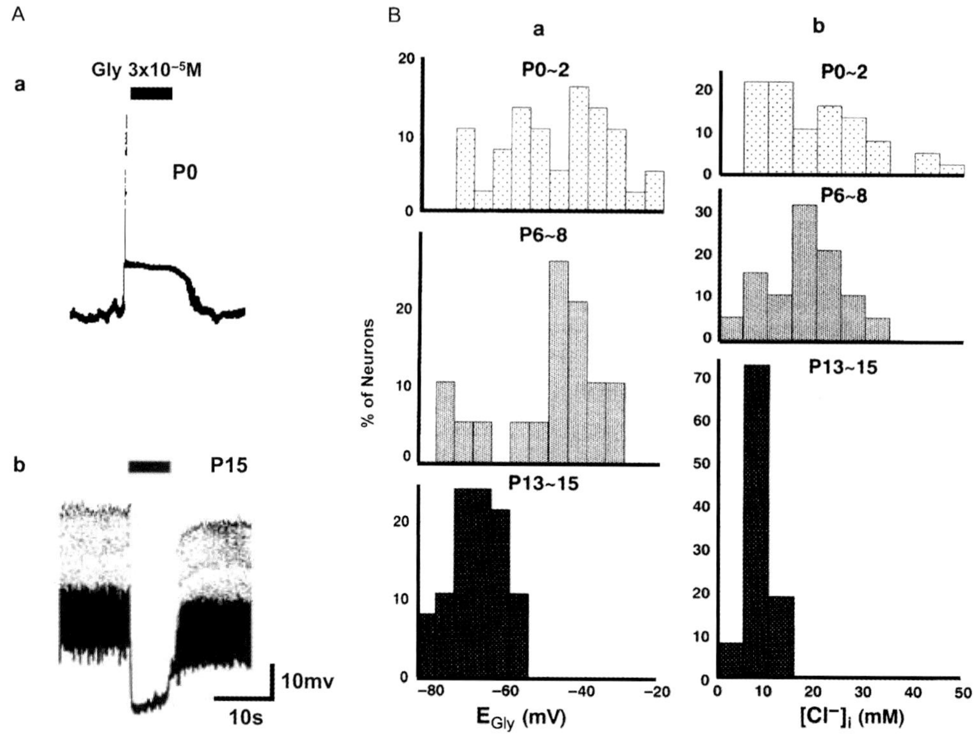

FIGURE 8.2 Developmental changes in glycine responses in lateral superior olive (LSO) neurons. **Aa.** At P0, 3×10^{-5}M glycine (*solid bar*) induced a depolarization that generated an action potential. The record was obtained though a gramicidin-perforated patch under current-clamp configuration. **Ab.** At P15, glycine induced a hyperpolarization, resulting in blockage of action potentials. Resting membrane potentials (E_m) were −53 and −60 mV in *a* and *b*, respectively. The reversal potentials of the glycine-induced currents (E_{Gly}) measured from the current/voltage (I/V) relationships in the LSO neurons of P0 and P15 rats were −35 and −70 mV, respectively, at a V_H of −50 mV. **B.** Developmental changes in intracellular Cl^- concentration ($[Cl^-]_i$) in LSO neurons. Histograms of the percentages of LSO neurons are indicated as a function of E_{Gly} (*a*) and $[Cl^-]_i$ (*b*) at P0–2, P6–8 and P13–15 rats. (From Kakazu et al., 1999, reproduced with permission.)

glycine-operated Cl^- channels, and thus E_{Gly} was considered to be equal to E_{Cl}. Under this assumption it was possible to calculate $[Cl^-]_i$ for each individual neuron using the Nernst equation. Figure 8.2B also shows significant differences in $[Cl^-]_i$ between P0–P2 and P13–P15 ($p < 0.01$), and between P6–P8 and P13–P15 ($p < 0.01$). These results clearly indicate that glycine's effect switches with maturation from depolarizing to hyperpolarizing due to a fall in $[Cl^-]_i$ in LSO neurons during the second week after birth.

Age-related changes in $[Cl^-]$i were also observed in the nucleus basalis of Meynert which is populated by large cholinergic neurons and is a major source of cholinergic input to the cerebral cortex. The Meynert neurons receive GABAergic inputs. The E_{GABA} measured from I/V curves became more negative with maturation. The mean $[Cl^-]_i$ calculated from the Nernst equation using both the known $[Cl^-]_o$ and the measured E_{GABA} had the following values: 34.8, 22.3 and 13.1 mM for P0, 2-week- and 6-week-old rat neurons, respectively (Akaike, 1996).

D. Effect of Neuronal Trauma

When rat embryonic 18-day hypothalamic neurons are cultured for 24–35 days, GABA application induces a hyperpolarization. In neurons traumatized by neurite transection, or osmotic challenges or excess heat, synaptically released or exogenously applied GABA was depolarizing (Fig. 8.3A) and this depolarization produced an increase in $[Ca^{2+}]_i$ (van den Pol, 1996). The depolarization after trauma lasted longer than a week.

E. Effect of H_2O_2 Treatment

During application of H_2O_2 for 1–6 h GABA-induced outward currents decreased with time and finally reversed to inward currents after 9 h (Fig. 8.3Ba). This phenomenon appears to be due to loss of tyrosine phosphorylation of KCC resulting in functional down-regulation of this cotransporter and consequent increase in $[Cl^-]_i$. Accordingly, during

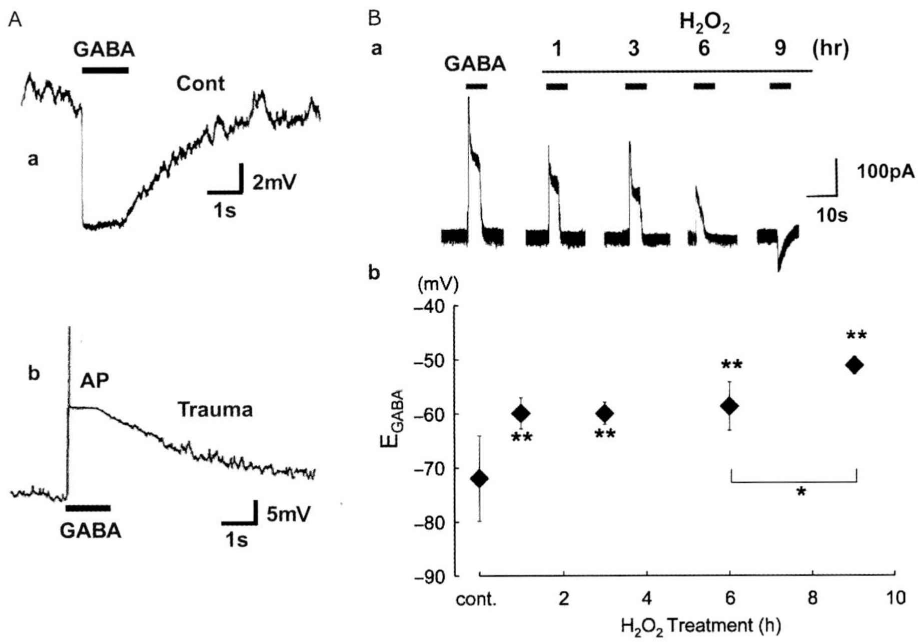

FIGURE 8.3 Change in polarity of GABA response induced by trauma or H_2O_2. **Aa.** 10^{-5}M GABA-induced membrane hyperpolarization from −60 to −72 mV in a rat hypothalamic neuron cultured for 33 days. **b.** In a traumatized neuron GABA depolarized the cell membrane from −60 to −38 mV and produced an action potential (AP). All recordings were made using gramicidin-perforated patch configuration. (Reproduced with permission from van den Pol et al., 1996). **B.** Changes in E_{GABA} in primary cultured hippocampal neurons during continuous incubation with H_2O_2. **a.** Representative whole-cell membrane currents recorded using the gramicidin-perforated patch technique, in response to application of GABA (3×10^{-5}M; thick horizontal bars), recorded at a V_H of −50 mV, before and at different times during H_2O_2 (5×10^{-5}M) incubation as indicated. The GABA-evoked outward currents gradually decreased in amplitude, and eventually after 9 h incubation with H_2O_2 they changed to an inward current. **b.** Averaged E_{GABA} measured in neurons incubated with H_2O_2 for up to 9 h as indicated. H_2O_2 incubation caused E_{GABA} to rapidly (<1 h) shift to a more depolarized level at a V_H of −50 mV. ** $p < 0.01$. (From Wake et al., 2007; reproduced with permission.)

continuous H_2O_2 incubation E_{GABA} is shifted towards more positive potentials and finally the GABA-induced current becomes inward (Fig. 8.3Ba). A more gradual depolarizing shift in E_{GABA} also occurred after successive application of H_2O_2 (Fig. 8.3Bb), presumably corresponding to a time when total KCC protein levels are decreased (Wake et al., 2007).

F. Effect of Furosemide on Intracellular $[Cl^-]_i$

Since Cl^- cannot permeate the gramicidin pores, the constancy of GABA- or glycine-induced currents reflect the existence of $[Cl^-]_i$ regulatory mechanisms that maintain a stable $[Cl^-]_i$. One of these mechanisms is KCC, which extrudes Cl^- actively under physiological conditions (Chapter 17). As shown in Fig. 8.4A, in dissociated neurons from the lateral superior olive of rats the glycine-induced Cl^- currents recorded with gramicidin patches gradually decrease in amplitude in the presence of 1 mM furosemide (Kakazu et al., 1999, 2000), a finding that is in agreement with previous observations made in hippocampal neurons (Thompson and Gahwiler, 1989). The $[Cl^-]_i$ inferred from E_{gly} indicated that $[Cl^-]_i$ increased approaching the values that would have if it was passively distributed across the plasma membrane (Fig. 8.4B). These results are in agreement with the notion that a furosemide-sensitive Cl^- extrusion mechanism plays an important role in maintaining $[Cl^-]_i$ below electrochemical equilibrium in mature CNS neurons. The furosemide-sensitive mechanism may be KCC. The electro-neutral operation of this cotransporter precludes direct observation of ion translocation by electrophysiological techniques. However, using the gramicidin-perforated patch method it is possible to indirectly assess its function by estimating $[Cl^-]_i$.

III. CONCLUSIONS

The development of the whole-cell patch technique has contributed enormously to our current understanding of membrane physiology. However, the disturbance of cell constituents by the artificial patch

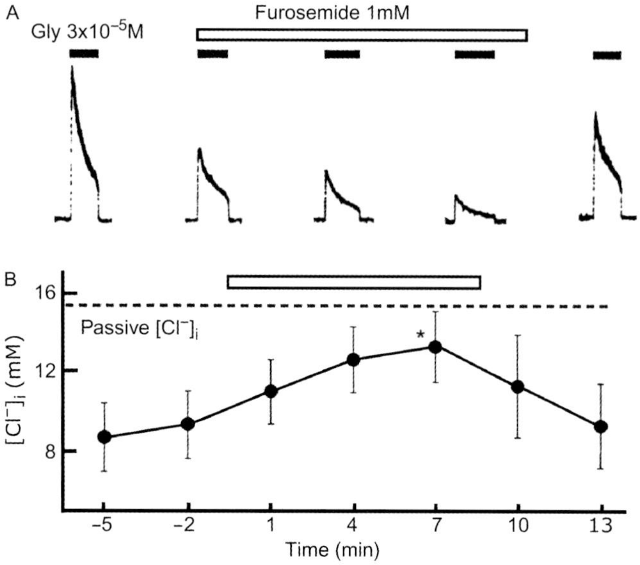

FIGURE 8.4 Neuronal $[Cl^-]_i$ maintained below electrochemical equilibrium by a furosemide-sensitive Cl^- transport mechanism. **A.** In the presence of 1 mM furosemide the amplitude of glycine-induced outward currents recorded in LSO neuron by using gramicidin-perforated patch method decreased gradually. In these experiments V_H was −50 mV. **B.** The $[Cl^-]_i$ values inferred from Egly were plotted as a function of time before, during and after the application of furosemide. The dashed line indicates the $[Cl^-]_i$ (15.7 mM) estimated for a passive distribution at V_h = −50 mV. The lower than electrochemical equilibrium $[Cl^-]_i$ was increased by furosemide, and returned to its original value after the removal of furosemide, suggesting that the furosemide-sensitive $[Cl^-]_i$ KCC actively extruded Cl^- maintaining it below electrochemical equilibrium. (Reproduced with permission from Kakazu et al., 2000.)

pipette solution limits the use of this method; it does not allow the study of native cell responses. The perforated patch recording mode using antibiotics such as nystatin, amphotericin B and gramicidin permits electrical access to the inside of the cell with minimal dialysis of the patch pipette solution and washout of intracellular contents. Gramicidin-perforated patch recording configuration allows the study of native GABA- or glycine-gated currents and at the same time it permits estimations of $[Cl^-]_i$. This is possible because gramicidin pores are permeable to cations but not to Cl^-. Further, gramicidin patches have allowed the study of cation-coupled Cl^- cotransporters such as KCC and NKCC, which play a major role in determining neuronal $[Cl^-]_i$. Gramicidin patch recording is a useful tool of broad applicability, which has provided important information on Cl^- channels and carriers in cells that preserve their physiological contents.

References

Abe, Y., Furukawa, K., Itoyama, Y., and Akaike, N. (1994). Glycine response in acutely dissociated ventromedial hypothalamic neuron of the rat: new approach with gramicidin perforated patch-clamp technique. *J. Neurophysiol.* **72**, 1530–1537.

Achilles, K., Okabe, A., Ikeda, M., Shimizu-Okabe, C., Yamada, J., Fukuda, A., Luhmann, H.J., and Kilb, W. (2007). Kinetic properties of Cl uptake mediated by Na^+-dependent K^+-$2Cl^-$ cotransport in immature rat neocortical neurons. *J. Neurosci.* **27**, 8616–8627.

Akaike, N. (1996). Gramicidin perforated patch recording and intracellular chloride activity in excitable cells. *Prog. Biophys. Mol. Biol.* **65**, 251–264.

Akaike, N., Fishman, H.M., Lee, K.S., Moore, L.E., and Brown, A. M. (1978). The units of calcium conduction in Helix neurones. *Nature* **274**, 379–382.

Akaike, N. and Harata, N. (1994). Nystatin perforated patch recording and its applications to analyses of intracellular mechanisms. *Jpn. J. Physiol.* **44**, 373–433.

Akaike, N. and Uneyama, H. (1994). ATP-induced K+ current oscillation in megakaryocytes: a unique purinoceptor. *News in Physiological Sciences* **9**, 49–53.

Ebihara, S., Shirato, K., Harata, N., and Akaike, N. (1995). Gramicidin-perforated patch recording: GABA response in mammalian neurones with intact intracellular chloride. *J. Physiol.* **484**, 77–86.

Ehrlich, I., Lohrke, S., and Friauf, E. (1999). Shift from depolarizing to hyperpolarizing glycine action in rat auditory neurones is due to age-dependent Cl^- regulation. *J. Physiol.* **520**, 121–137.

Hamill, O.P., Marty, A., Neher, E., Sakmann, B., and Sigworth, F.J. (1981). Improved patch-clamp techniques for high-resolution current recording from cells and cell-free membrane patches. *Pflugers Arch.* **391**, 85–100.

Harata, N. and Akaike, N. (1993). Metabotropic glutamate response in acutely dissociated hippocampal CA3 neurons of the rat. *Ann. NY Acad. Sci.* **707**, 511–514.

Higashijima, T., Ferguson, K.M., and Sternweis, P.C. (1987). Regulation of hormone-sensitive GTP-dependent regulatory proteins by chloride. *J. Biol. Chem.* **262**, 3597–3602.

Hladky, S.B. and Haydon, D.A. (1972). Ion transfer across lipid membranes in the presence of gramicidin A. I. Studies of the unit conductance channel. *Biochim. Biophys. Acta* **274**, 294–312.

Horn, R. and Marty, A. (1988). Muscarinic activation of ionic currents measured by a new whole-cell recording method. *J. Gen. Physiol.* **92**, 145–159.

Kakazu, Y., Akaike, N., Komiyama, S., and Nabekura, J. (1999). Regulation of intracellular chloride by cotransporters in developing lateral superior olive neurons. *J. Neurosci.* **19**, 2843–2851.

Kakazu, Y., Uchida, S., Nakagawa, T., Akaike, N., and Nabekura, J. (2000). Reversibility and cation selectivity of the K^+-Cl^- cotransport in rat central neurons. *J. Neurophysiol.* **84**, 281–288.

Kandler, K. and Friauf, E. (1995). Development of glycinergic and glutamatergic synaptic transmission in the auditory brainstem of perinatal rats. *J. Neurosci.* **15**, 6890–6904.

Korn, S.J. and Horn, R. (1989). Influence of sodium-calcium exchange on calcium current rundown and the duration of calcium-dependent chloride currents in pituitary cells, studied with whole cell and perforated patch recording. *J. Gen. Physiol.* **94**, 789–812.

Kostyuk, P.G., Krishtal, O.A., and Pidoplichko, V.I. (1975). Effect of internal fluoride and phosphate on membrane currents during intracellular dialysis of nerve cells. *Nature* **257**, 691–693.

Kurachi, Y., Asano, Y., Takikawa, R., and Sugimoto, T. (1989). Cardiac Ca current does not run down and is very sensitive to isoprenaline in the nystatin-method of whole cell recording. *Naunyn Schmiedebergs Arch. Pharmacol.* **340**, 219–222.

Lee, K.S., Weeks, T.A., Kao, R.L., Akaike, N., and Brown, A.M. (1979). Sodium current in single heart muscle cells. *Nature* **278**, 269–271.

Lynch, J.W. (2004). Molecular structure and function of the glycine receptor chloride channel. *Physiol. Rev.* **84**, 1051–1095.

Mistry, D.K. and Hablitz, J.J. (1990). Nystatin-perforated patch recordings disclose NMDA-induced outward currents in cultured neocortical neurons. *Brain Res.* **535**, 318–322.

Myers, V.B. and Haydon, D.A. (1972). Ion transfer across lipid membranes in the presence of gramicidin A. II. The ion selectivity. *Biochim. Biophys. Acta* **274**, 313–322.

Nabekura, J., Ueno, T., Okabe, A., Furuta, A., Iwaki, T., Shimizu-Okabe, C., Fukuda, A., and Akaike, N. (2002). Reduction of KCC2 expression and GABAA receptor-mediated excitation after in vivo axonal injury. *J. Neurosci.* **22**, 4412–4417.

Olsen, R.W. and Sieghart, W. (2009). GABA A receptors: subtypes provide diversity of function and pharmacology. *Neuropharmacology* **56**, 141–148.

Omura, T., Munakata, M., and Akaike, N. (1993). Nystatin-perforated patch recordings disclose KA-operated outward currents in rat cortical neurons. *Brain Res.* **627**, 345–348.

Rae, J., Cooper, K., Gates, P., and Watsky, M. (1991). Low access resistance perforated patch recordings using amphotericin. *B. J. Neurosci. Methods* **37**, 15–26.

Rhee, J.S., Ebihara, S., and Akaike, N. (1994). Gramicidin perforated patch-clamp technique reveals glycine-gated outward chloride current in dissociated nucleus solitarii neurons of the rat. *J. Neurophysiol.* **72**, 1103–1108.

Shirasaki, T., Harata, N., and Akaike, N. (1994). Metabotropic glutamate response in acutely dissociated hippocampal CA1 pyramidal neurones of the rat. *J. Physiol.* **475**, 439–453.

Sims, S.M., Lussier, B.T., and Kraicer, J. (1991). Somatostatin activates an inwardly rectifying K^+ conductance in freshly dispersed rat somatotrophs. *J. Physiol.* **441**, 615–637.

Tajima, Y., Ono, K., and Akaike, N. (1996). Perforated patch-clamp recording in cardiac myocytes using cation-selective ionophore gramicidin. *Am. J. Physiol.* **271**, C524–C532.

Thompson, S.M. and Gahwiler, B.H. (1989). Activity-dependent disinhibition. II. Effects of extracellular potassium, furosemide, and membrane potential on ECl^- in hippocampal CA3 neurons. *J. Neurophysiol.* **61**, 512–523.

Ueno, S., Ishibashi, H., and Akaike, N. (1992). Perforated-patch method reveals extracellular ATP-induced K^+ conductance in dissociated rat nucleus solitarii neurons. *Brain Res.* **597**, 176–179.

Uneyama, H., Uneyama, C., and Akaike, N. (1993). Intracellular mechanisms of cytoplasmic Ca^{2+} oscillation in rat megakaryocyte. *J. Biol. Chem.* **268**, 168–174.

van den Pol, A.N., Obrietan, K., and Chen, G. (1996). Excitatory actions of GABA after neuronal trauma. *J. Neurosci.* **16**, 4283–4292.

Wake, H., Watanabe, M., Moorhouse, A.J., Kanematsu, T., Horibe, S., Matsukawa, N., Asai, K., Ojika, K., Hirata, M., and Nabekura, J. (2007). Early changes in KCC2 phosphorylation in response to neuronal stress result in functional downregulation. *J. Neurosci.* **27**, 1642–1650.

Wendt, D.J., Starmer, C.F., and Grant, A.O. (1992). Na channel kinetics remain stable during perforated-patch recordings. *Am. J. Physiol.* **263**, C1234–C1240.

Ye, J.H. and Akaike, N. (1993). Calcium currents in pyramidal neurons acutely dissociated from the rat frontal cortex: a study by the nystatin perforated patch technique. *Brain Res.* **606**, 111–117.

CHAPTER

9

Measuring Electroneutral Chloride-dependent Ion Fluxes in Mammalian Cells and in Heterologous Expression Systems

Kenneth Gagnon

OUTLINE

I. INTRODUCTION

Physiologists have developed several methods to measure the concentration of intracellular Cl^- ($[Cl^-]_i$) in living cells. These techniques include Cl^--selective microelectrodes (Alvarez-Leefmans et al., 1988, 1990; Coles, 1995), gramicidin-perforated patch recordings (Akaike, 1996; Lu et al., 2008), Cl^--sensitive fluorescent dyes (Jayaraman and Verkman, 2000; Munkonge et al., 2004; Stern et al., 1995), and Cl^--sensitive fluorescent proteins (Kuner and Augustine, 2000). Some of these methods are discussed in more detail in Chapters 6–8 in this volume. Besides measurement of basal $[Cl^-]_i$, these methods can be used to determine *net* Cl^- fluxes from changes in $[Cl^-]_i$ (Rocha-Gonzalez et al., 2008). However, these methods assume that the changes in $[Cl^-]_i$ are solely due to transmembrane Cl^- fluxes, and that no buffering, sequestration or consumption/production of Cl^- occurs. Flux, in membrane transport, is defined as the amount of solute moving across a unit area of plasma membrane (typically 1 cm^2) per unit time (typically 1 minute). The flux is directly proportional to the permeability, surface area and solute concentration gradient (or difference) across the plasma membrane (Schultz, 1980). The *net* Cl^- flux is the algebraic sum of the unidirectional fluxes (i.e. influx and efflux). By measuring the two components of *net* flux separately, it is possible to study the kinetics of *one-way* movement of Cl^- by a given transporter. Moreover, unidirectional flux measurements are particularly powerful when there are multiple pathways capable of effecting transport of the solute of interest, which could obscure results obtained from techniques relying upon *net* flux measurements.

The permeability of the plasma membrane to solutes that could not normally cross the hydrophobic core of the phospholipid bilayer is increased by the presence of integral membrane transport proteins. Solutes are carried across membranes through pumps, channels, cotransporters, and exchangers. In general, organic molecules (e.g. sugars, amino acids) transported across biological membranes are often accompanied by inorganic ions such as sodium (Na^+), potassium (K^+), chloride (Cl^-) and bicarbonate (HCO_3^-). Some examples are the Na^+-glucose cotransporter (Hediger and Rhoads, 1994), monoamine transporters (e.g. serotonin, dopamine) (Torres and Caron, 2003), and the amino acid cotransporters (e.g. glutamate, glycine) (Kanai and Hediger, 2004). In the latter case, Cl^- movement is also associated with the movement of these amino acids. The electrogenic movement of solutes (i.e. the transfer of a net charge across the membrane) can be followed using various electrophysiological methods by measuring the transmembrane currents carried by the charged solute that is transported (i.e. *net* flux) (Alvarez-Leefmans et al., 1990).

When solute movement occurs with an electroneutral stoichiometry via membrane transporters (e.g. Cl^-/HCO_3^- exchanger, Na^+-HCO_3^- cotransporters, Na^+-K^+-$2Cl^-$ cotransporters) unidirectional flux measurements are an invaluable tool (Hebert et al., 2003; Romero, 2005; Soleimani, 2002). Unidirectional solute movement through these transport proteins can be measured either as an influx or an efflux. Influx experiments are a direct measure of solute movement into the cell from the extracellular space per unit time, whereas efflux experiments are an indirect measure as they are calculated by multiplying the rate constant of solute movement (k) by the intracellular concentration of the particular solute. In this chapter, using the electroneutral cation-chloride cotransporters as examples, we will describe the methodology of unidirectional ion fluxes (i.e. influxes and effluxes).

Although measurements of ion fluxes are regularly performed using native cells, there are often advantages in using "foreign" cell systems to express a transporter of interest. Advances in molecular biology have provided investigators with the tools (e.g. restriction enzymes, expression vectors, DNA polymerases) to clone and mutate genes and their encoded proteins for the study of their function and dysfunction. Two factors to consider for the characterization and study of electroneutral chloride-dependent ion flux are the choice of heterologous expression system (cultured cells versus frog oocytes), and the type of tracer ion utilized (i.e. radiolabeled isotope versus non-labeled or non-native solute).

II. HETEROLOGOUS EXPRESSION SYSTEMS

Heterologous expression is the introduction of either complementary DNA (cDNA) or RNA (cRNA) encoding for a protein of interest from one species into the cell of another species, such that the hosts' cellular machinery expresses the foreign protein. Cultured immortalized cells can be transfected with cDNA short term (transiently), or long term (stable), depending on whether the foreign DNA is integrated into the host genome. Transient DNA expression typically lasts 24–72 hours, whereas stable DNA expression potentially allows permanent overexpression of the protein.

Cells from bacteria (*Escherichia coli*), yeast (*Saccharomyces cerevisiae*), insects (*Spodoptera frugiperda*) and frogs (*Xenopus laevis*) can be used as hosts to heterologously express foreign proteins. However, mammalian cell lines are more appropriate for the expression of membrane proteins like transporters and channels, as the proper function of these proteins often requires post-translational modifications that can only fully occur in mammalian cells (e.g. glycosylation, deamindation, isoprenylation).

There are several well-established mammalian cell lines that have been used for heterologous expression of membrane proteins: the COS-7 cell line derived from the kidney of the African green monkey (*Cercopithecus aethiops*) (Yasumura and Kawakita, 1963), the CHO cell line derived from the ovaries of the Chinese hamster (*Cricetulus griseus*) (Tjio and Puck, 1958), and the HEK293 cell line generated by transformation of normal human embryonic kidney cells with sheared adenoviral DNA (Graham et al., 1977). HEK293 cells have been used as a heterologous expression system for both the transient and stable generation of functional mammalian and non-mammalian proteins. The amenability to a wide variety of transfection techniques, the high efficiency and accuracy in the translation and processing of proteins, and the easy reproduction and maintenance of this cell line are attributes which have made the HEK293 cell so popular among investigators for structural and functional analyses (Thomas and Smart, 2005).

In addition to mammalian cell lines, there is another widely used system to express membrane proteins: the unfertilized oocyte from the South African clawed frog (*Xenopus laevis*). The frog oocyte has become a popular choice for heterologous expression since it was first used in the expression of acetylcholine receptors (Barnard et al., 1982; Kusano et al., 1982; Miledi et al., 1982). Five factors which make frog oocytes an important heterologous expression system for modern molecular physiology are: (i) female *Xenopus laevis* produce oocytes year round; (ii) the low endogenous expression of many transport systems in oocytes; (iii) the high fidelity of translation in oocytes; (iv) the ease of microinjection of mRNA; and (v) the ability to measure membrane transport in a single cell (Wagner et al., 2000). However, experimental results obtained from overexpressing mammalian proteins in *Xenopus laevis* oocytes must be interpreted with caution. Heterologous proteins may interact with unique *Xenopus* oocytes' endogenous proteins and act much differently than when expressed in mammalian cells. For instance, it has been reported that some human genes are not correctly translated in *X. laevis* oocytes giving rise to heterogeneous N-terminal proteins (Fernandez et al., 2003; Fiers et al., 1982). Another concern is that nearly all mammals are diploid with two homologous copies of each chromosome, whereas approximately 30 million years ago, the *Xenopus* lineage underwent an allotetraploidization event and has four homologous copies of each chromosome (Bisbee et al., 1977; Evans et al., 2004). This extra chromosomal material may not have faced the same selection pressure as the original, and proteins derived from this material may have evolved to serve a different function. Therefore, if the mammalian proteins overexpressed in the frog oocyte have specific regulatory signaling cascades, then extra controls need to be added to these studies to validate the results.

III. TRACER VERSUS NON-TRACER FLUX MEASUREMENTS

Cl^- flux across the plasma membrane through channels and electrogenic transporters can be measured electrophysiologically. However, when Cl^- moves through transporters without carrying a net charge, the electroneutral movement requires another measurement technique. George Karl von Hevesy, a Hungarian radiochemist and Nobel laureate, was recognized in 1943 for his key role in the development of the tracer method, which uses radioactive tracers, to study chemical processes (Levi, 1976). Tracers can be used to measure the speed of chemical processes and to track the movement of a substance through a natural system such as a cell or a tissue (Rennie, 1999). Tracers may be artificially induced (i.e. radioisotope) or may be naturally occurring (i.e. K^+ efflux from a cell in an extracellular solution with zero K^+).

Radiolabeled tracers move through membrane pumps, transporters and exchangers along with the ions they mimic. They can be the ion themselves (e.g. ^{22}Na for sodium, ^{42}K for potassium, or ^{36}Cl for chloride) (Alper et al., 2006; Gamba et al., 1993; Hawke et al., 2001) or another ion with a similar ionic radius (e.g. ^{86}Rb as a congener for potassium) (Gagnon et al., 2007). In the case of a cotransport mechanism, the movement of Cl^- can also be mathematically determined from the movement of another cotransported ion species (e.g. Na^+ for the Na^+-Cl^- cotransporter) (Gamba et al., 1993). Non-labeled tracers (i.e. naturally occurring ions or solutes) also move through pumps, transporters and exchangers, but because of the zero trans conditions of the experiment, the amount of solute moved across the plasma membrane to the zero trans side can be directly measured (Delpire et al., 1991; Lauf et al., 2006). Investigators have also used both cation and anion replacements for sodium, potassium and chloride that are not native to the cell being studied (Armsby et al., 1995; De Franceschi et al., 1995; Gusev et al., 1999). Given a known extracellular or intracellular ion concentration, influx or efflux of these non-native ions into or out of cells can be determined by measuring the concentration differential across the plasma membrane.

The most frequent method used to measure non-labeled ionic concentrations is flame absorption spectrophotometry for Na^+, K^+ and Rb^+ (Gagnon et al., 2007; Lauf et al., 2006, 2008) and chloridometry for Cl^- (Ng et al., 1985; Slaunwhite et al., 1977). Another non-radioactive method for detecting the flux of solutes across cell membranes involves the use of fluorescent dyes. One example of this uses thallium (Tl^+) as a surrogate for K^+. Cell dye loading is accomplished by removing the culture medium and replacement with the AM ester of a thallium-sensitive dye. Cells loaded with dye are then excited using an argon laser to measure influx of Tl^+ as an increase in fluorescence above baseline (established prior to addition of Tl^+). The duration of Tl^+ flux experiments (e.g. 3–5 minutes) combined with the significant extracellular to intracellular Tl^+ concentration gradient make the amount of Tl^+ backflux negligible. Therefore, in this particular situation, increased fluorescence indicates both the *unidirectional* and *net* movement of Tl^+ through a membrane pump, transporter, channel, or exchanger (Weaver et al., 2004). This application has the advantage of rapid, facile, precise measurements of solute flux in isolated cells under a microscope (Geng et al., 2009), as well as confluent cells for screening large chemical libraries in multi-well plates using robotic high-throughput screening assays (Delpire et al., 2009).

IV. ELECTROCHEMICAL DRIVING FORCE OF ION FLUX

The electrochemical potential difference is the driving force for carrier-mediated passive transport of a charged solute x across the membrane. The electrochemical potential difference, $\Delta\mu_x$, is the sum of the chemical potential difference of solute x across the membrane (determined by the concentration gradient of the charged solute) and the electrical potential difference across the membrane (determined by the sign of the charge of the solute and the voltage difference across the membrane). The $\Delta\mu_x$ for a charged solute is defined by the following equation:

$$\Delta\mu_x = RT\ln([x]_i/[x]_o) + z_x FV_m$$

where R is the universal gas constant; T is the absolute temperature; z_x is the valence of solute x, F is the Faraday constant, and V_m, the membrane potential (electrical potential difference across the membrane).

The $\Delta\mu_x$ for an uncharged solute is solely dependent on the chemical component ($RT\ \ln([x]_i/[x]_o)$) as the electrical component (Z_xFV_m) reduces to zero. Whether charged or uncharged, carrier-mediated passive transport transfers solutes across the plasma membrane from areas of higher concentration to areas of lower concentration.

Cotransporters and exchangers are secondary active transport mechanisms that couple the "uphill" movement of one solute to the "downhill" movement of one or more solutes for which a favorable electrochemical potential difference exists, as discussed in detail in Chapter 5 in this volume. For example, the *electroneutral* Na^+-K^+-$2Cl^-$ cotransporter derives its energy mainly from the inwardly directed Na^+ electrochemical gradient (generated by the Na^+/K^+-ATPase) to drive the "uphill" movement of Cl^- and K^+ against their respective electrochemical potential gradients. This results in active uptake and accumulation of Cl^- above its electrochemical equilibrium potential (Chapter 5).

V. TECHNIQUES

Regardless of the model system involved (e.g. suspended cells, attached cells, or individual oocytes) measurement of unidirectional influx and efflux must be determined from the linear components of the flux. Influx experiments are initiated by replacing the extracellular solution with an osmotically equivalent salt solution (i.e. isosmotic with the control extracellular solution) containing the tracer compound. Removal of the extracellular medium and rinsing by an ice-cold $MgCl_2$ solution terminates the influx. The ice-cold wash also removes any exogenous extracellular tracer, allowing for an exact measurement of the amount of tracer that was transported into the cell. An important consideration when determining the unidirectional influx is to measure uptake during the non-saturating phase of solute uptake. Figure 9.1 illustrates the results of a hypothetical experiment for Rb^+ uptake versus time where during the first 10 minutes the amount of rubidium uptake is linear, then saturates over the next 10 minutes. As illustrated, the amount of rubidium uptake divided by time at either of the short dashed lines in the light gray shaded area result in equivalent solute influx (i.e. ~6 nmol $Rb^+ \times$ mg protein$^{-1} \times$ minute^{-1}). However, determining the rubidium uptake at two different time points in the saturating phase (longer dashed lines in the dark gray shaded area) results in both a considerable underestimation and different values for influx (i.e. ~4 vs. 3.1 nmol $Rb^+ \times$ mg protein$^{-1} \times$ minute^{-1}).

Efflux experiments, on the other hand, involve either preloading the tracer into the cells until steady-state equilibrium of the tracer is achieved (i.e. no apparent increase or decrease of tracer inside the cell), or using the native ion transported under zero-*trans* conditions (i.e. absence of the native ion at the *trans* side of the membrane). The amount of solute efflux measured per a given unit of time is added to an earlier measurement to reconstruct the solute efflux. The slope of the linear portions is the rate constant (k) of

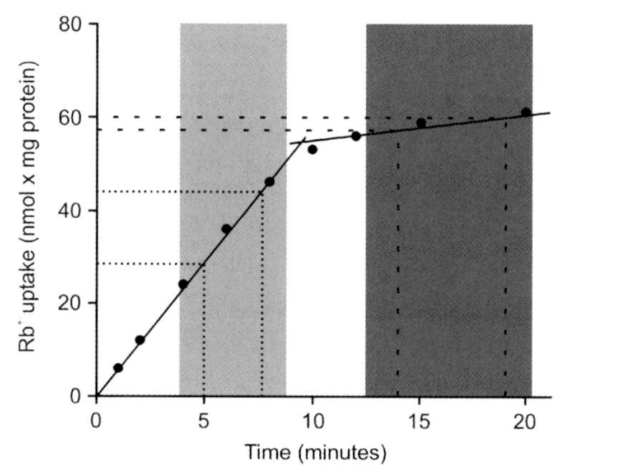

FIGURE 9.1 Tracer uptake at linear versus saturating phase. Plot of ^{86}Rb uptake at different time points demonstrating both initial velocity of rubidium uptake (light gray area) and saturated uptake (dark gray area). Small dashed lines demonstrate an equivalent Rb influx (~6 nmol $Rb^+ \times$ mg protein$^{-1} \times$ minute^{-1}) at both 5 and 7 minutes, respectively. The large dashed lines demonstrate an underestimation of unidirectional flux, and a non-equivalent amount of Rb influx (~4 vs. 3.1 nmol $Rb^+ \times$ mg protein$^{-1} \times$ minute^{-1}) in the saturating phase at 14 and 19 minutes, respectively.

solute efflux. Multiplying the rate constant (k) by the internal ion concentration (measured at the beginning of the experiment) gives the total solute efflux.

A. Suspended Cells

Let us consider red blood cells as an example of unidirectional flux measurement in a cell suspension. In this example we want to assess the activity of the red blood cell K^+-Cl^- cotransporter by measuring the chloride-dependent Rb^+ uptake using non-radioactive Rb^+ (surrogate for K^+) as a tracer.

Whole blood is first collected in heparinized syringes and centrifuged to separate the plasma and buffy coat layers from red blood cells. The isolated red blood cells are then washed several times in an isotonic solution by re-suspension/centrifugation. Next, equal amounts of red blood cells are incubated in either isotonic RbCl or $RbNO_3$ for 1 hour (linear phase). Rb^+ uptake is halted by re-suspension/centrifugation of the red blood cells in ice-cold isotonic $MgCl_2$ solution. The washed cells are then hemolyzed with a cation-free detergent so the cell hemoglobin and $[Rb^+]$ can be measured. Cell hemoglobin is determined by flame absorption spectrophotometry at 527 nm. The cell hematocrit (or packed cell volume) is then estimated as a percentage by tripling the hemoglobin concentration in g/dL and dropping the units. $[Rb^+]_i$ is measured by atomic emission spectrophotometry at 780.1 nm. The packed cells $OD^{527\,nm}$ is obtained by dividing the $OD^{527\,nm}$ of a known dilution of whole blood by its hematocrit value (see Fig. 9.2). The volume of cells is calculated from the ratio of sample $OD^{527\,nm}$ divided by packed cells $OD^{527\,nm}$ multiplied by total sample volume. Cl^--dependent Rb^+ influx (difference in Rb^+ influx in Cl^- and NO_3^-) is expressed in mmoles $\times$ liter cells^{-1} $\times$ hour^{-1}.

The greatest advantage to using the mature red cell for structure and function studies of plasma membrane transport proteins is its simplicity and elegance. The non-nucleated erythrocyte is unique among human cells in that the plasma membrane accounts for all of its antigenic, transport and mechanical characteristics (Mohandas and Gallagher, 2008). Examples of red cell membrane transport proteins include band 3 (anion transporter), aquaporin 1 (water transporter), Glut1 (glucose and L-dehydroascorbic acid transporter), Kidd antigen protein (urea transporter), RhAG (gas transporter, probably of carbon dioxide), Na^+-K^+-ATPase, Ca^{+2}-ATPase, Na^+-K^+-$2Cl^-$ cotransporter, Na^+-Cl^- cotransporter, K^+-Cl^- cotransporter and the "Gardos Channel" (see reviews in Brugnara, 1997; Lew and Bookchin, 2005; Milanick and Hoffman, 1986). Ironically, the greatest advantage of using mature red cells for membrane transport studies is also its greatest disadvantage. The lack of a nucleus prevents genetic manipulation and thus eliminates the use of molecular biology to manipulate signaling cascades or protein expression at the genetic level.

B. Attached Cell Influx (Mammalian Cultured Cells)

Cultured mammalian cells can also be grown to confluent monolayers in 6-well plates (or 35 mm dishes) to measure ion fluxes. Cell cultures are initially washed three times with a balanced salt solution. The monolayer is then pre-incubated in a saline solution containing different pharmacological agents. Following aspiration of the pre-incubation solution, a similar solution containing either a radiolabeled tracer (i.e. $^{42}K^+$, $^{86}Rb^+$, $^{22}Na^+$, $^{36}Cl^-$), or a non-radioactive cation (i.e. $^{85}Rb^+$) is added to the culture dishes for a specific amount of time. Unidirectional influx is then halted and remaining extracellular tracer is removed by several washes with an ice-cold solution. Cells are then lysed in sodium hydroxide, neutralized with acid and the ion flux is determined either by scintillation

$$\text{Rb influx} = \frac{e^{572\,nm}_{sample}}{e^{572\,nm}_{standard}} \times \frac{OD^{572\,nm}_{packed\ cells} \times DF_{Rb^+}}{OD^{572\,nm}_{sample} \times DF_{Hb}} - [Rb^+]^{cell}_{native}$$

where:

$e^{572\,nm}_{sample}$ is the Rb^+ emission for the sample

$e^{572\,nm}_{standard}$ is the Rb^+ emission for the 1 mM Rb^+ standard

$OD^{572\,nm}_{packed\ cells}$ is sample $OD^{572\,nm}$ divided by hematocrit

$OD^{572\,nm}_{sample}$ is sample $OD^{572\,nm}$

DF_{Rb^+} is the necessary dilution factor for flame photometry

DF_{Hb} is the necessary dilution factor for hemoglobinometry

$[Rb^+]^{cell}_{native}$ is the Rb^+ concentration natively occuring in red blood cells

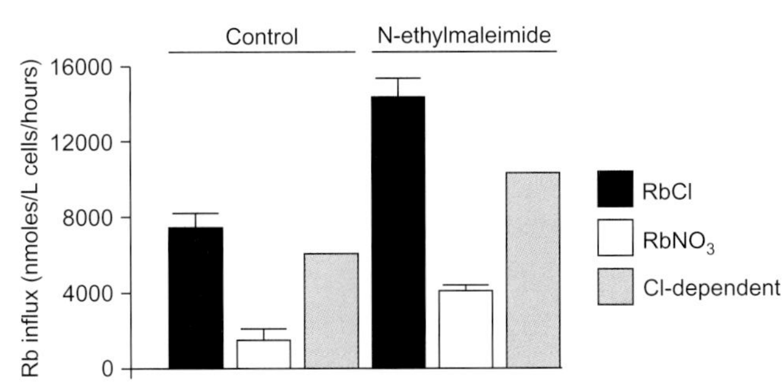

FIGURE 9.2 Chloride-dependent rubidium tracer influx in suspended red blood cells. Black and white bars represent the total Rb influx in control and N-ethylmaleimide treated red blood cells in the presence of chloride and nitrate, respectively. Gray bars are the chloride-dependent difference in each condition. Adapted from Lauf et al. (1984).

Sample volume : 300 μl **Total volume** : 750 μl **Influx time:** 15 minutes

[K⁺] : 5 mM or 5,000 pmoles/μl or 25,000 pmoles/5 μl **Rb Standard:** 5 μl

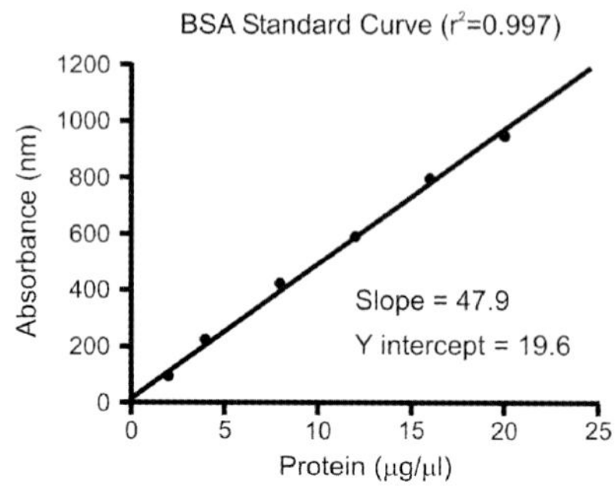

	Protein Absorbance		
	1	2	3
Control	369	389	378
Furosemide	383	390	386

Protein/dish = ((Absorbance - 19.6) / 47.9) / 300 x 750

	Scintillation Counter (cpm)				
	1	2	3	Std	Blank
Control	17119	16755	17671	11178	45
Furosemide	2025	1957	2024	11639	42

$$^{86}\text{Rb flux} = \frac{(\text{cpm}_{\text{sample}} - \text{cpm}_{\text{blank}}) \times ([\text{K}^+] / \text{cpm}_{\text{Rb standard}})}{(\text{Protein} \times \text{influx time})}$$

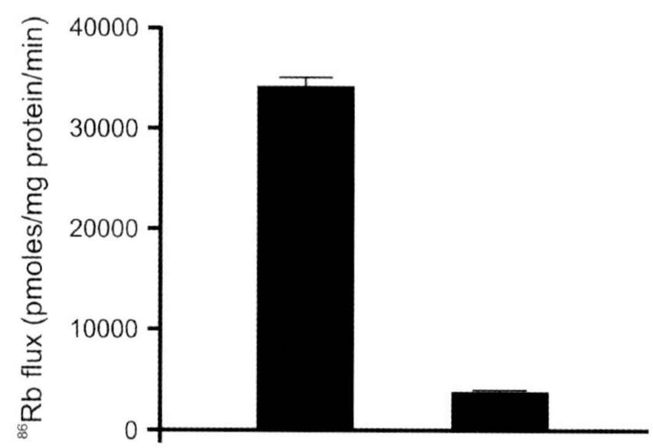

FIGURE 9.3 Rubidium tracer influx in cultured cells. Rb influx in untreated (control) and treated (2 mM furosemide) mammalian cell cultures. Adapted from Gagnon et al. (2007).

counter if radioactivity is used, or by flame absorption spectrophotometry if non-radiolabeled cations are employed. When Rb^+ is used to follow K^+ movement, the K^+ flux is calculated in nmol $K^+ \times$ mg protein$^{-1} \times$ min^{-1} (see Fig. 9.3).

C. Efflux Measurement in Cells Attached to a Substrate

Mammalian cells are first grown to confluency on plastic cell culture dishes. Cells are then pre-loaded with either a radiolabeled tracer (i.e. $^{42}K^+$, $^{86}Rb^+$, $^{22}Na^+$, $^{36}Cl^-$) or a non-radioactive cation (i.e. $^{85}Rb^+$) for 2 hours at 37°C. In the example illustrated in Fig. 9.4, the intracellular K^+ concentration of both media and cells was measured by atomic absorption spectrophotometry. For the washout kinetics, the cells are moved every minute into a fresh scintillation vial. At the end of the washout, the remainder of tracer in the cells is measured and the content of tracer at each time point is then reconstructed based on the amount of tracer lost. The slope of the linear components represents the rate constant of K^+ efflux (k). Efflux values can be calculated by multiplying the rate constant by the intracellular concentration of the ion (i.e. efflux = $k \times [K^+]_i$).

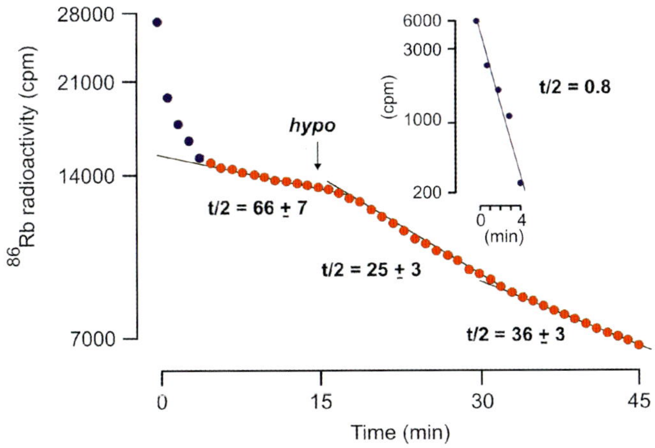

FIGURE 9.4 Radiolabeled tracer efflux in cultured pheochromocytoma cells (PC12). Cultured PC12 cells were pre-loaded with ^{86}Rb for 2 hours at 37°C then moved every minute into a fresh scintillation vial of isosmotic or hyposmotic (indicated by arrow) saline solution. After 45 minutes the remainder of ^{86}Rb in the cells is measured and the content of ^{86}Rb at each time point is reconstructed. The slope of the linear components represents the rate constant (*k*). The efflux can be calculated by multiplying the rate constant *k* by the intracellular concentration of potassium. Inset: the first five measurements of ^{86}Rb efflux were normalized (by subtracting the linear component of efflux) and the exponential component became linear. The extremely fast rate constant indicates that this initial component represents washout of radiolabeled tracer.

D. Flux Measurements in *Xenopus* Oocytes

Harvesting of *Xenopus laevis* oocytes involves surgical removal of unfertilized eggs through a small diagonal incision (~1–1.5 cm long) in the lower quadrant of the frog abdomen. The frog is first anesthesized with 0.2% tricaine methane sulfonate in deionized water. Frog skin is quite tough and therefore skin and abdominal muscle incisions should be made in two stages. After the initial incision through the skin, the muscle layer should be lifted with surgical forceps before creating the second incision to avoid inadvertently wounding any internal organs. Lobes of oocytes are then exteriorized, and the desired numbers of oocytes can be removed. The remainder of the oocyte mass is then carefully replaced into the coelomic cavity, and the incision is closed with silk sutures.

Stage V–VI oocytes require either manual or enzymatic defolliculation prior to injection with *in vitro* transcribed cRNA. Heterologous expression of cRNA in frog oocytes requires the use of a frog-specific expression vector. Before the cDNA vector can be transcribed into cRNA, it needs to be linearized with an enzyme (typically *Mlu*I) that leaves an adenine (A) at the end of an existing polyadenine stretch. Following linearization, the cDNA is transcribed into cRNA, purified, quantitated and checked by gel electrophoresis (1% agarose/0.693% formaldehyde).

Sample size: 1 oocyte **Influx time:** 1 hour

[K$^+$] : 5 mM or 5 nmoles/μl or 25 nmoles/5μl **Rb Standard:** 5 μl

[Cl$^-$] : 109 mM or 109 nmoles/μl or 545 nmoles/5 μl **Cl Standard: 5** μl

	^{86}Rb (cpm)	^{36}Cl (cpm)
Sample	570	1021
	582	1090
	324	1095
	550	1021
	670	1001
	564	1066
	763	1564
	609	1342
	510	1271
	410	1571
	349	998
	661	939
	604	926
	756	887
	612	1364
	426	906
	626	1283
	581	1780
	594	1080
	524	975
Standard	54129	53566
Blank	36	39

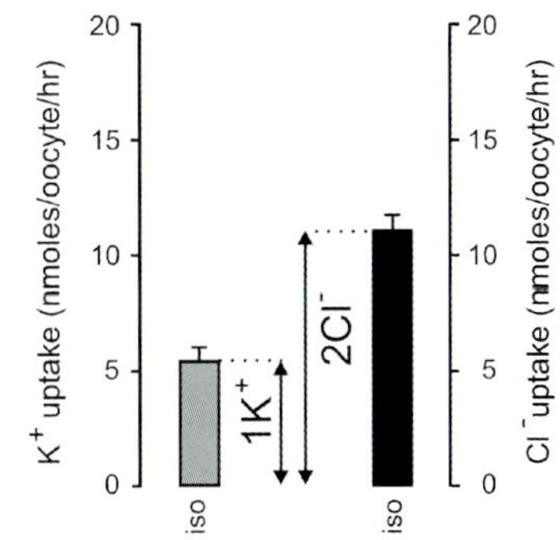

$$^{86}\text{Rb flux} = \frac{(\text{cpm}_{\text{sample}} - \text{cpm}_{\text{blank}}) \times ([\text{K}^+] / \text{cpm}_{\text{Rb standard}})}{(\text{Protein} \times \text{influx time})}$$

$$^{36}\text{Cl flux} = \frac{(\text{cpm}_{\text{sample}} - \text{cpm}_{\text{blank}}) \times ([\text{Cl}^-] / \text{cpm}_{\text{Cl standard}})}{(\text{Protein} \times \text{influx time})}$$

FIGURE 9.5 Radiolabeled tracer influx in *Xenopus laevis* oocytes. Tracer flux experiments with either ^{86}Rb (gray bar) or ^{36}Cl (black bar) were performed on two groups of stage VI frog oocytes (n = 20) injected with Na-K-2Cl cotransporter cRNA to measure ion influx. Note the 1:2 ratio of K to Cl ion influx, consistent with the potassium/chloride stoichiometry of the cotransporter.

The ion flux procedure involves groups of ~20–30 oocytes in 35 mm dishes. First, oocytes are washed with isosmotic saline, and then pre-incubated in isosmotic saline containing different pharmacological inhibitory agents. Next, the pre-incubation solution is replaced with a flux solution (isotonic or otherwise) containing a radiolabeled tracer (e.g. $^{36}Cl^-$ or $^{86}Rb^+$). After one hour of uptake, the radioactive solution is aspirated and the oocytes are washed several times with ice-cold solution. Single oocytes are then transferred into glass vials, lysed with sodium hydroxide, neutralized with glacial acetic acid, and tracer activity is measured by scintillation counter. Ion flux is expressed in nanomoles $K^+ \times \text{oocyte}^{-1} \times \text{hour}^{-1}$ (see Fig. 9.5).

VI. CONCLUSION

Measurement of $[Cl^-]_i$ in living cells can be achieved through the use of microelectrodes, fluorescent dyes and tracer fluxes. Every method has advantages and limitations depending on the experimental model and objective. In this chapter, I have presented the pros and cons of different heterologous systems (i.e. cultured mammalian cells versus frog oocytes) and different techniques for the measurement of unidirectional movement (both influx and efflux) of solutes across the plasma membrane. When multiple pathways capable of transporting a specific solute are present in the cell membrane (e.g. channels, exchangers and transporters), then unidirectional flux measurements are particularly powerful for identifying and isolating individual solute transport pathways.

References

Akaike, N. (1996). Gramicidin perforated patch recording and intracellular chloride activity in excitable cells. *Prog. Biophys. Molec. Biol.* **65**, 251–264.

Alper, S.L., Stewart, A.K., Chernova, M.N., Zolotarev, A.S., Clark, J.S., and Vandorpe, D.H. (2006). Anion exchangers in flux: functional differences between human and mouse SLC26A6 polypeptides. *Novartis Found Symp.* **273**, 107–119 discussion 119–125, 261–264.

Alvarez-Leefmans, F.J., Gamiño, S.M., Giraldez, F., and Nogueron, I. (1988). Intracellular chloride regulation in amphibian dorsal root ganglion neurons studied with ion-selective microelectrodes. *J. Physiol. (Lond.)* **406**, 225–246.

Alvarez-Leefmans, F.J., Giraldez, F., and Russell, J.M. (1990). Methods for measuring chloride transport across nerve, muscle, and glial cells. In *Chloride Channels and Carriers in Nerve, Muscle, and Glial Cells* (Alvarez-Leefmans, F.J. and Russell, J.M., eds), pp. 3–66. Plenum Press, New York.

Armsby, C.C., Brugnara, C., and Alper, S.L. (1995). Cation transport in mouse erythrocytes: role of K(+)-Cl− cotransport in regulatory volume decrease. *Am. J. Physiol.* **268**, C894–C902.

Barnard, E.A., Miledi, R., and Sumikawa, K. (1982). Translation of exogenous messenger RNA coding for nicotinic acetylcholine receptors produces functional receptors in Xenopus oocytes. *Proc. R. Soc. Lond. B Biol. Sci.* **215**, 241–246.

Bisbee, C.A., Baker, M.A., Wilson, A.C., Irandokht, H.A., and Fischberg, M. (1977). Albumin phylogeny for clawed frogs. *Science* **195**, 785–787.

Brugnara, C. (1997). Erythrocyte membrane transport physiology. *Curr. Opin. Hematol.* **4**, 122–127.

Coles, J.A. (1995). Measurement of cytosolic chloride activity by ion-selective microelectrodes. In *Measurement and Manipulation of Intracellular Ions Methods in Neurosicence* (Kraicer, J. and Dixon, S.J., eds), **27**, pp. (340–(360). Academic Press, San Diego.

De Franceschi, L., Beuzard, Y., and Brugnara, C. (1995). Sulfhydryl oxidation and activation of red cell K(+)-Cl− cotransport in the transgenic SAD mouse. *Am. J. Physiol.* **269**, C899–C906.

Delpire, E., Cornet, M., and Gilles, R. (1991). Volume regulation in rat pheochromocytoma cultured cells submitted to hypoosmotic conditions. *Arch. Inter. Physiol. Biochem.* **99**, 71–76.

Delpire, E., Days, E., Lewis, L.M., Mi, D., Kim, K., Lindsley, C.W., and Weaver, C.D. (2009). Small molecule screen identifies inhibitors of the neuronal K-Cl cotransporter KCC2. *Proc Natl Acad Sci USA.* **106**, 5383–5388.

Evans, B.J., Kelley, D.B., Tinsley, R.C., Melnick, D.J., and Cannatella, D.C. (2004). A mitochondrial DNA phylogeny of African clawed frogs: phylogeography and implications for polyploid evolution. *Mol. Phylogenet. Evol.* **33**, 197–213.

Fernandez, F.R., Morales, E., Rashid, A.J., Dunn, R.J., and Turner, R.W. (2003). Inactivation of Kv3.3 potassium channels in heterologous expression systems. *J. Biol. Chem.* **278**, 40890–40898.

Fiers, W., Remaut, E., Devos, R., Cheroutre, H., Contreras, R., Gheysen, D., Degrave, W., Stanssens, P., Tavernier, J., and Content, J. (1982). The human fibroblast and human immune interferon genes and their expression in homologous and heterologous cells. *Philos. Trans. R. Soc. Lond. B Biol. Sci.* **299**, 29–38.

Gagnon, K.B., Adragna, N.C., Fyffe, R.E., and Lauf, P.K. (2007). Characterization of glial cell K-Cl cotransport. *Cell. Physiol. Biochem.* **20**, 121–130.

Gamba, G., Saltzberg, S.N., Lombardi, M., Miyanoshita, A., Lytton, J., Hediger, M.A., Brenner, B.M., and Hebert, S.C. (1993). Primary structure and functional expression of a cDNA encoding the thiazide-sensitive, electroneutral sodium-chloride cotransporter. *Proc. Natl. Acad. Sci. USA* **90**, 2749–2753.

Geng, Y., Hoke, A., and Delpire, E. (2009). The Ste20 Kinases Ste20-related Proline-Alanine-rich Kinase and Oxidative-stress Response 1 Regulate NKCC1 function in Sensory Neurons. *J Biol Chem.* **284**, 14020–14028.

Graham, F.L., Smiley, J., Russell, W.C., and Nairn, R. (1977). Characteristics of a human cell line transformed by DNA from human adenovirus type 5. *J. Gen. Virol.* **36**, 59–74.

Gusev, G.P., Agalakova, N.I., and Lapin, A.V. (1999). Kinetics of K-Cl cotransport in frog erythrocyte membrane: effect of external sodium. *J. Membr. Biol.* **172**, 203–213.

Hawke, T.J., Lessard, S., Vickery, L., Lipskie, S.L., and Lindinger, M.I. (2001). Ouabain stimulates unidirectional and net potassium efflux in resting mammalian skeletal muscle. *Can. J. Physiol. Pharmacol.* **79**, 932–941.

Hebert, S.C., Mount, D.B., and Gamba, G. (2003). Molecular physiology of cation-coupled Cl(−) cotransport: the SLC12 family. *Pflugers Arch.*

Hediger, M.A. and Rhoads, D.B. (1994). Molecular physiology of sodium-glucose cotransporters. *Physiol. Rev.* **74**, 993–1026.

Jayaraman, S. and Verkman, A.S. (2000). Quenching mechanism of quinolinium-type chloride sensitive fluorescent indicators. *Biophys. Chem.* **85**, 49–57.

Kanai, Y. and Hediger, M.A. (2004). The glutamate/neutral amino acid transporter family SLC1: molecular, physiological and pharmacological aspects. *Pflugers Arch.* **447**, 469–479.

Kuner, T. and Augustine, G.J. (2000). A genetically encoded ratiometric indicator for chloride: capturing chloride transients in cultured hippocampal neurons. *Neuron* **27**, 447–459.

Kusano, K., Miledi, R., and Stinnakre, J. (1982). Cholinergic and catecholaminergic receptors in the Xenopus oocyte membrane. *J. Physiol.* **328**, 143–170.

Lauf, P.K., Adragna, N.C., and Garay, R.P. (1984). Activation by N-ethylmale-imide of a latent K + -C1− flux in human red blood cells. *Am. J. Physiol. Cell Physiol.* **246**, C385–C390.

Lauf, P.K., Adragna, N.C., Dupre, N., Bouchard, J.P., and Rouleau, G.A. (2006). K-Cl cotransport in red blood cells from patients with KCC3 isoform mutants. *Biochem. Cell. Biol.* **84**, 1034–1044.

Lauf, P.K., Chimote, A.A., and Adragna, N.C. (2008). Lithium fluxes indicate presence of Na-Cl cotransport (NCC) in human lens epithelial cells. *Cell Physiol. Biochem.* **21**, 335–346.

Levi, H. (1976). George von Hevesy memorial lecture. George Hevesy and his concept of radioactive indicators – in retrospect. *Eur. J. Nucl. Med.* **1**, 3–10.

Lew, V.L. and Bookchin, R.M. (2005). Ion transport pathology in the mechanism of sickle cell dehydration. *Physiol. Rev.* **85**, 179–200.

Lu, Y., Zheng, J., Xiong, L., Zimmermann, M., and Yang, J. (2008). Spinal cord injury-induced attenuation of GABAergic inhibition in spinal dorsal horn circuits is associated with down-regulation of the chloride transporter KCC2 in rat. *J. Physiol.* **586**, 5701–5715.

Milanick, M.A. and Hoffman, J.F. (1986). Ion transport and volume regulation in red blood cells. *Ann. NY Acad. Sci.* **488**, 174–186.

Miledi, R., Parker, I., and Sumikawa, K. (1982). Properties of acetylcholine receptors translated by cat muscle mRNA in Xenopus oocytes. *EMBO J.* **1**, 1307–1312.

Mohandas, N. and Gallagher, P.G. (2008). Red cell membrane: past, present, and future. *Blood* **112**, 3939–3948.

Munkonge, F., Alton, E.W., Andersson, C., Davidson, H., Dragomir, A., Edelman, A., Farley, R., Hjelte, L., McLachlan, G., Stern, M., and Roomans, G.M. (2004). Measurement of halide efflux from cultured and primary airway epithelial cells using fluorescence indicators. *J. Cyst. Fibros.* **3** (Suppl. 2), 171–176.

Ng, R.H., Altaffer, M., Ito, R., and Statland, B.E. (1985). The Technicon RA-1000 evaluated for measuring sodium, potassium, chloride, and carbon dioxide. *Clin. Chem.* **31**, 435–438.

Rennie, M.J. (1999). An introduction to the use of tracers in nutrition and metabolism. *Proc. Nutr. Soc.* **58**, 935–944.

Rocha-Gonzalez, H.I., Mao, S., and Alvarez-Leefmans, F.J. (2008). Na + , K + , 2Cl− cotransport and intracellular chloride regulation in rat primary sensory neurons: thermodynamic and kinetic aspects. *J. Neurophysiol.* **100**, 169–184.

Romero, M.F. (2005). Molecular pathophysiology of SLC4 bicarbonate transporters. *Curr. Opin. Nephrol. Hypertens.* **14**, 495–501.

Schultz, S.G. (1980). *Basic Principles of Membrane Transport*. Cambridge University Press.

Slaunwhite, D., Clements, J.C., and Reynoso, G. (1977). Clinical evaluation of the Technico Stat/Ion system. *Clin. Biochem.* **10**, 44–46.

Soleimani, M. (2002). Na + :HCO3− cotransporters (NBC): expression and regulation in the kidney. *J. Nephrol.* **15** (Suppl. 5), S32–S40.

Stern, M., Munkonge, F.M., Caplen, N.J., Sorgi, F., Huang, L., Geddes, D.M., and Alton, E.W. (1995). Quantitative fluorescence measurements of chloride secretion in native airway epithelium from CF and non-CF subjects. *Gene Ther.* **2**, 766–774.

Thomas, P. and Smart, T.G. (2005). HEK293 cell line: a vehicle for the expression of recombinant proteins. *J. Pharmacol. Toxicol. Methods* **51**, 187–200.

Tjio, J.H. and Puck, T.T. (1958). Genetics of somatic mammalian cells. II. Chromosomal constitution of cells in tissue culture. *J. Exp. Med.* **108**, 259–271.

Torres, G.E. and Caron, M.G. (2003). Center stage for the serotonin transporter: a gain-of-function polymorphism in persons with obsessive-compulsive disorder. *Mol. Pharmacol.* **64**, 196–198.

Wagner, C.A., Friedrich, B., Setiawan, I., Lang, F., and Broer, S. (2000). The use of Xenopus laevis oocytes for the functional characterization of heterologously expressed membrane proteins. *Cell Physiol. Biochem.* **10**, 1–12.

Weaver, C.D., Harden, D., Dworetzky, S.I., Robertson, B., and Knox, R.J. (2004). A thallium-sensitive, fluorescence-based assay for detecting and characterizing potassium channel modulators in mammalian cells. *J. Biomol. Screen.* **9**, 671–677.

Yasumura, Y. and Kawakita, M. (1963). The research for the SV40 by means of tissue culture technique. *Nippon Rinsho.* **21**, 1201–1219.

CHAPTER

10

Knockout Models of Cation-Chloride Cotransporters

Nicole Garbarini and Eric Delpire

OUTLINE

I. INTRODUCTION

Over the past three decades, the biological sciences have undergone a true revolution with the development of highly sophisticated molecular tools. These advances now allow investigators to easily modulate protein expression in individual cells and/or whole animals, and to address many questions related to protein function using novel approaches. Change in protein expression can be obtained through a variety of methods, such as overexpression of wild-type or mutant proteins, suppression of genes using antisense oligonucleotides and small interference RNAs, and random genome mutagenesis. One of the best ways to study the function of a known protein in the context of the whole organism is directed gene manipulation in embryonic stem cells. In this chapter, we will review implementation of this technique in mouse ES cells, and specifically focus on the creation of mouse models to study the function of cation-chloride cotransporters.

Every model organism has a particular set of advantages and disadvantages which factor into assessing its usefulness for particular scientific studies. Published genomic sequences through "shotgun" sequencing initiatives have rapidly advanced genetic manipulation of these fully sequenced organisms. Additionally, such genetic tractability has allowed the development of a wide variety of genetics informatics tools. Of course, other considerations in choosing a model organism include cost, time of reproductive cycle and, perhaps most importantly, the type of research that will be pursued in the laboratory. While invertebrates such as *C. elegans* and *Drosophila* have the advantage of quick generation cycles and inexpensive maintenance, there are several areas of research which benefit from studies of vertebrate and mammalian model organisms. Thus, the mouse emerges as a premier candidate for a mammalian model organism. It shares 99% of human genes, and can be used to study diseases present in humans, such as hypertension, diabetes, cancer, heart disease and neurological illnesses. Compared to mammals such as dogs or primates, the mouse is advantageous due to its small size, relatively shorter gestation period and cost of upkeep (Peters et al., 2007; Waterston and Consortium, 2002). Compared to rats, mouse strains have a longer history of genetic study and ES cell manipulation. Additionally, a wide variety of resources on mouse genetics are rapidly growing and publicly available (for review, see Peters et al., 2007).

Before we examine the various methods of gene manipulation in mice, a few general comments and definitions will be made. First, all the genetic work relevant to this chapter is performed in one specific murine species: *Mus musculus*. Within that species, there are many strains available, both outbred and inbred. An inbred strain, such as the widely used C57BL/6 ("C57 black 6" also called "black 6" or "B6") mouse, is typically genetically pure, i.e. all individuals share the same identical genetic make-up. Although this is true in principle, in reality, there has been some drifting between large populations of C57BL/6 mice (as well as within other strains) bred for many generations by large vendors. Second, it is important to note that genetic differences between inbred mouse strains can result in divergent phenotypes. However, since this is a known phenomenon, mouse geneticists have been able to use longitudinal studies of these strains to their advantage. Using inbred lines such as C57BL/6 (B) and DBA/2J (D), geneticists have created a panel of recombinant inbreds (homogeneous populations of mice descending from two parental inbred strains) that can be used to map traits and phenotypes to specific chromosomal regions (Crabbe et al., 1994; Hsu et al., 2007). BxD recombinant inbreds were first used to map genes to specific chromosomal locations. Prior to the sequencing of the entire mouse genome, this approach represented a solid method for chromosomal localization of genes. It was using this method that we localized the secretory Na^+-K^+-$2Cl^-$ cotransporter, NKCC1, on mouse chromosome 18 (Delpire et al., 1994).

For clarity in the next few sections, we define a transgenic mouse as a mouse carrying a specific and foreign transgene. Insertion of the transgene in the mouse genome is typically random, although there is some technology under development allowing the insertion of a transgene into a specific locus. Additionally, we define knockout and knock-in mice as mutant mice in which a specific gene has been mutated by homologous recombination in embryonic stem cells, with a knockout resulting in the absence of a functional RNA transcript and protein, and a knock-in resulting in the expression of a mutant protein. Note that a knock-in mutation can result in a phenotype similar to that of a knockout if the mutation also results in impaired protein function.

II. STRAIGHT KNOCKOUT VERSUS CONDITIONAL KNOCKOUT MOUSE MODELS

In creating a knockout animal, the first consideration is making a straight knockout versus a conditional knockout. This distinction rests on whether the deletion of the exon occurs in the germline: that is, prior to the development of mouse (for the straight knockout), versus during mouse development or later (for the conditional knockout). There are two main subtypes of conditional knockouts: a tissue-specific knockout where protein disruption is targeted to a specific tissue or cell type, or an inducible knockout where protein disruption can be induced, generally through diet manipulation. In other words, conditional knockouts can be designed in order to allow either spatial control or temporal control of genetic modification. Additionally, these subtypes can be combined such that protein disruption can be induced for temporal control of expression in a limited subset of cell/tissue types. Design of these targeting constructs will be discussed in detail in a later section.

III. KNOCK-IN MOUSE MODELS

Investigators may sometimes seek to study the effect of a particular protein mutation on mouse physiology. In order to "knock in" the expression of an altered protein, the mutation can be introduced into the mouse genome, again using homologous recombination to integrate the mutant protein into the mouse genome. In this case, rather than target an exon for deletion, the wild-type exon is substituted with the mutated exon. The exon can also be flanked by loxP sites to allow for swapping of DNA cassettes. There are many ways to use recombination sites that allow sophisticated exchanges of DNA fragments through recombination (for review, see García-Otín and Guillou, 2006). These recombination events are done in embryonic stem cells with very high rates of recombination, streamlining the process of ES cell clone selection.

IV. TARGETING A GENE BY HOMOLOGOUS RECOMBINATION

The creation of a knockout (or knock-in) mouse starts with the design of a cDNA construct for electroporation in embryonic stem cells. Historically, libraries of mouse genomic clones inserted in lambda phages were screened with ^{32}P-labeled cDNA fragments, positive clones were mapped by digestion with restriction enzymes, and exons then located using Southern blot analysis. This long process allowed investigators to draw rough restriction maps of genes and plan their targeting construct based on available and rare restriction sites. Today, with the sequencing of the whole genome, restriction maps and sequences are just a few mouse clicks away, and constructs can be designed in record time. Once designed, the constructs can be created from fragments of genomic DNA, either from relatively small lambda phage clones, or more recently from larger bacterial artificial chromosome (BAC) clones. An alternative is to amplify a large genomic DNA fragment by using polymerase chain reaction (PCR). The danger with this method is the possible insertion of multiple PCR errors that can affect the rate of recombination. As the insertion of the mutation occurs through homologous recombination, it is also important that the genomic fragments are derived from the same mouse strain as the ES cells. Using a different strain might reduce the rate of recombination due to differences in strain sequences. After identifying the exon target, constructs can be created by standard molecular subcloning procedures (restriction digests followed by ligation into plasmids), or more recently through recombineering methods in bacteria (reviews for further detailed reading: Copeland et al., 2001; Sawitzke et al., 2007).

Whichever way the construct is assembled, it will typically contain a large genomic fragment that drives the recombination (long arm), the region being mutated, and a shorter homologous arm. In our first example (Fig. 10.1A), one exon is omitted in the construct and recombination results in the exchange of the construct with one wild-type allele (straight knockout). In our second example shown in Fig. 10.1A, one exon is flanked by Cre recombinase recognition sites, or loxP sites, which will allow for deletion at a later time (inducible/conditional knockout).

Many exons within a specific gene can be targeted for deletion. However, it is advantageous to target a 5′ exon, as expression of a truncated protein might have undesirable effects, such as partial protein activity as opposed to truly complete protein deletion. Furthermore, to avoid deletion of an exon that results in a smaller but otherwise identical protein, it is important to avoid targeting an exon that undergoes alternative splicing, or an exon that disrupts the translation frame of the protein when omitted. Finally, it is also judicious to stay away from the first exon, as many genes have alternative promoters and protein translation may start at a downstream transcription initiation site.

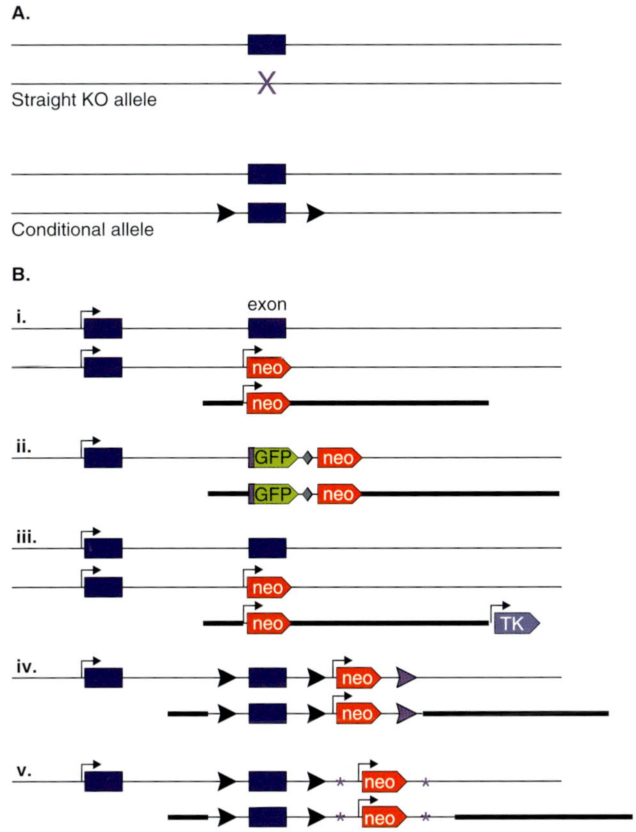

FIGURE 10.1 **A.** Schematic representation of the embryonic stem cell alleles for a straight knockout and conditional knockout mouse. **B.** Different strategies to create constructs for straight and conditional knockout mice: (1) The exon is substituted by a neomycin resistance gene cassette driven by its own promoter. (2) Promoter-less cassettes are fused to an exon of the gene of interest, and expression is driven by the native promoter of the gene. In this example, both GFP and neomycin are inserted with an internal ribosomal entry site (IRES, blue diamond) inserted between the two open reading frames. (3) Variation of (1) with thymidine kinase killer (TK) gene cassette placed after the long arm of recombination, to enhance selection of proper recombination. Errant recombination/random insertion of the construct therefore results in TK integration into the genome, and abrogates ES cell viability. (4) Placement of 3 loxP sites around targeted exon and neomycin. (5) Placement of loxP sites around the targeted exon and placement of FRT sites around the neomycin cassette. Thin arrows represent promoters, large arrowheads represent loxP sites, and stars represent FRT sites.

Because recombination is a rare event, a selection marker is added to the construct to eliminate embryonic stem cells that have not incorporated the construct. Although many selection markers can be used, the neomycin resistance gene cassette constitutes the most commonly used selection marker (Fig. 10.1B, i). As construct integration into the genome can occur by random insertion, the neomycin resistance gene cassette can be designed without a promoter as a fusion protein driven by the promoter of the targeted gene. This strategy can be more efficient than the traditional method of having a stand-alone promoter for neomycin, but can only be used if the gene of interest is expressed in embryonic stem cells. Furthermore, as the neomycin cassette is fused to the protein of interest, this method can also be combined with the use of a foreign expression marker such as β-galactosidase, luciferase, or as shown in the figure, green fluorescent protein (GFP) (Fig. 10.1B, ii). This method of fusing the neomycin cassette to the protein of interest was successfully used by our laboratory to generate the knockouts of NKCC1 and KCC3 (Delpire et al., 1999; Howard et al., 2002).

As an alternative to the promoter-less neomycin, many constructs include a killer gene which is located outside the region of recombination (Fig. 10.1B, iii). If the construct inserts itself randomly in the genome rather than recombining in the proper locus, the killer gene is inserted and affects the viability of the ES cell. The most commonly used system uses the thymidine kinase (TK) gene. It is our experience that this step can be eliminated, as it does not increase significantly the percentage of properly recombined neomycin-resistant clones.

The presence of the neomycin resistance gene cassette in the gene locus is without consequence for a straight knockout as the transcript is often unstable. In contrast, as a conditional knockout starts with a mutant mouse normally expressing the protein of interest, the neomycin resistance gene cassette is eliminated prior to the creation of an animal carrying two copies of the mutant allele (homozygous). There are multiple strategies to eliminate the neomycin resistance gene cassette. The first strategy is to insert a third loxP site in the construct, selectively flanking the exon and neomycin (Fig. 10.1B, iv). This allows for the loss of neomycin through CRE recombination. The second strategy is to use a different set of recombinase/recombinase recognition sites flanking the neomycin cassette, for example Flp ("flip") recombinase with FRT recombination sites, as depicted in Fig. 10.1B, v. Recombinases can be transfected in embryonic stem cells, or injected in the mouse embryo. However, the most convenient and efficient method of introducing the recombinase is through mating the conditional mouse with a recombinase-expressing transgenic mouse. This requires that the recombinase be expressed in the zygote. If three loxP sites are present, multiple combinations of recombination occur leading to four possible genotypes, as shown in Fig. 10.2: three loxP or no recombination (parental genotype), two loxP with loss of neomycin (desirable line), two loxP with loss of the exon, and one loxP (straight knockout). Genotypes can be sorted out using multiple PCR genotyping strategies. If FRT sites are used in combination of loxP sites, elimination of neomycin can be done efficiently without affecting the "floxed" exon (Fig. 10.1B, v).

Following creation of the targeting construct, this cDNA is then linearized and incorporated into embryonic stem cells through electroporation. Next, ES cells are plated on a fibroblast feeder layer in the presence of recombinant leukemia inhibitory factor (LIF, 1000 U/ml). After several days of antibiotic selection, surviving clones are picked and grown in 96-well plates again on feeder cells. The plates are then duplicated and DNA is isolated from a set of plates for analysis. The most reliable method to confirm proper recombination is through Southern blot analysis (see Fig. 10.3A). PCR can also be used, but requires the amplification of rather large fragments, as the region amplified needs to encompass both arms of recombination in entirety. Note that partial recombination in the proper locus is also possible, especially when the exon (or middle region) can serve as the short arm of recombination. This is particularly true with recombination sites such as loxP or FRT sites that can easily be omitted from recombination events (Fig. 10.3B). While the Southern blot strategy guarantees that the construct is recombined in its proper locus, it is nevertheless wise to follow up with PCR and sequencing to ensure the proper placement of the recombination sites. The site located closer to the long arm is generally inserted, but the presence of the

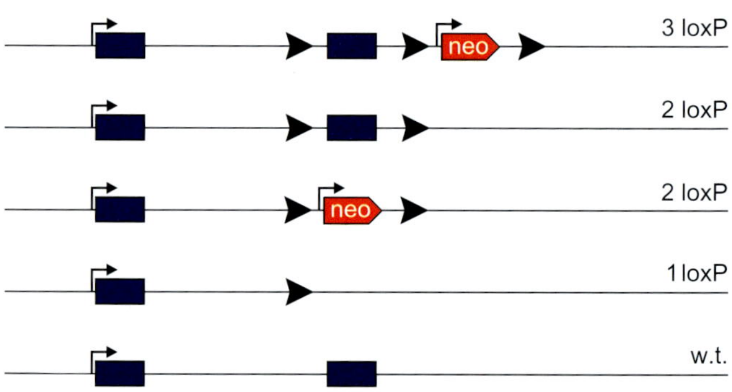

FIGURE 10.2 Schematic representation of all allele products generated by random recombination events after a mouse carrying the 3 loxP allele is mated to a transgenic mouse carrying the CRE recombinase in the zygote. Four genotypes can be observed: 3 loxP (parental allele), loss of neomycin (2 loxP), loss of the exon (alternative 2 loxP) and loss of both neomycin and exon (1 loxP). Note that the wild-type allele is also present.

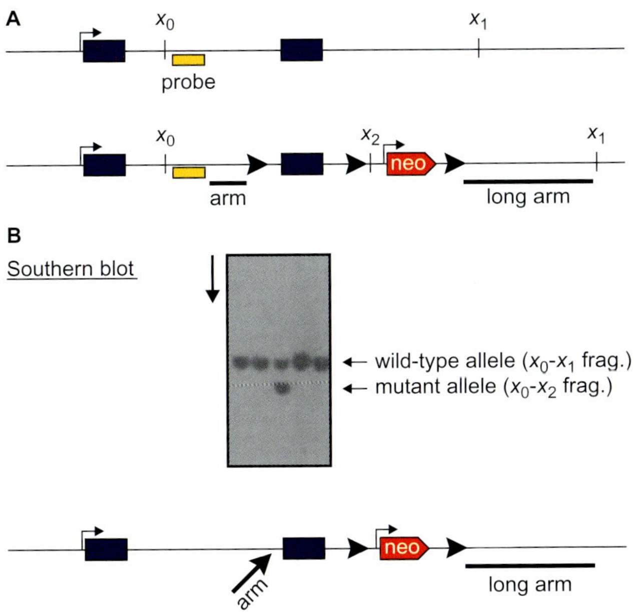

FIGURE 10.3 **A.** The wild-type (top) and mutant (bottom) alleles can be separated on Southern blot analysis after a restriction digest. In this example, a restriction site (X) is located upstream of the short arm of recombination and at the end of the long arm. A cDNA probe outside the recombination region labels the X_0–X_1 genomic DNA fragment. Note that in the mutant allele, insertion of neomycin introduces an additional X site (X_2), which decreases the size of the X_0–X_2 genomic fragment labeled by the probe. ES cell clones with proper recombination will demonstrate two fragments on Southern blot, whereas wild-type ES cells will display only the shorter fragment. **B.** As the long arm of the construct targets the construct to recombine in its proper location, the flanked exon can serve as a short arm, leaving the loxP site and short arm outside the recombination event. Thin arrows represent promoters, large arrowheads represent loxP sites.

recombination site further away from the long arm needs to be validated.

Historically, one of the favorite ES cell lines utilized is derived from a 129/Sv inbred mouse. Once the ES cell is injected in a pseudo-pregnant C57BL/6 female, contribution of 129/Sv ES cell and parental C57BL/6 cells to the skin will result in a chimeric mouse with patches of brown (agouti) and black hair. Thus, the contribution of the 129/Sv ES cells can be easily visualized in the chimeric mouse progeny – the browner the chimeric offspring, the larger is the contribution of the 129/Sv ES cells, and the more likely that these 129/Sv cells will also produce gametes. Germline transmission is obtained at the next generation, when the chimeric mouse is mated to a C57BL/6 mouse and an agouti pup is generated. Note that although the agouti pup is derived from the injected ES cells, it may or may not carry the mutant allele.

Once germline transmission is obtained, the animal is of mixed background C57BL/6:129/Sv. The chimeric mouse can then be re-mated to 129/Sv mice to obtain progeny that are close to 100% 129/Sv. However, this mouse strain is difficult to work with, as the litters are of very small sizes (1–4 pups). Typically, investigators first study the phenotype of the mouse in a mixed background while backcrossing to C57BL/6. This takes more than ten generations or typically two years of backcrossing. Many laboratories are now using C57BL/6 ES cell lines to circumvent this problem. However, it is worth noting that mice are usually stronger in a mixed background and that the phenotype can be "lost" in a pure background, especially if it results in weaker pups.

Homozygous mice can be easily made from mating two heterozygous animals. If the viability of the mice is not affected during embryogenesis, homozygous mice will be found at a 1:4 ratio (Mendelian distribution). It is relatively rare that a strong phenotype such as lethality is seen in heterozygous animals, although there is plenty of evidence showing a gene dosage effect. In the KCC2 knockout mouse for instance, heterozygous mice with 50% KCC2 expression levels show brain hyperexcitability in a behavioral test using phenyltetrazole (Woo et al., 2002), as well as in untreated brain slices, or slices treated with 4-amino pyridine (Zhu et al., 2008).

V. KNOCKOUTS OF CATION-CHLORIDE COTRANSPORTERS

All seven members of the cation-chloride cotransporter family have been knocked out (see Table 10.1), and with the exception of KCC1, and all of them exhibit remarkable phenotypes. Two knockouts result in postnatal lethality due to severe impairments: the NKCC2 knockout exhibits volume depletion, weight loss and a failure to thrive, as well as severe anion-gap metabolic acidosis, renal insufficiency, hypernatremia and hyperkalemia (Takahashi et al., 2000). The KCC2 knockout dies shortly after birth due to respiratory failure (Hubner et al., 2001). Interestingly, by targeting exon 1 of the mouse KCC2 gene, we obtained not a complete knockout of the cotransporter, but 95% reduction in its expression (Woo et al., 2002). The presence of 5% expression is likely due to the presence of an alternative promoter (Uvarov et al., 2007). The KCC2 hypomorph mice survive for up to 17 days after birth and experience frequent and massive seizures triggered by movement in the cage. The seizures could be prevented from birth up to 12 days by an injection twice a day of phenytoin (Dilantin), but they surprisingly reappeared at P12 leading to death by days P15–P17 (Woo et al.,

TABLE 10.1 Knockouts of cation-chloride cotransporters

Gene	Transporter	Year	Targeted exon	KO	Citation
Slc12a1	NKCC2	2000	exons 1–3	Full	Takahashi et al. (2000)
	NKCC2A	2007	exon 4A	partial	Oppermann et al. (2007)
	NKCC2B	2006	exon 4B	partial	Oppermann et al. (2006)
Slc12a2	NKCC1	1999	exon 18	Full	Delpire et al. (1999)
		1999	natural*	Full	Dixon et al. (1999)
		1999	exon 6	Full	Flagella et al. (1999)
Slc12a3	NCC	1998	exon 12	Full	Schultheis et al. (1998)
Slc12a4	KCC1	2007	exons 4,5	Full	Rust et al. (2007)
Slc12a5	KCC2	2001	exon 5	Full	Hubner et al. (2001)
			exon 1	95%	Woo et al. (2002)
Slc12a6	KCC3	2002	exon 6	Full	Howard et al. (2002)
		2003	exons 3,4	Full	Boettger et al. (2003)
Slc12a7	KCC4	2002	exons 4,5	Full	Boettger et al. (2002)

*Natural mutation in the NKCC1 gene isolated at the Jackson laboratories.

2002). As expected, based on the human phenotype of Gittleman syndrome (Simon et al., 1996), knockout of NCC results in hypokalemic alkalosis with hypomagnesemia and hypocalciuria (Morris et al., 2006; Schultheis et al., 1998). Consistent with chronic thiazide treatment associating with high bone mass density, NCC knockout mice exhibit higher renal tubular Ca^{2+} reabsorption, higher bone mass density, and lower bone remodeling than control mice (Nicolet-Barousse et al., 2005). Disruption of NKCC1 is interesting as it results in a wide variety of phenotypes, mirroring the wide expression pattern of the cotransporter. First and foremost, the NKCC1 knockout is deaf and exhibits a waltzer/shaker phenotype characteristic of inner ear defect (Delpire et al., 1999). The knockout suffers from intestinal transit problems (Flagella et al., 1999; Wouters et al., 2006), deficits in salivary production (Evans et al., 2000), deficits in sperm formation (Pace et al., 2000), blood pressure reduction (Castrop et al., 2005; Wall et al., 2006), deficits in sensory perception (Sung et al., 2000) and changes in brain excitability (Dzhala et al., 2005; Zhu et al., 2008). The phenotype of the KCC3 knockout greatly mirrors the phenotype of patients carrying truncating mutations in KCC3. Both the knockout mouse and humans with KCC3 truncation mutations have a severe early onset locomotor phenotype, nerve pathologies and decreased nerve signaling velocity (Boettger et al., 2003; Byun and Delpire, 2007; Howard et al., 2002). Additionally, KCC3 knockout mice exhibit age-related hearing loss (Boettger et al., 2003), high blood pressure (Adragna et al., 2004; Rust et al., 2006) and impaired acid/base balance in the kidney proximal tubule (Wang et al., 2003). Finally, the knockout of KCC4 results in renal tubular acidosis as well as deafness (Boettger et al., 2002).

To this date, no study has been published taking advantage of recombination sites or driving the knockout of cation-chloride cotransporters in specific tissues. Given the importance of the cation-chloride transporter family, as demonstrated by phenotypes of the full knockouts described above, studies of conditional cation-chloride cotransporter knockouts would be beneficial to understanding their contribution to physiological regulation. This is particularly true regarding CCCs with early death phenotypes. Inducible conditional knockout mice will allow investigators to study acute effects of cotransporter disruption, apart from the role of the cation-chloride cotransporters during development. Additionally, there has been a great increase of information regarding both alternative splice isoforms of the cation-chloride cotransporters and the role of different domains in the cotransporter. Knockout models of one specific cation-chloride cotransporter isoform can help determine function particular to that subset of cotransporter. Knock-in of mutant cotransporter can add a physiological context to newly identified interaction and phosphorylation domains. While the standard full knockout mice already generated continue to yield important information, such additional varieties of mouse genetic engineering will continue to complement the growing body of knowledge of the cation chloride-cotransporter family.

References

Adragna, N.C., Chen, Y., Delpire, E., Lauf, P.K., and Morris, M. (2004). Hypertension in K-Cl cotransporter-3 knockout mice. *Adv. Exp. Med. Biol.* **559**, 379–385.

Boettger, T., Hubner, C.A., Maier, H., Rust, M.B., Beck, F.X., and Jentsch, T.J. (2002). Deafness and renal tubular acidosis in mice lacking the K-Cl co-transporter Kcc4. *Nature* **416**, 874–878.

Boettger, T., Rust, M.B., Maier, H., Seidenbecher, T., Schweizer, M., Keating, D.J., Faulhaber, J., Ehmke, H., Pfeffer, C., Scheel, O., Lemcke, B., Horst, J., Leuwer, R., Pape, H.C., Volkl, H., Hubner, C.A., and Jentsch, T.J. (2003). Loss of K-Cl co-transporter KCC3 causes deafness, neurodegeneration and reduced seizure threshold. *Embo. J.* **22**, 5422–5434.

Byun, N. and Delpire, E. (2007). Axonal and periaxonal swelling precede peripheral neurodegeneration in KCC3 knockout mice. *Neurobiol. Dis.* **28**, 39–51.

Castrop, H., Lorenz, J.N., Hansen, P.B., Friis, U., Mizel, D., Oppermann, M., Jensen, B.L., Briggs, J., Skøtt, O., and Schnermann, J. (2005). Contribution of the basolateral isoform of the Na-K-2Cl-cotransporter (NKCC1/BSC2) to renin secretion. *Am. J. Physiol. Renal Physiol.* **289**, F1185–F1192.

Copeland, N.G., Jenkins, N.A., and Court, D.L. (2001). Recombineering: a powerful new tool for mouse functional genomics. *Nat. Rev. Genet.* **2**, 769–779.

Crabbe, J.C., Belknap, J.K., Buck, K.J., and Metten, P. (1994). Use of recombinant inbred strains for studying genetic determinants of responses to alcohol. *Alcohol Alcohol Suppl.* **2**, 67–71.

Delpire, E., Lu, J., England, R., Dull, C., and Thorne, T. (1999). Deafness and imbalance associated with inactivation of the secretory Na-K-2Cl co-transporter. *Nat. Genet.* **22**, 192–195.

Delpire, E., Rauchman, M.I., Beier, D.R., Hebert, S.C., and Gullans, S.R. (1994). Molecular cloning and chromosome localization of a putative basolateral Na-K-2Cl cotransporter from mouse inner medullary collecting duct (mIMCD-3) cells. *J. Biol. Chem.* **269**, 25677–25683.

Dixon, M.J., Gazzard, J., Chaudhry, S.S., Sampson, N., Schulte, B.A., and Steel, K.P. (1999). Mutation of the Na-K-Cl co-transporter gene Slc12a2 results in deafness in mice. *Hum. Mol. Genet.* **8**, 1579–1584.

Dzhala, V.I., Talos, D.M., Sdrulla, D.A., Brumback, A.C., Mathews, G.C., Benke, T.A., Delpire, E., Jensen, F.E., and Staley, K.J. (2005). NKCC1 transporter facilitates seizures in the developing brain. *Nat. Med.* **11**, 1205–1213.

Evans, R.L., Park, K., Turner, R.J., Watson, G.E., Nguyen, H.-V., Dennett, M.R., Hand, A.R., Flagella, M., Shull, G.E., and Melvin, J.E. (2000). Severe impairment of salivation in $Na^+/K^+/2Cl^-$ cotransporter (NKCC1)-deficient mice. *J. Biol. Chem.* **275**, 26720–26726.

Flagella, M., Clarke, L.L., Miller, M.L., Erway, L.C., Giannella, R.A., Andringa, A., Gawenis, L.R., Kramer, J., Duffy, J.J., Doetschman, T., Lorenz, J.N., Yamoah, E.N., Cardell, E.L., and Shull, G.E. (1999). Mice lacking the basolateral Na-K-2Cl cotransporter have impaired epithelial chloride secretion and are profoundly deaf. *J. Biol. Chem.* **274**, 26946–26955.

García-Otín, A.-L. and Guillou, F. (2006). Mammalian genome targeting using site-specific recombinases. *Front. in Biosci.* **11**, 1108–1136.

Howard, H.C., Mount, D.B., Rochefort, D., Byun, N., Dupré, N., Lu, J., Fan, X., Song, L., Rivière, J.-B., Prévost, C., Welch, R., England, R., Zhan, F.Q., Mercado, A., Siesser, W.B., George, A.L., Horst, J., Simonati, A., McDonald, M.P., Bouchard, J.-P., Mathieu, J., Delpire, E., and Rouleau, G.A. (2002). Mutations in the K-Cl cotransporter KCC3 cause a severe peripheral neuropathy associated with agenesis of the corpus callosum. *Nat. Genet.* **32**, 384–392.

Hsu, H.C., Lu, L., Yi, N., Van Zant, G., Williams, R.W., and Mountz, J.D. (2007). Quantitative trait locus (QTL) mapping in aging systems. *Methods Mol. Biol.* **371**, 321–348.

Hubner, C.A., Stein, V., Hermans-Borgmeyer, I., Meyer, T., Ballanyi, K., and Jentsch, T.J. (2001). Disruption of KCC2 reveals an essential role of K-Cl cotransport already in early synaptic inhibition. *Neuron* **30**, 515–524.

Morris, R.G., Hoorn, E.J., and Knepper, M.A. (2006). Hypokalemia in a mouse model of Gitelman's syndrome. *Am. J. Physiol. Renal Physiol.* **290**, F1416–F1420.

Nicolet-Barousse, L., Blanchard, A., Roux, C., Pietri, L., Bloch-Faure, M., Kolta, S., Chappard, C., Geoffroy, V., Morieux, C., Jeunemaitre, X., Shull, G.E., Meneton, P., Paillard, M., Houillier, P., and De Vernejoul, M.C. (2005). Inactivation of the Na-Cl co-transporter (NCC) gene is associated with high BMD through both renal and bone mechanisms: analysis of patients with Gitelman syndrome and Ncc null mice. *J. Bone Miner. Res.* **20**, 799–808.

Oppermann, M., Mizel, D., Huang, G., Li, C., Deng, C., Theilig, F., Bachmann, S., Briggs, J., Schnermann, J., and Castrop, H. (2006). Macula densa control of renin secretion and preglomerular resistance in mice with selective deletion of the B isoform of the Na,K,2Cl co-transporter. *J. Am. Soc. Nephrol.* **17**, 2143–2152.

Oppermann, M., Mizel, D., Kim, S.M., Chen, L., Faulhaber-Walter, R., Huang, Y., Li, C., Deng, C., Briggs, J., Schnermann, J., and Castrop, H. (2007). Renal function in mice with targeted disruption of the A isoform of the Na-K-2Cl co-transporter. *J. Am. Soc. Nephrol.* **18**, 440–448.

Pace, A.J., Lee, E., Athirakul, K., Coffman, T.M., O'Brien, D.A., and Koller, B.H. (2000). Failure of spermatogenesis in mouse lines deficient in the Na + -K + -2Cl− cotransporter. *J. Clin. Invest.* **105**, 441–450.

Peters, L.L., Robledo, R.F., Bult, C.J., Churchill, G.A., Paigen, B.J., and Svenson, K.L. (2007). The mouse as a model for human biology: a resource guide for complex trait analysis. *Nat. Rev. Genet.* **8**, 58–69.

Rust, M.B., Faulhaber, J., Budack, M.K., Pfeffer, C., Maritzen, T., Didie, M., Beck, F.X., Boettger, T., Schubert, R., Ehmke, H., Jentsch, T.J., and Hubner, C.A. (2006). Neurogenic mechanisms contribute to hypertension in mice with disruption of the K-Cl cotransporter KCC3. *Circ. Res.* **98**, 549–556.

Rust, M.B., Alper, S.L., Rudhard, Y., Shmukler, B.E., Vicente, R., Brugnara, C., Trudel, M., Jentsch, T.J., and Hübner, C.A. (2007). Disruption of erythroid K-Cl cotransporters alters erythrocyte volume and partially rescues erythrocyte dehydration in SAD mice. *J. Clin. Invest.* **117**, 1708–1717.

Sawitzke, J.A., Thomason, L.C., Costantino, N., Bubunenko, M., Datta, S., and Court, D.L. (2007). Recombineering: in vivo genetic engineering in E. coli, S. enterica, and beyond. *Methods Enzymol.* **421**, 171–199.

Schultheis, P.J., Lorenz, J.N., Meneton, P., Nieman, M.L., Riddle, T.M., Flagella, M., Duffy, J.J., Doetschman, T., Miller, M.L., and Shull, G.E. (1998). Phenotype resembling Gitelman's syndrome in mice lacking the apical Na + -Cl− cotransporter of the distal convoluted tubule. *J. Biol. Chem.* **273**, 29150–29155.

Simon, D.B., Nelson-Williams, C., Johnson Bia, M., Ellison, D., Karet, F.E., Morey Molina, A., Vaara, I., Iwata, F., Cushner, H.M., Koolen, M., Gainza, F.J., Gitelman, H.J., and Lifton, R.P. (1996). Gitelman's variant of Bartter's syndrome, inherited hypokalaemic alkalosis, is caused by mutations in the thiazide-sensitive Na-Cl cotransporter. *Nature Genetics* **12**, 24–30.

Sung, K.-W., Kirby, M., McDonald, M.P., Lovinger, D.M., and Delpire, E. (2000). Abnormal $GABA_A$-receptor mediated currents in dorsal root ganglion neurons isolated from Na-K-2Cl cotransporter null mice. *J. Neurosci.* **20**, 7531–7538.

Takahashi, N., Chernavvsky, D.R., Gomez, R.A., Igarashi, P., Gitelman, H.J., and Smithies, O. (2000). Uncompensated polyuria in a mouse model of Bartter's syndrome. *Proc. Natl. Acad. Sci. USA* **97**, 5434–5439.

Uvarov, P., Ludwig, A., Markkanen, M., Pruunsild, P., Kaila, K., Delpire, E., Timmusk, T., Rivera, C., and Airaksinen, M.S. (2007). A novel N-terminal isoform of the neuron-specific K-Cl cotransporter KCC2. *J. Biol. Chem.* **282**, 30570–30576.

Wall, S.M., Knepper, M.A., Hassell, K.A., Fischer, M.P., Shodeinde, A., Shin, W., Pham, T.D., Meyer, J.W., Lorenz, J.N., Beierwaltes, W.H., Dietz, J.R., Shull, G.E., and Kim, Y.H. (2006). Hypotension in NKCC1 null mice: role of the kidneys. *Am. J. Physiol. Renal Physiol.* **290**, F409–F416.

Wang, T., Delpire, E., Giebisch, G., Hebert, S.C., and Mount, D.B. (2003). Impaired fluid and bicarbonate absorption in proximal tubules (PT) of KCC3 knockout mice. *FASEB J.* **17**, A464.

Waterston, R.H. and Consortium, M.G.S. (2002). Initial sequencing and comparative analysis of the mouse genome. *Nature* **420**, 520–562.

Woo, N.-S., Lu, J., England, R., McClellan, R., Dufour, S., Mount, D.B., Deutch, A.Y., Lovinger, D.M., and Delpire, E. (2002). Hyperexcitability and epilepsy associated with disruption of the mouse neuronal-specific K-Cl cotransporter gene. *Hippocampus* **12**, 258–268.

Wouters, M., De Laet, A., Ver Donck, L., Delpire, E., van Bogaert, P.P., Timmermans, J.P., de Kerchove d'Exaerde, A., Smans, K., and Vanderwinden, J.M. (2006). Subtractive hybridization unravels a role for the ion co-transporter NKCC1 in the murine intestinal pacemaker. *Am. J. Physiol. Gastrointest. Liver Physiol.* **290**, G1219–G1227.

Zhu, L., Polley, N., Mathews, G.C., and Delpire, E. (2008). NKCC1 and KCC2 prevent hyperexcitability in the mouse hippocampus. *Epilepsy Res.* **79**, 201–212.

PART III

FROM CLONING TO STRUCTURE, FUNCTION AND REGULATION OF CHLORIDE TRANSPORTERS AND CHANNELS

CHAPTER

11

The NKCC and NCC Genes: An *in Silico* View

Mauricio Di Fulvio and Francisco Javier Alvarez-Leefmans

OUTLINE

I. AN *IN SILICO* VIEW OF THE NKCC AND NCC GENES

A. Databases and Definitions

Genes, genomes, transcripts and protein sequences are organized in and maintained by several publicly available databases. For instance, the *Reference Sequence* (Pruitt et al., 2007) and *GenBank* (Benson et al., 2008) databases are built and distributed by the National Center for Biotechnology Information (NCBI). Both databases together with the European Molecular Biology Laboratory Nucleotide Sequence Database (EMBL) (Cochrane et al., 2008) and the DNA DataBank of Japan (DDBJ) (Sugawara et al., 2008) form part of the International Nucleotide Sequence Database Collaboration (INSDC) (Galperin, 2008).

The NCBI's *GenBank* database annotates and organizes nucleotide sequences and their *predicted* protein translations through direct submissions of nucleotide sequences from individual laboratories and batch submissions of expressed sequence tags (*ESTs*), sequence tagged sites (*STS*), genome survey sequences (*GSS*) and high-throughput genome sequences (*HTGS*) from large-scale sequencing projects.

The Reference Sequence (*RefSeq*) Database annotates and identifies, with a unique number, DNA, RNA and protein sequences derived from records in the *GenBank* database. However, *RefSeqs* differ from the *GenBank* accession numbers in that each *RefSeq* is a synthesis of information, *not* related to primary research data. *RefSeq* accession numbers are distinguished from the *GenBank* ones by their distinct prefix format of two characters followed by an underscore character (i.e. NC_, NM_, NP_, XM_, XP_, etc). The *RefSeq* NC_$(n)_{6-9}$ format, where *n* is any number with six to nine digits, represents a complete genomic molecule whereas the NM_$(n)_{6-9}$ makes reference to transcript products, mature messenger RNA (mRNA) transcripts. A detailed description of the *RefSeq* database was recently published (Pruitt et al., 2007) and a full definition of the *RefSeq* is available on the *RefSeq* website (www.ncbi.nlm.nih.gov/RefSeq/key.html#accessions).

All members of the SLC12A family, as DNA, cDNA, mRNA or protein sequences, are clearly identified and cross-referenced in the *GenBank*, *RefSeq* and other databases (Table 11.1).

B. Nomenclature and Members of the SLC12A Family

The nomenclature of the electroneutral cation-chloride cotransporters (CCCs) gene family was proposed and approved by the Human Genome Organization (HUGO) Gene Nomenclature Committee (HGNC) (www.genenames.org) (Eyre et al., 2006; Ren et al., 2007; Wain et al., 2004). The accepted nomenclature for membrane transporter proteins have the SLC*n*X*m* format where SLC stands for solute carrier, *n* is the number that represents the family (12, defines the CCC family), X is any letter representing the subfamily (A, first subfamily) and *m* represents the actual member of the family. Each symbol is unique and identifies the gene and each gene has a unique accession number in *RefSeq*.

C. Homology, Identity and Similarity

When referring to amino acid or nucleotide sequences there are two concepts that are often confused: *identity* and *similarity*. Percentage of sequence *identity* is defined based on the number of places where two sequences have the *exact* same residue or nucleotide. For instance, when the aligned sequences match exactly the same residues or nucleotides, they are 100% identical. *Similarity* is a concept that goes beyond *identity*. It means that the aligned amino acid pair may consist of an amino acid that is not the same but has similar chemical, physicochemical and thermodynamical properties (e.g. hydrophobicity, charge, partition coefficient, etc.). Clearly, all identities are similarities. As different amino acids may share identical thermodynamical properties, similarity will always be higher than identity. Two protein sequences with low percentage of identity may exhibit similarities in their physiochemical properties, tertiary structure and/or biological properties.

The nine CCC gene members of the SLC12A family (SLC12A1-9) are organized and classified by *homology*, not by similarity or degree of identity (see below, section I.D). By definition, two nucleotide or protein sequences are homologous *if and only if* the two sequences evolved or acquired that state *directly* from the same ancestor gene. As shown in Fig. 11.1, phylogenetic analysis of aligned human SLC12A protein sequences using *PHYLIP* (evolution.genetics.washington.edu/phylip.html) demonstrate that all SLC12A proteins are evolutionary related or share a common ancestor. The phylogenetic distance between genes is defined as the fraction of mismatches at aligned positions, with gaps either ignored or counted as mismatches (Mount, 2004). It is important to note that phylogenetic distance values do not represent the degree of identity or similarity among genes, as highly similar protein sequences may arise

TABLE 11.1 Database accession numbers for all members of the human SLC12A family. Each human SLC12A gene is identified with an NC_ number indicating the complete genomic molecule (chromosome) followed by two numbers separated by two dots which indicate the nucleotide position in the chromosome. Accession numbers for transcripts or proteins of reference for each SLC12A gene are indicated with the prefix NM_ or NP_, respectively. Note that only one reference transcript may be annotated when only one full-length transcript cDNA of unique sequence is known, regardless of inferred splicing variants or the number of cloned cDNAs. Each SLC12A gene is also identified with mapped data supplied by NCBI (*GeneID*) providing descriptive information (i.e. genetic loci, official nomenclature, aliases, sequence accessions, phenotypes, homology, map locations, related Web sites, etc.). The *RefSeq* status indicates if the gene is classified as *Validated* (the *RefSeq* record has undergone an initial review to provide the preferred sequence standard), *Reviewed* (the *RefSeq* record has been reviewed to provide the preferred sequence standard and to add additional functional descriptive information and feature annotation, as relevant) or *Provisional* (an automatically provided record based on *GenBank* sequence data; there is support for the transcript and protein. This is the default status code applied to some genomes for which there is no clear information about the method used to define the sequence). The *PubMed* identification number (PMID) for each SLC12A gene links to published articles relevant to the entry in the NCBIs whereas the *Ensembl* accession number links to the current gene build of the Ensembl database. The *Online Mendelian Inheritance in Man* (OMIM) identification number provides links between human genes and genetic disorders. OMIM also contains textual information and references, copious links to Medline and sequence records. The *UniProt* identifier is provided by the EBI and is described as a curated protein sequence database. The HUGO Gene Nomenclature Committee (HGNC) assigns a unique and ideally meaningful name and symbol to every human gene. *GenBank* accession numbers of cDNAs relevant to the SLC12A gene are identified in the far left column.

Human Gene	Acronym	Reference Transcripts	Reference Protein	GeneID	RefSeq status	PMID	Ensembl	OMIM	UniProt ID	HGNC	GenBank
SLC12A1 (NC_000015) 46285790..46383567	NKCC2	NM_000338	NP_000329	6557	Validated	8640224	ENSG00000074803	600839	Q13621	10910	BX647067 BX647484 EF559316 U58130
SLC12A2 (NC_000005) 127447382..127553279	NKCC1	NM_001046	NP_001037	6558	Validated	7629105	ENSG00000064651	600840	P55011	10911	AY280459 BC033003 U30246
SLC12A3 (NC_000016) 55456643..55504850	NCC	NM_00339	NP_000330	6559	Reviewed	8812482	ENSG00000070915	600968	P55017	10912	AK315298 AK223133
SLC12A4 (NC_000016) 66560026..665356731	KCC1	NM_005072	NP_005063	6560	Provisional	8663127	ENSG00000124067	604119	Q9UP95	10913	AF047338 AF054505 AF054506 AK097808 BC021193
SLC12A5 (NC_000020) 44091245..44122196	KCC2	NM_020708	NP_065759	57468	Reviewed	–	ENSG00000124140	606726	Q9H2X9	13818	AB033002 AF208159 AK289758 BC132668 BC132670 U79245

(*Continued*)

TABLE 11.1 Continued

Human Gene	Acronym	Reference Transcripts	Reference Protein	GeneID	RefSeq status	PMID	Ensembl	OMIM	UniProt ID	HGNC	GenBank
SLC12A6 (NC_000015) 32417557..32309489	KCC3	NM_1042495 (KCC3c) NM_1042496 (KCC3d) NM_1042494 (KCC3c) NM_1042497 (KCC3e) NM_133647 (KCC3a) NM_005135 (KCC3b)	NP_01035960 NP_01035961 NP_01035959 NP_01035962 NP_598408 NP_005126	9990	Reviewed	10187864 10347194	ENSG00000140199	604878	Q9UHW9	10914	AF105366 AF108831 AF116242 AF477977 AF531258 AF531259 AF531260 AK128133 AK292550 AK315283 BC033894 BC051709 BC051744 BC070107 BC126241 BC126243 BX648195 DQ138323
SLC12A7 (NC_000005) 1165172..1103489	KCC4	NM_006598	NP_006589	10723	Validated	10347194	ENSG00000113504	604879	Q9Y666	10915	AF105365 BC007760 BC018982 BC098390
SLC12A8 (NC_000003) 126414299..126284170	CCC9	NM_024628	NP_078904	84561	Validated	11863360	ENSG00000163848	611316	Q96RF9	15595	AF345197 AF390442 AK123001 BC126158 BC126160 AK026841
SLC12A9 (NC_000007) 100288294..100302569	CIP	NM_020246	NP_064231	56996	Validated	10871601 11239002	ENSG00000146828	–	Q9BXP2	17435	AB033284 AF284422 AK024420 AK024421 AK024466 AK024494 AK090458 AK128873 BC000154 CR594298

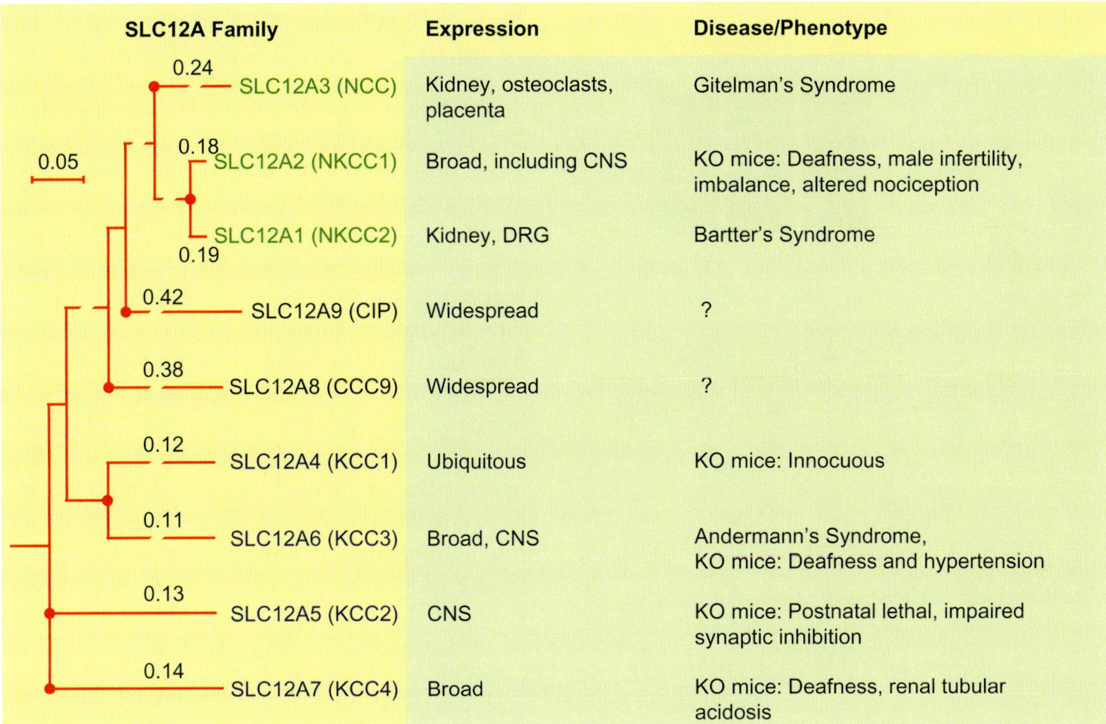

FIGURE 11.1 Unrooted phylogenetic cladogram tree of the SLC12A2 protein family. Cladogram drawn with *Phylodendron* (iubio.bio.indiana.edu/treeapp/) using phylogenetic data obtained with PHYLIP (evolution.genetics.washington.edu/phylip.html), a software package that calculates phylogenetic distances between aligned protein sequences. The human SLC12A protein sequences were aligned using *ClustalW* (www.ebi.ac.uk/clustalw/). The length of each branch (red lines) to the next node (red circles) is scaled in terms of expected numbers of nucleotide substitutions that occurred prior to the next level of separation (e.g. a branch of length 0.14 is 14 times as long as the one which would show a 1% difference between the nucleotide sequences at the beginning and the end of the branch). Gene expression patterns are also summarized together with their associated diseases/phenotypes in humans and/or mice.

from different ancestor genes. Homology between genes of different species is *orthology* (SLC12A (human) vs. *slc12a* (mouse, rat, etc.)), whereas any difference within a specie is *paralogy* (*slc12a1* (rat) vs. *slc12a2* (rat)). The SLC12A gene family members are paralogs among themselves since they derive from a same ancestral gene. A detailed phylogenetic analysis for SLC12A members can be found in the *Tree Families Database* (TreeFam, www.treefam.org), whereas the genetic alignment of all known SLC12A gene sequences are in *Ensembl* (www.ensembl.org, acc. number).

Estimation of the phylogenetic distances between SLC12A proteins allows the subclassification of the SLC12A gene family into three subgroups, as shown in Fig. 11.1: SLC12A4-7 (KCC1-KCC4), SLC12A1-3 (NKCC2, NKCC1 and NCC) and SLC12A8-9 (CCC9 and CIP) which in the figure are labeled in blue, green and black, respectively. The protein sequences of each member of these subgroups share a high degree of identity and may also share similar functions, although the latter needs further experimental demonstration.

D. How Similar are the Members of the SLC12A Family?

Homologous genes may share very different degrees or percentages of identity (Doolittle, 1981). The percentage of identity (*%ID*), defined as the number of aligned positions where the matching amino acids or nucleotides are identical (May, 2004), is calculated in many ways depending on the alignment algorithm used (Raghava and Barton, 2006). The SLC12A family of proteins shares a wide range of *%ID*. For instance, SLC12A1 (NKCC2) is ~62% and 52% identical to SLC12A2 (NKCC1) and SLC12A3 (NCC), respectively; but less than 25% identical to the other members of the family (Fig. 11.2). However, the SLC12A4 (KCC1) protein is 69%, 76% and 68% identical to SLC12A5 (KCC2), SLC12A 6 (KCC3) and SLC12A 7 (KCC4), respectively (inset, Fig. 11.2).

What is the meaning of the percentage of identity? In general terms, protein sequences belonging to the same homologous family (e.g. SLC12A proteins, which are evolutionary related (Fig. 11.1)) can be aligned to produce a *%ID* equal or greater than 30% (Murzin et al.,

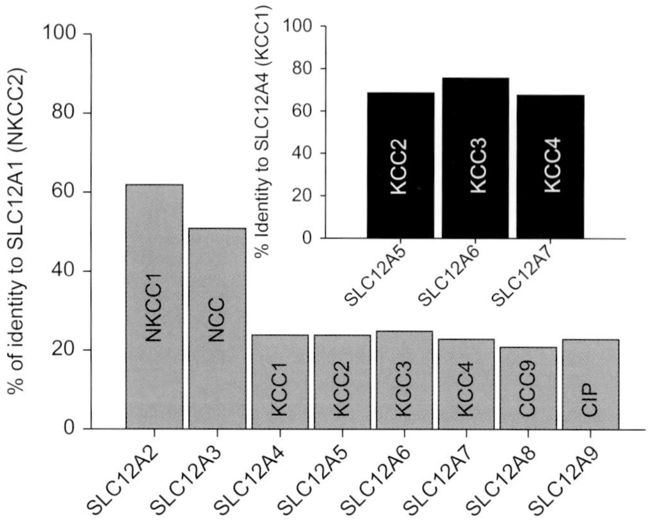

FIGURE 11.2 Alignment scores of the predicted protein sequences of SLC12A members expressed as percentage of identity. Gray bars represent the percentage of identity of members of the human SLC12A family (from SLC12A2 through 9) with respect to SLC12A1 (hNKCC2F (U58130)) whereas black bars indicate the percentage of identity of SLC12A5-7 (hKCC2, hKCC3 and hKCC4) with respect to SLC12A4 (KCC1). Progressive multiple alignments of SLC12A protein sequences were performed using *ClustalW*.

1995). This is clearly the case for SLC12A1-7 proteins (Figs 11.1 and 11.2). However, when this threshold is not reached (i.e. $\%ID < 30\%$) the probable common evolutionary origin of the proteins within a family must be *inferred* on the basis of shared structural and functional features (Murzin et al., 1995). This is the case for SLC12A8 (CCC9) and SLC12A9 (CIP). These two proteins with less than 30% identity with respect to the other members are still part of the SLC12 family because their predicted structural properties are *similar* to the ones predicted for the other SLC12A members, although their function remains to be elucidated.

The degree of identity permits the *inference* of structural and functional relationships and also identifies critical residues involved in functional properties as well as motifs, profiles, patterns, domains, etc. (Blundell et al., 1987). For instance, multiple sequence alignments of the SLC12A family have revealed common sequence patterns named "signature sequences" (i.e. the most evolutionary conserved regions). These sequences appear to be in the form of TX [L,I,V] SX [S,C] A [L,I,V] XT [N,D] X_2VX_2GG and SRXLG [P,A] $EXGX_2$ [L,I,V] G $[L,I,V]_2$ F (Park and Saier, 1996). Further, amino acid sequence analyses shows the presence of three different regions in the SLC12A proteins, two less conserved hydrophilic regions (i.e. N-terminal and C-terminal) and a most conserved hydrophobic core, predicted to be organized in several transmembrane α-helixes (Park and Saier, 1996).

E. Predicted but not Demonstrated Topologies of SLC12A1, 2 and 3 Proteins

A protein *topology* predicted *in silico* is halfway from the peptide sequence to the real three-dimensional structure of the protein (von Heijne, 2006). Hence, computer algorithms developed to predict protein topology or structure based on physicochemical properties of amino acid sequences as well as by comparison with known protein structures (e.g. threading and homology modeling) are invaluable tools to *infer* topology and/or function–structure relationships.

Most of the SLC12A proteins appear to share similar predicted structures with several transmembrane domains and long intracellular N- or C-termini. This assumption is based on the estimated hydrophilicity/hydrophobicity profiles of deduced SLC12A protein sequences according to Kyte-Doolittle's algorithm (Kyte and Doolittle, 1982). A key feature of this algorithm is the so-called "window size", i.e. the number of amino acids examined at a time to determine a point of hydrophobic character (Kyte and Doolittle, 1982). Hence, it is critical to choose a window size that corresponds to the expected size of the structural motif under investigation (i.e. a window size of 19–21 (about the size of a membrane spanning α-helix) will make hydrophobic, membrane-spanning domains stand out on the Kyte-Doolittle scale (typically >1.6)). However, windows sizes ranging from 11 to 15 amino acids were used to generate hydropathy plots predicting 12 transmembrane (TM) domains in mammalian members of the SLC12A family (Caron et al., 2000; Delpire et al., 1994; Gamba et al., 1994; Gillen et al., 1996; Hiki et al., 1999; Moore-Hoon and Turner, 1998; Payne and Forbush, 1994; Payne et al., 1996; Yerby et al., 1997). Although alternative topological models for members of the SLC12A family have been proposed (Park and Saier, 1996) and several transport protein families include members that probably have more or less than 12 TM domains (Espanol and Saier, 1995; Paulsen and Skurray, 1993), it is accepted that the SLC12A family are proteins of 12 TM domains.

It is now clear that the most important factor in determining membrane insertion is the hydrophobicity of 19–21 amino acid sequences (Zhao and London, 2006). This concept is better represented by using the experimentally determined transfer-free energies (ΔG) for each amino acid (i.e. a thermodynamic scale of hydrophobicity) originally proposed by Wimley and White (Wimley and White, 1996). Hence, the hydrophobicity plot of Wimley-White (also known as the *octanol* plot) identifies the position of transmembrane α-helices in protein sequences with less

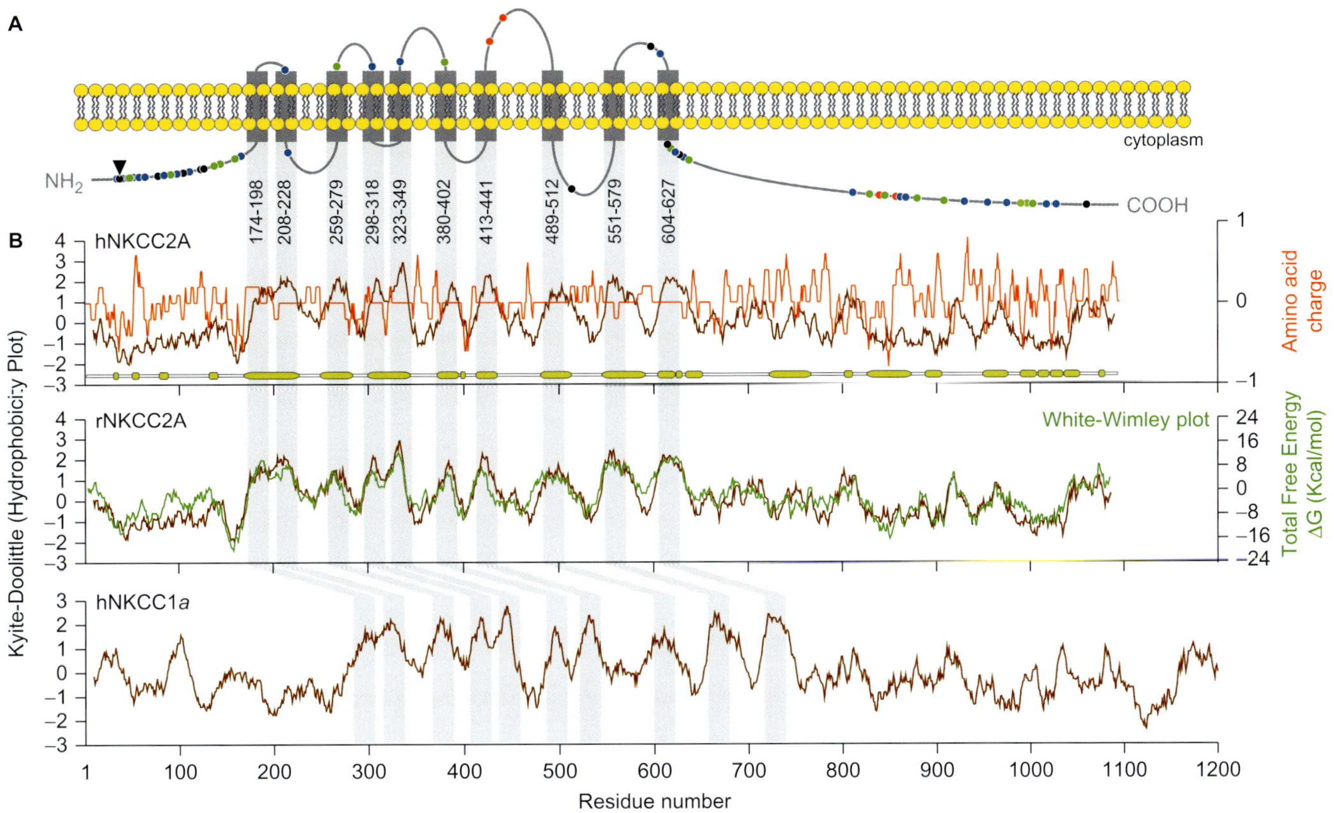

FIGURE 11.3 Kyte-Doolittle and White-Wimley plots of NKCC2 and NKCC1 protein sequences. **A.** Predicted NKCC protein topology. Putative transmembrane domains (TM) are indicated as gray boxes across the lipid bilayer. The position of NKCC2 amino acids predicted to be localized at the TM domains is numbered underneath each potential TM domain. The continuous gray line represents the amino acid chain of the NKCC2 proteins. Colored dots located at the cytoplasmic N-terminal and C-terminal portions of NKCC2s represent the location of residues predicted to be phosphorylated (blue: Ser, green: Thr and black: Tyr) and potential N-glycosylation sites (red dots). The potential site for tyrosine sulfination at the N-terminus of NKCC2 is indicated with an arrowhead. Phosphorylation and sulfination sites on NKCC2 proteins were predicted using *NetPhos* (www.cbs.dtu.dk/services/NetPhos) and *Sulfinator* (www.expasy.ch/tools/sulfinator), respectively. **B.** Hydropathy plots of hNKCC2A (ABU69043) (top), rNKCC2A (ABU63482) (center) and hNKCC1a (AAC50561) (bottom). These analyses were performed using a window size of 19 residues. Window sizes of 19 or 21 make hydrophobic, membrane-spanning domains stand out clearly (typically, values >1.6 on the Kyte and Doolittle scale). Under these conditions, hNKCC2 proteins are predicted to have 10 TM regions: 174–198, 208–228, 259–279, 298–318, 323–349, 380–402, 413–441, 489–512, 551–579 and 604–627. Each TM is ~20 residues in length and highly identical among species. All predicted TMs in NKCC2s have energetic preferences for being in the lipid environment as characterized by the total free energy (ΔG) above zero in the White-Wimley interface hydropathy plot. The mean charges of the amino acids are calculated by giving the residues D (Asp) and E (Glu) a charge of −1, K (Lys) and R (Arg) a charge of +1, and the residue H (His) a charge of +0.5. The represented data were obtained using *jEMBOSS* for Linux (emboss.sourceforge.net/Jemboss), *TMap*, *TMPredProtScale* (at the ExPASy molecular biology server) and *PROTEUS Structure Prediction Server v2.0* (wks16338.biology.ualberta.ca/proteus).

ambiguity than the Kyte-Doolittle plot. As shown in Fig. 11.3, *octanol* plots obtained for SLC12A1 (NKCC2), SLC12A2 (NKCC1) and SLC12A3 (NCC) are different to the ones originally proposed for these gene products using the Kyte-Doolittle algorithm with a window size of 11–15 (Delpire et al., 1994; Gamba et al., 1994; Payne and Forbush, 1994; Yerby et al., 1997). However, the *octanol* plot correlates very well with the Kyte-Doolittle plot if the latter is constructed using a window size of 19–21 amino acids (Fig. 11.3).

Prediction algorithms based solely on hydrophobicity plots (Kyte and Doolittle, 1982) or thermodynamic scales of hydrophobicity (Wimley and White, 1996) are somewhat incomplete and innacurate. The fact that ~5% of the transmembrane α-helices in the known structures are very short (<15 residues) and only partially span the membrane, together with the lack of critical thermodynamic data, have made transmembrane prediction algorithms somewhat unsatisfactory. It was not until recently that the free energy contributions from individual amino acids in different positions along the membrane was reported (Hessa et al., 2007). Hence, the accuracy of algorithms predicting TM helixes has been recently improved by the development of new tools such as *MemBrain* (Shen and Chou, 2008), *TopPred ΔG* (Hessa et al., 2007), *SCAMPI* (Bernsel et al., 2008), *ZPRED* (Granseth et al., 2006) and *PRO/PRODIV-TMHMM* (Viklund and Elofsson, 2004). Most of these algorithms are part of *TOPCONS* protein topology prediction server (topcons.cbr.su.se). By using *MemBrain* or *SCAMPI*, human SLC12A1, SLC12A2 and SLC12A3 proteins (i.e. NKCC2, NKCC1 and NCC) it can be predicted that these proteins may have 13 TM domains, whereas *PRODIV*, *PRO* or *OCTUPUS* predicts 12 TM domains (Fig. 11.4). It should be mentioned that the model having 13 TM domains places the N- and C-termini in different compartments (inside and outside,

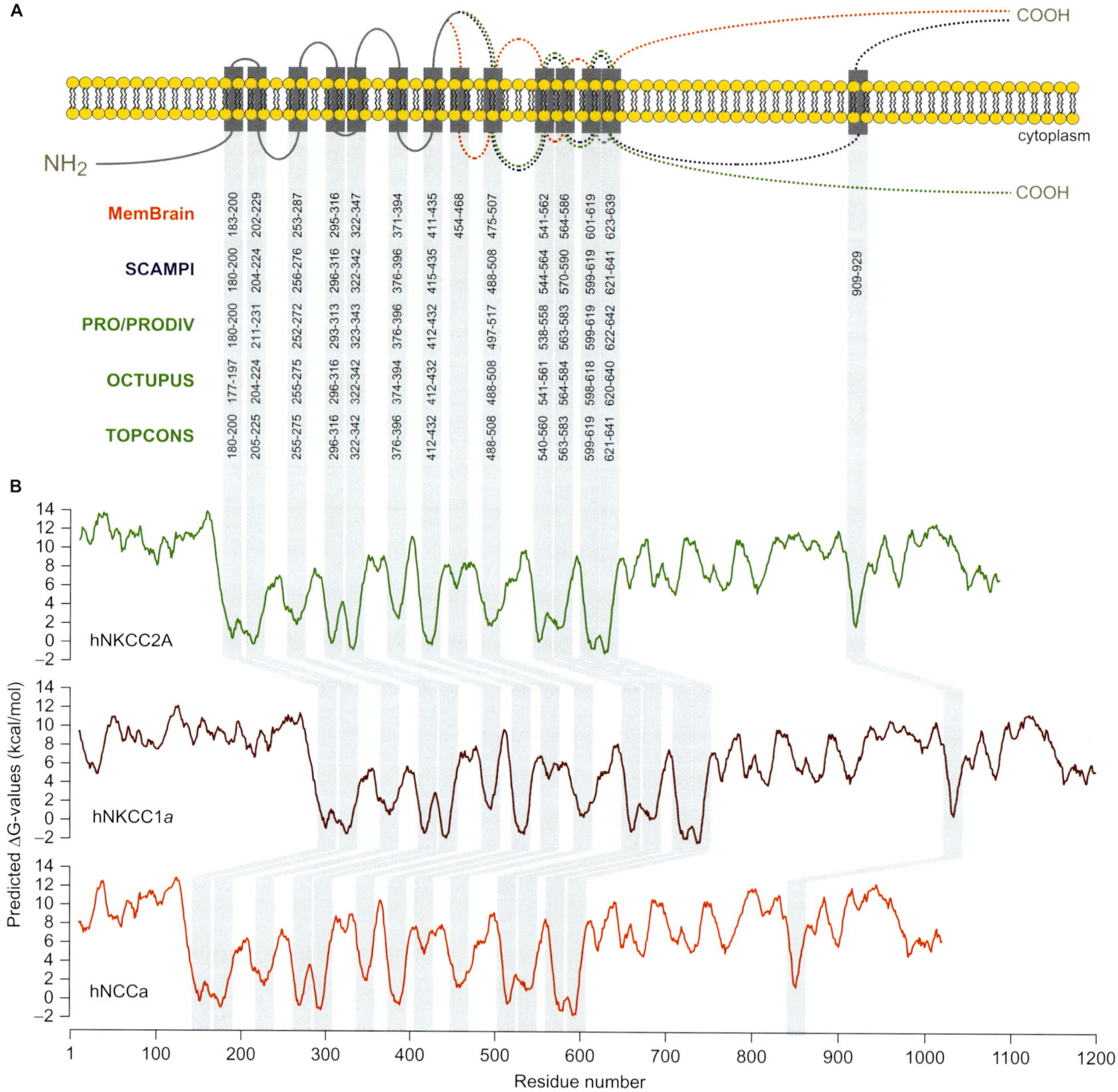

FIGURE 11.4 Consensus prediction of membrane protein topology. The topological information of hNKCC2A, hNKCC1a and hNCCa proteins (GenBank ABU69043, AAC50561 and AAC50355, respectively) was generated by using five different algorithms: *SCAMPI*, *OCTUPUS*, *ZPRED*, *PRO/PRODIV-TMHMM* (topcons.cbr.su.se/) and *MemBrain*, an algorithm used for predicting ends of TM domains that are shorter than 15 residues. **A.** Predicted topology of NKCC and NCC proteins according to the algorithms used (*MemBrain* (red), *SCAMPI* (blue), *PRO/PRODIV* and *TOPCONS* (green)). Predicted TM domains are indicated as gray boxes across the lipid bilayer. The NKCC/NCC amino acid location predicted to be located in each TM is numbered underneath each transmembrane domain and varies according to the algorithm used. The continuous gray line represents the amino acid chain of the NKCC/NCC proteins whereas dotted lines represent the potential topologies according to the algorithm used. The cytoplasmic N-terminal and C-terminal portions of NKCCs/NCC are indicated. **B.** Predicted total free energy (ΔG) values of each residue in hNKCC2A (top), hNKCC1a (center) and hNCCa (bottom) protein sequences.

respectively), which is not supported by current experimental evidence.

II. THE SLC12A1 GENE ENCODING FOR Na^+-K^+-$2Cl^-$ COTRANSPORTER 2 (NKCC2)

A. Genomic Organization of Mammalian SLC12A1 (NKCC2) Gene

The genomic organization of the mammalian NKCC2 gene is known for several species (*Homo sapiens* (NC_000015), *Pan troglodytes* (NC_006482), *Canis lupus familiaris* (NC_006612), *Mus musculus* (NC_000068), *Rattus norvegicus* (NC_005102), *Macaca mulatta* (NC_007864), *Bos taurus* (NC_007308) and *Equus caballus* (NC_009144)). The mammalian NKCC2 genes are represented in Fig. 11.5A. The putative mammalian NKCC2 homologs are genes of ~80–120 kb that encode for proteins of 1095–1100 residues. Bioinformatic DNA sequence analysis and molecular alignment with EST/cDNA clones show that the human NKCC2 gene contains 35 different GT-AG introns and the potential to transcribe 12 different transcripts (nine alternatively spliced variants and three unspliced forms). Further, the human NKCC2 gene contains five non-overlapping alternative last exons and ten validated alternative polyadenylation sites.

Comparative genomic analysis of mammalian NKCC2 genes using *ElDorado* (Genomatix, www.genomatix.de) reveals conserved sequences and highly similar structures in all NKCC2 gene orthologs (Fig. 11.5). The 5′ distal genomic region (i.e. ~2 kb upstream the putative A^1TG) of most mammalian NKCC2 genes contains at least two predicted promoters and a myriad of discrete nucleotide sequences that are >80% identical to consensus binding sequences for a wide variety of transcription factors (see Table 11.2 and Fig. 11.6). It is important to note that *in silico* predicted individual transcription binding sites in a gene promoter are *never* sufficient to indicate transcriptional function (Cartharius et al., 2005). However, all known mammalian NKCC2 promoter regions share nucleotide sequences which are 100% identical to the core consensus sites for many transcription factors. This appears to be the case for CREB (cAMP-responsive element binding proteins), HIFF (hypoxia inducible factor), HEAT (heat shock factors), GRE (glucocorticoid responsive and related elements), E2FF (E2F-myc activator/cell cycle regulator), HNF-1 (hepatic nuclear factor 1) to name a few of them. These findings suggest common and conserved regulatory mechanisms for the transcriptional regulation of the NKCC2 gene.

B. Expression of NKCC2 Genes

When some of the NKCC2 cDNAs were first cloned, Northern blot analysis of rat, rabbit and mouse tissues revealed expression of NKCC2 transcripts only in kidney (Gamba et al., 1994; Igarashi et al., 1995; Payne and Forbush, 1994). Moreover, cloning and characterization of several different full length mouse NKCC2 cDNAs obtained from kidney have provided conclusive evidence that at least four mNKCC2 isoforms are expressed in mouse kidney (Igarashi et al., 1995; Mount et al., 1999).

Today, it is widely accepted that NKCC2, regardless of species and/or variant, is expressed exclusively in the apical membrane of the thick ascending loop of Henle (TALH) in the kidney (Gamba, 2005). This tissue-restricted expression has been attributed to the presence of a tissue-specific promoter located upstream of the mouse NKCC2 gene. Partial functional characterization of this genomic region (*nt* -3585 to -1300 (numbered with respect to A^1TG)) revealed the presence of a transcription initiation site located at *nt* -1330 bp in the mouse NKCC2 gene (Igarashi et al., 1996). This mNKCC2 promoter region exhibited a strong transcriptional activity in kidney cells but not in 3T3 mouse fibroblasts (Igarashi et al., 1996). This observation was interpreted as proof of tissue specificity of NKCC2 expression in kidney.

However, several lines of evidence suggest that NKCC2 may not be restricted to kidney. For instance, two cloned full-length human NKCC2 cDNAs, one from kidney (U58130) and another from rat and human dorsal root ganglia (EF577032 and EF559316, respectively) argue against kidney-specific expression of NKCC2. Moreover, the hNKCC2 gene is supported and defined by 199 *GenBank* accessions from 195 cDNA clones, most of which were obtained from kidney but others came from brain tissues, pooled germ cell tumors, skeletal muscle and other tissues. Similarly, the mouse *slc12a1* gene (NC_000068) is supported by more that 200 EST/cDNA clones obtained from a variety of mouse tissues and protein expression by immunolabeling of non-renal tissues using antibodies against NKCC2 (Gavrikov et al., 2006; Kakigi et al., 2008). Admittedly, these immunolocalization studies in non-renal tissues have not been validated following rigorous criteria like, for example, showing antibody specificity in NKCC2 knockout mice (Rhodes and Trimmer, 2006; Saper, 2005). NKCC2 transcripts have also been detected in olfactory epithelium (Nickell et al., 2007).

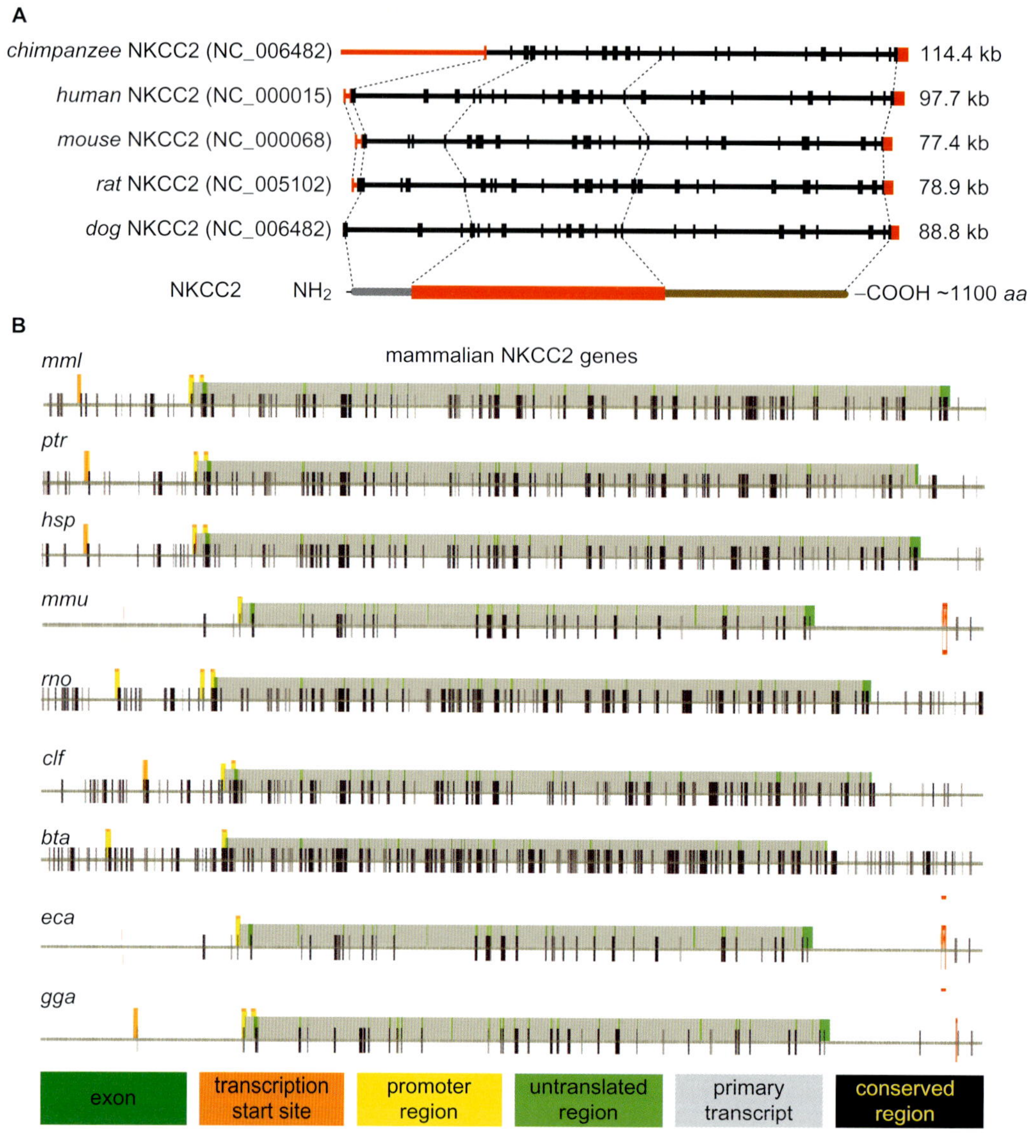

FIGURE 11.5 Comparative genomics of mammalian NKCC2 genes. **A.** In-scale exon-intron organization of known mammalian NKCC2 genes. Coding exons are represented as black boxes superimposed in horizontal lines representing introns. Left or right red boxes and lines represent sequences encoding for 5′- or 3′-UTRs, respectively. Dotted lines mark the exonic regions involved in coding three main protein regions (bottom): N-terminal (gray lane), hydrophobic core (horizontal red box) and C-terminal (brown line). The N- and C-termini are predicted to be intracellular whereas the hydrophobic core is predicted to contain 10–12 TM α-helixes. **B.** Genomic comparison between known mammalian NKCC2 genes using *ElDorado* suite for Comparative Genomics (www.genomatix.com). Genomic sequences are represented for *Macaca mulatta* (mml), *Pan troglodytes* (ptr), *Homo sapiens* (hsp), *Mus musculus* (mmu), *Rattus norvegicus* (rno), *Cannis lupus familiaris* (cfa), *Bos taurus* (bta), *Equus caballus* (eca) and *Gallus gallus* (gga). Each NKCC2 gene shows exons (green), predicted start sites (orange), putative promoter regions (yellow), 5′- and 3′-UTRs (dark green), the longest predicted primary transcript (gray) and the conserved regions (black).

The concept that the NKCC2 gene may be expressed in extra-renal tissues is further supported by a published database which provides microarray expression data from more than 12,000 probes and hybridized to hundreds of different samples from a wide variety of normal human tissues and cell lines (Su et al., 2002). This dataset provides a description of the normal mammalian transcriptome and is posted in the *GEO Profiles* under the accession number GDS596 (www.ncbi.nlm.nih.gov/geo). This database stores individual gene expression and molecular abundance profiles assembled from the *Gene Expression Omnibus* (GEO) repository. The hNKCC2 expression data is summarized in Fig. 11.7. This profiling clearly demonstrates that hNKCC2 transcripts are detected in tissues outside the kidney, although these extra renal NKCC2 transcripts appear to be expressed at low levels when compared to kidney. Although the presence of part of a transcript in a given tissue is not proof of NKCC2 protein expression,

TABLE 11.2 *In silico* analysis of transcription binding sites for the promoter region of the human NKCC2 gene using MatInspector 7.7.3 (Genomatix). The SLC12A1 genomic sequence analyzed encompasses ~620 bp located ~1.3 kb upstream the proposed initiation codon A[1]TG (exon 1) and~100 bp downstream the potential initiation sequence (see Fig. 11.6). The sequences in hNKCC2 gene promoter predicted to bind particular transcription factors are at least 80% identical to the described core consensus sequence for the transcription factor.

Transcription factor	Description	Sequence	Promoter's Position
PAX6	PAX-4/PAX-6 paired domain binding sites	attccACTCttaatctgct	1931
HOXF	Factors with moderate activity to homeo domain consensus sequence	ctctTAATctgcttact	1925
PTF1	Pancreas transcription factor 1, heterotrimeric transcription factor	ttaaTCTGcttacttcccatg	1922
RBPF	RBPJ – kappa	tgcaTGGGaagtaag	1930
OCT1	Octamer binding protein	cccATGCaaaagagt	1907
SF1F	Vertebrate steroidogenic factor	gagtCAAGgtaat	1896
EVI1	EVI1-myleoid transforming protein	caagcaAGATaaaatca	1875
GATA	GATA binding factors	gcaaGATAaaatc	1872
EVI1	EVI1-myleoid transforming protein	aagatAAAAtcatttca	1870
GATA	GATA binding factors	aaaTGATtttatc	1864
GFI1	Growth factor independence transcriptional repressor	taaAATCatttcaga	1866
HEAT	Heat shock factors	tactctgtaactAGAAagtagtgtc	1865
ATBF	AT-binding transcription factor	actactttctAGTTaca	1850
BCL6	POZ domain zinc finger expressed in B-Cells	ctgtaacTAGAaagtag	1842
OCT1	Octamer binding protein	agtatagaAATTgtc	1825
OCT1	Octamer binding protein	cacATGCtaaacaat	1788
FKHD	Fork head domain factors	acatgcTAAAcaatagg	1787
SORY	SOX/SRY-sex/testis determinig and related HMG box factors	gctaaaCAATaggaggt	1783
RORA	v-ERB and RAR-related orphan receptor alpha	acaataggaGGTCactaggactc	1778
EREF	Estrogen response elements	ggagGTCActaggactcat	1772
AP1R	MAF and AP1 related factors	tacagaTGAGtcctagtgacc	1769
AP4R	AP4 and related proteins	tcttacaGATGagtcct	1762
EVI1	EVI1-myleoid transforming protein	tcttacaGATGagtcct	1762
HNF6	Onecut homeodomain factor HNF6	ggtggaacTCAAtttat	1744
SORY	SOX/SRY-sex/testis determinig and related HMG box factors	ataAATTgagttccacc	1744
EVI1	EVI1-myleoid transforming protein	gggaataGATAaattga	1736
CLOX	CLOX and CLOX homology (CDP) factors	caatttATCTattcccaaatgac	1735
GATA	GATA binding factors	aataGATAaattg	1735
PIT1	GHF-1 pituitary specific pou domain transcription factor	caatTTATctatt	1735
PERO	Peroxisome proliferator-activated receptor	agtcatttgggaatAGATaaatt	1734
IKRS	Ikaros zinc finger family	atttGGGAataga	1728
PAX6	PAX-4/PAX-6 paired domain binding sites	tatTCCCaaatgactaatg	1726
AP1R	MAF and AP1 related factors	ttgccaTTAGtcatttgggaa	1726
BRNF	Brn POU domain factors	cccaaatgacTAATggcaa	1722
TCFF	TCF11 transcription factor	GTCAttt	1719
HOXF	Factors with moderate activity to homeo domain consensus sequence	tgacTAATggcaatgct	1716
NR2F	Nuclear receptor subfamily 2 factors	tgactaatggcaatGCTCatatttt	1716
THAP	THAP domain containing protein	actaatGGCAa	1714
OCT1	Octamer binding protein	ctAATGgcaatgctc	1713
SRFF	Serum response element binding factor	atgctcaTATTttttggac	1704
NKXH	NKX homeodomain factors	tcccTTAAcagcaat	1684
NKXH	NKX homeodomain factors	gctgTTAAgggatcc	1681
ATBF	AT-binding transcription factor	tctccttatgAATTaaa	1663
SORY	SOX/SRY-sex/testis determinig and related HMG box factors	tttAATTcataaggaga	1663
BRNF	Brn POU domain factors	tccttatgaATTAaagaaa	1661
HOXF	Factors with moderate activity to homeo domain consensus sequence	tcttTAATtcataagga	1661

(Continued)

TABLE 11.2 Continued

Transcription factor	Description	Sequence	Promoter's Position
BRNF	Brn POU domain factors	gtttcttTAATtcataagg	1660
LHXF	Lim homeodomain factors	agtttctTTAAttcata	1657
NKX6	NK6 homeobox transcription factors	tTTAAttcata	1657
RBIT	Regulator of B-Cell IgH transcription	tatgaATTAaaga	1657
HOMF	Homeodomain transcription factors	gagtttcttTAATtcat	1656
HNF6	Onecut homeodomain factor HNF6	aaagaaacTCTAtgtta	1649
OCT1	Octamer binding protein	tctatgttACTTcct	1641
ETSF	Human and murine ETS1 factors	tttagacAGGAagtaacatag	1640
CLOX	CLOX and CLOX homology (CDP) factors	ccatTAATcctttagacaggaa	1631
HOXF	Factors with moderate activity to homeo domain consensus sequence	ccatTAATcctttaga	1625
BRNF	Brn POU domain factors	aaaccatTAATcctttag	1624
HOXC	HOX - PBX complexes	aaaaGGATtaatggttt	1622
HOXF	Factors with moderate activity to homeo domain consensus sequence	ggatTAATggtttacta	1618
HNF1	Hepatic Nuclear Factor 1	gATTAatggtttactac	1617
NKX6	NK6 homeobox transcription factors	aTTAAtggttt	1616
HOXF	Factors with moderate activity to homeo domain consensus sequence	ctagtagTAAAccatta	1614
ATBF	AT-binding transcription factor	gtttactactAGTTacc	1609
NKXH	NKX homeodomain factors	aaaCTTAagtaccac	1585
NFAT	Nuclear factor of activated T-cells	tgaGGAAacttaagtacca	1584
TBPF	TATA-binding protein factor	tggtactTAAGtttcct	1584
RU49	Zinc finger transcription factor RU49, zinc finger proliferation 1 - Zipro1	aAGTAcc	1583
NKXH	NKX homeodomain factors	gtaCTTAagtttcct	1582
RORA	v-ERB and RAR-related orphan receptor alpha	gccccaccagGTCAttagccctg	1567
FXRE	Farnesoid X - activated receptor response elements	AGGTcattagccc	1565
GREF	Glucocorticoid responsive and related elements	gtggggcttacaGTACtat	1551
PLZF	C2H2 zinc finger protein PLZF	gctTACAgtactatt	1546
RU49	Zinc finger transcription factor RU49, zinc finger proliferation 1 - Zipro1	cAGTAct	1541
COMP	Factors which cooperate with myogenic proteins	caatgATTGgcctggaatagtac	1539
ZF35	Zinc finger protein ZNF35	gcctggAATAgta	1538
CLOX	CLOX and CLOX homology (CDP) factors	ttccaggcCAATcattgcacagt	1533
PBXC	PBX1 - MEIS1 complexes	gcaaTGATtggcctgga	1532
CAAT	CCAAT binding factors	caggCCAAtcattgc	1530
GFI1	Growth factor independence transcriptional repressor	gccAATCattgcaca	1527
AP1R	MAF and AP1 related factors	aagagaTGACtgtgcaatgat	1523
HEAT	Heat shock factors	tcatctctttcaAGAAagtttctaa	1511
BCL6	POZ domain zinc finger expressed in B-Cells	aactttcTTGAaagaga	1508
MYT1	MYT1 C2HC zinc finger protein	agaaAGTTtctaa	1499
CLOX	CLOX and CLOX homology (CDP) factors	tttctaacCAATaagcagaagcc	1493
CAAT	CCAAT binding factors	ctaaCCAAtaagcag	1490
CLOX	CLOX and CLOX homology (CDP) factors	ccaaTAAGcagaagcctaacttt	1488
HEAT	Heat shock factors	tagctttcagccAGAAagttaggct	1474
MYT1	MYT1 C2HC zinc finger protein	agaaAGTTaggct	1474
NF1F	Nuclear factor 1	tttcTGGCtgaaagctaagtg	1466
BRAC	Brachyury gene, mesoderm developmental factor	gggacacttAGCTttcagcca	1462
BRAC	Brachyury gene, mesoderm developmental factor	ctgaaagctAAGTgtcccata	1459
MYT1	MYT1 C2HC zinc finger protein	tgaaAGCTaagtg	1458
NKXH	NKX homeodomain factors	aaagctaAGTGtccc	1456
CLOX	CLOX and CLOX homology (CDP) factors	cgtcacccaggaatatATGGgac	1446
RUSH	SWI/SNF related nucleophosphoproteins with a RING finger DNA binding motif	tcCCATatatt	1445

(Continued)

TABLE 11.2 Continued

Transcription factor	Description	Sequence	Promoter's Position
TEAF	TEA/ATTS DNA binding domain factors	ataTATTcctggg	1441
ZF35	Zinc finger protein ZNF35	cccaggAATAtat	1441
AP1R	MAF and AP1 related factors	tcagcgTCACccaggaatata	1440
CREB	cAMP-responsive element binding proteins	attcctgggTGACgctgaagc	1437
CREB	cAMP-responsive element binding proteins	cctgggTGACgctgaagcagg	1434
WHNF	Winged helix binding sites	gtgACGCtgaa	1429
NFAT	Nuclear factor of activated T-cells	tgaGGAAatgtaacctgct	1419
PARF	PAR/bZIP family	tgaggaaatGTAAcctg	1417
ETSF	Human and murine ETS1 factors	tcttctgaGGAAatgtaacct	1416
STAT	Signal transducer and activator of transcription	cttcTTCTgaggaaatgta	1412
STAT	Signal transducer and activator of transcription	acatTTCCtcagaagaagg	1411
HEAT	Heat shock factors	catttcctcagaAGAAggctccttg	1410
BRAC	Brachyury gene, mesoderm developmental factor	aaggctcctTGGTggtaagtt	1396
CAAT	CCAAT binding factors	accaCCAAggagcct	1395
SNAP	snRNA-activating protein complex	ttcaaCTTAccaccaagga	1391
P53F	p53 tumor suppressor	tttacCCTGcccagaaatgttca	1376
MEF2	MEF2, myocyte-specific enhancer binding factor	gggcagggTAAAaatcatactta	1365
FKHD	Fork head domain factors	gcagggTAAAaatcata	1363
HOXC	HOX - PBX complexes	agtaTGATttttaccct	1361
BRNF	Brn POU domain factors	aaatCATActtaatattat	1354
NKXH	NKX homeodomain factors	aatattAAGTatgat	1352
NKXH	NKX homeodomain factors	atactTAATattatg	1349
OCT1	Octamer binding protein	gaCATAatattaagt	1347
CREB	cAMP-responsive element binding proteins	actttaTGACataatattaag	1346
FKHD	Fork head domain factors	tatgacATAAtattaag	1346
PARF	PAR/bZIP family	taatattatGTCAtaaa	1344
PARF	PAR/bZIP family	ctttatgacATAAtatt	1343
HOXF	Factors with moderate activity to homeo domain consensus sequence	tatgtcaTAAAgtagat	1338

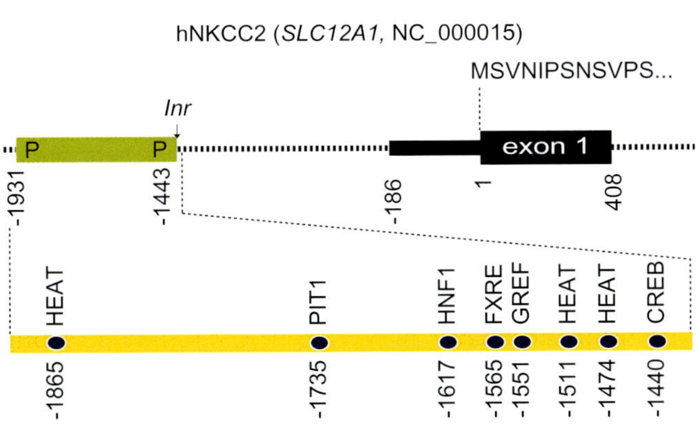

FIGURE 11.6 The promoter region of the human NKCC2 gene. Two promoter sequences (P) in the form of TATA boxes are shown in the 5′ distal genomic region (i.e. -1931 to -1443, upstream A[1]TG). Gene transcription is predicted to start at *Inr*, the initiator sequence located ~1440 bp upstream A[1]TG. Potential binding sites for selected transcriptions sites (e.g. CREB, PIT1, HEAT, GREF, FXRE and HNF1) are represented as blue circles over the promoter region comprising sequences -2000 to -1400. A detailed list of potential binding sites for transcription factors is shown in Table 11.2.

the functional sigificance of NKCC2 outside the kidney is unknown.

C. How Many NKCC2 Variants?

1. Mouse NKCC2 Variants

The presence of more than one mouse NKCC2 transcript is supported by cloning and functional characterization of several mNKCC2 cDNAs. Comparative analysis of full-length mouse NKCC2 cDNAs (U20973, U20974, U20975 (Igarashi et al., 1995), U61381 (Mount et al., 1999), U94518 and BC016888 (Strausberg et al., 2002)) with respect to the mouse NKCC2 genomic sequence demonstrates that these transcripts are the result of an alternative combination of three exons (exon 4 (NKCC2B); exon 5 (NKCC2A); and exon 6, (NKCC2F)) and an independent alternative stop codon expressed in exon 18, which would encode for a shorter NKCC2 protein at the C-terminus (Mount et al., 1999) (Fig. 11.8A).

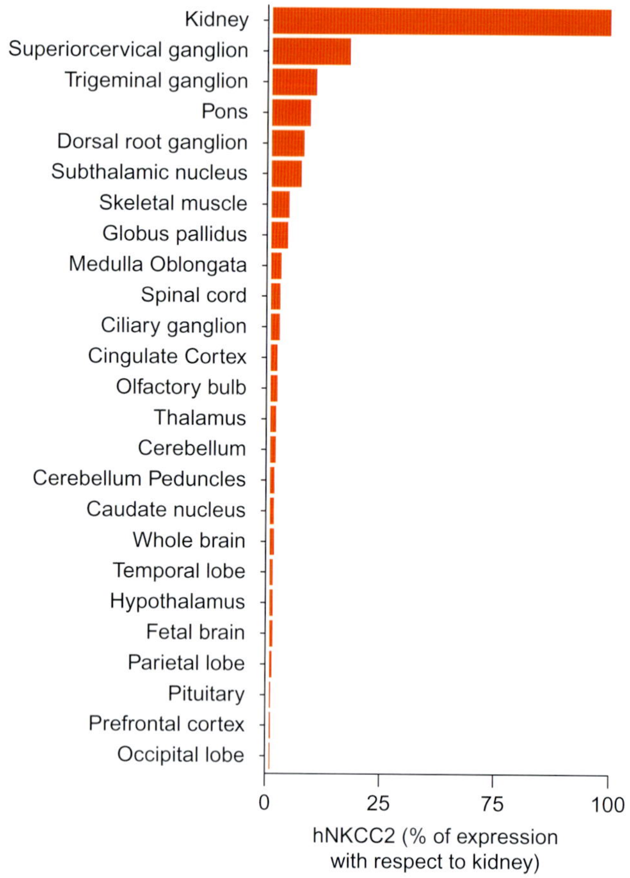

FIGURE 11.7 Expression profile of hNKCC2 transcripts. Expression results were obtained from GEO Profiles (record GDS596). The gene expression profile of hNKCC2 transcripts was established by using the *Affymetrix* probe *220281_at*, created from the hNKCC2 EST AI632015 which sequence corresponds to the 3′-UTR sequences of all known hNKCC2 transcripts. Results are shown as percentage of expression with respect to kidney (100%).

The mNKCC2A mRNA (U20973 (Igarashi et al., 1995)) can be reconstructed from 66 cDNA clones obtained from kidney and one from eye/retina (EL606477). This transcript covers 28 exons (when counting exon -1, see below section II.D) and is predicted to encode for a protein of 1095 residues (~120 kDa). Comparative *in silico* analysis of this cDNA sequence with the mouse gene reveals that U20973 (mNKCC2A) encodes some amino acid variations (e.g. A30S, D48G, M97F, D98G, A99H and others) that could be the result of polymorphic nucleotides present in the mice gene. An additional cDNA obtained from mouse kidney also encodes for mNKCC2A: BC016888 (Strausberg et al., 2002). This transcript contains some amino acid variations when compared with the mouse genomic sequence (e.g. M97F, D98G, A99H, V100N, P101T) and is 25 bp shorter than U20973′ 5′-UTR (210 bp vs. 232 bp). The 3′-UTR of the cDNAs encoding for mNKCC2A contains a typical polyadenylation signal (AATAAA) located 31 bp before the polyA tail (only present in BC016888).

Two shorter variants of mNKCC2A (here named mNKCC2A-*s1* (U61381) (Mount et al., 1999) and mNKCC2A-*s2* (AK052750)) were cloned from mouse kidney (Fig. 11.8A). The mNKCC2A-*s1* transcript starts with 200 bp of 5′-UTR, continues with a sequence encompassing exons 1 to 18 and ends with ~600 bp of 3′-UTR encoded by intron 17. This last portion of the transcript encodes for 56 unique residues and two potential polyadenylation signals located 97 bp and 420 bp downstream of the stop codon. The polyA tail is located ~30 bp from the last putative polyA signal. The cDNA coding for mNKCC2A-*s2* starts with a 5′-UTR of 252 bp and ends with exon A (exon 5). The 3′-UTR of this transcript is ~0.9 kb long and matches exactly intron 5, exon F (exon 6) and intron 6 of the mouse NKCC2 gene. Interestingly, this 3′-UTR region does not appear to encode for potential polyA signals. It is not known if this transcript encodes for a protein *in vivo* (predicted to be of 247 residues, ~27 kDa).

Only one full-length cDNA clone encoding for mNKCC2B has been cloned: U20974 (Igarashi et al., 1995) (Fig. 11.8A). This cDNA can be entirely reconstructed from 37 kidney cDNA clones as well as three clones from liver (i.e. AW259785, AW259766 and AW260385). The mNKCC2B transcript covers 28 exons of the gene and predicts encoding of a protein of 1095 residues (~120 kDa) with some amino acid variations that could reflect genetic polymorphisms of the mice NKCC2 gene. The 5′-UTR of the mNKCC2B mRNA is 230 bp long whereas the 3′-UTR is ~1.1 kb which encodes for a standard polyadenylation signal at 31 bp before the polyA.

The mNKCC2F is encoded by two cDNAs cloned from mouse kidney: U20975 (Igarashi et al., 1995) and U94518 (Fig. 11.8A). Both transcripts encompass 27 exons and can be reconstructed from ten cDNA clones obtained from kidney. Their nucleotide sequences only differ in a few base pairs. However, when compared to the mouse NKCC2 genomic sequence both transcripts encode amino acid variations (i.e. U94518: M97F, D98G, A99H, V100N, P101T and U20975: A30S, D48G, M97F, D98G, A99H). The 5′- and 3′-UTRs of both transcripts are ~233 bp and ~1.1 kb long, respectively. Both 3′-UTRs encode for a classical polyadenylation signal.

A non-coding mNKCC2 cDNA (named here mNKCC2A-S (Fig. 11.8A), probably incomplete at its 5′-end) was cloned from kidney (BC051100 (Strausberg et al., 2002)). There is no recognizable 5′-UTR in this transcript and sequences comprising exons 1–8 are missing. The BC051100 transcript starts from the 75th nucleotide of exon 9 and ends at mNKCC2A, B or F

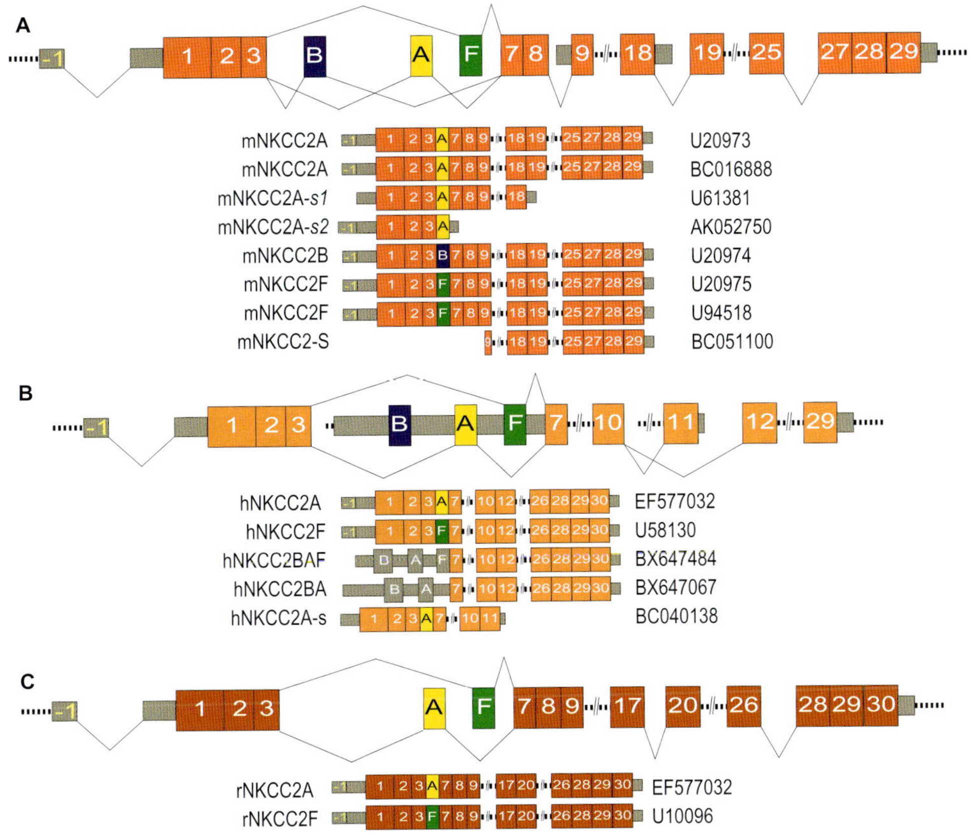

FIGURE 11.8 The transcripts encoded by mouse, human and rat NKCC2 genes. Diagramatic representation of full-length NKCC2 transcripts cloned from mouse **A**. human **B**. and rat **C**. sources. Each exon is represented as a colored box when coding whereas gray boxes represent 5′- or 3′-UTRs. Each transcript is identified by its unique GenBank accession number which is indicated.

transcripts. As it is, this transcript is not predicted to encode for a protein. The functional role of this transcript as well as whether the region comprising intron 8-exon 9 has promoter activities are unknown.

The mNKCC2 splice variants A, B, F and A-*s1* appear to be differentially expressed in the mouse kidney (Mount et al., 1999; Payne and Forbush, 1994) as discussed in Chapter 16 in this volume. They differ among themselves in the amino acid sequence of most of the predicted transmembrane (TM) domain 2 and in the intracellular loop interconnecting TM2 and TM3 (Figs 11.3 and 11.4). This region is encoded by exons A, B or F, which has been proposed to be involved in ion binding and translocation of the transporter (Plata et al., 2002).

2. *Human NKCC2 variants*

Five full-length human NKCC2 cDNAs have been cloned from different sources (adult and fetal kidney and dorsal root ganglion cells). Two of them, U58130 (Simon et al., 1996a) and EF559316, differ only in the presence of a unique 96 bp encoded by exon 6 (exon F) or 5 (exon A), respectively, whereas the cDNA clones BX647067 (hNKCC2BA) and BX647484 (hNKCC2BAF, from human fetal kidney) are two variants that may be produced by the election of two potential alternative promoters located in the third intron of the human NKCC2 gene (Fig. 11.8B).

The hNKCC2F splice variant appears to be encoded by the U58130 cDNA (Simon et al., 1996a). It can be reconstructed from more than 100 cDNA clones, most of them obtained from kidney and others from nervous tissue (brain (AL036755), corpus callosum (DA286101), and hippocampus (DA324269)) and liver (DB182625). The hNKCC2F mRNA expresses 26 exons and is predicted to encode for a protein of 1099 residues (~120 kDa). When the coding region of U58130 cDNA is compared to the genomic sequence, it can be seen that the cDNA encodes for an hNKCC2F protein with some amino acids variations (e.g. Q75H, G211L, A212G, Y213V, Y214I, www.uniprot.org/uniprot/Q13621). This could reflect genetic polymorphisms in the human NKCC2 gene. The 5′-UTR of hNKCC2F transcript is 216 bp long whereas its 3′-UTR contains ~0.5 kb followed by the polyA tail. There is no standard polyadenylation signal in the transcript. However, there are several canonical polyadenylation sites located in the last exon of the hNKCC2 gene, suggesting that the U58130 cDNA clone may be incomplete at its 3′-UTR.

There are two additional cDNAs encoding for hNKCC2: hNKCC2BAF and hNKCC2BA which contain very long (~5 kb long) 5′-UTRs (Fig. 11.8B). They are among the 5% longest UTRs posted in *GenBank* and, as such, they may be poorly translated (Davuluri et al., 2000). The 3′-UTR region of hNKCC2BAF and hNKCC2BA contain ~1.1 kb followed by a polyA tail implying that they are full-length cDNA clones. A classical polyadenylation signal in these cDNAs is expressed ~20 bp before the polyA. Both hNKCC2BAF (BX647067) and hNKCC2BA (BX647484) mRNAs are predicted to encode for proteins of 870 and 826 residues, respectively.

The potential of the hNKCC2 gene to generate additional transcripts, besides the ones mentioned above with alternative stop codons, is supported by a full-length hNKCC2A-s cDNA (BC040138), obtained from kidney (Strausberg et al., 2002). The 5′-UTR of this hNKCC2A cDNA is 186 bp long whereas its 3′-UTR is 285 bp followed by the polyA. The standard AATAAA polyadenylation signal is seen about 21 bp before the polyA tail. This cDNA would encode for a shorter protein of 430 amino acids as a result of the inclusion of an alternative stop codon located in exon 11. The function of all these additional hNKCCs, if any, is unknown.

Full-length hNKCC2A and hNKCC2B cDNAs have not yet been cloned from kidney. However, analysis of human EST database using exon 5 (exon A) or 4 (exon B) of the hNKCC2 gene reveals the existence of three "A" and two "B" clones from kidney (A: DC328468, DB471453 and DB479842, B: DC362459 and BG433923) suggesting that hNKCC2A and hNKCC2B, as expected, may be expressed in the human kidney (Fig. 11.8B). As mentioned above a full length hNKCC2A cDNA has been cloned from human dorsal root ganglion cells (EF559316 (Di Fulvio et al.)). The function of hNKCC2A remains to be elucidated.

3. *Rat NKCC2 Variants*

Two full-length rat NKCC2 cDNAs have been cloned from two different sources, one from kidney (U10096 (Gamba et al., 1994)) and another from the dorsal root ganglia (EF577032 (Di Fulvio et al., 2007)) (Fig. 11.8C). When these two cDNAs are compared *in silico* with the genomic sequence of the rat NKCC2 gene it is found that: U10096 encodes for rNKCC2F, whereas EF577032 encodes for rNKCC2A. The transcript EF577032 includes a 5′-UTR of 199 bp, a full open reading frame encoded by the following exons: 1-3, A, 7–17, 20–26 and 28–30, and ~700 bp of 3′-UTR. Although rNKCC2A and rNKCC2F cDNA sequences have a high percentage of identity (~98%), they show differences in their primary nucleotide sequence beyond the presence/absence of exon A or F, respectively. Indeed, three CG > GC changes are found in rNKCC2A which affect codons R^{261}, R^{267} and R^{430} (R > A) as known for the rNKCC2F sequence (U10096) (Gamba et al., 1994). These dinucleotide changes are also found in rat and mouse kidney ESTs (e.g. AI411202, AW012023, BF781835, CK482708, CK481236 and CK481074). It is not known if these changes are polymorphic. Interestingly, a GT dinucleotide insertion in position 3565 of rNKCC2A which modifies a GU-rich region in the 3′-UTR of the NKCC2A transcript (5′-UGUAGAUGU-3′), is located 32 bp upstream of a potential polyA signal region. This raises the possibility that additional polyA sites exist giving 3′-UTR heterogeneity to NKCC2 transcripts. The rNKCC2F has been extensively characterized by functional expression in oocytes (Gamba, 2005). The functional properties of rNKCC2A have not yet been studied.

D. Are there More NKCC2 Variants?

The answer is yes. The 5′-UTR of mouse, human and rat NKCC2 transcripts may be subjected to additional alternative splicing events. The transcription initiation site of the mouse NKCC2 gene was mapped -1330 bp from A^1TG (Igarashi et al., 1996). However, full-length NKCC2 cDNAs cloned from mouse, rat or rabbit kidneys contain ~200–280 bp of 5′-UTR (Di Fulvio et al., 2007; Gamba et al., 1994; Igarashi et al., 1995; Payne and Forbush, 1994). This suggests that ~1.1 kb of genomic DNA is absent in 5′-ends of mammalian NKCC2 transcripts. *In silico* analysis of ~2 kb genomic sequence upstream of the mammalian NKCC2 gene using several algorithms (*GeneSplicer* (Pertea et al., 2001), *NetGene2* (Brunak et al., 1991), *TwinScan* (Korf et al., 2001), *GenScan* (Burge and Karlin, 1997), *FirstEF* (Davuluri et al., 2001) and *NetUTR* (Eden and Brunak, 2004)) predicts the existence of a potential non-coding first exon (here referred to as *exon -1*, of ~80 bp in the rat, and its *intron -1* of ~1 kb) located ~70 bp from the putative TATA box and ~40 bp from a potential initiator sequence (Fig. 11.9A). Although the existence of this exon was previously recognized in mNKCC2 gene (Igarashi et al., 1996), its potential implications have not been addressed (see below). Evidence for this 5′-UTR splicing comes from *in silico* alignment of full-length NKCC2 transcripts cloned from rodent kidney cDNA libraries and its comparison with their respective genomic sequences. As shown in Fig. 11.9A, *in silico* results demonstrate that rat, mouse and rabbit NKCC2F cDNAs lack ~1.1 kb of genomic sequence due to alternative splicing. A similar result was obtained when aligning human ESTs (DC362018, DA636429, DA636076, DA629993, DA629479, DA629375, AI792946, BX092569, BP276654 and BP274908). Furthermore, evidence that exon -1 (i.e. splicing of intron -1, from *nt* -1159 to -194 in the rat NKCC2 gene) is expressed *in vivo* is supported by the existency of a rat EST clone (CK483746).

Interestingly, the genomic sequence encompassing *nt* -1270 to -194 (i.e. the sequence containing exon -1 and intron -1) in rat NKCC2 gene contains several regulatory sequences that resemble a typical promoter. This region contains a potential TATA-binding protein (TBP) element at *nt* -246, and binding sites for transcription factors such as the ubiquitous SP-1, USF, YY1, AP-1, NF κ-B, tissue-restricted transcription factors such as C/EBPα, HNF-1, Oct-6, HNF-3, SRY, as well as hormone-regulated transcription factors (e.g. GRE, GR, PR, ER). Most of these binding elements are conserved in mouse, rat and human; they are found in their corresponding regions of the NKCC2 genes (i.e. *nt* -1295 to -198 for mouse and -1389 to 186 for human) (Fig. 11.9B). The presence of conserved and highly identical TBP sites, together with several binding sites for regulatory factors in the genomic region comprising exon/intron -1 of the NKCC2 genes may confer promoter activity to this region or regulatory properties to transcribed mRNAs.

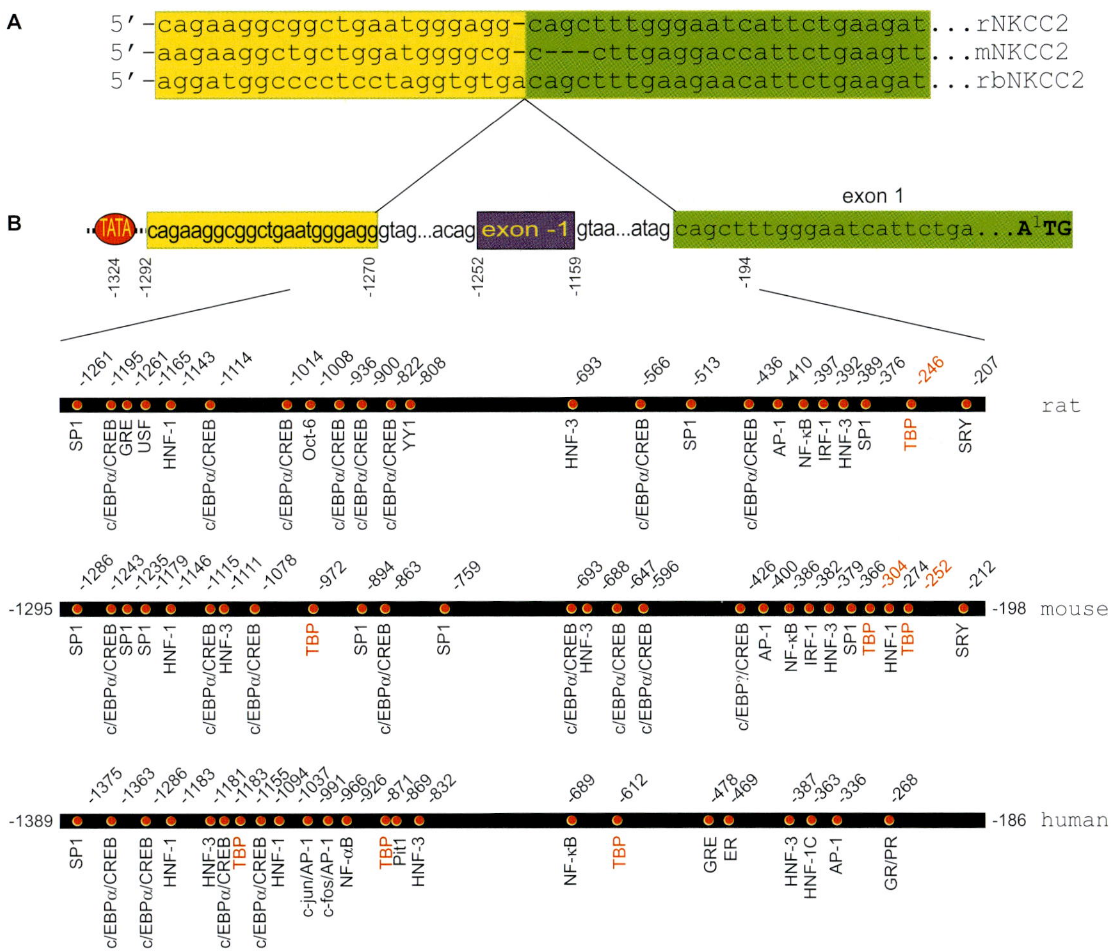

FIGURE 11.9 The 5′-UTR splicing of NKCC2 cDNAs. **A.** 5′-UTR of rat, mouse and rabbit NKCC2s (rNKCC2F (U10096), mNKCC2F (U20975) and rbNKCC2 (U07549), respectively) compared against the genomic sequence of rat or mouse NKCC2 genes. The nt positions of TATA box and the first codon ATG are indicated based on the rat NKCC2 gene sequence (NC_005102). Exon -1 and 1 (blue and green boxes, respectively) are shown, as well as the consensus alternative splicing sites. **B.** Potential binding sites for transcription factors predicted to be located between *nt* -1270 and -194 of the rat NKCC2 gene (equivalent positions in the mouse and human NKCC2 gene are indicated). The TATA box-binding protein element (TBP) is indicated in red. Information obtained at high astringency (minimum matrix conservation <80%, minimum number of sites = 3, matrix with in bp = 10 and pairsim to known sites = 50) by using *AliBaba v2.1* server and TRANSFAC database v2.3.1.0 (www.gene-regulation.com).

E. Regulators of NKCC2 Function

The regulation of NKCC2 protein function at the molecular level has been studied in detail in *Xenopus laevis* oocytes expression system. However, most of what we know about NKCC2 function comes from expression of rNKCC2F in *Xenopus* oocytes (Gamba, 2005), as discussed in Chapter 16 in this volume. It is important to note that there are many studies on regulation of NKCC2 done prior to the "cloning age", but discussing them is beyond the scope of the present chapter.

Pharmacological manipulations of phosphatases or kinases modulate NKCC2 activity in oocytes over expressing rNKCC2F. This has been the basis for proposing that phosphorylation of NKCC2 modulates its cotransport activity. Unlike other SLC12A members, attempts to heterologously express NKCC2 cDNAs in mammalian cells have failed (Isenring et al., 1998; Payne and Forbush, 1994), until recently (Benziane et al., 2007). Although *Xenopus* oocytes are still widely used as heterologous expression system, results obtained from this kind of cell must be interpreted with caution; proteins expressed in oocytes can act very differently when expressed in mammalian cells (Hendriks et al., 2004) as discussed in Chapter 9 in this volume.

F. Predicted Regulators of NKCC2 Gene Expression

Little is known about the mechanisms controlling NKCC2 gene expression (Gamba, 2005). Regulation of gene expression can be effected at many levels in the pathway from DNA to RNA to protein: (1) transcriptional, (2) post-transcriptional, e.g. splicing, translation, etc. and (3) post-translational, e.g. N-glycosylation, phosphorylation, etc. (Watson, 2004). The initiation of RNA transcription is the most important point of control. As shown in Table 11.2, the promoter of the human NKCC2 gene contains several consensus binding sites for a wide variety of transcription factors. While all of the binding sites are >85% identical to canonical consensus sites for transcription factors, some of them are conserved in all mammalian NKCC2 gene promoters (e.g. HNF-1, CREB, SP-1, AP-1) whereas others are not, suggesting that ortholog genes may not be identically regulated.

The HNF-1, a transcription factor involved in the regulation of several genes in various tissues, was proposed to participate in the "tissue-specific" expression of NKCC2 in mouse kidney (Igarashi et al., 1996). However, deletion of the putative HNF-1 responsive element located ~160 bp downstream of the TATA box in the mNKCC2 gene promoter (*nt* ~1520 upstream of A^1TG) did not affect its basal transcriptional activity (Igarashi et al., 1996). Further, mice with homozygous deletions of HNF-1 present a Lignac-de Toni-Debré-Fanconi syndrome characterized by polyuria, glucosuria, phosphaturia and generalized aminoaciduria. This syndrome results from multiple transport dysfunctions mainly restricted to the renal proximal tubules (Pontoglio et al., 1996) but it is not related to NKCC2, which is expressed in the thick ascending loop of Henle (Payne and Forbush, 1994).

Mammalian NKCC2 promoters share a conserved and highly identical 21 bp nucleotide sequence that may be recognized by the cAMP-responsive element binding protein (CREB). Two overlapping CREB sites are located very close to the TATA box in the human NKCC2 gene promoter. However, the role of cAMP in transcriptional regulation of NKCC2 gene is unknown. Interestingly, a predicted CREB binding site is located at -1332 bp (with respect to A^1TG), but ~80 bp downstream of the TATA box of the human NKCC2 gene (Figs 11.6 and 11.9B). Indirect evidence suggest that NKCC2 protein and activity increase after chronic administration of synthetic antidiuretic hormone, a known upregulator of cAMP in the kidney (Hebert et al., 1981; Molony et al., 1987). Interestingly, the CREB site appears to be spliced out in human, mouse, rat and rabbit NKCC2 5′-UTRs raising the possibility that cAMP may regulate this splicing. Taken together these observations suggest that CREB may have differential post-transcriptional and/or translational regulatory properties at the mRNA level. In line with this idea is the finding that only ~3% of total NKCC2 protein is at the cell surface of the thick ascending loop of Henle (TALH) under basal conditions and that surface-accessible NKCC2 increases dramatically in response to brief incubation with dibutyryl-cAMP or forskolin/IBMX (Ortiz, 2006).

Database analysis reveals that only ~15% of all human promoters contain consensus binding sequences for the farnesoid X activated receptor (FXR, ~110 bp downstream TATA), the TCF11 transcription factor (TCFF, ~260 bp downstream TATA) and THAP domain containing protein (THAP, ~260 bp downstream TATA), and the mammalian NKCC2 gene promoter appears to be one of them. Indeed, FXR, TCFF and THAP are predicted to bind human NKCC2 promoter sequences at ~110 bp and ~260 bp downstream TATA box, respectively (ElDorado, Genomatix.com) This suggests that the NKCC2 gene may be transcriptionally regulated by lipids.

G. Predicted Modulators of NKCC2 Function: Phosphorylation

Regulation of NKCC2 transport activity is associated with phosphorylation and/or dephosphorylation events as discussed in detail in Chapter 18 in this volume. Several consensus phosphorylation sites in human, rat, mouse and rabbit NKCC2A proteins are predicted *in silico* (Fig. 11.10). However, some consensus sites may not be conserved in all homologs or are predicted to be expressed within potential transmembrane or extracellular domains (see Fig. 11.3A). A few phosphoresidues (T^{96}, T^{101} and T^{111}) located in the N-terminus of rNKCC2F cotransporter have been implicated in the modulation of NKCC2 function (Ponce-Coria et al., 2008).

When expressed in *Xenopus laevis* oocytes, rNKCC2F, rabbit and mouse NKCC2s typically exhibit a high basal level of activity that can be enhanced by oocyte shrinkage and pharmacological maneuvers aimed at increasing intracellular phosphorylation reactions (Gamba, 2005). The first evidence of a role of phosphorylation in the regulation of NKCC2s function was obtained using whole kidneys of mice treated with vasopressin (Gimenez and Forbush, 2003) and an antibody (R5) predicted to recognize the phosphorylated threonines in hNKCC1 located within residues 208–223 (Y Y L R T F G H N T M D A V P R) (Flemmer et al., 2002). This region contains two conserved threonine residues: T^{212}/T^{217}, corresponding to T^{206}/T^{211}, T^{203}/T^{208} and T^{184}/T^{186} in mouse, rat and shark NKCC1s, respectively, as well as to T^{96}/T^{101}, T^{100}/T^{105} and T^{99}/T^{104} in rat/mouse, human and rabbit NKCC2s, respectively. These T-residues in NKCC2 are indicated with red arrowheads in Fig. 11.10. Since this hNKCC1 region is 93% identical to the corresponding region of NKCC2 proteins, and residue T^{189} was found in the shark homolog NKCC1 (skNKCC1) to be necessary for the protein to function (Darman and Forbush, 2002), it was reasonable to hypothesize that the R5 antibody cross-reacted with NKCC1s and NKCC2s and that similar phosphoregulatory mechanisms could be implicated in the regulation of NKCC2 function (Gimenez and Forbush, 2005).

WNK3 (WNK, with no lysine (K) kinase (Verissimo and Jordan, 2001)) appears to be a potent activator of rNKCC2F-mediated transport when coexpressed in *Xenopus laevis* oocytes (Rinehart et al., 2005). WNK3 has been proposed to regulate the activity of rNKCC2F by altering its expression at the plasma membrane. WNK3 increases rNKCC2F phosphorylation presumably at T^{96}/T^{101}, as assessed by R5 antibody (Rinehart et al., 2005). However, in this study, the impact of phophorylation of T^{96}/T^{101} on surface expression was not specifically demonstrated with mutagenized rNKCC2F constructs.

It has been shown that the metabolic sensing kinase AMPK (AMP-activated protein kinase) phosphorylates rbNKCC2A on its S^{129} *in vitro*. Further, AMPK and the N-terminus of rbNKCC2A appear to be physically associated in rat kidney, and in the cell line MMDD1. The involvement of S^{129} on rbNKCC2A activity was suggested based on the fact that rbNKCC2A S129A mutant shows reduced transport activity when expressed in *Xenopus* oocytes, regardless of basal/isotonic or hypertonic conditions. Interestingly, coexpression of AMPK with rbNKCC2A in *Xenopus* oocytes did not have a measurable effect on transport activity (Fraser et al., 2007). These results suggest that either phosphorylation of S^{129} and/or the residue itself are implicated in basal/isotonic rbNKCC2A activity. Interestingly, the rbNKCC2A protein sequence containing S^{129} (i.e. NRPS^{129}LLE) conforms a consensus site for casein kinase I/II and PKG/PKA which are also present in mouse, human and rat NKCC2s (Fig. 11.10). Further, in rat TALH membrane-permeant cGMP analogs decreased NKCC2 surface expression without affecting the total pool of NKCC2 (Ares et al., 2008). It would be interesting to determine if residue S^{129} is involved in NKCC2 trafficking.

H. Predicted Modulators of NKCC2 Function: N-glycosylation

N-linked glycosylation occurs at asparagines (N) within the consensus (*sequon*) NXS/T, where X is any amino acid except proline (Ben-Dor et al., 2004; Gavel and von Heijne, 1990; Kornfeld and Kornfeld, 1985). Immunoblots of oocyte membranes in which human or rat NKCC2F were transiently expressed yielded two bands corresponding to non-glycosylated (~115–125 kDa) and N-glycosylated (~150–160 kDa) proteins (Gamba et al., 1994; Starremans et al., 2003). On the other hand, *in vitro* translation of rat NKCC2F yields a non-glycosylated protein of ~110–120 kDa (Gamba et al., 1994), suggesting that the higher electrophoretic mobility (i.e. lower MW) of NKCC2 bands in immunoblots may represent non-glycosylated proteins.

N-glycosylation impacts rNKCC2F function when expressed in oocytes (Gamba, 2005). Protein N-glycosylation is a co- and post-translational modification crucial for folding, trafficking, insertion and function of many proteins (Jones et al., 2005; Kornfeld and Kornfeld, 1985). NKCC2s are glycoproteins (Haas and Forbush, 2000; Paredes et al., 2006; Payne and Forbush, 1994; Payne et al., 1995), thus N-glycosylation is expected to impact their functional and

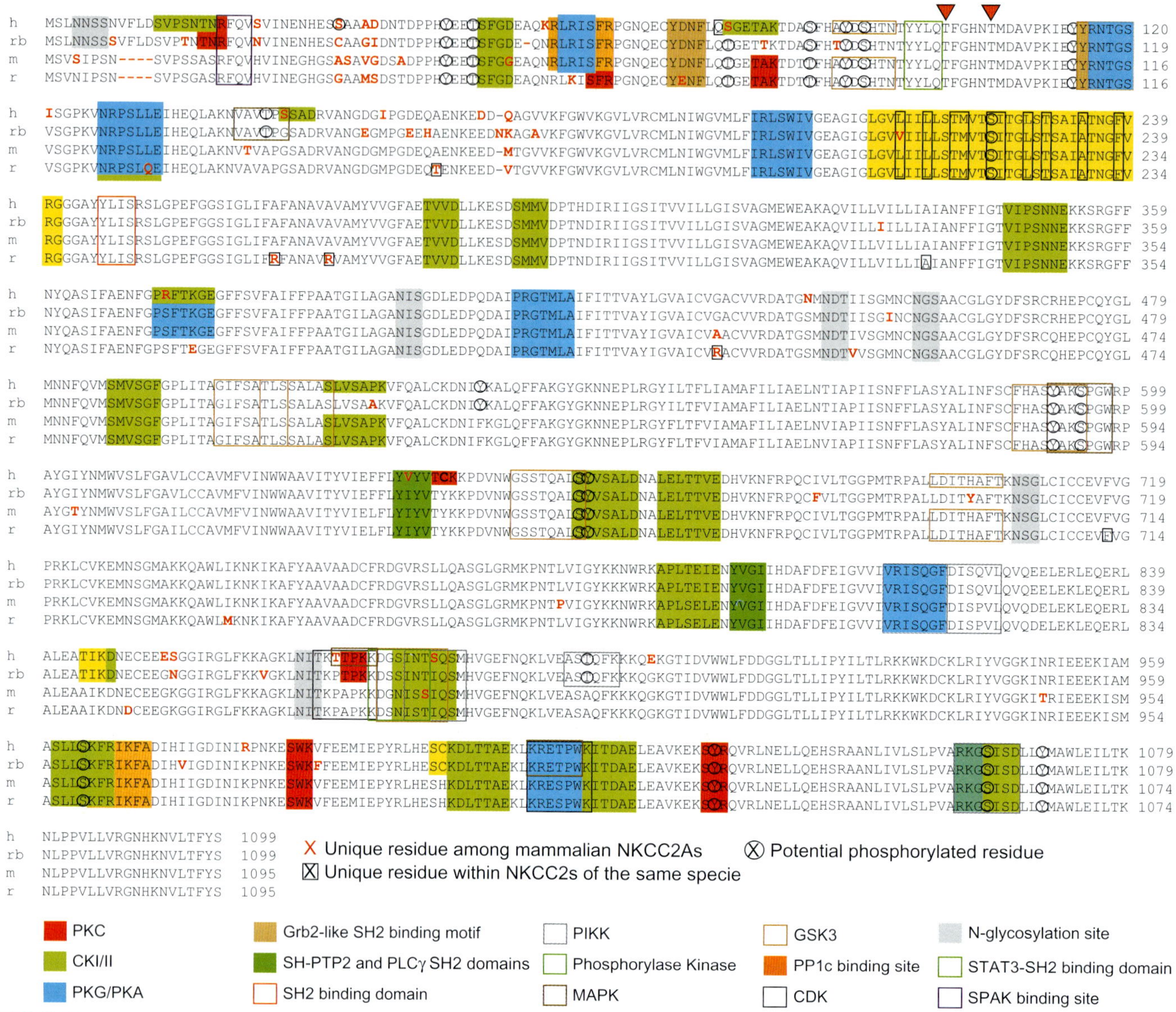

FIGURE 11.10 Alignment analysis of and predicted phosphorylation sites in the NKCC2A protein sequences. Predicted protein sequences corresponding to human (h), rabbit (rb), mouse (m) or rat (r) NKCC2As were obtained by translating the corresponding nucleotide sequences posted in the GenBank (EF559316 (hNKCC2A), U07547 (rbNKCC2A), U20973 (mNKCC2A) and EF577032 (rNKCC2A)). NKCC2A protein sequences were aligned by using *ClustalW*. The amino acid sequences predicted to be encoded by exon "A" are shown inside the yellow box. Unique residues found only in the mammalian NKCC2A variant are indicated in red letters, whereas boxed letters represent amino acids found in NKCC2A, but absent in NKCC2B or F splice variants of the same specie. The presence of consensus phosphorylation sites in the NKCC2A protein sequences was analyzed by using *ScanSite* (scansite.mit.edu), *KinasePhos* (kinasephos.mbc.nctu.edu.tw), *NetPhos* (cbs.dtu.dk/services/NetPhos) and *NetPhosK* (cbs.dtu.dk/services/NetPhosK) tools. N-glycosylation sites were analyzed by using the *NetNGlyc* server (cbs.dtu.dk/services/NetNGlyc). Amino acids located within colored squares represent predicted phosporylation sites or binding motifs, as indicated. Additional predicted features were collected from the ELM database server (elm.eu.org) and *InterPro* (www.ebi.ac.uk/interpro), which are also indicated below the alignment.

pharmacological properties. Prevention of N-glycosylation in rNKCC2F by site-directed mutagenesis produces a protein with reduced transport activity and decreased bumetanide sensitivity (Paredes et al., 2006). The impact of N-glycosylation on transport activity appears to be related to plasma membrane targeting (Benziane et al., 2007; Paredes et al., 2006). Interestingly, non-glycosylatable mutants of the rat *slc12a3* cotransporter (rNCC) cannot be efficiently targeted to the oocyte plasma membrane (Hoover et al., 2003). The effect of N-glycosylation on the functional activity and substrate affinity of other membrane transporters has been studied. For instance, prevention of N-glycosylation resulted in decreased substrate affinity of glucose transporter GLUT1 (Asano et al., 1991) and an organic cation transporter (Ott et al., 1992). Although the molecular mechanisms underlying these effects are not understood, changes in N-glycosylation may result in impaired protein targeting to the plasma membrane and/or changes in substrate affinitites. N-glycosylation has also been shown to reduce transport velocity, but not the *Km* in some transporters (e.g. GLYT1 (Olivares et al., 1995), Na^+/H^+ exchanger (Yusufi et al., 1988), Na/P*i*-2 transporter (Hayes et al., 1994) and serotonin transporter (Tate and Blakely, 1994)). Hence, it is conceivable that N-glycosylation of NKCC2s may affect basal transport activity due to altered plasma membrane insertion and/or modifications in ion affinities.

Five potential N-glycosylation sites are present in NKCC2s, which are conserved among species (N^{395}, N^{442}, N^{452}, N^{579}, N^{864} and N^{875}). Site-directed mutagenesis of rNKCC2F suggests that N^{442} and/or N^{452} are N-glycosylated (Paredes et al., 2006). Although this is consistent with the idea that only extracellular sites may be N-glycosylated (Hart, 1992; Jenkins et al., 1996; Kornfeld and Kornfeld, 1985), mutation of one of these residues renders rNKCC2F totally N-deglycosylated (Paredes et al., 2006). This suggests that N^{442} may interfere with N-glycosylation of N^{452} or vice versa. Although N-linked glycosylation occurs at the sequon NXS/T (Gavel and von Heijne, 1990; Kornfeld and Kornfeld, 1985), many NXS/T sequons are inefficiently or not glycosylated, even among different molecules of the same protein (Jenkins et al., 1996; Jones et al., 2005). This may be related to the amino acid composition of the NXS/T neighborhood and/or structural constraints (Jones et al., 2005). Several N-glycosylation consensus sites have been found to be poor oligosacharide acceptors (e.g. NWS, NDS, NGS or NLS); they are rarely N-glycosylated (Kasturi et al., 1997). Interestingly, residues N^{442} and N^{452} in rNKCC2s are encoded within the context N^{442}DT and N^{452}GS, respectively. This suggests that N^{452} in rNKCC2s may not be N-glycosylated. In line with this argument, rNKCC2F N442Q, N452Q or N442/452Q mutants are completely N-deglycosylated proteins (Paredes et al., 2006). It is tempting to speculate that N-glycosylation of N^{442} of NKCC2s may be influenced by N^{452} and/or distal residues on the amino acid sequence. This idea is based on the following facts: (i) Ns less than 12–14 residues apart are poorly glycosylated (Gavel and von Heijne, 1990); (ii) NXT sequons are preferred over NXS ones (Gavel and von Heijne, 1990); (iii) the presence of cysteine residues (e.g. C^{451} and C^{457} in rNKCC2) close to the NXS/T sequon may inhibit glycosylation (Bause et al., 1982; Bulleid et al., 1992; Hasegawa et al., 1992; Kane, 1993); (iv) some of the hNKCC2F mutants in Bartter's syndrome (A267S, G319R, A508T, D526N, Y998X) are not N-glycosylated when expressed in *Xenopus* oocytes (Starremans et al., 2003). Although the mechanisms involved in N-glycosylation of NKCC2 are still not known, differences in N-glycosylation among closely related NKCC2 proteins may determine their functional and pharmacological diversity.

I. What Happens if NKCC2 is not Functionally Expressed?

Inactivating homozygous mutations of the human NKCC2 gene product are the cause of antenatal Bartter syndrome type 1 (Simon et al., 1996a). This syndrome is characterized by renal tubular hypokalemic alkalosis, hypercalciuria, profound hydroelectrolytic imbalance and polyuria leading to polyhydramnios (Deschenes et al., 1993; Proesmans et al., 1985; Seyberth et al., 1985). Similarly, mice with targeted disruption of the promoter (3.5 kb upstream A^1TG) and first 3 exons (8.5 kb downstream A^1TG) of the NKCC2 gene results in a phenotype with many similarities to human Bartter's syndrome (Takahashi et al., 2000). However, homozygous deletions of exons A or B of the mNKCC2 gene do not result in mice with Bartter's syndrome (Oppermann et al., 2006, 2007). Indeed, NKCC2B-deficient mice are viable and fertile with a mild impairment of renal diluting function which is accompanied by a slight decrease of osmotic urine concentration and a significant right shift of the tubuloglomerular feedback function curve (Oppermann et al., 2006), as discussed in detail in Chapter 18 in this volume. The NKCC2A-deficient mice are also viable and fertile with no apparent abnormalities beyond a mild impairment of renal function after acute intravenous saline loading (Oppermann et al., 2007). It is not known if the phenotypes of these knockout mice are related to the particular localization of mNKCC2A and mNKCC2B within the nephron and/or if mNKCC2F can supply most if not all of the NKCC2 function.

The Human Gene Mutation Database (HGMD, www.hgmd.cf.ac.uk/ac/index.php) posts all known mutations of the hNKCC2 gene. A total of 22 mutations (15 missense, six deletions and one insertion) are known. Some of these mutations detected in patients with Bartter's syndrome were tested by heterologous expression in *Xenopus* oocytes of *in vitro* mutagenized hNKCC2F . These mutations were proven to be inactivating of hNKCC2F protein function (Simon et al., 1996a; Starremans et al., 2003; Vargas-Poussou et al., 1998). For instance, hNKCC2F G193R, A267S, G319R, A508T, del526N, and Y998X resulted in low expression of functionally impaired hNKCC2F proteins when expressed in *X. laevis* oocytes. Some of these mutants were poorly N-glycosylated and/or did not localize to the oocyte plasma membrane (Starremans et al., 2003). Although the impact of the above-mentioned mutations on the function of hNKCC2A or hNKCC2B isoforms is unknown, they are predicted to cause similar defects on these protein variants, although they may not play a role in the phenotypic outcome of Bartter's syndrome.

III. THE SLC12A2 GENE ENCODING FOR Na^+-K^+-$2Cl^-$ COTRANSPORTER 1 (NKCC1)

A. Genomic Organization of Mammalian NKCC1 Genes

The SLC12A2 gene encodes for the bumetanide-sensitive sodium-potassium-chloride cotransporter 1, also known as the basolateral Na-K-Cl symporter, BSC2 or NKCC1. Several NKCC1 genes were identified as putative orthologs of one another during the construction of *HomoloGene* (acc #20283). This database is a system for automated detection of gene homologs (including paralogs and orthologs) among the annotated genes of completely sequenced eukaryotic genomes (www.ncbi.nlm.nih.gov/sites/entrez?db = homologene).

The hNKCC1 gene is located on chromosome 5 at 5q23.3 and covers ~106 kb of genomic sequence (*nt* position 127,447,382 to 127,553,278 (*Ensembl* ENSG00000064651)). The nucleotide sequence of the hNKCC1 gene is defined by more than 250 cDNA clones obtained from colon, trachea, brain, stomach, eye, testis, lung and other tissues. *AceView*, a database that coalignes experimentally determined cDNA sequences (from *GenBank*, *dbEST* and *Trace*) with genomes (Thierry-Mieg and Thierry-Mieg, 2006), annotates the hNKCC1 gene with 32 different introns (31 GT-AG and 1 GC-AG splicing consensus) and the transcriptional potential to generate 13 different mRNAs, eight alternatively spliced variants and five non-spliced variants. The hNKCC1 gene has at least four probable alternative promoters, three non-overlapping alternative last exons and three validated alternative polyadenylation sites (Fig. 11.11A). Although the functional properties of the hNKCC1 promoter were not experimentally tested, a potential transcription start site was proposed to be located at -190 bp upstream A^1TG (Ibla et al., 2006).

As posted in *AceView*, the mouse NKCC1 gene is located in chromosome 18 (NC_000084, *Ensembl* ENSMUSG00000024597) and is defined by 422 cDNA clones, some obtained from mammary tissue, small intestinal epithelial progenitors, colon epithelial progenitor cells, prostate and other tissues. This mNKCC1 gene contains 26 different introns (25 GT-AG, 1 GC-AG consensus slicing sites) and is predicted to produce four different mNKCC1 mRNAs, two alternatively spliced variants and two non-spliced variants, probably due to the presence of two validated alternative polyadenylation sites (Fig. 11.11B).

The rat NKCC1 gene is located in chromosome 18 (18q12.1) and has a provisional *RefSeq* status. By using *Gnomon* (www.ncbi.nlm.nih.gov/genome/guide/gnomon.shtml), the genomic organization of rNKCC1 gene covers ~75 kb and can be organized, annotated and modeled (model number *hmm76891*) as displayed in Fig. 11.11C. The series of cDNA clones that appears to be originated from the same rNKCC1 gene is represented by *UniGene* (UGID 11523) with a full-length cDNA (AF051561), three partial sequences (AF086758, AF071863 and AF061084) and 98 ESTs obtained from a wide variety of rat tissues such as kidney, brain, muscle, eye, ovary, placenta, heart, dorsal root ganglion, prostate, colon, lung, inner ear and others. According to *Ensembl*, rNKCC1 gene has 29 well-defined exons, three of which are *ab initio* predicted exons, and a single gene promoter located upstream of exon 1.

B. How Many Variants of NKCC1?

In spite of the potential of the hNKCC1 gene to transcribe many spliced mRNAs which may encode or not for functional NKCC1 proteins, only two different full-length hNKCC1 cDNAs have been cloned: hNKCC1*a* (U30246 (Payne et al., 1995) and AY280459 (Liedtke et al., 2005)), and hNKCC1*b* (BC033003 (Strausberg et al., 2002)). The first two hNKCC1*a* cDNA clones (i.e. U30246 and AY280459) encode for identical proteins of 1212 amino acids (~131 kDa), but differ from each other at their 5′-UTRs; AY280459 and U30246 are 190 bp and 164 bp long, respectively. The 3′-UTR of both hNKCC1*a* cDNA clones contains ~3 kb of sequence followed by the polyA tail. The standard AATAAA polyadenylation signal is not present, but the variant ATTAAA is located ~26 bp before the polyA tail. The functional properties of hNKCC1*a* proteins have been partially established (Payne et al., 1995).

hNKCC1*a* differs from hNKCC1*b* in that the "*a*" variant does not express the first 67 bp of exon 26, resulting in the exclusion of an alternative stop codon (Fig. 11.12). Thus, hNKCC1*b* is a slightly shorter

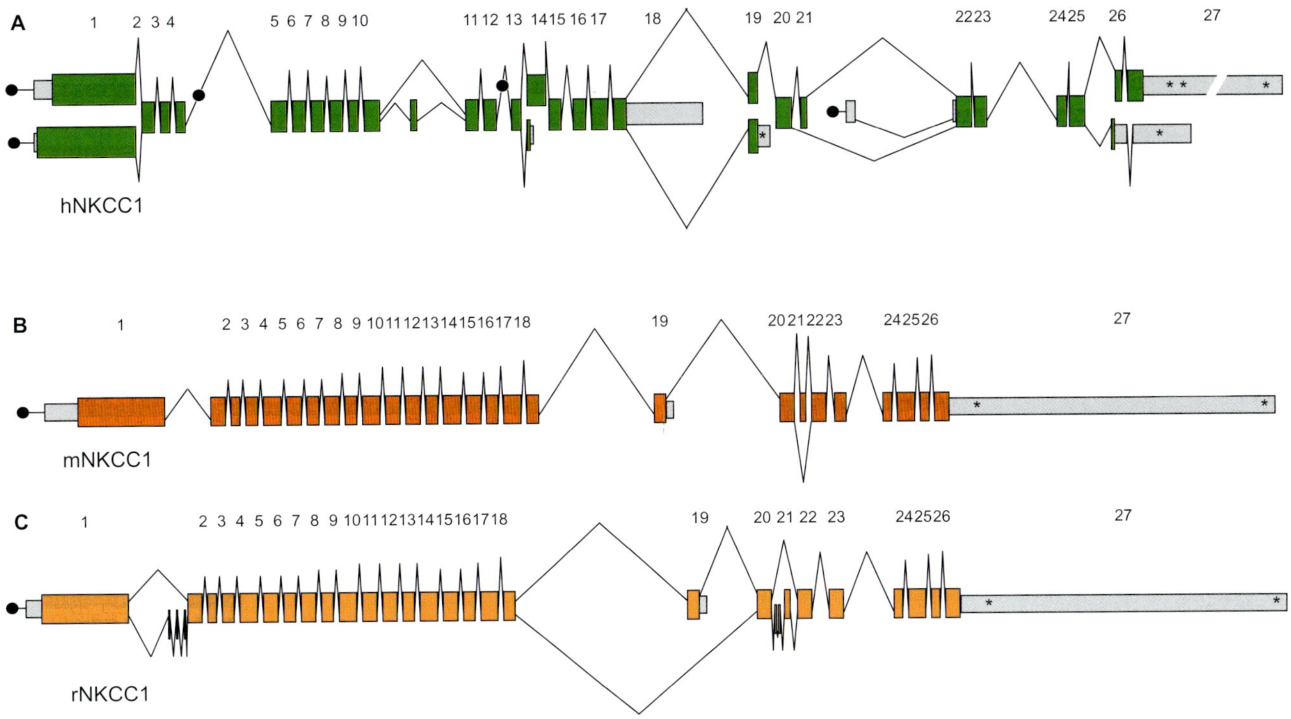

FIGURE 11.11 Promoters, exons, splicings and alternative ends of NKCC1 genes. Representation of human (**A**), mouse (**B**) and rat (**B**) NKCC1 genes (NC_000005, NC_005117 and NC_000084, respectively). Broken black lines represent introns with standard boundaries (GT-AG or GC-AG) that are exactly and experimentally supported (i.e. a cDNA sequence that exactly matches the genome over 16 bp, eight on both sides of the intron). Protein coding exons are represented by colored boxes whereas gray boxes represent the 5′-UTR (on the left) and 3′-UTR (on the right). Asterisks identify validated 3′ ends represented by polyadenylation sites in the form of AATAAA or similar and supported by cDNA accessions in the *GenBank*. Black circles represent potential promoters. Data obtained from *AceView* (www.ncbi.nlm.nih.gov/IEB/Research/Acembly).

hNKCC1 variant with the unique C-terminal sequence FYEPC. Further, hNKCC1*b* lacks the first 2.2 kb of exon 27, only expressed in hNKCC1*a*, resulting in a unique 3′-UTR sequence with an alternative polyadenylation site (Fig. 11.12). The hNKCC1*b* mRNA is a ~4.2 kb long mRNA with a probable incomplete 5′-UTR and thus predicted to be target for nonsense mediated mRNA decay. This hNKCC1*b* mRNA is predicted to encode for a protein of 1204 amino acids (~130 kDa). The 3′-UTR of this transcript contains ~0.65 kb of nucleotide sequence followed by the polyA tail. As for hNKCC1*a*, no standard polyadenylation signal exists in the 3′-UTR region of the hNKCC1*b*, but the variant TATAAA is expressed at ~22 bp before the polyA tail. The functional properties of hNKCC1*b* variant are unknown.

Several human cDNA clones in the form of hESTs, some from trachea (DB234759, DB241343 and DB217363), brain (AI124709), breast (BE084576), colon (AW859514), eye (AF439152), pterygium (DR421672) and other tissues (DB050295), suggest that exon 21 in hNKCC1 gene may also be spliced out in some cDNAs encoding for hNKCC1. On the other hand, inclusion of the 3′-end of exon 19 expressing an alternative stop codon supports the existence of an hNKCC1 variant lacking most of its C-terminus. This is supported by one cDNA clone from prostate (BP327886), which may be incomplete at its 5′ end.

So far, only one full-length mouse NKCC1 has been cloned and characterized (U13174 (Delpire et al., 1994)). This mNKCC1 is 87% identical to hNKCC1*a* (Fig. 11.12). Partial molecular characterization of an incomplete mNKCC1 cDNA variant (named here mNKCC1-*s*) suggests alternative exclusion of exon 21 of the mNKCC1 gene. This exon 21 would encode for 16 amino acids located in the C-terminal tail of the cotransporter protein (Randall et al., 1997). The expression of mNKCC1-*s* is also supported by mESTs encoding for mNKCC1 sequences where exon 21 is spliced out (e.g. CJ315514, DV071389, CN715009, AW412359 and BQ715209). The full-length cDNA encoding for mNKCC1-*s* has not been cloned and/or posted in the *GenBank* and the physiological significance of this particular mNKCC1 variant is unknown.

In silico analysis of the rat NKCC1 gene predicts the generation of several alternative spliced variants.

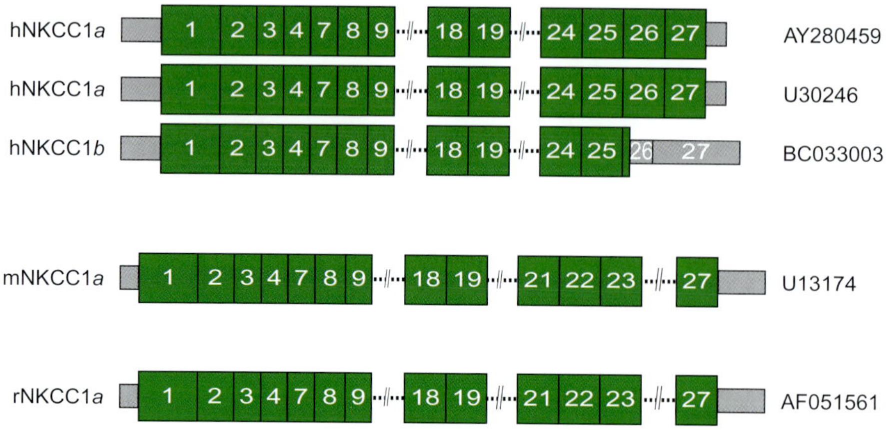

FIGURE 11.12 Human, mouse and rat full-length NKCC1 variant cDNAs. Protein coding exons are represented by filled and numbered green boxes whereas gray boxes represent the 5′-UTR (on the left) and 3′-UTR (on the right). The *GenBank* accession numbers for each cDNA are shown.

These spliced variants could result from the election of potential alternative stop codons or exons, as it may occur in human or mouse NKCC1 gene products. The existence of more than one rNKCC1 variant is supported, but not demonstrated, by the fact that several rNKCC1 mRNAs were detected by Northern blot analysis (Moore-Hoon and Turner, 1998). However, only one full-length rNKCC1 cDNA (AF051561) was cloned and partially characterized (Moore-Hoon and Turner, 1998). This rNKCC1 cDNA is 83% and 88% identical to human and mouse NKCC1*a* variants, respectively. Two incomplete rNKCC1 cDNAs (AF086758 (Fu et al., 1999) and AF086758 (Anzai et al., 1999)) were also cloned and posted in the *GenBank*. The expression of an rNKCC1-*s* variant (i.e. similar to mNKCC1-*s* lacking exon 21) in rat tissues is supported by the existence of rESTs encoding rNKCC1 sequences where exon 21 is not present. Some of these rESTs were obtained from dorsal root ganglion (BG666200, BG663902, CB717052 and BG672434) and others from brain (CO397329), colon (CB750614 and CB768404), prostate (BM385644) and lung (CF112886). The full-length rNKCC1-*s* variant has not yet been cloned, and its function is unknown.

Interestingly, the 5′-end of exon 26 (known to be included in the hNKCC1*b* variant) of rat, mouse and human NKCC1 genes have a high percentage of identity (>95%). However, there is no EST-based evidence supporting the expression of an analogous NKCC1*b* variant in rat or mice. Furthermore, in spite of the fact that exon 19 in human, mouse and rat NKCC1 genes is highly identical (>95%), there is no evidence in rats or mice for their 3′-end inclusion which encodes for an alternative stop codon and 3′-UTR in hNKCC1s. Hence, even though NKCC1 genes are highly similar among species, human, mice and rat NKCC1 genes may produce different variants.

C. Expression of NKCC1 Genes

The detection of several NKCC1 transcripts in mouse, rabbit and human tissues (Delpire et al., 1994; Payne et al., 1995) has been interpreted as proof of ubiquity of expression of NKCC1 gene products (Kaplan et al., 1996). Moreover, NKCC1 was introduced as the "housekeeping" isoform of the Na^+-K^+-$2Cl^-$ cotransporters (Payne et al., 1995) in spite of the fact that NKCC1 mRNAs were detected neither in liver nor in spleen (Delpire et al., 1994; Payne et al., 1995). However, tissue expression profiling of the hNKCC1 gene (*GeoProfile* record: GDS596 SLC12A2) supports the notion that hNKCC1 is indeed ubiquitously expressed. NKCC1 gene expression is extremely low in some tissues and varies between species (Fig. 11.13).

By definition, a *housekeeping gene* is a gene that is always constitutively expressed in any tissue, and that is involved in routine cellular metabolism. However, based on ubiquity, stability and relatively high levels of expression, no single gene may qualify as *housekeeping* since the expression levels of those genes may fluctuate (de Jonge et al., 2007). A list of suggested human housekeeping genes has been published (Eisenberg and Levanon, 2003). Interestingly, this list does not include NKCC1.

D. Is NKCC1 Gene Expression Transcriptionally Regulated?

This is a question that remains unanswered. Mammalian housekeeping genes show significantly lower promoter sequence conservation with respect to putative transcription start-sites of tissue-specific genes, e.g. NKCC1 vs. NKCC2 (Farre et al., 2007).

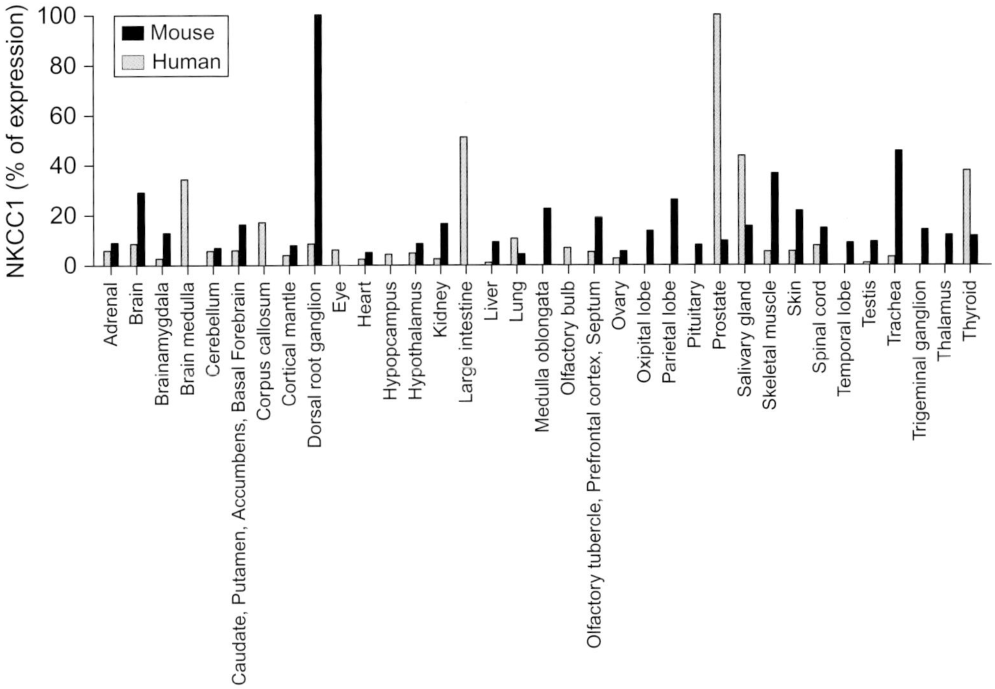

FIGURE 11.13 Expression profile of mouse and human NKCC1 transcripts. Expression results were obtained from *GEO Profiles* (Gene Expression Omnibus) records GDS596 and GDS868 from human or mouse databases, respectively. Expression levels of mNKCC1 transcripts (black bars) were assessed by using the mNKCC1 sequence GAT CCG AGT ATT CAT TGG TG corresponding to exon 23 whereas hNKCC1 expression levels (gray bars) were determined by using the *Affymetrix* probe *204404_at* specifically designed for hNKCC1 (transcript NM_001046, UniGene Hs.162585). Results are expressed as percentage relative to the NKCC1 mRNA levels in the corresponding tissue with the highest level of expression (dorsal root ganglion in mouse or prostate in human).

Interestingly, *in silico* analysis of human, rat and mouse NKCC1 distal sequences (~3 kb upstream the putative A¹TG) shows poor percentage of identity among them (~35%). However, the weak TATA box (TTTAAA) located at *nt* -208 bp from A¹TG in mouse NKCC1 gene promoter (Randall et al., 1997) is conserved in rat and human NKCC1 promoter sequences.

The NKCC1 promoter may be subject to transcriptional regulation. Using *ProScan* (www-bimas.cit.nih.gov/molbio/proscan) and *AliBaba v2.3* (www.gene-regulation.com) several conserved elements for SP-1 and AP2 binding factors are predicted in NKCC1 gene promoters suggesting that NKCC1 genes may be subjected to basal (constitutive) transcriptional regulation (Suske, 1999; Wierstra, 2008). The presence of GC-rich sequences in NKCC1 gene promoters predicts binding of negative modulators of basal transcription (e.g. GCF) (Faisst and Meyer, 1992). Further, EGF is a positive modulator of hNKCC1 gene transcription above basal levels. Increased expression by EGF is actinomycin D-dependent, supporting EGF-mediated transcriptional control of human NKCC1 mRNA expression (O'Mahony et al., 2008). Interestingly, hNKCC1 gene expression appears to be down-regulated in differentiated human colon adenocarcinoma cell line HT29. The mechanisms of this down-regulation are unknown (Matthews et al., 1998).

Decrease in hNKCC1 mRNA expression has been proposed to occur via functional binding of the hypoxia-inducible factor 1 (HIF-1) to responsive elements located at *nt* -38 (relative to A¹TG). The nature of HIF-mediated decrease of hNKCC1 mRNA was not studied (Ibla et al., 2006). However, the specific location of the HIF-1 responsive element (i.e. the 5′-UTR of hNKCC1 mRNAs) suggests that HIF-1 may be involved in the regulation of hNKCC1 mRNA expression at the post-transcriptional or translational level.

NKCC1 gene expression has been proposed to be developmentally regulated in the CNS (Clayton et al., 1998) as discussed in Chapters 19 and 24 in this volume. However, the mechanisms involved in NKCC1 gene expression during CNS development remain to be determined.

E. Post-transcriptional Regulation of NKCC1 Transcripts

In silico analysis of the 5′-UTR sequences of hNKCC1 mRNAs using *RegRNA* (regrna.mbc.nctu.

edu.tw/html/prediction.html) reveals interesting features regarding possible post-transcriptional regulatory mechanisms. The 190 bp long 5′-UTR of hNKCC1 mRNAs expresses two terminal oligopyrimidine tracts (TOPs) and a γIFN-activated inhibitor of mRNA translation (GAIT element). TOPs are known elements present in class II mRNAs that are regulated in a growth-dependent manner and that are involved in translational control according to the growth stage of the cell (Davuluri et al., 2000), whereas GAIT elements are involved in translational silencing (Sampath et al., 2003). The presence of these elements suggests that NKCC1 transcripts may be regulated at the translational level.

Interestingly, hNKCC1 activity is regulated via re-expression of the hNKCC1 transporter in response to some agonists in a cyclohexymide-dependent manner (Reynolds et al., 2007). This suggests a translational regulation of the transporter's transcript. Further, NKCC1 gene expression was reported to be involved in the control of normal cell proliferation and cell cycle (Panet et al., 2006, 2002, 2000) as discussed in detail in Chapter 27 in this volume.

F. Post-translational Control of NKCC1: Phosphorylation

Most of our knowledge regarding phosphorylation-mediated functional regulation of NKCC cotransporters comes from experiments performed in *Xenopus laevis* oocytes using skNKCC1, the NKCC1 isoform of shark (Gamba, 2005). It is accepted that phosphorylation of the N- and/or C-terminal domains of the NKCC1 protein modulates its transport activity (Flemmer et al., 2002; Gamba, 2005; Haas and Forbush, 2000; Matthews et al., 1998), as discussed in detail in Chapter 18 in this volume. To date, phosphorylation of duck, shark and human NKCC1s have been demonstrated to occur only on serine and threonine residues (Lytle, 1997; Lytle and Forbush, 1992). Although *in silico* analysis of predicted amino acid sequences of mammalian NKCC1 reveals consensus phosphorylation sites for tyrosine kinases, so far there is no evidence for NKCC1 phosphorylation on tirosine residues.

NKCC1 is phosphorylated in response to a number of stimuli including changes in cell volume, intracellular Cl^- concentration and pharmacological treatments, as discussed in detail in Chapter 18. At least three residues in the skNKCC1 protein (T^{184}, T^{189} and T^{202}) have been biochemically identified as phosphoacceptors (Darman and Forbush, 2002). It is worth noting that $T^{184}/T^{189}/T^{202}$ of skNKCC1 corresponds to $T^{212}/T^{217}/T^{230}$ in hNKCC1; $T^{206}/T^{211}/T^{224}$ in mNKCC1 and $T^{203}/T^{208}/T^{221}$ in rNKCC1. Interestingly, although the three threonine residues are conserved in the N-terminus of NKCC1 among species, the N-terminal region of skNKCC1 is only ~27% identical to that of NKCC1 from other species.

The N-terminal region of skNKCC1 contains a motif (RVXFXD) predicted to bind PP-1, protein phosphatase-1 (Darman et al., 2001). PP-1 was proposed as a modulator of skNKCC1 based on the pharmacological effects of calyculin A and okadaic acid (Lytle and Forbush, 1996). In line with this concept disruption or elimination of the PP-1/skNKCC1 interaction by site-directed mutagenesis resulted in skNKCC1 mutants with higher activity than wild type (Darman et al., 2001). Interestingly, the binding site for SPAK/OSR1 partially overlaps with the binding site for PP-1 on mNKCC1 (Piechotta et al., 2002). However, the functional significance of this overlap is unknown. For further discussion on phosphoregulation of NKCC1 see Chapters 16 and 18 in this volume.

G. Post-translational Control of NKCC1: N-glycosylation

In silico analysis of the mNKCC1 amino acid sequence reveals at least five potential N-glycosylation sites: N^{164}, N^{500}, N^{547}, N^{556} and N^{684}. These five N-glycosylation sites are conserved among species with the possible exception of skNKCC1, which has only four sites (it lacks the equivalent of N^{556}). Of the five conserved N-glycosylation sites two (N^{547} and N^{556}) are located in the large extracellular loop. Clearly, some of the N-glycosylation sites may fall within predicted intracellular or TM domains, and thus they may not be glycosylated *in vivo* (Hart, 1992; Jenkins et al., 1996; Kornfeld and Kornfeld, 1985). The impact of N-glycosylation on NKCC1 function has not been studied.

Evidence that NKCC1s are N-glycosylated *in vivo* comes from Western blot analyses in several species including frog, rat, cat and human brain and dorsal root ganglia (Alvarez-Leefmans et al., 2001; Marty et al., 2002; Muñoz et al., 2007), human cell lines (Payne et al., 1995), shark rectal gland (Lytle et al., 1992), rabbit parotid gland and kidney, and duck red cells (Lytle et al., 1995). These analyses show proteins with molecular weight higher than predicted for the core protein which decrease to the core molecular weight upon enzymatic deglycosylation. The functional role of N-glycosylation on NKCC1 expression and/or function is unknown.

H. What happens if NKCC1 is not Expressed?

So far, there are no reports of a disease related to NKCC1 gene abnormalities in humans. A critical role of NKCC1 in human deafness has been proposed (Delpire et al., 1999) as NKCC1 is essential for the production of endolymph in the inner-ear (Dixon et al., 1999). Several mice lacking NKCC1 gene expression were created by targeting different parts of the NKCC1 gene (Delpire et al., 1999; Dixon et al., 1999; Flagella et al., 1999; Pace et al., 2000, 2001). In all of these NKCC1 knockout mice, the most common phenotype is loss of hearing and inner ear dysfunction (Delpire and Mount, 2002). The impact of NKCC1 ablation in mice models is discussed in detail in Chapters 10, 16 and 21 in this volume.

IV. THE SLC12A3 GENE ENCODING FOR Na^+-Cl^- COTRANSPORTER (NCC)

A. Genomic Organization of Mammalian NCC Genes

The human SLC12A3 (hNCC) gene maps on chromosome 16 (NC_000016) at 16q13 (Mastroianni et al., 1996; Simon et al., 1996b) and spans ~50 kb from *nt* position 55,456,621 to 55,506,104 (*Ensembl* ENSG00000070915). This gene codifies for the thiazide-sensitive NaCl cotransporter (TSC), also known as NaCl cotransporter (NCC). The hNCC gene is defined by 54 *GenBank* cDNA sequences. Half of these cDNA sequences are derived from human kidney, whereas the other half is derived from other tissues and cell lines (e.g. cervix, placenta, liver, colon, stomach and cervical carcinoma cells). According to *AceView*, the hNCC gene contains at least 29 different introns with standard splicing sites (GT-AG). This gene is predicted to produce several different mRNAs, some alternatively spliced variants and other non-spliced forms. The hNCC gene contains two non-overlapping alternative last exons (exons 26 and 27), an overlapping exon with different boundaries (exon 24) and a unique exon (exon 20) with no similar counterpart either in rat or in the mouse NCC genes. The hNCC gene does not express an obvious consensus polyadenylation site on the 3′ side of exons 25–27 (see Fig. 11.14A).

The promoter region of the hNCC gene (~2 kb upstream A^1TG) has been cloned and partially characterized (MacKenzie et al., 2001). Transcription of the hNCC gene starts somewhere between *nt* -18 and -6 (with respect to A^1TG) and the hNCC promoter appears to be constitutively active in the Madin-Darby canine kidney (MDCK) cell line, but not in Chinese hamster ovary (CHO) cells. Two typical promoter elements are present in the hNCC gene (e.g. TATA and CCAAT boxes). The canonical TATA box located at *nt* -61 is responsible for most, but not all, the transcriptional activity of the hNCC gene (MacKenzie et al., 2001). Two inverted binding elements for SP-1 transcription factors are also present at *nt* -90 and -30. Several other elements are present in the hNCC gene promoter, some of them with the potential to bind transcription factors involved in tissue-specific expression or to be regulated by hormones or signaling pathways. Interestingly, a strong transcriptional repressing activity was mapped at position *nt* -1880 to -1000 in the hNCC gene promoter (MacKenzie et al., 2001). *In silico* analysis of the promoter region (2 kb downstream A^1TG) of human, mouse and rat NCC genes demonstrate a high degree of identity between them (~80% identical).

The mouse NCC gene covers 37 kb of genomic DNA on chromosome 8 (Pathak et al., 1996) (*Ensembl* ENSMUSG00000031766). According to *AceView*, the sequence of this gene is defined by more than 80 cDNA clones posted in *GenBank/dbEST*, most of them obtained from mouse kidney and some from inner ear, spleen and many other mouse tissues. The mNCC gene contains 25 well-defined GT-AG introns and contains at least three non-overlapping alternative last exons (exons 17, 19 and 25). Three possible alternative polyadenylation sites are located in exons 17, 19 and 25 (see Fig. 11.14B). The mNCC mRNAs appear to differ by truncation of the 3′ end, by overlapping exons with different boundaries, and by alternative splicing or retention of one intron. The mNCC gene is predicted to produce several different mRNAs (i.e. six alternatively spliced variants and one non-spliced form). Although the promoter region of the mNCC gene has not been cloned and characterized, a classical TATA-box element is located at *nt* -61 bp and transcription of this gene is predicted to start from at least *nt* -30.

The rat NCC gene (NC_005118) is located in chromosome 19 at 19p14-12 (Taniyama et al., 2001) and covers ~38 kb of genomic DNA sequence, from *nt* position 11,071,790 to 11,109,622 (*Ensembl* ENSRNOG00000018607). The rNCC and mNCC genes are ~70% identical. They contain 26 exons with the conventional GT-AG consensus splice sites and each rat or mouse NCC exon is >90% identical. Transcription of the rNCC gene is known to start from *nt* -18 (Taniyama et al., 2001) and is predicted to produce several variants according to the election of alternative exons.

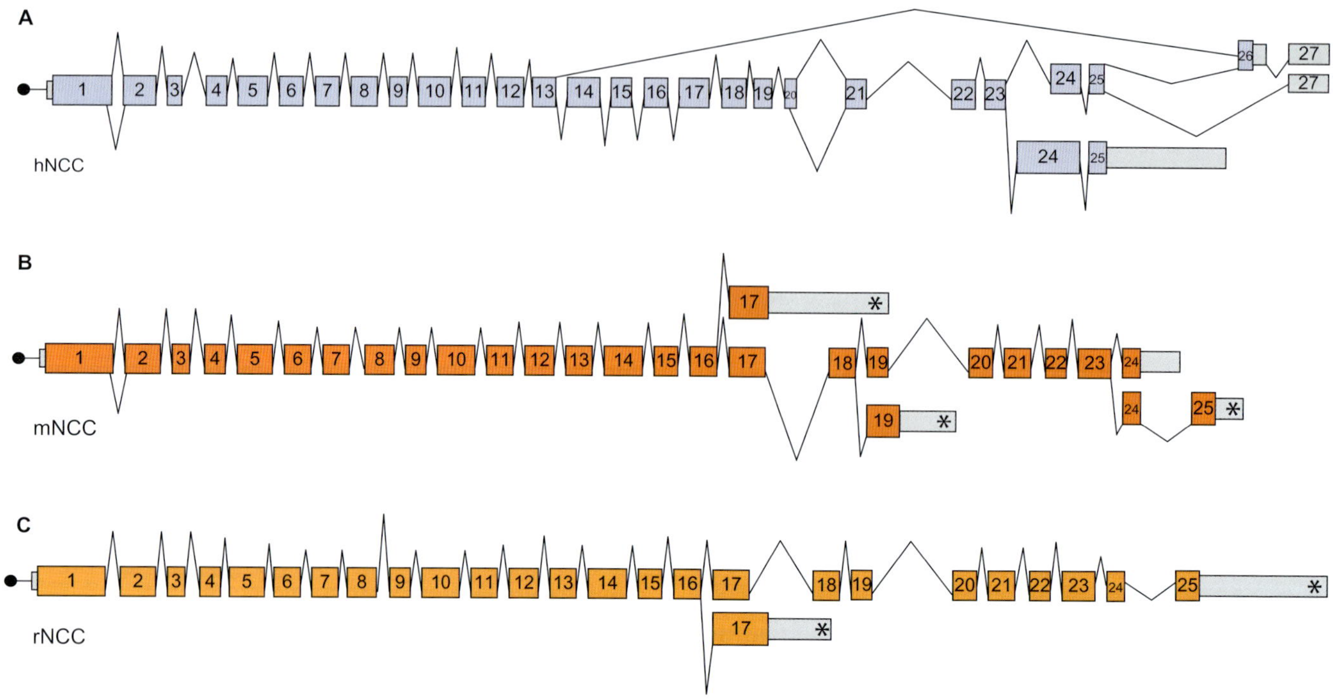

FIGURE 11.14 Promoters, exons, splicings and alternative ends of SLC12A3 genes. Representation of the human **A.**, mouse **B.** and rat **C.** NCC genes (NC_000016, NC_000074 and NC_005118, respectively). Broken black lines show introns with standard boundaries (GT-AG) that are exactly supported (i.e. a cDNA sequence exactly matches the genome over 16 bp, eight on both sides of the intron). Protein coding exons are represented by filled colored boxes, whereas gray boxes represent the 5′-UTR (on the left) and 3′-UTR (on the right). Asterisks identify validated or potential 3′ ends represented by polyadenylation sites and supported by cDNA accessions in the GenBank. Black circles represent potential promoters. Data obtained from *AceView* (www.ncbi.nlm.nih.gov/IEB/Research/Assembly).

The rNCC promoter region (~2 kb of distal genomic sequence, *GenBank* accession number AB024534), was cloned and partially characterized. The rNCC promoter is active in the human embryonic kidney-derived (HEK293) cell line, but not in the human hepatocellular carcinoma (HepG2) or the rat embryonic thoracic aorta vascular smooth muscle (A10) cell lines, respectively (Taniyama et al., 2001). This pattern of expression supports the notion that NCC is active in kidney cells.

B. How many Variants of NCC?

The potential of the hNCC gene to generate several variants is supported by at least four full-length human cDNAs cloned so far, more than 70 hESTs and incomplete cDNA clones (e.g. truncated, with no 5′- or 3′-UTRs). Three different full-length hNCC transcripts coming from the same gene are anotated in *RefSeq*: NM_000339, NM_001126107 and NM_001126108. Here we denote each of these clones as hNCCa, hNCCb and hNCCc, respectively. The hNCCa (*RefSeq* NM_000339) is supported by two full-length hNCC cDNAs (i.e. U44128 (Simon et al., 1996b) and BC111850). The predicted full-length mRNA for hNCCa can be reconstructed from 33 cDNA clones obtained from different human tissues. These hNCCa mRNAs correspond to the first 26 exons of the hNCC gene and are predicted to encode for a protein of 1030 amino acids (~114 kDa) (Figs 11.14A and 11.15A). *In silico* comparison of the hNCCa cDNAs with genomic sequences demonstrate that hNCC gene transcription starts at *nt* -6 (MacKenzie et al., 2001) and ends at exon 26. There is neither a standard (AATAAA) nor a variant polyadenylation signal in the last exons of the hNCC gene. Exon 27 is not included in hNCCa transcripts. This exon encodes for most of the 3′-UTRs in some hNCC transcripts. The hNCCa clone U44128 contains two changes when compared to BC111850 hNCCa cDNA and the human gene (i.e. A264G and D766E). Whether or not this represents genetic polymorphisms is unknown.

The *RefSeq* sequence NM_001126107 corresponds to hNCCb variant that lacks codon 95 (AGC, Q^{95}). The hNCCb transcripts appear to use an alternate in-frame splice site in the 5′ coding region of exon 2, resulting in a protein that is one residue (Q^{95}) shorter than variant hNCCa. The hNCCb variant is also

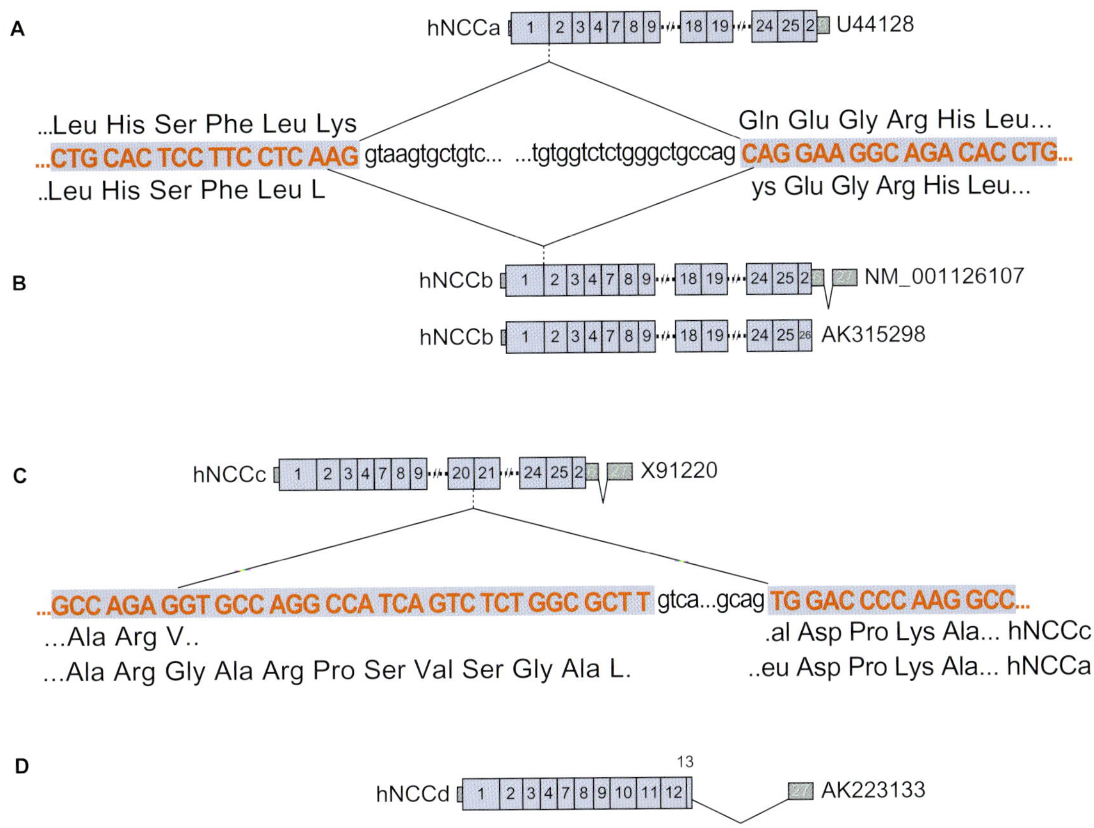

FIGURE 11.15 The hNCC variants. Representation of the four different human NCC variants encoding for hNCCa, hNCCb, hNCCc and hNCCd. (**A**) hNCCa (U44128), (**B**) hNCCb (AK315298 and NM_001126107), (**C**) hNCCc (X911220) and (**D**) hNCCd (AK223133). Protein coding exons are numbered and represented by colored boxes. The relevant amino acid sequences are shown.

characterized by alternative inclusion of 20 bp of intronic sequence connecting exons 26 and 27 (Fig. 11.15B). The hNCCb transcript NM_001126107 is supported by the hNCCb cDNA clone AK315298. This mRNA starts at *nt* -28 and ends at the stop codon in exon 26. The cDNA predicts the encoding of a protein having 1029 residues (~114 kDa). *In silico* analysis of hNCCa and hNCCb cDNAs demonstrate additional distinctive features. For instance, those found in codons 539 (TCC>CCC (S539P)), 778 (GAG>TAG (E778Y)) and 1017 (ATC>ATT (I^{1017})). It is not clear if these changes in hNCC cDNAs are polymorphic. A complete list of single nucleotide polymorphisms (SPNs) of the hNCC gene is available at www.ncbi.nlm.nih.gov/SNP.

The *RefSeq* NM_001126108 predicts the encoding of the third hNCC variant: hNCCc. This hNCC variant uses an alternate in-frame splice site in the 3′ coding region of exon 20 resulting in a transcript that lacks 27 bp in this exon (Figs 11.14A and 11.15C). Consequently, NM_001126108 encodes for an hNCCc protein which is nine amino acids shorter than hNCCa (1021 residues, ~113 kDa). The existence of this variant is supported by the full-length hNCCc cDNA X91220 clone (Mastroianni et al., 1996). The transcript for hNCCc starts at *nt* -25 and includes genomic sequences of exons 1–26, intron 26 and exon 27. This hNCCc cDNA contains several potential polymorphic changes in codons 445 (CAG>CGG (Q445R)), 459 (GCT>GTG (A459V)), 460 (GGC>GTC (G460V)) and 641 (TAC>TAT (Y^{641})). Its 3′-UTR is ~1.1 kb long and does not encode for a standard or a variant polyadenylation signal.

A fourth hNCC variant (AK223133), denoted here as hNCCd, was cloned from human kidney (Maruyama and Sugano, 1994). This transcript excludes hNCC genomic sequences encompassing the last 92 bp of exon 13 to exon 26 (Fig. 11.15D) and would encode for a protein of 534 amino acids (~57 kDa). Its function, if any, is unknown. Other transcript variants of the hNCC gene may exist. For instance, the hNCC EST clone BG428750 obtained from human kidney includes sequences corresponding to intron 24, exon 24–25 and part of intron 25.

The mouse NCC gene expresses at least four different transcripts as demonstrated by five full-length cDNA clones: U61085 (Kunchaparty et al., 1999), AK052691, AK085496, AK052755 (Katayama et al., 2005) and BC038612 (Strausberg et al., 2002). *In silico* alignment of these full-length cDNAs reveals different lengths in their 5-UTRs (Fig. 11.16A), supporting the notion that mNCC gene transcription may start after *nt* -33. The mNCC clone U61085 and the hNCCb clone AK315298 (see Fig. 11.15B) are 85% identical, they do not encode for Q^{95} (Q^{93} in mouse) and do not contain the 3′-UTR encoded by the last exon of mNCC gene (exon 25, Fig. 11.14B). Hence, U61085 cDNAs would encode for mNCCb. Interestingly, mouse NCCs are distinguished from hNCCs by the absence of a sequence encoded by exon 20, which is only present in the human NCC gene.

Mouse NCC clones AK052691 and BC038612 are almost identical in nucleotide sequence and differ only in two bps at the 3′-UTR. Both transcripts AK052691 and BC038612 can be reconstructed from 19 cDNA clones obtained from mouse kidney. These mNCC mRNAs start at *nt* -33, continue with a nucleotide sequence corresponding to the first 25 exons of

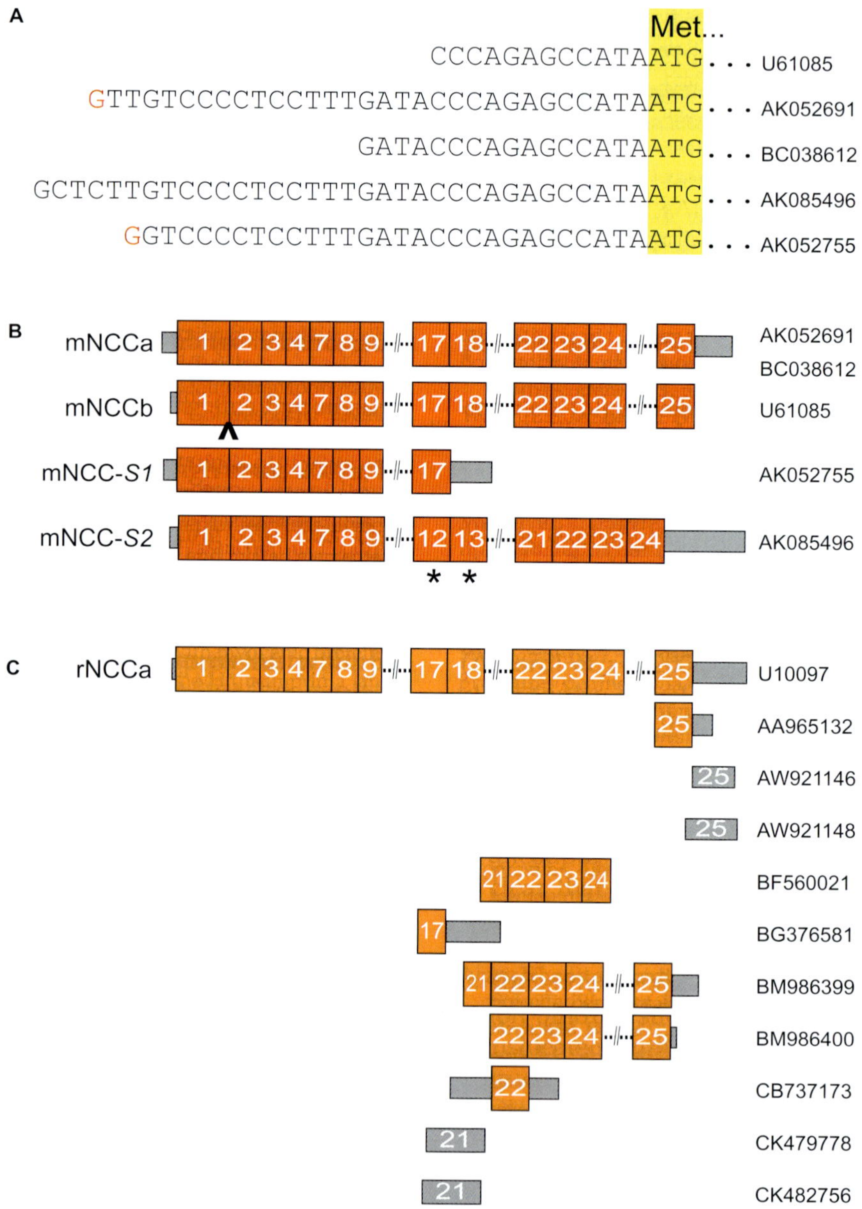

FIGURE 11.16 The mouse and rat NCC transcripts. **A.** Representation of the 5′-UTRs of mouse NCC transcripts suggesting variable initiation of transcription. **B.** Four full-length mouse cDNAs encoding for different mNCCs: mNCCa, mNCCb, mNCC-S1 and mNCC-S2. Protein coding exons are represented by filled colored boxes, whereas gray boxes represent the 5′-UTR (on the left) and 3′-UTR (on the right). The symbol ^ shown between exons 1 and 2 of mNCCb (U61085) indicates the site of single-codon splicing, as shown in Fig. 11.15A, whereas asterisks (*) under exon 12/13 of mNCC-S2 (AK085496) show single nucleotide insertions with respect to all other mNCC cDNAs. **C.** Shown is the only rNCCa full-length cDNA clone (U10097) in alignment with rEST clones.

the mNCC gene and end with ~261 bp of 3′-UTR. The last exon of these cDNAs encodes for a potential polyadenylation site (AAAAAA) (Fig. 11.14B). Moreover, *in silico* comparison between AK052691/BC038612 and U44128 (hNCCa) reveals ~85% identity among them, suggesting that these transcripts encode for the mNCCa variant and a protein of 1002 amino acids (~111 kDa).

The mNCC cDNA clone AK052755 may be the result of the election of an alternative stop codon and/or polyadenylation site located in exon 17 (see Fig. 11.16B). This mRNA starts at *nt* -29, contains 17 exons and ends with a 3′-UTR of ~700 bp. There are no evident signs for a standard or variant polyadenylation signal in the last exon (i.e. exon 17). Although mNCC AK052755 is ~85% identical in nucleotide sequence to AK223133 (hNCCd), these two cDNAs come from splicing occurring at different exons. Therefore, the mNCC cDNA AK052755 does not correspond to hNCCd, in spite of its high degree of identity (~85%). The mNCC isoform encoded by AK052755 may represent a unique variant of mNCC expressed only in rodents. The AK052755 transcript predicts the encoding of a protein of 726 amino acids (~80 kDa) of unknown function. Because of its relatively short length it is here denoted as mNCC-*S1* (Fig. 11.16B), to distinguish it from another short length mNCC variant (see below).

The clone AK085496 is another full-length mNCC cDNA that, in this case, results from the election of a stop codon located immediately after the last codon of exon 24 in the mNCC gene. This mRNA can be reconstructed from two additional cDNA clones (AI132066 and CB599768) obtained from mouse kidney. This mRNA starts at *nt* -34, includes genomic sequences corresponding to exons 1–24 and ends with ~1.4 kb of 3′-UTR. When compared to the genomic sequence of the mNCC gene and all full-length mNCC cDNAs, the AK085496 clone has three single nucleotide insertions, T^{1479}, C^{1618} and G^{2685} (numbered with respect to $A^{1}TG$), in exons 12, 13 and 23, respectively. It remains to be determined if these changes are polymorphisms, mutations, or simple sequencing errors. The AK085496 clone has neither a standard nor a variant polyadenylation signal in its last exon and intron 24. The mNCC cDNA AK085496 predicts the encoding of a shorter protein (i.e. 956 amino acids, ~106 kDa) because of the lack of exon 25, which encodes for the last 46 amino acids in the full-length mNCCa and mNCCb. This short mNCC cDNA, here denoted as mNCC-*S2*, does not appear to have a human counterpart (Figs 11.14A and 11.15). Finally, there is another mNCC cDNA (clone AW701155), probably incomplete, that nevertheless supports the idea that there are more than four variants of mNCCs. This mNCC clone AW701155 would encode for a mNCC transcript starting in exon 16 and ending at intron 19, which may contain an additional polyadenylation signal.

The rat *slc12a3* gene predicts the encoding of at least two rNCC transcripts (see Fig. 11.14C). This is supported by one full-length rNCC cDNA (U10097 (Gamba et al., 1994)) and nine rESTs cloned from rat kidney and one rEST (CB737173) cloned from hypothalamus (Fig. 11.16C). Two full-length rNCC cDNA clones of ~4.4 kb (rNCCa) and ~3.3 kb (rNCCb) were originally isolated from a rat kidney cDNA library. Both rNCC transcripts differ at their 3′-UTRs; rNCCb is shorter than rNCCa in 231 bp (Gamba et al., 1994). The rNCC cDNA posted in the *GenBank* with accession number U10097 corresponds to rNCCa. According to Gamba, the unique 3′-UTR sequence of rNCC1b appears to be the result of an alternative splicing (Gamba, 2005); however, the sequence of this alternatively spliced variant is not available in the *GenBank* database and thus it cannot be analyzed *in silico*. The rNCCa cDNA U10097 starts at *nt* -12 and ends with a long 3′-UTR (~1.4 kb) encoding a single polyadenylation site (see Figs 11.14C and 11.16C). Two additional polyadenylation sites are present ~1.4 kb downstream exon 25 in the rNCC gene. Their significance is unknown.

Comparative *in silico* analysis of rat, human and mouse NCC cDNAs sequences demonstrate ~85% identity among them. The existence of a shorter rNCC version similar to mNCC-S1 (AK052755) (Fig. 11.16B) has not yet been demonstrated, but can be inferred based on the rESTs BG376581, CK479778 and CK482756 which encode for exon 17 and/or its intron (Fig. 11.16C).

C. Expression of the NCC Gene

Tissue distribution analysis by Northern blot has shown that hNCC is expressed in the kidney (Mastroianni et al., 1996). The kidney-specific expression of hNCC transcripts is supported by the presence of hNCC promoter activity in HEK cells (MacKenzie et al., 2001) and the successful heterologous expression of the hNCC transcript in MDCK cells (de Jong et al., 2003b). However, low expression levels of hNCC transcripts were detected in other tissues such as small intestine, placenta, prostate, colon, spleen and leukocytes (Abuladze et al., 1998; Chang et al., 1996). Furthermore, the full-length cDNA clone AK315298 encoding for hNCCa was obtained from placenta (Wakamatsu et al., 2008). Interestingly, the tissue expression profiling of the hNCC gene (*GeoProfile* record: GDS596 SLC12A3) using a probe derived from the 3′-UTR of hNCCa, hNCCb and hNCCc, shows that the hNCC gene expression is not restricted to the human kidney (see Fig. 11.17A). Interestingly, hNCCs are expressed more or equally abundantly in sensory ganglia such as trigeminal and DRG, as well as in the superior cervical ganglia. The functional significance of NCC expression in or outside kidney is discussed in detail in Chapter 18.

Although the pattern of tissue expression of mNCC transcripts has not been reported, the EST database (www.ncbi.nlm.nih.gov/dbEST), *UniGene* (Mm.25804) and the *Geo Profiles* contain accessible information

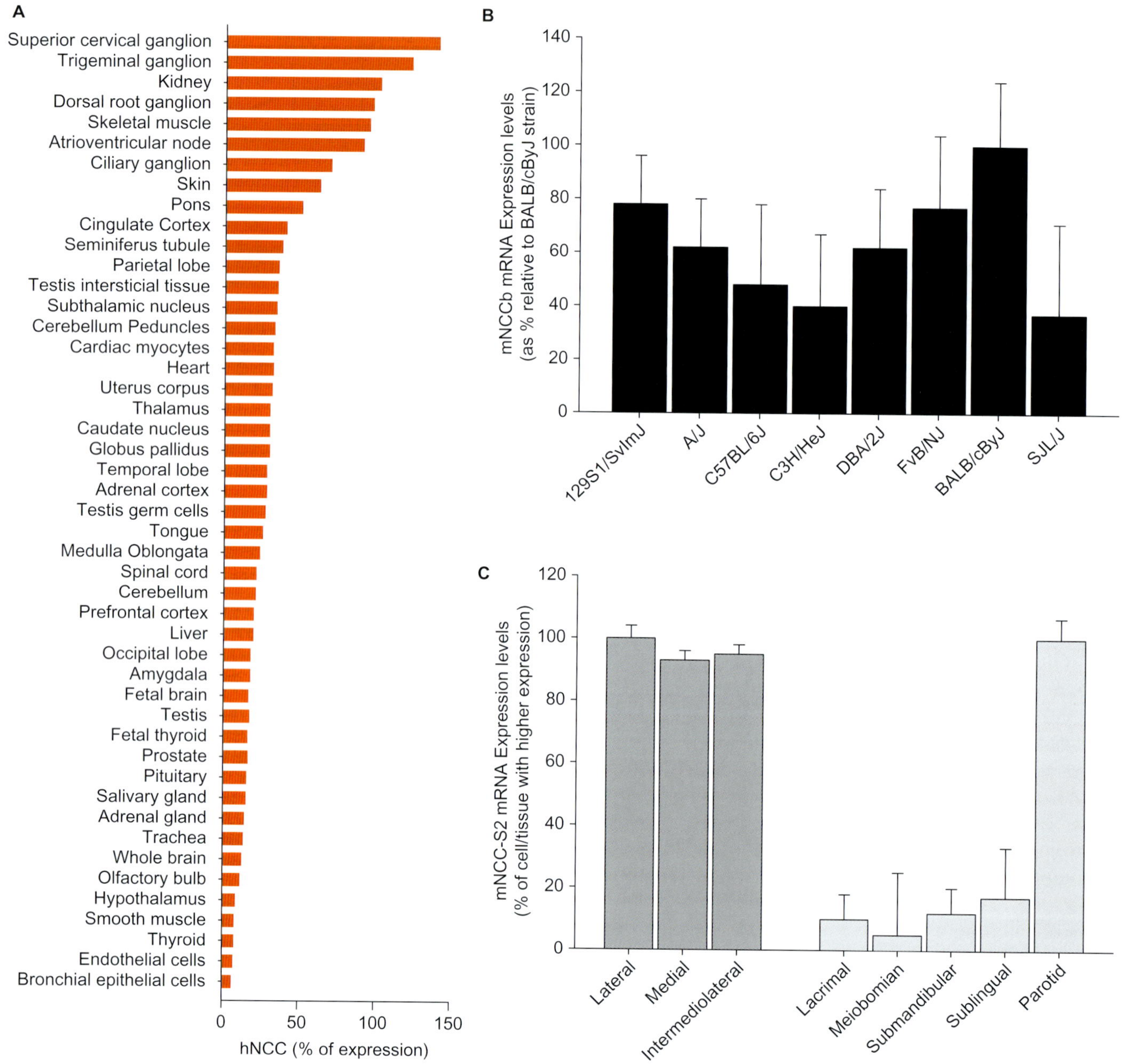

FIGURE 11.17 Expression profile of human NCC transcripts. Expression results were obtained from *GEO Profiles* (records GDS596, GDS2226, GDS1455 and GDS1009). **A.** Expression levels of hNCC transcripts were assessed by using the *Affymetrix* probe *215274_at* derived from hEST AI627943 encoding for the 3′-UTR of hNCCa, hNCCb and hNCCc mRNAs. Results are expressed as percentage with respect to kidney (100%). **B.** mNCCb mRNA expression analysis of hippocampi from eight inbred mice strains. Results obtained from platform GPL81 and by using the *Affimetrix* probe *93570_at* designed based on the mNCCb cDNA sequence (U61085). **C.** mNCC mRNA expression profile in mouse motoneurons and glandular tissues. Results obtained by using GPL81/93570_at (dark gray bars) or platform GPL1310 with probe 7544 (light gray bars) designed to recognize exon 24 and its intron in mNCC-S2.

about tissue distribution of mNCC transcripts. For instance, several mESTs encoding for mNCCs have been cloned from kidney, inner ear and spleen whereas the mouse *Geo Profile* records GDS2226 and GDS1455 (platform GPL81 *93570_at*), and GDS1009 (GPL1310/7544) show that mNCCb or mNCC-S2 mRNAs are expressed in extra-renal tissues (Figs 11.17B and 11.17C).

Definitive proof of rNCC expression in the rat kidney has been provided by using transgenic rats harboring a transgene consisting in region -2077 to -1 of the rNCC gene fused upstream of the LacZ (β-galactosidase) gene. β-galactosidase immunoreactivity was detected in these rats at the distal collecting tubules in the renal cortical region. However, β-galactosidase expression in other tissues of the same animal was not reported (Taniyama et al., 2001). The tissue expression pattern of the rNCC gene was ascertained by Northern blot analysis of several rat tissues (Gamba et al., 1994). Two rNCC transcripts were detected in kidney but not in other tissues.

D. Is NCC Gene Expression Transcriptionally Regulated?

Human and rat NCC gene promoters have been partially characterized (MacKenzie et al., 2001; Taniyama et al., 2001); however, virtually nothing is known about the molecular mechanisms involved in NCC gene transcription. Whole-animal studies have implicated aldosterone in the regulation of NCC expression. In rats, aldosterone significantly increases the number of thiazide-binding sites and NCC protein expression in renal cortical membranes (Chen et al., 1994; Kim et al., 1998; Velazquez et al., 1996). However, in other studies, aldosterone decreases rNCC abundance without altering its mRNA levels (Wang et al., 2001). The rNCC promoter contains two putative glucocorticoid-responsive elements (GRE), the mechanisms used by glucocorticoids and mineralocorticoids to stimulate rNCC are unknown. The functional significance of aldosterone on NCC activity is discussed in Chapter 18.

A putative cAMP-responsive element (CRE) has also been identified in the rNCC gene promoter, suggesting that rNCC transcription may be regulated by cAMP-inducing factors and/or drugs. The rNCC promoter region contains several binding elements for general transcription factors (Taniyama et al., 2001). *In silico* analysis of the rNCC promoter region (*GenBank* acc #AB024534, corresponding to *nt* -2077 to 35, relative to $A^{1}TG$) using *AliBaba v2.1* (www.gene-regulation.com) suggests the presence of additional binding elements for a wide arrange of transcription factors. For instance, binding sites for two retinoid X receptors α (RXRα) at *nt* -30 and *nt* -991, four NFκ-B sites (*nt* -475, -856, -942, -1563), ten TBP sites at *nt* -483, -1229, -1369, -1383, -1410, -1412, -1428, -1435, -1445, -1595, one activating transcription factor (ATF) at *nt* -1040, one cAMP-responsive element-binding protein 1 (CRE-BP1) at *nt* -1044, one *c*-jun transcription factor at *nt* -1046, one progesterone receptor (PR) at *nt* -1112, two thyroid hormone receptors β (T3Rβ) at *nt* -1188 and -1523, three hepatocyte nuclear factors-3 (HNF-3) at *nt* -1258, -1751, -1935, one estrogen receptor (ER) at *nt* -1523 and ten sites for the ubiquitously expressed transcription factor SP-1 (*nt* -91 -149, -173, -655, -803, -839, -865, -969, -1107 and -1662).

The initiation of transcription of the rNCC gene has been proposed to start at *nt* -20, very close to the classical TATA box (TATAA) located at *nt* -62 upstream of $A^{1}TG$ (Taniyama et al., 2001). However, potential additional TATAA sequences located at *nt* -1440 and -1223 which overlap with predicted binding sites for TBP may suggest additional regulatory sites for the initiation of transcription. Deletion of *nts* between -2077 and -1123 (predicted to contain at least two TATA boxes and two TBPs) or -125 and -1 (containing a TATA box and a potential transcription initiation site) in the rNCC promoter gene results in ~50% and ~75% reduction of basal transcriptional activity, respectively (Taniyama et al., 2001). These results suggest that half of the rNCC gene promoter activity is located between *nts* -2077 and -1123 whereas a major promoter is located between *nts* -125 and -1. Interestingly, deletion of *nts* -2077 to -1007 or -2077 to -563 resulted in ~25% and ~50% reduction of basal transcription, respectively, suggesting the presence of repressor/silencer activity between *nts* -1007 and -563 of the rNCC gene promoter.

The rNCC gene promoter also contains an HNF-3/homolog of the fork head-3 (HFH-3) binding site at *nt* -419. HFH-3 (also known as *Foxi1*, *Fkh10*, or *FREAC6*) expression appears to be restricted to the epithelium of the renal distal tubule, at least at the transcript level (Overdier et al., 1997). This HFH-3 binding site is conserved in mouse but not in human NCC gene promoters. The kidney-specific expression of rNCC, in terms of gene transcription, was attributed to the presence of a functional HFH-3 binding site in the promoter region of the rNCC gene. Mutagenesis analysis of the HFH-3 binding (*nt* -419) dramatically reduced transcriptional activity of the rNCC gene promoter region encompassing *nts* -564 to -1 as well as binding of an *in vitro* synthesized HFH-3 protein (Taniyama et al., 2001).

Although the role of the HFH-3 binding site on the entire promoter sequence has not been studied, mice lacking HFH-3 are deaf and display a phenotype very similar to that reported for the anion transporter pendrin knockout mice (Everett et al., 2001; Hulander et al., 2003) and the K^+-Cl^- cotransporter KCC4 knockout mice (Boettger et al., 2002). Interestingly, HFH-3 knockout mice do not express pendrin in the endolymphatic inner ear epithelium and anion exchangers 1 and 4 (AE1 and AE4) in kidney. However, HFH-3 knockout do not affect KCC4 expression, thus establishing HFH-3 as a direct or indirect upstream regulator of pendrin and AE1 and AE4 expression, in at least the inner-ear and kidney (Blomqvist et al., 2004; Hulander et al., 2003). Surprisingly, the possible link between HFH-3 and NCC expression *in vivo* has not been studied.

E. Post-translational Regulation of NCC Proteins: Phosphorylation

The regulation of NCC function has been mainly studied in *Xenopus* oocytes (Gamba, 2005). Although these studies have provided important new insights into the regulation of NCCs, a major drawback of the oocyte expression system is that it lacks the physiological background and polarized structure of native epithelial kidney cell. It is widely accepted, however, that Na^+-driven cotransporters (all NCCs and NKCC1s and NKCC2s) are activated by phosphorylation. Point and deletion mutants of NCC expressed in *Xenopus* oocytes have been extensively used as a model to elucidate the molecular mechanisms of NCC regulation by phosphorylation, as discussed in detail in Chapter 18 in this volume. These studies suggest that WNK3 kinase increases NCC function and NCC regulation may result from antagonism between WNK3 and WNK4; while WNK3 stimulates NCC activity via the C-terminal domain of the cotransporter whereas WNK4 inhibits. A note of caution in this kind of study is that detecting an effect of mutations using the *X. laevis* oocyte expression system may not recapitulate or reproduce with fidelity a genetic defect that occurs *in vivo*. Furthermore, the C-terminal portion of human, rat and mouse NCC proteins is not identical, as discussed, suggesting that there could be species differences between NCC regulatory mechanisms.

F. Post-translational Regulation of NCC Proteins: N-glycosylation

In silico analysis of hNCCa protein sequence using *NetGlyc v1.0* server (www.cbs.dtu.dk/services/NetNGlyc) predicts several N-glycosylation sequons which are conserved among species (i.e. N^{72}, N^{359}, N^{406}, N^{426}, N^{544}, N^{668} and N^{781}). However, some of these N residues may not be N-glycosylated *in vivo*. N^{406} and N^{426} are predicted to be located in the large extracellular loop between the TM7 and TM8 (see Fig. 11.4).

Western blot analyses of hNCC or rNCC kidney extracts (De Jong et al., 2002, 2003a) show bands at 135 – 150 kDa molecular weight, which is higher than predicted (~110 kDa) for the core molecular weight of NCC proteins. Deglycosylation with N-glycosidase F shifts the high molecular weight bands down to ~110 kDa (Hoover et al., 2003; Plotkin et al., 1996). Moreover, the molecular weight of rNCCa (U10097) and mNCCb (U61085) are reduced to ~115–120 kDa after enzymatic deglycosylation, when expressed in *X. laevis* oocytes (Hoover et al., 2003; Kunchaparty et al., 1999). Interestingly, the molecular size of mNCCb expressed in oocytes (~115–120 kDa) is different to that of the proteins expressed in native renal tissue, suggesting that the glycosylation machinery of oocytes may be different to that of distal convoluted tubular cells.

Definitive proof that rNCCa is N-glycosylated when heterologously expressed in *X. laevis* oocytes was provided by experiments of site-directed mutagenesis where N^{404} and/or N^{424} (equivalent to hNCC1′ N^{406} and N^{426}) were changed to glutamine (i.e. N404Q, N424Q and N404/424Q) (Hoover et al., 2003). Expression of the single or double rNCCa mutants in *X. laevis* oocytes resulted in proteins of ~110 kDa, suggesting that N^{404} or N^{424} are mutually and exclusively N-glycosylated (Hoover et al., 2003). Interestingly, the single rNCCa mutants (N404Q and N424Q) were half active when compared to wild type, whereas the double mutant (N404/424Q) was completely inactive. Taken together, these results suggest that N-glycosylation of N^{404} and N^{424} are necessary for a fully active rNCCa protein when expressed in *X. laevis* oocytes. Although the actual impact of N-glycosylation on the transport capacity of the rNCCa protein is unknown, the lack of activity in the double mutant rNCCa N404/424Q appears to be related to a deficient protein delivery to the oocyte plasma membrane (Hoover et al., 2003). However, single and double rNCCa mutants had similar expression levels at the plasma membrane, although reduced (~50%), when compared to wild type rNCCa. Taken together, these experiments suggested that N-glycosylation of rNCCa plays a critical role in its function when expressed in *X. laevis* oocytes. However, the mechanisms involved are yet not clear. Further discussion on this topic can be found in Chapter 18.

G. What Happens when NCC is not Expressed?

Mutations of the hNCC gene cause Gitelman's syndrome (GS, OMIM #263800, 600968), a condition similar to Bartter's syndrome (see section II.I above) but characterized by hypokalemic alkalosis combined with hypomagnesemia, low urinary calcium, and increased renin activity associated with normal blood pressure. These metabolic abnormalities are secondary to deficient NaCl reabsortion in the kidney (Simon et al., 1996b). Known hNCC gene mutations related or linked to Gitelman's syndrome are posted in the HGMD (www.hgmd.cf.ac.uk/ac/index.php). Out of 110 mutations of the hNCC gene, 81 are missense (~74%), ten splicing mutants (~9%), 11 small deletions (10%), five small insertions (~4%), a small insertion-deletion (*indel*, ~1%) and two gross deletions (~2%). Interestingly, GS mutations have not been described within exon 20, the genomic region absent in mouse or rat NCC genes. Only 18% of the reported GS cases are homozygous whereas half are double heterozygous. Interestingly, ~30% of the patients with GS had only one affected allele, or no mutation at all in the coding portion of the NCC gene in spite of clinical and biochemical diagnosis of GS (Riveira-Muñoz et al., 2007; Simon et al., 1996b). Mutations in the promoter region of the hNCC gene as well as 5′- and/or 3′-UTRs may explain some of the GS cases that do not show mutations in the coding portion of the NCC gene but exhibit all the clinical and biochemical signs of the disease. These putative mutations in the promoter region of the hNCC gene may affect the efficacy of transcription of the NCC gene. Mutations in the 5′- and/or 3′-UTRs may destabilize NCC transcripts. Clearly, this is an interesting avenue for studies.

NCC knockout mice were created by introducing the neomycin gene inside exon 12 in the mNCC gene (Schultheis et al., 1998). Interestingly, loss of NCC mRNA expression in the mouse causes mild perturbations of sodium and fluid volume homeostasis that appear to be largely compensated. However, knockout mice exhibit morphological changes in cells of the distal convoluted tubule. Besides hypocalciuria and hypomagnesemia, the NCC knockout mice lack the characteristic features of GS, suggesting that either mNCC function may not be identical to that in the human homolog and/or that the ablation of mNCC expression does not reproduce the impaired functional properties of the point mutations observed in GS.

The emerging picture of these studies is the decrease in cotransport activity when the mutant NCC proteins are expressed in *X. laevis* oocytes. It is not clear if the decrease in cotransport activity reflects decrease surface expression and/or a molecular defect in the translocation machinery. Some of the mutant NCC proteins are not glycosylated and do not appear to reach the oocyte plasma membrane. Because the NCC mutants may not be fully expressed in the plasma membrane, their transport activity cannot be determined.

Acknowledgements

This work was partially supported by the National Institute of Neurological Disorders and Stroke, Grant NS-29227 and Wright State University Boonshoft School of Medicine Seed Grant Program, Grant 229099 To FJA-L. We thank Dr. Dan Halm for his helpful comments in an early version of the manuscript.

References

Abuladze, N., Yanagawa, N., Lee, I., Jo, O.D., Newman, D., Hwang, J., Uyemura, K., Pushkin, A., Modlin, R.L., and Kurtz, I. (1998). Peripheral blood mononuclear cells express mutated NCCT mRNA in Gitelman's syndrome: evidence for abnormal thiazide-sensitive NaCl cotransport. *J. Am. Soc. Nephrol.* **9**, 819–826.

Alvarez-Leefmans, F.J., León-Olea, M., Mendoza-Sotelo, J., Alvarez, F.J., Antón, B., and Garduño, R. (2001). Immunolocalization of the Na^+-K^+-$2Cl^-$ cotransporter in peripheral nervous tissue of vertebrates. *Neuroscience* **104**, 569–582.

Anzai, N., Izumida, I., Kobayashi, Y., and Kawahara, K. (1999). Roles of vasopressin and hypertonicity in basolateral Na/K/2Cl cotransporter expression in rat kidney inner medullary collecting duct cells. *Jpn. J. Physiol.* **49**, 201–206.

Ares, G.R., Caceres, P., Alvarez-Leefmans, F.J., and Ortiz, P. A. (2008). cGMP decreases surface NKCC2 levels in the thick ascending limb: role of phosphodiesterase 2 (PDE2). *Am. J. Physiol. Renal Physiol.* **295**, F877–F887.

Asano, T., Katagiri, H., Takata, K., Lin, J.L., Ishihara, H., Inukai, K., Tsukuda, K., Kikuchi, M., Hirano, H., Yazaki, Y. et al. (1991). The role of N-glycosylation of GLUT1 for glucose transport activity. *J. Biol. Chem.* **266**, 24632–24636.

Bause, E., Hettkamp, H., and Legler, G. (1982). Conformational aspects of N-glycosylation of proteins Studies with linear and cyclic peptides as probes. *Biochem J.* **203**, 761–768.

Ben-Dor, S., Esterman, N., Rubin, E., and Sharon, N. (2004). Biases and complex patterns in the residues flanking protein N-glycosylation sites. *Glycobiology* **14**, 95–101.

Benson, D.A., Karsch-Mizrachi, I., Lipman, D.J., Ostell, J., and Wheeler, D.L. (2008). GenBank. *Nucleic Acids Res.* **36**, D25–D30.

Benziane, B., Demaretz, S., Defontaine, N., Zaarour, N., Cheval, L., Bourgeois, S., Klein, C., Froissart, M., Blanchard, A., Paillard, M., Gamba, G., Houillier, P., and Laghmani, K. (2007). NKCC2 surface expression in mammalian cells: down-regulation by novel interaction with aldolase B. *J. Biol. Chem.* **282**, 33817–33830.

Bernsel, A., Viklund, H., Falk, J., Lindahl, E., von Heijne, G., and Elofsson, A. (2008). Prediction of membrane-protein topology from first principles. *Proc. Natl. Acad. Sci. USA* **105**, 7177–7181.

Blomqvist, S.R., Vidarsson, H., Fitzgerald, S., Johansson, B.R., Ollerstam, A., Brown, R., Persson, A.E., Bergstrom, G.G., and Enerback, S. (2004). Distal renal tubular acidosis in mice that lack the forkhead transcription factor Foxi1. *J. Clin. Invest* **113**, 1560–1570.

Blundell, T.L., Sibanda, B.L., Sternberg, M.J., and Thornton, J.M. (1987). Knowledge-based prediction of protein structures and the design of novel molecules. *Nature* **326**, 347–352.

Boettger, T., Hubner, C.A., Maier, H., Rust, M.B., Beck, F.X., and Jentsch, T.J. (2002). Deafness and renal tubular acidosis in mice lacking the K-Cl co-transporter Kcc4. *Nature* **416**, 874–878.

Brunak, S., Engelbrecht, J., and Knudsen, S. (1991). Prediction of human mRNA donor and acceptor sites from the DNA sequence. *J. Mol. Biol.* **220**, 49–65.

Bulleid, N.J., Bassel-Duby, R.S., Freedman, R.B., Sambrook, J.F., and Gething, M.J. (1992). Cell-free synthesis of enzymically active tissue-type plasminogen activator Protein folding determines the extent of N-linked glycosylation. *Biochem. J.* **286**, 275–280.

Burge, C. and Karlin, S. (1997). Prediction of complete gene structures in human genomic DNA. *J. Mol. Biol.* **268**, 78–94.

Caron, L., Rousseau, F., Gagnon, E., and Isenring, P. (2000). Cloning and functional characterization of a cation-Cl^- cotransporter-interacting protein. *J. Biol. Chem.* **275**, 32027–32036.

Cartharius, K., Frech, K., Grote, K., Klocke, B., Haltmeier, M., Klingenhoff, A., Frisch, M., Bayerlein, M., and Werner, T. (2005). MatInspector and beyond: promoter analysis based on transcription factor binding sites. *Bioinformatics* **21**, 2933–2942.

Chang, H., Tashiro, K., Hirai, M., Ikeda, K., Kurokawa, K., and Fujita, T. (1996). Identification of a cDNA encoding a thiazide-sensitive sodium-chloride cotransporter from the human and its mRNA expression in various tissues. *Biochem. Biophys. Res. Commun.* **223**, 324–328.

Chen, Z., Vaughn, D.A., Blakely, P., and Fanestil, D.D. (1994). Adrenocortical steroids increase renal thiazide diuretic receptor density and response. *J. Am. Soc. Nephrol.* **5**, 1361–1368.

Clayton, G.H., Owens, G.C., Wolff, J.S., and Smith, R.L. (1998). Ontogeny of cation-Cl^- cotransporter expression in rat neocortex. *Brain Res. Dev. Brain Res.* **109**, 281–292.

Cochrane, G., Akhtar, R., Aldebert, P., Althorpe, N., Baldwin, A., Bates, K., Bhattacharyya, S., Bonfield, J., Bower, L., Browne, P., Castro, M., Cox, T., Demiralp, F., Eberhardt, R., Faruque, N., Hoad, G., Jang, M., Kulikova, T., Labarga, A., Leinonen, R., Leonard, S., Lin, Q., Lopez, R., Lorenc, D., McWilliam, H., Mukherjee, G., Nardone, F., Plaister, S., Robinson, S., Sobhany, S., Vaughan, R., Wu, D., Zhu, W., Apweiler, R., Hubbard, T., and Birney, E. (2008). Priorities for nucleotide trace, sequence and annotation data capture at the Ensembl Trace Archive and the EMBL Nucleotide Sequence Database. *Nucleic Acids Res.* **36**, D5–D12.

Darman, R.B., Flemmer, A., and Forbush, B. (2001). Modulation of ion transport by direct targeting of protein phosphatase type 1 to the Na-K-Cl cotransporter. *J. Biol. Chem.* **276**, 34359–34362.

Darman, R.B. and Forbush, B. (2002). A regulatory locus of phosphorylation in the N terminus of the Na-K-Cl cotransporter, NKCC1. *J. Biol. Chem.* **277**, 37542–37550.

Davuluri, R.V., Grosse, I., and Zhang, M.Q. (2001). Computational identification of promoters and first exons in the human genome. *Nat. Genet.* **29**, 412–417.

Davuluri, R.V., Suzuki, Y., Sugano, S., and Zhang, M.Q. (2000). CART classification of human 5′ UTR sequences. *Genome Res.* **10**, 1807–1816.

de Jong, J.C., Van Der Vliet, W.A., Van Den Heuvel, L.P., Willems, P. H., Knoers, N.V., and Bindels, R.J. (2002). Functional expression of mutations in the human NaCl cotransporter: evidence for impaired routing mechanisms in Gitelman's syndrome. *J. Am. Soc. Nephrol.* **13**, 1442–1448.

de Jong, J.C., Willems, P.H., Mooren, F.J., van den Heuvel, L.P., Knoers, N.V., and Bindels, R.J. (2003a). The structural unit of the thiazide-sensitive NaCl cotransporter is a homodimer. *J. Biol. Chem.* **278**, 24302–24307.

de Jong, J.C., Willems, P.H., van den Heuvel, L.P., Knoers, N.V., and Bindels, R.J. (2003b). Functional expression of the human thiazide-sensitive NaCl cotransporter in Madin-Darby canine kidney cells. *J. Am. Soc. Nephrol.* **14**, 2428–2435.

de Jonge, H.J., Fehrmann, R.S., de Bont, E.S., Hofstra, R.M., Gerbens, F., Kamps, W.A., de Vries, E.G., van der Zee, A.G., te Meerman, G.J., and ter Elst, A. (2007). Evidence based selection of housekeeping genes. *PLoS ONE* **2**, e898.

Delpire, E., Lu, J., England, R., Dull, C., and Thorne, T. (1999). Deafness and imbalance associated with inactivation of the secretory Na-K-2Cl co-transporter. *Nat. Genet.* **22**, 192–195.

Delpire, E. and Mount, D.B. (2002). Human and murine phenotypes associated with defects in cation-chloride cotransport. *Annu. Rev. Physiol.* **64**, 803–843.

Delpire, E., Rauchman, M.I., Beier, D.R., Hebert, S.C., and Gullans, S.R. (1994). Molecular cloning and chromosome localization of a putative basolateral $Na(^+)$-$K(^+)$-$2Cl^-$ cotransporter from mouse inner medullary collecting duct (mIMCD-3) cells. *J. Biol. Chem.* **269**, 25677–25683.

Deschenes, G., Burguet, A., Guyot, C., Hubert, P., Garabedian, M., Dechaux, M., Loirat, C., and Broyer, M. (1993). [Antenatal form of Bartter's syndrome]. *Ann. Pediatr. (Paris)* **40**, 95–101.

Di Fulvio, M., Garzon-Muvdi, T., and Alvarez-Leefmans, F.J. (2007). Molecular cloning and characterization of NKCC2A in non-renal tissues. *EF559316 and EF577032*, http://www.ncbi.nlm.nih.gov/sites/entrez.

Dixon, M.J., Gazzard, J., Chaudhry, S.S., Sampson, N., Schulte, B.A., and Steel, K.P. (1999). Mutation of the Na-K-Cl co-transporter gene Slc12a2 results in deafness in mice. *Hum. Mol. Genet.* **8**, 1579–1584.

Doolittle, R.F. (1981). Similar amino acid sequences: chance or common ancestry? *Science* **214**, 149–159.

Eden, E. and Brunak, S. (2004). Analysis and recognition of 5′ UTR intron splice sites in human pre-mRNA. *Nucleic Acids Res.* **32**, 1131–1142.

Eisenberg, E. and Levanon, E.Y. (2003). Human housekeeping genes are compact. *Trends Genet.* **19**, 362–365.

Espanol, M.J. and Saier, M.H., Jr. (1995). Topological and segmental phylogenetic analyses of the anion exchanger (band 3) family of transporters. *Mol. Membr. Biol.* **12**, 193–200.

Everett, L.A., Belyantseva, I.A., Noben-Trauth, K., Cantos, R., Chen, A., Thakkar, S.I., Hoogstraten-Miller, S.L., Kachar, B., Wu, D.K., and Green, E.D. (2001). Targeted disruption of mouse Pds provides insight about the inner-ear defects encountered in Pendred syndrome. *Hum. Mol. Genet.* **10**, 153–161.

Eyre, T.A., Ducluzeau, F., Sneddon, T.P., Povey, S., Bruford, E. A., and Lush, M.J. (2006). The HUGO Gene Nomenclature Database, 2006 updates. *Nucleic Acids Res.* **34**, D319–D321.

Faisst, S. and Meyer, S. (1992). Compilation of vertebrate-encoded transcription factors. *Nucleic Acids Res.* **20**, 3–26.

Farre, D., Bellora, N., Mularoni, L., Messeguer, X., and Alba, M.M. (2007). Housekeeping genes tend to show reduced upstream sequence conservation. *Genome Biol.* **8**, R140.

Flagella, M., Clarke, L.L., Miller, M.L., Erway, L.C., Giannella, R.A., Andringa, A., Gawenis, L.R., Kramer, J., Duffy, J.J., Doetschman, T., Lorenz, J.N., Yamoah, E.N., Cardell, E.L., and Shull, G.E. (1999). Mice lacking the basolateral Na-K-2Cl cotransporter have impaired epithelial chloride secretion and are profoundly deaf. *J. Biol. Chem.* **274**, 26946–26955.

Flemmer, A.W., Gimenez, I., Dowd, B.F., Darman, R.B., and Forbush, B. (2002). Activation of the Na-K-Cl cotransporter NKCC1 detected with a phospho-specific antibody. *J. Biol. Chem.* **277**, 37551–37558.

Fraser, S.A., Gimenez, I., Cook, N., Jennings, I., Katerelos, M., Katsis, F., Levidiotis, V., Kemp, B.E., and Power, D.A. (2007).

Regulation of the renal-specific Na^+-K^+-$2Cl^-$ co-transporter NKCC2 by AMP-activated protein kinase (AMPK). *Biochem. J.* **405**, 85–93.

Fu, L., Wong, J.A., Schneider, E.G., and Thomason, D.B. (1999). Unique 5′-end of a $Na(^+)$-$K(^+)$-$2Cl^-$ cotransporter-like mRNA expressed in rat skeletal muscle. *DNA Seq.* **10**, 127–132.

Galperin, M.Y. (2008). The Molecular Biology Database Collection: 2008 update. *Nucleic Acids Res.* **36**, D2–D4.

Gamba, G. (2005). Molecular physiology and pathophysiology of electroneutral cation-chloride cotransporters. *Physiol. Rev.* **85**, 423–493.

Gamba, G., Miyanoshita, A., Lombardi, M., Lytton, J., Lee, W. S., Hediger, M.A., and Hebert, S.C. (1994). Molecular cloning, primary structure, and characterization of two members of the mammalian electroneutral sodium-(potassium)-chloride cotransporter family expressed in kidney. *J. Biol. Chem.* **269**, 17713–17722.

Gavel, Y. and von Heijne, G. (1990). Sequence differences between glycosylated and non-glycosylated Asn-X-Thr/Ser acceptor sites: implications for protein engineering. *Protein Eng* **3**, 433–442.

Gavrikov, K.E., Nilson, J.E., Dmitriev, A.V., Zucker, C.L., and Mangel, S.C. (2006). Dendritic compartmentalization of chloride cotransporters underlies directional responses of starburst amacrine cells in retina. *Proc. Natl. Acad. Sci. USA* **103**, 18793–18798.

Gillen, C.M., Brill, S., Payne, J.A., and Forbush, B., 3rd (1996). Molecular cloning and functional expression of the K-Cl cotransporter from rabbit, rat, and human A new member of the cation-chloride cotransporter family. *J. Biol. Chem.* **271**, 16237–16244.

Gimenez, I. and Forbush, B. (2003). Short-term stimulation of the renal Na-K-Cl cotransporter (NKCC2) by vasopressin involves phosphorylation and membrane translocation of the protein. *J. Biol. Chem.* **278**, 26946–26951.

Gimenez, I. and Forbush, B. (2005). Regulatory phosphorylation sites in the NH2 terminus of the renal Na-K-Cl cotransporter (NKCC2). *Am. J. Physiol. Renal Physiol.* **289**, F1341–F1345.

Granseth, E., Viklund, H., and Elofsson, A. (2006). ZPRED: predicting the distance to the membrane center for residues in alpha-helical membrane proteins. *Bioinformatics* **22**, e191–e196.

Haas, M. and Forbush, B., 3rd (2000). The Na-K-Cl cotransporter of secretory epithelia. *Annu. Rev. Physiol.* **62**, 515–534.

Hart, G.W. (1992). Glycosylation. *Curr. Opin. Cell Biol.* **4**, 1017–1023.

Hasegawa, M., Orita, T., Kojima, T., Tomonoh, K., Hirata, Y., and Ochi, N. (1992). Improvement in the heterogeneous N-termini and the defective N-glycosylation of human interleukin-6 by genetic engineering. *Eur. J. Biochem.* **210**, 9–12.

Hayes, G., Busch, A., Lotscher, M., Waldegger, S., Lang, F., Verrey, F., Biber, J., and Murer, H. (1994). Role of N-linked glycosylation in rat renal Na/Pi-cotransport. *J. Biol. Chem.* **269**, 24143–24149.

Hebert, S.C., Culpepper, R.M., and Andreoli, T.E. (1981). NaCl transport in mouse medullary thick ascending limbs II. ADH enhancement of transcellular NaCl cotransport; origin of transepithelial voltage. *Am. J. Physiol.* **241**, F432–F442.

Hendriks, G., Koudijs, M., van Balkom, B.W., Oorschot, V., Klumperman, J., Deen, P.M., and van der Sluijs, P. (2004). Glycosylation is important for cell surface expression of the water channel aquaporin-2 but is not essential for tetramerization in the endoplasmic reticulum. *J. Biol. Chem.* **279**, 2975–2983.

Hessa, T., Meindl-Beinker, N.M., Bernsel, A., Kim, H., Sato, Y., Lerch-Bader, M., Nilsson, I., White, S.H., and von Heijne, G. (2007). Molecular code for transmembrane-helix recognition by the Sec61 translocon. *Nature* **450**, 1026–1030.

Hiki, K., D'Andrea, R.J., Furze, J., Crawford, J., Woollatt, E., Sutherland, G.R., Vadas, M.A., and Gamble, J.R. (1999). Cloning, characterization, and chromosomal location of a novel human K^+-Cl^- cotransporter. *J. Biol. Chem.* **274**, 10661–10667.

Hoover, R.S., Poch, E., Monroy, A., Vazquez, N., Nishio, T., Gamba, G., and Hebert, S.C. (2003). N-Glycosylation at two sites critically alters thiazide binding and activity of the rat thiazide-sensitive Na^+:Cl^- cotransporter. *J. Am. Soc. Nephrol.* **14**, 271–282.

Hulander, M., Kiernan, A.E., Blomqvist, S.R., Carlsson, P., Samuelsson, E.J., Johansson, B.R., Steel, K.P., and Enerback, S. (2003). Lack of pendrin expression leads to deafness and expansion of the endolymphatic compartment in inner ears of Foxi1 null mutant mice. *Development* **130**, 2013–2025.

Ibla, J.C., Khoury, J., Kong, T., Robinson, A., and Colgan, S.P. (2006). Transcriptional repression of Na-K-2Cl cotransporter NKCC1 by hypoxia-inducible factor-1. *Am. J. Physiol. Cell Physiol.* **291**, C282–C289.

Igarashi, P., Vanden Heuvel, G.B., Payne, J.A., and Forbush, B., 3rd (1995). Cloning, embryonic expression, and alternative splicing of a murine kidney-specific Na-K-Cl cotransporter. *Am. J. Physiol.* **269**, F405–F418.

Igarashi, P., Whyte, D.A., Li, K., and Nagami, G.T. (1996). Cloning and kidney cell-specific activity of the promoter of the murine renal Na-K-Cl cotransporter gene. *J. Biol. Chem.* **271**, 9666–9674.

Isenring, P., Jacoby, S.C., Payne, J.A., and Forbush, B., 3rd (1998). Comparison of Na-K-Cl cotransporters. NKCC1, NKCC2, and the HEK cell Na-L-Cl cotransporter. *J. Biol. Chem.* **273**, 11295–11301.

Jenkins, N., Parekh, R.B., and James, D.C. (1996). Getting the glycosylation right: implications for the biotechnology industry. *Nat. Biotechnol.* **14**, 975–981.

Jones, J., Krag, S.S., and Betenbaugh, M.J. (2005). Controlling N-linked glycan site occupancy. *Biochim. Biophys. Acta* **1726**, 121–137.

Kakigi, A., Nishimura, M., Takeda, T., Taguchi, D., and Nishioka, R. (2008). Expression of aquaporin1, 3, and 4, NKCC1, and NKCC2 in the human endolymphatic sac. *Auris Nasus Larynx*.

Kane, S.E. (1993). Mouse procathepsin L lacking a functional glycosylation site is properly folded, stable, and secreted by NIH 3T3 cells. *J. Biol. Chem.* **268**, 11456–11462.

Kaplan, M.R., Mount, D.B., and Delpire, E. (1996). Molecular mechanisms of NaCl cotransport. *Annu. Rev. Physiol.* **58**, 649–668.

Kasturi, L., Chen, H., and Shakin-Eshleman, S.H. (1997). Regulation of N-linked core glycosylation: use of a site-directed mutagenesis approach to identify Asn-Xaa-Ser/Thr sequons that are poor oligosaccharide acceptors. *Biochem. J.* **323**, 415–419.

Katayama, S., Tomaru, Y., Kasukawa, T., Waki, K., Nakanishi, M., Nakamura, M., Nishida, H., Yap, C.C., Suzuki, M., Kawai, J., Suzuki, H., Carninci, P., Hayashizaki, Y., Wells, C., Frith, M., Ravasi, T., Pang, K.C., Hallinan, J., Mattick, J., Hume, D.A., Lipovich, L., Batalov, S., Engstrom, P.G., Mizuno, Y., Faghihi, M.A., Sandelin, A., Chalk, A.M., Mottagui-Tabar, S., Liang, Z., Lenhard, B., and Wahlestedt, C. (2005). Antisense transcription in the mammalian transcriptome. *Science* **309**, 1564–1566.

Kim, G.H., Masilamani, S., Turner, R., Mitchell, C., Wade, J.B., and Knepper, M.A. (1998). The thiazide-sensitive Na-Cl cotransporter is an aldosterone-induced protein. *Proc. Natl. Acad. Sci. USA* **95**, 14552–14557.

Korf, I., Flicek, P., Duan, D., and Brent, M.R. (2001). Integrating genomic homology into gene structure prediction. *Bioinformatics* **17** (1), S140–S148.

Kornfeld, R. and Kornfeld, S. (1985). Assembly of asparagine-linked oligosaccharides. *Annu. Rev. Biochem.* **54**, 631–664.

Kunchaparty, S., Palcso, M., Berkman, J., Velazquez, H., Desir, G. V., Bernstein, P., Reilly, R.F., and Ellison, D.H. (1999). Defective processing and expression of thiazide-sensitive Na-Cl cotransporter as a cause of Gitelman's syndrome. *Am. J. Physiol.* **277**, F643–F649.

Kyte, J. and Doolittle, R.F. (1982). A simple method for displaying the hydropathic character of a protein. *J. Mol Biol.* **157**, 105–132.

Liedtke, C.M., Wang, X., and Smallwood, N.D. (2005). Role for protein phosphatase 2A in the regulation of Calu-3 epithelial Na^+-K^+-$2Cl^-$, type 1 co-transport function. *J. Biol. Chem.* **280**, 25491–25498.

Lytle, C. (1997). Activation of the avian erythrocyte Na-K-Cl cotransport protein by cell shrinkage, cAMP, fluoride, and calyculin-A involves phosphorylation at common sites. *J. Biol. Chem.* **272**, 15069–15077.

Lytle, C. and Forbush, B., 3rd (1992). The Na-K-Cl cotransport protein of shark rectal gland II. Regulation by direct phosphorylation. *J. Biol. Chem.* **267**, 25438–25443.

Lytle, C. and Forbush, B., 3rd (1996). Regulatory phosphorylation of the secretory Na-K-Cl cotransporter: modulation by cytoplasmic Cl. *Am. J. Physiol.* **270**, C437–C448.

Lytle, C., Xu, J.C., Biemesderfer, D., and Forbush, B., 3rd (1995). Distribution and diversity of Na-K-Cl cotransport proteins: a study with monoclonal antibodies. *Am. J. Physiol.* **269**, C1496–C1505.

Lytle, C., Xu, J.C., Biemesderfer, D., Haas, M., and Forbush, B., 3rd (1992). The Na-K-Cl cotransport protein of shark rectal gland I. Development of monoclonal antibodies, immunoaffinity purification, and partial biochemical characterization. *J. Biol. Chem.* **267**, 25428–25437.

MacKenzie, S., Vaitkevicius, H., and Lockette, W. (2001). Sequencing and characterization of the human thiazide-sensitive Na-Cl cotransporter (SLC12A3) gene promoter. *Biochem. Biophys. Res. Commun.* **282**, 991–1000.

Marty, S., Wehrle, R., Alvarez-Leefmans, F.J., Gasnier, B., and Sotelo, C. (2002). Postnatal maturation of Na^+, K^+, $2Cl^-$ cotransporter expression and inhibitory synaptogenesis in the rat hippocampus: an immunocytochemical analysis. *Eur. J. Neurosci.* **15**, 233–245.

Maruyama, K. and Sugano, S. (1994). Oligo-capping: a simple method to replace the cap structure of eukaryotic mRNAs with oligoribonucleotides. *Gene* **138**, 171–174.

Mastroianni, N., De Fusco, M., Zollo, M., Arrigo, G., Zuffardi, O., Bettinelli, A., Ballabio, A., and Casari, G. (1996). Molecular cloning, expression pattern, and chromosomal localization of the human Na-Cl thiazide-sensitive cotransporter (SLC12A3). *Genomics* **35**, 486–493.

Matthews, J.B., Hassan, I., Meng, S., Archer, S.Y., Hrnjez, B.J., and Hodin, R.A. (1998). Na-K-2Cl cotransporter gene expression and function during enterocyte differentiation Modulation of Cl^- secretory capacity by butyrate. *J. Clin. Invest.* **101**, 2072–2079.

May, A.C. (2004). Percent sequence identity; the need to be explicit. *Structure* **12**, 737–738.

Molony, D.A., Reeves, W.B., Hebert, S.C., and Andreoli, T.E. (1987). ADH increases apical Na^+, K^+, $2Cl^-$ entry in mouse medullary thick ascending limbs of Henle. *Am. J. Physiol.* **252**, F177–F187.

Moore-Hoon, M.L. and Turner, R.J. (1998). Molecular and topological characterization of the rat parotid Na^+-K^+-$2Cl^-$ cotransporter1. *Biochim. Biophys. Acta* **1373**, 261–269.

Mount, D.B., Baekgaard, A., Hall, A.E., Plata, C., Xu, J., Beier, D.R., Gamba, G., and Hebert, S.C. (1999). Isoforms of the Na-K-2Cl cotransporter in murine TAL I. Molecular characterization and intrarenal localization. *Am. J. Physiol* **276**, F347–F358.

Mount, D.W. (2004). *Bioinformatics: Sequence and Genome Analysis.* Cold Spring Harbor Laboratory Press, Cold Spring Harbor, NY.

Muñoz, A., Mendez, P., DeFelipe, J., and Alvarez-Leefmans, F.J. (2007). Cation-chloride cotransporters and GABA-ergic innervation in the human epileptic hippocampus. *Epilepsia* **48**, 663–673.

Murzin, A.G., Brenner, S.E., Hubbard, T., and Chothia, C. (1995). SCOP: a structural classification of proteins database for the investigation of sequences and structures. *J. Mol. Biol.* **247**, 536–540.

Nickell, W.T., Kleene, N.K., and Kleene, S.J. (2007). Mechanisms of neuronal chloride accumulation in intact mouse olfactory epithelium. *J. Physiol.* **583**, 1005–1020.

Olivares, L., Aragon, C., Gimenez, C., and Zafra, F. (1995). The role of N-glycosylation in the targeting and activity of the GLYT1 glycine transporter. *J. Biol. Chem.* **270**, 9437–9442.

O'Mahony, F., Toumi, F., Mroz, M.S., Ferguson, G., and Keely, S. J. (2008). Induction of $Na^+/K^+/2Cl^-$ cotransporter expression mediates chronic potentiation of intestinal epithelial Cl^- secretion by EGF. *Am. J. Physiol. Cell Physiol.* **294**, C1362–C1370.

Oppermann, M., Mizel, D., Huang, G., Li, C., Deng, C., Theilig, F., Bachmann, S., Briggs, J., Schnermann, J., and Castrop, H. (2006). Macula densa control of renin secretion and preglomerular resistance in mice with selective deletion of the B isoform of the Na K, 2Cl co-transporter. *J. Am. Soc. Nephrol.* **17**, 2143–2152.

Oppermann, M., Mizel, D., Kim, S.M., Chen, L., Faulhaber-Walter, R., Huang, Y., Li, C., Deng, C., Briggs, J., Schnermann, J., and Castrop, H. (2007). Renal function in mice with targeted disruption of the A isoform of the Na-K-2Cl co-transporter. *J. Am. Soc. Nephrol.* **18**, 440–448.

Ortiz, P.A. (2006). cAMP increases surface expression of NKCC2 in rat thick ascending limbs: role of VAMP. *Am. J. Physiol. Renal Physiol.* **290**, F608–F616.

Ott, R.J., Hui, A.C., and Giacomini, K.M. (1992). Inhibition of N-linked glycosylation affects organic cation transport across the brush border membrane of opossum kidney (OK) cells. *J. Biol. Chem.* **267**, 133–139.

Overdier, D.G., Ye, H., Peterson, R.S., Clevidence, D.E., and Costa, R.H. (1997). The winged helix transcriptional activator HFH-3 is expressed in the distal tubules of embryonic and adult mouse kidney. *J. Biol. Chem.* **272**, 13725–13730.

Pace, A.J., Lee, E., Athirakul, K., Coffman, T.M., O'Brien, D.A., and Koller, B.H. (2000). Failure of spermatogenesis in mouse lines deficient in the Na^+-K^+-$2Cl^-$ cotransporter. *J. Clin. Invest* **105**, 441–450.

Pace, A.J., Madden, V.J., Henson, O.W., Jr., Koller, B.H., and Henson, M.M. (2001). Ultrastructure of the inner ear of NKCC1-deficient mice. *Hear Res.* **156**, 17–30.

Panet, R., Eliash, M., and Atlan, H. (2006). $Na^+/K^+/Cl^-$ cotransporter activates MAP-kinase cascade downstream to protein kinase C, and upstream to MEK. *J. Cell Physiol.* **206**, 578–585.

Panet, R., Eliash, M., Pick, M., and Atlan, H. (2002). $Na(^+)/K(^+)/Cl(^-)$ cotransporter activates mitogen-activated protein kinase in fibroblasts and lymphocytes. *J. Cell Physiol.* **190**, 227–237.

Panet, R., Marcus, M., and Atlan, H. (2000). Overexpression of the $Na^+/K^+/Cl^-$ cotransporter gene induces cell proliferation and phenotypic transformation in mouse fibroblasts. *J. Cell Physiol.* **182**, 109–118.

Paredes, A., Plata, C., Rivera, M., Moreno, E., Vazquez, N., Muñoz-Clares, R., Hebert, S.C., and Gamba, G. (2006). Activity of the renal Na^+-K^+-$2Cl^-$ cotransporter is reduced by mutagenesis of N-glycosylation sites: role for protein surface charge in Cl^- transport. *Am. J. Physiol. Renal Physiol.* **290**, F1094–F1102.

Park, J.H. and Saier, M.H., Jr. (1996). Phylogenetic, structural and functional characteristics of the Na-K-Cl cotransporter family. *J. Membr. Biol.* **149**, 161–168.

Pathak, B.G., Shaughnessy, J.D., Jr., Meneton, P., Greeb, J., Shull, G. E., Jenkins, N.A., and Copeland, N.G. (1996). Mouse chromosomal location of three epithelial sodium channel subunit genes and an apical sodium chloride cotransporter gene. *Genomics* **33**, 124–127.

Paulsen, I.T. and Skurray, R.A. (1993). Topology, structure and evolution of two families of proteins involved in antibiotic and antiseptic resistance in eukaryotes and prokaryotes – an analysis. *Gene* **124**, 1–11.

Payne, J.A. and Forbush, B., 3rd (1994). Alternatively spliced isoforms of the putative renal Na-K-Cl cotransporter are differentially distributed within the rabbit kidney. *Proc. Natl. Acad. Sci. USA* **91**, 4544–4548.

Payne, J.A., Stevenson, T.J., and Donaldson, L.F. (1996). Molecular characterization of a putative K-Cl cotransporter in rat brain A neuronal-specific isoform. *J. Biol. Chem.* **271**, 16245–16252.

Payne, J.A., Xu, J.C., Haas, M., Lytle, C.Y., Ward, D., and Forbush, B., 3rd (1995). Primary structure, functional expression, and chromosomal localization of the bumetanide-sensitive Na-K-Cl cotransporter in human colon. *J. Biol. Chem.* **270**, 17977–17985.

Pertea, M., Lin, X., and Salzberg, S.L. (2001). GeneSplicer: a new computational method for splice site prediction. *Nucleic Acids Res.* **29**, 1185–1190.

Piechotta, K., Lu, J., and Delpire, E. (2002). Cation chloride cotransporters interact with the stress-related kinases Ste20-related proline-alanine-rich kinase (SPAK) and oxidative stress response 1 (OSR1). *J. Biol. Chem.* **277**, 50812–50819.

Plata, C., Meade, P., Vazquez, N., Hebert, S.C., and Gamba, G. (2002). Functional properties of the apical Na^+-K^+-$2Cl^-$ cotransporter isoforms. *J. Biol. Chem.* **277**, 11004–11012.

Plotkin, M.D., Kaplan, M.R., Verlander, J.W., Lee, W.S., Brown, D., Poch, E., Gullans, S.R., and Hebert, S.C. (1996). Localization of the thiazide sensitive Na-Cl cotransporter, rTSC1 in the rat kidney. *Kidney Int.* **50**, 174–183.

Ponce-Coria, J., San-Cristobal, P., Kahle, K.T., Vazquez, N., Pacheco-Alvarez, D., de Los Heros, P., Juarez, P., Muñoz, E., Michel, G., Bobadilla, N.A., Gimenez, I., Lifton, R.P., Hebert, S.C., and Gamba, G. (2008). Regulation of NKCC2 by a chloride-sensing mechanism involving the WNK3 and SPAK kinases. *Proc. Natl. Acad. Sci. USA* **105**, 8458–8463.

Pontoglio, M., Barra, J., Hadchouel, M., Doyen, A., Kress, C., Bach, J.P., Babinet, C., and Yaniv, M. (1996). Hepatocyte nuclear factor 1 inactivation results in hepatic dysfunction, phenylketonuria, and renal Fanconi syndrome. *Cell* **84**, 575–585.

Proesmans, W., Devlieger, H., Van Assche, A., Eggermont, E., Vandenberghe, K., Lemmens, F., Sieprath, P., and Lijnen, P. (1985). Bartter syndrome in two siblings – antenatal and neonatal observations. *Int. J. Pediatr. Nephrol.* **6**, 63–70.

Pruitt, K.D., Tatusova, T., and Maglott, D.R. (2007). NCBI reference sequences (RefSeq): a curated non-redundant sequence database of genomes, transcripts and proteins. *Nucleic Acids Res.* **35**, D61–D65.

Raghava, G.P. and Barton, G.J. (2006). Quantification of the variation in percentage identity for protein sequence alignments. *BMC Bioinformatics* **7**, 415.

Randall, J., Thorne, T., and Delpire, E. (1997). Partial cloning and characterization of Slc12a2: the gene encoding the secretory Na^+-K^+-$2Cl^-$ cotransporter. *Am. J. Physiol* **273**, C1267–C1277.

Ren, Q., Chen, K., and Paulsen, I.T. (2007). TransportDB: a comprehensive database resource for cytoplasmic membrane transport systems and outer membrane channels. *Nucleic Acids Res.* **35**, D274–D279.

Reynolds, A., Parris, A., Evans, L.A., Lindqvist, S., Sharp, P., Lewis, M., Tighe, R., and Williams, M.R. (2007). Dynamic and differential regulation of NKCC1 by calcium and cAMP in the native human colonic epithelium. *J. Physiol.* **582**, 507–524.

Rhodes, K.J. and Trimmer, J.S. (2006). Antibodies as valuable neuroscience research tools versus reagents of mass distraction. *J. Neurosci.* **26**, 8017–8020.

Rinehart, J., Kahle, K.T., de Los Heros, P., Vazquez, N., Meade, P., Wilson, F.H., Hebert, S.C., Gimenez, I., Gamba, G., and Lifton, R.P. (2005). WNK3 kinase is a positive regulator of NKCC2 and NCC, renal cation-Cl^- cotransporters required for normal blood pressure homeostasis. *Proc. Natl. Acad. Sci. USA* **102**, 16777–16782.

Riveira-Muñoz, E., Chang, Q., Godefroid, N., Hoenderop, J.G., Bindels, R.J., Dahan, K., and Devuyst, O. (2007). Transcriptional and functional analyses of SLC12A3 mutations: new clues for the pathogenesis of Gitelman syndrome. *J. Am. Soc. Nephrol.* **18**, 1271–1283.

Sampath, P., Mazumder, B., Seshadri, V., and Fox, P.L. (2003). Transcript-selective translational silencing by gamma interferon is directed by a novel structural element in the ceruloplasmin mRNA 3′ untranslated region. *Mol. Cell Biol.* **23**, 1509–1519.

Saper, C.B. (2005). An open letter to our readers on the use of antibodies. *J. Comp. Neurol.* **493**, 477–478.

Schultheis, P.J., Lorenz, J.N., Meneton, P., Nieman, M.L., Riddle, T. M., Flagella, M., Duffy, J.J., Doetschman, T., Miller, M.L., and Shull, G.E. (1998). Phenotype resembling Gitelman's syndrome in mice lacking the apical Na^+-Cl^- cotransporter of the distal convoluted tubule. *J. Biol. Chem.* **273**, 29150–29155.

Seyberth, H.W., Rascher, W., Schweer, H., Kuhl, P.G., Mehls, O., and Scharer, K. (1985). Congenital hypokalemia with hypercalciuria in preterm infants: a hyperprostaglandinuric tubular syndrome different from Bartter syndrome. *J. Pediatr.* **107**, 694–701.

Shen, H. and Chou, J.J. (2008). MemBrain: improving the accuracy of predicting transmembrane helices. *PLoS ONE* **3**, e2399.

Simon, D.B., Karet, F.E., Hamdan, J.M., DiPietro, A., Sanjad, S.A., and Lifton, R.P. (1996a). Bartter's syndrome, hypokalaemic alkalosis with hypercalciuria, is caused by mutations in the Na-K-2Cl cotransporter NKCC2. *Nat. Genet.* **13**, 183–188.

Simon, D.B., Nelson-Williams, C., Bia, M.J., Ellison, D., Karet, F.E., Molina, A.M., Vaara, I., Iwata, F., Cushner, H.M., Koolen, M., Gainza, F.J., Gitleman, H.J., and Lifton, R.P. (1996b). Gitelman's variant of Bartter's syndrome, inherited hypokalaemic alkalosis, is caused by mutations in the thiazide-sensitive Na-Cl cotransporter. *Nat. Genet.* **12**, 24–30.

Starremans, P.G., Kersten, F.F., Knoers, N.V., van den Heuvel, L. P., and Bindels, R.J. (2003). Mutations in the human Na-K-2Cl cotransporter (NKCC2) identified in Bartter syndrome type I consistently result in nonfunctional transporters. *J. Am. Soc. Nephrol.* **14**, 1419–1426.

Strausberg, R.L., Feingold, E.A., Grouse, L.H., Derge, J.G., Klausner, R.D., Collins, F.S., Wagner, L., Shenmen, C.M., Schuler, G.D., Altschul, S.F., Zeeberg, B., Buetow, K.H., Schaefer, C.F., Bhat, N.K., Hopkins, R.F., Jordan, H., Moore, T., Max, S.I., Wang, J., Hsieh, F., Diatchenko, L., Marusina, K., Farmer, A.A., Rubin, G.M., Hong, L., Stapleton, M., Soares, M.B., Bonaldo, M.F., Casavant, T.L., Scheetz, T.E., Brownstein, M.J., Usdin, T.B., Toshiyuki, S., Carninci, P., Prange, C., Raha, S.S., Loquellano, N.A., Peters, G.J., Abramson, R.D., Mullahy, S.J., Bosak, S. A., McEwan, P.J., McKernan, K.J., Malek, J.A., Gunaratne, P. H., Richards, S., Worley, K.C., Hale, S., Garcia, A.M., Gay, L. J., Hulyk, S.W., Villalon, D.K., Muzny, D.M., Sodergren, E.J., Lu, X., Gibbs, R.A., Fahey, J., Helton, E., Ketteman, M., Madan, A., Rodrigues, S., Sanchez, A., Whiting, M., Young, A.C., Shevchenko, Y., Bouffard, G.G., Blakesley, R.W., Touchman, J. W., Green, E.D., Dickson, M.C., Rodriguez, A.C., Grimwood, J., Schmutz, J., Myers, R.M., Butterfield, Y.S., Krzywinski, M.I.,

Skalska, U., Smailus, D.E., Schnerch, A., Schein, J.E., Jones, S.J., and Marra, M.A. (2002). Generation and initial analysis of more than 15,000 full-length human and mouse cDNA sequences. *Proc. Natl. Acad. Sci. USA* **99**, 16899–16903.

Su, A.I., Cooke, M.P., Ching, K.A., Hakak, Y., Walker, J.R., Wiltshire, T., Orth, A.P., Vega, R.G., Sapinoso, L.M., Moqrich, A., Patapoutian, A., Hampton, G.M., Schultz, P.G., and Hogenesch, J.B. (2002). Large-scale analysis of the human and mouse transcriptomes. *Proc. Natl. Acad. Sci. USA* **99**, 4465–4470.

Sugawara, H., Ogasawara, O., Okubo, K., Gojobori, T., and Tateno, Y. (2008). DDBJ. with new system and face. *Nucleic Acids Res.* **36**, D22–D24.

Suske, G. (1999). The Sp-family of transcription factors. *Gene* **238**, 291–300.

Takahashi, N., Chernavvsky, D.R., Gomez, R.A., Igarashi, P., Gitelman, H.J., and Smithies, O. (2000). Uncompensated polyuria in a mouse model of Bartter's syndrome. *Proc. Natl. Acad. Sci. USA* **97**, 5434–5439.

Taniyama, Y., Sato, K., Sugawara, A., Uruno, A., Ikeda, Y., Kudo, M., Ito, S., and Takeuchi, K. (2001). Renal tubule-specific transcription and chromosomal localization of rat thiazide-sensitive Na-Cl cotransporter gene. *J. Biol. Chem.* **276**, 26260–26268.

Tate, C.G. and Blakely, R.D. (1994). The effect of N-linked glycosylation on activity of the Na(+)- and Cl(−)-dependent serotonin transporter expressed using recombinant baculovirus in insect cells. *J. Biol. Chem.* **269**, 26303–26310.

Thierry-Mieg, D. and Thierry-Mieg, J. (2006). AceView: a comprehensive cDNA-supported gene and transcripts annotation. *Genome Biol.* **7** (Suppl. 1) S12.1–S14.

Vargas-Poussou, R., Feldmann, D., Vollmer, M., Konrad, M., Kelly, L., van den Heuvel, L.P., Tebourbi, L., Brandis, M., Karolyi, L., Hebert, S.C., Lemmink, H.H., Deschenes, G., Hildebrandt, F., Seyberth, H.W., Guay-Woodford, L.M., Knoers, N.V., and Antignac, C. (1998). Novel molecular variants of the Na-K-2Cl cotransporter gene are responsible for antenatal Bartter syndrome. *Am. J. Hum. Genet.* **62**, 1332–1340.

Velazquez, H., Bartiss, A., Bernstein, P., and Ellison, D.H. (1996). Adrenal steroids stimulate thiazide-sensitive NaCl transport by rat renal distal tubules. *Am. J. Physiol* **270**, F211–F219.

Verissimo, F. and Jordan, P. (2001). WNK kinases, a novel protein kinase subfamily in multi-cellular organisms. *Oncogene* **20**, 5562–5569.

Viklund, H. and Elofsson, A. (2004). Best alpha-helical transmembrane protein topology predictions are achieved using hidden Markov models and evolutionary information. *Protein Sci.* **13**, 1908–1917.

von Heijne, G. (2006). Membrane-protein topology. *Nat. Rev. Mol. Cell Biol.* **7**, 909–918.

Wain, H.M., Lush, M.J., Ducluzeau, F., Khodiyar, V.K., and Povey, S. (2004). Genew: the Human Gene Nomenclature Database, 2004 updates. *Nucleic Acids Res.* **32**, D255–D257.

Wakamatsu, A., Yamamoto, J., Kimura, K., Kaida, T., Tsuchiya, K., Iida, Y., Takayama, Y., Murakawa, K., Kanehori, K., Andoh, T., Kagawa, N., Sato, R., Kawamura, Y., Tanaka, S., Kisu, Y., Sugano, S., Goshima, N., Nomura, N., and Isogai, T. (2008). NEDO functional analysis of protein and research application project. http://www.ncbi.nlm.nih.gov/entrez/viewer.fcgi?db = nuccore&id = 164694894.

Wang, X.Y., Masilamani, S., Nielsen, J., Kwon, T.H., Brooks, H.L., Nielsen, S., and Knepper, M.A. (2001). The renal thiazide-sensitive Na-Cl cotransporter as mediator of the aldosterone-escape phenomenon. *J. Clin. Invest* **108**, 215–222.

Watson, J.D. (2004). *Molecular Biology of the Gene*. Pearson/Benjamin Cummings, San Francisco.

Wierstra, I. (2008). Sp1: emerging roles–beyond constitutive activation of TATA-less housekeeping genes. *Biochem. Biophys. Res. Commun.* **372**, 1–13.

Wimley, W.C. and White, S.H. (1996). Experimentally determined hydrophobicity scale for proteins at membrane interfaces. *Nat. Struct. Biol.* **3**, 842–848.

Yerby, T.R., Vibat, C.R., Sun, D., Payne, J.A., and O'Donnell, M.E. (1997). Molecular characterization of the Na-K-Cl cotransporter of bovine aortic endothelial cells. *Am. J. Physiol.* **273**, C188–C197.

Yusufi, A.N., Szczepanska-Konkel, M., and Dousa, T.P. (1988). Role of N-linked oligosaccharides in the transport activity of the Na^+/H^+ antiporter in rat renal brush-border membrane. *J. Biol. Chem.* **263**, 13683–13691.

Zhao, G. and London, E. (2006). An amino acid "transmembrane tendency" scale that approaches the theoretical limit to accuracy for prediction of transmembrane helices: relationship to biological hydrophobicity. *Protein Sci.* **15**, 1987–2001.

CHAPTER

12

The CLC Family of Chloride Channels and Transporters

Tobias Stauber, Gaia Novarino and Thomas J. Jentsch

OUTLINE

I. INTRODUCTION

CLC chloride channels and transporters are expressed from bacteria to humans. This highly conserved family comprises nine members in mammals (reviewed in Jentsch, 2008; Zifarelli and Pusch, 2007), which were identified by homology cloning subsequent to the family's founding member ClC-0 from *Torpedo marmorata* (Jentsch et al., 1990). According to their sequence homologies, they can be grouped into three subfamilies (Fig. 12.1). All members of the first subfamily, ClC-1, -2, -Ka and -Kb, have been shown to reside in the plasma membrane and function as Cl^- channels that mediate, e.g. transepithelial Cl^- transport or are involved in stabilizing the resting potential of the plasma membrane. By contrast, the members of the second (ClC-3, -4 and -5) and third (ClC-6 and -7) subfamilies localize predominantly to intracellular organelles and vesicles. Here, they may play a role in the acidification of these compartments by providing the shunt current for the vesicular H^+-ATPase (Jentsch, 2007). Intriguingly, just like the *E. coli* CLC, ecClC-1, the endosomal ClC-4 and -5 have been shown to function as Cl^-/H^+ antiporters rather than as Cl^- channels. This, however, does not change their capability of electrically neutralizing endosomal acidification. Due to the presence of a critical amino acid that is conserved specifically between the Cl^-/H^+ exchangers, this modus operandi has also been proposed for ClC-3 and the members of the third subfamily, but this has yet to be demonstrated experimentally.

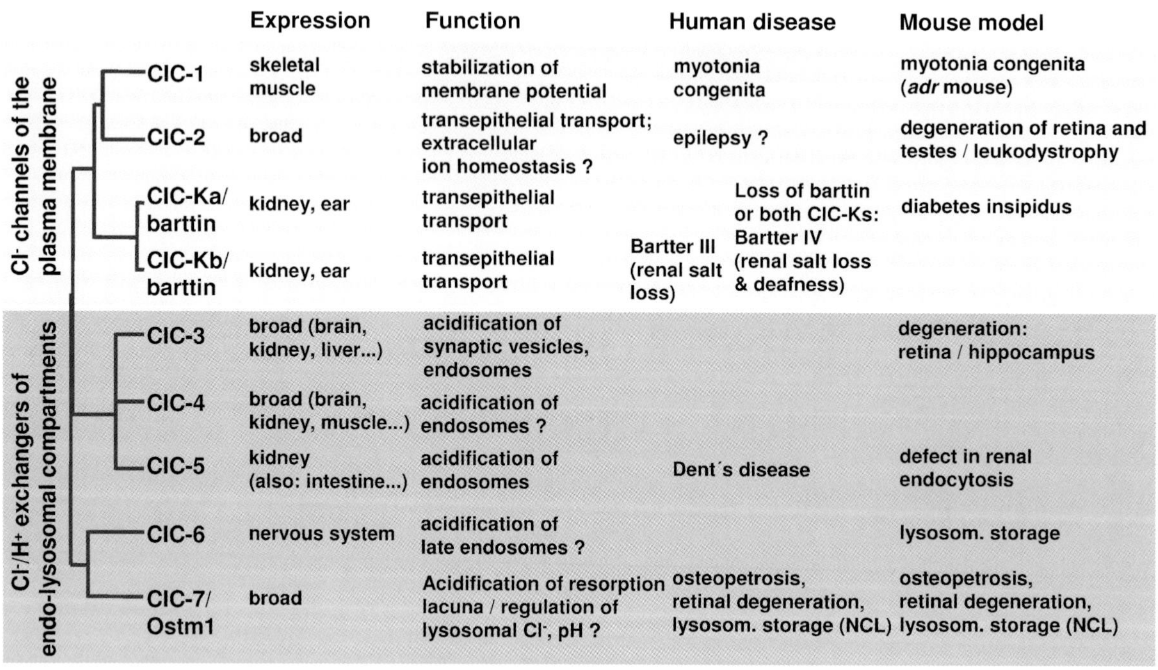

FIGURE 12.1 Overview of the mammalian CLC family of chloride channels and transporters. For the nine mammalian CLC proteins (where applicable with their β-subunits), the tissue expression and cell function are given. Known human and mouse pathologies are also listed. The first branch of the CLC proteins localizes to the plasma membrane, while the second and third subfamilies are localized predominantly to intracellular compartments of the endolysosomal pathway. Although not all vesicular CLC proteins have been characterized biophysically, they may all be Cl^-/H^+ exchangers, in contrast to the plasma membrane Cl^- channels of the first branch.

Members of all three branches of the CLC family are expressed in the nervous system. Their physiological importance has been demonstrated by the phenotypes of knockout mice and by identification of mutations in their respective genes in patients with neurological or other diseases (Jentsch, 2008; Jentsch et al., 2005b). The loss of ClC-2 in mice leads to retinal degeneration and leukoencephalopathy, a widespread vacuolation of the white matter in the CNS, as discussed below in Section III.A. ClC-3 knockout (KO) mice exhibit hippocampal and retinal degeneration (Section III.B). Mice lacking ClC-6 develop a lysosomal storage disease with symptoms resembling Kuf's disease, an adult-onset form of neuronal ceroid lipofuscinosis (NCL), as discussed below in Section III.C. Likewise, loss of ClC-7 or its essential β-subunit Ostm1 results in a severe NCL-like neurodegeneration (besides osteopetrosis), which is also accompanied by retinal degeneration (Section III.D).

In the first part of this chapter we discuss the general features of CLC proteins, such as their structure, mode of operation (Section II.A) and their roles in cell biology and physiology (Section II.B). In the second part of this chapter we cover in more detail the physiology and pathology of mammalian neuronal CLC proteins.

II. GENERAL PROPERTIES OF THE CLC PROTEIN FAMILY

A. Structure of CLC Proteins and Mechanism of Ion Translocation

1. The "Double-barreled" Channel

CLC proteins form rhombus-like dimers of twofold symmetry (Dutzler et al., 2002) (Fig. 12.2A). Although they function mostly as homodimers, heterodimerization has been shown within the first (Lorenz et al., 1996; Scholl et al., 2006; Weinreich and Jentsch, 2001) and second (Suzuki et al., 2006) subfamilies upon heterologous expression. ClC-4 and -5 have been reported to heteromerize *in vivo* as well (Mohammad-Panah et al., 2003), yet the physiological function of this remains unclear.

A CLC dimer works as a "double-barreled" translocation pathway in which each subunit provides its own pore (Dutzler et al., 2002; Weinreich and Jentsch, 2001). The independent operation of the two ion translocation pathways was deduced first from the analysis of the reconstituted *Torpedo* channel (Miller, 1982), years before the crystal structure of the bacterial ortholog became available (Dutzler et al., 2002) or even before the first CLC was cloned (Jentsch et al., 1990).

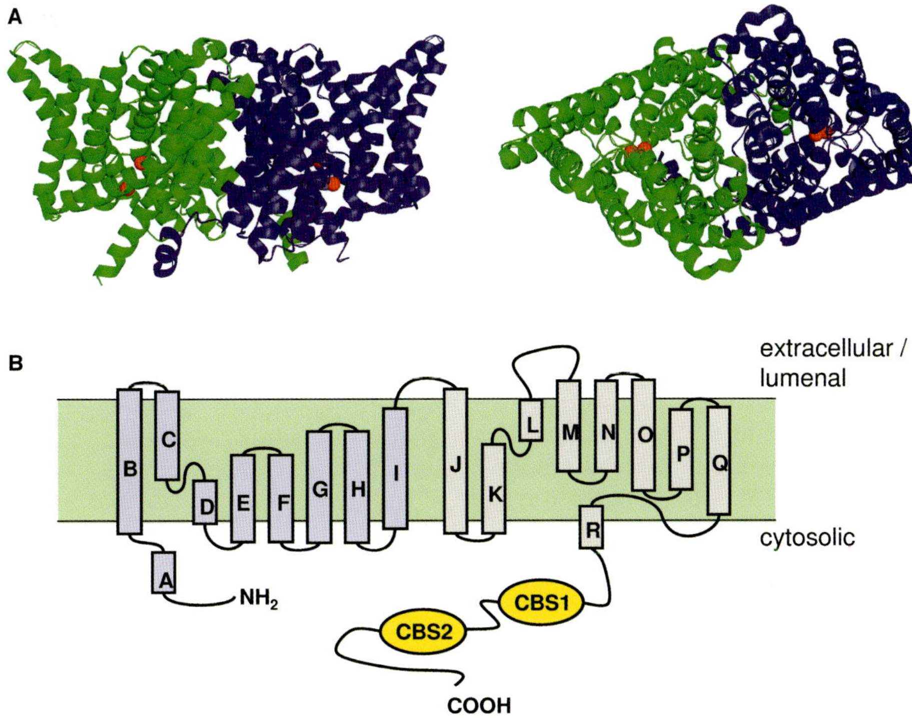

FIGURE 12.2 Structure and topology of CLC proteins. **A.** Ribbon presentation of the ecClC-1 crystal structure as seen from within the membrane with the extracellular side up (left) and from the extracellular face (right). The two subunits are in blue and green, respectively. The two red spheres in each subunit indicate the co-crystallized chloride ions in the permeation pathway. The pictures were generated with PyMOL using the coordinate 1OTS from the Protein Data Bank without the co-crystallized antibodies, and are based on the work of (Dutzler et al., 2003). **B.** Topological presentation of a single CLC subunit. The different shadings of the 18 α-helices (A–R) indicate the two similar halves. The two cytoplasmic CBS subdomains are shown in yellow.

The observed two equidistant conductance steps in combination with long periods of no channel activity suggested a fast opening and closing of two identical pores (called "protopore gate" or "fast gate" in the case of ClC-0) which is superimposed by a slower common gate, which closes both pores simultaneously. These characteristics were also observed when the by-then-cloned ClC-0 was expressed in *Xenopus* oocytes (Bauer et al., 1991). As a final proof of the "double-barrel" model, artificial dimers of wildtype (WT) ClC-0 subunits with mutant ClC-0 (Ludewig et al., 1996; Middleton et al., 1996) or with ClC-2 (Weinreich and Jentsch, 2001) were generated. As the mutant ClC-0 and ClC-2 subunits possess lower conductance levels than WT ClC-0, the independent conductance levels could be observed in single-channel recordings. Later on it was also shown that coupling of Cl^- flux to H^+ countertransport is mediated by each subunit individually (Nguitragool and Miller, 2007; Zdebik et al., 2008). Due to this largely independent function of the two pores, which contrasts sharply with the situation for voltage-dependent K^+ channels, mutations that affect single-channel conductance or the "fast gating" will generally not exert a dominant effect on WT subunits. This explains why dominantly inherited CLC-linked diseases are generally less severe than the recessive variants, in which both subunits are mutated.

2. *Subunit Topology and the Ion Translocation Pathway*

The crystal structures of prokaryotic CLC proteins (Dutzler et al., 2002, 2003) did not only confirm the dimeric structure with two translocation pathways, but also revealed the complex transmembrane topology. Each subunit possesses 18 α-helices of variable lengths, 17 of which (helices B–R) penetrate the membrane, often not completely spanning the lipid bilayer (Fig. 12.2B). This complex arrangement explains the difficulties encountered in previous biochemical analysis of the transmembrane topology of the transmembrane domain (Schmidt-Rose and Jentsch, 1997). The amino termini (including the amino-terminal α-helix A) and carboxy termini protrude into the cytosol.

The transmembrane domain of each subunit exhibits an antiparallel repeat of two similar halves (α-helices B–I and J–Q). Three Cl^--binding sites were identified on each subunit. The high-resolution crystal structure revealed an "internal" Cl^--binding site in the vestibule of the channel that opens to the cytoplasm (Dutzler et al., 2003). The second Cl^-, a centrally located ion, is coordinated by four amino acids from different α-helices of the subunit, two contributing with their main-chain nitrogen atoms and the other two, a serine and a tyrosine, with their side chains (Dutzler et al., 2002). Interestingly, the coordinating tyrosine lies in the beginning of α-helix R, which links the transmembrane domain with the large cytoplasmic carboxy-terminus of most CLC proteins (see below). Thus it might mediate the regulation of transport activity by the cytoplasmic domain. As no positive charges take part in the coordination of the centrally located Cl^-, it is proposed that this is accomplished by the dipoles of the four α-helices (Dutzler et al., 2002). Offering an alternative possibility, model calculations showed that a strictly conserved lysine in helix E, which is buried within the protein, could stabilize Cl^- binding by a long-range electrostatic interaction with this anion (Corry et al., 2004; Faraldo-Gómez and Roux, 2004). Supporting this hypothesis, mutation of this lysine in ClC-0 affected the channel's anion selectivity and fast gating (Engh et al., 2007a, b; Zhang et al., 2006).

The third, "external" Cl^- binding site was created when a glutamate that blocked the access of Cl^- from the extracytosolic side (Dutzler et al., 2002) was mutated to glutamine (Dutzler et al., 2003). This glutamate is conserved in most CLC proteins, with the exception of ClC-K channels, which possess a valine at this position and show little gating (Estévez et al., 2001; Waldegger and Jentsch, 2000). Voltage dependence of channel activity was shown to be affected by introducing this glutamate in ClC-K1 (Waldegger and Jentsch, 2000) or by neutralizing this position in other CLC proteins (Dutzler et al., 2003; Fahlke et al., 1997; Friedrich et al., 1999). The opening of the access of anions from the extracellular side upon the glutamate-to-glutamine mutation offers a beautiful explanation as to how this glutamate is involved in the "gating by the permeant ion". In this model, the external Cl^-, which itself acts as a voltage sensor (Chen and Miller, 1996; Pusch et al., 1995), competes for an anion binding side with the negatively charged side chain of the "gating glutamate". Thus, protonation of this glutamate and the turning away of the side chain opens the otherwise closed channel. Along this line, low pH has been shown to promote channel opening (Chen and Chen, 2001; Pusch, 2004; Rychkov et al., 1996; Traverso et al., 2006). The mechanism of Cl^--dependent opening of CLC channels (Niemeyer et al., 2003; Pusch et al., 1999; Rychkov et al., 1998; Schriever et al., 1999) may actually be mediated by "chloride-induced proton gating" (Bostick and Berkowitz, 2004).

Just as potassium channels have multi-ion pores, the three Cl^- binding sites in CLC proteins can be occupied simultaneously with affinities in the physiologically relevant range of 4–40 mM (Lobet and Dutzler, 2006). As the mammalian cytosolic Cl^- concentration is relatively low (5–40 mM), the internal binding site, which has the lowest binding affinity might only be occupied when the channel is open towards the higher external Cl^- concentration (of about 100 mM). This might determine the directionality of ion-dependent gating.

3. Cl^- Channels and Cl^-/H^+ Exchangers

When ecClC-1 was reconstituted into lipid bilayers, it yielded the unexpected finding that, in contrast to the well-studied ClC-0 Cl^- channel, it actually mediates Cl^-/H^+ exchange, most likely by transporting one proton in exchange for two Cl^- (Accardi and Miller, 2004). Fluxes of these ions are obligatory coupled (Nguitragool and Miller, 2006) and the gradient of one ion can drive the transport of the other ion against its electrochemical gradient (Accardi and Miller, 2004).

The exchanger can be converted into a pure chloride conductance by neutralizing either of two glutamates, the above-described "gating glutamate" (Accardi and Miller, 2004) and the "proton glutamate" (Accardi et al., 2005). The latter glutamate is located close to the cytoplasmic surface of the CLC protein, away from the Cl^--conducting funnel. Thus, the two ions have likely different pathways at the "internal" side that meet at the central "gating glutamate". As neutralizing the "proton glutamate" in ecClC-1 uncouples Cl^- conductance from H^+ transport, it is clear that there is no obligatory Cl^-/H^+ exchange site (Accardi et al., 2005). Two lines of evidence show that the coupling to proton countertransport is somehow related to the occupation of the central anion binding site. First, some polyatomic anions are translocated without exchange for protons and they only occupy the internal binding site as determined crystallographically (Nguitragool and Miller, 2006). Second, mutations in the anion-coordinating tyrosine that abrogated occupation of the central binding site led to an uncoupling of anion transport (Accardi et al., 2006).

The mammalian plasma membrane CLC channels of the first subfamily (and *Torpedo* ClC-0) have a valine at the position homologous to the ecClC-1 "proton glutamate". They have all been shown not to couple Cl^- flux to H^+ countertransport. By contrast, the intracellular CLC proteins of the second and third subfamilies possess a glutamate at this position, raising the possibility that they may be Cl^-/H^+ exchangers as well. Indeed, this has been shown for ClC-4 and -5 (Picollo and Pusch, 2005; Scheel et al., 2005), while it has not yet been determined for the others due to technical difficulties. In contrast to the situation with ecClC-1, neutralization of the "proton glutamate" in ClC-4 and -5 does not just abrogate the coupling of Cl^- flux from proton transport, but abolishes transport of both ions (Zdebik et al., 2008). The combined neutralization of both the "gating glutamate" and the "proton glutamate" allowed for Cl^- transport, again uncoupled from proton exchange (Zdebik et al., 2008), showing that these exchangers (when WT) have an obligatory Cl^-/H^+ exchange site. Due to the strong outward rectification of their currents, which could also be overcome by neutralizing the "gating glutamate" (Friedrich et al., 1999), proton transport by ClC-4 and -5 only takes place at positive cytoplasmic voltages (Picollo and Pusch, 2005; Scheel et al., 2005). The mechanism of coupling still remains to be elucidated and there is more to it than simply the presence of a glutamate at the "proton glutamate" position, as introduction of this amino acid into channels does not convert them into Cl^-/H^+ exchangers (Zdebik et al., 2008).

4. *Function of the Cytoplasmic Domains*

In contrast to ecClC-1, most CLC proteins, including the nine mammalian family members, possess long cytosolic carboxy-terminal portions of about 150–400 amino acids (Fig. 12.2B). The carboxy-terminus contains a pair of cystathionine β-synthetase (CBS) subdomains. These CBS subdomains are linked and flanked by sequences of variable length between the individual CLC proteins (Estévez et al., 2004). In biochemical and electrophysiological studies on ClC-1 (Estévez et al., 2004) and later in crystallographical studies of the carboxy-terminal regions of ClC-Ka (Markovic and Dutzler, 2007), it was shown that the CBS motifs dimerize not only intra-molecularly, as also shown in the crystal structure of the carboxy-terminal regions of ClC-0 (Meyer and Dutzler, 2006) and -5 (Meyer et al., 2007), but also inter-molecularly, supporting the idea that the carboxy-terminal domains are involved in dimerization.

A well-known function of the CBS domain is its role in channel gating (reviewed in detail in Chen, 2005; Zifarelli and Pusch, 2007). Large movements of the carboxy-termini of ClC-0 have been observed during slow (common) gating (Bykova et al., 2006), in agreement with the large temperature dependence of the common gate (Pusch et al., 1997). Recently, some CBS domains, those of ClC-2 and -5, have been found to bind adenosine nucleotides *in vitro* (Meyer et al., 2007; Scott et al., 2004). It is a matter of debate if these nucleotides affect the common gating of ClC-1 (Bennetts et al., 2005) and possibly ClC-2 (Niemeyer et al., 2004). Contradicting previous studies, ClC-1, for which nucleotide-mediated regulation of gating has even been reported to be pH dependent (Bennetts et al., 2007; Tseng et al., 2007), has recently been shown to be completely independent of intracellular ATP at all tested pH values (Zifarelli and Pusch, 2008). Since ClC-0, a close homolog of ClC-1, does not bind ATP (Meyer and Dutzler, 2006), it would be interesting to see whether the ClC-1 CBS domain binds ATP. That the CBS domain of ClC-5 could not be crystallized in the absence of nucleotides and the almost equal binding affinities of ATP, ADP and AMP (Meyer et al., 2007) argue against a regulatory role and suggest a structural function. Mutations in ClC-5 that render it incapable of nucleotide binding have no effect on its current/voltage relation, unless in the background of a neutralized gating glutamate when the mutations strongly decrease the currents at negative voltages (Meyer et al., 2007). The physiological meaning of this observation remains to be determined.

The sequence amino-terminally adjoining the CLC transmembrane domain consists of up to 120 amino acids (in ClC-7) and contains α-helix A. For ClC-2, this region is involved in gating. Its deletion leads to a constitutively open channel (Gründer et al., 1992). That it retains its function after "transplantation" into the carboxy-terminal domain (Gründer et al., 1992) suggests that ClC-2 possesses an internal binding site for it. A candidate for such an intramolecular receptor is a positively charged region between α-helices J and K (Jordt and Jentsch, 1997).

The amino- and carboxy-terminal domains are likely to bear sites for interaction with cytosolic proteins that may modulate them or regulate their trafficking, simply because they loom into the cytosol. Known transport signals include the PY ubiquitin–ligase binding motif between the two CBS subdomains of ClC-5 (Schwake et al., 2001), a dileucine motif in CBS2 of ClC-2 (Pena-Munzenmayer et al., 2005) and an amino-terminal dileucine cluster in ClC-3 (Zhao et al., 2007).

5. *Essential β-subunits*

For three CLC proteins, essential β-subunits have been identified, namely barttin for ClC-Ka and -Kb (Estévez et al., 2001) and Ostm1 for ClC-7 (Lange et al., 2006).

The ~40 kDa protein barttin (Birkenhäger et al., 2001) spans the membrane twice and possesses a short amino- and a long carboxy-terminal cytosolic part. It interacts with either of the ClC-K channels, presumably by binding separately to α-helices B and J in either subunit of the dimer (Tajima et al., 2007). Barttin stabilizes the ClC-K channels; it is involved in their trafficking (Estévez et al., 2001; Scholl et al., 2006) and alters their biophysical properties, such as their Ca^{2+} sensitivity (Waldegger et al., 2002) and gating (Scholl et al., 2006).

Ostm1 is a type-I transmembrane protein with a large, heavily glycosylated luminal and a short cytoplasmic domain that specifically interacts with the lysosomal ClC-7 (Lange et al., 2006). In mice lacking Ostm1, the level of ClC-7 protein (but not that of the mRNA) is reduced by more than 90% (Lange et al., 2006). As the late endosomal/lysosomal ClC-7 is the only mammalian CLC protein that lacks consensus sites for N-linked glycosylation, it is assumed that Ostm1 protects ClC-7 from lysosomal degradation. Whether the binding of Ostm1 alters the biophysical properties of ClC-7 is unknown; it has not been possible to measure the currents carried by this protein because of its lack of cell surface expression.

B. Cellular and Physiological Functions of CLC Proteins

CLC proteins exhibit a vast range of functions depending on their subcellular localization (plasma membrane versus intracellular compartments such as vesicles, endosomes and lysosomes). Their functions have become evident not only from their subcellular localizations but from the various pathologies resulting from gene mutations in humans and from the phenotypes of various knockout mice (Fig. 12.1) (Jentsch, 2008; Jentsch et al., 2005b; Zifarelli and Pusch, 2007).

1. *Plasma Membrane Channels*

Cl^- transport by CLC channels residing in the plasma membrane of various cell types serve key functions such as stabilization of transmembrane voltage, thereby affecting electrical excitability; cell volume control, counteracting cell swelling; and in epithelial cells making possible the transport of water and electrolytes.

ClC-1 is expressed almost exclusively in skeletal muscle (Steinmeyer et al., 1991b), providing the unusually high Cl^- conductance characteristic of skeletal muscle fibers, which accounts for about 80% of their resting membrane conductance (Rüdel and Lehmann-Horn, 1985; see also Chapter 1 in this volume). ClC-1 is important for transmembrane voltage stabilization and for membrane repolarization following muscle activity. Consistent with this view, it was found that in the myotonic mouse strain *adr* the open reading frame of ClC-1 was destroyed (Steinmeyer et al., 1991a) and that ClC-1 is the gene affected in humans suffering from myotonia congenita (Koch et al., 1992).

Multiple roles have been assigned to the broadly expressed ClC-2, which is described in more detail in Section III.A. These include the regulation of cell volume and the pH of the extracellular space. The latter function may explain the testicular and retinal degeneration observed in ClC-2 knockout mice (Bösl et al., 2001).

The highly homologous channels ClC-Ka and -Kb (in rodent -K1 and -K2, respectively) and their β-subunit barttin are expressed predominantly in kidney and the inner ear where they mediate transepithelial transport. In the kidney, ClC-Kb/barttin localizes to the basolateral membrane of cells of the thick ascending limb of Henle's loop (Estévez et al., 2001). Here it enables Cl^- exit to allow for salt reabsoption, as also seen from mutations that underlie the severe renal salt loss in Bartter syndrome type III (Simon et al., 1997). Mice lacking the ortholog of ClC-Ka, which is predominantly expressed in the thick limb of Henle's loop (Uchida et al., 1995; Vandewalle et al., 1997), develop a defect in urinary concentration ability resembling human nephrogenic diabetes insipidus (Matsumura et al., 1999). Mutations in the barttin gene *BSDN* (Birkenhäger et al., 2001), which should abrogate the function of both ClC-K channels, and the combined mutations of both ClC-K channels (Nozu et al., 2008; Schlingmann et al., 2004) cause the severe Bartter syndrome type IV, which is also accompanied by deafness (Rickheit et al., 2008). Thus, only abrogation of both ClC-K homologs results in deafness. This is consistent with their coexpression in the marginal cells of the stria vascularis and in the dark cells of the vestibular organ (Ando and Takeuchi, 2000; Estévez et al., 2001). Here, their channel activity indirectly supports potassium secretion into the cochlear endolymph of the scala media to sustain the high level of K^+ (150 mM), a prerequisite for hearing (see Section III.D in Chapter 21 in this volume).

2. *Intracellular CLC Proteins*

All intracellular CLC transporters are presumed to fulfill similar subcellular functions, namely mediating Cl^- flux to support luminal acidification of their respective intracytoplasmic compartment (Faundez and Hartzell, 2004; Jentsch, 2007). The members of the second and third branch of mammalian CLC proteins localize predominantly to endosomal/lysosomal organelles and vesicles that are acidified by the vesicular H^+-ATPase (V-ATPase) (Jefferies et al., 2008; Mellman et al., 1986) (Fig. 12.3). Proton import is electrogenic and requires a shunt current to prevent the build-up of a too high inside-positive potential that would prevent further acidification. From ion-substitution and channel-blocking experiments it was deduced that this shunt current is mediated by a Cl^- conductance (e.g. Bae and Verkman, 1990; Fuchs et al., 1989; Mellman et al., 1986; Schmid et al., 1989). Additional evidence supporting this view was the demonstration that pH decrease along the endocytic pathway is accompanied by an increase in luminal [Cl^-] (Sonawane and Verkman, 2003).

Consistent with their subcellular localization, the members of the second CLC subfamily have been shown to be required for the fast acidification of endosomes (ClC-4; Mohammad-Panah et al., 2003), of early (but not of late) endosomes (ClC-5 (Hara-Chikuma et al., 2005a)), and of early and late endosomes (ClC-3; Hara-Chikuma et al., 2005b) by examining cultured cells loaded with pH-sensitive dyes. In the latter two studies, the luminal [Cl^-] was also measured. In the respective knockout cells, an impairment of Cl^- accumulation was observed that correlated with a decreased fall in pH (Hara-Chikuma et al., 2005a, 2005b). In addition, an endosome-enriched preparation from liver of ClC-3-deficient mice showed an increased steady-state pH in comparison to that of WT mice (Yoshikawa et al., 2002), and purified synaptic vesicles (to which ClC-3 also localizes) from a different ClC-3 knockout line (Stobrawa et al., 2001) as well as renal cortical membrane preparations from ClC-5 knockout mice (Günther et al., 2003) showed impaired acidification kinetics. For the more late endosomal/lysosomal members of the third CLC subfamily, ClC-6 and -7 which will be discussed in Sections III.C and III.D, respectively, no effect on steady-state lysosomal pH was observed in mice lacking either of them (Kasper et al., 2005; Poet et al., 2006).

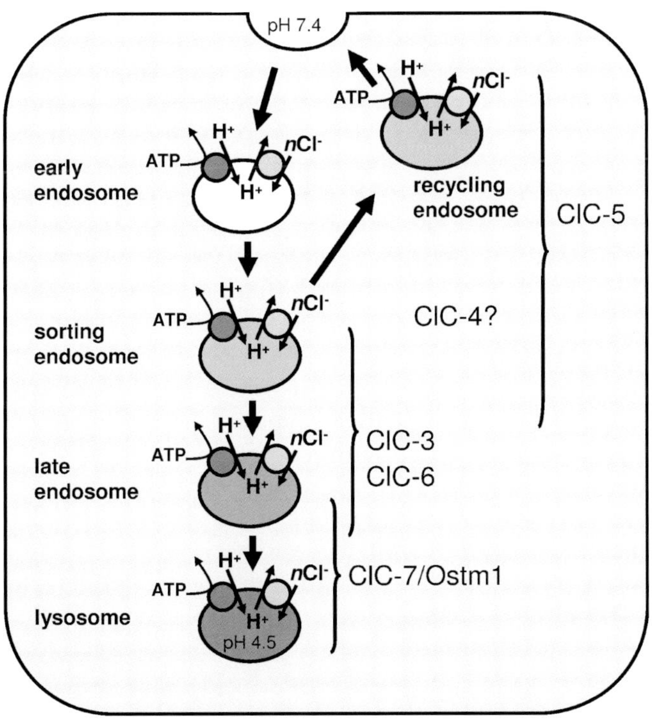

FIGURE 12.3 Subcellular localization of the intracellular CLC proteins. The scheme shows the proposed subcellular localizations of the intracellular CLC proteins. ClC-5 localizes to "earlier" compartments of the endocytic pathway, while ClC-3 and -6 localize to "later" compartments. ClC-4 is thought to localize "between" these, and ClC-7/Ostm1 is the only one that resides on lysosomes. The drop in luminal pH (from extracellularly pH 7.4 to lysosomal pH 4.5) is generated by the ATP-consuming proton pump. The shunt current is provided by the CLC proteins. Although only for ClC-4 and -5 it is known that they exchange Cl^- for H^+ (with an unknown stoichiometry), this function is also depicted for the other CLC isoforms.

ClC-4 and -5 have been shown to be Cl^-/H^+ exchangers rather than Cl^- channels (Picollo and Pusch, 2005; Scheel et al., 2005) and this may also be true for the other intracellular CLC proteins (see Section II.A). What difference would this make? When protons are exchanged for Cl^-, with a stoichiometry of $2Cl^-/1H^+$ this would obviously be electrogenic and could clearly supply the required shunt current for the V-ATPase (Fig. 12.4). With a $2Cl^-/1H^+$ stoichiometry, one out of three protons pumped by the ATPase would leave the vesicle to achieve electrical neutrality. Apart from the requirement of higher energy consumption by the V-ATPase to pump more protons into the lumen, there are two possible consequences. First, it would open the possibility that early in endocytosis, fast acidification is actually mediated by a CLC Cl^-/H^+ exchanger that allows for the initial proton influx driven by the efflux of chloride. The high extracellular [Cl^-] (>100 mM) is much larger than the relatively low cytoplasmic [Cl^-]. Within one minute after internalization, luminal [Cl^-] drops down to ~20 mM (Sonawane and Verkman, 2003). There is evidence that the negative Donnan potential, which results from negative charges of proteins inside the newly formed vesicle, promotes Cl^- efflux (Sonawane and Verkman, 2003). But whatever the driving force, if Cl^- leaves through a CLC Cl^-/H^+ exchanger, this efflux would inevitably lead to a fast, V-ATPase-independent acidification. Second, Cl^- gradients across the vesicular membrane will be coupled directly to the pH gradient, and not be accumulated only during acidification. This aspect is

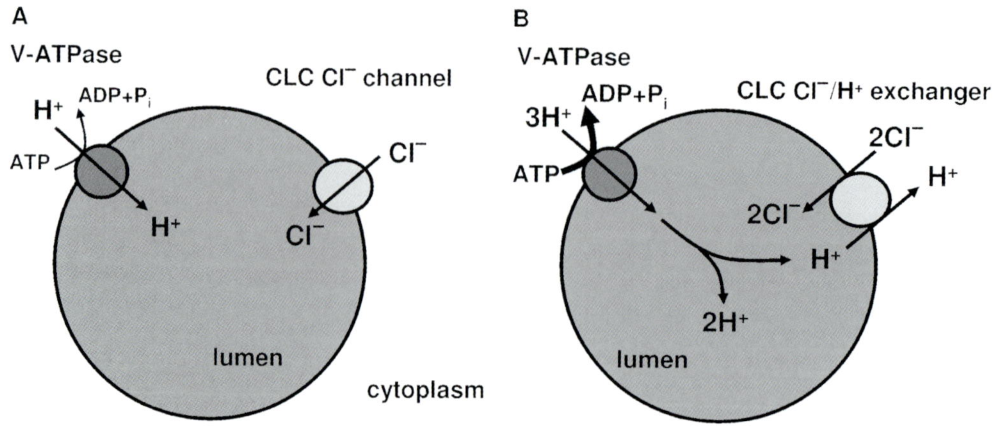

FIGURE 12.4 Electrical neutralization of acidification by CLC channels or transporters. The schematic drawing illustrates vesicular acidification by the V-ATPase. Under ATP consumption, it pumps protons into the lumen. Electrical neutralization could be achieved by a CLC chloride channel, which allows for a passive chloride influx (**A**). **B.** A CLC Cl^-/H^+ exchanger will also provide shunt current. While the stoichiometry for mammalian CLC Cl^-/H^+ exchangers is not known, the model depicts a $2Cl^-/1H^+$ stoichiometry (as known for ec-ClC-1).

especially interesting as the luminal [Cl^-] itself might be involved in the regulation of membrane trafficking, translocation of various molecules across the membrane, and enzymatic activities (Faundez and Hartzell, 2004; Jentsch, 2007).

ClC-3 is expressed in a broad range of tissues and cells, where it localizes to endosomes and endosome-related compartments, such as synaptic vesicles in neurons. Accordingly, diverse physiological functions, which will be discussed in Section III.B, have been assigned to it.

Very little is known about the physiological role of the endosomal ClC-4. When ectopically co-expressed with its close relatives ClC-3 and -5, they co-localize and can be co-immunoprecipitated (Suzuki et al., 2006). Due to its co-localization and interaction with ClC-5 in renal proximal tubules (PT) (Mohammad-Panah et al., 2003) a role for endosomal acidification in PT cells was implicated. But ClC-4 is less important (if at all) than ClC-5 (see below) in this process, as ClC-4 cannot compensate for the loss of ClC-5. ClC-4-deficient mice showed no obvious phenotype that could be attributed to the lack of ClC-4 (Rugarli et al., 1995). Furthermore, an unpublished ClC-4 knockout mouse from our laboratory does not show proteinuria (unpublished data).

ClC-5 is the best-studied member of the second CLC subfamily. It is predominantly expressed in kidney and intestine (Devuyst et al., 1999; Steinmeyer et al., 1995; Vandewalle et al., 2001). ClC-5 is expressed in endosomes where it co-localizes with the V-ATPase and endocytic cargo (Günther et al., 1998; Sakamoto et al., 1999). ClC-5 is important for apical endocytosis in renal proximal tubule cells (Piwon et al., 2000; Wang et al., 2000), likely because it is required for the acidification of early endosomes (Günther et al., 2003; Hara-Chikuma et al., 2005a). The consequences of ClC-5 deficiency in KO mice and mutations in humans include the development of Dent's disease with proteinuria and (indirectly via impaired vitamin D metabolism (Maritzen et al., 2006)) hyperphosphaturia, hypercalciuria and kidney stones (recently reviewed in Jentsch, 2008; Jentsch et al., 2005a).

ClC-6 and -7 (which will be covered in Sections III.C and III.D, respectively) reside on late endosomal/lysosomal compartments (Kasper et al., 2005; Poet et al., 2006). The disruption of either Cl^- transporter leads to neuronal lysosomal storage disease (Kasper et al., 2005; Poet et al., 2006). ClC-6 is highly expressed in dorsal root ganglia. In ClC-6 KO mice, these sensory neurons exhibit high levels of storage material, which was correlated with a decrease in pain sensitivity as assayed by tail-flick analysis (see below). ClC-7 expression is not restricted to the nervous system; it is also inserted into the acid-secreting membrane of osteoclasts. ClC-7 KO mice develop osteopetrosis (Kornak et al., 2001).

III. PHYSIOLOGY AND PATHOLOGY OF THE NEURONAL CLC PROTEINS

A. ClC-2: a Role in Extracellular Ion Homeostasis?

ClC-2 is a broadly expressed plasma membrane Cl^- channel (Thiemann et al., 1992). It mediates slowly activating inwardly rectifying currents that, depending on the expression system, activate at voltages between −40 and −80 mV (Thiemann et al., 1992). This voltage dependence implies that only a small proportion of ClC-2 would be open at physiological resting membrane voltages.

ClC-2 is inhibited by Cd^{2+} and Zn^{2+} (Clark et al., 1998), but is relatively insensitive to DIDS (4,4′-diisothiocyanatostilbene-2,2′-disulfonic acid disodium salt) (Thiemann et al., 1992). When expressed in *Xenopus* oocytes, ClC-2 currents are activated by cell swelling (Furukawa et al., 1998; Gründer et al., 1992; Jordt and Jentsch, 1997), suggesting that they may play a role in cell volume regulation. Swelling activation of ClC-2 has also been observed in native cells (Clark et al., 1998; Comes et al., 2006; Huber et al., 2004). However, ClC-2 does not underlie the swelling-activated currents (VRAC or $I_{Cl,swell}$) typically seen in most cells (Nilius et al., 1997; Sardini et al., 2003). VRAC current can be distinguished from ClC-2-mediated currents by its anion selectivity ($I^- > Cl^-$ for VRAC and $Cl^- > I^-$ for ClC-2) and by its outward rectification (Gründer et al., 1992; Nilius and Droogmans, 2001). Although ClC-2-mediated currents can be activated by swelling, salivary acinar cells from ClC-2-deficient mice do not exhibit impaired volume regulation (Nehrke et al., 2002). Thus, the physiological importance of the activation of ClC-2 by cell swelling is still an unresolved issue.

Like other CLC channels, ClC-2 currents are sensitive to pH and Cl^-. ClC-2 current amplitude increases upon extracellular acidification up to pH 6.5 (Jordt and Jentsch, 1997). However, further acidification leads to a decrease in amplitude of native ClC-2 currents in parotid acinar cells, suggesting the presence of two different proton-binding sites in the channel (Arreola et al., 2002). ClC-2 gating also depends on the $[Cl^-]_i$ (Haug et al., 2003; Niemeyer et al., 2003; Pusch et al., 1999); a rise in $[Cl^-]_i$ increases ClC-2 open probability towards positive voltages.

In contrast to its close homolog ClC-1, which displays a restricted tissue distribution, ClC-2 is broadly expressed. Its mRNA has been found in all tissues and cell lines investigated (Thiemann et al., 1992). It is most highly expressed in brain and in epithelia. ClC-2 has been reported to be present in apical (Mohammad-Panah et al., 2001; Murray et al., 1995) or basolateral (Catalan et al., 2004; Lipecka et al., 2002) membranes of lung and intestinal epithelia. Knockout-controlled immunocytochemistry strongly supports a basolateral localization in mouse colon (Zdebik and Jentsch, unpublished data), which is also supported by functional data (Zdebik et al., 2004) (see below). ClC-2 is also expressed in the kidney (Morales et al., 2001; Obermüller et al., 1998; Thiemann et al., 1992), but its intrarenal expression pattern is poorly defined.

The broad tissue distribution of ClC-2 gave rise to several speculations on its physiological roles. Its expression in lung epithelia, pancreas and colon (Thiemann et al., 1992), all tissues affected in cystic fibrosis, led to the suggestion that ClC-2 may function as an alternative pathway for Cl^- secretion in this disease (Schwiebert et al., 1998). However, in the lung ClC-2 is expressed from early on and its expression declines after birth, whereas expression of the cystic fibrosis transmembrane conductance regulator (CFTR) increases postnatally (Zdebik et al., 2004). In the intestine, colonic cells take up Cl^- via the basolateral Na-K-2Cl cotransporter, NKCC1, and secrete it through apical CFTR channels (Strong et al., 1994). If ClC-2 localized together with CFTR to the apical membrane of the same colonic epithelial cells and mediated a parallel Cl^- efflux, the additional impairment of ClC-2 in CFTR-deficient mice would be expected to worsen the cystic fibrosis phenotype. However, this is not the case, and ClC-2/CFTR double-deficient mice survive even better than CFTR-deficient mice (Zdebik et al., 2004). This, together with data from Ussing chamber experiments, suggests that ClC-2 does not reside in the apical, but rather in the basolateral membrane (Zdebik et al., 2004).

ClC-2 has also been proposed to regulate $[Cl^-]_i$ in neurons, an important determining factor for synaptic inhibition mediated through $GABA_A$ and glycine receptors (Clark et al., 1998; Smith et al., 1995; Staley et al., 1996). Since ClC-2 gating depends on $[Cl^-]_i$ (Niemeyer et al., 2003; Pusch et al., 1999), a rise in intraneuronal $[Cl^-]$ may open ClC-2 leading to depolarizing Cl^- efflux, if E_{Cl} is less negative than E_m. Indeed, transfection of dorsal root ganglion neurons with ClC-2 shifted E_{Cl} (estimated from E_{GABA}) to values close to electrochemical equilibrium (Staley et al., 1996). Haug and colleagues identified three different mutations in *CLCN2* by screening families with idiopathic generalized epilepsies (IGE), which had been previously mapped to 3q26, a chromosomal region close to the locus of *CLCN2* (3q27) (Haug et al., 2003). One mutation, reported to have a dominant-negative effect, truncates the ClC-2 protein at helix F (Gins597), leading to a loss of function. Although Haug and co-workers reported that the truncating mutation (Gins597) strongly suppressed currents when co-expressed with wildtype ClC-2, the mutant lacked a dominant-negative effect when expressed in a 1:1 ratio with wildtype ClC-2 in *Xenopus* oocytes, resembling the situation in heterozygous patients (Blanz et al., 2007). The second mutation leads to the deletion of 11 nucleotides of an intron (IVS2-14del11). This was reported to alter splicing, thereby increasing the quantity of a protein that lacks 44 amino acids in the transmembrane portion. When expressed heterologously, this construct displayed a strong dominant-negative effect (Haug et al., 2003). Finally, Haug and colleagues identified a missense mutation (G715E) between CBS1 and CBS2 in the carboxy-terminus of ClC-2. This mutation was reported to shift the voltage dependence of ClC-2 towards more positive voltages in a Cl^--dependent manner. This would represent a gain of function, in contrast with the loss of function caused by the truncating mutations. However, work from other groups could not reproduce the functional data of Haug et al. (Blanz et al., 2007; Niemeyer et al., 2004). The group of Sepúlveda failed to detect an effect on splicing of the deletion IVS2-14del11 (Niemeyer et al., 2004). Instead of a gain of function for the mutant G715E, they only found slower deactivation kinetics of ClC-2 currents. Two additional screens for ClC-2 mutations in epileptic patients identified only two polymorphisms (D'Agostino et al., 2004; Marini et al., 2004) which do not alter the biophysical properties of ClC-2 expressed in *Xenopus* oocytes and that were also found in non-epileptic persons (Blanz et al., 2007).

To elucidate the physiological role of ClC-2, its gene was disrupted in mice (Bösl et al., 2001). ClC-2 KO mice were viable, grew normally and showed no immediately visible physical or behavioral abnormalities. Contrary to previous speculations (Blaisdell et al., 2000; Huber et al., 1998; Malinowska et al., 1995; Murray et al., 1995), disruption of ClC-2 did not lead to lung or kidney abnormalities, a gastric acid secretion defect, or increased sensitivity to seizure-inducing agents (Blanz et al., 2007; Bösl et al., 2001). Altogether these observations warrant skepticism towards the proposed role of ClC-2 in the etiology of epilepsy.

Unexpectedly, disruption of ClC-2 in mice resulted in blindness, male infertility (Bösl et al., 2001; Nehrke et al., 2002) and leukoencephalopathy (Blanz et al., 2007). *Clcn2*$^{-/-}$ males were infertile due to degeneration of seminiferous tubules. Germ cells of any stage were missing in males older than six months and spermatids could not be identified in *Clcn2*$^{-/-}$ mice (Bösl et al., 2001). Seminiferous tubules were filled with abnormal Sertoli cells. Knockout-controlled patch-clamp analysis of Sertoli cells showed that they indeed expressed a ClC-2 current.

Histological analysis of the retina of *Clcn2*$^{-/-}$ mice showed a dramatic and early loss of photoreceptors. Starting from P14 in the outer nuclear layer (ONL), a great number of shady, pyknotic nuclei were observed and the number of cells decreased to 50% with respect to control (WT) mice. The width of the ONL was reduced further with time until only some scattered nuclei remained in old mice (Bösl et al., 2001).

Both photoreceptors and germ cells depend on transepithelial transport, mediated by the retinal pigment epithelium (RPE) and Sertoli cells, respectively. Ussing chamber experiments revealed a decreased short-circuit current across the *Clcn2*$^{-/-}$ RPE. It is therefore tempting to speculate that ClC-2 is important for ion homeostasis of the subretinal space, which is formed between the RPE and photoreceptors. Similarly, ClC-2 may be an important determinant of ion concentrations in the extracellular cleft between Sertoli cells and germ cells. Changes in the extracellular ion concentration surrounding germ cells and photoreceptors might cause their degeneration (Bösl et al., 2001).

ClC-2 knockout mice also show a widespread, progressive spongiform vacuolation of the white matter of the brain and spinal cord, accompanied by glial activation (Blanz et al., 2007) (Fig. 12.5). The CNS vacuolation was found within fiber tracts and developed during the period of myelination. Vacuole-like holes became evident at P28 in the white matter of the cerebellum and in fiber tracts of the brainstem of *Clcn2*$^{-/-}$ mice. The area covered by vacuole-like structures increases with age primarily due to enlargement in vacuole size rather than an increase in their number. Unlike the CNS, peripheral nerves were not affected. Electron microscopy showed that vacuoles developed within the compact myelin sheets. The cell bodies of oligodendrocytes and axons did not show anomalous morphology even when they were in close proximity to vacuoles. Besides the above changes, gene expression profiling with whole-cerebellum RNA demonstrated an increase in several gene transcripts involved in inflammatory processes.

Although ClC-2 knockout mice showed severe vacuolation of the cerebellum, they did not have obvious defects in motor performance, probably because of the lack of neuronal degeneration. As expected for a demyelinating disorder, there was a decrease in nerve conduction velocity determined by measuring auditory brainstem responses. Similar results are found in other demyelinating mouse models (Roncagliolo et al., 2000) and in humans with subclinical forms of CNS myelopathies (Elidan et al., 1982; Robinson and Rudge, 1975; Starr, 1978).

In the brain, ClC-2 is expressed in neurons and glia (Sik et al., 2000). In glial cells it is localized to the end-feet of astrocytes surrounding blood vessels in the hippocampus, and in Bergmann glia in the cerebellum. ClC-2-positive puncta have also been observed surrounding oligodendrocytic somata. Double-immunolabeling with Cx47, a connexin selectively expressed in oligodendrocytes (Altevogt and Paul, 2004), showed co-localization of ClC-2 along the circumference of oligodendrocyte cell bodies (Blanz et al., 2007).

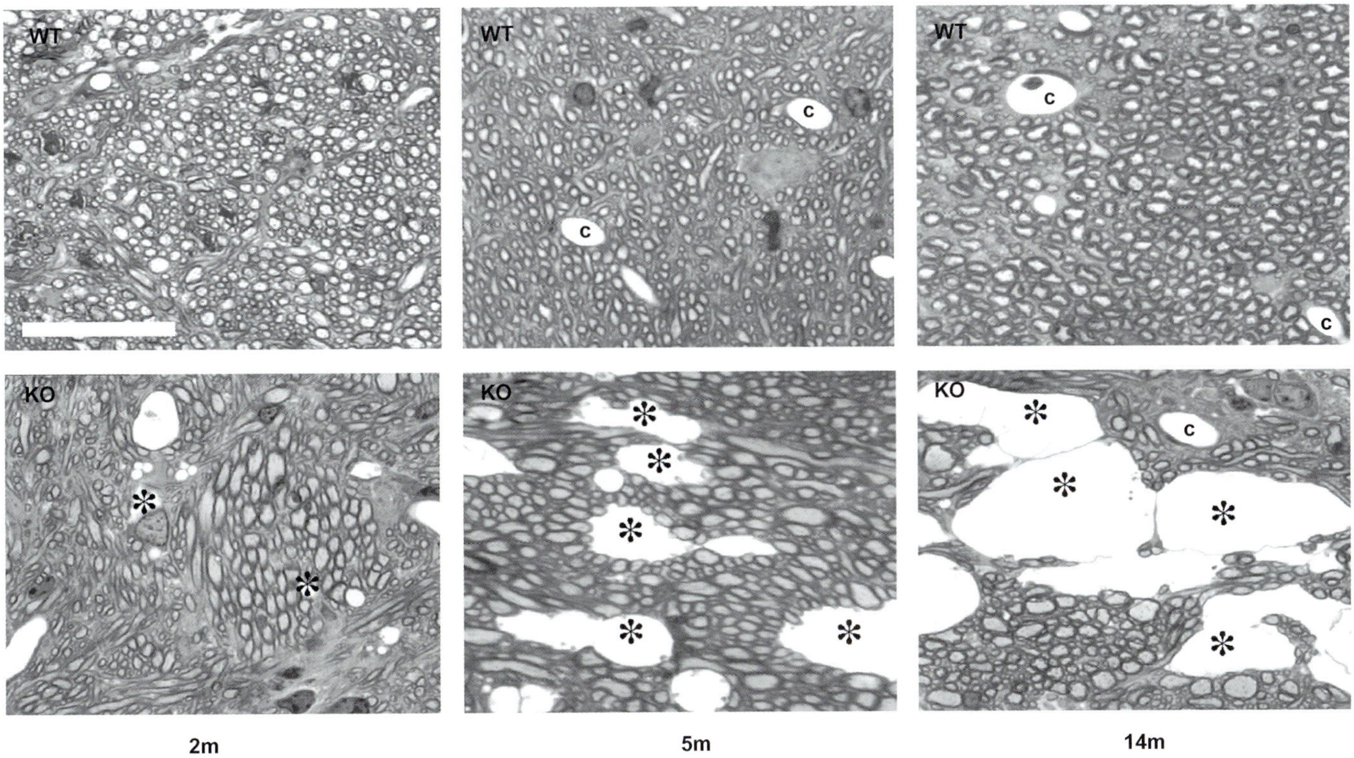

FIGURE 12.5 Spongiform vacuolation in the white matter of ClC-2 knockout mice. Semi-thin sections of the middle cerebellar peduncle of 2-, 5-, and 14-month-old WT and ClC-2 knockout mice. Abundant vacuoles (asterisks) are visible in cerebellar white matter of 5- and 14-month-old KO but not WT animals, in which capillaries (c) are indicated. Scale bar: 20 µm. Image from Blanz et al. (2007).

The widespread and extensive vacuolation observed in old mice lacking ClC-2 is reminiscent of the one observed in mice lacking the inwardly rectifying K^+ channel Kir4.1, which is also localized in astrocytic endfeet and on the cell body of oligodendrocytes (Butt and Kalsi, 2006; Neusch et al., 2001), and to the one observed in the double knockout of Cx47 and Cx32, two connexins thought to be involved in K^+ homeostasis in astrocytic and oligodendrocityc syncytiae (Menichella et al., 2003). Although these knockout mice differ from ClC-2-deficient mice to some degree, for instance *Kir4.1*$^{-/-}$ mice display neuronal cell loss that resulted in severe neurological deficits, altogether these observations suggest that ClC-2 may participate in buffering of extracellular ions, particularly K^+. This attractive hypothesis has also been proposed for the ClC-2 function in retina and testis. Interestingly, the vacuolation observed in the optic nerve of Cx47/32 double knockout mice began after eye opening and could be suppressed by inhibiting optic nerve activity with tetrodotoxin, suggesting that these connexins are needed to buffer extracellular K^+ that increases during neuronal activity (Menichella et al., 2006). As ClC-2-deficient mice are blind, there is no activity in their optic nerves. Indeed, in analogy to the experiments of Menichella and collaborators, this nerve in particular did not degenerate in *Clcn2*$^{-/-}$ mice (Blanz et al., 2007). This suggests that ClC-2 is needed for oligodendrocyte integrity when neuronal activity leads to dynamic changes of extracellular ion concentrations.

The CNS pathology of *Clcn2*$^{-/-}$ mice resembles the clinical phenotype of some human leukodystrophy; several of these patients had vacuolation of the CNS, but only mild neurological deficits. However, none of the patients analyzed in the study of Blanz and collaborators displayed pathogenic *CLCN2* mutations (Blanz et al., 2007). Further, there are no reports of human leukodystrophy accompanied with retinal degeneration.

B. ClC-3: Important for the Acidification of Endosomes and Synaptic Vesicles

ClC-3 is a member of the second branch of the CLC family (Fig. 12.1). Members of this branch localize to the membranes of intracellular organelles, particularly those along the endocytic pathway (Jentsch, 2008). Cell fractionation and immunofluorescence studies showed that ClC-3 localizes to late endosomes and synaptic vesicles (Salazar et al., 2004; Stobrawa et al., 2001; Yoshikawa et al., 2002). This localization on endosomes and synaptic vesicles is consistent with the fact that synaptic vesicles often recycle through an endosomal compartment (Südhof, 2004). In ClC-3-overexpressing cells, the protein localizes mainly to intracellular compartments, and less than 6% of the total protein is found at the cell surface (Zhao et al., 2007). Further, ClC-3 overexpression results in the formation of large intracellular ClC-3-positive vesicular structures (Li et al., 2002).

Several interaction partners have been proposed for ClC-3. For instance, the interaction between an amino terminal dileucine motive and clathrin was suggested to be required for ClC-3 endocytosis (Zhao et al., 2007). Additionally, the sorting of ClC-3 to synaptic vesicles is thought to be mediated by the adaptor protein AP-3 (Salazar et al., 2004; Seong et al., 2005). The human ClC-3 splice variant, ClC-3B, which possesses an additional PDZ-binding motif at the extreme carboxy-terminus, can bind the PDZ domains of NHERF-1, PDZK1 and GOPC (Gentzsch et al., 2003; Ogura et al., 2002). These interactions might be important for the correct trafficking of ClC-3B. When ClC-3B alone is heterologously expressed in C127 cells, it localizes diffusely to intracellular compartments, which may represent Golgi and ER. But when ClC-3B is co-expressed with NHERF-1, it localizes to the leading edge of membrane ruffles (Ogura et al., 2002).

The biophysical properties of ClC-3 currents are still controversial. It has been proposed that ClC-3 mediates time-independent, slightly outwardly-rectifying Cl^- currents, which are inhibited upon activation of protein kinase C (Kawasaki et al., 1994). In addition, it has been claimed for several years that activation of ClC-3 triggers swelling-activated Cl^- currents (Duan et al., 1997; von Weikersthal et al., 1999; Yamazaki et al., 1998). Another group proposed ClC-3 as a Ca^{2+}-dependent, CamKII-activated channel (Huang et al., 2001). These currents do not resemble those observed for ClC-4 and ClC-5, with which ClC-3 shares 80% sequence identity, and they are also observed in cells from ClC-3 knockout mice (Arreola et al., 2002; Gong et al., 2004; Stobrawa et al., 2001). In 2000, Weinman and co-workers recorded ClC-3 currents that seem to be truly mediated by ClC-3 (Li et al., 2000). In ClC-3-transfected cells, they observed currents that displayed a conductivity sequence with $Cl^- > I^-$, which is typical of CLC proteins (Li et al., 2000). A mutation of the gating glutamate (see Section II.A) in ClC-4 and -5 abolished their strong outward rectification (Friedrich et al., 1999), and exactly the same effect was observed when the corresponding ClC-3 mutant was expressed (Li et al., 2002). Recently, another study described outwardly rectifying Cl^- currents upon ClC-3 overexpression in HEK-293 cells (Matsuda et al., 2008). In this work, mutations in the gating glutamate also led to a loss of rectification. However, many groups have been trying extensively to measure ClC-3-mediated currents (Friedrich et al., 1999; Picollo and Pusch, 2005; Weylandt et al., 2001), without reproducibly yielding currents. It now seems very likely that ClC-3, just like ClC-4 and -5, mediates Cl^-/H^+ exchange.

ClC-3 was detected in brain, retina, pancreas, kidney, adrenal gland, liver, skeletal muscle and heart. Within the brain, ClC-3 is abundant in hippocampus, cerebellum and olfactory bulb (Stobrawa et al., 2001). To elucidate the cellular and physiological role of ClC-3, three groups have independently disrupted its gene in mice. All three groups observed that ClC-3-deficient mice exhibit severe neurodegeneration that leads to a complete loss of the hippocampus (Fig. 12.6) and of photoreceptors (Dickerson et al., 2002; Stobrawa et al., 2001; Yoshikawa et al., 2002). In spite of the severe neurodegeneration, ClC-3 knockout mice are viable and survive for more than a year. However, there is an overall higher mortality and a reduction in weight.

Our laboratory showed that hippocampal degeneration starts in the CA1 region of the hippocampus two weeks after birth (Stobrawa et al., 2001). At

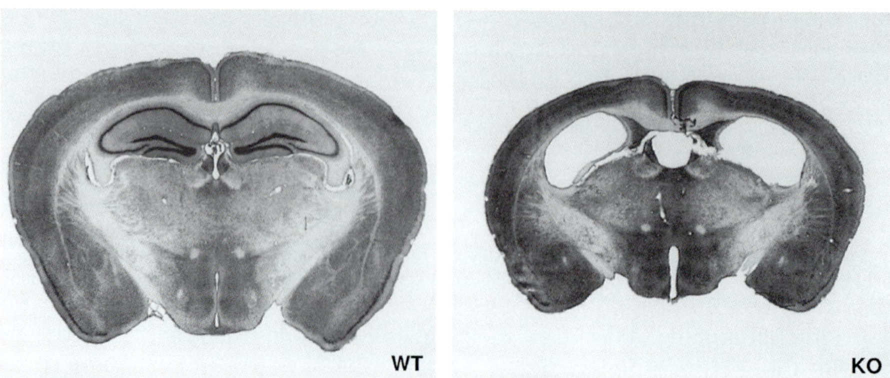

FIGURE 12.6 The hippocampus is degenerated in ClC-3 knockout mice. Nissl-stained frontal sections of 7-month-old wildtype (WT) and ClC-3 knockout (KO) mice show a clear degeneration of the hippocampus in ClC-3 knockout mice. Images from Stobrawa et al. (2001).

three months of age, the hippocampus is replaced by large cavities contiguous with the ventricular system. Degeneration is not restricted to the hippocampus, and is accompanied by an activation of microglia and astrogliosis. Interestingly, in the ClC-3 knockout line generated by Dickerson and colleagues, the hippocampal degeneration has a slower time course developing over months, instead of weeks, and progresses in a different sequence among neuronal subpopulations (Dickerson et al., 2002). These authors also reported that *Clcn3*$^{-/-}$ mice are particularly sensitive to midazolam, a benzodiazepine that enhances activity of $GABA_A$ receptors, pointing to an abnormality in GABA neurotransmission (Dickerson et al., 2002). ClC-3 knockout mice display increased basal locomotion accompanied by an overall reduced resting time. They have also difficulties in tests for motor skills (Stobrawa et al., 2001; Yoshikawa et al., 2002). However, they are able to improve their performances, showing that motor learning is not abolished (Stobrawa et al., 2001).

Yoshikawa and colleagues reported that their ClC-3 knockout line displayed some features of neuronal ceroid lipofuscinosis (NCL), a subtype of lysosomal storage disease. They observed accumulation of subunit c of the mitochondrial ATP synthase, a hydrophobic protein that is normally degraded in lysosomes, and a mild accumulation of intracellular electron-dense material (Yoshikawa et al., 2002). Comparison of ClC-7 knockout mice, which display severe lysosomal storage (see Section III.D), with mice disrupted for ClC-3 in our laboratory (Stobrawa et al., 2001) showed that the latter mice did not exhibit intraneuronal storage material and only minimally accumulated subunit c (Kasper et al., 2005). The reasons for these subtle differences between phenotypes of the three ClC-3 knockout lines are not clear, but might be related to the genetic background. Indeed, back-crossing ClC-3 knockout mice to C57/Bl6 in our laboratory resulted in increased mortality (unpublished data).

Photoreceptors in ClC-3-deficient mice degenerate, leading to complete blindness. Histological analysis of the retina showed a dramatic and selective loss of the outer nuclear layer (ONL) and of the outer segments (OS) starting from P12 (Stobrawa et al., 2001; Yoshikawa et al., 2002). Electroretinograms showed that most of the ClC-3-deficient mice do not respond to flash light stimuli. The CNS degeneration affects most drastically the hippocampus, a region where ClC-3 is highly expressed. In the retina, the highest ClC-3 immunoreactivity was found in the inner and outer plexiform layers (IPL and OPL), consistent with the localization of ClC-3 to synaptic vesicles, whereas degeneration takes place in the ONL (Stobrawa et al., 2001). The selective degeneration observed in *Clcn3*$^{-/-}$ mice may be due to the high vulnerability of both the hippocampus and photoreceptors. The hippocampal CA1 region is the brain area that first degenerates in knockout mice, consistent with its well-known sensitivity to other insults, e.g. cell death following anoxia, glutamate toxicity, or epileptic seizures (Grooms et al., 2000; Nunn et al., 1994).

The reason for the observed neurodegeneration is not clear but may be secondary to a defective acidification of intracellular organelles. The V-type H^+-ATPase, which is involved in vesicle acidification, needs a parallel anion conductance for efficient H^+ pumping (Jefferies et al., 2008), which may be mediated by ClC-3. *In vitro* acidification of synaptic vesicles from ClC-3-deficient mice was indeed reduced (Stobrawa et al., 2001), and the intra-vesicular steady-state pH of an endosome-enriched preparation from liver of ClC-3 knockout mice was slightly increased (Yoshikawa et al., 2002). The electrochemical H^+ gradient across the membrane of synaptic vesicles is an important driving force for the uptake of neurotransmitters. Hence, neurotransmitter uptake might be altered in *Clcn3*$^{-/-}$ mice, although initial experiments did not support this hypothesis (Stobrawa et al., 2001).

C. ClC-6: a Late Endosomal Chloride Transporter in Neurons

Like ClC-3 and -7, ClC-6 localizes to the endo-lysosomal pathway in neurons and its absence leads to neuronal pathology (Poet et al., 2006). ClC-6 was cloned in parallel with ClC-7 (Brandt and Jentsch, 1995). These two proteins share about 45% sequence identity and form the third subfamily of mammalian CLC proteins. In contrast to the members of the second subfamily, ClC-3 through -5, ClC-6 and -7 could not be co-immunoprecipitated with each other or with any other intracellular CLC protein (Suzuki et al., 2006). The fact that ClC-6 can complement the phenotype of the yeast gef1 mutant, in which the single yeast CLC gene is disrupted (Greene et al., 1993), indicates that ClC-6 mediates Cl^- flux (Kida et al., 2001). Nevertheless, no ClC-6 currents could be measured when the protein was expressed ectopically in *Xenopus* oocytes (Brandt and Jentsch, 1995; Buyse et al., 1997) or COS cells (Buyse et al., 1998). This is likely due to its intracellular localization (see below), which so far has prevented the biophysical characterization of ClC-6. Therefore it is not known whether ClC-6 is a chloride channel or a Cl^-/H^+ exchanger, although the presence of a "proton glutamate" (see Section II.A) argues for the latter.

The mRNA encoding for ClC-6 shows broad tissue expression, including brain, kidney, testes, lung, skeletal muscle, thymus, intestine and pancreas (Brandt and Jentsch, 1995; Eggermont et al., 1997; Kida et al., 2001). Splice variants of the ClC-6 mRNA have been identified by RT-PCR (Eggermont et al., 1997), but their physiological relevance is doubtful, as they encode severely truncated proteins. In spite of its ubiquitous transcription, the ClC-6 protein is almost exclusively expressed in the nervous system (Poet et al., 2006).

Using the overexpression of an epitope-tagged version, ClC-6 was at first reported to be an endoplasmic reticulum (ER)-resident protein (Buyse et al., 1998). Later, however, it was shown to co-localize with markers of both early and late endosomes by another group (Suzuki et al., 2006). From immunocytochemistry and subcellular fractionation of native tissues, using knockout mice as control, it is now clear that native ClC-6 is an endosomal protein (Poet et al., 2006). As it partially co-localizes with lamp-1 (lysosome-associated membrane protein-1) and is partially shifted into lysosomal fractions in ClC-7 knockout mice, it is believed to reside predominantly on late endosomes (Poet et al., 2006). This localization has also been described for endogenous ClC-6 in the human SH-SY5Y neuroblastoma cell line, while ectopically (over-)expressed ClC-6 was mislocalized to early and recycling endosomes (Ignoul et al., 2007). ClC-6 is incorporated into detergent-resistant membrane domains, depending on a stretch of basic amino acids. This incorporation seems to play a role in the localization of ClC-6, since alanine substitution of this stretch of amino acids shifts the overexpressed ClC-6 from early endosomes to late endosomes/lysosomes, where it partially co-localizes with ClC-7 (Ignoul et al., 2007).

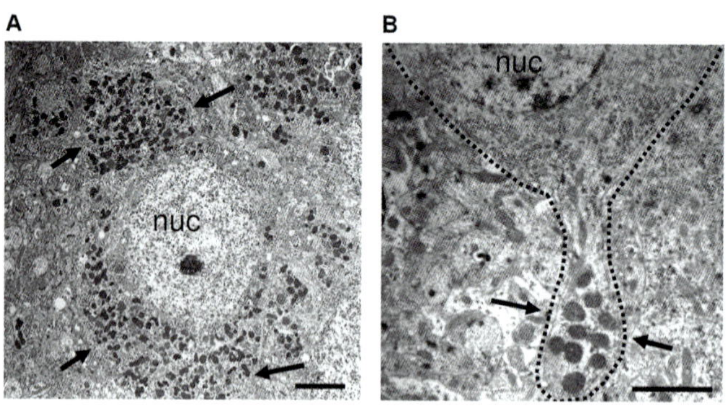

FIGURE 12.7 Lysosomal storage material in neurons from ClC-6 and ClC-7 knockout mice. Electron micrographs of ClC-7-deficient cortical neurons (**A**) and ClC-6-deficient hippocampal neurons (**B**) clearly show the accumulation of electron-dense storage material (arrows). Note the difference in distribution, with the storage material scattered through the entire cell body in ClC-7 knockout and the restricted localization to the initial axon segment in ClC-6 knockout. Modified from Kasper et al. (2005) (**A**) and Poet et al. (2006) (**B**). nuc = nucleus; scale bars: 2 μm.

ClC-6 knockout mice show no immediately apparent changes in phenotype; they are of normal size and have a normal lifespan (Poet et al., 2006). However, these mice accumulate lysosomal storage material in central and peripheral neurons that becomes visible at four weeks of age (Poet et al., 2006). This autofluorescent storage material does not only include lysosomal markers, such as lamp-1 and cathepsin D, but also stains positive for saposin D and subunit c of the mitochondrial ATPase, both typical components of storage material in human neuronal ceroid lipofuscinosis (NCL) (Palmer et al., 1992; Tyynelä et al., 1993). In contrast to the lysosomal storage material in ClC-7 knockout mice, which accumulates in neuronal somata (Kasper et al., 2005; Poet et al., 2006) (see Section III.D), these deposits are found in the initial axonal segment in ClC-6 knockout neurons (Fig. 12.7). It often (e.g. about 20% of cortical neurons) leads to a swelling of the proximal axon (Poet et al., 2006).

Tail-flick latencies (as a measure of pain threshold) are doubled in the ClC-6 knockout mouse. This might be due to impairment of dorsal root ganglion neurons that accumulate large amounts of storage material in ClC-6 KO animals (Poet et al., 2006).

The phenotype of the ClC-6 knockout mouse is less severe than that of the ClC-3 and -7 knockout mice (see Sections III.B and III.D, respectively); ClC-6-deficient mice display neither neurodegeneration nor blindness. The phenotypes observed in the ClC-6 knockout mouse resemble those of Kuf's disease, a mild, adult-onset form of NCL. Two (out of 75 tested) patients were found to carry heterozygous missense mutations (Poet et al., 2006). Interestingly, one of them displayed the mutation (V580M) within α-helix R, which lines the anion permeation pathway (see Section II.A). But as no other *CLCN6* mutation was found in the other allele, there is no proof for a causal relation between the ClC-6 mutations and Kuf's disease (Poet et al., 2006). Due to the lack of an appropriate expression system, the biophysical properties of these mutants could not yet be tested.

In spite of the lysosomal pathology, the steady-state lysosomal pH of cultured hippocampal neurons from ClC-6 knockout mice is not altered (Poet et al., 2006). This does not exclude an altered pH of late endosomes (to which ClC-6 predominantly localizes), which in the long run (over several months) could lead to lysosomal storage, possibly by a slight trafficking defect in the endosomal/lysosomal system.

D. ClC-7/Ostm1: Role in Osteoclasts, Neuronal Lysosomes and Retina

ClC-7, together with its β-subunit Ostm1, has been shown to be engaged in various physiological processes such as bone resorption by osteoclasts and lysosomal function in neurons, depending on its tissue-specific subcellular localization. Accordingly, its loss or dysfunction leads to diseases as diverse as osteopetrosis, neuronal lysosomal storage with neurodegeneration and blindness (Jentsch, 2008).

Together with ClC-6, with which it shares 45% sequence homology (Brandt and Jentsch, 1995) (see Section III.C), ClC-7 forms the third CLC subfamily (Fig. 12.1). Just like ClC-6, ClC-7 possesses the "proton glutamate" common to CLC Cl^-/H^+ exchangers (see Section II.A), indicating that ClC-7 might also be an exchanger. But as for ClC-6, no currents could be detected when ClC-7 was overexpressed in *Xenopus* oocytes (Brandt and Jentsch, 1995), and no biophysical characterization has been reported so far.

ClC-7 mRNA is broadly expressed. By Northern blot it was detected in all tested murine tissues including brain, liver, heart, placenta, lung, thymus, colon, testis and kidney (Brandt and Jentsch, 1995). By *in situ* hybridization of mouse embryos, ClC-7 was shown to be highly expressed in the nervous system (dorsal root and trigeminal ganglia and brain) (Kornak et al., 2001). Hybridization of adult mouse sections also yielded a strong signal in cerebellar Purkinje cells as well as in other non-nervous tissues (e.g. renal proximal tubules) (Kida et al., 2001). Expression of X-Gal under the endogenous ClC-7 promoter in ClC-7 knockout mice showed additionally expression in the eye (lens, retinal pigment epithelium and neuroretina) and in osteoclasts (Kornak et al., 2001). A broad neuronal expression of ClC-7 protein was confirmed by immunocytochemistry (Kasper et al., 2005).

Endogenous ClC-7 co-localizes strongly with the late endosomal/lysosomal marker lamp-1 in mouse fibroblasts (Kornak et al., 2001) and hippocampal neurons (Kasper et al., 2005). Immunoblotting following subcellular fractionation revealed the presence of ClC-7 in late endosomal and lysosomal fractions (Kornak et al., 2001). Its presence on lysosomes has been demonstrated additionally by electron microscopy of immunogold-labeled cortical neurons (Kasper et al., 2005). ClC-7 is also targeted to these compartments when heterologously expressed (Lange et al., 2006; Suzuki et al., 2006), hence likely impeding its biophysical characterization. ClC-7 is the only member of the intracellular CLC proteins that localizes so "late" in the endo-lysosomal pathway. Only when it is lacking in ClC-7 knockout tissue are the late endosome-localized ClC-3 and -6 partially shifted into the lysosomal fraction (Poet et al., 2006). Just like ClC-6, ClC-7 did not co-immunoprecipitate with any other CLC protein of the second or third subfamily (Suzuki et al., 2006). In osteoclasts, ClC-7 localizes not only to intracellular compartments but also to the ruffled border at the interface between these bone-resorbing cells and the bone (Kornak et al., 2001), a localization important for the function of ClC-7 in bone resorption (see below).

The stability of ClC-7 depends on its association with its β-subunit Ostm1 (Lange et al., 2006). Ostm1 was identified as a membrane protein of unknown function, the deletion or mutations of which lead to the severe osteopetrotic phenotype of the spontaneous grey-lethal mouse and to rare cases of human recessive, malignant infantile osteopetrosis (Chalhoub et al., 2003) – hence the name Ostm1, short for osteopetrosis-associated membrane protein 1. Independently, Ostm1 was cloned as GIPN (GAIP-interacting protein N-terminus), and a ubiquitin ligase activity was attributed to it (Fischer et al., 2003). However, such an activity is virtually excluded in the light of our current knowledge about the topology of Ostm1 as a type-I transmembrane protein with a cleavable amino-terminal signal peptide, a long, heavily glycosylated extracytoplasmic domain (which contains a putative RING-finger domain; Fischer et al., 2003) and a short, cytoplasmic, carboxy-terminal amino acid stretch (Lange et al., 2006). Just as for ClC-7, transcripts of Ostm1 could be detected in all examined tissues (Chalhoub et al., 2003; Fischer et al., 2003; Lange et al., 2006). In native cells, ClC-7 and Ostm1 perfectly co-localize on late endosomes/lysosomes, on the way to (or in) which the glycosylated luminal part of Ostm1 is proteolytically cleaved, and in the ruffled border of osteoclasts (Lange et al., 2006). Trafficking of heterologously expressed Ostm1 to lamp-1-positive compartments depends specifically on ClC-7, whereas heterologously expressed ClC-7 reaches those structures also in the absence of Ostm1 (Lange et al., 2006). The two proteins can be co-immunoprecipitated with each other and the protein level of either is strongly reduced in the mouse lacking the other (ClC-7 knockout and grey-lethal, respectively) (Lange et al., 2006). As ClC-7 is not glycosylated, one can speculate that its interaction with the heavily glycosylated Ostm1 stabilizes it against degradation in lysosomes. At least in osteoclasts, transcription of the ClC-7-encoding gene, *Clcn7*, and that of its β-subunit, *Ostm1*, are co-regulated by the same (namely, microphthalmia) transcription factor (Meadows et al., 2007). As expected from the relationship of ClC-7 and Ostm1, the grey-lethal mouse, which lacks Ostm1, and the ClC-7 knockout mouse show highly similar phenotypes, such as gray fur in an *agouti* background (indicating a role of ClC-7/Ostm1 in melanocyte physiology) and the short lifespan of just a few weeks, as well as osteopetrosis, retina degeneration, lysosomal storage and neurodegeneration (see below) (Lange et al., 2006). Accordingly, mutations in both human genes have been identified which cause the same diseases (see below).

1. Role in Bone Resorption by Osteoclasts

The ruffled border of osteoclasts is a specialized plasma membrane domain that faces the resorption lacuna, an acidified space between the osteoclast and the bone (Teitelbaum, 2000). Massive exocytosis delivers lysosomal enzymes into the resorption lacuna and the ruffled border itself gains features similar to the late endosomal/lysosomal membrane. Also the V-ATPase with its a3 subunit (Nishi and Forgac, 2000) and ClC-7/Ostm1 (Kornak et al., 2001; Lange et al., 2006) localize to it. The V-ATPase actively acidifies the resorption lacuna (Blair et al., 1989) and ClC-7/Ostm1 is believed to mediate the shunt current, in a manner equivalent to the general role of intracellular CLC proteins in enabling luminal acidification by charge compensation (see Section II.B). The ruffled border of osteoclasts from ClC-7 knockout mice is admittedly underdeveloped (raising the possibility that the lack of ClC-7/Ostm1 rather leads to an impaired build-up of this membrane), and these cells do not significantly degrade bone material (Kornak et al., 2001). By staining ivory-attached osteoclasts with the pH-sensitive dye acridine orange, it could indeed be demonstrated that acidification of the resorption lacuna is greatly abrogated in ClC-7 knockout osteoclasts (Kornak et al., 2001).

Deficient bone resorption leads to osteopetrosis, which is characterized by dense, fragile bones that lack the bone marrow (Tolar et al., 2004). Mutations in the a3 subunit of the V-ATPase cause this disease in mice (Li et al., 1999; Scimeca et al., 2000) and men (Frattini et al., 2000; Kornak et al., 2000). Consistent with the hypothesis that Cl^- transport by ClC-7/Ostm1 mediates the electrical neutralization for the acidification by the proton pump, the ClC-7 knockout and the grey-lethal mice also display severe osteopetrotic phenotypes with secondary effects like a lack of teeth eruption (Kornak et al., 2001; Lange et al., 2006). Furthermore, numerous mutations in ClC-7 (Cleiren et al., 2001; Frattini et al., 2003; Kornak et al., 2001; Waguespack et al., 2007, 2003) and Ostm1 (Chalhoub et al., 2003; Maranda et al., 2008; Pangrazio et al., 2006; Quarello et al., 2004; Ramírez et al., 2004; Souraty et al., 2007) have been found that cause various types of osteopetrosis.

2. Function in Lysosomes of the Nervous System

In brain, ClC-7 is present in all cell types but is most prominently expressed in neurons (Kasper et al., 2005). Consistent with ClC-7 and Ostm1 being localized to neuronal late endosomes/lysosomes (Kasper et al., 2005; Lange et al., 2006), both ClC-7 knockout (Kasper et al., 2005) and Ostm1-deficient grey-lethal (Lange et al., 2006) mice develop a neuronal pathology with clear symptoms of lysosomal dysfunction. Just as in ClC-6 knockout mice (Poet et al., 2006) (see Section III.C), neurons of ClC-7- and Ostm1-deficient mice accumulate electron-dense lysosomal storage material that stains positive for subunit c of the mitochondrial ATPase. This indicates that also loss of ClC-7/Ostm1 leads to NCL, which in this case is accompanied by microglial activation, astrogliosis and a widespread degeneration of the CNS. Also human osteopetrosis patients with mutations in either *CLCN7* (Frattini et al., 2003) or *OSTM1* (Maranda et al., 2008; Pangrazio et al., 2006) have been described that show neuronal symptoms. The NCL-typical accumulation of subunit c of the mitochondrial ATPase is much more pronounced in the ClC-7 knockout mouse (Kasper et al., 2005) than in a ClC-3 knockout mouse for which NCL-like phenotypes have been reported (Yoshikawa et al., 2002) (see Section III.B). In contrast to the restriction of lysosomal storage material to the initial axon segments in ClC-6 knockout mice (Poet et al., 2006) (see Section III.C), electron-dense deposits were scattered throughout the cell body of ClC-7 knockout and grey-lethal mice (Kasper et al., 2005; Lange et al., 2006) (Fig. 12.7A). Storage material is not only found in neurons, but also in renal proximal tubules of ClC-7 knockout mice (Kasper et al., 2005).

The exact molecular mechanism underlying the accumulation of storage material in these mice is not clear. Although the phenotype suggests a lysosomal dysfunction, no impaired activity of the lysosomal enzyme TPP I in ClC-7-deficient neurons or fibroblasts was detected (Kasper et al., 2005). Moreover, the steady-state lysosomal pH in ClC-7 knockout and grey-lethal neurons is not altered (Kasper et al., 2005; Lange et al., 2006), just as in the case of ClC-6 deficiency (Poet et al., 2006). Again, a defect in the acidification kinetics, maybe during the transition from late endosomes to lysosomes, cannot be excluded. Likewise it is unknown whether the lysosomal chloride concentration is normal. Conductive pathways negligible in the presence of ClC-7/Ostm1 could potentially neutralize the charge generated during acidification in ClC-7/Ostm1-deficient lysosomes. In such a scenario, the altered luminal $[Cl^-]$ might lead to the observed phenotypes (Jentsch, 2007).

3. *Retinal Degeneration*

Human osteopetrosis is frequently accompanied by blindness, which often results from an osteopetrotic narrowing of the optical canal and a compression of the optic nerve (Steward, 2003). Indeed, as seen with either ClC-2 and ClC-3 knockout mice (Bösl et al., 2001; Stobrawa et al., 2001) (see Sections III.A and III.B), ClC-7 knockout and grey-lethal mice display retinal degeneration (Kornak et al., 2001; Lange et al., 2006). In both ClC-7- and Ostm1-deficient mice, retinal degeneration starts at about two weeks after birth and rapidly progresses until there are hardly any photoreceptors left at four weeks of age. By this time, the number of ganglion cells is only weakly reduced, indicating that retinal degeneration is independent from the narrowing of the optical canal, which actually occurs in ClC-7 knockout mice (Kornak et al., 2001). Additional proof that in these animal models retinal degeneration is tissue-intrinsic and not a secondary effect of osteopetrosis was provided by examining the retinas of mice lacking the a3 subunit of the V-ATPase and of ClC-7-deficient mice whose osteopetrosis was rescued by osteoclast-specific expression of ClC-7 under the TRAP promoter (Kasper et al., 2005). The former mice, despite exhibiting severe osteopetrosis, did not show retinal degeneration, whereas in the latter mice of the same age, the rescue of the bone phenotype did not affect retinal cell death. Also electroretinograms of ClC-7 knockout mice showed a primary reduction in photoreceptor function, whereas no dysfunction was detected in a3-deficient mice (Kasper et al., 2005).

Thus, ClC-7 is involved in at least two independent mechanisms that can lead to blindness. One is indirect via its role in bone resorption, as also osteopetrosis associated with mutations in the a3 subunit of the V-ATPase can be accompanied by visual impairment (Steward, 2003). The other is related to the NCL phenotype, as visual impairment is also observed in most subtypes of NCL (Goebel and Wisniewski, 2004). While the former mechanism, which plays a minor role (Kasper et al., 2005), is well understood, namely, constriction of the optic nerve, it is not clear how NCL leads to retinal degeneration. Finally, it cannot be excluded that the degeneration of the retina starts in the non-neuronal cells of the retinal pigment epithelium, in which ClC-7 is also expressed (Kornak et al., 2001), and that due to the fast cell death no accumulation of storage material can be found in these cells.

IV. OUTLOOK

Due to the intensive work in the field of CLC research, much knowledge about the CLC proteins has been gained since the first CLC was cloned less than 20 years ago. Within just five years, all nine mammalian members have been cloned. Numerous electrophysiological studies, mutageneses, and the rather recent determination of the crystal structures of bacterial channels and various cytoplasmic domains have largely contributed to elucidating the structure and operation mechanism of these proteins. Genetic surveys and importantly knockout mice models for the different CLC transporters and channels have yielded insight into their tremendously broad physiological roles.

Nonetheless, many open questions are yet to be addressed. In the structure–function field, it is still unclear what exactly makes the difference between Cl^- channels and Cl^-/H^+ exchangers. It would be of great interest to see whether the members of the third CLC subfamily, ClC-6 and -7, are also Cl^-/H^+ exchangers. On a more cell biological level, the search for interaction partners of the CLC proteins is certainly not over. Maybe even more β-subunits will be identified, or proteins that regulate the activity or the trafficking of CLC proteins. It still remains to be brought to light if providing a shunt for the proton pump is the only function of the intracellular CLC proteins or if luminal Cl^- also plays a major role. Along that line, one may ask what leads to the lysosomal storage in ClC-6 and -7 deficiencies. Even though huge progress has been made in the investigation of their physiological roles, there is still ignorance, e.g. as to how the lack of ClC-3 or -7 leads to retinal degeneration. Finally, the discovery or development of specific inhibitors advances research and most importantly might help cure patients suffering from CLC-related diseases.

References

Accardi, A., Lobet, S., Williams, C., Miller, C., and Dutzler, R. (2006). Synergism between halide binding and proton transport in a CLC-type exchanger. *J. Mol. Biol.* **362**, 691–699.

Accardi, A. and Miller, C. (2004). Secondary active transport mediated by a prokaryotic homologue of ClC Cl^- channels. *Nature* **427**, 803–807.

Accardi, A., Walden, M., Nguitragool, W., Jayaram, H., Williams, C., and Miller, C. (2005). Separate ion pathways in a Cl^-/H^+ exchanger. *J. Gen. Physiol.* **126**, 563–570.

Altevogt, B.M. and Paul, D.L. (2004). Four classes of intercellular channels between glial cells in the CNS. *J. Neurosci.* **24**, 4313–4323.

Ando, M. and Takeuchi, S. (2000). mRNA encoding "ClC-K1, a kidney Cl^- channel" is expressed in marginal cells of the stria vascularis of rat cochlea: its possible contribution to Cl^- currents. *Neurosci. Lett.* **284**, 171–174.

Arreola, J., Begenisch, T., Nehrke, K., Nguyen, H.V., Park, K., Richardson, L., Yang, B., Schutte, B.C., Lamb, F.S., and Melvin, J.E. (2002). Secretion and cell volume regulation by salivary acinar cells from mice lacking expression of the Clcn3 Cl^- channel gene. *J. Physiol.* **545.1**, 207–216.

Bae, H.R. and Verkman, A.S. (1990). Protein kinase A regulates chloride conductance in endocytic vesicles from proximal tubule. *Nature* **348**, 637–639.

Bauer, C.K., Steinmeyer, K., Schwarz, J.R., and Jentsch, T.J. (1991). Completely functional double-barreled chloride channel expressed from a single *Torpedo* cDNA. *Proc. Natl. Acad. Sci. USA* **88**, 11052–11056.

Bennetts, B., Parker, M.W., and Cromer, B.A. (2007). Inhibition of skeletal muscle ClC-1 chloride channels by low intracellular pH and ATP. *J. Biol. Chem.* **282**, 32780–32791.

Bennetts, B., Rychkov, G.Y., Ng, H.L., Morton, C.J., Stapleton, D., Parker, M.W., and Cromer, B.A. (2005). Cytoplasmic ATP-sensing domains regulate gating of skeletal muscle ClC-1 chloride channels. *J. Biol. Chem.* **280** (37), 32452–32458.

Birkenhäger, R., Otto, E., Schürmann, M.J., Vollmer, M., Ruf, E.M., Maier-Lutz, I., Beekmann, F., Fekete, A., Omran, H., Feldmann, D., Milford, D.V., Jeck, N., Konrad, M., Landau, D., Knoers, N. V.A.M., Antignac, C., Sudbrack, R., Kispert, A., and Hildebrandt, F. (2001). Mutation of *BSND* causes Bartter syndrome with sensorineural deafness and kidney failure. *Nat. Genet.* **29**, 310–314.

Blair, H.C., Teitelbaum, S.L., Ghiselli, R., and Gluck, S. (1989). Osteoclastic bone resorption by a polarized vacuolar proton pump. *Science* **245**, 855–857.

Blaisdell, C.J., Edmonds, R.D., Wang, X.T., Guggino, S., and Zeitlin, P.L. (2000). pH-regulated chloride secretion in fetal lung epithelia. *Am. J. Physiol. Lung Cell Mol. Physiol.* **278**, L1248–L1255.

Blanz, J., Schweizer, M., Auberson, M., Maier, H., Muenscher, A., Hubner, C.A., and Jentsch, T.J. (2007). Leukoencephalopathy upon disruption of the chloride channel ClC-2. *J. Neurosci.* **27**, 6581–6589.

Bösl, M.R., Stein, V., Hübner, C., Zdebik, A.A., Jordt, S.E., Mukhophadhyay, A.K., Davidoff, M.S., Holstein, A.F., and Jentsch, T.J. (2001). Male germ cells and photoreceptors, both depending on close cell–cell interactions, degenerate upon ClC-2 Cl^--channel disruption. *EMBO J.* **20**, 1289–1299.

Bostick, D.L. and Berkowitz, M.L. (2004). Exterior site occupancy infers chloride-induced proton gating in a prokaryotic homolog of the ClC chloride channel. *Biophys J.* **87**, 1686–1696.

Brandt, S. and Jentsch, T.J. (1995). ClC-6 and ClC-7 are two novel broadly expressed members of the CLC chloride channel family. *FEBS Lett.* **377**, 15–20.

Butt, A.M. and Kalsi, A. (2006). Inwardly rectifying potassium channels (Kir) in central nervous system glia: a special role for Kir4.1 in glial functions. *J. Cell Mol. Med.* **10**, 33–44.

Buyse, G., Trouet, D., Voets, T., Missiaen, L., Droogmans, G., Nilius, B., and Eggermont, J. (1998). Evidence for the intracellular location of chloride channel (ClC)-type proteins: co-localization of ClC-6a and ClC-6c with the sarco/endoplasmic-reticulum Ca^{2+} pump SERCA2b. *Biochem. J.* **330**, 1015–1021.

Buyse, G., Voets, T., Tytgat, J., De Greef, C., Droogmans, G., Nilius, B., and Eggermont, J. (1997). Expression of human pICln and ClC-6 in Xenopus oocytes induces an identical endogenous chloride conductance. *J. Biol. Chem.* **272**, 3615–3621.

Bykova, E.A., Zhang, X.D., Chen, T.Y., and Zheng, J. (2006). Large movement in the C terminus of CLC-0 chloride channel during slow gating. *Nat. Struct. Mol. Biol.* **13**, 1115–1119.

Catalan, M., Niemeyer, M.I., Cid, L.P., and Sepulveda, F.V. (2004). Basolateral ClC-2 chloride channels in surface colon epithelium: regulation by a direct effect of intracellular chloride. *Gastroenterology* **126**, 1104–1114.

Chalhoub, N., Benachenhou, N., Rajapurohitam, V., Pata, M., Ferron, M., Frattini, A., Villa, A., and Vacher, J. (2003). Grey-lethal mutation induces severe malignant autosomal recessive osteopetrosis in mouse and human. *Nat. Med.* **9**, 399–406.

Chen, M.F. and Chen, T.Y. (2001). Different fast-gate regulation by external Cl^- and H^+ of the muscle-type ClC chloride channels. *J. Gen. Physiol.* **118**, 23–32.

Chen, T.Y. (2005). Structure and function of clc channels. *Annu. Rev. Physiol* **67**, 809–839.

Chen, T.Y. and Miller, C. (1996). Nonequilibrium gating and voltage dependence of the ClC-0 Cl^- channel. *J. Gen. Physiol.* **108**, 237–250.

Clark, S., Jordt, S.E., Jentsch, T.J., and Mathie, A. (1998). Characterization of the hyperpolarization-activated chloride current in dissociated rat sympathetic neurons. *J. Physiol. (Lond.)* **506**, 665–678.

Cleiren, E., Benichou, O., Van Hul, E., Gram, J., Bollerslev, J., Singer, F.R., Beaverson, K., Aledo, A., Whyte, M.P., Yoneyama, T., deVernejoul, M.C., and Van Hul, W. (2001). Albers-Schönberg disease (autosomal dominant osteopetrosis, type II) results from mutations in the ClCN7 chloride channel gene. *Hum. Mol. Genet.* **10**, 2861–2867.

Comes, N., Abad, E., Morales, M., Borras, T., Gual, A., and Gasull, X. (2006). Identification and functional characterization of ClC-2 chloride channels in trabecular meshwork cells. *Exp. Eye Res.* **83**, 877–889.

Corry, B., O'Mara, M., and Chung, S.H. (2004). Conduction mechanisms of chloride ions in ClC-type channels. *Biophys. J.* **86**, 846–860.

D'Agostino, D., Bertelli, M., Gallo, S., Cecchin, S., Albiero, E., Garofalo, P.G., Gambardella, A., St Hilaire, J.M., Kwiecinski, H., Andermann, E., and Pandolfo, M. (2004). Mutations and polymorphisms of the CLCN2 gene in idiopathic epilepsy. *Neurology* **63**, 1500–1502.

Devuyst, O., Christie, P.T., Courtoy, P.J., Beauwens, R., and Thakker, R.V. (1999). Intra-renal and subcellular distribution of the human chloride channel, CLC-5, reveals a pathophysiological basis for Dent's disease. *Hum. Mol. Genet.* **8**, 247–257.

Dickerson, L.W., Bonthius, D.J., Schutte, B.C., Yang, B., Barna, T. J., Bailey, M.C., Nehrke, K., Williamson, R.A., and Lamb, F.S. (2002). Altered GABAergic function accompanies hippocampal degeneration in mice lacking ClC-3 voltage-gated chloride channels. *Brain Res.* **958**, 227–250.

Duan, D., Winter, C., Cowley, S., Hume, J.R., and Horowitz, B. (1997). Molecular identification of a volume-regulated chloride channel. *Nature* **390**, 417–421.

Dutzler, R., Campbell, E.B., Cadene, M., Chait, B.T., and MacKinnon, R. (2002). X-ray structure of a ClC chloride channel at 3.0 A reveals the molecular basis of anion selectivity. *Nature* **415**, 287–294.

Dutzler, R., Campbell, E.B., and MacKinnon, R. (2003). Gating the selectivity filter in ClC chloride channels. *Science* **300**, 108–112.

Eggermont, J., Buyse, G., Voets, T., Tytgat, J., De Smedt, H., Droogmans, G., and Nilius, B. (1997). Alternative splicing of ClC-6 (a member of the CIC chloride-channel family) transcripts generates three truncated isoforms one of which, ClC-6c, is kidney-specific. *Biochem. J.* **325**, 269–276.

Elidan, J., Sohmer, H., Gafni, M., and Kahana, E. (1982). Contribution of changes in click rate and intensity on diagnosis of multiple sclerosis by brainstem auditory evoked potentials. *Acta Neurol. Scand.* **65**, 570–585.

Engh, A.M., Faraldo-Gomez, J.D., and Maduke, M. (2007a). The mechanism of fast-gate opening in ClC-0. *J. Gen. Physiol.* **130**, 335–349.

Engh, A.M., Faraldo-Gomez, J.D., and Maduke, M. (2007b). The role of a conserved lysine in chloride- and voltage-dependent ClC-0 fast gating. *J. Gen. Physiol.* **130**, 351–363.

Estévez, R., Boettger, T., Stein, V., Birkenhäger, R., Otto, M., Hildebrandt, F., and Jentsch, T.J. (2001). Barttin is a Cl^--channel β-subunit crucial for renal Cl^--reabsorption and inner ear K^+-secretion. *Nature* **414**, 558–561.

Estévez, R., Pusch, M., Ferrer-Costa, C., Orozco, M., and Jentsch, T.J. (2004). Functional and structural conservation of CBS domains from CLC chloride channels. *J. Physiol.* **557**, 363–378.

Fahlke, C., Yu, H.T., Beck, C.L., Rhodes, T.H., and George, A.L., Jr. (1997). Pore-forming segments in voltage-gated chloride channels. *Nature* **390**, 529–532.

Faraldo-Gómez, J.D. and Roux, B. (2004). Electrostatics of ion stabilization in a ClC chloride channel homologue from Escherichia coli. *J. Mol. Biol.* **339**, 981–1000.

Faundez, V. and Hartzell, H.C. (2004). Intracellular chloride channels: determinants of function in the endosomal pathway. *Sci. STKE*, **2004** (233), re8.

Fischer, T., De Vries, L., Meerloo, T., and Farquhar, M.G. (2003). Promotion of Gαi3 subunit down-regulation by GIPN, a putative E3 ubiquitin ligase that interacts with RGS-GAIP. *Proc. Natl. Acad. Sci. USA* **100**, 8270–8275.

Frattini, A., Orchard, P.J., Sobacchi, C., Giliani, S., Abinun, M., Mattsson, J.P., Keeling, D.J., Andersson, A.K., Wallbrandt, P., Zecca, L., Notarangelo, L.D., Vezzoni, P., and Villa, A. (2000). Defects in TCIRG1 subunit of the vacuolar proton pump are responsible for a subset of human autosomal recessive osteopetrosis. *Nat. Genet.* **25**, 343–346.

Frattini, A., Pangrazio, A., Susani, L., Sobacchi, C., Mirolo, M., Abinun, M., Andolina, M., Flanagan, A., Horwitz, E.M., Mihci, E., Notarangelo, L.D., Ramenghi, U., Teti, A., Van Hove, J., Vujic, D., Young, T., Albertini, A., Orchard, P.J., Vezzoni, P., and Villa, A. (2003). Chloride channel *ClCN7* mutations are responsible for severe recessive, dominant, and intermediate osteopetrosis. *J. Bone Miner. Res.* **18**, 1740–1747.

Friedrich, T., Breiderhoff, T., and Jentsch, T.J. (1999). Mutational analysis demonstrates that ClC-4 and ClC-5 directly mediate plasma membrane currents. *J. Biol. Chem.* **274**, 896–902.

Fuchs, R., Male, P., and Mellman, I. (1989). Acidification and ion permeabilities of highly purified rat liver endosomes. *J. Biol. Chem.* **264**, 2212–2220.

Furukawa, T., Ogura, T., Katayama, Y., and Hiraoka, M. (1998). Characteristics of rabbit ClC-2 current expressed in Xenopus oocytes and its contribution to volume regulation. *Am. J. Physiol.* **274**, C500–C512.

Gentzsch, M., Cui, L., Mengos, A., Chang, X.B., Chen, J.H., and Riordan, J.R. (2003). The PDZ-binding chloride channel ClC-3B localizes to the Golgi and associates with CFTR-interacting PDZ proteins. *J. Biol. Chem.* **278**, 6440–6449.

Goebel, H.H. and Wisniewski, K.E. (2004). Current state of clinical and morphological features in human NCL. *Brain Pathol.* **14**, 61–69.

Gong, W., Xu, H., Shimizu, T., Morishima, S., Tanabe, S., Tachibe, T., Uchida, S., Sasaki, S., and Okada, Y. (2004). ClC-3-independent, PKC-dependent activity of volume-sensitive Cl channel in mouse ventricular cardiomyocytes. *Cell Physiol. Biochem.* **14**, 213–224.

Greene, J.R., Brown, N.H., DiDomenico, B.J., Kaplan, J., and Eide, D.J. (1993). The *GEF1* gene of *Saccharomyces cerevisiae* encodes an integral membrane protein; mutations in which have effects on respiration and iron-limited growth. *Mol. Gen. Genet.* **241**, 542–553.

Grooms, S.Y., Opitz, T., Bennett, M.V., and Zukin, R.S. (2000). Status epilepticus decreases glutamate receptor 2 mRNA and protein expression in hippocampal pyramidal cells before neuronal death. *Proc. Natl. Acad. Sci. USA* **97**, 3631–3636.

Gründer, S., Thiemann, A., Pusch, M., and Jentsch, T.J. (1992). Regions involved in the opening of CIC-2 chloride channel by voltage and cell volume. *Nature* **360**, 759–762.

Günther, W., Lüchow, A., Cluzeaud, F., Vandewalle, A., and Jentsch, T.J. (1998). ClC-5, the chloride channel mutated in Dent's disease, colocalizes with the proton pump in endocytotically active kidney cells. *Proc. Natl. Acad. Sci. USA* **95**, 8075–8080.

Günther, W., Piwon, N., and Jentsch, T.J. (2003). The ClC-5 chloride channel knock-out mouse – an animal model for Dent's disease. *Pflügers Arch.* **445**, 456–462.

Hara-Chikuma, M., Wang, Y., Guggino, S.E., Guggino, W.B., and Verkman, A.S. (2005a). Impaired acidification in early endosomes of ClC-5 deficient proximal tubule. *Biochem. Biophys. Res. Commun.* **329**, 941–946.

Hara-Chikuma, M., Yang, B., Sonawane, N.D., Sasaki, S., Uchida, S., and Verkman, A.S. (2005b). ClC-3 chloride channels facilitate endosomal acidification and chloride accumulation. *J. Biol. Chem.* **280**, 1241–1247.

Haug, K., Warnstedt, M., Alekov, A.K., Sander, T., Ramírez, A., Poser, B., Maljevic, S., Hebeisen, S., Kubisch, C., Rebstock, J., Horvath, S., Hallmann, K., Dullinger, J.S., Rau, B., Haverkamp, F., Beyenburg, S., Schulz, H., Janz, D., Giese, B., Muller-Newen, G., Propping, P., Elger, C.E., Fahlke, C., Lerche, H., and Heils, A. (2003). Mutations in *CLCN2* encoding a voltage-gated chloride channel are associated with idiopathic generalized epilepsies. *Nat. Genet.* **33**, 527–532.

Huang, P., Liu, J., Robinson, N.C., Musch, M.W., Kaetzel, M.A., and Nelson, D.J. (2001). Regulation of human ClC-3 channels by multifunctional Ca^{2+}/calmodulin dependent protein kinase. *J. Biol. Chem.* **276**, 20093–20100.

Huber, S., Braun, G., Schroppel, B., and Horster, M. (1998). Chloride channels ClC-2 and ICln mRNA expression differs in renal epithelial ontogeny. *Kidney Int. Suppl.* **67**, S149–S151.

Huber, S.M., Duranton, C., Henke, G., Van De Sand, C., Heussler, V., Shumilina, E., Sandu, C.D., Tanneur, V., Brand, V., Kasinathan, R. S., Lang, K.S., Kremsner, P.G., Hubner, C.A., Rust, M.B., Dedek, K., Jentsch, T.J., and Lang, F. (2004). Plasmodium induces swelling-activated ClC-2 anion channels in the host erythrocyte. *J. Biol. Chem.* **279**, 41444–41452.

Ignoul, S., Simaels, J., Hermans, D., Annaert, W., and Eggermont, J. (2007). Human ClC-6 is a late endosomal glycoprotein that associates with detergent-resistant lipid domains. *PLoS ONE* **2**, e474.

Jefferies, K.C., Cipriano, D.J., and Forgac, M. (2008). Function, structure and regulation of the vacuolar (H(+))-ATPases. *Arch. Biochem. Biophys.* **476** (1), 33–42.

Jentsch, T.J. (2007). Chloride and the endosomal–lysosomal pathway: emerging roles of CLC chloride transporters. *J. Physiol.* **578**, 633–640.

Jentsch, T.J. (2008). CLC chloride channels and transporters: from genes to protein structure, pathology and physiology. *Crit. Rev. Biochem. Mol. Biol.* **43**, 3–36.

Jentsch, T.J., Maritzen, T., and Zdebik, A.A. (2005a). Chloride channel diseases resulting from impaired transepithelial transport or vesicular function. *J. Clin. Invest* **115**, 2039–2046.

Jentsch, T.J., Poet, M., Fuhrmann, J.C., and Zdebik, A.A. (2005b). Physiological functions of CLC Cl^- channels gleaned from human genetic disease and mouse models. *Ann. Rev. Physiol.* **67**, 779–807.

Jentsch, T.J., Steinmeyer, K., and Schwarz, G. (1990). Primary structure of *Torpedo marmorata* chloride channel isolated by expression cloning in *Xenopus* oocytes. *Nature* **348**, 510–514.

Jordt, S.E. and Jentsch, T.J. (1997). Molecular dissection of gating in the ClC-2 chloride channel. *EMBO J.* **16**, 1582–1592.

Kasper, D., Planells-Cases, R., Fuhrmann, J.C., Scheel, O., Zeitz, O., Ruether, K., Schmitt, A., Poet, M., Steinfeld, R., Schweizer, M., Kornak, U., and Jentsch, T.J. (2005). Loss of the chloride channel ClC-7 leads to lysosomal storage disease and neurodegeneration. *EMBO J.* **24**, 1079–1091.

Kawasaki, M., Uchida, S., Monkawa, T., Miyawaki, A., Mikoshiba, K., Marumo, F., and Sasaki, S. (1994). Cloning and expression of a protein kinase C-regulated chloride channel abundantly expressed in rat brain neuronal cells. *Neuron* **12**, 597–604.

Kida, Y., Uchida, S., Miyazaki, H., Sasaki, S., and Marumo, F. (2001). Localization of mouse CLC-6 and CLC-7 mRNA and their functional complementation of yeast CLC gene mutant. *Histochem. Cell Biol.* **115**, 189–194.

Koch, M.C., Steinmeyer, K., Lorenz, C., Ricker, K., Wolf, F., Otto, M., Zoll, B., Lehmann-Horn, F., Grzeschik, K.H., and Jentsch, T. J. (1992). The skeletal muscle chloride channel in dominant and recessive human myotonia. *Science* **257**, 797–800.

Kornak, U., Kasper, D., Bösl, M.R., Kaiser, E., Schweizer, M., Schulz, A., Friedrich, W., Delling, G., and Jentsch, T.J. (2001). Loss of the ClC-7 chloride channel leads to osteopetrosis in mice and man. *Cell* **104**, 205–215.

Kornak, U., Schulz, A., Friedrich, W., Uhlhaas, S., Kremens, B., Voit, T., Hasan, C., Bode, U., Jentsch, T.J., and Kubisch, C. (2000). Mutations in the a3 subunit of the vacuolar H^+-ATPase cause infantile malignant osteopetrosis. *Hum. Mol. Genet.* **9**, 2059–2063.

Lange, P.F., Wartosch, L., Jentsch, T.J., and Fuhrmann, J.C. (2006). ClC-7 requires Ostm1 as a beta-subunit to support bone resorption and lysosomal function. *Nature* **440**, 220–223.

Li, X., Shimada, K., Showalter, L.A., and Weinman, S.A. (2000). Biophysical properties of ClC-3 differentiate it from swelling-activated chloride channels in Chinese hamster ovary-K1 cells. *J. Biol. Chem.* **275**, 35994–35998.

Li, X., Wang, T., Zhao, Z., and Weinman, S.A. (2002). The ClC-3 chloride channel promotes acidification of lysosomes in CHO- K1 and Huh-7 cells. *Am. J. Physiol. Cell Physiol.* **282**, C1483–C1491.

Li, Y.P., Chen, W., Liang, Y., Li, E., and Stashenko, P. (1999). *Atp6i*-deficient mice exhibit severe osteopetrosis due to loss of osteoclast-mediated extracellular acidification. *Nat. Genet.* **23**, 447–451.

Lipecka, J., Bali, M., Thomas, A., Fanen, P., Edelman, A., and Fritsch, J. (2002). Distribution of ClC-2 chloride channel in rat and human epithelial tissues. *Am. J. Physiol. Cell Physiol.* **282**, C805–C816.

Lobet, S. and Dutzler, R. (2006). Ion-binding properties of the ClC chloride selectivity filter. *Embo. J.* **25**, 24–33.

Lorenz, C., Pusch, M., and Jentsch, T.J. (1996). Heteromultimeric CLC chloride channels with novel properties. *Proc. Natl. Acad. Sci. USA* **93**, 13362–13366.

Ludewig, U., Pusch, M., and Jentsch, T.J. (1996). Two physically distinct pores in the dimeric ClC-0 chloride channel. *Nature* **383**, 340–343.

Malinowska, D.H., Kupert, E.Y., Bahinski, A., Sherry, A.M., and Cuppoletti, J. (1995). Cloning, functional expression, and characterization of a PKA-activated gastric Cl- channel. *Am. J. Physiol.* **268**, C191–C200.

Maranda, B., Chabot, G., Decarie, J.C., Pata, M., Azeddine, B., Moreau, A., and Vacher, J. (2008). Clinical and cellular manifestations of OSTM1-related infantile osteopetrosis. *J. Bone Miner. Res.* **23**, 296–300.

Marini, C., Scheffer, I.E., Crossland, K.M., Grinton, B.E., Phillips, F. L., McMahon, J.M., Turner, S.J., Dean, J.T., Kivity, S., Mazarib, A., Neufeld, M.Y., Korczyn, A.D., Harkin, L.A., Dibbens, L.M., Wallace, R.H., Mulley, J.C., and Berkovic, S.F. (2004). Genetic architecture of idiopathic generalized epilepsy: clinical genetic analysis of 55 multiplex families. *Epilepsia* **45**, 467–478.

Maritzen, T., Rickheit, G., Schmitt, A., and Jentsch, T.J. (2006). Kidney-specific upregulation of vitamin D3 target genes in ClC-5 KO mice. *Kidney Int.* **70**, 79–87.

Markovic, S. and Dutzler, R. (2007). The structure of the cytoplasmic domain of the chloride channel ClC-Ka reveals a conserved interaction interface. *Structure* **15**, 715–725.

Matsuda, J.J., Filali, M.S., Volk, K.A., Collins, M.M., Moreland, J.G., and Lamb, F.S. (2008). Overexpression of CLC-3 in HEK293T cells yields novel currents that are pH dependent. *Am. J. Physiol. Cell Physiol.* **294**, C251–C262.

Matsumura, Y., Uchida, S., Kondo, Y., Miyazaki, H., Ko, S.B., Hayama, A., Morimoto, T., Liu, W., Arisawa, M., Sasaki, S., and Marumo, F. (1999). Overt nephrogenic diabetes insipidus in mice lacking the CLC-K1 chloride channel. *Nat. Genet.* **21**, 95–98.

Meadows, N.A., Sharma, S.M., Faulkner, G.J., Ostrowski, M.C., Hume, D.A., and Cassady, A.I. (2007). The expression of Clcn7 and Ostm1 in osteoclasts is coregulated by microphthalmia transcription factor. *J. Biol. Chem.* **282**, 1891–1904.

Mellman, I., Fuchs, R., and Helenius, A. (1986). Acidification of the endocytic and exocytic pathways. *Annu. Rev. Biochem.* **55**, 663–700.

Menichella, D.M., Goodenough, D.A., Sirkowski, E., Scherer, S.S., and Paul, D.L. (2003). Connexins are critical for normal myelination in the CNS. *J. Neurosci.* **23**, 5963–5973.

Menichella, D.M., Majdan, M., Awatramani, R., Goodenough, D.A., Sirkowski, E., Scherer, S.S., and Paul, D.L. (2006). Genetic and physiological evidence that oligodendrocyte gap junctions contribute to spatial buffering of potassium released during neuronal activity. *J. Neurosci.* **26**, 10984–10991.

Meyer, S. and Dutzler, R. (2006). Crystal structure of the cytoplasmic domain of the chloride channel ClC-0. *Structure* **14**, 299–307.

Meyer, S., Savaresi, S., Forster, I.C., and Dutzler, R. (2007). Nucleotide recognition by the cytoplasmic domain of the human chloride transporter ClC-5. *Nat. Struct. Mol. Biol.* **14**, 60–67.

Middleton, R.E., Pheasant, D.J., and Miller, C. (1996). Homodimeric architecture of a ClC-type chloride ion channel. *Nature* **383**, 337–340.

Miller, C. (1982). Open-state substructure of single chloride channels from *Torpedo* electroplax. *Philos. Trans. R. Soc. Lond. B Biol. Sci.* **299**, 401–411.

Mohammad-Panah, R., Gyomorey, K., Rommens, J., Choudhury, M., Li, C., Wang, Y., and Bear, C.E. (2001). ClC-2 contributes to native chloride secretion by a human intestinal cell line, Caco-2. *J. Biol. Chem.* **276**, 8306–8313.

Mohammad-Panah, R., Harrison, R., Dhani, S., Ackerley, C., Huan, L.J., Wang, Y., and Bear, C.E. (2003). The chloride channel ClC-4 contributes to endosomal acidification and trafficking. *J. Biol. Chem.* **278**, 29267–29277.

Morales, M.M., Nascimento, D.S., Capella, M.A., Lopes, A.G., and Guggino, W.B. (2001). Arginine vasopressin regulates CFTR and ClC-2 mRNA expression in rat kidney cortex and medulla. *Pflugers Arch.* **443**, 202–211.

Murray, C.B., Morales, M.M., Flotte, T.R., McGrath-Morrow, S.A., Guggino, W.B., and Zeitlin, P.L. (1995). CIC-2: a developmentally dependent chloride channel expressed in the fetal lung and downregulated after birth. *Am. J. Respir. Cell Mol. Biol.* **12**, 597–604.

Nehrke, K., Arreola, J., Nguyen, H.V., Pilato, J., Richardson, L., Okunade, G., Baggs, R., Shull, G.E., and Melvin, J.E. (2002). Loss of hyperpolarization-activated Cl^- current in salivary acinar cells from Clcn2 knockout mice. *J. Biol. Chem* **26**, 23604–23611.

Neusch, C., Rozengurt, N., Jacobs, R.E., Lester, H.A., and Kofuji, P. (2001). Kir4.1 potassium channel subunit is crucial for oligodendrocyte development and in vivo myelination. *J. Neurosci.* **21**, 5429–5438.

Nguitragool, W. and Miller, C. (2006). Uncoupling of a CLC Cl^-/ H^+ exchange transporter by polyatomic anions. *J. Mol. Biol.* **362**, 682–690.

Nguitragool, W. and Miller, C. (2007). Inaugural Article: CLC Cl/ H+ transporters constrained by covalent cross-linking. *Proc. Natl. Acad. Sci. USA* **104**, 20659–20665.

Niemeyer, M.I., Cid, L.P., Zuniga, L., Catalán, M., and Sepúlveda, F. V. (2003). A conserved pore-lining glutamate as a voltage- and chloride-dependent gate in the ClC-2 chloride channel. *J. Physiol.* **553**, 873–879.

Niemeyer, M.I., Yusef, Y.R., Cornejo, I., Flores, C.A., Sepúlveda, F. V., and Cid, L.P. (2004). Functional evaluation of human ClC-2 chloride channel mutations associated with idiopathic generalized epilepsies. *Physiol. Genomics* **19**, 74–83.

Nilius, B. and Droogmans, G. (2001). Ion channels and their functional role in vascular endothelium. *Physiol. Rev.* **81**, 1415–1459.

Nilius, B., Eggermont, J., Voets, T., Buyse, G., Manolopoulos, V., and Droogmans, G. (1997). Properties of volume-regulated anion channels in mammalian cells. *Prog. Biophys. Mol. Biol.* **68**, 69–119.

Nishi, T. and Forgac, M. (2000). Molecular cloning and expression of three isoforms of the 100-kDa a subunit of the mouse vacuolar proton-translocating ATPase. *J. Biol. Chem.* **275**, 6824–6830.

Nozu, K., Inagaki, T., Fu, X.J., Nozu, Y., Kaito, H., Kanda, K., Sekine, T., Igarashi, T., Nakanishi, K., Yoshikawa, N., Iijima, K., and Matsuo, M. (2008). Molecular analysis of digenic inheritance in Bartter syndrome with sensorineural deafness. *J. Med. Genet.* **45**, 182–186.

Nunn, J.A., LePeillet, E., Netto, C.A., Hodges, H., Gray, J.A., and Meldrum, B.S. (1994). Global ischaemia: hippocampal pathology and spatial deficits in the water maze. *Behav. Brain Res.* **62**, 41–54.

Obermüller, N., Gretz, N., Kriz, W., Reilly, R.F., and Witzgall, R. (1998). The swelling-activated chloride channel ClC-2, the chloride channel ClC- 3, and ClC-5, a chloride channel mutated in kidney stone disease, are expressed in distinct subpopulations of renal epithelial cells. *J. Clin. Invest* **101**, 635–642.

Ogura, T., Furukawa, T., Toyozaki, T., Yamada, K., Zheng, Y.J., Katayama, Y., Nakaya, H., and Inagaki, N. (2002). ClC-3B, a novel ClC-3 splicing variant that interacts with EBP50 and facilitates expression of CFTR-regulated ORCC. *FASEB J.* **16**, S63–S65.

Palmer, D.N., Fearnley, I.M., Walker, J.E., Hall, N.A., Lake, B.D., Wolfe, L.S., Haltia, M., Martinus, R.D., and Jolly, R.D. (1992). Mitochondrial ATP synthase subunit c storage in the ceroid-lipofuscinoses (Batten disease). *Am. J. Med. Genet.* **42**, 561–567.

Pangrazio, A., Poliani, P.L., Megarbane, A., Lefranc, G., Lanino, E., Di Rocco, M., Rucci, F., Lucchini, F., Ravanini, M., Facchetti, F., Abinun, M., Vezzoni, P., Villa, A., and Frattini, A. (2006). Mutations in OSTM1 (grey lethal) define a particularly severe form of autosomal recessive osteopetrosis with neural involvement. *J. Bone Miner. Res.* **21**, 1098–1105.

Pena-Munzenmayer, G., Catalan, M., Cornejo, I., Figueroa, C. D., Melvin, J.E., Niemeyer, M.I., Cid, L.P., and Sepulveda, F.V. (2005). Basolateral localization of native ClC-2 chloride channels in absorptive intestinal epithelial cells and basolateral sorting encoded by a CBS-2 domain di-leucine motif. *J. Cell Sci.* **118**, 4243–4252.

Picollo, A. and Pusch, M. (2005). Chloride/proton antiporter activity of mammalian CLC proteins ClC-4 and ClC-5. *Nature* **436**, 420–423.

Piwon, N., Günther, W., Schwake, M., Bösl, M.R., and Jentsch, T.J. (2000). ClC-5 Cl^--channel disruption impairs endocytosis in a mouse model for Dent's disease. *Nature* **408**, 369–373.

Poet, M., Kornak, U., Schweizer, M., Zdebik, A.A., Scheel, O., Hoelter, S., Wurst, W., Schmitt, A., Fuhrmann, J.C., Planells-Cases, R., Mole, S.E., Hubner, C.A., and Jentsch, T.J. (2006). Lysosomal storage disease upon disruption of the neuronal chloride transport protein ClC-6. *Proc. Natl. Acad. Sci. USA* **103**, 13854–13859.

Pusch, M. (2004). Structural insights into chloride and proton-mediated gating of CLC chloride channels. *Biochemistry* **43**, 1135–1144.

Pusch, M., Jordt, S.E., Stein, V., and Jentsch, T.J. (1999). Chloride dependence of hyperpolarization-activated chloride channel gates. *J. Physiol. (Lond.)* **515**, 341–353.

Pusch, M., Ludewig, U., and Jentsch, T.J. (1997). Temperature dependence of fast and slow gating relaxations of ClC-0 chloride channels. *J. Gen. Physiol.* **109**, 105–116.

Pusch, M., Ludewig, U., Rehfeldt, A., and Jentsch, T.J. (1995). Gating of the voltage-dependent chloride channel ClC-0 by the permeant anion. *Nature* **373**, 527–531.

Quarello, P., Forni, M., Barberis, L., Defilippi, C., Campagnoli, M.F., Silvestro, L., Frattini, A., Chalhoub, N., Vacher, J., and Ramenghi, U. (2004). Severe malignant osteopetrosis caused by a GL gene mutation. *J. Bone Miner. Res.* **19**, 1194–1199.

Ramírez, A., Faupel, J., Goebel, I., Stiller, A., Beyer, S., Stockle, C., Hasan, C., Bode, U., Kornak, U., and Kubisch, C. (2004). Identification of a novel mutation in the coding region of the grey-lethal gene *OSTM1* in human malignant infantile osteopetrosis. *Hum. Mutat.* **23**, 471–476.

Rickheit, G., Maier, H., Strenzke, N., Andreescu, C.E., De Zeeuw, C.I., Muenscher, A., Zdebik, A.A., and Jentsch, T.J. (2008). Endocochlear potential depends on Cl^- channels: mechanism underlying deafness in Bartter syndrome IV. *Embo. J.* **27**, 2907–2917.

Robinson, K. and Rudge, P. (1975). Auditory evoked responses in multiple sclerosis. *Lancet* **1**, 1164–1166.

Roncagliolo, M., Benitez, J., and Eguibar, J.R. (2000). Progressive deterioration of central components of auditory brainstem responses during postnatal development of the myelin mutant taiep rat. *Audiol. Neurootol.* **5**, 267–275.

Rüdel, R. and Lehmann-Horn, F. (1985). Membrane changes in cells from myotonia patients. *Physiol. Rev.* **65**, 310–356.

Rugarli, E.I., Adler, D.A., Borsani, G., Tsuchiya, K., Franco, B., Hauge, X., Disteche, C., Chapman, V., and Ballabio, A. (1995). Different chromosomal localization of the *Clcn4* gene in Mus spretus and C57BL/6J mice. *Nat. Genet.* **10**, 466–471.

Rychkov, G.Y., Pusch, M., Astill, D.S., Roberts, M.L., Jentsch, T.J., and Bretag, A.H. (1996). Concentration and pH dependence of skeletal muscle chloride channel ClC-1. *J. Physiol. (Lond.)* **497**, 423–435.

Rychkov, G.Y., Pusch, M., Roberts, M.L., Jentsch, T.J., and Bretag, A.H. (1998). Permeation and block of the skeletal muscle chloride channel, ClC-1, by foreign anions. *J. Gen. Physiol.* **111**, 653–665.

Sakamoto, H., Sado, Y., Naito, I., Kwon, T.H., Inoue, S., Endo, K., Kawasaki, M., Uchida, S., Nielsen, S., Sasaki, S., and Marumo, F. (1999). Cellular and subcellular immunolocalization of ClC-5 channel in mouse kidney: colocalization with H^+-ATPase. *Am. J. Physiol.* **277**, F957–F965.

Salazar, G., Love, R., Styers, M.L., Werner, E., Peden, A., Rodriguez, S., Gearing, M., Wainer, B.H., and Faundez, V. (2004). AP-3-dependent mechanisms control the targeting of a chloride channel (ClC-3) in neuronal and non-neuronal cells. *J. Biol. Chem.* **279**, 25430–25439.

Sardini, A., Amey, J.S., Weylandt, K.H., Nobles, M., Valverde, M.A., and Higgins, C.F. (2003). Cell volume regulation and swelling-activated chloride channels. *Biochim. Biophys. Acta* **1618**, 153–162.

Scheel, O., Zdebik, A., Lourdel, S., and Jentsch, T.J. (2005). Voltage-dependent electrogenic chloride proton exchange by endosomal CLC proteins. *Nature* **436**, 424–427.

Schlingmann, K.P., Konrad, M., Jeck, N., Waldegger, P., Reinalter, S. C., Holder, M., Seyberth, H.W., and Waldegger, S. (2004). Salt wasting and deafness resulting from mutations in two chloride channels. *N. Engl. J. Med.* **350**, 1314–1319.

Schmid, S., Fuchs, R., Kielian, M., Helenius, A., and Mellman, I. (1989). Acidification of endosome subpopulations in wild-type Chinese hamster ovary cells and temperature-sensitive acidification-defective mutants. *J. Cell Biol.* **108**, 1291–1300.

Schmidt-Rose, T. and Jentsch, T.J. (1997). Transmembrane topology of a CLC chloride channel. *Proc. Natl. Acad. Sci. USA* **94**, 7633–7638.

Scholl, U., Hebeisen, S., Janssen, A.G., Muller-Newen, G., Alekov, A., and Fahlke, C. (2006). Barttin modulates trafficking and function of ClC-K channels. *Proc. Natl. Acad. Sci. USA* **103**, 11411–11416.

Schriever, A.M., Friedrich, T., Pusch, M., and Jentsch, T.J. (1999). CLC chloride channels in Caenorhabditis elegans. *J. Biol. Chem.* **274**, 34238–34244.

Schwake, M., Friedrich, T., and Jentsch, T.J. (2001). An internalization signal in ClC-5, an endosomal Cl -channel mutated in Dent's disease. *J. Biol. Chem.* **276**, 12049–12054.

Schwiebert, E.M., Cid-Soto, L.P., Stafford, D., Carter, M., Blaisdell, C.J., Zeitlin, P.L., Guggino, W.B., and Cutting, G.R. (1998). Analysis of ClC-2 channels as an alternative pathway for chloride conduction in cystic fibrosis airway cells. *Proc. Natl. Acad. Sci. USA* **95**, 3879–3884.

Scimeca, J.C., Franchi, A., Trojani, C., Parrinello, H., Grosgeorge, J., Robert, C., Jaillon, O., Poirier, C., Gaudray, P., and Carle, G.F. (2000). The gene encoding the mouse homologue of the human osteoclast-specific 116-kDa V-ATPase subunit bears a deletion in osteosclerotic (oc/oc) mutants. *Bone* **26**, 207–213.

Scott, J.W., Hawley, S.A., Green, K.A., Anis, M., Stewart, G., Scullion, G.A., Norman, D.G., and Hardie, D.G. (2004). CBS domains form energy-sensing modules whose binding of adenosine ligands is disrupted by disease mutations. *J. Clin. Invest.* **113**, 274–284.

Seong, E., Wainer, B.H., Hughes, E.D., Saunders, T.L., Burmeister, M., and Faundez, V. (2005). Genetic analysis of the neuronal and ubiquitous AP-3 adaptor complexes reveals divergent functions in brain. *Mol. Biol. Cell* **16**, 128–140.

Sik, A., Smith, R.L., and Freund, T.F. (2000). Distribution of chloride channel-2-immunoreactive neuronal and astrocytic processes in the hippocampus. *Neuroscience* **101**, 51–65.

Simon, D.B., Bindra, R.S., Mansfield, T.A., Nelson-Williams, C., Mendonca, E., Stone, R., Schurman, S., Nayir, A., Alpay, H., Bakkaloglu, A., Rodriguez-Soriano, J., Morales, J.M., Sanjad, S. A., Taylor, C.M., Pilz, D., Brem, A., Trachtman, H., Griswold, W., Richard, G.A., John, E., and Lifton, R.P. (1997). Mutations in the chloride channel gene, *CLCNKB*, cause Bartter's syndrome type III. *Nat. Genet.* **17**, 171–178.

Smith, R.L., Clayton, G.H., Wilcox, C.L., Escudero, K.W., and Staley, K.J. (1995). Differential expression of an inwardly rectifying chloride conductance in rat brain neurons: a potential mechanism for cell-specific modulation of postsynaptic inhibition. *J. Neurosci.* **15**, 4057–4067.

Sonawane, N.D. and Verkman, A.S. (2003). Determinants of [Cl^-] in recycling and late endosomes and Golgi complex measured using fluorescent ligands. *J. Cell Biol.* **160**, 1129–1138.

Souraty, N., Noun, P., Djambas-Khayat, C., Chouery, E., Pangrazio, A., Villa, A., Lefranc, G., Frattini, A., and Megarbane, A. (2007). Molecular study of six families originating from the Middle-East and presenting with autosomal recessive osteopetrosis. *Eur. J. Med. Genet.* **50**, 188–199.

Staley, K., Smith, R., Schaack, J., Wilcox, C., and Jentsch, T.J. (1996). Alteration of GABAA receptor function following gene transfer of the CLC-2 chloride channel. *Neuron* **17**, 543–551.

Starr, A. (1978). Sensory evoked potentials in clinical disorders of the nervous system. *Annu. Rev. Neurosci.* **1**, 103–127.

Steinmeyer, K., Klocke, R., Ortland, C., Gronemeier, M., Jockusch, H., Gründer, S., and Jentsch, T.J. (1991a). Inactivation of muscle chloride channel by transposon insertion in myotonic mice. *Nature* **354**, 304–308.

Steinmeyer, K., Ortland, C., and Jentsch, T.J. (1991b). Primary structure and functional expression of a developmentally regulated skeletal muscle chloride channel. *Nature* **354**, 301–304.

Steinmeyer, K., Schwappach, B., Bens, M., Vandewalle, A., and Jentsch, T.J. (1995). Cloning and functional expression of rat CLC-5, a chloride channel related to kidney disease. *J. Biol. Chem.* **270**, 31172–31177.

Steward, C.G. (2003). Neurological aspects of osteopetrosis. *Neuropathol. Appl. Neurobiol.* **29**, 87–97.

Stobrawa, S.M., Breiderhoff, T., Takamori, S., Engel, D., Schweizer, M., Zdebik, A.A., Bösl, M.R., Ruether, K., Jahn, H., Draguhn, A., Jahn, R., and Jentsch, T.J. (2001). Disruption of ClC-3, a chloride channel expressed on synaptic vesicles, leads to a loss of the hippocampus. *Neuron* **29**, 185–196.

Strong, T.V., Boehm, K., and Collins, F.S. (1994). Localization of cystic fibrosis transmembrane conductance regulator mRNA in the human gastrointestinal tract by in situ hybridization. *J. Clin. Invest.* **93**, 347–354.

Südhof, T.C. (2004). The synaptic vesicle cycle. *Annu. Rev. Neurosci.* **27**, 509–547.

Suzuki, T., Rai, T., Hayama, A., Sohara, E., Suda, S., Itoh, T., Sasaki, S., and Uchida, S. (2006). Intracellular localization of ClC chloride channels and their ability to form hetero-oligomers. *J. Cell Physiol.* **206**, 792–798.

Tajima, M., Hayama, A., Rai, T., Sasaki, S., and Uchida, S. (2007). Barttin binds to the outer lateral surface of the ClC-K2 chloride channel. *Biochem. Biophys. Res. Commun.* **362**, 858–864.

Teitelbaum, S.L. (2000). Bone resorption by osteoclasts. *Science* **289**, 1504–1508.

Thiemann, A., Gründer, S., Pusch, M., and Jentsch, T.J. (1992). A chloride channel widely expressed in epithelial and non-epithelial cells. *Nature* **356**, 57–60.

Tolar, J., Teitelbaum, S.L., and Orchard, P.J. (2004). Osteopetrosis. *N. Engl. J. Med.* **351**, 2839–2849.

Traverso, S., Zifarelli, G., Aiello, R., and Pusch, M. (2006). Proton sensing of CLC-0 mutant E166D. *J. Gen. Physiol.* **127**, 51–65.

Tseng, P.Y., Bennetts, B., and Chen, T.Y. (2007). Cytoplasmic ATP inhibition of CLC-1 is enhanced by low pH. *J. Gen. Physiol.* **130**, 217–221.

Tyynelä, J., Palmer, D.N., Baumann, M., and Haltia, M. (1993). Storage of saposins A and D in infantile neuronal ceroid-lipofuscinosis. *FEBS Lett.* **330**, 8–12.

Uchida, S., Sasaki, S., Nitta, K., Uchida, K., Horita, S., Nihei, H., and Marumo, F. (1995). Localization and functional characterization of rat kidney-specific chloride channel, ClC-K1. *J. Clin. Invest* **95**, 104–113.

Vandewalle, A., Cluzeaud, F., Bens, M., Kieferle, S., Steinmeyer, K., and Jentsch, T.J. (1997). Localization and induction by dehydration of ClC-K chloride channels in the rat kidney. *Am. J. Physiol.* **272**, F678–F688.

Vandewalle, A., Cluzeaud, F., Peng, K.C., Bens, M., Lüchow, A., Günther, W., and Jentsch, T.J. (2001). Tissue distribution and subcellular localization of the ClC-5 chloride channel in rat intestinal cells. *Am. J. Physiol. Cell Physiol.* **280**, C373–C381.

von Weikersthal, S.F., Barrand, M.A., and Hladky, S.B. (1999). Functional and molecular characterization of a volume-sensitive chloride current in rat brain endothelial cells. *J. Physiol.* **516**, 75–84.

Waguespack, S.G., Hui, S.L., Dimeglio, L.A., and Econs, M.J. (2007). Autosomal dominant osteopetrosis: clinical severity and natural history of 94 subjects with a chloride channel 7 gene mutation. *J. Clin. Endocrinol. Metab.* **92**, 771–778.

Waguespack, S.G., Koller, D.L., White, K.E., Fishburn, T., Carn, G., Buckwalter, K.A., Johnson, M., Kocisko, M., Evans, W.E., Foroud, T., and Econs, M.J. (2003). Chloride channel 7 (CLCN7) gene mutations and autosomal dominant osteopetrosis, type II. *J. Bone Miner. Res.* **18**, 1513–1518.

Waldegger, S., Jeck, N., Barth, P., Peters, M., Vitzthum, H., Wolf, K., Kurtz, A., Konrad, M., and Seyberth, H.W. (2002). Barttin increases surface expression and changes current properties of ClC-K channels. *Pflügers Arch.* **444**, 411–418.

Waldegger, S. and Jentsch, T.J. (2000). Functional and structural analysis of ClC-K chloride channels involved in renal disease. *J. Biol. Chem.* **275**, 24527–24533.

Wang, S.S., Devuyst, O., Courtoy, P.J., Wang, X.T., Wang, H., Wang, Y., Thakker, R.V., Guggino, S., and Guggino, W.B. (2000). Mice lacking renal chloride channel, CLC-5, are a model for Dent's disease, a nephrolithiasis disorder associated with defective receptor-mediated endocytosis. *Hum. Mol. Genet.* **9**, 2937–2945.

Weinreich, F. and Jentsch, T.J. (2001). Pores formed by single subunits in mixed dimers of different CLC chloride channels. *J. Biol. Chem.* **276**, 2347–2353.

Weylandt, K.H., Valverde, M.A., Nobles, M., Raguz, S., Amey, J. S., Díaz, M., Nastrucci, C., Higgins, C.F., and Sardini, A. (2001). Human ClC-3 is not the swelling-activated chloride channel involved in cell volume regulation. *J. Biol. Chem.* **276**, 17461–17467.

Yamazaki, J., Duan, D., Janiak, R., Kuenzli, K., Horowitz, B., and Hume, J.R. (1998). Functional and molecular expression of volume-regulated chloride channels in canine vascular smooth muscle cells. *J. Physiol. (Lond.)* **507**, 729–736.

Yoshikawa, M., Uchida, S., Ezaki, J., Rai, T., Hayama, A., Kobayashi, K., Kida, Y., Noda, M., Koike, M., Uchiyama, Y., Marumo, F., Kominami, E., and Sasaki, S. (2002). CLC-3 deficiency leads to phenotypes similar to human neuronal ceroid lipofuscinosis. *Genes Cells* **7**, 597–605.

Zdebik, A.A., Cuffe, J.E., Bertog, M., Korbmacher, C., and Jentsch, T.J. (2004). Additional disruption of the ClC-2 Cl(−) channel does not exacerbate the cystic fibrosis phenotype of cystic fibrosis transmembrane conductance regulator mouse models. *J. Biol. Chem.* **279**, 22276–22283.

Zdebik, A.A., Zifarelli, G., Bergsdorf, E.Y., Soliani, P., Scheel, O., Jentsch, T.J., and Pusch, M. (2008). Determinants of anion-proton coupling in mammalian endosomal CLC proteins. *J. Biol. Chem.* **283**, 4219–4227.

Zhang, X.D., Li, Y., Yu, W.P., and Chen, T.Y. (2006). Roles of K149, G352, and H401 in the channel functions of ClC-0: testing the predictions from theoretical calculations. *J. Gen. Physiol.* **127**, 435–447.

Zhao, Z., Li, X., Hao, J., Winston, J.H., and Weinman, S.A. (2007). The ClC-3 chloride transport protein traffics through the plasma membrane via interaction of an N-terminal dileucine cluster with clathrin. *J. Biol. Chem.* **282**, 29022–29031.

Zifarelli, G. and Pusch, M. (2007). CLC chloride channels and transporters: a biophysical and physiological perspective. *Rev. Physiol. Biochem. Pharmacol.* **158**, 23–76.

Zifarelli, G. and Pusch, M. (2008). The muscle chloride channel ClC-1 is not directly regulated by intracellular ATP. *J. Gen. Physiol.* **131**, 109–116.

CHAPTER

13

Calcium-activated Chloride Channels

Fiona C. Britton, Normand Leblanc and James L. Kenyon

OUTLINE

I. INTRODUCTION

Ca^{2+}-activated chloride currents ($I_{Cl(Ca)}$) mediated by a Ca^{2+}-activated Cl^- conductance ($g_{Cl(Ca)}$) were first described in *Xenopus* oocytes (Miledi, 1982; Barish, 1983) and rod inner segments from salamander retinas (Bader et al., 1982). By 1991, $I_{Cl(Ca)}$ with similar properties had been described in central and peripheral neuronal cell bodies (Owen et al., 1984, 1986; Mayer, 1985; Bader et al., 1987; Scott et al., 1988; Schlichter et al., 1989), developing skeletal muscle (Hume and Thomas, 1989), cardiac muscle (Zygmunt and Gibbons, 1991), smooth muscle (Byrne and Large, 1987, 1988; Pacaud et al., 1989), lacrimal gland cells (Marty et al., 1984; Evans and Marty, 1986) and pituitary cells (Korn and Weight, 1987). These initial publications began investigations that continue today into the properties and function of $I_{Cl(Ca)}$. Although intriguing candidate genes have been recently identified, the molecular identity of the Ca^{2+}-activated Cl^- channels that mediate $I_{Cl(Ca)}$ is so far unresolved. Functional evidence points to the existence of at least two functional classes of channels that mediate

$I_{Cl(Ca)}$: Cl^- channels activated by Ca^{2+} without requirement for an intermediary kinase, i.e. Ca^{2+}-activated Cl^- channels (Cl_{Ca} channels), and Cl^- channels activated by Ca^{2+}/calmodulin-dependent kinase II (CaMKII) phosphorylation, i.e. CaMKII-dependent channels. Cl_{Ca} channels are typically small conductance channels (5pS or less) that are also commonly activated by depolarization. These channels, commonly found in oocytes, central and peripheral neurons, cardiac muscle, smooth muscle and exocrine secretory cells are the topic of this review. The CaMKII-activated Cl^- channels are best characterized in epithelial tissues where they are medium to large conductance channels (tens to hundreds of pS). Members of the CLCA gene family have been proposed to contribute to these channels and are the subject of recent reviews (Hartzell et al., 2005a; Loewen and Forsyth, 2005). They are not considered in detail here. The unknown molecular basis of the Cl_{Ca} channels leaves open the possibility of a molecular relationship between the Cl_{Ca} channels and the CaMKII-activated Cl^- channels, i.e. proteins encoded by CLCA may contribute to Cl_{Ca} channels. Our discussion of CLCA will be limited to consideration of this possibility.

Figures 13.1 and 13.2 illustrate two voltage clamp protocols used to elicit and characterize $I_{Cl(Ca)}$ mediated by Cl_{Ca} channels. Figure 13.1A shows recordings of membrane potential (upper traces) and membrane current (lower traces) and the activation of $I_{Cl(Ca)}$ by increases in intracellular Ca^{2+} ($[Ca^{2+}]_i$) mediated by voltage-gated Ca^{2+} channels in the plasma membrane of rat dorsal root ganglion (DRG) neurons in culture. As shown here, step depolarizations to -12 and $+35$mV activated voltage-gated inward Ca^{2+} current (I_{Ca}) that mediated a Ca^{2+} influx that activates a delayed current that was inward at -12mV (negative to E_{Cl}), outward at $+35$mV (positive to E_{Cl}), and reversed at E_{Cl} (not shown). Repolarizations from steps that activated I_{Ca} and the delayed current elicited large, slowly deactivating inward tail currents that also reversed at E_{Cl}. That the delayed current and the tail currents are Ca^{2+} dependent can be inferred from the observation that both are greatly diminished if the depolarization is to $+80$mV where Ca^{2+} entry is small due to a reduced driving force for Ca^{2+} (e.g. step to $+80$mV in Fig. 13.1A). Taken together, these observations establish the delayed current and the tail current as $I_{Cl(Ca)}$ based on their Cl^- selectivity and activation by increased $[Ca^{2+}]_i$. In addition to activation by Ca^{2+} influx, $I_{Cl(Ca)}$ can also be activated by release of Ca^{2+} from intracellular stores (Fig. 13.1B) including Ca^{2+}-induced Ca^{2+} release (Ayar and Scott, 1999; Gillo et al., 1989; Martínez-Pinna et al., 2000; Lancaster et al., 2002), caffeine-induced Ca^{2+} release (Currie and Scott, 1992; Ivanenko et al., 1993), or release activated by the endogenous agonists cyclic-ADP ribose (Crawford et al., 1997; Currie et al., 1993; Pollock et al., 1999) or IP_3 (Wang and Kotlikoff, 1997b). A more detailed consideration of the kinetics of activation and deactivation of $I_{Cl(Ca)}$ as determined by the rise and fall of $[Ca^{2+}]_i$, the change in membrane potential, and the kinetics of the response of the Cl_{Ca} channels to these two activators is given below. Lastly, we note that observations qualitatively similar to these have been made in each of the cell types cited above and more.

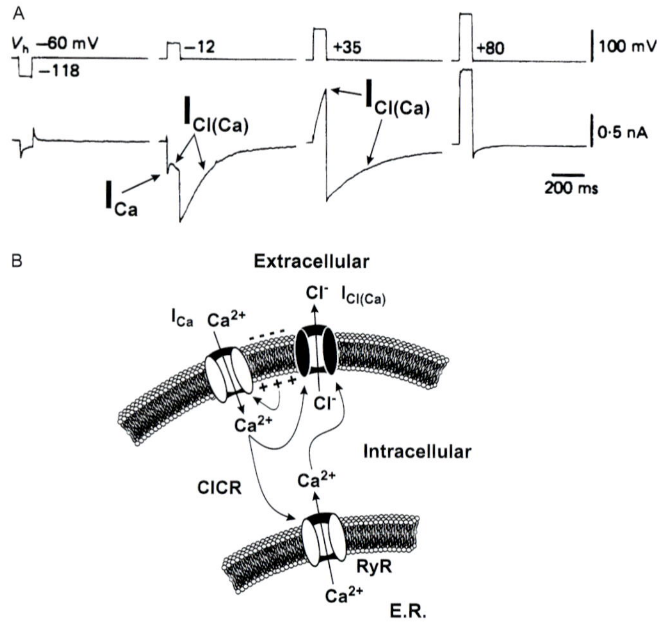

FIGURE 13.1 $I_{Cl(Ca)}$ recorded from a rat dorsal root ganglion neuron. **A.** Membrane currents elicited in whole-cell patch clamp recordings of sensory neurons isolated from neonatal rats and kept in culture for 2–11 weeks. From holding potential (V_h) of -60mV, four voltage clamp steps to potentials ranging from -118 to $+80$mV (upper traces) elicited whole-cell currents shown in the lower traces. The step to -118mV produced no time-dependent current. The step to -12mV evoked a brief transient inward Ca^{2+} current (I_{Ca}) that was immediately followed by the development of a delayed secondary inward current that is consistent with $I_{Cl(Ca)}$. Repolarization to V_h yielded a large inward tail current that reflects slow deactivation of $I_{Cl(Ca)}$ following termination of Ca^{2+} influx at negative potentials. The step to $+35$mV elicited a slowly developing outward $I_{Cl(Ca)}$. Adapted from Mayer (1985) with permission from the Physiological Society (UK). **B.** Cartoon illustrating the roles of voltage-dependent Ca^{2+} channels and Ca^{2+} release from endoplasmic reticulum (ER) Ca^{2+} stores in the activation of Ca^{2+}-activated Cl^- channels in neurons. The diagram shows that $I_{Cl(Ca)}$ can be stimulated by Ca^{2+} entering the cell through I_{Ca} or via Ca^{2+}-induced Ca^{2+} release (CICR) involving ryanodine receptors (RyR) located in the ER membrane.

Figure 13.2 illustrates the protocol used to elicit and characterize $I_{Cl(Ca)}$ in olfactory sensory neurons (OSN) where it is essential for olfaction, as discussed in more detail in Chapter 20 in this volume. This system is of particular interest because of the mounting evidence that they are mediated by the bestrophin family of proteins (Pifferi et al., 2006). In these cells, odorant

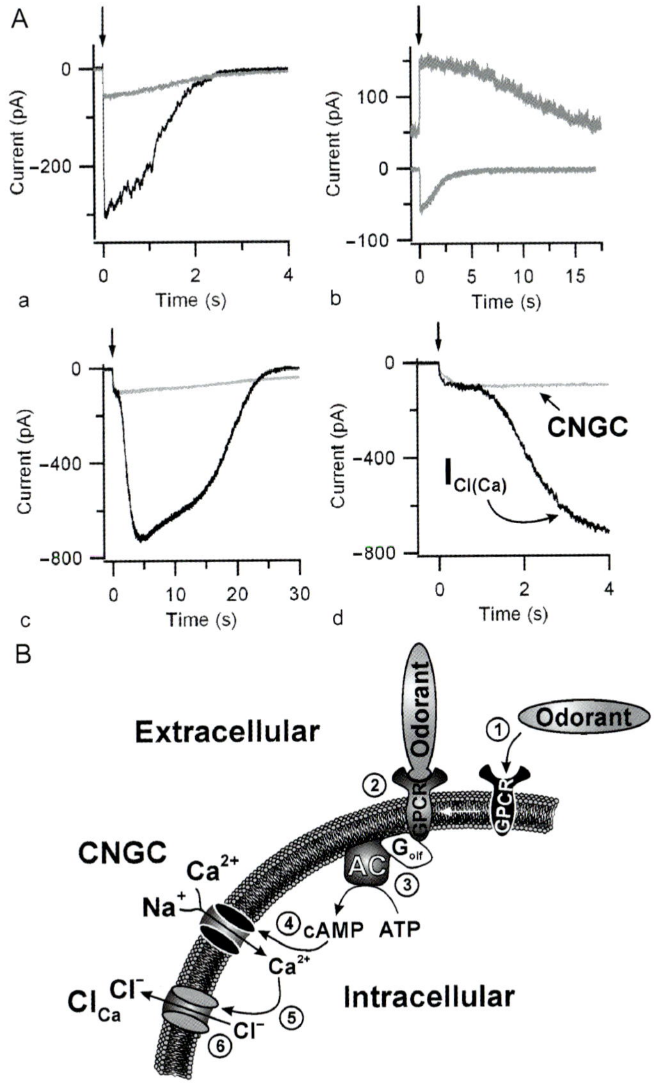

FIGURE 13.2 Ca^{2+}-activated Cl^- currents evoked by Ca^{2+} entry through cyclic nucleotide-gated channels in olfactory sensory neurons. **A.** Whole-cell patch clamp currents recorded from two mouse olfactory sensory neurons (OSNs). Flash photolysis was used to release cAMP (indicated by arrows) to activate cyclic nucleotide-gated cation channel (CNGC) currents. **Panel A–a:** The dark and light gray traces were recorded respectively in the absence and presence of 500 μM NFA, a blocker of Cl_{Ca} channels, at a holding potential of −50 mV. **Panel A–b:** In the same neuron as panel A–a, the top and bottom traces were recorded with 500 μM NFA at +50 and −50 mV, respectively. Notice the smaller inward current relative to the outward current which is consistent with the outwardly rectifying properties of CNGCs. **Panel A–c:** Similar experiment in a different OSN which highlights the biphasic nature of the currents activated by cAMP. **Panel A–d:** Same current traces on an expanded scale. Notice that NFA blocked the delayed secondary inward current but had no effect on the initial current. These results are consistent with the idea that photorelease of cAMP activates CNGCs first and in turn allows Ca^{2+} to enter the cell and activate NFA-sensitive $I_{Cl(Ca)}$. Adapted from Boccaccio and Menini (2007) with permission of the American Physiological Society. **B.** Cartoon showing the signal transduction pathway involved in the activation of Ca^{2+}-activated Cl^- channels (Cl_{Ca}) by CNGCs during olfaction. As shown, binding of an odorant molecule to a G-protein-coupled receptor (GPCR; step 1) stimulates the GTP-binding protein Golf (step 2) which stimulates membrane-bound adenylyl cyclase (AC) leading to the synthesis of the second messenger 3′–5′ cyclic adenosine monophosphate (cAMP; step 3) from ATP. Newly formed cAMP then stimulates CNGCs (step 4). Ca^{2+} entry through CNGCs in turn activates Cl_{Ca} channels (step 5) leading to Cl^- efflux (step 6) and membrane depolarization.

molecules bind to specific receptors in the cilia that are coupled to the G-protein G_{olf} that activates a cascade that includes activation of membrane-bound adenylyl cyclase, increased concentration of cAMP, and activation of cyclic nucleotide-gated channels permeable to Na^+, K^+ and Ca^{2+} (CNGC). Ca^{2+} entry through these channels increases $[Ca^{2+}]_i$ activating $I_{Cl(Ca)}$ thereby amplifying the excitatory process (Reisert et al., 2003; Boccaccio and Menini, 2007). This cascade as characterized in whole-cell patch clamp recordings from mouse OSNs is shown in Fig. 13.2A. In each panel, the arrow indicates a flash of UV light that photoreleased cAMP. The dark and light gray current traces were obtained in the absence and presence respectively of 500 μM niflumic acid (NFA), a non-selective Cl_{Ca} channel blocker. As shown in panel A–a, photorelease of cAMP in OSNs held at −50 mV induced a transient inward current inhibited by NFA thereby establishing the presence of $I_{Cl(Ca)}$. Panel A–b shows a comparison of the amplitude and time course of currents recorded in the presence of NFA from a neuron held at −50 mV (negative to E_{Cl}) and +50 mV (positive to E_{Cl}). The NFA-insensitive, cAMP-dependent current displays the typical outward rectification for CNGC channels (Kleene, 1993). Panels A–c and A–d show the biphasic activation of inward current at −50 mV at two different time scales, again in the absence and presence of NFA. These records illustrate the delayed nature of the NFA-sensitive component, i.e. $I_{Cl(Ca)}$, implying that Ca^{2+} entry through CNGCs activates $I_{Cl(Ca)}$. The elements of the signaling cascade linking odorant binding to activation of $I_{Cl(Ca)}$ are illustrated in Fig. 13.2B.

In addition to activation by $[Ca^{2+}]_i$, Cl_{Ca} channels are commonly activated by depolarization. Thus, the activation and deactivation of $I_{Cl(Ca)}$ depends on membrane potential, Ca^{2+} fluxes across the plasma membrane and intracellular membranes, cytoplasmic Ca^{2+} binding, and the activity of signaling cascades including kinases and phosphatases. Sorting out the individual contributions of these mechanisms is complicated because each will modify the others. A productive experimental approach to study the gating and regulation of $I_{Cl(Ca)}$ is to record the current in a voltage-clamp protocol in which $[Ca^{2+}]_i$ is held at a known value with a Ca^{2+} buffer (EGTA, HEDTA, BAPTA). This approach has been used extensively in smooth muscle myocytes (Ledoux et al., 1999, 2003, 2005; Piper et al., 2002; Angermann et al., 2006) and other cell types (Ishikawa and Cook, 1993; Arreola et al., 1996; Evans and Marty, 1986; Nilius et al., 1997; Ward and Kenyon, 2000). Results obtained using this approach are reviewed below.

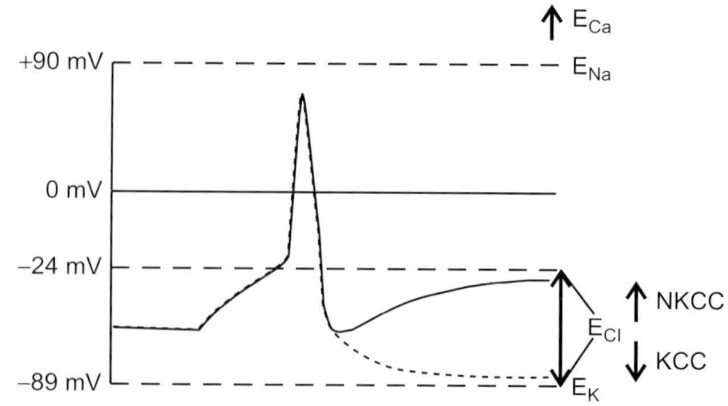

FIGURE 13.3 Equilibrium potentials of the physiologically relevant cations (E_{Na}, E_K, E_{Ca}) and E_{Cl} in relation to the physiological range of membrane potentials. The equilibrium potentials of the cations are positive to (E_{Na}, E_{Ca}) or negative to (E_K) physiological resting and action potentials. In contrast, physiological values of E_{Cl} lie within the range of the resting and action potentials. Two action potentials illustrate the effect of E_{Cl} on the afterpotential activated by Ca^{2+} entry during the depolarization. Increased activity of NKCC will shift E_{Cl} to less negative potentials (−24 mV in this illustration). In this case, increased activation of $g_{Cl(Ca)}$ will result in an afterdepolarization (solid line). Increased activity of KCC will shift E_{Cl} to more negative potentials (−89 mV in this illustration). In this case, increased $g_{Cl(Ca)}$ results in an afterhyperpolarization (dotted line). The equilibrium potentials were calculated using the Nernst equation for a temperature of 37°C and the following extra- and intra-cellular ion concentrations in mM: $[K^+]_i = 140$; $[K^+]_o = 5$; $[Na^+]_i = 5$; $[Na^+]_o = 145$; $[Ca^{2+}]_i = 0.0001$; $[Ca^{2+}]_o = 1$; $[Cl^-]_i = 4$ to 45; $[Cl^-]_o = 110$.

II. THE Cl^- EQUILIBRIUM POTENTIAL IN DEVELOPMENT AND INJURY

The effect of Cl^- currents on membrane potential depends on the amplitude of the Cl^- conductance (g_{Cl}) and the value of the Cl^- equilibrium potential (E_{Cl}). E_{Cl} differs from the equilibrium potentials of the physiologically relevant cations in two ways. First, the equilibrium potentials of the cations are at the extremes of the physiological range of membrane potentials (about −90 mV for E_K, +60 to +90 for E_{Na}, and +80 to +130 for E_{Ca}, at 37°C) whereas physiological intracellular Cl^- concentrations ($[Cl^-]_i$) range from 4 to 45 mM (Chapters 7, 19 and 22 in this volume) resulting in a broad spectrum of E_{Cl} values (from ~ −89 to −24 mV at 37°C) centered in the physiological range of membrane potentials (Fig. 13.3). Thus, an increase in membrane permeability to Na^+ or Ca^{2+} reliably depolarizes cells, an increase in membrane permeability to K^+ reliably hyperpolarizes cells, but an increase in g_{Cl} can cause hyperpolarization, depolarization, or no change in membrane potential. The second difference is that the equilibrium potentials for the cations are relatively constant over the life of the cell whereas E_{Cl} often varies during development and in response to injury as discussed below and in Chapters 7, 19, 22 and 23 in this volume. This is particularly true in the case of central and peripheral neurons where there is abundant evidence of developmental and pathological changes in E_{Cl}. Clearly, in order to understand the biological function of Cl_{Ca} channels (or any other Cl^- channel), one must first understand the transport mechanisms that set $[Cl^-]_i$ and E_{Cl} and how these mechanisms change during development and injury.

Two families of cation-chloride cotransporters actively accumulate or extrude Cl^-, thereby shifting E_{Cl} away from the resting potential (see Chapters 2–4, 16 and 17 in this volume; as well as the following reviews: Russell, 2000; Payne et al., 2003; Adragna et al., 2004; Price et al., 2005; Gamba, 2005) . K^+-Cl^- cotransporters (KCC) couple the extrusion of K^+ and Cl^- from the cytoplasm thereby lowering $[Cl^-]_i$ and shifting E_{Cl} to more negative potentials such that increases in g_{Cl} will stabilize or hyperpolarize resting membrane potential (Fig. 13.3). Thus, KCC activity makes Cl^- channels inhibitory.

Na^+-K^+-Cl^- (NKCC) cotransporters couple the entry of Na^+, K^+ and Cl^- into the cytoplasm thereby increasing $[Cl^-]_i$ above electrochemical equilibrium thereby shifting E_{Cl} to less negative potentials such that increases in g_{Cl} will depolarize the membrane potential (Fig. 13.3). The functional consequence of this depolarization can be inhibition, excitation, or both excitation and inhibition depending on the cell in question. Inhibitory mechanisms include depolarization-dependent inhibition of voltage-gated Na^+ or Ca^{2+} conductances with consequent reduction in neurotransmitter release like in presynaptic inhibition (Willis, 1999), or shunting excitatory currents (Purves et al., 2008, as discussed in detail in Chapters 5 and 22 in this volume). Excitatory effects include activation of voltage-gated Na^+ and Ca^{2+} channels triggering action potentials and activation of Ca^{2+} influx. A well-characterized example of NKCC-mediated excitation is the excitatory action of GABA in developing central nervous system (CNS) neurons, as discussed in Chapter 7 in this volume (reviewed also in Cherubini et al., 1991; Staley and Smith, 2001; Price et al., 2005; Ben-Ari et al., 2007). A complex mix of NKCC-mediated inhibition and excitation has been characterized in dorsal root ganglion (DRG) neurons where a $GABA_A$ receptor-mediated increase in g_{Cl} underlies both reduced neurotransmitter release (presynaptic inhibition) or enhancement of primary afferent depolarization that can trigger efferent action potentials leading to neurogenic inflammation and sensitization as discussed in Chapter 22 in this volume. Thus, the functional consequences of changes in NKCC and Cl^- channel activity are not easily predicted.

The pattern of expression and development of NKCC1 and KCC isoforms differs in the central and peripheral nervous systems (CNS and PNS). In embryonic CNS (Ganguly et al., 2001; Li et al., 1998; Owens et al., 1996; Reichling et al., 1994) and PNS neurons (Kenyon, 2000) active Cl^- accumulation by NKCC1 sets E_{Cl} between −40 and −24 mV such that an increase in g_{Cl} depolarizes the neurons. As discussed below, the fraction of neurons expressing a large $I_{Cl(Ca)}$ tends to be higher early in development raising the possibility that, like $GABA_A$ receptors, Cl_{Ca} channels may cause depolarization that activates Ca^{2+} entry which is important for development of the CNS. This possibility has not been investigated to our knowledge.

In late embryonic or early postnatal development, many CNS neurons (but not PNS neurons) reduce NKCC1 expression and activity while increasing KCC expression and activity. Thus, in adult CNS neurons E_{Cl} is typically shifted toward or negative to the resting potential (reviewed in Chapters 7, 19 and 22 in this volume; and in Cherubini et al., 1991; Staley and Smith, 2001; Price et al., 2005; Ben-Ari et al., 2007). As described above, in this situation an increase in g_{Cl} will stabilize or hyperpolarize the resting potential thereby reducing excitability and Ca^{2+} influx via voltage-gated Ca^{2+} channels. Here, again, the physiological impact of this developmental change in E_{Cl} is well documented in the context of Cl^--selective $GABA_A$ and glycine receptors but little is known about the potential function of $I_{Cl(Ca)}$ as E_{Cl} becomes more negative.

In addition to the developmental changes in cation-chloride-cotransporter function and E_{Cl} discussed above, studies of the responses by neurons in the dorsal horn of the spinal cord and in DRGs have identified changes in response to injury that are important in pain sensation as discussed in Chapters 20, 22 and 23 in this volume. For instance, KCC2 expression in dorsal horn neurons is reduced following injury favoring a depolarizing shift in E_{Cl} such that GABA and glycine responses become depolarizing and potentially excitatory, an effect that has been linked to increased pain sensation (Coull et al., 2003). In the PNS, effects consistent with the complex function of NKCC have been described. Morales-Aza et al. (2004) found that NKCC disappeared from DRG neurons in an experimental arthritis model. These authors suggested that a probable negative shift of E_{Cl} resulting from the decrease in NKCC could reduce presynaptic inhibition in the dorsal horn of the spinal cord leading to inflammatory pain. That the connection between NKCC activity and pain is complex is suggested by the data of Valencia-de Ita et al. (2006) who found that NKCC activity and depolarized E_{Cl} are necessary for the development of neurogenic inflammation in response to capsaicin. This response is mediated by a cascade including the activation of dorsal root reflexes and the generation of efferent action potentials leading to the release of inflammatory mediators in the periphery (Willis, 1999). Other authors have found that NKCC activity contributes to the development of inflammation and sensation of pain (Sung et al., 2000; Laird et al., 2004; Granados-Soto et al., 2005). Thus, NKCC activity can increase or decrease nociception depending on the neuron involved and the experimental model of pain, as discussed in Chapters 20, 22 and 23 in this volume.

In summary, there is a substantial literature describing developmental and pathological alterations in the expression and function of cation-chloride cotransporters and the functional roles of GABA and glycine receptor-mediated increases in g_{Cl}. This raises the possibility that Cl_{Ca} channel-mediated increases in g_{Cl} have similar functions (discussed in Chapters 22 and 23, and in Granados-Soto et al., 2005). Observations reviewed below that Cl_{Ca} channel expression and activity decrease during development and increase following injury support this idea.

III. EXPRESSION OF Cl_{Ca} CHANNELS IN DEVELOPMENT AND INJURY

Two observations point to a specific function of $I_{Cl(Ca)}$ in development and regeneration of peripheral neurons. First, Cl_{Ca} channels are preferentially expressed by peripheral neurons that are not in contact with peripheral tissues (i.e. in developing or regenerating neurons). Second, $I_{Cl(Ca)}$ in these neurons is often quite large (>100 pA/pF). We review these observations here.

The majority of the earliest observations of $I_{Cl(Ca)}$ were made with neurons and skeletal muscle myocytes isolated from embryonic birds and mammals (Bader et al., 1987; Hume and Thomas, 1989; Ivanenko et al., 1993; Mayer, 1985; Owen et al., 1984, 1986; Schlichter et al., 1989; Scott et al., 1988). Bernheim et al. (1989) found that the proportion of quail trigeminal neurons expressing $I_{Cl(Ca)}$ increased from 20% on the 5th day of embryonic development to near 100% by the 7th day and then fell to about 50% at the time of hatching at 17 days. The rise in the frequency of expression of $I_{Cl(Ca)}$ correlated with axon outgrowth whereas the fall in the frequency of expression correlated with the establishment of contacts with the target tissues. Bernheim et al. (1989) further found that expression was maintained or increased in culture in the absence of target tissue but declined if ganglia were incubated in the presence of target tissue. The implication is that a factor in the target tissue reduces expression or function of Cl_{Ca} channels. The identity of this factor remains unknown although Bernheim et al. (1989) ruled out the growth factors NGF, BDNF and CNTF individually or in combination. Mammalian DRG neurons undergo a qualitatively similar pattern of expression as the proportion of rat DRG neurons expressing $I_{Cl(Ca)}$ is about 50% in neurons isolated from neonatal rats and kept in culture <1 week and then rises to about 64% for neurons kept in culture for >1 week (Currie and Scott, 1992). In acutely isolated adult rat DRG neurons, $I_{Cl(Ca)}$ was undetected in small diameter neurons (i.e. a population enriched in nociceptive neurons) and found in fewer than 50% of medium and large diameter neurons (Abdulla and Smith, 1999). This last observation indicates that Cl_{Ca} channel expression correlates with sensory modality. This is consistent with the observation in quail trigeminal neurons that $I_{Cl(Ca)}$ is expressed less frequently in neurons positive for substance P-like immunoreactivity (Schlichter et al., 1989).

Thus, in the PNS $I_{Cl(Ca)}$ is found most often in embryonic sensory neurons that have not yet contacted their targets or have lost contact with their target as a result of cell isolation and culture. Similarly, the proportion of adult rat sympathetic and sensory neurons expressing $I_{Cl(Ca)}$ increases after the neurons have lost contact with their target as a result of axotomy (André et al., 2003; Hilaire et al., 2005; Lancaster et al., 2002; Sánchez-Vives and Gallego, 1994). In each of these studies, peripheral axons were cut and the animals allowed to recover for 4 to 10 days before the neurons were characterized electrophysiologically either in intact ganglia (Sánchez-Vives and Gallego, 1994) or following dispersal (André et al., 2003; Hilaire et al., 2005; Lancaster et al., 2002). In both protocols, a subset of neurons that do not express $I_{Cl(Ca)}$ in control were found to express the current following axotomy during regeneration. Further, in the studies of sensory neurons, there was a tendency for $I_{Cl(Ca)}$ to appear preferentially in large or medium diameter neurons (i.e. a population enriched in low threshold mechanosensory neurons with few nociceptive neurons) (Abdulla and Smith, 1999; André et al., 2003; Hilaire et al., 2005; Lancaster et al., 2002). These findings are additional evidence that expression of $I_{Cl(Ca)}$ correlates with sensory modality (Schlichter et al., 1989). Although these difficult and labor intensive studies are highly suggestive, it must be noted that they provide limited mechanistic information. In particular, only Bernheim et al. (1989) examined candidate signals that control $I_{Cl(Ca)}$ expression and there are no data on how $I_{Cl(Ca)}$ functions in neurons growing toward their target tissues.

The analyses of $I_{Cl(Ca)}$ expression reviewed above are largely limited to comparison of the percentage of neurons expressing $I_{Cl(Ca)}$ of magnitude over an arbitrary threshold. Thus, data that could be treated as continuous are instead treated as nominal (present vs. absent) with the loss of information about the amplitude distribution of the current, i.e. the number of neurons with a given current. The single published amplitude distribution (Fig. 8 in Bernheim et al., 1989) shows $I_{Cl(Ca)}$ amplitudes that do not follow a normal distribution but rather show a broad distribution from no current to very large current. In order to illustrate the benefit of an extended analysis of

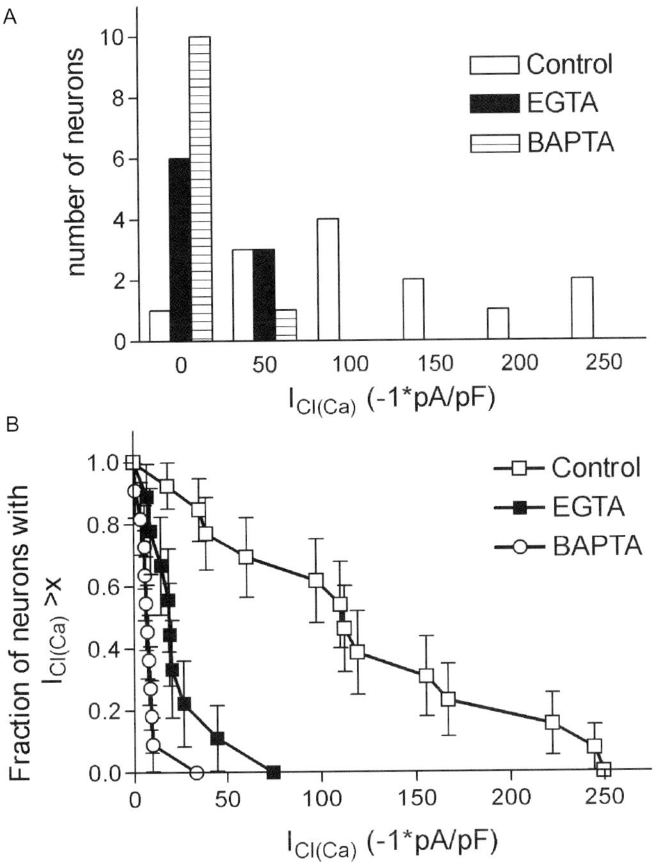

FIGURE 13.4 Distribution of $I_{Cl(Ca)}$ amplitudes in embryonic chick DRG neurons. Data are replotted from Kenyon and Ward (2000). Panel A shows the distribution of $I_{Cl(Ca)}$ amplitudes recorded in the presence of minimal exogenous intracellular Ca^{2+} buffer (control) or 2 mM of either EGTA or BAPTA. Panel B shows the same data plotted as a cumulative distribution (see text).

$I_{Cl(Ca)}$ expression we have reanalyzed measurements of $I_{Cl(Ca)}$ amplitude in embryonic chick DRG neurons that were dialyzed with minimal exogenous Ca^{2+} buffer (control) or 2 mM of either EGTA or BAPTA in order to establish the spatial relationship between the voltage-gated Ca^{2+} channels and Cl_{Ca} channels (Ward and Kenyon, 2000). The distributions of the amplitudes of $I_{Cl(Ca)}$ in the three conditions are plotted in Fig. 13.4A. Analysis using a threshold of 9 pA/pF finds that 13 of 13 control neurons expressed $I_{Cl(Ca)}$ with a mean amplitude of -125.6 ± 22 pA/pF, 8 of 8 neurons dialyzed with EGTA expressed $I_{Cl(Ca)}$ with a mean amplitude of -28.5 ± 7.5 pA/pF, and 3 of 11 neurons dialyzed with BAPTA expressed $I_{Cl(Ca)}$ with a mean amplitude of -17.5 ± 7.9 pA/pF. Thus, treating the data as nominative finds that dialysis with EGTA does not affect the fraction of neurons expressing $I_{Cl(Ca)}$ whereas dialysis with BAPTA reduces the fraction of neurons expressing $I_{Cl(Ca)}$ ($p < 0.0001$, Fisher's exact test). Both treatments reduce the amplitude of the current compared with control (EGTA $p < 0.001$; BAPTA $p < 0.05$, Dunnett's test). A complete picture of the data is shown in Fig. 13.4 as a conventional distribution (panel A) or a cumulative "survival" distribution (panel B). Panel A shows in control a broad, non-Gaussian distribution of $I_{Cl(Ca)}$ magnitudes ranging from −35 to −250 pA/pF reflecting our experience that from cell to cell one expected to record a measurable $I_{Cl(Ca)}$ but the magnitude was unpredictable. Panel A also shows the distributions of $I_{Cl(Ca)}$ amplitudes in neurons dialyzed with exogenous Ca^{2+} buffers EGTA and BAPTA. The distributions of $I_{Cl(Ca)}$ amplitudes are clearly shifted to smaller currents with a higher fraction of neurons showing very small currents. The cumulative distribution in panel B plots the same data in a format convenient for a log-rank analysis, a powerful method to determine whether one group has a greater tendency to have a larger value than another (McGehee and Oxford, 1991). A log-rank test finds that both Ca^{2+} buffers reduce $I_{Cl(Ca)}$ amplitude and that BAPTA significantly reduces $I_{Cl(Ca)}$ amplitude compared with EGTA ($p < 0.009$).

In summary, there are many separate observations of parallel changes in E_{Cl} and expression of $I_{Cl(Ca)}$ with development and injury. Although this is highly suggestive of a functional importance, the nature of that function remains unknown. The literature is limited to speculation that $I_{Cl(Ca)}$ contributes in some way to excitability or volume regulation or both (cf. Sánchez-Vives and Gallego, 1994). This is an area where more experiments and data are needed and where molecular data would be particularly valuable. Unfortunately, as described below, the molecular components of Cl_{Ca} channels are unclear and the available transcript level data do not match the expression pattern described above. That is, the message encoding two candidates for Cl_{Ca} channels in DRG neurons (CLCA and TTHY) falls following axotomy (Al-Jumaily et al., 2007), a treatment that increases $I_{Cl(Ca)}$ expression (André et al., 2003). Al-Jumaily et al. (2007) tested bestrophin expression but could not generate reliable results for mBest1 or mBest4. This is a particularly frustrating state of affairs.

IV. BIOPHYSICAL PROPERTIES OF Cl_{Ca} CHANNELS

A. Permeation and Selectivity

Information on the permeation and selectivity of Cl_{Ca} channels comes from measurements of shifts in the reversal potential of $I_{Cl(Ca)}$ in response to ion substitution in whole-cell or macropatch recordings.

Although there is substantial variability in the ability of Cl_{Ca} channels to select Cl^- over cations, there is a conserved selectivity sequence for Cl_{Ca} channels expressed by a range of cell types. The reversal potential of $I_{Cl(Ca)}$ in neurons and smooth muscle is little affected by the replacement of extracellular Na^+ with tetraethylammonium or choline (Amédée et al., 1990; Mayer, 1985; Owen et al., 1984; Pacaud et al., 1989). Similarly, replacement of extracellular Na^+ with N-methyl-D-glucamine has no effect on the reversal potential of $I_{Cl(Ca)}$ elicited by expression of mouse Best2 and TMEM16A (Qu et al., 2004; Schroeder et al., 2008; Yang et al., 2008). These observations imply that the Cl_{Ca} channels in these studies are highly selective for Cl^- over cations.

In contrast, Cl_{Ca} channels in cone photoreceptor cells (Barnes and Hille, 1989) and olfactory sensory neurons (Hallani et al., 1998) are somewhat less selective with $P_{Cations}/P_{Cl}$ of 0.071 and 0.035, respectively. Even less selective for Cl^- over cations are Cl_{Ca} channels in salivary gland acinar cells ($P_{Cations}/P_{Cl} = 0.25$; Martin, 1993) and in *Xenopus* oocytes ($P_{Cations}/P_{Cl} = 0.42$–$0.71$; Young et al., 1984; $P_{Cations}/P_{Cl} = 0.1$; Qu and Hartzell, 2000). The molecular basis for this range of selectivities is unclear and may reflect pores encoded by different gene products, different subunits, or different regulatory states of the channels.

There is less variability in the ability of Cl_{Ca} channels expressed in a number of cell types to select among anions; most investigations find a permeability sequence for anions matching the lyotropic sequence $SCN^- > I^- > Br^- \geq Cl^- > F^-$: smooth muscle myocytes from portal vein (Wang and Large, 1991), ear (Amédée et al., 1990) and pulmonary artery (Clapp et al., 1996), chick skeletal muscle myotubes (Hume and Thomas, 1989), lacrimal gland (Ishikawa and Cook, 1993), rat and mouse olfactory receptor neurons (Reisert et al., 2003), OSNs (Pifferi et al., 2006), epididymal cells (Huang et al., 1993), and T84 (Cliff and Frizzell, 1990) and B-TC3 (Kozak and Logothetis, 1997) lines. This sequence is also characteristic of Cl^- currents mediated by volume-regulated Cl^- channels (Jentsch et al., 2002), $GABA_A$ and $GABA_C$ receptors (Wotring et al., 1999), glycine receptors (Bormann et al., 1987) and CFTR channels (Anderson et al., 1991) as well as channels mediated by heterologous expression of CLCA, bestrophin and TMEM16A (Caputo et al., 2008; Hartzell et al., 2005a; Qu et al., 2004; Schroeder et al., 2008; Yang et al., 2008). The implication here is that these Cl^- channels share a common mechanism for ion permeation characterized by a low electric field profile determined mainly by the dehydration energy profiles of the permeant anions rather than by major differences in the ability of anions to interact with a binding site within the conductive pore. Measurements of Cl_{Ca} channel permeability to organic anions of various sizes has led investigators to conclude that the pore of Cl_{Ca} channels is quite large at 6Å (Qu and Hartzell, 2000). This is similar to pore diameters reported for ClC channels, $GABA_A$ receptors, $GABA_C$ receptors, glycine receptors and CFTR channels (4.5 to 6Å) (Anderson et al., 1991; Fahlke et al., 1997; Linsdell et al., 1997; McCarty and Zhang, 2001; Rychkov et al., 1998; Wotring et al., 1999). Determination of the mechanism whereby the low electric field profile and relatively large pore dimensions of Cl_{Ca} channels (and other Cl^- channels) relate to their significant permeability to cations will require structural information.

B. Single Channel Conductance

Measurements of single channel currents and fluctuation analysis of macroscopic currents have identified Cl_{Ca} channels with a wide range of unitary channel conductances. That being noted, observations of Cl_{Ca} channels with conductances greater than 50pS are relatively rare (Bajnath et al., 1993; Dixon et al., 1993; Fahmi et al., 1995; Hussy, 1992; Suzuki, 2006; Young et al., 1984). Most studies describe Cl_{Ca} channels with single channel conductances below 5pS, i.e. at or below the resolution of most single channel recordings: ventricular myocytes (1pS) (Collier et al., 1996), pituitary cells (2.5pS) (Taleb et al., 1988) and olfactory sensory neurons (0.5 to 1.6pS) (Kleene, 1997; Larsson et al., 1997; Pifferi et al., 2006; Reisert et al., 2003). These values are similar to unitary conductances of single Cl_{Ca} channels encoded by mBest2 (0.26pS) (Pifferi et al., 2006), the *Drosophila* ortholog of human Best1, dBest1 (2pS) (Chien et al., 2006), and slightly smaller than the one encoded by TMEM16A (8.3pS) (Yang et al., 2008). Although there is clear evidence for the expression of large conductance Ca^{2+}-activated Cl^- channels in some neurons and other cell types, a majority of studies suggest that the predominant Cl_{Ca} channel has a very small unitary conductance.

C. Steady-state Activation by Voltage and Ca^{2+}

The activation of Cl_{Ca} channels by depolarization and Ca^{2+} has been characterized using voltage-clamp measurements of $I_{Cl(Ca)}$ in experiments where $[Ca^{2+}]_i$ is controlled by EGTA, BAPTA or HEDTA. While smooth muscle myocytes have been examined most extensively (Angermann et al., 2006; Greenwood et al., 2001, 2004; Ledoux et al., 2005, 2003), other cell types have provided similar data (Arreola et al., 1996; Evans and Marty, 1986; Ishikawa and Cook, 1993; Kuruma and Hartzell, 2000; Nilius et al., 1997; Ward and Kenyon, 2000). These studies have established that the activation of Cl_{Ca} channels is qualitatively similar to the activation of large conductance Ca^{2+}-activated K^+ channels (BK channels). However, as we describe here, there are quantitative differences in the control of these two channel types. Further, the activation of $I_{Cl(Ca)}$ by depolarization and Ca^{2+} resembles the activation of TRP family channels by multiple mechanisms. Thus, concepts arising from studies of TRPV and TRPM channels may provide important insights into Cl_{Ca} channel gating.

The activation of BK channels has two defining characteristics. First, at a given [Ca^{2+}], depolarization increases BK channel conductance from 0 to the maximum value at which the open probability of the channels is near 1 (Barrett et al., 1982; Latorre and Brauchi, 2006). This voltage-dependent activation is well characterized by a Boltzmann function:

$$G = \frac{G_{max}}{\left\{1 + \exp\left(\frac{E_{0.5} - E_m}{s}\right)\right\}} \qquad \text{(Eq. 1)}$$

where G is the conductance, G_{max} is the maximum conductance, E_m is the membrane potential, $E_{0.5}$ is the membrane potential at which $G = 0.5 \cdot G_{max}$, and s is the change in mV that produces an e-fold change in G. The parameter s is determined by the number of effective charges that cross the membrane electric field as the channel opens or closes, i.e. $z = RT/sF$ where R, T and F have their usual meanings ($RT/F = 25\,mV$ at room temperature). For BK channels, $s = 13$ to $35\,mV$ and is only modestly dependent on [Ca^{2+}] over the range 0.5 to 500 μM (Barrett et al., 1982; Carl and Sanders, 1989; Cui et al., 1997; Latorre and Brauchi, 2006; Wei et al., 1994).

The second characteristic of BK channel gating is that Ca^{2+} increases BK channel conductance by shifting $E_{0.5}$ to more negative potentials for 0.1 to 10 μM [Ca^{2+}]. This effect is described empirically by a linear relationship between $E_{0.5}$ and the log of the Ca^{2+} concentration (log[Ca^{2+}]) (cf. Carl et al., 1996):

$$E_{0.5} = \Delta E_{0.5} \cdot \log\left(\frac{[Ca^{2+}]}{Ca_0}\right) \qquad \text{(Eq. 2)}$$

where $\Delta E_{0.5}$ is the shift in $E_{0.5}$ for a ten-fold change in [Ca^{2+}] and Ca_0 is the [Ca^{2+}] at which $G = 0.5$ at $E_m = 0\,mV$. Thus, $\Delta E_{0.5}$ is a measure of the Ca^{2+} sensitivity of channel gating. For BK channels $\Delta E_{0.5}$ values range from 2.5 to 297 mV with most determinations falling between 30 and 100 mV (Kapicka et al., 1994).

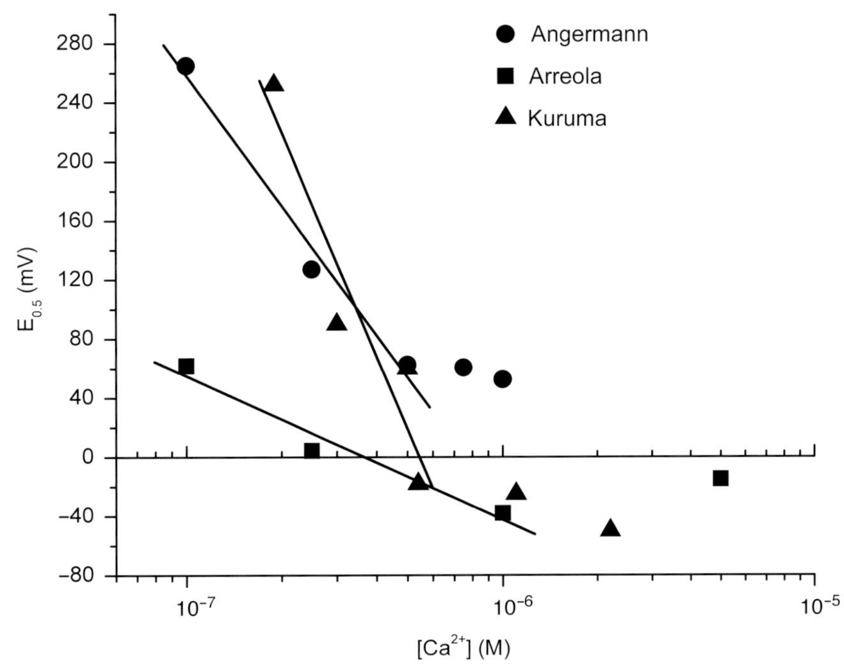

FIGURE 13.5 E0.5 values reported by Angermann et al. (2006), Arreola et al. (1996) and Kuruma and Hartzell (2000). Straight lines are least squares fits to the data representing the lowest 3 (Angermann et al., 2006; Arreola et al., 1996) or 4 (Kuruma and Hartzell, 2000) values of [Ca^{2+}] described in those publications. The $E_{0.5}$ value at 5 μM higher [Ca^{2+}] shown for Arreola et al. (1996) was obtained by reading the relative conductance values as a function of [Ca^{2+}]$_i$ from Fig. 7 of that paper and fitting them to a modified Boltzmann function Eq. 3. The $E_{0.5}$ values at 1.1 and 2.2 μM [Ca^{2+}] shown for Kuruma and Hartzell (2000) were obtained by reading the mean V½ values from Fig. 4B of that paper. The $\Delta E_{0.5}$ values from the straight lines shown here are 97 mV for Arreola, 288 mV for Kuruma and 293 mV for Angermann. Kuruma and Hartzell (2000) fit data from a "typical patch" finding $\Delta E_{0.5}$ of 295 mV.

Characterization of the voltage and Ca^{2+} dependence of Cl_{Ca} channel gating is difficult because the small single channel conductance of Cl_{Ca} channels precludes direct measurement of open probability and because the determination of a complete activation curve requires measurements spanning over 200 mV (see below). That being noted, data obtained in studies of $I_{Cl(Ca)}$ expressed by parotid acinar cells, *Xenopus* oocytes, and smooth muscle myocytes establish that across these cell types, Cl_{Ca} channels, like BK channels, are activated by depolarization and $[Ca^{2+}]$ (Angermann et al., 2006; Arreola et al., 1996; Kuruma and Hartzell, 2000). However, the activation of Cl_{Ca} channels does not share the defining characteristics of BK channel activation listed above. First, the slope parameter s increases with increasing $[Ca^{2+}]$ (Angermann et al., 2006; Kuruma and Hartzell, 2000). Second, the range of $[Ca^{2+}]$ over which Ca^{2+} activates Cl_{Ca} channels by shifting $E_{0.5}$ is limited to 0.1 to 1 μM $[Ca^{2+}]$, i.e. Eq. 2 does not hold for $[Ca^{2+}]$ greater than 1 μM. As shown in Fig. 13.5, there is a linear relationship between $E_{0.5}$ and the log$[Ca^{2+}]$ for $[Ca^{2+}]$ less than 1 μM observed in parotid acinar cells and *Xenopus* oocytes with $\Delta E_{0.5}$ values of 97 mV (acinar cells; Arreola et al., 1996) and 288 mV (*Xenopus* oocyte; Kuruma and Hartzell, 2000). However, for higher $[Ca^{2+}]$ $E_{0.5}$ is independent of $[Ca^{2+}]$. Similarly, Angermann et al. (2006) in a study of $I_{Cl(Ca)}$ in smooth muscle myocytes dialyzed with AMP-PNP to minimize protein phosphorylation found a linear relationship between $E_{0.5}$ and log$[Ca^{2+}]$ with $\Delta E_{0.5}$ equal to 293 mV for $[Ca^{2+}]$ less than 0.75 μM. Here, again, $E_{0.5}$ was independent of $[Ca^{2+}]$ for higher $[Ca^{2+}]$. That is, each of these studies found that for $[Ca^{2+}] \leq 1\,\mu M$, Ca^{2+} activates Cl_{Ca} channels by shifting $E_{0.5}$ to negative potentials and that the Ca^{2+} sensitivity of Cl_{Ca} channels is equal to or greater than that of BK channels.

Third, high $[Ca^{2+}]$ does not increase $g_{Cl(Ca)}$ to G_{max} at negative potentials in parotid acinar cells (Arreola et al., 1996) and in smooth muscle myocytes dialyzed with ATP to promote protein phosphorylation (Angermann et al., 2006). In contrast, Kuruma and Hartzell (2000) concluded that saturating $[Ca^{2+}]$ does increase $g_{Cl(Ca)}$ to G_{max} independent of voltage in *Xenopus* oocyte membranes if a Ca^{2+}-dependent "rundown" of the conductance was minimized. Interestingly, in smooth muscle myocytes dialyzed with AMP-PNP to minimize protein phosphorylation, the ability of high $[Ca^{2+}]$ to increase $g_{Cl(Ca)}$ toward G_{max} was enhanced (Angermann et al., 2006). Taken together, the data of Angermann et al. (2006) and Kuruma and Hartzell (2000) suggest that the ability of Ca^{2+} to maximally activate $g_{Cl(Ca)}$ is altered by channel regulation.

Fourth, negative potentials cannot shut down $g_{Cl(Ca)}$ for $[Ca^{2+}] \geq 1\,\mu M$ implying that high Ca^{2+} activates Cl_{Ca} channels by a voltage-independent mechanism (Angermann et al., 2006; Arreola et al., 1996; Kuruma and Hartzell, 2000). In this case, the activation of $g_{Cl(Ca)}$ is best described by a modified Boltzmann equation (cf. Angermann et al., 2006; Kuruma and Hartzell, 2000; Qu et al., 2003a):

$$G = G_v + G_c = \frac{G_{max} - G_c}{\left\{1 + \exp\left(\frac{E_{0.5} - E_m}{s}\right)\right\}} + G_c \qquad \text{(Eq.3)}$$

where G_v is the voltage-dependent component of conductance, G_{max} is the maximum conductance and G_c is the voltage-independent component of conductance. (Note that typographical errors are common in listings of this equation (Angermann et al., 2006; Kuruma and Hartzell, 2000).) In all studies where Cl_{Ca} channel gating has been quantified (Angermann et al., 2006; Arreola et al., 1996; Kuruma and Hartzell, 2000; Qu et al., 2003a), G_c increases with increasing Ca^{2+} such that G_c can be described by the Hill equation:

$$G_c = \frac{G_{c,max}}{1 + \left[\frac{[Ca^{2+}]}{Ca_{0.5}}\right]^{-n}} \qquad \text{(Eq.4)}$$

where $G_{c,max}$ is the maximum voltage-independent conductance, $Ca_{0.5}$ is $[Ca^{2+}]$ at which $G_c = 0.5 \cdot G_{c,max}$ and n is the number of Ca^{2+} ions that must bind to open the channel. Note that in this formalism $G_{c,max}$ and G_{max} will be the same if channels opened by the voltage-dependent mechanism and by the voltage-independent mechanism have the same single channel conductance and the same open probability. Observations described above that saturating Ca^{2+} did not activate $g_{Cl(Ca)}$ to G_{max} are consistent with either a lower single channel conductance or open probability or both for channels opened by the voltage-independent mechanism.

Because the commonly studied voltage-dependent Na_v, K_v, Ca_v, and BK channels do not have significant voltage-independent activation, there is little precedent for the analysis of combined voltage-dependent and voltage-independent activation of ion channels. Voltage-gated ion channels where voltage-independent activation has been considered is the transient receptor potential channel family, i.e. TRP. Matta and Ahern (2007) found that modulators of TRPV and TRPM activity (i.e. temperature, capsaicin, menthol) open the channels both by shifting $E_{0.5}$ and by a mechanism that is independent of voltage leading them to use the modified Boltzmann equation (Eq. 3) to describe their data. Matta and Ahern proposed an allosteric model in which voltage and modulators are independently coupled to TRP channel opening. We suggest that the conceptual framework developed to understand how capsaicin and lipids interact with voltage to open TRPV1 channels should also be applied to future studies of Cl_{Ca} channel gating by Ca^{2+} and voltage. More similarities between Cl_{Ca} and TRP channel gating are reviewed below.

The value of the parameter s in Eqs 1 and 3 shows an interesting progression in voltage-gated channels: $s = 4$ mV for Na_v and K_v channels (Aggarwal and MacKinnon, 1996; Hodgkin and Huxley, 1952), $s = 13$ to 35 mV for BK channels (Barrett et al., 1982; Carl and Sanders, 1989; Latorre and Brauchi, 2006), $s = 31$ to 35 mV TRPV and TRPM channels (Nilius et al., 2005; Voets et al., 2004, 2005), and $s = 42$ to 100 mV for Cl_{Ca} channels increasing with increasing $[Ca^{2+}]$ (Angermann et al., 2006; Arreola et al., 1996; Evans and Marty, 1986; Kuruma and Hartzell, 2000; Qu et al., 2003a). Thus, of the channel families listed here, Cl_{Ca} channels are among the most sensitive to changes in Ca^{2+} (i.e. large $\Delta E_{0.5}$ values) and the least sensitive to changes in voltage (i.e. large s values). A practical implication of the shallow voltage-dependence of Cl_{Ca} channels is that activation from 10% to 90% of full activation usually spans over 200 mV. This combines with the large $\Delta E_{0.5}$ and the technical difficulties of setting membrane potential outside +200 to −200 mV to limit the range of $[Ca^{2+}]$ where the minimum and maximum voltage-dependent activation can be directly determined. As a practical matter, the determination of $E_{0.5}$, k and G_{max} depends on extrapolating a Boltzmann or modified Boltzmann function from a limited portion of the activation curve. As a result, there is often considerable uncertainty in the estimations of these parameters (Angermann et al., 2006).

More interestingly, the shallow voltage-dependent activation of $I_{Cl(Ca)}$ implies that relatively few charges cross the membrane electric field as the channels open and close. Restating the s values in terms of $z = RT/sF$: $z \approx 4.5$ to 6 for Na_v and K_v channels (Hille, 2001, p.58), $z \approx 1.5$ for BK channels (Barrett et al., 1982; Carl and Sanders, 1989; Latorre and Brauchi, 2006), $z \approx 0.75$ for TRP channels (Nilius et al., 2005; Voets et al., 2004, 2005) and $z \approx 0.59$ to 0.25 for Cl_{Ca} channels (Angermann et al., 2006; Arreola et al., 1996; Evans and Marty, 1986; Kuruma and Hartzell, 2000). Thus, among the channels listed, Cl_{Ca} channel opening involves the movement of the smallest amount of gating charge. As pointed out by Voets, Nilius, and co-workers (Voets et al., 2005), but underappreciated (Zhu, 2007), a small gating charge means that a large shift in $E_{0.5}$ requires only a small change in the Gibbs free energy of channel gating, i.e. $\Delta\Delta G = z \cdot F \cdot \Delta E_{0.5}$. Thus, the shallow voltage dependence of gating and the high Ca^{2+} sensitivity (i.e. large $\Delta E_{0.5}$ values) of Cl_{Ca} channels are seen to be related. Further, Zhu (2007) points out in the context of the relationship between z and $\Delta\Delta G$ that "many physiological stimuli could take advantage of this unique property to activate TRPs". Cl_{Ca} channels share that "unique property" raising the possibility that, like TRP channels, Cl_{Ca} channels respond to many physiological stimuli in addition to Ca^{2+} and voltage. As we discuss below, this concept is easily reconciled with regulation by phosphorylation/dephosphorylation. It also suggests that investigation of other regulators (temperature, lipids) might be profitable.

D. Kinetics of Activation by Voltage and Ca^{2+}

Three groups (Angermann et al., 2006; Arreola et al., 1996; Kuruma and Hartzell, 2000) carried out systematic studies of the kinetics of Cl_{Ca} channel and developed similar models of channel gating that account for the following observations. First, the Ca^{2+} dependence of activation of Cl_{Ca} channels is characterized by voltage-dependent Hill coefficients ranging

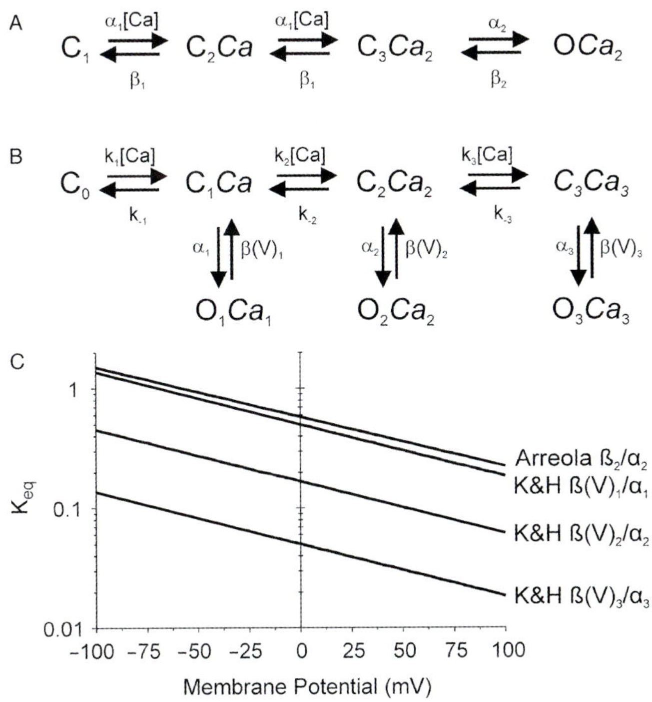

FIGURE 13.6 Kinetic schemes describing Cl_{Ca} channel gating. Panel A from Arreola et al. (1996). Panel B from Kuruma and Hartzell (2000). Panel C shows the voltage dependencies of the equilibrium constants of the closed to open transitions in the models in panels A and B.

from 1.2 to 3.5 implying that two or more Ca^{2+} ions bind during channel activation. Second, the inability of high $[Ca^{2+}]$ to activate $g_{Cl(Ca)}$ to G_{max} discussed above (i.e. $G_{max} > G_{c,max}$) led to models in which the transitions into and out of the open state (or states) do not involve Ca^{2+} binding. Third, the time courses of activation and deactivation of $I_{Cl(Ca)}$ are Ca^{2+} and voltage dependent implying Ca^{2+}- and voltage-dependent rate constants.

These considerations led Arreola et al. (1996) to propose the kinetic scheme shown in Fig. 13.6A in which Ca^{2+} binds to the channel in two sequential, moderately voltage-dependent steps to create an activated channel that opens via a more strongly voltage-dependent step that does not involve Ca^{2+} binding. Based on their data, these authors assigned the voltage dependence of the open–closed transition to the rate of channel closing (β_2). With parameters determined from measurements of activation and deactivation of $g_{Cl(Ca)}$ in response to voltage-clamp steps at various levels of Ca^{2+}, the model accurately simulates $g_{Cl(Ca)}$ activation over a wide range of Ca^{2+} and voltage but is less successful at simulating deactivation. The inability of saturating Ca^{2+} to activate $g_{Cl(Ca)}$ at negative potentials to levels achieved by Ca^{2+} plus depolarization arises in the model from the shallow voltage dependence of the equilibrium constant for the closed to open transition ($\approx$100 mV per e-fold change in $K_2 = \beta_2/\alpha_2$) and the relatively high values of K_2 for potentials between -50 and $+50$ mV. That is, the open probability at saturating Ca^{2+} rises from about 0.5 to 0.75 as voltage goes from -50 to $+50$ mV.

Kuruma and Hartzell (2000) used measurements of $g_{Cl(Ca)}$ responses to jumps of voltage and Ca^{2+} to develop the more complex gating scheme shown in Fig. 13.6B with three voltage-independent Ca^{2+}-binding steps and three open states. This model simulates channel activation well and is better at simulating

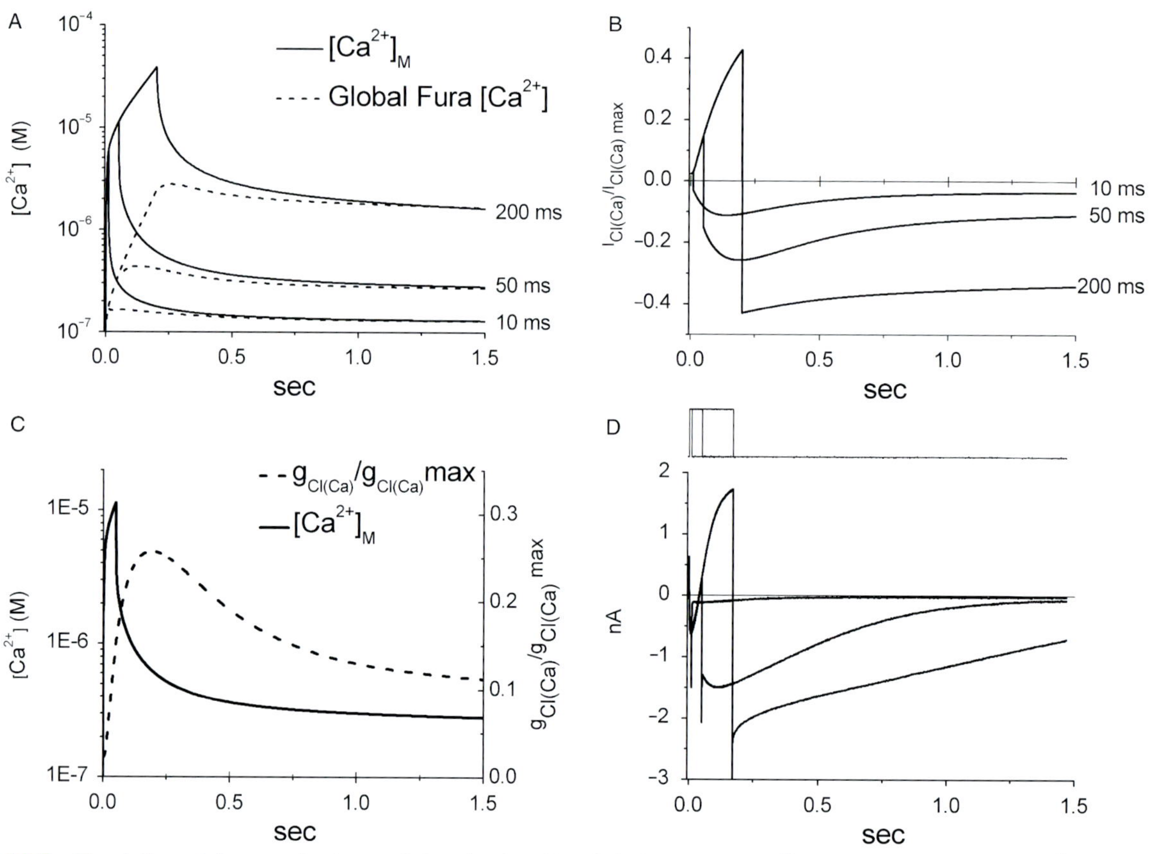

FIGURE 13.7 Simulation of $I_{Cl(Ca)}$ activation and deactivation. Panel A: Simulated $[Ca^{2+}]_i$ transients elicited by Ca^{2+} influx (10, 50 or 200 ms steps from –80 to 0 mV) calculated using parameters chosen to mimic diffusion and buffering in a rat DRG neuron (see McHugh and Kenyon (2003) for details). The solid lines are $[Ca^{2+}]$ in the cytoplasm next to the plasma membrane ($[Ca^{2+}]_M$). The dashed lines are simulated fura-2 measurements of global $[Ca^{2+}]_i$. Panel B: Simulated $I_{Cl(Ca)}$ calculated using $[Ca^{2+}]_M$ in panel A to drive the four-state model of Arreola et al. (1996) modified by setting $\alpha 1$ equal to $3.8 \times 10^4 M^{-1} msec^{-1}$ and β_1 equal to 0.019 $msec^{-1}$ at -80 mV and 0.013 $msec^{-1}$ at 0 mV. $I_{Cl(Ca)}/I_{Cl(Ca)max} = g_{Cl(Ca)}/g_{Cl(Ca)max} \cdot (E - E_{Cl})$ and E_{Cl} is -40 mV. Panel C: Time course of $[Ca^{2+}]_M$ and $g_{Cl(Ca)}/g_{Cl(Ca)max}$ corresponding to the 50 ms record in panels A and B. Panel D: Membrane currents dominated by Ca^{2+} current and $I_{Cl(Ca)}$ recorded from a chick DRG neuron at 20°C as described in Ward and Kenyon (2000). Membrane potential was stepped from –70 to 0 mV for durations of 10, 50 and 170 ms (top traces).

deactivation than the Arreola et al. model. Similar to the Arreola et al. model, the voltage dependence of channel gating is in the closing rate constants ($\beta(V)$) with a shallow voltage dependence of the equilibrium constants for the transitions to the open states (Fig. 13.6C). However, the absolute values of these equilibrium constants ($K_{eq1} = \alpha_1/\beta(V)_1$, $K_{eq2} = \alpha_2/\beta(V)_2$, and $K_{eq3} = \alpha_3/\beta(V)_3$) are smaller than K_2 in the Arreola et al. model with K_{eq2} and K_{eq3} being 0.3 and 0.09 of K_2. K_{eq3} falls from 0.08 to 0.03 as voltage is shifted from -50 to $+50$ mV such that at saturating Ca^{2+} over 90% of the channels will open into O_3Ca_3 at physiologically relevant potentials (the open probability increases from 0.92 to 0.97 as voltage goes from -50 to $+50$ mV). Thus, in this model the simulated $g_{Cl(Ca)}$ has a large, essentially voltage-independent component that can be described by Eq. 4 with $n = 3$, and $G_{max} = G_{c,max}$. Modification of the model replacing the C_3Ca_3 state with the O_3Ca_3 state has little effect on its behavior at physiological potentials.

In principle, measurements of the kinetics of Cl_{Ca} channel activation and deactivation in response to steps in potential and $[Ca^{2+}]$ and measurements of $[Ca^{2+}]$ in the cytoplasm near the plasma membrane ($[Ca^{2+}]_M$) are sufficient to model $I_{Cl(Ca)}$ responses elicited in voltage-clamp experiments. Although the work described above provides parameters for Cl_{Ca} channel gating, direct measurements of $[Ca^{2+}]_M$ are not available. Accordingly, Kenyon and Scott (2002) used a calculated time course of $[Ca^{2+}]_M$ from Sala and Hernández-Cruz (1990) to test the ability of a simple model in which the activation and deactivation of the Cl_{Ca} channels was rapid relative to changes in membrane potential and $[Ca^{2+}]_M$ to simulate the activation and deactivation of $I_{Cl(Ca)}$ in a voltage-clamp experiment. Figure 13.7 advances this work by combining the kinetic model of Cl_{Ca} channel gating developed by Arreola et al. (1996) with the model of Ca^{2+} buffering and diffusion developed by McHugh and Kenyon (2003). Panel A shows the predicted time course of $[Ca^{2+}]_M$ and the global $[Ca^{2+}]_i$ calculated using parameters chosen to mimic a rat dorsal root ganglion neuron (see McHugh and Kenyon, 2003 for details). The three pairs of traces correspond to responses predicted for 10, 50 and 200 ms step depolarizations each of which raises $[Ca^{2+}]_M$ into the micromolar range that will activate $I_{Cl(Ca)}$ based on properties of Cl_{Ca} channels described above. The calculated $[Ca^{2+}]_M$ was used as the activating Ca^{2+} in a modified version of the four-state Arreola et al. model in which α_1 and β_1 were chosen to set $K_1 = 500$ nM (compared to 320 nM in Arreola et al., 1996). The simulated $I_{Cl(Ca)}$ are shown in panel B. Recordings of $I_{Cl(Ca)}$ from an embryonic chick DRG neuron (from Kenyon and Scott, 2002) are shown in panel D. The predicted behavior of $I_{Cl(Ca)}$ differs from observations in chick and rat DRG neurons in that the model shows substantial $I_{Cl(Ca)}$ at resting $[Ca^{2+}]_i$ and -80 mV and relatively large $I_{Cl(Ca)}$ in response to the small $[Ca^{2+}]_i$ transient elicited by the 10 ms depolarization. However, the predicted $I_{Cl(Ca)}$ mimics the time courses of activation and deactivation of $I_{Cl(Ca)}$ during and after the 50 and 200 ms depolarizations quite well. In particular, unlike the simulation by Kenyon and Scott (2002), this model predicts that following depolarizations less than 200 ms in duration $I_{Cl(Ca)}$ tail current amplitudes increase before deactivating. As shown in panel C for the 50 msec depolarization, this time course results from the relatively slow activation kinetics (α_2) that cause the current to lag the $[Ca^{2+}]_i$ transient, i.e. the current is still activating as the $[Ca^{2+}]_i$ transient starts to decline.

V. REGULATION OF Cl_{Ca} CHANNELS BY KINASES AND PHOSPHATASES

Hormones and neurotransmitters modulate synaptic potentials and action potential firing rate by regulating the activity of various classes of ion channels. As discussed above, Cl_{Ca} channels identified in oocytes, central and peripheral neurons, cardiac muscle, smooth muscle and exocrine secretory cells are commonly activated directly by $[Ca^{2+}]_i$ and voltage without requirement for a Ca^{2+}-dependent enzyme. That being noted, signaling molecules such as kinases and Ca^{2+}/calmodulin (CaM) might be expected to modulate channel gating by altering the sensitivities to these direct activators. In a recent study, Kaneko et al. (2006) showed that the Ca^{2+}-activated Cl^- conductance in olfactory sensory neurons is stimulated by CaM by enhancing the Ca^{2+} sensitivity of the channels. The possibility that Cl_{Ca} channels in neuronal cells are modulated by kinases and phosphatases, G-protein-coupled receptor signaling, or protein–protein interactions has received little experimental attention. However, many studies have provided evidence that Cl_{Ca} channels in smooth muscle cells are regulated by phosphorylation and our discussion will thus focus on these data.

Early speculation about the possibility of Cl_{Ca} channel regulation in smooth muscle and other cell types stemmed from the observation that macroscopic and unitary $I_{Cl(Ca)}$ often ran down after seal rupture or patch excision. In the case of $I_{Cl(Ca)}$ rundown observed in whole-cell patch clamp experiments, the rundown could reflect downregulation of the Cl_{Ca} channels or the loss of activating Ca^{2+}. The first observation indicating that kinase activity regulated Cl_{Ca} function was reported by Wang and Kotlikoff (1997a) who recorded $[Ca^{2+}]_i$ and $I_{Cl(Ca)}$ in isolated tracheal smooth muscle myocytes during caffeine-mediated release of Ca^{2+} and during exposure to the Ca^{2+} ionophore ionomycin. In the caffeine experiments, $I_{Cl(Ca)}$ declined more rapidly than the activating $[Ca^{2+}]_i$. In the ionomycin experiments, $I_{Cl(Ca)}$ declined while $[Ca^{2+}]_i$ remained elevated. Further, inhibition of CaMKII slowed the rate of decay of $I_{Cl(Ca)}$ in the caffeine protocol and prevented the decay of $I_{Cl(Ca)}$ in the ionomycin protocol. Taken together, these data provided strong evidence that the Cl_{Ca} channels of tracheal smooth muscle cells are inactivated by CaMKII-dependent phosphorylation of the channel or associated regulatory protein. However, they do not provide evidence with regard to the mechanism of this inactivation that could be a block of the channel pore or shift in the channel sensitivity to voltage and calcium.

Greenwood, Leblanc, and co-workers carried out a systematic investigation of the regulation of smooth muscle $I_{Cl(Ca)}$ by phosphorylation and dephosphorylation that provides insight into the mechanism. Greenwood et al. (2001) compared $I_{Cl(Ca)}$ recorded from myocytes dialyzed with Ca^{2+} buffers to control $[Ca^{2+}]_i$ and compared currents and their regulation by CaMKII in rabbit pulmonary artery, coronary artery and portal vein myocytes. Although the currents in each of these cell types had qualitatively similar activation and deactivation kinetics and sensitivities to voltage and Ca^{2+}, their responses to inhibition of CaMKII differed markedly. Inhibition of CaMKII activity at 500 nM $[Ca^{2+}]_i$ increased $I_{Cl(Ca)}$ amplitude in pulmonary artery myocytes and coronary artery myocytes by shifting the activation curve to more negative potentials. The observation that CaMKII activity shifts the voltage sensitivity of the channel rules out a simple block of the channel as the mechanism underlying regulation of $I_{Cl(Ca)}$ by this enzyme. In portal vein myocytes, inhibition of CaMKII reduced $I_{Cl(Ca)}$. Thus, there are cell-specific mechanisms downstream of CaMKII-dependent phosphorylation that determine the effects of that phosphorylation on $I_{Cl(Ca)}$: (1) the channels are composed of different homomeric or heteromeric subunits, some of which are phosphorylated in a distinct fashion by CaMKII; (2) the pore-forming subunits are not directly phosphorylated by CaMKII; they are instead associated with a distinct array of accessory subunits exhibiting differences in their regulation by CaMKII; or (3) the target of CaMKII is remote from the Cl_{Ca} channel or regulatory subunit, e.g. another kinase or a phosphatase, which results in a change in the state of phosphorylation of the channel, translocation of a scaffolding protein, or another signal transduction event.

The regulatory actions of phosphorylation are typically balanced by phosphatase-dependent dephosphorylation. Ledoux et al. (2003) investigated the coordinate regulation of $I_{Cl(Ca)}$ in rabbit coronary artery myocytes at constant $[Ca^{2+}]_i$ by CaMKII and calcineurin. Both enzymes were expressed in these myocytes and translocated to or near the plasma membrane in response to elevated $[Ca^{2+}]_i$. Inhibition of CaMKII activity increased $I_{Cl(Ca)}$ amplitude at 1 μM $[Ca^{2+}]_i$ but not at $[Ca^{2+}]_i \leq 500$ nM. Inhibition of calcineurin activity reduced $I_{Cl(Ca)}$ amplitude at $[Ca^{2+}]_i$ levels up to 500 nM but not at 1 μM $[Ca^{2+}]_i$. Further, inhibition of calcineurin potentiated the effect of subsequent inhibition of CaMKII indicating that both enzymes are active in the control situation. Thus, the kinase and phosphatase have antagonistic actions on $I_{Cl(Ca)}$ amplitude but the relative activities of the two enzymes differ as a function of $[Ca^{2+}]_i$ with the balance tipping toward calcineurin at lower $[Ca^{2+}]_i$ and toward CaMKII at higher $[Ca^{2+}]_i$.

Greenwood et al. (2004) further characterized the regulation of $I_{Cl(Ca)}$ by calcineurin in pulmonary artery myocytes dialyzed with Ca^{2+} buffers. They found that the myocytes expressed both the α and β isoforms of calcineurin and that $I_{Cl(Ca)}$ amplitude was increased by dialysis with the α isoform of calcineurin but not the β isoform. Consistent with the later observation, the α isoform but not the β isoform translocated toward the plasma membrane when $[Ca^{2+}]_i$ was raised to ~500 nM Ca^{2+}. In addition, $I_{Cl(Ca)}$ amplitude was reduced by inhibition of calcineurin by cyclosporin with the effect being more reliable if the current was activated by Ca^{2+} influx vs. dialysis with 500 nM $[Ca^{2+}]_i$. This observation suggests that the activation and regulation of $I_{Cl(Ca)}$ by an increase in $[Ca^{2+}]_i$ depends on the location and kinetics of that increase. $I_{Cl(Ca)}$ elicited by an elevated fixed level of $[Ca^{2+}]_i$ runs down to less than ~30% of its initial level in pulmonary artery myocytes, an effect attributed to CaMKII-mediated phosphorylation because it is reversed by blocking CaMKII with KN-93 (Greenwood et al., 2001) or by replacing the ATP substrate with its non-hydrolysable form AMP-PNP (Angermann et al., 2006). This suggests that under these conditions, the regulation of Cl_{Ca} channels appears to be dominated by CaMKII activity. However, at lower levels of stimulation by transient and sporadic increases in $[Ca^{2+}]_i$ caused by activation of L-type Ca^{2+} channels, the regulation of $I_{Cl(Ca)}$ may be under the control of a more balanced contribution by CaMKII and calcineurin. This contrasts with coronary artery myocytes where the phosphorylation balance seems to be shifted toward the phosphatase (Ledoux et al., 2003). These findings suggest very fine and distinct tuning mechanisms of the same channel class in the two cell types due to local environmental differences determined by the pattern of expression of the two enzymes, their localization and translocation during signaling.

In order to clarify the biophysical mechanisms by which phosphorylation regulates Cl_{Ca} channels, Angermann et al. (2006) carried out a systemic analysis of the Ca^{2+} and voltage dependence of $I_{Cl(Ca)}$ in pulmonary artery myocytes dialyzed for 20 minutes with either 3 mM ATP or 3 mM AMP-PNP to induce general phosphorylation or dephosphorylation, respectively. At a given $[Ca^{2+}]_i$, phosphorylation decreased $I_{Cl(Ca)}$ amplitude and the rate of activation while speeding deactivation compared to currents recorded under the dephosphorylation condition. Over the range of potentials where the reduced amplitude of $I_{Cl(Ca)}$ in the phosphorylation condition permitted comparison, the K_D for $I_{Cl(Ca)}$ activation by Ca^{2+} was not different between the two protocols. In contrast, dephosphorylation shifted the voltage dependence of $I_{Cl(Ca)}$ activation to more negative potentials and caused the appearance of a sustained $I_{Cl(Ca)}$ that could not be shut off by membrane hyperpolarization to potentials as negative as −200 mV. Angermann et al. (2006) applied the kinetic model of Kuruma and Hartzell (2000) with parameters appropriate for their pulmonary artery myocytes and were able to simulate basic properties of $I_{Cl(Ca)}$ observed in their experiments. They found that the effects of general phosphorylation were simulated by converting the O2 and O3 states to closed states and increasing the values of the voltage-dependent rate constants ($\beta(V)$ in Fig. 13.6B) by shifting their $V_{0.5}$ values to more positive potentials and increasing a scaling factor (parameter a in their Table 1). Based on these findings, Angermann et al. (2006) proposed that phosphorylation resulted in a state-dependent block of Cl_{Ca} channels, especially at elevated intracellular Ca^{2+} levels.

VI. CLONING, STRUCTURE, AND FUNCTION OF Cl_{Ca} CHANNELS

The advance of understanding with regard to the function, expression and gating of Cl_{Ca} channels now requires molecular information. Unfortunately, determination of the molecular nature of Cl_{Ca} channels is difficult because, unlike voltage-gated channels that can be identified by their homology in the voltage-sensing S4 domain and pore loop domains, there are no conserved motifs or shared membrane topology among the chloride channel families that have been cloned and identified thus far (reviewed by Jentsch, 2002; see also Chapter 12 in this volume). Thus, candidate genes must be identified functionally and the bar is high to prove that a gene encodes an essential component of a Cl^- channel. In addition to Cl_{Ca} channels, this is true for ligand-gated GABA and glycine receptors, the cystic fibrosis transmembrane conductance regulator (CFTR) and the voltage-gated chloride channel (ClC) superfamily. Further, there are no selective pharmacological modulators of $I_{Cl(Ca)}$ that can be used to isolate ion channel proteins or to investigate the pore structure. That being noted, functional and mutational assays have identified four families of proteins that produce $I_{Cl(Ca)}$ when expressed in heterologous systems: CLCA, bestrophin, Tweety and TMEM16A/B. We review these here.

A. CLCA Proteins as Molecular Candidates for Cl_{Ca} Channels

Benos and co-workers isolated a protein from bovine tracheal epithelium that mediated a CaMKII-activated Cl^- channel when incorporated into lipid bilayers (Fuller et al., 1994; Ran et al., 1992). The native 140 kDa protein forms an anion-selective, 4,4′-diisothiocyanatostilbene-2,2′-disulfonic acid (DIDS) and dithiothreitol (DTT)-sensitive, 25–30 pS channel with an anion selectivity of $I^- > NO_3^- > Br^- > Cl^-$. The reconstituted channel is activated by CaMKII or by micromolar $[Ca^{2+}]$ (>3 μM) in the absence of CaMKII (Fuller et al., 1994). Further, DTT disrupted the 140 kDa protein yielding a 38 kDa subunit that was not associated with channel activity in bilayers. Using an antibody against the 38 kDa peptide, Cunningham et al. (1995) screened a bovine tracheal cDNA expression library and isolated a cDNA clone encoding a 903-amino acid protein that showed no homology to other previously reported Cl^- channel sequences. Expression of this clone, termed bCLCA1, in *Xenopus* oocytes and COS cells resulted in the appearance of an $I_{Cl(Ca)}$ that was not inhibited by NFA. Since the initial cloning of bCLCA1, numerous mammalian CLCA genes that mediate $I_{Cl(Ca)}$ and have distinct, tissue-specific expression patterns have been described (Loewen and Forsyth, 2005). The currents mediated by each of these CLCA genes differ from $I_{Cl(Ca)}$ found in oocytes, central and peripheral neurons, cardiac muscle, smooth muscle and exocrine secretory cells described above in requiring millimolar Ca^{2+} for activation (Fuller et al., 2001), are not activated by depolarization, and have different pharmacological profiles (Britton et al., 2002).

The differences between native $I_{Cl(Ca)}$ and currents generated by CLCA imply that if the CLCA protein forms a Cl_{Ca} channel, additional protein subunits are required to produce native currents. Greenwood et al. (2002) took a shot in the dark and tested the ability of the regulatory β1-subunit of the BK channel to modify CLCA channel function. Co-expression of CLCA and the β1-subunit resulted in $I_{Cl(Ca)}$ activated by 250 and 500 nM $[Ca^{2+}]$ with kinetics and voltage dependence of activation resembling native $I_{Cl(Ca)}$ expressed in oocytes, central and peripheral neurons, cardiac muscle, smooth muscle and exocrine secretory cells. This work provides a proof of concept that auxiliary subunits modulate CLCA functional properties.

In addition to the requirement for millimolar $[Ca^{2+}]$ for activation and lack of time- and voltage-dependent gating, there are other reasons to be skeptical about the role of the CLCA family in $I_{Cl(Ca)}$. Briefly, CLCA proteins are proteolytically processed and several findings identify CLCA proteins as secretory proteins with physiological functions in cell adhesion, mucus production in asthma and cystic fibrosis, cell division and tumor metastasis (reviewed in detail by Loewen, 2005 and by Hartzell, 2005a). Accordingly, here we will focus on the more recent candidates to encode Cl_{Ca} channels: bestrophin, Tweety and TMEM16A.

B. Bestrophins as Molecular Candidates for Cl_{Ca} Channels

Bestrophin was identified as the protein product of the *VMD2* gene mutated in Best vitelliform macular dystrophy, or Best disease (Marquardt et al., 1998; Petrukhin et al., 1998). Best disease is an autosomal dominant disorder characterized by an accumulation of lipofuscin-like material in the macular area. It has an early onset leading to a progressive loss of vision (Marquardt et al., 1998; Petrukhin et al., 1998). The bestrophin family currently includes four human genes: Best1, Best2, Best3 and Best4 (Stöhr et al., 2002). Highly conserved orthologs of the human genes have been identified in mouse (Burkin et al., 2003; Krämer et al., 2004; Qu et al., 2004), *Xenopus* (Qu et al., 2003b), *Drosophila* (Tavsanli et al., 2001) and *Caenorhabditis elegans* (Sun et al., 2002; Tsunenari et al., 2003). Although the sequence homology is high at the NH_2 terminus of each isoform, the carboxyl termini are unique providing regions for the design of isoform-specific antibodies and probes. Thus, bestrophin expression and localization in a number of cell types, including DRG neurons and OSNs, has been determined (reviewed by Hartzell et al., 2008).

Hydropathy analysis predicts either four (Bakall et al., 2003; Milenkovic et al., 2007; Sun et al., 2002) or six (Petrukhin et al., 1998; Qu et al., 2003b) hydrophobic, putative transmembrane domains (TMDs) within the highly conserved 350 amino acid N-terminal region. Tsunenari et al. (2003) provided evidence from structural analysis experiments of human Best1 that TMD-1, -2, -4 and -6 traverse the membrane, that TMD-3 is cytoplasmic and that TMD-5 is partially inserted into the extracellular side of the membrane. However, Milenkovic et al. (2007) suggested that TMD1, -2, -5 and -6 of hBest1 traverse the membrane and that a relatively hydrophobic segment (containing putative TMD-3 and -4) is cytoplasmic. Both of these studies located the NH_2 and COOH termini cytoplasmically. Further studies, including determination of the crystal structure of the bestrophin proteins, are required to establish the basis for the different proposed membrane topologies.

C. Biophysical Properties and Permeation of Bestrophin Channels

Expression of human, mouse, *Xenopus*, *C. elegans* or *Drosophila* bestrophin in HEK cells results in the appearance of $I_{Cl(Ca)}$ (Chien et al., 2006; O'Driscoll, 2007; Pifferi et al., 2006; Qu et al., 2004, 2003b; Srivastava et al., 2008; Sun et al., 2002; Tsunenari et al., 2003, 2006). Although the current/voltage relationships and the rectification of the currents differ among the bestrophins tested (Table 13.1), the currents have an anion permeability sequence of $I^- > Br^- > Cl^- > F^-$ similar to native Cl_{Ca} channels (O'Driscoll, 2007; Qu et al., 2003b; Qu and Hartzell, 2004; Sun et al., 2002). Similarly, the expressed currents are sensitive to the anion channel blockers that inhibit native $I_{Cl(Ca)}$: DIDS, NFA and SITS (O'Driscoll, 2007; Pifferi et al., 2006; Qu et al., 2003b; Qu and Hartzell, 2004; Sun et al., 2002) (Table 13.1). In the *Drosophila* S2 cell line, single channel analysis of endogenous bestrophin-dependent currents were mediated by ~2 pS single channels (Chien et al., 2006). The single channel conductance of mouse Best2 heterologously expressed in HEK cells is ~0.26 pS (Pifferi et al., 2006). The small single channel conductances of the bestrophin-mediated channels are consistent with single channel conductances of native currents described above and support the idea that bestrophin genes encode at least part of the native Cl_{Ca} channels in a variety of cell types.

All the disease-causing bestrophin1 mutations that have been examined in expression systems alter Cl^- channel function. These include the T6P, A10V, Y85H, R92C, W93C, N99K, D104E, R218S, Y227N, A243T, Q293K, G299E, E300D, D301E, T307I and A243V mutations, all of which produce currents with amplitudes <20% that of wildtype bestrophin (Hartzell et al., 2005b). Further, point mutations in bestrophin proteins

TABLE 13.1 Biophysical properties of heterologously expressed Bestrophins. From O'Driscoll (2007)

	Time and voltage dependence	I-V Relation	Permeability	Ca^{2+} Sensitivity	Pharmacology
hBest1*^	Time-independent	Weak outward rectification	$NO_3^- > I^- > Br^- > Cl^-$	Ca^{2+} sensitive; K_D n.d.	DIDS 500 μM
hBest2*^	Time-independent	Almost linear	$NO_3^- > I^- > Br^- > Cl^-$	Ca^{2+} sensitive; K_D n.d.	
hBest3^	Time dependent	Strong inward rectification		Ca^{2+} sensitive; K_D n.d.	
hBest4^#	Time & voltage independent	Almost linear		K_D of 230 nM	
dBestl*$	Time-independent	Outward rectification	$SCN^- > Cl^-$	Ca^{2+} sensitive; K_D n.d.	
cBestl*	Some time-dependence	Inward rectification		Ca^{2+} sensitive; K_D n.d.	
xBest2a%	Time & voltage independent	Linear	$I^- > Br^- > Cl^- >> Asp$	K_D of 210 nM	
xBest2b%	Time & voltage independent	Linear	$I^- > Br^- > Cl^- >> Asp$	K_D of 228 nM	
mBest1*	Time & voltage independent	Almost linear	$SCN^- > I^- > Cl^-$	Ca^{2+} sensitive; K_D n.d.	DIDS 100 μM NFA 100 μM
mBest2+ϕψ	Time & voltage independent	Almost linear	$SCN^- > I^- > Br^- > Cl^- > F^-$ $I^- > NO^-_3 > Br^- > Cl^- >> MeS$	K_D of 230 nM K_D of 400 nM	DIDS 300 μM NFA 10 μM SITS 400 μM
mBest3+ϖ	Time & voltage independent	Almost linear	$SCN^- > I^- > Cl^-$	K_D of 174 nm	DIDS 100 μM NFA 100 μM
mBest3 truncatedξ	some time dependence and voltage independent	Linear	$SCN^- > I^- > Br^- > Cl$	Ca^{2+} insensitive	

*(Sun et al., 2002), ^ (Tsunenari et al., 2003), # (Tsunenari et al., 2006), $ (Chien et al., 2006), % (Qu et al., 2003),+ (O'Driscoll, 2007), ϕ (Qu et al., 2004), ψ (Pifferi et al., 2006), ϖ (Srivastava et al., 2008), ξ (Qu et al., 2006).

produced currents that differ from wildtype currents in their relative conductance and permeability (O'Driscoll, 2007; Qu et al., 2003b; Qu and Hartzell, 2004). This is noteworthy as it is strong evidence that bestrophin proteins comprise pore forming components of the Cl_{Ca} channel. Extensive site directed mutagenesis of the highly conserved putative second transmembrane domain (TMD2) of bestrophin proteins indicate that TMD2 is important in the permeation pathway and may form the Cl_{Ca} channel pore (Qu et al., 2004, 2006; Qu and Hartzell, 2004). Cysteine scanning of amino acids between positions 69 and 104 revealed that, unlike K_v channels, the amino acids forming the mBest2 selectivity filter are not discretely localized but are distributed over approximately 20 amino acids within the transmembrane domain (Qu et al., 2006). In addition, the ability of RNAi targeting bestrophin to reduce native $I_{Cl(Ca)}$ in the *Drosophila* S2 cell line is further evidence for a role of bestrophins in Cl_{Ca} channels. In these experiments, native $I_{Cl(Ca)}$ was abolished by RNAi constructs targeting dBest1 and dBest2 but was not affected by RNAi constructs to dBest3 or dBest4 (Chien et al., 2006). Moreover, RNAi directed against hBest1 suppressed endogenous $I_{Cl(Ca)}$ in human airway epithelial cells (Barro Soria et al., 2006; Duta et al., 2004).

The Ca^{2+} sensitivity of expressed bestrophin currents has been examined in dose–response relationships with various Ca^{2+} concentrations (Table 13.1). The expression of mouse and Xenopus Best2 in HEK cells generated $I_{Cl(Ca)}$ activated by physiological levels of intracellular Ca^{2+} (K_D of 210–400 nM) (Pifferi et al., 2006; Qu et al., 2004, 2003b). Best3 cloned from mouse heart generated macroscopic $I_{Cl(Ca)}$ with a K_D of 174 nM (O'Driscoll, 2007). Tsunenari et al. (2006) demonstrated that hBest4 can be activated by Ca^{2+} in a cell-free environment. A calculated K_D of 230 nM is similar to that reported for Best2 (Qu et al., 2004, 2003b) and Best3 (O'Driscoll, 2007) and would indicate that Ca^{2+} binds directly to the channel without a requirement for an indirect protein.

All bestrophin proteins isolated to date have a conserved highly acidic region composed of five consecutive acidic residues ^{301}EDDDD^{305} and three nearby negative charges. This motif is located after putative TMD6 on the cytosolic region of the protein. Sixteen mutations of hBest1 identified in patients with Best disease are located in this region and heterologous expression of some of these mutants in HEK293 cells resulted in $I_{Cl(Ca)}$ substantially reduced compared to wildtype (Marmorstein and Kinnick, 2007). Tsunenari et al. (2006) pointed out that this conserved stretch of negatively charged residues is a candidate Ca^{2+} binding domain based on its similarity to the high affinity Ca^{2+} binding site of the BK channel, referred to as the "Ca^{2+} bowl", that contains ten negative charges of which five are consecutive (Schreiber et al., 1999).

Clearly, there is a great deal of interest in the expression and function of the bestrophin family. Ample data indicate that expression of bestrophin proteins is sufficient to generate $I_{Cl(Ca)}$ and that the bestrophin genes encode channels. However, the biological role or roles of this protein are unsettled. As pointed out by Hartzell (Hartzell et al., 2005a), "The troubled history of Cl^- channel identification (Clapham, 1998; Jentsch et al., 2002) should make one exceedingly circumspect about the conclusion that mBest2 forms the Cl^- channel pore, despite the fact that mBest2 selectivity can be changed by mutation." Indeed, functions other than encoding Cl_{Ca} channels have been proposed for bestrophins including modulation of voltage-dependent Ca^{2+} channels (Marmorstein et al., 2006; Rosenthal et al., 2006; Yu et al., 2008) and formation of volume-sensitive Cl^- channels (Chien and Hartzell, 2007; Fischmeister and Hartzell, 2005). Of particular importance is the lack of time and voltage dependence of most bestrophin-mediated $I_{Cl(Ca)}$ (Fig. 13.8). The exceptions being human Best3 and *C. elegans* Best1 currents (Sun et al., 2002). As summarized above, mBest2 is a strong candidate to mediate $g_{Cl(Ca)}$ in OSNs where the native $g_{Cl(Ca)}$ is similarly time and voltage independent. However, this gating contrasts sharply with the characteristic activation by Ca^{2+} and depolarization of $g_{Cl(Ca)}$ in oocytes, central and peripheral neurons, cardiac muscle, smooth muscle and exocrine secretory cells. Although bestrophins seem less likely to underlie these currents, several observations indicate that they cannot be ruled out. As discussed in the context of CLCA, time and voltage dependence might arise from the action of as yet unidentified subunit proteins. Similarly, some studies have found evidence that bestrophins form dimers, tetramers and pentamers. Heteromeric combinations are possible (Stanton et al., 2006; Sun et al., 2002; Yu et al., 2006, 2007). The time and voltage dependence of native currents might arise from the action of as yet unidentified subunit proteins. Lastly, the time and voltage dependence of native smooth muscle $g_{Cl(Ca)}$ is subject to regulation by kinases and phosphatases. As shown by Angermann et al. (2006) native $g_{Cl(Ca)}$ tends to lose time and voltage dependence in conditions that favor dephosphorylation. Thus, it may be that the time and voltage dependence of native currents requires appropriate regulatory contribution.

D. Tweety Channels as Molecular Candidates for Cl_{Ca} Channels

Tweety proteins are a novel group of membrane proteins recently identified as putative chloride channels (Suzuki, 2006). There are three members of the Tweety family in mammals, TTYH1 (Campbell et al., 2000), TTYH2 (Rae et al., 2001) and TTYH3 (Suzuki and Mizuno, 2004) that have been identified based on sequence homology to the protein encoded by the *Drosophila melanogaster tty* gene. The topological structure for Tweety proteins was proposed from hydrophobicity analysis and glycosylation site mutagenesis (He et al., 2008) to incorporate five

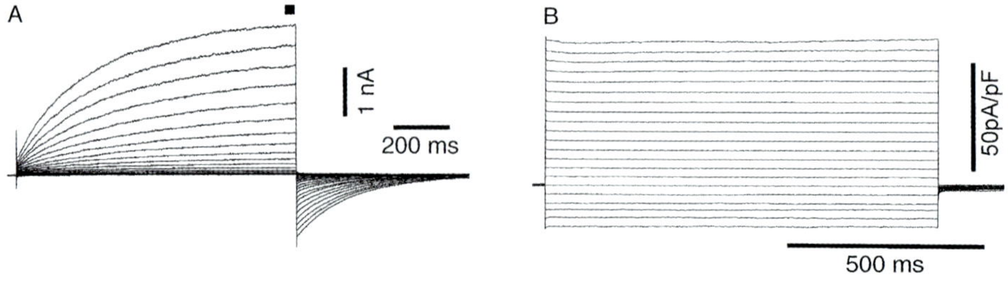

FIGURE 13.8 Different time- and voltage-dependent activation of native and bestrophin-induced $I_{Cl(Ca)}$. Whole-cell currents recorded from a rabbit pulmonary artery smooth muscle myocyte (panel A) and from a HEK-293 cell expressing mBest3 (panel B). Currents were activated with pipette solutions containing 250 nM free $[Ca^{2+}]_i$. Voltage steps were applied in 10 mV steps from a holding potential of −50 mV to voltages between −100 and +140 mV (A) and between −100 and +100 mV (B).

membrane-spanning domains in which the NH_2 terminus is located extracellularly and the COOH terminus is located cytoplasmically. This membrane topology is similar to BK channels (Suzuki and Mizuno, 2004).

Suzuki and Mizuno (2004) reported that human TTYH3 encodes a large-conductance Ca^{2+}-activated Cl^- channel. Expression of human TTYH3 in Chinese hamster ovary cells generated a unique Cl^- current activated by addition of a Ca^{2+} ionophore or with 100 μM Ca^{2+} in the pipette solution. The evoked current was completely inhibited by DIDS (10 μM), but was insensitive to NFA (300 μM), SITS (10 μM), $ZnCl_2$ (10 μM) and DTT (10 μM). The current/voltage relation was linear in symmetrical Cl^- with slope conductance of 260 pS. Like native large-conductance Ca^{2+}-activated Cl^- channels, hTTYH3 channels show complex gating kinetics, voltage-dependent inactivation, require micromolar $[Ca^{2+}]$ for activation, and have a similar permeability order of $I^- > Br^- > Cl^-$ determined from anion substitution experiments. Analysis of pore mutants suggested that the conserved positively charged amino acids (R366 and H370) contribute to anion selectivity. Similarly, hTTYH2 encodes a large-conductance Ca^{2+}-activated Cl^- channel. However, a splice variant of TTYH1 that lacks exon 11 acts as a volume-sensitive chloride channel (Suzuki, 2006). Taken together, the functional properties of TTYH3 and the expression and localization of Tweety channels in neurons suggest that TTYHs are strong candidates to encode large-conductance Cl^- channels. Further studies on the hTTYH family should lead to the elucidation of their physiological roles as novel Cl^- channels.

E. TMEM16 as Molecular Candidates for Cl_{Ca} Channels

During preparation of this review, three independent groups identified the TMEM16 family of genes as encoding a novel membrane protein associated with Ca^{2+}-dependent Cl^- channel activity (Caputo et al., 2008; Schroeder et al., 2008; Yang et al., 2008). This family consists of ten paralogs that encode highly conserved membrane spanning proteins with NH_2 and COOH termini located in the cytosol. When expressed in recombinant expression systems, human, mouse and Xenopus TMEM16A (Anoctamin 1 or ANO1; and mouse TMEM16B or ANO2) generate a chloride-selective Ca^{2+}-dependent conductance activated by agonists (endothelin-1, angiotensin II, carbachol, UTP, ATP, lysophosphatidic acid and histamine) (Yang et al., 2008), intracellular injection or photorelease of IP_3 (Schroeder et al., 2008; Yang et al., 2008), exposure to the Ca^{2+} ionophore ionomycin (Caputo et al., 2008; Yang et al., 2008), and fixed high $[Ca^{2+}]_i$ (Caputo et al., 2008; Schroeder et al., 2008; Yang et al., 2008). The biophysical and pharmacological properties of TMEM16A-mediated currents are consistent with native $I_{Cl(Ca)}$ recorded from many secretory and retinal epithelia, smooth muscle and sensory neurons including similar anion permeability ($SCN^- >> NO_3^- > I^- > Br^- > Cl^- >> F^- >>>$ gluconate) (Caputo et al., 2008; Schroeder et al., 2008; Yang et al., 2008), a voltage-sensitive K_d for activation by intracellular Ca^{2+} of 2.6 μM and 400 nM at −60 and +60 mV, respectively (Caputo et al., 2008), and a single-channel conductance of 8.3 pS (Yang et al., 2008). In contrast to currents mediated by the molecular candidates discussed above, TMEM16-induced currents are time and voltage dependent exhibiting outward rectification, slow activation upon depolarization and slow deactivation kinetics following repolarization (Caputo et al., 2008; Schroeder et al., 2008; Yang et al., 2008). TMEM16A-elicited currents were inhibited by the classical anion channel blockers niflumic acid, DIDS, NPPB and DPC at concentrations that inhibit native $I_{Cl(Ca)}$ (Caputo et al., 2008; Schroeder et al., 2008; Yang et al., 2008). In addition, two groups have mutated the channel at different locations within the putative transmembrane domains finding altered anion permeability sequence (Caputo et al., 2008; Yang et al., 2008) consistent with TMEM16A being a pore-forming subunit. Lastly, TMEM16A expression has been identified in a number of cell types where $I_{Cl(Ca)}$ has been recorded including sensory DRG neurons. Thus, we anticipate that the discovery of this family of ion channels will lead to exciting investigations into the expression and function of $I_{Cl(Ca)}$.

VII. CONCLUDING REMARKS

The wide variety of cells expressing large $I_{Cl(Ca)}$ which is regulated and often changes during development and in response to injury suggests that a Ca^{2+}-activated Cl^- conductance provides a broadly useful function (or functions). However, more than 25 years after the discovery of $I_{Cl(Ca)}$, the molecular identity of Cl_{Ca} channels is still unknown. We do not know whether the different cells obtain this function with the widespread expression of a single gene, expression of multiple members of a single gene family, or expression of multiple genes from multiple gene families. Although each of the first candidate genes (CLCA, bestrophin and Tweety) has merit (particularly the bestrophin family in the context of olfaction), none of them has so far cracked open the genomic door enabling molecular investigations. The latest candidate, TMEM16, appears to be stronger than any of these but remains to be tested. Until the molecular door can be opened, advances in a number of areas will be difficult as illustrated by the incremental advances in understanding Cl_{Ca} channel biology made since the last major reviews of Cl_{Ca} channels. These advances include interesting but observational studies in neurons that correlate $I_{Cl(Ca)}$ and cation-chloride cotransport expression in development and injury as well as the important characterization of the regulation of $I_{Cl(Ca)}$ in smooth muscle myocytes by kinases and phosphatases.

Acknowledgements

NL was supported by R01 HL075477. FB was supported by R01 HL091238. NL and FB were supported by 5P20 RR15581 from the National Center for Research Resources (NCRR), a component of the National Institutes of Health (NIH) supporting a Center of Biomedical Research Excellence (COBRE) at the University of Nevada School of Medicine, Reno, Nevada. JLK was supported by grants from the American Heart Association (Grant-in-Aid 98-257) and the NIH (R01 NS32144 and R01 NS041037). The contents of this work are solely the responsibility of the authors and do not necessarily represent the official views of NCRR or NIH. We thank Dr. K.E. O'Driscoll for providing Table 13.1 and the records in Fig. 13.8B.

References

Abdulla, F.A. and Smith, P.A. (1999). Neuropeptide Y actions and the distribution of Ca^{2+}-dependent Cl^- conductance in rat dorsal root ganglion neurons. *J. Auton. Nerv. Syst.* **78**, 24–29.

Adragna, N.C., DiFulvio, M., and Lauf, P.K. (2004). Regulation of K-Cl cotransport: from function to genes. *J. Membr. Biol.* **201**, 109–137.

Aggarwal, S.K. and MacKinnon, R. (1996). Contribution of the S4 segment to gating charge in the *Shaker* K^+ channel. *Neuron* **16**, 1169–1177.

Al-Jumaily, M., Kozlenkov, A., Mechaly, I., Fichard, A., Matha, V., Scamps, F., Valmier, J., and Carroll, P. (2007). Expression of three distinct families of calcium-activated chloride channel genes in the mouse dorsal root ganglion. *Neurosci. Bull.* **23**, 293–299.

Amédée, T., Large, W.A., and Wang, Q. (1990). Characteristics of chloride currents activated by noradrenaline in rabbit ear artery cells. *J. Physiol. (Lond.)* **428**, 501–516.

Anderson, M.P., Gregory, R.J., Thompson, S., Souza, D.W., Paul, S., Mulligan, R.C., Smith, A.E., and Welsh, M.J. (1991). Demonstration that CFTR is a chloride channel by alteration of its anion selectivity. *Science* **253**, 202–205.

André, S., Boukhaddaoui, H., Campo, B., Al-Jumaily, M., Mayeux, V., Greuet, D., Valmier, J., and Scamps, F. (2003). Axotomy-induced expression of calcium-activated chloride current in subpopulations of mouse dorsal root ganglion neurons. *J. Neurophysiol.* **90**, 3764–3773.

Angermann, J.E., Sanguinetti, A.R., Kenyon, J.L., Leblanc, N., and Greenwood, I.A. (2006). Mechanism of the inhibition of Ca^{2+}-activated Cl^- currents by phosphorylation in pulmonary arterial smooth muscle cells. *J. Gen. Physiol.* **128**, 73–87.

Arreola, J., Melvin, J.E., and Begenisich, T. (1996). Activation of calcium-dependent chloride channels in rat parotid acinar cells. *J. Gen. Physiol.* **108**, 35–47.

Ayar, A. and Scott, R.H. (1999). The actions of ryanodine on Ca^{2+}-activated conductances in rat cultured DRG neurones; evidence for Ca^{2+}-activated Ca^{2+} release. *Naunyn-Schmiedebergs Arch. Pharmacol.* **359**, 81–91.

Bader, C.R., Bertrand, D., and Schlichter, R. (1987). Calcium-activated chloride current in cultured sensory and parasympathetic quail neurones. *J. Physiol. (Lond.)* **394**, 125–148.

Bader, C.R., Bertrand, D., and Schwartz, E.A. (1982). Voltage-activated and calcium-activated currents studied in solitary rod inner segments from the salamander retina. *J. Physiol. (Lond.)* **331**, 253–284.

Bajnath, R.B., Groot, J.A., de Jonge, H.R., Kansen, M., and Bijman, J. (1993). Calcium ionophore plus excision induce a large conductance chloride channel in membrane patches of human colon carcinoma cells HT-29cl.19A. *Experientia* **49**, 313–316.

Bakall, B., Marmorstein, L.Y., Hoppe, G., Peachey, N.S., Wadelius, C., and Marmorstein, A.D. (2003). Expression and localization of bestrophin during normal mouse development. *Invest. Ophthalmol. Vis. Sci.* **44**, 3622–3628.

Barish, M.E. (1983). A transient calcium-dependent chloride current in the immature Xenopus oocyte. *J. Physiol. (Lond.)* **342**, 309–325.

Barnes, S. and Hille, B. (1989). Ionic channels of the inner segment of tiger salamander cone photoreceptors. *J. Gen. Physiol.* **94**, 719–743.

Barrett, J.N., Magleby, K.L., and Pallotta, B.S. (1982). Properties of single calcium-activated potassium channels in cultured rat muscle. *J. Physiol. (Lond.)* **331**, 211–230.

Barro Soria, R., Spitzner, M., Schreiber, R., and Kunzelmann, K. (2006). Bestrophin 1 enables Ca^{2+} activated Cl^- conductance in epithelia. *J. Biol. Chem.*.

Ben-Ari, Y., Gaiarsa, J.L., Tyzio, R., and Khazipov, R. (2007). GABA: a pioneer transmitter that excites immature neurons and generates primitive oscillations. *Physiol. Rev.* **87**, 1215–1284.

Bernheim, L., Bader, C.R., Bertrand, D., and Schlichter, R. (1989). Transient expression of a Ca^{2+}-activated Cl^- current during development of quail sensory neurons. *Dev. Biol.* **136**, 129–139.

Boccaccio, A. and Menini, A. (2007). Temporal development of cyclic nucleotide-gated and Ca^{2+}-activated Cl^- currents in isolated mouse olfactory sensory neurons. *J. Neurophysiol.* **98**, 153–160.

Bormann, J., Hamill, O.P., and Sakmann, B. (1987). Mechanism of anion permeation through channels gated by glycine and gamma-aminobutyric acid in mouse cultured spinal neurones. *J. Physiol. (Lond.)* **385**, 243–286.

Britton, F.C., Ohya, S., Horowitz, B., and Greenwood, I.A. (2002). Comparison of the properties of CLCA1 generated currents and $I_{Cl(Ca)}$ in murine portal vein smooth muscle cells. *J. Physiol. (Lond.)* **539**, 107–117.

Burkin, H.R., Horowitz, B., Hume, J.R., and Britton, F.C. (2003). Identification of novel murine bestrophin transcripts that are expressed in mouse heart. *J. Gen. Physiol.* **122** (1), 62.

Byrne, N.G. and Large, W.A. (1988). Membrane ionic mechanisms activated by noradrenaline in cells isolated from the rabbit portal vein. *J. Physiol.* **404**, 557–573.

Byrne, N.G. and Large, W.A. (1987). Action of noradrenaline on single smooth muscle cells freshly dispersed from the rat anococcygeus muscle. *J. Physiol.* **389**, 513–525.

Campbell, H.D., Kamei, M., Claudianos, C., Woollatt, E., Sutherland, G.R., Suzuki, Y., Hida, M., Sugano, S., and Young, I.G. (2000). Human and mouse homologues of the *Drosophila melanogaster* tweety (tty) gene: a novel gene family encoding predicted transmembrane proteins. *Genomics* **68**, 89–92.

Caputo, A., Caci, E., Ferrera, L., Pedemonte, N., Barsanti, C., Sondo, E., Pfeffer, U., Ravazzolo, R., Zegarra-Moran, O., and Galietta, L.J. (2008). TMEM16A, a membrane protein associated with calcium-dependent chloride channel activity. *Science* **322**, 590–594.

Carl, A., Lee, H.K., and Sanders, K.M. (1996). Regulation of ion channels in smooth muscles by calcium. *Am. J. Physiol. Cell Physiol.* **271**, C9–C34.

Carl, A. and Sanders, K.M. (1989). Ca^{2+}-activated K channels of canine colonic myocytes. *Am. J. Physiol. Cell Physiol.* **257**, C470–C480.

Cherubini, E., Gaiarsa, J.L., and Ben Ari, Y. (1991). GABA: an excitatory transmitter in early postnatal life. *Trends Neurosci.* **14**, 515–519.

Chien, L.T. and Hartzell, H.C. (2007). *Drosophila* bestrophin-1 chloride current is dually regulated by calcium and cell volume. *J. Gen. Physiol.* **130**, 513–524.

Chien, L.T., Zhang, Z.R., and Hartzell, H.C. (2006). Single Cl^- channels activated by Ca^{2+} in *Drosophila* S2 cells are mediated by bestrophins. *J. Gen. Physiol.* **128**, 247–259.

Clapham, D.E. (1998). The list of potential volume-sensitive chloride currents continues to swell (and shrink). *J. Gen. Physiol.* **111**, 623–624.

Clapp, L.H., Turner, J.L., and Kozlowski, R.Z. (1996). Ca^{2+}-activated Cl^- currents in pulmonary arterial myocytes. *Am. J. Physiol.* **270**, H1577–H1584.

Cliff, W.H. and Frizzell, R.A. (1990). Separate Cl^- conductances activated by cAMP and Ca^{2+} in Cl^--secreting epithelial cells. *Proc. Natl. Acad. Sci. USA* **87**, 4956–4960.

Collier, M.L., Levesque, P.C., Kenyon, J.L., and Hume, J.R. (1996). Unitary Cl^- channels activated by cytoplasmic Ca^{2+} in canine ventricular myocytes. *Circ. Res.* **78**, 936–944.

Coull, J.A., Boudreau, D., Bachand, K., Prescott, S.A., Nault, F., Sik, A., de Koninck, P., and de Koninck, Y. (2003). Trans-synaptic shift in anion gradient in spinal lamina I neurons as a mechanism of neuropathic pain. *Nature* **424**, 938–942.

Crawford, J.H., Wootton, J.F., Seabrook, G.R., and Scott, R.H. (1997). Activation of Ca^{2+}-dependent currents in dorsal root ganglion neurons by metabotropic glutamate receptors and cyclic ADP-ribose precursors. *J. Neurophysiol.* **77**, 2573–2584.

Cui, J., Cox, D.H., and Aldrich, R.W. (1997). Intrinsic voltage dependence and Ca^{2+} regulation of mslo large conductance Ca-activated K^+ channels. *J. Gen. Physiol.* **109**, 647–673.

Cunningham, S.A., Awayda, M.S., Bubien, J.K., Ismailov, I.I., Arrate, M.P., Berdiev, B.K., Benos, D.J., and Fuller, C.M. (1995). Cloning of an epithelial chloride channel from bovine trachea. *J. Biol. Chem.* **270**, 31016–31026.

Currie, K.P.M. and Scott, R.H. (1992). Calcium-activated currents in cultured neurones from rat dorsal root ganglia. *Br. J. Pharmacol.* **106**, 593–602.

Currie, K.P.M., Swann, K., Galione, A., and Scott, R.H. (1993). Activation of Ca^{2+}-dependent currents in cultured rat dorsal root ganglion neurones by a sperm factor and cyclic ADP-ribose. *Molec. Biol. Cell* **3**, 1415–1425.

Dixon, D.M., Valkanov, M., and Martin, R.J. (1993). A patch-clamp study of the ionic selectivity of the large conductance, Ca-activated chloride channel in muscle vesicles prepared from Ascaris suum. *J. Membr. Biol.* **131**, 143–149.

Duta, V., Szkotak, A.J., Nahirney, D., and Duszyk, M. (2004). The role of bestrophin in airway epithelial ion transport. *FEBS Lett.* **577**, 551–554.

Evans, M.G. and Marty, A. (1986). Calcium-dependent chloride currents in isolated cells from rat lacrimal glands. *J. Physiol. (Lond.)* **378**, 437–460.

Fahlke, C., Dürr, C., and George, A.L., Jr. (1997). Mechanism of ion permeation in skeletal muscle chloride channels. *J. Gen. Physiol.* **110**, 551–564.

Fahmi, M., Garcia, L., Taupignon, A., Dufy, B., and Sartor, P. (1995). Recording of a large-conductance chloride channel in normal rat lactotrophs. *Am. J. Physiol.* **269**, E969–E976.

Fischmeister, R. and Hartzell, H.C. (2005). Volume sensitivity of the bestrophin family of chloride channels. *J. Physiol.* **562**, 477–491.

Fuller, C.M., Ismailov, I.I., Keeton, D.A., and Benos, D.J. (1994). Phosphorylation and activation of a bovine tracheal anion channel by Ca2 + /calmodulin-dependent protein kinase II. *J. Biol. Chem.* **269**, 26642–26650.

Fuller, C.M., Ji, H.L., Tousson, A., Elble, R.C., Pauli, B.U., and Benos, D.J. (2001). Ca^{2+}-activated Cl^- channels: a newly emerging anion transport family. *Pflügers Arch.* **443** (Supplement 1), S107–S110.

Gamba, G. (2005). Molecular physiology and pathophysiology of electroneutral cation-chloride cotransporters. *Physiol. Rev.* **85**, 423–493.

Ganguly, K., Schinder, A.F., Wong, S.T., and Poo, M. (2001). GABA itself promotes the developmental switch of neuronal GABAergic responses from excitation to inhibition. *Cell* **105**, 521–532.

Gillo, B., Landau, E.M., Moriarty, T.M., Roberts, J.L., and Sealfon, S.C. (1989). A novel calcium-dependent chloride current in *Xenopus* oocytes injected with rat brain messenger RNA. *J. Physiol. (Lond.)* **417**, 47–61.

Granados-Soto, V., Arguelles, C.F., and Alvarez-Leefmans, F.J. (2005). Peripheral and central antinociceptive action of Na^+-K^+-$2Cl^-$ cotransporter blockers on formalin-induced nociception in rats. *Pain* **114**, 231–238.

Greenwood, I.A., Ledoux, J., and Leblanc, N. (2001). Differential regulation of Ca^{2+}-activated Cl^- currents in rabbit arterial and portal vein smooth muscle cells by Ca^{2+}-calmodulin-dependent kinase. *J. Physiol. (Lond.)*, **534**, 395–408.

Greenwood, I.A., Ledoux, J., Sanguinetti, A., Perrino, B.A., and Leblanc, N. (2004). Calcineurin Aα but not Aβ augments $I_{Cl(Ca)}$ in rabbit pulmonary artery smooth muscle cells. *J. Biol. Chem.* **279**, 38830–38837.

Greenwood, I.A., Miller, L.J., Ohya, S., and Horowitz, B. (2002). The large conductance potassium channel beta-subunit can interact with and modulate the functional properties of a calcium-activated chloride channel, CLCA1. *J. Biol. Chem.* **277**, 22119–22122.

Hallani, M., Lynch, J.W., and Barry, P.H. (1998). Characterization of calcium-activated chloride channels in patches excised from the dendritic knob of mammalian olfactory receptor neurons. *J. Membr. Biol.* **161**, 163–171.

Hartzell, C., Putzier, I., and Arreola, J. (2005a). Calcium-activated chloride channels. *Annu. Rev. Physiol.* **67**, 719–758.

Hartzell, C., Qu, Z., Putzier, I., Artinian, L., Chien, L.T., and Cui, Y. (2005b). Looking chloride channels straight in the eye: bestrophins, lipofuscinosis, and retinal degeneration. *Physiology (Bethesda)* **20**, 292–302.

Hartzell, H.C., Qu, Z., Yu, K., Xiao, Q., and Chien, L.T. (2008). Molecular physiology of bestrophins: multifunctional membrane proteins linked to best disease and other retinopathies. *Physiol. Rev.* **88**, 639–672.

He, Y., Ramsay, A.J., Hunt, M.L., Whitbread, A.K., Myers, S.A., and Hooper, J.D. (2008). N-glycosylation analysis of the human Tweety family of putative chloride ion channels supports a penta-spanning membrane arrangement: impact of N-glycosylation on cellular processing of Tweety homologue 2 (TTYH2). *Biochem. J.* **412**, 45–55.

Hilaire, C., Campo, B., Andre, S., Valmier, J., and Scamps, F. (2005). K^+ current regulates calcium-activated chloride current-induced after depolarization in axotomized sensory neurons. *Eur. J. Neurosci.* **22**, 1073–1080.

Hille, B. (2001). *Ion Channels of Excitable Membranes*. Sinauer Associates, Sunderland, MA.

Hodgkin, A.L. and Huxley, A.F. (1952). Currents carried by sodium and potassium ions through the membrane of the giant axon of. *Loligo. J. Physiol. (Lond.)* **116**, 449–472.

Huang, S.J., Fu, W.O., Chung, Y.W., Zhou, T.S., and Wong, P.Y. (1993). Properties of cAMP-dependent and Ca(2+)-dependent whole cell Cl^- conductances in rat epididymal cells. *Am. J. Physiol.* **264**, C794–C802.

Hume, R.I. and Thomas, S.A. (1989). A calcium- and voltage-dependent chloride current in developing chick skeletal muscle. *J. Physiol. (Lond.)* **417**, 241–261.

Hussy, N. (1992). Calcium-activated chloride channels in cultured embryonic Xenopus spinal neurons. *J. Neurophysiol.* **68**, 2042–2050.

Ishikawa, T. and Cook, D.I. (1993). A Ca^{2+}-activated Cl^- current in sheep parotid secretory cells. *J. Membr. Biol.* **135**, 261–271.

Ivanenko, A., Baring, M.D., Airey, J.A., Sutko, J.L., and Kenyon, J.L. (1993). A caffeine- and ryanodine-sensitive Ca^{2+} store in avian sensory neurons. *J. Neurophysiol.* **70**, 710–722.

Jentsch, T.J., Stein, V., Weinreich, F., and Zdebik, A.A. (2002). Molecular structure and physiological function of chloride channels. *Physiol. Rev.* **82**, 503–568.

Kaneko, H., Mohrlen, F., and Frings, S. (2006). Calmodulin contributes to gating control in olfactory calcium-activated chloride channels. *J. Gen. Physiol.* **127**, 737–748.

Kapicka, C.L., Carl, A., Hall, M.L., Percival, A.L., Frey, B.W., and Kenyon, J.L. (1994). Comparison of large conductance Ca^{2+}-activated K^+ channels in artificial bilayer and patch-clamp experiments. *Am. J. Physiol. Cell Physiol.* **266**, C601–C610.

Kenyon, J.L. (2000). The reversal potential of Ca^{2+}-activated Cl^- currents indicates that chick sensory neurons accumulate intracellular Cl^-. *Neurosci. Lett.* **296**, 9–12.

Kenyon, J.L. and Scott, R.H. (2002). Ca^{2+}-activated Cl^- channels as Ca^{2+}-sensors with particular reference to the modulation of neuronal excitability. In *Calcium-Activated Chloride Channels* (Fuller, C.M. ed.). Academic Press, San Diego.

Kleene, S.J. (1997). High-gain, low-noise amplification in olfactory transduction. *Biophys. J.* **73**, 1110–1117.

Kleene, S.J. (1993). Origin of the chloride current in olfactory transduction. *Neuron* **11**, 123–132.

Korn, S.J. and Weight, F.F. (1987). Patch-clamp study of the calcium-dependent chloride current in AtT-20 pituitary cells. *J. Neurophysiol.* **58**, 1431–1451.

Kozak, J.A. and Logothetis, D.E. (1997). A calcium-dependent chloride current in insulin-secreting beta TC-3 cells. *Pflügers Arch.* **433**, 679–690.

Krämer, F., Stohr, H., and Weber, B.H. (2004). Cloning and characterization of the murine Vmd2 RFP-TM gene family. *Cytogenet. Genome Res.* **105**, 107–114.

Kuruma, A. and Hartzell, H.C. (2000). Bimodal control of a Ca^{2+}-activated Cl^- channel by different Ca^{2+} signals. *J. Gen. Physiol.* **115**, 59–80.

Laird, J.M., Garcia-Nicas, E., Delpire, E.J., and Cervero, F. (2004). Presynaptic inhibition and spinal pain processing in mice: a possible role of the NKCC1 cation-chloride co-transporter in hyperalgesia. *Neurosci. Lett.* **361**, 200–203.

Lancaster, E., Oh, E.J., Gover, T.D., and Weinreich, D. (2002). Calcium and calcium-activated currents in vagotomized rat primary vagal afferent neurons. *J. Physiol. (Lond.)* **540**, 543–556.

Larsson, H.P., Kleene, S.J., and Lecar, H. (1997). Noise analysis of ion channels in non-space-clamped cables: estimates of channel parameters in olfactory cilia. *Biophys. J.* **72**, 1193–1203.

Latorre, R. and Brauchi, S. (2006). Large conductance Ca^{2+}-activated K^+ (BK) channel: activation by Ca^{2+} and voltage. *Biol. Res.* **39**, 385–401.

Ledoux, J., Chartier, D., and Leblanc, N. (1999). Inhibitors of calmodulin-dependent protein kinase are nonspecific blockers of voltage-dependent K+ channels in vascular myocytes. *J. Pharmacol. Exp. Ther.* **290**, 1165–1174.

Ledoux, J., Greenwood, I.A., and Leblanc, N. (2005). Dynamics of Ca2 + -dependent Cl− channel modulation by niflumic acid in rabbit coronary arterial myocytes. *Mol. Pharmacol.* **67**, 163–173.

Ledoux, J., Greenwood, I.A., Villeneuve, L.R., and Leblanc, N. (2003). Modulation of Ca^{2+}-dependent Cl^- channels by calcineurin in rabbit coronary arterial myocytes. *J. Physiol.* **552**, 701–714.

Li, Y.-X., Schaffner, A.E., Walton, M.K., and Barker, J.L. (1998). Astrocytes regulate developmental changes in the chloride ion gradient of embryonic rat ventral spinal cord neurones in culture. *J. Physiol. (Lond.)* **509**, 847–858.

Linsdell, P., Tabcharani, J.A., Rommens, J.M., Hou, Y.X., Chang, X.B., Tsui, L.C., Riordan, J.R., and Hanrahan, J.W. (1997). Permeability of wild-type and mutant cystic fibrosis transmembrane conductance regulator chloride channels to polyatomic anions. *J. Gen. Physiol.* **110**, 355–364.

Loewen, M.E. and Forsyth, G.W. (2005). Structure and function of CLCA proteins. *Physiol. Rev.* **85**, 1061–1092.

Marmorstein, A.D. and Kinnick, T.R. (2007). Focus on molecules: bestrophin (best-1). *Exp. Eye Res.* **85**, 423–424.

Marmorstein, L.Y., Wu, J., McLaughlin, P., Yocom, J., Karl, M.O., Neussert, R., Wimmers, S., Stanton, J.B., Gregg, R.G., Strauss, O., Peachey, N.S., and Marmorstein, A.D. (2006). The light peak of the electroretinogram is dependent on voltage-gated calcium channels and antagonized by bestrophin (best-1). *J. Gen. Physiol.* **127**, 577–589.

Marquardt, A., Stohr, H., Passmore, L.A., Kramer, F., Rivera, A., and Weber, B.H. (1998). Mutations in a novel gene, VMD2, encoding a protein of unknown properties cause juvenile-onset vitelliform macular dystrophy (Best's disease). *Hum. Mol. Genet.* **7**, 1517–1525.

Martin, D.K. (1993). Small conductance chloride channels in acinar cells from the rat mandibular salivary gland are directly controlled by a G-protein. *Biochem. Biophys. Res. Commun.* **192**, 1266–1273.

Martínez-Pinna, J., McLachlan, E.M., and Gallego, R. (2000). Distinct mechanisms for activation of Cl^- and K^+ currents by Ca^{2+} from different sources in mouse sympathetic neurones. *J. Physiol. (Lond.)* **527**, 249–264.

Marty, A., Tan, Y.P., and Trautmann, A. (1984). Three types of calcium-dependent channel in rat lacrimal glands. *J. Physiol. (Lond.)* **357**, 293–325.

Matta, J.A. and Ahern, G.P. (2007). Voltage is a partial activator of rat thermosensitive TRP channels. *J. Physiol. (Lond.)* **585**, 469–482.

Mayer, M.L. (1985). A calcium-activated chloride current generates the after-depolarization of rat sensory neurones in culture. *J. Physiol. (Lond.)* **364**, 217–239.

McCarty, N.A. and Zhang, Z.R. (2001). Identification of a region of strong discrimination in the pore of CFTR. *Am. J. Physiol. Lung Cell Mol. Physiol.* **281**, L852–L867.

McGehee, D.S. and Oxford, G.S. (1991). Bradykinin modulates the electrophysiology of cultured rat sensory neurons through a pertussis toxin-insensitive G protein. *Mol. Cell. Neurosci.* **2**, 21–30.

McHugh, J.M. and Kenyon, J.L. (2003). An Excel®-based model of Ca^{2+} diffusion and fura-2 measurements in a spherical cell. *Am. J. Physiol. Cell Physiol.* **286**, C342–C348.

Miledi, R. (1982). A calcium-dependent transient outward current in *Xenopus laevis* oocytes. *Proc. R. Soc. Lond. B* **215**, 491–497.

Milenkovic, V.M., Rivera, A., Horling, F., and Weber, B.H. (2007). Insertion and topology of normal and mutant bestrophin-1 in the endoplasmic reticulum membrane. *J. Biol. Chem.* **282**, 1313–1321.

Morales-Aza, B.M., Chillingworth, N.L., Payne, J.A., and Donaldson, L.F. (2004). Inflammation alters cation chloride cotransporter expression in sensory neurons. *Neurobiol. Dis.* **17**, 62–69.

Nilius, B., Prenen, J., Voets, T., Van den Bremt, K., Eggermont, J., and Droogmans, G. (1997). Kinetic and pharmacological properties of the calcium-activated chloride-current in macrovascular endothelial cells. *Cell Calcium* **22**, 53–63.

Nilius, B., Talavera, K., Owsianik, G., Prenen, J., Droogmans, G., and Voets, T. (2005). Gating of TRP channels: a voltage connection? *J. Physiol. (Lond.)* **567** (1), 35–44.

O'Driscoll, K.E. (2007). *Molecular and Functional Expression of the Murine Bestrophin Family from Cardiovascular Tissues* (pp. 1–244). University of Nevada.

Owen, D.G., Segal, M., and Barker, J.L. (1986). Voltage-clamp analysis of a Ca^{2+}- and voltage-dependent chloride conductance in cultured mouse spinal neurons. *J. Neurophysiol.* **55**, 1115–1135.

Owen, D.G., Segal, M., and Barker, J.L. (1984). A Ca-dependent Cl^- conductance in cultured mouse spinal neurones. *Nature* **311**, 567–570.

Owens, D.F., Boyce, L.H., Davis, M.B., and Kriegstein, A.R. (1996). Excitatory GABA responses in embryonic and neonatal cortical slices demonstrated by gramicidin perforated-patch recordings and calcium imaging. *J. Neurosci.* **16**, 6414–6423.

Pacaud, P., Loirand, G., Lavie, J.L., Mironneau, C., and Mironneau, J. (1989). Calcium-activated chloride current in rat vascular smooth muscle cells in short-term primary culture. *Pflügers Arch.* **413**, 629–636.

Payne, J.A., Rivera, C., Voipio, J., and Kaila, K. (2003). Cation-chloride co-transporters in neuronal communication, development and trauma. *Trends Neurosci.* **26**, 199–206.

Petrukhin, K., Koisti, M.J., Bakall, B., Li, W., Xie, G., Marknell, T., Sandgren, O., Forsman, K., Holmgren, G., Andreasson, S., Vujic, M., Bergen, A.A., Garty-Dugan, V., Figueroa, D., Austin, C.P., Metzker, M.L., Caskey, C.T., and Wadelius, C. (1998). Identification of the gene responsible for Best macular dystrophy. *Nat. Genet.* **19**, 241–247.

Pifferi, S., Pascarella, G., Boccaccio, A., Mazzatenta, A., Gustincich, S., Menini, A., and Zucchelli, S. (2006). Bestrophin-2 is a candidate calcium-activated chloride channel involved in olfactory transduction. *Proc. Natl. Acad. Sci. USA* **103**, 12929–12934.

Piper, A.S., Greenwood, I.A., and Large, W.A. (2002). Dual effect of blocking agents on Ca^{2+}-activated Cl^- currents in rabbit pulmonary artery smooth muscle cells. *J. Physiol.* **539**, 119–131.

Pollock, J., Crawford, J.H., Wootton, J.F., Seabrook, G.R., and Scott, R.H. (1999). Metabotropic glutamate receptor activation and intracellular cyclic ADP-ribose release Ca^{2+} from the same store in cultured DRG neurones. *Cell Calcium* **26**, 139–148.

Price, T.J., Cervero, F., and de Koninck, Y. (2005). Role of cation-chloride-cotransporters (CCC) in pain and hyperalgesia. *Curr. Top. Med. Chem.* **5**, 547–555.

Purves, D., Augustine, G.J., Fitzpatrick, D., Hall, W.C., LaMantia, A.-S., McNamara, J.O., and White, L.E. (2008). *Neuroscience* (pages 112–114). Sinauer Associates, Sunderland, MA.

Qu, Z., Chien, L.T., Cui, Y., and Hartzell, H.C. (2006). The anion-selective pore of the bestrophins, a family of chloride channels associated with retinal degeneration. *J. Neurosci.* **26**, 5411–5419.

Qu, Z., Fischmeister, R., and Hartzell, C. (2004). Mouse bestrophin-2 is a bona fide Cl^- channel: identification of a residue important in anion binding and conduction. *J. Gen. Physiol.* **123**, 327–340.

Qu, Z. and Hartzell, C. (2004). Determinants of anion permeation in the second transmembrane domain of the mouse bestrophin-2 chloride channel. *J. Gen. Physiol.* **124**, 371–382.

Qu, Z. and Hartzell, H.C. (2000). Anion permeation in Ca^{2+}-activated Cl^- channels. *J. Gen. Physiol.* **116**, 825–844.

Qu, Z., Wei, R.W., and Hartzell, H.C. (2003a). Characterization of Ca^{2+}-activated Cl^- currents in mouse kidney inner medullary collecting duct cells. *Am. J. Physiol. Renal Physiol.* **285**, F326–F335.

Qu, Z., Wei, R.W., Mann, W., and Hartzell, H.C. (2003b). Two bestrophins cloned from Xenopus laevis oocytes express Ca^{2+}-activated Cl^- currents. *J. Biol. Chem.* **278**, 49563–49572.

Rae, F.K., Hooper, J.D., Eyre, H.J., Sutherland, G.R., Nicol, D.L., and Clements, J.A. (2001). TTYH2, a human homologue of the Drosophila melanogaster gene tweety, is located on 17q24 and upregulated in renal cell carcinoma. *Genomics* **77**, 200–207.

Ran, S., Fuller, C.M., Arrate, M.P., Latorre, R., and Benos, D.J. (1992). Functional reconstitution of a chloride channel protein from bovine trachea. *J. Biol. Chem.* **267**, 20630–20637.

Reichling, D.B., Kyrozis, A., Wang, J., and MacDermott, A.B. (1994). Mechanisms of GABA and glycine depolarization-induced calcium transients in rat dorsal horn neurons. *J. Physiol. (Lond.)* **476**, 411–421.

Reisert, J., Bauer, P.J., Yau, K.W., and Frings, S. (2003). The Ca-activated Cl channel and its control in rat olfactory receptor neurons. *J. Gen. Physiol.* **122**, 349–363.

Rosenthal, R., Bakall, B., Kinnick, T., Peachey, N., Wimmers, S., Wadelius, C., Marmorstein, A., and Strauss, O. (2006). Expression of bestrophin-1, the product of the VMD2 gene, modulates voltage-dependent Ca2+ channels in retinal pigment epithelial cells. *FASEB J.* **20**, 178–180.

Russell, J.M. (2000). Sodium–potassium–chloride cotransport. *Physiol. Rev.* **80**, 211–276.

Rychkov, G.Y., Pusch, M., Roberts, M.L., Jentsch, T.J., and Bretag, A. H. (1998). Permeation and block of the skeletal muscle chloride channel, ClC-1, by foreign anions. *J. Gen. Physiol.* **111**, 653–665.

Sala, F. and Hernández–Cruz, A. (1990). Calcium diffusion modeling in a spherical neuron Relevance of buffering properties. *Biophys. J.* **57**, 313–324.

Sánchez-Vives, M.V. and Gallego, R. (1994). Calcium-dependent chloride current induced by axotomy in rat sympathetic neurons. *J. Physiol. (Lond.)* **475**, 391–400.

Schlichter, R., Bader, C.R., Bertrand, D., Dubois-Dauphin, M., and Bernheim, L. (1989). Expression of substance P and of a Ca^{2+}-activated Cl^- current in quail sensory trigeminal neurons. *Neuroscience* **30**, 585–594.

Schreiber, M., Yuan, A., and Salkoff, L. (1999). Transplantable sites confer calcium sensitivity to BK channels. *Nat. Neurosci.* **2**, 416–421.

Schroeder, B.C., Cheng, T., Jan, Y.N., and Jan, L.Y. (2008). Expression cloning of TMEM16A as a calcium-activated chloride channel subunit. *Cell* **134**, 1019–1029.

Scott, R.H., McGuirk, S.M., and Dolphin, A.C. (1988). Modulation of divalent cation-activated chloride ion currents. *Br. J. Pharmacol.* **94**, 653–662.

Srivastava, A., Romanenko, V.G., Gonzalez-Begne, M., Catalan, M. A., and Melvin, J.E. (2008). A variant of the Ca(2+)-activated Cl channel Best3 is expressed in mouse exocrine glands. *J. Membr. Biol.*.

Staley, K. and Smith, R. (2001). A new form of feedback at the $GABA_A$ receptor. *Nat. Neurosci.* **4**, 674–676.

Stanton, J.B., Goldberg, A.F., Hoppe, G., Marmorstein, L.Y., and Marmorstein, A.D. (2006). Hydrodynamic properties of porcine bestrophin-1 in Triton X-100. *Biochim. Biophys. Acta* **1758**, 241–247.

Stöhr, H., Marquardt, A., Nanda, I., Schmid, M., and Weber, B.H. (2002). Three novel human VMD2-like genes are members of the evolutionary highly conserved RFP-TM family. *Eur. J. Hum. Genet.* **10**, 281–284.

Sun, H., Tsunenari, T., Yau, K.W., and Nathans, J. (2002). The vitelliform macular dystrophy protein defines a new family of chloride channels. *Proc. Natl. Acad. Sci. USA* **99**, 4008–4013.

Sung, K.-W., Kirby, M., McDonald, M.P., Lovinger, D.M., and Delpire, E. (2000). Abnormal $GABA_A$-receptor mediated currents in dorsal root ganglion neurons isolated from Na-K-2Cl cotransporter null mice. *J. Neurosci.* **20**, 7531–7538.

Suzuki, M. (2006). The Drosophila tweety family: molecular candidates for large-conductance Ca^{2+}-activated Cl^- channels. *Exp. Physiol.* **91**, 141–147.

Suzuki, M. and Mizuno, A. (2004). A novel human Cl^- channel family related to Drosophila flightless locus. *J. Biol. Chem.* **279**, 22461–22468.

Taleb, O., Feltz, P., Bossu, J.L., and Feltz, A. (1988). Small-conductance chloride channels activated by calcium on cultured endocrine cells from mammalian pars intermedia. *Pflügers Arch.* **412**, 641–646.

Tavsanli, B.C., Pappu, K.S., Mehta, S.Q., and Mardon, G. (2001). Dbest1, a *Drosophila* homolog of human Bestrophin, is not required for viability or photoreceptor integrity. *Genesis* **31**, 130–136.

Tsunenari, T., Nathans, J., and Yau, K.W. (2006). Ca2+-activated Cl− current from human bestrophin-4 in excised membrane patches. *J. Gen. Physiol.* **127**, 749–754.

Tsunenari, T., Sun, H., Williams, J., Cahill, H., Smallwood, P., Yau, K.W., and Nathans, J. (2003). Structure–function analysis of the bestrophin family of anion channels. *J. Biol. Chem.* **278**, 41114–41125.

Valencia-de Ita, S., Lawand, N.B., Lin, Q., Castañeda–Hernandez, G., and Willis, W.D. (2006). Role of the Na^+-K^+-$2Cl^-$ cotransporter in the development of capsaicin-induced neurogenic inflammation. *J. Neurophysiol.* **95**, 3553–3561.

Voets, T., Droogmans, G., Wissenbach, U., Janssens, A., Flockerzi, V., and Nilius, B. (2004). The principle of temperature-dependent gating in cold- and heat-sensitive TRP channels. *Nature* **430**, 748–754.

Voets, T., Talavera, K., Owsianik, G., and Nilius, B. (2005). Sensing with TRP channels. *Nat. Chem. Biol.* **1**, 85–92.

Wang, Q. and Large, W.A. (1991). Noradrenaline-evoked cation conductance recorded with the nystatin whole-cell method in rabbit portal vein cells. *J. Physiol.* **435**, 21–39.

Wang, Y.X. and Kotlikoff, M.I. (1997a). Inactivation of calcium-activated chloride channels in smooth muscle by calcium/calmodulin-dependent protein kinase. *Proc. Natl. Acad. Sci. USA* **94**, 14918–14923.

Wang, Y.-X. and Kotlikoff, M.I. (1997b). Muscarinic signaling pathway for calcium release and calcium-activated chloride current in smooth muscle. *Am. J. Physiol.* **273**, C509–C519.

Ward, S.M. and Kenyon, J.L. (2000). The spatial relationships between Ca^{2+} channels and Ca^{2+}-activated channels and the function of Ca^{2+}-buffering in avian sensory neurons. *Cell Calcium* **28**, 233–246.

Wei, A., Solaro, C., Lingle, C., and Salkoff, L. (1994). Calcium sensitivity of BK-type K_{Ca} channels determined by a separable domain. *Neuron* **13**, 671–681.

Willis, W.D. (1999). Dorsal root potentials and dorsal root reflexes: a double-edged sword. *Exp. Brain Res.* **124**, 395–421.

Wotring, V.E., Chang, Y., and Weiss, D.S. (1999). Permeability and single channel conductance of human homomeric rho1 GABAC receptors. *J. Physiol.* **521** (Pt 2), 327–336.

Yang, Y.D., Cho, H., Koo, J.Y., Tak, M.H., Cho, Y., Shim, W.S., Park, S.P., Lee, J., Lee, B., Kim, B.M., Raouf, R., Shin, Y.K., and Oh, U. (2008). TMEM16A confers receptor-activated calcium-dependent chloride conductance. *Nature* **455**, 1210–1215.

Young, G.P., Young, J.D., Deshpande, A.K., Goldstein, M., Koide, S.S., and Cohn, Z.A. (1984). A Ca^{2+}-activated channel from Xenopus laevis oocyte membranes reconstituted into planar bilayers. *Proc. Natl. Acad. Sci. USA* **81**, 5155–5159.

Yu, K., Cui, Y., and Hartzell, H.C. (2006). The bestrophin mutation A243V, linked to adult-onset vitelliform macular dystrophy, impairs its chloride channel function. *Invest. Ophthalmol. Vis. Sci.* **47**, 4956–4961.

Yu, K., Qu, Z., Cui, Y., and Hartzell, H.C. (2007). Chloride channel activity of bestrophin mutants associated with mild or late-onset macular degeneration. *Invest. Ophthalmol. Vis. Sci.* **48**, 4694–4705.

Yu, K., Xiao, Q., Cui, G., Lee, A., and Hartzell, H.C. (2008). The best disease-linked Cl− channel hBest1 regulates Ca V 1 (L-type) Ca2+ channels via src-homology-binding domains. *J. Neurosci.* **28**, 5660–5670.

Zhu, M.X. (2007). Understanding the role of voltage gating of polymodal TRP channels. *J. Physiol. (Lond.)* **585**, 321–322.

Zygmunt, A.C. and Gibbons, W.R. (1991). Calcium-activated chloride current in rabbit ventricular myocytes. *Circ. Res.*, 424–437.

CHAPTER

14

$GABA_A$ Receptor Channels

Robert L. Macdonald and Emmanuel J. Botzolakis

OUTLINE

I. INTRODUCTION

Gamma-aminobutyric acid (GABA) is the most abundant inhibitory neurotransmitter in the CNS. Following release from presynaptic vesicles, GABA exerts fast inhibitory effects by interacting with GABA receptors, whose primary function is to hyperpolarize neuronal membranes in mature CNS neurons. GABA receptors are found both presynaptically, where they decrease the likelihood of neurotransmitter release, and postsynaptically, where they decrease the likelihood of neuronal firing. There are two types of GABA receptor, termed $GABA_A$ and $GABA_B$ receptors. $GABA_A$ receptors are fast-activating Cl^- channels from the Cys-loop family of ligand-gated ion channels (Olsen and Tobin, 1990; Macdonald and Olsen, 1994; Rabow et al., 1995). Activation of $GABA_A$ receptors causes membrane hyperpolarization by allowing Cl^- influx, reflecting the relatively low concentration of Cl^- found intracellularly in most adult CNS neurons (see Chapters 7, 8 and 17). $GABA_A$ receptors can also mediate depolarizing responses in most immature CNS neurons and in mature peripheral neurons (e.g. primary afferent neurons), as discussed in Chapters 7 and 22 in this volume. In these cases the intracellular Cl^- concentration is maintained above electrochemical equilibrium and activation of $GABA_A$ receptors causes a depolarizing Cl^- efflux that can be either excitatory (Chapter 7) or inhibitory (Chapter 22). In contrast, $GABA_B$ receptors are members of the

G protein-coupled receptor family and exert slow inhibitory effects either by activating outwardly rectifying voltage-gated potassium channels or by inhibiting voltage-gated calcium channels (Mintz and Bean, 1993; Wagner and Dekin, 1993; Ulrich and Bettler, 2007).

Of these receptor types, GABA$_A$ receptors mediate the majority of GABAergic signaling and are thus most important for maintaining inhibitory tone in the mammalian brain. Indeed, pharmacological blockade of GABA$_A$ receptors rapidly induces seizures in animals (Kapur and Macdonald, 1997; Poulter et al., 1999) and epileptiform activity in neuronal slice preparations (De Deyn and Macdonald, 1990; Schneiderman, 1997). Reduction of GABAergic inhibition due to mutation of GABA$_A$ receptors has been shown to be associated with several types of idiopathic generalized epilepsies, including childhood absence epilepsy, generalized epilepsy with febrile seizures plus, and juvenile myoclonic epilepsy (Macdonald et al., 2006). Enhancement of GABA$_A$ receptor function is the basis of action of a number of antiepileptic, sedating, and intoxicating drugs including benzodiazepines (BZDs), barbiturates, anesthetics, neurosteroids, and ethanol (Olsen and Tobin, 1990; Macdonald and Olsen, 1994; Rabow et al., 1995). Targeting GABA$_A$ receptors has thus been an effective approach for treating neuropsychiatric disorders characterized by neuronal hyperexcitability (e.g. epilepsy, anxiety and insomnia).

II. MOLECULAR BIOLOGY OF GABA$_A$ RECEPTORS

Like other members of the Cys-loop superfamily of ligand-gated ion channels (LGICs), which includes nicotinic acetylcholine receptors (nAChR) (Noda et al., 1983), 5-hydroxytryptamine type 3 serotonin receptors (5-HT$_3$R) (Maricq et al., 1991), zinc-activated channels (ZAC) (Davies et al., 2003) and glycine receptors (GlyR) (Langosch et al., 1988), GABA$_A$ receptors are heteropentameric ion channels assembled from a large family of homologous subunits (Table 14.1) (Schofield et al., 1987; Mamalaki et al., 1989; Nayeem et al., 1994; Knight et al., 1998; Barrera et al., 2008). Molecular cloning studies have identified eight subunit families (α, β, γ, δ, ε, θ, π and ρ), thus providing enormous potential for receptor heterogeneity. Some of these families contain multiple subunit subtypes (α1-6, β1-3, γ1-3, ρ1-3), splice variants (e.g. β2S and β2L; β3-v1 and β3-v2; γ2S and γ2L), and alternatively

TABLE 14.1 GABA$_A$ receptor subunit gene list (from Simon et al., 2004)

		Chromosome		
Subunit	**Gene**	**Human**	**Mouse**	**Rat**
α1	*GABRA1*	5q34	11	10
α2	*GABRA2*	4p12	5	14
α3	*GABRA3*	Xq28	X	X
α4	*GABRA4*	4p12	5	14
α5	*GABRA5*	15q13.2	7	1
α6	*GABRA6*	5q34	11	10
β1	*GABRB1*	4p12	5	14
β2	*GABRB2*	5q34	11	10
β3	*GABRB3*	15q13.2	7	1
γ1	*GABRG1*	4p12	5	14
γ2	*GABRG2*	5q34	11	10
γ3	*GABRG3*	15q13.2	7	1
δ	*GARBD*	1p36.3	4	5
ε	*GABRE*	Xq28	X	X
θ	*GABRQ*	Xq28	X	X
π	*GABRP*	5q35.1	11	10
ρ1	*GABRR1*	6q15	4	5
ρ2	*GABRR2*	6q15	4	5
ρ3	*GABRR3*	3q12.1	16	11

edited transcripts (e.g. α3I and α3M), further increasing the potential for heterogeneity (Levitan et al., 1988; Pritchett et al., 1989; Shivers et al., 1989; Ymer et al., 1989a, b; Whiting et al., 1990; Herb et al., 1992; Kirkness and Fraser, 1993; Hedblom and Kirkness, 1997; Davies et al., 1997; Bonnert et al., 1999; Simon et al., 2004; Ohlson et al., 2007). Sequence homology is ~60–80% among members of the same $GABA_A$ receptor family, ~20–40% among members of different families, and ~10–20% among members of the $GABA_A$ receptor and other Cys-loop receptor families (Olsen and Tobin, 1990).

Most genes encoding human $GABA_A$ receptor subunits are found in tight clusters comprising 1 or 2 α subunits, 1 β subunit and 1 γ subunit gene (in the order β-α-[α]-γ). Chromosome 4 contains α2, α4, β1 and γ1 subunit genes; chromosome 5 contains α1, α6, β2 and γ2 subunit genes; and chromosome 15 contains α5, β3 and γ3 subunit genes (Buckle et al., 1989; Wilcox et al., 1992; Russek and Farb, 1994; Hicks et al., 1994; McLean et al., 1995; Glatt et al., 1997; Simon et al., 2004). The gene encoding the α3 subunit is found on the X chromosome near the genes encoding θ and ε subunits (Bell et al., 1989; Wilke et al., 1997), which are most homologous to β and γ subunits, respectively (Wilke et al., 1997; Bonnert et al., 1999). Thus, it is believed that the incredible diversity of $GABA_A$ receptor subunits arose following duplication of a progenitor α-β-γ gene cluster (Russek and Farb, 1994; Hicks et al., 1994; McLean et al., 1995; Greger et al., 1995; Wilke et al., 1997; Darlison et al., 2005). The exceptions are the relatively isolated genes encoding the δ (chromosome 1) (Sommer et al., 1990), π (chromosome 5) (Bailey et al., 1999a) and ρ (chromosomes 3 and 6) (Cutting et al., 1992; Bailey et al., 1999b; Simon et al., 2004) subunits, which given their lower degree of homology to other subunit subtypes, likely resulted from earlier gene duplication and transposition (Fig. 14.1) (Simon et al., 2004).

While the functional significance of gene clustering remains uncertain, it may be important for coordinating transcription, and, consequently, for limiting receptor heterogeneity (Russek and Farb, 1994; McLean et al., 1995; Wilke et al., 1997; Darlison et al., 2005). Indeed, it is remarkable that the most abundant $GABA_A$ receptor isoforms in the cortex (α1β2γ2) (McKernan and Whiting, 1996; Pirker et al., 2000) and cerebellum (α6β2γ2) (McKernan and Whiting, 1996; Pirker et al., 2000) are composed of subunits encoded by genes clustered on chromosome 5, and, similarly, that brain regions known to express the α3 subunit also express the ε and θ subunits, which form the X chromosome cluster (Fritschy et al., 1992; Moragues et al., 2000). Additional support for this hypothesis

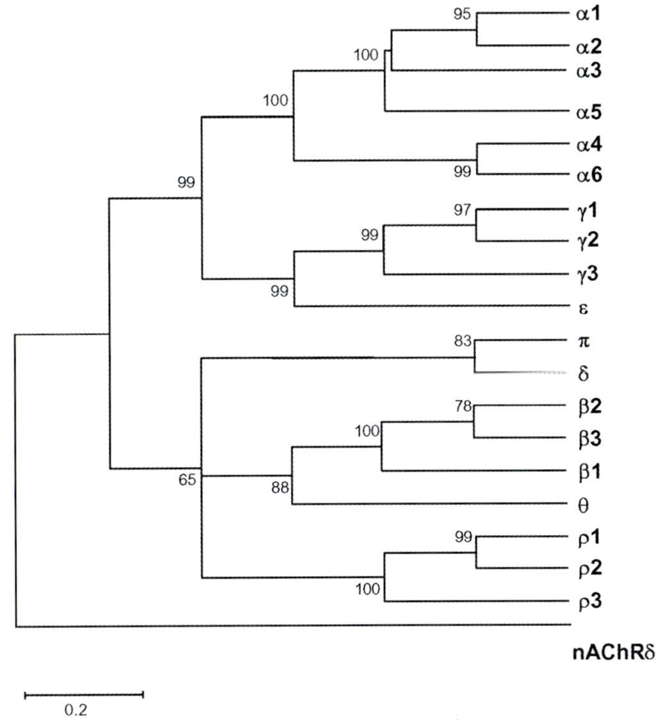

FIGURE 14.1 Dendrogram showing the sequence relationships of the 19 known $GABA_A$ receptor genes. The length of horizontal branches connecting any two $GABA_A$ receptor subunits represents the fractional divergence in their amino acid sequence. Sequences used were those of the mature protein after signal peptide removal. The scale bar corresponds to 20% sequence divergence. Numbers at the junctions are bootstrap values for 100,000 replications. (Reproduced from Simon et al., 2004, with permission.)

comes from the observation that genomic disruption of α6 subunit expression leads to decreased transcript levels for co-clustered subunits (Uusi-Oukari et al., 2000). There are, however, many examples of receptor isoforms (see below) comprised of subunits encoded by non-clustered genes (McKernan and Whiting, 1996), reflecting the fact that individual neurons can simultaneously express as many as ten different subunit subtypes (Brooks-Kayal et al., 2001).

III. ASSEMBLY OF $GABA_A$ RECEPTORS

$GABA_A$ receptors, like other Cys-loop receptors, are assembled from homologous subunits (Fig. 14.2A) as pseudo-symmetrical heteropentamers (Fig. 14.2B). Assembly occurs in the endoplasmic reticulum (ER) and is thought to be slow and relatively inefficient (Green and Millar, 1995), involving multiple assembly intermediates and relying heavily on luminal and cytoplasmic molecular chaperones (Bollan et al., 2003; Wanamaker and Green, 2007; Sarto-Jackson and

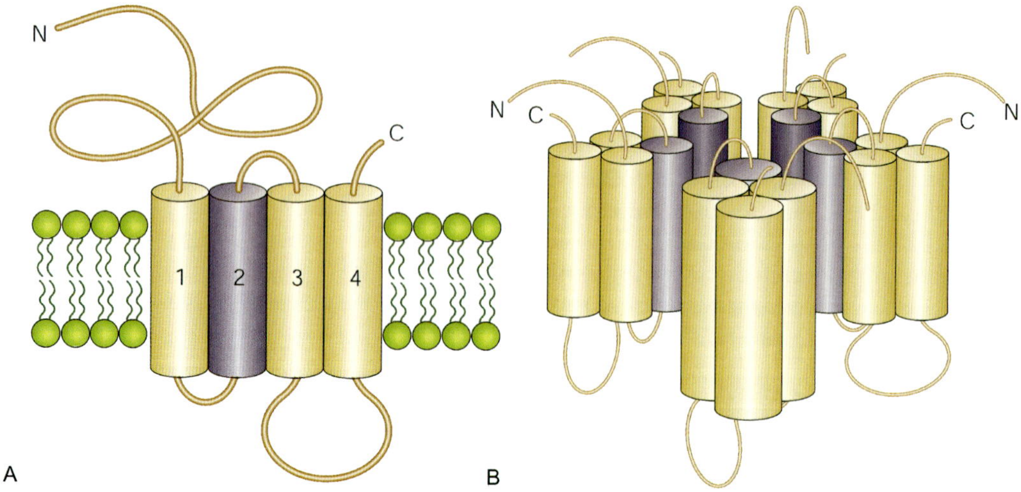

FIGURE 14.2 Structure of Cys-loop receptor ion channels. **A.** Individual subunits contain four hydrophobic transmembrane (TM) domains. The large N-terminal domain is located extracellularly and is believed to contain neurotransmitter and modulator binding sites. The C-terminal domain is also located extracellularly; however, it is typically only a few amino acids long. The intracellular domain between TM3 and TM4 is the most divergent region and contains numerous consensus sites for the action of both serine/threonine and tyrosine protein kinases. **B.** Receptors are assembled as pseudo-symmetrical pentamers. The ion channel is formed by a ring of TM2 domains (blue). This is surrounded by a second ring composed of TM1 and TM3 domains, which is in turn surrounded by a third ring composed of TM4 domains. (Reproduced from Moss and Smart, 2001, with permission.)

Sieghart, 2008). Luminal chaperones include calnexin, which recognizes immature glycans (Helenius and Aebi, 2004); immunoglobulin heavy-chain-binding protein (BiP), which recognizes exposed hydrophobic residues (Gething, 1999); and protein disulfide isomerase, which catalyzes the formation of appropriate disulfide bonds (Maattanen et al., 2006). Working together, these proteins facilitate folding and oligomerization of $GABA_A$ receptor subunits and provide a stringent quality control system (Bollan et al., 2003; Ellgaard and Frickel, 2003). Although less is known about the role of cytoplasmic chaperones in $GABA_A$ receptor assembly, their involvement has been inferred based on the known interaction between $\alpha4\beta2$ nAChRs and 14-3-3η (Jeanclos et al., 2001; Exley et al., 2006), a member of a large family of cytoplasmic regulatory, scaffolding and adaptor proteins (Wang and Shakes, 1996).

In addition to these cellular requirements, there are also specific subunit requirements for $GABA_A$ receptor assembly that serve to limit receptor heterogeneity (Angelotti and Macdonald, 1993). While all subunit combinations appear capable of oligomerization (Connolly et al., 1996), sucrose density centrifugation studies indicate that only a small subset can form pentamers, a prerequisite for receptor function and surface expression (lower molecular weight oligomers are retained in the ER and degraded) (Connolly et al., 1996, 1999; Gorrie et al., 1997; Taylor et al., 2000; Klausberger et al., 2001; Bollan et al., 2003; Sarto-Jackson and Sieghart, 2008; Lo et al., 2008). For example, when recombinant $\alpha1$, $\beta2$ or $\gamma2$ subunits were expressed individually in heterologous cells, primarily monomers and dimers were formed (Gorrie et al., 1997; Connolly et al., 1999; Taylor et al., 1999; Sarto-Jackson and Sieghart, 2008; Lo et al., 2008). Similarly, coexpression of either $\alpha1$ or $\beta3$ subunits with $\gamma2$ subunits yielded primarily dimers and trimers (Tretter et al., 1997; Sarto-Jackson and Sieghart, 2008). In contrast, coexpression of $\alpha1$ and $\beta2/3$ subunits formed pentamers, as did coexpression of $\alpha1$, $\beta2/3$, and $\gamma2$ subunits (Gorrie et al., 1997; Tretter et al., 1997; Connolly et al., 1999; Taylor et al., 2000; Klausberger et al., 2001; Sarto-Jackson and Sieghart, 2008; Lo et al., 2008). Thus, the presence of both α and β subunits appears to be required for pentameric assembly (Angelotti and Macdonald, 1993). Known exceptions are the $\beta3$ and $\rho1$ subunits, which efficiently form homopentameric receptors when expressed individually (Taylor et al., 1999; Pan et al., 2006).

Interestingly, ternary receptors appear to assemble with higher efficiency than binary receptors, suggesting existence of an assembly hierarchy within the subunit family that further limits surface receptor heterogeneity. For example, coexpression of α and β subunits with either γ, δ, ε, or π subunits yields a relatively homogeneous receptor population with kinetic properties distinct from receptors formed following coexpression of only α and β subunits (Angelotti and Macdonald, 1993; Saxena and Macdonald, 1994; Fisher and Macdonald, 1997b; Haas and Macdonald, 1999; Neelands and Macdonald, 1999; Neelands et al., 1999;

Lagrange et al., 2007). Similarly, while the functional and pharmacological signature of αβ receptors has been identified in a subset of hippocampal neurons (Mortensen and Smart, 2006), the overwhelming majority of native receptors are composed of ternary subunit combinations (McKernan and Whiting, 1996; Olsen and Sieghart, 2008). The most widely expressed are αβγ and αβδ receptors, though αβε, αβθ and αβπ receptors may be important in certain brain regions (see below). The α1β2γ2 isoform is the most abundant, accounting for ~40% of all $GABA_A$ receptors (McKernan and Whiting, 1996; Olsen and Sieghart, 2008).

For αβγ receptors, a subunit stoichiometry of 2α:2β:1γ (Chang et al., 1996; Tretter et al., 1997; Farrar et al., 1999) and a subunit arrangement of γ-β-α-β-α (counterclockwise when viewed top-down from the synaptic cleft) (Baumann et al., 2001, 2002; Baur et al., 2006) has been proposed (Fig. 14.3A). Atomic force microscopy studies have suggested a similar stoichiometry and arrangement for αβδ receptors, with the δ subunit taking the place of the γ subunit in the pentamer (i.e. δ-β-α-β-α) (Barrera et al., 2008). It should be noted, however, that there is evidence for alternate patterns of assembly. For example, multiple γ subunit subtypes have been identified in a subset of native (Quirk et al., 1994; Khan et al., 1994; Benke et al., 1996) and recombinant (Backus et al., 1993) receptors. While it is unclear if multiple δ subunits can also be incorporated in the same pentamer, recent evidence using concatenated subunits suggested at least two additional arrangements are possible: δ-α-β-β-α (GABA gated) and δ-α-β-α-β (THDOC gated) (Kaur et al., 2009). The ε subunit, which is highly homologous to the γ subunit (Wilke et al., 1997), appears particularly promiscuous, as it can not only form αβε receptors with multiple stoichiometries (2α:2β:1ε and 2α:1β:2ε), but can also coassemble with α, β and γ subunits to form αβγε receptors (Neelands et al., 1999; Davies et al., 2001; Wagner et al., 2005; Jones and Henderson, 2007; Bollan et al., 2008). The θ and π subunits, which are most homologous to β and δ subunits, respectively (Simon et al., 2004), exhibit similar promiscuity, having been shown to assemble as both ternary (αβθ and αβπ) and quaternary (αβγθ and αβγπ) receptors (Bonnert et al., 1999; Neelands and Macdonald, 1999).

IV. SPATIAL AND TEMPORAL REGULATION OF $GABA_A$ RECEPTORS

Expression of $GABA_A$ receptors is tightly regulated both spatially and temporally, with individual subunits having distinct, but often overlapping, distributions (Fig. 14.4). Based on *in situ* hybridization (Zhang et al., 1991; Wisden et al., 1992; Laurie et al., 1992a, b) and immunohistochemical (Fritschy et al., 1994; Fritschy and Mohler, 1995; Pirker et al., 2000) studies, α1 subunits appear to be widely expressed in the adult, but not the developing, brain. In contrast, α2 subunits are widely expressed during development (though not in cerebellum), but are limited to cortex, hippocampus, basal ganglia, amygdala, certain thalamic nuclei, and hypothalamus in the adult. α3 subunits have a similar spatiotemporal pattern, with high expression in most brain areas during development but limited expression in the adult, when it is found mainly in cortex and certain thalamic nuclei (most notably in the nucleus reticularis). Of note, α2 and α3 subunits are the predominant α subtypes in brainstem and spinal cord. α4 subunits have little expression during development; in the adult, their highest levels are found in cortex, hippocampus and thalamus. α5 and α6 subunits have the most restricted distributions of all α subtypes, being localized mainly to hippocampus and cerebellum, respectively. α5 subunits, however, are expressed highly during development (except in cerebellum, where α3 subunits predominate). Among β subunits, β2 subunits are most widely expressed in the adult, except in hippocampus, where β1 and β3 subunits predominate (particularly in the dentate gyrus). Conversely, β3 subunits are most widely expressed during development, though β2 subunits are also expressed in cortex. γ2 subunits are the most widely expressed of the γ subunits, both in developing and adult brain. γ1 subunits are expressed in basal ganglia, forebrain, amygdala and cerebellum, while γ3 subunits are expressed in cortex, basal ganglia and certain thalamic nuclei (Herb et al., 1992). δ subunits are virtually absent during early development, but in the adult, are found in cortex, hippocampus, basal ganglia, thalamus and cerebellum. Expression of ε subunits was initially undetectable in whole brain mRNA (Wilke et al., 1997), but upon closer evaluation, was identified in amygdala, thalamus and the subthalamic nucleus (Davies et al., 1997). θ subunits are found at low levels in cortex, intermediate levels in hypothalamus and high levels in amygdala, hippocampus and brainstem (Bonnert et al., 1999). Although π subunits are found mainly in peripheral tissues, low levels have been identified in hippocampus and temporal cortex (Hedblom and Kirkness, 1997). ρ subunits are expressed in retina on bipolar cells (Enz et al., 1996), cerebellum during development and adult brainstem and spinal cord (Boue-Grabot et al., 1998; Lopez-Chavez et al., 2005; Mejia et al., 2008).

Using information provided by mRNA colocalization studies, testable hypotheses were generated

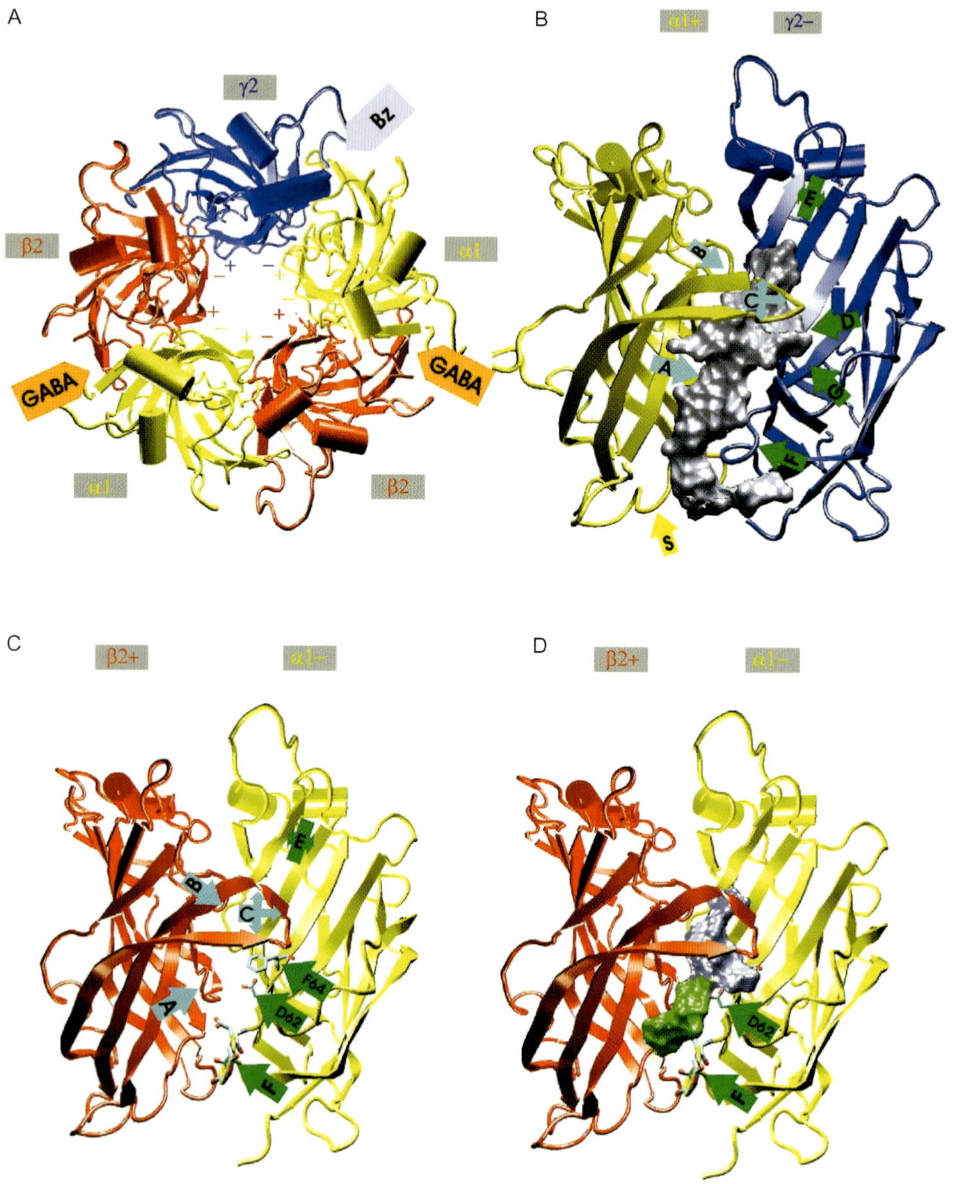

FIGURE 14.3 Structure of $GABA_A$ receptor extracellular domains based on homology modeling to the nAChR. **A.** The putative arrangement of α1β2γ2 $GABA_A$ receptors, as viewed from the synaptic cleft. The + (plus) and − (minus) sides of each subunit are labeled on the inner circumference of the channel. The subunit interfaces at which benzodiazepines (Bz) and GABA bind are labeled with lilac and orange arrows, respectively. **B.** The benzodiazepine binding pocket, composed of the interface between the extracellular domains of the α and γ subunits, as viewed from the side. The buried volume provided by the α^+/γ^- interface is shown in space filling representation. Binding site "loops" are labeled A through G. The eponymous Cys-loop (S) is labeled with a yellow arrow. **C.** The GABA binding pocket. The interface between α and β subunits, as viewed from the side. The muscimol photolabel target, "loop D" α1F64, and "loop D" D62, a residue whose accessibility has been shown to change upon antagonist binding, but not upon GABA binding (Holden and Czajkowski, 2002). Residues 182–184 in loop F of the α subunit, which change their accessibility upon agonist and antagonist action, are also shown (Newell and Czajkowski, 2003). **D.** The ligand docking area is shown in space filling representation. The blue, upper portion is confined by loops A, B, C, D and E and presumably contains the agonist binding subsite. The green, membrane-near portion in direct contact with loop F residues and the α1D62 residue is thought to be occupied by some antagonists. (Reproduced from Ernst et al., 2003, with permission.)

regarding possible subunit combinations that might exist *in vivo* (as opposed to what subunits could theoretically assemble *in vitro*). For example, transcripts encoding α1, β2 and γ2 subunits were coexpressed throughout the brain, while transcripts encoding α4, β2 and δ subunits were coexpressed in thalamus and hippocampus, suggesting they had a high probability of coassembly in those areas. However, many brain regions were found to simultaneously express a variety of subunit subtypes (dentate gyrus, for example,

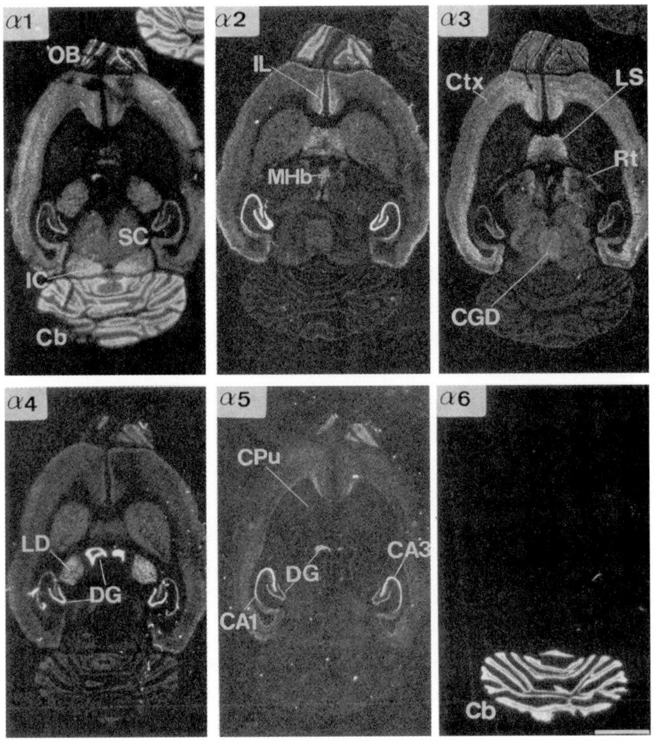

FIGURE 14.4 Anatomical distribution of GABA$_A$ receptor subunits. *In situ* hybridization studies illustrate the distribution of α subunit mRNAs in horizontal rat brain sections. OB, olfactory bulb; SC, superior colliculus; IC, inferior colliculus; Cb, cerebellum; IL, infralimbic cortex; MHb, medial habenular nucleus; Ctx, neocortex; LS, lateral septum; Rt, reticular thalamic nucleus; CGD, central dray, dorsal; LD, laterodorsal thalamic nucleus; DG, dentate gyrus; CPu, caudate, putamen; CA1, field 1 of Ammon's horn; CA3, field 3 of Ammon's horn; scale bar, 4.4 mm. (Reproduced from Wisden et al., 1992, with permission.)

expresses α1-5, β1-3, γ1-3 and δ subunits) (Wisden et al., 1992), making the potential for receptor heterogeneity tremendous. As a result, early estimates put the total number of different isoforms in the adult brain at nearly 800 (Barnard et al., 1998)! To address this issue more directly, immunoprecipitation and immunoaffinity chromatography studies were performed using subunit-specific antibodies (Quirk et al., 1995; Jechlinger et al., 1998; Sur et al., 1999; Sieghart and Sperk, 2002). These studies, combined with detailed electrophysiological and pharmacological profiling of native receptors, demonstrated that receptor heterogeneity *in vivo* is surprisingly limited (Sieghart and Sperk, 2002; Olsen and Sieghart, 2008). In total, only 11 receptor isoforms (α1β2γ2, α2β2γ2, α3β2γ2, α4β2γ2, α4β2δ, α4β3δ, α5βxγ2, α6βxγ2, α6β2δ, α6β3δ and ρx) were identified "unequivocally" (Olsen and Sieghart, 2008). An additional eight isoforms (α1β3γ2, α1βxδ, α5β3γ2, αxβ1γ, αxβ1δ, αxβx, α1α6βxγ and α1α6βxδx) were thought to exist with "high probability", and the existence of an additional nine isoforms (αxβxγ1, αxβxγ3, αxβxε, αxβxθ, αxβxπ, αxαyβxγ2, ρ1, ρ2 and ρ3) were considered "tentative" (Olsen and Sieghart, 2008). Note that some of the listed isoforms contain multiple α subunit subtypes, which confer unique functional and pharmacological properties upon receptors (Duggan et al., 1991; Luddens et al., 1991). The same is thought to be true for β subunit subtypes, but evidence for this is more equivocal (Li and De Blas, 1997; Olsen and Sieghart, 2008).

Of these, α1βxγ2, α1βxδ, α2βxγ2, α3βxγ2 and α4βxδ isoforms are the predominant cortical receptors; α1βxγ2, α4βxγ2 and α4βxδ isoforms are the predominant thalamic receptors (α3βxγ2 is also important in the nucleus reticularis); α1βxγ2, α2βxγ2 and α5βxγ2 isoforms are the predominant hypothalamic receptors; α1βxγ2, α1βxδ, α2βxγ2, α4βxδ and α5β3γ2 isoforms are the predominant hippocampal receptors; and α1βxγ2, α1βxδ, α1α6βxγ2, α1α6βxδ, α6βxγ2 and α6βxδ isoforms are the predominant cerebellar receptors (Wisden et al., 1992; McKernan and Whiting, 1996; Pirker et al., 2000; Brickley et al., 2001; Caraiscos et al., 2004; Farrant and Nusser, 2005; Glykys et al., 2007). Interestingly, subunit composition is an important determinant of receptor subcellular localization. For example, receptors containing γ subunits are typically found in synapses, while those containing δ subunits are typically found in peri- and/or extrasynaptic compartments (Nusser et al., 1998; Brickley et al., 2001; Farrant and Nusser, 2005). The known exceptions are α4 and α5 subunit-containing αβγ receptors, which are thought to be primarily peri- and extrasynaptic (Caraiscos et al., 2004; Glykys et al., 2007). Similarly, receptors assembled without γ and δ subunits (i.e. αβ receptors) are localized extrasynaptically (Mortensen and Smart, 2006). Interestingly, β subunit subtype was recently shown to be an important determinant of subcellular localization in dentate gyrus (Herd et al., 2008).

V. STRUCTURE OF GABA$_A$ RECEPTORS

Although a crystal structure of the GABA$_A$ receptor is unavailable, several experimental approaches have provided important insight into its tertiary and quaternary structure. These include imaging by electron and atomic force microscopy (Nayeem et al., 1994; Barrera et al., 2008), scanning for accessibility of substituted cysteine residues (Bera et al., 2002; Akabas, 2004; Jansen and Akabas, 2006), site-directed mutagenesis (Kash et al., 2003; Olsen et al.,

2004; Filippova et al., 2004; Jones-Davis et al., 2005; Padgett et al., 2007; Hanson and Czajkowski, 2008), photoaffinity labeling (Olsen et al., 2004; Li et al., 2006) and homology modeling based on high resolution structures of the *Torpedo marmorata* nAChR and its soluble molluscan cousin, the acetylcholine binding protein (AChBP) (Unwin, 1993; Brejc et al., 2001; Cromer et al., 2002; Miyazawa et al., 2003; Ernst et al., 2003; Trudell and Bertaccini, 2004; Unwin, 2005; O'Mara et al., 2005; Campagna-Slater and Weaver, 2007). The recent crystallization of the nAChR extracellular domain (Dellisanti et al., 2007) and a prokaryotic Cys-loop receptor (Hilf and Dutzler, 2008) has also provided important validation for the predicted $GABA_A$ receptor structure.

Hydropathy analysis suggests that individual $GABA_A$ receptor subunits are composed of a large (~200 amino acids) extracellular N-terminal domain, followed by four α-helical transmembrane domains (M1, M2, M3 and M4) that are connected by cytoplasmic (M1–M2 and M3–M4) and extracellular (M2–M3) linkers (Fig. 14.2A) (Olsen and Tobin, 1990; Macdonald and Olsen, 1994; Smith and Olsen, 1995). When viewed from the synaptic cleft (i.e. perpendicular to the plasma membrane), assembled $GABA_A$ receptors have a circular structure (~80 Å in diameter), with individual subunits arranged pseudo-symmetrically around a central ion-conducting pore (~20 Å in diameter) (Fig. 14.2B and 14.3A). The extracellular domain is composed of two sets of β-sheets joined by a highly conserved disulfide bridge (the Cys-loop), for which the receptor superfamily is named (Unwin, 2005). These β-sheets contain seven major "loops" that are important for agonist binding. The "principal" side of the extracellular domain (also referred to as the "+" side) is composed of the A, B and C loops, while the "complementary" side (the "−" side) is composed of the D, E, F and G loops (Ernst et al., 2003) (Fig. 14.3B). GABA binding occurs at the interface between the principal side of the β subunit and the complementary side of the α subunit (Fig. 14.3A and 14.3C), while BZDs bind at the interface between the principal side of the α subunit and the complementary side of the γ subunit (Fig. 14.3A and 14.3B). Interestingly, the BZD and GABA binding sites are highly homologous, suggesting that the BZD binding site evolved from a former GABA binding site (Galzi and Changeux, 1995). An endogenous ligand for the BZD binding site, however, has yet to be identified. Of note, in addition to containing binding sites for a variety of agonists, antagonists and allosteric modulators (Macdonald and Olsen, 1994; Olsen et al., 2004), the extracellular subunit domain is the primary determinant of receptor assembly (Hackam et al., 1997, 1998; Taylor et al., 1999; Enz and Cutting, 1999; Taylor et al., 2000; Klausberger et al., 2000).

The M1 domain contains residues important for determining receptor kinetic properties and sensitivity to allosteric modulators such as BZDs and neurosteroids (Bianchi et al., 2001; Bianchi and Macdonald, 2002; Engblom et al., 2002; Jones-Davis et al., 2005; Keramidas et al., 2006; Akk et al., 2008). The M2 domain lines the transmembrane portion of the pore (Fig. 14.2), determines ion selectivity and channel conductance, and contains the channel gate (Giraudat et al., 1986; Imoto et al., 1986; Akabas et al., 1994; Xu and Akabas, 1996; Serafini et al., 2000; Wilkins et al., 2002; Keramidas et al., 2004; Gonzales et al., 2008). A leucine residue at the 9′ position is highly conserved among members of the Cys-loop family and has therefore been the focus of active investigation. This residue is thought to line the narrowest part of the pore, and, consequently, mutation of this residue has profound effects on channel function (Chang and Weiss, 1998; Bianchi and Macdonald, 2001a; Scheller and Forman, 2002; Bianchi et al., 2007). The M3 and M4 domains are packed somewhat loosely with respect to the M1 and M2 domains, creating cavities where allosteric modulators such as volatile and intravenous anesthetics bind (Schofield and Harrison, 2005; Richardson et al., 2007). The lipid-facing M4 domain is farthest from the channel pore and is thought mainly to provide subunit stability in the plasma membrane. In nAChRs, for example, the M4 domain can be replaced by an unrelated transmembrane domain without abolishing channel function, whereas replacement of any other transmembrane domain renders channels nonfunctional (Tobimatsu et al., 1987).

The amino acid linkers connecting the transmembrane domains are also important determinants of receptor function. The intracellular M1–M2 linker extends the channel pore and contributes to charge selectivity (Jensen et al., 2002; Filippova et al., 2004; Wotring and Weiss, 2008). The extracellular M2–M3 linker is responsible for transducing ligand binding in the N-terminal domain to channel gating in the transmembrane domain. Based on charge swapping mutations, this appears to involve a salt bridge between loop 2 (and possibly loop 7) in the N-terminal domain and a highly conserved lysine residue in the M2–M3 linker (Kash et al., 2003). Of note, mutation of this conserved residue in the γ2 subunit of the $GABA_A$ receptor (K289M) has been associated with generalized epilepsy with febrile seizures plus (Baulac et al., 2001), and mutations of the same residue in the α1 subunit of the GlyR (K276E and K276Q) have been associated with hyperekplexia (Langosch et al., 1994;

Harvey et al., 2008). In each case, mutation of the conserved lysine substantially inhibited channel function (Bianchi et al., 2002b), consistent with a role for this domain in transducing binding to gating. Interestingly, the mechanisms underlying the coupling of binding and gating appear to be highly conserved within the Cys-loop superfamily. Functional chimeras have been generated between nAChR subunit extracellular domains and 5-HT_3R subunit transmembrane domains and, similarly, between $GABA_A$ receptors and GlyRs (Eisele et al., 1993; Mihic et al., 1997). These chimeric receptors are gated by ligands specific for the extracellular domain, but have functional properties that depend on the identity of the transmembrane domain. The large intracellular M3–M4 linker (~100–200 amino acids) is also thought to play a role in channel function, having recently been shown in 5-HT_3Rs and nAChRs to contain a motif (the "MA" stretch) that is an important determinant of single-channel conductance (Kelley et al., 2003; Peters et al., 2004; Hales et al., 2006).

The M3–M4 linker, however, has a variety of other functions. Indeed, this domain serves as the primary interface between the $GABA_A$ receptor and the intracellular milieu. It contains phosphorylation sites for PKA, PKC and PTK (Brandon et al., 2002; Kittler and Moss, 2003; Jacob et al., 2008), and protein binding domains that are important for receptor clustering, sorting, targeting and trafficking (Moss and Smart, 2001; Kittler and Moss, 2003; Chen and Olsen, 2007; Jacob et al., 2008). Known interacting proteins include AP2, which promotes endocytosis by recruiting receptors into clathrin-coated pits (Kittler et al., 2005; Smith et al., 2008); GRIF1, which regulates kinesin-mediated vesicular transport (Beck et al., 2002; Brickley et al., 2005); gephyrin, which clusters and stabilizes receptors in synapses (Essrich et al., 1998; Kneussel and Betz, 2000; Moss and Smart, 2001; Tretter et al., 2008); BIG2, which facilitates vesicular transport (Charych et al., 2004); GABARAP, which promotes receptor trafficking from the Golgi to the cell surface (Wang et al., 1999; Chen et al., 2000; Moss and Smart, 2001); PRIPs, which modulate phosphatase activity (Kanematsu and Hirata, 2002; Kanematsu et al., 2007); and PLIC1, which stabilizes receptors on the cell surface by preventing ubiquitination-mediated degradation (Fig.14.5) (Bedford et al., 2001). The M3–M4 linker is also thought to mediate the direct interaction between $GABA_A$ receptors and other neurotransmitter receptors such as the $GABA_B$, D_5 and P_2X receptors (Liu et al., 2000; Balasubramanian et al., 2004; Boue-Grabot et al., 2004). In addition, recent evidence suggests that motifs in the M3–M4 linker are required for complete pentameric assembly (Lo et al., 2008).

VI. BIOPHYSICAL AND KINETIC PROPERTIES OF $GABA_A$ RECEPTOR CHANNELS

$GABA_A$ receptor channels are relatively impermeant to cations (permeability ratio of K^+ to $Cl^- < 0.05$) but highly permeable to anions such as Cl^- and HCO_3^- (Bormann et al., 1987). The channel is considerably more permeable to Cl^- than to HCO_3^- (ratio of Cl^-:HCO_3^- permeability is ~5:1), and consequently, the majority of charge transfer that follows channel activation *in vivo* is Cl^- mediated. However, HCO_3^- may carry some of the current when the Cl^- gradient is collapsed (Grover et al., 1993; Perkins and Wong, 1997; Dallwig et al., 1999; Kim et al., 2009), as may be the case in immature neurons and in a subset of mature neurons (Chapters 7 and 22). Interestingly, GABA-mediated HCO_3^- efflux has been shown to trigger Ca^{2+} influx (Kulik et al., 2000; Chavas et al., 2004), which may contribute to the ability of GABA to serve as a trophic signal in maturing networks (Ben-Ari, 2002; Owens and Kriegstein, 2002).

The canonical $\alpha1\beta x\gamma2$ $GABA_A$ receptor opens to a main conductance level of 26–30 pS and to several less frequent subconductance levels (Bormann et al., 1987; Macdonald et al., 1989; Twyman et al., 1990; Newland et al., 1991; Fisher and Macdonald, 1997b; Haas and Macdonald, 1999; Burkat et al., 2001). To fully activate the receptor, binding of two molecules of GABA (to independent sites) is required. Once bound with GABA, the channel exhibits complex patterns of activity. Detailed kinetic analysis of native (Macdonald et al., 1989; Twyman et al., 1990; Newland et al., 1991) and recombinant (Fisher and Macdonald, 1997b; Haas and Macdonald, 1999; Burkat et al., 2001) receptors demonstrated the existence of three open states and at least five closed states (based on exponential fitting of open and closed time distributions). When activated by a saturating concentration of GABA (1 mM), open times of 0.3, 2.0 and 3.5 ms (referred to as O1, O2 and O3) were observed with relative amplitudes of 24, 48% and 28%, respectively, the overall mean open time being 2.1 ms (Haas and Macdonald, 1999) (Fig. 14.6). The overall mean closed time was 21.0 ms (with individual components ranging from 0.2 to 990 ms in duration), thus yielding an open probability of ~0.1 when activated by 1 mM GABA (note that this may be an overestimation, as recordings are rarely obtained from patches containing a "single" channel).

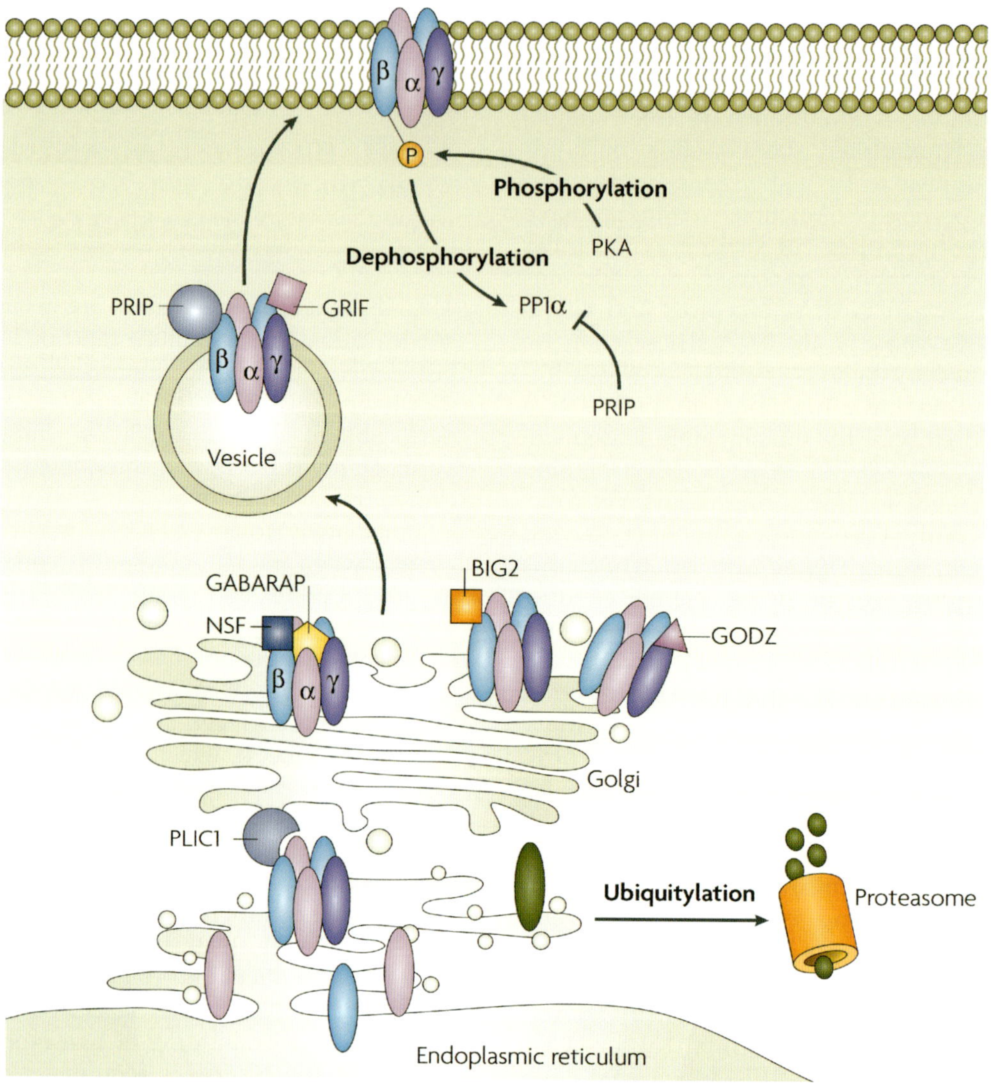

FIGURE 14.5 $GABA_A$ receptor trafficking. $GABA_A$ receptor subunits are synthesized and assembled into pentameric structures in the endoplasmic reticulum (ER). The fate of $GABA_A$ receptor subunits can be modulated by ubiquitylation and subsequent ER-associated degradation by the proteasome. Ubiquitylated subunits can be modulated by their association with the ubiquitin-like protein PLIC1, which facilitates receptor accumulation at the synapse by preventing degradation. Trafficking to the synapse is also facilitated by $GABA_A$ receptor-associated protein (GABARAP), which associates with the γ2 subunit and aids in trafficking from the Golgi network to the plasma membrane. N-ethylmaleimide-sensitive factor (NSF) and brefeldin-A-inhibited GDP/GTP exchange factor 2 (BIG2) are also localized to the Golgi network, where they bind to β subunits and modulate trafficking. Palmitoylation of γ subunits occurs in the Golgi apparatus as a result of an association with the palmitoyltransferase Golgi-specific DHHC zinc-finger-domain protein (GODZ), and is a critical step in the delivery of receptors to the plasma membrane. $GABA_A$ receptor interacting factor proteins (GRIFs) also have a role in membrane trafficking, as do phospholipase-C-related catalytically inactive proteins (PRIPs), which have essential roles in trafficking and modulating the phosphorylation state of $GABA_A$ receptors. PKA, protein kinase A; PP1α, protein phosphatase 1α. (Reproduced from Jacob et al., 2008, with permission.)

Channel openings tended to occur in bursts (a series of openings separated by brief closures), of which there were at least three types, each containing a single type of opening (i.e. either O1 or O2 or O3). This indicated that receptors could not transition directly from one open state to another. Interestingly, activating receptors with progressively lower GABA concentrations increased the relative contribution of O1 openings at the expense of both O2 and O3 openings, without altering individual open times (though the overall mean open time was decreased since the lifetime of O1 is short). Considering that receptors have negligible spontaneous activity, these data have been interpreted to mean that O1 and O2/O3 represent receptor sojourns in mono- and di-liganded open states, respectively (Twyman et al., 1990). What

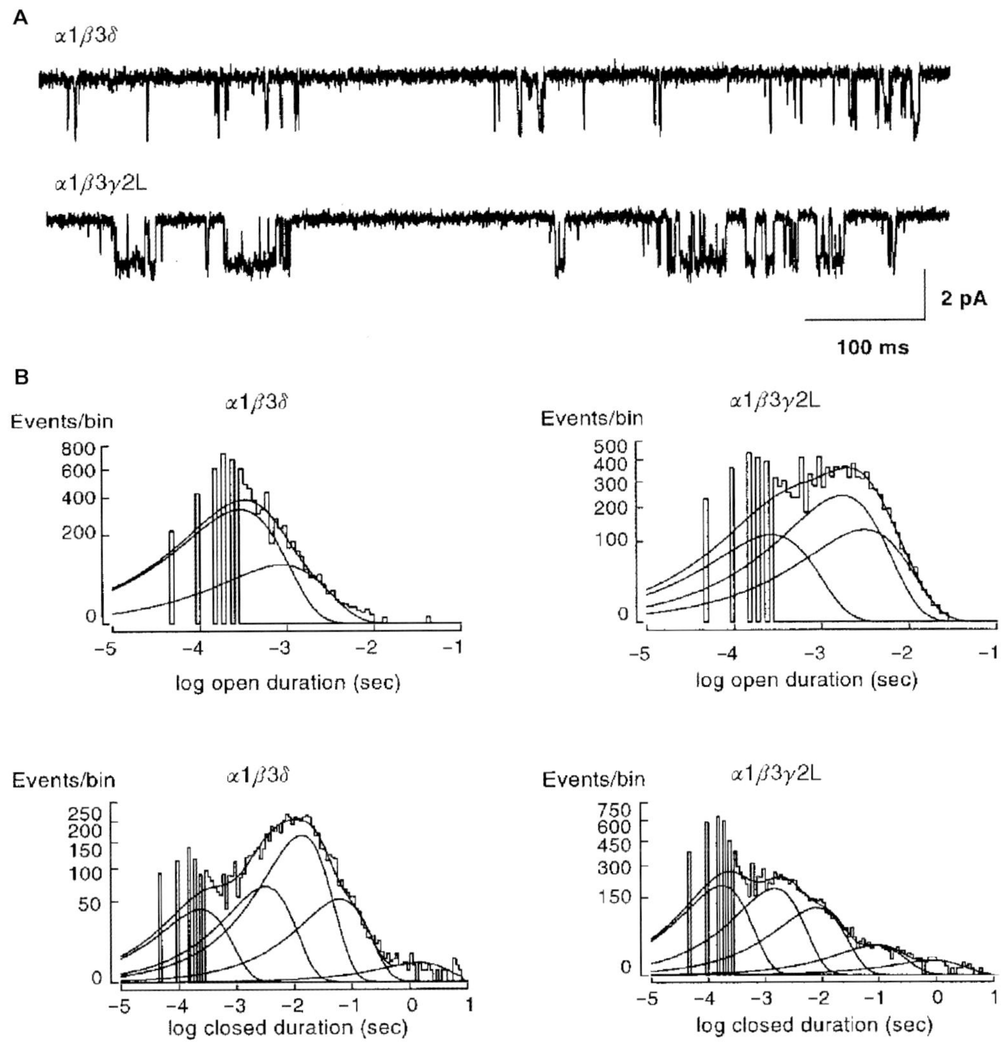

FIGURE 14.6 Single-channel properties of α1β3γ2L and α1β3δ GABA$_A$ receptors. **A**. Single-channel GABA$_A$ receptor currents recorded from outside-out membrane patches voltage-clamped at −75 mV. Traces shown were continuous 700 ms recordings in response to 1 mM GABA. Channel openings are downwards. **B**. Duration histograms of open and closed intervals were formed from steady-state single-channel data. Open duration histograms for α1β3γ2L GABA$_A$ receptors were fitted best with three components, while α1β3δ open duration histograms were fitted best with only two components. Closed interval histograms were fitted best by five components in both cases. (Reproduced from Haas and Macdonald, 1999, with permission.)

remains unclear, however, is why O1 openings were detectable at all in the context of a saturating GABA concentration, as this should have effectively driven receptor occupancy in mono-liganded states near zero. One possibility is that two O1 states exist with similar mean open times, one mono-liganded and the other di-liganded. In other words, there may actually be four open states, two of which are simply indistinguishable with classical exponential fitting of open time distributions (Lagrange et al., 2007).

Given the complex channel activity observed at the microscopic level, it is not surprising that GABA$_A$ receptor macroscopic current properties (i.e. the ensemble response of hundreds or thousands of channels) are also quite complex (Fig. 14.7) (Haas and Macdonald, 1999; Bianchi et al., 2001; Bianchi and Macdonald, 2002; Mozrzymas et al., 2003; Lagrange et al., 2007). Rapid application of a saturating GABA concentration to excised outside-out patches from hippocampal neurons or HEK293T cells transiently expressing α1βxγ2 receptors gives rise to large amplitude currents that activate in the submillisecond time domain. In the context of prolonged agonist exposure, these currents undergo extensive multi-phasic desensitization, typically with three to four time constants ranging from <10 ms to >1000 ms (Celentano and Wong, 1994; Haas and Macdonald, 1999; Bianchi and Macdonald, 2002; Lagrange et al., 2007). This phenomenon (also referred

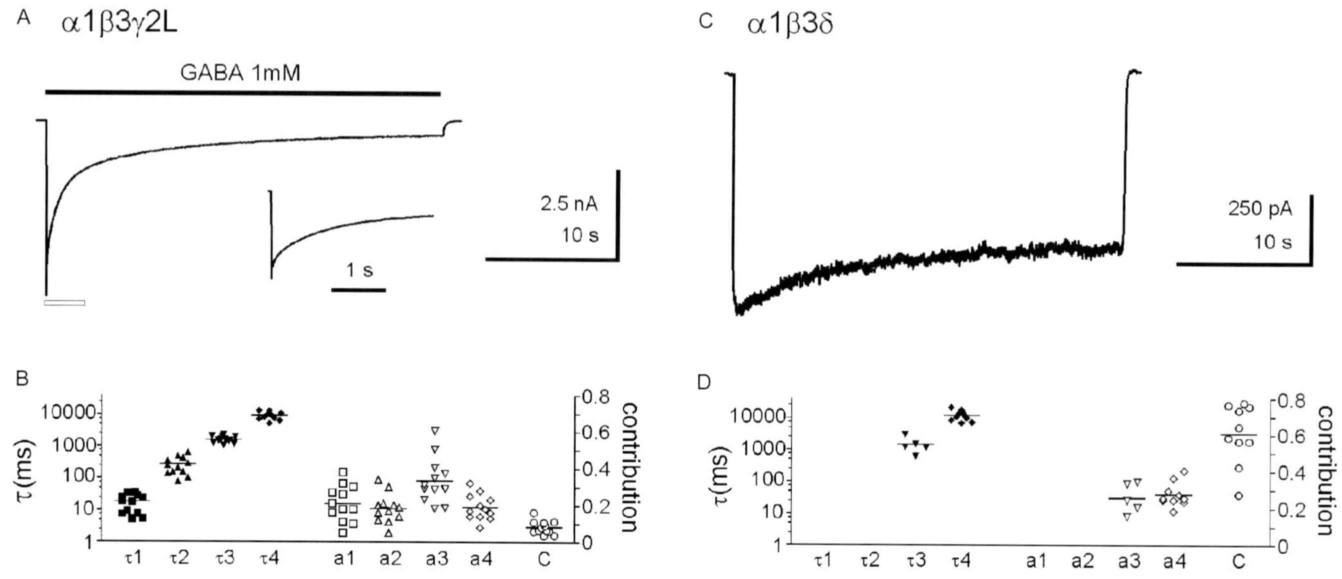

FIGURE 14.7 Macroscopic properties of α1β3γ2L and α1β3δ GABA$_A$ receptor currents. (a) Response of α1β3γ2L receptors transiently expressed in HEK293T cells to a 28 s concentration jump using 1 mM GABA (filled bar). The inset shows the first 3 s on an expanded time scale. (c) Response of α1β3δ receptors to the same protocol as in panel (a). The parameters used to fit α1β3γ2L and α1β3δ currents are shown as scatter plots in panels (b) and (d), respectively. The left ordinate indicates the time constants (τ1–τ4; note the logarithmic scale), and the right ordinate indicates the relative contribution of the corresponding time constants (a1–a4), as well as the constant term to account for incomplete desensitization. For each parameter, a horizontal line is drawn through the mean. (Reproduced with permission from Bianchi and Macdonald, 2002.)

to as current "sag") is not caused by receptor internalization or loss of the chloride gradient, but rather is an intrinsic property of the channel, reflecting the progressive accumulation of receptors in long-lived non-conducting (desensitized) states (Celentano and Wong, 1994; Bianchi and Macdonald, 2002). Interestingly, desensitization has a steep concentration dependence (Haas and Macdonald, 1999; Bianchi et al., 2007). While this has been suggested to reflect a concentration-dependent entry rate into the microscopic desensitized state, multiple modeling studies have demonstrated that this gating structure is not actually required (Haas and Macdonald, 1999; Bianchi et al., 2007). Instead, the loss of desensitization associated with application of low concentrations of GABA is most likely caused by failure to synchronously activate receptors, which masks the otherwise concentration-independent process (much like slow application of agonist) (Jones and Westbrook, 1995; Bianchi and Macdonald, 2002; Bianchi et al., 2007). Evidence for this comes from detailed kinetic analysis of macroscopic desensitization, which shows that while the relative contribution of each desensitization time constant changes with GABA concentration, the actual time constants do not (Haas and Macdonald, 1999).

Following GABA washout, currents typically deactivate bi-phasically (Jones and Westbrook, 1995; Haas and Macdonald, 1999), though multi-phasic deactivation has been described (Lagrange et al., 2007). The time course of deactivation depends strongly on the duration of GABA exposure prior to washout, with longer applications being associated with slower deactivation (Jones and Westbrook, 1995; Haas and Macdonald, 1999; Bianchi et al., 2007; Botzolakis et al., 2009). This reflects the inability of GABA to unbind from receptors in desensitized states (note that GABA is also "trapped" on receptors in open and pre-open states) (Bianchi and Macdonald, 2001b, 2002; Bianchi et al., 2002a, b), which represent an increasing fraction of receptors with longer GABA applications (Bianchi et al., 2007). While receptors in desensitized states are technically no different from other closed states, their long lifetimes provide channels with the opportunity to reopen long after GABA washout, the macroscopic correlate of which is prolonged deactivation (Jones and Westbrook, 1995, 1996). As a result, the phenomena of desensitization and deactivation are commonly referred to as being "coupled" (Jones and Westbrook, 1995; Bianchi et al., 2001, 2007).

Both the macroscopic and microscopic kinetic properties of GABA$_A$ receptors are highly influenced by subunit composition, thus providing a mechanism for neurons to fine tune their sensitivity to GABA (Angelotti and Macdonald, 1993; Saxena and Macdonald, 1994; Burgard et al., 1996; Fisher et al., 1997; Haas and Macdonald, 1999; Neelands and

Macdonald, 1999; Neelands et al., 1999; Bianchi and Macdonald, 2002; Bianchi et al., 2002a, b, 2007; Feng and Macdonald, 2004; Feng et al., 2004; Lagrange et al., 2007; Barberis et al., 2007; Picton and Fisher, 2007; Rula et al., 2008). For example, in contrast to αβγ receptors, αβδ receptors have only two open states, both of which are relatively short-lived (0.3 and 1.0 ms) (Fig. 14.6). The shorter of these accounts for 80% of all openings, yielding an overall mean open time of only 0.4 ms (Fisher et al., 1997; Fisher and Macdonald, 1997a, b; Haas and Macdonald, 1999). In addition, αβδ receptors have a longer mean closed time (~36 ms), reflecting their increased likelihood of entering long-lived closed states. This, combined with their decreased mean open time, makes their overall open probability when activated by 1 mM GABA much lower than that of αβγ receptors (~0.02 vs. ~0.1, respectively). Consequently, macroscopic currents evoked from αβδ receptors have kinetic properties that are different from those evoked from αβγ receptors (Fig. 14.7). Indeed, αβδ receptor currents are typically small, slowly activating, minimally desensitizing and rapidly deactivating (Saxena and Macdonald, 1996; Fisher and Macdonald, 1997b; Haas and Macdonald, 1999; Bianchi and Macdonald, 2002). Interestingly, while comparing αβδ and αβγ *macroscopic* currents provides support for the idea that desensitization and deactivation are coupled phenomena, it should be noted that at the *microscopic* level, αβδ receptors actually appear to have similar, if not increased, access to long-lived closed states (Fisher and Macdonald, 1997b; Haas and Macdonald, 1999). This suggests that macroscopic desensitization may not simply reflect receptor accumulation in any particular state or set of states, but rather may depend on a more complex interplay between all rate constants in the gating scheme (Mozrzymas et al., 2003; Bianchi et al., 2007). Similarly, while deactivation may be influenced by receptor trapping in desensitized states, other rate constants (such as the unbinding rate) undoubtedly also contribute to its time course.

A variety of mathematical approaches have been used to describe the behavior of ion channels; however, it is generally agreed that Markov models comprising multiple, reversibly connected states, each corresponding to a distinct receptor conformation (i.e. open or closed), provide the best fits of channel data (Fig. 14.8) (Korn and Horn, 1988; McManus et al., 1988; Sansom et al., 1989). Although applying Markov models to ion channels involves several assumptions (primarily, that state transitions are probabilistic and independent of previous channel activity), thus far, they have proven extremely useful for describing the behavior not only of $GABA_A$ receptors (Weiss and Magleby, 1989; Twyman et al., 1989a; Celentano and Wong, 1994; Jones and Westbrook, 1995; Haas and Macdonald, 1999; Lagrange et al., 2007), but also of numerous other ligand- and voltage-gated channels (Horn and Vandenberg, 1984; Zagotta et al., 1994; Schoppa and Sigworth, 1998; Rothberg and Magleby, 2000; Sigg and Bezanilla, 2003; Burzomato et al., 2004; Chakrapani et al., 2004; Lape et al., 2008). Indeed, Markov models have provided an important conceptual framework for interpreting the effects of disease-causing mutations and a variety of modulators (Twyman et al., 1989a; Twyman et al., 1989b, 1992; Twyman and Macdonald, 1992; Bianchi and Macdonald, 2001a; Feng et al., 2004; Mercik et al., 2006; Mozrzymas et al., 2007; Plested et al., 2007). However, for models to have any practical utility, they must account for both microscopic and macroscopic channel behavior (that being said, simple models are often quite valuable for systematically exploring the relationship between microscopic rate constants and macroscopic phenomena; Fig. 14.8B, 14.8C (Bianchi et al., 2007)). Considering receptor responses under both microscopic and macroscopic conditions is critical, as neither can independently provide enough information for a unique reaction scheme to be generated with certainty. Macroscopic currents, for example, constrain gating schemes by providing non-equilibrium kinetic data, which is typically unavailable from near-equilibrium single-channel studies. Single channel data, in contrast, provides direct information regarding channel open

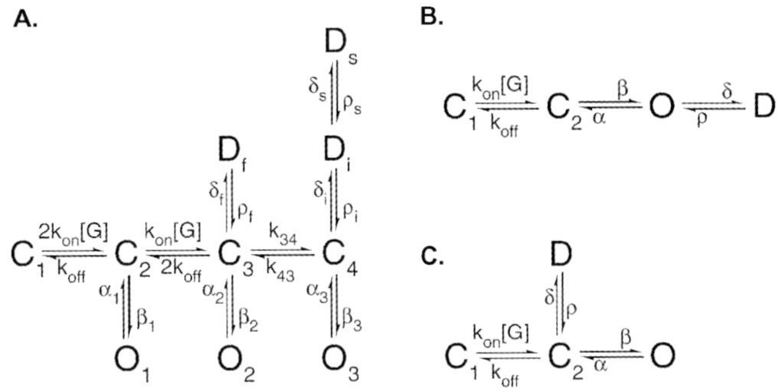

FIGURE 14.8 Kinetic models of ligand-gated ion channel function. **A.** Comprehensive kinetic model for the α1β3γ2L receptor isoform (Haas and Macdonald, 1999) illustrating reversibly connected closed (C), open (O) and desensitized (D_f, "fast"; D_i, "intermediate"; D_s, "slow") states. For simplicity, the two distal "intraburst" C states connected to each O state were omitted. The microscopic transitions associated with agonist binding were labeled k_{on} and k_{off} for association and dissociation rate constants, respectively. Each agonist binding step was taken to be equivalent and independent (the first binding and unbinding rates were therefore multiplied by 2). [G], concentration of GABA. **B.** Simple four-state kinetic model in "linear" arrangement. A single binding step connects C_1 (the unbound closed state) and C_2 (the ligand-bound closed state). The O state is accessed from C_2 and is arranged in series with the D state. **C.** Simple four-state kinetic model in "branched" arrangement. D and O states are arranged in parallel, each being directly accessible from C_2. (Reproduced from Bianchi et al., 2007, with permission.)

and closed states, including the number of each, their connectivity, and in the case of the open states, their approximate lifetimes. The first comprehensive model was developed by Haas and Macdonald (Haas and Macdonald, 1999), and was based on an earlier model by Twyman and Macdonald (Twyman et al., 1990) that stemmed from detailed single channel analysis (Fig. 14.8 A). This model contained two GABA binding steps and a total of 16 states – three open and 13 closed. Three of the closed states were given the special designation of "desensitized" states, as they allowed for the macroscopic phenomenon of desensitization to occur. Although this model has been updated recently to take into account several additional macroscopic and microscopic observations (Lagrange et al., 2007), its core gating structure has not changed in almost a decade.

VII. PHARMACOLOGICAL PROPERTIES OF $GABA_A$ RECEPTORS

$GABA_A$ receptors have a rich pharmacology (for a review, see Macdonald and Olsen, 1994, or Johnston, 1996). In addition to GABA, a number of GABA analogs can directly activate the receptor, including the plant alkaloid muscimol and its conformationally restricted analog tetrahydroisoxazolopyridinol (THIP). Endogenous agonists include taurine and β-alanine, both of which are found at relatively high concentrations in the brain (Lerma et al., 1986). Interestingly, while each of these agonists has a different microscopic affinity for GABA (muscimol > GABA > β-alanine ≈ THIP), when used at EC-equivalent concentrations, all give rise to similar microscopic and macroscopic currents (Jones et al., 1998; Bianchi et al., 2007). $GABA_A$ receptor currents can be competitively antagonized by the convulsant drug, bicuculline and by the pyridazinyl derivative of GABA, gabazine (SR95531). Of note, bicuculline can also block spontaneously active $GABA_A$ receptors (Bianchi and Macdonald, 2002), indicating it should also be classified as an inverse $GABA_A$ receptor agonist along with β-carbolines such as DMCM. Non-competitive antagonists include the convulsants picrotoxin, TBPS, pentylenetetrazole and penicillin, the latter being a classic open channel blocker of $GABA_A$ receptors (Twyman et al., 1992; Feng et al., 2009). The anticonvulsant barbiturates and BZDs allosterically enhance $GABA_A$ receptor currents, but through different binding sites and by different mechanisms (Twyman et al., 1989). Barbiturates increase the fraction of long (O3) openings at the expense of short (O1 and O2) openings, thus increasing channel mean open time. In contrast, BZDs increase the microscopic affinity of GABA for the receptor without altering channel mean open time, thus increasing channel opening frequency (at subsaturating concentrations of GABA). The BZD binding site is targeted by several agents, including the inverse agonist β-carbolines, the imidazolpyridines (zolpidem, alpidem), the BZD inverse agonist Ro 15-4513, and the BZD receptor antagonist flumazenil. Other positive allosteric modulators include ethanol, neurosteroids (THDOC and allopregnanolone) and several volatile and intravenous anesthetics (halothane, diethylether, enflurane, isoflurane, alphaxalone, ketamine and propofol). Negative allosteric modulators include pregnenolone sulfate, zinc and furosemide. In addition, there are several classes of allosteric modulators with mixed effects (i.e. positive and negative modulation depending on GABA concentration and context of receptor activation). These include the insecticides dieldrin and lindane, the anti-helminthic ivermectin, lanthanum and pH.

The pharmacological properties of $GABA_A$ receptors, much like their kinetic properties, are highly sensitive to subunit composition (for a review, see Hevers and Luddens, 1998). For example, BZD modulation requires the presence of a γ subunit; however, only α(1, 2, 3 or 5)βγ receptor isoforms are BZD sensitive. Conversely, α(1, 2, 3 or 5)βγ receptor isoforms are less sensitive to furosemide, while those containing α4 and α6 subunits are highly sensitive. Zolpidem has highest affinity for α1 subtype-containing receptors, low affinity for α2 and α3 subtype-containing receptors, and almost no affinity for α5 subtype-containing receptors. Receptors containing β2 or β3 subtypes are highly sensitive to loreclezole, while those containing the β1 subtype are relatively insensitive. Inclusion of a γ subunit dramatically reduces receptor sensitivity to zinc and neurosteroids. Receptor incorporation of δ subunits increases receptor sensitivity to neurosteroids and ethanol. In some cases, subunit composition determines the polarity of modulation (i.e. enhancement vs. inhibition). For example, while Ro 15-4513 is an inverse agonist at the BZD binding site for α(1, 2, 3 or 5)βγ receptor isoforms, it is an agonist of α4 and α6 subtype-containing receptors. Similarly, while lanthanum enhances α1βγ receptor currents, it blocks α6βγ receptor currents.

VIII. MODES OF GABAergic INHIBITION

$GABA_A$ receptors mediate two modes of inhibitory neurotransmission. The first, termed "phasic" inhibition, involves the transient activation of postsynaptic

$GABA_A$ receptors by nearly saturating concentrations of GABA released from presynaptic vesicles. This process gives rise to inhibitory postsynaptic currents (IPSCs) that activate rapidly (rise times of ~1 ms or less) but decay slowly (time constants of 10s to 100s of ms) (Maconochie et al., 1994; Jones and Westbrook, 1995). In the experimental setting, several types of IPSCs can be recorded, each having slightly different kinetic properties (Otis et al., 1994; Mody and Pearce, 2004). These include "miniature" IPSCs (mIPSCs), "spontaneous" IPSCs (sIPSCs) and "evoked" IPSCs (eIPSCs). mIPSCs are triggered by the spontaneous release of GABA from a single synaptic vesicle (i.e. action potential independent) (Fig. 14.9A). In contrast, sIPSCs are triggered by spontaneously occurring action potentials in presynaptic terminals and typically involve release of GABA from multiple synaptic vesicles (Fig. 14.9B). eIPSCs are triggered following experimentally induced action potentials, and, like sIPSCs, involve release of GABA from multiple synaptic vesicles. The properties of IPSCs are highly variable, depending on brain region, developmental stage and neuron type (Vicini et al., 2001; Mozrzymas and Barberis, 2004). While both pre- and postsynaptic factors influence IPSC shape, it is generally accepted that postsynaptic factors are the primary determinants. Indeed, the GABA transient is thought to reach nearly saturating concentrations in <100 μs and decay in <1 ms due to a combination of diffusion and reuptake (Clements, 1996; Glavinovic, 1999; Ventriglia and Di Maio, 2003). Thus, IPSC duration significantly outlasts the presence of GABA in the synaptic cleft, suggesting that IPSCs are shaped primarily by the intrinsic properties of postsynaptic receptors. In support of this hypothesis, application of ultra-brief pulses of nearly saturating GABA to membrane patches excised from mammalian cell lines expressing recombinant $GABA_A$ receptors gives rise to currents with kinetic properties that resemble IPSCs (Jones and Westbrook, 1995; Haas and Macdonald, 1999). Currents evoked from receptors containing the γ subunit are most similar to IPSCs, as they activate rapidly and deactivate slowly (Jones and Westbrook, 1995; Haas and Macdonald, 1999; Lagrange et al., 2007). This should not imply, however, that presynaptic factors cannot also influence IPSC shape. Several studies suggest that the GABA transient may actually decay with a time constant as brief as 100 μs (Mozrzymas, 2004). Kinetic modeling studies predict that such ultra-brief exposures prevent postsynaptic receptors from reaching maximal activation,

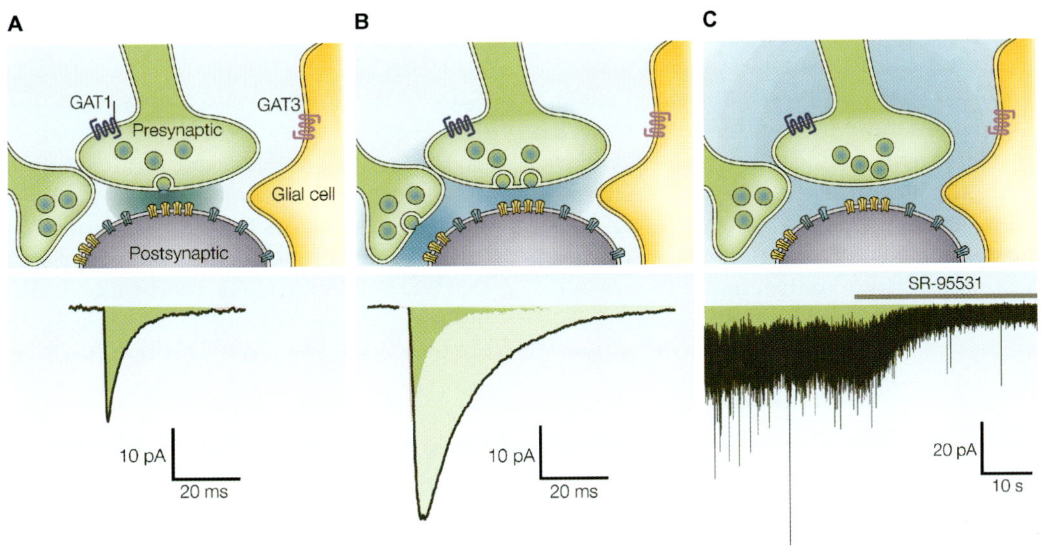

FIGURE 14.9 Modes of $GABA_A$ receptor activation. **A.** The release of GABA (blue shading) from a single presynaptic vesicle activates only those postsynaptic $GABA_A$ receptors clustered immediately beneath the release site (yellow). The current record shows an averaged waveform of the resulting miniature inhibitory postsynaptic currents (mIPSCs). **B.** Action potential-dependent release of multiple vesicles or evoked release from several terminals promotes GABA "spillover", which activates both synaptic and extrasynaptic receptors (blue). The current record shows the larger and much slower averaged waveform of the resulting eIPSCs. The area of the mIPSC shown in panel (a) is superimposed for comparison. **C.** A low concentration of ambient GABA, which persists despite the activity of the neuronal and glial GABA transporters (GAT1 and GAT3, respectively), tonically activates extrasynaptic $GABA_A$ receptors. The trace shows the tonic current that results from stochastic opening of these high-affinity receptors, with superimposed phasic currents. The $GABA_A$ receptor antagonist gabazine (SR-95531) blocks phasic IPSCs and tonic channel activity, causing a change in the "holding" current and a reduction in current variance. The shaded area beneath the current record before SR-95531 application represents the charge carried by tonically active $GABA_A$ receptors. The current records are from whole-cell patch-clamp recordings of granule cells in acute cerebellar slices from adult mice. (Reproduced from Farrant and Nusser, 2005, with permission.)

thus leading to smaller amplitude and more rapidly deactivating IPSCs (Mozrzymas, 2004; Lagrange et al., 2007). In addition, while the concentration of GABA in synapses is generally thought to be nearly saturating at ~1 mM (Edwards et al., 1990; Otis et al., 1994; Jones and Westbrook, 1995; Strecker et al., 1999), the observation that diazepam can increase IPSC amplitudes in some brain regions suggests this may not always be the case; if GABA is truly saturating in the synapse, then increasing receptor affinity for GABA with a BZD should not affect IPSC amplitude, as the postsynaptic receptors are already fully liganded (Frerking et al., 1995; Defazio and Hablitz, 1998; Hill et al., 1998). Activating receptors with subsaturating concentrations of GABA can significantly impact IPSC kinetics, as the time courses of $GABA_A$ receptor activation and deactivation are both highly sensitive to GABA concentration (slower and faster, respectively) (Lagrange et al., 2007; Bianchi et al., 2007).

In addition to mediating fast synaptic inhibition, there is now compelling evidence that $GABA_A$ receptors are also involved in slower forms of non-synaptic inhibition, a phenomenon termed "tonic" inhibition, illustrated in Fig. 14.9C. This phenomenon is mediated by peri- and extra-synaptic $GABA_A$ receptors that are persistently activated by subsaturating concentrations of ambient GABA. While the sources and precise concentration of ambient GABA are still uncertain (unlike other neurotransmitters such as dopamine and serotonin, the GABA concentration cannot be measured directly), it is generally believed to arise from a combination of synaptic overflow and non-vesicular release, and to reach concentrations of ~1 μM (Attwell et al., 1993; Zoli et al., 1999; Farrant and Nusser, 2005). Interestingly, the contribution of the tonic current to overall inhibitory tone may actually be greater than the summed charge transfer of phasic currents (Brickley et al., 1996; Hamann et al., 2002). It should also be noted that tonic and phasic currents are thought to be differentially modulated by various pharmacological agents (Feng and Macdonald, 2004; Feng et al., 2004, 2008) and to play distinct roles in the pathogenesis of neurological disorders such as epilepsy (Dibbens et al., 2004; Feng et al., 2006; Eugene et al., 2007). This reflects the fact that tonic currents are mediated by a different subset of receptor isoforms, which have kinetic properties distinct from those mediating phasic currents. α4βxδ and α6βxδ receptor isoforms are thought to be the primary mediators of tonic inhibition (Farrant and Nusser, 2005), though α1βxδ and α5βxγ2 receptors may also play an important role in the hippocampus (Glykys et al., 2008). Indeed, properties conferred by the δ subunit are consistent with a role in tonic inhibition; αβδ receptors desensitize much slower and less extensively than αβγ receptors and have a lower GABA EC_{50}, ideal for receptors that must respond to very low concentrations of GABA for extended periods of time (Saxena and Macdonald, 1996; Haas and Macdonald, 1999; Lagrange et al., 2007).

IX. INVOLVEMENT OF $GABA_A$ RECEPTORS IN EPILEPSY

Epilepsy is associated with abnormal, hypersynchronous activation of neuronal populations. Although the mechanistic basis for partial and generalized forms of epilepsy are uncertain, there is considerable evidence that impaired GABAergic inhibition underlies several types of epilepsy. Indeed, pharmacological blockade of GABAergic inhibition with $GABA_A$ receptor antagonists such as penicillin, picrotoxin or bicuculline produces paroxysmal bursting in isolated neurons and seizures in experimental animals (Schwartzkroin and Prince, 1980; Hotson and Prince, 1981; De Deyn et al., 1992). Similar experiments have shown that $GABA_A$ receptor blockade produces paroxysmal depolarization shifts (PDSs), which are the interictal manifestations of epileptiform events (Hwa et al., 1991; Schiller, 2002). This local paroxysmal bursting can spread to involve large areas of hippocampus or generalize to cortex when inhibition is further weakened and when other synchronizing factors occur, such as altered extracellular concentrations of potassium and calcium (Korn et al., 1987).

However, conclusive evidence that loss of GABAergic inhibition was involved in the pathogenesis of human epilepsy syndromes did not come until the recent discovery of mutations in genes encoding $GABA_A$ receptor subunits that were associated with idiopathic generalized epilepsies (Macdonald et al., 2006). The first α subunit mutation to be reported was the α1 subunit mutation, A322D, in a family with juvenile myoclonic epilepsy (JME). Mutant receptors had substantially reduced maximal currents (Cossette et al., 2002). This was caused by reduced levels of $GABA_A$ receptor surface expression, which resulted from ER retention and accelerated proteosomal and lysosomal degradation of subunits following failure of the M3 domain to insert properly in the plasma membrane (Gallagher et al., 2004, 2005, 2007; Bradley et al., 2008). Another α1 subunit mutation, a single base-pair deletion predicted to produce a frameshift and a premature translation-termination codon, 975delC, S326fs328X, was recently identified as a *de novo* mutation in an individual with childhood absence epilepsy

(CAE) (Maljevic et al., 2006). Little is known regarding the mechanisms by which this mutation causes disease, except that current was not detectable when the mutant subunit was coexpressed with β2 and γ2 subunits. Whether this was caused by impaired channel function or trafficking, however, remains unknown. Three β3 subunit mutations, β3(S15F), β3(P11S), and β3(G32R), were also reported recently in families with CAE (Tanaka et al., 2008). These mutations were found to decrease the amplitude of GABA-evoked currents, a finding likely attributable to altered channel trafficking since mutant subunits have altered glycosylation in the absence of altered total protein expression. In the δ subunit, two epilepsy susceptibility variants, E177A and R220H, were discovered in small GEFS+ families (Dibbens et al., 2004). When coexpressed with α1 and β2 subunits, these variants reduced single channel mean open time and the levels of $GABA_A$ receptor surface expression, thus substantially decreasing whole cell current amplitudes (Feng et al., 2006).

Interestingly, more mutations have been identified in the γ2 subunit than in all other subunit subtypes combined. A family with generalized epilepsy with febrile seizures plus (GEFS+) was found to have a γ2 subunit mutation, K289M, located in the extracellular M2–M3 linker (Baulac et al., 2001), a region implicated in transduction of ligand binding to channel gating (Kash et al., 2003). Consistent with the known importance of this protein domain to channel function, recordings from HEK293T cells expressing α1β2γ2(K289M) receptors were found to have defective channel gating (shortened mean open times) and accelerated deactivation (Bianchi et al., 2002b), as illustrated in Fig. 14.10. A family with CAE and febrile

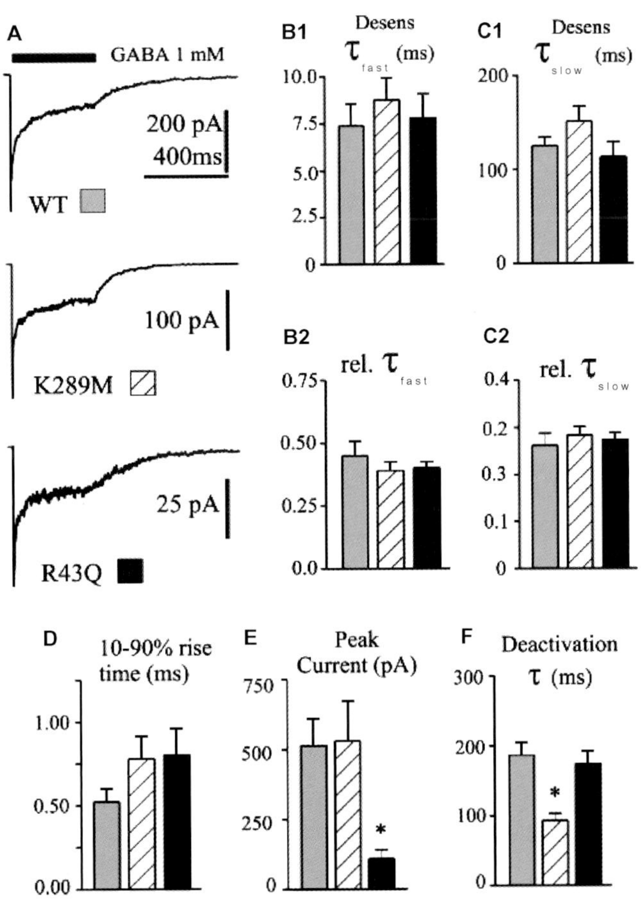

FIGURE 14.10 Macroscopic kinetic properties of wild-type and mutant $GABA_A$ receptor currents. **A.** Representative current traces obtained from wild-type α1β3γ2 or mutant α1β3γ2(K289M or R43Q) $GABA_A$ receptors expressed transiently in HEK293T cells during 400 msec jumps into 1 mM GABA. Time scale of top trace applies to all three traces. **B.** Neither the fast **B1** nor the slow **C1** time constant of desensitization, nor their relative contributions (b2, c2) were significantly altered by the mutations. **D.** Current activation rate, as indicated by the 10–90% rise time of the current, was not significantly altered by the mutations. **E.** Peak current amplitudes were significantly smaller for α1β3γ2(R43Q) $GABA_A$ receptors. *$p < 0.01$. **F.** Current deactivation after removal of GABA was significantly faster for α1β3γ2(K289M) $GABA_A$ receptors. *$p < 0.001$. Data were obtained from 8–13 patches. (Reproduced from Bianchi et al., 2002, with permission.)

seizures had a γ2 subunit mutation, R43Q, located in the N-terminal extracellular domain in the BZD binding domain (Wallace et al., 2001). As shown in Fig. 14.10, this mutation reduced peak current amplitudes without altering channel kinetics (Bianchi et al., 2002b). The basis for the reduced current was reduced surface expression due to retention of the receptor in the ER (Sancar and Czajkowski, 2004; Kang and Macdonald, 2004), possibly due to disruption of inter-subunit contacts at the γ-β subunit interface (Hales et al., 2005). However, it has been reported that γ2(R43Q) subunits may assemble with α3 subunits (Frugier et al., 2007), suggesting the effects of the mutation may depend on the specific α subunit subtype involved. The γ2 subunit mutation, R139G, altered BZD sensitivity and accelerated desensitization (Audenaert et al., 2006). A γ2 subunit mutation, Q351X, which introduced a PTC in the M3–M4 linker was identified in a family with GEFS+ (Harkin et al., 2002). Mutant receptors had no GABA sensitivity when expressed in oocytes, probably due to ER retention. A γ2 subunit splice donor site mutation, IVS6 + 2T-G, was identified in a family with CAE and febrile seizures (Kananura et al., 2002). The effect of this mutation is unknown but was predicted to lead to non-functional receptors. A γ2 subunit, mutation Q1X, that introduced a PTC (Q1X) between the signal peptide and the mature peptide was identified in a family with severe myoclonic epilepsy of infancy (Hirose, 2004). The functional consequence of the mutation is unknown but may be haploinsufficiency, since the mutation would likely trigger nonsense mediated mRNA decay, thus preventing production of even a signal peptide.

Acknowledgements

Funding was provided by NIH R01-NS33300 and R01-NS51590 to RLM and NIH T32-GM07347 to the Vanderbilt Medical Scientist Training Program (MSTP). The authors thank Katharine Gurba and Sandra Camp for assistance with references.

References

Akabas, M.H. (2004). $GABA_A$ receptor structure-function studies: a reexamination in light of new acetylcholine receptor structures. *Int. Rev. Neurobiol.* **62**, 1–43.

Akabas, M.H., Kaufmann, C., Archdeacon, P., and Karlin, A. (1994). Identification of acetylcholine receptor channel-lining residues in the entire M2 segment of the alpha subunit. *Neuron* **13**, 919–927.

Akk, G., Li, P., Bracamontes, J., Reichert, D.E., Covey, D.F., and Steinbach, J.H. (2008). Mutations of the GABA-A receptor alpha1 subunit M1 domain reveal unexpected complexity for modulation by neuroactive steroids. *Mol. Pharmacol.* **74**, 614–627.

Angelotti, T.P. and Macdonald, R.L. (1993). Assembly of $GABA_A$ receptor subunits: alpha 1 beta 1 and alpha 1 beta 1 gamma 2S subunits produce unique ion channels with dissimilar single-channel properties. *J. Neurosci.* **13**, 1429–1440.

Attwell, D., Barbour, B., and Szatkowski, M. (1993). Nonvesicular release of neurotransmitter. *Neuron* **11**, 401–407.

Audenaert, D., Schwartz, E., Claeys, K.G., Claes, L., Deprez, L., Suls, A., Van Dyck, T., Lagae, L., Van Broeckhoven, C., Macdonald, R.L., and De Jonghe, P. (2006). A novel GABRG2 mutation associated with febrile seizures. *Neurology* **67**, 687–690.

Backus, K.H., Arigoni, M., Drescher, U., Scheurer, L., Malherbe, P., Mohler, H., and Benson, J.A. (1993). Stoichiometry of a recombinant $GABA_A$ receptor deduced from mutation-induced rectification. *Neuroreport* **5**, 285–288.

Bailey, M.E., Albrecht, B.E., Johnson, K.J., and Darlison, M.G. (1999a). Genetic linkage and radiation hybrid mapping of the three human $GABA_C$ receptor rho subunit genes: GABRR1, GABRR2 and GABRR3. *Biochim. Biophys. Acta* **1447**, 307–312.

Bailey, M.E., Matthews, D.A., Riley, B.P., Albrecht, B.E., Kostrzewa, M., Hicks, A.A., Harris, R., Muller, U., Darlison, M.G., and Johnson, K.J. (1999b). Genomic mapping and evolution of human GABA(A) receptor subunit gene clusters. *Mamm. Genome* **10**, 839–843.

Balasubramanian, S., Teissere, J.A., Raju, D.V., and Hall, R.A. (2004). Hetero-oligomerization between $GABA_A$ and $GABA_B$ receptors regulates $GABA_B$ receptor trafficking. *J. Biol. Chem.* **279**, 18840–18850.

Barberis, A., Mozrzymas, J.W., Ortinski, P.I., and Vicini, S. (2007). Desensitization and binding properties determine distinct alpha1beta2gamma2 and alpha3beta2gamma2 $GABA_A$ receptor-channel kinetic behavior. *Eur. J. Neurosci.* **25**, 2726–2740.

Barnard, E.A., Skolnick, P., Olsen, R.W., Mohler, H., Sieghart, W., Biggio, G., Braestrup, C., Bateson, A.N., and Langer, S.Z. (1998). International Union of Pharmacology. XV. Subtypes of gamma-aminobutyric acidA receptors: classification on the basis of subunit structure and receptor function. *Pharmacol. Rev.* **50**, 291–313.

Barrera, N.P., Betts, J., You, H., Henderson, R.M., Martin, I.L., Dunn, S.M., and Edwardson, J.M. (2008). Atomic force microscopy reveals the stoichiometry and subunit arrangement of the alpha4beta3delta $GABA_A$ receptor. *Mol. Pharmacol.* **73**, 960–967.

Baulac, S., Huberfeld, G., Gourfinkel-An, I., Mitropoulou, G., Beranger, A., Prud'homme, J.F., Baulac, M., Brice, A., Bruzzone, R., and LeGuern, E. (2001). First genetic evidence of $GABA_A$ receptor dysfunction in epilepsy: a mutation in the gamma2-subunit gene. *Nat. Genet.* **28**, 46–48.

Baumann, S.W., Baur, R., and Sigel, E. (2001). Subunit arrangement of gamma-aminobutyric acid type A receptors. *J. Biol. Chem.* **276**, 36275–36280.

Baumann, S.W., Baur, R., and Sigel, E. (2002). Forced subunit assembly in alpha1beta2gamma2 $GABA_A$ receptors. Insight into the absolute arrangement. *J. Biol. Chem.* **277**, 46020–46025.

Baur, R., Minier, F., and Sigel, E. (2006). A $GABA_A$ receptor of defined subunit composition and positioning: concatenation of five subunits. *FEBS Lett.* **580**, 1616–1620.

Beck, M., Brickley, K., Wilkinson, H.L., Sharma, S., Smith, M., Chazot, P.L., Pollard, S., and Stephenson, F.A. (2002). Identification, molecular cloning, and characterization of a novel $GABA_A$ receptor-associated protein. *GRIF-1. J. Biol. Chem.* **277**, 30079–30090.

Bedford, F.K., Kittler, J.T., Muller, E., Thomas, P., Uren, J.M., Merlo, D., Wisden, W., Triller, A., Smart, T.G., and Moss, S.J. (2001). $GABA_A$ receptor cell surface number and subunit stability are regulated by the ubiquitin-like protein Plic-1. *Nat. Neurosci.* **4**, 908–916.

Bell, M.V., Bloomfield, J., McKinley, M., Patterson, M.N., Darlison, M.G., Barnard, E.A., and Davies, K.E. (1989). Physical linkage of a $GABA_A$ receptor subunit gene to the DXS374 locus in human Xq28. *Am. J. Hum. Genet.* **45**, 883–888.

Ben-Ari, Y. (2002). Excitatory actions of GABA during development: the nature of the nurture. *Nature Reviews. Neurosci.* **3**, 728–739.

Benke, D., Honer, M., Michel, C., and Mohler, H. (1996). $GABA_A$ receptor subtypes differentiated by their gamma-subunit variants: prevalence, pharmacology and subunit architecture. *Neuropharmacology* **35**, 1413–1423.

Bera, A.K., Chatav, M., and Akabas, M.H. (2002). $GABA_A$ receptor M2–M3 loop secondary structure and changes in accessibility during channel gating. *J. Biol. Chem.* **277**, 43002–43010.

Bianchi, M.T. and Macdonald, R.L. (2001a). Mutation of the 9′ leucine in the GABA(A) receptor gamma2L subunit produces an apparent decrease in desensitization by stabilizing open states without altering desensitized states. *Neuropharmacology* **41**, 737–744.

Bianchi, M.T. and Macdonald, R.L. (2001b). Agonist trapping by $GABA_A$ receptor channels. *J. Neurosci.* **21**, 9083–9091.

Bianchi, M.T., Haas, K.F., and Macdonald, R.L. (2001). Structural determinants of fast desensitization and desensitization-deactivation coupling in GABAa receptors. *J. Neurosci.* **21**, 1127–1136.

Bianchi, M.T. and Macdonald, R.L. (2002). Slow phases of $GABA_A$ receptor desensitization: structural determinants and possible relevance for synaptic function. *J. Physiol.* **544**, 3–18.

Bianchi, M.T., Haas, K.F., and Macdonald, R.L. (2002a). Alpha1 and alpha6 subunits specify distinct desensitization, deactivation and neurosteroid modulation of $GABA_A$ receptors containing the delta subunit. *Neuropharmacology* **43**, 492–502.

Bianchi, M.T., Song, L., Zhang, H., and Macdonald, R.L. (2002b). Two different mechanisms of disinhibition produced by $GABA_A$ receptor mutations linked to epilepsy in humans. *J. Neurosci.* **22**, 5321–5327.

Bianchi, M.T., Botzolakis, E.J., Haas, K.F., Fisher, J.L., and Macdonald, R.L. (2007). Microscopic kinetic determinants of macroscopic currents: insights from coupling and uncoupling of $GABA_A$ receptor desensitization and deactivation. *J. Physiol.* **584**, 769–787.

Bollan, K., Robertson, L.A., Tang, H., and Connolly, C.N. (2003). Multiple assembly signals in gamma-aminobutyric acid (type A) receptor subunits combine to drive receptor construction and composition. *Biochem. Soc. Trans* **31**, 875–879.

Bollan, K.A., Baur, R., Hales, T.G., Sigel, E., and Connolly, C.N. (2008). The promiscuous role of the epsilon subunit in $GABA_A$ receptor biogenesis. *Mol. Cell Neurosci.* **37**, 610–621.

Bonnert, T.P., McKernan, R.M., Farrar, S., le Bourdelles, B., Heavens, R.P., Smith, D.W., Hewson, L., Rigby, M.R., Sirinathsinghji, D.J., Brown, N., Wafford, K.A., and Whiting, P.J. (1999). Theta, a novel gamma-aminobutyric acid type A receptor subunit. *Proc. Natl. Acad. Sci. USA* **96**, 9891–9896.

Bormann, J., Hamill, O.P., and Sakmann, B. (1987). Mechanism of anion permeation through channels gated by glycine and gamma-aminobutyric acid in mouse cultured spinal neurones. *J. Physiol.* **385**, 243–286.

Botzolakis, E.J., Maheshwari, A., Feng, H.J., Lagrange, A.H., Shaver, J.H., Kassebaum, N.F., Venkataraman, R., Baudenbacher, F.J., and Macdonald, R.L. (2009). Achieving synaptically relevant pulses of neurotransmitter using PDMS microfluidics. *J. Neurosci. Methods* **177**, 294–302.

Boue-Grabot, E., Roudbaraki, M., Bascles, L., Tramu, G., Bloch, B., and Garret, M. (1998). Expression of GABA receptor rho subunits in rat brain. *J. Neurochem.* **70**, 899–907.

Boue-Grabot, E., Toulme, E., Emerit, M.B., and Garret, M. (2004). Subunit-specific coupling between gamma-aminobutyric acid type A and P2X2 receptor channels. *J. Biol. Chem.* **279**, 52517–52525.

Bradley, C.A., Taghibiglou, C., Collingridge, G.L., and Wang, Y.T. (2008). Mechanisms involved in the reduction of $GABA_A$ receptor alpha1-subunit expression caused by the epilepsy mutation A322D in the trafficking-competent receptor. *J. Biol. Chem.* **283**, 22043–22050.

Brandon, N., Jovanovic, J., and Moss, S. (2002). Multiple roles of protein kinases in the modulation of gamma-aminobutyric acid(A) receptor function and cell surface expression. *Pharmacol. Ther.* **94**, 113–122.

Brejc, K., van Dijk, W.J., Klaassen, R.V., Schuurmans, M., van Der, O.J., Smit, A.B., and Sixma, T.K. (2001). Crystal structure of an ACh-binding protein reveals the ligand-binding domain of nicotinic receptors. *Nature* **411**, 269–276.

Brickley, K., Smith, M.J., Beck, M., and Stephenson, F.A. (2005). GRIF-1 and OIP106, members of a novel gene family of coiled-coil domain proteins: association in vivo and in vitro with kinesin. *J. Biol. Chem.* **280**, 14723–14732.

Brickley, S.G., Cull-Candy, S.G., and Farrant, M. (1996). Development of a tonic form of synaptic inhibition in rat cerebellar granule cells resulting from persistent activation of $GABA_A$ receptors. *J. Physiol.* **497**, 753–759.

Brickley, S.G., Revilla, V., Cull-Candy, S.G., Wisden, W., and Farrant, M. (2001). Adaptive regulation of neuronal excitability by a voltage-independent potassium conductance. *Nature* **409**, 88–92.

Brooks-Kayal, A.R., Shumate, M.D., Jin, H., Rikhter, T.Y., Kelly, M.E., and Coulter, D.A. (2001). Gamma-aminobutyric acid(A) receptor subunit expression predicts functional changes in hippocampal dentate granule cells during postnatal development. *J. Neurochem.* **77**, 1266–1278.

Buckle, V.J., Fujita, N., Ryder-Cook, A.S., Derry, J.M., Barnard, P.J., Lebo, R.V., Schofield, P.R., Seeburg, P.H., Bateson, A.N., and Darlison, M.G. (1989). Chromosomal localization of $GABA_A$ receptor subunit genes: relationship to human genetic disease. *Neuron* **3**, 647–654.

Burgard, E.C., Tietz, E.I., Neelands, T.R., and Macdonald, R.L. (1996). Properties of recombinant gamma-aminobutyric acid A receptor isoforms containing the alpha 5 subunit subtype. *Mol. Pharmacol.* **50**, 119–127.

Burkat, P.M., Yang, J., and Gingrich, K.J. (2001). Dominant gating governing transient $GABA_A$ receptor activity: a first latency and Po/o analysis. *J. Neurosci.* **21**, 7026–7036.

Burzomato, V., Beato, M., Groot-Kormelink, P.J., Colquhoun, D., and Sivilotti, L.G. (2004). Single-channel behavior of heteromeric alpha1beta glycine receptors: an attempt to detect a conformational change before the channel opens. *J. Neurosci.* **24**, 10924–10940.

Campagna-Slater, V. and Weaver, D.F. (2007). Molecular modelling of the $GABA_A$ ion channel protein. *J. Mol. Graph. Model.* **25**, 721–730.

Caraiscos, V.B., Elliott, E.M., You, T., Cheng, V.Y., Belelli, D., Newell, J.G., Jackson, M.F., Lambert, J.J., Rosahl, T.W., Wafford, K.A., MacDonald, J.F., and Orser, B.A. (2004). Tonic inhibition in mouse hippocampal CA1 pyramidal neurons is mediated by alpha5 subunit-containing gamma-aminobutyric acid type A receptors. *Proc. Natl. Acad. Sci. USA* **101**, 3662–3667.

Celentano, J.J. and Wong, R.K. (1994). Multiphasic desensitization of the $GABA_A$ receptor in outside-out patches. *Biophys. J.* **66**, 1039–1050.

Chakrapani, S., Bailey, T.D., and Auerbach, A. (2004). Gating dynamics of the acetylcholine receptor extracellular domain. *J. Gen. Physiol.* **123**, 341–356.

Chang, Y., Wang, R., Barot, S., and Weiss, D.S. (1996). Stoichiometry of a recombinant $GABA_A$ receptor. *J. Neurosci.* **16**, 5415–5424.

Chang, Y. and Weiss, D.S. (1998). Substitutions of the highly conserved M2 leucine create spontaneously opening rho1 gamma-aminobutyric acid receptors. *Mol. Pharmacol.* **53**, 511–523.

Charych, E.I., Yu, W., Miralles, C.P., Serwanski, D.R., Li, X., Rubio, M., and De Blas, A.L. (2004). The brefeldin A-inhibited GDP/GTP exchange factor 2, a protein involved in vesicular trafficking, interacts with the beta subunits of the GABA receptors. *J. Neurochem.* **90**, 173–189.

Chavas, J., Forero, M.E., Collin, T., Llano, I., and Marty, A. (2004). Osmotic tension as a possible link between GABA(A) receptor activation and intracellular calcium elevation. *Neuron* **44**, 701–713.

Chen, L., Wang, H., Vicini, S., and Olsen, R.W. (2000). The gamma-aminobutyric acid type A ($GABA_A$) receptor-associated protein (GABARAP) promotes $GABA_A$ receptor clustering and modulates the channel kinetics. *Proc. Natl. Acad. Sci. USA* **97**, 11557–11562.

Chen, Z.W. and Olsen, R.W. (2007). $GABA_A$ receptor associated proteins: a key factor regulating $GABA_A$ receptor function. *J. Neurochem.* **100**, 279–294.

Clements, J.D. (1996). Transmitter timecourse in the synaptic cleft: its role in central synaptic function. *Trends Neurosci.* **19**, 163–171.

Connolly, C.N., Kittler, J.T., Thomas, P., Uren, J.M., Brandon, N.J., Smart, T.G., and Moss, S.J. (1999). Cell surface stability of gamma-aminobutyric acid type A receptors. Dependence on protein kinase C activity and subunit composition. *J. Biol. Chem.* **274**, 36565–36572.

Connolly, C.N., Krishek, B.J., McDonald, B.J., Smart, T.G., and Moss, S.J. (1996). Assembly and cell surface expression of heteromeric and homomeric gamma-aminobutyric acid type A receptors. *J. Biol. Chem.* **271**, 89–96.

Cossette, P., Liu, L., Brisebois, K., Dong, H., Lortie, A., Vanasse, M., Saint-Hilaire, J.M., Carmant, L., Verner, A., Lu, W.Y., Wang, Y.T., and Rouleau, G.A. (2002). Mutation of GABRA1 in an autosomal dominant form of juvenile myoclonic epilepsy. *Nat. Genet.* **31**, 184–189.

Cromer, B.A., Morton, C.J., and Parker, M.W. (2002). Anxiety over GABA(A) receptor structure relieved by AChBP. *Trends Biochem. Sci.* **27**, 280–287.

Cutting, G.R., Curristin, S.Z., oghbi, H., O'Hara, B., Seldin, M.F., and Uhl, G.R. (1992). Identification of a putative gamma-aminobutyric acid (GABA) receptor subunit rho2 cDNA and colocalization of the genes encoding rho2 (GABRR2) and rho1 (GABRR1) to human chromosome 6q14-q21 and mouse chromosome 4. *Genomics* **12**, 801–806.

Dallwig, R., Deitmer, J.W., and Backus, K.H. (1999). On the mechanism of GABA-induced currents in cultured rat cortical neurons. *Pflugers Arch.* **437**, 289–297.

Darlison, M.G., Pahal, I., and Thode, C. (2005). Consequences of the evolution of the $GABA_A$ receptor gene family. *Cell Mol. Neurobiol.* **25**, 607–624.

Davies, P.A., Hanna, M.C., Hales, T.G., and Kirkness, E.F. (1997). Insensitivity to anaesthetic agents conferred by a class of GABA(A) receptor subunit. *Nature* **385**, 820–823.

Davies, P.A., Kirkness, E.F., and Hales, T.G. (2001). Evidence for the formation of functionally distinct alphabetagammaepsilon GABA(A) receptors. *J. Physiol.* **537**, 101–113.

Davies, P.A., Wang, W., Hales, T.G., and Kirkness, E.F. (2003). A novel class of ligand-gated ion channel is activated by $Zn2+$. *J. Biol. Chem.* **278**, 712–717.

De Deyn, P.P., D'Hooge, R., Marescau, B., and Pei, Y.Q. (1992). Chemical models of epilepsy with some reference to their applicability in the development of anticonvulsants. *Epilepsy Res.* **12**, 87–110.

De Deyn, P.P. and Macdonald, R.L. (1990). Guanidino compounds that are increased in cerebrospinal fluid and brain of uremic patients inhibit GABA and glycine responses on mouse neurons in cell culture. *Ann. Neurol.* **28**, 627–633.

Defazio, T. and Hablitz, J.J. (1998). Zinc and zolpidem modulate mIPSCs in rat neocortical pyramidal neurons. *J. Neurophysiol.* **80**, 1670–1677.

Dellisanti, C.D., Yao, Y., Stroud, J.C., Wang, Z.Z., and Chen, L. (2007). Crystal structure of the extracellular domain of nAChR alpha1 bound to alpha-bungarotoxin at 1.94 A resolution. *Nat. Neurosci.* **10**, 953–962.

Dibbens, L.M., Feng, H.J., Richards, M.C., Harkin, L.A., Hodgson, B.L., Scott, D., Jenkins, M., Petrou, S., Sutherland, G.R., Scheffer, I.E., Berkovic, S.F., Macdonald, R.L., and Mulley, J.C. (2004). GABRD encoding a protein for extra- or peri-synaptic $GABA_A$ receptors is a susceptibility locus for generalized epilepsies. *Hum. Mol. Genet.* **13**, 1315–1319.

Duggan, M.J., Pollard, S., and Stephenson, F.A. (1991). Immunoaffinity purification of $GABA_A$ receptor alpha-subunit iso-oligomers. Demonstration of receptor populations containing alpha 1 alpha 2, alpha 1 alpha 3, and alpha 2 alpha 3 subunit pairs. *J. Biol. Chem.* **266**, 24778–24784.

Edwards, F.A., Konnerth, A., and Sakmann, B. (1990). Quantal analysis of inhibitory synaptic transmission in the dentate gyrus of rat hippocampal slices: a patch-clamp study. *J. Physiol.* **430**, 213–249.

Eisele, J.L., Bertrand, S., Galzi, J.L., Devillers-Thiery, A., Changeux, J.P., and Bertrand, D. (1993). Chimaeric nicotinic-serotonergic receptor combines distinct ligand binding and channel specificities. *Nature* **366**, 479–483.

Ellgaard, L. and Frickel, E.M. (2003). Calnexin, calreticulin, and ERp57: teammates in glycoprotein folding. *Cell Biochem. Biophys.* **39**, 223–247.

Engblom, A.C., Carlson, B.X., Olsen, R.W., Schousboe, A., and Kristiansen, U. (2002). Point mutation in the first transmembrane region of the beta 2 subunit of the gamma-aminobutyric acid type A receptor alters desensitization kinetics of gamma-aminobutyric acid- and anesthetic-induced channel gating. *J. Biol. Chem.* **277**, 17438–17447.

Enz, R., Brandstatter, J.H., Wassle, H., and Bormann, J. (1996). Immunocytochemical localization of the GABAc receptor rho subunits in the mammalian retina. *J. Neurosci.* **16**, 4479–4490.

Enz, R. and Cutting, G.R. (1999). Identification of 70 amino acids important for $GABA_C$ receptor rho1 subunit assembly. *Brain Res.* **846**, 177–185.

Ernst, M., Brauchart, D., Boresch, S., and Sieghart, W. (2003). Comparative modeling of $GABA_A$ receptors: limits, insights, future developments. *Neuroscience* **119**, 933–943.

Essrich, C., Lorez, M., Benson, J.A., Fritschy, J.M., and Luscher, B. (1998). Postsynaptic clustering of major $GABA_A$ receptor subtypes requires the gamma 2 subunit and gephyrin. *Nat. Neurosci.* **1**, 563–571.

Eugene, E., Depienne, C., Baulac, S., Baulac, M., Fritschy, J.M., Le Guern, E., Miles, R., and Poncer, J.C. (2007). $GABA_A$ receptor gamma 2 subunit mutations linked to human epileptic syndromes differentially affect phasic and tonic inhibition. *J. Neurosci.* **27**, 14108–14116.

Exley, R., Moroni, M., Sasdelli, F., Houlihan, L.M., Lukas, R.J., Sher, E., Zwart, R., and Bermudez, I. (2006). Chaperone protein 14-3-3 and protein kinase A increase the relative abundance of low agonist sensitivity human alpha 4 beta 2 nicotinic acetylcholine receptors in Xenopus oocytes. *J. Neurochem.* **98**, 876–885.

Farrant, M. and Nusser, Z. (2005). Variations on an inhibitory theme: phasic and tonic activation of $GABA_A$ receptors. *Nat. Rev. Neurosci.* **6**, 215–229.

Farrar, S.J., Whiting, P.J., Bonnert, T.P., and McKernan, R.M. (1999). Stoichiometry of a ligand-gated ion channel determined by fluorescence energy transfer. *J. Biol. Chem.* **274**, 10100–10104.

Feng, H.J., Bianchi, M.T., and Macdonald, R.L. (2004). Pentobarbital differentially modulates alpha1beta3delta and alpha1beta3gamma2L $GABA_A$ receptor currents. *Mol. Pharmacol.* **66**, 988–1003.

Feng, H.J., Kang, J.Q., Song, L., Dibbens, L., Mulley, J., and Macdonald, R.L. (2006). Delta subunit susceptibility variants E177A and R220H associated with complex epilepsy alter channel gating and surface expression of alpha4beta2delta $GABA_A$ receptors. *J. Neurosci.* **26**, 1499–1506.

Feng, H.J. and Macdonald, R.L. (2004). Multiple actions of propofol on alphabetagamma and alphabetadelta $GABA_A$ receptors. *Mol. Pharmacol.* **66**, 1517–1524.

Feng, H.J., Mathews, G.C., Kao, C., and Macdonald, R.L. (2008). Alterations of $GABA_A$-receptor function and allosteric modulation during development of status epilepticus. *J. Neurophysiol.* **99**, 1285–1293.

Feng, H.J., Botzolakis, E.J., and Macdonald, R.L. (2009). Context-dependent modulation of alphabetagamma and alphabetadelta GABA(A) receptors by penicillin: implications for phasic and tonic inhibition. *Neuropharmacology* **56**, 161–173.

Filippova, N., Wotring, V.E., and Weiss, D.S. (2004). Evidence that the TM1-TM2 loop contributes to the rho1 GABA receptor pore. *J. Biol. Chem.* **279**, 20906–20914.

Fisher, J.L. and Macdonald, R.L. (1997a). Functional properties of recombinant $GABA_A$ receptors composed of single or multiple beta subunit subtypes. *Neuropharmacology* **36**, 1601–1610.

Fisher, J.L. and Macdonald, R.L. (1997b). Single channel properties of recombinant $GABA_A$ receptors containing gamma 2 or delta subtypes expressed with alpha 1 and beta 3 subtypes in mouse L929 cells. *J. Physiol.* **505**, 283–297.

Fisher, J.L., Zhang, J., and Macdonald, R.L. (1997). The role of alpha1 and alpha6 subtype amino-terminal domains in allosteric regulation of gamma-aminobutyric acida receptors. *Mol. Pharmacol.* **52**, 714–724.

Frerking, M., Borges, S., and Wilson, M. (1995). Variation in GABA mini amplitude is the consequence of variation in transmitter concentration. *Neuron* **15**, 885–895.

Fritschy, J.M., Benke, D., Mertens, S., Oertel, W.H., Bachi, T., and Mohler, H. (1992). Five subtypes of type A gamma-aminobutyric acid receptors identified in neurons by double and triple immunofluorescence staining with subunit-specific antibodies. *Proc. Natl. Acad. Sci. USA* **89**, 6726–6730.

Fritschy, J.M. and Mohler, H. (1995). $GABA_A$-receptor heterogeneity in the adult rat brain: differential regional and cellular distribution of seven major subunits. *J. Comp. Neurol.* **359**, 154–194.

Fritschy, J.M., Paysan, J., Enna, A., and Mohler, H. (1994). Switch in the expression of rat $GABA_A$-receptor subtypes during postnatal development: an immunohistochemical study. *J. Neurosci.* **14**, 5302–5324.

Frugier, G., Coussen, F., Giraud, M.F., Odessa, M.F., Emerit, M.B., Boue-Grabot, E., and Garret, M. (2007). A gamma 2(R43Q) mutation, linked to epilepsy in humans, alters $GABA_A$ receptor assembly and modifies subunit composition on the cell surface. *J. Biol. Chem.* **282**, 3819–3828.

Gallagher, M.J., Ding, L., Maheshwari, A., and Macdonald, R.L. (2007). The $GABA_A$ receptor alpha1 subunit epilepsy mutation A322D inhibits transmembrane helix formation and causes proteasomal degradation. *Proc. Natl. Acad. Sci. USA* **104**, 12999–13004.

Gallagher, M.J., Shen, W., Song, L., and Macdonald, R.L. (2005). Endoplasmic reticulum retention and associated degradation of a $GABA_A$ receptor epilepsy mutation that inserts an aspartate in the M3 transmembrane segment of the alpha1 subunit. *J. Biol. Chem.* **280**, 37995–38004.

Gallagher, M.J., Song, L., Arain, F., and Macdonald, R.L. (2004). The juvenile myoclonic epilepsy $GABA_A$ receptor alpha1 subunit mutation A322D produces asymmetrical, subunit position-dependent reduction of heterozygous receptor currents and alpha1 subunit protein expression. *J. Neurosci.* **24**, 5570–5578.

Galzi, J.L. and Changeux, J.P. (1995). Neuronal nicotinic receptors: molecular organization and regulations. *Neuropharmacology* **34**, 563–582.

Gething, M.J. (1999). Role and regulation of the ER chaperone BiP. *Semin. Cell Dev. Biol.* **10**, 465–472.

Giraudat, J., Dennis, M., Heidmann, T., Chang, J.Y., and Changeux, J.P. (1986). Structure of the high-affinity binding site for noncompetitive blockers of the acetylcholine receptor: serine 262 of the delta subunit is labeled by [3H]chlorpromazine. *Proc. Natl. Acad Sci. USA* **83**, 2719–2723.

Glatt, K., Glatt, H., and Lalande, M. (1997). Structure and organization of GABRB3 and GABRA5. *Genomics* **41**, 63–69.

Glavinovic, M.I. (1999). Monte Carlo simulation of vesicular release, spatiotemporal distribution of glutamate in synaptic cleft and generation of postsynaptic currents. *Pflugers Arch.* **437**, 462–470.

Glykys, J., Mann, E.O., and Mody, I. (2008). Which GABA(A) receptor subunits are necessary for tonic inhibition in the hippocampus? *J. Neurosci.* **28**, 1421–1426.

Glykys, J., Peng, Z., Chandra, D., Homanics, G.E., Houser, C.R., and Mody, I. (2007). A new naturally occurring $GABA_A$ receptor subunit partnership with high sensitivity to ethanol. *Nat. Neurosci.* **10**, 40–48.

Gonzales, E.B., Bell-Horner, C.L., Dibas, M.I., Huang, R.Q., and Dillon, G.H. (2008). Stoichiometric analysis of the TM2 6′ phenylalanine mutation on desensitization in alpha1beta2 and alpha1beta2gamma2 $GABA_A$ receptors. *Neurosci. Lett.* **431**, 184–189.

Gorrie, G.H., Vallis, Y., Stephenson, A., Whitfield, J., Browning, B., Smart, T.G., and Moss, S.J. (1997). Assembly of $GABA_A$ receptors composed of alpha1 and beta2 subunits in both cultured neurons and fibroblasts. *J. Neurosci.* **17**, 6587–6596.

Green, W.N. and Millar, N.S. (1995). Ion-channel assembly. *Trends Neurosci.* **18**, 280–287.

Greger, V., Knoll, J.H., Woolf, E., Glatt, K., Tyndale, R.F., DeLorey, T.M., Olsen, R.W., Tobin, A.J., Sikela, J.M., and Nakatsu, Y. (1995). The gamma-aminobutyric acid receptor gamma 3 subunit gene (GABRG3) is tightly linked to the alpha 5 subunit gene (GABRA5) on human chromosome 15q11-q13 and is transcribed in the same orientation. *Genomics* **26**, 258–264.

Grover, L.M., Lambert, N.A., Schwartzkroin, P.A., and Teyler, T.J. (1993). Role of HCO_3^- ions in depolarizing $GABA_A$ receptor-mediated responses in pyramidal cells of rat hippocampus. *J. Neurophysiol.* **69**, 1541–1555.

Haas, K.F. and Macdonald, R.L. (1999). $GABA_A$ receptor subunit gamma2 and delta subtypes confer unique kinetic properties on recombinant $GABA_A$ receptor currents in mouse fibroblasts. *J. Physiol.* **514**, 27–45.

Hackam, A.S., Wang, T.L., Guggino, W.B., and Cutting, G.R. (1997). The N-terminal domain of human GABA receptor rho1 subunits contains signals for homooligomeric and heterooligomeric interaction. *J. Biol. Chem.* **272**, 13750–13757.

Hackam, A.S., Wang, T.L., Guggino, W.B., and Cutting, G.R. (1998). Sequences in the amino termini of GABA rho and $GABA_A$ subunits specify their selective interaction in vitro. *J. Neurochem.* **70**, 40–46.

Hales, T.G., Dunlop, J.I., Deeb, T.Z., Carland, J.E., Kelley, S.P., Lambert, J.J., and Peters, J.A. (2006). Common determinants of single channel conductance within the large cytoplasmic loop of

5-hydroxytryptamine type 3 and alpha4beta2 nicotinic acetylcholine receptors. *J. Biol. Chem.* **281**, 8062–8071.

Hales, T.G., Tang, H., Bollan, K.A., Johnson, S.J., King, D.P., McDonald, N.A., Cheng, A., and Connolly, C.N. (2005). The epilepsy mutation, gamma2(R43Q) disrupts a highly conserved inter-subunit contact site, perturbing the biogenesis of GABA$_A$ receptors. *Mol. Cell Neurosci.* **29**, 120–127.

Hamann, M., Rossi, D.J., and Attwell, D. (2002). Tonic and spillover inhibition of granule cells control information flow through cerebellar cortex. *Neuron* **33**, 625–633.

Hanson, S.M. and Czajkowski, C. (2008). Structural mechanisms underlying benzodiazepine modulation of the GABA$_A$ receptor. *J. Neurosci.* **28**, 3490–3499.

Harkin, L.A., Bowser, D.N., Dibbens, L.M., Singh, R., Phillips, F., Wallace, R.H., Richards, M.C., Williams, D.A., Mulley, J.C., Berkovic, S.F., Scheffer, I.E., and Petrou, S. (2002). Truncation of the GABA(A)-receptor gamma2 subunit in a family with generalized epilepsy with febrile seizures plus. *Am. J. Hum. Genet.* **70**, 530–536.

Harvey, R.J., Topf, M., Harvey, K., and Rees, M.I. (2008). The genetics of hyperekplexia: more than startle!. *Trends Genet.* **24**, 439–447.

Hedblom, E. and Kirkness, E.F. (1997). A novel class of GABA$_A$ receptor subunit in tissues of the reproductive system. *J. Biol. Chem.* **272**, 15346–15350.

Helenius, A. and Aebi, M. (2004). Roles of N-linked glycans in the endoplasmic reticulum. *Annu. Rev. Biochem.* **73**, 1019–1049.

Herb, A., Wisden, W., Luddens, H., Puia, G., Vicini, S., and Seeburg, P.H. (1992). The third gamma subunit of the gamma-aminobutyric acid type A receptor family. *Proc. Natl. Acad. Sci. USA* **89**, 1433–1437.

Herd, M.B., Haythornthwaite, A.R., Rosahl, T.W., Wafford, K.A., Homanics, G.E., Lambert, J.J., and Belelli, D. (2008). The expression of GABA$_A$ beta subunit isoforms in synaptic and extrasynaptic receptor populations of mouse dentate gyrus granule cells. *J. Physiol.* **586**, 989–1004.

Hevers, W. and Luddens, H. (1998). The diversity of GABA$_A$ receptors. Pharmacological and electrophysiological properties of GABA$_A$ channel subtypes. *Mol. Neurobiol.* **18**, 35–86.

Hicks, A.A., Bailey, M.E., Riley, B.P., Kamphuis, W., Siciliano, M.J., Johnson, K.J., and Darlison, M.G. (1994). Further evidence for clustering of human GABA$_A$ receptor subunit genes: localization of the alpha 6-subunit gene (GABRA6) to distal chromosome 5q by linkage analysis. *Genomics* **20**, 285–288.

Hilf, R.J. and Dutzler, R. (2008). X-ray structure of a prokaryotic pentameric ligand-gated ion channel. *Nature* **452**, 375–379.

Hill, M.W., Reddy, P.A., Covey, D.F., and Rothman, S.M. (1998). Contribution of subsaturating GABA concentrations to IPSCs in cultured hippocampal neurons. *J. Neurosci.* **18**, 5103–5111.

Hirose, S. (2004). [Genetic approach to the pathogeneses of epilepsy]. *Rinsho Shinkeigaku* **44**, 975–978.

Holden, J.H. and Czajkowski, C. (2002). Different residues in the GABA(A) receptor alpha 1T60-alpha 1K70 region mediate GABA and SR-95531 actions. *J. Biol. Chem.* **277**, 18785–18792.

Horn, R. and Vandenberg, C.A. (1984). Statistical properties of single sodium channels. *J. Gen. Physiol.* **84**, 505–534.

Hotson, J.R. and Prince, D.A. (1981). Penicillin- and barium-induced epileptiform bursting in hippocampal neurons: actions on Ca^{++} and K^+ potentials. *Ann. Neurol.* **10**, 11–17.

Hwa, G.G., Avoli, M., Oliver, A., and Villemure, J.G. (1991). Bicuculline-induced epileptogenesis in the human neocortex maintained in vitro. *Exp. Brain Res.* **83**, 329–339.

Imoto, K., Methfessel, C., Sakmann, B., Mishina, M., Mori, Y., Konno, T., Fukuda, K., Kurasaki, M., Bujo, H., and Fujita, Y. (1986). Location of a delta-subunit region determining ion transport through the acetylcholine receptor channel. *Nature* **324**, 670–674.

Jacob, T.C., Moss, S.J., and Jurd, R. (2008). GABA$_A$ receptor trafficking and its role in the dynamic modulation of neuronal inhibition. *Nat. Rev. Neurosci.* **9**, 331–343.

Jansen, M. and Akabas, M.H. (2006). State-dependent cross-linking of the M2 and M3 segments: functional basis for the alignment of GABA$_A$ and acetylcholine receptor M3 segments. *J. Neurosci.* **26**, 4492–4499.

Jeanclos, E.M., Lin, L., Treuil, M.W., Rao, J., DeCoster, M.A., and Anand, R. (2001). The chaperone protein 14-3-3eta interacts with the nicotinic acetylcholine receptor alpha 4 subunit. Evidence for a dynamic role in subunit stabilization. *J. Biol. Chem.* **276**, 28281–28290.

Jechlinger, M., Pelz, R., Tretter, V., Klausberger, T., and Sieghart, W. (1998). Subunit composition and quantitative importance of hetero-oligomeric receptors: GABA$_A$ receptors containing alpha6 subunits. *J. Neurosci.* **18**, 2449–2457.

Jensen, M.L., Timmermann, D.B., Johansen, T.H., Schousboe, A., Varming, T., and Ahring, P.K. (2002). The beta subunit determines the ion selectivity of the GABA$_A$ receptor. *J. Biol. Chem.* **277**, 41438–41447.

Johnston, G.A. (1996). GABA$_A$ receptor pharmacology. *Pharmacol. Ther.* **69**, 173–198.

Jones, B.L. and Henderson, L.P. (2007). Trafficking and potential assembly patterns of epsilon-containing GABA$_A$ receptors. *J. Neurochem.* **103**, 1258–1271.

Jones, M.V., Sahara, Y., Dzubay, J.A., and Westbrook, G.L. (1998). Defining affinity with the GABA$_A$ receptor. *J. Neurosci.* **18**, 8590–8604.

Jones, M.V. and Westbrook, G.L. (1995). Desensitized states prolong GABA$_A$ channel responses to brief agonist pulses. *Neuron* **15**, 181–191.

Jones, M.V. and Westbrook, G.L. (1996). The impact of receptor desensitization on fast synaptic transmission. *Trends Neurosci.* **19**, 96–101.

Jones-Davis, D.M., Song, L., Gallagher, M.J., and Macdonald, R.L. (2005). Structural determinants of benzodiazepine allosteric regulation of GABA$_A$ receptor currents. *J. Neurosci.* **25**, 8056–8065.

Kananura, C., Haug, K., Sander, T., Runge, U., Gu, W., Hallmann, K., Rebstock, J., Heils, A., and Steinlein, O.K. (2002). A splice-site mutation in GABRG2 associated with childhood absence epilepsy and febrile convulsions. *Arch. Neurol.* **59**, 1137–1141.

Kanematsu, T., Fujii, M., Mizokami, A., Kittler, J.T., Nabekura, J., Moss, S.J., and Hirata, M. (2007). Phospholipase C-related inactive protein is implicated in the constitutive internalization of GABA$_A$ receptors mediated by clathrin and AP2 adaptor complex. *J. Neurochem.* **101**, 898–905.

Kanematsu, T. and Hirata, M. (2002). [The analysis of protein–protein interaction with special reference to PRIP-1]. *Nippon Yakurigaku Zasshi* **119**, 241–246.

Kang, J.Q. and Macdonald, R.L. (2004). The GABA$_A$ receptor gamma2 subunit R43Q mutation linked to childhood absence epilepsy and febrile seizures causes retention of alpha1beta2-gamma2S receptors in the endoplasmic reticulum. *J. Neurosci.* **24**, 8672–8677.

Kapur, J. and Macdonald, R.L. (1997). Rapid seizure-induced reduction of benzodiazepine and Zn^{2+} sensitivity of hippocampal dentate granule cell GABA$_A$ receptors. *J. Neurosci.* **17**, 7532–7540.

Kash, T.L., Jenkins, A., Kelley, J.C., Trudell, J.R., and Harrison, N.L. (2003). Coupling of agonist binding to channel gating in the GABA(A) receptor. *Nature* **421**, 272–275.

Kaur, K.H., Baur, R., and Sigel, E. (2009). Unanticipated structural and functional properties of delta subunit containing GABA$_A$ receptors. *J. Biol. Chem.* **284**, 7889–7896.

Kelley, S.P., Dunlop, J.I., Kirkness, E.F., Lambert, J.J., and Peters, J.A. (2003). A cytoplasmic region determines single-channel conductance in 5-HT3 receptors. *Nature* **424**, 321–324.

Keramidas, A., Kash, T.L., and Harrison, N.L. (2006). The pre-M1 segment of the alpha1 subunit is a transduction element in the activation of the $GABA_A$ receptor. *J. Physiol.* **575**, 11–22.

Keramidas, A., Moorhouse, A.J., Schofield, P.R., and Barry, P.H. (2004). Ligand-gated ion channels: mechanisms underlying ion selectivity. *Prog. Biophys. Mol. Biol.* **86**, 161–204.

Khan, Z.U., Gutierrez, A., and De Blas, A.L. (1994). Short and long form gamma 2 subunits of the $GABA_A$/benzodiazepine receptors. *J. Neurochem.* **63**, 1466–1476.

Kim, D.Y., Fenoglio, K.A., Kerrigan, J.F., and Rho, J.M. (2009). Bicarbonate contributes to $GABA_A$ receptor-mediated neuronal excitation in surgically resected human hypothalamic hamartomas. *Epilepsy Research* **83**, 89–93.

Kirkness, E.F. and Fraser, C.M. (1993). A strong promoter element is located between alternative exons of a gene encoding the human gamma-aminobutyric acid-type A receptor beta 3 subunit (GABRB3). *J. Biol. Chem.* **268**, 4420–4428.

Kittler, J.T., Chen, G., Honing, S., Bogdanov, Y., McAinsh, K., Arancibia-Carcamo, I.L., Jovanovic, J.N., Pangalos, M.N., Haucke, V., Yan, Z., and Moss, S.J. (2005). Phospho-dependent binding of the clathrin AP2 adaptor complex to $GABA_A$ receptors regulates the efficacy of inhibitory synaptic transmission. *Proc. Natl. Acad. Sci. USA* **102**, 14871–14876.

Kittler, J.T. and Moss, S.J. (2003). Modulation of $GABA_A$ receptor activity by phosphorylation and receptor trafficking: implications for the efficacy of synaptic inhibition. *Curr. Opin. Neurobiol.* **13**, 341–347.

Klausberger, T., Ehya, N., Fuchs, K., Fuchs, T., Ebert, V., Sarto, I., and Sieghart, W. (2001). Detection and binding properties of $GABA_A$ receptor assembly intermediates. *J. Biol. Chem.* **276**, 16024–16032.

Klausberger, T., Fuchs, K., Mayer, B., Ehya, N., and Sieghart, W. (2000). $GABA_A$ receptor assembly Identification and structure of gamma(2) sequences forming the intersubunit contacts with alpha(1) and beta(3) subunits. *J. Biol. Chem.* **275**, 8921–8928.

Kneussel, M. and Betz, H. (2000). Receptors, gephyrin and gephyrin-associated proteins: novel insights into the assembly of inhibitory postsynaptic membrane specializations. *J. Physiol.* **525**, 1–9.

Knight, A.R., Hartnett, C., Marks, C., Brown, M., Gallager, D., Tallman, J., and Ramabhadran, T.V. (1998). Molecular size of recombinant alpha1beta1 and alpha1beta1gamma2 $GABA_A$ receptors expressed in Sf9 cells. Receptors. *Channels* **6**, 1–18.

Korn, S.J., Giacchino, J.L., Chamberlin, N.L., and Dingledine, R. (1987). Epileptiform burst activity induced by potassium in the hippocampus and its regulation by GABA-mediated inhibition. *J. Neurophysiol.* **57**, 325–340.

Korn, S.J. and Horn, R. (1988). Statistical discrimination of fractal and Markov models of single-channel gating. *Biophys. J.* **54**, 871–877.

Kulik, A., Nishimaru, H., and Ballanyi, K. (2000). Role of bicarbonate and chloride in GABA- and glycine-induced depolarization and [Ca2+]i rise in fetal rat motoneurons in situ. *J. Neurosci.* **20**, 7905–7913.

Lagrange, A.H., Botzolakis, E.J., and Macdonald, R.L. (2007). Enhanced macroscopic desensitization shapes the response of alpha4 subtype-containing $GABA_A$ receptors to synaptic and extrasynaptic GABA. *J. Physiol.* **578**, 655–676.

Langosch, D., Laube, B., Rundstrom, N., Schmieden, V., Bormann, J., and Betz, H. (1994). Decreased agonist affinity and chloride conductance of mutant glycine receptors associated with human hereditary hyperekplexia. *EMBO J.* **13**, 4223–4228.

Langosch, D., Thomas, L., and Betz, H. (1988). Conserved quaternary structure of ligand-gated ion channels: the postsynaptic glycine receptor is a pentamer. *Proc. Natl. Acad. Sci. USA* **85**, 7394–7398.

Lape, R., Colquhoun, D., and Sivilotti, L.G. (2008). On the nature of partial agonism in the nicotinic receptor superfamily. *Nature* **454**, 722–727.

Laurie, D.J., Seeburg, P.H., and Wisden, W. (1992a). The distribution of 13 $GABA_A$ receptor subunit mRNAs in the rat brain. II. Olfactory bulb and cerebellum. *J. Neurosci.* **12**, 1063–1076.

Laurie, D.J., Wisden, W., and Seeburg, P.H. (1992b). The distribution of thirteen $GABA_A$ receptor subunit mRNAs in the rat brain. III. Embryonic and postnatal development. *J. Neurosci.* **12**, 4151–4172.

Lerma, J., Herranz, A.S., Herreras, O., Abraira, V., and Martin, D.R. (1986). In vivo determination of extracellular concentration of amino acids in the rat hippocampus. A method based on brain dialysis and computerized analysis. *Brain Res.* **384**, 145–155.

Levitan, E.S., Schofield, P.R., Burt, D.R., Rhee, L.M., Wisden, W., Kohler, M., Fujita, N., Rodriguez, H.F., Stephenson, A., and Darlison, M.G. (1988). Structural and functional basis for $GABA_A$ receptor heterogeneity. *Nature* **335**, 76–79.

Li, G.D., Chiara, D.C., Sawyer, G.W., Husain, S.S., Olsen, R.W., and Cohen, J.B. (2006). Identification of a $GABA_A$ receptor anesthetic binding site at subunit interfaces by photolabeling with an etomidate analog. *J. Neurosci.* **26**, 11599–11605.

Li, M. and De Blas, A.L. (1997). Coexistence of two beta subunit isoforms in the same gamma-aminobutyric acid type A receptor. *J. Biol. Chem.* **272**, 16564–16569.

Liu, F., Wan, Q., Pristupa, Z.B., Yu, X.M., Wang, Y.T., and Niznik, H.B. (2000). Direct protein–protein coupling enables cross-talk between dopamine D5 and gamma-aminobutyric acid A receptors. *Nature* **403**, 274–280.

Lo, W.Y., Botzolakis, E.J., Tang, X., and Macdonald, R.L. (2008). A conserved Cys-loop receptor aspartate residue in the M3–M4 cytoplasmic loop is required for $GABA_A$ receptor assembly. *J. Biol. Chem.* **283**, 29740–29752.

Lopez-Chavez, A., Miledi, R., and Martinez-Torres, A. (2005). Cloning and functional expression of the bovine $GABA_C$ rho2 subunit. Molecular evidence of a widespread distribution in the CNS. *Neurosci. Res.* **53**, 421–427.

Luddens, H., Killisch, I., and Seeburg, P.H. (1991). More than one alpha variant may exist in a $GABA_A$/benzodiazepine receptor complex. *J. Recept. Res.* **11**, 535–551.

Maattanen, P., Kozlov, G., Gehring, K., and Thomas, D.Y. (2006). ERp57 and PDI: multifunctional protein disulfide isomerases with similar domain architectures but differing substrate-partner associations. *Biochem. Cell Biol.* **84**, 881–889.

Macdonald, R.L., Kang, J.Q., Gallagher, M.J., and Feng, H.J. (2006). GABA(A) receptor mutations associated with generalized epilepsies. *Adv. Pharmacol.* **54**, 147–169.

Macdonald, R.L. and Olsen, R.W. (1994). $GABA_A$ receptor channels. *Annu. Rev. Neurosci.* **17**, 569–602.

Macdonald, R.L., Rogers, C.J., and Twyman, R.E. (1989). Kinetic properties of the $GABA_A$ receptor main conductance state of mouse spinal cord neurones in culture. *J. Physiol.* **410**, 479–499.

Maconochie, D.J., Zempel, J.M., and Steinbach, J.H. (1994). How quickly can $GABA_A$ receptors open? *Neuron* **12**, 61–71.

Maljevic, S., Krampfl, K., Cobilanschi, J., Tilgen, N., Beyer, S., Weber, Y.G., Schlesinger, F., Ursu, D., Melzer, W., Cossette, P., Bufler, J., Lerche, H., and Heils, A. (2006). A mutation in the $GABA_A$ receptor alpha(1)-subunit is associated with absence epilepsy. *Ann. Neurol.* **59**, 983–987.

Mamalaki, C., Barnard, E.A., and Stephenson, F.A. (1989). Molecular size of the gamma-aminobutyric acidA receptor purified from mammalian cerebral cortex. *J. Neurochem.* **52**, 124–134.

Maricq, A.V., Peterson, A.S., Brake, A.J., Myers, R.M., and Julius, D. (1991). Primary structure and functional expression of the 5HT3 receptor, a serotonin-gated ion channel. *Science* **254**, 432–437.

McKernan, R.M. and Whiting, P.J. (1996). Which GABA$_A$-receptor subtypes really occur in the brain? *Trends Neurosci.* **19**, 139–143.

McLean, P.J., Farb, D.H., and Russek, S.J. (1995). Mapping of the alpha 4 subunit gene (GABRA4) to human chromosome 4 defines an alpha 2-alpha 4-beta 1-gamma 1 gene cluster: further evidence that modern GABA$_A$ receptor gene clusters are derived from an ancestral cluster. *Genomics* **26**, 580–586.

McManus, O.B., Weiss, D.S., Spivak, C.E., Blatz, A.L., and Magleby, K.L. (1988). Fractal models are inadequate for the kinetics of four different ion channels. *Biophys. J.* **54**, 859–870.

Mejia, C., Garcia-Alcocer, G., Berumen, L.C., Rosas-Arellano, A., Miledi, R., and Martinez-Torres, A. (2008). Expression of GABArho subunits during rat cerebellum development. *Neurosci. Lett.* **432**, 1–6.

Mercik, K., Pytel, M., Cherubini, E., and Mozrzymas, J.W. (2006). Effect of extracellular pH on recombinant alpha1beta2gamma2 and alpha1beta2 GABA$_A$ receptors. *Neuropharmacology* **51**, 305–314.

Mihic, S.J., Ye, Q., Wick, M.J., Koltchine, V.V., Krasowski, M.D., Finn, S.E., Mascia, M.P., Valenzuela, C.F., Hanson, K.K., Greenblatt, E.P., Harris, R.A., and Harrison, N.L. (1997). Sites of alcohol and volatile anaesthetic action on GABA$_A$ and glycine receptors. *Nature* **389**, 385–389.

Mintz, I.M. and Bean, B.P. (1993). GABA$_B$ receptor inhibition of P-type Ca2+ channels in central neurons. *Neuron* **10**, 889–898.

Miyazawa, A., Fujiyoshi, Y., and Unwin, N. (2003). Structure and gating mechanism of the acetylcholine receptor pore. *Nature* **423**, 949–955.

Mody, I. and Pearce, R.A. (2004). Diversity of inhibitory neurotransmission through GABA$_A$ receptors. *Trends Neurosci.* **27**, 569–575.

Moragues, N., Ciofi, P., Lafon, P., Odessa, M.F., Tramu, G., and Garret, M. (2000). cDNA cloning and expression of a gamma-aminobutyric acid A receptor epsilon-subunit in rat brain. *Eur. J. Neurosci.* **12**, 4318–4330.

Mortensen, M. and Smart, T.G. (2006). Extrasynaptic alphabeta subunit GABA$_A$ receptors on rat hippocampal pyramidal neurons. *J. Physiol.* **577**, 841–856.

Moss, S.J. and Smart, T.G. (2001). Constructing inhibitory synapses. *Nat. Rev. Neurosci.* **2**, 240–250.

Mozrzymas, J.W. (2004). Dynamism of GABA$_A$ receptor activation shapes the "personality" of inhibitory synapses. *Neuropharmacology* **47**, 945–960.

Mozrzymas, J.W. and Barberis, A. (2004). Changes of GABA$_A$ receptor activation kinetics in hippocampal neurons cultured for different periods of time. *Cell Mol. Biol. Lett.* **9**, 61–67.

Mozrzymas, J.W., Barberis, A., Mercik, K., and Zarnowska, E.D. (2003). Binding sites, singly bound states, and conformation coupling shape GABA-evoked currents. *J. Neurophysiol.* **89**, 871–883.

Mozrzymas, J.W., Barberis, A., and Vicini, S. (2007). GABAergic currents in RT and VB thalamic nuclei follow kinetic pattern of alpha3- and alpha1-subunit-containing GABA$_A$ receptors. *Eur. J. Neurosci.* **26**, 657–665.

Nayeem, N., Green, T.P., Martin, I.L., and Barnard, E.A. (1994). Quaternary structure of the native GABA$_A$ receptor determined by electron microscopic image analysis. *J. Neurochem.* **62**, 815–818.

Neelands, T.R., Fisher, J.L., Bianchi, M., and Macdonald, R.L. (1999). Spontaneous and gamma-aminobutyric acid (GABA)-activated GABA$_A$ receptor channels formed by epsilon subunit-containing isoforms. *Mol. Pharmacol.* **55**, 168–178.

Neelands, T.R. and Macdonald, R.L. (1999). Incorporation of the pi subunit into functional gamma-aminobutyric Acid(A) receptors. *Mol. Pharmacol.* **56**, 598–610.

Newell, J.G. and Czajkowski, C. (2003). The GABA$_A$ receptor alpha 1 subunit Pro174-Asp191 segment is involved in GABA binding and channel gating. *J. Biol. Chem.* **278**, 13166–13172.

Newland, C.F., Colquhoun, D., and Cull-Candy, S.G. (1991). Single channels activated by high concentrations of GABA in superior cervical ganglion neurones of the rat. *J. Physiol.* **432**, 203–233.

Noda, M., Takahashi, H., Tanabe, T., Toyosato, M., Kikyotani, S., Furutani, Y., Hirose, T., Takashima, H., Inayama, S., Miyata, T., and Numa, S. (1983). Structural homology of Torpedo californica acetylcholine receptor subunits. *Nature* **302**, 528–532.

Nusser, Z., Sieghart, W., and Somogyi, P. (1998). Segregation of different GABA$_A$ receptors to synaptic and extrasynaptic membranes of cerebellar granule cells. *J. Neurosci.* **18**, 1693–1703.

O'Mara, M., Cromer, B., Parker, M., and Chung, S.H. (2005). Homology model of the GABA$_A$ receptor examined using Brownian dynamics. *Biophys. J.* **88**, 3286–3299.

Ohlson, J., Pedersen, J.S., Haussler, D., and Ohman, M. (2007). Editing modifies the GABA$_A$ receptor subunit alpha3. *RNA* **13**, 698–703.

Olsen, R.W., Chang, C.S., Li, G., Hanchar, H.J., and Wallner, M. (2004). Fishing for allosteric sites on GABA$_A$ receptors. *Biochem. Pharmacol.* **68**, 1675–1684.

Olsen, R.W. and Sieghart, W. (2008). GABA$_A$ receptors: subtypes provide diversity of function and pharmacology. *Neuropharmacology*.

Olsen, R.W. and Tobin, A.J. (1990). Molecular biology of GABA$_A$ receptors. *FASEB J.* **4**, 1469–1480.

Otis, T.S., De Koninck, Y., and Mody, I. (1994). Lasting potentiation of inhibition is associated with an increased number of gamma-aminobutyric acid type A receptors activated during miniature inhibitory postsynaptic currents. *Proc. Natl. Acad. Sci. USA* **91**, 7698–7702.

Owens, D.F. and Kriegstein, A.R. (2002). Is there more to GABA than synaptic inhibition? *Nat. Rev. Neurosci* **3**, 715–727.

Padgett, C.L., Hanek, A.P., Lester, H.A., Dougherty, D.A., and Lummis, S.C. (2007). Unnatural amino acid mutagenesis of the GABA$_A$ receptor binding site residues reveals a novel cation-pi interaction between GABA and beta 2Tyr97. *J. Neurosci.* **27**, 886–892.

Pan, Y., Ripps, H., and Qian, H. (2006). Random assembly of GABA rho1 and rho2 subunits in the formation of heteromeric GABA(C) receptors. *Cell Mol. Neurobiol.* **26**, 289–305.

Perkins, K.L. and Wong, R.K. (1997). The depolarizing GABA response. *Can. J. Physiol. Pharmacol.* **75**, 516–519.

Peters, J.A., Kelley, S.P., Dunlop, J.I., Kirkness, E.F., Hales, T.G., and Lambert, J.J. (2004). The 5-hydroxytryptamine type 3 (5-HT3) receptor reveals a novel determinant of single-channel conductance. *Biochem. Soc. Trans.* **32**, 547–552.

Picton, A.J. and Fisher, J.L. (2007). Effect of the alpha subunit subtype on the macroscopic kinetic properties of recombinant GABA$_A$ receptors. *Brain Res.* **1165**, 40–49.

Pirker, S., Schwarzer, C., Wieselthaler, A., Sieghart, W., and Sperk, G. (2000). GABA(A) receptors: immunocytochemical distribution of 13 subunits in the adult rat brain. *Neuroscience* **101**, 815–850.

Plested, A.J., Groot-Kormelink, P.J., Colquhoun, D., and Sivilotti, L.G. (2007). Single-channel study of the spasmodic mutation alpha1A52S in recombinant rat glycine receptors. *J. Physiol.* **581**, 51–73.

Poulter, M.O., Brown, L.A., Tynan, S., Willick, G., William, R., and McIntyre, D.C. (1999). Differential expression of alpha1, alpha2, alpha3, and alpha5 GABA$_A$ receptor subunits in seizure-prone and seizure-resistant rat models of temporal lobe epilepsy. *J. Neurosci.* **19**, 4654–4661.

Pritchett, D.B., Sontheimer, H., Shivers, B.D., Ymer, S., Kettenmann, H., Schofield, P.R., and Seeburg, P.H. (1989). Importance of a novel GABA$_A$ receptor subunit for benzodiazepine pharmacology. *Nature* **338**, 582–585.

Quirk, K., Gillard, N.P., Ragan, C.I., Whiting, P.J., and McKernan, R.M. (1994). Gamma-aminobutyric acid type A receptors in the rat brain can contain both gamma 2 and gamma 3 subunits, but gamma 1 does not exist in combination with another gamma subunit. *Mol. Pharmacol.* **45**, 1061–1070.

Quirk, K., Whiting, P.J., Ragan, C.I., and McKernan, R.M. (1995). Characterisation of delta-subunit containing $GABA_A$ receptors from rat brain. *Eur. J. Pharmacol.* **290**, 175–181.

Rabow, L.E., Russek, S.J., and Farb, D.H. (1995). From ion currents to genomic analysis: recent advances in $GABA_A$ receptor research. *Synapse* **21**, 189–274.

Richardson, J.E., Garcia, P.S., O'Toole, K.K., Derry, J.M., Bell, S.V., and Jenkins, A. (2007). A conserved tyrosine in the beta2 subunit M4 segment is a determinant of gamma-aminobutyric acid type A receptor sensitivity to propofol. *Anesthesiology* **107**, 412 418.

Rothberg, B.S. and Magleby, K.L. (2000). Voltage and Ca2+ activation of single large-conductance Ca^{2+}-activated K^+ channels described by a two-tiered allosteric gating mechanism. *J. Gen. Physiol.* **116**, 75–99.

Rula, E.Y., Lagrange, A.H., Jacobs, M.M., Hu, N., Macdonald, R.L., and Emeson, R.B. (2008). Developmental modulation of $GABA_A$ receptor function by RNA editing. *J. Neurosci.* **28**, 6196–6201.

Russek, S.J. and Farb, D.H. (1994). Mapping of the beta 2 subunit gene (GABRB2) to microdissected human chromosome 5q34-q35 defines a gene cluster for the most abundant $GABA_A$ receptor isoform. *Genomics* **23**, 528–533.

Sancar, F. and Czajkowski, C. (2004). A $GABA_A$ receptor mutation linked to human epilepsy (gamma2R43Q) impairs cell surface expression of alphabetagamma receptors. *J. Biol. Chem.* **279**, 47034–47039.

Sansom, M.S., Ball, F.G., Kerry, C.J., McGee, R., Ramsey, R.L., and Usherwood, P.N. (1989). Markov, fractal, diffusion, and related models of ion channel gating. A comparison with experimental data from two ion channels. *Biophys. J.* **56**, 1229–1243.

Sarto-Jackson, I. and Sieghart, W. (2008). Assembly of $GABA_A$ receptors (Review). *Mol. Membr. Biol.* **25**, 302–310.

Saxena, N.C. and Macdonald, R.L. (1994). Assembly of $GABA_A$ receptor subunits: role of the delta subunit. *J. Neurosci.* **14**, 7077–7086.

Saxena, N.C. and Macdonald, R.L. (1996). Properties of putative cerebellar gamma-aminobutyric acid A receptor isoforms. *Mol. Pharmacol.* **49**, 567–579.

Scheller, M. and Forman, S.A. (2002). Coupled and uncoupled gating and desensitization effects by pore domain mutations in GABA(A) receptors. *J. Neurosci.* **22**, 8411–8421.

Schiller, Y. (2002). Inter-ictal- and ictal-like epileptic discharges in the dendritic tree of neocortical pyramidal neurons. *J. Neurophysiol.* **88**, 2954–2962.

Schneiderman, J.H. (1997). The role of long-term potentiation in persistent epileptiform burst-induced hyperexcitability following $GABA_A$ receptor blockade. *Neuroscience* **81**, 1111–1122.

Schofield, C.M. and Harrison, N.L. (2005). Transmembrane residues define the action of isoflurane at the $GABA_A$ receptor alpha-3 subunit. *Brain Res.* **1032**, 30–35.

Schofield, P.R., Darlison, M.G., Fujita, N., Burt, D.R., Stephenson, F.A., Rodriguez, H., Rhee, L.M., Ramachandran, J., Reale, V., and Glencorse, T.A. (1987). Sequence and functional expression of the $GABA_A$ receptor shows a ligand-gated receptor super-family. *Nature* **328**, 221–227.

Schoppa, N.E. and Sigworth, F.J. (1998). Activation of Shaker potassium channels. III. An activation gating model for wild-type and V2 mutant channels. *J. Gen. Physiol.* **111**, 313–342.

Schwartzkroin, P.A. and Prince, D.A. (1980). Changes in excitatory and inhibitory synaptic potentials leading to epileptogenic activity. *Brain Res.* **183**, 61–76.

Serafini, R., Bracamontes, J., and Steinbach, J.H. (2000). Structural domains of the human $GABA_A$ receptor 3 subunit involved in the actions of pentobarbital. *J. Physiol.* **524**, 649–676.

Shivers, B.D., Killisch, I., Sprengel, R., Sontheimer, H., Kohler, M., Schofield, P.R., and Seeburg, P.H. (1989). Two novel $GABA_A$ receptor subunits exist in distinct neuronal subpopulations. *Neuron* **3**, 327–337.

Sieghart, W. and Sperk, G. (2002). Subunit composition, distribution and function of $GABA_A$ receptor subtypes. *Curr. Top. Med. Chem.* **2**, 795–816.

Sigg, D. and Bezanilla, F. (2003). A physical model of potassium channel activation: from energy landscape to gating kinetics. *Biophys. J.* **84**, 3703–3716.

Simon, J., Wakimoto, H., Fujita, N., Lalande, M., and Barnard, E.A. (2004). Analysis of the set of $GABA_A$ receptor genes in the human genome. *J. Biol. Chem.* **279**, 41422–41435.

Smith, G.B. and Olsen, R.W. (1995). Functional domains of $GABA_A$ receptors. *Trends Pharmacol. Sci.* **16**, 162–168.

Smith, K.R., McAinsh, K., Chen, G., Arancibia-Carcamo, I.L., Haucke, V., Yan, Z., Moss, S.J., and Kittler, J.T. (2008). Regulation of inhibitory synaptic transmission by a conserved atypical interaction of $GABA_A$ receptor beta- and gamma-subunits with the clathrin AP2 adaptor. *Neuropharmacology*.

Sommer, B., Poustka, A., Spurr, N.K., and Seeburg, P.H. (1990). The murine $GABA_A$ receptor delta-subunit gene: structure and assignment to human chromosome 1. DNA. *Cell Biol.* **9**, 561–568.

Strecker, G.J., Park, W.K., and Dudek, F.E. (1999). Zinc and flunitrazepam modulation of GABA-mediated currents in rat suprachiasmatic neurons. *J. Neurophysiol.* **81**, 184–191.

Sur, C., Farrar, S.J., Kerby, J., Whiting, P.J., Atack, J.R., and McKernan, R.M. (1999). Preferential coassembly of alpha4 and delta subunits of the gamma-aminobutyric acidA receptor in rat thalamus. *Mol. Pharmacol.* **56**, 110–115.

Tanaka, M., Olsen, R.W., Medina, M.T., Schwartz, E., Alonso, M.E., Duron, R.M., Castro-Ortega, R., Martinez-Juarez, I.E., Pascual-Castroviejo, I., Machado-Salas, J., Silva, R., Bailey, J.N., Bai, D., Ochoa, A., Jara-Prado, A., Pineda, G., Macdonald, R.L., and Delgado-Escueta, A.V. (2008). Hyperglycosylation and reduced GABA currents of mutated GABRB3 polypeptide in remitting childhood absence epilepsy. *Am. J. Hum. Genet.* **82**, 1249–1261.

Taylor, P.M., Connolly, C.N., Kittler, J.T., Gorrie, G.H., Hosie, A., Smart, T.G., and Moss, S.J. (2000). Identification of residues within $GABA_A$ receptor alpha subunits that mediate specific assembly with receptor beta subunits. *J. Neurosci.* **20**, 1297–1306.

Taylor, P.M., Thomas, P., Gorrie, G.H., Connolly, C.N., Smart, T.G., and Moss, S.J. (1999). Identification of amino acid residues within $GABA_A$ receptor beta subunits that mediate both homomeric and heteromeric receptor expression. *J. Neurosci.* **19**, 6360–6371.

Tobimatsu, T., Fujita, Y., Fukuda, K., Tanaka, K., Mori, Y., Konno, T., Mishina, M., and Numa, S. (1987). Effects of substitution of putative transmembrane segments on nicotinic acetylcholine receptor function. *FEBS Lett.* **222**, 56–62.

Tretter, V., Ehya, N., Fuchs, K., and Sieghart, W. (1997). Stoichiometry and assembly of a recombinant $GABA_A$ receptor subtype. *J. Neurosci.* **17**, 2728–2737.

Tretter, V., Jacob, T.C., Mukherjee, J., Fritschy, J.M., Pangalos, M.N., and Moss, S.J. (2008). The clustering of $GABA_A$ receptor subtypes at inhibitory synapses is facilitated via the direct binding of receptor alpha 2 subunits to gephyrin. *J. Neurosci.* **28**, 1356–1365.

Trudell, J.R. and Bertaccini, E. (2004). Comparative modeling of a $GABA_A$ alpha1 receptor using three crystal structures as templates. *J. Mol. Graph. Model.* **23**, 39–49.

Twyman, R.E., Green, R.M., and Macdonald, R.L. (1992). Kinetics of open channel block by penicillin of single $GABA_A$ receptor channels from mouse spinal cord neurones in culture. *J. Physiol.* **445**, 97–127.

Twyman, R.E. and Macdonald, R.L. (1992). Neurosteroid regulation of $GABA_A$ receptor single-channel kinetic properties of mouse spinal cord neurons in culture. *J. Physiol.* **456**, 215–245.

Twyman, R.E., Rogers, C.J., and Macdonald, R.L. (1989a). Differential regulation of gamma-aminobutyric acid receptor channels by diazepam and phenobarbital. *Ann. Neurol.* **25**, 213–220.

Twyman, R.E., Rogers, C.J., and Macdonald, R.L. (1989b). Pentobarbital and picrotoxin have reciprocal actions on single $GABA_A$ receptor channels. *Neurosci. Lett.* **96**, 89–95.

Twyman, R.E., Rogers, C.J., and Macdonald, R.L. (1990). Intraburst kinetic properties of the $GABA_A$ receptor main conductance state of mouse spinal cord neurones in culture. *J. Physiol.* **423**, 193–220.

Ulrich, D. and Bettler, B. (2007). $GABA_B$ receptors: synaptic functions and mechanisms of diversity. *Curr. Opin. Neurobiol.* **17**, 298–303.

Unwin, N. (1993). Nicotinic acetylcholine receptor at 9 A resolution. *J. Mol. Biol.* **229**, 1101–1124.

Unwin, N. (2005). Refined structure of the nicotinic acetylcholine receptor at 4A resolution. *J. Mol. Biol.* **346**, 967–989.

Uusi-Oukari, M., Heikkila, J., Sinkkonen, S.T., Makela, R., Hauer, B., Homanics, G.E., Sieghart, W., Wisden, W., and Korpi, E.R. (2000). Long-range interactions in neuronal gene expression: evidence from gene targeting in the $GABA_A$ receptor beta2-alpha6-alpha1-gamma2 subunit gene cluster. *Mol. Cell Neurosci.* **16**, 34–41.

Ventriglia, F. and Di Maio, V. (2003). Stochastic fluctuations of the quantal EPSC amplitude in computer simulated excitatory synapses of hippocampus. *Biosystems* **71**, 195–204.

Vicini, S., Ferguson, C., Prybylowski, K., Kralic, J., Morrow, A.L., and Homanics, G.E. (2001). $GABA_A$ receptor alpha1 subunit deletion prevents developmental changes of inhibitory synaptic currents in cerebellar neurons. *J. Neurosci.* **21**, 3009–3016.

Wagner, D.A., Goldschen-Ohm, M.P., Hales, T.G., and Jones, M.V. (2005). Kinetics and spontaneous open probability conferred by the epsilon subunit of the $GABA_A$ receptor. *J. Neurosci.* **25**, 10462–10468.

Wagner, P.G. and Dekin, M.S. (1993). $GABA_b$ receptors are coupled to a barium-insensitive outward rectifying potassium conductance in premotor respiratory neurons. *J. Neurophysiol.* **69**, 286–289.

Wallace, R.H., Marini, C., Petrou, S., Harkin, L.A., Bowser, D.N., Panchal, R.G., Williams, D.A., Sutherland, G.R., Mulley, J.C., Scheffer, I.E., and Berkovic, S.F. (2001). Mutant $GABA_A$ receptor gamma2-subunit in childhood absence epilepsy and febrile seizures. *Nat. Genet.* **28**, 49–52.

Wanamaker, C.P. and Green, W.N. (2007). Endoplasmic reticulum chaperones stabilize nicotinic receptor subunits and regulate receptor assembly. *J. Biol. Chem.* **282**, 31113–31123.

Wang, H., Bedford, F.K., Brandon, N.J., Moss, S.J., and Olsen, R. W. (1999). GABA(A)-receptor-associated protein links $GABA_A$ receptors and the cytoskeleton. *Nature* **397**, 69–72.

Wang, W. and Shakes, D.C. (1996). Molecular evolution of the 14-3-3 protein family. *J. Mol. Evol.* **43**, 384–398.

Weiss, D.S. and Magleby, K.L. (1989). Gating scheme for single GABA-activated Cl^- channels determined from stability plots, dwell-time distributions, and adjacent-interval durations. *J. Neurosci.* **9**, 1314–1324.

Whiting, P., McKernan, R.M., and Iversen, L.L. (1990). Another mechanism for creating diversity in gamma-aminobutyrate type A receptors: RNA splicing directs expression of two forms of gamma 2 phosphorylation site. *Proc. Natl. Acad. Sci. USA* **87**, 9966–9970.

Wilcox, A.S., Warrington, J.A., Gardiner, K., Berger, R., Whiting, P., Altherr, M.R., Wasmuth, J.J., Patterson, D., and Sikela, J.M. (1992). Human chromosomal localization of genes encoding the gamma 1 and gamma 2 subunits of the gamma-aminobutyric acid receptor indicates that members of this gene family are often clustered in the genome. *Proc. Natl. Acad. Sci. USA* **89**, 5857–5861.

Wilke, K., Gaul, R., Klauck, S.M., and Poustka, A. (1997). A gene in human chromosome band Xq28 (GABRE) defines a putative new subunit class of the $GABA_A$ neurotransmitter receptor. *Genomics* **45**, 1–10.

Wilkins, M.E., Hosie, A.M., and Smart, T.G. (2002). Identification of a beta subunit TM2 residue mediating proton modulation of GABA type A receptors. *J. Neurosci.* **22**, 5328–5333.

Wisden, W., Laurie, D.J., Monyer, H., and Seeburg, P.H. (1992). The distribution of 13 $GABA_A$ receptor subunit mRNAs in the rat brain. I. Telencephalon, diencephalon, mesencephalon. *J. Neurosci.* **12**, 1040–1062.

Wotring, V.E. and Weiss, D.S. (2008). Charge scan reveals an extended region at the intracellular end of the GABA receptor pore that can influence ion selectivity. *J. Gen. Physiol.* **131**, 87–97.

Xu, M. and Akabas, M.H. (1996). Identification of channel-lining residues in the M2 membrane-spanning segment of the $GABA_A$ receptor alpha1 subunit. *J. Gen. Physiol.* **107**, 195–205.

Ymer, S., Schofield, P.R., Draguhn, A., Werner, P., Kohler, M., and Seeburg, P.H. (1989a). $GABA_A$ receptor beta subunit heterogeneity: functional expression of cloned cDNAs. *EMBO J.* **8**, 1665–1670.

Ymer, S., Schofield, P.R., Shivers, B.D., Pritchett, D.B., Luddens, H., Kohler, M., Werner, P., Sontheimer, H., Kettenmann, H., and Seeburg, P.H. (1989b). Molecular studies of the $GABA_A$ receptor. *J. Protein Chem.* **8**, 352–355.

Zagotta, W.N., Hoshi, T., and Aldrich, R.W. (1994). Shaker potassium channel gating. III: Evaluation of kinetic models for activation. *J. Gen. Physiol.* **103**, 321–362.

Zhang, J.H., Araki, T., Sato, M., and Tohyama, M. (1991). Distribution of $GABA_A$-receptor alpha 1 subunit gene expression in the rat forebrain. *Brain Res. Mol. Brain Res.* **11**, 239–247.

Zoli, M., Jansson, A., Sykova, E., Agnati, L.F., and Fuxe, K. (1999). Volume transmission in the CNS and its relevance for neuropsychopharmacology. *Trends Pharmacol. Sci.* **20**, 142–150.

CHAPTER

15

The Puzzles of Volume-activated Anion Channels

Yasunobu Okada, Kaori Sato, Abduqodir H. Toychiev, Makoto Suzuki, Amal K. Dutta, Hana Inoue and Ravshan Z. Sabirov

OUTLINE

I. INTRODUCTION: ANION CHANNELS AND THEIR MULTIPLE FUNCTIONS

Anion channels (ACs) are present both in the plasma membrane and in the membranes of intracellular organelles. Animal cells express a large variety of anion channels in their plasma membrane (Table 15.1). ACs are involved in a wide range of functions, such as inhibitory synaptic transmission through plasma membrane hyperpolarization, epithelial Cl^- transport, as well as transport of other organic anions such as glutamate and anionic forms of ATP (mainly ATP^{4-} and Mg-ATP^{2-}).

In contrast to cation channels, ACs are not directly involved in the initiation and termination of action potentials in nerves and muscles. In neurons and other cell types, membrane potential is determined by the relative electromotive forces and the conductances of each ion permeation pathway, as illustrated in the equivalent circuit shown in Fig. 15.1. Opening of voltage-gated Na^+ and/or Ca^{2+} channels results in inward depolarizing currents underlying action potentials whereas voltage-gated K^+ channels counteract the depolarization; the equilibrium potentials for Na^+ and Ca^{2+} (E_{Na} and E_{Ca}) are far positive to the resting potential (E_m), whereas the equilibrium potential for K^+ (E_K) is more negative than E_m. In neurons, the opening of Cl^- channels results in either membrane hyperpolarization or depolarization depending on the E_{Cl} value, which is in turn determined by the intracellular Cl^- concentration ($[Cl^-]_i$), as discussed in several chapters throughout this book (e.g. Chapters 1, 7, 13, 20 and 22). The $[Cl^-]_i$ level is determined by the relative contribution of plasmalemmal Cl^- transporters such as the Na^+-K^+-$2Cl^-$ cotransporter (NKCC), Na^+-Cl^- cotransporter (NCC), K^+-Cl^- cotransporter (KCC), Cl^-/HCO_3^- exchanger (AE) and Cl^-/H^+ exchanger (ClC-3, -4, -5, -7), as well as the Na^+- and Cl^--dependent neurotransmitter transporters of the SLC6 family (Fig. 15.2). In most mature central neurons and skeletal muscle cells, $[Cl^-]_i$ is at or below electrochemical equilibrium. Thus, opening of neuronal ligand-gated anion channels, such as the glycine receptor (GlyR) and the $GABA_A$ receptor ($GABA_AR$), as well as muscle voltage-dependent anion channels, ClC-1, leads to either stabilization of the membrane potential or hyperpolarization. The latter often results in inhibition. Dysfunctional mutations of the genes encoding for these Cl^- channels cause anion channelopathies coupled to hyperexcitability in neurons and in skeletal muscle (see Chapter 1). Table 15.2 lists examples such as startle disease for GlyR, Angelman syndrome, Prader–Willi syndrome and idiopathic epilepsies for $GABA_AR$, as well as Becker's disease and Thomsen's disease for ClC-1 (see Chapters 12 and 14 in this volume).

By promoting net Cl^- transport, ACs participate in transepithelial transport (secretion or absorption) of electrolytes and fluids. For instance, the epithelial cAMP/PKA-activated Cl^- channel, CFTR, is involved in NaCl and fluid secretion accomplished by respiratory, pancreatic and intestinal epithelial cells and in NaCl reabsorption by sweat gland duct cells. Hereditary dysfunction of CFTR causes cystic fibrosis (Table 15.2) characterized by airway, pancreatic and intestinal obstructions as well as by elevation of electrolyte contents of the sweat. ClC-Kb participates in

TABLE 15.1 Mammalian types of plasmalemmal anion channels

Type	Name	Subtype or abbreviation
Ligand-gated	Glycine receptor	GlyR
	GABA receptor	$GABA_AR$
		$GABA_CR$
Voltage-dependent	ClC member	ClC-1
		ClC-2
		ClC-Ka(K1)
		ClC-Kb(K2)
cAMP/PKA-activated	CFTR	CFTR
Ca^{2+}-activated		CaCC
Volume-activated	ClC member	ClC-2
	Outwardly rectifying	VSOR
	Large conductance	Maxi-anion
Acid-activated	Outwardly rectifying	ASOR

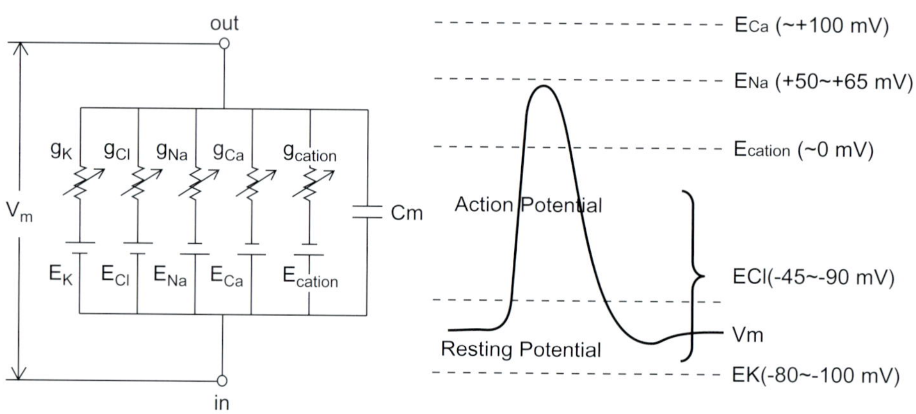

FIGURE 15.1 Electrical equivalent circuit model of the plasma membrane and the cell membrane potential (resting or action potentials). V_m, C_m, g_i, and E_i represent the membrane potential, the membrane capacitance, the conductance of each ion (i) permeation pathway and the electromotive force (=equilibrium potential) for each ion, respectively. The equilibrium potential of each ion can be calculated using the Nernst equation.

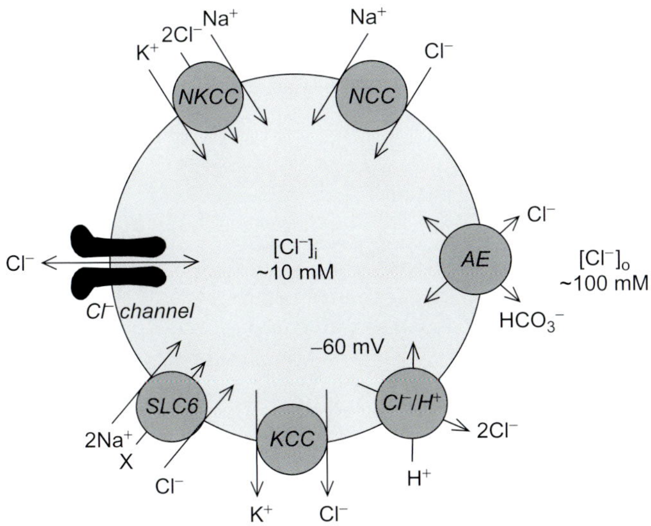

FIGURE 15.2 Intracellular Cl^- homeostasis is determined by the functional activities of various Cl^- transporters expressed in the plasma membrane of a given cell. These Cl^- transporter molecules include the cation-chloride cotransporters NKCC, NCC, KCC, members of the SLC6 family, Cl^-/HCO_3^- antiporters (AEs), and Cl^-/H^+ antiporters (ClC-3, -4 , -5 or -7). X represents a neurotransmitter such as GABA, dopamine, glycine, etc. (Chen et al., 2004). Direction of Cl^- transport depends on the chemical or the electrochemical potential gradients across the plasma membrane.

TABLE 15.2 Examples of anion channelopathies

Anion channel	Subunit	Disease	Abbreviation
GlyR	α1	Hereditary hyperekplexia (Startle disease)	STHE
$GABA_AR$	β2	Angelman syndrome	AS
		Prader–Willi syndrome	PWS
	γ3	Autosonal dominant idiopathic epilepsies	ADIE
ClC-1		Recessive myotonia congenita (Becker's disease)	RMC
		Dominant myotonia congenita (Thomsen's disease)	DMC
ClC-Kb		Bartter syndrome (Type III)	BS (III)
CFTR		Cystic fibrosis	CF

Cl^- reabsorption across the basolateral membrane of the thick ascending limb, the distal convoluted tubule and the collecting duct in the human kidney. Thus, its hereditary dysfunction results in Bartter syndrome (type III) which is associated with severe renal salt loss (Table 15.2).

In addition to these well-established roles, plasmalemmal ACs participate in many other functions or processes such as cell volume regulation (Okada, 1997, 2004; Hoffmann et al., 2009), cell proliferation (Shen et al., 2000; Wondergem et al., 2001), cell differentiation (Voets et al., 1997b), cell death induction (Okada et al., 2001, 2004), acquired cisplatin resistance (Lee et al., 2007; Shimizu et al., 2008), ATP release (Sabirov and Okada, 2004a, 2005) and glutamate release (Kimelberg and Mongin, 1998; Liu et al., 2006; Malarkey and Parpura, 2008). These functions depend on the activities of particular types of plasmalemmal ACs characterized by their sensitivity to changes in cell volume. In the present chapter we review current knowledge on volume-activated anion channels (VAACs) by focusing on their nomenclature, permeability to organic osmolytes and molecular identity.

II. WHAT ARE VOLUME-ACTIVATED ANION CHANNELS (VAACs)?

Upon exposure to hyposmotic media, cells initially swell, and then return to their initial volume, a phenomenon termed regulatory volume decrease (RVD). This regulatory response is accomplished through the loss of intracellular solutes along with osmotically obligated water. RVD is typically accomplished by the net efflux of KCl and osmotically committed water. This is achieved by activation of K^+ efflux pathways in conjunction with available anion channels. These anion channels, the focus of this review, include Cl^- channels with basal activity (background Cl^- conductance), Ca^{2+}-activated Cl^- channels, cAMP-activated Cl^- channels, ClC-2-like Cl^- channels, large-conductance Cl^- channels (Maxi-anion) and volume-sensitive outwardly rectifying (VSOR) Cl^- channels (Okada, 1997). Among these volume-activated anion channels, only ClC-2, Maxi-anion and VSOR are ubiquitously expressed. Thus, these three channel types belong to a group tentatively termed volume-activated anion channels (VAACs) or volume-regulated anion channels (VRACs). Table 15.3 summarizes the physiological, biophysical and pharmacological properties of these three VAACs.

A. ClC-2: Inwardly Rectifying, Small-conductance VAAC

To date, the CLC proteins are probably the best studied Cl^- channels/transporters (Jentsch et al., 1999, 2002, 2005a, b; see Chaper 12 in this volume). Among the nine members of the mammalian CLC family (see

TABLE 15.3 Comparison of physiological, biophysical and pharmacological properties of ClC-2, Maxi-anion and VSOR channels

	ClC-2	Maxi-anion	VSOR
Physiological properties			
volume sensitivity	swelling-enhanced	swelling-activated	swelling-activated
intracellular ATP dependence	upregulated by ATP depletion	upregulated by ATP depletion	non-hydrolytic ATP requirement
Biophysical properties			
pore radius	?	1.28–1.34 nm	0.58–0.65 nm
single-channel conductance	small (2–3 pS)	large (300–400 pS)	intermediate (10–80 pS)
rectification	inward rectification	no rectification	outward rectification
voltage-dependent kinetics	depolarization-induced inactivation	depolarization-induced inactivation	depolarization-induced inactivation
	hyperpolarization-induced activation	hyperpolarization-induced inactivation	
halide permeability sequence	$Cl^- \geq Br^- > I^-$	$I^- > Br^- > Cl^-$	$I^- > Br^- > Cl^-$
glutamate/aspartate permeability	small	large	intermediate
Pharmacological properties			
DIDS sensitivity	−+	+	++
NPPB sensitivity	+	+	++
heavy metal sensitivity	Cd^{2+}, Zn^{2+}-sensitive	Gd^{3+}-sensitive	−
arachidonate sensitivity	?	++	++

Fig. 12.1 in Chapter 12 in this volume), ClC-2 and ClC-3 have been found to be sensitive to cell volume changes, whereas the remaining CLCs do not seem to be affected by cell swelling. ClC-3 was originally found to be activated by cell swelling and the ClC-3 protein was proposed to be the molecular counterpart of VSOR. However, it is now known that ClC-3 is distinct from VSOR (see below). In fact, ClC-3 is not a channel but a Cl^-/H^+ antiporter that resides in endosomes and synaptic vesicles as discussed in detail in Chapter 12 in this volume. In conclusion, among the mammalian CLC channels/transporters, only ClC-2 belongs to the category of VAACs (see Table 15.3).

ClC-2 is activated not only by cell swelling (Grunder et al., 1992) but also by strong membrane hyperpolarization (Thiemann et al., 1992) and extracellular acidification (Jordt and Jentsch, 1997). Activation of ClC-2 does not depend on intracellular ATP (Park et al., 1998), in fact, its activation is induced by depletion of intracellular ATP (Rutledge et al., 2002). The channel pore size of ClC-2 has not been determined, but its single channel conductance (Weinreich and Jentsch, 2001) is known to be relatively small (2–3 pS). The macroscopic ClC-2 current exhibits inward rectification and depolarization-induced inactivation (Kajita et al., 2000; Fahlke, 2001). The halide selectivity of ClC-2 is of the strong electric-field type (Eisenman's sequence IV or V) with a permeability sequence $Cl^- > Br^- > I^-$ (Thiemann et al., 1992). The ClC-2 currents are poorly blocked by the stilbene-derivative anion channel blocker DIDS, but are somewhat effectively blocked by 5-nitro-2-(3-phenylpropylamino)-benzoate (NPPB) (Thiemann et al., 1992; Furukawa et al., 1998). The most effective blockers of ClC-2 are the heavy metals Cd^{2+} and Zn^{2+} (Schwiebert et al., 1995; Bond et al., 1998; Clark et al., 1998; Duan et al., 2000; Kajita et al., 2000).

Swelling-induced activation of ClC-2 has been linked to RVD process in *Xenopus laevis* oocytes (Furukawa et al., 1998), insect Sf9 cells (Xiong et al., 1999) and mammalian hepatoma cells (Roman et al., 2001). In human colonic T84 cells, however, the involvement of ClC-2 as an RVD mechanism has been challenged (Bond et al., 1998). In neurons, ClC-2 has been proposed to play a role stabilizing the resting membrane potential (Staley et al., 1996). However, ClC-2 knockout mice do not exhibit epileptic seizures but unexpectedly show degeneration of photoreceptors and male germ cells (Bosl et al., 2001), as well as vacuolation of the white matter in brain and spinal cord (Blanz et al., 2007). In sum, although ClC-2 is activated by cell swelling its physiological role in volume regulation is still puzzling. A detailed discussion of other functional and molecular aspects of ClC-2 can be found in Chapter 12 in this volume.

B. Maxi-anion: An Ohmic Large-conductance VAAC

In a large variety of cells, osmotic swelling or membrane patch excision activates a VAAC channel, called Maxi-anion channel (Sabirov and Okada, 2009). The single-channel conductance of Maxi-anion channels is large (300 to 400 pS), whether determined in cell-attached or inside-out patch configurations (Falke and Misler, 1989; Jalonen, 1993; Schwiebert et al., 1994; Sabirov et al., 2001; Dutta et al., 2004; Liu et al., 2006, 2008a) (see Table 15.3). In cell-attached membrane patches, the channel is also activated in response to ATP depletion induced by ischemia or hypoxia both in cardiomyocytes (Dutta et al., 2004, 2008) and in astrocytes (Liu et al., 2006, 2008a). Since another type of VAAC, VSOR, was frequently co-activated in these experiments, swelling-activated Maxi-anion currents could only be observed in isolation by blocking VSOR activity with phloretin, a relatively specific VSOR blocker, or by removing ATP from the pipette (intracellular) solution since ATP is required for VSOR activity (Sabirov et al., 2001). Figure 15.3 shows sample traces of macro-patch currents recorded from Maxi-anion channels in rat cardiomyocytes, and Fig. 15.4 shows sample traces of Maxi-anion single-channel currents recorded from rat fibroblasts. In many studies, it has been observed that Maxi-anion currents exhibit multiple subconductance states ranging from 15 to 200 pS (Schwarze and Kolb, 1984; Olesen and Bundgaard, 1992; Sun et al., 1993; Dutta et al., 2004).

Using the non-electrolyte partitioning method, we estimated the radii of the intracellular and extracellular entrances to the channels pore, and found them to be 1.16 nm and 1.42 nm, respectively (Sabirov and Okada, 2004b). Measurements by symmetrical two-sided application of polyethylene glycol molecules yielded an average functional pore radius of approximately 1.3 nm. As shown in Fig. 15.4, the current/voltage (I/V) relationship of the fully open state of the single channel is linear, with no rectification. The open channel probability exhibits a bell-shaped dependence on voltages with the maximum at around 0 mV (see Sabirov et al., 2001; Dutta et al., 2002). The macroscopic Maxi-anion current shows time-dependent inactivation when the transmembrane voltage exceeds ±15 to ±20 mV (Fig. 15.3). The halide selectivity of the Maxi-anion channel measured in many cell types is consistent with that of a weak electric-field type (Eisenman's sequence I) with a permeability sequence $I^- > Br^- > Cl^-$ (Soejima and Kokubun, 1988; Schlichter et al., 1990; Bajnath et al., 1993; Kajita et al., 1995; Sabirov et al., 2001). The Maxi-anion current is only partially suppressed by a number of conventional anion channel blockers such as DIDS, SITS, NPPB and DPC (Schmid et al., 1988; Groschner and Kukovetz, 1992; Pahapill and Schlichter, 1992; Sun et al., 1992; Kajita et al., 1995; Vaca, 1999; Sabirov et al., 2001). Arachidonic acid at micromolar concentrations effectively blocks the Maxi-anion channel (Zachar and Hurnak, 1994; Riquelme and Parra, 1999; Dutta et al., 2002, 2004; Liu et al., 2006). Gd^{3+} is the most effective blocker for this channel; however, it acts only from

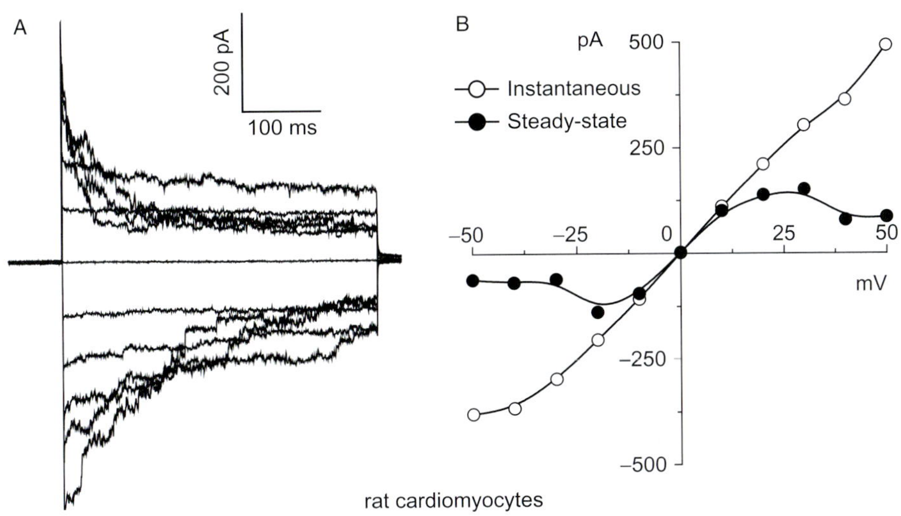

FIGURE 15.3 Voltage-dependent inactivation of Maxi-anion currents recorded in macro-patches excised from a neonatal rat ventricular myocyte in primary culture. **A.** Representative current traces recorded upon application of step pulses form 0 to ±50 mV in 10 mV increments. **B.** I/V relationships of instantaneous (open circles) and steady-state (filled circles) currents measured at the beginning and end of the current responses shown in A. The methods employed in this experiment are same as those described in Dutta et al. (2004).

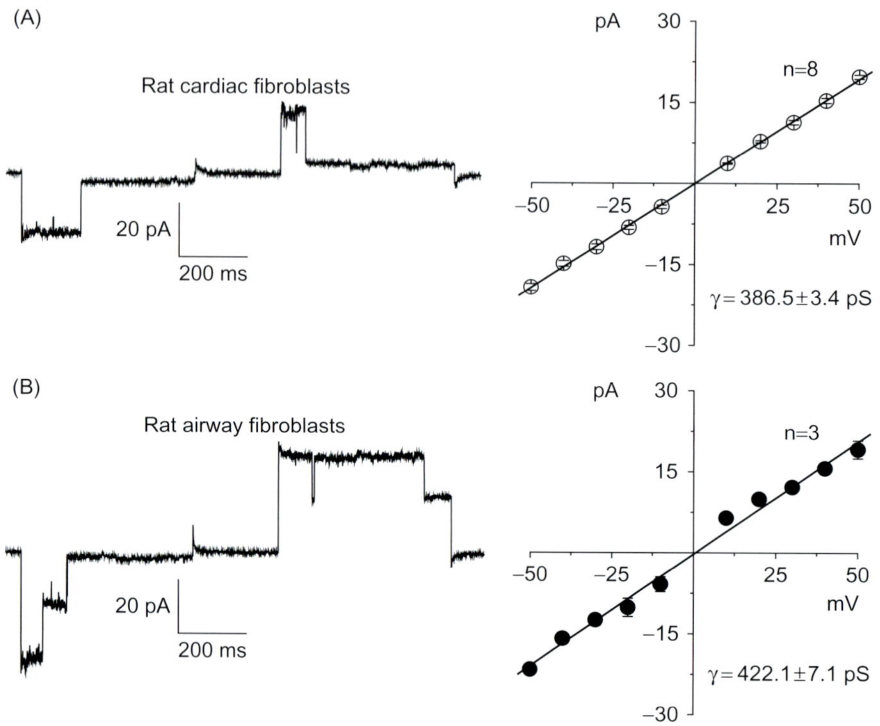

FIGURE 15.4 Voltage-dependent inactivation and I/V relationships of single-channel currents recorded in inside-out patches excised from neonatal rat cardiac fibroblasts (**A**) and airway fibroblasts in primary culture (**B**). *Left panels*: Representative unitary current traces recorded upon application of alternating ±50 mV pulses from a holding potential of 0 mV. *Right panels*: Unitary I/V relationships of instantaneous currents. For detailed methods see Dutta et al. (2004).

the extracellular side (Sabirov et al., 2001; Dutta et al., 2004, 2008; Liu et al., 2006, 2008a).

In view of its volume sensitivity, the Maxi-anion channel has been considered a major route for volume-regulatory Cl^- efflux during RVD in many cell types, including epithelial cells (Schwiebert et al., 1992; Mitchell et al., 1997a; Bernucci et al., 2003), lymphocytes (Cahalan and Lewis, 1988; Schlichter et al., 1990), myoblasts (Hurnak and Zachar, 1992), neuroblastoma (Falke and Misler, 1989) and astrocytes (Jalonen, 1993).

C. VSOR: An Outwardly Rectifying Intermediate-conductance VAAC

To distinguish this channel from the inwardly rectifying ClC-2 and the non-rectifying Maxi-anion, we call this channel the volume-sensitive outwardly rectifying anion channel (VSOR) (Okada, 1997, 2006). Among all VAACs, VSOR is likely the most typical swelling-activated anion channel found in most animal cells (Okada, 1997) (see Table 15.3). At physiological intracellular ATP levels, VSOR contributes a major part to the swelling-activated whole-cell conductance. This channel has also been referred to as the swelling-activated Cl^- channel, $I_{Cl,swell}$ (Ackerman et al., 1994a, b; Nilius et al., 1997; d'Anglemont de Tassigny et al., 2003); the volume-activated Cl^- current, $I_{Cl,vol}$ (Nilius et al., 1994), the volume-sensitive organic osmolyte anion channel, VSOAC (Strange and Jackson, 1995; Strange et al., 1996); and the volume-regulated anion current, VRAC (Levitan and Garber, 1998; Nilius and Droogmans, 2003).

In addition to its volume sensitivity and moderate outward rectification, VSOR exhibits a number of phenotypic properties that are summarized in Table 15.3. This channel has an intermediate unitary conductance when compared with ClC-2 or the Maxi-anion channel; it ranges between 10 and 80 pS (Worrell et al., 1989; Solc and Wine, 1991; Banderali and Roy, 1992; Grygorczyk and Bridges, 1992; Weiss and Lang, 1992; Kelly et al., 1994; Okada et al., 1994; Petersen et al., 1994; Jackson and Strange, 1995; Boese et al., 1996; Meyer and Korbmacher, 1996; Tsumura et al., 1996; Duan et al., 1997a; Voets et al., 1997a; Inoue et al., 2005). Figure 15.5 shows typical whole-cell and single-channel currents simultaneously recorded from mouse cortical neurons using a double-patch clamp approach, i.e. simultaneous whole-cell and cell-attached recordings (Okada et al., 1994; Petersen et al., 1994; Inoue et al., 2005). This methodological

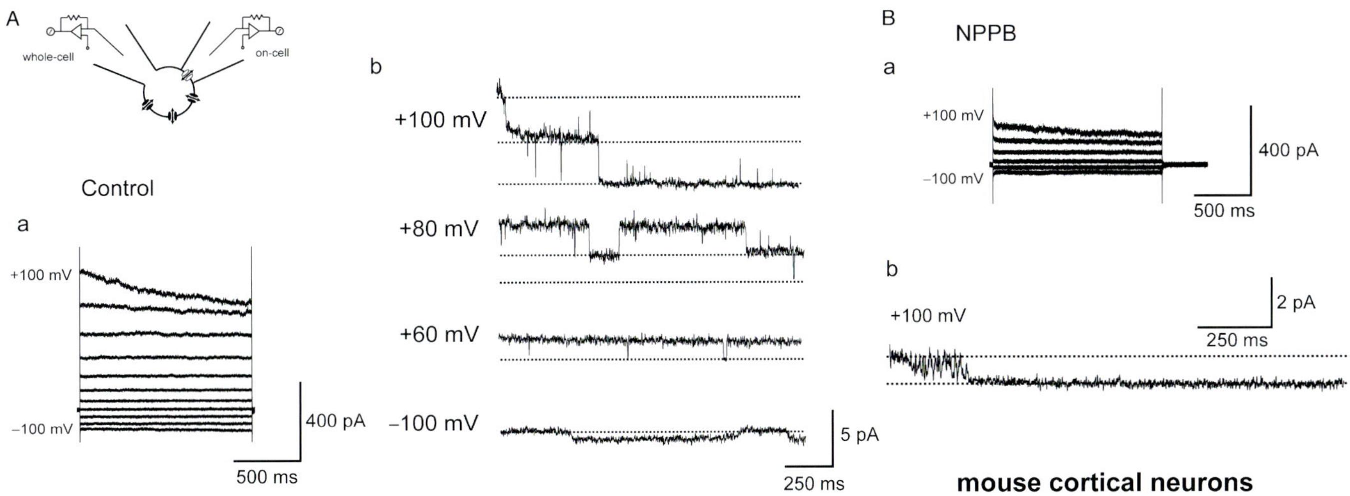

FIGURE 15.5 Double-patch recordings of VSOR currents in embryonic mouse cortical neurons in primary culture under hypotonic conditions in the absence (**A**) and presence (**B**) of 80 μM NPPB in the on-cell pipette and bath solutions. *Inset*: Scheme of the double-patch configuration. a: Whole-cell currents in response to step pulses from −100 mV to +100 mV in 20 mV increments applied from a holding potential of −40 mV. b: On-cell single-channel currents in response to given step pulses. Dotted lines designate the closed-state and open-state levels. For methods see Inoue et al. (2005).

approach and ensemble average currents from inside-out recordings (Tsumura et al., 1996) clearly demonstrate that the moderate outward rectification of the macroscopic currents is mainly caused by an outwardly rectifying property, which is inherent to the single-channel conductance (Grygorczyk and Bridges, 1992; Okada et al., 1994; Meyer and Korbmacher, 1996; Tsumura et al., 1996; Inoue et al., 2005). When using the double-patch recording method, upon application of large positive voltages, time-dependent inactivation of whole-cell VSOR currents are observed together with stepwise closing unitary VSOR events (see Fig. 15.5A, as well as Okada et al., 1994; Inoue et al., 2005).

The halide permeability sequence of VSOR is $I^- > Br^- > Cl^-$, which corresponds to Eisenman's sequence I (Strange et al., 1996; Nilius et al., 1997; Okada, 1997). The most salient physiological property of VSOR is its dependence on cytosolic free ATP (Okada, 1997). This property distinguishes VSOR from ClC-2 and Maxi-anion (Table 15.3). In contrast to intracellular free ATP, intracellular free Mg^{2+} is an inhibitor of VSOR channel activity (Oiki et al., 1994). Intracellular protons also potently decrease VSOR activity, with an effective p*K* value of ~6.3 (Sabirov et al., 2000). Consistent with this feature, acidification due to lactacidosis causes inhibition of RVD (Mori et al., 2002; Nabekura et al., 2003). VSOR exhibits a broad sensitivity to conventional anion channel blockers, including the stilbene derivatives SITS and DIDS, the carboxylate analogs NPPB, DPC and 9-AC, and the DPC-derivative arylaminobenzoates niflumic acid and flufenamic acid (Strange et al., 1996; Nilius et al., 1997; Okada, 1997, 2006). VSOR activity is also inhibited by various structurally unrelated chemicals, such as some inhibitors or substrates of P-glycoprotein, like arachidonic acid, and the sulfonylurea receptor inhibitor glibenclamide (Okada, 2006). Tamoxifen, a competitive antagonist of estrogen receptors, known to inhibit P-glycoprotein, has also been used as an effective VSOR blocker. However, it must be noted that the effect of tamoxifen is not universal inasmuch as it is ineffective in blocking VSOR in mouse muscle cells (Voets et al., 1997b), bovine PCE cells (Mitchell et al., 1997b), rat sympathetic neurons (Leaney et al., 1997) and mouse cortical neurons (Inoue et al., 2005). A new generation of relatively specific VSOR blockers include the bisphenol phloretin (Fan et al., 2001), the ethacrynic-acid derivative indanone compound DCPIB (Decher et al., 2001) and the acidic di-aryl-urea derivative NS3728 (Helix et al., 2003). In contrast to ClC-2, VSOR activity is insensitive to the extracellular application of Cd^{2+} and Zn^{2+}, as shown in Fig. 15.6. In addition, VSOR currents have also been reported to be poorly sensitive to Gd^{3+} (Hazama et al., 1999; Sabirov et al., 2001).

In contrast to NPPB (Fig. 15.5B), phloretin and arachidonic acid, the stilbene derivatives SITS or DIDS inhibit VSOR currents in a voltage-dependent manner (Kubo and Okada, 1992; Okada, 1997). The stilbene derivatives suppress outward currents more prominently than inward currents, facilitating their inactivation kinetics at positive potentials. This suggests that SITS and DIDS block the open-state channel at

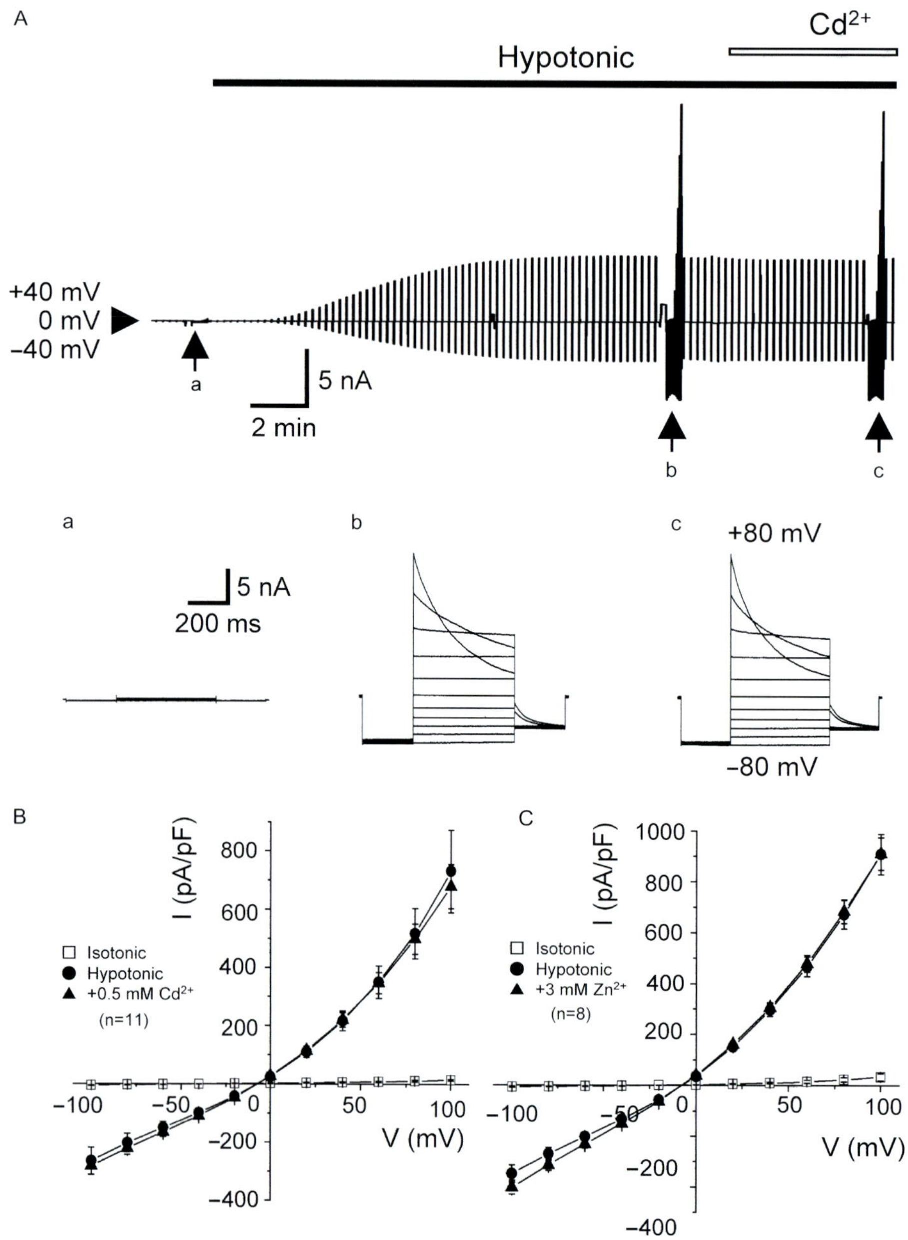

FIGURE 15.6 Effects of Cd^{2+} and Zn^{2+} on whole-cell VSOR currents recorded from human epithelial HeLa cells. **A.** Representative recordings of currents before and during hypotonic challenge, and with 0.5 mM Cd^{2+} absent or present in the bath solution. Alternating $\pm$40 mV pulses (2 s duration, every 15 s) or step pulses form $-$80 mV to $+$80 mV in 20 mV increments (at arrows) were applied to elicit currents. a, b and c: Expanded traces of current responses to step pulses applied at arrows in the upper trace under isotonic conditions (a) and hypotonic (85.9% osmolality) conditions (b, c) in the absence (a, b) and presence (c) of Cd^{2+}. **B.** *I*/*V* relationships before (open symbols) and during (filled symbols) hypotonic stimulation in the absence (circles) and presence (triangles) of 0.5 mM Cd^{2+}. **C.** *I*/*V* relationships before and during hypotonic stimulation in the absence and presence of 3 mM Zn^{2+}. Each symbol represents the mean current$\pm$SEM (vertical bar) for 11 (B) and 8 (C) observations. The isotonic and hypotonic bath solutions contained (mM): 110 CsCl, 5 $MgSO_4$, 10 HEPES and 90 and 45 mannitol, respectively. The pipette solution contained (mM): 110 CsCl, 2 $MgSO_4$, 1 EGTA, 2 Na_2ATP, 0.3 Na_3GTP and 50 mannitol. Detailed methods can be found in Shimizu et al. (2000, 2004).

the external entrance of the VSOR pore. Essentially, similar open-channel blocks were observed upon extracellular application of some anionic forms of glibenclamide (Liu et al., 1998) and ATP (Jackson and Strange, 1995; Tsumura et al., 1996; Poletto Chaves and Varanda, 2008). Suramin also exhibits open-channel block in bovine endothelial cells (Droogmans et al., 1999), mouse Leydig cells (Poletto Chaves and

Varanda, 2008) and human epithelial HeLa cells (Fig. 15.7). Taken together, these observations suggest that VSOR possesses an outer vestibule larger than the sizes of SITS, glibenclamide, suramin and ATP, but a pore size that is smaller than these molecules. In fact, recent non-electrolyte partition studies (Ternovsky et al., 2004) revealed that the cut-off radius of the VSOR pore is 0.63 nm, a value close to the effective radii of $SITS^{2-}$ (~0.55 nm) and ATP^{4-} (0.58 nm), and slightly smaller than the effective radii of $glibenclamide^{-}$ (~0.60 nm) and $suramin^{6-}$ (~0.91 nm), as illustrated in Fig. 15.8.

In sum, three members of VAAC – ClC-2, Maxi-anion and VSOR – share similarity as far as their volume sensitivity is concerned, but are distinct from each other in their unitary conductances, I/V relationships, halide selectivity, heavy metal sensitivity and intracellular ATP dependence (Table 15.3).

III. WHICH VAACs ARE VOLUME-SENSITIVE ORGANIC OSMOLYTE ANION CHANNELS (VSOACs)?

Upon osmotic swelling, cells downregulate their volume by releasing not only inorganic osmolytes, such as K^+ and Cl^-, but also organic ones, like amino acids, polyols, methylamines and nucleotides. Since volume-induced release of organic osmolytes is inhibited by a number of anion channel blockers, it has been suggested that some anion channels serve as pathways for swelling-induced efflux of organic solutes (Kimelberg et al., 1990; Kirk et al., 1992; Sanchez-Olea et al., 1993; Strange et al., 1996; Kirk, 1997). The pore radius of Maxi-anion (1.16–1.42 nm) is larger than the effective radii of the anionic amino acids such as glutamate, gluconate, or aspartate (0.34–0.35 nm) and the effective radii of the nucleotides: ATP^{4-}, $HATP^{3-}$, $MgATP^{2-}$, ADP^{3-} and UTP^{4-} (0.53–0.62 nm) (see Table 2 in Sabirov and Okada, 2005). In contrast, the pore radius of VSOR, ~0.63 nm, is large enough to conduct negatively charged amino acids, but not nucleotides. Because of its small unitary conductance, it is reasonable to assume the pore of ClC-2 to be much smaller than the pore of VSOR. Therefore, most likely ClC-2 is not permeable to nucleotides, and is only poorly permeable to anionic amino acids. In fact, the relative permeability of ClC-2 to aspartate has been measured in cardiac myocytes and found to be quite low ($P_{aspartate}/P_{Cl}$~0.04; Duan et al., 2000). This value is in contrast with the $P_{aspartate}/P_{Cl}$ values reported for VSOR: 0.20 (Banderali and Roy, 1992), 0.42 (Lewis et al., 1993), 0.10 (Vandenberg et al., 1994), 0.48 (Jackson et al., 1996) and 0.63 (Yamazaki et al., 1998). It is also in contrast with the values determined for Maxi-anion: 0.23 (Sabirov et al., 2001; Bell et al., 2003) and 0.42 (Vallejos and Riquelme, 2007). Based on these measurements, it is possible to conclude that Maxi-anion and VSOR, but presumably not ClC-2, are volume-sensitive organic anion channels (VSOACs).

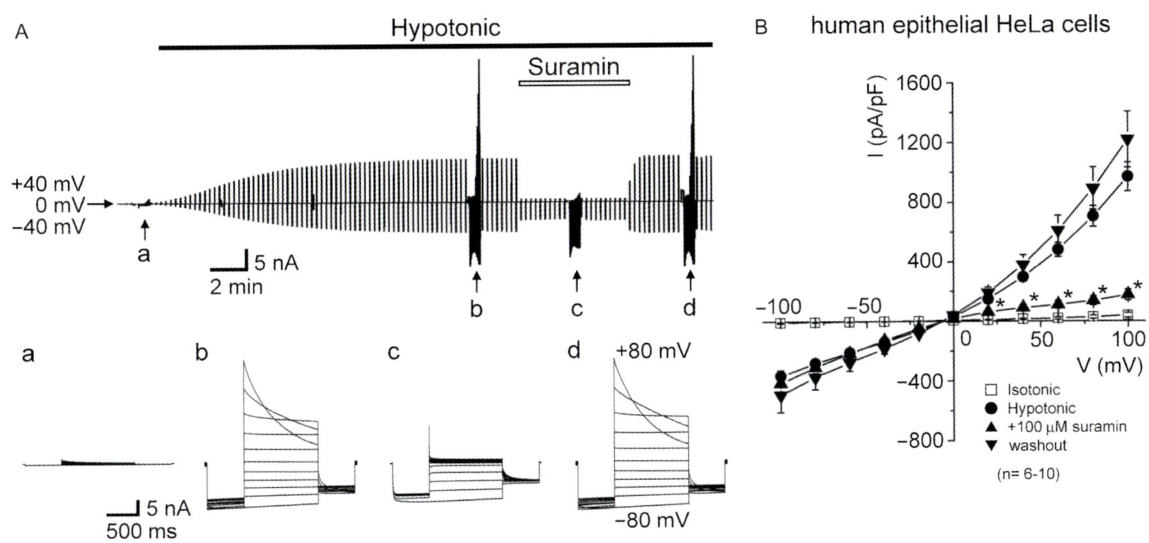

FIGURE 15.7 Effects of suramin on whole-cell VSOR currents recorded from human epithelial HeLa cells. **A.** Representative recordings of currents before and during a hypotonic challenge in the presence or absence of 0.1 mM suramin in the bathing solution. Pulse conditions, arrows, a and b are same as those in Fig. 15.6. c and d: Traces of current responses under hypotonic conditions during application of suramin (c) and after washout of suramin (d). **B.** I/V relationships before (open symbols) and during (filled symbols) hypotonic stimulation in the absence (circles) and presence (triangles) or 0.1 mM suramin as well as after washout of suramin (inverted triangles). Each symbol represents the mean current±SEM (vertical bar) for 6–10 observations. For details on the methods see Shimizu et al. (2000, 2004).

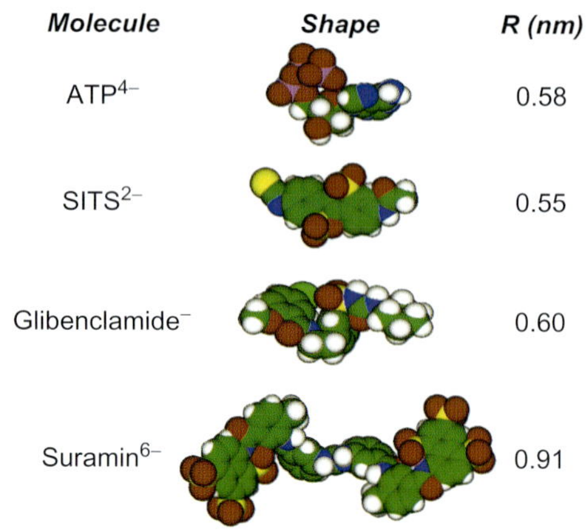

FIGURE 15.8 Shapes and dimensions of open-channel blockers for VSOR. The unhydrated radii were calculated as a geometric mean of three dimensions according to the formula: $R_x = (1/2)(l_1\, l_2\, l_3)^{1/3}$, where l_1, l_2 and l_3 are ion dimensions estimated from space-filling models. The dimension data were calculated using Molecular Modeling Pro computer software (Norgwyn Montgomery Software Inc., North Wales, PA) after energy minimization by bond rotation. Color coding: C – green, H – white, O – red, N – blue, P – purple, S – yellow.

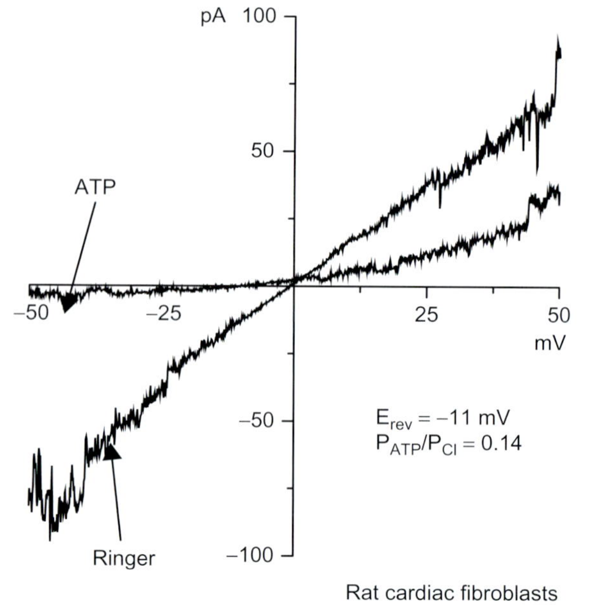

FIGURE 15.9 ATP permeability of Maxi-anion. Ramp I/V recordings from an inside-out patch excised from neonatal rat cardiac fibroblasts in primary culture in standard electrolyte saline solution (Ringer) and in a 100 mM Na_2ATP bath solution (ATP). Pipette solution was a standard Ringer's solution. The inward current recorded in the ATP-exposed patch represents an ATP^{4-} current. The amplitude of outward current became smaller when ATP was present in the bath, suggesting an open-channel blocking of currents by ATP^{4-}, as observed in mammary C127 cells (Sabirov et al., 2001) and astrocytes (Liu et al., 2008a). For detailed methods see the two papers cited above.

A. Maxi-anion: An ATP- and Glutamate-releasing VSOAC

ATP is not only a universal energy source, but also an extracellular second messenger for autocrine and paracrine signaling (Dubyak and el-Moatassim, 1993; Fields and Stevens, 2000; Bodin and Burnstock, 2001; Forrester, 2003; Sabirov and Okada, 2004a, 2005, 2009). When stimulated, cells release ATP which then binds to P2 purinergic receptors located in their own membrane, or in the membranes of neighboring cells, thereby signaling in autocrine or paracrine fashions (Burnstock, 2004). Non-lytic release of ATP is mediated by exocytosis, transporters and channels. Since most ATP molecules exist in the cytoplasm in anionic forms (~87% as $MgATP^{2-}$ and ~11% as ATP^{4-}; Sabirov and Okada, 2005) at physiological pH, anionic channels may constitute the most effective efflux pathways for these molecules which are driven by an outwardly directed electrochemical potential gradient of the order of 10^8 and 10^{10} for $MgATP^{2-}$ and ATP^{4-}, respectively. This gradient was calculated based on the assumption that intracellular and extracellular ATP levels are at mM and nM levels, respectively; and that the membrane potential is around −60 mV (Sabirov and Okada, 2005).

It has long been considered that a conductive pathway for ATP release is provided by CFTR (Reisin et al., 1994; Schwiebert et al., 1995). However, several lines of evidence argue against this hypothesis (Abraham et al., 1997; Sabirov and Okada, 2004a, 2005). Instead, recent studies have provided ample supporting evidence that Maxi-anion is an ATP release channel. First, Maxi-anion channels have a large pore size, with an effective radius of ~1.3 nm (Sabirov and Okada, 2004b). Such a wide pore provides sufficient room to accommodate ATP^{4-} and $MgATP^{2-}$ (0.57–0.65 nm). Second, in a variety of cell types we have been able to show inward currents carried by ATP^{4-} and $MgATP^{2-}$ flowing through Maxi-anion channels, when all anions in the intracellular solution are replaced with 100 mM ATP or MgATP (Fig. 15.9). The permeability ratio P_{ATP}/P_{Cl} estimated by the reversal potential under bi-ionic conditions was around 0.1 for the Maxi-anion channel of mouse mammary C127 cells (Sabirov et al., 2001; Dutta et al., 2002), rabbit kidney macula densa cells (Bell et al., 2003), rat ventricular cardiomyocytes (Dutta et al., 2004, 2008), mouse astrocytes (Liu et al., 2008a) and rat cardiac fibroblasts (Fig. 15.9). In rat cardiomyocytes, the P_{MgATP}/P_{Cl} value (0.16) was even higher than P_{ATP}/P_{Cl} (Dutta et al., 2004). ATP currents were sensitive to the blockers SITS, NPPB and Gd^{3+} (Sabirov et al., 2001) and sensitive to arachidonic acid (Dutta et al., 2002). Third, stimuli that are effective in activating Maxi-anion were also effective in provoking massive release of ATP. For instance, osmotic stress elicited the release of ATP from C127 cells (Sabirov et al., 2001), cardiomyocytes (Dutta et al., 2004, 2008) and astrocytes (Liu et al., 2008b); salt stress released ATP from macula densa cells (Bell et al., 2003); and ischemic or hypoxic stress resulted in the release of ATP from astrocytes (Liu et al., 2008a). Fourth, in all these studies, the typical Maxi-anion blockers (Gd^{3+}, SITS, NPPB) effectively suppressed the stimulated ATP release, whereas the VSOR blockers (phloretin and glibenclamide) failed to inhibit the ATP release. Fifth, Dutta et al. (2008) found that the spatial distribution of the Maxi-anion expression at the cell surface of both neonatal and adult rat cardiomyocytes coincided with the spatial distribution of their ATP releasing sites determined by patch-clamp under scanning ion conductance microscopy, the so-called "smart-patch" technique (Gu et al., 2002), coupled to a biosensor ATP detection method (Hazama et al., 1998a; Hayashi et al., 2004).

Both ATP and glutamate represent important extracellular second messengers released from mammalian cells, and they are sometimes called gliotransmitters because brain astrocytes release them in response to a number of physiological and pathological stimuli (Haydon, 2001; Hansson and Ronnback, 2003; Volterra and Meldolesi, 2005; Fellin et al., 2006). Astrocytes express the Maxi-anion channel which is permeable to glutamate with $P_{glutamate}/P_{Cl}$ of 0.21 (Liu et al., 2006). Therefore, it is possible that the astrocytic Maxi-anion channel serves as a pathway for glutamate release from stimulated astrocytes. Consistent with this inference, swelling- and ischemia-induced glutamate release from mouse astrocytes has been shown to be largely suppressed by a number of Maxi-anion blockers (Liu et al., 2006). However, even the most powerful blocker of Maxi-anion, Gd^{3+}, only partially (half at most) inhibited swelling- and ischemia-induced release of glutamate, indicating the existence of an additional releasing pathway in astrocytes (Liu et al., 2006). As discussed below, this second pathway is VSOR.

In summary, Maxi-anion is a volume-sensitive organic anion channel (VSOAC) serving as a release pathway for two major extracellular second messengers or gliotransmitters, ATP and glutamate. Since activation of Maxi-anion is facilitated by ATP depletion (Table 15.3), it is likely that its pore may represent a predominant pathway for ATP and glutamate under stress/emergency conditions, when the cell metabolism is somehow reduced or impaired (Fig. 15.10).

B. VSOR: A Glutamate-releasing VSOAC

Swelling-induced efflux of organic osmolytes has long been known to occur in a variety of organisms (Strange et al., 1996; Kirk, 1997) including mammalian cells (Pasantes-Morales and Schousboe, 1989; Siebens and Spring, 1989). Kimelberg and collaborators (Kimelberg et al., 1990) observed swelling-induced release of glutamate, aspartate and taurine in cultured astrocytes, which was sensitive to anion transport blockers. Roy and collaborators provided the first evidence for swelling-induced amino acid release via

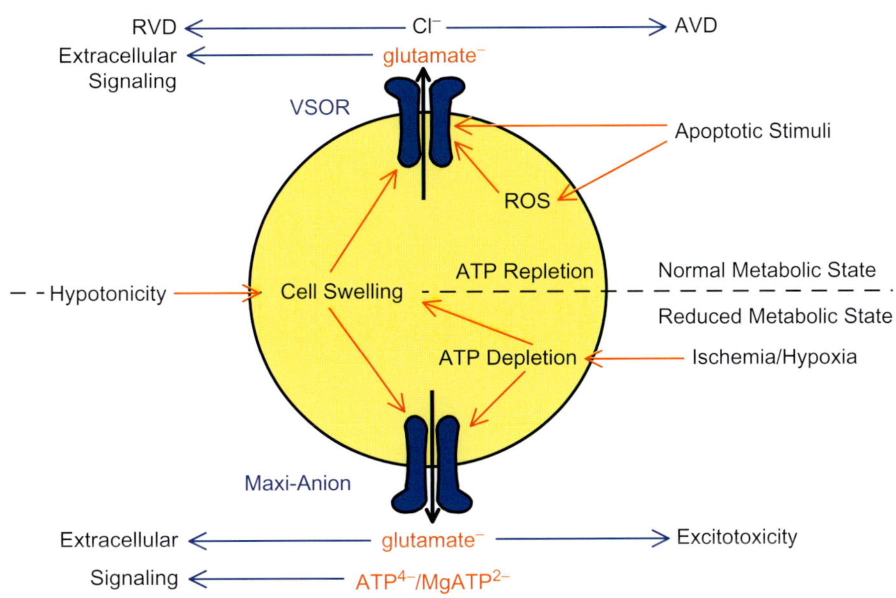

FIGURE 15.10 Schematic drawing representing the roles of VSOR and Maxi-anion in glutamate and ATP release. VSOR-mediated release of Cl^- is known to be involved in regulatory volume decrease (RVD) under hypotonic conditions (see Okada, 1997), and in apoptotic volume decrease (AVD) under isotonic stimulation with apoptotic inducers (see Okada et al., 2001, 2006 as well as text for details).

VSOR from MDCK cells (Banderali and Roy, 1992; Roy and Malo, 1992). They showed a relatively high permeability not only to anionic amino acids such as glutamate ($P_{glutamate}/P_{Cl} = 0.18$) and aspartate ($P_{aspartate}/P_{Cl} = 0.2$), but also to a zwitterionic amino acid, taurine ($P_{taurine}/P_{Cl} = 0.49$). Subsequent electrophysiological studies performed in a variety of cells also showed significant permeabilities of VSORs to glutamate ($P_{glutamate}/P_{Cl}$~0.06 to 0.20 (Chan et al., 1994; Arreola et al., 1995; Roy, 1995; Basavappa et al., 1996; Boese et al., 1996; Levitan and Garber, 1998; Schmid et al., 1998; Carpaneto et al., 1999; Sakai et al., 1999; Liu et al., 2006) and to taurine ($P_{taurine}/P_{Cl}$~0.15 to 0.26 (Jackson and Strange, 1993; Roy, 1995; Lewis et al., 1996).

Glutamate is released from astrocytes during increased electrical activity in normal brain tissue and upon pathologic insults such as ischemia and trauma. Astroglial swelling is currently considered to be the main cause of glutamate release in the brain (Pasantes-Morales et al., 2002; Phillis and O'Regan, 2003; Kimelberg et al., 2004). However, stimulation of bradykinin receptors has also been shown to induce the release of glutamate from astrocytes (Parpura et al., 1994), even without concomitant cell swelling (Liu, 2007). Putative pathways for astrocytic glutamate release are currently considered to include exocytosis, reverse-mode operation of some Na^+-dependent glutamate transporters, gap junction hemichannels and chloride channels (Phillis and O'Regan, 2003; Evanko et al., 2004; Parpura et al., 2004). Since the astrocytic VSOR is also considerably permeable to glutamate (with a $P_{glutamate}/P_{Cl}$~0.15 (Liu et al., 2006)), this channel is likely to serve as a pathway for glutamate release from swollen astrocytes. Indeed, two effective blockers of the astrocytic VSOR, phloretin and tamoxifen, significantly suppressed the Gd^{3+}-insensitive (i.e. Maxi-anion-independent) component of glutamate release in mouse astrocytes exposed to hypotonic or ischemic conditions (Liu et al., 2006). This observation is in good agreement with the observations of Mongin and collaborators (Abdullaev et al., 2006) showing that the VSOR blockers, tamoxifen and DCPIB, effectively inhibited swelling-activated D-[^{3}H] aspartate release from rat astrocytes. Taken together, these data demonstrate that glutamate is released from swollen astrocytes, not only via the Maxi-anion channels but also through VSOR channels (Fig. 15.10). Further, in the absence of osmotic cell swelling, a VSOR current is activated by a number of apoptotic stimuli (Shimizu et al., 2004) and reactive oxygen species, ROS (Browe and Baumgarten, 2004; Shimizu et al., 2004; Varela et al., 2004). Thus, it is likely that VSOR-mediated glutamate release is also induced by apoptotic stimuli and ROS (Fig. 15.10). In fact, Mongin and collaborators (Harrigan et al., 2008) have recently confirmed ROS-induced, VSOR-mediated release of D-[^{3}H] aspartate from rat microglia.

The β-amino acid taurine is one of major organic osmolytes in mammalian tissues, and it is mostly (96% at pH 7.4) present as a zwitterion at physiological pH due to its pK_1' of 1.5 and pK_2' of 8.82. In whole-cell experiments performed at pH $\geq$ 8.2, using very high concentrations of taurine to ensure a sufficient amount of anionic form of this amino acid, VSOR was shown in various cell types to exhibit a relatively high permeability to this osmolyte. The following values for $P_{taurine}/P_{Cl}$ have been measured: ~0.49 in MDCK cells (Banderali and Roy, 1992), ~0.2 in rat C6 glioma cells (Jackson and Strange, 1993), ~0.31 to 0.36 in human glial U-138MG cells (Roy, 1995), and ~0.34 to 0.40 in calf endothelial cells (Manolopoulos et al., 1997b). Although this fact alone does not warrant that VSOR serves as a pathway for the neutral, zwitterionic form of taurine, the "VSOR = taurine channel hypothesis" was put forward mainly based on pharmacological observations showing that a number of VSOR blockers inhibit swelling-induced taurine release (Jackson and Strange, 1993; Lewis et al., 1996; Chou et al., 1997; Manolopoulos et al., 1997a, b; Shen et al., 2001). However, there are several other pieces of evidence at variance with this hypothesis. First, the pharmacological properties of swelling-induced taurine release and swelling-activated Cl^- currents (or I^- efflux), although similar, are not precisely identical (Lambert and Hoffmann, 1994; Van Winkle et al., 1994; Moran et al., 1997; Stutzin et al., 1997; Tomassen et al., 2004). Second, the activation time course of taurine and Cl^- or I^- fluxes do not parallel (Moran et al., 1997; Stutzin et al., 1999; Tomassen et al., 2004). Third, the osmolarity thresholds for taurine release and Cl^- current activation are different (Ordaz et al., 2004). Fourth, swelling activates taurine release without activating Cl^- currents or I^- efflux and vice versa in some cell types (Shennan et al., 1994; Kirk, 1997; Stegen et al., 2000).

Given that the effective radius of the VSOR pore (0.63 nm) is very close to that of ATP (0.57–0.65 nm) (Ternovsky et al., 2004), there remains the possibility that VSOR can permeate ATP, albeit not very efficiently; ATP could be freed from the voltage-dependent open-channel blockade. Indeed, at very large positive potentials, ATP block is released and the molecule squeezed out of the VSOR pore into the external cell surfaces (Hisadome et al., 2002). However, there is no evidence for actual exit of cytosolic ATP through the VSOR pore. Furthermore, most of the results obtained so far argue against the "VSOR = ATP channel hypothesis". First, swelling-induced ATP release is not suppressed by the VSOR blockers glibenclamide (Hazama et al., 1998b, 1999, 2000b; Mitchell et al., 1998), DPC (Mitchell et al., 1998), DIDS (Mitchell et al., 1998) and SITS (Hazama et al., 1999). Second, the most effective blocker of swelling-induced ATP release, Gd^{3+}, failed to inhibit VSOR activity in various cells types (Hazama et al., 1999, 2000b; Sabirov et al., 2001). Third, monoclonal antibodies raised against membrane proteins from swollen Intestine 407 cells blocked swelling-induced ATP release without affecting VSOR (Hazama et al., 1999, 2000a). Finally, heterologous expression of CFTR upregulates swelling-induced ATP release (Ando-Akatsuka et al., 2002) but downregulates VSOR activity (Vennekens et al., 1999; Ando-Akatsuka et al., 2002). Taken together, it appears that VSOR is not involved in ATP release within the normal range of transmembrane voltages, although the extracellular VSOR entrance can accommodate extracellular ATP and might even translocate it to the intracellular compartment at very high positive potentials.

In summary, VSOR is a volume-sensitive organic anion channel (VSOAC), which serves as a pathway for glutamate release from animal cells. This pathway is especially important in astrocytes. It is, however, unlikely that VSOR functions as a pathway for the release of ATP and taurine. Since VSOR activation requires intracellular ATP (Table 15.3), it is likely that the VSOR channel may represent a predominant pathway for glutamate release from cells containing normal intracellular ATP concentrations under physiological metabolic states (Fig. 15.10).

IV. WHAT ARE THE MOLECULAR IDENTITIES OF MAXI-ANION AND VSOR?

Among plasmalemmal anion channels expressed in animal cells (Table 15.1), so far, only the molecular entities of ligand-gated anion channels (GLyR, $GABA_AR$, $GABA_CR$), voltage-dependent anion channel members of the ClC superfamily (ClC-1, ClC-2, ClC-Ka/K1, ClC-Kb/K2) and a cAMP/PKA-activated anion channel (CFTR) have been identified. However, the molecular nature of other anion channels (CaCC, VSOR, Maxi-anion and ASOR) remains unknown despite the fact that they are ubiquitously expressed and functionally well defined (see Chapters 12 to 14 in this volume). Biochemical purification approaches for these ACs are hampered by the relatively scarce protein expression and by the lack of specific, high-affinity ligands or blockers. Further, molecular biological approaches such as the expression cloning strategy cannot be applied for these ubiquitously and constitutively expressed membrane-spanning proteins because of lack of good negative controls. For the case of Ca^{2+}-activated Cl^- channel (CaCC), several candidates have been recently identified, but the story remains inconclusive. They include the bestrophins (Chien et al., 2006; Hartzell et al., 2008) and other proteins, as discussed in detail in Chapters 1, 13 and 20 in this volume. For VSOR and Maxi-anion, things are not very different; all candidate molecules so far proposed have been judged to be unacceptable (see below).

A. Maxi-anion Channel Protein is Distinct from the Voltage-dependent Anion Channel Porin and Tweety

Some electrophysiological properties of Maxi-anion, such as the very large single-channel conductance and the bell-shaped voltage dependence of the open probability, are similar to those of the voltage-dependent anion channel (VDAC or porin) expressed in the outer membrane of mitochondria (Mannella, 1997; Colombini, 2004), and in the plasma membrane of several cell types (Thinnes, 1992; Buettner et al., 2000). Further, the activity of the Maxi-anion channel is blocked by anti-VDAC1 monoclonal antibody in rat astrocytes (Dermietzel et al., 1994) and by antisense oligonucleotide-mediated knockdown of pl-VDAC in mouse neuroblastoma C1300 cells (Bahamonde et al., 2003). Based on these similar properties, it has been proposed that the plasmalemmal VDAC (pl-VDAC) is the most likely candidate of the Maxi-anion protein. However, gene silencing experiments have refuted this hypothesis. Indeed, silencing all three isoforms of VDAC (VDAC1-3), for instance by silencing VDAC2 in fibroblasts derived from VDAC1/VDAC3 double-deficient mice failed to affect the Maxi-anion activity (Sabirov et al., 2006).

Recently, Suzuki and Mizuno (2004) reported that a gene called *tweety*, which is found in the *Drosophila flightless* locus, has a structure similar to that of known ion channels. The human homolog of *tweety*, hTTYH1, has been proposed to encode for a large-conductance, Ca^{2+}-activated anion channel (Suzuki, 2006). We tested the "Maxi-anion = TTYH1 hypothesis" by transfecting two splice variants of the TTYH1 gene (TTYH1-E and TTYH1-SV) into HEK293T cells. As shown in Fig. 15.11, both TTYH1-E- and TTYH1-SV-transfected cells failed to demonstrate the unitary current that is phenotypic of the Maxi-anion current. In conclusion, for the time being, there is no sufficient evidence backing the hypothesis that TTYH1 encodes the Maxi-anion channel.

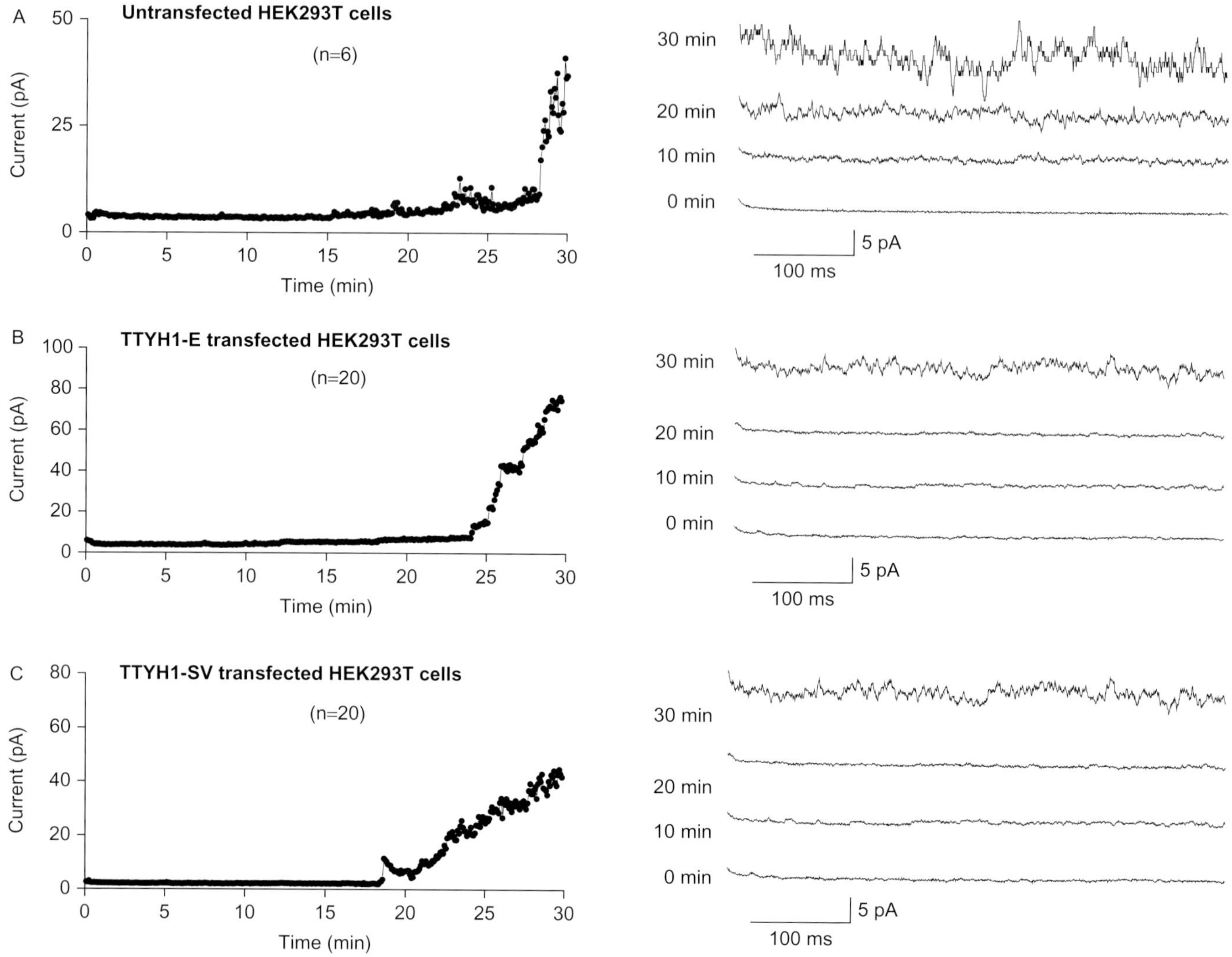

FIGURE 15.11 Non-Maxi-anion currents activated in inside-out patches at +25 mV after excision from non-transfected control HEK293T cells (**A**) and HEK293T cells transfected with TTYH1-E (**B**) and TTYH1-SV (**C**) cDNAs. *Left panels*: Time courses of mean patch currents observed after patch excision (at 0 min). *Right panels*: Expanded traces of the currents. The pipette and bath solutions contained standard Ringer solutions.

B. VSOR is a Protein Distinct from P-glycoprotein, pI_{Cln}, ClC-3 and Best

So far, three proteins have been proposed as candidates for the molecular counterpart of VSOR. They are the multidrug resistance ABC transporter P-glycoprotein (PGP), the 26 kDa protein pI_{Cln} and ClC-3 (Paulmichl et al., 1992; Valverde et al., 1992; Duan et al., 1997b). All three proteins were described as VSOR candidates based on the observation that their over-expression induced or enhanced functional expression of the VSOR current.

The first hypothesis assumed that PGP switches from a multidrug-resistant drug pump to a chloride channel upon osmotic cell swelling (Gill et al., 1992; Valverde et al., 1992). This "transporter/channel switching hypothesis" was initially taken to one's heart, because it had been established that another ABC transporter, CFTR, represented the molecular entity of the cAMP/PKA-activated anion channel. As summarized in several review articles (Nilius et al., 1997; Okada, 1997; Okada et al., 1998; Idriss et al., 2000), this attractive hypothesis had to be discarded in the face of mounting evidence showing that P-glycoprotein is distinct from VSOR. The evidence includes the following results obtained in our laboratory: (1) VSOR activity was not affected by antisense-induced abolition of PGP expression in human epithelial Intestine 407 cells (Tominaga et al., 1995); (2) VSOR currents were not blocked in Intestine 407 cells by three types of anti-PGP monoclonal antibodies (Tominaga et al., 1995); (3) neither the PGP inhibitor verapamil nor the PGP drug pump substrates vincristine and daunomycin prevented the swelling-induced VSOR activation in Intestine 407 cells (Tominaga et al., 1995); (4) the maximal VSOR current amplitude was indistinguishable between PGP-transfected KB-G2 cells and their parental PGP-lacking KB-3-1 cells (Miwa et al., 1997). The original observation showing enhancement of the VSOR current by PGP overexpression (Valverde et al., 1992) may have been caused by some special experimental conditions (e.g. protein–protein interaction) because PGP overexpression was found to upregulate VSOR activity (Valverde et al., 1996; Miwa et al., 1997) by augmenting the channel's sensitivity to cell volume changes or to cell membrane expansion (Miwa et al., 1997), an effect that was independent of PGP's ATPase activity (Okada et al., 1997).

The second hypothesis proposed that an intracellular 235-amino acid protein pI_{Cln}, which has no putative transmembrane domain, is involved in forming the VSOR pore after its insertion into the plasma membrane upon osmotic swelling, thereby conferring sensitivity of VSOR to extracellular ATP, cAMP and cGMP by exposing the glycine-rich nucleotide-binding site to the extracellular solution (Paulmichl et al., 1992). However, VSOR currents were not suppressed by extracellular cAMP and cGMP in Intestine 407 cells (Tsumura et al., 1996). Moreover, overexpression of pI_{Cln} in *Xenopus laevis* oocytes resulted in activation of anion currents with channel properties (rectification, anion selectivity and blocking) that were clearly different from those of VSOR (Voets et al., 1996; Buyse et al., 1997). Furthermore, reconstitution of recombinant pI_{Cln} into lipid bilayers often yielded cationic, rather than anionic currents which exhibited totally different rectification and single-channel conductance properties (Li et al., 1998; Strange, 1998; Furst et al., 2000), and which sometimes gave anionic currents with properties markedly different from those of VSOR (Garavaglia et al., 2002). Finally, it must be noted that pI_{Cln} has been recently identified as a constitutive cytosolic and nuclear protein which is a component of spliceosomal nuclear ribonucleoprotein (Pu et al., 1999, 2000).

The third hypothesis proposes that ClC-3 is the molecular counterpart of VSOR. This was proposed by Duan and co-workers (Duan et al., 1997a, b, 1999), based on the following observations. (1) Transfection of the cDNA coding the guinea-pig cardiac ClC-3 in NIH3T3 cells conferred a basal outward-rectifiying anion conductance to these cells (Duan et al., 1997b). This anion current was enhanced by osmotic swelling and was inhibited by a PKC activator (Duan et al., 1997b). The current also exhibited an intermediate unitary conductance, a $I^- > Cl^-$ selectivity, inactivation kinetics at large positive potentials, and sensitivity to DIDS, tamoxifen and extracellular ATP (Duan et al., 1997b). These properties, with the exception of the PKC sensitivity (see below) are similar to those of VSOR currents. (2) A single amino acid replacement (N579K) was found to change the halide selectivity from $I^- > Cl^-$ to $Cl^- > I^-$ and the channel rectification from outward to inward (Duan et al., 1997b). (3) The S51A mutation abolished the volume sensitivity and the PKC sensitivity (Duan et al., 1999). (4) Dialysis with synthetic C-termini peptides, which block ClC-3 interaction with F-actin, partially suppressed swelling-activated anion currents in NIH3T3 cells transfected with the short isoform of ClC-3 (McCloskey et al., 2007). (5) Anti-ClC-3 antibodies suppressed the VSOR-like swelling-activated anion currents endogenously expressed in guinea-pig atrial cells and canine smooth muscle cells (Duan et al., 2001; Wang et al., 2003; Yamamoto-Mizuma et al., 2004). (6) Knock-down of ClC-3 by antisense oligonucleotides or antisense cRNAs also suppressed the swelling-activated chloride currents native to HeLa cells and *Xenopus* oocytes (Hermoso et al., 2002). Some of these observations have been supported by other laboratories: With regard to the first, fifth and sixth points, Zhou and collaborators showed that transfection of guinea-pig ClC-3 cDNA in embryonic rat smooth muscle (A10) cells also augmented swelling-activated anion currents and that both antisense oligonucleotides and antibodies to ClC-3 suppressed the swelling-activated anion current (Zhou et al., 2005). Furthermore, sensitivity of endogenous VSOR-like currents to anti-ClC-3 antibodies has also been observed in AGS cells (Jin et al., 2003), LNCaP cells (Lemonnier et al., 2004) and NPE cells (Do et al., 2005). In addition, it has been recently shown that knockdown of ClC-3 by short-hairpin RNA reduced (but not abolished) anionic currents activated by lysophosphatidic acid or osmotic swelling in human corneal keratinocytes and lung fibroblasts (Yin et al., 2008). However, it must be stressed that despite all these observations, we cannot exclude the possibility that ClC-3 is not the VSOR channel *per se* but rather a regulator of its activity.

An alternative possibility could be that ClC-3 represents only one subtype of VSOR, or ClC3 in combination with (an)other channel(s) mediate VSOR currents. This was pointed out by Wang and collaborators on the basis of their observations showing that antisense ClC-3 knockdown only partially reduced swelling-activated anion currents, and that the sensitivity of the remaining current became altered by a PKC activator and by extracellular ATP (Wang et al., 2000). This possibility is also supported by recent data from Hume's group. They found that in ClC-3-deficient atrial cells, swelling-activated Cl^- currents are devoid of some of the properties inherent to the wildtype atrial cells, such as intracellular ATP dependence, as well as sensitivity to PKC and to anti-ClC-3 antibodies (Yamamoto-Mizuma et al., 2004). Although the authors reached the conclusion that ClC-3 constitutes one subtype of VSORs by assuming compensation by some other ClC member(s) in ClC-3-deficient cells, a simplest interpretation of the data would be that ClC-3 in atrial myocytes plays a regulating role in the activity and properties of VSOR. In this regard, it must be pointed out that some properties of ClC-3-associated anion currents observed by the Hume's group are markedly different from the phenotypic properties of VSOR observed in a wide variety of cells. First, their ClC-3-associated current was constitutively very active in non-swollen cells (Duan et al., 1997b), whereas the VSOR activity becomes prominent only after osmotic swelling (or other maneuvers such as ROS, GTPγS or reduced intracellular ionic strength; see Okada et al., 1997, 2006). Second, the open probability (P_o~0.6) of a ClC-3-associated single-channel event (Duan et al., 1997b) is much lower than that of VSOR (P_o~1 at negative potentials; Okada, 1997). Third, the ClC-3-associated current is inhibited by cAMP (Nagasaki et al., 2000), whereas cAMP was found to have no effect on the VSOR current in many cell types (Okada, 1997) or even enhance the VSOR current in dog atrial myocytes (Du and Sorota, 1997), mouse ventricular myocytes (Gong et al., 2004) and human epithelial cells (Shimizu et al., 2000; Shuba et al., 2000). Further, at variance with the hypothesis that ClC3 mediates VSOR currents is the observation by two independent groups that ClC-3 overexpression elicits anion currents that are completely independent of VSOR currents (Li et al., 2000; Huang et al., 2001). Finally, recent studies using ClC-3 knockout mice have provided the most critical evidence that native VSOR currents are totally independent of ClC-3 expression. Indeed, currents that were identical to those carried through VSOR in whole-cell mode were observed in hepatocytes and pancreatic acinar cells (Stobrawa et al., 2001) and parotid salivary acinar cells (Arreola et al., 2002) isolated from ClC-3 knockout mice. Furthermore, biophysical and pharmacological properties of VSOR currents in ventricular myocytes of ClC-3-deficient and wildtype mice were found to be indistinguishable, either at the macroscopic whole-cell current level (Gong et al., 2004), or at the microscopic single-channel level (Wang et al., 2005).

In summary, it is clear that PGP, pI_{Cln} and ClC-3 are not the molecular entities behind VSOR currents, although there remains the possibility that ClC-3 may participate in the regulation of the VSOR activity in some special situations.

Recently, Hartzell and co-workers have provided evidence that bestrophins are a new family of anion channels. Bestrophins are represented by four members in *Drosophila* and humans, and three members in mice (Hartzell et al., 2005, 2008). Among the members, hBest1, hBest2, hBest4, mBest2, dBest1 and dBest2 have been shown to function as CaCCs in heterologous expression systems (Sun et al., 2002; Qu et al., 2003, 2004; Tsunenari et al., 2003, 2006; Chien et al., 2006). Some hBest1 mutants have been associated with macular degeneration called Best vitelliform macular dystrophy (Marquardt et al., 1998; Petrukhin et al., 1998; White et al., 2000; Yu et al., 2007). Recently, Chien and Hartzell (2007) have found that not only a rise in the intracellular Ca^{2+} level, but also osmotic cell swelling are activators for dBest1-associated anion currents in *Drosophila* S2 cells. The properties of both the swelling-activated and Ca^{2+}-activated dBest1 currents are, however, distinct from those of VSOR and characterized by a sigmoidal whole-cell I/V relationship, a linear single-channel I/V relationship, no discernible voltage-dependent inactivation kinetics at positive potentials, and a small unitary conductance of ~2 pS (Chien et al., 2006; Chien and Hartzell, 2007). Also, human HEK293 cells do not express hBest1 and hBest2 (Chien et al., 2006) and mouse C127 cells lack expression of mBest1, mBest2 and mBest3 (Toychiev, Sabirov and Okada, unpublished RT-PCR observations), whereas both cell lines have native VSOR activities. Furthermore, transfection with the dominant negative mutant (G299E) of hBest1 failed to affect endogenous VSOR currents in HEK293 cells (Fischmeister and Hartzell, 2005). Thus, it appears that dBest1 is one subset of VAACs, but dBest1 and hBest1 are not related to VSOR.

Taken together, we must admit that for the time being, we have no good candidate molecules for VSOR currents. One of the difficulties encountered in the VSOR identification has been its ubiquitous expression. We now have a cell line (the cisplatin-resistant K-CP4 cell line), which lacks endogenous VSOR activity (Lee et al., 2007; Shimizu et al., 2008). This cell line may be useful for testing molecular candidates for VSOR in the near future.

V. ADDENDUM: THOUGHTS ON WHETHER ClC-3 IS A Cl^- CHANNEL OR A Cl^-/H^+ ANTIPORTER

Although it is now unlikely that ClC-3 mediates VSOR currents, one cannot exclude the possible involvement of ClC-3 in the formation or regulation of other type(s) of Cl^- channel. In fact, Weinman and collaborators reported that transfection of the short form of rat ClC-3 in CHO-K1 cells resulted in the activation of swelling-insensitive anion currents, without affecting endogenous VSOR currents (Li et al., 2000). These ClC-3-associated currents were characterized by very sharp outward rectification, a $Cl^- > I^-$ permeability sequence, lack of inactivation kinetics at positive potentials, and insensitivity to DIDS and NPPB. These properties are distinct from the properties of VSOR currents. In addition, Nelson and collaborators reported that human ClC-3 transfected into tsA cells induced swelling-insensitive, Ca^{2+}/CaMKII-activated Cl^- currents (Huang et al., 2001). In a more recent study, they indicated that ClC-3 is expressed mainly on the synaptic membrane in mouse hippocampal neurons and functions as a Ca^{2+}/CaMKII-activated Cl^- channel (Wang et al., 2006). However, previous molecular expression studies indicated that native ClC-3 localizes exclusively in intracellular compartments such as synaptic vesicles, endosomes and secretory granules (Stobrawa et al., 2001; Gentzsch et al., 2003; Moreland et al., 2006; Miller et al., 2007). Furthermore when heterologously overexpressed, ClC-3 was also found to be mainly localized intracellularly (Weylandt et al., 2001; Li et al., 2002; Zhao et al., 2007). Consistent with this observation of intracellular expression, Cl^--transporting functions of ClC-3 have been found to be required for respiratory burst mediated by NADPH oxidase (NOX2) at secretory granules in polymorphonuclear leukocytes (PMNs) (Moreland et al., 2006; Moreland et al., 2007) and for NOX1-mediated ROS production in vascular smooth muscle cells (Miller et al., 2007). Also, it must be stressed that halide permeability sequence for stimulated ROS production in PMNs ($Cl^- > Br^- > I^-$) was found to be distinct from that for VSOR by Moreland et al. (2006).

Since ClC-3, -4, and -5 belong to a common branch of the CLC gene family with ~80% sequence identity (Jentsch et al., 2005b), it has been assumed that ClC-3 may not be a Cl^- channel, but a Cl^-/H^+ antiporter like ClC-4 and ClC-5 (Picollo and Pusch, 2005). Matsuda et al. have recently provided direct evidence that ClC-3, overexpressed in HEK293 cells, functions as an electrogenic Cl^-/H^+ antiporter with sharp outward rectification properties (Matsuda et al., 2008). Further details on ClC-3 function can be found in Chapter 12 in the present volume.

Acknowledgements

We are grateful to T. Okayasu for the preparation of this manuscript. The work from our laboratory which is cited in this chapter was supported by Grants-in-Aid for Scientific Research (A) and (C) to YO and RZS, from the Ministry of Education, Culture, Sports, Science and Technology of Japan.

References

Abdullaev, I.F., Rudkouskaya, A., Schools, G.P., Kimelberg, H.K., and Mongin, A.A. (2006). Pharmacological comparison of swelling-activated excitatory amino acid release and Cl^- currents in cultured rat astrocytes. *J. Physiol.* **572**, 677–689.

Abraham, E.H., Okunieff, P., Scala, S., Vos, P., Oosterveld, M.J.S., Chen, A.Y., Shrivastav, B., Guidotti, G., Reddy, M.M., Quinton, P.M., Haws, C., Wine, J., Grygorczyk, R., Tabcharani, J.A., Hanrahan, J.W., Gunderson, K.L., Kopito, R., and Grygorczyk, R. (1997). Cystic fibrosis transmembrane conductance regulator and adenosine triphosphate. *Science* **275**, 1324–1326.

Ackerman, M.J., Krapivinsky, G.B., Gordon, E., Krapivinsky, L., and Clapham, D.C. (1994a). Characterization of a native swelling-induced chloride current, ICl.swell, and its regulatory protein, pICln, in Xenopus oocytes. *Jpn. J. Physiol.* **44** (Suppl. 2), S17–S24.

Ackerman, M.J., Wickman, K.D., and Clapham, D.E. (1994b). Hypotonicity activates a native chloride current in *Xenopus* oocytes. *J. Gen. Physiol.* **103**, 153–179.

Ando-Akatsuka, Y., Abdullaev, I.F., Lee, E.L., Okada, Y., and Sabirov, R.Z. (2002). Down-regulation of volume-sensitive Cl^- channels by CFTR is mediated by the second nucleotide-binding domain. *Pflugers Arch.* **445**, 177–186.

Arreola, J., Begenisich, T., Nehrke, K., Nguyen, H.-V., Park, K., Richardson, L., Yang, B., Schutte, B.C., Lamb, F.S., and Melvin, J.E. (2002). Secretion and cell volume regulation by salivary acinar cells from mice lacking expression of the *Clcn3* Cl^- channel gene. *J. Physiol.* **545**, 207–216.

Arreola, J., Melvin, J.E., and Begenisich, T. (1995). Volume-activated chloride channels in rat parotid acinar cells. *J. Physiol.* **484**, 677–687.

Bahamonde, M.I., Fernandez-Fernandez, J.M., Guix, F.X., Vazquez, E., and Valverde, M.A. (2003). Plasma membrane voltage-dependent anion channel mediates antiestrogen-activated maxi Cl^- currents in C1300 neuroblastoma cells. *J. Biol. Chem.* **278**, 33284–33289.

Bajnath, R.B., Groot, J.A., de Jonge, H.R., Kansen, M., and Bijman, J. (1993). Calcium ionophore plus excision induce a large conductance chloride channel in membrane patches of human colon carcinoma cells HT-29cl.19A. *Experientia* **49**, 313–316.

Banderali, U. and Roy, G. (1992). Anion channels for amino acids in MDCK cells. *Am. J. Physiol. Cell. Physiol.* **263**, C1200–C1207.

Basavappa, S., Huang, C.C., Mangel, A.W., Lebedev, D.V., Knauf, P.A., and Ellory, J.C. (1996). Swelling-activated amino acid efflux in the human neuroblastoma cell line CHP-100. *J. Neurophysiol.* **76**, 764–769.

Bell, P.D., Lapointe, J.Y., Sabirov, R., Hayashi, S., Peti-Peterdi, J., Manabe, K., Kovacs, G., and Okada, Y. (2003). Macula densa cell signaling involves ATP release through a maxi anion channel. *Proc. Natl. Acad. Sci. USA* **100**, 4322–4327.

Bernucci, L., Umana, F., Llanos, P., and Riquelme, G. (2003). *Large chloride* channel from pre-eclamptic human placenta. *Placenta* **24**, 895–903.

Blanz, J., Schweizer, M., Auberson, M., Maier, H., Muenscher, A., Hubner, C.A., and Jentsch, T.J. (2007). Leukoencephalopathy upon disruption of the chloride channel ClC-2. *J. Neurosci.* **27**, 6581–6589.

Bodin, P. and Burnstock, G. (2001). Purinergic signalling: ATP release. *Neurochem. Res.* **26**, 959–969.

Boese, S.H., Kinne, R.K., and Wehner, F. (1996). Single-channel properties of swelling-activated anion conductance in rat inner medullary collecting duct cells. *Am. J. Physiol. Renal Physiol.* **271**, F1224–F1233.

Bond, T.D., Ambikapathy, S., Mohammad, S., and Valverde, M.A. (1998). Osmosensitive Cl^- currents and their relevance to regulatory volume decrease in human intestinal T84 cells: outwardly vs. inwardly rectifying currents. *J. Physiol.* **511**, 45–54.

Bosl, M.R., Stein, V., Hubner, C., Zdebik, A.A., Jordt, S.E., Mukhopadhyay, A.K., Davidoff, M.S., Holstein, A.F., and Jentsch, T.J. (2001). Male germ cells and photoreceptors, both dependent on close cell–cell interactions, degenerate upon ClC-2 Cl^- channel disruption. *EMBO J.* **20**, 1289–1299.

Browe, D.M. and Baumgarten, C.M. (2004). Angiotensin II (AT1) receptors and NADPH oxidase regulate Cl^- current elicited by β1 integrin stretch in rabbit ventricular myocytes. *J. Gen. Physiol.* **124**, 273–287.

Buettner, R., Papoutsoglou, G., Scemes, E., Spray, D.C., and Dermietzel, R. (2000). Evidence for secretory pathway localization of a voltage-dependent anion channel isoform. *Proc. Natl. Acad. Sci. USA* **97**, 3201–3206.

Burnstock, G. (2004). Introduction: P2 receptors. *Curr. Top. Med. Chem.* **4**, 793–803.

Buyse, G., Voets, T., Tytgat, J., De Greef, C., Droogmans, G., Nilius, B., and Eggermont, J. (1997). Expression of human pI_{Cln} and ClC-6 in *Xenopus* oocytes induces an identical endogenous chloride conductance. *J. Biol. Chem.* **272**, 3615–3621.

Cahalan, M.D. and Lewis, R.S. (1988). Role of potassium and chloride channels in volume regulation by T lymphocytes. In *Cell Physiology of Blood* (Gunn, R. and Parker, J., eds), pp. 282–301. Rockefeller University Press, NY.

Carpaneto, A., Accardi, A., Pisciotta, M., and Gambale, F. (1999). Chloride channels activated by hypotonicity in N2A neuroblastoma cell line. *Exp. Brain Res.* **124**, 193–199.

Chan, H.C., Fu, W.O., Chung, Y.W., Huang, S.J., Chan, P.S., and Wong, P.Y. (1994). Swelling-induced anion and cation conductances in human epididymal cells. *J. Physiol.* **478**, 449–460.

Chen, N.H., Reith, M.E., and Quick, M.W. (2004). Synaptic uptake and beyond: the sodium- and chloride-dependent neurotransmitter transporter family SLC6. *Pflugers Arch.* **447**, 519–531.

Chien, L.-T. and Hartzell, H.C. (2007). *Drosophila* bestrophin-1 chloride current is dually regulated by calcium and cell volume. *J. Gen. Physiol.* **130**, 513–524.

Chien, L.-T., Zhang, Z.-R., and Hartzell, H.C. (2006). Single Cl^- channels activated by Ca^{2+} in drosophila S2 cells are mediated by bestrophins. *J. Gen. Physiol.* **128**, 247–259.

Chou, C.Y., Shen, M.R., Chen, T.M., and Huang, K.E. (1997). Volume-activated taurine transport is differentially activated in human cervical cancer HT-3 cells but not in human papillomavirus-immortalized Z183A and normal cervical epithelial cells. *Clin. Exp. Pharmacol. Physiol.* **24**, 935–939.

Clark, S., Jordt, S.-E., Jentsch, T.J., and Mathie, A. (1998). Characterization of the hyperpolarization-activated chloride current in dissociated rat sympathetic neurons. *J. Physiol.* **506**, 665–678.

Colombini, M. (2004). VDAC: the channel at the interface between mitochondria and the cytosol. *Mol. Cell. Biochem.* **256/257**, 107–115.

d'Anglemont de Tassigny, A., Souktani, R., Ghaleh, B., Henry, P., and Berdeaux, A. (2003). Structure and pharmacology of swelling-sensitive chloride channels, $I_{Cl,swell}$. *Fundam. Clin. Pharmacol.* **17**, 539–553.

Decher, N., Lang, H.J., Nilius, B., Bruggemann, A., Busch, A.E., and Steinmeyer, K. (2001). DCPIB is a novel selective blocker of $I_{Cl,swell}$ and prevents swelling-induced shortening of guinea-pig atrial action potential duration. *Br. J. Pharmacol.* **134**, 1467–1479.

Dermietzel, R., Hwang, T.K., Buettner, R., Hofer, A., Dotzler, E., Kremer, M., Deutzmann, R., Thinnes, F.P., Fishman, G.I., and Spray, D.C. (1994). Cloning and in situ localization of a brain-derived porin that constitutes a large-conductance anion channel in astrocytic plasma membranes. *Proc. Natl. Acad. Sci. USA* **91**, 499–503.

Do, C.W., Lu, W., Mitchell, C.H., and Civan, M.M. (2005). Inhibition of swelling-activated Cl^- currents by functional anti-ClC-3 antibody in native bovine non-pigmented ciliary epithelial cells. *Invest. Ophthalmol. Vis. Sci.* **46**, 948–955.

Droogmans, G., Maertens, C., Prenen, J., and Nilius, B. (1999). Sulphonic acid derivatives as probes of pore properties of volume-regulated anion channels in endothelial cells. *Br. J. Pharmacol.* **128**, 35–40.

Du, X.Y. and Sorota, S. (1997). Modulation of dog atrial swelling-induced chloride current by cAMP: protein kinase A-dependent and -independent pathways. *J. Physiol.* **500**, 111–122.

Duan, D., Cowley, S., Horowitz, B., and Hume, J.R. (1999). A serine residue in ClC-3 links phosphorylation-dephosphorylation to chloride channel regulation by cell volume. *J. Gen. Physiol.* **113**, 57–70.

Duan, D., Hume, J.R., and Nattel, S. (1997a). Evidence that outwardly rectifying Cl^- channels underlie volume-regulated Cl^- currents in heart. *Circ. Res.* **80**, 103–113.

Duan, D., Winter, C., Cowley, S., Hume, J.R., and Horowitz, B. (1997b). Molecular identification of a volume-regulated chloride channel. *Nature* **390**, 417–421.

Duan, D., Ye, L., Britton, F., Horowitz, B., and Hume, J.R. (2000). A novel anionic inward rectifier in native cardiac myocytes. *Circ. Res.* **86**, e63–e71.

Duan, D., Zhong, J., Hermoso, M., Satterwhite, C.M., Rossow, C.F., Hatton, W.J., Yamboliev, I., Horowitz, B., and Hume, J.R. (2001). Functional inhibition of native volume-sensitive outwardly rectifying anion channels in muscle cells and *Xenopus* oocytes by anti-ClC-3 antibody. *J. Physiol.* **531**, 437–444.

Dubyak, G.R. and el-Moatassim, C. (1993). Signal transduction via P2-purinergic receptors for extracellular ATP and other nucleotides. *Am. J. Physiol. Cell. Physiol.* **265**, C577–C606.

Dutta, A.K., Korchev, Y.E., Shevchuk, A.I., Hayashi, S., Okada, Y., and Sabirov, R.Z. (2008). Spatial distribution of maxi-anion channel on cardiomyocytes detected by smart-patch technique. *Biophys. J.* **94**, 1646–1655.

Dutta, A.K., Okada, Y., and Sabirov, R.Z. (2002). Regulation of an ATP-conductive large-conductance anion channel and swelling-induced ATP release by arachidonic acid. *J. Physiol.* **542**, 803–816.

Dutta, A.K., Sabirov, R.Z., Uramoto, H., and Okada, Y. (2004). Role of ATP-conductive anion channel in ATP release from neonatal rat cardiomyocytes in ischaemic or hypoxic conditions. *J. Physiol.* **559**, 799–812.

Evanko, D.S., Zhang, Q., Zorec, R., and Haydon, P.G. (2004). Defining pathways of loss and secretion of chemical messengers from astrocytes. *Glia* **47**, 233–240.

Fahlke, C. (2001). Ion permeation and selectivity in ClC-type chloride channels. *Am. J. Physiol. Renal Physiol.* **280**, F748–F757.

Falke, L.C. and Misler, S. (1989). Activity of ion channels during volume regulation by clonal N1E115 neuroblastoma cells. *Proc. Natl. Acad. Sci. USA* **86**, 3919–3923.

Fan, H.-T., Morishima, S., Kida, H., and Okada, Y. (2001). Phloretin differentially inhibits volume-sensitive and cyclic AMP-activated, but not Ca-activated, Cl^- channels. *Br. J. Pharmacol.* **133**, 1096–1106.

Fellin, T., Pascual, O., and Haydon, P.G. (2006). Astrocytes coordinate synaptic networks: balanced excitation and inhibition. *Physiology (Bethesda)* **21**, 208–215.

Fields, R.D. and Stevens, B. (2000). ATP: an extracellular signaling molecule between neurons and glia. *Trends Neurosci.* **23**, 625–633.

Fischmeister, R. and Hartzell, H.C. (2005). Volume sensitivity of the bestrophin family of chloride channels. *J. Physiol.* **562**, 477–491.

Forrester, T. (2003). A purine signal for functional hypertemia in skeletal and cardiac muscle. In *Current Topics in Membranes: Extracellular Nucleotides and Nucleosides: Release, Receptors, and Physiological and Pathophysiological Effects* (Schwiebert, E.M., ed.), pp. 269–305. Academic Press, Amsterdam.

Furst, J., Bazzini, C., Jakab, M., Meyer, G., Konig, M., Gschwentner, M., Ritter, M., Schmarda, A., Botta, G., Benz, R., Deetjen, P., and Paulmichl, M. (2000). Functional reconstitution of ICln in lipid bilayers. *Pflugers Arch.* **440**, 100–115.

Furukawa, T., Ogura, T., Katayama, Y., and Hiraoka, M. (1998). Characteristics of rabbit ClC-2 current expressed in *Xenopus* oocytes and its contribution to volume regulation. *Am. J. Physiol. Cell Physiol.* **274**, C500–C512.

Garavaglia, M.L., Rodighiero, S., Bertocchi, C., Manfredi, R., Furst, J., Gschwentner, M., Ritter, M., Bazzini, C., Botta, G., Jakab, M., Meyer, G., and Paulmichl, M. (2002). ICln channels reconstituted in heart-lipid bilayer are selective to chloride. *Pflugers Arch.* **443**, 748–753.

Gentzsch, M., Cui, L., Mengos, A., Chang, X.-B., Chen, J.-H., and Riordan, J.R. (2003). The PDZ-binding chloride channel ClC-3B localizes to the golgi and associates with cystic fibrosis transmembrane conductance regulator-interacting PDZ proteins. *J. Biol. Chem.* **278**, 6440–6449.

Gill, D.R., Hyde, S.C., Higgins, C.F., Valverde, M.A., Mintenig, G. M., and Sepulveda, F.V. (1992). Separation of drug transport and chloride channel functions of the human multidrug resistance P-glycoprotein. *Cell* **71**, 23–32.

Gong, W., Xu, H., Shimizu, T., Morishima, S., Tanabe, S., Tachibe, T., Uchida, S., Sasaki, S., and Okada, Y. (2004). ClC-3-independent, PKC-dependent activity of volume-sensitive Cl channel in mouse ventricular cardiomyocytes. *Cell. Physiol. Biochem.* **14**, 213–224.

Groschner, K. and Kukovetz, W.R. (1992). Voltage-sensitive chloride channels of large conductance in the membrane of pig aortic endothelial cells. *Pflugers Arch.* **421**, 209–217.

Grunder, S., Thiemann, A., Pusch, M., and Jentsch, T.J. (1992). Regions involved in the opening of CIC-2 chloride channel by voltage and cell volume. *Nature* **360**, 759–762.

Grygorczyk, R. and Bridges, M.A. (1992). Whole-cell chloride conductances in cultured brushed human nasal epithelial cells. *Can. J. Physiol. Pharmacol.* **70**, 1134–1141.

Gu, Y., Gorelik, J., Spohr, H.A., Shevchuk, A., Lab, M.J., Harding, S.E., Vodyanoy, I., Klenerman, D., and Korchev, Y.E. (2002). High-resolution scanning patch-clamp: new insights into cell function. *FASEB J.* **16**, 748–750.

Hansson, E. and Ronnback, L. (2003). Glial neuronal signaling in the central nervous system. *FASEB J.* **17**, 341–348.

Harrigan, T.J., Abdullaev, I.F., Jourd'heuil, D., and Mongin, A.A. (2008). Activation of microglia with zymosan promotes excitatory amino acid release via volume-regulated anion channels: the role of NADPH oxidases. *J. Neurochem.* **106**, 2449–2462.

Hartzell, C., Putzier, I., and Arreola, J. (2005). Calcium-activated chloride channels. *Annu. Rev. Physiol.* **67**, 719–758.

Hartzell, H.C., Qu, Z., Yu, K., Xiao, Q., and Chien, L.-T. (2008). Molecular physiology of bestrophins: multifunctional membrane proteins linked to best disease and other retinopathies. *Physiol. Rev.* **88**, 639–672.

Hayashi, S., Hazama, A., Dutta, A.K., Sabirov, R.Z., and Okada, Y. (2004). Detecting ATP release by a biosensor method. *Sci. STKE* **2004** (258), pl14.

Haydon, P.G. (2001). GLIA: listening and talking to the synapse. *Nat. Rev. Neurosci.* **2**, 185–193.

Hazama, A., Ando-Akatsuka, Y., Fan, H.-T., Tanaka, S., and Okada, Y. (2000a). CFTR-dependent and -independent ATP release induced by osmotic swelling. In *Control and Disease of Sodium Dependent Transportation Proteins and Ion Channels* (Suketa, Y., Carafoli, E., Lazdunski, M., Mikoshiba, K., Okada, Y., and Wright, E.M., eds), pp. 429–431. Elsevier, Amsterdam.

Hazama, A., Fan, H.-T., Abdullaev, I., Maeno, E., Tanaka, S., Ando-Akatsuka, Y., and Okada, Y. (2000b). Swelling-activated, cystic fibrosis transmembrane conductance regulator-augmented ATP release and Cl^- conductances in murine C127 cells. *J. Physiol.* **523**, 1–11.

Hazama, A., Hayashi, S., and Okada, Y. (1998a). Cell surface measurements of ATP release from single pancreatic β cells using a novel biosensor technique. *Pflugers Arch.* **437**, 31–35.

Hazama, A., Miwa, A., Miyoshi, T., Shimizu, T., and Okada, Y. (1998b). ATP release from swollen or CFTR-expressing epithelial cells. In *Cell Volume Regulation: The Molecular Mechanism and Volume Sensing Machinery* (Okada, Y., ed.), pp. 93–98. Elsevier, Amsterdam.

Hazama, A., Shimizu, T., Ando-Akatsuka, Y., Hayashi, S., Tanaka, S., Maeno, E., and Okada, Y. (1999). Swelling-induced, CFTR-independent ATP release from a human epithelial cell line: lack of correlation with volume-sensitive Cl^- channels. *J. Gen. Physiol.* **114**, 525–533.

Helix, N., Strobaek, D., Dahl, B.H., and Christophersen, P. (2003). Inhibition of the endogenous volume-regulated anion channel (VRAC) in HEK293 cells by acidic di-aryl-ureas. *J. Membr. Biol.* **196**, 83–94.

Hermoso, M., Satterwhite, C.M., Andrade, Y.N., Hidalgo, J., Wilson, S.M., Horowitz, B., and Hume, J.R. (2002). ClC-3 is a fundamental molecular component of volume-sensitive outwardly rectifying Cl^- channels and volume regulation in HeLa cells and Xenopus laevis oocytes. *J. Biol. Chem.* **277**, 40066–40074.

Hisadome, K., Koyama, T., Kimura, C., Droogmans, G., Ito, Y., and Oike, M. (2002). Volume-regulated anion channels serve as an auto/paracrine nucleotide release pathway in aortic endothelial cells. *J. Gen. Physiol.* **119**, 511–520.

Hoffmann, E.K., Lambert, I.H., and Pedersen, S.F. (2009). Physiology of cell volume regulation in vertebrates. *Physiol. Rev.* **89**, 193–277.

Huang, P., Liu, J., Di, A., Robinson, N.C., Musch, M.W., Kaetzel, M.A., and Nelson, D.J. (2001). Regulation of human CLC-3 channels by multifunctional Ca^{2+}/calmodulin-dependent protein kinase. *J. Biol. Chem.* **276**, 20093–20100.

Hurnak, O. and Zachar, J. (1992). Maxi chloride channels in L6 myoblasts. *Gen. Physiol. Biophys.* **11**, 389–400.

Idriss, H.T., Hannun, Y.A., Boulpaep, E., and Basavappa, S. (2000). Regulation of volume-activated chloride channels by P-glycoprotein: phosphorylation has the final say! *J. Physiol.* **524**, 629–636.

Inoue, H., Mori, S., Morishima, S., and Okada, Y. (2005). Volume-sensitive chloride channels in mouse cortical neurons: characterization and role in volume regulation. *Eur. J. Neurosci.* **21**, 1648–1658.

Jackson, P.S., Churchwell, K., Ballatori, N., Boyer, J.L., and Strange, K. (1996). Swelling-activated anion conductance in skate hepatocytes: regulation by cell Cl^- and ATP. *Am. J. Physiol. Cell Physiol.* **270**, C57–C66.

Jackson, P.S. and Strange, K. (1993). Volume-sensitive anion channels mediate swelling-activated inositol and taurine efflux. *Am. J. Physiol. Cell Physiol.* **265**, C1489–C1500.

Jackson, P.S. and Strange, K. (1995). Characterization of the voltage-dependent properties of a volume-sensitive anion conductance. *J. Gen. Physiol.* **105**, 661–676.

Jalonen, T. (1993). Single-channel characteristics of the large-conductance anion channel in rat cortical astrocytes in primary culture. *Glia* **9**, 227–237.

Jentsch, T.J., Friedrich, T., Schriever, A., and Yamada, H. (1999). The CLC chloride channel family. *Pflugers Arch.* **437**, 783–795.

Jentsch, T.J., Neagoe, I., and Scheel, O. (2005a). CLC chloride channels and transporters. *Curr. Opin. Neurobiol.* **15**, 319–325.

Jentsch, T.J., Poet, M., Fuhrmann, J.C., and Zdebik, A.A. (2005b). Physiological functions of CLC Cl^- channels gleaned from human genetic disease and mouse models. *Annu. Rev. Physiol.* **67**, 779–807.

Jentsch, T.J., Stein, V., Weinreich, F., and Zdebik, A.A. (2002). Molecular structure and physiological function of chloride channels. *Physiol. Rev.* **82**, 503–568.

Jin, N.G., Kim, J.K., Yang, D.K., Cho, S.J., Kim, J.M., Koh, E.J., Jung, H.C., So, I., and Kim, K.W. (2003). Fundamental role of ClC-3 in volume-sensitive Cl^- channel function and cell volume regulation in AGS cells. *Am. J. Physiol. Gastrointest. Liver Physiol.* **285**, G938–G948.

Jordt, S.E. and Jentsch, T.J. (1997). Molecular dissection of gating in the ClC-2 chloride channel. *EMBO J.* **16**, 1582–1592.

Kajita, H., Kotera, T., Shirakata, Y., Ueda, S., Okuma, M., Oda-Ohmae, K., Takimoto, M., Urade, Y., and Okada, Y. (1995). A maxi Cl^- channel coupled to endothelin B receptors in the basolateral membrane of guinea-pig parietal cells. *J. Physiol.* **488**, 65–75.

Kajita, H., Omori, K., and Matsuda, H. (2000). The chloride channel ClC-2 contributes to the inwardly rectifying Cl^- conductance in cultured porcine choroid plexus epithelial cells. *J. Physiol.* **523**, 313–324.

Kelly, M.E., Dixon, S.J., and Sims, S.M. (1994). Outwardly rectifying chloride current in rabbit osteoclasts is activated by hyposmotic stimulation. *J. Physiol.* **475**, 377–389.

Kimelberg, H.K., Goderie, S.K., Higman, S., Pang, S., and Waniewski, R.A. (1990). Swelling-induced release of glutamate, aspartate, and taurine from astrocyte cultures. *J. Neurosci.* **10**, 1583–1591.

Kimelberg, H.K. and Mongin, A.A. (1998). Swelling-activated release of excitatory amino acids in the brain: relevance for pathophysiology. *Contrib. Nephrol.* **123**, 240–257.

Kimelberg, H.K., Nestor, N.B., and Feustel, P.J. (2004). Inhibition of release of taurine and excitatory amino acids in ischemia and neuroprotection. *Neurochem. Res.* **29**, 267–274.

Kirk, K. (1997). Swelling-activated organic osmolyte channels. *J. Membr. Biol.* **158**, 1–16.

Kirk, K., Ellory, J.C., and Young, J.D. (1992). Transport of organic substrates via a volume-activated channel. *J. Biol. Chem.* **267**, 23475–23478.

Kubo, M. and Okada, Y. (1992). Volume-regulatory Cl^- channel currents in cultured human epithelial cells. *J. Physiol.* **456**, 351–371.

Lambert, I.H. and Hoffmann, E.K. (1994). Cell swelling activates separate taurine and chloride channels in Ehrlich mouse ascites tumor cells. *J. Membr. Biol.* **142**, 289–298.

Leaney, J.L., Marsh, S.J., and Brown, D.A. (1997). A swelling-activated chloride current in rat sympathetic neurones. *J. Physiol.* **501**, 555–564.

Lee, E.L., Shimizu, T., Ise, T., Numata, T., Kohno, K., and Okada, Y. (2007). Impaired activity of volume-sensitive Cl^- channel is involved in cisplatin resistance of cancer cells. *J. Cell. Physiol.* **211**, 513–521.

Lemonnier, L., Shuba, Y., Crepin, A., Roudbaraki, M., Slomianny, C., Mauroy, B., Nilius, B., Prevarskaya, N., and Skryma, R. (2004). Bcl-2-dependent modulation of swelling-activated Cl^- current and ClC-3 expression in human prostate cancer epithelial cells. *Cancer Res.* **64**, 4841–4848.

Levitan, I. and Garber, S.S. (1998). Anion competition for a volume-regulated current. *Biophys. J.* **75**, 226–235.

Lewis, R.A., Bursell, J.D., and Kirk, K. (1996). Anion-selectivity of the swelling-activated osmolyte channel in eel erythrocytes. *J. Membr. Biol.* **149**, 103–111.

Lewis, R.S., Ross, P.E., and Cahalan, M.D. (1993). Chloride channels activated by osmotic stress in T lymphocytes. *J. Gen. Physiol.* **101**, 801–826.

Li, C., Breton, S., Morrison, R., Cannon, C.L., Emma, F., Sanchez-Olea, R., Bear, C., and Strange, K. (1998). Recombinant pI_{Cln} forms highly cation-selective channels when reconstituted into artificial and biological membranes. *J. Gen. Physiol.* **112**, 727–736.

Li, X., Shimada, K., Showalter, L.A., and Weinman, S.A. (2000). Biophysical properties of ClC-3 differentiate it from swelling-activated chloride channels in Chinese hamster ovary-K1 cells. *J. Biol. Chem.* **275**, 35994–35998.

Li, X., Wang, T., Zhao, Z., and Weinman, S.A. (2002). The ClC-3 chloride channel promotes acidification of lysosomes in CHO-K1 and Huh-7 cells. *Am. J. Physiol. Cell Physiol.* **282**, C1483–C1491.

Liu, H.-T. (2007). Bradykinin-induced communication between astrocytes and neurons is mediated by glutamate released via volume-sensitive outwardly rectifying anion channels. *J. Physiol. Sci.* **57**, S126.

Liu, H.-T., Sabirov, R.Z., and Okada, Y. (2008a). Oxygen-glucose deprivation induces ATP release via maxi-anion channels in astrocytes. *Purinergic Signal* **4**, 147–154.

Liu, H.-T., Tashmukhamedov, B.A., Inoue, H., Okada, Y., and Sabirov, R.Z. (2006). Roles of two types of anion channels in glutamate release from mouse astrocytes under ischemic or osmotic stress. *Glia* **54**, 343–357.

Liu, H.-T., Toychiev, A.H., Takahashi, N., Sabirov, R.Z., and Okada, Y. (2008b). Maxi-anion channel as a candidate pathway for osmosensitive ATP release from mouse astrocytes in primary culture. *Cell Res.* **18**, 558–565.

Liu, Y., Oiki, S., Tsumura, T., Shimizu, T., and Okada, Y. (1998). Glibenclamide blocks volume-sensitive Cl^- channels by dual mechanisms. *Am. J. Physiol. Cell Physiol.* **275**, C343–C351.

Malarkey, E.B. and Parpura, V. (2008). Mechanisms of glutamate release from astrocytes. *Neurochem. Int.* **52**, 142–154.

Mannella, C.A. (1997). Minireview: on the structure and gating mechanism of the mitochondrial channel, VDAC. *J. Bioenerg. Biomembr.* **29**, 525–531.

Manolopoulos, V.G., Droogmans, G., and Nilius, B. (1997a). Hypotonicity and thrombin activate taurine efflux in BC3H1 and C2C12 myoblasts that is down regulated during differentiation. *Biochem. Biophys. Res. Commun.* **232**, 74–79.

Manolopoulos, V.G., Voets, T., Declercq, P.E., Droogmans, G., and Nilius, B. (1997b). Swelling-activated efflux of taurine and other organic osmolytes in endothelial cells. *Am. J. Physiol. Cell Physiol.* **273**, C214–C222.

Marquardt, A., Stohr, H., Passmore, L.A., Kramer, F., Rivera, A., and Weber, B.H. (1998). Mutations in a novel gene, *VMD2*, encoding a protein of unknown properties cause juvenile-onset vitelliform macular dystrophy (Best's disease). *Hum. Mol. Genet.* **7**, 1517–1525.

Matsuda, J.J., Filali, M.S., Volk, K.A., Collins, M.M., Moreland, J.G., and Lamb, F.S. (2008). Overexpression of CLC-3 in HEK293T cells yields novel currents that are pH dependent. *Am. J. Physiol. Cell Physiol.* **294**, C251–C262.

McCloskey, D.T., Doherty, L., Dai, Y.-P., Miller, L., Hume, J.R., and Yamboliev, I.A. (2007). Hypotonic activation of short ClC3 isoform is modulated by direct interaction between its cytosolic C-terminal tail and subcortical actin filaments. *J. Biol. Chem.* **282**, 16871–16877.

Meyer, K. and Korbmacher, C. (1996). Cell swelling activates ATP-dependent voltage-gated chloride channels in M-1 mouse cortical collecting duct cells. *J. Gen. Physiol.* **108**, 177–193.

Miller, F.J., Jr., Filali, M., Huss, G.J., Stanic, B., Chamseddine, A., Barna, T.J., and Lamb, F.S. (2007). Cytokine activation of nuclear factor κB in vascular smooth muscle cells requires signaling endosomes containing Nox1 and ClC-3. *Circ. Res.* **101**, 663–671.

Mitchell, C.H., Wang, L., and Jacob, T.J. (1997a). A large-conductance chloride channel in pigmented ciliary epithelial cells activated by GTPgammaS. *J. Membr. Biol.* **158**, 167–175.

Mitchell, C.H., Zhang, J.J., Wang, L., and Jacob, T.J. (1997b). Volume-sensitive chloride current in pigmented ciliary epithelial cells: role of phospholipases. *Am. J. Physiol. Cell Physiol.* **272**, C212–C222.

Mitchell, D.M., Zhou, M., Pariyarath, R., Wang, H., Aitchison, J.D., Ginsberg, H.N., and Fisher, E.A. (1998). Apoprotein B100 has a prolonged interaction with the translocon during which its lipidation and translocation change from dependence on the microsomal triglyceride transfer protein to independence. *Proc. Natl. Acad. Sci. USA* **95**, 14733–14738.

Miwa, A., Ueda, K., and Okada, Y. (1997). Protein kinase C-independent correlation between P-glycoprotein expression and volume sensitivity of Cl^- channel. *J. Membr. Biol.* **157**, 63–69.

Moran, J., Miranda, D., Pena-Segura, C., and Pasantes-Morales, H. (1997). Volume regulation in NIH/3T3 cells not expressing P-glycoprotein. II. Chloride and amino acid fluxes. *Am. J. Physiol. Cell Physiol.* **272**, C1804–C1809.

Moreland, J.G., Davis, A.P., Bailey, G., Nauseef, W.M., and Lamb, F.S. (2006). Anion channels, including ClC-3, are required for normal neutrophil oxidative function, phagocytosis, and transendothelial migration. *J. Biol. Chem.* **281**, 12277–12288.

Moreland, J.G., Davis, A.P., Matsuda, J.J., Hook, J.S., Bailey, G., Nauseef, W.M., and Lamb, F.S. (2007). Endotoxin priming of neutrophils requires NADPH oxidase-generated oxidants and is regulated by the anion transporter ClC-3. *J. Biol. Chem.* **282**, 33958–33967.

Mori, S., Morishima, S., Takasaki, M., and Okada, Y. (2002). Impaired activity of volume-sensitive anion channel during lactacidosis-induced swelling in neuronally differentiated NG108-15 cells. *Brain Res.* **957**, 1–11.

Nabekura, T., Morishima, S., Cover, T.L., Mori, S.-I., Kannan, H., Komune, S., and Okada, Y. (2003). Recovery from lactacidosis-induced glial cell swelling with the aid of exogenous anion channels. *Glia* **41**, 247–259.

Nagasaki, M., Ye, L., Duan, D., Horowitz, B., and Hume, J.R. (2000). Intracellular cyclic AMP inhibits native and recombinant volume-regulated chloride channels from mammalian heart. *J. Physiol.* **523**, 705–717.

Nilius, B. and Droogmans, G. (2003). Amazing chloride channels: an overview. *Acta Physiol. Scand.* **177**, 119–147.

Nilius, B., Eggermont, J., Voets, T., Buyse, G., Manolopoulos, V., and Droogmans, G. (1997). Properties of volume-regulated anion channels in mammalian cells. *Prog. Biophys. Mol. Biol.* **68**, 69–119.

Nilius, B., Oike, M., Zahradnik, I., and Droogmans, G. (1994). Activation of a Cl^- current by hypotonic volume increase in human endothelial cells. *J. Gen. Physiol.* **103**, 787–805.

Oiki, S., Kubo, M., and Okada, Y. (1994). Mg^{2+} and ATP-dependence of volume-sensitive Cl^- channels in human epithelial cells. *Jpn. J. Physiol.* **44** (Suppl. 2), S77–S79.

Okada, Y. (1997). Volume expansion-sensing outward-rectifier Cl^- channel: fresh start to the molecular identity and volume sensor. *Am. J. Physiol. Cell Physiol.* **273**, C755–C789.

Okada, Y. (2004). Ion channels and transporters involved in cell volume regulation and sensor mechanisms. *Cell Biochem. Biophys.* **41**, 233–258.

Okada, Y. (2006). Cell volume-sensitive chloride channel: phenotypic properties and molecular identity. In *Mechanisms and Significance of Cell Volume Regulation* (Lang, F., ed.), pp. 9–24. Karger, Basel.

Okada, Y., Maeno, E., Shimizu, T., Dezaki, K., Wang, J., and Morishima, S. (2001). Receptor-mediated control of regulatory volume decrease (RVD) and apoptotic volume decrease (AVD). *J. Physiol.* **532**, 3–16.

Okada, Y., Maeno, E., Shimizu, T., Manabe, K., Mori, S., and Nabekura, T. (2004). Dual roles of plasmalemmal chloride channels in induction of cell death. *Pflugers Arch.* **448**, 287–295.

Okada, Y., Oiki, S., Tominaga, M., Kubo, M., Miwa, A., Tominaga, T., Tsumura, T., and Ueda, K. (1997). Volume-sensitive Cl^- channel in human epithelial cells: regulation by ATP and relation to P-glycoprotein. *Jpn. J. Physiol.* **47** (Suppl. 1), S19–S20.

Okada, Y., Petersen, C.C., Kubo, M., Morishima, S., and Tominaga, M. (1994). Osmotic swelling activates intermediate-conductance Cl^- channels in human intestinal epithelial cells. *Jpn. J. Physiol.* **44**, 403–409.

Okada, Y., Shimizu, T., Maeno, E., Tanabe, S., Wang, X., and Takahashi, N. (2006). Volume-sensitive chloride channels involved in apoptotic volume decrease and cell death. *J. Membr. Biol.* **209**, 21–29.

Okada, Y., Tominaga, M., Tominaga, T., Hazama, A., Tsumura, T., Morishima, S., Oiki, S., Ueda, K., Miyoshi, T., and Miwa, A. (1998). Volume sensor anion channel. In *Cell Volume Regulation: The Molecular Mechanism and Volume Sensing Machinery* (Okada, Y., ed.), pp. 15–22. Elsevier, Amsterdam.

Olesen, S.P. and Bundgaard, M. (1992). Chloride-selective channels of large conductance in bovine aortic endothelial cells. *Acta Physiol. Scand.* **144**, 191–198.

Ordaz, B., Vaca, L., Franco, R., and Pasantes-Morales, H. (2004). Volume changes and whole cell membrane currents activated during gradual osmolarity decrease in C6 glioma cells: contribution of two types of K^+ channels. *Am. J. Physiol. Cell Physiol.* **286**, C1399–C1409.

Pahapill, P.A. and Schlichter, L.C. (1992). Cl^- channels in intact human T lymphocytes. *J. Membr. Biol.* **125**, 171–183.

Park, K., Arreola, J., Begenisich, T., and Melvin, J.E. (1998). Comparison of voltage-activated Cl^- channels in rat parotid acinar cells with ClC-2 in a mammalian expression system. *J. Membr. Biol.* **163**, 87–95.

Parpura, V., Basarsky, T.A., Liu, F., Jeftinija, K., Jeftinija, S., and Haydon, P.G. (1994). Glutamate-mediated astrocyte-neuron signalling. *Nature* **369**, 744–747.

Parpura, V., Scemes, E., and Spray, D.C. (2004). Mechanisms of glutamate release from astrocytes: gap junction "hemichannels", purinergic receptors and exocytotic release. *Neurochem. Int.* **45**, 259–264.

Pasantes-Morales, H., Franco, R., Ochoa, L., and Ordaz, B. (2002). Osmosensitive release of neurotransmitter amino acids: relevance and mechanisms. *Neurochem. Res.* **27**, 59–65.

Pasantes-Morales, H. and Schousboe, A. (1989). Release of taurine from astrocytes during potassium-evoked swelling. *Glia* **2**, 45–50.

Paulmichl, M., Li, Y., Wickman, K., Ackerman, M., Peralta, E., and Clapham, D. (1992). New mammalian chloride channel identified by expression cloning. *Nature* **356**, 238–241.

Petersen, C.C., Kubo, M., Morishima, S., Tominaga, M., and Okada, Y. (1994). Single-channel recordings of volume-sensitive Cl^- channels in human intestinal epithelial cells. *Jpn. J. Physiol.* **44** (Suppl. 2), S73–S75.

Petrukhin, K., Koisti, M.J., Bakall, B., Li, W., Xie, G., Marknell, T., Sandgren, O., Forsman, K., Holmgren, G., Andreasson, S., Vujic, M., Bergen, A.A., McGarty-Dugan, V., Figueroa, D., Austin, C.P., Metzker, M.L., Caskey, C.T., and Wadelius, C. (1998).

Identification of the gene responsible for Best macular dystrophy. *Nat. Genet.* **19**, 241–247.

Phillis, J.W. and O'Regan, M.H. (2003). Characterization of modes of release of amino acids in the ischemic/reperfused rat cerebral cortex. *Neurochem. Int.* **43**, 461–467.

Picollo, A. and Pusch, M. (2005). Chloride/proton antiporter activity of mammalian CLC proteins ClC-4 and ClC-5. *Nature* **436**, 420–423.

Poletto Chaves, L.A. and Varanda, W.A. (2008). Volume-activated chloride channels in mice Leydig cells. *Pflugers Arch.* **457**, 493-504.

Pu, W.T., Krapivinsky, G.B., Krapivinsky, L., and Clapham, D.E. (1999). pICln inhibits snRNP biogenesis by binding core spliceosomal proteins. *Mol. Cell. Biol.* **19**, 4113–4120.

Pu, W.T., Wickman, K., and Clapham, D.E. (2000). *ICln* is essential for cellular and early embryonic viability. *J. Biol. Chem.* **275**, 12363–12366.

Qu, Z., Fischmeister, R., and Hartzell, C. (2004). Mouse bestrophin-2 is a bona fide Cl^- channel: identification of a residue important in anion binding and conduction. *J. Gen. Physiol.* **123**, 327–340.

Qu, Z., Wei, R.W., Mann, W., and Hartzell, H.C. (2003). Two bestrophins cloned from Xenopus laevis oocytes express Ca^{2+}-activated Cl^- currents. *J. Biol. Chem.* **278**, 49563–49572.

Reisin, I.L., Prat, A.G., Abraham, E.H., Amara, J.F., Gregory, R.J., Ausiello, D.A., and Cantiello, H.F. (1994). The cystic fibrosis transmembrane conductance regulator is a dual ATP and chloride channel. *J. Biol. Chem.* **269**, 20584–20591.

Riquelme, G. and Parra, M. (1999). Regulation of human placental chloride channel by arachidonic acid and other *cis* unsaturated fatty acids. *Am. J. Obstet. Gynecol.* **180**, 469–475.

Roman, R.M., Smith, R.L., Feranchak, A.P., Clayton, G.H., Doctor, R.B., and Fitz, J.G. (2001). ClC-2 chloride channels contribute to HTC cell volume homeostasis. *Am. J. Physiol. Gastrointest. Liver Physiol.* **280**, G344–G353.

Roy, G. (1995). Amino acid current through anion channels in cultured human glial cells. *J. Membr. Biol.* **147**, 35–44.

Roy, G. and Malo, C. (1992). Activation of amino acid diffusion by a volume increase in cultured kidney (MDCK) cells. *J. Membr. Biol.* **130**, 83–90.

Rutledge, E., Denton, J., and Strange, K. (2002). Cell cycle- and swelling-induced activation of a *Caenorhabditis elegans* ClC channel is mediated by CeGLC-7 α/β phosphatases. *J. Biol. Chem.* **158**, 435–444.

Sabirov, R.Z., Dutta, A.K., and Okada, Y. (2001). Volume-dependent ATP-conductive large-conductance anion channel as a pathway for swelling-induced ATP release. *J. Gen. Physiol.* **118**, 251–266.

Sabirov, R.Z. and Okada, Y. (2004a). ATP-conducting maxi-anion channel: a new player in stress-sensory transduction. *Jpn. J. Physiol.* **54**, 7–14.

Sabirov, R.Z. and Okada, Y. (2004b). Wide nanoscopic pore of maxi-anion channel suits its function as an ATP-conductive pathway. *Biophys. J.* **87**, 1672–1685.

Sabirov, R.Z. and Okada, Y. (2005). ATP release via anion channels. *Purinergic Signal.* **1**, 311–328.

Sabirov, R.Z. and Okada, Y. (2009). The maxi-anion channel: a classical channel playing novel roles through an unidentified molecular entity. *J. Physiol. Sci.* **59**, 3–21.

Sabirov, R.Z., Prenen, J., Droogmans, G., and Nilius, B. (2000). Extra- and intracellular proton-binding sites of volume-regulated anion channels. *J. Membr. Biol.* **177**, 13–22.

Sabirov, R.Z., Sheiko, T., Liu, H., Deng, D., Okada, Y., and Craigen, W.J. (2006). Genetic demonstration that the plasma membrane maxianion channel and voltage-dependent anion channels are unrelated proteins. *J. Biol. Chem.* **281**, 1897–1904.

Sakai, H., Nakamura, F., and Kuno, M. (1999). Synergetic activation of outwardly rectifying Cl^- currents by hypotonic stress and external Ca^{2+} in murine osteoclasts. *J. Physiol.* **515**, 157–168.

Sanchez-Olea, R., Pena, C., Moran, J., and Pasantes-Morales, H. (1993). Inhibition of volume regulation and efflux of osmoregulatory amino acids by blockers of Cl^- transport in cultured astrocytes. *Neurosci. Lett.* **156**, 141–144.

Schlichter, L.C., Grygorczyk, R., Pahapill, P.A., and Grygorczyk, C. (1990). A large, multiple-conductance chloride channel in normal human T lymphocytes. *Pflugers Arch.* **416**, 413–421.

Schmid, A., Blum, R., and Krause, E. (1998). Characterization of cell volume-sensitive chloride currents in freshly prepared and cultured pancreatic acinar cells from early postnatal rats. *J. Physiol.* **513**, 453–465.

Schmid, A., Gogelein, H., Kemmer, T.P., and Schulz, I. (1988). Anion channels in giant liposomes made of endoplasmic reticulum vesicles from rat exocrine pancreas. *J. Membr. Biol.* **104**, 275–282.

Schwarze, W. and Kolb, H.A. (1984). Voltage-dependent kinetics of an anionic channel of large unit conductance in macrophages and myotube membranes. *Pflugers Arch.* **402**, 281–291.

Schwiebert, E.M., Egan, M.E., Hwang, T.-H., Fulmer, S.B., Allen, S.S., Cutting, G.R., and Guggino, W.B. (1995). CFTR regulates outwardly rectifying chloride channels through an autocrine mechanism involving ATP. *Cell* **81**, 1063–1073.

Schwiebert, E.M., Karlson, K.H., Friedman, P.A., Dietl, P., Spielman, W.S., and Stanton, B.A. (1992). Adenosine regulates a chloride channel via protein kinase C and a G protein in a rabbit cortical collecting duct cell line. *J. Clin. Invest* **89**, 834–841.

Schwiebert, E.M., Mills, J.W., and Stanton, B.A. (1994). Actin-based cytoskeleton regulates a chloride channel and cell volume in a renal cortical collecting duct cell line. *J. Biol. Chem.* **269**, 7081–7089.

Shen, M.-R., Droogmans, G., Eggermont, J., Voets, T., Ellory, J.C., and Nilius, B. (2000). Differential expression of volume-regulated anion channels during cell cycle progression of human cervical cancer cells. *J. Physiol.* **529**, 385–394.

Shen, M.R., Chou, C.Y., and Ellory, J.C. (2001). Swelling-activated taurine and K^+ transport in human cervical cancer cells: association with cell cycle progression. *Pflugers Arch.* **441**, 787–795.

Shennan, D.B., McNeillie, S.A., and Curran, D.E. (1994). The effect of a hyposmotic shock on amino acid efflux from lactating rat mammary tissue: stimulation of taurine and glycine efflux via a pathway distinct from anion exchange and volume-activated anion channels. *Exp. Physiol.* **79**, 797–808.

Shimizu, T., Lee, E.L., Ise, T., and Okada, Y. (2008). Volume-sensitive Cl^- channel as a regulator of acquired cisplatin resistance. *Anticancer Res.* **28**, 75–83.

Shimizu, T., Morishima, S., and Okada, Y. (2000). Ca^{2+}-sensing receptor-mediated regulation of volume-sensitive Cl^- channels in human epithelial cells. *J. Physiol.* **528**, 457–472.

Shimizu, T., Numata, T., and Okada, Y. (2004). A role of reactive oxygen species in apoptotic activation of volume-sensitive Cl^- channel. *Proc. Natl. Acad. Sci. USA* **101**, 6770–6773.

Shuba, Y.M., Prevarskaya, N., Lemonnier, L., Van Coppenolle, F., Kostyuk, P.G., Mauroy, B., and Skryma, R. (2000). Volume-regulated chloride conductance in the LNCaP human prostate cancer cell line. *Am. J. Physiol. Cell Physiol.* **279**, C1144–C1154.

Siebens, A.W. and Spring, K.R. (1989). A novel sorbitol transport mechanism in cultured renal papillary epithelial cells. *Am. J. Physiol. Renal Physiol.* **257**, F937–F946.

Soejima, M. and Kokubun, S. (1988). Single anion-selective channel and its ion selectivity in the vascular smooth muscle cell. *Pflugers Arch.* **411**, 304–311.

Solc, C.K. and Wine, J.J. (1991). Swelling-induced and depolarization-induced C1-channels in normal and cystic fibrosis epithelial cells. *Am. J. Physiol. Cell Physiol.* **261**, C658–C674.

Staley, K., Smith, R., Schaack, J., Wilcox, C., and Jentsch, T.J. (1996). Alteration of GABAA receptor function following gene transfer of the CLC-2 chloride channel. *Neuron* **17**, 543–551.

Stegen, C., Matskevich, I., Wagner, C.A., Paulmichl, M., Lang, F., and Broer, S. (2000). Swelling-induced taurine release without chloride channel activity in *Xenopus laevis* oocytes expressing anion channels and transporters. *Biochim. Biophys. Acta* **1467**, 91–100.

Stobrawa, S.M., Breiderhoff, T., Takamori, S., Engel, D., Schweizer, M., Zdebik, A.A., Bösl, M.R., Ruether, K., Jahn, H., Draguhn, A., Jahn, R., and Jentsch, T.J. (2001). Disruption of ClC-3, a chloride channel expressed on synaptic vesicles, leads to a loss of the hippocampus. *Neuron* **29**, 185–196.

Strange, K. (1998). Perspective: molecular identity of the outwardly rectifying, swelling-activated anion channel: time to reevaluate pICln. *J. Gen. Physiol.* **111**, 617–622.

Strange, K., Emma, F., and Jackson, P.S. (1996). Cellular and molecular physiology of volume-sensitive anion channels. *Am. J. Physiol. Cell Physiol.* **270**, C711–C730.

Strange, K. and Jackson, P.S. (1995). Swelling-activated organic osmolyte efflux: a new role for anion channels. *Kidney Int.* **48**, 994–1003.

Stutzin, A., Eguiguren, A.L., Cid, L.P., and Sepulveda, F.V. (1997). Modulation by extracellular Cl^- of volume-activated organic osmolyte and halide permeabilities in HeLa cells. *Am. J. Physiol. Cell Physiol.* **273**, C999–C1007.

Stutzin, A., Torres, R., Oporto, M., Pacheco, P., Eguiguren, A.L., Cid, L.P., and Sepulveda, F.V. (1999). Separate taurine and chloride efflux pathways activated during regulatory volume decrease. *Am. J. Physiol. Cell Physiol.* **277**, C392–C402.

Sun, H., Bristow, B.N., Qu, G., and Wasserman, S.A. (2002). A heterotrimeric death domain complex in Toll signaling. *Proc. Natl. Acad. Sci. USA* **99**, 12871–12876.

Sun, X.P., Supplisson, S., and Mayer, E. (1993). Chloride channels in myocytes from rabbit colon are regulated by a pertussis toxin-sensitive G protein. *Am. J. Physiol. Gastrointest. Liver Physiol.* **264**, G774–G785.

Sun, X.P., Supplisson, S., Torres, R., Sachs, G., and Mayer, E. (1992). Characterization of large-conductance chloride channels in rabbit colonic smooth muscle. *J. Physiol.* **448**, 355–382.

Suzuki, M. (2006). The Drosophila tweety family: molecular candidates for large-conductance Ca^{2+}-activated Cl^- channels. *Exp. Physiol.* **91**, 141–147.

Suzuki, M. and Mizuno, A. (2004). A novel human Cl^- channel family related to Drosophila flightless locus. *J. Biol. Chem.* **279**, 22461–22468.

Ternovsky, V.I., Okada, Y., and Sabirov, R.Z. (2004). Sizing the pore of the volume-sensitive anion channel by differential polymer partitioning. *FEBS Lett.* **576**, 433–436.

Thiemann, A., Grunder, S., Pusch, M., and Jentsch, T.J. (1992). A chloride channel widely expressed in epithelial and non-epithelial cells. *Nature* **356**, 57–60.

Thinnes, F.P. (1992). Evidence for extra-mitochondrial localization of the VDAC/porin channel in eucaryotic cells. *J. Bioenerg. Biomembr.* **24**, 71–75.

Tomassen, S.F.B., Fekkes, D., de Jonge, H.R., and Tilly, B.C. (2004). Osmotic swelling-provoked release of organic osmolytes in human intestinal epithelial cells. *Am. J. Physiol. Cell Physiol.* **286**, C1417–C1422.

Tominaga, M., Tominaga, T., Miwa, A., and Okada, Y. (1995). Volume-sensitive chloride channel activity does not depend on endogenous P-glycoprotein. *J. Biol. Chem.* **270**, 27887–27893.

Tsumura, T., Oiki, S., Ueda, S., Okuma, M., and Okada, Y. (1996). Sensitivity of volume-sensitive Cl^- conductance in human epithelial cells to extracellular nucleotides. *Am. J. Physiol. Cell Physiol.* **271**, C1872–C1878.

Tsunenari, T., Nathans, J., and Yau, K.-W. (2006). Ca^{2+}-activated Cl^- current from human bestrophin-4 in excised membrane patches. *J. Gen. Physiol.* **127**, 749–754.

Tsunenari, T., Sun, H., Williams, J., Cahill, H., Smallwood, P., Yau, K.-W., and Nathans, J. (2003). Structure-function analysis of the bestrophin family of anion channels. *J. Biol. Chem.* **278**, 41114–41125.

Vaca, L. (1999). SITS blockade induces multiple subconductance states in a large conductance chloride channel. *J. Membr. Biol.* **169**, 65–73.

Vallejos, C. and Riquelme, G. (2007). The maxi-chloride channel in human syncytiotrophoblast: a pathway for taurine efflux in placental volume regulation? *Placenta* **28**, 1182–1191.

Valverde, M.A., Bond, T.D., Hardy, S.P., Taylor, J.C., Higgins, C.F., Altamirano, J., and Alvarez-Leefmans, F.J. (1996). The multidrug resistance P-glycoprotein modulates cell regulatory volume decrease. *EMBO J.* **15**, 4460–4468.

Valverde, M.A., Diaz, M., Sepulveda, F.V., Gill, D.R., Hyde, S.C., and Higgins, C.F. (1992). Volume-regulated chloride channels associated with the human multidrug-resistance P-glycoprotein. *Nature* **355**, 830–833.

Van Winkle, L.J., Patel, M., Wasserlauf, H.G., Dickinson, H.R., and Campione, A.L. (1994). Osmotic regulation of taurine transport via system beta and novel processes in mouse preimplantation conceptuses. *Biochim. Biophys. Acta* **1191**, 244–255.

Vandenberg, J.I., Yoshida, A., Kirk, K., and Powell, T. (1994). Swelling-activated and isoprenaline-activated chloride currents in guinea pig cardiac myocytes have distinct electrophysiology and pharmacology. *J. Gen. Physiol.* **104**, 997–1017.

Varela, D., Simon, F., Riveros, A., Jorgensen, F., and Stutzin, A. (2004). NAD(P)H oxidase-derived H_2O_2 signals chloride channel activation in cell volume regulation and cell proliferation. *J. Biol. Chem.* **279**, 13301–13304.

Vennekens, R., Trouet, D., Vankeerberghen, A., Voets, T., Cuppens, H., Eggermont, J., Cassiman, J.-J., Droogmans, G., and Nilius, B. (1999). Inhibition of volume-regulated anion channels by expression of the cystic fibrosis transmembrane conductance regulator. *J. Physiol.* **515**, 75–85.

Voets, T., Buyse, G., Tytgat, J., Droogmans, G., Eggermont, J., and Nilius, B. (1996). The chloride current induced by expression of the protein pI_{Cln} in *Xenopus* oocytes differs from the endogenous volume-sensitive chloride current. *J. Physiol.* **495**, 441–447.

Voets, T., Droogmans, G., and Nilius, B. (1997a). Modulation of voltage-dependent properties of a swelling-activated Cl^- current. *J. Gen. Physiol.* **110**, 313–325.

Voets, T., Wei, L., De Smet, P., Van Driessche, W., Eggermont, J., Droogmans, G., and Nilius, B. (1997b). Downregulation of volume-activated Cl^- currents during muscle differentiation. *Am. J. Physiol. Cell. Physiol.* **272**, C667–C674.

Volterra, A. and Meldolesi, J. (2005). Astrocytes, from brain glue to communication elements: the revolution continues. *Nat. Rev. Neurosci.* **6**, 626–640.

Wang, G.-X., Hatton, W.J., Wang, G.L., Zhong, J., Yamboliev, I., Duan, D., and Hume, J.R. (2003). Functional effects of novel anti-ClC-3 antibodies on native volume-sensitive osmolyte and anion channels in cardiac and smooth muscle cells. *Am. J. Physiol. Heart Circ. Physiol.* **285**, H1453–H1463.

Wang, J., Xu, H., Morishima, S., Tanabe, S., Jishage, K., Uchida, S., Sasaki, S., Okada, Y., and Shimizu, T. (2005). Single-channel properties of volume-sensitive Cl^- channel in ClC-3-deficient cardiomyocytes. *Jpn. J. Physiol.* **55**, 379–383.

Wang, L., Chen, L., and Jacob, T.J.C. (2000). The role of ClC-3 in volume-activated chloride currents and volume regulation in bovine epithelial cells demonstrated by antisense inhibition. *J. Physiol.* **524**, 63–75.

Wang, X.Q., Deriy, L.V., Foss, S., Huang, P., Lamb, F.S., Kaetzel, M.A., Bindokas, V., Marks, J.D., and Nelson, D.J. (2006). CLC-3 channels modulate excitatory synaptic transmission in hippocampal neurons. *Neuron* **52**, 321–333.

Weinreich, F. and Jentsch, T.J. (2001). Pores formed by single subunits in mixed dimers of different CLC chloride channels. *J. Biol. Chem.* **276**, 2347–2353.

Weiss, H. and Lang, F. (1992). Ion channels activated by swelling of Madin Darby canine kidney (MDCK) cells. *J. Membr. Biol.* **126**, 109–114.

Weylandt, K.-H., Valverde, M.A., Nobles, M., Raguz, S., Amey, J.S., Diaz, M., Nastrucci, C., Higgins, C.F., and Sardini, A. (2001). Human ClC-3 is not the swelling-activated chloride channel involved in cell volume regulation. *J. Biol. Chem.* **276**, 17461–17467.

White, K., Marquardt, A., and Weber, B.H. (2000). VMD2 mutations in vitelliform macular dystrophy (Best disease) and other maculopathies. *Hum. Mutat.* **15**, 301–308.

Wondergem, R., Gong, W., Monen, S.H., Dooley, S.N., Gonce, J.L., Conner, T.D., Houser, M., Ecay, T.W., and Ferslew, K.E. (2001). Blocking swelling-activated chloride current inhibits mouse liver cell proliferation. *J. Physiol.* **532**, 661–672.

Worrell, R.T., Butt, A.G., Cliff, W.H., and Frizzell, R.A. (1989). A volume-sensitive chloride conductance in human colonic cell line T84. *Am. J. Physiol. Cell. Physiol.* **256**, C1111–C1119.

Xiong, H., Li, C., Garami, E., Wang, Y., Ramjeesingh, M., Galley, K., and Bear, C.E. (1999). ClC-2 activation modulates regulatory volume decrease. *J. Membr. Biol.* **167**, 215–221.

Yamamoto-Mizuma, S., Wang, G.-X., Liu, L.L., Schegg, K., Hatton, W.J., Duan, D., Horowitz, T.L.B., Lamb, F.S., and Hume, J.R. (2004). Altered properties of volume-sensitive osmolyte and anion channels (VSOACs) and membrane protein expression in cardiac and smooth muscle myocytes from *Clcn3*$^{-/-}$ mice. *J. Physiol.* **557**, 439–456.

Yamazaki, J., Duan, D., Janiak, R., Kuenzli, K., Horowitz, B., and Hume, J.R. (1998). Functional and molecular expression of volume-regulated chloride channels in canine vascular smooth muscle cells. *J. Physiol.* **507**, 729–736.

Yin, Z., Tong, Y., Zhu, H., and Watsky, M.A. (2008). ClC-3 is required for LPA-activated Cl^- current activity and fibroblast-to-myofibroblast differentiation. *Am. J. Physiol. Cell Physiol.* **294**, C535–C542.

Yu, K., Qu, Z., Cui, Y., and Hartzell, H.C. (2007). Chloride channel activity of bestrophin mutants associated with mild or late-onset macular degeneration. *Invest. Ophthalmol. Vis. Sci.* **48**, 4694–4705.

Zachar, J. and Hurnak, O. (1994). Arachidonic acid blocks large-conductance chloride channels in L6 myoblasts. *Gen. Physiol. Biophys.* **13**, 193–213.

Zhao, Z., Li, X., Hao, J., Winston, J.H., and Weinman, S.A. (2007). The ClC-3 chloride transport protein traffics through the plasma membrane via interaction of an N-terminal dileucine cluster with clathrin. *J. Biol. Chem.* **282**, 29022–29031.

Zhou, J.-G., Ren, J.-L., Qiu, Q.-Y., He, H., and Guan, Y.-Y. (2005). Regulation of intracellular Cl^- concentration through volume-regulated ClC-3 chloride channels in A10 vascular smooth muscle cells. *J. Biol. Chem.* **280**, 7301–7308.

CHAPTER

16

The Sodium-dependent Chloride Cotransporters

Gerardo Gamba

OUTLINE

I. INTRODUCTION

Transcellular ion transport in epithelial cells requires specific plasma membrane proteins that mediate ion influx and efflux. The electrochemical gradient that promotes Na^+ influx and K^+ efflux is generated by the Na^+/K^+-ATPase, which, with the so far known exceptions of the retinal pigment epithelium and the choroid plexus, is expressed at the basolateral membrane of epithelia. In the apical membrane, transport of Na^+ or K^+ following their gradients occurs through ion channels, or by secondary active transporters in which Na^+ or K^+ transport is coupled with that of ions or other solutes. Among these secondary active transport mechanisms are the Na^+-coupled cotransporters, which translocate anions and cations with electroneutral stoichiometry and thus ion movement occurs without producing changes in transmembrane potential. Because of these features, this family of solute carriers is known as the electroneutral cation-Cl^--coupled cotransporters. These cotransporters are not only critical for ion absorption and secretion in epithelial cells, but also play a key role in the maintenance and regulation of cell volume in both epithelial and non-epithelial cells. Since Na^+ influx and K^+ efflux by electroneutral cotransporters are offset by the Na^+/K^+-ATPase, it is believed that the net effect of these cotransporters' activity is the inward movement of Cl^-. Besides transepithelial ion movements and cell volume regulation, a variety of physiological roles have been proposed for electroneutral cotransporters in the regulation of intracellular chloride concentration ($[Cl^-]_i$). A key example is the modulation of $[Cl^-]_i$ in neurons, a process that is critical in the postsynaptic membrane response to neurotransmitters acting on receptors coupled with Cl^- channels, as discussed in several chapters throughout this book.

Four types of electroneutral cotransporter systems have been identified based on a number of criteria, including the type and number of cations coupled to Cl^-, the stoichiometry of the process and the sensitivity of the cotransporters to various inhibitors. The four identified systems include: (1) the benzothiadiazine (or thiazide)-sensitive Na^+-Cl^- cotransporter, (2) the sulfamoylbenzoic (or bumetanide)-sensitive Na^+-K^+-$2Cl^-$ cotransporter, (3) the sulfamoylbenzoic (or bumetanide)-sensitive Na^+-Cl^- cotransporter and (4) the dihydroindenyloxy-alkanoic acid (DIOA)-sensitive K^+-Cl^- cotransporter.

Conclusive physiological evidence for the existence of the electroneutral cation-chloride-coupled cotransporters became available at the end of the 1970s. First, there was the observation of an Na^+-Cl^- interdependent electroneutral pathway in the winter flounder urinary bladder (Renfro, 1977). This was followed by the observation of the Na^+-K^+-$2Cl^-$ mechanism in Ehrlich ascites cells (Geck et al., 1980), and then a K^+-Cl^- pathway in red blood cells (Lauf and Theg, 1980; Dunham and Ellory, 1981). Since then, a remarkable amount of information has been generated on several aspects of these transport systems in many different cells and experimental conditions. As will be discussed, identification of the cotransporters at the molecular level revealed that the electroneutral cation-Cl^--coupled cotransporter family can be divided into two branches (Gamba, 2005): the Na^+-driven branch, which encompasses the cotransporters using Na^+ (with or without K^+) as the cation(s) coupled to Cl^-, and the K^+-driven branch, composed of cotransporters that use only K^+ as the Cl^--coupled cation. The present chapter discusses the Na^+-dependent Cl^- cotransporters and concentrates on the information that became available as a consequence of cloning the cDNAs encoding for some of their members. The major subjects for discussion for each transporter in this chapter are: (1) molecular biology, (2) functional properties of the recombinant proteins, (3) structure–function analysis, (4) physiological role and (5) participation of Na^+-Cl^--coupled cotransporters in pathophysiology of monogenic and polygenic diseases. The K^+-Cl^- cotransporter branch is discussed in detail in Chapters 3 and 17 in this volume.

II. MOLECULAR BIOLOGY OF THE SODIUM-DEPENDENT CHLORIDE COTRANSPORTERS

Soon after the discovery of the Na^+-dependent Cl^- cotransporters (Geck et al., 1980; Renfro, 1975), the molecular identification of these transport systems was attempted by many laboratories (Dunham et al., 1990; Feit et al., 1988; Jorgensen et al., 1984; Turner and Geroge, 1990; Luo et al., 1990; Ellison et al., 1990). A major breakthrough in this field occurred in the early 1990s when researchers were able to identify the genes encoding Na^+-Cl^- (Gamba et al., 1993, 1994) and Na^+-K^+-$2Cl^-$ cotransporters (Xu et al., 1994; Payne and Forbush, 1994; Gamba et al., 1994; Delpire et al., 1994). As eloquently predicted by Homer Smith in his book *From Fish to Philosopher* (Smith, 1953), cDNAs encoding these proteins were first identified in fish sources and then in mammalian sources using homology approaches with fish cDNAs.

A. The Thiazide-sensitive Na^+-Cl^- Cotransporter

The first electroneutral cotransporter protein identified at the molecular level was the thiazide-sensitive Na^+-Cl^- cotransporter (NCC) from *Pseudopleuronectes americanus* (winter flounder) urinary bladder, following an expression cloning strategy using *Xenopus laevis* oocytes as a heterologous expression system (Gamba et al., 1993). The 3.7 kb cDNA clone exhibited an open reading frame of 3069 bp and predicted a protein comprised of 1023 amino acid residues with a core molecular mass of 112 kDa. The hydropathy analysis (Kyte and Doolittle, 1982) suggested for the first time the basic two-dimensional topology of the Na^+-coupled-Cl^- cotransporters shown in Fig. 16.1. These

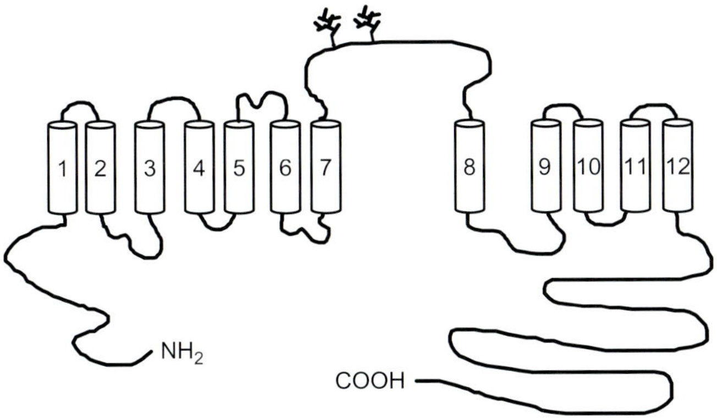

FIGURE 16.1 Proposed topologies for the electroneutral cation-chloride cotransporters. A central hydrophobic domain contains 12 putative transmembrane segments divided in two by a long hydrophilic loop that is presumably located outside the cell since it is gylcosylated (Hoover et al., 2003; Paredes et al., 2006). The hydrophobic domain is flanked by a short amino terminal and a long carboxyl-terminal domain located within the cell.

results were consistent with a central hydrophobic domain containing 12 α-helices compatible with putative transmembrane-spanning segments, flanked by a short amino terminal and long carboxyl terminal domains of intracellular localization. The algorithm predicted a long hydrophilic loop interconnecting transmembrane segments 7 and 8 (see Chapter 11). Later studies using some members of the SLC12A family revealed that this loop is indeed glycosylated (Hoover et al., 2003; Moreno et al., 2006; Paredes et al., 2006). Northern blot analysis of winter flounder tissues revealed expression of a 3.7 kb transcript in the urinary bladder and a shorter 3.0 kb message due to alternative splicing in several tissues including the gonads, intestine, eye, brain, skeletal muscle and heart (Gamba et al., 1993). The functional consequence of this variant has not been resolved (Merino et al., 1999). Primary sequences of the thiazide-sensitive cotransporter (NCC) were later reported from four mammalian species, including *Rattus norvegicus* (rat) (Gamba et al., 1994), *Mus musculus* (mouse) (Kunchaparty et al., 1999), *Oryctolagus cuniculus* (rabbit) (Velazquez et al., 1998) and *Homo sapiens* (human) (Simon et al., 1996b; Mastroianni et al., 1996b). The extent of identity among mammalian NCCs is ~90%, and of any mammalian with flounder is ~60%. Interestingly, rabbit and human NCCs are longer than rat and mouse orthologs due to the presence of 17–26 amino acid residues in the carboxyl terminal domain. These extra residues were shown to be encoded in humans by a separate exon (exon 20), which is not present in mouse or rat. It is noteworthy that in humans, there is a putative protein kinase A (PKA) site (RPS) within the extra fragment that is not present in rabbit, mouse or rat NCC.

Tissue distribution analysis by Northern blot in the rat revealed renal specific expression of NCC (Gamba et al., 1994). However, recent studies have also revealed its presence in intestine (Bazzini et al., 2005) and bone cells (Dvorak et al., 2007) as discussed in detail in Chapter 11.

B. Isoform 2 of the Bumetanide-sensitive Na^+-K^+-$2Cl^-$ Cotransporter (NKCC2)

Two genes encoding bumetanide-sensitive Na^+-K^+-$2Cl^-$ cotransporters were identified as part of the SLC12A family. These genes are known as *SLC12A1* and *SLC12A2*. The *SLC12A1* gene encodes the Na^+-K^+-$2Cl^-$ cotransporter (NKCC2) which is characteristically expressed in the apical membrane of the thick ascending limb of Henle (TALH) (see Chapter 11). NKCC2 cDNA was simultaneously identified from mammalian kidney in 1994 by Payne and Forbush (Payne and Forbush, 1994) and Gamba et al. (Gamba et al., 1994) after screening cDNA libraries from rabbit and rat renal outer medulla, respectively. Later, human and mouse NKCC2 sequences were reported (Simon et al., 1996a; Igarashi et al., 1995). Isolated cDNA clones were about 4.5 kb in size, with an open reading frame of 3,285 bp encoding a 1,095 residue protein. As shown in Fig. 16.2, the predicted NKCC2 general topology is similar to NCC, featuring a central hydrophobic domain of ~475 residues containing 12 putative membrane-spanning segments flanked by two predominantly hydrophilic domains: a short amino terminal domain of ~165 amino acids and a long carboxyl terminal domain of ~450 residues. The long extracellular loop between transmembrane segments 7 and 8 is N-glycosylated (Paredes et al., 2006). Functional expression analysis in *X. laevis* oocytes demonstrated that the isolated clones encode a bumetanide-sensitive Na^+-K^+-$2Cl^-$ cotransporter (Gamba et al., 1994). Tissue distribution by Northern blot analysis showed that transcripts were present only in total RNA from kidney; all other tissues tested were negative.

At least six variants of NKCC2 may be expressed in the mouse kidney due to the combination of two alternative splicing mechanisms (Fig. 16.2) (Mount et al., 1999; Gamba, 2001). The first splicing event was described in rabbit kidney (Payne and Forbush, 1994) and is also present in mouse (Igarashi et al., 1995), rat (Yang et al., 1996) and human (Simon et al., 1996) kidney. It is secondary to the expression of three mutually exclusive cassette exons of 96 bp designated A, B and F which encode 32 amino acid residues corresponding to the second half of the putative transmembrane domain TM2, as well as part of the interconnecting

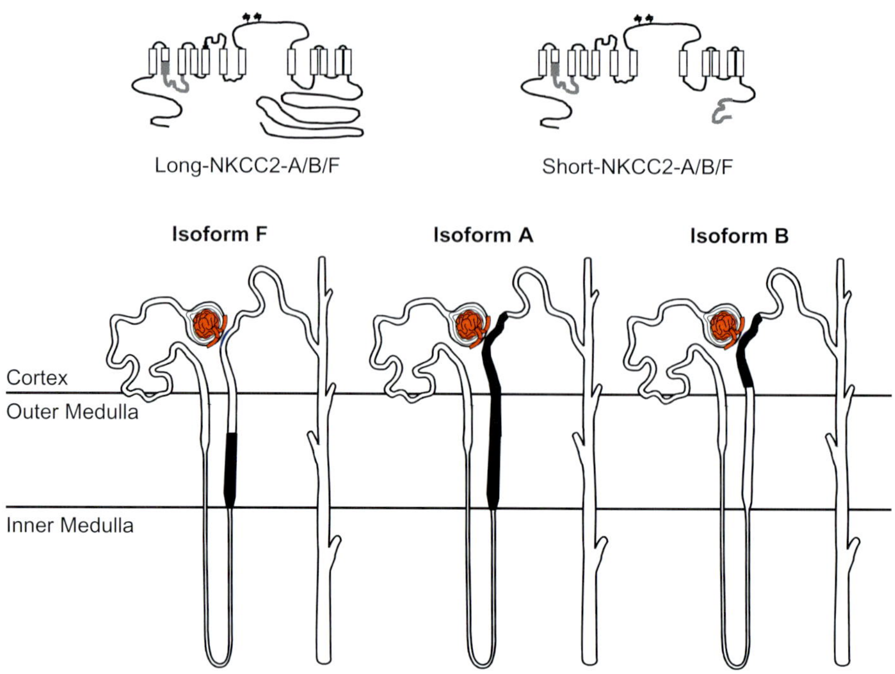

FIGURE 16.2 Splice variants of mouse. The region of the transmembrane domain 2 and the interconnecting segment between transmembrane domains 2 and 3, shown in gray, depict the mutually exclusive cassette exons A, B or F. Long isoform L-NKCC2 contains 329 amino acid residues in carboxyl terminal domain that are not present in shorter isoform S-NKCC2, which contains 55 unique residues at the end of carboxyl terminal domain (highlighted in gray and shown by an arrow). The localization of the F, A and B isoforms along TALH is shown. Isoform F is only expressed in inner strip of the outer medulla, isoform A is present all along TALH and isoform B is exclusively expressed in cortical portion of the TALH.

loop between TM2 and TM3 (Fig. 16.2). Thus, three NKCC2 proteins are produced differing only in the 32 amino acid residues encoded by cassettes A, B or F. NKCC2 variants A and F have also been identified from *Squalus acanthias* (shark) kidney (Gagnon et al., 2002).

The second splicing of the *SLC12A1* gene has been observed in the mouse kidney. It is the consequence of a poly-adenylation site within the intron between exons 16 and 17, predicting a protein with a shorter C-terminal domain (Mount et al., 1999). Thus, two NKCC2 proteins are predicted, which are identical in the amino terminus and transmembrane domains, but different in length and in the sequence of the carboxyl terminal domain. The first 74 residues of the carboxyl terminal domain are identical. After that, the longer isoform contains 383 residues that are not present in the shorter isoform, which, in turn, contains 55 residues not present in the longer variant (Fig. 16.2). Since the two splicing events are independent of each other, a total of six isoforms are predicted to be produced in mouse kidney: three long NKCC2 isoforms (A, B and F), and three short NKCC2 isoforms (A, B and F) (Mount et al., 1999). The possible functional significance of spliced isoforms is discussed below.

C. Isoform 1 of the Bumetanide-sensitive Na^+-K^+-$2Cl^-$ Cotransporter (NKCC1)

The *SLC12A2* gene encodes the ubiquitously expressed Na^+-K^--$2Cl^-$ cotransporter isoform NKCC1. This cotransporter is present in epithelial and non-epithelial cells. In epithelial cells its expression is restricted to the basolateral membrane with the so far known exception of the choroid plexus (Plotkin et al., 1997), where NKCC1 resides in the apical membrane (Chapter 29). In 1994, the same two independent research teams that cloned NKCC2 also identified cDNA encoding NKCC1 from fish (Xu et al., 1994) and mammalian sources (Delpire et al., 1994). The predicted two-dimensional topology of NKCC1 is similar to NCC and NKCC2 (see Fig. 16.1 and Chapter 11). In fact, NKCC1 is the only member of the family to date for which the proposed topology is to some extent supported by experimental data (Gerelsaikhan and Turner, 2000). The first eight transmembrane segments exhibit the classical ~20 residue helices. In contrast, TMs 9 and 10, and 11 and 12 together are ~36 residues in length, suggesting a hairpin-like structure in the membrane or a non-helical or partial-helical structure. NKCC1 has been identified at the molecular level in several species, including mouse (Delpire et al., 1994), rat (Moore-Hoon and Turner, 1998), human (Payne et al., 1995), *Bos taurus* (bovine) (Yerby et al., 1997), shark (Xu et al., 1994), *Anguilla anguilla* (eel) (Cutler and Cramb, 2002) and *Dicentrarchus labrax* (sea bass) (Lorin-Nebel et al., 2006), and even from the plant *Arabidopsis thaliana* (Colmenero-Flores et al., 2007). There is evidence suggesting the existence of one alternatively spliced isoform of NKCC1. A fragment of this splice variant was detected in mouse-brain total RNA; it lacks 48bp that correspond to the entire exon 21. Thus, this predicts that 16 residues of the carboxy terminal are not present in this short NKCC1 variant (Randall et al., 1997). The existence of a splice transcript was supported by an RNAse protection assay. Distribution analysis within the brain showed that an NKCC1 transcript lacking exon 21 is present in all areas examined except in the choroid plexus, where only the full-length isoform containing exon 21 is expressed. When reconstructed in expression vectors and expressed in heterologous systems the splice variant performs as an Na^+-K^+-$2Cl^-$ cotransporter (Vibat et al., 2001). Further, it has been suggested that exon 21 of NKCC1 is implicated in differential sorting in polarized epithelial cells (Carmosino et al., 2008). It is also worth noting that the absence of the exon 21 sequence removes a potential PKA phosphorylation site present in the full-length NKCC1 sequence.

D. Genes, Promoters and Phylogenetic Analysis

SLC12A1, the gene encoding NKCC2 has been mapped in humans to chromosome 15 (Simon et al., 1996), in rat to chromosome 3 (Wang et al., 1997) and in mouse to chromosome 2 (Quaggin et al., 1995). *SLC12A1* in humans encompasses at least 80kb and contains at least 26 exons (see Chapter 11 for a detail description of the gene). An *SLC12A1* promoter region has been cloned from mouse genomic DNA (Igarashi et al., 1996). The NKCC2 transcript starts with a first exon of 34bp that is non-coding, followed by a first intron of 1101bp and a second exon containing the translation start codon. The cloned promoter is composed of 2255bp and sequence analysis revealed a TATA box located at position -29 and consensus recognition sites for several transcription factors, of which the most interesting could be a binding site for HNF-1 at -211bp. In developing mouse kidney, the expression of HNF-1 precedes the expression of NKCC2 (Lazzaro et al., 1992). HNF-1 has been implicated in the regulation of tissue-specific expression of genes in liver, pancreas, kidney and intestine. Using TALH-derived cells, it was demonstrated that deletion of -2255 to -1529bp produced a ~ three-fold increase in luciferase activity, suggesting that this region contains negative regulatory elements. Deletion from -1529 to -469bp had no further effect, but deletion from -469 to -190 resulted in a 76% reduction of promoter activity, suggesting that this region contains positive regulatory elements. An HFN-1 binding site is located in this region. A cAMP response-element binding protein is located at nucleotide -1111. It is also known that Na^+-K^+-$2Cl^-$ cotransporter activity in TALH is increased by vasopressin (Hebert et al., 1987) and that chronic administration of vasopressin DDAVP to Sprague-Dawley and Brattleboro rats increases NKCC2 abundance at the protein level (Kim et al., 1999). For further discussion on the NKCC2 potential promoters and transcription factors, see Chapter 11 in this volume.

The gene *SLC12A2* encodes NKCC1 and is located to chromosome 5q23 in humans (Payne et al., 1995) and to chromosome 18 in mouse (Delpire et al., 1994). A complete gene has been cloned from mouse DNA, covering a region of 75kb (Randall et al., 1997) and composed of 28 exons. Sequencing of the promoter region revealed the presence of numerous SP1 consensus sites and binding sites for different transcription factors, including MEF2, a CACCC binding, OTF/1-2A, $NF_{kappa}B$ and AP-2. Transfection of mouse IMDC3 cells with a 2063bp promoter region ligated to a luciferase reporter gene (pGL3) yielded significant luciferase activity. Deletions of >1kb which reduced the promoter region to 702 or 516bp resulted in a significant increase of luciferase activity, suggesting the existence of silencer sequences in the deleted bases. Additional deletions resulted in progressive reduction of luciferase activity, suggesting the presence of enhancer elements.

The gene encoding the thiazide sensitive Na^+-Cl^- cotransporter (*SLC12A3*) has been mapped to chromosome 16q13 in humans (Simon et al., 1996; Mastroianni et al., 1996), to chromosome 19p12-14 in rat (Taniyama et al., 2001) and to chromosome 8 in mouse (Pathak et al., 1996). Human *SLC12A3* is 55kb long and contains 26 exons (Simon et al., 1996). Transcription initiation is confined to an area from -18 to -6bp upstream of the translation start codon. The promoter activity observed in the mouse distal convoluted cell line MDCT (MacKenzie et al., 2001), with a construct containing 1019bp of the 5′ flanking region was reduced only 25% by eliminating the first 885bp. Sequence analysis of the promoter revealed the presence of a TATA element, two Sp binding sites and potential binding sites for NF-1/CTF or NY-I/CP-I. Interestingly, the promoter activity of the rat NCC gene is inhibited by acidosis. This is consistent with a marked fall in renal cortical abundance of NCC

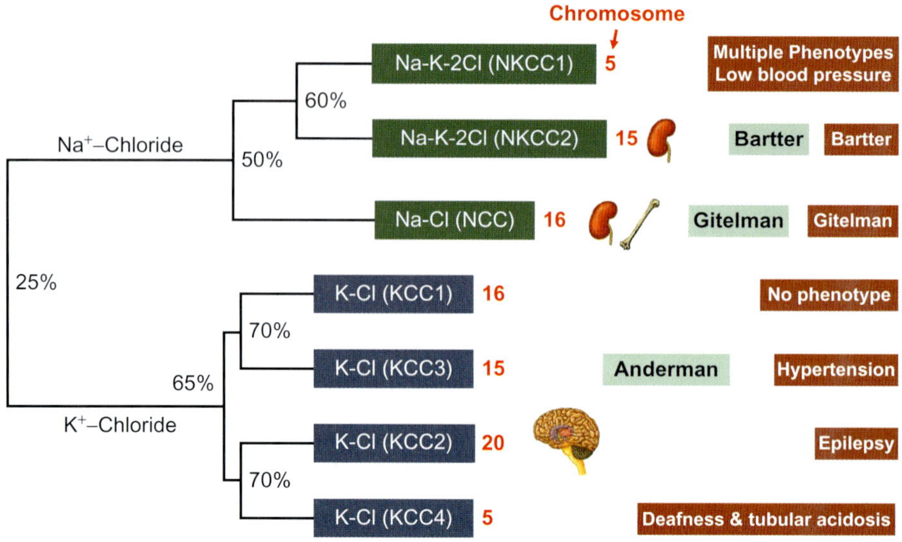

FIGURE 16.3 Phylogenetic tree of the electroneutral cation-coupled chloride cotransporter family SLC12. Numbers indicate the degree of identity. Na-chloride-coupled cotransporters are shown in green and K-coupled chloride cotransporters are shown in blue. Human chromosome location is shown by numbers in red. Tissue-specific or exclusive expression is shown for NKCC2, NCC and KCC2. Inherited diseases due to inactivating mutations are shown in blue for NKCC2, NCC and KCC3. Phenotypes of knockout mice are shown in red boxes for each cotransporter.

protein assessed by Western blot (Kim et al., 2000), and by Metolazone binding to plasma membranes from renal cortex (Fanestil et al., 1997) of rats exposed to chronic NH_4Cl loading.

Figure 16.3 shows a phylogenetic tree of all members of the *SLC12A* family for which functional properties are known. Human protein sequences were used for the alignment analysis. Figure 16.3 also shows the chromosome which each gene has been mapped to in humans, the known tissue distribution, the inherited disease due to inactivating mutations and the consequence of knocking out each gene in the mouse. Two main branches are identified, one composed of Na^+-dependent cotransporters which includes NKCC1, NKCC2 and NCC. The overall degree of identity between NKCC1 and NKCC2 is ~60%, while among these two proteins and NCC it is ~50%. The degree varies, however, within specific domains of the cotransporters; identity is >80% within the central hydrophobic membrane-spanning domain, ~50% in the carboxyl terminal domain and less than 10% in the amino terminal domain. The other branch is composed of the cotransporters that use K^+. These include KCC1, KCC2, KCC3 and KCC4. Extensive discussion of these cotransporters is presented in Chapters 3 and 17. The degree of identity between the two branches is only ~25%. Although not shown in Fig. 16.3, there are other genes, *SLC12A8* and *SLC12A9*, which exhibit some degree of identity (~20%) with all members of the family. The functional properties and roles of *SLC12A8* and *SLC12A9* are unknown.

III. FUNCTIONAL PROPERTIES

NKCC1 is a member of the *SLC12A* family which has been extensively characterized in cells that express it (for an in-depth review see Russell, 2000). In contrast, functional characterization of NKCC2 and NCC has been relatively scarce (O'Grady et al., 1987; Rose, 1991; Bachmann et al., 1999). However, with the cloning of cDNAs encoding some variants of these proteins, in-depth characterization of their major functional, pharmacologic and some regulatory properties has been made possible.

A. The Thiazide-sensitive Na^+-Cl^- Cotransporter

Heterologous expression of teleosts, rat, mouse and human NCC has been achieved in *X. laevis* oocytes (Gamba et al., 1993, 1994; Monroy et al., 2000; Kunchaparty et al., 1999; Sabath et al., 2004; De Jong et al., 2002). This amphibian expression system is a useful tool for robust and reproducible NCC expression. Expression of NCC in mammalian cells transfected with NCC cDNAs has been reported by two independent groups in MDCK (De Jong et al., 2003b) and HEK-293 (Richardson et al., 2008). However, in both cases, NCC expression was low and insufficient for kinetic characterization, when compared with oocytes. As shown in Table 16.1, a number of interesting

TABLE 16.1 Ion transport and thiazide-sensitive kinetics of mammalian and flounder NCC expressed in *Xenopus laevis* oocytes

	Rat TSC Monroy et al. (2000) Moreno et al. (2006)	**Mouse TSC Sabath et al. (2004)**	**Flounder TSC Monroy et al. (2000) Moreno et al. (2006)**
Na^+ Km (mM)	5.5 ± 1.0	7.2 ± 0.4	30.0 ± 6.0
Cl^- Km (mM)	2.6 ± 0.6	5.6 ± 0.6	15.2 ± 2.0
Metolazone IC_{50} μM	0.3 ± 0.001	0.4 ± 0.001	4.0 ± 0.08

differences have been observed between fish (flounder) and mammalian (rat and mouse) NCC. The apparent Km values for Na^+ and Cl^- in rat (Monroy et al., 2000) or mouse NCC proteins (Sabath et al., 2004) are significantly lower than the Km values observed in the flounder NCC (Vazquez et al., 2002). In addition, the affinity for Na^+ and Cl^- in mammalian NCCs is similar, whereas in flounder the affinity for extracellular Cl^- is higher than the affinity for Na^+. In all cases, NCC activity is inhibited by thiazide-type diuretics with the following profile: polythiazide > metolazone > bendroflumethiazide > trichloromethiazide > chlorthalidone. However, flounder NCC exhibited lower affinity for each thiazide. In fact, at a concentration of 10^{-4}M, the less potent thiazides, trichloromethiazide and chlorthalidone, reduced flounder NCC activity by only 68% and 46%, respectively (Vazquez et al., 2002), whereas the same concentration of all thiazides inhibited rat TSC by >95% (Monroy et al., 2000).

Two different proposals for the order of ion binding to cotransporter have been advanced for NCC. One was obtained before NCC cDNAs were available by assessing the metolazone binding to membranes extracted from rat renal cortex. In this study Tran and collaborators (Tran et al., 1990) observed an Na^+ binding increase in the putative thiazide-sensitive transport protein. Cl^- decreased the affinity for metolazone, while Na^+ increased the affinity for Cl^-. Thus, the authors proposed that the thiazide receptor contains two binding sites: one that is selective for Na^+, and the other that can recognize either Cl^- or metolazone in a competitive fashion. In this model, occupancy of the Na^+ site increases the affinity of the second site for Cl^- and/or metolazone (Chang and Fujita, 1999).

The other model was proposed by Monroy and collaborators (Monroy et al., 2000) based on functional expression of rat NCC in *X. laevis* oocytes. It was observed that affinity for Na^+ or Cl^- changed as a function of counter ion concentration. That is, the lower the extracellular Na^+ concentration, the lower the Cl^- affinity, a relationship that supports the initial conclusions of Tran et al. (Tran et al., 1990). However, it was also observed that lower extracellular Cl^- concentrations correlate with lower Na^+ affinity. In addition, it was observed that the IC_{50} for metolazone inhibition was shifted to the left when the diuretic dose–response curves were performed in low Na^+ or Cl^- conditions, suggesting that both ions compete with metolazone for binding to the cotransporter. Thus, the proposed model included a random order of binding with both ions affecting affinity for the counter ion and competing with thiazide diuretics (Monroy et al., 2000). Data discussed below (see section IV) suggest that affinity-defining domains or residues for Cl^- and thiazide binding are located in different parts of the protein (Moreno et al., 2006).

B. The Na^+-K^+-$2Cl^-$ Cotransporter 2 (NKCC2)

Several NKCC2 splice variants from several species have been analyzed at the functional level. These species include: shark (Gagnon et al., 2002), rat (Gamba et al., 1994), mouse (Plata et al., 1999), rabbit (Payne and Forbush, 1994; Gimenez et al., 2002) and human (Starremans et al., 2003a). As discussed above, three variants of NKCC2 named A, B and F have been identified. They differ in the sequence of part of TM segment 2 and an interconnecting sequence between segments 2 and 3 (Fig. 16.2). It was first observed that the three mouse NKCC2 variants A, B and F behave as Na^+-K^+-$2Cl^-$, suggesting that the difference between variants could be in the kinetics of ion transport or bumetanide affinity (Plata et al., 1999). This hypothesis was also supported by intra-renal localization studies that demonstrated axial distribution of these variants along TALH (Payne and Forbush, 1994; Igarashi et al., 1995; Yang et al., 1996). As shown in Fig. 16.2, while the A isoform is present in both cortical and medullary TALH, the B isoform is present only in cortical TALH and the F isoform is expressed only in medullary TALH. In this regard, early studies (Burg, 1982; Rocha and Kokko, 1973; Reeves et al., 1988) indicated that NaCl transport rate in mTALH is significantly more rapid than in cTALH, but with greater diluting power in the later segment, suggesting the existence of transport heterogeneity properties along TALH. Plata et al. (Plata et al., 2002) and Giménez et al. (Gimenez et al., 2002) simultaneously obtained evidence in support of this hypothesis using mouse and rabbit NKCC2 variants, respectively. The NKCC2F variant exhibits the lowest affinity for cotransported ions and is more sensitive to changes in extracellular osmolarity. As shown in Fig. 16.2, this is the isoform that is predominantly expressed in the inner stripe of the outer medulla, where the salt concentration is very high and where greater changes in extracellular osmolarity occur. The NKCC2A variant exhibits the highest transport capacity and is expressed throughout the TALH. Finally, the mouse NKCC2B variant is the one with the highest affinity for the three cotransported ions, and is located in a region of TALH in which Na^+ and Cl^- concentration of tubular fluid has been greatly reduced. Thus, the dilution power of cTALH may be explained by the presence and functional characteristics of the three alternatively spliced variants of NKCC2. Data discussed below (see section IV) suggest that specific residues encoded by exons A, B or F are responsible for the observed differences in functional properties. Data from these studies have been recently used to estimate a mathematical model for all three NKCC2 isoforms (Marcano et al., 2009).

A shorter variant (due to carboxyl terminal truncation) of NKCC2 has been identified in mouse (Fig. 16.2). The expression of this shorter NKCC2 variant was demonstrated in mouse outer medulla, using a rabbit polyclonal antibody directed against 55 unique residues of this protein (Mount et al., 1999). The short NKCC2 variant is present in some, but not all TALH cells. The proportion of immunostained cells decreased along individual medullary rays towards the cortex. When heterologously expressed in *X. laevis* oocytes, the short NKCC2 variant cRNA encodes a K^+-independent, but nevertheless, loop diuretic-sensitive, Na^+-Cl^- cotransporter that requires exposure to hypotonicity for activation (Plata et al., 2001). Activation is apparently achieved by trafficking the cotransporter from cytosolic pools to the plasma membrane. In addition, cotransporter activity in hypotonic media is inhibited by cAMP and further stimulated by blocking PKA activity with H89. Thus, the shorter NKCC2 variant may provide a molecular explanation for previous physiological studies that suggested a switch from Na^+-Cl^- to Na^+-K^+-Cl^- cotransporter mode in TALH by extracellular osmolarity or by the presence of vasopressin. Eveloff and co-workers (Eveloff and Calamia, 1986; Alvo et al., 1985) observed in rabbit TALH cells a K^+-independent, furosemide-sensitive Na^+-Cl^- cotransporter in hypotonic conditions. It became K^+ dependent, constituting the Na^+-K^+-Cl^- cotransporter, when cells were changed to isotonic medium. In addition, Sun et al. (Sun et al., 1991) found that in mouse TALH cells vasopressin (i.e. cAMP) shifts the mode of apical cotransport from Na^+-Cl^- to Na^+-K^+-$2Cl^-$. Thus, the large NKCC2 is an Na^+-K^+-$2Cl^-$ cotransporter that is activated by increased tonicity or the presence of vasopressin, while the shorter NKCC2 is an Na^+-Cl^- cotransporter that is activated in hypotonicity or by inhibition of protein kinase A activity.

Functional expression of the short NKCC2 in *X. laevis* oocytes has suggested additional functional roles for this variant; it exerts a dominant-negative effect upon the activity of the longer NKCC2 that can be abrogated by cAMP (Plata et al., 1999). Using confocal microcopy, in oocytes injected with both the long (NKCC2F) and the short variant of mouse NKCC2, Meade and collaborators (Meade et al., 2003) observed that the short NKCC2 reduced the activity of the long NKCC2. This was correlated with the observation that the short NKCC2 variant prevented cotransporter trafficking and surface membrane expression of the long NKCC2. This effect of the short NKCC2 variant could be prevented by cAMP. These studies suggest that in mouse medullary TALH, activation of Na^+-K^+-$2Cl^-$ cotransporter by hormones that increase intracellular cAMP (e.g. vasopressin, PTH) requires the presence of the short NKCC2 protein. The absence of cAMP allows the short form of NKCC2 to reduce cotransporter activity, whereas in the presence of cAMP, the negative effect of the short isoform on NKCC2 is inhibited. In this regard, Mount and collaborators (Mount et al., 1999) observed that expression of the short NKCC2 is axially distributed along TALH, as cortical TALH appears to lack this isoform. This heterogeneity may explain the observation that in mouse the vasopressin effect is present only in the medullary TALH (Hebert et al., 1981a).

C. Isoform 1 of the Bumetanide-sensitive Na^+-K^+-$2Cl^-$ Cotransporter (NKCC1)

NKCC1 has been identified at the molecular level from several sources, but detailed functional expression analysis has been performed only for human (Payne et al., 1995) and shark orthologs (Xu et al., 1994; Isenring and Forbush, 1997). These studies were done in HEK-293 cells transfected with the corresponding cDNAs and showed that, similar to what we discussed for the case of the flounder and mammalian NCCs, the human NKCC1 exhibits significantly higher affinity for Na^+, K^+ and Cl^-, as well as for the inhibitor bumetanide when compared with shark

TABLE 16.2 Ion transport kinetics and bumetanide affinity of the Na^+-K^+-$2Cl^-$ cotransporter NKCC1 from human and shark

	Sodium Km (mM)	Rubidium Km (mM)	Chloride Km (mM)	Bumetanide Ki (μM)	References
hNKCC1	19.6 ± 4.9	2.68 ± 0.72	26.5 ± 1.3	0.16	Payne et al. (1995)
hNKCC1	15.2 ± 1.5	1.6–2.5	31 ± 1.0	0.044–0.079	Isenring and Forbush (1997)
HEK-293	22	12	110	0.054	Xu et al. (1994)
sNKCC1	42	12	110	0.54	Xu et al. (1994)
sNKCC1	165 ± 34	14 ± 8.0	101 ± 24	0.57	Payne et al. (1995)
sNKCC1	113 ± 11	9.6–11.6	102 ± 7	0.22–0.30	Isenring and Forbush (1997)

hNKCC1 = human NKCC1, sNKCC1 = shark NKCC1.

NKCC1 (Table 16.2). These differences were then exploited to define major affinity modifier regions within the cotransporters (Isenring and Forbush, 2001) as discussed below. In addition, the endogenous NKCC1 cotransporter has been extensively characterized *in situ* in many different cells and tissues, and Chapter 18 is devoted to the discussion of these data. Finally, two alternatively spliced variants of mouse NKCC1 have been found (Randall et al., 1997). A recent study suggests that the presence or absence of exon 21 which characterizes these two variants may account for differential sorting of NKCC1 in polarized cells (Carmosino et al., 2008).

IV. STRUCTURE–FUNCTION RELATIONSHIPS

As mentioned above, the only published study on the proposed topology of one member of the SLC12A family was performed by Gerelsaikhan and Turner using *in vitro* translation of human NKCC1 constructs (Gerelsaikhan and Turner, 2000). They concluded that NKCC1 may be composed of 12 membrane-spanning segments flanked by amino and carboxyl terminal domains located within the cell. TM segments 1–8 exhibit the classical ~20 residue α-helices and TMs 9–10 and 11–12 are ~36 residues in length. These apparently form tight hairpin-like structures in the membrane or take up either a non-helical or a partial-helical structure (see Chapter 11 for further discussion).

A. Ion or Diuretic Affinity Modifier Domains or Residues

Several attempts to begin to understand structure–function relationship issues in the Na^+-coupled Cl^- cotransporters have been made on NKCC1 (for review see Isenring and Forbush, 2001), NKCC2 (Plata et al., 2002; Tovar-Palacio et al., 2004; Gimenez et al., 2002; Gagnon et al., 2005; Gimenez and Forbush, 2007) and NCC (Moreno et al., 2006). In the cases of NKCC1 and NCC, differences in functional properties between fish and mammalian transporters were used to design chimeric proteins to begin to define domains, regions or residues that are critical for defining ion transport or diuretic binding affinities. In contrast, most of the information on NKCC2 has come from in-depth analysis of the amino acid residues comprising exons A, B and F.

Shark and human NKCC1 exhibit 74% identity and the functional differences that are shown in Table 16.2. Isenring and Forbush constructed several chimeric proteins by interchanging fragments between human and shark NKCC1. By exchanging the amino- and carboxyl-terminal domains between human and shark NKCC1 they provided evidence suggesting that the protein domains that define ion affinity are located within the central hydrophobic regions of the NKCC1 protein, which contains the 12 transmembrane segments (Isenring and Forbush, 1997). They observed that chimeras containing the central domain of shark NKCC1 with amino- and carboxyl-terminal domains of human NKCC1 exhibit ion transport kinetics similar to wild-type shark NKCC1, and vice versa. Then, additional chimeras were constructed and analyzed by in-depth kinetic analysis (Isenring and Forbush, 1997; Isenring et al., 1998a, b). The chimeric design took into account the transmembrane segments that are identical in NKCC1 of both species. The conclusions from these series of studies are depicted in Fig. 16.4, and suggest that three TM segments play an important role in defining ion transport kinetics in NKCC1. Accordingly, TM2 is involved in Na^+ and Rb^+ kinetics, TM4 in Rb^+ and Cl^- kinetics and TM7 in Na^+, Rb^+ and Cl^- kinetics. Interestingly, the behavior of several chimeric proteins with respect to bumetanide inhibition was completely different from that observed in

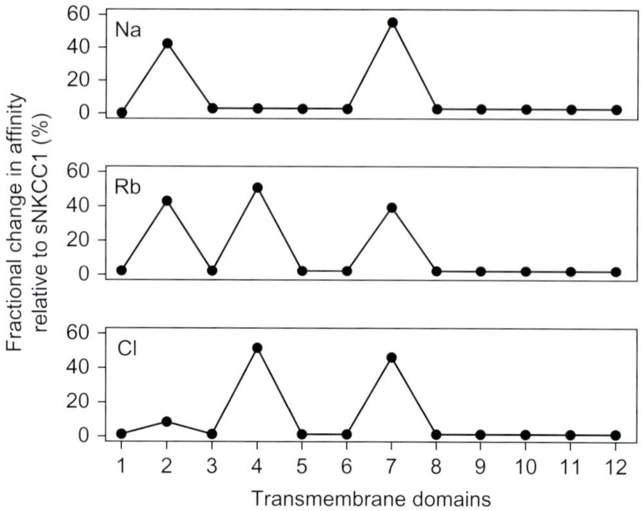

FIGURE 16.4 Contribution of each transmembrane segment of NKCC1 to fractional changes in apparent affinity for Na^+, K^+ and Cl^- in chimeric cotransporters from human and shark NKCC1. (Modified from Isenring and Forbush, 2001.)

FIGURE 16.5 Proposed model of the TM1 to TM3 region of the NKCC2 after analysis of the contribution of several residues of exons A, B or F on defining ion transport affinity. (Modified from Gimenez and Forbush, 2007. Figure kindly provided by Professor Biff Forbush.)

ion transport kinetics. In the case of bumetanide, data suggested that TM segments 2–6 and 10–12 play a role in defining affinity for loop diuretics. Thus, results from studies using shark and human NKCC1 chimeric cotransporters argued against a previous hypothesis that loop diuretics and Cl^- compete for the same binding site (Forbush and Palfrey, 1983).

As discussed above, analysis of the three spliced variants A, B and F of rabbit NKCC2 revealed a clear role of the segments encoded by these exons (TM2) in transport kinetics properties, not only for Na^+ and Rb^+, but also for Cl^- and bumetanide (Fig. 16.2) (Plata et al., 2002; Gimenez et al., 2002). Further studies (Gagnon et al., 2002) showed that NKCC2 variants A and F, but not B, are present in the shark kidney, and also that the NKCC2A variant exhibits higher affinity for transported ions than the NKCC2F variant. Then, additional chimeric proteins were constructed: chimera A/F contained the TM2 sequence of variant A followed by an interconnecting sequence of variant F, and vice versa (Gagnon et al., 2003). The results suggested that Cl^- transport affinity defining domains between isoforms A and F were located within the predicted interconnecting segment between TM2 and TM3, rather than within the TM2 segment. Interestingly, the chimeras' behavior with respect to bumetanide affinity was opposite that of Cl^- affinity, suggesting that residues located within TM2 and not within the TM2–TM3 interconnecting segment play a role in defining bumetanide affinity. Finally, in one study (Gimenez and Forbush, 2007), a series of single or multiple point mutations were performed between rabbit NKCC2 variants B and F, or A and B, in order to switch one or few residues between variants. The results demonstrated that replacement of six residues in NKCC2B resulted in a variant with Na^+ and Cl^- affinities identical to those of NKCC2F. Three of these residues are located within the TM2 segment (ATG in B isoform were switched to TAY) and the other three within the interconnecting segment between TM2 and TM3 (SVT in B isoform were changed to MCV). Thus, with respect to Na^+ and Cl^- affinity, the six mutations (ATG-TAY/SVT-MCV) turned NKCC2B into the F variant. These same residues turned NKCC2B into the A variant for differences in Cl^- affinity. Based on these observations it was suggested that the interconnecting loop between TM2 and TM3 may actually be part of a membrane-embedded domain, as illustrated in Fig. 16.5.

All studies of NKCC1 and NKCC2 discussed in the previous paragraphs were performed in native, chimeric and mutant proteins that were expected to perform as Na^+-K^+-$2Cl^-$ cotransporters. Thus, no information was obtained concerning structural requirements to define specificity for ions or diuretics. In this regard, as discussed above, a short isoform of NKCC2 in mouse behaves as a K^+-independent, but loop diuretic-sensitive Na^+-Cl^- cotransporter (Plata et al., 2001), suggesting that sequences within the carboxyl terminal domain could be critical to endow NKCC2 with K^+ transport ability. However, Tovar-Palacio and collaborators (Tovar-Palacio et al., 2004) constructed chimeras mixing rat NCC and rat NKCC2F in which both amino- and carboxyl-terminal domains were switched between cotransporters. As shown in Fig. 16.6, the chimera with the central hydrophobic domain of NKCC2 flanked by amino- and carboxyl-terminal domains of rat NCC exhibits NKCC2 behavior. That is, it acted as a bumetanide-sensitive Na^+-K^+-$2Cl^-$ cotransporter, indicating that the residues that endow rat NKCC2F with its functional properties are located within the central hydrophobic domain. Thus, there is no clear explanation so far on how the short NKCC2 variant performs as a K^+-independent transporter that is bumetanide sensitive, i.e. a bumetanide-sensitive Na^+-Cl^- cotransporter. The unique 55 amino acid segment at the end of this isoform contains several negatively charged residues, suggesting that this fragment could interact with the cotransporter core to prevent K^+ translocation, like the ball and chain mechanism proposed for some membrane channels (Malysiak and Grzywna, 2008).

Similar chimera construction and directed point mutation approaches have been used to gain insight into the structure–function relationships of NCC. As discussed above, it was observed that rat (Monroy et al., 2000) and mouse (Sabath et al., 2004) NCCs exhibit similar functional properties that are significantly different from those documented for flounder NCC (Vazquez et al., 2002). Since identity degree between mammalian and flounder NCC is about 60%, differences in functional properties and primary sequences between rat and flounder NCCs were exploited to define affinity modifying domains (Moreno et al., 2006). In this study, NCCs were divided into five fragments: amino terminal domain, TM segments 1–7, extracellular glycosylated loop, TM segments 8–12 and the entire carboxyl terminal domain that was swapped between rat and flounder NCC. The results indicated that the functional characteristics of rat and flounder NCC are defined by their central hydrophobic domains. As shown in Fig. 16.7, however, the authors were able to define different segments within the central domain defining Cl^- or thiazide affinity. Chimeras in which the TM segments 1–7 were interchanged between flounder and rat NCCs demonstrated that affinity defining residues for Cl^- are located within these segments. Similarly, it was observed that affinity defining residues for thiazide inhibition are located within TM segments 8–12. This information argued against a previous proposal that thiazide diuretics and Cl^- compete for the same binding site (Tran et al., 1990). The observation that Cl^- affinity is defined by residues located within TM segments 1–7 is supported by another study from Moreno et al. (Moreno et al., 2004). The authors analyzed the functional consequences of single nucleotide polymorphisms located within the exons of the human *SLC12A3* gene that changed the

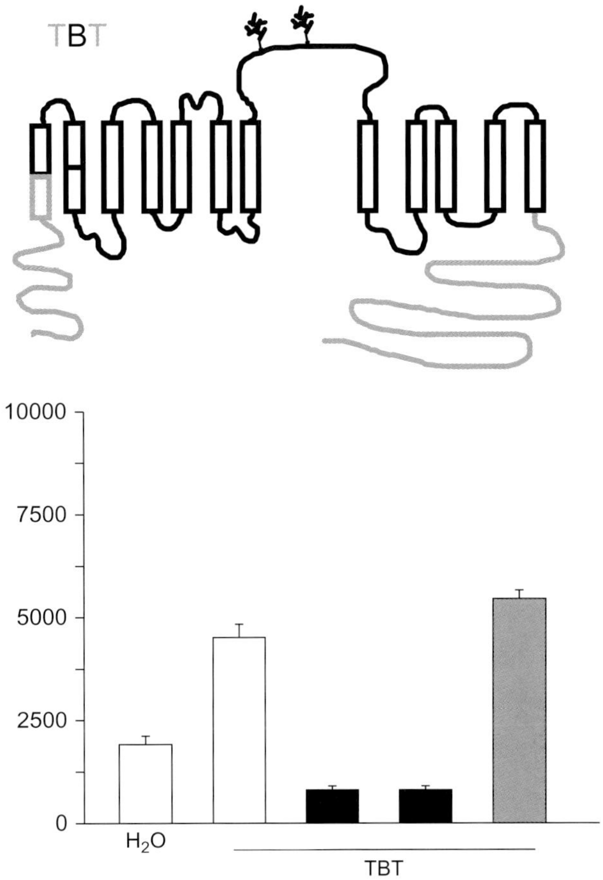

FIGURE 16.6 Functional characteristics of a chimeric transporter containing the amino- and carboxyl-terminal domains of NCC (gray) fused into the central hydrophobic transmembrane domain of NKCC2 (black). This chimera was named TBT. Injection of TBT cRNA into *Xenopus laevis* oocytes resulted in an increase in $^{86}Rb^+$ uptake mechanisms that is evident in the presence of Na^+, K^+ and Cl^- (open bars). Increased uptake is reduced in the absence of extracellular Cl^- (black bar), or in the presence of 100 μM bumetanide (black bar), but is not affected by 100 μM metolazone (gray bar). (Modified from Tovar-Palacio et al., 2004.)

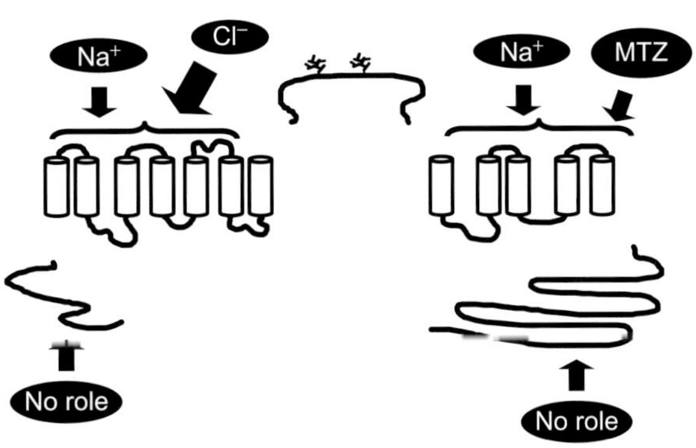

FIGURE 16.7 Affinity modifying domains in the thiazide-sensitive Na^+-Cl^- cotransporter, NCC.

primary structure of NCC. The results revealed that a highly conserved glycine within the fourth transmembrane domain plays a critical role in defining the level of cotransporter activity and the affinity for Cl^-. Accordingly, it was observed that a glycine to alanine SNP at position 264 resulted in decreased activity of the cotransporter by 50%, with increased affinity for Cl^- by one order of magnitude.

Analysis of the putative glycosylation sites in rat NCC by means of mutagenesis revealed that glycosylation itself is another component of the cotransporter that affects thiazide affinity. In one study, Hoover et al. (Hoover et al., 2003) eliminated the two glycosylation sites located within the extracellular loop between TM segments 7 and 8. It was observed that the elimination of one site (either N404 or N424) resulted in a 50% reduction of NCC activity and the elimination of both sites was associated with a 95% reduction. Interestingly, prevention of glycosylation was associated with increased affinity for extracellular Cl^- and metolazone. This effect of glycosylation upon thiazide affinity seems to be unique to rat NCC since it was not observed after preventing glycosylation of flounder NCC (Moreno et al., 2006) or rat NKCC2F (Paredes et al., 2006). In this last cotransporter, elimination of glycosylation by means of site-directed mutagenesis was associated with increased affinity for extracellular chloride, but no effect was observed on the affinity for the loop diuretic bumetanide.

B. Residues Involved in Regulation of the Cotransporter Activity

Chapter 18 in this volume focuses on the analysis of the regulation of cotransporter's activity, a field that has experienced significant progress during the last decade. Thus, here we just highlight the structure–function relationship aspects. Accordingly, we discuss the residues that have been found to be critical for the regulation of the Na-coupled Cl^- cotransporters. It is well known that activity of NKCC1/2 and NCC is tightly regulated; activity of these cotransporters should be coordinated with that of Cl^- extrusion mechanisms. It has been established that, at least for NKCC1 and NKCC2 proteins, cotransporter regulation is associated with their phosphorylation/dephosphorylation. Further, several lines of evidence strongly suggest that $[Cl^-]_i$ is a common pathway of NKCCs regulation. When $[Cl^-]_i$ falls, the cotransporter becomes phosphorylated, and when $[Cl^-]_i$ rises, the cotransporter is dephosphorylated and inhibited. These effects of $[Cl^-]_i$ on cotransporter activity have been demonstrated for NKCC1 (Lytle and McManus, 2002; Dowd and Forbush, 2003; Flemmer et al., 2002), NKCC2 (Ponce-Coria et al., 2008) and NCC (Pacheco-Alvarez et al., 2006). In all cases, it was observed that activation of the cotransporter by intracellular Cl^- depletion is associated with increased phosphorylation of the cotransporter in several conserved threonine residues located within the amino terminal domain. The first study that presented evidence suggesting that activation of NKCC1 cotransporter was associated with phosphorylation of a threonine residue was carried out by Lytle and Forbush (Lytle and Forbush, 1992), who analyzed suspensions of shark rectal-gland tubules. The peptide FGHNTIDAVP, which became phosphorylated, was isolated in this study. When shark NKCC1 cDNA was identified at the molecular level (Xu et al., 1994) it was shown that this peptide corresponds to amino acid residues 184–194, which are located within the amino terminal domain. It was later demonstrated by mass spectrophotometry that phosphoacceptor amino acid sites within the cotransporter amino terminal domain corresponded to the threonines located at positions 184, 189 and 202 of NKCC1. Of these, threonine 189 is absolutely required for NKCC1 to be functional. Then, the same group performed studies *in vivo* using a specific antiphospho-NKCC1 antibody denominated R5, which was raised against a synthetic peptide of the amino terminal domain containing threonines 212 and 217 of human NKCC1 (corresponding to threonines 184 and 189 of shark NKCC1) (Flemmer et al., 2002; Darman and Forbush, 2002). The results demonstrated that activation of NKCC1 by intracellular Cl^- depletion is associated with phosphorylation of these threonines. Later, the same R5 phosphoantibody was used to demonstrate that both in rat NKCC2 and NCC, the conserved threonines (96/101 and 53/58, respectively) become phosphorylated when intracellular Cl^- is depleted (Moriguchi et al., 2005; Ponce-Coria et al., 2000; Pacheco-Alvarez et al., 2006). In addition, a recent publication using human NCC also deemed the corresponding sites (threonines 55 and 60) to be critical for NCC regulation by intracellular Cl^- and suggested the serine 91 as a third potential residue to be involved (Richardson et al., 2008).

C. The Na-coupled Chloride Cotransporters Form Homodimers

There is evidence suggesting that Na^+-K^+-$2Cl^-$ and Na^+-Cl^- cotransporters form homodimers in the plasma membrane. First, Moore-Hoon and Turner (Moore-Hoon and Turner, 2000) used rat parotid plasma membrane to analyze NKCC1 using the reversible chemical cross-linker DTSSP (3,3′-dithiobis-[sulfosuccinimidyl propionate]). They observed that NKCC1 migrates at ~335 kDa. After protein denaturation, single monomers of approximately ~170 kDa were obtained, in which the investigators were unable to detect the presence of any additional protein. Similar results were later observed on human NKCC2 by Starremans et al. (Starremans et al., 2003b) and human NCC by De Jong et al. (De Jong et al., 2003a). These studies were performed in *X. laevis* oocytes injected with *in vitro*-transcribed cRNA from FLAG-and/or HA-tagged wild-type cotransporters and concatamer constructions. These experiments revealed that FLAG-NKCC2 and HA-NKCC2 are physically linked (Starremans et al., 2003b). Similar observations were obtained for FLAG-NCC and HA-NCC constructs (De Jong et al., 2003a). Thus, NKCC1, NKCC2 and NCC are able to form homodimers and might be functional in this conformation. Interestingly, some data suggest that members of the *SLC12A* family can build heterocomplexes comprised of different members of the family. For example, an orphan gene of the family known as CIP does not appear to transport ions itself, but does appear to inhibit transport activity of NKCC1 or activate KCC2 (Wenz et al., 2009). However, it does not appear to inhibit the activity of other family members, raising the possibility that CIP may form an activating heterocomplex specifically with NKCC1 (Caron et al., 2000). In this regard, a recent study has shown, by means of yeast two-hybrid and pull-down assays, that K^+-Cl^- cotransporter isoforms can interact among themselves and even with NKCC1 (Simard et al., 2007). The functional significance of such interactions is still to be determined. In this context the study of Wenz and co-workers (2009) adds a novel facet to the regulation of cation-coupled Cl^- cotransporters by identifying CIP as a possible reciprocal regulator of NKCC1 and KCC2.

V. PHYSIOLOGICAL ROLES

Electroneutral cation-chloride cotransporters are membrane proteins that translocate Cl^- ions together with a cation that can be Na^+, K^+ or both. Their stoichiometries – $1Na^+$-$1Cl^-$, $1K^+$-$1Cl^-$ or $1Na^+$-$1K^+$-$2Cl^-$ – are consistent with the transport process being electroneutral. The cation gradient is the main determinant of the direction in which Cl^- ions are moved by members of the *SLC12A* gene family, as discussed in Chapter 5. Thus, Na-coupled cotransporters translocate ions inwardly across the plasma membrane. Because Na^+ is quickly returned to the extracellular space, the activity of NKCC1, NKCC2 and NCC primarily serves to regulate $[Cl^-]_i$. This is one of the major functions of these cotransporters at the cell physiology level as discussed in Chapter 5 and in Alvarez-Leefmans (2001). This role of the SLC12A family members is particularly critical in setting intra neuronal $[Cl^-]_i$, either above or below its electrochemical equilibrium potential. Thus, the activity of these cotransporters is critical in determining the polarity and magnitude of the effect of neurotransmitters that gate Cl^- channel in postsynaptic membranes, such as GABA, as discussed in several chapters throughout this book. Another major function of electroneutral cation-chloride cotransporters is their well-known role as regulators of cell volume (Russell, 2000). Na-coupled-Cl^- cotransporters also participate in transepithelial movement of ions. NCC and NKCC2 are polarized to the apical membrane in certain nephron segments in which they participate in renal salt reabsorption (see below). In kidney, NKCC1 is expressed in the basolateral membrane and is critical for salt secretion.

A. The Thiazide-sensitive Na^--Cl^- Cotransporter

NCC is the major salt transport pathway in the apical membrane of mammalian distal convoluted tubule (DCT) cells (Kunau et al., 1975; Velazquez et al., 1984; Costanzo, 1985; Ellison et al., 1987; Plotkin et al., 1996), a nephron region that mediates reabsorption of 5–10% of glomerular filtrate. The molecular mechanism of salt reabsorption in DCT is illustrated in Fig. 16.8. The majority of studies in human, rabbit, mouse and rat kidneys agree that NCC is the only sodium reabsorption pathway along the early DCT and shares this work with the Na^+ channel ENaC in late DCT (Loffing and Kaissling, 2003; Loffing et al., 2001; Ellison et al., 1987; Obermuller et al., 1995; Campean et al., 2001; Reilly and Ellison, 2000). The Na^+ gradient that drives transport from the lumen to the interstitium is generated and maintained by very intense activity of Na^+/K^+ ATPase which is polarized to the basolateral membrane (Doucet, 1988). Potassium entering the cell through the Na^+/K^+ pump is secreted by the luminal membrane via ROMK K^+ channels (Xu et al., 1997) and by an apical K^+-Cl^- cotransporter (Amorim et al., 2003). Thus, the rate of Na^+-Cl^- reabsorption determines, in part, the rate of K^+ secretion. In addition, NCC modulates magnesium and calcium reabsorption, the latter in an inversely related fashion with sodium reabsorption. The higher the sodium reabsorption, the lower the calcium reabsorption and vice versa (Costanzo, 1985).

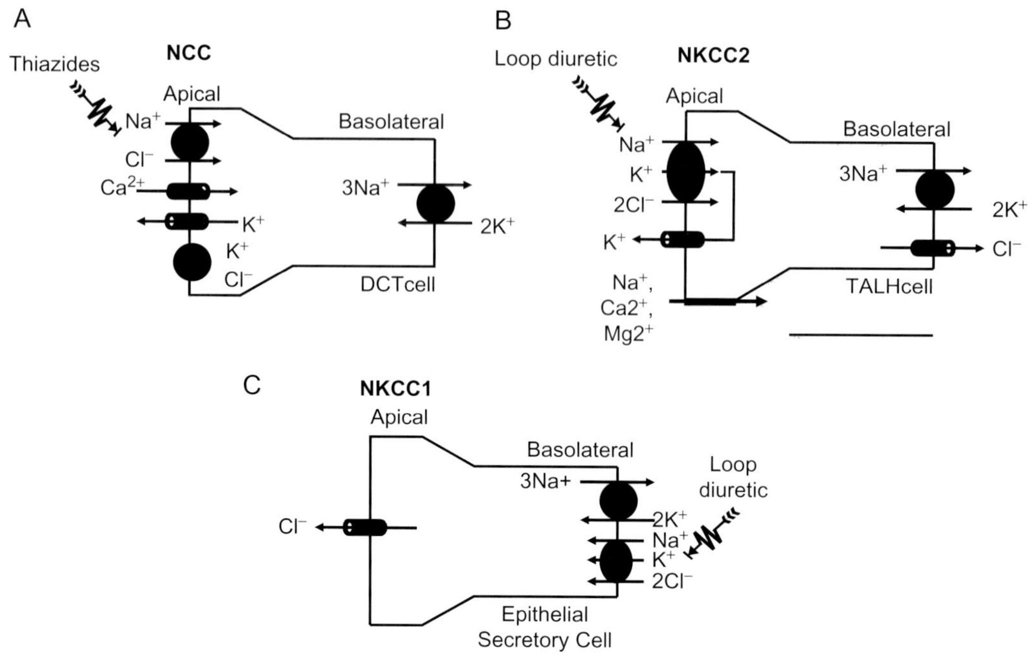

FIGURE 16.8 Representation of sodium-coupled chloride cotransporters in transepithelial ion transport. **A.** Distal convoluted tubule. Salt is transported in apical membrane by NCC. **B.** Thick ascending limb of Henle's loop. Salt is transported in apical membrane by NKCC2. **C.** Secretory epithelial cell from trachea, gills, intestine, salivary gland, etc. Salt is transported from interstitial space into cell by means of NKCC1.

Consistent with its role in renal salt reabsorption, NCC is regulated by the mineralocorticoid hormone aldosterone that is produced by the adrenal glands (Masilamani et al., 1999; Kim et al., 1998; Nielsen et al., 2002). Despite positive regulation by aldosterone, NCC is the only transport protein along the nephron that is decreased during the aldosterone escape phenomenon (Wang et al., 2001), indicating that NCC is a major target of pressure natriuresis. This conclusion was suggested many years ago by Majid and Navar based on measurements of pressure natriuresis in intact dogs (Majid and Navar, 1994). As illustrated in Fig. 16.8A, the apical Na^+-Cl^- cotransporter in DCT is the major target for thiazide-type diuretics (chlorthalidone, hydrochlorothiazide, bendroflumethiazide, metolazone) (Stokes et al., 1984; Ellison et al., 1987; Gamba et al., 1994), which are among the most commonly prescribed drugs in the world due to their beneficial effects in the treatment of arterial hypertension, edematous states and renal stone disease (Rose, 1991). Thus, pharmacologic inhibition of NCC is frequently used in the clinical setting.

Although NCC was considered to be a kidney-specific gene for many years, recent studies have clearly demonstrated its expression in intestine (Bazzini et al., 2005) and bone (Dvorak et al., 2007). In addition, a thiazide-sensitive Na^+-Cl^- cotransporter has been postulated to exist in the brain (Deng and Chen, 2003), blood vessels (Clader et al., 1992), pancreas (Bernstein et al., 1995), peripheral blood mononuclear cells (Abuladze et al., 1998), bone (Barry et al., 1997), gallbladder (Cremaschi et al., 2000) and heart (Drewnowska and Baumgarten, 1991) (see also Chapter 11). In the intestine, the role of NCC is not clear, but it may be important in salt and calcium absorption. In bone, NCC activity is associated with the rate of bone formation. It has been shown in many clinical studies that use of thiazide diuretics in elderly subjects promotes an increase in bone mineral density and helps to prevent pathological fractures (Jones et al., 1995; Schoofs et al., 2003). Consistent with this beneficial effect of thiazides, it was recently demonstrated that NCC is expressed in osteoblasts of rat and human bones and that addition of thiazides to osteoblast in culture increases the formation of mineralized nodules. This effect of thiazides was not present after decreasing NCC expression by transfecting cells with an NCC antisense plasmid (Dvorak et al., 2007).

B. The Na^+-K^+-$2Cl^-$ Cotransporter 2 (NKCC2)

The bumetanide-sensitive Na^+-K^+-$2Cl^-$ cotransporter NKCC2 is the major salt transport pathway in mammalian TALH, a nephron segment in which 15–20% of glomerular filtrate is reabsorbed. The activity of NKCC2, which is located in the apical membrane of the epithelial cells of the TALH (Kaplan, 1996), is also critical for the maintenance of the countercurrent multiplication mechanism, and thus for our ability to dilute or concentrate urine. It is also crucial for divalent cation (Ca^{2+} and Mg^{2+}) and ammonium (NH_4^+) reabsorption (for reviews see Greger, 1985; Hebert, 2003). The molecular mechanisms of salt reabsorption by TALH are illustrated in Fig. 16.8B. NKCC2 is the major pathway for sodium reabsorption in TALH apical membranes (Greger et al., 1981, 1983; Greger and Schlatter, 1983; Hebert et al., 1981a, Greger et al., 2003), but for this to occur, the simultaneous operation of other membrane transport proteins is required. Na^+ and Cl^- leave the cells via the basolateral membrane through Na^+/K^+ ATPase and Cl^- channels (CLC-Kb), respectively, while K^+ ions return to the tubular lumen via an inwardly rectifying K^+ channel known as ROMK. This recycling of K^+ is fundamental for maintaining a sufficient concentration of this ion in the lumen and generating a positive voltage within the TALH lumen. The positive voltage in the lumen drives the reabsorption of a second cation, Ca^{2+}, Mg^{2+} or Na^+, through a paracellular pathway.

NKCC2 also plays an important role in cardiovascular and renal pharmacology because this cotransporter is the main pharmacologic target of loop diuretics, such as furosemide or bumetanide, which are the most potent natriuretic agents available for clinical use. By inhibiting NKCC2, loop diuretics reduce the rate of salt reabsorption in TALH, producing significant natriuresis and diuresis. Under physiological conditions, when the macula densa cells at the end of TALH sense an increase in NaCl delivery, a compensatory decrease in glomerular filtration rate occurs thereby preventing the loss of salt in urine (Schnermann, 1998). This compensation is absent, however, when loop diuretics are present, because the salt-sensing protein in macula densa is also NKCC2 (Nielsen et al., 1998). Since the B variant of NKCC2 is the most abundant variant in macula densa cells (Gimenez et al., 2002), and it is not expressed in shark kidney, which also lacks a macula densa (Gagnon et al., 2002), it has been proposed that NKCC2B is the Cl^- sensor in TALH cells. However, variant-specific knockout mice lacking NKCC2B developed a compensatory increase in NKCC2A expression in the TALH cells and only exhibited a very slight shift to the right of the tubuloglomerular feedback curve (Oppermann et al., 2006). Isoform-specific null mice for the A variant also showed slight changes in tubuloglomerular feedback (Oppermann et al., 2007). These observations suggest that both isoforms working together compose the Cl^- sensor in the macula densa.

Increasing net NaCl reabsorption in TALH by hormones generating cAMP via Gs-coupled receptors such as vasopressin, glucagon, PTH, β-adrenergic and calcitonin is a fundamental mechanism for regulating salt transport in this nephron segment (Hebert et al., 1981b, c). Among these hormones, the most important is the antidiuretic hormone vasopressin, which increases NaCl absorption by TALH through a mechanism that appears to involve trafficking of NKCC2 to the apical plasma membrane (Gimenez and Forbush, 2003; Plata et al., 1999; Meade et al., 2003). Consistent with this notion is the observation in medullary TAL suspensions that most NKCC2 protein is located in intracellular vesicles, and that addition of cAMP increases their concentration by more than 100% when compared with basal conditions (Ortiz, 2006). Long-term increases in vasopressin are associated with increased expression of NKCC2 and maximal urinary concentration ability (Kim et al., 1999).

C. The Na^+-K^+-$2Cl^-$ Cotransporter 1 (NKCC1)

NKCC1 is a salt transport pathway present in several epithelial and non-epithelial cells. As mentioned above, at the cellular physiology level, NKCC1 plays important roles in the regulation of cell volume (Lang et al., 1998; Hoffmann, et al., 2009) and $[Cl^-]_i$ (Russell, 2000; Alvarez-Leefmans, 2001; Gamba, 2005; Rocha-González et al., 2008). As illustrated in Fig. 16.8C, by virtue of its localization in the basolateral membrane of most epithelial cells, NKCC1 plays a critical role in providing cells with K^+ or Cl^- ions that are secreted through the apical membrane. Most of our knowledge about the role of NKCC1 in specific organ or system physiology has come from analyzing the consequences of eliminating NKCC1 expression by traditional knockout models (Delpire and Mount, 2002). A full chapter in this book is devoted to the analysis of knockout mice in the study of the SLC12A family of membrane transporters (Chapter 10).

Mice lacking NKCC1 expression were produced simultaneously by Delpire et al. (Delpire et al., 1999) and Flagella et al. (Flagella et al., 1999) through targeted disruption of exons 9 and 6, respectively. NKCC1-null mice exhibited growth retardation apparent from the first day of life and were deaf due to a sensorineural defect with dysfunction of epithelial secretion in the labyrinth (see Chapter 21 in this volume). Thus, it is likely that NKCC1 plays an important role in providing epithelial cells with enough K^+ to be secreted into the cochlear chamber (Sakaguchi et al., 1998). Supporting this conclusion, a similar phenotype occurs in knockout mice lacking the K^+ channel (*minK*), which is present in the apical membrane of the stria vascularis, and through which K^+ ions are secreted (Vetter et al., 1996).

NKCC1-null mice also exhibit an intestinal phenotype. About 30% of the colony died between days 18 and 26 of life, due to cecum bleeding and severe blockade of the colon. This phenotype is similar to that of CFTR knockout mice (Snouwaert et al., 1992), indicating that NKCC1 may play a role in providing epithelial cells with ions that will then be secreted through the apical membrane. In addition, NKCC1 knockout mice also exhibited severe salivation impairment (Evans et al., 2000). The reduced secretory capacity of epithelial cells in NKCC1 null mice is also evidenced by the fact that males are infertile due to a deficit in spermatocytes production accompanied by architectural disruption of the testis and epididymis (Pace et al., 2000).

Although initial studies reported no changes in blood pressure in NKCC1-null mice (Pace et al., 2000), other groups, using more direct analyses, demonstrated that blood pressure in these mice is decreased (Flagella et al., 1999; Meyer et al., 2002; Wall et al., 2006), probably due to a decreased vascular tone. In agreement with this hypothesis, it has been recently shown that bumetanide decreased blood pressure in normal mice by inhibiting the activity of NKCC1 in vascular beds, an effect that was not observed in NKCC1-null mice (Garg et al., 2007).

The absence of NKCC1 also has consequences for the function of primary sensory neurons. The cell bodies of these neurons are located in brain stem and dorsal root ganglia. These neurons are depolarized by GABA in the adult, and this depolarization plays an important role in generating presynaptic inhibition, as discussed in Chapter 22 in this volume. This depolarization is mediated by an inward current due to efflux of Cl^-. This is possible because primary sensory neurons maintain a higher than passive $[Cl^-]_i$ in adult animals due to the functional expression of a $Na^+,K^+,2Cl^-$ cotransporter, as first demonstrated by Alvarez-Leefmans and collaborators (Alvarez-Leefmans et al., 1988; Rocha-González et al., 2008). Later studies confirmed these observations and identified NKCC1 as the isoform responsible for keeping the outward Cl^- gradient in these cells (Plotkin et al., 1997; Sung et al., 2000). In NKCC1-null mice the GABA-evoked inward current is decreased, abolished or even reversed in polarity. Further, it was also shown that NKCC1-null animals exhibited increased nociceptive threshold in the hot plate test and in touch-evoked pain (allodynia), suggesting that NKCC1 is involved in pain perception (Sung et al., 2000; Laird et al., 2004). Consistent with this notion, it has been demonstrated that bumetanide either applied peripherally or intrathecally in the spinal cord of rats significantly reduces nociception in the formalin test (Granados-Soto et al., 2005). Finally, the absence of NKCC1 has been shown to be a protective factor against cerebral ischemia (Chen et al., 2005) as discussed in Chapter 25 in this volume, and bacteremia due to bacterial pneumonia (Nguyen et al., 2007).

VI. THE ROLE OF THE SODIUM-COUPLED CATION CHLORIDE COTRANSPORTERS IN INHERITED DISEASE

Given the important roles of NCC and NKCC2 in renal physiology, it was expected that mutations in these cotransporters could be associated with monogenic disease. In fact, to date, inactivating mutations of NCC and NKCC2 have been shown to be linked with development of inherited conditions such as Gitelman's and Bartter's diseases, respectively. In addition, by affecting the way in which NCC is modulated by certain kinases, increased activity of this cotransporter has been associated with another monogenic disease known as pseudohypoaldosteronims type II, or Gordon's disease.

A. Gitelman's Disease

This is an inherited monogenic autosomic recessive disease. The clinical presentation usually becomes evident by the second decade of life. It is characterized by hypokalemic metabolic alkalosis, arterial hypotension, hypocalciuria and hypomagenesemia. The clinical presentation resembles that observed in patients intoxicated with thiazide-type diuretics. Because of this, NCC became a strong candidate for the cause of the disease and linkage between Gitelman's disease and the locus for *SLC12A3* in human chromosome 16 was found by several groups (Simon et al., 1996b; Pollak et al., 1996; Mastroianni et al., 1996a,b; Lemmink et al., 1996). At present, more than 70 families have been studied and more than 100 different mutations of NCC have been reported (http://archive.uwcm.ac.uk/uwcm/mg/hgmd0.html). In addition, a phenotype resembling Gitelman's disease was obtained in mice by targeted disruption of the NCC gene (Schultheis et al., 1998) and functional analysis of human or rat NCC containing some of the reported point mutations revealed that mutant NCC proteins are non-functional (Kunchaparty et al., 1999; De Jong et al., 2002; Sabath et al., 2004).

Although Gitelman's disease is an autosomal recessive disorder, surprisingly, many Gitelman patients have been diagnosed with mutations in only one allele. These patients represent up to 30% of kindreds (Cruz et al., 2001; Maki et al., 2004; Lemmink et al., 1998; Lin et al., 2004; Syren et al., 2002; Simon et al., 1996). Because inheritance of Gitelman's disease is clearly recessive and heterozygous relatives of patients with Gitelman's disease are clinically and metabolically asymptomatic, it is likely that undetected mutations in the other allele also existed. Up to five potential mechanisms by which mutations reduce or abolish transporter activity have been described: (1) impairment of protein synthesis, (2) impairment protein processing, (3) impairment of insertion of a functional protein into the plasma membrane, (4) impairment of functional properties of the cotransporter and (5) acceleration of protein removal from the membrane or degradation. Using heterologous expression system in *X. laevis* oocytes, initial reports of missense mutations (a single nucleotide change resulting in a codon encoding for a different amino acid) reported along NCC protein revealed that the cotransporter is synthesized, but not properly glycosylated and thus not expressed at plasma membrane. Similarly, a study showed that elimination of NCC glycosylation remarkably reduces the activity of the cotransporter due to a deficient insertion into the plasma membrane (Hoover et al., 2003). Later studies of functional properties and surface expression of more mutants of NCC (De Jong et al., 2002; Sabath et al., 2004) revealed that some Gitelman's missense mutations resulting in partially functional proteins are explained by the third mechanism mentioned above, that is a cotransporter with apparently normal processing and normal functional and kinetic properties, but in which insertion into plasma membrane is partially impaired.

B. Bartter's Disease

Bartter's disease is a monogenic, but nevertheless heterogeneous, disease in which at least five different genes in TALH are implicated. Three of these genes are directly involved in salt reabsorption and two genes regulate salt transporters. Thus, Bartter's disease is classified into five types: the first four (I to IV) are due to inactivating mutations of NKCC2, the potassium channel ROMK, the basolateral Cl^- channel CLC-KB and the CLC-KB chaperon protein Barttin, whereas type V is the result of activating mutations in the calcium sensing receptor located in the basolateral membrane (Hebert, 2003). The characteristic clinical picture includes a salt wasting state with low blood pressure, metabolic alkalosis with hypokalemia, hypercalciuria, hypereninemia and secondary aldosteronism (Shaer, 2001). The disease is usually apparent at birth or even earlier, due to excessive accumulation of amniotic fluid (polyhydramnios). A striking difference with Gitelman's disease is that patients exhibit hypercalciuria and some develop nephrocalcinosis.

As seen in Table 16.3, there are similarities and differences in the clinical features of patients with Bartter's disease based on the gene causing the disease; therefore, there is a clinical/molecular correlation. All patients present hypokalemia and metabolic alkalosis, which are less severe in type III patients. Antenatal presentation and polyhydramnios are more often seen in Bartter's type I, II and IV, than in type III disease. In addition, patients with type III disease do not develop nephrocalcinosis. Type IV is associated with deafness and with development of chronic renal failure (Meade et al., 2003; Hebert, 2003). Therefore,

TABLE 16.3 Clinical presentation of patients with Bartter's disease according to genomic classification in type I to IV

	Type I	Type II	Type III	Type IV
Affected gene	SLC12A1	KCNJ1/ROMK	ClCNKB	BSND
Age of onset	Neonatal	Neonatal	First decade	Neonatal
Polyhydramnios	Yes	Yes	No	Yes
Sensorineural deafness	rare	rare	No	Yes
Poliuria	Yes	Yes	Uncommon	Yes
End stage renal disease	No	No	No	Yes
Hypokalemia	Severe	Severe	Moderate	Severe
Metabolic alkalosis	Moderate	Moderate	Mild	Moderate
Nephrocalcinosis	Yes	Yes	No	Yes

age of clinical presentation and certain features such as nephrocalcinosis, deafness and chronic renal failure help the clinician in the identification of the proper molecular target.

C. Gordon's Disease

A disease known as Gordon's disease, pseudohypoaldosteronism type II (PHAII), familial hypertension with hyperkalemia is the mirror image of Gitelman's disease. It is an inherited disease with a dominant pattern of inheritance. PHAII patients feature arterial hypertension with hyperkalemia, despite a normal glomerular filtration rate, and decreased urinary H^+ excretion with the consequent hyperchloremic metabolic acidosis. Hypertension appears in all patients until adult life if untreated. Thus, metabolic abnormalities develop first and hypertension develops later in life. In addition, while Gitelman patients exhibit increased bone mineral density (Cruz, 2001; Nicolet-Barousse et al., 2005), PHAII patients develop a significant decrease (Mayan et al., 2002). Interestingly, all clinical features in PHAII patients are quite sensitive to treatment with low-dose, thiazide-type diuretics (Gordon and Hodsman, 1986; Mayan et al., 2002). Thus, it was hypothesized that PHAII could be due to gain-of-function mutations in the *SLC12A3* gene encoding for NCC. However, no significant linkage was found between PHAII and the *SLC12A3* locus on chromosome 16 (Simon et al., 1995). Instead, three different loci for PHAII have been observed in the human genome located on chromosomes 1q31-42 (Mansfield et al., 1997), 12p13 (Disse-Nicodeme et al., 2000) and 17p11-q21 (Mansfield et al., 1997). In addition, the existence of PHAII kindreds not linked to these chromosome regions indicates the existence of another locus (Disse-Nicodeme et al., 2001). The gene responsible in families linked to chromosome 1 remains a mystery, but in chromosomes 12 and 17, it was found that Gordon's disease is associated with mutation of two kinases known as WNK1 and WNK4, respectively (Wilson et al., 2001). In the first case, nucleotide deletions in the first intron of the WNK1 gene correlated with increased expression of a normal WNK1 protein, while in the second case, missense mutations in a conserved acidic region of WNK4 are responsible for the disease.

Most of the pathophysiology of the Gordon's disease seems to be explained by the effects of WNKs upon the activity of salt transport proteins in the distal nephron, particularly NCC. WNK4 is a negative regulator of NCC activity. This effect of WNK4 upon NCC is prevented by introducing the PHAII-type mutations (Wilson et al., 2003; Yang et al., 2003; Cai et al., 2006; Golbang et al., 2006). Thus, under normal conditions, WNK4 inhibits NCC activity and in patients with PHAII mutations this inhibition is not present leading to a chronic increase in the activity of NCC that produces arterial hypertension and hyperkalemia. Experiments in transgenic animals support this hypothesis. A clear PHAII syndrome has been reproduced in BAC transgenic or knock-in mouse harboring WNK4 with PHAII-type mutations (Lalioti et al., 2006; Yang et al., 2007). In these studies, PHAII clinical features were rescued by treatment with a thiazide-type diuretic or by crossing the mouse with the NCC null mice. A more detailed discussion is provided in Chapter 18.

VII. POTENTIAL ROLE IN POLYGENIC DISEASES

The known physiological roles of cation-coupled chloride cotransporters and their involvement in monogenic diseases strongly suggest that these cotransporters could be implicated in development of complex polygenic diseases, such as arterial hypertension and osteoporosis.

High blood pressure is one of the most common diseases in adults, with a prevalence of ~20%. Arterial hypertension is diagnosed by chronic sustained increase in diastolic blood pressure >90 mmHg, usually accompanied with systolic blood pressure >140 mmHg. Arterial hypertension increases the risk of developing acute myocardial infarction, stroke, chronic renal failure and congestive cardiac failure. Hypertension is a disease with an important genetic component. NCC and NKCC2 are strong candidates for defining blood pressure levels and involvement in the genesis of arterial hypertension. This is due to their role in renal salt handling, in the treatment of hypertension and the fact that inherited diseases caused by inactivating mutations in the genes encoding these cotransporters feature low blood pressure. Supporting this hypothesis, in a recent study based on genetic analysis of the Framingham population study, it was observed that people exhibiting mutations in one allele of the *SLC12A1* (NKCC2) or *SLC12A3* (NCC) genes tend to have lower blood pressure and a lower risk of hypertension (Ji et al., 2008). Additionally, a recent study demonstrated strong genome-wide association in several populations of SPAK or STK39, a kinase that regulates NCC function (Chapter 18) and systolic and diastolic blood pressure (Wang et al., 2009).

Osteoporosis is a polygenic disease of the skeleton that is secondary to a decrease in bone mineral density, with a disruption of the normal bone architecture and the consequent increase in the risk of fractures. Due to functional interaction of NCC with calcium-transport mechanisms, this cotransporter is involved in renal calcium absorption. Inactivating mutations of NCC in Gitelman's disease are associated with high bone mineral density (Cruz, 2001; Nicolet-Barousse et al., 2005), whereas activation of NCC in PHAII subjects is accompanied by decreased bone mineral density (Mayan et al., 2002). In addition, there is strong epidemiological evidence that hypertensive patients treated with thiazides are at a lower risk for osteoporosis (Ray et al., 1989; Jones et al., 1995; Sebastian, 2000; Reid et al., 2000; Schoofs et al., 2003). In this regard, it has been shown recently that the NCC protein is expressed in osteoblasts in rat and human bones and that addition of thiazides to osteoblast cells in culture increases the formation of mineralized nodules by a mechanism that may involve NCC (Dvorak et al., 2007). Therefore, osteoporosis is another polygenic disease in which this cotransporter could be implicated.

Acknowledgements

I dedicate this chapter with eternal gratitude to the memory of my mentor Steven C. Hebert MD for his invaluable help and collaboration of over 20 years. Experimental work performed in the author's laboratory has been supported by the Mexican Council of Science and Technology (CONACYT), NIH Grants DK36803 and DK064635, the Wellcome Trust, The Foundation Leducq for the Transatlantic Network on Hypertension – Renal Salt Handling in the Control of Blood Pressure.

References

Abuladze, N., Yanagawa, N., Lee, I., Jo, O.D., Newman, D., Hwang, J., Uyemura, K., Pushkin, A., Modlin, R.L., and Kurtz, I. (1998). Peripheral blood mononuclear cells express mutated NCCT mRNA in Gitelman's syndrome: evidence for abnormal thiazide-sensitive NaCl cotransport. *J. Am. Soc. Nephrol.* **9**, 819–826.

Alvarez-Leefmans, F.J., Gamino, S.M., Giraldez, F., and Nogueron, I. (1988). Intracellular chloride regulation in amphibian dorsal root ganglion neurones studied with ion-selective microelectrodes. *J. Physiol.* **406**, 225–246.

Alvarez-Leefmans, F.J. (2001). Intracellular chloride regulation. In *Cell Physiology Source Book. A Molecular Approach* (Sperelakis, N., ed.), pp. 301–318. Academic Press, San Diego, CA.

Alvo, M., Calamia, J., and Eveloff, J. (1985). Lack of potassium effect on Na-Cl cotransport in the medullary thick ascending limb. *Am. J. Physiol. (Renal Fluid Electrolyte Physiology)* **249**, F34–F39.

Amorim, J.B., Bailey, M.A., Musa-Aziz, R., Giebisch, G., and Malnic, G. (2003). Role of luminal anion and pH in distal tubule potassium secretion. *Am. J. Physiol. Renal Physiol.* **284**, F381–F388.

Bachmann, S., Bostanjoglo, M., Schmitt, R., and Ellison, D.H. (1999). Sodium transport-related proteins in the mammalian distal nephron – distribution, ontogeny and functional aspects. *Anat. Embryol. (Berl.)* **200**, 447–468.

Barry, E.L.R., Gesek, F.A., Kaplan, M.R., Hebert, S.C., and Friedman, P.A. (1997). Expression of the sodium-chloride cotransporter in osteoblast-like cells: effects of thiazide diuretics. *Am. J. Physiol. (Cell Physiology)* **272**, C109–C116.

Bazzini, C., Vezzoli, V., Sironi, C., Dossena, S., Ravasio, A., Debiasi, S., Garavaglia, M., Rodighiero, S., Meyer, G., Fascio, U., Furst, J., Ritter, M., Botta, G., and Paulmichl, M. (2005). Thiazide-sensitive NaCl cotransporter in the intestine: possible role of HCTZ in the intestinal Ca2+ uptake. *J. Biol. Chem.* **280**, 19902–19910.

Bernstein, P.L., Zawalach, W., Bartiss, A., Reilly, R., Palcso, M., and Ellison, D.H. (1995). The thiazide-sensitive Na-Cl cotransporter is expressed in rat endocrine pancreas. *J. Am. Soc. Nephrol.* **6**, 732.

Burg, M.B. (1982). Thick ascending limb of Henle's loop. *Kidney International* **22**, 454–464.

Cai, H., Cebotaru, V., Wang, Y.H., Zhang, X.M., Cebotaru, L., Guggino, S.E., and Guggino, W.B. (2006). WNK4 kinase regulates surface expression of the human sodium chloride cotransporter in mammalian cells. *Kidney International* **69**, 2162–2170.

Campean, V., Kricke, J., Ellison, D., Luft, F.C., and Bachmann, S. (2001). Localization of thiazide-sensitive Na(+)-Cl(−) cotransport and associated gene products in mouse DCT. *Am. J. Physiol. Renal Physiol.* **281**, F1028–F1035.

Carmosino, M., Giménez, I., Caplan, M., and Forbush, B. (2008). Exon loss accounts for differential sorting of Na-K-Cl cotransporters in polarized epithelial cells. *Mol. Biol. Cell.* **19**, 4341–4351.

Caron, L., Rousseau, F., Gagnon, E., and Isenring, P. (2000). Cloning and functional characterization of a cation C1− cotransporter interacting protein. *J. Biol. Chem.* **275**, 32027–32036.

Chang, H. and Fujita, T. (1999). A kinetic model of the thaizide-sensitive Na-Cl cotransporter. *Am. J. Physiol. (Renal Physiol.)* **276**, F952–F959.

Chen, H., Luo, J., Kintner, D.B., Shull, G.E., and Sun, D. (2005). Na(+)-dependent chloride transporter (NKCC1)-null mice exhibit less gray and white matter damage after focal cerebral ischemia. *J. Cereb. Blood Flow Metab.* **25**, 54–66.

Clader, J.A., Schacheter, M., and Sever, P.S. (1992). Direct vascular actions of hydrochlorothaizide and indapamide in isolated small vessels. *Eur. J. Pharmacol.* **220**, 19–26.

Colmenero-Flores, J.M., Martinez, G., Gamba, G., Vazquez, N., Iglesias, D.J., Brumos, J., and Talon, M. (2007). Identification and functional characterization of cation-chloride cotransporters in plants. *Plant J.* **50**, 278–292.

Costanzo, L.S. (1985). Localization of diuretic action in microperfused rat distal tubules: Ca and Na transport. *Am. J. Physiol. (Renal Fluid Electrolyte Physiology)* **248**, F527–F535.

Cremaschi, D., Porta, C., Botta, G., Bazzini, C., Baroni, M.D., and Garavaglia, M. (2000). Apical Na(+)-Cl(−) symport in rabbit gallbladder epithelium: a thiazide-sensitive cotransporter (TSC). *J. Membr. Biol.* **176**, 53–65.

Cruz, D.N. (2001). The renal tubular Na-Cl co-transporter (NCCT): a potential genetic link between blood pressure and bone density? *Nephrol. Dial. Transplant.* **16**, 691–694.

Cruz, D.N., Shaer, A.J., Bia, M.J., Lifton, R.P., and Simon, D.B. (2001). Gitelman's syndrome revisited: an evaluation of symptoms and health- related quality of life. *Kidney International* **59**, 710–717.

Cutler, C.P. and Cramb, G. (2002). Two isoforms of the $Na^+/K^+/2Cl^-$ cotransporter are expressed in the European eel (Anguilla anguilla). *Biochim. Biophys. Acta* **1566**, 92–103.

Darman, R.B. and Forbush, B. (2002). A regulatory locus of phosphorylation in the N terminus of the Na-K-Cl cotransporter, NKCC1. *J. Biol. Chem.* **277**, 37542–37550.

De Jong, J.C., Van Der Vliet, W.A., van den Heuvel, L.P., Willems, P.H., Knoers, N.V., and Bindels, R.J. (2002). Functional expression of mutations in the human NaCl cotransporter: evidence for impaired routing mechanisms in Gitelman's syndrome. *J. Am. Soc. Nephrol.* **13**, 1442–1448.

De Jong, J.C., Willems, P.H., Mooren, F.J., van den Heuvel, L.P., Knoers, N.V., and Bindels, R.J. (2003a). The structural unit of the thiazide-sensitive NaCl cotransporter is a homodimer. *J. Biol. Chem.* **278**, 24302–24307.

De Jong, J.C., Willems, P.H., van den Heuvel, L.P., Knoers, N.V., and Bindels, R.J. (2003b). Functional expression of the human thiazide-sensitive NaCl cotransporter in Madin-Darby canine kidney cells. *J. Am. Soc. Nephrol.* **14**, 2428–2435.

Delpire, E., Rauchman, M.I., Beier, D.R., Hebert, S.C., and Gullans, S.R. (1994). Molecular cloning and chromosome localization of a putative basolateral Na^+-K^+-$2Cl^-$ cotransporter from mouse inner medullary collecting duct (mIMCD-3) cells. *J. Biol. Chem.* **269**, 25677–25683.

Delpire, E., Lu, J., England, R., Dull, C., and Thorne, T. (1999). Deafness and imbalance associated with inactivation of the secretory Na-K-2Cl co-transporter. *Nat. Genet.* **22**, 192–195.

Delpire, E. and Mount, D.B. (2002). Human and murine phenotypes associated with defects in cation-chloride cotransport. *Annu. Rev. Physiol* **64**, 803–843.

Deng, L. and Chen, G. (2003). Cyclothiazide potently inhibits gamma-aminobutyric acid type A receptors in addition to enhancing glutamate responses. *Proc. Natl. Acad. Sci. USA* **100**, 13025–13029.

Disse-Nicodeme, S., Achard, J.M., Desitter, I., Houot, A.M., Fournier, A., Corvol, P., and Jeunemaitre, X. (2000). A new locus on chromosome 12p13.3 for pseudohypoaldosteronism type II, an autosomal dominant form of hypertension. *Am. J. Hum. Genet.* **67**, 302–310.

Disse-Nicodeme, S., Desitter, I., Fiquet-Kempf, B., Houot, A.M., Stern, N., Delahousse, M., Potier, J., Ader, J.L., and Jeunemaitre, X. (2001). Genetic heterogeneity of familial hyperkalaemic hypertension. *J. Hypertens.* **19**, 1957–1964.

Doucet, A. (1988). Function and control of Na-K-ATPase in single nephron segments of the mammalian kidney. *Kidney International* **34**, 749–760.

Dowd, B.F. and Forbush, B. (2003). PASK (proline-alanine-rich STE20-related kinase), a regulatory kinase of the Na-K-Cl cotransporter (NKCC1). *J. Biol. Chem.* **278**, 27347–27353.

Drewnowska, K. and Baumgarten, C.M. (1991). Regulation of cellular volume in rabbit ventricular myocytes: bumetanide, chlorthiazide, and ouabain. *Am. J. Physiol. (Cell Physiology)* **260**, C122–C131.

Dunham, P., Jessen, F., and Hoffman, E.K. (1990). Inhibition of Na-K-Cl cotransport in Ehrlich ascites cells by antiserum against purified proteins of the cotransporter. *Proc. Nat. Acad. Sci. USA* **87**, 6823–6828.

Dunham, P.B. and Ellory, J.C. (1981). Passive potassium transport in low potassium sheep red cells: dependence upon cell volume and chloride. *J. Physiol* **318**, 511–530.

Dvorak, M.M., De Joussineau, C., Carter, D.H., Pisitkun, T., Knepper, M.A., Gamba, G., Kemp, P.J., and Riccardi, D. (2007). Thiazide diuretics directly induce osteoblast differentiation and mineralized nodule formation by interacting with a sodium chloride co-transporter in bone. *J. Am. Soc. Nephrol.* **18**, 2509–2516.

Ellison, D.H., Morrisey, J., and Desir, G.V. (1990). Rabbit thiazide diuretic receptors: Solubilization, characterization and purification. *J. Am. Soc. Nephrol.* **1**, 683.

Ellison, D.H., Velazquez, H., and Wright, F.S. (1987). Thiazide-sensitive sodium chloride cotransport in early distal tubule. *Am. J. Physiol. (Renal Fluid Electrolyte Physiology)* **253**, F546–F554.

Evans, R.L., Park, K., Turner, R.J., Watson, G.E., Nguyen, H.V., Dennett, M.R., Hand, A.R., Flagella, M., Shull, G.E., and Melvin, J.E. (2000). Severe impairment of salivation in $Na^+/K^+/2Cl^-$ cotransporter (NKCC1)-deficient mice. *J. Biol. Chem.* **275**, 26720–26726.

Eveloff, J. and Calamia, J. (1986). Effect of osmolarity on cation fluxes in medullary thick ascending limb cells. *Am. J. Physiol. (Renal Fluid Electrolyte Physiology)* **250**, F176–F180.

Fanestil, D.D., Vaughan, D.A., and Blakely, P. (1997). Metabolic acid–base influences on renal thiazide receptor density. *Am. J. Physiol. (Regulatory Integrative Comparative Physiology)* **272**, R2004–R2008.

Feit, P.W., Hoffman, E.K., Schiodt, M., Kristensen, P., Jessen, F., and Dunham, P.B. (1988). Purification of proteins of the Na/Cl cotransporter from membranes of Ehrilch ascites cells using a bumetanide-sepharose affinity column. *J. Membr. Biol.* **103**, 135–147.

Flagella, M., Clarke, L.L., Miller, M.L., Erway, L.C., Giannella, R.A., Andringa, A., Gawenis, L.R., Kramer, J., Duffy, J.J., Doetschman, T., Lorenz, J.N., Yamoah, E.N., Cardell, E.L., and Shull, G.E. (1999). Mice lacking the basolateral Na-K-2Cl cotransporter have impaired epithelial chloride secretion and are profoundly deaf. *J. Biol. Chem.* **274**, 26946–26955.

Flemmer, A.W., Gimenez, I., Dowd, B.F., Darman, R.B., and Forbush, B. (2002). Activation of the Na-K-Cl cotransporter NKCC1 detected with a phospho-specific antibody. *J. Biol. Chem.* **277**, 37551–37558.

Forbush, B., III and Palfrey, H.C. (1983). [^{3}H]Bumetanide binding to membranes isolated from dog kidney outer medulla. Relationship to the Na, K,Cl co-transport system. *J. Biol. Chem.* **258**, 11787–11792.

Gagnon, E., Bergeron, M.J., Brunet, G.M., Daigle, N.D., Simard, C.F., and Isenring, P. (2003). Molecular mechanisms of Cl transport by the renal Na-K-Cl cotransporter: identification of an intracellular locus that may form part of a high affinity Cl-binding site. *J. Biol. Chem.* **279**, 5648–5654.

Gagnon, E., Bergeron, M.J., Daigle, N.D., Lefoll, M.H., and Isenring, P. (2005). Molecular Mechanisms of cation transport by the renal Na^+-K^+-Cl^- cotransporter: Structural insight into the operating characteristics of the ion transport sites. *J. Biol. Chem.* **280**, 32555–32563.

Gagnon, E., Forbush, B., Flemmer, A.W., Gimenez, I., Caron, L., and Isenring, P. (2002). Functional and molecular characterization of the shark renal Na-K-Cl cotransporter: novel aspects. *Am. J. Physiol. Renal Physiol.* **283**, F1046–F1055.

Gamba, G., Miyanoshita, A., Lombardi, M., Lytton, J., Lee, W.S., Hediger, M.A., and Hebert, S.C. (1994). Molecular cloning, primary structure and characterization of two members of the mammalian electroneutral sodium-(potassium)-chloride cotransporter family expressed in kidney. *J. Biol. Chem.* **269**, 17713–17722.

Gamba, G., Saltzberg, S.N., Lombardi, M., Miyanoshita, A., Lytton, J., Hediger, M.A., Brenner, B.M., and Hebert, S.C. (1993). Primary structure and functional expression of a cDNA encoding the thiazide-sensitive, electroneutral sodium-chloride cotransporter. *Proc. Nat. Acad. Sci. USA* **90**, 2749–2753.

Gamba, G. (2001). Alternative splicing and diversity of renal transporters. *Am. J. Physiol. Renal Physiol.* **281**, F781–F794.

Gamba, G. (2005). Molecular physiology and pathophysiology of the electroneutral cation-chloride cotransporters. *Physiol. Rev.* **85**, 423–493.

Garg, P., Martin, C.F., Elms, S.C., Gordon, F.J., Wall, S.M., Garland, C.J., Sutliff, R.L., and O'Neill, W.C. (2007). Effect of the Na-K-2Cl cotransporter NKCC1 on systemic blood pressure and smooth muscle tone. *Am. J. Physiol. Heart Circ. Physiol.* **292**, H2100–H2105.

Geck, P., Pietrzyk, C., Burckhardt, B.C., Pfeiffer, B., and Heinz, E. (1980). Electrically silent cotransport of Na^+, K^+ and Cl^- in Ehrlich cells. *Biochim. Biophys. Acta,* **600**, 432–447.

Gerelsaikhan, T. and Turner, R.J. (2000). Transmembrane topology of the secretory Na^+-K^+-$2Cl^-$ cotransporter (NKCC1) studied by in vitro translation. *J. Biol. Chem.* **275**, 40471–40477.

Gimenez, I., Isenring, P., and Forbush, B., III (2002). Spatially distributed alternative splice variants of the renal Na-K-Cl cotransporter exhibit dramatically different affinities for the transported ions. *J. Biol. Chem.* **277**, 8767–8770.

Gimenez, I. and Forbush, B. (2003). Short-term stimulation of the renal Na-K-Cl cotransporter (NKCC2) by vasopressin involves phosphorylation and membrane translocation of the protein. *J. Biol. Chem.* **278**, 26946–26951.

Gimenez, I. and Forbush, B. (2007). The residues determining differences in ion affinities among the alternative splice variants F, A, and B of the mammalian renal Na–K-Cl cotransporter (NKCC2). *J. Biol. Chem.* **282**, 6540–6547.

Golbang, A.P., Cope, G., Hamad, A., Murthy, M., Liu, C.H., Cuthbert, A.W., and O'Shaughnessy, K.M. (2006). Regulation of the expression of the Na/Cl cotransporter (NCCT) by WNK4 and WNK1: evidence that accelerated dynamin-dependent endocytosis is not involved. *Am. J. Physiol. Renal Physiol.* **291**, F1369–F1376.

Gordon, R.D. and Hodsman, G.P. (1986). The syndrome of hypertension and hyperkalaemia without renal failure: long term correction by thiazide diuretic. *Scott. Med. J.* **31**, 43–44.

Granados-Soto, V., Arguelles, C.F., and Alvarez-Leefmans, F.J. (2005). Peripheral and central antinociceptive action of Na^+-K^+-$2Cl^-$ cotransporter blockers on formalin-induced nociception in rats. *Pain* **114**, 231–238.

Greger, R. (1981). Chloride reabsorption in the rabbit cortical thick ascending limb of the loop of Henle. A sodium dependent process. *Pflugers Archiv.* **390**, 38–43.

Greger, R. and Schlatter, E. (1981). Presence of luminal K^+, a prerequisite for active NaCl transport in the cortical thick ascending limb of Henle's loop of rabbit kidney. *Pflugers Archiv.* **392**, 92–94.

Greger, R. and Schlatter, E. (1983). Properties of the lumen membrane of the cortical thick ascending limb of Henle's loop of rabbit kidney. *Pflugers Archiv.* **396**, 315–324.

Greger, R., Schlatter, E., and Lang, F. (1983). Evidence for electroneutral sodium chloride cotransport in the cortical thick ascending limb of Henle's loop of rabbit kidney. *Pflugers Archiv.* **396**, 308–314.

Greger, R. (1985). Ion transport mechanisms in thick ascending limb of Henle's loop of mammalian nephron. *Physiol Rev.* **65**, 760–797.

Hebert, S.C., Culpepper, R.M., and Andreoli, T.E. (1981a). NaCl transport in mouse medullary thick ascending limbs. I. Functional nephron heterogeneity and ADH-stimulated NaCl cotransport. *Am. J. Physiol. (Renal Fluid Electrolyte Physiology)* **241**, F412–F431.

Hebert, S.C., Culpepper, R.M., and Andreoli, T.E. (1981b). NaCl transport in mouse medullary thick ascending limbs. II. ADH enhancement of transcellular NaCl cotrasport; origin of transepithelial voltage. *Am. J. Physiol. (Renal Fluid Electrolyte Physiology)* **241**, F432–F442.

Hebert, S.C., Culpepper, R.M., and Andreoli, T.E. (1981c). NaCl transport in mouse medullary thick ascending limbs. III. Modulation of ADH efect by peritubular osmolality. *Am. J. Physiol. (Renal Fluid Electrolyte Physiology)* **241**, F443–F451.

Hebert, S.C. (2003). Bartter syndrome. *Curr. Opin. Nephrol. Hypertens.* **12**, 527–532.

Hoffmann, E.K., Lambert, I.H., and Pedersen, S.F. (2009). Physiology of cell volume regulation in vertebrates. *Physiol. Rev.* **89**, 193–277.

Hoover, R.S., Poch, E., Monroy, A., Vazquez, N., Nishio, T., Gamba, G., and Hebert, S.C. (2003). N-Glycosylation at two sites critically alters thiazide binding and activity of the rat thiazide-sensitive Na^+:Cl^- cotransporter. *J. Am. Soc. Nephrol.* **14**, 271–282.

Igarashi, P., Vanden Heuver, G.B., Payne, J.A, and Forbush, B., III (1995). Cloning, embryonic expression, and alternative splicing of a murine kidney-specific Na-K-Cl cotransporter. *Am. J. Physiol. (Renal Fluid Electrolyte Physiology)* **269**, F406–F418.

Igarashi, P., Whyte, D.A., Kui, L., and Nagami, G.T. (1996). Cloning and kidney cell-specific activity of the promoter of the murine renal Na-K-Cl cotransporter gene. *J. Biol. Chem.* **271**, 9666–9674.

Isenring, P. and Forbush, B., III (1997). Ion and bumetanide binding by the Na-K-Cl cotransporter. Importance of transmembrane domains. *J. Biol. Chem.* **272**, 24556–24562.

Isenring, P., Jacoby, S.C., Chang, J., and Forbush, B., III (1998a). Mutagenic mapping of the Na-K-Cl cotransporter for domains involved in ion transport and bumetanide binding. *J. Gen. Physiol.* **112**, 549–558.

Isenring, P., Jacoby, S.C., and Forbush, B., III (1998b). The role of transmembrane domain 2 in cation transport by the Na-K-Cl cotransporter. *Proc. Nat. Acad. Sci. USA* **95**, 7179–7184.

Isenring, P. and Forbush, B. (2001). Ion transport and ligand binding by the Na-K-Cl cotransporter, structure–function studies. *Comp. Biochem. Physiol. A Mol. Integr. Physiol.* **130**, 487–497.

Ji, W., Foo, J.N., O'Roak, B.J., Zhao, H., Larson, M.G., Simon, D.B., Newton-Cheh, C., State, M.W., Levy, D., and Lifton, R.P. (2008). Rare independent mutations in renal salt handling genes contribute to blood pressure variation. *Nat. Genet.* **40**, 592–599.

Jones, G., Nguyen, T., Sambrook, P.N., and Eisman, J.A. (1995). Thiazide diuretics and fractures: can meta-analysis help? *J. Bone Miner. Res.* **10**, 106–111.

Jorgensen, P.L., Petersen, J., and Rees, W.D. (1984). Identification of a Na^+, K^+, Cl^- cotransport protein of Mr 34000 from kidney by photolabeling with [3H]bumetanide. *Biochim. Biophys. Acta,* **775**, 105–110.

Kaplan, M.R., Plotkin, M.D., Lee, W.-S., Xu, Z.-C., Lytton, J., and Hebert, S.C. (1996). Apical localization of the Na-K-Cl cotransporter, rBSC1, on rat thick ascending limbs. *Kidney International* **49**, 40–47.

Kim, G.-H., Ecelbarger, C.A., Mitchell, C., Packer, R.K., Wade, J.B., and Knepper, M.A. (1999). Vasopressin increases Na-K-2Cl cotransporter expression in thick ascending limb of Henle's loop. *Am. J. Physiol. (Renal Physiol.)* **276**, F96–F103.

Kim, G.-H., Masilamani, S., Turner, R., Mitchell, C., Wade, J.B., and Knepper, M.A. (1998). The thiazide-sensitive Na-Cl cotransporter is an aldosterone-induced protein. *Proc. Nat. Acad. Sci. USA* **95**, 14552–14557.

Kim, G.-H., Martin, S.W., Fernandez-Llama, P., Masilamani, S., Packer, R.K., and Knepper, M.A. (2000). Long-term regulation of renal Na-dependent cotransporters and ENaC: response to altered acid-base intake. *Am. J. Physiol. Renal Physiol.* **279**, F459–F467.

Kunau, R.T., Weller, D.R., and Webb, H.L. (1975). Clarification of the site of action of chlorothiazide in the rat nephron. *J. Clin. Invest* **56**, 401–407.

Kunchaparty, S., Palcso, M., Berkman, J., Zquez, H., Desir, G.V., Bernstein, P., Reilly, R.F., and Ellison, D.H. (1999). Defective processing and expression of thiazide-sensitive Na-Cl cotransporter as a cause of Gitelman's syndrome. *Am. J. Physiol.* **277**, F643–F649.

Kyte, J. and Doolittle, R.F. (1982). A simple method for displaying the hydropathic character of a protein. *J. Mol. Biol.* **157**, 105–132.

Laird, J.M., Garcia-Nicas, E., Delpire, E.J., and Cervero, F. (2004). Presynaptic inhibition and spinal pain processing in mice: a possible role of the NKCC1 cation-chloride co-transporter in hyperalgesia. *Neurosci. Lett.* **361**, 200–203.

Lalioti, M.D., Zhang, J., Volkman, H.M., Kahle, K.T., Hoffmann, K. E., Toka, H.R., Nelson-Williams, C., Ellison, D.H., Flavell, R., Booth, C.J., Lu, Y., Geller, D.S., and Lifton, R.P. (2006). Wnk4 controls blood pressure and potassium homeostasis via regulation of mass and activity of the distal convoluted tubule. *Nat. Genet.* **38**, 1124–1132.

Lang, F., Busch, G.L., Ritter, M., Volkl, H., Waldegger, S., Gulbins, E., and Haussinger, D. (1998). Functional significance of cell volume regulatory mechanisms. *Physiol. Rev.* **78**, 247–306.

Lauf, P.K. and Theg, B.E. (1980). A chloride dependent K+ flux induced by N-ethylmaleimide in genetically low K+ sheep and goat erythrocytes. *Biochem. Biophys. Res. Commun.* **92**, 1422–1428.

Lazzaro, D., De Simone, V., De Magistris, L., Lehtonen, E., and Cortese, R. (1992). LFB1 and LFB3 homeoproteins are sequentially expressed during kidney development. *Development* **114**, 469–479.

Lemmink, H.H., Knoers, N.V., Karolyi, L., van Dijk, H., Niaudet, P., Antignac, C., Guay-Woodford, L.M., Goodyer, P.R., Carel, J. C., Hermes, A., Seyberth, H.W., Monnens, L.A., and van den Heuvel, L.P. (1998). Novel mutations in the thiazide-sensitive NaCl cotransporter gene in patients with Gitelman syndrome with predominant localization to the C-terminal domain. *Kidney International* **54**, 720–730.

Lemmink, H.H., van den Heuvel, L.P., van Dijk, H.A., Merkx, G. F., Smilde, T.J., Taschner, P.E., Monnens, L.A., Hebert, S.C., and Knoers, N.V. (1996). Linkage of Gitelman syndrome to the thiazide-sensitive sodium-chloride cotransporter gene with identification of mutations in Dutch families. *Pediatr. Nephrol.* **10**, 403–407.

Lin, S.H., Cheng, N.L., Hsu, Y.J., and Halperin, M.L. (2004). Intrafamilial phenotype variability in patients with Gitelman syndrome having the same mutations in their thiazide-sensitive sodium/chloride cotransporter. *Am. J. Kidney Dis.* **43**, 304–312.

Loffing, J. and Kaissling, B. (2003). Sodium and calcium transport pathways along the mammalian distal nephron: from rabbit to human. *Am. J. Physiol. Renal Physiol.* **284**, F628–F643.

Loffing, J., Loffing-Cueni, D., Valderrabano, V., Klausli, L., Hebert, S.C., Rossier, B.C., Hoenderop, J.G., Bindels, R.J., and Kaissling, B. (2001). Distribution of transcellular calcium and sodium transport pathways along mouse distal nephron. *Am. J. Physiol. Renal Physiol.* **281**, F1021–F1027.

Lorin-Nebel, C., Boulo, V., Bodinier, C., and Charmantier, G. (2006). The $Na^+/K^+/2Cl^-$ cotransporter in the sea bass Dicentrarchus labrax during ontogeny: involvement in osmoregulation. *J. Exper. Biol.* **209**, 4908–4922.

Luo, H., Beaumont, K., Vaughn, D.A., and Fanestil, D.D. (1990). Solubilization of thiazide diuretic receptors from rat kidney membranes. *Biochim. Biophys. Acta* **1052**, 119–122.

Lytle, C. and Forbush, B., III (1992). The Na-K-Cl cotransport protein of shark rectal gland. II Regulation by direct phosphorylation. *J. Biol. Chem.* **267**, 25438–25443.

Lytle, C. and McManus, T. (2002). Coordinate modulation of Na-K-2Cl cotransport and K-Cl cotransport by cell volume and chloride. *Am. J. Physiol. Cell Physiol.* **283**, C1422–C1431.

MacKenzie, S., Vaitkevicius, H., and Lockette, W. (2001). Sequencing and characterization of the human thiazide-sensitive Na-Cl cotransporter (SLC12A3) gene promoter. *Biochem. Biophys. Res. Commun.* **282**, 991–1000.

Majid, D.S.A. and Navar, G.L. (1994). Blockade of distal nephron sodium transport attenuates pressure natriuresis in dogs. *Hypertension* **23**, 1040–1045.

Maki, N., Komatsuda, A., Wakui, H., Ohtani, H., Kigawa, A., Aiba, N., Hamai, K., Motegi, M., Yamaguchi, A., Imai, H., and Sawada, K.I. (2004). Four novel mutations in the thiazide-sensitive Na-Cl co-transporter gene in Japanese patients with Gitelman's syndrome. *Nephrol. Dial. Transplant.* [Incomplete].

Malysiak, K. and Grzywna, Z.J. (2008). On the possible methods for the mathematical description of the ball and chain model of ion channel inactivation. *Cell Mol. Biol. Lett.* **13**, 535–552.

Mansfield, T.A., Simon, D.B., Farfel, Z., Bia, M., Tucci, J.R., Lebel, M., Gutkin, M., Vialettes, B., Christofilis, M.A., Kauppinen-Makelin, R., Mayan, H., Risch, N., and Lifton, R.P. (1997). Multilocus linkage of familial hyperkalaemia and hypertension, pseudohypoaldosteronism type II, to chromosomes 1q31-42 and 17p11-q21. *Nat. Genet.* **16**, 202–205.

Marcano, M., Yang, H.M., Nieves-González, A., Clausen, C., and Moore, L.C (2009). Parameter estimation for mathematical models of NKCC2 cotransporter isoforms. *Am. J. Physiol. Renal Physiol.* **296**, F369–F381.

Masilamani, S., Kim, G.H., Mitchell, C., Wade, J.B., and Knepper, M.A. (1999). Aldosterone-mediated regulation of ENaC alpha, beta, and gamma subunit proteins in rat kidney. *J. Clin. Invest* **104**, R19–R23.

Mastroianni, N., Bettinelli, A., Bianchetti, M., Colussi, G., de Fusco, M., Sereni, F., Ballabio, A., and Casari, G. (1996a). Novel molecular variants of the Na-Cl cotransporter gene are responsible for Gitelman syndrome. *Am. J. Hum. Genet.* **59**, 1019–1026.

Mastroianni, N., DeFusco, M., Zollo, M., Arrigo, G., Zuffardi, O., Bettinelli, A., Ballabio, A., and Casari, G. (1996b). Molecular cloning, expression pattern, and chromosomal localization of the human Na-Cl thiazide-sensitive cotransporter (SLC12A3). *Genomics* **35**, 486–493.

Mayan, H., Vered, I., Mouallem, M., Tzadok-Witkon, M., Pauzner, R., and Farfel, Z. (2002). Pseudohypoaldosteronism type II: marked sensitivity to thiazides, hypercalciuria, normomagnesemia, and low bone mineral density. *J. Clin. Endocrinol. Metab* **87**, 3248–3254.

Meade, P., Hoover, R.S., Plata, C., Vazquez, N., Bobadilla, N.A., Gamba, G., and Hebert, S.C. (2003). cAMP-dependent activation of the renal-specific Na^+-K^+-$2Cl^-$ cotransporter is mediated by regulation of cotransporter trafficking. *Am. J. Physiol. Renal Physiol.* **284**, F1145–F1154.

Meade, P., Sabath, E., and Gamba, G. (2003). Fisiopatología molecular del sìndrome de Bartter. *Rev. Invest. Clin.* **55**, 74–83.

Merino, A., Hebert, S.C., and Gamba, G. (1999). Correlation between water salinity and tissue expression of the thiazide-sensitive cotransporter (TSC) in teleost. *J. Am. Soc. Nephrol.* **10**, 39A.

Meyer, J.W., Flagella, M., Sutliff, R.L., Lorenz, J.N., Nieman, M.L., Weber, C.S., Paul, R.J., and Shull, G.E. (2002). Decreased blood pressure and vascular smooth muscle tone in mice lacking basolateral Na^+-K^+-$2Cl^-$ cotransporter. *Am. J. Physiol. Heart Circ. Physiol.* **283**, H1846–H1855.

Molony, D.A., Reeves, W.B., Hebert, S.C., and Andreoli, T.E. (1987). ADH increases apical Na^+, K^+, $2Cl^-$ entry in mouse medullary thick ascending limbs of Henle. *Am. J. Physiol. (Renal Fluid Electrolyte Physiology)*, **252**, F177–F187.

Monroy, A., Plata, C., Hebert, S.C., and Gamba, G. (2000). Characterization of the thiazide-sensitive Na(+)-Cl(−) cotransporter: a new model for ions and diuretics interaction. *Am. J. Physiol. Renal Physiol.* **279**, F161–F169.

Moore-Hoon, M.L. and Turner, R.J. (1998). Molecular and topological characterization of the rat parotid Na^+-K^+-$2Cl^-$ cotransporter1. *Biochim. Biophys. Acta*, **1373**, 261–269.

Moore-Hoon, M.L. and Turner, R.J. (2000). The structural unit of the secretory Na^+-K^+-$2Cl^-$ cotransporter (NKCC1) is a homodimer. *Biochemistry* **39**, 3718–3724.

Moreno, E., San Cristobal, P., Rivera, M., Vazquez, N., Bobadilla, N.A., and Gamba, G. (2006). Affinity defining domains in the Na-Cl cotransporter: different location for Cl^- and thiazide binding. *J. Biol. Chem.* **281**, 17266–17275.

Moreno, E., Tovar-Palacio, C., De Los, H.P., Guzman, B., Bobadilla, N.A., Vazquez, N., Riccardi, D., Poch, E., and Gamba, G. (2004). A single nucleotide polymorphism alters the activity of the renal Na^+:Cl^- cotransporter and reveals a role for transmembrane segment 4 in chloride and thiazide affinity. *J. Biol. Chem.* **279**, 16553–16560.

Moriguchi, T., Urushiyama, S., Hisamoto, N., Iemura, S., Uchida, S., Natsume, T., Matsumoto, K., and Shibuya, H. (2005). WNK1 regulates phosphorylation of cation-chloride-coupled cotransporters via the STE20-related kinases, SPAK and OSR1. *J. Biol. Chem.* **280**, 42685–42693.

Mount, D.B., Baekgard, A., Hall, A.E., Plata, C., Xu, J., Beier, D.R., Gamba, G., and Hebert, S.C. (1999). Isoforms of the Na-K-2Cl transporter in murine TAL I. Molecular characterization and intrarenal localization. *Am. J. Physiol. (Renal Physiol.)* **276**, F347–F358.

Nguyen, M., Pace, A.J., and Koller, B.H. (2007). Mice lacking NKCC1 are protected from development of bacteremia and hypothermic sepsis secondary to bacterial pneumonia. *J. Exper. Med.* **204**, 1383–1393.

Nicolet-Barousse, L., Blanchard, A., Roux, C., Pietri, L., Bloch-Faure, M., Kolta, S., Chappard, C., Geoffroy, V., Morieux, C., Jeunemaitre, X., Shull, G.E., Meneton, P., Paillard, M., Houillier, P., and De Vernejoul, M.C. (2005). Inactivation of the Na-Cl co-transporter (NCC) gene is associated with high BMD through both renal and bone mechanisms: analysis of patients with Gitelman syndrome and Ncc null mice. *J. Bone Miner. Res.* **20**, 799–808.

Nielsen, J., Kwon, T.H., Masilamani, S., Beutler, K., Hager, H., Nielsen, S., and Knepper, M.A. (2002). Sodium transporter abundance profiling in the kidney: effect of spironolactone. *Am. J. Physiol. Renal Physiol.* **283**, F923–F933.

Nielsen, S., Maunsbach, A.B., Ecelbarger, C.A., and Knepper, M.A. (1998). Ultrastructural localization of Na-K-2Cl cotransporter in thick ascending limb and macula densa of rat kidney. *Am. J. Physiol.* **275**, F885–F893.

O'Grady, S.M., Palfrey, H.C., and Field, M. (1987). Characteristics and function of Na-K-Cl cotransport in epithelial tissues. *Am. J. Physiol. (Cell Physiology)* **253**, C177–C192.

Obermuller, N., Bernstein, P., Velázquez, H., Reilly, R., Moser, D., Ellison, D.H., and Bachman, S. (1995). Expression of the thiazide-sensitive Na-Cl cotransporter in rat and human kidney. *Am. J. Physiol. (Renal Fluid Electrolyte Physiology)* **269**, F900–F910.

Oppermann, M., Mizel, D., Huang, G., Li, C., Deng, C., Theilig, F., Bachmann, S., Briggs, J., Schnermann, J., and Castrop, H. (2006). Macula densa control of renin secretion and preglomerular resistance in mice with selective deletion of the B isoform of the Na,K,2Cl co-transporter. *J. Am. Soc. Nephrol.* **17**, 2143–2152.

Oppermann, M., Mizel, D., Kim, S.M., Chen, L., Faulhaber-Walter, R., Huang, Y., Li, C., Deng, C., Briggs, J., Schnermann, J., and Castrop, H. (2007). Renal function in mice with targeted disruption of the A isoform of the Na-K-2Cl co-transporter. *J. Am. Soc. Nephrol.* **18**, 440–448.

Ortiz, P.A. (2006). cAMP increases surface expression of NKCC2 in rat thick ascending limbs: role of VAMP. *Am. J. Physiol. Renal Physiol.* **290**, F608–F616.

Pace, A.J., Lee, E., Athirakul, K., Coffman, T.M., O'Brien, D.A., and Koller, B.H. (2000). Failure of spermatogenesis in mouse lines deficient in the Na^+-K^+-$2Cl^-$ cotransporter. *J. Clin. Invest* **105**, 441–450.

Pacheco-Alvarez, D., San Cristobal, P., Meade, P., Moreno, E., Vazquez, N., Munoz, E., Diaz, A., Juarez, M.E., Gimenez, I., and Gamba, G. (2006). The Na-Cl cotransporter is activated and phosphorylated at the amino terminal domain upon intracellular chloride depletion. *J. Biol. Chem.* **281**, 28755–28763.

Paredes, A., Plata, C., Rivera, M., Moreno, E., Vazquez, N., Munoz-Clares, R., Hebert, S.C., and Gamba, G. (2006). Activity of the renal Na^+:K^+:$2Cl^-$ cotransporter is reduced by mutagenesis of N-glycosylation sites: role for protein surface charge in Cl^- transport. *Am. J. Physiol. Renal Physiol.* **290**, F1094–F1102.

Pathak, B.G., Shaughnessy, J.D., Jr., Meneton, P., Greeb, J., Shull, G.E., Jenkins, N.A., and Copeland, N.G. (1996). Mouse chromosomal location of three epithelial sodium channel subunit genes and an apical sodium chloride cotransporter gene. *Genomics* **33**, 124–127.

Payne, J.A. and Forbush, B., III (1994). Alternatively spliced isoforms of the putative renal Na-K-Cl cotransporter are differentially distributed within the rabbit kidney. *Proc. Nat. Acad. Sci. USA* **91**, 4544–4548.

Payne, J.A., Xu, J.-C., Haas, M., Lytle, C.Y., Ward, D., and Forbush, B., III (1995). Primary structure, functional expression, and chromosomal localization of the bumetanide-sensitive Na-K-Cl cotransporter in human colon. *J. Biol. Chem.* **270**, 17977–17985.

Plata, C., Meade, P., Hall, A.E., Welch, R.C., Vazquez, N., Hebert, S.C., and Gamba, G. (2001). Alternatively spliced isoform of the apical Na-K-Cl cotransporter gene encodes a furosemide sensitive Na-Cl cotransporter. *Am. J. Physiol. Renal Physiol.* **280**, F574–F582.

Plata, C., Mount, D.B., Rubio, V., Hebert, S.C., and Gamba, G. (1999). Isoforms of the Na-K-2Cl cotransporter in murine TAL. II. Functional characterization and activation by cAMP. *Am. J. Physiol. (Renal Physiol.)* **276**, F359–F366.

Plata, C., Meade, P., Vazquez, N., Hebert, S.C., and Gamba, G. (2002). Functional properties of the apical Na+-K+-2Cl− cotransporter isoforms. *J. Biol. Chem.* **277**, 11004–11012.

Plotkin, M.D., Kaplan, M.R., Peterson, L.N., Gullans, S.R., Hebert, S.C., and Delpire, E. (1997). Expression of the Na^+-K^+-$2Cl^-$ cotransporter BSC2 in the nervous system. *Am. J. Physiol. (Cell Physiology),* **272**, C173–C183.

Plotkin, M.D., Kaplan, M.R., Verlander, J.M., Lee, W.-S., Brown, D., Poch, E., Gullans, S.R., and Hebert, S.C. (1996). Localization of the thiazide sensitive Na-Cl cotransporter, rTSC1, in the rat kidney. *Kidney International* **50**, 174–183.

Pollak, M.R., Delaney, V.B., Graham, R.M., and Hebert, S.C. (1996). Gitelman's syndrome (Bartter's variant) maps to the thiazide-sensitive cotransporter gene locus on chromosome 16q13 in a large kindred. *J. Am. Soc. Nephrol.* **7**, 2244–2248.

Ponce-Coria, J., San Cristobal, P., Kahle, K.T., Vazquez, N., Pacheco-Alvarez, D., De Los, H.P., Juarez, P., Munoz, E., Michel, G., Bobadilla, N.A., Gimenez, I., Lifton, R.P., Hebert, S.C., and Gamba, G. (2008). Regulation of NKCC2 by a chloride-sensing mechanism involving the WNK3 and SPAK kinases. *Proc. Natl. Acad. Sci. USA* **105**, 8458–8463.

Quaggin, S.E., Payne, J.A., Forbush, B., III, and Igarashi, P. (1995). Localization of the renal Na-K-Cl cotransporter gene (Slc12a1) on mouse chromosome 2. *Mamm.Genome* **6**, 557–558.

Randall, J., Thorne, T., and Delpire, E. (1997). Partial cloning and characterization of Slc12a2: the gene encoding the secretory Na^+-K^+-$2Cl^-$ cotransporter. *Am. J. Physiol. (Cell Physiology),* **273**, C1267–C1277.

Ray, W.A., Griffin, M.R., Downey, W., and Melton, L.J., III (1989). Long-term use of thiazide diuretics and risk of hip fracture. *The Lancet* **I**, 687–690.

Reeves, W.B., Molony, D.A., and Andreoli, T.E. (1988). Diluting power of thick limbs of Henle. III. Modulation of in vitro diluting power. *Am. J. Physiol.* **255**, F1145–F1154.

Reid, I.R., Ames, R.W., Orr-Walker, B.J., Clearwater, J.M., Horne, A. M., Evans, M.C., Murray, M.A., McNeil, A.R., and Gamble, G.D. (2000). Hydrochlorothiazide reduces loss of cortical bone in normal postmenopausal women: a randomized controlled trial. *Am. J. Med.* **109**, 362–370.

Reilly, R.F. and Ellison, D.H. (2000). Mammalian distal tubule: physiology, pathophysiology, and molecular anatomy. *Physiol. Rev.* **80**, 277–313.

Renfro, J.L. (1975). Water and ion transport by the urinary bladder of the teleost Pseudopleuronectus americanus. *Am. J. Physiol.* **228**, 52–61.

Renfro, J.L. (1977). Interdependence of active Na+ and Cl− transport by the isolated urinary bladder of the teleost, Pseudopleuronectes americanus. *J. Exper. Zool.* **199**, 383–390.

Richardson, C., Rafiqi, F.H., Karlsson, H.K., Moleleki, N., Vandewalle, A., Campbell, D.G., Morrice, N.A., and Alessi, D. R. (2008). Activation of the thiazide-sensitive Na^+-Cl^- cotransporter by the WNK-regulated kinases SPAK and OSR1. *J. Cell Sci.* **121**, 675–684.

Rocha, A.S. and Kokko, J.P. (1973). Sodium chloride and water transport in the medullary thick ascending limb of Henle. Evidence for active chloride transport. *J. Clin. Invest* **52**, 612–623.

Rocha-González, H.I., Mao, S., and Alvarez-Leefmans, F.J. (2008). Na^+,K^+,2Cl^- cotransport and intracellular chloride regulation in rat primary sensory neurons: thermodynamic and kinetic aspects. *J. Neurophysiol.* **100**, 169–184.

Rose, B.D. (1991). Diuretics. *Kidney International* **39**, 336–352.

Russell, J.M. (2000). Sodium-potassium-chloride cotransport. *Physiol Rev.* **80**, 211–276.

Sabath, E., Meade, P., Berkman, J., De Los, H.P., Moreno, E., Bobadilla, N.A., Vazquez, N., Ellison, D.H., and Gamba, G. (2004). Pathophysiology of functional mutations of the thiazide-sensitive Na-Cl cotransporter in Gitelman disease. *Am. J. Physiol. Renal Physiol.* **287**, F195–F203.

Sakaguchi, N., Crouch, J.J., Lytle, C., and Schulte, B.A. (1998). Na-K-Cl cotransporter expression in the developing and senescent gerbil cochlea. *Hear. Res.* **118**, 114–122.

Schnermann, J. (1998). Juxtaglomerular cell complex in the regulation of renal salt excretion. *Am. J. Physiol. (Regulatory Integrative Comparative Physiology)* **274**, R263–R279.

Schoofs, M.W., van der, K.M., Hofman, A., de Laet, C.E., Herings, R.M., Stijnen, T., Pols, H.A., and Stricker, B.H. (2003). Thiazide diuretics and the risk for hip fracture. *Ann. Intern. Med.* **139**, 476–482.

Schultheis, P.J., Lorenz, J.N., Meneton, P., Nieman, M.L., Riddle, T.M., Flagella, M., Duffy, J.J., Doetschman, T., Miller, M.L., and Shull, G.E. (1998). Phenotype resembling Gitelman's syndrome in mice lacking the apical Na^+-Cl^- cotransporter of the distal convoluted tubule. *J. Biol. Chem.* **273**, 29150–29155.

Sebastian, A. (2000). Thiazides and bone. *Am. J. Med.* **109**, 429–430.

Shaer, A.J. (2001). Inherited primary renal tubular hypokalemic alkalosis: a review of Gitelman and Bartter syndromes. *Am. J. Med. Sci.* **322**, 316–332.

Simard, C.F., Bergeron, M.J., Frenette-Cotton, R., Carpentier, G.A., Pelchat, M.E., Caron, L., and Isenring, P. (2007). Homooligomeric and heterooligomeric associations between K^+-Cl^- cotransporter isoforms and between K^+-Cl^- and Na^+-K^+-Cl^- cotransporters. *J. Biol. Chem.* **282**, 18083–18093.

Simon, D.B., Karet, F.E., Hamdan, J.M., Di Pietro, A., Sanjad, S.A., and Lifton, R.P. (1996a). Bartter's syndrome, hypokalaemic alkalosis with hypercalciuria, is caused by mutations in the Na-K-2Cl cotransporter NKCC2. *Nature Genetics* **13**, 183–188.

Simon, D.B., Nelson-Williams, C., Johnson-Bia, M., Ellison, D., Karet, F.E., Morey-Molina, A., Vaara, I., Iwata, F., Cushner, H.M., Koolen, M., Gainza, F.J., Gitelman, H.J., and Lifton, R.P. (1996b). Gitelman's variant of Bartter's syndrome, inherited hypokalaemic alkalosis, is caused by mutations in the thiazide-sensitive Na-Cl cotransporter. *Nature Genetics* **12**, 24–30.

Simon, D.B., Farfel, Z., Ellison, D., Bia, M., Tucci, J., and Lifton, R.P. (1995). Examination of the thiazide-sensitive Na-Cl cotransporter as a candidate gene in Gordon's syndrome. *J. Am. Soc. Nephrol.* **6**, 632.

Smith, H.W. (1953). *From Fish to Philosopher*. Little Brown and Company, Boston.

Snouwaert, J.N., Brigman, K.K., Latour, A.M., Malouf, N.N., Boucher, R.C., Smithies, O., and Koller, B.H. (1992). An animal model for cystic fibrosis made by gene targeting. *Science* **257**, 1083–1088.

Starremans, P.G., Kersten, F.F., Knoers, N.V., van den Heuvel, L.P., and Bindels, R.J. (2003a). Mutations in the human Na-K-2Cl cotransporter (NKCC2) identified in Bartter syndrome type I consistently result in nonfunctional transporters. *J. Am. Soc. Nephrol.* **14**, 1419–1426.

Starremans, P.G., Kersten, F.F., van den Heuvel, L.P., Knoers, N.V., and Bindels, R.J. (2003b). Dimeric architecture of the human bumetanide-sensitive Na-K-Cl Co-transporter. *J. Am. Soc. Nephrol.* **14**, 3039–3046.

Stokes, J.B., Lee, I., and D'Amico, M. (1984). Sodium chloride absorption by the urinary bladder of the winter flounder. A thiazide-sensitive, electrically neutral transport system. *J. Clin. Invest* **74**, 7–16.

Sun, A., Grossman, E.B., Lombardi, M., and Hebert, S.C. (1991). Vasopressin alters the mechanism of apical Cl^- entry from Na^+:Cl^- to Na^+:K^+:2Cl^- cotransport in mouse medullary thick ascending limb. *J. Membr. Biol.* **120**, 83–94.

Sung, K.W., Kirby, M., McDonald, M.P., Lovinger, D.M., and Delpire, E. (2000). Abnormal $GABA_A$ receptor-mediated currents in dorsal root ganglion neurons isolated from Na-K-2Cl cotransporter null mice. *J. Neurosci.* **20**, 7531–7538.

Syren, M.L., Tedeschi, S., Cesareo, L., Bellantuono, R., Colussi, G., Procaccio, M., Ali, A., Domenici, R., Malberti, F., Sprocati, M., Sacco, M., Miglietti, N., Edefonti, A., Sereni, F., Casari, G., Coviello, D.A., and Bettinelli, A. (2002). Identification of fifteen novel mutations in the SLC12A3 gene encoding the Na-Cl co-transporter in Italian patients with Gitelman syndrome. *Hum. Mutat.* **20**, 78.

Taniyama, Y., Sato, K., Sugawara, A., Uruno, A., Ikeda, Y., Kudo, M., Ito, S., and Takeuchi, K. (2001). Renal tubule-specific transcription and chromosomal localization of rat thiazide-sensitive Na-Cl cotransporter gene. *J. Biol. Chem.* **276**, 26260–26268.

Tovar-Palacio, C., Bobadilla, N.A., Cortes, P., Plata, C., De Los, H.P., Vazquez, N., and Gamba, G. (2004). Ion and diuretic specificity of chimeric proteins between apical Na^+:K^+:2Cl^- and Na^+:Cl^- cotransporters. *Am. J. Physiol. Renal Physiol.* **287**, F570–F577.

Tran, J.M., Farrell, M.A., and Fanestil, D.D. (1990). Effect of ions on binding of the thiazide-type diuretic metolazone to kidney membrane. *Am. J. Physiol. (Renal Fluid Electrolyte Physiology)* **258**, F908–F915.

Turner, R.J. and Geroge, J.N. (1990). Solubilization and partial purification of the rabbit parotid Na/K/Cl-dependent bumetanide binding site. *J. Membr. Biol.* **113**, 203–210.

Vazquez, N., Monroy, A., Dorantes, E., Munoz-Clares, R.A., and Gamba, G. (2002). Functional differences between flounder and rat thiazide-sensitive Na-Cl cotransporter. *Am. J. Physiol. Renal Physiol.* **282**, F599–F607.

Velazquez, H., Good, D.W., and Wright, F.S. (1984). Mutual dependence of sodium and chloride absorption by renal distal tubule. *Am. J. Physiol. (Renal Fluid Electrolyte Physiology)* **247**, F904–F911.

Velazquez, H., Naray-Fejes-Toth, A., Silva, T., Andujar, E., Reilly, R.F., Desir, G.V., and Ellison, D.H. (1998). Rabbit distal convoluted tubule coexpresses NaCl cotransporter and 11 beta-hydroxysteroid dehydrogenase II mRNA. *Kidney International* **54**, 464–472.

Vetter, D.E., Mann, J.R., Wangemann, P., Liu, J., McLaughlin, K.J., Lesage, F., Marcus, D.C., Lazdunski, M., Heinemann, S.F., and

Barhanin, J. (1996). Inner ear defects induced by null mutation of the isk gene. *Neuron* **17**, 1251–1264.

Vibat, C.R., Holland, M.J., Kang, J.J., Putney, L.K., and O'Donnell, M.E. (2001). Quantitation of Na^+-K^+-$2Cl^-$ cotransport splice variants in human tissues using kinetic polymerase chain reaction. *Anal. Biochem.* **298**, 218–230.

Wall, S.M., Knepper, M.A., Hassell, K.A., Fischer, M.P., Shodeinde, A., Shin, W., Pham, T.D., Meyer, J.W., Lorenz, J.N., Beierwaltes, W.H., Dietz, J.R., Shull, G.E., and Kim, Y.H. (2006). Hypotension in NKCC1 null mice: role of the kidneys. *Am. J. Physiol. Renal Physiol.* **290**, F409–F416.

Wang, J., Pravenec, M., Kren, V., and Kurtz, T.W. (1997). Linkage mapping of the Na-K-2Cl cotransporter gene (Slc12a1) to rat chromosome 3. *Mamm. Genome* **8**, 379.

Wang, X.Y., Masilamani, S., Nielsen, J., Kwon, T.H., Brooks, H. L., Nielsen, S., and Knepper, M.A. (2001). The renal thiazide-sensitive Na-Cl cotransporter as mediator of the aldosterone-escape phenomenon. *J. Clin. Invest* **108**, 215–222.

Wang, Y., O'Connell, J.R., McArdle, P.F., Wade, J.B., Dorff, S.E., Shah, S.J., Shi, X., Pan, L., Rampersaud, E., Shen, H., Kim, J.D., Subramanya, A.R., Steinle, N.I., Parsa, A., Ober, C.C., Welling, P.A., Chakravarti, A., Weder, A.B., Cooper, R.S., Mitchell, B.D., Shuldiner, A.R., and Chang, Y.P. (2009). From the cover: whole-genome association study identifies STK39 as a hypertension susceptibility gene. *Proc. Natl. Acad. Sci. USA* **106**, 226–231.

Wenz, M., Hartmann, A.M., Friauf, E., and Nothwang, H.G. (2009). CIP1 is an activator of the K^+-Cl^- cotransporter KCC2. *Biochem. Biophys. Res. Commun.* **381**:388-392.

Wilson, F.H., Disse-Nicodeme, S., Choate, K.A., Ishikawa, K., Nelson-Williams, C., Desitter, I., Gunel, M., Milford, D.V., Lipkin, G.W., Achard, J.M., Feely, M.P., Dussol, B., Berland, Y., Unwin, R.J., Mayan, H., Simon, D.B., Farfel, Z., Jeunemaitre, X., and Lifton, R.P. (2001). Human hypertension caused by mutations in WNK kinases. *Science* **293**, 1107–1112.

Wilson, F.H., Kahle, K.T., Sabath, E., Lalioti, M.D., Rapson, A.K., Hoover, R.S., Hebert, S.C., Gamba, G., and Lifton, R.P. (2003). Molecular pathogenesis of inherited hypertension with hyperkalemia: the Na-Cl cotransporter is inhibited by wild-type but not mutant WNK4. *Proc. Natl. Acad. Sci. USA* **100**, 680–684.

Xu, J.-C., Lytle, C., Zhu, T.T., Payne, J.A., Benz, E., Jr., and Forbush, B., III (1994). Molecular cloning and functional expression of the bumetanide-sensitive Na-K-Cl cotransporter. *Proc. Nat. Acad. Sci. USA* **91**, 2201–2205.

Xu, J.Z., Hall, A.E., Peterson, L.N., Bienkowski, M.J., Eessalu, T. E., and Hebert, S.C. (1997). Localization of the ROMK protein on apical membranes of rat kidney nephron segments. *Am. J. Physiol. Renal Physiol.* **273**, F739–F749.

Yang, T., Huang, Y.G., Singh, I., Schnermann, J., and Briggs, J.P. (1996). Localization of bumetanide- and thiazide-sensitive Na-K-Cl cotransporters along the rat nephron. *Am. J. Physiol. (Renal Fluid Electrolyte Physiology)* **271**, F931–F939.

Yang, C.L., Angell, J., Mitchell, R., and Ellison, D.H. (2003). WNK kinases regulate thiazide-sensitive Na-Cl cotransport. *J. Clin. Invest* **111**, 1039–1045.

Yang, S.S., Morimoto, T., Rai, T., Chiga, M., Sohara, E., Ohno, M., Uchida, K., Lin, S.H., Moriguchi, T., Shibuya, H., Kondo, Y., Sasaki, S., and Uchida, S. (2007). Molecular pathogenesis of pseudohypoaldosteronism type II: generation and analysis of a Wnk4 (D561A/+) knockin mouse model. *Cell Metab.* **5**, 331–344.

Yerby, T.R., Vibat, C.R.T., Sun, D., Payne, J.A., and O'Donnell, M.E. (1997). Molecular characterization of the Na-K-Cl cotransporter of bovine aortic endothelial cells. *Am. J. Physiol. (Cell Physiology)* **273**, C188–C197.

CHAPTER

17

The Potassium-Chloride Cotransporters: from Cloning to Structure and Function

John A. Payne

OUTLINE

I. OVERVIEW OF THE K^+-Cl^- COTRANSPORTERS: A HISTORICAL PERSPECTIVE

K^+-Cl^- cotransporters are carrier proteins that mediate the coupled, electroneutral transport of one K^+ ion and one Cl^- ion across the plasma membrane. This coupled transport is *obligatory* as it absolutely requires the simultaneous presence of both K^+ and Cl^- ions on the same side of the membrane for net transport to proceed. The K^+-Cl^- cotransporters are members of a larger gene family called the cation-chloride cotransporters (CCCs) or solute carrier family 12 (SLC12). This gene family also includes the Na^+-K^+-Cl^- cotransporter (NKCC) and the Na^+-Cl^- cotransporter (NCC). Research on the cation chloride cotransporters now spans over four decades (see Chapter 3). Much of the early work characterizing the CCCs was performed using vertebrate red blood cells. In the 1960s and 1970s, numerous red cell studies characterized movements of K^+ as well as Na^+ that were clearly the result of the processes we now know as KCC and NKCC (e.g. Hoffman and Kregenow, 1966; Funder and Wieth, 1967; Kregenow, 1971a, b, 1973; Sachs, 1971; Wiley and Cooper, 1974; Schmidt and McManus, 1977a–c). In these early studies, however, the absolute dependency of these cation fluxes on the presence of Cl^- was not clearly established. There are a number of reasons why it was difficult to link the obligatory nature of chloride's involvement in cation transport in these early red cell studies. First, there was the "cationocentric point of view" (as so aptly put by Russell, 2000) that was so pervasive at the time, which held that Cl^- was passively distributed and moved simply in response to changes in membrane potential secondary to cation transport. This view stemmed largely from the fact that many of the early studies on Cl^- permeability were performed on cells that turned out to be exceptions to the rule, i.e. vertebrate red cells and skeletal muscle cells. These two cell types, which were widely accepted experimental models for ion transport, were shown to exhibit remarkably high passive Cl^- permeability and an apparent equilibrium distribution of Cl^- across their plasma membranes (Chapter 1). Researchers simply extended findings from red cells and skeletal muscle cells to all other cell types, and consequently, Cl^- was thought to be of little importance and uninteresting! Relative to cation permeability, however, Cl^- permeability is uniquely high in vertebrate red cells and skeletal muscle cells due to a robust anion exchanger (AE1 in red cells) or high Cl^- conductance (ClC in skeletal muscle cells). We now know that most cells do not have high Cl^- permeability, actively maintain intracellular Cl^- away from its electrochemical potential equilibrium and exhibit tight regulatory control of intracellular $[Cl^-]$. In performing this important cellular function, the CCCs permit energy stored in the Cl^- electrochemical gradient to be used by other transport pathways to perform useful electrical, chemical and even osmotic work. A second and more important reason for the difficulty in linking Cl^- to cation transport in red cells was the simple presence of a robust anion exchanger. This Cl^-/HCO_3^- exchanger, which is necessary for the red cell's physiological role in CO_2 transport, mediates such large Cl^- fluxes across the red cell plasma membrane that it is difficult to demonstrate clearly the Cl^- fluxes via other much less robust transport systems, like KCC and NKCC. Hence, most early red cell researchers overlooked the significance of Cl^- involvement in the cation fluxes and simply assumed that Cl^- was important only as a counter ion to cations and important in charge conservation, i.e. they assumed Cl^- was only *functionally coupled* to the cation fluxes through the membrane potential and not obligatorily coupled.

While there were clear difficulties of studying cation chloride cotransport in vertebrate red cells, they were not insurmountable with time. The use of disulfonic stilbene derivatives to inhibit the red cell anion exchanger proved helpful. Furthermore, researchers using other cell systems which exhibit low Cl^- permeability, such as the Ehrlich ascites tumor cell and the squid giant axon, began to address the issue of Cl^- transport. A pivotal time period in the history of the cation-chloride cotransporters came in 1979–1980 when a number of publications appeared in which chloride was shown clearly to be coupled to cation transport (Russell, 1979; Dunham et al., 1980; Ellory and Dunham, 1980; Geck et al., 1980; Lauf and Theg, 1980). Using Ehrlich cells, Geck et al. (1980) provided the most lucid description of an Na^+-K^+-Cl^- cotransporter by presenting evidence for a transport process that was electroneutral, inhibited by furosemide, not inhibited by ouabain and exhibited tight coupling of Na^+, K^+ and Cl^- ion fluxes with a stoichiometry of $1Na^+$:$1K^+$:$2Cl^-$ (Chapters 2 and 5). Although Russell (1979) had clearly demonstrated the coupling of Cl^- to Na^+ transport in the squid giant axon, the role of K^+ in this process was not apparent until a subsequent publication where he elegantly characterized an NKCC in this cell type (Russell, 1983). Early descriptions of K^+-Cl^- cotransport were provided by Ellory and Dunham (1980), Lauf and Theg (1980), and Dunham et al. (1980) using either swollen or N-ethylmaleimide (NEM)-treated sheep red cells or untreated human red cells. All three of these reports described a Cl^--dependent K^+ transport process, but it was not entirely clear at the time whether this apparent KCC was a distinct process/protein from NKCC as described by Geck et al. (1980). The most important aspect of all of these early CCC reports is that they collectively set the stage for the idea of obligatory coupling of chloride to cation transport. Shortly after these papers appeared, numerous reports began to distinguish the three separate and distinct cation-chloride cotransporters (i.e. NKCC, NCC and KCC) along with their respective functional properties and their remarkable diversity of expression, as discussed in detail in Chapters 2 and 3.

During the period of functional characterization of KCC in the 1980s and 1990s, no other cell type provided a better model system than the vertebrate red cell. It is in the red cell that KCC's sensitivity to volume, to NEM and to loop diuretics as well as its kinetic and regulatory properties were clearly identified and thoroughly studied. Not surprisingly, the knowledge gained about KCC in vertebrate red cells permitted rapid functional characterization of the KCCs once they were cloned and sequenced. The purpose of this chapter is to provide a review of the KCCs from their initial cloning to studies of structure and function in neurons. The regulation of the KCCs will be covered in a separate chapter, so only some important regulatory aspects of KCC operation will be touched upon.

II. MOLECULAR BIOLOGY OF THE K^+-Cl^- COTRANSPORTERS

Among the cation-chloride cotransporter gene family, the K^+-Cl^- cotransporters exhibit the greatest genetic diversity. Four distinct isoforms of the K^+-Cl^- cotransporter have been identified, KCC1 (Slc12a4), KCC2 (Slc12a5), KCC3 (Slc12a6) and KCC4 (Slc12a7). Additionally, these KCC isoforms exhibit alternative exon splicing and the use of alternative promoters and first exons, leading to an even greater diversity (discussed further below). In this section I describe the molecular characterization of the K^+-Cl^- cotransporters along with their diverse expression patterns.

A. Molecular Characterization

The K^+-Cl^- cotransporters were the last of the functionally distinct CCCs to be identified at a molecular level. Since the Na-dependent cation chloride cotransporters, NKCC and NCC, had previously been cloned and characterized (Gamba et al., 1993, 1994; Payne and Forbush, 1994; Xu et al., 1994; Delpire et al., 1994), one might assume that KCC would be quite easily identified using standard homology-based screening techniques (e.g. low stringency PCR and library screening). Unfortunately, this proved difficult and indicated that there must be low identity between KCC and the Na-dependent cation-chloride cotransporters. The breakthrough in the molecular identification of KCC came from the use of a new and rapidly developing GeneBank database, composed of human expressed sequence tags (hESTs). Two independent groups identified a number of hESTs with low but significant identity (<50%) with the human NKCC1 using BLASTN searches of the hEST database. This sequence information along with corresponding clones from the IMAGE Consortium was then coupled with standard molecular cloning techniques to identify two distinct gene products, KCC1 and KCC2 (Gillen et al., 1996; Payne et al., 1996). Gillen et al. (1996) identified KCC1 from rabbit and human, whereas Payne et al. (1996) identified KCC1 and KCC2 from rat. After these initial reports, KCC1 (Holtzman et al., 1998; Pellegrino et al., 1998) and KCC2 (Song et al., 2002) were cloned from other mammalian species (see Tables 17.1 and 17.2).

The molecular identification of KCC3 and KCC4 came a few years after the initial cloning of KCC1 and KCC2. Remarkably, three independent groups concurrently reported the cloning of human KCC3 (Hiki et al., 1999; Mount et al., 1999; Race et al., 1999). Using differential display PCR, Hiki et al. (1999) cloned human KCC3 in their attempts to identify transcripts that were up-regulated in primary cultured human umbilical vein endothelial cells following treatment with vascular endothelial cell growth factor. Race et al. (1999) reported the cloning of human KCC3, and Mount et al. (1999) reported the cloning of human KCC3 and human and mouse KCC4. In cloning KCC3 and KCC4, both of these latter groups used a strategy very similar to that employed in the original identification of KCC1 and KCC2, i.e. BLASTN searches of the hEST database. Interestingly, the KCC3 cDNA cloned by Hiki et al. (1999) predicted a protein that was shorter (KCC3b: 1099 amino acids) and significantly different at the amino terminus than that identified by Mount et al. (1999) and Race et al. (1999) (KCC3a: 1150 amino acids). Because tissue Northern blots with KCC3 probes showed the presence of several bands, these two different KCC3s were predicted at the time to be the result of alternative splicing. This prediction has since proved to be correct as KCC3a and KCC3b result from the use of alternative first exons in the KCC3 gene (Fig. 17.1C and see section II.C: KCC3).

Since the initial cloning of all four KCC isoforms, additional vertebrate and non-vertebrate orthologs of KCC1-4 have been reported (Tables 17.1 and 17.2). All of the vertebrate KCCs code for proteins over 1000 amino acids with core molecular weights of 120–128 kDa (Table 17.1). Each has a similar predicted structure with 12 transmembrane segments and intracellular amino- and carboxy-terminal ends with the latter making up nearly half the protein size (Fig. 17.2). Not surprisingly, this molecular structure fits well with that predicted for NKCC and NCC. Over the entire protein length, the KCCs are 25–30% identical to the Na-dependent cation chloride cotransporters NCC and NKCC [N(K)CCs]. The greatest identity between the KCCs and the N(K)CCs is present within the predicted transmembrane (TM) region and a short intracellular loop between putative TM2 and TM3. The one major structural difference between the KCCs and the N(K)CCs is the position of the extracellular loop containing potential N-linked glycosylation sites. In each of the KCCs, this loop is between putative TM5 and TM6, whereas with NKCC and NCC the loop is between putative TM7 and TM8. Just as with NKCC and NCC (Lytle et al., 1992; Kunchaparty et al., 1999), a number of studies have confirmed that the KCCs are heavily glycosylated with apparent molecular weights of 120–180 kDa (Gillen et al., 1996; Payne, 1997; Hiki et al., 1999; Williams et al., 1999; Pearson et al., 2001; Muñoz et al., 2007).

TABLE 17.1 Potassium-chloride cotransporter transcript and protein

Organism	Common name	ORF (bp)	Amino acids	Core MW (kDa)	cDNA Accession No.	Protein Accession No.
KCC1 (Slc12a4)						
Homo sapiens	human	3258	1085	120.7	NM_005072	NP_005063
Sus scrofa	pig	3261	1086	120.8	NM_213949	NP_999114
Mus musculus	mouse	3258	1085	120.6	NM_009195	NP_033221
Rattus norvegicus	rat	3258	1085	120.6	NM_019229	NP_062102
Canis lupus familiaris	dog	3261	1086	120.7	NM_001010952	NP_001010952
Oryctolagus cuniculus	rabbit	3258	1085	120.8	NM_001082703	NP_001076172
Ovis aries	sheep	3261	1086	120.7	AF_515770	AA_P47188
KCC2 (Slc12a5)						
Homo sapiens	human	3351	1116	123.5	NM_020708	NP_065759
Mus musculus	mouse	3348	1115	123.6	NM_020333	NP_065066
Rattus norvegicus	rat	3351	1116	123.6	NM_134363	NP_599190
KCC3 variant a (Slc12a6)						
Homo sapiens	human	3453	1150	127.6	NM_133647	NP_598408
Pan troglodytes	chimpanzee	3453	1150	127.6	XM_001173480	XP_001173480
Mus musculus	mouse	3453	1150	127.5	NM_133649	NP_598410
Rattus norvegicus	rat	3453	1150	127.6	NM_001109630	NP_001103100
KCC3 variant b (Slc12a6)						
Homo sapiens	human	3300	1099	122.1	NM_005135	NP_005126
Mus musculus	mouse	3300	1099	122.0	NM_133648	NP_598409
Rattus norvegicus	rat	3300	1099	122.0	XM_001074289	XP_001074289
KCC4 (Slc12a7)						
Homo sapiens	human	3252	1083	119.1	NM_006598	NP006589
Mus musculus	mouse	3252	1083	119.5	NM_011390	NP035520
Rattus norvegicus	rat	3252	1083	119.4	XM_001071999	XP_001071999
Oryctolagus cuniculus	rabbit	3321	1106	120.9	NM_001082123	NP_001075592
Non-vertebrate putative KCCs						
Ciona intestinalis	tunicate	3483	1160	127.6	NM_001033830	NP_001029002
Strongylocentrotus purpuratus	sea urchin	3069	1022	112.8	XM_778532	XP_783625
Drosophila melanogaster	fruit fly	3225	1074	119.3	NM_166630	NP_726377
Caenorhabditis elegans	nematode	2934	977	108.4	NM_068509	NP_500910

Predicted KCC sequences are displayed for *Canis* KCC1, *Pan* KCC1, and *Rattus* KCC4. There are multiple putative KCC sequences in the database for *Drosophila* and *Caenorhabditis*, but most represent splice variants from the same gene. The accesssion number for only the longest variant is listed.

B. Tissue, Cell and Subcellular Expression of Transcript and Protein

Both human and rat KCC1 exhibited a broad tissue expression pattern on Northern blots. A prominent 3.8 kb transcript was observed in all tissues along with a much less abundant 4.4 kb transcript in some rat tissues (Gillen et al., 1996). This ubiquitous expression pattern of KCC1 was further supported by the presence of KCC1 hESTs identified from a number of different tissue libraries. KCC1 mRNA and protein have been reported in numerous epithelial cells (Di Stefano et al., 2001; Lee et al., 2003, 2005b; Misri et al., 2006; Ubels et al., 2006) and osteoclasts (Kajiya et al., 2006) as well as neurons and glial cells (Kanaka et al., 2001; Mikawa et al., 2002; Becker et al., 2003). Extensive tissue localization with anti-KCC1 antibodies has not been performed, but it is widely considered to be the "housekeeping" isoform of the KCCs. In contrast to KCC1, Northern blots of rat KCC2 exhibited a single

TABLE 17.2 Potassium-chloride cotransporter genes

Organism	Common name	Chromosome	Position	Exons	Gene ID
KCC1 (Slc12a4)					
Homo sapiens	human	16	16q22.1	24	6560
Sus scrofa	pig	NA	NA		396992
Mus musculus	mouse	8	8 D3 | 8 53.0 cM		20498
Rattus norvegicus	rat	19	19q12		29501
Canis lupus familiaris	dog	5	NA		479679
Oryctolagus cuniculus	rabbit	NA	NA		100009441
KCC2 (Slc12a5)					
Homo sapiens	human	20	20q13.12	27	57468
Mus musculus	mouse	2*	2 G2-G3*		57138
Rattus norvegicus	rat	3	3q42		171373
KCC3 (Slc12a6)					
Homo sapiens	human	15	15q13-q15	28	9990
Pan troglodytes	chimpanzee	15	NA		750207
Mus musculus	mouse	2	2 E3		107723
Rattus norvegicus	rat	3	3q34		691209
KCC4 (Slc12a7)					
Homo sapiens	human	5	5p15	24	10723
Mus musculus	mouse	13	13 C1 | 13 43.0 cM		20499
Rattus norvegicus	rat	1	1p11		308069
Oryctolagus cuniculus	rabbit	NA	NA		100008847

*Note that Sallinen et al. (2001) have reported a disruption of the well-established homology between human chromosome 20 and mouse chromosome 2, and they have assigned the murine homolog of KCC2 (*Slc12a5*) to chromosome 5 (5G2-G3).

5.6 kb transcript that was exclusively expressed in brain tissue at exceptionally high levels (Payne et al., 1996). This restricted distribution of KCC2 was further supported by the finding that all KCC2 hESTs identified from the database were derived exclusively from brain or retinal tissue. Evidence for an exclusively neuronal distribution of KCC2 in rat brain was provided by reverse transcriptase PCR (RT-PCR) and *in situ* hybridization studies (Payne et al., 1996; Rivera et al., 1999). Using total RNA from a cell line of glial origin as well as rat primary cultured astrocytes in RT-PCR experiments, a product of the correct size was observed with specific primers for KCC1 but not with specific primers for KCC2 (Payne et al., 1996). In contrast to glial cells, RT-PCR experiments with total RNA from primary cultured neurons (e.g. amacrine cells of the retina) exhibited products of appropriate size and were observed with specific primers for both KCC1 and KCC2 (Payne et al., 1996). *In situ* hybridization studies in rat brain using a KCC2-specific antisense cRNA probe showed intense expression of mRNA throughout the brain, including cortex, hippocampus, cerebellum and brainstem (Payne et al., 1996; Rivera et al., 1999). The fact that the white matter of the corpus callosum was devoid of any hybridization signal supported the view that the distribution is exclusively neuronal. Numerous subsequent protein localization studies with KCC2-specific antibodies have fully corroborated the neuronal specificity of KCC2 (e.g. Lu et al., 1999; Williams et al., 1999; Gulyas et al., 2001; Hubner et al., 2001). Immunolocalization reveals a remarkably punctate distribution of KCC2 along dendrites (Zhu et al. 2005), indicating that the transporter is primarily targeted to the postsynaptic membrane and closely associated with synapses. Remarkably, KCC2 appears to be localized at both inhibitory and excitatory synapses (Williams et al., 1999; Gulyas et al., 2001). While its localization at inhibitory synapses correlates well with its function in intracellular [Cl^-] regulation of neurons and hence functional coupling with $GABA_A$ and glycine receptors, KCC2 localization at excitatory synapses was surprising (Gulyas et al., 2001). Recent work suggests that KCC2 localization at excitatory synapses may play a unique non-transporting, morphogenic role in neurons (Li et al., 2007).

As mentioned above, KCC3 was shown to exhibit rather extensive transcript heterogeneity in the initial cloning reports (Hiki et al., 1999; Mount et al., 1999;

A

KCC1 Amino-terminal Variants

```
               1        10        20        30        40        50        60
               |        |         |         |         |         |         |
        hKCC1  MPHFTVVPVDGPRRGDYDNLEGLSWVDYGERAELDDS--DGHGNHRESSPFLSPLEASRG
  mouse KCC1a  ---------------------------------MGDTLHPGQGNHRENSPFLCPLDASRG
```

KCC1 Carboxy-terminal Variants

```
               965      975       985       995       1005      1015      1025
               |        |         |         |         |         |         |
        hKCC1  YSDEEDESAVGADKIQMTWTRDKYMTETWDPSHAPDNFRELVHIKPDQSNVRRMHTAVKL
hKCC1 variant1 YSDEEDESAVGADKIQMTWTRDKYMTETWDPSHAPDNFRELVHIKPDQSNVRRMHTAVKL
hKCC1 variant2 YSDEEDESAVGADKIQMTWTRDKYMTETWDPSHAPDNFRELVHIKP--------------

               1026     1035      1045      1055      1065      1075      1085
               |        |         |         |         |         |         |
        hKCC1  NEVIVTRSHDARLVLLNMPGPPRNSEGDENYMEFLEVLTEGLERVLLVRGGGREVITIYS
hKCC1 variant1 NEVIVTRSHDARLVLLNMPGPPRNSEGDENCIPLWRGRQLGGG-----------------
hKCC1 variant2 ------------------------------------------------------------
```

B

KCC2 N-terminal Variants

```
               1        10        20        30        40        50        60
               |        |         |         |         |         |         |
      hKCC2a   MSRRFTVTSLPPAGPARSPDPESRRHSVADPRHLPGEDVKGDGNPKESSPFINSTDTEKG
      hKCC2b   -----------------------MLNNLTDCEDGDGGANPGDGNPKESSPFINSTDTEKG
```

C

KCC3 N-terminal Variants

```
               1        10        20        30        40        50        60
               |        |         |         |         |         |         |
      hKCC3a   MHPPETTTKMASVRFMVTPTKIDDIPGLSDTSPDLSSRSSSRVRFSSRESVPETSRSEPM
  hKCC3a-x2M   MHPPETTTKMASVRFMVTPTKIDDIPGLSDTSPDLSSRSSSRVRFSSRESVPETSRSEPM
      hKCC3b   --------------------------------------------------MPHFTVTKVE
  hKCC3b-x2M   --------------------------------------------------MPHFTVTKVE
 hKCC3a-S1/2   -----------------------------------------------------------M

               61       70        80        90        100       110       120
               |        |         |         |         |         |         |
      hKCC3a   SEMSGATTSLATVALDPP-SDRTSHPQDVIE-DLSQNSITGEHSQLLDDGHKKARNAYLN
  hKCC3a-x2M   SEMSGATTSLATVALDPP-SDRTSHPQDVIE-----------------DDGHKKARNAYLN
      hKCC3b   DPEEGAA---ASISQEPSLADIKARIQDSDEPDLSQNSITGEHSQLLDDGHKKARNAYLN
  hKCC3b-x2M   DPEEGAA---ASISQEPSLADIKARIQDSDEP---------------DDGHKKARNAYLN
 hKCC3a-S1/2   SEMSGATTSLATVALDPP-SDRTSHPQDVIE-DLSQNSITGEHSQLLDDGHKKARNAYLN
```

D

KCC4 N-terminal Variants

```
               1        10        20        30        40        50        60
               |        |         |         |         |         |         |
       hKCC4   -------------------------MPTNFTVVPVEAHADGGGDETAERTEAPGTPEGPE
 rabbit KCC4   MVYTALTWQRLHGTAAGLVPSHLPQEGEKGSTHPTPRPLGTTPRVTAHIEPPRPWAAGAE

               61       70        80        90        100       110       120
               |        |         |         |         |         |         |
       hKCC4   PERPSPGDGNPRENSPFLNNVEVEQESFFEGKNMALFEEEMDSNPMVSSLLNKLANYTNL
 rabbit KCC4   PPLPA-GDGSARESSPFIGSAAADGDSLLEGKNMALFEEEMDSNPMVSSLLNKLANYTNL
```

FIGURE 17.1 Alignment of alternative first exons and splicing variants that have been proposed for each of the KCC isoforms. **A.** KCC1 amino-terminal and carboxy-terminal variants as published by Pellegrino et al. (1998) and Crable et al. (2005). **B.** KCC2 amino-terminal variants as published by Uvarov et al. (2007). **C.** KCC3 amino-terminal variants as published by Mercado et al. (2005). **D.** KCC4 amino-terminal variant published for rabbit as published by Valezquez and Silva (2003). The human (h) sequence is shown except where noted (mouse and rabbit).

Race et al., 1999). At least three mRNA transcripts sizes were observed on Northern blots with KCC3 probes, ranging in size from ~4.5 kb to ~8 kb. We now know that much of this variation is due to alternative splicing of the KCC3 gene (see section II.C: KCC3). A ~4.5 and ~7 kb transcript were observed in kidney, whereas ~7 and ~8 kb transcripts were more broadly expressed in heart placenta, lung, liver and pancreas (Race et al., 1999). Skeletal muscle and brain expressed primarily the ~8 kb transcript of KCC3. On Northern blots, Mercado et al. (2005) showed that KCC3a-specific probes hybridized with the ~8 kb transcript were observed in brain, heart, kidney and skeletal muscle, whereas KCC3b-specific probes hybridized

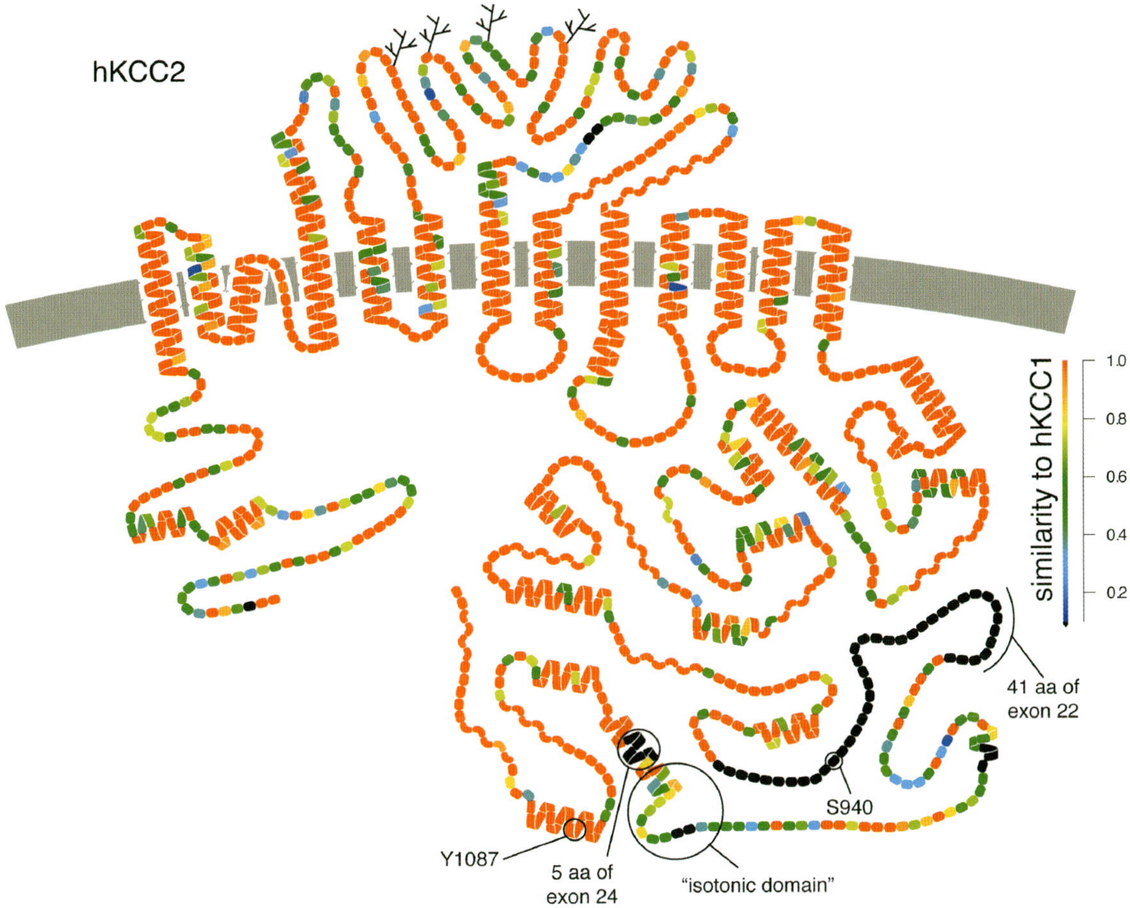

FIGURE 17.2 Hypothetical structural model of the K^+-Cl^- cotransporter. Human KCC2 (hKCC2) and human KCC1 (hKCC1) are compared to one another by colors that indicate the degree of similarity on a per residue basis: red residues are identical and black residues are absent from hKCC1 (alignment performed using CLUSTAL X ver. 1.8). The branched lines are potential N-linked glycosylation sites between putative transmembrane segments 5 and 6. Predicted secondary structural elements are shown as helices for predicted α-helix and wavy lines for predicted β-sheet. A reentrant loop between putative transmembrane segments 2 and 3 has been predicted for NKCC2 (see Gimenez and Forbush, 2007) and is shown for KCC2. See text for description of carboxy-terminal notations of hKCC2. This figure was kindly provided by Bliss Forbush using DNAPLOT.

with the ~7 kb transcript were observed only in kidney. Hence, KCC3b appears to be restricted to the kidney, and its mRNA transcript is shorter by ~1 kb than that of KCC3a. Antibodies to KCC3 have been prepared, and protein localization studies have been reported in nervous system, kidney, inner ear and gastric mucosa (Pearson et al., 2001; Boettger et al., 2003; Mercado et al., 2005; Fujii et al., 2008). KCC3 exhibited broad expression in most brain regions, being found in large neurons of the hippocampus, cortex, brainstem and cerebellum. In the nervous system, KCC3 was most highly expressed in the spinal cord primarily along myelin sheaths that surround axons (Pearson et al., 2001). KCC3 exhibited colocalization with CNPase, an oligodendrocyte marker, confirming its presence in these cells and its close association with myelin sheaths (Pearson et al., 2001). Interestingly, even though KCC3 knockout mice exhibited a peripheral neuropathy associated with locomotor defects (Howard et al., 2002), no KCC3 immunoreactivity is observed in peripheral nerves of wildtype animals (Pearson et al., 2001). The base of the choroid plexus epithelium also exhibited strong KCC3 immunoreactivity, indicating that KCC3 was expressed either at the basolateral membrane of the choroid plexus epithelium or possibly in the vasculature just underneath the epithelium (Pearson et al., 2001). KCC3 immunoreactivity within rodent kidney was most prominent in basolateral membranes of proximal tubules (Mercado et al., 2005). In the mouse inner ear, KCC3 was expressed in many of the supporting cells of the hair cells along with type 1 and type III fibrocytes underneath the stria vascularis (Boettger et al., 2003). In gastric mucosa, KCC3 localized to the basolateral membrane of luminal parietal cells (Fujii et al., 2008).

On Northern blots, probes of KCC4 detected a single 5.3 kb transcript in numerous tissues with the most prominent found in heart and kidney (Mount et al., 1999). Immunolocalization studies with KCC4 antibodies have been performed in inner ear, kidney and nervous system (Boettger et al., 2002; Velazquez and Silva, 2003; Karadsheh et al., 2004). KCC4 exhibited a developmental expression pattern in the mouse inner ear. One week after birth, KCC4 immunoreactivity was found in the stria vascularis, the developing supporting cells (organ of Corti) and outer hair cells. Two weeks after birth, KCC4 immunoreactivity became restricted to a subset of supporting cells, i.e. Deiter's cells. KCC4 immunoreactivity was found in the basolateral membranes of numerous nephron segments, including the proximal tubule, medullary thick ascending limb, distal convoluted tubule, connecting segment and alpha-intercalated cells of the collecting duct (Boettger et al., 2002; Velazquez and Silva, 2003). In the nervous system, KCC4 immunoreactivity was observed primarily in brainstem and peripheral nerves. KCC4 immunoreactivity was absent from the cortex, hippocampus and cerebellum. KCC4 staining in the brainstem was observed only in the cranial nerves and nuclei, where it colocalized with CNPase and MAP2, indicating expression in both oligodendrocytes and neurons. In peripheral nerves, KCC4 exhibited immunoreactivity that was consistent with localization in axons and Schwann cells (Karadsheh et al., 2004). Additionally, KCC4 antibodies stained the apical membrane of the choroid plexus (Karadsheh et al., 2004). Interestingly, KCC3 and KCC4 appear to be expressed together in a number of cell types, including Deiter's cells of the inner ear, renal proximal tubule cells, oligodendrocytes and choroid plexus epithelium.

C. Transcript and Protein Heterogeneity: Alternative Splicing, Alternative Exons and Alternative Transcriptional Start Sites

All of the KCC isoforms have been reported to exhibit some form of transcript variation, yet little is known about the physiological significance of the predicted protein heterogeneity. Most of the variation reported for the KCCs is due to the use of alternative first exons and would lead to alteration in potential regulatory sites. In the following section, I describe some of the reported variation for each of the KCC isoforms. In some cases, it is not known if the reported variation in transcript translates into variation at the protein level and hence functional level.

1. *KCC1*

Pellegrino et al. (1998) and Crable et al. (2005) have reported the identification of a number of alternative transcripts for KCC1 which would lead to variations in the amino-terminus and carboxy-terminus of the protein. The predicted proteins from some of these alternative transcripts are displayed in Fig. 17.1A. Most of the amino-terminal splice variants of KCC1 appear to be due to alternative first exons but this has not been extensively analyzed. Using the terminology of Crable et al. (2005), exon 1 (this is the variant identified in the initial cloning) encodes 39 amino acids, whereas exon 1a encodes only 7 amino acids. The alternative transcripts that lead to carboxy-terminal changes appear to be "readthrough" transcripts that incorporate intronic sequence, leading to either premature termination of the protein or incorporation of additional coding sequence prior to termination (Crable et al., 2005). Most of these carboxy-terminal variants appear to be minor transcripts compared to the full-length original KCC1 (Crable et al., 2005). Other than the initial cloned KCC1, none of the other reported variants of KCC1 have been functionally characterized. Hence, the significance of these variants remains uncertain. It is important to point out that relatively minor truncations of the carboxy-terminus of the KCC1 are known to lead to non-functional proteins (Casula et al., 2001).

2. *KCC2*

Uvarov et al. (2007) have recently shown that the neuronal K^+-Cl^- cotransporter, KCC2, exhibits amino terminal heterogeneity due to the use of alternative promoters and alternative first exons (Fig. 17.1B). Exon 1a encodes 40 amino acids (KCC2a) and exon 1b encodes 17 amino acids (KCC2b; this is the variant identified in the initial cloning). Both KCC2a and KCC2b are exclusively expressed in neurons of the central nervous system; however, they differ in the developmental timing and level of expression. In brains of newborn mice, transcript expression of KCC2a and KCC2b are similar. Upon postnatal development, however, KCC2b is steeply up-regulated whereas KCC2a is not (Uvarov et al., 2007). KCC2b appears to be the predominant variant in the mature rodent cortex. KCC2a exhibits significant levels of mRNA expression in the brainstem and spinal cord during prenatal and early postnatal development, indicating it may play an important developmental role in these CNS regions. This latter supposition is supported by the finding that mice lacking KCC2b but retaining normal levels of KCC2a survive up to 2 weeks after birth (Woo et al., 2002), whereas mice lacking both KCC2 variants die at birth because of severe motor deficits, including the inability to breath (Hubner et al., 2001). Flux analysis of KCC2a and KCC2b transiently expressed in HEK 293 cells demonstrated that both variants exhibit comparable K^+-Cl^- cotransport activity. Importantly, both KCC2a and KCC2b exhibit constitutive activity under isotonic conditions (often termed "isotonic" activity) which is a unique functional feature of KCC2 and likely important with regard to the control of cell $[Cl^-]$ in neurons (Payne, 1997), as discussed below in section III.

3. *KCC3*

Transcript heterogeneity of KCC3 was evident during its initial cloning. The KCC3 cDNA cloned by Hiki et al. (1999) predicted a protein of 1099 amino acids which was shorter than the 1150 amino acid protein predicted from the cDNA cloned by Race et al. (1999) and Mount et al. (1999). These two KCC3 variants termed KCC3a (1150aa) and KCC3b (1099aa) were identical except for their amino-termini, indicating that they may have resulted from alternative splicing events. The identification of two distinct KCC3 cDNAs was consistent with the presence of at least two transcripts observed on tissue Northern blots (Mount et al., 1999). The fact that KCC3a and KCC3b were both predicted to be on human chromosome 15 (15q13) also supported the hypothesis that these were splice variants of the same gene, SLC12A6 (Hiki et al., 1999; Race et al., 1999). A subsequent report clearly demonstrated that KCC3a and KCC3b are produced by the use of alternative first exons in the KCC3 gene (Mercado et al., 2005) (Fig. 17.1C). Exon 1a of the KCC3 gene encodes 90 amino acids, whereas exon 1b of the KCC3 gene encodes 39 amino acids, and this accounts for the 51 amino acid difference in the two predicted proteins. In addition to the use of alternative first exons in KCC3, exon 2 (encoding 15 amino acids) is subject to alternative splicing, resulting in two additional KCC3 transcripts, KCC3a-x2M (KCC3a minus exon 2) and KCC3b-x2M (KCC3b minus exon 2). Mercado et al. (2005) have identified two additional KCC3 transcripts that differ only by the presence of alternative upstream non-coding exons (KCC3a-S1 and KCC3a-S2). Hence, they have identical open reading frames and both code for the same KCC3 variant (KCC3a-S). Interestingly, KCC3a-S differs from KCC3a because of the use of an alternative internal splicing acceptor site in exon 1a, which removes the first initiating methionine of exon 1a. This leads to the use of a downstream initiating methionine in KCC3a-S and removal of the first 51 amino acids. As there are three in-frame methionines in exon 1a, there remains some uncertainty about which is used, but the Fig. 17.1C follows the description provided by Mercado et al. (2005). In summary, six different KCC3 transcripts are known to exist, and they code for five different KCC3 variants. All of these KCC3 variants function as K^+-Cl^- cotransporters based on flux analysis of protein expressed in *Xenopus* oocytes (Mercado et al., 2005). Whether this KCC3 amino-terminus heterogeneity leads to differences in physiological function is not clear. Since exons 1a, 1b and 2 of the KCC3 gene encode a number of potential phospho-acceptor sites it seems reasonable that such KCC3 heterogeneity would lead to differences in regulation.

4. *KCC4*

Transcript heterogeneity has not been extensively studied in KCC4. However, Valezquez and Silva have reported 5′-variation in the transcript cloned from rabbit (Velazquez and Silva, 2003). The rabbit KCC4 transcript predicts a longer amino-terminus which exhibits little identity to that of the human, rat, or mouse orthologs (Fig. 17.1D). Rabbit KCC4 is longer at the amino-terminus by 24 amino acids. Furthermore, the first 89 amino acids of the rabbit KCC4 predicted protein show no identity to any known KCC isoform. It is currently unclear if this variation is due to some type of alternative splicing event or due to a cloning artifact. No functional characterization of the rabbit KCC4 was reported (Velazquez and Silva, 2003).

III. STRUCTURE–FUNCTION STUDIES WITH THE K⁺-Cl⁻ COTRANSPORTERS

After the cloning of the KCCs, information on the basic operation of the different KCC isoforms, including their transport kinetics, ion selectivity, inhibitor sensitivity, regulation and other functional differences, emerged rapidly, mainly from studies using heterologous expression systems such as *Xenopus* oocytes and culture mammalian cells. The findings reported by a number of research groups for each of the KCCs expressed in various expression systems are nicely summarized in a recent review by Gamba (2005). All of the KCC isoforms exhibited hallmark characteristics of K^+-Cl^- cotransport that were so well documented by previous red cell studies. These include "loop" diuretic (furosemide and bumetanide) sensitivity, activation by N-ethylmaleimide (NEM), and activation by hypotonic swelling. With regards to activation by hypotonic swelling, it is important to point out that KCC expression studies using human embryonic kidney-293 (HEK-293) cells or Chinese hamster ovary (CHO) cells often exhibited little swelling activation of the expressed KCC (Payne, 1997; Hiki et al., 1999; Strange et al., 2000). This appears to be due largely to a deficit in the signaling mechanism present in the cell system rather than to an unresponsive KCC to swelling. For example, while KCC2 was not activated by hypotonic swelling in HEK-293 cells (Payne, 1997) or CHO cells (Strange et al., 2000), it was clearly activated by hypotonic swelling when expressed in *Xenopus* oocytes (Strange et al., 2000; Song et al., 2002) or MDCK cells (Payne, unpublished data). These findings bring to light the importance of understanding the limitations of studying regulation of expressed protein in heterologous expression systems.

In contrast to an abundance of structure–function studies performed with the sodium transporting members of the CCC family (NKCC and NCC), only a handful of structure–function studies have been performed with the KCC isoforms. Early structure–function studies on the KCCs examined simple truncation mutants of KCC1 expressed either in *Xenopus* oocytes or HEK-293 cells (Casula et al., 2001; Lauf et al., 2001b). In general, these studies demonstrated that while short truncations (46 amino acids) of the amino-terminus may diminish function, carboxy terminal truncations as small as 8 amino acids lead to total loss of function. Interestingly, the carboxy terminal truncations reported by Casula et al. (2001) exhibited no function despite expression of the protein and its apparent presence at the plasma membrane. These data are consistent with the findings that additions of epitope tags at the amino-terminus of the KCCs are well tolerated functionally but that such additions at the carboxy-terminus lead to non-functional protein (Gillen et al., 1996; Payne, 1997). The functional importance of the carboxy-terminus of the KCCs may be related to recent findings that the CCCs appear to form oligomeric carrier assemblies by interacting at the carboxy-terminus of the proteins (Simard et al., 2007).

Additional structure–function studies have examined some unique operational and structural features of KCC2. In alignments of the KCC isoforms, it became immediately apparent that KCC2 had a large extended region in the carboxy-terminus, exhibiting little identity with the other KCCs (Figs 17.2 and 17.3). This unique region of KCC2 is coded by two extra exons (exon 22 and 24) which are apparently absent from the other KCC genes. Exon 22 of KCC2 codes for 41 amino acids and exon 24 of KCC2 codes for 5 amino acids. Exon 23 of the KCC2 gene is conserved among the KCC genes; however, the identity of the predicted amino acids in exon 23 of KCC2 is quite low when compared to the other KCC isoforms. Overall, the predicted protein spanning these three exons of KCC2 is unique among the KCCs in containing a large number of serines and negatively charged residues, and it was originally predicted to be an important regulatory region (Payne et al., 1996). This prediction has been fully supported by recent structure–function studies.

As mentioned above, one unique functional feature of KCC2 that has been noted in numerous heterologous expression studies is the presence of a significant basal constitutive activity under isotonic conditions, often

Alignment of KCC Isoforms

```
                       --|--       *       exon 22              --|--
hKCC2    QILKQMHLTKNEREREIQSITDESRGSIRRKNPANTRLRLNVPEETAGDSEEKPEEEVQL
hKCC1    QMLRQMRLTKTERERE-----------------------------------------AQL
hKCC3    QMLRHMRLSKTERDRE-----------------------------------------AQL
hKCC4    QMLKQMQLSKNEQERE-----------------------------------------AQL


                                 exon 23
hKCC2    IHDQSAPSCPSSSPSPGEEPEGEGETDPEKVHLTWTKDKSVAEKNKGPSPVSSEGIKDFF
hKCC1    VKDRHSALRLESLYSDEEDES---AVGADKIQMTWTRDKYMTETWDPSH--APDNFRELV
hKCC3    VKDRNSMLRLTSIGSDEDEET---ETYQEKVHMTWTKDKYMASRGQKAK--SMEGFQDLL
hKCC4    IHDRNTASHTAAAARTQAPP-----T-PDKVQMTWTREKLIAEKYRSRD-TSLSGFKDLF


            exon
          --| 24 |--
hKCC2    SMKPEWENLNQSNVRRMHTAVRLN
hKCC1    HIKP-----DQSNVRRMHTAVKLN
hKCC3    NMRP-----DQSNVRRMHTAVKLN
hKCC4    SMKP-----DQSNVRRMHTAVKLN
```

FIGURE 17.3 Carboxy-terminal alignment of the KCC isoforms. Additional KCC2 exons (exon 22 and 24) are noted in boldface. The 15 amino acid region identified by Mercado et al. (2006) as required for constitutive "isotonic" activity of KCC2 is underlined. Serine-940 (S940) of KCC2 is noted with an asterisk. Lee et al. (2007) have reported that protein kinase C directly phosphorylates S940 of KCC2 and that S940 phosphorylation is required for activation of KCC2 by protein kinase C. Human (h) sequences are displayed.

referred to as "isotonic" activity (e.g. Payne, 1997; Strange et al., 2000; Song et al., 2002). This constitutive activity of KCC2 likely plays an important role in establishing the steady state intracellular [Cl^-] of neurons. One structure–function study tested the hypothesis that the unique structural region in the carboxy-terminal domain of KCC2 conferred its constitutive "isotonic" activity. Mercado et al. (2006) tested this hypothesis using chimeric constructs in which portions of the carboxy-terminal domains were switched between KCC2 and KCC4. Two generations of KCC2-4 chimeras were used to narrow down the region conferring constitutive "isotonic" activity to a 15 amino acid region within exon 23 of KCC2 (Figs 17.2 and 17.3). Interestingly, they found that the constitutive "isotonic" activity of KCC2 was unaffected by serine/threonine (S/T) phosphatase inhibitor calyculin A whereas the swelling-induced activity of KCC2 was completely abolished by pre-incubation with calyculin A. Based upon these findings, they concluded that the constitutive "isotonic" activity and the swelling-induced activity of KCC2 are mediated by two distinct mechanisms.

Red cell studies have clearly shown the importance of phosphorylation–dephosphorylation events in the regulation of the K^+-Cl^- cotransporter, and this has been confirmed in numerous KCC expression studies (Jennings and Al-Rohil, 1990). Both S/T and tyrosine phosphorylation–dephosphorylation events have been implicated in the regulation of the red cell KCC. Specifically, it has been shown that inhibition of S/T protein phosphatases (e.g. calyculin A, okadaic acid) prevent activation of K^+-Cl^- cotransport whereas inhibition of S/T protein kinases (e.g. staurosporine) activate K^+-Cl^- cotransport (Jennings and Schulz, 1991; Kaji and Tsukitani, 1991; Starke and Jennings, 1993; Bize and Dunham, 1994; Bize et al., 1999). Additionally, tyrosine kinase inhibitors (genistein) prevent activation and tyrosine phosphatase inhibitors promote activation of the K^+-Cl^- transporter (Flatman et al., 1996). While little is known about the exact mechanism, some of the phosphorylation substrate proteins involved in KCC regulation are beginning to emerge, including the S/T kinases, Ste20-related proline alanine-rich kinases (SPAK), oxidative stress response kinase (OSR1) and the "with no lysine" kinases (WNK). One early structure–function study examined the regulatory role of a well-conserved tyrosine residue present in the carboxy-terminus of all KCC isoforms (Strange et al., 2000; see Y1087 in Fig. 17.2). In KCC2 and KCC4, this particular tyrosine was identified to be a consensus tyrosine protein kinase phosphorylation site. When this tyrosine residue in KCC2 was mutated to aspartate (Y1087D) to mimic phosphorylation and then expressed in *Xenopus* oocytes, constitutive "isotonic" activity as well as swelling-induced activity of the transporter were dramatically inhibited despite normal trafficking of the mutated protein to the plasma membrane. Remarkably, replacement of the tyrosine residue in KCC2 with phenylalanine (Y1087F) did not alter the constitutive or swelling-induced activity from that observed with non-mutated controls. Similar data were obtained when they expressed equivalent mutants of KCC1 (Y1056D and Y1056F) in *Xenopus* oocytes. Furthermore, Strange et al. (2000) demonstrated that neither tyrosine phosphatase (dephostatin, pervanadate) nor kinase (genistein) inhibitors altered basal KCC2 activity expressed in CHO cells. They concluded that tyrosine phosphorylation of KCC2 protein did not play an important role in regulating KCC2 activity, but that this well-conserved tyrosine does play an important functional role in the KCCs.

While the importance of phosphorylation–dephosphorylation events in the regulation of the K^+-Cl^- cotransporter has been inferred for many years, it has not been clear if changes in its activity were due to changes in the phosphorylation state of the transporter itself. This issue was recently examined by Lee et al. (2007) in their study of the regulation of KCC2 by protein kinase C (PKC), an S/T protein kinase. First, using an *in vitro* kinase assay and fusion proteins of KCC2 containing point mutations of potential PKC phosphorylation sites (S/T for alanine), they demonstrated that purified PKC directly phosphorylated serine 940 (S940) within the carboxyl-terminal domain of KCC2 (Figs 17.2 and 17.3). Second, they showed that S940 is a major site for PKC-dependent phosphorylation of KCC2 protein expressed in HEK-293 cells and that phosphorylation of S940 increases KCC2 activity by increasing its cell surface stability. Third, they confirmed their recombinant studies by demonstrating that PKC-dependent phosphorylation of KCC2 increases cell surface expression of KCC2 in cultured hippocampal neurons. The fact that phosphorylation of S940 leads to KCC2 activation seems contrary to the dogma that dephosphorylation of K^+-Cl^- cotransport activates it. However, it must be remembered that S940 is one of many phosphorylation sites in KCC2 that are altered by PKC – some phosphorylated and some dephosphorylated. In fact, in other studies we have found that activation of PKC with phorbol esters leads to a net dephosphorylation of KCC2 expressed in MDCK cells. Interestingly, we found phosphoserine levels to increase but phospho-threonine levels to decrease in greater proportion. These findings are consistent with a net dephosphorylation of KCC2 being associated with its activation.

IV. PHYSIOLOGICAL FUNCTION OF K^+-Cl^- COTRANSPORTER IN NEURONS

In mediating the coupled transport of K^+ and Cl^-, the KCCs carry no net charge across the plasma membrane (i.e. they are electroneutral), and they do not directly hydrolyze ATP (i.e. they are secondary active). In most mammalian cells, the KCCs derive their energy for net transport from the asymmetric distribution of potassium across the plasma membrane which is generated by the Na,K-ATPase (i.e. the K^+ chemical potential difference, $\Delta\mu_K$). Using the energy in the K^+ chemical concentration gradient, the KCCs can extrude Cl^- from a cell against the Cl^- chemical concentration gradient (i.e. Cl^- chemical potential difference, $\Delta\mu_{Cl}$) – hence, secondary active transport of Cl^-. As cotransporters, however, the KCCs are, by definition, bi-directional and can carry out net ion influx as well as efflux, depending upon the magnitude and sign of energy stored in the chemical concentration gradients of the transported ions (i.e. thermodynamic driving force for KCC, $\Delta\mu_{KCC} = \Delta\mu_K + \Delta\mu_{Cl}$; for more information on thermodynamics of KCC, see Chapter 5). This later property is of immense importance to understanding the physiology of the KCCs. Because this transporter is poised near thermodynamic equilibrium, it has the potential to mediate net ion flux in either direction and thus serve as a "dynamic buffer" of the transported ions. To illustrate this point consider the well-described and much studied Cl-HCO_3 exchanger of red cells which with its robust constitutive activity can mitigate changes in intracellular Cl^- and HCO_3^-. As we will see, K^+-Cl^- cotransport in neurons operates close to thermodynamic equilibrium and KCC2, like the red cell anion exchanger, exhibits all of the transport properties required of a "dynamic buffer". In this section, I provide a review of the physiological functions of K^+-Cl^- cotransport in neurons and specifically discuss the operation of KCC2 and KCC3, the two KCC isoforms that perform significant neuronal functions. Additionally, I have attempted to highlight some unique aspects of K^+-Cl^- cotransporter operation in neurons.

A. Regulation of Intracellular $[Cl^-]$

1. *Pump–Leak Model of Cl^- Homeostasis*

The "pump–leak" model of cellular ion homeostasis can be applied to Cl^- just as it was applied to Na^+ and K^+ of sheep red cells some 50 years ago by Tosteson and Hoffman (1960). Here, steady state $[Cl^-]_i$ will be determined by the relative contributions of active (conservative) "pumps" and passive (dissipative) "leaks". Since most animal cells maintain steady state $[Cl^-]_i$ well above electrochemical equilibrium, there must be a conservative transport pathway that mediates Cl^- accumulation in the cell interior. The one major exception to this general rule of Cl^- accumulation in animal cells is observed in neurons that exhibit hyperpolarizing synaptic inhibition mediated by ligand-gated anion channels. Because $[Cl^-]_i$ in such neurons must be below electrochemical equilibrium to permit Cl^- influx to carry hyperpolarizing current, a conservative transport pathway that extrudes Cl^- from the neuron is necessary. The most important and clearly identifiable Cl^- "pumps" or conservative Cl^- transport pathways are the secondary active cation-chloride cotransporters. By maintaining intracellular $[Cl^-]$ either above or below electrochemical equilibrium, these transport proteins store energy in the Cl^- electrochemical gradient to be used by dissipative "leak" pathways to perform useful electrical and chemical work, i.e. Cl^- secretion in epithelial cells and Cl^--dependent signaling in neurons. Of the CCCs, NKCC1 and KCC2 are of particular interest with regard to neuronal Cl^- homeostasis. The NKCC1 isoform is prominently expressed in the central nervous system where it is found not only in neurons but also in vascular endothelial cells, glial cells, and epithelial cells of the choroid plexus (O'Donnell et al., 1995; Plotkin et al., 1997; Yan et al., 2001) as discussed in Chapters 25, 26, 29 and 30 in this book. NKCC1 is also highly expressed in adult primary afferent neurons in the peripheral nervous system (Alvarez-Leefmans et al., 1988; Plotkin et al, 1997), where it is involved in maintaining $[Cl^-]$ above electrochemical equilibrium, a key phenomenon in gating of somatosensory information through GABAergic mechanisms, as discussed in detail in Chapter 22. In contrast, KCC2 is expressed exclusively and at high levels in mature central neurons (Payne et al., 1996; Lu et al., 1999; Rivera et al., 1999; Williams and Payne, 2004), and it is believed to be the primary neuronal Cl^- extrusion mechanism that is responsible for keeping $[Cl^-]$i lower than equilibrium (E_{Cl} more negative than E_m), a unique feature of this cell type.

2. *Is KCC2 the Neuronal Cl^- Extrusion Mechanism?*

Years before the KCCs were identified at a molecular level, a K^+-Cl^- cotransport mechanism was suspected to function in neurons, maintaining the low intracellular $[Cl^-]$ essential for hyperpolarizing synaptic inhibition (see Alvarez-Leefmans, 1990). This mechanism was often referred to in the early literature as a 'Cl⁻-pump" (e.g. Llinas et al., 1974). While a number of reports suggested the involvement of a primary active Cl^- extrusion mechanism in neurons (Shiroya et al., 1989; Inoue et al., 1991), the participation of electroneutral, secondary active transporters (i.e., Na^+-K^+-Cl^- cotransporter) in "active Cl^- transport" had been identified in dorsal root ganglion neurons (Alvarez-Leefmans et al., 1988), and numerous non-nervous tissues. Evidence supporting the role of a K^+-dependent Cl^- extrusion mechanism of neurons was provided by studies using both vertebrate and invertebrate systems. The crayfish stretch receptor neuron (Aickin et al., 1982, 1984), insect neurosecretory cells (Dubreil et al., 1995) and vertebrate cortical and hippocampal neurons (Thompson et al., 1988a, b; Thompson and Gahwiler, 1989a, b) were reported to use the favorable outwardly directed K^+ chemical concentration gradient to drive Cl^- extrusion. These studies essentially characterized a neuronal K^+-Cl^- cotransport by examining shifts in E_{Cl} as measured by E_{GABA} or E_{IPSP} following alterations in extracellular $[K^+]$ or furosemide treatment. Recognizing the uniqueness of neuronal Cl^- homeostasis (i.e. E_{Cl} more negative than E_m) among animal cells, the simple fact that KCC2 exhibited an exclusive and robust expression pattern in neurons throughout the central nervous system provided me with enough circumstantial evidence to hypothesize that KCC2 was the neuronal "Cl^- pump" responsible for low cell $[Cl^-]$ (Payne et al., 1996). This hypothesis has been fully supported by an abundance of research from my laboratory and those of others.

After the development of specific antibodies against KCC2, immunolocalization studies confirmed the neuron-specific expression of KCC2 protein (e.g. Lu et al., 1999; Williams et al., 1999; Vardi et al., 2000; Vu et al., 2000; Gulyas et al., 2001). While KCC2 was found at the somata of neurons, it exhibited distinct punctate localization along dendrites where it often colocalized with $GABA_A$ receptors, providing more circumstantial evidence consistent with the role of KCC2 in neuronal Cl^- extrusion (Williams et al., 1999). One of the more intriguing findings from the localization studies was the fact that KCC2 was present near excitatory inputs in rat hippocampus which was certainly not expected given its proposed role in Cl^- homeostasis (Gulyas et al., 2001). At the time, it was suggested that KCC2 could be playing a role in volume regulation, but we now have an additional explanation to this perplexing finding (discussed below). In addition to information about specific localization of KCC2, antibodies also established that KCC2 protein expression in the brain correlated temporally with the well-established appearance of hyperpolarizing $GABA_A$ receptor-mediated responses in the developing brain (Lu et al., 1999; Vu et al., 2000; Stein et al., 2004). In the rat hippocampus, fast GABAergic transmission is depolarizing (and often excitatory) at birth, yet becomes hyperpolarizing and inhibitory by the second week of postnatal development (see discussion below). This implies that the neuronal Cl^- extrusion mechanism of hippocampal neurons does not become functionally expressed until the end of the first week after birth. Indeed, KCC2 mRNA and protein expression in the rat hippocampus was absent at birth yet appeared by the end of the first week of postnatal development (Rivera et al., 1999; Stein et al., 2004).

The most convincing data to support the major role of KCC2 in active neuronal Cl^- extrusion and regulation of intracellular $[Cl^-]$ have come from functional studies of neurons in which the expression level of KCC2 has been manipulated (Rivera et al., 1999; Chudotvorova et al., 2005; Zhu et al, 2005; Lee et al., 2005a; Akerman and Cline, 2006; Cancedda et al., 2007) and from the production of KCC2 knockout mice (Hubner et al., 2001; Woo et al., 2002). Using cultured rat hippocampal slices, Rivera et al. (1999) demonstrated that KCC2 protein was greatly reduced after 15 h exposure of the slices to KCC2 antisense oligonucleotides. Measurements of E_{GABA-A} in slices treated with KCC2 antisense oligonucleotides were near the resting membrane potential or depolarizing, indicating that neuronal Cl^- was at or above equilibrium (E_{Cl} equal or more positive than E_m). Control slices, however, always exhibited hyperpolarizing responses to the $GABA_A$ receptor agonist muscimol, indicating that Cl^- was below equilibrium (E_{Cl} more negative than E_m). These data clearly demonstrated that the Cl^- extrusion mechanism was significantly impaired in the KCC2 antisense treated slices, and they provided the first functional data supporting the hypothesis that KCC2 was essential for hyperpolarizing $GABA_A$-receptor-mediated responses. In contrast to the KCC2 knockdown experiments of Rivera et al. (1999), a number of more recent studies have taken a "gain of function" approach to examining the role of KCC2 as the main Cl^- extrusion pathway of neurons (Chudotvorova et al., 2005; Lee et al., 2005a; Akerman and Cline, 2006; Cancedda et al., 2007). Each of these studies demonstrated that transfection and overexpression of KCC2 in neurons at an early developmental stage can convert normally depolarizing $GABA_A$ receptor-mediated responses into hyperpolarizing ones by lowering neuronal $[Cl^-]$ prematurely. Collectively, these studies demonstrated the key role that KCC2 plays in Cl^- homeostasis as the primary Cl^- extrusion pathway of neurons. These studies also demonstrated the importance of KCC2 in development of the nervous system (discussed further below).

The importance of KCC2 in neuronal function has also come from *Kcc2*$^{-/-}$ knockout mice. Here, it is important to remember that KCC2 is subject to alternative splicing that results from the use of alternative first exons, leading to the production of two different KCC2 variants, KCC2a and KCC2b (see section II.C above). These two KCC2 variants were not discovered until after the development of the *Kcc2*$^{-/-}$ knockout mice. Two *Kcc2*$^{-/-}$ mouse strains have been developed each by a different research group. Hubner et al. (2001) targeted exon 4 of *Slc12a5*, which we now know inhibited expression of both KCC2a and KCC2b. These true *Kcc2*$^{-/-}$ mice died immediately after birth because of excitatory actions of GABA and glycine which prevent motor function, including the control of respiration. In contrast, Woo et al. (2002) targeted the disruption of exon 1b of *Slc12a5*, thus their mice lacked KCC2b expression but retained KCC2a expression. While it was not known at the time, the expression of KCC2a in these particular knockout mice could account for the small amount of KCC2 protein (5% of normal) that was expressed after the first week of postnatal development. Remarkably, some of the mice produced by Woo et al. (2002) survived for up to 2–3 weeks, but they exhibited significant difficulties with their posture and gait as well as recurrent seizures. The latter being the likely cause of their demise. Since cultured cortical neurons prepared from the KCC2b knockout mice exhibit impaired Cl^- regulation and depolarizing GABAergic responses (Zhu et al., 2005), it appears that KCC2b is the variant responsible for neuronal Cl^- regulation at least in the cortex. Additional studies will be required to identify the exact roles of KCC2a and KCC2b in neurons of other brain regions.

3. *Role of KCC2 in Development of the Nervous System*

GABAergic transmission exhibits many unique features that permit it to serve important roles during the life of a neuron – from controlling the maturation of neurons and neuronal networks during development to operating as the main inhibitory system in mature central neurons. This versatility of GABAergic transmission lies partly in the plasticity of the system, i.e. the ability of the neuron to change subtly the magnitude and even the polarity of the driving force of $GABA_A$ receptor-mediated responses. For example, in the immature brain GABA acting on $GABA_A$ receptors is depolarizing and some times excitatory, yet in the mature brain $GABA_A$ receptor-mediated responses are inhibitory and often hyperpolarizing. This developmental polarity change in GABAergic transmission, often termed the developmental "shift", is due primarily to alterations in the temporal and spatial expression of the main neuronal Cl^- transporters, NKCC1 and KCC2 (Payne et al., 2003). Immature neurons predominantly express NKCC1 which maintains high, above equilibrium intracellular $[Cl^-]$, and hence these cells exhibit depolarizing $GABA_A$ receptor-mediated responses. With the appearance of KCC2 expression later in neuronal development, intracellular $[Cl^-]$ is lowered (see Chapter 7), and $GABA_A$ receptor-mediated responses become hyperpolarizing. This developmental "shift" has been observed in a wide variety of different neurons from various different regions of the central nervous system, and hence, it is believed to be a fundamental characteristic in the development of central neurons. It should be emphasized, however, that this developmental "shift" of $GABA_A$ receptor-mediated responses is tightly linked to the maturation of a particular neuron, and therefore the exact timing of its appearance exhibits significant regional differences (Rivera et al., 1999; Li et al., 2002; Vinay and Jean-Xavier, 2008), animal species differences (altricial vs. precocious young; see Rivera et al., 1999), and even sex differences (Kyrozis et al., 2006; Perrot-Sinal et al., 2007). Significantly, the temporal expression of KCC2 correlates well with each of these observed alterations in timing of the developmental "shift", providing support for the key role that KCC2 plays in this phenomenon.

GABA and $GABA_A$ receptors are present and functional well before synapse formation, indicating that they are involved in non-synaptic processes early in development. Since depolarizing GABAergic signaling occurs during a period when neurons exhibit much of their morphological development and synaptogenesis, these depolarizations likely provide an important developmental signal. Indeed, $GABA_A$ receptor-induced depolarizations can generate action potentials, open voltage-gated Ca^{+2} channels, or activate N-methyl-D-aspartate (NMDA) receptors by removing their Mg^{+2} block. Each of these responses can lead to transient elevations of intracellular Ca^{+2} and trigger Ca^{+2}-dependent mechanisms involved in a wide range of developmental processes, including neuronal proliferation, migration and maturation (Owens and Kriegstein, 2002). A number of recent studies have provided evidence identifying $GABA_A$ receptor-mediated depolarizations as a key signal in early development of neurons and neuronal networks. These studies manipulated the expression of KCC2 in order to prematurely decrease intracellular $[Cl^-]$ in immature neurons and therefore alter the timing of the developmental "shift" of $GABA_A$ receptor-mediated responses (Chudotvorova et al., 2005;

Akerman and Cline, 2006; Cancedda et al., 2007). Two of these studies examined premature expression of KCC2 *in vivo*. Electroporation of immature tectal cells of the *Xenopus* tadpole with KCC2 prevented normal development of glutamatergic synapses and increased the development of inhibitory GABAergic inputs (Akerman and Cline, 2006). Cancedda et al. (2007) used *in utero* electroporation of rat embryos to express KCC2 and found dramatic alterations in the maturation of cortical neurons. Early expression of KCC2 has also been examined in cultured hippocampal neurons and resulted in a significant increase in the density of GABA receptors and formation of GABAergic synapses (Chudotvorova et al., 2005). Another study has taken a slightly different approach to shift $GABA_A$ receptor-mediated responses from depolarizing to hyperpolarizing by knocking down expression of NKCC1 in newly generated neurons of the adult hippocampus (Ge et al., 2006). Remarkably, this caused significant defects in synapse formation and dendritic development of the immature neurons and prevented their integration into the mature network. It is clear from these studies that depolarizing $GABA_A$ receptor-mediated responses in early development provide an important signal for the development of neurons and their networks and that KCC2 functions not only in the transition of GABAergic transmission (from depolarizing to hyperpolarizing) but also as an integral part of the entire developmental process of the nervous system. An important question that remains is how KCC2 mediates its effect on neuronal development. Is it the cotransporter itself that provides the signal or is it acting indirectly through its effect on intracellular $[Cl^-]$ (Chudotvorova et al., 2005)? If it is the latter it will elevate the status of Cl^- as an important signaling molecule.

The role of KCC2 in neuronal development has taken a new and exciting turn recently with the finding that it plays a morphogenic role in the formation of excitatory synapses (Li et al., 2007). Remarkably, this function of KCC2 does not require its transport activity, but rather KCC2 appears to act as a key cytoskeletal scaffolding protein involved in dendritic spine morphogenesis. Li et al. (2007) noted that cultured cortical neurons from *Kcc2*$^{-/-}$ mice exhibited dendritic protrusions that were long and had aberrant "filopodia-like" spines. Moreover, many of these protrusions showed significantly reduced expression of pre- and postsynaptic marker proteins, indicating reduced synapse formation in cortical neurons from *Kcc2*$^{-/-}$ mice. Indeed, the frequency of miniature excitatory postsynaptic currents was reduced in these cells from *KCC2*$^{-/-}$ mice, confirming a significant reduction in the number of functional excitatory synapses. Most remarkable was the finding that normal spine phenotype could be restored by transfecting the *Kcc2*$^{-/-}$ cortical neurons with a non-transporting construct of KCC2 (i.e. lacking the amino-terminus of the protein). Clearly, the morphogenic role of KCC2 in excitatory synapse formation must be structural rather than related to its transport operation. Li et al. (2007) went on to show using immunoprecipitation assays that KCC2 interacts with 4.1N protein, a previously characterized postsynaptic protein which in turn interacts with cytoskeletal proteins, including spectrin and actin. This study helps to explain the earlier finding by Gulyas et al. (2001) that KCC2 was highly expressed at excitatory inputs of the rat hippocampus. Thus, it now appears that KCC2 plays a key role in the development of both inhibitory and excitatory synapses.

4. *How does KCC2 Help Regulate Neuronal [Cl⁻]?*

In any homeostatic process, the effector mechanism must be sensitive to changes in the parameter being controlled. This implies that KCC2 must be sensitive to changes in $[Cl^-]_i$. This sensitivity of KCC2 to changes in $[Cl^-]_i$ could be direct and/or indirect through allosteric regulation. After considering the typical ion gradients that exist in neurons, I have reasoned that KCC2 might be capable of sensing and regulating $[Cl^-]_i$ directly through the effect that changes in $[Cl^-]_i$ have on the thermodynamic driving force of K^+-Cl^- cotransport (Payne, 1997). Such a mechanism is rapid and efficient as the effector is also the sensor. Precedence for a transporter operating in such a manner is provided by the C^--HCO_3^- exchanger in vertebrate red cells (Cala and Grinstein, 1988) and in gastric parietal cells during histamine-induced secretion (Thomas and Machen, 1991). This type of transporter operation has been previously referred to as "dynamic buffering" or "thermodynamic activation" (Cala and Grinstein, 1988; Thomas and Machen, 1991).

In order for a transporter to operate as a "dynamic buffer", a number of conditions must be met. First, the transporter must be constitutively active and must have a high transport capacity. Second, the transporter must be near thermodynamic equilibrium. Third, its ion transport affinities must be near physiological ion concentrations so as to permit maximal alterations in transporter velocity to respond to incipient and modest changes in ion concentration. Given these conditions, a Cl^- transporter would appear to be "activated" by a rapid increase in $[Cl^-]_i$, and it would operate to mitigate that change by eliciting a compensatory net Cl^- efflux. Does KCC2 of neurons meet these conditions? In mature neurons where $[Cl^-]_i$ is low (5–10 mM), i.e. below equilibrium, the outwardly directed K^+ chemical gradient is closely balanced by the inwardly directed Cl^- chemical gradient, i.e. KCC2 is operating near thermodynamic equilibrium (zero point on ordinate of Fig. 17.4). Thus, one of the conditions is met. Significantly, $[Cl^-]_i$ and $[K^+]_o$ are small compared to $[Cl^-]_o$ and $[K^+]_i$ (100–130 mM); thus, small concentration changes such as those occurring with $GABA_AR$ activation (elevated $[Cl^-]_i$) or with high excitatory activity (elevated $[K^+]_o$) will dramatically alter the *magnitude* and possibly the *direction* of the thermodynamic force driving net K^+-Cl^- cotransport. Hence, KCC2 can lower neuronal $[Cl^-]$ to levels that permit fast hyperpolarizing inhibition, but because KCC2 is poised near its reversal point, it can also mediate net K^+ and Cl^- uptake in response to elevated extracellular $[K^+]$. This latter property permits KCC2 to operate as a dynamic buffer of extracellular $[K^+]$ (Payne, 1997). The role of KCC2 in regulation of extracellular $[K^+]$ is important to consider given the fact that the extracellular space in the brain is relatively small (~20% of total) and continually threatened by the reservoir of K^+ held within neurons by the membrane voltage. What about the other conditions required for a significant thermodynamic effect on KCC2 activity? Remarkably, KCC2 appears to meet each one: constitutive activity, proper kinetics (transport affinities for extracellular K^+ and intracellular Cl^- are in the physiological range), and high transport capacity. Using heterologous expression systems, we and others have shown that membrane-resident KCC2 is constitutively active and exhibits high basal transport activity (Payne, 1997; Strange et al., 2000; Song et al., 2002). Constitutive activity and high transport activity for KCC2 have also been inferred by studies using native protein in cultured neurons and brain slices (e.g. Jarolimek et al., 1999; DeFazio et al., 2000). Significantly, we demonstrated that KCC2 has a high transport affinity for external Rb^+/K^+, K_m~5 mM (Payne, 1997; Williams and Payne, 2004) – close to the physiological level of $[K^+]_o$ in the brain interstitium (3–4 mM). The neuronal Cl^- extrusion pathway is often reported as being robust, and immunolocalization studies, showing that KCC2 is highly expressed througout most of the CNS, is consistent with a high transport capacity (Lu et al., 1999; Williams et al., 1999; Vardi et al., 2000; Vu et al., 2000; Gulyas et al., 2001).

The apparent "control" of KCC2 as described above does not preclude additional allosteric regulation either through alteration in unitary transport rate and/or surface protein density. In this regard, a study by Kelsch et al. (2001) demonstrated that KCC2 activity in cultured hippocampal neurons required protein tyrosine kinase (PTK) activity. Within minutes after genistein or lavendustin A application (tyrosine kinase

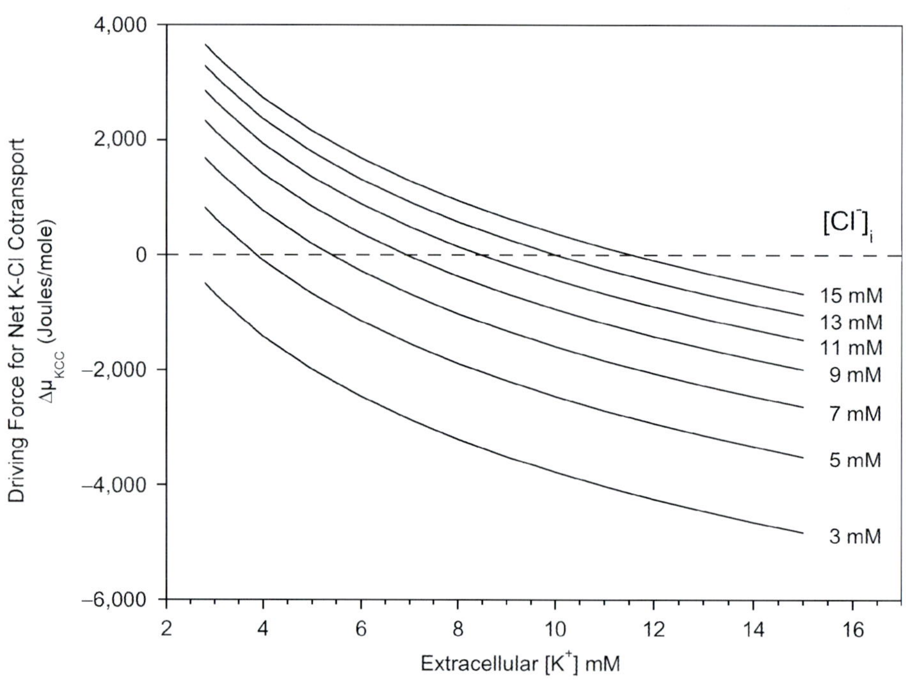

FIGURE 17.4 Thermodynamic driving force for K^+-Cl^- cotransport (J/mole) as a function of extracellular $[K^+]$ at various intracellular $[Cl^-]$ ($[Cl^-]_i$). Driving force was calculated using constant intracellular $[K^+]$ of 100 mM and extracellular $[Cl^-]$ of 130 mM at 37°C using the following equation: $\Delta\mu_{KCC} = \Delta\mu_K + \Delta\mu_{Cl} = RT \ln[K^+]_i/[K^+]_o + RT \ln[Cl^-]_i/[Cl^-]_o$, where the subscripts i and o denote intracellular and extracellular, respectively, R is the gas constant (8.314 J mol − 1 K − 1) and T is absolute temperature. This figure is modified from Payne (1997).

inhibitors), E_{GABA} shifted to more positive values in the dendrites of hippocampal neurons, indicating a rapid inhibition of Cl^- extrusion via KCC2 and concomitant elevation of dendritic $[Cl^-]_i$. Furthermore, their study indicated that KCC2 activity could be rapidly induced (10 min) if insulin or IGF-1 was coapplied with a cytoplasmic PTK (c-Src via recording pipette). As tyrosine phosphorylation is known to increase amplitudes of $GABA_A$ receptor Cl^- currents (Moss et al., 1995; Wan et al., 1997) allosteric regulation of KCC2 by PTKs would permit efficient functional coupling between the Cl^- "leak" and Cl^- "pump" pathways. Although the study by Kelsch et al. (2001) indicated that KCC2 was biochemically regulated via a tyrosine kinase cascade in hippocampal neurons, the mechanism by which KCC2 transport capacity was altered could not be determined. Recently, Lee and coworkers (2007) reported that activation of KCC2 by protein kinase C (PKC) required the phosphorylation of a specific serine on KCC2 (S940). This PKC phosphorylation of S940 appeared to be direct as *in vitro* experiments showed that purified PKC could specifically phosphorylate this site of a carboxy-terminal fusion protein of KCC2. They also provided evidence that the activation of KCC2 induced by PKC was due to an increase in surface stability of the transporter. In addition to PTK and PKC, a number of other kinases have been shown to alter KCC2 activity, including brain-type creatine kinase (Inoue et al., 2004, 2006), with no lysine kinases (WNKs; Kahle et al., 2005; Gagnon et al., 2006), and Ste-20 related proline alanine-rich kinase (SPAK; Gagnon et al., 2006). Collectively, these studies demonstrate that KCC2 is subject to significant allosteric regulation by a myriad of cell signaling mechanisms.

B. Regulation of Cell Volume

The most widely stated physiological function of the K^+-Cl^- cotransporter is a role in cell volume regulation. This KCC function is long-standing because much of the early work on the KCCs was in vertebrate red cells stimulated by swelling in hyposmotic media. Cell volume regulation refers to the process whereby a cell is able to restore volume to normal following an osmotic challenge. As animal cells are incapable of withstanding any significant hydrostatic pressure difference across their plasma membranes, volume regulation results from the activation of transporters that mediate net solute movement with concurrent flow of osmotically obliged water. As a net efflux mechanism, KCC can participate in the restoration of cell volume following osmotic swelling. Indeed, a role for KCC in cell volume regulation has been demonstrated for some vertebrate red cells (e.g. Kregenow, 1971b, 1981; Lauf, 1982) as well as non-erythroid cells (Thornhill and Laris, 1984), as discussed in Chapter 3. The particular KCC isoform(s) that is/are responsible for cell volume regulation has/have been the subject of much interest. While each KCC isoform has been shown to be activated by hyposmotic swelling, one must be careful in assigning a volume regulatory role to a transporter when one has only demonstrated volume sensitivity. A transporter should be considered volume regulatory only after it has been demonstrated that the transport rate is a graded function of the stimulus and that *net* ion transport is large enough to restore cell volume to normal. In the absence of such measurements, it is best to refer to the transporter as volume sensitive until more extensive experimentation is conducted.

1. *Is KCC3 the Primary Volume Regulatory KCC?*

It has generally been assumed that KCC1 with its wide tissue expression is the "housekeeping" isoform responsible for cell volume regulation. However, vertebrate red cells where K^+-Cl^- cotransport has been clearly shown to be involved in cell volume regulation express both KCC1 and KCC3 isoforms (Pellegrino et al., 1998; Su et al., 1999; Lauf et al., 2001a; Crable et al., 2005). In a more recent study, Rust et al. (2007) have provided convincing data that volume-sensitive K^+-Cl^- cotransporter activity in mouse red cells is largely mediated by KCC3. In this study, Rust et al. (2007) examined K^+-Cl^- cotransporter activity in the red cells of three different knockout mice, $Kcc1^{-/-}$, $Kcc3^{-/-}$ and the double mutant, $Kcc1^{-/-}Kcc3^{-/-}$. Compared to red cells from wildtype mice cotransporter activity following hyposmotic swelling was unaffected in $Kcc1^{-/-}$ mice, whereas it was dramatically reduced in $Kcc3^{-/-}$ mice and completely abolished in the double mutant, $Kcc1^{-/-}Kcc3^{-/-}$. A significant role of KCC3 in volume homeostasis of other cell types is supported by a preceding study which examined the volume regulatory response of neurons and renal tubule cells of $Kcc3^{-/-}$ mice (Boettger et al., 2003). In this study, volume regulation after hyposmotic swelling was dramatically reduced in hippocampal pyramidal neurons cultured from $Kcc3^{-/-}$ mice compared to those cultured from wildtype mice. Additionally, isolated perfused proximal straight tubules from the $Kcc3^{-/-}$ mice exhibited impaired cell volume regulation following hyposmotic swelling. From these knockout studies, it appears that KCC3 is the primary isoform involved in cell volume regulation following hyposmotic swelling.

One peculiar finding of the *Kcc3*$^{-/-}$ mice was that as they aged a distinct neurodegeneration was observed in both the central and peripheral nervous systems. This neurodegeneration was often accompanied by vascuolization and significant axonal swelling (Boettger et al., 2003). Furthermore, while the *Kcc3*$^{-/-}$ mice exhibited peripheral neuropathy (Howard et al., 2002), strangely enough KCC3 protein was not observed in the sciatic nerve or axons of the mature nervous system. A recent study has helped to clarify these findings. Byun and Delpire (2007) demonstrated that KCC3 is expressed in the sciatic nerve of juvenile mice but is absent in the sciatic nerve of adult mice. They also showed that the axonal swelling observed in *Kcc3*$^{-/-}$ mice progressed throughout postnatal development and was the likely cause of the neurodegeneration. These data support a key role of KCC3 in volume regulation of peripheral axons and revealed the importance of this cell function to the maintenance of neurons.

2. *Hyposmotic vs. Isosmotic Swelling*

At this point it seems appropriate to take time to discuss some issues of significant importance to the volume regulatory function of K^+-Cl^- cotransport in neurons. The first issue concerns the nature of volume perturbation most commonly incurred by neurons and by mammalian cells in general, and how this directly affects the thermodynamics of K^+-Cl^- cotransport. Such alterations in KCC thermodynamics following cell swelling places significant constraints on cotransporter operation and may dictate cellular localization of a particular K^+-Cl^- cotransporter based on its specific cellular function.

Most studies examining swelling activation of the KCCs have used hyposmotic solutions to mediate the volume perturbation. While experimentally expedient, this mechanism of cell swelling is not the most physiological for mammalian systems. Most mammalian cells and neurons in particular gain or lose volume as a result of the activation of ion transport mechanisms under conditions of constant osmolarity, i.e. isosmotic swelling/shrinkage (for review see, O'Neill, 1999). Good examples of isosmotic swelling/shrinkage are evident both in transporting epithelia and in neurons. In epithelial cells, significant changes in cell volume can occur whenever there is a subtle mismatch in the large apical and basolateral fluxes activated during secretion or absorption. In neurons, isosmotic swelling occurs in dendrites following intense synaptic activity, leading to significant channel-mediated Na^+ and Cl^- uptake along with osmotically obliged water. While the difference between hyposmotic and isosmotic swelling may seem small at first, the nature of the swelling and the mechanism by which the K^+-Cl^- cotransporter becomes activated can be significantly different. Briefly, with hyposmotic swelling, the volume increase is due exclusively to water movement. Hence, cell volume increases and intracellular $[Cl^-]$ decreases. In contrast, isosmotic swelling is the result of osmotically obliged water uptake secondary to solute uptake (typically Na^+ and Cl^-). This is a subtle but important point. Hyposmotic swelling reduces $[Cl^-]_i$ yet isosmotic swelling increases $[Cl^-]_i$, and this is the key to understanding the difference between the two types of cell swelling. With either hyposmotic or isosmotic swelling, the K^+-Cl^- cotransporter could be allosterically regulated by a truly volume-sensitive mechanism (e.g. volume-sensitive kinase/phosphatase system). Unlike hyposmotic swelling, however, isosmotic swelling could potentially use two additional mechanisms for activation of K^+-Cl^- cotransport: an increase in intracellular $[Cl^-]$ could allosterically activate K^+-Cl^- cotransport (e.g. Cl^--sensitive kinase/phosphatase system) and/or the increase in intracellular $[Cl^-]$ could simply increase the thermodynamic force driving net KCC efflux. Significantly, studies have shown that elevation of intracellular $[Cl^-]$ can be a potent activator of K^+-Cl^- cotransport (Lytle and McManus, 2002; Williams and Payne, 2004). Furthermore, in neurons with a robust constitutively active K^+-Cl^- cotransporter, like KCC2, the alteration in the driving force favoring net efflux is an appealing mechanism for volume regulation since it would occur simply as a consequence of KCC2's primary function in neuronal Cl^- homeostasis. Because KCC2 exhibits all of the necessary transport properties of a neuronal "dynamic buffer" (see IV.A; Payne, 1997), it is poised to respond rapidly to any incipient increase in intracellular $[Cl^-]$ with a net K^+, Cl^- and water efflux. It follows from this discussion that while KCC3 may be the major KCC isoform responsible for volume regulation following hyposmotic swelling in neurons and other cells, KCC2 can operate as a significant volume regulatory process following the more physiological isosmotic swelling of neurons.

The second issue to discuss deals with the thermodynamics of K^+-Cl^- cotransport in neurons, and the constraints it places on the operation of K^+-Cl^- cotransport in neurons. It is now widely accepted that KCC2 plays a key role in Cl^- homeostasis of mature neurons (see section IV.A). As KCC2 is the primary neuronal Cl^- extrusion mechanism, it is largely responsible for the low intracellular $[Cl^-]$ needed for hyperpolarizing $GABA_A$ and glycine receptor-mediated responses observed in mature neurons. Most estimates of intracellular $[Cl^-]$ of neurons have been obtained using the gramicidin perforated-patch clamp technique which is believed to preserve the endogenous neuronal intracellular Cl^- levels because the gramicidin pore is impermeable to Cl^- (see Chapter 8). Using this technique, one can monitor current changes with voltage ramps following the application of a $GABA_A$ receptor agonist and obtain a measurement of the reversal potential of $GABA_A$ receptor-mediated responses (E_{GABA}). Since the $GABA_A$ receptor is primarily permeable to Cl^-, E_{GABA} provides an estimate of the Cl^- equilibrium potential (E_{Cl}) and from knowledge of extracellular $[Cl^-]$, one can easily calculate intracellular $[Cl^-]$. It should be emphasized, however, that the $GABA_A$ receptor is also permeable to HCO_3^- which carries depolarizing current through the channel (Kaila and Voipio, 1987). Therefore, measurements of E_{GABA} will exhibit more positive values than E_{Cl} and therefore calculations based on E_{GABA} will overestimate intracellular $[Cl^-]$ (see Chapter 31). Estimates of intracellular $[Cl^-]$ based upon E_{GABA} using perforated-patch clamp range between 5 and 10 mM for a wide variety of different mature neurons (Chapter 7). This level of intracellular $[Cl^-]$ is quite low compared to the typical $[Cl^-]$ level found in virtually all other animal cells (>40 mM; see below section IV.A, and Chapter 22). With such a low level of Cl^- in mature neurons, the ability of K^+-Cl^- cotransport to aid in the recovery of volume following cell swelling will be limited by substrate (Cl^-) availability as well as by a steep Cl^- chemical concentration gradient against which the carrier must transport. As discussed above, this is where the difference between isosmotic and hyposmotic swelling becomes most evident. The increase in intracellular $[Cl^-]$ with isosmotic swelling provides much needed substrate (i.e. Cl^-) and a more favorable thermodynamic force driving net ion and water efflux. In stark contrast, hyposmotic swelling causes intracellular $[Cl^-]$ to decrease further and the thermodynamic force driving net KCC efflux becomes less favorable. Since intracellular $[Cl^-]$ is at such low levels, subtle changes in its value will have dramatic effects on the thermodynamic force driving net KCC (note broadening between lines in Fig. 17.4 as $[Cl^-]_i$ decreases). Based upon considerations of substrate availability and thermodynamic driving force, it is difficult to see how any cell with an intracellular $[Cl^-]$ of <5 mM could use KCC to move meaningful volumes of water absent a mechanism for Cl^- recycling. Yet, as mentioned above, Boettger et al. (2003) have reported that cultured hippocampal pyramidal neurons (with normal $[Cl^-]_i$ of 5.6 mM by their measurement using gramicidin perforated patch clamp) mediated full recovery of cell volume within 20 min following a ~25% hyposmotic swelling (300 mOsm/l to 230 mOsm/l) apparently via KCC3 (since neurons of *Kcc3*$^{-/-}$ mice exhibited no volume recovery from swelling). How can this conundrum be explained? The most reasonable explanations require a means of Cl^- recycling and/or intracellular compartmentalization of Cl^- and the distinct subcellular localization of KCC2 and KCC3 in those neurons where these two proteins are coexpressed. One must realize that measurements of intracellular $[Cl^-]$ using the perforated patch-clamp technique provide a very localized estimate of intracellular $[Cl^-]$ (e.g. soma vs. dendrite), and it is not appropriate to generalize intracellular $[Cl^-]$ for the entire neuron from one localized measurement. Recent research indicates that there exist significant steady state intracellular Cl^- gradients in neurons with axo-somato-dendritic gradients as large as ~20 mM in certain neurons (Duebel et al., 2006; Price and Trussell, 2006; Khirug et al., 2008). By all accounts, these gradients are maintained primarily (but not exclusively; see Khirug et al., 2008) by members of the CCC family, i.e. KCC2 as a Cl^- extrusion pathway maintaining low $[Cl^-]_i$ and NKCC1 as a Cl^- accumulation pathway maintaining high $[Cl^-]_i$. By necessity, KCC2 and NKCC1 must also exhibit distinct neuronal subcellular localizations, and the available data support this conclusion (Vardi et al., 2000; Szabadics et al., 2006; Khirug et al., 2008). Based on thermodynamics and the transport characteristics of KCC3, it follows that in order for this KCC isoform to play a significant role in regulation of neuronal volume, KCC3 must be localized in neuronal subcellular regions where intracellular $[Cl^-]$ is high enough (>20 mM) to provide the necessary substrate and energy for efficient volume recovery. This would require KCC3 to follow a distribution that is more closely associated with NKCC1 than with KCC2. Immunolocalization studies which examine the colocalization of these transporters at the cellular and subcellular level will be required to test this hypothesis.

V. CONCLUDING REMARKS

The molecular characterization of the different K^+-Cl^- cotransporter isoforms has highlighted a much greater functional diversity than would have ever been predicted from the early red cell studies. K^+-Cl^- cotransporters participate not only in cell volume regulation as first studied in vertebrate red cells but also in transepithelial transport, in intracellular Cl^- homeostasis and possibly in regulation of extracellular $[K^+]$. The interest in these transporters by neuroscientists has increased dramatically over the years, and this is no doubt due to the many key roles that the neuronal K^+-Cl^- cotransporter, KCC2, plays in the central nervous system. Many of these roles are directly linked to KCC2's transport operation, and its contribution to intracellular [Cl^-] regulation. By controlling intracellular [Cl^-], KCC2 dictates the electrical consequences of the ligand-gated anion channels, $GABA_A$ and glycine receptors in central neurons. As we have discussed, this function can have significant effects on neuronal maturation, proliferation and targeting, and the development of both inhibitory and excitatory synapses as well as the development of neuronal networks. Additionally, recent work on KCC2 has pointed to a function that goes beyond the process of transporting ions. KCC2 appears to be a scaffolding protein interacting with the neuronal cytoskeleton, where it can have a morphogenic role in the development of excitatory synapses (Li et al., 2007).

While I have centered the discussion on the operation of K^+-Cl^- cotransport in neurons under physiological conditions, alterations in neuronal K^+-Cl^- cotransport during pathological states has become a significant area of research (see Chapters 23 and 24 in this book). KCC2 expression in neurons decreases dramatically in certain pathological states (e.g. trauma, epilepsy, inflammation and ischemia/hypoxia), and this can result in the appearance of depolarizing $GABA_A$ receptor-mediated responses. Understanding the cellular mechanisms that control KCC2 expression and activity under both physiological and pathological states is an active area of research. But beyond this, researchers must determine why KCC2 expression changes so dramatically and rapidly in certain pathological states. It seems possible that reduction in KCC2 expression and appearance of depolarizing $GABA_A$ receptor-mediated responses following a traumatic event may be an adaptive response that provides neurons with greater flexibility in recovering from the insult. This may provide neurons with a key signal necessary for neuronal sprouting, retargeting and reconnecting within the circuit. Clearly, this is an exciting time for studying the functional roles of the K^+-Cl^- cotransporters in neurons.

Acknowledgements

The author is indebted to Drs Peter Cala, Christian Lytle and Kai Kaila for many thoughtful discussions about the function of K-Cl cotransport in neurons. Their kindness in reading and providing comments on the manuscript was much appreciated. Dr Bliss Forbush kindly provided Fig. 17.2. The author's research has been funded by National Institute of Health (NS-36296) and American Heart Association (National and Western States Affiliate).

References

Aickin, C.C., Deisz, R.A., and Lux, H.D. (1982). Ammonium action on post-synaptic inhibition in crayfish neurones: implications for the mechanism of chloride extrusion. *J. Physiol. (Lond.)* **329**, 319–339.

Aickin, C.C., Deisz, R.A., and Lux, H.D. (1984). Mechanisms of chloride transport in crayfish stretch receptor neurones and guinea pig vas deferens: implications for inhibition mediated by GABA. *Neurosci. Lett.* **47**, 239–244.

Akerman, C.J. and Cline, H.T. (2006). Depolarizing GABAergic conductances regulate the balance of excitation to inhibition in the developing retinotectal circuit in vivo. *J. Neurosci.* **26**, 5117–5130.

Alvarez-Leefmans, F.J., Gamino, S.M., Giraldez, F., and Nogueron, I. (1988). Intracellular chloride regulation in amphibian dorsal root ganglion neurones studied with ion-selective microelectrodes. *J. Physiol.* **406**, 225–246.

Alvarez-Leefmans, F.J. (1990). Intracellular Cl^- regulation and synaptic inhibition in vertebrate and invertebrate neurons. In *Chloride Channels and Carriers in Nerve, Muscle, and Glial Cells* (Alvarez-Leefmans, F.J. and Russell, J.M., eds), pp. 109–158. Plenum Press, New York.

Becker, M., Nothwang, H.G., and Friauf, E. (2003). Differential expression pattern of chloride transporters NCC, NKCC2, KCC1, KCC3, KCC4, and AE3 in the developing rat auditory brainstem. *Cell Tissue Res.* **312**, 155–165.

Bize, I. and Dunham, P.B. (1994). Staurosporine, a protein kinase inhibitor, activates K-Cl cotransport in LK sheep erythrocytes. *Am. J. Physiol. Cell Physiol.* **266**, C759–C770.

Bize, I., Guvenc, B., Robb, A., Buchbinder, G., and Brugnara, C. (1999). Serine/threonine protein phosphatases and regulation of K-Cl cotransport in human erythrocytes. *Am. J. Phsyiol. Cell Physiol.* **277**, C926–C936.

Boettger, T., Hubner, C.A., Maier, H., Rust, M.B., Beck, F.X., and Jentsch, T.J. (2002). Deafness and renal tubular acidosis in mice lacking the K-Cl cotransporter, KCC4. *Nature* **416**, 874–878.

Boettger, T., Rust, M.B., Maier, H., Seidenbecher, T., Schweizer, M., Keating, D.J., Faulhaber, J., Ehmke, H., Pfeffer, C., Scheel, O., Lemcke, B., Horst, J., Leuwer, R., Pape, H.C., Völkl, H., Hübner, C.A., and Jentsch, T.J. (2003). Loss of K-Cl co-transporter KCC3 causes deafness, neurodegeneration and reduced seizure threshold. *Embo J.* **22**, 5422–5434.

Byun, N. and Delpire, E. (2007). Axonal and periaxonal swelling precede peripheral neurodegeneration in KCC3 knockout mice. *Neurobiol. Dis.* **28**, 39–51.

Cala, P.M. and Grinstein, S. (1988). Coupling between Na/H and Cl/HCO_3 exchange in pH and volume regulation. In *Na/H Exchange* (Grinstein, S., ed.), pp. 201–208. CRC Press, Boca Raton.

Cancedda, L., Fiumelli, H., Chen, K., and Poo, M.-M. (2007). Excitatory GABA action is essential for morphological maturation of cortical neurons in vivo. *J. Neurosci.* **27**, 5224–5235.

Casula, S., Shmukler, B.E., Wilhelm, S., Stuart-Tilley, A.K., Su, W., Chernova, M.N., Brugnara, C., and Alper, S.L. (2001). A dominant negative mutant of the KCC1 K-Cl cotransporter: both N- and C-terminal cytoplasmic domains are required for K-Cl cotransport activity. *J. Biol. Chem.* **276**, 41870–41878.

Chudotvorova, I., Ivanov, A., Rama, S., Hübner, C.A., Pellegrino, C., Ben-Ari, Y., and Medina, I. (2005). Early expression of KCC2 in rat hippocampal cultures augments expression of functional GABA synapses. *J. Physiol. (Lond.)* **566**, 671–679.

Crable, S.C., Hammond, S.M., Papes, R., Rettig, R.K., Zhou, G.-P., Gallagher, P.G., Joiner, C.H., and Anderson, K. (2005). Multiple isoforms of the KCl cotransporter are expressed in sickle and normal erythroid cells. *Exp. Hematol.* **33**, 624–631.

DeFazio, R.A., Keros, S., Quick, M.W., and Hablitz, J.J. (2000). Potassium-coupled chloride cotransport controls intracellular chloride in rat neocoritical pyramidal neurons. *J. Neurosci.* **20**, 8069–8076.

Delpire, E., Rauchman, M.I., Beier, D.R., Hebert, S.C., and Gullans, S.R. (1994). Molecular cloning and chromosome localization of a putative basolateral Na^+-K^+-$2Cl^-$ cotransporter from mouse inner medullary collecting duct (mIMCD-3) cells. *J. Biol. Chem.* **269** (41), 25677–25683.

Di Stefano, A., Jounier, S., and Wittner, M. (2001). Evidence supporting a role for KCl cotransporter in the thick ascending limb of Henle's loop. *Kid. Int.* **60**, 1809–1823.

Dubreil, V., Hue, B., and Pelhate, M. (1995). Outward chloride/potassium co-transport in insect neurosecretory cells (DUM neurones). *Comp. Biochem. Physiol.* **111A**, 263–270.

Duebel, J., Haverkamp, S., Schleich, W., Feng, G., Augustine, G.J., Kuner, T., and Euler, T. (2006). Two-photon imaging reveals somatodendritic chloride gradient in retinal ON-type bipolar cells expressing the biosensor Clomeleon. *Neuron* **49**, 81–94.

Dunham, P.B., Stewart, G.W., and Ellory, J.C. (1980). Chloride-activated passive potassium transport in human erythrocytes. *Proc. Natl. Acad. Sci. USA* **77**, 1711–1715.

Ellory, J.C. and Dunham, P. (1980). Volume-dependent passive potassium transport in LK sheep red cells. In *Membrane Transport in Erythrocytes, Alfred Benzon Symposium 14* (Lassen, U.V., Ussing, H.H., and Wieth, J.O., eds), pp. 409–427. Munksgaard.

Flatman, P.W., Adragna, N.C., and Lauf, P.K. (1996). Role of protein kinases in regulating sheep erythrocyte K-Cl cotransport. *Am. J. Physiol. Cell Physiol.* **271**, C255–C263.

Fujii, T., Takahashi, Y., Itomi, Y., Fujita, K., Morii, M., Tabuchi, Y., Asano, S., Tsukada, K., Takeguchi, N., and Sakai, H. (2008). K^+-Cl^- cotransporter-3a up-regulates Na^+,K^+ ATPase in lipid rafts of gastric luminal parietal cells. *J. Biol. Chem.* **283**, 6869–6877.

Funder, J. and Wieth, J.O. (1967). Effects of some monovalent anions on fluxes of Na and K, and on glucose metabolism of ouabain-treated human red cells. *Acta Physiol. Scand.* **71**, 168–185.

Gagnon, K.B.E., England, R., and Delpire, E. (2006). Volume sensitivity of cation-Cl^- cotransporters is modulated by the interaction of two kinases: Ste20-related proline–alanine-rich kinase and WNK4. *Am. J. Physiol. Cell Physiol.* **290**, C134–C142.

Gamba, G. (2005). Molecular physiology and pathophysiology of electroneutral cation-chloride cotransporters. *Physiol. Rev.* **85**, 423–493.

Gamba, G., Miyanoshita, A., Lombardi, M., Lytton, J., Lee, W.-S., Hediger, M., and Hebert, S.C. (1994). Molecular cloning, primary structure, and characterization of two members of the mammalian electroneutral sodium-(potassium)-chloride cotransporter family expressed in kidney. *J. Biol. Chem.* **269**, 17713–17722.

Gamba, G., Saltzberg, S.N., Lambardi, M., Miyanoshita, A., Lytton, J., Hediger, M.A., Brenner, B.M., and Hebert, S.C. (1993). Primary structure and functional expression of a cDNA encoding the thiazide-sensitive, electroneutral sodium-chloride cotransporter. *Proc. Natl. Acad. Sci. USA* **90**, 2749–2753.

Ge, S., Goh, E.L., Sailor, K.A., Kitabatake, Y., Ming, G.L., and Song, H. (2006). GABA regulates synaptic integration of newly generated neurons in the adult brain. *Nature* **439**, 589–593.

Geck, P., Pietrzyk, C., Burckhardt, B.-C., Pfeiffer, B., and Heinz, E. (1980). Electrically silent cotransport of Na, K, and Cl in Ehrlich cells. *Biochim. Biophys. Acta* **600**, 432–447.

Gillen, C.M., Brill, S., Payne, J.A., and Forbush, B., III (1996). Molecular cloning and functional expression of the K-Cl cotransport from rabbit, rat, and human: a new member of the cation-chloride cotransporter family. *J. Biol. Chem.* **271**, 16237–16244.

Gimenez, I. and Forbush, B., III (2007). The residues determining differences in ion affinities among the alternative splice variants F, A, and B of the mammalian renal Na-K-Cl cotransporter (NKCC2). *J. Biol. Chem.* **282**, 6540–6547.

Gulyas, A.I., Sik, A., Payne, J.A., Kaila, K., and Freund, T.F. (2001). The K-Cl cotransporter, KCC2, is highly expressed in the vicinity of excitatory synapses in the rat hippocampus. *Eur. J. Neurosci.* **13**, 2205–2217.

Hiki, K., D'Andrea, R.J., Furze, J., Crawford, J., Woollatt, E., Sutherland, G.R., Vadas, M.A., and Gamble, J.R. (1999). Cloning, characterization, and chromosomal localization of a novel human K-Cl cotransporter. *J. Biol. Chem.* **274**, 10661–10667.

Hoffman, J.F. and Kregenow, F.M. (1966). The characterization of new energy dependent cation transport processes in red blood cells. *Ann. NY Acad. Sci.* **137**, 566–576.

Holtzman, E.J., Kumar, S., Faaland, C.A., Warner, F., Logue, P.J., Erickson, S.J., Ricken, G., Waldman, J., Kumar, S., and Dunham, P.B. (1998). Cloning, characterization, and gene organization of K-Cl cotransporter from pig and human kidney and *C. elegans*. *Am. J. Physiol.* **275**, F550–F564.

Howard, H.C., Mount, D.B., Rochefort, D., Byun, N., Dupre, N., Lu, J., Fan, X., Song, L., Riviere, J.-B., Prevost, C., Horst, J., Simonati, A., George, A.L., McDonald, M.P., Bouchard, J.-P., Mathieu, J., Delpire, E., and Rousleau, G.A. (2002). The K-Cl cotransporter KCC3 is mutant in a severe peripheral neuropathy associated with agenesis of the corpus callosum. *Nature Genet.* **32**, 384–392.

Hubner, C.A., Stein, V., Hermans-Borgmeyer, I., Meyer, T., Ballanyi, K., and Jentsch, T.J. (2001). Disruption of KCC2 reveals an essential role of K-Cl cotransport already in early synaptic inhibition. *Neuron* **30**, 515–524.

Inoue, K., Ueno, S., and Fukuda, A. (2004). Interaction of neuron-specific K^+-Cl^- cotransporter, KCC2 with brain-type creatine kinase. *FEBS Lett.* **564**, 131–135.

Inoue, K., Yamada, J., Ueno, S., and Fukuda, A. (2006). Brain-type creatine kinase activates neuron-specific K^+-Cl^- co-transporter KCC2. *J. Neurochem.* **96**, 598–608.

Inoue, M., Hara, M., Zeng, X.-T., Hirose, T., Ohnishi, S., Yasukura, T., Uriu, T., Omori, K., Minato, A., and Inagaki, C. (1991). An ATP-driven Cl^- pump regulates Cl^- concentrations in rat hippocampal neurons. *Neurosci. Lett.* **134**, 75–78.

Jarolimek, W., Lewen, A., and Misgeld, U. (1999). A furosemide-sensitive K-Cl cotransporter counteracts intracellular Cl^- accumulation and depletion in cultured rat midbrain neurons. *J. Neurosci.* **19**, 4695–4704.

Jennings, M.L. and Al-Rohil, N. (1990). Kinetics of activation and inactivation of swelling-stimulated K^+/Cl^--trasport. The volume-sensitive parameter is the rate constant for inactivation. *J. Gen. Physiol.* **95**, 1021–1040.

Jennings, M.L. and Schulz, R.K. (1991). Okadaic acid inhibition of KCl cotransport. Evidence that protein dephosphorylation is necessary for activation of transport by either cell selling or N-ethylmaleimide. *J. Gen. Physiol.* **97**, 799–818.

Kahle, K.T., Rinehart, J., de Los Heros, P., Louvi, A., Meade, P., Vazquez, N., Hebert, S.C., Gamba, G., Gimenez, I., and Lifton, R.P. (2005). WNK3 modulates transport of Cl^- in and out of cells: implications for control of cell volume and neuronal excitability. *Proc. Natl. Acad. Sci.* **102**, 16783–16788.

Kaila, K. and Voipio, J. (1987). Postsynaptic fall in intracellular pH induced by GABA-activated bicarbonate conductance. *Nature* **330**, 163–165.

Kaji, D.M. and Tsukitani, Y. (1991). Role of protein phosphatase in activation of KCl cotransport in human erythrocytes. *Am. J. Physiol. Cell Physiol.* **260**, C178–C182.

Kajiya, H., Okamoto, F., Li, J.-P., and Okabe, K. (2006). Expression of mouse osteoclast K-Cl cotransporter-1 and its role during bone resorption. *J. Bone Mineral Res.* **21**, 984–992.

Kanaka, C., Ohno, K., Kuriyama, K., Itoh, K.T., Fukuda, A., and Sato, K. (2001). The differential expression patterns of messenger RNAs encoding K-Cl cotransporters (KCC1,2) and Na-K-2Cl cotransporter (NKCC1) in the rat nervous system. *Neurosci.* **104**, 933–946.

Karadsheh, M.F., Byun, N., Mount, D.B., and Delpire, E. (2004). Localization of the KCC4 potassium-chloride cotransporter in the nervous system. *Neurosci.* **123**, 381–391.

Kelsch, W., Hormuzdi, S., Straube, E., Lewen, A., Monyer, H., and Misgeld, U. (2001). Insulin-like growth factor 1 and a cytosolic tyrosine kinase activate chloride outward transport during maturation of hippocampal neurons. *J. Neurosci.* **21**, 8339–8347.

Khirug, S., Yamada, J., Afzalov, R., Voipio, J., Khiroug, L., and Kaila, K. (2008). GABAergic depolarization of the axon initial segment in cortical principal neurons is caused by the Na-K-2Cl cotransporter NKCC1. *J. Neurosci.* **28**, 4635–4639.

Kregenow, F.M. (1971a). The response of duck erythrocytes to hypertonic media. Further evidence for a volume-controlling mechanism. *J. Gen. Physiol.* **58**, 396–412.

Kregenow, F.M. (1971b). The response of duck erythrocytes to nonhemolytic hypotonic media: evidence of a volume controlling mechanism. *J. Gen. Physiol.* **58**, 372–395.

Kregenow, F.M. (1973). The response of duck erythrocytes to norepinephrine and an elevated extacellular potassium: volume regulation in isotonic media. *J. Gen. Physiol.* **61**, 509–527.

Kregenow, F.M. (1981). Osmoregulatory salt transporting mechanisms: control of cell volume in anisotonic media. *Ann. Rev. Physiol.* **43**, 493–505.

Kunchaparty, S., Palcso, M., Berkman, J., Velazquez, H., Desir, G.V., Bernstein, P., Reilly, R.F., and Ellison, D.H. (1999). Defective processing and expression of thiazide-sensitive Na-Cl cotransporter as a cause of Gitelman's syndrome. *Am. J. Phsyiol. Renal Physiol.* **277**, F643–F649.

Kyrozis, A., Chudomel, O., Moshé, S.L., and Galanopoulou, A.S. (2006). Sex-dependent maturation of $GABA_A$ receptor-mediated synaptic events in rat substantia nigra reticulata. *Neurosci. Lett.* **398**, 1–5.

Lauf, P.K. (1982). Evidence for chloride dependent potassium and water transport induced by hyposmotic stress in erythrocytes of the marine teleost, *Opsanus tau*. *J. Comp. Physiol. B* **146**, 9–16.

Lauf, P.K. and Theg, B.E. (1980). A chloride dependent K^+ flux induced by N-ethylmaleimide in genetically low K^+ sheep and goat erythrocytes. *Biochem. Biophys. Res. Commun.* **92**, 1422–1428.

Lauf, P.K., Zhang, J., Delpire, E., Fyffe, R.E.W., Mount, D.B., and Adragna, N.C. (2001a). K-Cl cotransport: immunocytochemical and functional evidence for more than one KCC isoform in high K and low K sheep erythrocytes. *Comp. Biochem. Physiol.* **130**, 499–509.

Lauf, P.K., Zhang, J., Gagnon, K.B.E., Delpire, E., Fyffe, R.E.W., and Adragna, N.C. (2001b). K-Cl cotransport: immunohistochemical and ion flux studies in human embryonic kidney (HEK-293) cells transfected with full-length and c-terminal domain-truncated KCC1 cDNAs. *Cell Physiol. Biochem.* **11**, 143–160.

Lee, H., Chen, C.X., Liu, Y.J., Aizenman, E., and Kandler, K. (2005a). KCC2 expression in immature rat cortical neurons is sufficient to switch the polarity of GABA responses. *Eur. J. Neurosci.* **21**, 2593–2599.

Lee, H.H.C., Walker, J.A., Williams, J.R., Goodier, R.J., Payne, J.A., and Moss, S.J. (2007). Direct protein kinase C-dependent phosphorylation regulates the cell surface stability and activity of the potassium chloride cotransporter, KCC2. *J. Biol. Chem.* **282**, 29777–29784.

Lee, S., Park, J., Jung, H., Lee, S., Oh, J., Lee, H., Jun, H., Cho, W., and Lee, J. (2005b). Expression and distribution of ion transport mRNAs in human nasal mucosa and nasal polyps. *Acta Oto-Laryngol.* **125**, 745–752.

Lee, S.Y., Maniak, P.J., Rhodes, R., Ingbar, D.H., and O'Grady, S.M. (2003). Basolateral Cl^- transport is stimulated by terbutaline in adult rat alveolar epithelial cells. *J. Membr. Biol.* **191**, 133–139.

Li, H., Khirug, S., Cai, C., Ludwig, A., Blaesse, P., Kolikova, J., Afzalov, R., Coleman, S.K., Lauri, S., Airaksinen, M.S., Keinanen, K., Khiroug, L., Saarma, M., Kaila, K., and Rivera, C. (2007). KCC2 interacts with the dendritic cytoskeleton to promote spine development. *Neuron* **56**, 1019–1033.

Li, H., Tornberg, J., Kaila, K., Airaksinen, M.S., and Rivera, C. (2002). Patterns of cation-chloride cotransporter expression during emryonic rodent CNS development. *Eur. J. Neurosci.* **16**, 2358–2370.

Llinas, R., Baker, R., and Precht, W. (1974). Blockage of inhibition by ammonium acetate action on chloride pump in cat trochlear motoneurons. *J. Neurophysiol.* **37**, 522–532.

Lu, J., Karadsheh, M., and Delpire, E. (1999). Developmental regulation of the neuronal-specific isoform of K-Cl cotransporter KCC2 in postnatal rat brains. *J. Neurobiol.* **39**, 558–568.

Lytle, C. and McManus, T.J. (2002). Coordinate modulation of Na-K-2Cl cotransport and K-Cl cotransport by cell volume and chloride. *Am. J. Physiol. Cell Physiol.* **283**, C1422–C1431.

Lytle, C., Xu, J.-C., Biemesderfer, D., Haas, M., and Forbush, B., III (1992). The Na-K-Cl cotransporter protein of shark rectal gland. I. Development of monoclonal antibodies, immunoaffinity purification, and partial biochemical characterization. *J. Biol. Chem.* **267**, 25428–25437.

Mercado, A., Broumand, V., Zandi-Nejad, K., Enck, A., and Mount, D.B. (2006). A C-terminal domain in KCC2 confers constitutive K-Cl cotransport. *J. Biol. Chem.* **281**, 1016–1026.

Mercado, A., Vazquez, N., Song, L., Cortes, R., Enck, A.H., Welch, R., Delpire, E., Gamba, G., and Mount, D.B. (2005). NH_2-terminal heterogeneity in the KCC3 K^+-Cl^- cotransporter. *Am. J. Phsyiol. Renal Physiol.* **289**, F1246–F1261.

Mikawa, S., Wang, C., Shu, F., Wang, T., Fukuda, A., and Sato, K. (2002). Developmental changes in KCC1, KCC2 and NKCC1 mRNAs in the rat cerebellum. *Dev. Brain Res.* **136**, 93–100.

Misri, S., Chimote, A.A., Adragna, N.C., Warwar, R., Brown, T.L., and Lauf, P.K. (2006). KCC isoforms in a human lens epithelial cell line (B3) and lens tissue extracts. *Exp. Eye Res.* **83**, 1287–1294.

Moss, S.J., Gorrie, G.H., Amato, A., and Smart, T.G. (1995). Modulation of $GABA_A$ receptors by tyrosine phosphorylation. *Nature* **377**, 344–348.

Mount, D.B., Mercado, A., Song, L., Xu, J., George, A.L., Delpire, E., and Gamba, G. (1999). Cloning and characterization of KCC3 and KCC4, new members of the cation-chloride cotransporter gene family. *J. Biol. Chem.* **274**, 16355–16362.

Muñoz, A., Mendez, P., DeFelipe, J., and Alvarez-Leefmans, F.J. (2007). Cation-chloride cotransporters and GABA-ergic innervation in the human epileptic hippocampus. *Epilepsia* **48**, 663–673.

O'Donnell, M.E., Martinez, A., and Sun, D. (1995). Cerebral microvascular endothelial cell Na-K-Cl cotransport: regulation by astrocyte-conditioned medium. *Am. J. Physiol. Cell Physiol.* **268**, C747–C754.

O'Neill, W.C. (1999). Physiological significance of volume-regulatory transporters. *Am. J. Physiol. Cell Physiol.* **276**, C995–C1011.

Owens, D.F. and Kriegstein, A.R. (2002). Is there more to GABA than synaptic inhibition? *Nature Rev. Neurosci.* **3**, 715–727.

Payne, J.A. (1997). Functional characterization of the neuronal-specific K-Cl cotransporter: implications for $[K^+]_o$ regulation. *Am. J. Physiol. Cell Physiol.* **273**, C1516–C1525.

Payne, J.A. and Forbush, B., III (1994). Alternatively spliced isoforms of the putative renal Na-K-Cl cotransporter are differentially distributed within the rabbit kidney. *Proc. Natl. Acad. Sci. USA* **91**, 4544–4548.

Payne, J.A., Rivera, C., Voipio, J., and Kaila, K. (2003). Cation-chloride cotransporters in neuronal communication, development, and trauma. *Trends Neurosci.* **26**, 199–206.

Payne, J.A., Stevenson, T.J., and Donaldson, L.F. (1996). Molecular characterization of a putative K-Cl cotransporter in rat brain: a neuronal-specific isoform. *J. Biol. Chem.* **271**, 16245–16252.

Pearson, M.M., Lu, J., Mount, D.B., and Delpire, E. (2001). Localization of the K-Cl cotranspoter, KCC3, in the central and peripheral nervous systems: expression in the choroid plexus, large neurons, and white matter tracts. *Neurosci.* **103**, 481–491.

Pellegrino, C.M., Rybicki, A.C., Musto, S., Nagel, R.L., and Schwartz, R.S. (1998). Molecular identification and expression of erythroid K-Cl cotransporter in human and mouse erythroleukemic cells. *Blood Cells, Molecules, and Diseases* **24**, 31–40.

Perrot-Sinal, T.S., Sinal, C.J., Reader, J.C., Speert, D.B., and McCarthy, M.M. (2007). Sex differences in the chloride cotransporters, NKCC1 and KCC2, in the developing hypothalamus. *J. Neuroendocrinol.* **19**, 302–308.

Plotkin, M.D., Kaplan, M.R., Peterson, L.N., Gullans, S.R., Hebert, S.C., and Delpire, E. (1997). Expression of the Na^+-K^+-$2Cl^-$ cotransporter BSC2 in the nervous system. *Am. J. Physiol. Cell Physiol.* **272**, C173–C183.

Price, G.D. and Trussell, L.O. (2006). Estimate of the chloride concentration in a central glutamatergic terminal: a gramicidin perforated-patch study on the calyx of Held. *J. Neurosci.* **26**, 11432–11436.

Race, J.E., Makhlouf, F.N., Logue, P.J., Wilson, F.H., Dunham, P.B., and Holtzman, E.J. (1999). Molecular cloning and functional characterization of KCC3, a new K-Cl cotransporter. *Am. J. Physiol. Cell Physiol.* **277**.

Rivera, C., Voipio, J., Payne, J.A., Ruusuvuori, E., Lahtinen, H., Lamsa, K., Pirvola, U., Saarma, M., and Kaila, K. (1999). The K^+-Cl^- cotransporter KCC2 renders GABA hyperpolarizing during neuronal maturation. *Nature* **397**, 251–255.

Russell, J.M. (1979). Chloride and sodium influx: a coupled uptake mechanism in the squid giant axon. *J. Gen. Physiol.* **73**, 801–818.

Russell, J.M. (1983). Cation-coupled chloride influx in squid axon. Role of potassium and stoichiometry of the transport process. *J. Gen. Physiol.* **81**, 909–925.

Russell, J.M. (2000). Sodium-potassium-chloride cotransport. *Physiol. Rev.* **80**, 211–276.

Rust, M.B., Alper, S.L., Rudhard, Y., Shmukler, B.E., Vicente, R., Brugnara, C., Trudel, M., Jentsch, T.J., and Hübner, C.A. (2007). Disruption of erythroid K-Cl cotransporters alters erythrocyte volume and partially rescues erythrocyte dehydration in SAD mice. *J. Clin. Invest.* **117**, 1708–1717.

Sachs, J.R. (1971). Ouabain-insensitive sodium movements in the human red blood cell. *J. Gen. Physiol.* **57**, 259–282.

Sallinen, R., Thornberg, J., Putkiranta, M., Horelli-Kuitunen, N., Airaksinen, M.S., and Wessman, M. (2001). Chromosomal localization of SLC12A5/Slc12a5, the human and mouse genes for the neuron-specific K^+-Cl^- cotransporter (KCC2) defines a new region of conserved homology. *Cytogenet. Cell Genet.* **94**, 67–70.

Schmidt, W.F. and McManus, T.J. (1977a). Ouabain-insensitive salt and water movements in duck red cells. I. Kinetics of cation transport under hypertonic conditions. *J. Gen. Physiol.* **70**, 59–79.

Schmidt, W.F. and McManus, T.J. (1977b). Ouabain-insensitive salt and water movements in duck red cells. II. Norepinephrine stimulation of sodium plus potassium cotransport. *J. Gen. Physiol.* **70**, 81–97.

Schmidt, W.F. and McManus, T.J. (1977c). Ouabain-insensitive salt and water movements in duck red cells. III. The role of chloride in the volume response. *J. Gen. Physiol.* **70**, 99–121.

Shiroya, T., Fukunaga, R., Akashi, K., Shimada, N., Takagi, Y., Nishino, T., Hara, M., and Inagaki, C. (1989). An ATP-driven Cl^- pump in the brain. *J. Biol. Chem.* **264**, 17416–17421.

Simard, C.F., Bergeron, M.J., Frenette-Cotton, R., Carpentier, G.A., Pelchat, M.-E., Caron, L., and Isenring, P. (2007). Homooligomeric and heterooligomeric associations between K^+-Cl^- cotransporter isoforms and between K^+-Cl^- and Na^+-K^+-Cl^- cotransporters. *J. Biol. Chem.* **282**, 18083–18093.

Song, L., Mercado, A., Vazquez, N., Xie, Q., Desai, R., George, A.L., Gamba, G., and Mount, D.B. (2002). Molecular, functional, and genomic characterization of human KCC2, the neuronal K-Cl cotransporter. *Mol. Brain Res.* **103**, 91–105.

Starke, L.C. and Jennings, M.L. (1993). K-Cl cotransport in rabbit red cells: further evidence for regulation by protein phosphatase type 1. *Am. J. Physiol. Cell Physiol.* **264**, C118–C124.

Stein, V., Hermans-Borgmeyer, I., Jentsch, T.J., and Hubner, C.A. (2004). Expression of the KCl cotransporter KCC2 parallels neuronal maturation and the emergence of low intracellular chloride. *J. Comp. Neurol.* **468**, 57–64.

Strange, K., Singer, T.D., Morrison, R., and Delpire, E. (2000). Dependence of KCC2 K-Cl cotransport activity on a conserved carboxy terminus tyrosine residue. *Am. J. Physiol. Cell Physiol.* **279**, C860–C867.

Su, W., Shmukler, B.E., Chernova, M.N., Stuart-Tilley, A.K., de Franceschi, L., Brugnara, C., and Alper, S.L. (1999). Mouse K-Cl cotransporter KCC1: cloning, mapping, pathological expression, and functional regulation. *Am. J. Physiol. Cell Physiol.* **277**, C899–C912.

Szabadics, J., Varga, C., Molnár, G., Oláh, S., Barzó, P., and Tamás, G. (2006). Excitatory effect of GABAergic axo-axonic cells in cortical microcircuits. *Science* **311**, 233–235.

Thomas, H.A. and Machen, T.E. (1991). Regulation of Cl/HCO_3 exchange in gastric parietal cells. *Cell Reg.* **2**, 727–737.

Thompson, S.M., Deisz, R.A., and Prince, D.A. (1988a). Outward chloride/cation co-transport in mammalian coritcal neurons. *Neurosci. Lett.* **89**, 49–54.

Thompson, S.M., Deisz, R.A., and Prince, D.A. (1988b). Relative contributions of passive equilibrium and active transport to the distribution of chloride in mammalian cortical neurons. *J. Neurophysiol.* **60**, 105–124.

Thompson, S.M. and Gahwiler, B.H. (1989a). Activity-dependent disinhibition. II. Effects of extracellular potassium, furosemide, and membrane potential on E_{Cl} in hippocampal CA3 neurons. *J. Neurophysiol.* **61**, 512–523.

Thompson, S.M. and Gahwiler, B.H. (1989b). Activity-dependent disinhibition. III. Desensitization and $GABA_B$ receptor-mediated presyaptic inhibition in the hippocampus in vitro. *J. Neurophysiol.* **61**, 524–533.

Thornhill, W.B. and Laris, P.C. (1984). KCl loss and cell shrinkage in the Ehrlich ascites tumor cell induced by hypotonic media, 2-deoxyglucose and propranolol. *Biochim. Biophys. Acta* **773**, 207–218.

Tosteson, D.C. and Hoffman, J.F. (1960). Regulation of cell volume by active cation transport in high and low potassium sheep red cells. *J. Gen. Physiol.* **44**, 169–194.

Ubels, J.L., Hoffman, H.M., Srikanth, S., Resau, J.H., and Webb, C.P. (2006). Gene expression in rat lacrimal gland duct cells collected using laser capture microdissection: evidence for K+ secretion by duct cells. *Invest. Opthalmol. Vis. Sci.* **47**, 1876–1885.

Uvarov, P., Ludwig, A., Markkanen, M., Pruunsild, P., Kaila, K., Delpire, E., Timmusk, T., Rivera, C., and Airaksinen, M.S. (2007). A novel N-terminal isoform of the neuronal-specific K-Cl cotransporter KCC2. *J. Biol. Chem.* **282**, 30570–30576.

Vardi, N., Zhang, L., Payne, J.A., and Sterling, P. (2000). Evidence that different cation chloride cotransporters in retinal neurons allow opposite responses to GABA. *J. Neurosci.* **20**, 7657–7663.

Velazquez, H. and Silva, T. (2003). Cloning and localization of KCC4 in rabbit kidney: expression in the distal convoluted tubule. *Am. J. Physiol. Renal Physiol.* **285**, F49–F58.

Vinay, L. and Jean-Xavier, C. (2008). Plasticity of spinal cord locomotor networks and contribution of cation-chloride cotransporters. *Brain Res. Rev.* **57**, 103–110.

Vu, T.Q., Payne, J.A., and Copenhagen, D.R. (2000). Localization and developmental expression patterns of the neuronal K-Cl cotransporter (KCC2) in the rat retina. *J. Neurosci.* **20**, 1414–1423.

Wan, Q., Man, H.Y., Braunton, J., Wang, W., Salter, M.W., Becker, L.E., and Wang, Y.T. (1997). Modulation of $GABA_A$ receptor function by tyrosine phosphorylation of β-subunits. *J. Neurosci.* **17**, 5062–5069.

Wiley, J.S. and Cooper, R.A. (1974). A furosemide-sensitive cotransport of sodium plus potassium in the human red cell. *J. Clin. Invest.* **53**, 745–755.

Williams, J.R. and Payne, J.A. (2004). Cation transport by the neuronal K-Cl cotransporter, KCC2: thermodynamics and kinetics of alternate transport modes. *Am. J. Physiol. Cell Physiol.* **287**, C919–C931.

Williams, J.R., Sharp, J.W., Kumari, V.G., Wilson, M., and Payne, J.A. (1999). The neuronal-specific K-Cl cotransporter, KCC2: antibody development and initial characterization of the protein. *J. Biol. Chem.* **274**, 12656–12664.

Woo, N.S., Lu, J., England, R., McClellan, R., Dufour, S., Mount, D.B., Deutch, A.Y., Lovinger, D.M., and Delpire, E. (2002). Hyperexcitability and epilepsy associated with disruption of the mouse neuronal-specific K-Cl cotransporter gene. *Hippocampus* **12**, 258–268.

Xu, J.C., Lytle, C., Zhu, T., Payne, J.A., Benz, E., and Forbush, B., III (1994). Molecular cloning and functional expression of the bumetanide-sensitive Na-K-Cl cotransporter. *Proc. Natl. Acad. Sci. USA* **91**, 2201–2205.

Yan, Y., Dempsey, R.J., and Sun, D. (2001). Expression of Na-K-Cl cotransporter in rat brain during development and its localization in mature astrocytes. *Brain Res.* **91**, 43–55.

Zhu, L., Lovinger, D.M., and Delpire, E. (2005). Cortical neurons lacking KCC2 expression show impaired regulation of intracellular chloride. *J. Neurophysiol.* **93**, 1557–1568.

CHAPTER

18

Regulation of Cation-Chloride Cotransporters

Gerardo Gamba, Nicole Garbarini and Eric Delpire

OUTLINE

I. INTRODUCTION

Ion transporters, as well as other membrane-bound transport proteins, are regulated at multiple levels (Fig. 18.1). This regulation starts at the transcriptional level, where the broadly defined promoter of a specific transporter gene determines spatial (tissue- or cell-type specificity) and temporal expression. The second level of regulation occurs post-translationally, with trafficking of the protein to the plasma membrane, its insertion, and later its retrieval from the membrane. The overall turnover of transporters in the membrane is rather short, measured in terms of hours. Thus, there is a constant supply of newly synthesized transporters ready for transit to the plasma membrane. The amount of transporters at the cell surface, or steady-state level, is in itself certainly not constant, but often can vary widely based on conditions surrounding the tissue. In fact, it is becoming increasingly clear that tissues are highly dynamic, remodeling themselves in response to physiological (or pathophysiological) conditions, and changing transport properties are an important part of this process. Finally, over their lifetime at the cell surface, transporters may undergo conformational changes that affect their activity. These changes are often directly caused by intracellular cytoplasmic signaling cascades that have been triggered by a wide variety of extracellular stimuli. This chapter will cover most of these mechanisms of regulation and will focus more specifically on the modulation of transporter activity during expression at the cell surface.

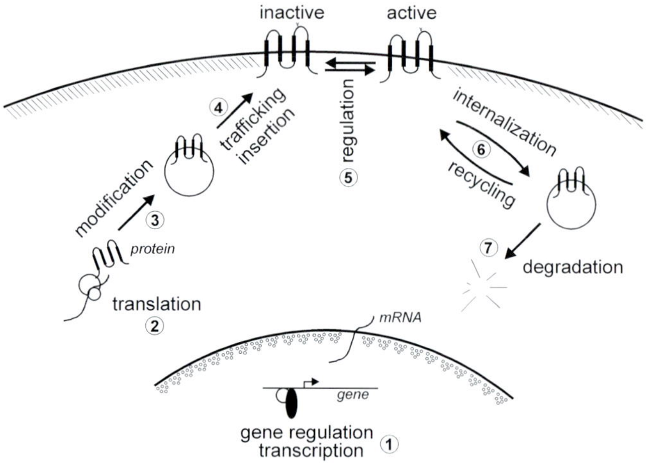

FIGURE 18.1 Schematic representation of the life of a membrane protein, starting from gene transcription to degradation. The scheme identifies seven major steps of regulation: (1) transcription, (2) translation, (3) maturation, (4) trafficking/insertion, (5) regulation, (6) internalization and (7) degradation.

II. REGULATION OF PROTEIN EXPRESSION AND ABUNDANCE

The literature provides ample evidence that the amount of cation-chloride cotransporters at the cell membrane is not fixed or constant, but modulated in response to changing physiological conditions. Indeed, multiple studies have highlighted significant alteration in cotransporter expression in the distal nephron upon dietary changes. For instance, dietary salt restriction results in a significant increase in expression of the thiazide-sensitive Na^+-Cl^- cotransporter in the distal convoluted tubule (Masilamani et al., 2002; Sandberg et al., 2006). This increase is mediated by aldosterone, which is known to activate gene expression by binding to regulatory elements in the 5′ flanking region of genes. However, concomitant to the level of NCC increasing at the apical membrane, there was no change in mRNA abundance, as demonstrated by RNA protection assay (Masilamani et al., 2002). Abundance of NKCC2 in the thick ascending limb of Henle is also modulated by a variety of factors including vasopressin, angiotensin, prostaglandins, acidosis and salt loading (Attmane-Elakeb et al., 1998; Ecelbarger et al., 2001; Knepper et al., 2003; Kwon et al., 2003).

Cotransporter expression is also regulated during development. The most studied example is the up-regulation of the neuronal-specific K^+-Cl^- cotransporter KCC2 during postnatal development (Clayton et al., 1998; Lu et al., 1999; Rivera et al., 1999). This increase is mediated at least in part by brain-derived neurotrophic factor (BDNF) (Aguado et al., 2003; Rivera et al., 2002). This aspect of KCC2 regulation will be covered in greater detail in Chapter 17. The literature also provides many hints that cotransporter expression levels are affected under pathophysiological conditions. For example, changes in KCC2 expression are observed in central neurons after epileptic seizures (Palma et al., 2006) or following nerve injury (Coull et al., 2003). Similarly, changes in KCC3 and KCC4 expression have been observed in tumor cells (Shen et al., 2003; Shiozaki et al., 2006). While these changes in expression levels under physiological or pathological conditions are well documented, the specific mechanism(s) by which these changes occur are not always understood.

A. Promoter Analysis

There are only a few studies which have examined the promoter and 5′ flanking region of cation-chloride cotransporter genes. The promoter and 5′ flanking region of the two Na^+-K^+-$2Cl^-$ cotransporters and the Na^+-Cl^- cotransporter have received some attention (Igarashi et al., 1996; Randall et al., 1997; Taniyama et al., 2001). As expected for a protein with restricted expression pattern, the renal specificity of NKCC2 is determined at the gene transcription level, as demonstrated by nuclear run-off assays, as well as luciferase assays of promoter deletion mutants in kidney and fibroblast cells. The promoter contains the binding site for the transcription factor HNF-1 or hepatocyte nuclear factor-1, and this likely confers kidney specificity to NKCC2 expression (Igarashi et al., 1996). Expression of the thiazide-sensitive Na^+-Cl^- cotransporter is thought to be driven in the distal convoluted tubule of the kidney by HFH-3 hepatocyte nuclear factor-3/folk head homolog-3, a transcription factor expressed in this segment of the renal tubule (Taniyama et al., 2001). However, it has recently been demonstrated that NCC is also expressed outside the kidney, in intestine (Bazzini et al., 2005), bone (Dvorak et al., 2007) and eye (Lauf et al., 2008).

There is some indication that the KCCs are likewise regulated by transcription factors. Analysis of the murine KCC2 gene has identified a neuronal restrictive silencing element (NRSE) within intron 1, possibly confining expression of the cotransporter to central neurons (Karadsheh and Delpire, 2001). Indeed, in non-neuronal cells, binding of the transcription factors NRSF/REST to NRSE represses gene transcription, whereas in neuronal cells which lack the NRSF/REST transcription factors, transcription and expression are allowed (Schoenherr and Anderson, 1995). However, KCC2's NRSE alone is not necessary for the neuron-specific expression, as a 6.8 kb region covering 1.4 kb of the promoter and KCC2 intron 1 lacking the NRSE can still sufficiently replicate the pattern of endogenous KCC2 expression in the brain (Uvarov et al., 2005).

Analysis of the KCC2 gene has also identified several potential transcription factor binding sites in the KCC2 promoter region. One of these proposed transcription factors, early growth response 4 (Egr4) has been demonstrated in cultured neurons as having a functional effect on KCC2 transcription (Uvarov et al., 2006, 2005).

B. Isoform Variants

The repertoire of cotransporter function is expanded through the existence of multiple cotransporter variants or isoforms. Such isoforms are generated by alternative promoters or alternative splicing. To date, there is no report of RNA editing events regulating cotransporter expression or function. The differential expression of cotransporter isoforms adds another level of regulation by altering intrinsic transport properties. The best example of this is the existence of three cassette exons (A, B and F) encoding the second transmembrane domain, TM2, and part of the intracellular loop between TM2 and TM3 in NKCC2 (Igarashi et al., 1995; Payne and Forbush, 1994). These alternatively spliced cotransporters are differentially expressed along the thick ascending limb of the kidney nephron, and the specific exon sequences confer distinct transport properties to the cotransporter at these locations (Oppermann et al., 2006, 2007; Payne and Forbush, 1994). The physiological relevance of this splicing is covered in greater detail in Chapter 16.

In 1997, Randall and colleagues reported a splice variant of NKCC1 which lacks exon 21 (Randall et al., 1997). The addition of exon 21 generates a consensus site for PKA phosphorylation which is absent in the isoform lacking the exon. The functional significance of this has not yet been determined. Exon 21 also contains a dileucine motif, and inclusion of this exon has recently been shown to direct the expression of the cotransporter to the basolateral membrane (Carmosino et al., 2008). In the absence of exon 21, the transporter is localized on the apical membrane. These splice variants were quantified in a variety of human tissues by Vibat and O'Donnell and demonstrated to be highly variable from tissue to tissue, indicating that this differential splicing may play an important regulatory role (Vibat et al., 2001). Importantly, the region encoded by exon 21 has no corresponding sequence within the NKCC2 gene, likely accounting for the apical localization of NKCC2 (Carmosino et al., 2008).

The K^+-Cl^- cotransporters likewise are reported to undergo modification through alternative splicing. Four isoforms of KCC1 have been reported in erythrocytes, differentiated by a variable first exon. Interestingly, there is a trend towards increased expression of exon 1b in red blood cells carrying sickle hemoglobin. The 1b isoform displays an elongated N′-terminus and deleted phosphorylation sites, as compared to the KCC1 expressed in wild-type erythrocytes (Crable et al., 2005).

Although there was only one known isoform of KCC2 for over a decade, we now know the existence of at least two gene products derived from alternate promoters. In 2007, Uvarov and colleagues identified KCC2a which is encoded by an alternative promoter and first exon located upstream of KCC2b, the isoform cloned in 1996. These two isoforms differ by 40 amino acids in their N-terminal region. The KCC2a isoform makes up almost 50% of the low KCC2 levels present after birth, but only between 5 and 10% of the total KCC2 in the mature cortex (Uvarov et al., 2007). The second isoform, KCC2b, has low expression levels early in life, but is up-regulated during postnatal development. It is the most abundant isoform in the mature brain (Lu et al., 1999; Payne et al., 1996). More studies are needed to understand the functional significance of these individually transcribed KCC2 transporters.

Several splice isoforms of KCC3 also exist. KCC3a was cloned from human tissues and reported by two groups (Mount et al., 1999; Race et al., 1999), whereas a second isoform, KCC3b, was reported by a third group in the same year (Hiki et al., 1999). Differences between these isoforms exist in the amino termini due to an alternative exon 1 generated by alternative transcription initiation within independent promoter regions (Pearson et al., 2001). Further analysis in later studies of KCC3 demonstrated that this cotransporter also undergoes alternative splicing to generate three additional isoforms which also differ in the N′-terminal tail (Mercado et al., 2005). Removal of the 15-residue exon 2 occurs in both KCC3a and KCC3b, generating KCC3a-x2M and KCC3b-x2M, respectively. Additionally, an alternative acceptor site within exon 1 results in a KCC3 isoform 60 residues shorter in length, termed KCC3a-S.

Several studies have begun to point towards the functional differences between KCC3 isoforms. In vascular smooth muscle cells, the nitric oxide signaling pathway appears to preferentially up-regulate mRNA transcripts of KCC3b over KCC3as (Di Fulvio et al., 2003). Also, Mercado and co-workers report that the five isoforms do not differ significantly in their sensitivity to extracellular osmolarity, although KCC3a-x2M and in KCC3a-S were shown to activate more rapidly than other KCC3 isoforms under hypotonic conditions (Mercado et al., 2005). Further indication of the physiological role of this KCC3 heterogeneity comes from isoform localization and sequence analysis. Transcript for KCC3a is widely expressed in the brain, lung, kidney, heart and muscle, whereas transcript for KCC3b is mainly located in the kidney (Mercado et al., 2005; Pearson et al., 2001). Several PKC phosphorylation sites and serine/threonine residues exist in the KCC3a-specific region, sites which are absent in KCC3b. Additionally, the deletion of exon 2 removes several putative phospho-regulatory sites (Hiki et al., 1999; Mercado et al., 2005; Race et al., 1999). In 2003, Velasquez and Silva reported the cloning of KCC4 from rabbit. In this species, the sequence had an additional 23 residues not found in mouse or human sequences. Whether or not this difference is due to alternative splicing is currently not known.

III. POST-TRANSLATIONAL PROCESSING

During and following the synthesis of the polypeptide chain, several modifications are made to the protein to allow proper trafficking and assembly of functional units. One such modification is glycosylation, in which glycan groups are enzymatically linked to proteins. Glycosylation is required in the trafficking of many membrane-bound transport proteins, including the cation-chloride cotransporters. There is increasing evidence that points to dimerization and/or oligomerization as another important post-translational mechanism of transport regulation. Oligomerization of proteins can be critical for release from the ER, trafficking to the cell surface and functional activity at the membrane. Furthermore, transport proteins are known to expand their functional repertoire via changes in hetero-oligomeric assembly.

In this section, we will describe post-translational modifications of the cation-chloride cotransporters. We will focus on studies of the glycosylation state of the CCCs, and studies that suggest CCC functional regulation is dependent on multimeric assembly.

A. Glycosylation

The molecular cloning of cation-chloride cotransporters identified key conserved asparagine residues in the large extracellular loop that connect transmembrane domains TM7 and TM8 in the Na^+-Cl^- and the two Na^+-K^+-$2Cl^-$ cotransporters (Delpire et al., 1994; Gamba et al., 1994, 1993; Payne et al., 1995; Xu et al., 1994). These asparagine residues are putative sites of N-linked glycosylation, a post-translational event in which an oligosaccharide chain is transferred to the nascent polypeptide at asparagine residues within the specific consensus sequence Asn-Xaa-Ser/Thr, where Xaa is any amino acid except a proline (Landolt-Marticorena and Reithmeier, 1994). Evidence that these asparagines are sites of glycosylation was later provided for NCC by enzymatic deglycosylation of isolated membrane proteins, and by incubation of NCC-expressing *Xenopus laevis* oocytes with tunicamicin (a compound that blocks the synthesis of all N-linked glycoproteins). Both experiments resulted in significant protein size reduction on Western blots, collapsing the size to the core molecular weight of the cotransporter (Hoover et al., 2003). As expected, mutagenesis of these residues led to decreased cotransporter function ensuing from decreased cell surface expression. Interestingly, it was also shown that elimination of glycosylation altered the binding of thiazide to the few cotransporters that made it to the plasma membrane, indicating that the glycan groups somehow hindered the access of the protein to the inhibitor. Similarly, the activity of NKCC2 is reduced by exchanging Asn residues (N442, N452) with Gln residues. Although cell surface expression was not affected by one single residue substitution, there was a significant reduction of NKCC2 at the membrane when the two asparagine residues were mutated (Paredes et al., 2006).

Very little direct information exists on the role of glycosylation of K^+-Cl^- cotransporters, with the exception of the identification of four conserved sites in the loop between transmembrane segments TM5 and TM6 (Gillen et al., 1996; Hiki et al., 1999; Mount et al., 1999; Payne et al., 1996).

B. Oligomerization

Many membrane-bound proteins exist as oligomers, and proper multimeric assembly of these proteins is required for protein stability, endoplasmic reticulum release, trafficking to the cell membrane and/or function at the cell surface. Additionally, formation of hetero-oligomers can also change protein function dependent on the multimeric makeup, as seen in G-protein-coupled receptors (Milligan and Smith, 2007). Studies of cation-chloride cotransporters suggest that this family of proteins likewise exists as oligomers in their native state.

The first indications of homo-oligomeric cation-chloride cotransporters come from Western blot experiments revealing the presence of immunoreactive bands approximately twice the core molecular weight of the cotransporter. Crosslinking studies with full-length and truncated NKCC1 clones demonstrated that NKCC1 forms homodimers (Moore-Hoon and Turner, 2000). Analysis of high-molecular weight protein complexes excreted in the urine, followed by crosslinking of membranes from kidney tissue, likewise demonstrated that both NCC and NKCC2 exist as multimers (McKee et al., 2000). Further studies demonstrated that homodimers formed by NCC and NKCC2 are functional units (de Jong et al., 2003; Starremans et al., 2003). Oligomerization determinants were mapped using yeast 2-hybrid and GST pull-down experiments. In both NKCC1 and NKCC2, there are two self-interacting domains located in the carboxy termini (Brunet et al., 2005; Simard et al., 2004).

Western blotting for the KCCs has likewise revealed the presence of immunoreactive bands at double the cotransporter's molecular weight. The first indications of KCC oligomerization came from experiments by Casula et al., which demonstrated that truncated KCC1 could exert a dominant negative effect when coexpressed with full-length KCC1 (Casula et al., 2001). More recently, Simard and colleagues demonstrated that the KCCs, as well as NKCC1, can form hetero-oligomeric structures when heterologously expressed in *Xenopus laevis* oocytes (Simard et al., 2007). Whether these associations occur *in vivo*, or have physiological and functional significance, is, however, unknown.

A recent study has, however, demonstrated that homo-oligomerization of KCC2 occurs *in vivo* and likely has functional effects. Analysis of rat brain tissues shows that in young immature brainstem neurons, KCC2 is expressed at the membranes, but exists mostly as a monomeric protein. This is in contrast with mature neurons, which express more multimers than monomers. Functional activation of KCC2 develops during brain maturation concomitant with oligomerization of the cotransporter, thus suggesting that oligomerization is a critical and required mechanism for KCC2 function (Blaesse et al., 2006).

IV. MEMBRANE RETRIEVAL AND DEGRADATION

Very little is known about the specific processes and stimuli which contribute to retrieval of cation-chloride transporters from the membrane. However, based on other fields of ion transport, it is clear that this regulated process should receive more attention. For both KCC2 and NKCC1, it was shown that PKC phosphorylation affects surface expression (membrane stability) of the cotransporters. However, PKC activity has opposite effects on the two cotransporters: in the case of KCC2, the rate of internalization from the plasma membrane was decreased thereby inducing an increased cell surface stability (Lee et al., 2007), whereas PKC activity triggered internalization of NKCC1 reducing cell surface expression (Del Castillo et al., 2005). The retrieval mechanisms of KCC2 were shown to be dynamin- and clathrin-dependent mechanisms, and shown to involve a non-canonical di-leucine motif LLXXEE located in the cytoplasmic carboxyl terminus of the cotransporter (Zhao et al., 2008).

V. REGULATION OF TRANSPORT RATE

So far in this chapter, we have addressed some of the mechanisms that impact the number of "functional" transporters expressed at the membrane of cells at any given time (an important factor behind the V_{max} in transport kinetics). These mechanisms involved the synthesis, trafficking and post-translational modification of transporters, as well as the retrieval of proteins from the membrane and their degradation. In this section, we will discuss the modulation of cotransporter activity, strictly speaking, the increase and decrease in the function of transporters already existing in the plasma membrane. These functional changes are related to alteration in ion concentrations, allosteric modifications and protein modification by phosphorylation/dephosphorylation.

A. Ionic Concentration

The concentration of ions obviously plays an important role in determining how much substrate crosses the plasma membrane. Transport is first dependent upon the ion concentrations on both sides (*cis* and *trans*) of the membrane. While inward and outward transport occurs through all cation-chloride cotransporters, the net movement is determined by the ion gradients. Thermodynamic considerations of transport through cation-chloride cotransporters are covered in detail in other chapters. However, ions can have effects on transport different from those related to gradients. Indeed, ion binding at sites different from the sites of transport can cause an allosteric change in transport rate. One such indication that this may occur comes via kinetic analysis of K^+-Cl^- cotransport in sheep red blood cells. KCC1 has a second Cl^- binding site on the transporter (Delpire and Lauf, 1992) and furosemide binding is altered by K^+ and Cl^- ions, which are transported across the plasma membrane, as well as by Na^+ ions, which are not transported (Lauf, 1984). Similar ionic effects have also been shown with the Na^+-K^+-$2Cl^-$ cotransporter (for review, see Russell, 2000). Thus, changes in ion composition, both physiological and pathophysiological, at either side of the membrane will have profound effects on transport properties.

B. Cell Volume and Osmotic Stress

The early studies that have identified the cotransporters as functional transport units and first described their basic transport properties have also demonstrated cation-chloride cotransporter sensitivity to cell volume changes (Dunham and Ellory, 1981; Geck et al., 1980). Since this original observation was made almost 30 years ago, many studies have attempted to understand the relationship between cell volume and ion transport. While it is clear that Na^+-K^+-$2Cl^-$ and K^+-Cl^- cotransporters are oppositely affected by cell volume changes, and that phosphorylation of the transporters is one of the last steps leading to changes in cotransporter activity, the signaling events between cell volume and the final phosphorylation steps are still not well understood.

C. Growth and Trophic Factors

Many circulating factors have been shown to modulate transport properties of epithelia and cells. These transport properties are in part due to changes in cation-chloride cotransporter activity. We have already provided some examples of factors that lead to altered cotransporter expression at the plasma membrane in different tubule segments and that are dependent upon dietary changes and hormonal control. In Table 18.1, we summarize several studies reporting short-term effects of growth factors and other trophic factors on Na^+-K^+-$2Cl^-$ cotransporter activity.

Studies on KCC2 have identified several growth factors and trophic factors which may be involved with regulation; however, it is unknown if this is a direct effect, or more likely part of a larger signaling cascade. Insulin growth factor-1 (IGF-1) as well as the cytoplasmic protein tyrosine kinase c-src impacts KCC2 activity. Coapplication of IGF-1 and c-src has been shown to rapidly activate KCC2 in immature neurons (Kelsch et al., 2001), and as further des-cribed in other chapters, affect KCC2 turnover (Rivera et al., 2005, 2004). However, it is unknown if BDNF signaling through the TrkB receptors directly causes a change in KCC2

TABLE 18.1 Factors affecting Na^+-K^+-$2Cl^-$ cotransporter activity

NKCC1 activation		
Deoxygenation	turkey erythrocytes[(2)]	Muzyamba et al. (1999)
Shear stress	vascular endothelium	Topper et al. (1997)
IL-6	brain endothelial cells	Sun et al. (1997)
IL-1b, TNF	vascular endothelium	Topper et al. (1997)
Arsenite	ferret erythrocytes[(1)]	Flatman and Creanor (1999)
Norepinephrine	H29 human colonic adenocarc.	Kim et al. (1988)
	turkey erythrocytes	Ueberschär and Bakker-Grunwald (1983)
	rabbit tracheal epithelial	Liedtke (1990)
	bovine retinal pigment epithelium	Edelman and Miller (1991)
Isoproterenol	turkey erythrocytes	Muzyamba et al. (1999)
	simian eccrine cells	Toyomoto et al. (1997)
acethylcholine	hen trachea	Winding and Bindslev (1990)
β-adrenergic	rat parotid acinar cells	Paulais and Turner (1992)
	skeletal muscle	Gosmanov et al. (2002)
cAMP[(3)]	simian eccrine cells	Toyomoto et al. (1997)
	shark rectal gland	Palfrey et al. (1984)
	osteoblasts	Whisenant et al. (1991)
	ocular ciliary body non-pigmented EC Crook et al. (1992)	
Ca^{2+}	vascular smooth muscle	Smith and Smith (1987)
	rabbit tracheal epithelial	Liedtke (1990)
	vascular endothelial cells	O'Donnell (1991)
	osteoblasts	Whisenant et al. (1991)
Ca^{2+}, TPA	Ehrlich ascites	Jensen et al. (1993)
PMA	human airway epithelial cells	Liedtke (1995)

VIP	T84 colonic cells	Dharmsathaphorn et al. (1985)
	H29 human colonic adenocarc.	Kim et al. (1988)
	small intestinal crypts	O'Brien et al. (1993)
PGE1	T84 colonic cells	Weymer et al. (1985)
	H29 human colonic adenocarc.	Kim et al. (1988)
PGE2	osteoblasts	Whisenant et al. (1991)
	fetal lung	McCray and Bettencourt (1993)
α-thrombin	hamster fibroblasts	Paris and Pouyssegur (1986)
EGF	hamster fibroblasts	Paris and Pouyssegur (1986)
	retinal pigment epithelium	Arrindell et al. (1992)
Bradykinin	endothelial cells	Brock et al. (1986); Klein and O'Neill (1990)
	Erhlich ascites	Jensen et al. (1993)
Vasopressin	endothelial cells	Brock et al. (1986)
		O'Donnell et al. (2005)
Angiotensin II	vascular smooth muscle	Smith and Smith (1987)
Parathyroid h.	osteoblasts	Whisenant et al. (1991)
Calcitonin (cAMP)	renal tubule cells	Vuillemin et al. (1992)
NGF	PC12 cells	Leung et al. (1994)
Insulin	3T2-L1 adipocytes	Sargeant et al. (1995)
NKCC1 inactivation		
oxidative stress	vascular endothelial cells	Elliott and Schilling (1992)
	skeletal muscle	Sen et al. (1995)
TPA	Balb/3T3	Guo and O'Brien (1996)
	simian eccrine cells	Toyomoto et al. (1997)
PMA	ocular ciliary body non-pigmented EC Crook et al. (1992)	
	T84 colonic cells	Matthews et al. (1993)
	fetal human pigmented ciliary EC	Von Brauchitsch and Crook (1993)
PBDu	rabbit non-pigmented ciliary epithelium	Dong and Delamere (1994)
Isoproterenol	vascular smooth muscle	Smith and Smith (1987)
cAMP	vascular smooth muscle	Smith and Smith (1987)
ANP, cGMP	rabbit myocytes	Clemo and Baumgarten (1991); Clemo et al. (1992)

Notes: (1) inactivated by staurosporine, PP1, genestein, A23187 + EGTA
(2) inactivated by staurosporine
(3) or shown through forskolin, theophyllin

activity at the cell surface via direct interaction or phosphoregulation of the cotransporter.

D. Phosphoregulation of the Cation-Chloride Cotransporters

Like many other proteins, cation-chloride cotransporters are directly regulated by phosphorylation and dephosphorylation events (Fig. 18.2). A role for kinases and phosphatases in the regulation of Cl-dependent K^+ fluxes was first proposed in the very early studies that described the properties of K^+-Cl^- cotransport (Hall and Ellory, 1986; Lauf, 1983; Sachs, 1988) and Na^+-K^+-$2Cl^-$ cotransport (Altamirano et al., 1988; Geck et al., 1980; Hall and Ellory, 1985; Hoffmann et al., 1983; Lytle and Forbush, 1992; Ueberschär and Bakker-Grunwald, 1985). Evidence to support these ideas came first from studies using ATP depletion, followed by studies using vanadate and/or fluoride as unspecific protein phosphatase inhibitors (Altamirano et al., 1988; Paulais and Turner, 1992). Based on kinetics of transport activation, and the observation of a time delay for activation, Jennings and al-Rohil suggested in 1991 that swelling-induced activation of K^+-Cl^- cotransport was due to net dephosphorylation (Jennings and al-Rohil, 1990).

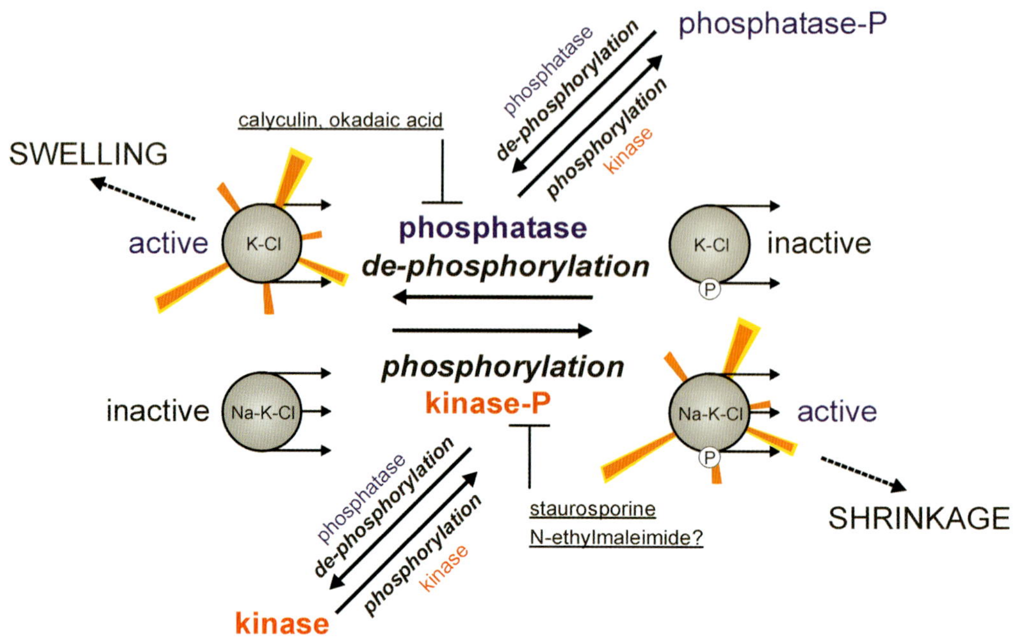

FIGURE 18.2 Mirror regulation of Na^+-K^+-$2Cl^-$ and K^+-Cl^- cotransporters by phosphorylation. Model shows that phosphorylation of NKCC leads to its activation, whereas dephosphorylation of KCC leads to its activation. Model also illustrates that shrinkage activates NKCC, while swelling activates KCC. The major kinases and phosphatases are also regulated by phosphorylation/dephosphorylation events. The putative mode of action of some pharmacological compounds is also illustrated.

They argued that this dephosphorylation was likely the result of kinase inhibition rather than stimulation of a phosphatase. One year later, they provided evidence showing that PP1 inactivation resulted in the activation of the cotransporter (Jennings and Schultz, 1991). A role for PP1 was also reported for the modulation of Na^+-K^+-$2Cl^-$ cotransporter (Palfrey and Pewitt, 1993; Pewitt et al., 1990). In fact, the cotransporter was demonstrated to be a PP1 binding partner, as it contains a x[F/W] consensus binding motif (Darman and Forbush, 2002).

Na^+-K^+-$2Cl^-$ and K^+-Cl^- cotransporters hold an inverse, reciprocal activation–deactivation relationship where phosphorylation activates the Na^+-dependent cotransporters and dephosphorylation activates the Na^+-independent cotransporters (Gagnon et al., 2006; Jennings and Schultz, 1991; Lauf et al., 1992; Mercado et al., 2000; Starke and Jennings, 1993). This indicates a coordinated balance between cation-chloride influx/efflux and suggests coexpression of the two types of cotransporters. Whereas the identity of the protein phosphatase was recognized in the early 1990s (Jennings and Schultz, 1991; Kaji and Tsukitani, 1991; Krarup and Dunham, 1996; Krarup et al., 1998; Palfrey and Pewitt, 1993; Pewitt et al., 1990), the identity of the kinases regulating NKCC1 have remained elusive.

Early studies in secretory epithelium of the shark rectal gland showed increased Cl^- transport by vasoactive intestinal peptide, or VIP (Stoff et al., 1979). The peptide, via a mechanism involving activation of adenyl cyclase, increases cAMP levels. Similar effects on Cl^- transport can be elicited by forskolin, an activator of the adenyl cyclase, or by theophylline or IBMX, inhibitors of the phosphodiesterase. In the avian salt gland, VIP and forskolin treatment result in increased phosphorylation of NKCC1 (Torchia et al., 1992). These data strongly suggest a role for the cAMP-dependent protein kinase in the activation of the cotransporter. Parallel to these studies showing transport activation by cAMP, there were also many studies reporting inhibition of Na^+-K^+-$2Cl^-$ cotransport by cAMP (Garay and Ciccone, 1982; Guandalini et al., 1982; Owen, 1984). This cell-type-dependent effect of cAMP indicates that the cyclic nucleotide-dependent protein kinase does not directly affect the cotransporter, but rather acts somewhere upstream of the direct mechanism of cotransporter activation.

A similar argument can be made for protein kinase C (PKC). Early studies have established a relationship between activation of muscarinic receptors, protein kinase C and NKCC1 (Table 18.1). In a variety of cell types, activation of PKC by phorbol-12-myristate-13-acetate (PMA/TPA) leads, however, to inhibition of cotransporter activity (Table 18.1).

Phosphorylation of NKCC1 has been examined in a variety of systems. The first technique utilized NKCC1-specific immunoprecipitation from protein lysates exposed to ^{32}P-orthophosphate. These experiments established that NKCC1 phosphorylation mirrors NKCC1 activity (Lytle, 1997; Lytle and Forbush, 1992). Furthermore, hormones such as bradykinin and vasopressin in endothelial cells increased NKCC1 phosphorylation signals (O'Donnell et al., 1995), whereas cGMP but not cAMP increased phosphorylation of flounder intestine brush border membrane NKCC1 (Suvitayavat et al., 1994). The phosphorylation of NKCC1 at specific sites is now demonstrated through the use of mass spectrometry (Darman and Forbush, 2002; Vitari et al., 2006) and phosphopeptide-specific antibodies (Flemmer et al., 2002; Vitari et al., 2006). Immunoreactivity mirrors closely the activity of the cotransporter (Flemmer et al., 2002).

1. *Ste20 Kinases*

Between 1997 and 1999, two kinases related to the yeast Sterile20p (Ste20p) kinase were identified in mammalian cells by three independent groups (DeAizpurua et al., 1997; Tamari et al., 1999; Ushiro et al., 1998). Typical for this competitive era of cloning novel cDNAs, the groups were unaware of each others' discovery and the names of the kinases PASK (proline-alanine-rich sterile 20p kinase), SPAK (Ste20p-related proline-alanine-rich kinase), and OSR1 (oxidative stress response 1) reflect this lack of coordination. To add to the nomenclature debacle, GenBank assigned the name STK39 to human SPAK/PASK, seemingly unaware of the existence of OSR1, which today stands alone in its database without clear connection to SPAK. For clarity, from this point, we will refer to these two kinases as SPAK and OSR1. Models of SPAK and OSR1 kinase are illustrated in Fig. 18.4.

Within one species (mouse or human), SPAK and OSR1 share a high degree of homology, both in their N-terminal catalytic domain and their C-terminal regulatory domain. No other protein kinase shares homology in the C-terminal domain, making SPAK and OSR1 unique kinases, likely to serve unique functions. Close examination of the genomes from a wide variety of organisms identifies OSR1 as the original kinase, with SPAK likely arising from gene duplication during late vertebrate evolution (Delpire and Gagnon, 2008 and Fig. 18.4).

In a 2002 study searching for proteins that interact with the K^+-Cl^- cotransporter KCC3, Piechotta et al. found a direct interaction between SPAK/OSR1 and the cytosolic amino terminus of KCC3. The kinase–cotransporter interaction was extended to NKCC1 and NKCC2 (Piechotta et al., 2002). Yeast 2-hybrid screening and immunoprecipitation experiments have identified a number of additional proteins that bind to SPAK and OSR1. These proteins include other kinases such as

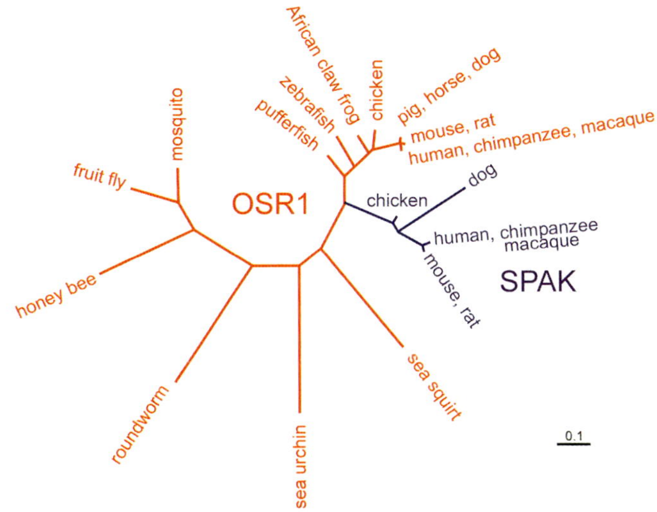

FIGURE 18.3 Representation of SPAK and OSR1 kinases. The kinase domain follows a relatively short amino terminus. The extreme carboxyl terminus (~90 residues) forms a protein fold which interacts with other proteins containing RFxV motifs. Sequence of a caspase cleavage site is also highlighted.

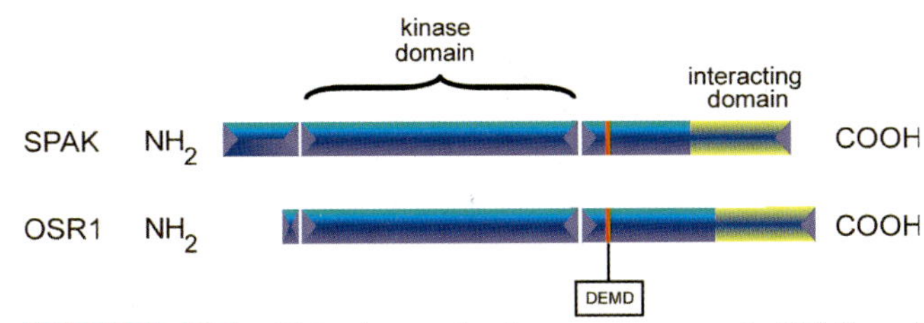

FIGURE 18.4 This cluster dendrogram represents SPAK and OSR1 kinases from a variety of organisms. It provides evidence for OSR1 evolving first, as it is found from roundworm to mammals. As SPAK is only found in birds and mammals, the gene likely originated from OSR1 gene duplication during late vertebrate evolution. Adapted from Delpire and Gagnon (2008).

WNK1, WNK2, WNK4, AATYK, AATYK3, heat shock and cytoskeletal proteins such as HSP105 and gelsolin, and other membrane proteins such as the tumor necrosis factor receptors RELT and RELT-2 and the *C. elegans* chloride channel CLH3 (Denton et al., 2005; Moriguchi et al., 2006; Piechotta et al., 2003, 2002; Polek et al., 2006).

Extensive work has been done to characterize the binding between these kinases and the cotransporters. An RFx[V/I] motif was identified within several cation-chloride cotransporters and specific residues were tested through mutagenesis using yeast 2-hybrid assays (Piechotta et al., 2002). These experiments identified several details about the nature of this motif. First, the arginine at position 1 of the motif can be substituted with a lysine residue but not an alanine, indicating the importance of the positive charge. Second, the phenylalanine residue at position 2 cannot be substituted by tyrosine, valine or alanine residues. Third, there is no conservation of the residue at position 3. Fourth, the valine residue at position 4 can be substituted by an isoleucine, but not leucine or alanine. Finally, the RFxV motif requires five additional residues at its 3′ end, at least for yeast 2-hybrid interaction. Although mutagenesis of these residues affected protein–protein interaction, no consensus could be extracted from these experiments. Interestingly, a large segment of the kinase regulatory domain, consisting of the 90 most C-terminal residues, was shown to be necessary for protein interaction with the RFxV sequence, suggesting that this rather large region of the kinase formed a protein fold that could interact with an RFxVxxxxx peptide (Piechotta et al., 2002). In fact, this domain was recently crystallized in the presence of a WNK4 peptide RFQVTSSKE, and the structure reveals the presence of a hydrophobic groove or cavity that can accommodate the hydrophobic phenylalanine residue at position 2, and the presence of aspartic and glutamic acid residues which can form salt bridges with the arginine residue at position 1 (Villa et al., 2007). The cavity is followed by a larger pocket for which function has not yet been determined. It could conceivably accommodate the additional five residues shown to be necessary for interaction, and by extension could conceivably provide a more specific binding site for a particular protein. It was noted by the authors that a threonine residue directly follows the RFxV motif. Because the possibility exists that this residue might be phosphorylated *in vivo*, they proceeded to test whether the phosphorylated peptide could bind to the kinase. They found that once phosphorylated, the peptide was no longer recognized by the kinase. This intriguing observation leaves the possibility that phosphorylation of this residue might constitute a mechanism by which SPAK binding can be regulated in cells. This exciting possibility has not yet received experimental support. Interestingly, following this threonine residue are two serines, further raising the possibility that phosphorylation of additional residues within the 9 residue peptide might constitute a mechanism of binding/unbinding between kinase and substrate. This theory has even greater relevance since a genome-wide search of proteins with putative SPAK binding motifs revealed that serine residues are by far the most frequent amino acids found at positions 5, 6, 7 and 8 of the motif (Delpire and Gagnon, 2007).

Among the proteins containing an RFxV motif, the with-no-lysine kinase 4 (WNK4) has been best studied for its relationship with SPAK. In heterologous expression systems, WNK1 and WNK4 were demonstrated to be upstream effectors of SPAK (Gagnon et al., 2006; Moriguchi et al., 2006). Binding of WNK4 to SPAK is

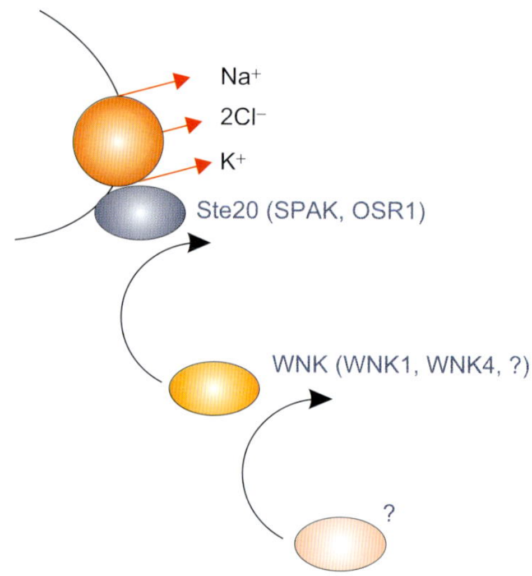

FIGURE 18.5 Model of NKCC1 regulation by kinases. Binding of the Ste20 kinases SPAK and OSR1 is necessary for NKCC1 phosphorylation and activation. WNK kinases lie upstream of the Ste20 kinases. WNKs bind to SPAK and OSR1 and phosphorylate and activate them. WNKs are themselves phosphorylated and activated by upstream kinase(s) responsive to osmolarity changes.

indeed required for SPAK phosphorylation and activation. In a model of NKCC1 regulation, Gagnon and Delpire postulated a chain of kinase reactions involving SPAK/OSR1 as kinases directly phosphorylating NKCC1, a WNK kinase phosphorylating SPAK, and additional upstream kinases activating WNK (Fig. 18.5). As binding of the C-terminal domain of SPAK is required for both SPAK and WNK4, it is likely that there is a hierarchical order to the interaction; first, WNK binds and activates SPAK, and next SPAK binds and activates the cotransporter (Gagnon et al., 2006). It has been demonstrated that WNK4 phosphorylates SPAK on two specific threonine and serine residues: T243 and S383 (for mouse NKCC1 sequence) (Vitari et al., 2005). *In vivo* experiments have shown that residue T243 is critical to SPAK function. In addition, *in vitro* phosphorylation experiments have identified T243 as one of the four threonine residues in the activation loop which is phosphorylated by SPAK through an intramolecular autophosphorylation reaction (Gagnon et al., 2006). Whether autophosphorylation of this threonine along the other threonines of the activation loop was due to the presence of manganese, rather than magnesium, in the *in vitro* phosphorylation reaction is still unresolved. Interestingly, mutation of S383, the other residue targeted by WNK4, into an alanine has no functional effect on the activation of NKCC1 in *Xenopus laevis* oocytes coinjected with WNK4-SPAK-NKCC1 cRNA (unpublished data). However, the fact that the S383A mutant is functional, but still requires WNK4 activation, suggests that WNK4 phosphorylation of S383 has a similar effect as mutation of this residue into an alanine. The exact mechanism by which S383 affects SPAK function is still unresolved.

Phosphorylation of NKCC1 is also not completely understood. In 2002, Darman and Forbush used mass spectrometry analysis and Edman sequencing to identify three threonine residues (T184, T189 and T202) located in the N-terminus of shark NKCC1 which are phosphorylated upon forskolin treatment (Darman and Forbush, 2002). These residues correspond to T206, T211 and T224 in the mouse NKCC1 sequence. In 2006, Vitari et al. used similar techniques on recombinant NKCC1 which had been phosphorylated by SPAK *in vitro*. They identified residues T197, T201 and T206 as targets of SPAK phosphorylation, whereas residue T211 was not phosphorylated (Vitari et al., 2006). These data were confirmed by site-directed mutagenesis experiments (Gagnon et al., 2007). It is worth noting that T206 and T211 are the most significant residues, as their substitution to alanine residues completely abrogates NKCC1 function (Darman and Forbush, 2002; Gagnon et al., 2007). Interestingly, T206 was common to both SPAK and forskolin-induced phosphorylation. In contrast, substitution of residues T197, T199 (not a target of phosphorylation), T201 and T224 has minimal effect of NKCC1 function. Whether there are truly two independent kinases, as the data suggest, still needs to be resolved. As the kinase acting downstream of the forskolin stimulation is still unknown, work is clearly needed both to identify and determine if these kinases act dependently or independently. One possibility is that phosphorylation of specific residues by one kinase could prime the phosphorylation of different residues by the other kinase. Finally, it is important to note that the phosphorylation events discussed in this section do not preclude the possibility that additional kinases modulate NKCC1 at other serine/threonine sites within the amino or carboxyl termini of the cotransporter.

NKCC2 regulation by SPAK and WNK3 has been assessed under low Cl^- stimulation conditions (Ponce-Coria et al., 2008). When expressed in *Xenopus laevis* oocytes, the renal-specific cotransporter was unaffected by coinjection of SPAK alone, but activated by the coinjection of both SPAK and WNK3. Furthermore, just as in NKCC1, NKCC2 was also inactivated by expressing dominant negative kinases: SPAK (K104R) or WNK3 (D294A). The study also demonstrated that WNK3 lies upstream of SPAK and the cotransporter, and that binding between SPAK and WNK3 is required for cotransporter regulation (Ponce-Coria et al., 2008). Taken together, these results demonstrate similar regulatory pathways for the two Na^+-K^+-$2Cl^-$ cotransporters.

Based on the high conservation between the phospho-threonine residues in the amino terminus of NKCC1, NKCC2 and NCC, and the presence of SPAK binding sites on the three cotransporters, regulation of the Na^+-Cl^- cotransporter by SPAK and OSR1 has also been recently investigated (Richardson et al., 2008). The Ste20 kinases were shown to phosphorylate NCC at three conserved threonines within the N-terminal tail of the cotransporter. This phosphorylation could be triggered by exposing the cells to a low Cl^- hypotonic medium which activated a WNK1-SPAK/OSR1-dependent signaling pathway.

Although not much information is known about phosphorylation sites on K^+-Cl^- cotransporters, heterologous expression of dominant negative SPAK together with KCC2 in *Xenopus laevis* oocytes revealed a role for the kinase in the inactivation of the cotransporter (Gagnon et al., 2006). This observation, reproduced in another study, seemed to be KCC2 specific, as the other K^+-Cl^- cotransporters were seemingly directly regulated by WNK4 independent of SPAK (Garzón-Muvdi et al., 2007).

Certainly more work is needed to understand fully the relationship between SPAK/OSR1 and WNKs. The work described here points to an important and complex interplay between SPAK/OSR1 and WNK taking place *in vivo* to impact cation-chloride cotransport function. The contribution of each of these kinases likely adds additional levels of regulation and specificity to CCC modulation. However, there is a significant body of work on the independent role of the WNK kinases in CCC regulation, and this will be discussed in the next section.

3. *WNK Kinases*

The WNK family of serine/threonine kinases was discovered in 2000 by the group of Melanie Cobb during a search for novel members of the mitogen-activated protein (MAP) kinase and extracellular-signal regulated kinase (ERK) families in the central nervous system. Functional and sequence analyses revealed that a newly isolated kinase was both closely related to MAP kinases, yet also fundamentally different. The kinase had high homology to serine/threonine kinases in the catalytic domain, yet lacked the conserved catalytic lysine (K) which most of the serine/threonine kinases possess in subdomain II, and which is required for ATP binding. Thus, the new kinase was named WNK1 for "with no K". Since WNK1 was able to autophosphorylate and phosphorylate myelin basic protein, a non-specific kinase substrate, it was evident that WNK1 possessed catalytic activity (Xu et al., 2000). The crystal structure of the WNK1 catalytic domain revealed that the "missing" lysine is located within subdomain I (Min et al., 2004). In the initial study, it was observed that among many stimuli tested, only the NaCl concentration of the extracellular medium affected WNK1 autophosphorylation. In addition, some specific mutations, like D368A were also shown to eliminate WNK1 kinase activity (Xu et al., 2000).

The initial identification of WNK1 was quickly followed by the isolation of three additional members of the family: WNK2, WNK3 and WNK4 (Verissimo and Jordan, 2001). WNKs can be divided in three domains: a short amino terminal domain of 146 to 220 amino acid residues in length, a serine/threonine kinase domain of 294 residues and finally a carboxyl terminal domain of variable size ranging from 796 residues in WNK4 to 1888 in WNK1 (Fig. 18.6). The degree of conservation among the four isoforms is very high within the kinase domain (>80%), but extremely low within the amino or carboxyl terminal domains (<15%). There are two coiled-coil domains in the carboxyl terminal domain conserved among the four WNKs, as well as a conserved auto-inhibitory motif shown to inhibit autophosphorylation in WNK1 (Lenertz et al., 2005; Xu et al., 2000).

The importance of WNKs could have passed unnoticed to the transport physiology world, but for Lifton's group (Wilson et al., 2001) which demonstrated in a landmark 2001 paper that intronic deletions of WNK1 or missense mutations in an acidic conserved region of WNK4 were the cause of a human form of inherited arterial hypertension accompanied by hyperkalemia (increased K^+ serum levels) and metabolic acidosis, known as pseudo-hypo-aldosteronism type II (PHAII). Two main reasons led to the hypothesis that PHAII was the result of increased activity of the distal nephron Na^+-Cl^- cotransporter, NCC. First, PHAII patients feature a clinical condition which is the mirror image of Gitelman's syndrome, an inherited monogenic disease which causes arterial hypotension, hypokalemia and metabolic alkalosis, and which is due to inactivating mutations along the *SLC12A3* gene encoding NCC (Simon et al., 1996). Second, clinicians had long observed that PHAII patients were particularly sensitive to very low doses of thiazide diuretics, a pharmacological agent which targets the Na^+-Cl^- cotransporter (Costanzo, 1985; Gamba et al., 1993; Stokes et al., 1984). Merely 20% of the typical thiazide dose was enough to improve hypertension and the other clinical features of PHAII. Furthermore, PHAII kindreds did not exhibit positive linkage to the *SLC12A3* region of chromosome 16, eliminating the possibility of gain of function mutations in the cotransporter to account for the disease (Simon et al., 1995). Since WNK1 and WNK4 were both shown to be expressed in the distal nephron (Wilson et al., 2001), WNKs were proposed to modulate the activity of NCC and a few months later this was indeed demonstrated simultaneously by two independent groups (Wilson et al., 2003; Yang et al., 2003).

a. Modulation of NCC Activity by WNK4

In 2003, Wilson and co-workers (Wilson et al., 2003) showed that wild-type WNK4 reduces the activity of NCC when coexpressed in *Xenopus laevis* oocytes

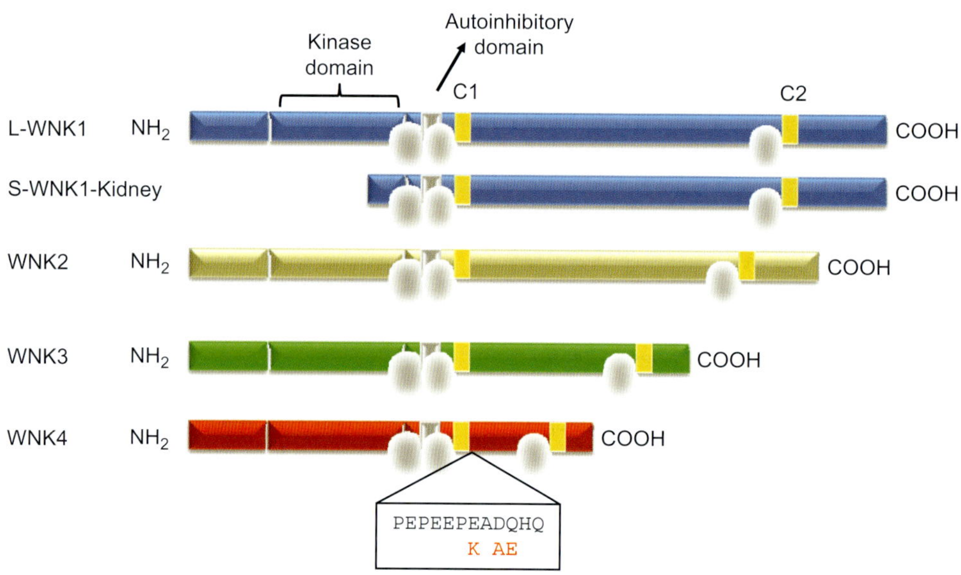

FIGURE 18.6 Proposed topological model of WNK proteins. The central kinase domain is shown by bracket. The auto-inhibitory domain and the coiled/coiled domains are highlighted. Locations of the negatively charged residues stretch in which WNK4 mutations occur are shown below WNK4.

(Table 18.2). This decrease was due, at least in part, to a decrease in the amount of NCC present at the cell surface. In contrast, the inhibitory effect of WNK4 was not observed when using mutant WNK4 cRNAs, either harboring the D318A mutation which eliminates WNK4 catalytic activity, or containing one of the PHAII-type mutations (E562K) reported in at least one family with PHAII (Wilson et al., 2001). Interestingly, other PHAII-type mutations tested were still able to inhibit NCC (Yang et al., 2003). These observations suggest that wild-type WNK4 is a natural inhibitor of NCC, and that inhibition requires the catalytic activity of the kinase (Table 18.2). PHAII-type mutations result in a partial or complete loss of WNK4 ability to inhibit the cotransporter, thereby resulting in arterial hypertension, hypokalemia and metabolic acidosis. The inhibitory effect of WNK4 upon NCC activity was confirmed by other groups, using not only *Xenopus laevis* oocytes, but also mammalian cells in culture (Cai et al., 2006; Golbang et al., 2006). In addition to NCC, other transport systems are regulated by WNK4 that help to explain the phenotype in PHAII patients. For instance, the activity of the epithelial potassium channel ROMK, located in the distal nephron and responsible for K^+ secretion into the urine, is also inhibited by WNK4. Thus, PHAII-type WNK4 mutations dis-inhibit ROMK leading to hyperkalemia (Kahle et al., 2003). In addition, similar to NCC, it has also been observed that WNK4 inhibits the activity of the epithelial sodium channel, ENaC, and that this effect is prevented by PHAII-type mutations (Ring et al., 2007). This demonstrates the presence of another component in the development of hypertension and hyperkalemia in PHAII patients, this being the loss of ENaC inhibition by mutant WNK4.

The proposed roles for wild-type WNK4 and PHAII-mutant WNK4 upon NCC have also been corroborated by two independent groups using transgenic animals as *in vivo* models. These studies further suggest that the consequences of WNK4 mutations upon NCC were sufficient to explain the whole PHAII phenotype. By introducing genomic segments harboring wild-type or PHAII-mutant WNK4 into the mouse genome, Lalioti et al. produced BAC transgenic mice (Lalioti et al., 2006) with extra activity of wild-type WNK4, Tg(WNK4^{WT}), as well as transgenic mice overexpressing the PHAII-type mutant WNK4, Tg(WNK4^{PHAII}) (see Fig. 18.7). When compared to normal mice, the arterial pressure of Tg(WNK4^{WT}) mice was significantly lower, whereas the arterial pressure of Tg(WNK4^{PHAII}) mice was significantly higher. These observations correlate well with the transgene effect upon the distal convoluted tubule (DCT). Furthermore, Tg(WNK4^{WT}) mice exhibited a decreased size and number of DCTs, while a clear hyperplasia and hypertrophy was observed in Tg(WNK4^{PHAII}) mice. Additionally, serum potassium levels were reduced in Tg(WNK4^{WT}) mice, while increased in Tg(WNK4^{PHAII}) mice. Taken together, the extra activity of wild-type WNK4 produces a Gitelman-like syndrome in which the animals exhibit arterial hypotension, hypokalemia and decreased number of DCTs, consistent with wild-type WNK4 inhibiting NCC activity. In contrast, the presence of the PHAII mutation in Tg(WNK4^{PHAII}) mice was associated with a PHAII-like syndrome featuring arterial hypertension, hyperkalemia and increased number of DCTs, again consistent with the idea that a single PHAII-type mutation on WNK4 switched its effect on NCC from inhibition to activation. Similar observations were obtained using a PHAII-type mutant knock-in mouse in which homologous recombination was used to create a mutant allele of WNK4 harboring the D561A mutation (Yang et al., 2007). The BAC transgenic animals and

TABLE 18.2 Effects of WNKs and their catalytically inactive forms on SLC12A cotransporters

	NCC	NKCC1	NKCC2	KCC1	KCC2	KCC3	KCC4
WNK1	↔	↔	↔	↔	↔	↔	↔
WNK1-DA	—	—	—	—	—	—	—
WNK3	↑↑	↑↑	↑↑	↓↓	↓↓	↓↓	↓↓
WNK3-DA	↓↓	↓↓	↓↓	↑↑	↑↑	↑↑	↑↑
WNK4	↓	↓ * ↑	↓	↓	↓	↓	↓
WNK4-DA	↔	—	—	↔	↑	↑	↔

↔ —*One study shows inhibition and one activation (see text for explanation)

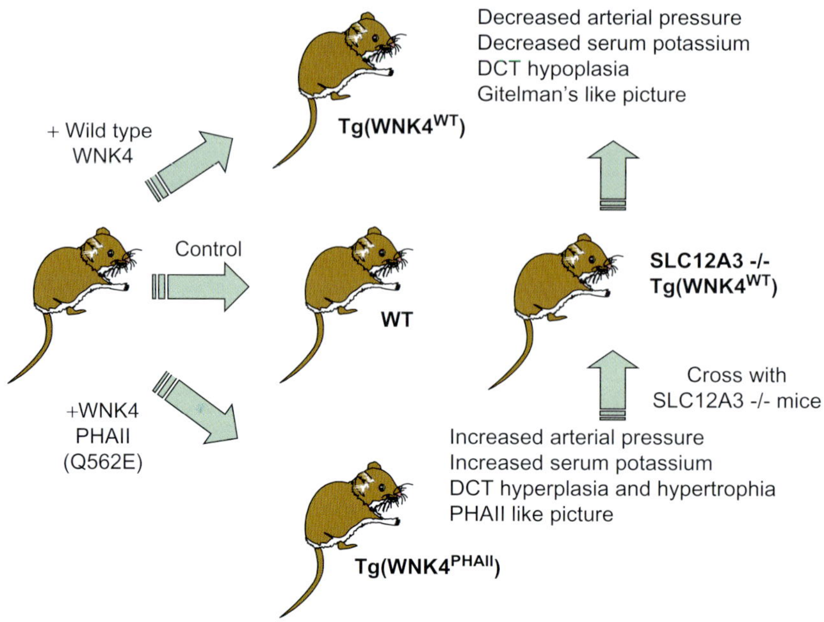

FIGURE 18.7 BAC transgenic mice demonstrate that wild-type WNK4 is a natural inhibitor of NCC and mutant PHAII-type WNK4 is an activator. Based on the results of Lalioti et al. (Lalioti et al., 2006).

knock-in mice are somewhat different models: in the first model, the mice have two normal WNK4 alleles plus extra copies of wild-type or PHAII mutant WNK4 DNAs. In contrast, in the second model, there are only two WNK4 alleles, one normal and one mutated. The clinical consequences of both models were, however, similar. Interestingly, in both studies, eliminating the activity of NCC was enough to reverse the entire phenotype. Indeed, crossing the Tg(WNK4^{PHAII}) mice with NCC null mice prevented the appearance of arterial hypertension and of hyperkalemia, even in mice exposed to high potassium diet (Lalioti et al., 2006). Similarly, treatment of WNK4-D561A$^{+/-}$ knock-in mice with thiazide also reversed all clinical features (Yang et al., 2007).

A recent study has shown that WNK4 effect towards NCC is modulated by the hypertension-induced hormone angiotensin II and that this effect involves SPAK and NCC phosphorylation (San Cristobal P., 2009). Reconstitution experiments in oocytes using combinations of NCC, WNK4, SPAK and the angiotensin II AT1 receptor cRNAs demonstrated that wild-type WNK4 inhibition on NCC is switched to activation in the presence of angiotensin II. The effect required the presence of AT1 receptor and was completely abrogated with the specific AT1 receptor blocker losartan. In mammalian cells from distal convoluted tubule the angiotensin II effect was accompanied by SPAK and NCC phosphorylation in key residues known to be associated with their activation (Pacheco-Alvarez et al., 2006; Richardson et al., 2008). Thus, angiotensin II modulates the WNK4-SPAK-NCC interaction to switch WNK4 from an inhibitor to an activator of NCC. Interestingly, WNK4 harboring PHAII-type mutations no longer inhibited NCC and also did not respond to angiotensin II providing constitutive activation of the signaling pathway between AT1 receptor and NCC.

WNK4 also regulates other members of the SLC12 family (Table 18.2). Using *Xenopus laevis* oocytes as a heterologous expression system, WNK4 was observed to inhibit NKCC1 by decreasing its expression at the cell surface (Kahle et al., 2004). However, using the same expression system, it was later demonstrated that the effect of WNK4 on NKCC1 activity is completely opposite in the presence of SPAK (Gagnon et al., 2006). Indeed, while coinjection of NKCC1 cRNA with either WNK4 cRNA or SPAK cRNA alone had no significant effect on NKCC1 activity, the coinjection of the cotransporter with both kinases resulted in a significant activation of the cotransporter. These observations were consistent with data showing that phosphorylation of the amino terminal domain of NKCC1 was only achieved when coincubated with WNK1 or WNK4 together with SPAK or OSR1 (Moriguchi et al., 2006; Vitari et al., 2005, 2006). In support of these observations, it was also demonstrated that WNK1 is activated by hyperosmotic stress which is known to significantly increase the activity of NKCC1 (Lenertz et al., 2005; Zagórska et al., 2007), and that WNK1 modulates the activity of NKCC1 in HeLa cells (Anselmo et al., 2006). Finally, WNK4 is a negative regulator of the K^+-Cl^- cotransporters; this effect is dependent on WNK4 catalytic activity but independent of SPAK (Garzón-Muvdi et al., 2007).

b. Modulation of NCC Activity by WNK1

Deletions within the first intron of WNK1 result in an increased expression of WNK1 protein. An association between these mutations and PHAII suggests that increased expression of WNK1 is another possible cause of the disease. Consistent with this hypothesis, single nucleotide polymorphisms within the WNK1 gene are associated with variations in blood pressure levels (Newhouse et al., 2005), and heterozygous WNK1$^{-/+}$ mice exhibit reduced blood pressure (Zambrowicz et al., 2003). A direct effect of WNK1 on NCC or any other member of the SLC12A family has yet to be demonstrated using *Xenopus laevis* oocytes (Table 18.2) as an expression system (Yang et al., 2003). Thus, it is currently unknown if WNK1 alone can modulate NCC function. It is possible that WNK1 affects NCC activity by interacting with WNK4 and/or interacting with a renal-specific spliced isoform of WNK1. Indeed, Yang and co-workers did not observe an effect of WNK1 on NCC-mediated $^{22}Na^+$ uptake in *Xenopus laevis* oocytes, while WNK4 inhibited NCC activity (Yang et al., 2003). However, coinjection of WNK1, WNK4 and NCC cRNAs resulted in the abrogation of WNK4-induced inhibition of the cotransporter, indicating that WNK1 was able to prevent WNK4 inhibition of NCC through a mechanism that seemed to include physical interaction between the two kinases (Lenertz et al., 2005; Yang et al., 2003, 2005). This observation could suggest that increased WNK1 protein in PHAII patients could prevent WNK4-induced inhibition of NCC, thus increasing the activity of this cotransporter. This explanation would be valid if WNK1 was expressed in the distal convoluted tubule. However, WNK1 is mainly expressed as a shorter spliced isoform in the kidney (Delaloy et al., 2003; O'Reilly et al., 2003) (Fig. 18.6). This variant is the result of alternative splicing of exons 1–4. The transcription of this isoform is under the control of an intron 4 alternative promoter and contains a sequence from an extra exon designated exon 4a, located between exons 4 and 5. As shown in Fig. 18.6, the shorter WNK1 isoform (S-WNK1) lacks the first 437 amino acid residues, including almost the entire kinase domain. In the kidney, the L-WNK1 isoform is present along the entire nephron, while S-WNK1 is exclusively expressed in the aldosterone-sensitive distal nephron, and is particularly abundant in the distal convoluted tubule (DCT) and connecting tubule (CNT). Therefore, there are more copies of S-WNK1 than L-WNK1 in the aldosterone-sensitive distal nephron. Interestingly, it was observed that S-WNK1, by interacting with L-WNK1 in a dominant negative fashion, eliminated the L-WNK1-induced inhibition of WNK4 (Subramanya et al., 2006). Our current understanding of WNK1-induced hypertension in PHAII patients is that in normal subjects, the ratio of S-WNK1/L-WNK1 expression in DCT and CNT is in favor of S-WNK1. Prevention of L-WNK1 inhibition of WNK4 allows WNK4 to inhibit NCC. Conversely, intronic deletions of *PRKWNK1* in PHAII patients result in increased expression of L-WNK1. As a consequence, the ratio of S-WNK1 to L-WNK1 is reduced and there are many copies of L-WNK1 that are not affected by S-WNK1. These copies can inhibit WNK4 which in turn results in increased NCC activity, increased DCT salt reabsorption, and increased arterial pressure. In support of this model, it has also been shown that S-WNK1 and L-WNK1 interact and regulate the activity of the potassium channel ROMK in a similar fashion (Lazrak et al., 2006).

c. Modulation of Cation-Chloride Cotransporters by WNK3

As shown in Fig. 18.6, the WNK family is composed of four genes, two of which (WNK1 and WNK4) have been extensively studied due to their causative role in PHAII. However, WNK2 and WNK3 are also important kinases. In fact, analysis of WNK3 coexpression with cation-chloride cotransporters in *Xenopus laevis* oocytes revealed a series of effects that positioned WNK3 as a strong candidate, along with SPAK and OSR1, as the intracellular Cl^--sensitive kinase (Lytle and McManus, 2002).

WNK3 cDNA was first cloned by Holden, Cox and Raymond from a human fetal brain cDNA library (Holden et al., 2004). WNK3 is an 1,800 amino acid long serine/threonine kinase from which at least two alternatively spliced variants have been identified. One variant originates from an alternative splice donor that introduces 47 amino acid residues into exon 18 and is expressed solely in brain. The other variant results from alternative splicing of exon 22 and eliminates nine residues at the end of the carboxyl terminal domain. RT-PCR analysis demonstrated that WNK3 transcripts are present in all tissues (Holden et al., 2004) and immunohistochemistry has demonstrated WNK3 protein present in epithelial cells of the intestine, stomach, pancreas, bile duct and along the entire nephron. Interestingly, WNK3 is also expressed in neurons containing GABA receptors, suggesting a direct role for the kinase in cation-chloride cotransport regulation, and indirect regulation of GABAergic function (Kahle et al., 2005; Rinehart et al., 2005).

The effect of WNK3 on the SLC12 cotransporters has been examined using the *Xenopus laevis* oocyte expression system (Table 18.2). WNK3 was shown to be a powerful activator of the sodium-coupled chloride cotransporters NCC, NKCC1 and NKCC2 (Kahle et al., 2005; Rinehart et al., 2005) (Fig. 18.8). The activity of these cotransporters is increased by more than two-fold when coexpressed with WNK3. Activation of NCC by WNK3 is associated with increased amounts of cell-surface NCC, suggesting that WNK3 promotes insertion of NCC into the plasma membrane (Rinehart et al., 2005). Activation of NKCC1 and NKCC2 is accompanied by phosphorylation of threonines in the amino terminal domain. These phospho-acceptor sites were previously shown to become phosphorylated in NKCC1 when activated by depletion of $[Cl^-]_i$ (Darman et al., 2001; Darman and Forbush, 2002; Dowd and Forbush, 2003) and in NKCC2 when activated by cell shrinkage (Gimenez and Forbush, 2005) or by vasopressin *in vivo* (Giménez and Forbush, 2003). Interestingly, the WNK3-induced activation of NKCC1 and NKCC2 is observed even in oocytes incubated in hypotonic medium to induce cell swelling. For example, NKCC1 activity in shrunken oocytes injected with NKCC1 cRNA alone was similar to NKCC1 activity in swollen oocytes coinjected with NKCC1 and WNK3 cRNA. This situation is similar to NKCC1's loss of osmotic sensitivity seen in oocytes coinjected with SPAK and WNK4 (Gagnon et al., 2006).

It is known that eliminating an aspartic acid in WNK kinase domains renders these proteins catalytically inactive (Xu et al., 2000). In WNK3, this aspartate corresponds to amino acid residue 294. Thus, the WNK3-D294A mutant has been extensively used to define the extent to which WNK3 requires the catalytic activity of the kinase. Interestingly, catalytically inactive WNK3 (WNK3-D294A) is not able to activate NKCC1, NKCC2 and NCC. Instead, WNK3-D294A became a powerful inhibitor of these cotransporters (Fig. 18.8). When NCC, NKCC1 or NKCC2 are coinjected with WNK3-D294A cRNA, the basal activity of the cotransporters is reduced by more than 70%. The inhibitory effect was observed even in shrunken oocytes in which NKCC1 or NKCC2 are maximally active (Kahle et al., 2005). In addition, inactivation of NKCC1 and NKCC2 by catalytically inactive WNK3 was accompanied by dephosphorylation of the amino terminal threonines. Again, this situation is similar to the requirement of WNK4 catalytic activity for NKCC1 stimulation (Gagnon et al., 2006).

As discussed above, the SLC12 family is composed of two branches that are completely opposite

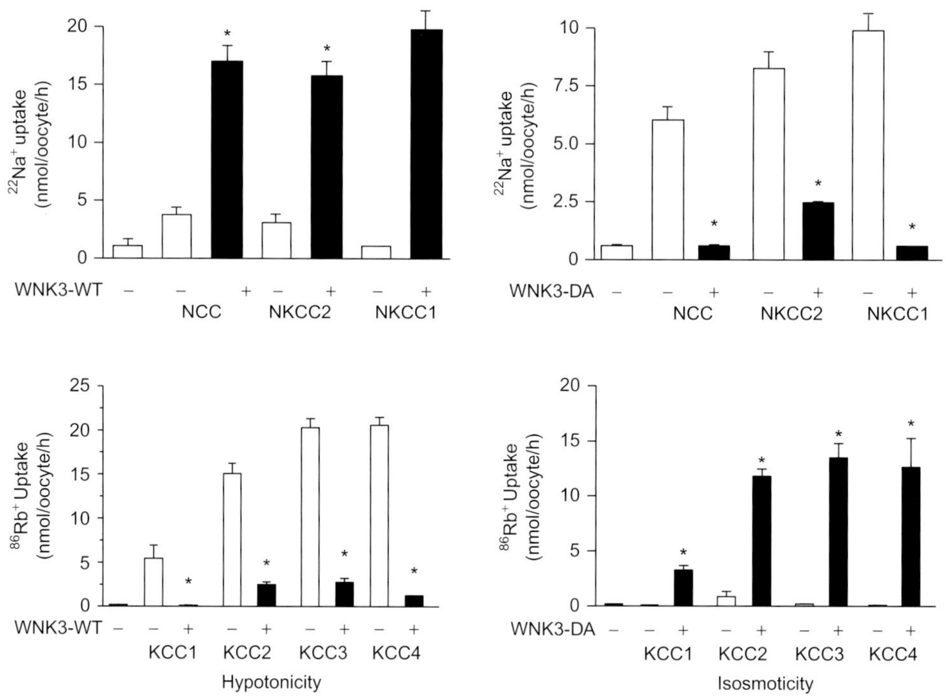

FIGURE 18.8 Effect of wild-type (WNK3-WT) and catalytically inactive WNK3 (WNK3-DA) on SLC12A cotransporters. A and B depict effect on Na^+-driven cotransporters. These experiments were performed in isotonic conditions. C and D depict effect on K^+-driven cotransporters. Experiments were performed in hypotonic conditions for C and isotonic conditions for D. The figure is a compilation of experiments from several studies (de Los Heros et al., 2006; Garzón-Muvdi et al., 2007; Kahle et al., 2005; Rinehart et al., 2005; Wilson et al., 2003).

to each other. One branch includes cotransporters that use sodium as a cation coupled to chloride: NKCC1, NKCC2 and NCC. These cotransporters follow the sodium gradient established by the Na^+/K^+-ATPase, and move chloride ions from outside to inside. Their activity is increased by cell shrinkage through mechanisms that are associated with phosphorylation processes. In contrast, the other branch consists of cotransporters that are sodium independent and move K^+ together with Cl^-. Thus, following the gradient established by the Na^+/K^+-ATPase for K^+, KCCs move chloride ions from inside the cell to outside. These cotransporters are activated by cell swelling via mechanisms that involve dephosphorylation processes. Interestingly, the WNK3 effect upon KCCs is completely opposite to that for NKCCs. When expressed in *Xenopus laevis* oocytes, the K^+-Cl^- cotransporters KCC1, KCC3 and KCC4 are completely inactive under isotonic conditions (Mercado et al., 2000, 2005), while KCC2 exhibits a small, but significant activity (Song et al., 2002; Strange et al., 2000). All four cotransporters are activated by cell swelling (Mercado et al., 2000, 2005; Song et al., 2002; Strange et al., 2000); however, when coexpressed with wild-type WNK3, activation by cell swelling is completely prevented (de Los Heros et al., 2006) (Fig. 18.8). These results suggest that wild-type WNK3 is a powerful inhibitor of the Na^+-independent branch of the SLC12 family. The effect of catalytically inactive WNK3 (WNK3-D294A) is also significant and opposite to the effect of wild-type WNK3. Indeed, coinjection of *Xenopus laevis* oocytes with any KCC and WNK3-D294A results in substantial activation of the cotransporters (Fig. 18.8), even under isotonic conditions in which these cotransporters are usually inactive. Moreover, it was observed that activation of K^+-Cl^- cotransporters by WNK3-D294A was prevented by incubation of oocytes in the presence of calyculin A and/or cyclosporine A, indicating that the increased cotransporter activity was associated with the activity of protein phosphatases 1 and 2B (de Los Heros et al., 2006). Thus, for the activation or inhibition of all members of the SLC12 family, WNK3 and WNK3-D294A are bypassing the normal changes in cell volume and/or $[Cl^-]_i$ that are usually required for regulation. A recent study shows that activation of NKCC2 by intracellular chloride depletion requires the presence and activity of WNK3 and SPAK, and that WNK3 lies upstream of SPAK (Ponce-Coria et al., 2008).

3. *Apoptosis-Associated Tyrosine Kinase*

AATYK (apoptosis-associated tyrosine kinase), a protein with dual serine/threonine and tyrosine kinase activity, was identified and cloned in 1997 using subtractive hybridization techniques from myeloid precursor cells triggered to undergo apoptosis (Gaozza et al., 1997). The kinase was identified based on its up-regulation upon apoptotic induction. Later studies showed that AATYK is highly expressed in neurons and is also up-regulated during neuronal differentiation (Raghunath et al., 2000). AATYK and a closely related protein (known today as AATYK3 (Tomomura et al., 2007)) were also identified as interactors of SPAK through a yeast 2-hybrid screen (Piechotta et al., 2003). As SPAK regulates NKCC1 activity, the function of AATYK was also assessed in relationship to the cotransporter. Coinjection of AATYK with NKCC1 resulted in a complete loss of NKCC1 function (Gagnon et al., 2007). Several experiments were performed to understand the mechanism by which the kinase affects NKCC1 function. First, it was shown that AATYK does not affect cotransporter expression at the cell surface, thus indicating that the kinase does not impact NKCC1 activity via insertion or removal of NKCC1 into the plasma membrane. Second, and more puzzling, was the observation that AATYK exerted its effect even when its catalytic kinase activity was abolished through mutagenesis. This observation pointed to AATYK acting as a scaffolding protein rather than a phosphoregulator. Third, the need for SPAK binding to the regulatory domain was also tested through mutagenesis of the two RFxV SPAK binding motifs which were mutated to prevent SPAK interaction (Fig. 18.9). In this case, inactivating mutations of the SPAK binding motifs completely abolished the AATYK effect, indicating that AATYK was exerting its effect on NKCC1 through the Ste20 kinase. Interestingly, there was an additional requirement for the AATYK effect: the interaction of the kinase with PP1. The possibility of PP1's involvement was considered after the realization that the regulatory domain of AATYK also contained a PP1 binding motif (Fig. 18.9). When oocytes were injected with an AATYK mutant that was unable to interact with PP1, there was no inhibition of NKCC1 function by the mutant. This observation led to a model where both kinase (SPAK) and phosphatase (PP1) needed to be brought together to promote NKCC1 inhibition (Gagnon et al., 2007). This model proposes that PP1 is held in proximity of SPAK to dephosphorylate and inactivate the kinases. Interestingly, SPAK and PP1 binding motifs are also present on the amino terminus of NKCC1, itself suggesting that the cotransporter itself might serve as a scaffold (Darman et al., 2001; Piechotta et al., 2002). AATYK might play a similar role by scaffolding PP1 and the kinase Cdk5 at its carboxyl terminus (Honma et al., 2003). Finally, for AATYK to be relevant to NKCC1 function, it needs to be expressed in the

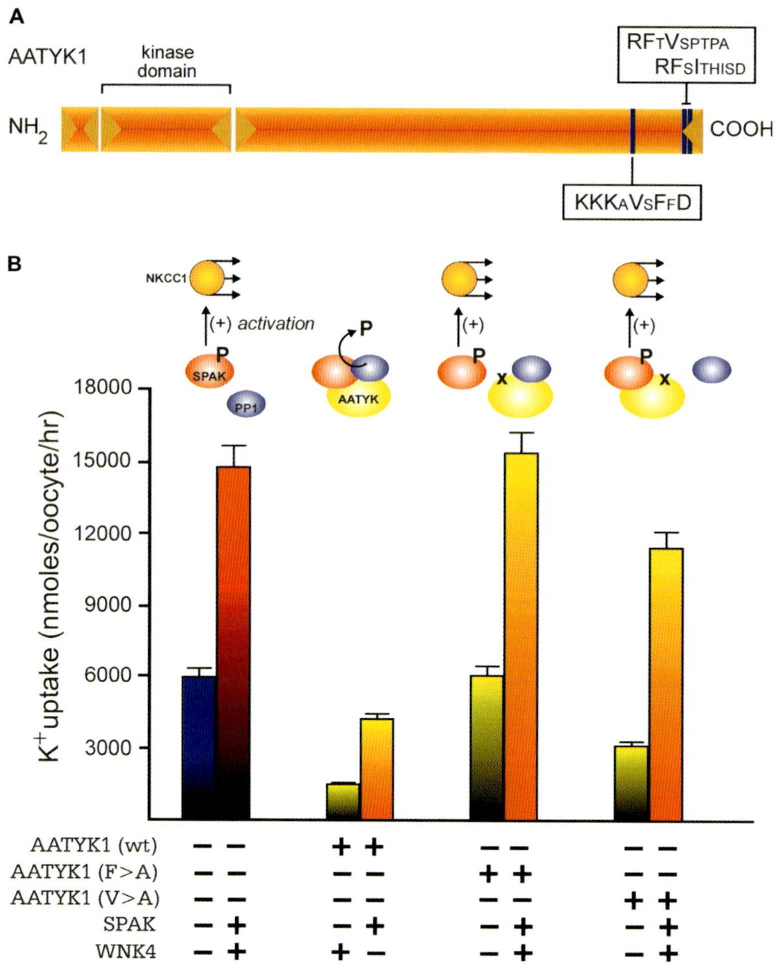

FIGURE 18.9 Putative role of AATYK in the modulation of NKCC1 function. **A.** Representation of the 1,800 amino acid long AATYK1 protein showing the kinase domain and long downstream regulatory domain. One PP1 binding site and two SPAK binding sites are highlighted. Binding of respective proteins to these sites was confirmed by yeast 2-hybrid analysis (Gagnon et al., 2007). **B.** Summary of data from Gagnon et al. (2007) showing inhibition of NKCC1 occurring upon expression of AATYK. Mutations of the SPAK binding domains or the PP1 binding domain prevent this inhibition.

same cells as the cotransporter. Such colocalization has not yet been tested in detail, but it is important to note both proteins are expressed in neurons, and have particularly high expression in sensory dorsal root ganglion neurons (Alvarez-Leefmans et al., 2002; Sung et al., 2000; Tomomura et al., 2007).

4. *AMP-activated Kinase*

In 2007, Fraser and co-workers demonstrated that AMP-activated protein kinase also regulated NKCC2 function (Fraser et al., 2007). It was demonstrated that (1) AMP-activated kinase was coexpressed with NKCC2 on the apical membrane of the thick ascending limb of Henle (Fraser et al., 2005); (2) the kinase coimmunoprecipitated with the N-terminus of NKCC2; (3) the kinase phosphorylated NKCC2 at residue S126 (rabbit NKCC2); and (4) mutation of S126 resulted in a decrease in basal NKCC2 transport, as measured in *Xenopus laevis* oocytes through ^{86}Rb influx. Interestingly, residue S126 is highly conserved among the Na^+-dependent cation-chloride cotransporters, indicating a broader function of the AMP-activated kinase on ion transport. The addition of this new player makes the study of cotransporter phosphoregulation even more challenging.

VI. CONCLUSIONS

As evidenced in this chapter, the regulation of cation-chloride cotransporters is very complex and involves many processes. Work of the past 30 years has only begun highlighting the complexity of regulation, and many more studies are required to fill the many remaining blanks and unanswered questions. Because of its important role in kidney function and the existence of potent pharmacological agents, the physiology of cation-chloride cotransporters is better known in this organ, and the impact of its regulation by complex phosphorylation/dephosphorylation mechanisms (and its players) has received greater attention. The development of specific phosphopeptide antibodies and specific kinase inhibitors should facilitate our understanding of cation-chloride cotransporter regulation. As the role of cation-chloride cotransporters expands to other systems, including the nervous system (a topic covered in many chapters of this book), particular attention to the regulatory pathways will certainly constitute the focus of future efforts.

References

Aguado, F., Carmona, M.A., Pozas, E., Aguilo, A., Martinez-Guijarro, F.J., Alcantara, S., Borrell, V., Yuste, R., Ibanez, C. F., and Soriano, E. (2003). BDNF regulates spontaneous correlated activity at early developmental stages by increasing synaptogenesis and expression of the K^+-Cl^- co-transporter KCC2. *Development* **130**, 1267–1280.

Altamirano, A.A., Breitwieser, G.E., and Russel, J.M. (1988). Vanadate and fluoride effects on Na-K-Cl cotransport in squid giant axon. *Am. J. Physiol.* **254**, C582–C586.

Alvarez-Leefmans, F.J., Leon-Olea, M., Mendoza-Sotelo, J., Alvarez, F.J., Anton, B., and Garduno, R. (2002). Immunolocalization of the Na^+-K^+-$2C^-$ cotransporter in peripheral nervous tissue of vertebrates. *Neuroscience* **104**, 569–582.

Anselmo, A.N., Earnest, S., Chen, W., Juang, Y.C., Kim, S.C., Zhao, Y., and Cobb, M.H. (2006). WNK1 and OSR1 regulate the Na^+, K^+, $2Cl^-$ cotransporter in HeLa cells. *Proc. Natl. Acad Sci. USA* **103**, 10883–10888.

Arrindell, E.L., McKay, B.S., Jaffe, G.J., and Burke, J.M. (1992). Modulation of potassium transport in cultured retinal pigment epithelium and retinal glial cells by serum and epidermal growth factor. *Exp. Cell Res.* **203**, 192–197.

Attmane-Elakeb, A., Mount, D.B., Sibella, V., Vernimmen, C., Hebert, S.C., and Bichara, M. (1998). Stimulation by in vivo and in vitro metabolic acidosis of expression of rBSC-1, the Na^+-$K^+(NH_4^+)$-$2Cl^-$ cotransporter of the rat medullary thick ascending limb. *J. Biol. Chem.* **273**, 33681–33691.

Bazzini, C., Vezzoli, V., Sironi, C., Dossena, S., Ravasio, A., De Biasi, S., Garavaglia, M., Rodighiero, S., Meyer, G., Fascio, U., Fürst, J., Ritter, M., Bottà, G., and Paulmichl, M. (2005). Thiazide-sensitive NaCl− cotransporter in the intestine: possible role of hydrochlorothiazide in the intestinal Ca^{2+} uptake. *J. Biol. Chem.* **280**, 19902–19910.

Blaesse, P., Guillemin, I., Schindler, J., Schweizer, M., Delpire, E., Khiroug, L., Friauf, E., and Nothwang, H.G. (2006). Oligomerization of KCC2 correlates with development of inhibitory neurotransmission. *J. Neurosci.* **26**, 10407–10419.

Brock, T.A., Brugnara, C., Canessa, M., and Gimbrone, M.A.J. (1986). Bradykinin and vasopressin stimulate Na^+-K^+-Cl^- cotransport in cultured endothelial cells. *Am. J. Physiol.* **250**, C888–C895.

Brunet, G.M., Gagnon, E., Simard, C.F., Daigle, N.D., Caron, L., Noël, M., Lefoll, M.H., Bergeron, M.J., and Isenring, P. (2005). Novel insights regarding the operational characteristics and teleological purpose of the renal Na+-K+-Cl2 cotransporter (NKCC2s) splice variants. *J. Gen. Physiol.* **126**, 325–337.

Cai, H., Cebotaru, V., Wang, Y.H., Zhang, X.M., Cebotaru, L., Guggino, S.E., and Guggino, W.B. (2006). WNK4 kinase regulates surface expression of the human sodium chloride cotransporter in mammalian cells. *Kidney Int.* **69**, 2162–2170.

Carmosino, M., Giménez, I., Caplan, M., and Forbush, B. (2008). Exon loss accounts for differential sorting of Na-K-Cl cotransporters in polarized epithelial cells. *Mol. Biol. Cell.*

Casula, S., Shmukler, B.E., Wilhelm, S., Stuart-Tilley, A.K., Su, W., Chernova, M.N., Brugnara, C., and Alper, S.L. (2001). A dominant negative mutant of the KCC1 K-Cl cotransporter: both N- and C-terminal cytoplasmic domains are required for K-Cl cotransport activity. *J. Biol. Chem.* **276**, 41870–41878.

Clayton, G.H., Owens, G.C., Wolf, J.S., and Smith, R.L. (1998). Ontogeny of cation-Cl^- cotransporter expression in rat neocortex. *Brain Research. Developmental Brain Research* **109**, 281–292.

Clemo, H.F. and Baumgarten, C.M. (1991). Atrial natriuretic factor decreases cell volume of rabbit atrial and ventricular myocytes. *Am. J. Physiol. (Cell Physiol.)* **260**, C681–C690.

Clemo, H.F., Feher, J.J., and Baumgarten, C.M. (1992). Modulation of rabbit ventricular cell volume and Na+/K+/2Cl− cotransport by cGMP and atrial natriuretic factor. *J. Gen. Physiol.* **100**, 89–114.

Costanzo, L.S. (1985). Localization of diuretic action in microperfused rat distal tubules: Ca and Na transport. *Am. J. Physiol. (Renal Physiol.)* **248**, F527–F535.

Coull, J.A., Boudreau, D., Bachand, K., Prescott, S.A., Nault, F., Sik, A., De Koninck, P., and De Koninck, Y. (2003). Trans-synaptic shift in anion gradient in spinal lamina I neurons as a mechanism of neuropathic pain. *Nature* **424**, 938–942.

Crable, S.C., Hammond, S.M., Papes, R., Rettig, R.K., Zhou, G.P., Gallagher, P.G., Joiner, C.H., and Anderson, K.P. (2005). Multiple isoforms of the KC1 cotransporter are expressed in sickle and normal erythroid cells. *Exp. Hematol.* **33**, 624–631.

Crook, R.B., von Brauchitsch, D.K., and Polansky, J.R. (1992). Potassium transport in nonpigmented epithelial cells of ocular ciliary body: inhibition of a Na^+, K^+, Cl^- cotransporter by protein kinase C. *J. Cell. Physiol.* **153**, 214–220.

Darman, R.B., Flemmer, A., and Forbush, B. (2001). Modulation of ion transport by direct targeting of PP1 to the Na-K-Cl cotransporter. *J. Biol. Chem.* **276**, 34359–34362.

Darman, R.B. and Forbush, B. (2002). A regulatory locus of phosphorylation in the N terminus of the Na-K-Cl cotransporter, NKCC1. *J. Biol. Chem.* **277**, 37542–37550.

de Jong, J.C., Willems, P.H., Mooren, F.J., van den Heuvel, L.P., Knoers, N.V., and Bindels, R.J. (2003). The structural unit of the thiazide-sensitive NaCl cotransporter is a homodimer. *J. Biol. Chem.* **278**, 24302–24307.

de Los Heros, P., Kahle, K.T., Rinehart, J., Bobadilla, N.A., Vazquez, N., San Cristobal, P., Mount, D.B., Lifton, R.P., Hebert, S.C., and Gamba, G. (2006). WNK3 bypasses the tonicity requirement for K-Cl cotransporter activation via a phosphatase-dependent pathway. *Proc. Natl. Acad Sci. USA* **103**, 1976–1981.

DeAizpurua, H.J., Cram, D.S., Naselli, G., Devereux, L., and Dorow, D.S. (1997). Expression of mixed lineage kinase-1 in pancreatic beta-cell lines at different stages of maturation and during embryonic pancreas development. *J. Biol. Chem.* **272**, 16364–16373.

Del Castillo, I.C., Fedor-Chaiken, M., Song, J.C., Starlinger, V., Yoo, J., Matlin, K.S., and Matthews, J.B. (2005). Dynamic regulation of Na^+-K^+-$2Cl^-$ cotransporter surface expression by PKC-{epsilon} in Cl^--secretory epithelia. *Am. J. Physiol. Cell Physiol.* **289**, C1332–C1342.

Delaloy, C., Lu, J., Houot, A.M., Disse-Nicodeme, S., Gasc, J.M., Corvol, P., and Jeunemaitre, X. (2003). Multiple promoters in the

WNK1 gene: one controls expression of a kidney-specific kinase-defective isoform. *Mol. Cell Biol.* **23**, 9208–9221.

Delpire, E. and Gagnon, K.B. (2007). Genome-wide analysis of SPAK/OSR1 binding motifs. *Physiol. Genomics* **28**, 223–231.

Delpire, E. and Gagnon, K.B. (2008). SPAK and OSR1: STE20 kinases involved in the regulation of ion homoeostasis and volume control in mammalian cells. *Biochem. J.* **409**, 321–331.

Delpire, E. and Lauf, P.K. (1992). Kinetics of DIDS inhibition of swelling-activated K-Cl cotransport in low K sheep erythrocytes. *J. Membrane Biol.* **126**, 89–96.

Delpire, E., Rauchman, M.I., Beier, D.R., Hebert, S.C., and Gullans, S.R. (1994). Molecular cloning and chromosome localization of a putative basolateral Na-K-2Cl cotransporter from mouse inner medullary collecting duct (mIMCD-3) cells. *J. Biol. Chem.* **269**, 25677–25683.

Denton, J., Nehrke, K., Yin, X., Morrison, R., and Strange, K. (2005). GCK-3, a newly identified Ste20 kinase, binds to and regulates the activity of a cell cycle-dependent ClC anion channel. *J. Gen. Physiol.* **125**, 113–125.

Dharmsathaphorn, K., Mandel, K.G., Masui, H., and McRoberts, J.A. (1985). Vasoactive intestinal polypeptide-induced chloride secretion by a colonic epithelial cell line. Direct participation of a basolaterally localized Na^+-K^+-Cl^- cotransport system. *J. Clin Invest.* **75**, 462–471.

Di Fulvio, M., Lauf, P.K., and Adragna, N.C. (2003). The NO signaling pathway differentially regulates KCC3a and KCC3b mRNA expression. *Nitric Oxide* **9**, 165–171.

Dong, J. and Delamere, N.A. (1994). Protein kinase C inhibits Na(+)-K(+)-2Cl− cotransporter activity in cultured rabbit nonpigmented ciliary epithelium. *Am. J. Physiol.* **276**, C1553–C1560.

Dowd, B.F. and Forbush, B. (2003). PASK (proline-alanine-rich STE20-related kinase), a regulatory kinase of the Na-K-Cl cotransporter (NKCC1). *J. Biol. Chem.* **278**, 27347–27353.

Dunham, P.B. and Ellory, J.C. (1981). Passive potassium transport in low potassium sheep red cells: dependence upon cell volume and chloride. *J. Physiol. (Lond.)* **318**, 511–530.

Dvorak, M.M., De Joussineau, C., Carter, D.H., Pisitkun, T., Knepper, M.A., Gamba, G., Kemp, P.J., and Riccardi, D. (2007). Thiazide diuretics directly induce osteoblast differentiation and mineralized nodule formation by interacting with a sodium chloride co-transporter in bone. *J. Am. Soc. Nephrol.* **18**, 2509–2516.

Ecelbarger, C.A., Kim, G.H., Wade, J.B., and Knepper, M.A. (2001). Regulation of the abundance of renal sodium transporters and channels by vasopressin. *Exp. Neurol.* **171**, 227–234.

Edelman, J.L. and Miller, S.S. (1991). Epinephrine stimulates fluid absorption across bovine retinal pigment epithelium. *Invest. Ophthalmol. Vis. Sci.* **32**, 3033–3040.

Elliott, S.J. and Schilling, W.P. (1992). Oxidant stress alters Na+ pump and Na^+-K^+-Cl^- cotransporter activities in vascular endothelial cells. *Am. J. Physiol.* **263**, H96–H102.

Flatman, P.W. and Creanor, J. (1999). Stimulation of Na^+-K^+-$2Cl^-$ cotransport by arsenite in ferret erythrocytes. *J. Physiol.* **519**, 143–152.

Flemmer, A.W., Gimenez, I., Dowd, B.F., Darman, R.B., and Forbush, B. (2002). Activation of the Na-K-Cl cotransporter NKCC1 detected with a phospho-specific antibody. *J. Biol. Chem.* **277**, 37551–37558.

Fraser, S., Mount, P., Hill, R., Levidiotis, V., Katsis, F., Stapleton, D., Kemp, B.E., and Power, D.A. (2005). Regulation of the energy sensor AMP-activated protein kinase in the kidney by dietary salt intake and osmolality. *Am. J. Physiol. Renal Physiol.* **288**, F578–F586.

Fraser, S.A., Gimenez, I., Cook, N., Jennings, I., Katerelos, M., Katsis, F., Levidiotis, V., Kemp, B.E., and Power, D.A. (2007). Regulation of the renal-specific Na^+-K^+-$2Cl^-$ co-transporter NKCC2 by AMP-activated protein kinase (AMPK). *Biochem. J.* **405**, 85–93.

Gagnon, K.B., England, R., and Delpire, E. (2006). Characterization of SPAK and OSR1, regulatory kinases of the Na-K-2Cl cotransporter. *Mol. Cell Biol.* **26**, 689–698.

Gagnon, K.B., England, R., and Delpire, E. (2006). Volume sensitivity of cation-chloride cotransporters is modulated by the interaction of two kinases: SPAK and WNK4. *Am. J. Physiol. Cell Physiol.* **290**, C134–C142.

Gagnon, K.B., England, R., and Delpire, E. (2007). A single binding motif is required for SPAK activation of the Na-K-2Cl cotransporter. *Cell Physiol. Biochem.* **20**, 131–142.

Gagnon, K.B., England, R., Diehl, L., and Delpire, E. (2007). Apoptosis associated tyrosine kinase scaffolding of protein phosphatase 1 and SPAK reveals a novel pathway for Na-K-2C1 cotransporter regulation. *Am. J. Physiol. (Cell Physiol.)* **292**, C1809–C1815.

Gamba, G., Miyanoshita, A., Lombardi, M., Lytton, J., Lee, W.-S., Hediger, M., and Hebert, S.C. (1994). Molecular cloning, primary structure, and characterization of two members of the mammalian electroneutral sodium-(potassium)-chloride cotransporter family expressed in kidney. *J. Biol. Chem.* **269**, 17713–17722.

Gamba, G., Saltzberg, S.N., Lombardi, M., Miyanoshita, A., Lytton, J., Hediger, M.A., Brenner, B.M., and Hebert, S.C. (1993). Primary structure and functional expression of a cDNA encoding the thiazide-sensitive, electroneutral sodium-chloride cotransporter. *Proc. Natl. Acad. Sci. USA* **90**, 2749–2753.

Gaozza, E., Baker, S.J., Vora, R.K., and Reddy, E.P. (1997). AATYK: a novel tyrosine kinase induced during growth arrest and apoptosis of myeloid cells. *Oncogene* **15**, 3127–3135.

Garay, R.P. and Ciccone, J. (1982). Inhibition of the Na/K cotransport system by cAMP and intracellular Ca^{2+} in human red cells. *Biochim. Biophys. Acta* **688**, 786–792.

Garzón-Muvdi, T., Pacheco-Alvarez, D., Gagnon, K.B., Vázquez, N., Ponce-Coria, J., Moreno, E., Delpire, E., and Gamba, G. (2007). WNK4 kinase is a negative regulator of K^+-Cl^- cotransporters. *Am. J. Physiol. Renal Physiol.* **292**, F1197–F1207.

Geck, P., Pietrzyk, C., Burckhardt, B.-C., Pfeiffer, B., and Heinz, E. (1980). Electrically silent cotransport of Na^+, K^+ and Cl− in Ehrlich cells. *Biochim. Biophys. Acta* **600**, 432–447.

Gillen, C.M., Brill, S., Payne, J.A., and Forbush, B.I. (1996). Molecular cloning and functional expression of the K-Cl cotransporter from rabbit, rat, and human. A new member of the cation-chloride cotransporter family. *J. Biol. Chem.* **271**, 16237–16244.

Giménez, I. and Forbush, B. (2005). Regulatory phosphorylation sites in the NH2 terminus of the renal Na-K-Cl cotransporter (NKCC2). *Am. J. Physiol. Renal Physiol.* **289**, F1341–F1345.

Giménez, I. and Forbush, B. (2003). Short-term stimulation of the renal Na-K-Cl cotransporter (NKCC2) by vasopressin involves phosphorylation and membrane translocation of the protein. *J. Biol. Chem.* **278**, 26946–26951.

Golbang, A.P., Cope, G., Hamad, A., Murthy, M., Liu, C.H., Cuthbert, A.W., and O'Shaughnessy, K.M. (2006). Regulation of the expression of the Na/Cl cotransporter by WNK4 and WNK1: evidence that accelerated dynamin-dependent endocytosis is not involved. *Am. J. Physiol. Renal Physiol.* **291**, F1369–F1376.

Gosmanov, A.R., Nordtvedt, N.C., Brown, R., and Thomason, D.B. (2002). Exercise effects on muscle beta-adrenergic signaling for MAPK-dependent NKCC activity are rapid and persistent. *J. Appl. Physiol.* **93**, 1457–1465.

Guandalini, S., Rao, M.C., Smith, P.L., and Field, M. (1982). cGMP modulation of ileal ion transport: in vitro effects of Escherichia coli heat-stable enterotoxin. *Am. J. Physiol.* **243**, G36–G41.

Guo, Y. and O'Brien, T. (1996). Restoration of responsiveness to phorbol ester by reconstitution of a functional Na/K/Cl cotransporter in cotransporter-deficient BALB/c 3T3 cells. *Mol. Carcinogen.* **17**, 35–40.

Hall, A.C. and Ellory, J.C. (1985). Measurement and stoichiometry of bumetanide-sensitive (2Na:1K:3Cl) cotransport in ferret red cells. *J. Membrane Biol.* **85**, 205–213.

Hall, A.C. and Ellory, J.C. (1986). Evidence for the presence of volume-sensitive KCl transport in "young" human red cells. *Biochim. Biophys. Acta* **858**, 317–320.

Hiki, K., D'Andrea, R.J., Furze, J., Crawford, J., Woollatt, E., Sutherland, G.R., Vadas, M.A., and Gamble, J.R. (1999). Cloning, characterization, and chromosomal location of a novel human K^+-Cl^- cotransporter. *J. Biol. Chem.* **274**, 10661–10667.

Hoffmann, E.K., Sjoholm, C., and Simonsen, L.O. (1983). Na^+, Cl^- cotransport in Ehrlich asites tumor cells activated during volume regulation (regulatory volume increase). *J. Membrane Biol.* **76**, 269–280.

Holden, S., Cox, J., and Raymond, F.L. (2004). Cloning, genomic organization, alternative splicing and expression analysis of the human gene WNK3 (PRKWNK3). *Gene* **335**.

Honma, N., Asada, A., Takeshita, S., Enomoto, M., Yamakawa, E., Tsutsumi, K., Saito, T., Satoh, T., Itoh, H., Kaziro, Y., Kishimoto, T., and Hisanaga, S. (2003). Apoptosis-associated tyrosine kinase is a Cdk5 activator p35 binding protein. *Biochem. Biophys. Res. Commun.* **310**, 398–404.

Hoover, R.S., Poch, E., Monroy, A., Vazquez, N., Nishio, T., Gamba, G., and Hebert, S.C. (2003). N-Glycosylation at two sites critically alters thiazide binding and activity of the rat thiazide-sensitive Na(+):Cl(−) cotransporter. *J. Am. Soc. Nephrol.* **14**, 271–282.

Igarashi, P., Vanden Heuvel, G.B., Payne, J.A., and Forbush, B. (1995). Cloning, embryonic expression, and alternative splicing of a murine kidney-specific Na-K-Cl cotransporter. *Am. J. Physiol.* **269**, F405–F418.

Igarashi, P., Whyte, D.A., Li, K., and Nagami, G.T. (1996). Cloning and kidney cell-specific activity of the promoter of the murine renal Na-K-Cl cotransporter gene. *J. Biol. Chem.* **271**, 9666–9674.

Jennings, M.L. and al-Rohil, N. (1990). Kinetics of activation and inactivation of swelling-stimulated K^+/Cl^- transport. The volume-sensitive parameter is the rate constant for inactivation. *J. Gen. Physiol.* **95**, 1021–1040.

Jennings, M.L. and Schultz, R.K. (1991). Okadaic acid inhibition of KCl cotransport. Evidence that protein dephosphorylation is necessary for activation of transport by either swelling or N-ethylmaleimide. *J. Gen. Physiol.* **97**, 799–817.

Jensen, B.S., Jessen, F., and Hoffmann, E.K. (1993). Na^+-K^+-Cl^- cotransport and its regulation in Ehrlich ascites tumor cells. $Ca2^+$/calmodulin and protein kinase C dependent pathways. *J. Membrane Biol.*, **131**, 161–178.

Kahle, K.T., Gimenez, I., Hassan, H., Wilson, F.H., Wong, R.D., Forbush, B., Aronson, P.S., and Lifton, R.P. (2004). WNK4 regulates apical and basolateral Cl^- flux in extrarenal epithelia. *Proc. Natl. Acad. Sci. (USA)* **101**, 2064–2069.

Kahle, K.T., Rinehart, J., de Los Heros, P., Louvi, A., Meade, P., Vazquez, N., Hebert, S.C., Gamba, G., Gimenez, I., and Lifton, R.P. (2005). WNK3 modulates transport of Cl^- in and out of cells: implications for control of cell volume and neuronal excitability. *Proc. Natl. Acad. Sci. USA* **102**, 16783–16788.

Kahle, K.T., Wilson, F.H., Leng, Q., Lalioti, M.D., O'Connell, A.D., Dong, K., Rapson, A.K., MacGregor, G.G., Giebisch, G., Hebert, S.C., and Lifton, R.P. (2003). WNK4 regulates the balance between renal NaCl reabsorption and K^+ secretion. *Nat. Genet.* **35**, 372–376.

Kaji, D. and Tsukitani, Y. (1991). Role of protein phosphatase in activation of KCl cotransport in human erythrocytes. *Am. J. Physiol.* **260**, C176–C182.

Karadsheh, M.F. and Delpire, E. (2001). A neuronal restrictive silencing element is found in the KCC2 gene: molecular basis for KCC2 specific expression in neurons. *J. Neurophysiol.* **85**, 995–997.

Kelsch, W., Hormuzdi, S., Straube, E., Lewen, A., Monyer, H., and Misgeld, U. (2001). Insulin-like growth factor 1 and a cytosolic tyrosine kinase activate chloride outward transport during maturation of hippocampal neurons. *J. Neurosci.* **21**, 8339–8347.

Kim, H.D., Tsai, Y.-S., Franklin, C.C., and Turner, J.T. (1988). Characterization of $Na^+/K^+/Cl^-$ cotransport in cultured HT29 human colonic adenocarcinoma cells. *Biochim. Biophys. Acta* **946**, 397–404.

Klein, J.D. and O'Neill, W.C. (1990). Effect of bradykinin on Na-K-2Cl cotransport and bumetanide binding in aortic endothelial cells. *J. Biol. Chem.* **265**, 22238–22242.

Knepper, M.A., Kim, G.H., and Masilamani, S. (2003). Renal tubule sodium transporter abundance profiling in rat kidney: response to aldosterone and variations in NaCl intake. *Ann. NY Acad. Sci.* **986**, 562–569.

Krarup, T. and Dunham, P.B. (1996). Reconstitution of calyculin-inhibited K-Cl cotransport in dog erythrocyte ghosts by exogenous PP-1. *Am. J. Physiol. (Cell Physiol.)* **270**, C898–C902.

Krarup, T., Jakobsen, L.D., Jensen, B.S., and Hoffmann, E.K. (1998). Na^+-K^+-$2Cl^-$ cotransport in Ehrlich cells: regulation by protein phosphatases and kinases. *Am. J. Physiol. Cell Physiol.* **275**, C239–C250.

Kwon, T.H., Nielsen, J., Kim, Y.H., Knepper, M.A., Frøkiaer, J., and Nielsen, S. (2003). Regulation of sodium transporters in the thick ascending limb of rat kidney: response to angiotensin II. *Am. J. Physiol. Renal Physiol.* **285**, F152–F165.

Lalioti, M.D., Zhang, J., Volkman, H.M., Kahle, K.T., Hoffmann, K.E., Toka, H.R., Nelson-Williams, C., Ellison, D.H., Flavell, R., Booth, C.J., Lu, Y., Geller, D.S., and Lifton, R.P. (2006). Wnk4 controls blood pressure and potassium homeostasis via regulation of mass and activity of the distal convoluted tubule. *Nat. Genet.* **38**, 1124–1132.

Landolt-Marticorena, C. and Reithmeier, R.A. (1994). Asparagine-linked oligosaccharides are localized to single extracytosolic segments in multi-span membrane glycoproteins. *Biochem. J.* **302**, 253–260.

Lauf, P.K. (1983). Thiol-dependent passive K/Cl transport in sheep red blood cells: V. Dependence on metabolism. *Am. J. Physiol.* **245**, C445–C448.

Lauf, P.K. (1984). Thiol-dependent passive K/Cl transport in sheep red cells. IV. Furosemide inhibition as a function of external Rb^+, Na^+, and Cl^-. *J. Membrane Biol.* **77**, 57–62.

Lauf, P.K., Bauer, J., Adragna, N.C., Fujise, H., Zade-Oppen, A.M.M., Ryu, K., and Delpire, E. (1992). Erythrocyte K-Cl cotransport: properties and regulation. *Am. J. Physiol.* **263**, C917–C932.

Lauf, P.K., Chimote, A.A., and Adragna, N.C. (2008). Lithium fluxes indicate presence of Na-Cl cotransport (NCC) in human lens epithelial cells. *Cell. Physiol. Biochem.* **21**, 335–346.

Lazrak, A., Liu, Z., and Huang, C.L. (2006). Antagonistic regulation of ROMK by long and kidney-specific WNK1 isoforms. *Proc. Natl. Acad. Sci. USA* **103**, 1615–1620.

Lee, H.L.H., Walker, J.A., Williams, J.R., Goodier, R.J., Payne, J.A., and Moss, S.J. (2007). Direct protein kinase C-dependent phosphorylation regulates the cell surface stability and activity of the potassium chloride cotransporter KCC2. *J. Biol. Chem.* **282**, 29777–29784.

Lenertz, L.Y., Lee, B.H., Min, X., Xu, B.E., Wedin, K., Earnest, S., Goldsmith, E.J., and Cobb, M.H. (2005). Properties of WNK1 and implications for other family members. *J. Biol. Chem.* **280**, 26653–26658.

Leung, S., O'Donnell, M.E., Martinez, A., and Palfrey, H.C. (1994). Regulation by nerve growth factor and protein phosphorylation of Na/K/2Cl cotransport and cell volume in PC12 cells. *J. Biol. Chem.* **269**, 10581–10589.

Liedtke, C.M. (1990). Calcium and alpha-adrenergic regulation of Na-Cl(K) cotransport in rabbit tracheal epithelial cells. *Am. J. Physiol.* **259**, L66–L72.

Liedtke, C.M. (1995). The role of protein kinase C in alpha-adrenergic regulation of NaCl(K) cotransport in human airway epithelial cells. *Am. J. Physiol* **268**, L414–L423.

Lu, J., Karadsheh, M., and Delpire, E. (1999). Developmental regulation of the neuronal-specific isoform of K-Cl cotransporter KCC2 in postnatal rat brains. *J. Neurobiol.* **39**, 558–568.

Lytle, C. (1997). Activation of the avian erythrocyte Na-K-Cl cotransport protein by cell shrinkage, cAMP, fluoride, and calyculin-A involves phosphorylation at common sites. *J. Biol. Chem.* **272**, 15069–15077.

Lytle, C. and Forbush, B.I. (1992). The Na-K-Cl cotransport protein of shark rectal gland. II. Regulation by direct phosphorylation. *J. Biol. Chem.* **267**, 25438–25443.

Lytle, C. and McManus, T. (2002). Coordinate modulation of Na-K-2Cl cotransport and K-Cl cotransport by cell volume and chloride. *Am. J. Physiol. Cell Physiol.* **283**, C1422–C1431.

Masilamani, S., Wang, X., Kim, G.H., Brooks, H., Nielsen, J., Nielsen, S., Nakamura, K., Stokes, J.B., and Knepper, M.A. (2002). Time course of renal Na-K-ATPase, NHE3, NKCC2, NCC, and ENaC abundance changes with dietary NaCl restriction. *Am. J. Physiol. Renal Physiol.* **283**, F648–F657.

Matthews, J.B., Awtrey, C.S., Hecht, G., Tally, K.J., Thompson, R.S., and Madara, J.L. (1993). Phorbol ester sequentially downregulates cAMP-regulated basolateral and apical Cl− transport pathways in T84 cells. *Am. J. Physiol. (Cell Physiol.)* **265**, C1109–C1117.

McCray, P.B.J. and Bettencourt, J.D. (1993). Prostaglandins stimulate fluid secretion in human fetal lung. *J. Dev. Physiol.* **19**, 29–36.

McKee, J.A., Kumar, S., Ecelbarger, C.A., Fernández-Llama, P., Terris, J., and Knepper, M.A. (2000). Detection of Na(+) transporter proteins in urine. *J. Am. Soc. Nephrol.* **11**, 2128–2132.

Mercado, A., Song, L., Vazquez, N., Mount, D.B., and Gamba, G. (2000). Functional comparison of the K^+-Cl^- cotransporters KCC1 and KCC4. *J. Biol. Chem.* **275**, 30326–30334.

Mercado, A., Vazquez, N., Song, L., Cortes, R., Enck, A.H., Welch, R., Delpire, E., Gamba, G., and Mount, D.B. (2005). NH2-terminal heterogeneity in the KCC3 K^+-Cl^- cotransporter. *Am. J. Physiol. Renal Physiol.* **289**, F1246–F1261.

Milligan, G. and Smith, N.J. (2007). Allosteric modulation of heterodimeric G-protein-coupled receptors. *Trends Pharmacol. Sci.* **28**, 615–620.

Min, X., Lee, B.H., Cobb, M.H., and Goldsmith, E.J. (2004). Crystal structure of the kinase domain of WNK1, a kinase that causes a hereditary form of hypertension. *Structure* **12**, 1303–1311.

Moore-Hoon, M.L. and Turner, R.J. (2000). The structural unit of the secretory Na^+-K^+-$2Cl^-$ cotransporter (NKCC1) is a homodimer. *Biochemistry* **39**, 3718–3724.

Moriguchi, T., Urushiyama, S., Hisamoto, N., Iemura, S.I., Uchida, S., Natsume, T., Matsumoto, K., and Shibuya, H. (2006). WNK1 regulates phosphorylation of cation-chloride-coupled cotransporters via the STE20-related kinases, SPAK and OSR1. *J. Biol. Chem.* **280**, 42685–42693.

Mount, D.B., Mercado, A., Song, L., Xu, J., George, J.A.L., Delpire, E., and Gamba, G. (1999). Cloning and characterization of KCC3 and KCC4, new members of the cation-chloride cotransporter gene family. *J. Biol. Chem.* **274**, 16355–16362.

Muzyamba, M.C., Cossins, A.R., and Gibson, J.S. (1999). Regulation of Na+-K+-2Cl− cotransport in turkey red cells: the role of oxygen tension and protein phosphorylation. *J. Physiol. (Lond.)* **517**, 421–429.

Newhouse, S.J., Wallace, C., Dobson, R., Mein, C., Pembroke, J., Farrall, M., Clayton, D., Brown, M., Samani, N., Dominiczak, A., Connell, J.M., Webster, J., Lathrop, G.M., Caulfield, M., and Munroe, P.B. (2005). Haplotypes of the WNK1 gene associate with blood pressure variation in a severely hypertensive population from the British Genetics of Hypertension study. *Hum. Mol. Genet.* **14**, 1805–1814.

O'Brien, J.A., Walters, R.J., Valverde, M.A., and Sepúlveda, F.V. (1993). Regulatory volume increase after hypertonicity- or vasoactive-intestinal-peptide-induced cell-volume decrease in small-intestinal crypts is dependent on Na^+-K^+-$2Cl^-$ cotransport. *Pflugers Arch.* **423**, 67–73.

O'Donnell, M.E. (1991). Endothelial cell sodium-potassium-chloride cotransport. Evidence of regulation by Ca^{2+} and protein kinase C. *J. Biol. Chem.* **266**, 11559–11566.

O'Donnell, M.E., Duong, V., Suvatne, J., Foroutan, S., and Johnson, D.M. (2005). Arginine vasopressin stimulation of cerebral microvascular endothelial cell Na-K-Cl cotransporter activity is V1 receptor and [Ca] dependent. *Am. J. Physiol. (Cell Physiol.)* **289**, C283–C292.

O'Donnell, M.E., Martinez, A., and Sun, D. (1995). Endothelial Na-K-Cl cotransport regulation by tonicity and hormones: phosphorylation of cotransport protein. *Am. J. Physiol. (Cell Physiol.)* **269**, C1513–C1523.

Oppermann, M., Mizel, D., Huang, G., Li, C., Deng, C., Theilig, F., Bachmann, S., Briggs, J., Schnermann, J., and Castrop, H. (2006). Macula densa control of renin secretion and preglomerular resistance in mice with selective deletion of the B isoform of the Na,K,2Cl co-transporter. *J. Am. Soc. Nephrol.* **17**, 2143–2152.

Oppermann, M., Mizel, D., Kim, S.M., Chen, L., Faulhaber-Walter, R., Huang, Y., Li, C., Deng, C., Briggs, J., Schnermann, J., and Castrop, H. (2007). Renal function in mice with targeted disruption of the A isoform of the Na-K-2Cl co-transporter. *J. Am. Soc. Nephrol.* **18**, 440–448.

O'Reilly, M., Marshall, E., Speirs, H.J., and Brown, R.W. (2003). WNK1, a gene within a novel blood pressure control pathway, tissue-specifically generates radically different isoforms with and without a kinase domain. *J. Am. Assoc. Nephrol.* **14**, 2447–2456.

Owen, N.E. (1984). Regulation of Na/K/Cl cotransport in vascular smooth muscle cells. *Biochem. Biophys. Res. Comm.* **125**, 500–508.

Pacheco-Alvarez, D., San Cristóbal, P., Meade, P., Moreno, E., Vázquez, N., Muñoz, E., Díaz, A., Juárez, M.E., Giménez, I., and Gamba, G. (2006). The Na-Cl cotransporter is activated and phosphorylated at the amino terminal domain upon intracellular chloride depletion. *J. Biol. Chem.* **281**, 28755–28763.

Palfrey, H.C. and Pewitt, E.B. (1993). The ATP and Mg2+ dependence of Na^+-K^+-$2Cl^-$ cotransport reflects a requirement for protein phosphorylation: studies using calyculin A. *Pflugers Archiv.* **425**, 321–328.

Palfrey, H.C., Silva, P., and Epstein, F.H. (1984). Sensitivity of cAMP-stimulated salt secretion in shark rectal gland to "loop" diuretics. *Am. J. Physiol.* **246**, C242–C246.

Palma, E., Amici, M., Sobrero, F., Spinelli, G., Di Angelantonio, S., Ragozzino, D., Mascia, A., Scoppetta, C., Esposito, V., Miledi, R., and Eusebi, F. (2006). Anomalous levels of Cl^- transporters in the hippocampal subiculum from temporal lobe epilepsy patients make GABA excitatory. *Proc. Natl. Acad. Sci. USA* **103**, 8465–8468.

Paredes, A., Plata, C., Rivera, M., Moreno, E., Vázquez, N., Muñoz-Clares, R., Hebert, S.C., and Gamba, G. (2006). Activity of the renal Na^+-K^+-$2Cl^-$ cotransporter is reduced by mutagenesis of N-glycosylation sites: role for protein surface charge in Cl^- transport. *Am. J. Physiol. Renal Physiol.* **290**, F1094–F1102.

Paris, S. and Pouyssegur, J. (1986). Growth factors activate the bumetanide sensitive $Na^+/K^+/Cl^-$ cotransport in hamster fibroblasts. *J. Biol. Chem.* **261**, 6177–6183.

Paulais, M. and Turner, R.J. (1992). Activation of the Na^+-K^+-$2Cl^-$ cotransporter in rat parotid acinar cells by aluminum fluoride and phosphatase inhibitors. *J. Biol. Chem.* **267**, 21558–21563.

Paulais, M. and Turner, R.J. (1992). Beta-adrenergic upregulation of the Na^+-K^+-$2Cl^-$ cotransporter in rat parotid acinar cells. *J. Clin. Invest* **89**, 1142–1147.

Payne, J.A. and Forbush, B.I. (1994). Alternatively spliced isoforms of the putative renal Na-K-Cl cotransporter are differentially distributed within the rabbit kidney. *Proc. Natl. Acad. Sci. USA* **91**, 4544–4548.

Payne, J.A., Stevenson, T.J., and Donaldson, L.F. (1996). Molecular characterization of a putative K-Cl cotransporter in rat brain. A neuronal-specific isoform. *J. Biol. Chem.* **271**, 16245–16252.

Payne, J.A., Xu, J.-C., Haas, M., Lytle, C.Y., Ward, D., and Forbush, B.I. (1995). Primary structure, functional expression, and chromosome localization of the bumetanide sensitive Na-K-Cl cotransporter in human colon. *J. Biol. Chem.* **270**, 17977–17985.

Pearson, M., Lu, J., Mount, D.B., and Delpire, E. (2001). Localization of the K-Cl cotransporter, KCC3, in the central and peripheral nervous systems: expression in choroid plexus, large neurons, and white matter tracts. *Neuroscience* **103**, 483–493.

Pewitt, E.B., Hegde, R.S., Haas, M., and Palfrey, H.C. (1990). The regulation of Na/K/2Cl cotransport and bumetanide binding in avian erythrocytes by protein phosphorylation and dephosphorylation. Effects of kinase inhibitors and okadaic acid. *J. Biol. Chem.* **265**, 20747–20756.

Piechotta, K., Garbarini, N.J., England, R., and Delpire, E. (2003). Characterization of the interaction of the stress kinase SPAK with the Na^+-K^+-$2Cl^-$ cotransporter in the nervous system: evidence for a scaffolding role of the kinase. *J. Biol. Chem.* **278**, 52848–52856.

Piechotta, K., Lu, J., and Delpire, E. (2002). Cation-chloride cotransporters interact with the stress-related kinases SPAK and OSR1. *J. Biol. Chem.* **277**, 50812–50819.

Polek, T.C., Talpaz, M., and Spivak-Kroizman, T. (2006). The TNF receptor, RELT, binds SPAK and uses it to mediate p38 and JNK activation. *Biochem. Biophys. Res. Commun.* **343**, 125–134.

Ponce-Coria, J., San-Cristobal, P., Kahle, K.T., Vazquez, N., Pacheco-Alvarez, D., de Los Heros, P., Juárez, P., Muñoz, E., Michel, G., Bobadilla, N.A., Gimenez, I., Lifton, R.P., Hebert, S.C., and Gamba, G. (2008). Regulation of NKCC2 by a chloride-sensing mechanism involving the WNK3 and SPAK kinases. *Proc. Natl. Acad. Sci. USA* **105**, 8458–8463.

Race, J.E., Makhlouf, F.N., Logue, P.J., Wilson, F.H., Dunham, P.B., and Holtzman, E.J. (1999). Molecular cloning and functional characterization of KCC3, a new K-Cl cotransporter. *Am. J. Physiol. (Cell Physiol.)* **277**, C1210–C1219.

Raghunath, M., Patti, R., Bannerman, P., Lee, C.M., Baker, S., Sutton, L.N., Phillips, P.C., and Damodar Reddy, C. (2000). A novel kinase, AATYK induces and promotes neuronal differentiation in a human neuroblastoma (SH-SY5Y) cell line. *Mol. Brain Res.* **77**, 151–162.

Randall, J., Thorne, T., and Delpire, E. (1997). Partial cloning and characterization of *Slc12a2*: the gene encoding the secretory Na^+-K^+-$2Cl^-$ cotransporter. *Am. J. Physiol. (Cell Physiol.)* **273**, C1267–C1277.

Richardson, C., Rafiqi, F.H., Karlsson, H.K., Moleleki, N., Vandewalle, A., Campbell, D.G., Morrice, N.A., and Alessi, D. R. (2008). Activation of the thiazide-sensitive Na^+-Cl^- cotransporter by the WNK-regulated kinases SPAK and OSR1. *J. Cell Sci.* **121**, 675–684.

Rinehart, J., Kahle, K.T., de Los Heros, P., Vazquez, N., Meade, P., Wilson, F.H., Hebert, S.C., Gimenez, I., Gamba, G., and Lifton, R.P. (2005). WNK3 kinase is a positive regulator of NKCC2 and NCC, renal cation-Cl^- cotransporters required for normal blood pressure homeostasis. *Proc. Natl. Acad Sci. USA* **102**, 16777–16782.

Ring, A.M., Cheng, S.X., Leng, Q., Kahle, K.T., Rinehart, J., Lalioti, M.D., Volkman, H.M., Wilson, F.H., Hebert, S.C., and Lifton, R.P. (2007). WNK4 regulates activity of the epithelial Na+ channel in vitro and in vivo. *Proc. Natl. Acad Sci. USA* **104**, 4020–4024.

Rivera, C., Li, H., Thomas-Crusells, J., Lahtinen, H., Viitanen, T., Nanobashvili, A., Kokaia, Z., Airaksinen, M.S., Voipio, J., Kaila, K., and Saarma, M. (2002). BDNF-induced TrkB activation down-regulates the K^+-Cl^- cotransporter KCC2 and impairs neuronal Cl− extrusion. *J. Cell Biol.* **159**, 747–752.

Rivera, C., Voipio, J., and Kaila, K. (2005). Two developmental switches in GABAergic signalling: the K^+-Cl^- cotransporter KCC2 and carbonic anhydrase CAVII. *J. Physiol. (Lond.)* **562**, 27–36.

Rivera, C., Voipio, J., Payne, J.A., Ruusuvuori, E., Lahtinen, H., Lamsa, K., Pirvola, U., Saarma, M., and Kaila, K. (1999). The K^+/Cl^- co-transporter KCC2 renders GABA hyperpolarizing during neuronal maturation. *Nature* **397**, 251–255.

Rivera, C., Voipio, J., Thomas-Crusells, J., Li, H., Emri, Z., Sipila, S., Payne, J.A., Minichiello, L., Saarma, M., and Kaila, K. (2004). Mechanism of activity-dependent downregulation of the neuron-specific K-Cl cotransporter KCC2. *J. Neurosci.* **24**, 4683–4691.

Russell, J.M. (2000). Sodium-potassium-chloride cotransport. *Physiol. Rev.* **80**, 211–276.

Sachs, J.R. (1988). Volume-sensitive K influx in human red cell ghosts. *J. Gen. Physiol.* **92**, 685–711.

San-Cristobal, P., Pacheco-Alvarez, D., Richardson, C., Ring, A.M., Vazquez, N., Rafiqi, F.H., Chari, D., Kahle, K.T., Leng, Q., Bobadilla, N.A., Hebert, S.C., Alessi, D.R., Lifton, R.P., and Gamba, G. (2009). Angiotensin II signaling increases activity of the renal Na-Cl cotransporter through a WNK4-SPAK – dependent pathway. *Proc. Natl. Acad. Sci. USA* **106**, 4384–4389.

Sandberg, M.B., Maunsbach, A.B., and McDonough, A.A. (2006). Redistribution of distal tubule Na^+-Cl^- cotransporter (NCC) in response to a high-salt diet. *Am. J. Physiol. Renal Physiol.* **291**, F503–F508.

Sargeant, R.J., Liu, Z., and Klip, A. (1995). Action of insulin on Na^+-K^+-ATPase and the Na^+-K^+-$2Cl^-$ cotransporter in 3T3-L1 adipocytes. *Am. J. Physiol.* **269**, C217–C225.

Schoenherr, C.J. and Anderson, D.J. (1995). The neuron-restrictive silencer factor (NRSF): a coordinate repressor of multiple neuron-specific genes. *Science* **267**, 1360–1363.

Sen, C.K., Kolosova, I., Hanninen, O., and Orlov, S.N. (1995). Inward potassium transport systems in skeletal muscle derived cells are highly sensitive to oxidant exposure. *Free Radic. Biol. Med.* **18**, 795–800.

Shen, M.R., Chou, C.Y., Hsu, K.F., Hsu, Y.M., Chiu, W.T., Tang, M.J., Alper, S.L., and Ellory, J.C. (2003). KCl cotransport is an important modulator of human cervical cancer growth and invasion. *J. Biol. Chem.* **278**, 39941–39950.

Shiozaki, A., Miyazaki, H., Niisato, N., Nakahari, T., Iwasaki, Y., Itoi, H., Ueda, Y., Yamagishi, H., and Marunaka, Y. (2006). Furosemide, a blocker of $Na^+/K^+/2Cl^-$ cotransporter, diminishes proliferation of poorly differentiated human gastric cancer cells by affecting G0/G1 state. *J. Physiol. Sci.* **56**, 401–406.

Simard, C.F., Bergeron, M.J., Frenette-Cotton, R., Carpentier, G.A., Pelchat, M.E., Caron, L., and Isenring, P. (2007). Homooligomeric and heterooligomeric associations between K^+-Cl^- cotransporter isoforms and between K^+-Cl^- and Na^+-K^+-Cl^- cotransporters. *J. Biol. Chem.* **282**, 18083–18093.

Simard, C.F., Brunet, G.M., Daigle, N.D., Montminy, V., Caron, L., and Isenring, P. (2004). Self-interacting domains in the C terminus of a cation-Cl^- cotransporter described for the first time. *J. Biol. Chem.* **279**, 40769–40777.

Simon, D.B., Farfel, Z., Ellison, D., Bia, M., Tucci, J., and Lifton, R.P. (1995). Examination of the thiazide-sensitive Na-Cl cotransporter as a candidate gene in Gordon's syndrome. *J. Am. Soc. Nephrol.* **6**, 632.

Simon, D.B., Nelson-Williams, C., Johnson Bia, M., Ellison, D., Karet, F.E., Morey Molina, A., Vaara, I., Iwata, F., Cushner, H.M.,

Koolen, M., Gainza, F.J., Gitelman, H.J., and Lifton, R.P. (1996). Gitelman's variant of Bartter's syndrome, inherited hypokalaemic alkalosis, is caused by mutations in the thiazide-sensitive Na-Cl cotransporter. *Nature Genetics* **12**, 24–30.

Smith, J.B. and Smith, L. (1987). $Na^+/K^+/Cl^-$ cotransport in cultured vascular smooth muscle cells: stimulation by angiotensin II and calcium ionophores, inhibition by cyclic AMP and calmodulin antagonists. *J. Membr. Biol.* **99**, 51–63.

Song, L., Mercado, A., Vazquez, N., Xie, Q., Desai, R., George, A.L., Jr., Gamba, G., and Mount, D.B. (2002). Molecular, functional, and genomic characterization of human KCC2, the neuronal K-Cl cotransporter. *Brain Res. Mol. Brain Res.* **103**, 91–105.

Starke, L.C. and Jennings, M.L. (1993). K-Cl cotransport in rabbit red cells: further evidence for regulation by protein phosphatase. *Am. J. Physiol. (Cell Physiol.)* **264**, C118–C124.

Starremans, P.G., Kersten, F.F., Knoers, N.V., van den Heuvel, L.P., and Bindels, R.J. (2003). Mutations in the human Na-K-2Cl cotransporter (NKCC2) identified in Bartter syndrome type I consistently result in nonfunctional transporters. *J. Am. Soc. Nephrol.* **14**, 1419–1426.

Stoff, J.S., Rosa, R., Hallac, R., Silva, P., and Epstein, F.H. (1979). Hormonal regulation of active chloride transport in the dogfish rectal gland. *Am. J. Physiol.* **237**, F138–F144.

Stokes, J.B., Lee, I., and D'Amico, M. (1984). Sodium chloride absorption by the urinary bladder of the winter flounder: a thiazide-sensitive, electrically neutral transport system. *J. Clin. Invest.* **74**, 7–16.

Strange, K., Singer, T.D., Morrison, R., and Delpire, E. (2000). Dependence of KCC2 K-Cl cotransporter activity on a conserved carboxy terminus tyrosine residue. *Am. J. Physiol. (Cell Physiol.)* **279**, C860–C867.

Subramanya, A.R., Yang, C.L., Zhu, X., and Ellison, D.H. (2006). Dominant-negative regulation of WNK1 by its kidney-specific kinase-defective isoform. *Am. J. Physiol. Renal Physiol.* **290**, F619–F624.

Sun, D., Lytle, C., and O'Donnell, M.E. (1997). IL-6 secreted by astroglial cells regulates Na-K-Cl cotransport in brain microvessel endothelial cells. *Am. J. Physiol. Cell Physiol.* **272**, C1829–C1835.

Sung, K.-W., Kirby, M., McDonald, M.P., Lovinger, D.M., and Delpire, E. (2000). Abnormal $GABA_A$-receptor mediated currents in dorsal root ganglion neurons isolated from Na-K-2Cl cotransporter null mice. *J. Neurosci.* **20**, 7531–7538.

Suvitayavat, W., Dunham, P.B., Haas, M., and Rao, M.C. (1994). Characterization of the proteins of the intestinal Na^+-K^+-$2Cl^-$ cotransporter. *Am. J. Physiol. (Cell Physiol.)* **267**, C375–C384.

Tamari, M., Daigo, Y., and Nakamura, Y. (1999). Isolation and characterization of a novel serine threonine kinase gene on chromosome 3p22-21.3. *J. Hum. Genet.* **44**, 116–120.

Taniyama, Y., Sato, K., Sugawara, A., Uruno, A., Ikeda, Y., Kudo, M., Ito, S., and Takeuchi, K. (2001). Renal tubule-specific transcription and chromosomal localization of rat thiazide-sensitive Na-Cl cotransporter gene. *J. Biol. Chem.* **276**, 26260–26268.

Tomomura, M., Morita, N., Yoshikawa, F., Konishi, A., Akiyama, H., Furuichi, T., and Kamiguchi, H. (2007). Structural and functional analysis of the apoptosis-associated tyrosine kinase (AATYK) family. *Neuroscience* **148**, 510–521.

Topper, J.N., Wasserman, S.M., Anderson, K.R., Cai, J., Falb, D., and Gimbrone, M.A.J. (1997). Expression of the bumetanide-sensitive Na-K-Cl cotransporter BSC2 is differentially regulated by fluid mechanical and inflammatory cytokine stimuli in vascular endothelium. *J. Clin. Invest.* **99**, 2941–2949.

Torchia, J., Lytle, C., Pon, D.J., Forbush, B.I., and Sen, A.K. (1992). The Na-K-Cl cotransporter of avian salt gland. Phosphorylation in response to cAMP-dependent and calcium-dependent secretogogues. *J. Biol. Chem.* **267**, 25444–25450.

Toyomoto, T., Knutsen, D., Soos, G., and Sato, K. (1997). Na-K-2Cl cotransporters are present and regulated in simian eccrine clear cells. *Am. J. Physiol* **273**, R270–R277.

Ueberschär, S. and Bakker-Grunwald, T. (1983). Bumetanide-sensitive potassium transport and volume regulation in turkey erythrocytes. *Biochim. Biophys. Acta* **731**, 243–250.

Ueberschär, S. and Bakker-Grunwald, T. (1985). Effects of ATP and cAMP on the (Na^++K^++$2Cl^-$)-cotransport system in turkey erythrocytes. *Biochim. Biophys. Acta* **818**, 260–266.

Ushiro, H., Tsutsumi, T., Suzuki, K., Kayahara, T., and Nakano, K. (1998). Molecular cloning and characterization of a novel Ste20-related protein kinase enriched in neurons and transporting epithelia. *Arch. Biochem. Biophys.* **355**, 233–240.

Uvarov, P., Ludwig, A., Markkanen, M., Pruunsild, P., Kaila, K., Delpire, E., Timmusk, T., Rivera, C., and Airaksinen, M.S. (2007). A novel N-terminal isoform of the neuron-specific K-Cl cotransporter KCC2. *J. Biol. Chem.* **282**, 30570–30576.

Uvarov, P., Ludwig, A., Markkanen, M., Rivera, C., and Airaksinen, M.S. (2006). Upregulation of the neuron-specific K^+/Cl^- cotransporter expression by transcription factor early growth response 4. *J. Neurosci.* **26**, 13463–13473.

Uvarov, P., Pruunsild, P., Timmusk, T., and Airaksinen, M.S. (2005). Neuronal K^+/Cl^- co-transporter (KCC2) transgenes lacking neurone restrictive silencer element recapitulate CNS neurone-specific expression and developmental up-regulation of endogenous KCC2 gene. *J. Neurochem.* **95**, 1144–1155.

Verissimo, F. and Jordan, P. (2001). WNK kinases, a novel protein kinase subfamily in multi-cellular organisms. *Oncogene* **20**, 5562–5569.

Vibat, C.R., Holland, M.J., Kang, J.J., Putney, L.K., and ME, O.D. (2001). Quantitation of Na^+-K^+-$2Cl^-$ cotransport splice variants in human tissues using kinetic polymerase chain reaction. *Anal Biochem.* **298**, 218–230.

Villa, F., Goebel, J., Rafiqi, F.H., Deak, M., Thastrup, J., Alessi, D.R., and van Aalten, D.M.F. (2007). Structural insights into the recognition of substrates and activators by the OSR1 kinase. *EMBO Report* **8**, 839–845.

Vitari, A.C., Deak, M., Morrice, N.A., and Alessi, D.R. (2005). The WNK1 and WNK4 protein kinases that are mutated in Gordon's hypertension syndrome, phosphorylate and active SPAK and OSR1 protein kinases. *Biochem. J.* **391**, 17–24.

Vitari, A.C., Thastrup, J., Rafiqi, F.H., Deak, M., Morrice, N.A., Karlsson, H.K., and Alessi, D.R. (2006). Functional interactions of the SPAK/OSR1 kinases with their upstream activator WNK1 and downstream substrate NKCC1. *Biochem. J.* **397**, 223–231.

Von Brauchitsch, D.K. and Crook, R.B. (1993). Protein kinase C regulation of a Na^+, K^+, Cl^- cotransporter in fetal human pigmented ciliary epithelial cells. *Exp. Eye Res.* **57**, 699–708.

Vuillemin, T., Teulon, J., Geniteau-Legendre, M., Baudouin, B., Estrade, S., Cassingena, R., Ronco, P., and Vandewalle, A. (1992). Regulation by calcitonin of Na^+-K^+-Cl^- cotransport in a rabbit thick ascending limb cell line. *Am. J. Physiol. (Cell Physiol.)* **263**, C563–C572.

Weymer, A., Huott, P., Liu, W., McRoberts, J.A., and Dharmsathaphorn, K. (1985). Chloride secretory mechanism induced by prostaglandin E1 in a colonic epithelial cell line. *J. Clin. Invest.* **76**, 1826–1836.

Whisenant, N., Zhang, B.-X., Khademazad, M., Loessberg, P., and Muallem, S. (1991). Regulation of Na-K-2Cl cotransport in osteoblasts. *Am. J. Physiol.* **261**, C433–C440.

Wilson, F.H., Disse-Nicodeme, S., Choate, K.A., Ishikawa, K., Nelson-Williams, C., Desitter, I., Gunel, M., Milford, D.V., Lipkin, G.W., Achard, J.M., Feely, M.P., Dussol, B., Berland, Y., Unwin, R.J., Mayan, H., Simon, D.B., Farfel, Z., Jeunemaitre, X., and Lifton, R.P. (2001). Human hypertension caused by mutations in WNK kinases. *Science* **293**, 1107–1112.

Wilson, F.H., Kahle, K.T., Sabath, E., Lalioti, M.D., Rapson, A.K., Hoover, R.S., Hebert, S.C., Gamba, G., and Lifton, R.P. (2003). Molecular pathogenesis of inherited hypertension with hyperkalemia: the Na-Cl cotransporter is inhibited by wild-type but not mutant WNK4. *Proc. Natl. Acad. Sci. USA* **100**, 680–684.

Winding, B. and Bindslev, N. (1990). Characterization of a muscarinic receptor controlling Cl^- secretion in hen trachea. *Am. J. Physiol.* **258**, C982–C987.

Xu, B., English, J.M., Wilsbacher, J.L., Stippec, S., Goldsmith, E.J., and Cobb, M.H. (2000). WNK1, a novel mammalian serine/threonine protein kinase lacking the catalytic lysine in subdomain II. *J. Biol. Chem.* **275**, 16795–16801.

Xu, J.-C., Lytle, C., Zhu, T.T., Payne, J.A., Benz, E.J., and Forbush, B.I. (1994). Molecular cloning and functional expression of the bumetanide-sensitive Na-K-2Cl cotransporter. *Proc. Natl. Acad. Sci.* **91**, 2201–2205.

Yang, C.L., Angell, J., Mitchell, R., and Ellison, D.H. (2003). WNK kinases regulate thiazide-sensitive Na-Cl cotransport. *J. Clin. Invest.* **111**, 1039–1045.

Yang, C.L., Zhu, X., Wang, Z., Subramanya, A.R., and Ellison, D.H. (2005). Mechanisms of WNK1 and WNK4 interaction in the regulation of thiazide-sensitive NaCl cotransport. *J. Clin. Invest.* **115**, 1379–1387.

Yang, S.S., Morimoto, T., Rai, T., Chiga, M., Sohara, E., Ohno, M., Uchida, K., Lin, S.H., Moriguchi, T., Shibuya, H., Kondo, Y., Sasaki, S., and Uchida, S. (2007). Molecular pathogenesis of pseudohypoaldosteronism type II: generation and analysis of a Wnk4(D561A/+) knockin mouse model. *Cell Metabolism* **5**, 331–344.

Zagórska, A., Pozo-Guisado, E., Boudeau, J., Vitari, A.C., Rafiqi, F.H., Thastrup, J., Deak, M., Campbell, D.G., Morrice, N.A., Prescott, A.R., and Alessi, D.R. (2007). Regulation of activity and localization of the WNK1 protein kinase by hyperosmotic stress. *J. Cell Biol.* **176**, 89–100.

Zambrowicz, B.P., Abuin, A., Ramirez-Solis, R., Richter, L.J., Piggott, J., BeltrandelRio, H., Buxton, E.C., Edwards, J., Finch, R.A., Friddle, C.J., Gupta, A., Hansen, G., Hu, Y., Huang, W., Jaing, C., Key, B.W.J., Kipp, P., Kohlhauff, B., Ma, Z.Q., Markesich, D., Payne, R., Potter, D.G., Qian, N., Shaw, J., Schrick, J., Shi, Z.Z., Sparks, M.J., Van Sligtenhorst, I., Vogel, P., Walke, W., Xu, N., Zhu, Q., Person, C., and Sands, A.T. (2003). Wnk1 kinase deficiency lowers blood pressure in mice: a gene-trap screen to identify potential targets for therapeutic intervention. *Proc. Natl. Acad Sci. USA* **100**, 14109–14114.

Zhao, B., Wong, A.Y., Murshid, A., Bowie, D., Presley, J.F., and Bedford, F.K. (2008). Identification of a novel di-leucine motif mediating K^+-Cl^- cotransporter KCC2 constitutive endocytosis. *Cell Signal* **20**, 1769–1779.

PART IV

CATION-CHLORIDE COTRANSPORTERS IN NEURAL FUNCTION AND DYSFUNCTION

CHAPTER

19

GABA, Glycine and Cation-Chloride Cotransporters in Retinal Function and Development

Noga Vardi and Ling-Li Zhang

OUTLINE

I. INTRODUCTION

Chloride ions (Cl^-) are pivotal in neuronal signaling; they permeate through anion channels thereby regulating membrane potential and excitability in neurons. A large proportion of Cl^- permeable channels are gated by the neurotransmitters GABA and glycine. Both neurotransmitters act by opening anion channels that are mainly, but not exclusively, permeable to Cl^-. However, since the permeability to Cl^- is significantly greater than that to other anions, the reversal potential for GABA (E_{GABA}) is determined primarily by the equilibrium potential for chloride (E_{Cl}). Therefore E_{GABA} can be considered close to E_{Cl}. Thus, for practical purposes, in this chapter we will use E_{GABA}, $E_{glycine}$ and E_{Cl} interchangeably even though they are not necessarily equal. Whether Cl^- flows into or out of a neuron critically depends on the value of E_{Cl} with respect to E_m, the resting membrane potential, as discussed in detail in Chapters 5 and 13 in this volume. When E_{Cl} is more negative than E_m, Cl^- flows into the cell and hyperpolarizes it; conversely, when E_{Cl} is less negative than E_m, Cl^- flows out and depolarizes the cell.

Relatively small changes in intracellular $[Cl^-]$ are reflected in fluctuations in E_{Cl} around the resting potential and thus can switch GABA's action from hyperpolarizing to depolarizing. The ease of switching a transmitter from being hyperpolarizing and inhibitory into depolarizing and excitatory contributes to the versatility and diversity of neuronal circuits. The first part of this chapter discusses the diversity of GABA and glycine's actions in mature nervous system. It focuses mainly on the retina because, as explained below, the wealth of information regarding specific cell types and their functions in this tissue greatly facilitates the understanding of computations and performance of neuronal circuits. The second part of the chapter discusses the function of inhibitory neurotransmitters and cation-coupled Cl^- cotransporters in retinal development.

II. FUNCTION OF GABA, GLYCINE AND CHLORIDE COTRANSPORTERS IN THE ADULT RETINA

A. Retinal Cell Types and Circuits

The retina is a thin sheet of brain tissue (100 to 250 μm thick) that grows out into the eye to provide neural processing for image processing. The retina includes photoreceptors and two stages of neural processing. Its output cells project centrally and the information they convey is analyzed by about half of the cerebral cortex (Van Essen et al., 1992). Neuronal processes in the cerebral cortex can span millimeters to centimeters while in the retina the two synaptic layers span only 60 μm, and most lateral processes span only several hundred μm. Therefore, it is relatively easy to identify retinal neurons and quantify their synaptic connections. This revealed that the retina comprises about 75 discrete neuron types connected in specific and highly stereotyped patterns. Each of these neuronal types also has a unique set of neurochemicals (e.g. transmitters, transporters, receptors) and physiological response to light stimulation. Further, the known circuits can explain both intrinsic retinal mechanisms and visual performance to a high degree. For these reasons, the retina is an excellent model of brain function. What follows is a brief account of retinal structure and function, a necessary prelude to understand the workings of inhibitory circuits.

1. *Feedforward Pathways (also Called Vertical Pathways)*

The general design of the retina encompasses several parallel feedforward pathways and many intricate feedback circuits. The feedforward pathways comprise three orders of hierarchical neurons that transmit information in two stages, using glutamate as their neurotransmitter. The feedback circuits modulate this information at each stage of processing via interneurons that use GABA or glycine as their main neurotransmitters (Fig. 19.1). The first order neurons are the photoreceptors, which detect light using a special compartment called the outer segment. The outer segment is packed with membranous disks that provide a very large surface area housing molecules of rhodopsin (the chromophore that captures photons and transduces its energy into electrical energy) and its signaling cascade elements. In mammals, there are two classes of photoreceptors: rods and cones. The rods are characterized for having a thin and long outer segment. The rods detect very dim light, down to a single photon event. These photoreceptors are essential for nocturnal vision. The cones, in primates, include three spectral types, whose outer segment is thicker and shorter than that of the rods. These photoreceptors have a threshold for light that is four orders of magnitude higher than that of rods. Cones are essential for diurnal color vision. The outer segments of rods and cones packed in the first layer of the retina thereby maximize light capture. Light is transduced into electrical signals by a G-protein-coupled cascade that greatly amplifies the signal. This transduction occurs in the photoreceptor outer segment, and the resulting membrane depolarization modulates the glutamate released from photoreceptor terminals. Photoreceptors do not generate action potentials because information needs to be transmitted only over a short distance of ~100 μm from the outer segment to the photoreceptor synaptic terminal. Unique to photoreceptors is the fact that the "unstimulated" dark membrane potential is maintained depolarized at ~45 mV, and light hyperpolarizes it. Consequently, in total darkness, photoreceptors continuously release glutamate, and light reduces this release.

Glutamate released from photoreceptors acts on second order neurons called bipolar cells. At its apex, a bipolar cell's soma extends a primary dendrite toward the first synaptic layer (also called the outer plexiform layer) that collects information from the photoreceptors. At its base the soma emits an axon that branches at specific levels within the second synaptic layer (also called the inner plexiform layer). There are about ten different types of bipolar cells, which are classified by their morphology. The

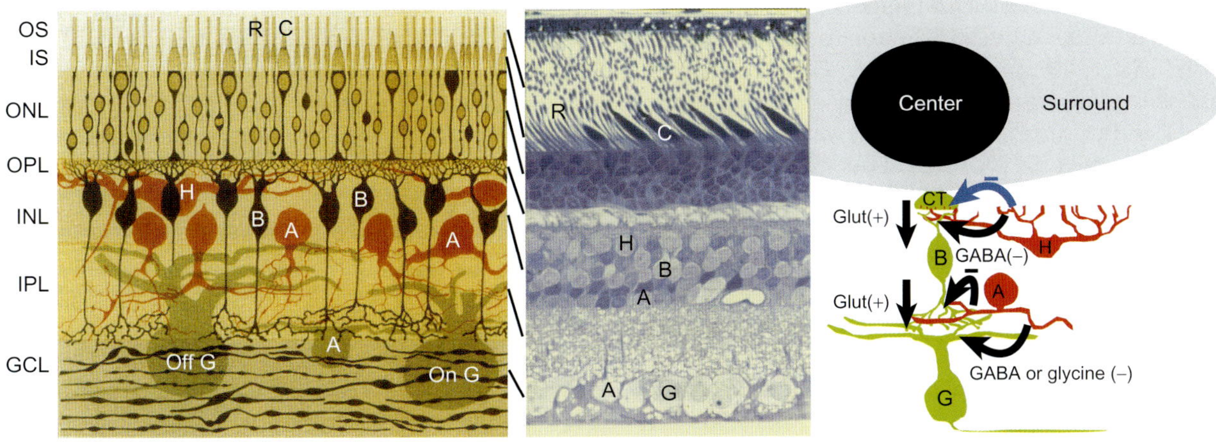

FIGURE 19.1 Basic organization of the retina and its circuits. **Left,** schematic of retinal cell classes (from Tartuferi, 1887). R, rod; C, cone; H, horizontal cell; B, bipolar cell; A, amacrine cell; G, ganglion cell. **Middle,** 1 μm thick plastic radial section of monkey retina stained with toluidine blue. Diagonal lines point to the middle of each layer in the two panels. **Right,** schematic of cell connections and a typical receptive field of a ganglion cell with its excitatory center and inhibitory surround. The receptive field looks oval (due to space constraint) but is actually circular. Forward pathways: cone terminals (CT) release glutamate onto horizontal and bipolar cells. Bipolar cells release glutamate onto amacrine and ganglion cells (straight black arrows). Feedback circuits: horizontal cells provide feedback to photoreceptor terminals (blue curved arrow) and GABAergic feedforward to bipolar cell dendrites (black curved arrow). Amacrine cells provide GABAergic or glycinergic input to bipolar axon terminals and to ganglion cell dendrites (curved black arrows). The synaptic layers: OS, outer segment; IS, inner segment; ONL, outer nuclear layer; OPL, outer plexiform layer; INL, inner nuclear layer; IPL, inner plexiform layer; GCL, ganglion cell layer.

most important morphological feature that distinguishes each of these cell types is the stratification level (or sublamina) of their axon terminals within the inner plexiform layer. Roughly, half of the bipolar cells stratify in the sublaminas that reside closer to the somata of the bipolar cells. These constitute the OFF class. The other half of bipolar cells, the ON bipolar cells, stratify in the laminas that are distal to the bipolar somas. The OFF bipolar cells include about five cell types, all of which contact cones and thus are called cone bipolar cells. The ON bipolar cells include about five types that contact cones, and a single type, the rod bipolar cell, that contacts rods. The different types of cone bipolar cells are thought to divide the time domain of the image into separate information channels such that each bipolar type, or channel, processes a different range of temporal frequencies. The single rod bipolar cell type is a key link in transmitting information at low light levels (see below).

Like photoreceptors, the OFF bipolar cells respond to light with a hyperpolarization. This is because these cells express AMPA/kainite receptor channels that, in response to glutamate released in the dark, stay open and keep the cells depolarized. When the synaptic release of glutamate decreases with light, the receptor channels close, and the cells hyperpolarize. In contrast, the ON bipolar cells respond to light with a depolarization. This is because they express a specific type of G-protein-coupled glutamate receptor, mGluR6, that initiates a signaling cascade that eventually closes cation channels (Nawy and Jahr, 1990; Shiells and Falk, 1990; de la Villa et al., 1995). Thus, in darkness, when synaptic release of glutamate from the photoreceptors is high, these cells are hyperpolarized. When light reduces synaptic glutamate release, the cation channels are allowed to open and the ON bipolar cells depolarize. Like photoreceptors, bipolar cells do not generate action potentials. Rather, information is transmitted by graded potentials that modulate the release of glutamate from their axon terminals in a graded fashion.

Glutamate released from bipolar cells is sensed by the third order neurons, the ganglion cell. Ganglion cells also come in about ten different morphological types that are roughly divided into two classes: ON and OFF ganglion cells. ON ganglion cells receive their input from ON bipolar cells and increase their spiking activity in response to an increment of light intensity whereas OFF ganglion cells receive their input from OFF bipolar cells and increase their activity in response to a decrement of light. Different types of ganglion cells convey different aspects or features of the light image. For example, certain ganglion cells convey small spatial details while others blur spatial information but are highly sensitive to contrast and motion, and yet others are sensitive to the direction of motion. Different types of ganglion cells emit dendrites that stratify in different sublaminas of the inner plexiform layer, and thus contact bipolar cells that mostly stratify in the same sublamina. Thus parallel pathways are formed, each conveying a different attribute of the light image. The different types of ganglion cells code image attributes as spike trains that are sent in parallel to the lateral geniculate nucleus for further processing of the visual image. Thus the retina "fragments" the light information for efficient analysis, processing and transmission, and the brain must put back together the image attributes to perceive a coherent picture. In addition to these "image-forming" ganglion cell types, there are certain types of ganglion cell that have intrinsic phototransduction mechanisms, and their output is sent to the suprachiasmatic nucleus to entrain the circadian clock.

2. *Feedback Circuits and Lateral Interactions*

The information that flows forward is intercepted at each synaptic layer by lateral elements. At the first synaptic layer, one (in the case of rodents) or two (in cats and primates) type(s) of horizontal cell modulate the information that flows from photoreceptors to bipolar cells. At the second synaptic layer, about 30 types of amacrine cells modulate the information that flows from bipolar cells to ganglion cells (Fig. 19.1 right). In addition to inhibitory circuits mediated by horizontal cells within the outer plexiform layer and by amacrine cells within the inner plexiform layer, there exists a cell type that communicates *between* the two synaptic layers. This neuron is appropriately termed the inter-plexiform cell, and it is GABAergic in cats and monkeys. Inter-plexiform cells synapse primarily onto the dendrites of bipolar cells. The function of these cells in mammals is unknown.

Horizontal cells have a large dendritic field and are electrically coupled by gap junctions, so they collect and average light information from a large area. This averaged signal is fed back onto the synaptic terminals of the photoreceptors and onto the dendrites of bipolar cells. The mechanism of this feedback is still controversial. In some but not all species studied, horizontal cells express GAD and immunostain for GABA (Vardi et al., 1994; Zhang et al., 2006b). In rabbit and monkey, this expression varies along retinal eccentricity (Grünert and Wässle, 1990; Johnson and Vardi, 1998). GABA receptors located postsynaptic to horizontal cells were found by immunostaining on bipolar dendrites, but not on photoreceptors (Vardi and Sterling, 1994). Furthermore, although some studies report $GABA_A$ responses by photoreceptors, most do not. Yet, horizontal cells of all types, in all species, feed back onto photoreceptors as evidenced by recordings from photoreceptors, bipolar cells and ganglion cells. The variability in GABA content between species, combined with failure to detect GABA receptors on photoreceptors, suggests that the feedback onto photoreceptors is not GABAergic. Current theories attribute this feedback to an ephaptic effect or a modulation of pH in the synaptic cleft (Kamermans and Fahrenfort, 2004; Davenport et al., 2008).

Amacrine cells receive light information from bipolar cells and feed this information back to bipolar cell's axon terminals, and forward to other amacrine cells and to ganglion cell dendrites. The synapses from amacrine cells are mostly inhibitory and account for over 60% of all synapses in the inner plexiform layer. The 30 types of amacrine cells are classified mainly by their morphology (dendritic area and stratification level in the inner plexiform layer) (Masland, 2004). Roughly half of the amacrine cells are GABAergic and the other half are glycinergic. Amacrine cells do not corelease GABA and glycine, but many GABAergic types do release an additional neurotransmitter, such as dopamine, acetylcholine, NO, or a variety of peptides (Vardi and Auerbach, 1995). Thus, amacrine cell circuits are highly diverse and contribute in many ways to all aspects of signal processing and its modulation.

3. *Basic Physiological Responses to Light: the Concept of Receptive Field*

The output of a ganglion cell does not depend on the absolute energy that strikes its photoreceptors, but on the local contrast. Thus, both the light/dark pattern that strikes a particular cell and the pattern that strikes its neighbors affect the cell's response. Furthermore, the response is also greatly modulated by the history of the light stimulus since the system adapts, i.e. it modulates its gain to enable high sensitivity around the ambient light intensities. The key feature of the system that makes possible these response properties is the receptive field. The receptive field of a retinal neuron is defined as the area in the visual field (in degrees) or on the retina (in μm) whose light stimulation affects the neuron's response. A typical receptive field is circular and is composed of an excitatory (or inhibitory) center and an inhibitory (or excitatory) surround (Fig. 19.1 right).

For an ON ganglion cell, a spot of light at the cell's center increases spike rate; a larger spot further increases this response but only up to a certain spot size; this defines the receptive field center. A further increase of spot size decreases the light response. The annulus, whose light stimulation decreases spike frequency, defines the inhibitory antagonistic surround. While a light increment in the surround reduces spike frequency, a light decrement in this area increases spike frequency. For OFF ganglion cells, the reverse occurs (i.e. inhibitory center and excitatory surround); the cell increases its spike rate when the light intensity striking its center decreases or when the intensity striking its surround increases. The sensitivity of the ganglion cell to illuminating a small spot in its receptive field depends on the distance of the spot from the receptive field center. Plotting the cell response as a function of distance from the center for either the center or the surround gives a "Mexican hat" shape, which is modeled by subtracting two Gaussians functions, the center minus the surround.

This center/surround structure of the receptive field has several consequences. First, it defines an optimal stimulus size and thus an optimal spatial frequency for a particular cell. Different ganglion cell types have different receptive field sizes and therefore different functions. Ganglion cells with small dendritic areas have small receptive field center sizes and they can resolve high spatial frequencies (e.g. primate midget ganglion cells). Ganglion cells with larger dendritic areas are optimized to resolve lower spatial frequencies (e.g. primate parasol ganglion cell). The second consequence of a receptive field is that it increases the contrast sensitivity of the ganglion cell, thus converting the cell into a contrast detector rather than a light detector. This is because the strongest stimulus paradigm for an ON ganglion cell is not light stimulation of the center alone, but a light increment in the center simultaneous with a decrement in the surround; i.e. the cell responds best to contrast presented within its receptive field. Larger cells (such as the primate parasol cell) have higher contrast sensitivity (lower threshold) because they integrate information from a larger area. The third consequence of a receptive field is that the center-surround structure increases the dynamic range of the cell. By subtracting the mean luminance, a cell's voltage response can be mapped around the mean luminance thus readjusting for different mean intensities.

The center-surround organization was first discovered in ganglion cells, but it arises already at the first synapse. Thus, while a photoreceptor hyperpolarizes in response to light striking its outer segment, it depolarizes in response to light striking its neighbors. The basis for this antagonistic surround is the horizontal cell network that provides a shallow surround by collecting input from a large area and feeding it back to the photoreceptor terminals. This center-surround structure is then transmitted to bipolar cells. These cells show a more significant surround due to the summation of the output from several cones, and to the additional surround inhibition from horizontal cells to bipolar dendrites and from amacrine cells to their axons. Like horizontal cells, amacrine cells that contribute to the formation of the receptive field surround collect information from a large area, greater than the bipolar cell's center or the ganglion cell's center area. At the next stage of processing, the center-surround structure of bipolar cells is summed by the ganglion cells and again this summation deepens the surround. Further extension of the surround results from amacrine cell inhibition of ganglion cell dendrites. This inhibition is typically GABAergic or glycinergic, but additional neuromodulatory transmitters are coreleased, thus providing further computation abilities.

4. *Contribution of inhibitory interneurons to specific retinal functions*

In addition to contributing to the receptive field structure, GABAergic and/or glycinergic interneurons actively take part in feeding information forward at low light levels, contributing to adaptation processes, and shaping the temporal waveforms of bipolar and ganglion cell responses. Three examples will be discussed below.

a. Glycinergic Amacrine Cell (AII) Participates in the Rod Bipolar Feedforward Pathway

As described above in section II.A.1, low-level light is captured by rods, and the signal is transmitted to the rod bipolar cell (recall that this cell is an ON type). Unlike cone bipolar cells, which directly contact ganglion cells, the rod bipolar cell sends its signal to ganglion cells indirectly. The rod bipolar cell contacts a glycinergic amacrine cell, called the "AII amacrine cell", that forms gap junctions with neighboring AII cells and with ON cone bipolar cells. These cone bipolar cells then contact ON ganglion cells (Fig. 19.2). There is no OFF rod bipolar cell, yet OFF ganglion cells do respond to low light levels with a reduction of their spontaneous spike rate (Mastronarde, 1983). Information reaches the OFF ganglion cell from the rod bipolar cells through the AII amacrine cell that makes glycinergic synapses both onto OFF cone bipolar and OFF ganglion cells. Thus glycinergic amacrine cells are key linkers in the rod bipolar pathway and not merely modulators. Since the electrical synapses from AII to ON bipolar cells retain the polarity of the response, and the glycinergic synapses to OFF bipolar cells reverse the polarity, the AII amacrine cells provide information of opposite polarities to both ON and OFF ganglion cells. Consequently, a photon absorbed by a rod

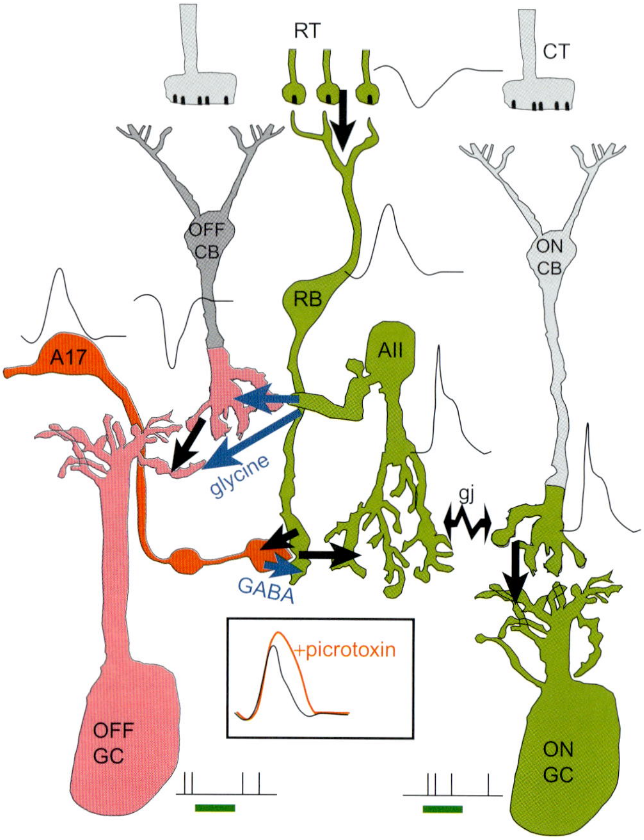

FIGURE 19.2 The rod bipolar pathway. At low light levels, below cone threshold, the information flows from rod terminals to rod bipolar cells to AII amacrine cells. From here the information splits. In the ON sublamina, it flows to the axon terminals of cone bipolar cells (via gap junctions) and then on to the ON ganglion cells (green pathway). In the OFF sublamina, it flows from the AII to the axons of OFF bipolar cells and to the OFF ganglion cells (via glycinergic synapses). Information also flows from the OFF cone bipolar cells to the OFF ganglion cells (pink pathway). Black arrows show glutamatergic transmission; blue arrows show glycinergic and GABAergic transmission. Schematized response is shown for each cell type. Note inversion of polarity from rod to rod bipolar and from AII amacrine to OFF cone bipolar and OFF ganglion cells. Also note that the response becomes faster and more transient as it proceeds from rod to rod bipolar cell and to AII amacrine cell. Responses prior to the ganglion cells are graded while ganglion cell responses are spikes; green bars under ganglion cell responses represent the light stimulus. Inset shows a schematized response of a rod bipolar cell before (black) and after (red) blocking inhibition. This inhibition arrives from the A17 amacrine cell that receives input from the rod bipolar cell and releases GABA back onto it. RT, rod terminal; CT, cone terminal; RB, rod bipolar cell; CB, cone bipolar cell; GC, ganglion cell; gj, gap junction.

would increase spiking frequency in the ON ganglion cell while decreasing spiking rate in the OFF ganglion cell. Thus, the AII amacrine negatively correlates the responses of neighboring ON and OFF ganglion cells (Murphy and Rieke, 2008). The "bridge" from the ON system to the OFF system is also manifested at higher light levels. In daylight, when light decreases, the cones hyperpolarize the ON cone bipolar cells, and this is relayed to the AII amacrine cells through the cone bipolar to AII gap junctions (in the opposite direction to night vision). The AII then releases less glycine and releases the OFF cell from its inhibition. Thus, this disinhibition contributes to excitation of the OFF ganglion cells, and by doing so it extends the dynamic range of the OFF ganglion cell in daylight (Manookin et al., 2008).

b. The GABAergic Amacrine Cells (A17) quicken the Time Course of the Responses to Light

An important function of inhibition in the rod bipolar cell pathway is to speed up the responses to light (Fig. 19.2, boxed inset). The light response of the AII amacrine cell is faster and more transient than that of the rod bipolar cell. The response of the bipolar cell is in turn faster than the light-evoked response of the rod. There are multiple mechanisms that speed up the rod-to-bipolar light-evoked response (Armstrong-Gold and Rieke, 2003; Xu et al., 2008) and the AII response (Nelson, 1982; Smith and Vardi, 1995; Bloomfield and Volgyi, 2004a; Bloomfield and Volgyi, 2004b). Relevant to this chapter, an important mechanism that accomplishes this speeding up of the responses to light is the feedback inhibition from a certain type of amacrine cell called A17. This cell type radiates from its soma many dendrites that extend for long distances. The dendrites are varicose, and the varicosities are connected with slender dendrites from other cells. This morphological feature causes the electrical signals to decay fast, thus rendering each varicosity electrically isolated from each other. The A17 varicosities receive input from the rod bipolar axon terminal and feed back to the same terminal via a reciprocal GABAergic synapse (Nelson and Kolb, 1985; Chavez et al., 2006). Since this feedback signal is delayed by two synapses relative to the rod-to-bipolar signal, the A17 cell starts to inhibit the rod bipolar's light response after it has reached its peak. Consequently, the response of the rod bipolar cell to a long-lasting light stimulus has two phases, a transient one that peaks around 150 msec after stimulation, and a sustained phase with lower amplitude that remains for the duration of the stimulus. When the light stimulus is briefer than the response rise time, the response has only one phase. Blocking GABAergic inhibition or both GABAergic and glycinergic inhibition greatly increases the decay time, suggesting that the feedback accelerates the light response (Dong and Werblin, 1998; Dong and Hare, 2002; Xu et al., 2008). Thus, the A17's GABAergic inhibition of rod bipolar cells facilitates its light response and enables this cell to resolve higher temporal frequencies than the rod. Interestingly, rod bipolar cells express at least three different ligand-gated chloride channels, $GABA_A$, $GABA_C$ and glycine receptors, each of which displays a different response time course. The sharpening of the light-evoked response is attributed mainly to the relatively slow $GABA_C$ receptor (Eggers and Lukasiewicz, 2006).

c. Asymmetric Inhibition Computes Directional Selectivity

A fascinating example of computation by inhibitory circuits is the extraction of motion direction (highly developed in rabbits). The retina has two types of directionally selective ganglion cells, the ON type and the ON-OFF type. Most work has been performed on the more dense ON-OFF cell type (see reviews by Taylor and Vaney, 2002; Demb, 2007; Zhou and Lee, 2008). A directionally selective ganglion cell responds best to an object moving in one (preferred) of four directions (up, down, left or right) and it does not respond to an object moving in the opposite direction (null). Blocking $GABA_A$ receptors results in the loss of direction selectivity, i.e. the cell now responds equally to an object moving in any direction.

A classical model to explain directional selectivity was suggested by Barlow and Levick (1965). In this model (Fig. 19.3A), inhibition arises from the null direction with a delay. When the object moves from point 1 toward point 2, the delayed inhibition from point 1 coincides with the excitation from point 2 and nulls it. A similar process occurs when the object continues to move from point 2 to 3. In contrast, when the object moves from point 3 to point 2, the inhibition from 2 arrives too late to nullify the excitation from point 3, so the cell responds. It is now known that the GABAergic neuron that provides this asymmetric input is the "starburst" amacrine cell. The starburst cell itself exhibits a directionally selective response to moving objects, but this selectivity is for radial movement; the cell's soma shows a greater response to centrifugal movement (objects moving away from the soma) than to centripetal movement (objects moving toward the soma). However, the starburst cell has no axon, and its soma's response is practically irrelevant to the circuit. The dendrites of the starburst cell work as independent units; they receive bipolar input throughout their length, but release neurotransmitter only from their ends. Calcium imaging makes it possible to record the dendrite's response and to assess the output of the cell. It has been found that each dendrite is sensitive to motion away from the soma. Thus each dendrite is selective for linear motion in one direction, but a cell with 4–5 dendrites radiating from its soma is endowed with the capacity to integrate and transmit information about all directions of motion of the stimulus. The mechanism for this selectivity to stimulus motion direction is thought to arise from reciprocal inhibition between starburst cells (Fig. 19.3C). This inhibition creates the surround of the starburst cell that in turn creates the directional selectivity of its dendrites (Lee and Zhou, 2006). An additional mechanism that may explain the radial direction selectivity of the starburst cells is attributed to differential distribution of KCC2 and NKCC transporters along the dendrites of these cells (Gavrikov et al., 2003, 2006).

Starburst cells also release acetylcholine and GABA onto the ganglion cells. The function of acetylcholine is probably to enhance the excitatory input that ganglion cells receive from bipolar cells, but its precise contribution is not clear. The function of GABA is clearly to shape the directional selectivity of the ganglion cell. Since all cells distribute evenly across the retina, each directionally selective ganglion cell is surrounded by several starburst cells. However, only the starburst cells on the null side provide inhibition to the directionally selective ganglion cell. Therefore, when an object moves in the null direction, the inhibition coincides with the excitation (Fig. 19.3B, C) and nulls the response (Fried et al., 2002; Lee and Zhou, 2006). In contrast, when the object moves in the preferred direction, the inhibition is small and lags behind the excitation, so it does not greatly interfere with the response. Thus, in many respects, the model shown in Fig. 19.3C supports the general idea proposed by Barlow and Levick.

B. Diversity of GABA Actions in Different Types of Mature Neurons and its Relation to the Expression of Chloride Cotransporters

As noted at the beginning of this chapter, E_{GABA} is relatively close to E_m and thus the action of GABA in neuronal circuits may be modulated by changing E_{Cl} and/or E_m. Indeed, as discussed in several chapters throughout this book, GABA and glycine can be hyperpolarizing and inhibitory in some neurons, depolarizing and even excitatory in others, or have a dual excitatory and inhibitory action in the same neuron. In the latter case E_{Cl} is thought to have a different value with respect to E_m in different cell compartments. In these cells, opposing responses to GABA or glycine can be spatially segregated. Moreover, besides being spatially segregated, the responses can also be

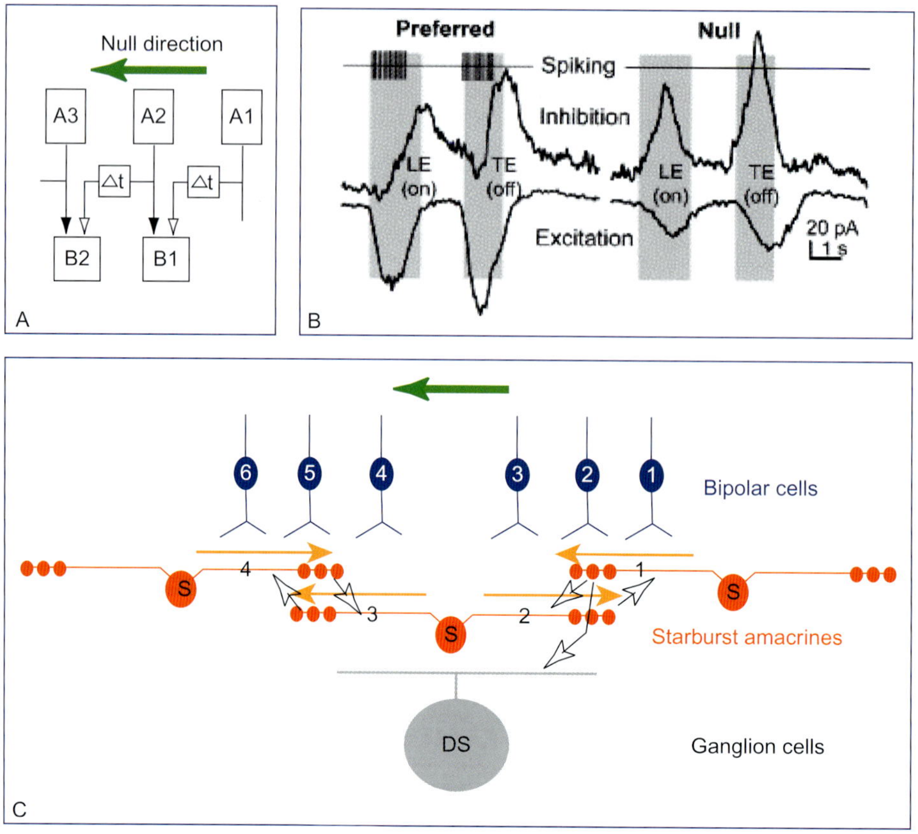

FIGURE 19.3 Asymmetric inhibitory input creates directionally selective cells. **A.** A "concept" model modified from Barlow and Levick (1965). "A" units provide excitatory input to "B" units (filled arrows). "A" units also send a delayed inhibitory output leftward but not rightward (empty arrows). When the stimulus moves in the null direction, inhibition from location A1 coincides with excitation from A2 and nulls the response at B1. When the stimulus moves in the preferred direction, excitation from A2 arrives before inhibition from A1, and B1 can respond. Thus B units are directionally selective and summing these units will retain directional selectivity (not shown). **B.** A summary (reproduced with permission from Fried et al., 2005) of excitatory and inhibitory currents in the directionally selective ON-OFF ganglion cell. This cell gives action potentials (top traces) both when the leading edge (LE) and when the trailing edge (TE) of the object enter the receptive field center and move in the preferred direction. It does not give action potentials when the object moves in the null direction. Clamping the cell at 0 mV, to record inhibitory currents (middle traces), and at E_{Cl}, to record excitatory currents (bottom traces), shows that in the preferred direction, inhibitory current lags excitatory current while, in the null direction, it coincides or even precedes the excitatory current, just as predicted by Barlow and Levick. **C.** A model (modified from Zhou and Lee, 2008) showing the interaction between cells that participate in computing directional selectivity. Arrays of bipolar cells are blue, starburst cells are red, and the directionally selective (DS) ganglion cell is gray. Bipolar cells provide excitation both to the starbursts and the ganglion cell. Starburst dendrites inhibit each other in a symmetrical manner (open arrows) and this provides them with a receptive field surround. In the direction of movement shown (null for the example DS), starburst dendrite 1 (dendrite numbers shown in black) responds before starburst dendrite 2 and inhibits it, thus nulling the response of dendrite 2. Since the dendrites are electrically isolated, i.e. inhibition from dendrite 1 does not affect the response of dendrite 3, dendrite 3 does respond to this direction, which is centrifugal. Thus reciprocal inhibition endows the starburst dendrites with responses that are selective for movement away from the soma as shown by the orange arrows. Starburst cells from the null side (right), but not from the preferred side, also provide inhibitory input to the ganglion cell (open bent arrow) and possibly to the bipolar cell (not shown). The starburst cells therefore create the asymmetry shown in A. Therefore, inhibition resulting from bipolar 1's output (which is in the surround of the DS cell) coincides with excitation from bipolar 2 (which is in the center). In summary, reciprocal inhibition between starburst cells, and asymmetric inhibition between these cells and the directionally selective ganglion cell vetoes the response to movement in the null direction, and thus computes direction of motion.

temporally segregated by dynamic changes in $[Cl^-]_i$ and therefore in E_{Cl} that switch the responses to GABA or glycine from inhibitory to excitatory (or vice versa), as illustrated in Fig. 19.4. In almost all mature CNS neurons studied to date the level of intracellular Cl^- is determined by the relative functional expression of the active Cl^- extruder KCC2 and active Cl^- accumulators such as NKCC1. Diverse patterns of expression of these transporters may result in a variety of signal processing modes in different networks. The pattern of expression of cation-coupled Cl^- cotransporters in the retina is illustrated in the diagram of Fig. 19.5.

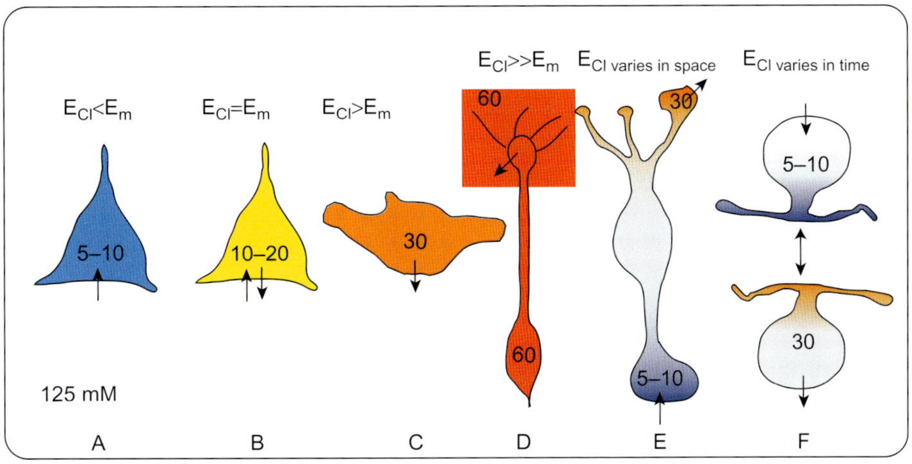

FIGURE 19.4 Diversity of intracellular chloride concentrations in neurons and the consequences of activating chloride channels. Numbers within the cells are approximate $[Cl^-]_i$ and in the analysis below we assume a membrane resting potential (E_m) of −65 mV. Arrows show direction of Cl^- flux when chloride channels are open. **A.** Cells that are hyperpolarized by GABA: ganglion cells and most mature neurons. GABA causes chloride influx. **B.** Cells that are shunted by GABA: certain amacrine cells, neocortical neurons and many mature neurons. Net Cl^- flux is null, but due to conductance increase, GABA shunts the excitatory current and inhibits the cell. **C.** Cells that are depolarized by GABA e.g. immature neurons, horizontal cells, DRG neurons, and hippocampal granule cells. In these cell types GABA causes chloride efflux. **D.** Cells in which the chloride current produces a depolarizing receptor potential e.g.: olfactory receptor neurons (in these cells E_{Cl} is more positive than E_m and opening of calcium-activated chloride channels produces a depolarizing chloride efflux). **E.** Cells with compartmentalized E_{Cl}: ON bipolar cells (E_{Cl} is more positive than E_m in dendrites, and more negative than E_m in axon); possibly starburst amacrine cells (E_{Cl} is more positive than E_m in proximal dendrites, and more negative than E_m in distal); principal neurons (E_{Cl} is high in axon initial segment and low in the rest of the cell). **F.** Cells with dynamic E_{Cl}: certain amacrine cells (E_{Cl} shifts positive with NO); some hippocampal cells (E_{Cl} shifts positive with NO); gonadotropin-releasing hormone neuron (E_{Cl} shifts negative with long application GABA); many hippocampal cells (E_{Cl} shifts negative with repetitive application of GABA); certain neurons in dorsal suprachiasmatic nucleus (E_{Cl} shifts positive at night).

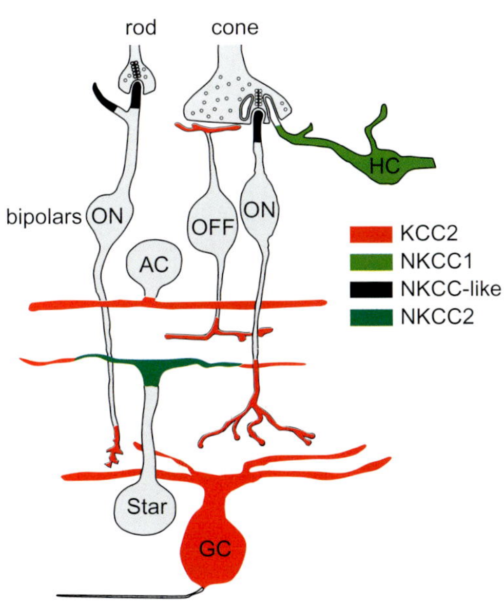

FIGURE 19.5 Expression of chloride cotransporters in adult retinal cells. HC, horizontal cell; AC, amacrine cell; GC, ganglion cell; star, starburst amacrine cell; ON, ON bipolar cell; OFF, OFF bipolar cell. ON bipolar dendritic tips express a protein that is not NKCC1, but might be similar to it (black). Among different types of cone bipolar cells only certain types express KCC2. Rods and cones express KCC2 transiently, but probably not in the adult.

1. *GABA as a Hyperpolarizing Inhibitory Neurotransmitter: Neurons with Lower than Passive Intracellular Cl^- Levels*

Most mature brain neurons hyperpolarize to GABA and glycine. In retina, this strict and clear type of inhibition is exemplified in the multiple inputs from GABAergic and glycinergic amacrine cells onto ganglion cells. Retinal ganglion cells receive glutamatergic excitatory input from bipolar cells and inhibitory input from GABAergic and glycinergic amacrine cells. There are about 30 types of amacrine cells whose primary function is to shape the receptive field of ganglion cells. They modulate ganglion cells' temporal response, control their gain and sensitize them to motion. Realizing the importance of Cl^- currents to information processing in the retina, Miller and Dacheux (1983) first measured $[Cl^-]_i$ in different types of neurons from mudpuppy retina, using double-barrel Cl^--sensitive microelectrodes. This approach allowed them to measure E_m and intracellular Cl^- activity in single cells. They found that Ganglion cells have an E_{Cl} more negative than the resting E_m in darkness. Specifically, the average E_{Cl} in these neurons was −49 mV whereas the "dark" membrane potential was

−39 mV. Subsequent studies in mature mammalian ganglion cells using gramicidin-perforated patches and the Cl^- indicator MEQ showed much lower values: 60 to −80 mV for E_{Cl} and around −60 mV for the resting E_m (Demb et al., 2001; Murphy and Rieke, 2006; Zhang et al., 2006b), but E_{Cl} is always close to or more negative than resting E_m; thus GABA and glycine produce inhibition by membrane shunting and/or hyperpolarization.

The lower than passive $[Cl^-]_i$ of ganglion cells suggests that they are endowed with an active Cl^- extruder like KCC2. Immunolabeling studies using a specific antibody against KCC2 revealed that mature ganglion cells indeed express KCC2 (Figs 19.5 and 19.6A). Immunofluorescence and electron microscopy revealed that KCC2 immunolabeling was associated with the plasma membrane. KCC2 immunolabeling was found to be relatively weak at the somas and much stronger at the dendrites, in consonance with the fact that most inhibitory synapses contact the dendrites. This proximity of KCC2 to synaptic regions suggests that the transporter tightly regulates $[Cl^-]_i$ in the postsynaptic membranes.

2. *GABA as a Depolarizing and Excitatory Neurotransmitter: Neurons with Higher than Passive Intracellular Cl⁻ Levels*

It is unusual for mature neurons to maintain an $[Cl^-]_i$ higher than predicted for a passive Cl^- distribution across their plasma membrane. Primary sensory neurons and olfactory sensory neurons are well-documented examples of *mature* nerve cells in which $[Cl^-]_i$ is maintained above electrochemical equilibrium (Alvarez-Leefmans et al., 1988; Kaneko et al., 2004; Rocha-González et al., 2008). The functional significance of the outward Cl^- gradients in these sensory neurons is discussed in detail in Chapters 20 and 22 in this volume. In mature brain, there are fast spiking interneurons that maintain an E_{Cl} close to the spike threshold and therefore GABA excites them (Martina et al., 2001). Another example is the gonadotropin-releasing hormone neurons in the hypothalamus, which are consistently depolarized by GABA. Further, immunostaining shows that most of these cells express NKCC1 (DeFazio et al., 2002). In the mature retina, a well-studied example where GABA acts to depolarize the cell is provided by the horizontal cell (see below).

a. Retinal Horizontal Cells

It has long been known that Cl^- gradients play key roles in light-evoked signaling and image processing in the retina because replacing Cl^- with an impermeant anion blocks ON ganglion cell responses (Miller and Dacheux, 1973). Cl^--dependent mechanisms were studied by systematically measuring intra- and extracellular Cl^- *activities* throughout the retina using Cl^--selective microelectrodes. In horizontal cells of mudpuppy and fish, average intracellular Cl^- *activity* was ~42 mM. Thus, taking an activity coefficient of ~0.76 (Parsons, 1959), the average intracellular Cl^- concentration is ~55 mM. The average E_{Cl} was −17 mV, which is about 10 mV more depolarized than

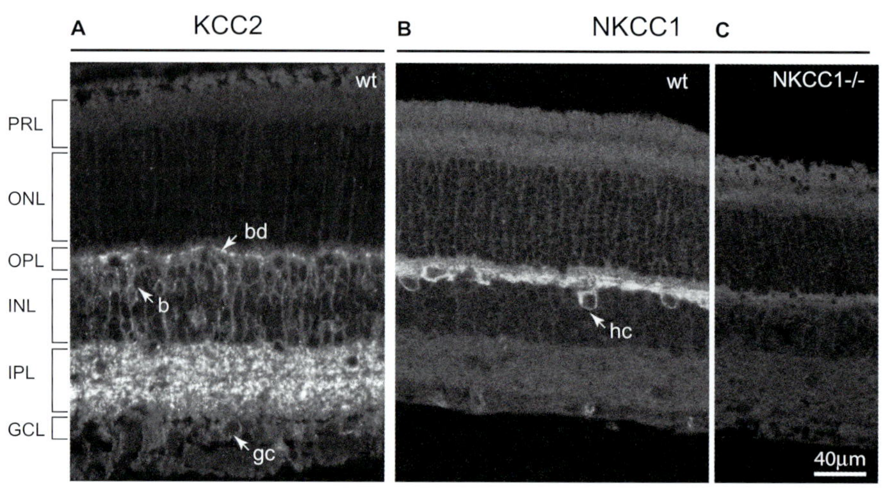

FIGURE 19.6 Immunostaining of mouse retina for KCC2 and NKCC1 (radial sections). **A.** Staining for KCC2 shows that the strongest stain lies in the synaptic layers. Stained puncta in the outer plexiform layer (OPL) belong to the dendrites of OFF bipolar cells (bd). Staining in bipolar cell somas (b) is strong, in ganglion cell somas (gc) is weak, and in amacrine cell somas is too weak to be seen in this image. **B, C.** Staining for NKCC1 shows that in the wild type (wt) labeling in horizontal cell somas (hc) and their processes is strong, and there is no staining anywhere else. The faint staining elsewhere is at background level, same as in NKCC1 knockout mouse ($NKCC^{-/-}$). Staining of horizontal cells in NKCC1 knockout retina is absent. PRL, photoreceptor layer; remaining retinal layers are as labeled in Fig. 19.1. Reproduced with permission from Zhang et al. 2007.

the dark resting potential (Miller and Dacheux, 1983; Djamgoz and Laming, 1987). Subsequent intracellular recordings with sharp microelectrodes from skate rod-driven dissociated horizontal cells and from rabbit horizontal cells in the eye-cup preparation showed that GABA depolarized these cells, and this depolarization was mediated by $GABA_A$ receptors, further suggesting that $[Cl^-]_i$ is higher than electrochemical equilibrium in these cells (Lasater et al., 1984; Blanco et al., 1996). Additional observations consistent with the hypothesis that retinal horizontal cells maintain a higher than passive $[Cl^-]_i$ came from experiments using gramicidin-perforated patch microelectrodes. In these experiments, the authors recorded GABA responses of dissociated rabbit horizontal cells, thus eliminating network influences. E_{Cl} and $[Cl^-]_i$ were estimated to be $-29\,mV$ and 44 mM, respectively (Varela et al., 2005b).

Active Cl^- accumulation in horizontal cells is most likely mediated by NKCC1. However, the evidence that supports this hypothesis is based only on immunostaining. The commonly used monoclonal T4 antibody weakly immunolabels horizontal cells in several mammalian retinas and strongly immunolabels horizontal cells of lower vertebrates and mouse horizontal cells treated with SDS (Vardi et al., 2000, 2002; Li and Shen, 2007; Li et al., 2008). However, this antibody is not NKCC isoform specific; it recognizes both NKCC1 and NKCC2. To determine the NKCC isoform expressed in horizontal cells we used an antibody prepared against the C-terminus of mouse NKCC1 (Plotkin et al., 1997a). This antibody showed strong immunostaining in horizontal cells of mouse retina, which was absent in control NKCC1-null mice, as shown in Fig. 19.6B, C (Zhang et al., 2007). In these studies, KCC2 was not detected in mouse horizontal cells but an earlier study by a different group reported KCC2 immunolocalization in rat horizontal cells labeled with calbindin (Vu et al., 2000). In our opinion, this apparent colocalization results from a relatively low image resolution and the proximity of horizontal cell processes to structures that do express KCC2. Still missing are functional experiments that test the contribution and regulation of NKCC1 and possibly KCC2 to Cl^- accumulation in horizontal cells.

While the evidence for $[Cl^-]_i$ being higher than electrochemical equilibrium and that the $GABA_A$-mediated depolarization in horizontal cells is strong, the function of this depolarization remains unclear. The main source of GABA within the outer plexiform layer, the layer at which horizontal cells arborize, is the horizontal cell itself (Vardi et al., 1994; Johnson and Vardi, 1998). Another source is a unique interneuron called the interplexiform cell, but this cell does not seem to make synaptic contacts with horizontal cells (Nakamura et al., 1980; Chun and Wässle, 1989). Thus, the $GABA_A$ receptors on horizontal cells are autoreceptors that could provide positive feedback onto the horizontal cell network (see below). We think this feedback is positive because horizontal cells do not spike and their signaling is transmitted via changes in graded membrane potential. Interestingly, the pharmacological properties of the $GABA_A$ receptors on horizontal cells of several species often differ from that of the classical $GABA_A$ or $GABA_C$ receptors, suggesting that they may express a different type of $GABA_{A/C}$ receptor. Consistent with this, receptor subunits of $GABA_{A/C}$ have not yet been detected by immunostaining. Furthermore, certain types of horizontal cells in zebra fish do not express GABA receptors, and yet they are depolarized in response to GABA. It appears that this depolarization is mediated by the Na^+-dependent GABA transporter (Nelson et al., 2008). Thus, the GABAergic feedback loop provided by horizontal cells may be an important design aspect, but how the feedback occurs may vary between species and between horizontal cell types.

Two hypotheses have been proposed to explain the function of the positive GABAergic feedback loop: (a) the loop couples horizontal cells chemically and (b) it slows down their light-evoked response. As explained at the beginning of this chapter, the main function of horizontal cells is to collect input from a large area, compute the mean local luminance and feed it back to the photoreceptors and bipolar cells. A horizontal cell's dendritic diameter spans $60\,\mu m$ (in mouse), but since they are electrically coupled by gap junctions, their receptive field (i.e. collection area) is doubled (Shelley et al., 2006). It is possible that the feedback depolarization by GABA provides another way to increase this receptive field. This GABA-mediated depolarization would occur particularly in darkness when photoreceptors constantly release glutamate, and thus depolarize the horizontal cells via AMPA/kainate receptors. This depolarization would then increase GABA release and that would further depolarize the same cell and its neighbors. Thus, the positive feedback by GABA can be viewed as chemical coupling that enhances the cell's response to glutamate and extends its receptive field much like electrical coupling does (Varela et al., 2005b). Alternatively, GABA's positive feedback may function to slow down the kinetics of the light response. It has been shown that the horizontal cell light response in the dark is much slower than that in the light, and this effect can be mimicked by GABA application. Thus, while GABA released in darkness helps to depolarize the cells, it also increases cell conductance, so when glutamate decreases in light, the horizontal cell hyperpolarizes and GABA release decreases with a delay. Consequently, the hyperpolarization of the horizontal cell is slowed (Kamermans and Werblin, 1992).

3. *Spatial Gradient of E_{Cl} and Compartmentalization of GABA Function*

Cl^- has long been thought to distribute evenly within a cell but there is mounting evidence showing that, within many neuron types, Cl^- concentration does vary along different parts of the cell. It is tempting to draw analogies between Cl^- and Ca^{2+}. The latter is known to be compartmentalized and there are numerous proteins that bind Ca^{2+} and buffer its local concentration. It would be exciting if future research identifies a protein that binds Cl^- and restricts its intracellular diffusion. However, fast transport of Cl^- across the plasma membrane in one region and lack of transport in another may be sufficient to generate and maintain a gradient within the cell. In addition, intracellular organelles, which have been found to have high Cl^- concentrations and CLC channels and transporters, may regulate and dynamically modulate Cl^- concentration within a local area that could define a compartment. In the following section, we will discuss the functional significance of Cl^- gradients in ON bipolar cells and describe evidence that supports it.

a. Intracellular Cl^- Gradients in Retinal ON Bipolar Cells

We have already discussed the function of horizontal cells in forming the receptive field surround of photoreceptors and bipolar cells. The principle of this action is to antagonize the effect of glutamate released from photoreceptors. Glutamate depolarizes OFF bipolar cells and hyperpolarizes ON bipolar cells. To antagonize glutamate's action, GABA released from horizontal cells should hyperpolarize OFF bipolar dendrites and depolarize ON bipolar dendrites. Since both classes of bipolar cells express $GABA_A$ receptors on their distal dendrites, this opposite polarity can only be achieved if E_{Cl} in OFF bipolar dendrites is close to or more negative than the resting E_m, whereas in ON bipolar dendrites E_{Cl} is more depolarized than E_m (Vardi and Sterling, 1994). Yet, at the axon terminal, ON bipolar cells receive GABAergic and glycinergic feedback from amacrine cells, and this feedback is known to hyperpolarize the cell (Suzuki et al., 1990). Therefore, theory predicts that ON bipolar cells should maintain relatively high $[Cl^-]_i$ in the dendrites and relatively low $[Cl^-]_i$ in the axon terminal, i.e. ON bipolar neurons maintain a Cl^- gradient within the cell. This hypothesis was tested by several approaches as discussed below.

b. Immunostaining

KCC2 immunolabeling in a variety of mammalian retinas clearly shows that the axon terminals of rod bipolar cells, a type of ON bipolar cell, are immunopositive and that this immunostaining concentrates near the GABA receptors (Vardi et al., 2000). On the other hand, the somas of these cells are weakly immunoreactive for KCC2 and the dendritic tips are KCC2 immunonegative. In contrast, OFF bipolar cells show clear KCC2 immunoreactivity both in their axon terminals and their dendritic tips. Immunolabeling for NKCC using T4 monoclonal antibody shows that NKCC immunoreactivity is restricted to the dendritic tips of the ON bipolar cell. Neither the ON bipolar axon terminal nor any compartment of the OFF bipolar cell are NKCC immunoreactive. Similar differential NKCC immunolabeling between ON and OFF bipolar cells and between dendrites and axons in ON bipolar cells has also been shown in goldfish retina (Li et al., 2008). These observations suggest that the same cell expresses both an active Cl^- extruder and an active Cl^- uptake cotransporter protein, but that they are located at different endings of the cell, just as predicted by theory. Thus, KCC2 extrudes Cl^- from the axon terminal permitting GABA and glycine released by amacrine cells to hyperpolarize the cell, while NKCC1 accumulates Cl^- at the tips of the dendrites permitting GABA released by horizontal cells to depolarize it.

A complication of this clear picture arose when the staining obtained with the T4 antibody was retained in NKCC1-null mice retina (Zhang et al., 2007). Furthermore, staining with an antibody that is specific to NKCC1 revealed staining in the horizontal cells, but not in the ON bipolar dendritic tips. These results suggest that ON bipolar dendritic tips (at least in mouse) do not express NKCC1, but a closely related protein. NKCC2 could be a candidate. However, available antibodies against this protein used in Western blotting do not reveal a band with the mobility described for the kidney protein NKCC2.

c. Whole-cell Recording with Gramicidin

The hypotheses supported by immunostaining predict two important differences in E_{Cl}: a difference between ON and OFF bipolar cells (where E_{Cl} in ON dendrites have a more depolarized value than that of OFF dendrites) and a difference within the ON bipolar cell (where the dendritic tips have a more depolarized value than at the axon). Several labs have tested these hypotheses by focal application of GABA to either the dendrites or the axon. The results are consistent with the predictions but the details are still somewhat controversial, as discussed below.

Comparing E_{Cl} values between classes of bipolar cells by puffing GABA onto their dendrites (in mouse slice preparations) revealed that the E_{Cl} of the rod bipolar cell was significantly more depolarized than that of the ON cone bipolar cell which in turn was more depolarized than that of the OFF cone bipolar cell (−39, −55 and −67 mV, respectively). The dark membrane potential of these cells is estimated to be −45 mV; therefore GABA is expected to depolarize the rod bipolar cell but hyperpolarize both types of cone bipolar cells. Comparing the E_{Cl} of rod bipolar dendrites to that of axon terminals by selectively puffing GABA either on the dendrite or the axon showed that there was no difference (Satoh et al., 2001). However, these experiments were done without blocking network signaling. Similar experiments were performed in rat slices where cells were pharmacologically isolated from the network. For each rod bipolar cell, E_{Cl} in the dendrites was more depolarized than that in the axon. Although small (~4 mV) the difference was significant (Billups and Attwell, 2002). Nonetheless, the authors concluded that this difference was too small to contribute to GABA antagonism. Interestingly, in these experiments, although E_{Cl} changed when the holding membrane potential was changed for more than a minute, E_{Cl} in the dendrites always remained more depolarized than that in the axon, favoring the idea of compartmentalization of intracellular Cl^-.

The possible E_{Cl} gradient in ON bipolar cells was re-examined in mouse dissociated rod bipolar cells (Varela et al., 2005a). Dissociation has two advantages: first, complex network effects are eliminated; second, visualization of the processes allows more precise focal application of GABA. Also, using mouse as experimental model is convenient because its rod bipolar cells express mainly $GABA_A$ receptors on their dendritic tips and $GABA_C$ receptors at their axon terminals. Thus GABA-gated Cl^- channels on the dendrites can be activated by puffing GABA and TPMPA (to block the residual $GABA_C$ receptors) and those on the axons by puffing GABA and bicuculin (to block the residual $GABA_A$ receptors). Their analysis revealed that E_{Cl} in dendrites is significantly more depolarized than in axon terminals (−33 vs. −60 mV, respectively). Current clamp experiments verified that, as suggested by immunocytochemistry, GABA applied onto these cells' dendrites has a depolarizing action whereas when applied to the axon terminals has a hyperpolarizing action.

d. Imaging with Clomeleon

A breakthrough in estimating $[Cl^-]_i$ within different parts of a cell came with the development of the Cl^- sensor Clomeleon , as discussed in detail in Chapter 7 in this volume. A mouse expressing Clomeleon under control of thy1 promoter has been developed. Several retinal cell types in this genetically modified mouse, including three types of ON cone bipolar cells (but not the rod bipolar cell), express Clomeleon. This provides a direct method for testing the theory that ON bipolar cells maintain a Cl^- gradient. Furthermore, since the cells in these mice fluoresce, they can be identified by their morphology (Duebel et al., 2006). These results are discussed in detail in Chapter 7 of this book. Briefly, it was found that $[Cl^-]_i$ in ON bipolar cells is 35 mM in the dendrites, 27 mM in their soma and 25 mM in their axon, i.e. E_{Cl} is −33 mV, −40 mV and −42 mV, respectively. Since the average resting E_m of these cells is −42 mV, GABA applied to the dendrites depolarizes them, while at the soma or the axon causes shunting inhibition. When cell type was considered, Duebel and collaborators found that one type of bipolar cell (type 9) exhibits a significant (20 mV) E_{Cl} gradient with respect to the soma ($[Cl^-]_i$ is 44 mM in dendrite vs. 24 mM in soma) while another type of bipolar cell (type 7) has only a ~4 mV difference. Accordingly, GABA puffed onto the dendrites of type 9 decreased $[Cl^-]_i$ but when puffed onto their axon increased $[Cl^-]_i$. Simultaneous application of furosemide and bumetanide to maximally block the Cl^- gradients reduced, but did not abolish them. One possible interpretation of this finding is that the Cl^- uptake mechanism in ON bipolar dendritic tips is not sufficiently sensitive to loop diuretics.

In summary, most studies to date support the notion that ON bipolar cells exhibit an intracellular Cl^- gradient, but the data still suffer from high variability. Although some of this variability may reflect true differences between cells, much of it is probably due to limitations of the methods used to measure $[Cl^-]_i$ and E_{Cl} and insufficient resolution to discriminate between different parts of a single cell. Antibodies used for immunostaining are often nonspecific, may have low sensitivity, and do not prove function. Gramicidin-perforated membrane patch may retain the native anion content of a cell, but the recordings suffer from high access resistance and inadequate voltage clamp. Further, the reversal of GABA or glycine currents does not necessarily correspond to E_{Cl} because the anion channels opened by these neurotransmitters are often non-selective for Cl^-. Clomeleon imaging may be a promising technique for Cl^- imaging. However, it is inherently difficult to reliably image thin processes; slight deviations from the focal plane introduce large variability. In addition, Clomeleon is difficult to calibrate reliably and it is highly sensitive to intracellular pH. Some of these limitations are discussed in Chapter 7. New chemical and GFP-based fluorescent Cl^- indicators are discussed in Chapter 6.

4. *Temporal Change of E_{Cl}: Modulation by Neurotransmitters*

In the previous sections, we discussed examples of cells in which E_{Cl} is fixed in the whole cell or in some cell compartments. However, $[Cl^-]_i$ and thus E_{Cl} may change over time. The best known and studied examples come from developing neurons that experience a dramatic shift in E_{Cl} after maturation (see below). Shifts in E_{Cl} may also occur during a circuit's malfunctioning, as discussed in Chapters 22 to 24. In healthy adult neurons, E_{Cl} can also change, with a relatively fast time course and in response to neurotransmitter stimulation. In the sections that follow, we discuss modulation of E_{Cl} by GABA, by membrane potential and by nitric oxide (NO).

a. Modulation of E_{Cl} by GABA

Neurons typically maintain E_{Cl} constant even in the face of continuous inhibition. There are, however, numerous examples of neurons that undergo positive shifts in E_{Cl} under repetitive GABAergic inhibition. These shifts are typically more pronounced in the dendrites than in the soma. The vulnerability of dendrites to such changes is probably due to their relatively small volume so that $GABA_A$ receptor-mediated Cl^- influx exceeds Cl^- extruder capacity leading to intracellular Cl^- accumulation. Under certain conditions, this positive shift in E_{Cl} can be of sufficient magnitude to cause a depolarizing response to GABA. Two explanations have been given for this phenomenon. One hypothesis suggests that the positive shift in E_{GABA} is due to the contribution of HCO_3^- flux through GABA receptor channels becoming evident. When E_{Cl} lies below the resting membrane potential, HCO_3^- efflux through the anion channels gated by GABA is outweighed by Cl^- influx. However, following Cl^- accumulation, E_{Cl} becomes close to E_m and the HCO_3^- efflux becomes evident thereby producing a depolarization (Staley and Proctor, 1999). Another possible explanation, based on measurements of $[Cl^-]_i$, attributes the positive shift in E_{GABA} only to a rise in $[Cl^-]_i$ resulting in a positive shift in E_{Cl}. In cultured hippocampal neurons, focal application of GABA to the soma can elevate somatic $[Cl^-]_i$ (measured with Clomeleon) by 10–40 mM, potentially shifting E_{Cl} by >30 mV above E_m (Kuner and Augustine, 2000). Further, Cl^- accumulated in the soma diffuses to proximal dendrites elevating their $[Cl^-]$. Similarly, in hippocampal slices tetanic stimulation of GABAergic basket cells synapsing onto CA1 pyramidal cell's soma leads to a postsynaptic depolarization that lasts for 1 to 5 sec. This depolarization correlates with an increase in $[Cl^-]_i$ (measured with the Cl^- indicator MEQ) and it decreases but does not completely disappear with bicuculline, a $GABA_A$ receptor-blocker (Isomura et al., 2003). It has been suggested that a rise in extracellular potassium increases the activity of NKCC1 and decreases or reverses the activity of KCC2, leading to a rise in $[Cl^-]_i$ to levels higher than would be expected from a passive distribution. For further discussion on the effects of increasing extracellular potassium on KCCs see Chapter 17 and the review by Voipio and Kaila (2000). It is possible that the changes in the GABA response are mediated by multiple mechanisms, including changes in the regulation of cation Cl^- cotransporter's (CCCs) surface expression (a mechanism with a relatively slow time course), CCCs phosphorylation (a mechanism with a relatively fast time course), and by environmental intra- and extracellular ions including HCO_3^- (fastest process) (reviewed by Kaila, 1994; Voipio and Kaila, 2000). None of these mechanisms are mutually exclusive.

b. Modulation of E_{Cl} by NO

An intriguing modulation of E_{Cl} with a short time constant (1–3 sec), which appears to be independent of CCCs, was demonstrated in chick amacrine cells and rat hippocampal neurons (Hoffpauir et al., 2006). When cultured amacrine cells are stimulated with low micromolar concentrations of NO, GABA-gated currents are enhanced. This enhancement is due to a substantial positive shift (15–35 mV) in the reversal potential of the current. A similar NO effect is seen on glycine-gated currents, indicating that the positive shift in E_{GABA} is due to an increase in cytosolic $[Cl^-]$. The change in E_{GABA} is sufficiently large that it shifts from values below the resting membrane potential (resting $E_m = \sim -60$ mM) to levels above it. Consequently, NO inverts the sign of GABA- and glycine-gated voltage responses from hyperpolarizing to depolarizing. This NO effect can also be demonstrated by recording from a postsynaptic amacrine cell while stepping up the voltage of a presynaptic amacrine cell that releases GABA or glycine. Thus, E_{GABA} can be shifted by network activity and not only by exogenous application of GABA. Furthermore, current clamp recordings show that the NO-induced depolarizing response to GABA can elicit action potentials, suggesting that GABA responses are turned from inhibitory to excitatory.

All shifts in E_{Cl} so far discussed were due to a change in the balance between the expression and activity of different CCCs and Cl^- conductances. The effect of NO appears to be independent of CCCs since inhibitors of known transporters (furosemide, bumetanide, DIDS, ethacrynic acid) do not block the shift. Furthermore, this shift does not require extracellular Cl^-, as it still occurs when extracellular chloride concentration is decreased to virtually 0 mM. These experiments suggest that the transient elevation of $[Cl^-]_i$ in response to NO is due to Cl^- release from intracellular organelles such as endosomal compartments. This movement of Cl^- across endosomal compartments may be mediated by CLC channels (see Chapter 12). The mechanism by which NO stimulates Cl^- efflux from endosomes is not clear. It probably does not involve soluble guanylyl cyclase since blocking this enzyme does not block the shift in E_{GABA}. The effect of NO on the GABA response underscores the importance of Cl^- in neural processing and raises the possibility that there exist Cl^--binding proteins/organelles that buffer Cl^- concentration in different cell compartments.

c. Modulation of E_{Cl} by Membrane Potential

We have already discussed experiments that investigated if retinal rod bipolar cells maintain Cl^- gradients within different parts of the same cell. In one of these studies, the authors clamped the membrane voltage for 2 min before determining E_{Cl} and found that E_{Cl} correlated with the value of the clamped membrane voltage (Billups and Attwell, 2002). In darkness, resting membrane potential and E_{Cl} were similar (−40 mV). When membrane voltage was clamped to a depolarized potential, E_{Cl} became more depolarized; when clamped to a hyperpolarized potential, E_{Cl} became more hyperpolarized. However, the changes in E_{Cl} never matched with those in E_m, confirming that Cl^- is not passively distributed across the plasma membrane. In these experiments, the dynamic change in E_{Cl} occurred under conditions in which Ca^{2+}-dependent neurotransmitter release was blocked and no exogenous neurotransmitters were applied. It was concluded that bipolar cells have a significant "resting" permeability to Cl^-. Although ClC2 is expressed by these cells and may tune their E_{Cl} (Enz et al., 1999), it probably does not contribute to this behavior since this channel is activated by hyperpolarization. Alternatively, the Cl^- permeability may be due to calcium-independent tonic release of GABA, a result that is reminiscent of E_{Cl} changes due to repetitive GABA applications. A similar phenomenon has been described in neurons of the lateral superior olive (Ehrlich et al., 1999). Whatever the cause of E_{Cl}'s dynamic changes, they may be viewed as an adapting behavior of the rod bipolar cell providing it with an inherently transient voltage response (albeit over a time course of seconds) to a continuous GABA release. When the rod bipolar cell is depolarized by a prolonged light stimulus, GABA released from amacrine cells would initially produce a strong hyperpolarizing response. With continued GABA release, E_{Cl} would shift towards E_m thereby decreasing the driving force for Cl^- and consequently decreasing the amplitude of the response. When the cell is hyperpolarized in constant darkness, GABA released from horizontal cells would initially depolarize the dendrites. With time, E_{Cl} would move in the negative direction so that subsequent depolarizations by GABA would decrease in amplitude.

III. FUNCTION OF GABA, GLYCINE AND CATION-CHLORIDE COTRANSPORTERS IN RETINAL DEVELOPMENT

As discussed in Chapter 17, a dramatic event in neural maturation is the shift (also referred to as "the switch") in the action of GABA and glycine from excitatory to inhibitory. This switch occurs amidst a complicated but well-orchestrated sequence of events where multiple neuron types are born each at its own time and integrated into a functional neural circuit. It is therefore important to ask how does this apparently general rule of the "GABA switch" relate to this sequence of developmental processes. The retina provides an excellent model to study these questions because of the wealth of information regarding the intermediate developmental steps of different defined cell types (Fig. 19.7). In considering the "GABA switch" during development, several specific questions come to mind and are addressed in the sections that follow. (1) Where are GABA and glycine produced and released, and where are their receptors expressed at each developmental stage? (2) When during the maturation process does GABA switch its action for each cell type? (3) What functions do GABA and glycine serve at cellular and circuit level at each developmental stage? (4) What sets E_{GABA} at each developmental stage for each cell type?

A. Content of GABA and Glycine during Development

Localization of GABA during development has been addressed in numerous studies in many species (reviewed by Redburn, 1992; Nag and Wadhwa, 1997; Lee et al., 1999; Ferreiro-Galve et al., 2008). In mouse, immunostaining for GABA at postnatal (P) 0 (birth)

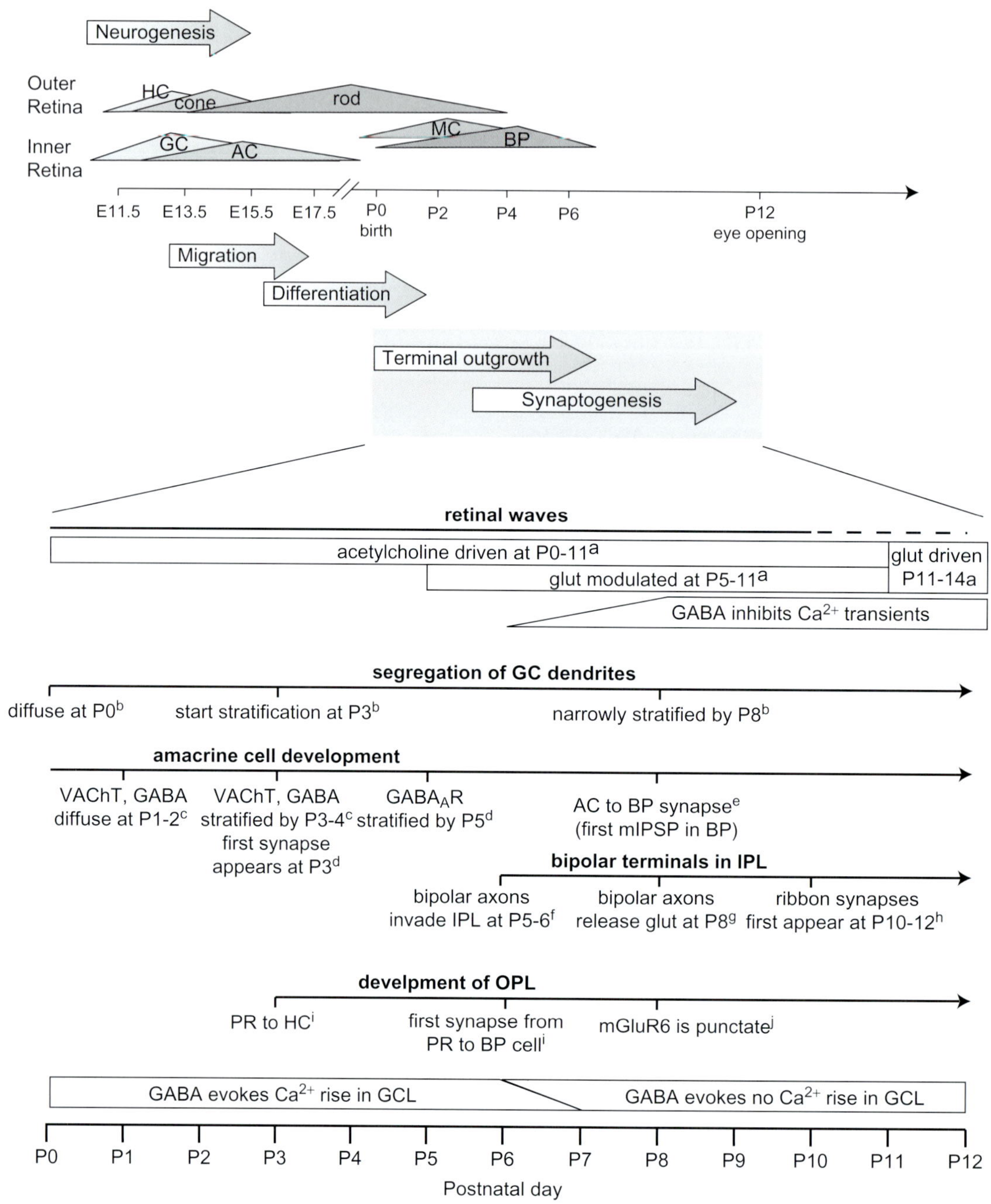

FIGURE 19.7 Temporal sequence of GABA's action and timing of major developmental events in developing mouse retina. Where mouse data are not available, rat data are quoted. **Top,** time line of neurogenesis of different cell classes. **Bottom,** expanded time line for terminal outgrowth and synaptogenesis. Glut, glutamate; GC, ganglion cells; HC, horizontal cells; AC, amacrine cells; MC, Müller cells; BP, bipolar cells; PR, photoreceptors; VAChT: vesicular acetylcholine transporter; IPL, inner plexiform layer; GCL, ganglion cell layer. Sources of data: [a] Bansal et al. (2000); [b] Diao et al. (2004); [c] Stacy and Wong (2003); [d] Fisher (1979); Sassoè-Pognetto and Wässle (1997, rat); [e] Schubert et al. (2008); [f] Ueda et al. (1997); [g] Sherry et al. (2003); [h] Olney (1968); Fisher (1979); [i] Olney (1968); Nomura et al. (1994); [j] Nomura et al. (1994).

to P3 is positive throughout the retinal layers (Fig. 19.8). Similarly in turtle, cells expressing GABA and GAD (glutamic acid decarboxylase), the enzyme that catalyzes the conversion of glutamate to GABA, are both present long before hatching (Nguyen and Grzywacz, 2000). As with GABA, glycine immunoreactivity already appears at birth throughout the retina including the neuroblast layer (Fletcher and Kalloniatis, 1997; Pow and Hendrickson, 2000). The content of GABA has been measured in rat retina at different time points during development. During the first two postnatal weeks, GABA content is about 7 nmol/kg protein.

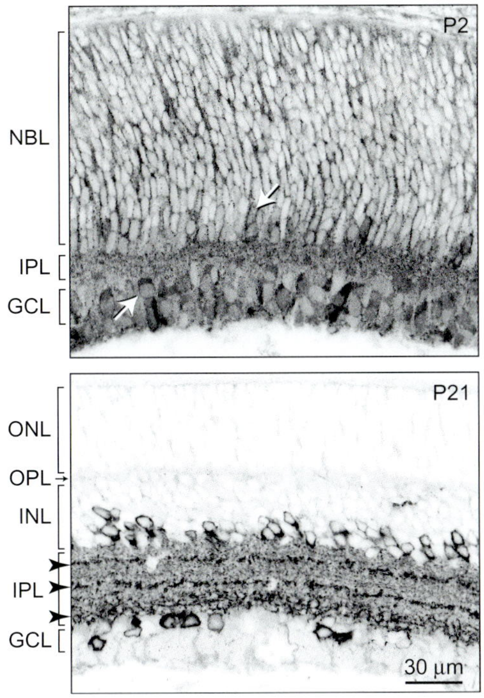

FIGURE 19.8 GABA is present in all layers of the developing retina. Immunostaining for GABA in P2 and P21 mouse retina. At P2, GABA is present throughout the neuroblastic layer (NBL), in the developing inner plexiform layer (IPL) and in all or most somas of the ganglion cell layer (GCL). Cells that stain strongly (arrows) are probably amacrine cells. In contrast, at P21, when the retina is mature, GABA is present only in amacrine cells. Somas of amacrine cells flank the IPL, and their processes distribute throughout the IPL, but most strongly in 3 strata (arrowheads). Layers are as in previous figures. Reproduced with permission from Zhang et al. 2006.

While the concentration of GABA is constant during this time, the activity of GAD increases progressively, suggesting an additional source of GABA production right after birth. Indeed, GABA may be produced from putrescine since both putrescine and ornithine decarboxylase decrease progressively as GAD increases (de Mello et al., 1993). Thus, before GAD expression is adequate to produce sufficient GABA, GABA is synthesized by an alternative pathway, underscoring its importance as a trophic factor.

At P4 in mouse retina (and at the corresponding developmental age in other mammals), most of the somas in the neuroblast layer are already negative for GABA immunostaining, but some of them, located at the position of newly migrated horizontal cells, are weakly stained. Interestingly, horizontal cells increase the expression of their GABAergic markers (GABA, GAD) during the first postnatal week, and thereafter decrease their expression. In some species, including mouse, immunostaining for GABA or GAD cannot be detected in the adult horizontal cells (reviewed by Redburn, 1992). At that time, many progenitor cells have already differentiated, the inner nuclear layer starts to emerge and many amacrine cells assume their final position in the tiers of somas that border the inner plexiform layer. Many of these amacrine cells are now strongly stained for GABA and the labeling pattern begins to resemble that of the adult. Thus, before differentiation, most progenitor cells have the potential to release GABA, but as soon as the cells differentiate, depending on their type and destination, some cells lose this potential and others strengthen it. At maturity, only half of the amacrine cells release GABA while the other half release glycine.

What is the mechanism of neurotransmitter release at an early age? Neither conventional synapses nor synaptic vesicles can be seen at birth. The release mechanism has been studied in hippocampus where release of GABA and glutamate can be demonstrated in response to an electrical stimulus, suggesting that it is voltage dependent. Furthermore, similar to conventional vesicular release, this release can be evoked by exposure to a hyperosmotic solution (ACSF + 500 mM sucrose). This evidence supports the notion that neurotransmitters are released by neuronal precursors. However, this release is clearly unconventional since it is Ca^{2+} independent and does not require the SNARE complex. The release mechanism is still unknown. At this time during development, several synaptic proteins, including synaptophysin, different isoforms of SV2, and the vesicular GABA and glycine transporter VGAT, are expressed diffusely (de Mello et al., 1993; Wang et al., 2003). The transition to conventional synaptic release is gradual. In the inner plexiform layer, immature synapses between amacrine cells and ganglion cells in the first postnatal week have electron dense pre- and postsynaptic membranes, but no aggregates of synaptic vesicles (Olney, 1968a; Weidman and Kuwabara, 1969). The latter appear in the second postnatal week. Since immature synapses exist, it is possible that they use some elements of the release machinery that have already been developed thereby releasing the neurotransmitter in an unconventional manner. Alternatively, neurotransmitters may leak or be released via gap junction hemichannels.

B. Expression of GABA and Glycine Receptors during Development

The expression of GABA and glycine receptors during development was studied with immunostaining using antibodies specific for a variety of receptor subunits. In all vertebrates tested so far, GABA and glycine receptors are already expressed prior to birth. In mammals, receptor distribution and staining patterns were carefully studied mainly in postnatal retinas. The receptor subunits that are expressed at P0–P1, distribute diffusely over the plasma membrane with only a few clusters that correspond to conventional synapses (Sassoe-Pognetto and Wässle, 1997). As development progresses, staining becomes punctate with each puncta representing a cluster of receptors, typically at a postsynaptic site.

C. Functions of GABA during Development

1. *Timeline of GABA's Action on Amacrine and Ganglion Cells: Function at the Cellular Level*

The time course of GABA's action has been studied in mouse, rat, rabbit, ferret, turtle and chick. The most complete data set was provided for mouse, and most of this work was performed on ganglion and amacrine cells (partially because their location at the surface of the retina allows easy accessibility). The changes during development were determined by imaging the effect of GABA on intracellular Ca^{2+} and Cl^- with the fluorescent dyes fura-2 and MEQ, respectively (Zhang et al., 2006b). At P0 to P5, GABA or the GABA agonist muscimol increases intracellular Ca^{2+} concentration ($[Ca^{2+}]_i$) in over 99% of the cells, and this increase is reversibly blocked by the $GABA_A$ receptor-channel blocker picrotoxin (Fig. 19.9). The main carrier of this $GABA_A$ response is thought to be Cl^- since GABA consistently evokes an intracellular Ca^{2+} rise in nominally CO_2/HCO_3^--free (HEPES-based) medium. Further, the rise in $[Ca^{2+}]_i$ correlates with an increase in MEQ fluorescence, suggesting that $[Cl^-]_i$ decreases during the response due to Cl^- efflux. This efflux should depolarize the membrane and activate voltage-gated Ca^{2+} channels. Indeed the increase in $[Ca^{2+}]_i$ in response to GABA must be due to Ca^{2+} influx since it does not occur in Ca^{2+}-free medium. Thus, during the first postnatal week, GABA in mouse retinal neurons is depolarizing and promotes voltage-sensitive Ca^{2+} entry, and this phenomenon is mediated by ionotropic GABA receptors.

At P6, the effect of GABA starts to change. On this transitional day, GABA evokes an increase in $[Ca^{2+}]_i$ in only about half of the retinas. By P7, GABA no longer elicits a rise in $[Ca^{2+}]_i$. Concomitant with this phenomenon at P7–P12 GABA or muscimol causes an *increase* in $[Cl^-]_i$ most likely due to Cl^- influx because $[Cl^-]_i$ is lower than the electrochemical equilibrium.

The above-described effects of GABA on $[Ca^{2+}]_i$ strongly suggest that $[Cl^-]_i$ in amacrine and ganglion cells is higher than the electrochemical equilibrium in

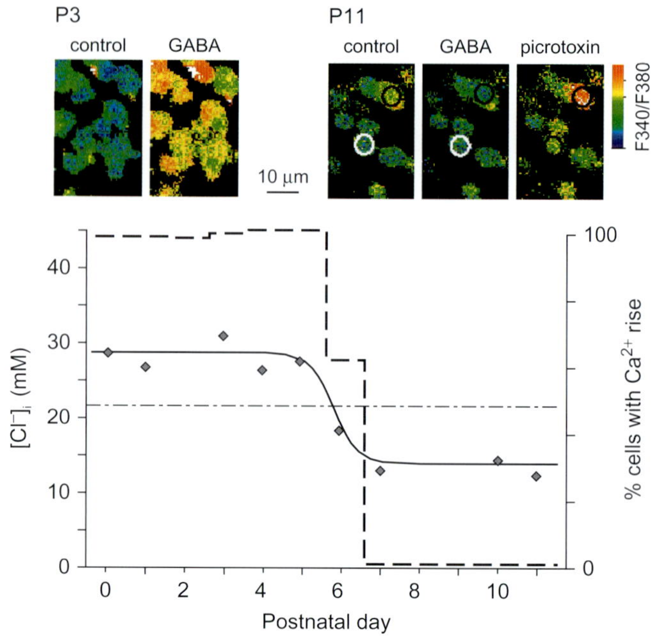

FIGURE 19.9 Timeline of GABA's action in the developing mouse retina. **Top,** calcium imaging of the ganglion cell layer before and during application of GABA at P3 and P11. At P3, all cells increased their $[Ca^{2+}]_i$ in response to GABA. At P11, one cell (circled in white) decreased its $[Ca^{2+}]_i$, but most cells did not change their concentrations. When picrotoxin is applied, $[Ca^{2+}]_i$ is elevated, suggesting tonic GABA is inhibitory. **Bottom,** diamonds and solid line graph the average $[Cl^-]_i$ during development; data are averages from the "reference method" and the gramicidin-perforated patch recordings. Dashed line shows the % cells that respond to GABA with a calcium increase. The dashed-dotted line indicates the intracellular chloride level required for GABA to activate low-voltage-activated calcium channels. Data points above this line indicate a potential for calcium to rise with GABA application. Reproduced with permission from Zhang et al. 2006.

the first postnatal week, and declines during the second postnatal week. This hypothesis was tested with gramicidin-perforated patch recording and MEQ fluorescent imaging during different phases of development. Consistent with the actions of GABA observed with fura-2 and MEQ, gramicidin-perforated patch recording at P1–P3 reveals that E_{GABA} is ~ -45 mV, a value that remains constant irrespective of whether the extracellular medium was buffered with bicarbonate or with HEPES that is nominally free of HCO_3^-. At P7–P11, E_{GABA} becomes significantly more negative, with a mean value of -60 mV.

In these experiments $[Cl^-]_i$ was also measured directly by fluorescence imaging using "the reference method" (Zhang et al., 2006b). This method compares the MEQ fluorescence elicited by the native $[Cl^-]_i$ (in Ames medium) to that elicited by a specific reference Cl^- solution as described in detail in the study by Zhang et al. (2006b). The developmental profile of chloride concentration was measured relative to reference concentrations of 22 and 30 mM. These reference Cl^- concentrations were chosen because similar intracellular Cl^- levels would set E_{Cl} to the threshold membrane voltage for low- and high-voltage-activated calcium channels. The results showed that neighboring cells respond in a similar manner, suggesting that amacrine and ganglion cells have similar chloride concentrations. At P0–P5, when GABA evokes a rise in $[Ca^{2+}]_i$ in virtually all cells, over 75% of the cells exhibit $[Cl^-]_i$ above 22 mM ($E_{Cl} > -46$ mV) potentially allowing GABA to activate low-voltage-activated Ca^{2+} channels. Over 50% of the cells have $[Cl^-]_i$ above 30 mM ($E_{Cl} > -38$ mV), potentially allowing GABA to activate high-voltage-activated Ca^{2+} channels. At P6, when GABA's action is "switching", the percentage of cells with over 30 mM $[Cl^-]_i$ drops sharply to 22%. By P8, when the developmental GABA switch has occurred, rarely any cell has over 30 mM $[Cl^-]_i$, yet about half the cells still have concentrations over 22 mM. At P10 – P11, $[Cl^-]_i$ drops below 22 mM in all cells ($E_{Cl} < -46$ mV), and thus GABA cannot depolarize the cells to the point of activating low-voltage-activated Ca^{2+} channels.

Specific average values for $[Cl^-]_i$ were obtained by modifying the Stern-Volmer equation (Zhang et al., 2006b). The average $[Cl^-]_i$ was: 29 mM (20–40 mM range) at P0–P5, 18 mM (10–30 mM range) at P6, and 14 mM (5–20 mM range) after P6 (Fig. 19.9). The E_{Cl} values were estimated to be ~ -40 mV in the first postnatal week and ~ -60 mV in the second postnatal week, thus allowing only about one day (P6) for the voltage drop to occur.

2. *Effects of GABA on Neuronal and Synaptic Development: Role at the Level of Local Circuits*

Excitatory activity is important in a variety of developmental processes, most notably for the refinement and stabilization of synaptic connections. In the CNS, the GABAergic system develops prior to the glutamatergic system and becomes inhibitory when synaptic glutamate starts to be released. Therefore, it seems reasonable to hypothesize that GABA depolarizations in early development provides the necessary level of excitatory drive (reviewed by Ben Ari, 2001). The coincidence of the GABA's "switch" with glutamate release also occurs in the retina, particularly in the inner retina, since bipolar cells develop later than amacrine cells. In mouse, bipolar cells invade the inner plexiform layer and start to release glutamate at P7, the time at which the GABA "switch" occurs (Fischer et al., 1998; Zhou, 2001; Catsicas and Mobbs, 2001). In the outer retina, the excitatory action of GABA during the early phases of cell development varies, and it can be separated into two: the effects on the rod's genesis and those on the cone's maintenance.

a. Effect of Ligand-gated Chloride Channels on Rod's Genesis

The highest rate of rod genesis occurs at P0 (mouse), and this rate is critically affected by the neuroactive sulfonic acid, taurine. Adding taurine to an *in vitro* developing retina at P0 or blocking its uptake dramatically increases rod production (Young and Cepko, 2004). Application of GABA alone or glycine alone does not mimic the effect of taurine, but simultaneous application of these transmitters does. It was suggested that the effect of taurine is mediated through the glycine receptor subunit GlyRα2, which is only expressed during development. This was inferred because overexpressing GlyRα2 increased rod production and knocking it down with RNA interference decreased it (surprisingly, knocking GlyRα2 out had no effect) (Young-Pearse et al., 2006). Further, the effect of taurine could be blocked by strychnine, a glycine receptor antagonist. However, the effect could also be blocked by the $GABA_A$ receptor-channel antagonists bicuculin or picrotoxin, suggesting that $GABA_A$ receptors are also involved. Indeed, the α6 subunit of $GABA_A$ receptors is expressed early on during cone and amacrine cell genesis even before the expression of GlyRα2. Since GABA, glycine and taurine are all released during rod genesis, they may all affect rod production via ligand-gated chloride channels.

b. Effect of GABA on Cone's Maintenance

At about the same time that rods are produced, cones grow their axon and form synapses with horizontal cells (Fig. 19.7). Prior to this, horizontal cells assume their final position in the inner nuclear layer and their processes define the outer plexiform layer. At this stage of their development, horizontal cells express GABAergic markers. Also at this time, GABA is excitatory, as suggested by the fact that it increases intracellular Ca^{2+} in cone photoreceptors (rabbit; Huang and Redburn, 1996), and that it is required for cone development. Peanut agglutinin staining (a cone marker) shows that blocking $GABA_A$ receptors disrupts cone distribution and synaptogenesis. This disruption can be rescued by applying GABA (Messersmith and Redburn, 1993; Huang et al., 2000). This effect appears to be mediated by horizontal cells since ablating these cells with kainic acid diminishes the array of cone terminals normally formed at P5 (Redburn, 1992). Since horizontal cells transiently express GABA during cone growth, and cones transiently express GABA receptors during this time

(Schnitzer and Rusoff, 1984; Mitchell et al., 1999), it is inferred that horizontal cells stabilize cones and their terminals by releasing GABA onto the developing cones. However, horizontal cells also transiently immunoreact for taurine (Pow et al., 1994; Lake, 1994), and since taurine affects $GABA_A$ receptors, both of these transmitters may contribute to cone maturation.

c. Effect of GABA Transmission on Retinal Waves: Function in Global Network Properties

A retinal wave is a spontaneous burst of action potentials accompanied by elevation of $[Ca^{2+}]_i$ which is highly correlated between neighboring cells and which propagates across the retina (Meister et al., 1991). Retinal waves appear early in development, as soon as the first differentiated cells (amacrine and ganglion cells) express voltage-gated sodium and calcium channels and thus become excitable. In rabbit, this occurs at embryonic day 22, about 20 days prior to eye opening (Syed et al., 2004). Consequently, these waves are a major type of activity in the developing retina, and they are critical for the development of the visual system upstream of retina (Katz and Shatz, 1996). Naturally, the study of GABA's function in early development was geared mainly toward determining its effect on these waves. The dynamic properties of these waves, their mechanism of generation and the modulation of their properties have been divided into three stages. In rabbit, chick and turtle retinas, the earliest waves (stage I) seem to be generated by cAMP via adenosine receptors. At this stage, $GABA_{A/C}$ and $GABA_B$ receptor agonists block the waves whereas $GABA_B$ antagonists enhance them, suggesting that GABA is used to reduce excitability (Catsicas and Mobbs, 2001; Syed et al., 2004). In stage II, waves are generated by nicotinic receptors. At this stage, $GABA_B$ receptors no longer function and the effect of $GABA_{A/C}$ receptors is subtle. Depending on the study and possibly on the methodology, application of $GABA_{A/C}$ antagonists led to slightly different results. In one study, these antagonists reduced the bursting activity of early waves (ferret, P1–P10), as would be expected from a depolarizing action of GABA and as found in the hippocampus (Fischer et al., 1998; Marchionni et al., 2007). In other studies, when waves were observed in mouse and rabbit with intracellular Ca^{2+} imaging at low magnification, blocking $GABA_{A/C}$ receptors did not affect the global spatial and temporal properties of the waves (Syed et al., 2004; Wang et al., 2007). However, when observed at high magnification, $GABA_{A/C}$ blockers increased the frequency of spontaneous intracellular Ca^{2+} transients in some cells and decreased it in others. Another source of variability between results arose from the blockers concentration used in each study. Low concentrations of antagonists applied to turtle retina at pre-hatching stage 25 (when GABA was depolarizing) blocked the waves, supporting GABA's excitatory role; but a higher concentration of the antagonists enhanced the speed of propagation (Sernagor et al., 2003). Analyzing the activity of ganglion cells by a multi-electrode array revealed that blocking $GABA_{A/C}$ receptors correlated the spiking activity of these cells (Wang et al., 2007). Whole-cell recordings showed that blocking these receptors did not affect the excitability of ganglion cells directly, but it did affect the conductance and excitability of a specific type of amacrine cell that is known to be crucial for wave generation, the starburst amacrine cells. Thus, tonic or phasic release of GABA (likely from the starburst cells themselves) modulates the activity of the starburst cells and "decorrelates" spiking activity across ganglion cells.

Although the effect of blocking $GABA_A$ receptors is variable between studies, that of applying GABA itself is less variable. GABA increases $[Ca^{2+}]_i$ due to its depolarizing nature, yet it blocks the waves. This seemingly paradoxical effect can be explained if the continuous excitation by applied GABA may inactivate sodium and calcium channels, thereby inhibiting the waves. This in turn can explain the variability of results obtained by blocking $GABA_A$ receptors. As discussed above, the depolarizing effect of GABA when transiently released depends on the precise timing and magnitude of the increase in membrane conductance and the residual depolarizing phase relative to other modulated conductances. Since, in all studies, the investigators apply GABA or $GABA_A$ receptor blockers globally, to the whole retina, the intricate interactions of the evoked conductances and voltages are not obvious and data interpretation should be cautious. Furthermore, in most experimental paradigms, the retina is perfused with a medium that may wash away tonically released GABA. Thus, these *in vitro* experiments may artificially reduce the effect of endogenously released GABA, leading to underestimation and possibly misinterpretation of GABA's effect.

While the effect of endogenous GABA on stage II waves is subtle, its effect on stage III waves and afterwards (after the "switch") is dramatic. Applying $GABA_{A/C}$ receptor antagonists at P6–P9 (mouse) increases the frequency of spontaneous Ca^{2+} transients. At P10–P11, when spontaneous Ca^{2+} transients are absent, $GABA_{A/C}$ receptor antagonists still elicit waves that are indistinguishable from those naturally occurring at P0–P7. Thus, at stage III and afterwards, GABA gradually inhibits retinal waves laying down the groundwork for ganglion cells to respond, as they do, to visual stimuli.

D. Mechanisms that Determine E_{GABA} during Development

1. *Contribution of KCC2*

The evidence so far discussed suggests that the GABA switch is caused primarily by a change in E_{Cl} and thus in $[Cl^-]_i$ which is set by the concerted activity of Cl^- transporters and channels in the plasma membrane of neurons. In the retina, as in central brain regions, KCC2 appears to be the main active Cl^- extrusion mechanism. Western blots of mouse and rat retinas do not detect KCC2 during the first postnatal week, but show a quick rise in expression level after P7, the time at which the "switch" occurs (Vu et al., 2000; Zhang et al., 2007). Similarly, in ferret retina, KCC2 is hardly detectable before the "switch" time, but starts to rise afterwards (P18). However, since the ratio of KCC2 relative to synaptophysin remains constant (Fig. 19.10A), the rise in KCC2 expression reflects an increase in synaptic density rather than evidence for a correlation with the "switch" (Zhang et al., 2006a). By immunostaining, a method more sensitive than Western blotting, KCC2 can be detected already at birth, but the expression level is very low and is restricted to the inner retina. The early detection of KCC2 is consistent with several studies performed in brain that examined the issue physiologically (Delpy et al., 2008). Expression level of KCC2 continues to rise in parallel with the rate of synapse formation (Fig. 19.10A and 19.10B). Indeed, the dendrites of ganglion and amacrine cells, which house most of the synapses, consistently stain more strongly for KCC2 than their somas, which house only a few synapses.

2. *Contribution of NKCC1*

While the contribution of KCC2 to the mature E_{Cl} level and GABA response is not in doubt, the mechanisms that set the E_{Cl} at depolarizing levels in immature retina are still unclear. The negative value of E_{Cl} with respect to that of the resting E_m implies that there must be a Cl^- transporter that accumulates Cl^- against its thermodynamic equilibrium. The prime candidate for neuronal Cl^- accumulation is NKCC1, a transporter that has been shown to contribute to

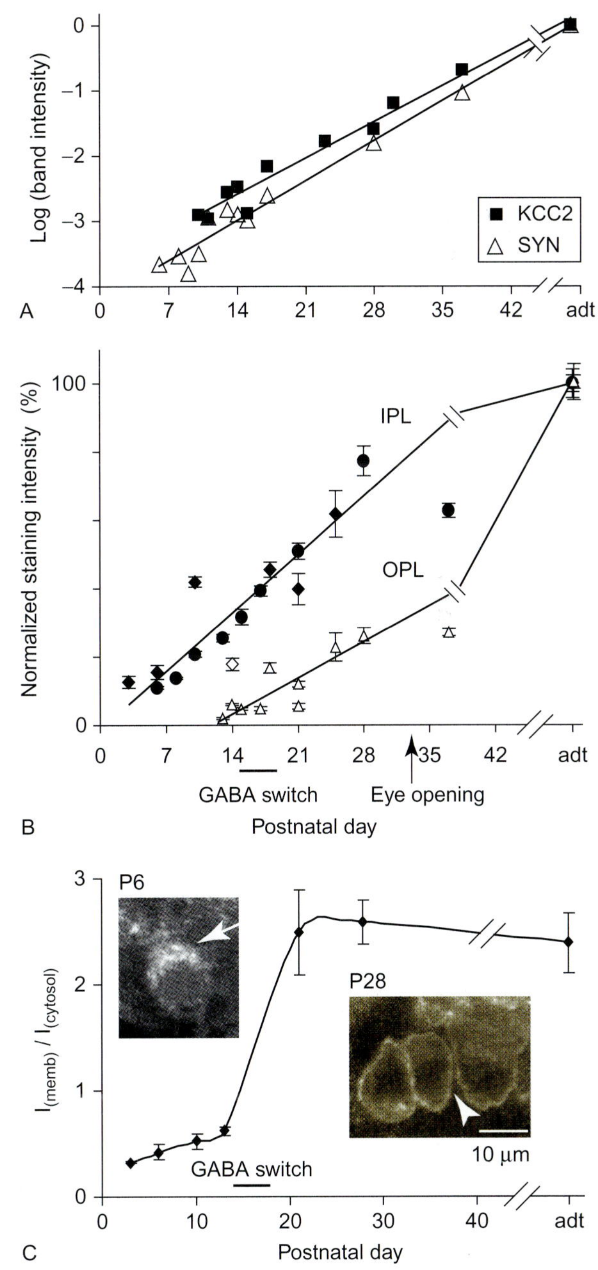

FIGURE 19.10 KCC2 expression parallels synaptophysin and KCC2's targeting to the membrane may determine switch time (ferret). **A.** Western blots of membrane fractions from different-aged ferret retinas were probed, first for KCC2, then stripped and re-probed for synaptophysin (SYN). Log of band intensity was normalized to adult (adt) intensity levels. The slopes of the KCC2 and synaptophysin graphs are similar, so expression of KCC2 parallels that of synaptophysin. Horizontal bar below *x*-axis in (B) shows time of GABA's switch and arrow points to time of eye opening. **B.** KCC2 expression during development was quantified from immunocytochemistry as the average staining intensity per unit area in IPL or OPL normalized to the adult level. Data was fitted with a regression line (without using the adult data point). The intensity in the OPL is initially lower than that in the IPL because the OFF bipolar dendrites in the OPL (which express KCC2 in the adult) develop more slowly than the amacrine and ganglion cells' processes in the IPL. **C.** Ratio of average KCC2 staining intensity in the plasma membrane to that in the cytosol. There is a sharp increase in the proportion of KCC2 localized to the membrane relative to the cytosol around switch time. Insets: examples of somas stained for KCC2 at P6 and P28. Reproduced with permission from Zhang et al. 2006.

higher than passive E_{Cl} values in mature sensory neurons, as discussed above in section II.B and in Chapter 22 in this volume. Initial evidence supporting this candidacy relied on the observed down-regulation of NKCC1 mRNA and NKCC1 protein during development of CNS neurons (Plotkin et al., 1997b; Kanaka et al., 2001; Li et al., 2002; Mikawa et al., 2002; Ikeda et al., 2003) (see also section III in Chapter 24). Further support for this notion came from estimates of E_{Cl} and measurement of NKCC1 mRNA transcripts from the same cells in neocortex. This showed a good correlation between NKCC1 expression; the estimated E_{Cl}, and the degree to which bumetanide shifted the cells' E_{Cl} (Yamada et al., 2004). Additionally, immature cortical neurons that were electroporated with shRNA to knockdown NKCC1 showed a more hyperpolarized E_{Cl} (Wang and Kriegstein, 2008). In retina, careful examination of the functional expression and localization of NKCC1 disagrees with this notion, as detailed below.

Immunostaining for NKCC1 with the T4 monoclonal antibody, which recognizes both NKCC1 and NKCC2 (Lytle et al., 1992), and with an antibody against the C-terminus of NKCC1 (Kaplan et al., 1996) reveals that in early developmental stages of mouse retina NKCC1 is expressed throughout the retina (Li et al., 2008). However, high magnification and double labeling with known cell type markers reveals that the stain does not localize to immature neurons (Zhang et al., 2007) but to Müller cells – the retinal radial glial cells that extend processes from one end of the retina (outer limiting membrane) to the other (inner limiting membrane).

Since immunostaining often gives variable results and highly depends on fixation protocol, we carried out an exhaustive functional study using NKCC1-null mice (see Chapter 10 in this volume). At P0–P5, GABA consistently evokes an intracellular Ca^{2+} rise in amacrine and ganglion cells in NKCC1$^{-/-}$ that exhibits the same rate and magnitude as that in the wild type (WT), suggesting that GABA has a depolarizing action in these NKCC1$^{-/-}$ neurons. Further, $[Cl^-]_i$ in amacrine and ganglion cells of NKCC1-null retinas is 30 mM, same as in WT mice of this age. These results are not due to compensatory mechanisms in NKCC1$^{-/-}$ retinas since applying high concentrations of bumetanide (40 μM) to wild-type and NKCC1-null mice even for long periods does not abolish the GABA-induced Ca^{2+} increase. Furthermore, neither bumetanide nor low extracellular $[Na^+]$ lowers $[Cl^-]_i$ in these immature mouse ganglion and amacrine cells. We conclude that in immature retina, NKCC1 is expressed in Müller cells but not in neurons, and that there must be a different active Cl^- uptake transport mechanism that increases and maintains $[Cl^-]_i$ above electrochemical equilibrium, and that this mechanism is bumetanide insensitive and Na^+ independent. Müller cells may use NKCC1 to remove extracellular K^+, consistent with the function of these cells.

3. *Mechanisms that Regulate Expression of Chloride Transporters*

a. Possible Basis for the Developmental GABA Switch

In most brain regions in which it has been studied, including retina, the time of GABA's switch coincides with the release of glutamate. In retina, the switch occurs at P7, exactly when bipolar cells start to develop and release glutamate. In the previous section, we stated that in ferret retina KCC2 expression increases at the same rate as synaptophysin. This gradual increase in KCC2 expression cannot easily explain the relatively fast "switch" of GABA from excitatory to inhibitory. These two facts can be reconciled by at least two possible mechanisms. First, down-regulation of the Cl^- uptake mechanism by itself could provide the basis for the "switch". However, down-regulation of NKCC1 in ferret retina, where it was measured relative to synaptophysin, does not provide a clue since NKCC1 is up-regulated in horizontal cells at the time in which the GABA switch occurs. Also, since NKCC1 probably does not accumulate Cl^- in immature retinal neurons, its hypothetical down-regulation may be irrelevant to the process underlying the "switch". However, the idea may be applicable to Cl^- accumulator mechanisms whose nature is to be determined in future studies.

The second possibility is that in addition to the mechanisms that regulate protein expression, there are mechanisms that regulate the transport activity of the proteins. This idea was examined in ferret retina (a common model to study development) because its ganglion cell somas are relatively large and permit subcellular localizations. At P6–P14, when KCC2 expression is low and E_{Cl} is still maintained at depolarizing values, KCC2 accumulates in the cytosol. As ganglion cells mature, this protein is trafficked to the plasma membrane, where it resides in the adult (Zhang et al., 2006a). Plotting immunostaining intensity at the plasma membrane relative to that in the cytosol reveals a sharp increase in membrane staining around the time of the switch, suggesting that both expression and trafficking are regulated during development (Fig. 19.10C).

The idea that additional regulatory mechanisms control the "switch" was also examined in cultured hippocampal neurons where KCC2 is expressed at early developmental stages but it is inactive. Upon maturation, KCC2 becomes active, and this activity can be enhanced by phosphorylation mediated by a growth factor (IGF-1) (Kelsch et al., 2001). Unlike cultured neurons, the developing brain does not seem to use phosphorylation as a means to trigger the "switch" (Stein et al., 2004; Khirug et al., 2005). Possibly other mechanisms, such as increased membrane insertion, take part. Either way, it appears that the switch of GABA's action occurs via several regulatory mechanisms, which are different between *in vivo* and cultured preparations, and possibly between different cell types (see below).

b. KCC2 Expression Parallels Neuronal Differentiation, but there are Exceptions

Brain regions that differentiate earlier express KCC2 earlier, e.g., hindbrain versus forebrain. In the retina we extend this idea to different cell types. Amacrine and ganglion cells differentiate before bipolar cells, and they indeed express KCC2 before bipolar cells (Zhang et al., 2006a). The polarity of bipolar cell responses to GABA has not been investigated so far. The progression of KCC2 expression in ganglion and bipolar cells and its predicted outcome is illustrated in Fig. 19.11. However, many cell types do not follow this rule. Horizontal cells that do not express KCC2 in the adult never seem to express this protein. Rods, which exhibit relatively high $[Cl^-]_i$ in the adult, also do not express KCC2 at maturity (Vardi et al., 2000; Thoreson et al., 2002, 2003; Markova et al., 2008; but see Li et al., 2008). However, they do transiently express KCC2 during development (at least in ferret; Zhang et al., 2006a). Therefore, in addition to differentiation and general up-regulation of KCC2, other factors must exist that regulate expression of KCC2 and allow it to down-regulate.

c. Regulation of KCC2 Expression and Activity may be Compartmentalized

Since the GABA "switch" occurs when glutamate starts to be released, one can hypothesize that the development of glutamate regulates the expression of KCC2. However, in brain slices and in hippocampal cultures, glutamate does not regulate KCC2 expression or affects the GABA "switch" (Ganguly et al., 2001; Ludwig et al., 2003). In cultured hippocampal neurons, it is believed that GABA itself up-regulates the expression of KCC2 by depolarizing the cell and elevating $[Ca^{2+}]_i$. When KCC2 expression is sufficient to shift E_{Cl} below the resting E_m, GABA no longer elevates $[Ca^{2+}]_i$ and no longer up-regulates KCC2 expression (Ganguly et al., 2001). A similar conclusion was reached using *in vivo* turtle retina. In these experiments, chronically blocking $GABA_A$ receptors prevents the shift of GABA from depolarizing to hyperpolarizing and the retinal waves continue to propagate at the developmental age in which they usually stop. However, chronically depolarizing E_m with KCl did not affect GABA action or the retinal waves (Leitch et al., 2005). Since glutamate or KCl, which also elevates intracellular Ca^{2+}, does not regulate KCC2 expression, it appears that GABA does not regulate KCC2 via global depolarization or global Ca^{2+} elevation. Alternatively, $GABA_A$ receptors may regulate KCC2 locally at the synaptic level rather than at the cell level. This idea is supported by physiological measurements of wave activity performed on rabbit retina. These measurements showed that when GABA had already become inhibitory, glycine was still excitatory, suggesting that KCC2 expression varies across different compartments of the same developing neuron (Zhou, 2001). In this case, a compartment that expresses $GABA_A$ receptors may express KCC2 before a compartment that expresses glycine receptors. We propose that as soon as a postsynaptic compartment receives a mature synaptic contact from a GABAergic (or a glycinergic) neuron, this compartment expresses KCC2 (see Fig. 19.11). Possibly the glycinergic amacrine cells in rabbit retina form contacts with ganglion cells after the GABAergic cells do so. Prior to the formation of synaptic contacts between the GABAergic (or glycinergic) amacrine cells with the ganglion cells, the ganglion cells respond to GABA or glycine via extrasynaptic receptors. This hypothesis suggests that developing neurons are compartmentalized; i.e. the nature of their response to GABA or glycine is not homogeneous along the dendrites. Regions of the dendrites that receive a mature synaptic contact express KCC2 and may hyperpolarize in response to GABA while regions that express receptors but did not receive contacts may still depolarize. Such fine details cannot be detected by recording from the cell's soma, since this averages responses from many synapses. Future experiments can test this hypothesis using voltage sensitive dyes.

d. Effect of Exogenous Factors on Regulation of KCC2 Expression

In addition to the ongoing normal mechanisms that regulate KCC2 expression during development (whatever they might be) expression of KCC2 in retina is modulated by light. In turtle retina, dark rearing slows down the normal developmental increase in expression of this protein (Sernagor et al., 2003) and this in turn

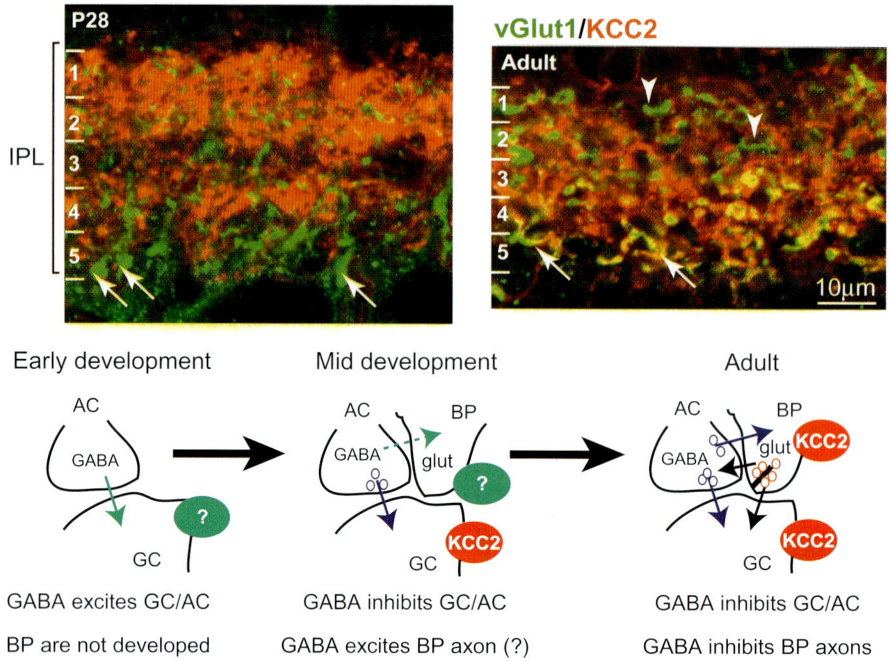

FIGURE 19.11 Expression of KCC2 in bipolar cells lags behind that in amacrine and ganglion cells. Images from ferret retinas that are double stained for KCC2 (red) and glutamate vesicular transporter vGlut1 (green) which specifically labels bipolar axon terminals. At P28, KCC2 expression is strong in amacrine and ganglion cells, but is absent in bipolar axon terminals. In the adult, rod bipolar terminals (identified by their large size and their location in stratum 5 of the IPL; arrows) express KCC2, but certain other bipolar terminals remain unstained (arrowheads). Numbers in the IPL designate stratum number. The schematic shows development of the synaptic diad in the IPL. In early development (before the GABA switch in ganglion cells), bipolar axon terminals are undeveloped and amacrine cells (AC) make few synapses, mainly immature, onto ganglion cells (GC); at that time the ganglion cells probably express a chloride accumulator, but its identity is still unknown. At about switch time (mid development), ganglion and amacrine cells express KCC2 (shown only for GC) and their response to GABA has switched to hyperpolarizing. In contrast, bipolar axon terminals that have started to arborize in the IPL and release glutamate, still do not express KCC2. We predict that bipolar cells at this stage depolarize in response to GABA. In the adult, both bipolar terminals and ganglion cells express KCC2 and receive mature synapses that hyperpolarize to GABA. Reproduced with permission from Zhang et al. 2006.

extends the time period of retinal waves. Since light modulates GABA release, the effect of light on KCC2 expression may be mediated by GABA. Expression of KCC2 may also be modulated by light after maturation, in which case such regulation could contribute to signal processing and dark adaptation. Thus, reduction of GABA in the dark might decrease KCC2 expression and thereby reduce inhibitory input in the dark.

In conclusion, GABA, glycine and taurine appear early in development and they act mainly through ionotropic receptors whose subunits and properties change during different stages of development. During the first postnatal week of the developing mouse retina, these transmitters are depolarizing and may be excitatory, and they contribute to rod production, cone growth and tuning the properties of retinal waves. At some point GABA (most likely at synaptic sites) up-regulate the expression of KCC2 and its action becomes inhibitory.

References

Alvarez-Leefmans, F.J., Gamino, S.M., Giraldez, F., and Nogueron, I. (1988). Intracellular chloride regulation in amphibian dorsal root ganglion neurones studied with ion-selective microelectrodes. *J. Physiol.* **406**, 225–246.

Armstrong-Gold, C.E. and Rieke, F. (2003). Bandpass filtering at the rod to second-order cell synapse in salamander (*Ambystoma tigrinum*) retina. *J. Neurosci.* **23**, 3796–3806.

Bansal, A., Singer, J.H., Hwang, B.J., Xu, W., Beaudet, A., and Feller, M.B. (2000). Mice lacking specific nicotinic acetylcholine receptor subunits exhibit dramatically altered spontaneous activity patterns and reveal a limited role for retinal waves in forming ON and OFF circuits in the inner retina. *J. Neurosci.* **20**, 7672–7681.

Ben Ari, Y. (2001). Developing networks play a similar melody. *Trends Neurosci.* **24**, 353–360.

Billups, D. and Attwell, D. (2002). Control of intracellular chloride concentration and GABA response polarity in rat retinal ON bipolar cells. *J. Physiol.* **545**, 183–198.

Blanco, R., Vaquero, C.F., and de la Villa, P. (1996). The effects of GABA and glycine on horizontal cells of the rabbit retina. *Vision Res.* **36**, 3987–3995.

Bloomfield, S.A. and Volgyi, B. (2004a). Function and plasticity of homologous coupling between AII amacrine cells. *Vision Res.* **44**, 3297–3306.

Bloomfield, S.A. and Volgyi, B. (2004b). Function and plasticity of homologous coupling between AII amacrine cells. *Vision Res.* **44**, 3297–3306.

Catsicas, M. and Mobbs, P. (2001). GABAb receptors regulate chick retinal calcium waves. *J. Neurosci.* **21**, 897–910.

Chavez, A.E., Singer, J.H., and Diamond, J.S. (2006). Fast neurotransmitter release triggered by Ca^{2+} influx through AMPA-type glutamate receptors. *Nature* **443**, 705–708.

Chun, M.-H. and Wässle, H. (1989). GABA-like immunoreactivity in the cat retina: electron microscopy. *J. Comp. Neurol.* **279**, 55–67.

Davenport, C.M., Detwiler, P.B., and Dacey, D.M. (2008). Effects of pH buffering on horizontal and ganglion cell light responses in primate retina: evidence for the proton hypothesis of surround formation. *J. Neurosci.* **28**, 456–464.

De la Villa, P., Kurahashi, T., and Kaneko, A. (1995). L-Glutamate-induced responses and cGMP-activated channels in three subtypes of retinal bipolar cells dissociated from the cat. *J. Neurosci.* **15**, 3571–3582.

De Mello, M.C., Guerra-Peixe, R., and de Mello, F.G. (1993). Excitatory amino acid receptors mediate the glutamate-induced release of GABA synthesized from putrescine in cultured cells of embryonic avian retina. *Neurochem. Int.* **22**, 249–253.

DeFazio, R.A., Heger, S., Ojeda, S.R., and Moenter, S.M. (2002). Activation of A-type gamma-aminobutyric acid receptors excites gonadotropin-releasing hormone neurons. *Mol. Endocrinol.* **16**, 2872–2891.

Delpy, A., Allain, A.E., Meyrand, P., and Branchereau, P. (2008). NKCC1 cotransporter inactivation underlies embryonic development of chloride-mediated inhibition in mouse spinal motoneuron. *J. Physiol.* **586**, 1059–1075.

Demb, J.B. (2007). Cellular mechanisms for direction selectivity in the retina. *Neuron* **55**, 179–186.

Demb, J.B., Zaghloul, K., and Sterling, P. (2001). Cellular basis for the response to second-order motion cues in Y retinal ganglion cells. *Neuron* **32**, 711–721.

Diao, L., Sun, W., Deng, Q., and He, S. (2004). Development of the mouse retina: emerging morphological diversity of the ganglion cells. *J. Neurobiol.* **61**, 236–249.

Djamgoz, M.B. and Laming, P.J. (1987). Micro-electrode measurements and functional aspects of chloride activity in cyprinid fish retina: extracellular activity and intracellular activities of L- and C-type horizontal cells. *Vision Res.* **27**, 1481–1489.

Dong, C.J. and Hare, W.A. (2002). GABAc feedback pathway modulates the amplitude and kinetics of ERG b-wave in a mammalian retina in vivo. *Vision Res.* **42**, 1081–1087.

Dong, C.J. and Werblin, F.S. (1998). Temporal contrast enhancement via $GABA_C$ feedback at bipolar terminals in the tiger salamander retina. *J. Neurophysiol.* **79**, 2171–2180.

Duebel, J., Haverkamp, S., Schleich, W., Feng, G., Augustine, G. J., Kuner, T., and Euler, T. (2006). Two-photon imaging reveals somatodendritic chloride gradient in retinal ON-type bipolar cells expressing the biosensor Clomeleon. *Neuron* **49**, 81–94.

Eggers, E.D. and Lukasiewicz, P.D. (2006). Receptor and transmitter release properties set the time course of retinal inhibition. *J. Neurosci.* **26**, 9413–9425.

Ehrlich, I., Lohrke, S., and Friauf, E. (1999). Shift from depolarizing to hyperpolarizing glycine action in rat auditory neurons is due to age-dependent Cl^- regulation. *J. Physiol.* **520** (Pt 1), 121–137.

Enz, R., Ross, B.J., and Cutting, G.R. (1999). Expression of the voltage-gated chloride channel ClC-2 in rod bipolar cells of the rat retina. *J. Neurosci.* **19**, 9841–9847.

Ferreiro-Galve, S., Candal, E., Carrera, I., Anadon, R., and Rodriguez-Moldes, I. (2008). Early development of GABAergic cells of the retina in sharks: an immunohistochemical study with GABA and GAD antibodies. *J. Chem. Neuroanat.* **36**, 6–16.

Fischer, K.F., Lukasiewicz, P.D., and Wong, R.O. (1998). Age-dependent and cell class-specific modulation of retinal ganglion cell bursting activity by GABA. *J. Neurosci.* **18**, 3767–3778.

Fisher, L.J. (1979). Development of synaptic arrays in the inner plexiform layer of neonatal mouse retina. *J. Comp. Neurol.* **187**, 359–372.

Fletcher, E.L. and Kalloniatis, M. (1997). Localisation of amino acid neurotransmitters during postnatal development of the rat retina. *J. Comp. Neurol.* **380**, 449–471.

Fried, S.I., Munch, T.A., and Werblin, F.S. (2002). Mechanisms and circuitry underlying directional selectivity in the retina. *Nature* **420**, 411–414.

Ganguly, K., Schinder, A.F., Wong, S.T., and Poo, M. (2001). GABA itself promotes the developmental switch of neuronal GABAergic responses from excitation to inhibition. *Cell* **105**, 521–532.

Gavrikov, K.E., Dmitriev, A.V., Keyser, K.T., and Mangel, S.C. (2003). Cation-chloride cotransporters mediate neural computation in the retina. *Proc. Natl. Acad. Sci. USA* **100**, 16047–16052.

Gavrikov, K.E., Nilson, J.E., Dmitriev, A.V., Zucker, C.L., and Mangel, S.C. (2006). Dendritic compartmentalization of chloride cotransporters underlies directional responses of starburst amacrine cells in retina. *Proc. Natl. Acad. Sci. USA* **103**, 18793–18798.

Grünert, U. and Wässle, H. (1990). GABA-like immunoreactivity in the macaque monkey retina: a light and electron microscopic study. *J. Comp. Neurol.* **297**, 509–524.

Hoffpauir, B., McMains, E., and Gleason, E. (2006). Nitric oxide transiently converts synaptic inhibition to excitation in retinal amacrine cells. *J. Neurophysiol.* **95**, 2866–2877.

Huang, B., Mitchell, C.K., and Redburn-Johnson, D.A. (2000). GABA and $GABA_A$ receptor antagonists alter developing cone photoreceptor development in neonatal rabbit retina. *Vis. Neurosci.* **17**, 925–935.

Huang, B.O. and Redburn, D.A. (1996). GABA-induced increases in $[Ca^{2+}]_i$ in retinal neurons of postnatal rabbits. *Vis. Neurosci.* **13**, 441–447.

Ikeda, M., Toyoda, H., Yamada, J., Okabe, A., Sato, K., Hotta, Y., and Fukuda, A. (2003). Differential development of cation-chloride cotransporters and Cl^- homeostasis contributes to differential GABAergic actions between developing rat visual cortex and dorsal lateral geniculate nucleus. *Brain Res.* **984**, 149–159.

Isomura, Y., Sugimoto, M., Fujiwara-Tsukamoto, Y., Yamamoto-Muraki, S., Yamada, J., and Fukuda, A. (2003). Synaptically activated Cl^- accumulation responsible for depolarizing GABAergic responses in mature hippocampal neurons. *J. Neurophysiol.* **90**, 2752–2756.

Johnson, M.A. and Vardi, N. (1998). Regional differences in GABA and GAD immunoreactivity in rabbit horizontal cells. *Vis. Neurosci.* **15**, 743–753.

Kaila, K. (1994). Ionic basis of $GABA_A$ receptor channel function in the nervous system. *Prog. Neurobiol.* **42**, 489–537.

Kamermans, M. and Fahrenfort, I. (2004). Ephaptic interactions within a chemical synapse: hemichannel-mediated ephaptic inhibition in the retina. *Curr. Opin. Neurobiol.* **14**, 531–541.

Kamermans, M. and Werblin, F. (1992). GABA-mediated positive autofeedback loop controls horizontal cell kinetics in tiger salamander retina. *J. Neurosci.* **12**, 2451–2463.

Kanaka, C., Ohno, K., Okabe, A., Kuriyama, K., Itoh, T., Fukuda, A., and Sato, K. (2001). The differential expression patterns of messenger RNAs encoding K-Cl cotransporters (KCC1,2) and

Na-K-2Cl cotransporter (NKCC1) in the rat nervous system. *Neuroscience* **104**, 933–946.

Kaneko, H., Putzier, I., Frings, S., Kaupp, U.B., and Gensch, T. (2004). Chloride accumulation in mammalian olfactory sensory neurons. *J. Neurosci.* **24**, 7931–7938.

Kaplan, M.R., Plotkin, M.D., Brown, D., Hebert, S.C., and Delpire, E. (1996). Expression of the mouse Na-K-2Cl cotransporter, mBSC2, in the terminal inner medullary collecting duct, the glomerular and extraglomerular mesangium, and the glomerular afferent arteriole. *J. Clin. Invest.* **98**, 723–730.

Katz, L.C. and Shatz, C.J. (1996). Synaptic activity and the construction of cortical circuits. *Science* **274**, 1133–1138.

Kelsch, W., Hormuzdi, S., Straube, E., Lewen, A., Monyer, H., and Misgeld, U. (2001). Insulin-like growth factor 1 and a cytosolic tyrosine kinase activate chloride outward transport during maturation of hippocampal neurons. *J. Neurosci.* **21**, 8339–8347.

Khirug, S., Huttu, K., Ludwig, A., Smirnov, S., Voipio, J., Rivera, C., Kaila, K., and Khiroug, L. (2005). Distinct properties of functional KCC2 expression in immature mouse hippocampal neurons in culture and in acute slices. *Eur. J. Neurosci.* **21**, 899–904.

Kuner, T. and Augustine, G.J. (2000). A genetically encoded ratiometric indicator for chloride: capturing chloride transients in cultured hippocampal neurons. *Neuron* **27**, 447–459.

Lake, N. (1994). Taurine and GABA in the rat retina during postnatal development. *Vis. Neurosci.* **11**, 253–260.

Lasater, E.M., Dowling, J.E., and Ripps, H. (1984). Pharmacological properties of isolated horizontal and bipolar cells from the skate retina. *J. Neurosci.* **4**, 1966–1975.

Lee, M.Y., Shin, S.L., Han, S.H., and Chun, M.H. (1999). The birthdates of GABA-immunoreactive amacrine cells in the rat retina. *Exp. Brain Res.* **128**, 309–314.

Lee, S. and Zhou, Z.J. (2006). The synaptic mechanism of direction selectivity in distal processes of starburst amacrine cells. *Neuron* **51**, 787–799.

Leitch, E., Coaker, J., Young, C., Mehta, V., and Sernagor, E. (2005). GABA type-A activity controls its own developmental polarity switch in the maturing retina. *J. Neurosci.* **25**, 4801–4805.

Li, B., McKernan, K., and Shen, W. (2008). Spatial and temporal distribution patterns of Na-K-2Cl cotransporter in adult and developing mouse retinas. *Vis. Neurosci.* **25**, 109–123.

Li, B. and Shen, W. (2007). Cation Cl^- cotransporters in the dendrites of goldfish bipolar cells. *NeuroReport* **18**, 625–629.

Li, H., Tornberg, J., Kaila, K., Airaksinen, M.S., and Rivera, C. (2002). Patterns of cation-chloride cotransporter expression during embryonic rodent CNS development. *Eur. J. Neurosci.* **16**, 2358–2370.

Ludwig, A., Li, H., Saarma, M., Kaila, K., and Rivera, C. (2003). Developmental up-regulation of KCC2 in the absence of GABAergic and glutamatergic transmission. *Eur. J. Neurosci.* **18**, 3199–3206.

Lytle, C., Xu, J.C., Biemesderfer, D., Haas, M., and Forbush, B., III (1992). The Na-K-Cl cotransport protein of shark rectal gland. I. Development of monoclonal antibodies, immunoaffinity purification, and partial biochemical characterization. *J. Biol. Chem.* **267**, 25428–25437.

Manookin, M.B., Beaudoin, D.L., Ernst, Z.R., Flagel, L.J., and Demb, J.B. (2008). Disinhibition combines with excitation to extend the operating range of the OFF visual pathway in daylight. *J. Neurosci.* **28**, 4136–4150.

Manookin, M.B. and Demb, J.B. (2006). Presynaptic mechanism for slow contrast adaptation in mammalian retinal ganglion cells. *Neuron* **50**, 453–464.

Marchionni, I., Omrani, A., and Cherubini, E. (2007). In the developing rat hippocampus a tonic $GABA_A$-mediated conductance selectively enhances the glutamatergic drive of principal cells. *J. Physiol.* **581**, 515–528.

Markova, O., Mukhtarov, M., Real, E., J.acob, Y., and Bregestovski, P. (2008). Genetically encoded chloride indicator with improved sensitivity. *J. Neurosci. Methods* **170**, 67–76.

Martina, M., Royer, S., and Pare, D. (2001). Cell-type-specific GABA responses and chloride homeostasis in the cortex and amygdala. *J. Neurophysiol.* **86**, 2887–2895.

Masland, R.H. (2004). Neuronal cell types. *Curr. Biol.* **14**, R497–R500.

Mastronarde, D.N. (1983). Correlated firing of cat retinal ganglion cells. II. Responses of X- and Y-cells to single quantal events. *J. Neurophysiol.* **49**, 325–349.

Meister, M., Wong, R.O., Baylor, D.A., and Shatz, C.J. (1991). Synchronous bursts of action potentials in ganglion cells of the developing mammalian retina. *Science* **252**, 939–943.

Messersmith, E.K. and Redburn, D.A. (1993). The role of GABA during development of the outer retina in the rabbit. *Neurochem. Res.* **18**, 463–470.

Mikawa, S., Wang, C., Shu, F., Wang, T., Fukuda, A., and Sato, K. (2002). Developmental changes in KCC1, KCC2 and NKCC1 mRNAs in the rat cerebellum. *Brain Res. Dev. Brain Res.* **136**, 93–100.

Miller, R.F. and Dacheux, R.F. (1973). Information processing in the retina: importance of chloride ions. *Science* **181**, 266–268.

Miller, R.F. and Dacheux, R.F. (1983). Intracellular chloride in retinal neurons: measurement and meaning. *Vision Res.* **23**, 399–411.

Mitchell, C.K., Huang, B., and Redburn-Johnson, D.A. (1999). $GABA_A$ receptor immunoreactivity is transiently expressed in the developing outer retina. *Vis. Neurosci.* **16**, 1083–1088.

Murphy, G.J. and Rieke, F. (2006). Network variability limits stimulus-evoked spike timing precision in retinal ganglion cells. *Neuron* **52**, 511–524.

Murphy, G.J. and Rieke, F. (2008). Signals and noise in an inhibitory interneuron diverge to control activity in nearby retinal ganglion cells. *Nat. Neurosci.* **11**, 318–326.

Nag, T.C. and Wadhwa, S. (1997). Expression of GABA in the fetal, postnatal, and adult human retinas: an immunohistochemical study. *Vis. Neurosci.* **14**, 425–432.

Nakamura, Y., McGuire, B.A., and Sterling, P. (1980). Interplexiform cell in cat retina: identification by uptake of gamma-[^{3}H]aminobutyric acid and serial reconstruction. *Proc. Natl. Acad. Sci. USA* **77** (1), 658–661.

Nawy, S. and Jahr, C.E. (1990). Suppression by glutamate of cGMP-activated conductance in retinal bipolar cells. *Nature* **346**, 269–271.

Nelson, R. (1982). AII amacrine cells quicken time course of rod signals in the cat retina. *J. Neurophysiol.* **47**, 928–947.

Nelson, R., Bender, A.M., and Connaughton, V.P. (2008). Transporter-mediated GABA responses in horizontal and bipolar cells of zebrafish retina. *Vis. Neurosci.* **25**, 155–165.

Nelson, R. and Kolb, H. (1985). A17: A broad-field amacrine cell in the rod system of the cat retina. *J. Neurophysiol.* **54**, 592–614.

Nguyen, L.T. and Grzywacz, N.M. (2000). Colocalization of choline acetyltransferase and gamma-aminobutyric acid in the developing and adult turtle retinas. *J. Comp. Neurol.* **420**, 527–538.

Nomura, A., Shigemoto, R., Nakamura, Y., Okamoto, N., Mizuno, N., and Nakanishi, S. (1994). Developmentally-regulated postsynaptic localization of a metabotropic glutamate-receptor in rat rod bipolar cells. *Cell* **77**, 361–369.

Olney, J.W. (1968a). An electron microscopic study of synapse formation, receptor outer segment development, and other aspects of developing mouse retina. *Invest. Ophthal.* **7**, 250–268.

Olney, J.W. (1968b). Centripetal sequence of appearance of receptor-bipolar synaptic structures in developing mouse retina. *Nature* **218**, 281–282.

Parsons, R. (1959). *Handbook of Electrochemical Constants.* Butterworths Scientific Publications, London.

Plotkin, M.D., Kaplan, M.R., Peterson, L.N., Gullans, S.R., Hebert, S.C., and Delpire, E. (1997a). Expression of the Na^+-K^+-$2Cl^-$ cotransporter BSC2 in the nervous system. *Am. J. Physiol.* **272**, C173–C183.

Plotkin, M.D., Snyder, E.Y., Hebert, S.C., and Delpire, E. (1997b). Expression of the Na-K-2Cl cotransporter is developmentally regulated in postnatal rat brains: a possible mechanism underlying GABA's excitatory role in immature brain. *J. Neurobiol.* **33**, 781–795.

Pow, D.V. and Hendrickson, A.E. (2000). Expression of glycine and the glycine transporter Glyt-1 in the developing rat retina. *Vis. Neurosci.* **17**, 1R–9R.

Pow, D.V., Wright, L.L., and Vaney, D.I. (1994). The immunocytochemical detection of amino-acid neurotransmitters in paraformaldehyde-fixed tissues. *J. Neurosci. Meth.* **56**, 115–123.

Redburn, D.A. (1992). Development of GABAergic neurons in the mammalian retina. *Prog. Brain Res.* **90**, 133–147.

Rocha-González, H.I., Mao, S., and Alvarez-Leefmans, F.J. (2008). Na^+, K^+,$2Cl^-$ cotransport and intracellular chloride regulation in rat primary sensory neurons: thermodynamic and kinetic aspects. *J. Neurophysiol.* **100**, 169–184.

Sassoè-Pognetto, M. and Wässle, H. (1997). Synaptogenesis in the rat retina: subcellular localization of glycine receptors, $GABA_A$ receptors, and the anchoring protein gephyrin. *J. Comp. Neurol.* **381**, 158–174.

Satoh, H., Kaneda, M., and Kaneko, A. (2001). Intracellular chloride concentration is higher in rod bipolar cells than in cone bipolar cells of the mouse retina. *Neurosci. Lett.* **310**, 161–164.

Schnitzer, J. and Rusoff, A.C. (1984). Horizontal cells of the mouse retina contain glutamic acid decarboxylase-like immunoreactivity during early developmental stages. *J. Neurosci.* **4**, 2948–2955.

Schubert, T., Kerschensteiner, D., Eggers, E.D., Misgeld, T., Kerschensteiner, M., Lichtman, J.W., Lukasiewicz, P.D., and Wong, R.O. (2008). Development of presynaptic inhibition onto retinal bipolar cell axon terminals is subclass-specific. *J. Neurophysiol.* **100**, 304–316.

Sernagor, E., Young, C., and Eglen, S.J. (2003). Developmental modulation of retinal wave dynamics: shedding light on the GABA saga. *J. Neurosci.* **23**, 7621–7629.

Shelley, J., Dedek, K., Schubert, T., Feigenspan, A., Schultz, K., Hombach, S., Willecke, K., and Weiler, R. (2006). Horizontal cell receptive fields are reduced in connexin57-deficient mice. *Eur. J. Neurosci.* **23**, 3176–3186.

Sherry, D.M., Wang, M.M., Bates, J., and Frishman, L.J. (2003). Expression of vesicular glutamate transporter 1 in the mouse retina reveals temporal ordering in development of rod vs. cone and ON vs. OFF circuits. *J. Comp. Neurol.* **465**, 480–498.

Shiells, R.A. and Falk, G. (1990). Glutamate receptors of rod bipolar cells are linked to a cyclic GMP cascade via a G-protein. *Proc. R. Soc. Lond. B* **242**, 91–94.

Smith, R.G. and Vardi, N. (1995). Simulation of the AII amacrine cell of mammalian retina: functional consequences of electrical coupling and regenerative membrane properties. *Vis. Neurosci.* **12**, 851–860.

Stacy, R.C. and Wong, R.O. (2003). Developmental relationship between cholinergic amacrine cell processes and ganglion cell dendrites of the mouse retina. *J. Comp. Neurol.* **456**, 154–166.

Staley, K.J. and Proctor, W.R. (1999). Modulation of mammalian dendritic $GABA_A$ receptor function by the kinetics of Cl^- and $HCO3^-$ transport. *J. Physiol.* **519**, 693–712.

Stein, V., Hermans-Borgmeyer, I., J.entsch, T.J., and Hubner, C.A. (2004). Expression of the KCl cotransporter KCC2 parallels neuronal maturation and the emergence of low intracellular chloride. *J. Comp. Neurol.* **468**, 57–64.

Suzuki, S., Tachibana, M., and Kaneko, A. (1990). Effects of glycine and GABA on isolated bipolar cells of the mouse retina. *J. Physiol.* **421**, 645–662.

Syed, M.M., Lee, S., Zheng, J., and Zhou, Z.J. (2004). Stage-dependent dynamics and modulation of spontaneous waves in the developing rabbit retina. *J. Physiol.* **560**, 533–549.

Taylor, W.R. and Vaney, D.I. (2002). Diverse synaptic mechanisms generate direction selectivity in the rabbit retina. *J. Neurosci.* **22**, 7712–7720.

Thoreson, W.B., Bryson, E.J., and Rabl, K. (2003). Reciprocal interactions between calcium and chloride in rod photoreceptors. *J. Neurophysiol.* **90**, 1747–1753.

Thoreson, W.B., Stella, S.L., Jr., Bryson, E.I., Clements, J., and Witkovsky, P. (2002). D2-like dopamine receptors promote interactions between calcium and chloride channels that diminish rod synaptic transfer in the salamander retina. *Vis. Neurosci.* **19**, 235–247.

Ueda, Y., Iwakabe, H., Masu, M., Suzuki, M., and Nakanishi, S. (1997). The mGluR6 5′ upstream transgene sequence directs a cell-specific and developmentally regulated expression in retinal rod and ON-type cone bipolar cells. *J. Neurosci.* **17**, 3014–3023.

Van Essen, D.C., Anderson, C.H., and Felleman, D.J. (1992). Information processing in the primate visual system: an integrated systems perspective. *Science* **255**, 419–423.

Vardi, N. and Auerbach, P. (1995). Specific cell types in cat retina express different forms of glutamic acid decarboxylase. *J. Comp. Neurol.* **351**, 374–384.

Vardi, N., Dhingra, A., Zhang, L.L., Lyubarsky, A., Wang, T.L., and Morigiwa, K. (2002). Neurochemical organization of the first visual synapse. *Keio J. Med.* **51**, 154–164.

Vardi, N., Kaufman, D.L., and Sterling, P. (1994). Horizontal cells in cat and monkey retina express different isoforms of glutamic acid decarboxylase. *Vis. Neurosci.* **11**, 135–142.

Vardi, N. and Sterling, P. (1994). Subcellular localization of $GABA_A$ receptor on bipolar cells in macaque and human retina. *Vision Res.* **34**, 1235–1246.

Vardi, N., Zhang, L.L., Payne, J.A., and Sterling, P. (2000). Evidence that different cation chloride cotransporters in retinal neurons allow opposite responses to GABA. *J. Neurosci.* **20**, 7657–7663.

Varela, C., Blanco, R., and de la Villa, P. (2005a). Depolarizing effect of GABA in rod bipolar cells of the mouse retina. *Vision Res.* **20**, 7657–7663.

Varela, C., Rivera, L., Blanco, R., and de la Villa, P. (2005b). Depolarizing effect of GABA in horizontal cells of the rabbit retina. *Neurosci. Res.* **53**, 257–264.

Voipio, J. and Kaila, K. (2000). GABAergic excitation and K(+)-mediated volume transmission in the hippocampus. *Prog. Brain Res.* **125**, 329–338 [Review] [53 refs].

Vu, T.Q., Payne, J.A., and Copenhagen, D.R. (2000). Localization and developmental expression patterns of the neuronal K-Cl cotransporter (KCC2) in the rat retina. *J. Neurosci.* **20**, 1414–1423.

Wang, C.T., Blankenship, A.G., Anishchenko, A., Elstrott, J., Fikhman, M., Nakanishi, S., and Feller, M.B. (2007). $GABA_A$ receptor-mediated signaling alters the structure of spontaneous activity in the developing retina. *J. Neurosci.* **27**, 9130–9140.

Wang, D.D. and Kriegstein, A.R. (2008). GABA regulates excitatory synapse formation in the neocortex via NMDA receptor activation. *J. Neurosci.* **28**, 5547–5558.

Wang, M.M., Janz, R., Belizaire, R., Frishman, L.J., and Sherry, D.M. (2003). Differential distribution and developmental expression of synaptic vesicle protein 2 isoforms in the mouse retina. *J. Comp. Neurol.* **460**, 106–122.

Weidman, T.A. and Kuwabara, T. (1969). Development of the rat retina. *Invest. Ophthalmol.* **8**, 60–69.

Williams, J.R., Sharp, J.W., Kumari, V.G., Wilson, M., and Payne, J.A. (1999). The neuron-specific K-Cl cotransporter, KCC2. Antibody development and initial characterization of the protein. *J. Biol. Chem.* **274**, 12656–12664.

Xu, Y., Sulaiman, P., Feddersen, R.M., Liu, J., Smith, R.G., and Vardi, N. (2008). Retinal ON bipolar cells express a new PCP2 splice variant that accelerates the light response. *J. Neurosci.* **28**, 8873–8884.

Yamada, J., Okabe, A., Toyoda, H., Kilb, W., Luhmann, H.J., and Fukuda, A. (2004). Cl^- uptake promoting depolarizing GABA actions in immature rat neocortical neurones is mediated by NKCC1. *J. Physiol.* **557**, 829–841.

Young-Pearse, T.L., Ivic, L., Kriegstein, A.R., and Cepko, C.L. (2006). Characterization of mice with targeted deletion of glycine receptor alpha 2. *Mol. Cell Biol.* **26**, 5728–5734.

Young, T.L. and Cepko, C.L. (2004). A role for ligand-gated ion channels in rod photoreceptor development. *Neuron* **41**, 867–879.

Zhang, L.L., Delpire, E., and Vardi, N. (2007). NKCC1 does not accumulate chloride in developing retinal neurons. *J. Neurophysiol.* **98**, 266–277.

Zhang, L.L., Fina, M.E., and Vardi, N. (2006a). Regulation of KCC2 and NKCC during development: membrane insertion and differences between cell types. *J. Comp. Neurol.* **499**, 132–143.

Zhang, L.L., Pathak, H.R., Coulter, D.A., Freed, M.A., and Vardi, N. (2006b). Shift of intracellular chloride concentration in ganglion and amacrine cells of developing mouse retina. *J. Neurophysiol.* **95**, 2404–2416.

Zhou, Z.J. (2001). A critical role of the strychnine-sensitive glycinergic system in spontaneous retinal waves of the developing rabbit. *J. Neurosci.* **21**, 5158–5168.

Zhou, Z.J. and Lee, S. (2008). Synaptic physiology of direction selectivity in the retina. *J. Physiol.* **586**, 4371–4376.

CHAPTER

20

Chloride-based Signal Amplification in Olfactory Sensory Neurons

Stephan Frings

OUTLINE

I. ODOR-INDUCED CHLORIDE CURRENTS IN OLFACTORY SENSORY NEURONS

Sensory cells detect their adequate stimuli by specialized structures which are localized in sensory organelles such as microvillae or cilia. Stimulus detection is followed by a transduction step that leads to a change in membrane potential. This electrical signal, the receptor potential, is ultimately translated into a change of afferent nerve impulse activity which can be read out by the brain and interpreted in terms of sensory information. Receptor potentials are generated by the gating of transduction channels which, in many cases, are cation channels of the TRP (transient receptor potential) or CNG (cyclic nucleotide-gated) families. With seven subfamilies of TRP channels (Venkatachalam and Montell, 2007), a modular system of six CNG-channel subunits (Bradley et al., 2004) and an assortment of cation channels from the degenerin family (ASIC, ENaC, MEC, and others; Kellenberger and Schild, 2002), the variety of transduction channels is large. The functional properties of these channels are, however, similar. Once activated by their specific gating mechanism, they conduct a depolarizing inward current, carried by Ca^{2+} and monovalent cations. Thus, stimulation of a sensory cell invariably has two consequences: the generation of an intracellular Ca^{2+} signal, and a change in membrane voltage. In this chapter, I discuss a special mechanism that employs the stimulus-induced Ca^{2+} signal to boost the change of the membrane voltage ten-fold or more. This mechanism operates with Ca^{2+} dependent, depolarizing Cl^- currents and is used by olfactory sensory neurons (OSNs) to detect weak olfactory stimuli.

OSNs have to deal with weak stimuli because they work with low-sensitivity odorant receptors. The logic of the olfactory system is that the brain reads out the combined response of many OSNs, each of which can

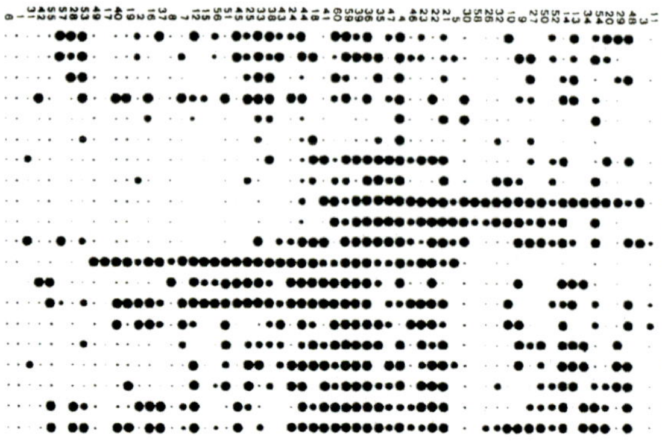

FIGURE 20.1 The logic of odor coding. The response of 60 frog OSNs to 20 different odorants is displayed using spot sizes to represent response intensities. Larger spots indicate higher rates of odor-induced action potentials. The OSNs are numbered 1–60 on the top, odorant names are given on the right. A vertical column in this matrix represents the response spectrum of an individual OSN. It can be seen that most OSNs respond to several odorants and, therefore, provide ambiguous – fuzzy – information to the brain. In contrast, the horizontal rows provide unambiguous identification of odorants, as each row contains a unique pattern of activity across all 60 OSNs. The nose represents olfactory information by a precise combinatorial code to which many OSNs contribute their own fuzzy signal. Altered from Sicard and Holley (1984). With permission of Elsevier Ltd.

be activated by many different odorants at different concentrations. While the information contained in the response of a single OSN is fuzzy, the information contained in the ensemble activity pattern is precise. An experiment by Sicard and Holley (1984) illustrates this point (Fig. 20.1). It shows the response intensity of 60 different OSNs to a panel of 25 different odorants. Since each individual OSN responds to 2–24 odorants (a column in Fig. 20.1), it is not possible to identify any odorant by any single OSN activity. However, each odorant produces a unique activity pattern over all 60 OSNs (a row in Fig. 20.1) and can be identified by it. This way of combinatorial encoding of odor information makes it possible to detect a vast number of different odorants and odorant mixtures, for even subtle differences between activity patterns can be analyzed. In fact, the olfactory system presents to the inhaled air, hundreds of different odorant receptors (about 400 in humans, 900 in dogs, 1200 in mice), the combinations of which amount to a variety of activity patterns which, to all intents and purposes, is unlimited. Each OSN expresses only one type of receptor, and all OSNs with the same receptor type converge onto the same point on the surface of the olfactory bulb (Mombaerts, 2006). It is this surface (the glomerular layer) where a two-dimensional activity map represents each odor. Information is coded both in the spatial distribution of activity, and also in the way the activity pattern develops during the first few hundred milliseconds of smelling (Friedrich, 2006). Thus, spatial and temporal pattern analysis extracts accurate information which is supplied by a large number of inaccurate sensors. More specific information about olfactory coding can be obtained from some excellent recent reviews (Elsaesser and Paysan, 2007; Mori et al., 2006; Schaefer and Margrie, 2007; Vosshall and Stocker, 2007).

Working with fuzzy odorant receptors creates problems for signal transduction in OSNs, resulting from the low efficiency of the initial transduction step. This step, the activation of adenylyl cyclase and the production of the second messenger cAMP (Fig. 20.2), can only work efficiently if the odorant receptor binds a ligand with high affinity. This, however, is unlikely. As a consequence of their low selectivity, the receptors are mostly activated by non-perfect ligands, that is, by odorants for which they have a relatively low affinity. *In vitro* experiments have consistently shown that odorant concentrations in the range of 1–1000 μM are necessary to obtain robust activation of OSNs (e.g. Firestein et al., 1993). Compared to the sensitivity of olfactory organs observed in behavioral assays, this concentration range appears very high indeed. It suggests that olfactory receptor proteins are only weakly activated during normal operation. Bhandawat et al. (2005) estimated that the mean dwell time of an odorant at its receptor is less than 1 ms, a consequence of low-affinity binding. Such a brief contact is hardly sufficient to trigger the transduction cascade, and the synthesis of cAMP proceeds at a low rate. Measurements with amphibian OSNs revealed that the maximal total rate of cAMP synthesis in an OSN at saturating stimulus intensity is about 200,000 cAMP molecules per second (Takeuchi and Kurahashi, 2005). This corresponds to the approximate rate of cGMP hydrolysis in a rod photoreceptor induced by a single photon. OSNs thus stand at the bottom end of the scale of metabotropic transduction efficiencies, while photoreceptors arguably hold the top position.

It appears that OSNs solve this problem by utilizing a chloride-based signal-amplification mechanism. The rising cAMP concentration in the sensory cilia opens cAMP-gated channels which conduct Ca^{2+} influx and give rise to a ciliary Ca^{2+} signal (Leinders-Zufall et al., 1997; Fig. 20.2). Triggered by this Ca^{2+} signal, Cl^- channels open in the ciliary membrane, allow Cl^- to flow out of the cilia and to amplify the depolarizing receptor potential (Kleene and Gesteland, 1991; Kurahashi and Yau, 1993; Lowe and Gold, 1993). Signal amplification is rapid and efficient: Within about 50 ms, the receptor current is amplified roughly ten-fold (Boccaccio and Menini, 2007) so that even small primary cation currents are converted into large

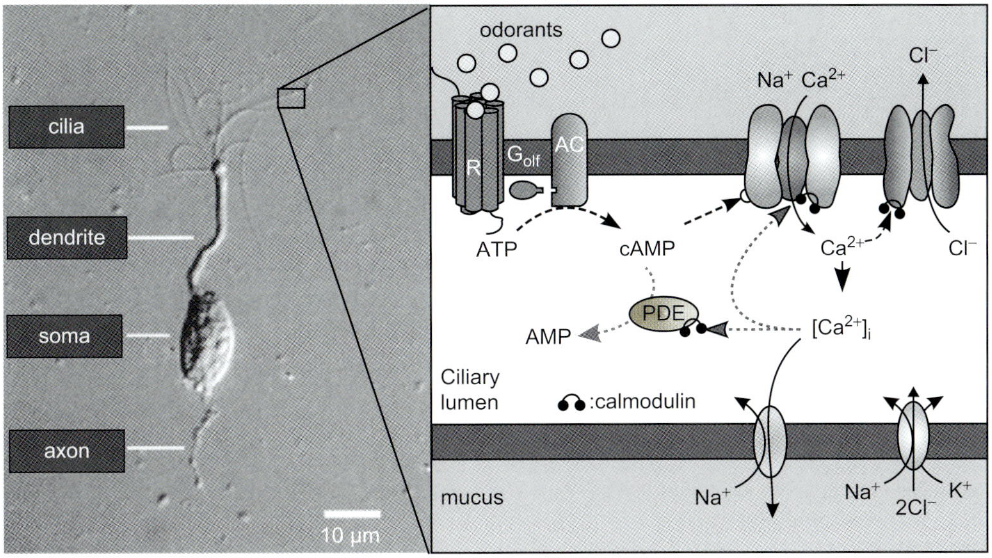

FIGURE 20.2 Olfactory signal transduction and chloride-based signal amplification. The sensory cilia contain all components of olfactory signal transduction. The initial step consists of the activation of odorant receptor proteins (R) which trigger cAMP synthesis by adenylyl cyclase typ III (AC) through a GTP-binding protein (G_{olf}). The second step is the generation of the excitatory receptor potential. Depolarization of the ciliary membrane by cAMP-gated, Ca^{2+}-permeable channels is amplified by Ca^{2+}-activated Cl^- channels. Termination of this process involves phosphodiesterase (PDE) and a feedback inhibition of cAMP-gated channels by Ca^{2+}/calmodulin. Ca^{2+} extrusion from the cilia is coupled to Na^+ uptake, and Cl^- accumulation is mediated in part by an electroneutral cation-chloride cotransporter. OSN from Kleene and Gesteland (1981). With permission from Elsevier Ltd.

depolarizations. Following its discovery in frog OSNs (Kleene and Gesteland, 1991), it soon became clear that this amplification mechanism is a general feature of vertebrate OSNs, operating in fish (Sato and Suzuki, 2000), amphibia (Kleene, 1993; Kurahashi and Yau, 1993) and mammals (Lowe and Gold, 1993). A remarkable feature of this amplification mechanism was revealed when Lowe and Gold (1993) examined the relationship between stimulus intensity and stimulus-induced currents in rat OSNs. It turned out that this relationship is highly non-linear, characterized by a Hill coefficient of 3 to 5 (Fig. 20.3A). This non-linearity results from the activation of the Cl^- current and basically introduces an activation threshold into the OSN transduction cascade. Current noise below this threshold is not amplified, while current signals that exceed the threshold only by a small increment are strongly amplified by eliciting the Cl^- current. A further key property of the signal amplification was gleaned from variance analysis of ciliary currents. Kleene (1997) found out that the Ca^{2+}-dependent Cl^- current amplified the signal amplitude without amplifying its noise (Fig. 20.3B). These findings revealed a novel mode of physiological signal amplification that works fast, with high gain, and with low noise. Indeed, the coordinated activation of cAMP-gated cation channels and Ca^{2+}-gated Cl^- channels in sensory cilia enables OSNs to produce excitatory receptor potentials even with fuzzy odorant receptors and their weak activation. In the sections that follow, I summarize the current understanding of the molecular organization of this remarkable mechanism.

II. CALCIUM-ACTIVATED CHLORIDE CHANNELS IN OLFACTORY SENSORY CILIA

During odor detection, the concentration of cAMP rises inside the cilia. Takeuchi and Kurahashi (2005) derived the ciliary cAMP concentration in newt OSNs from the cAMP-dependent current (Fig. 20.4A) and estimated that the concentration rises at a rate of 10–100 μM/s, depending on stimulus intensity. Considering that a single sniff in mammals can be as short as 100 ms (Verhagen et al., 2007), the cAMP concentration is expected to rise to roughly 1–10 μM during a sniff, assuming similar cAMP production rates in mammals. This is enough to open the cAMP-gated channel and to admit Ca^{2+} influx into the cilia. The cAMP-gated channels are highly cAMP sensitive. Their two modulatory subunits CNGA4 and CNGB1b increase the cAMP sensitivity of the primary channel subunit CNGA2 and enable channel opening by only a few μM of cAMP (Fig. 20.4B; Bönigk et al., 1999;

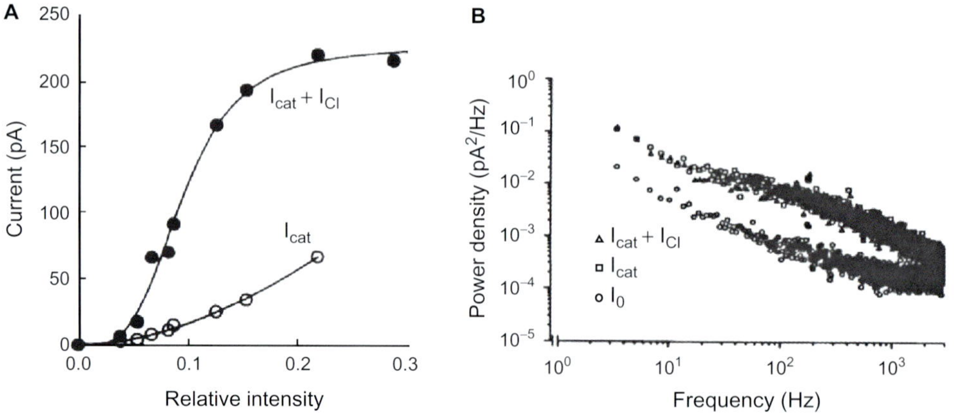

FIGURE 20.3 High-gain, low-noise amplification. **A.** cAMP-induced current in a newt OSN. The current was elicited by photorelease of cAMP from caged cAMP. Increasing duration of photolysis produced increasing concentrations of free cAMP and is indicated on the graph as relative intensity. The intensity dependence was described by a Hill function with a Hill coefficient of 3.7 reflecting the high cooperativity of channel gating. A chloride contribution to this current of 70% is revealed by blocking I_{Cl} with 5 mM SITS (4-acetoamido-4′-isothiocyanatostilbene-2,2′disulphonic acid). A gain, $(I_{cat} + I_{Cl})/I_{cat}$, of 2–3 is typical for amphibia. In mammalian OSNs the gain is in the range of 10. Altered from Lowe and Gold (1993). With permission of Nature Publishing Group. **B.** Power spectral analysis of cAMP-induced current noise in a single frog OSN cilium. Three power density spectra illustrate the noise in the absence of cAMP (I_0, circles), after opening of cAMP-gated channels (I_{cat}, squares), and after Ca^{2+} influx did fully activate the Ca^{2+}-dependent Cl^- channels ($I_{cat} + I_{Cl}$, triangles). This analysis shows that the noise spectrum of the amplified current ($I_{cat} + I_{Cl}$) is not significantly different from that of the non-amplified current (I_{cat}). It illustrates the special low-noise performance of chloride-based signal amplification. Altered from Kleene (1997). With permission of Biophysical Society.

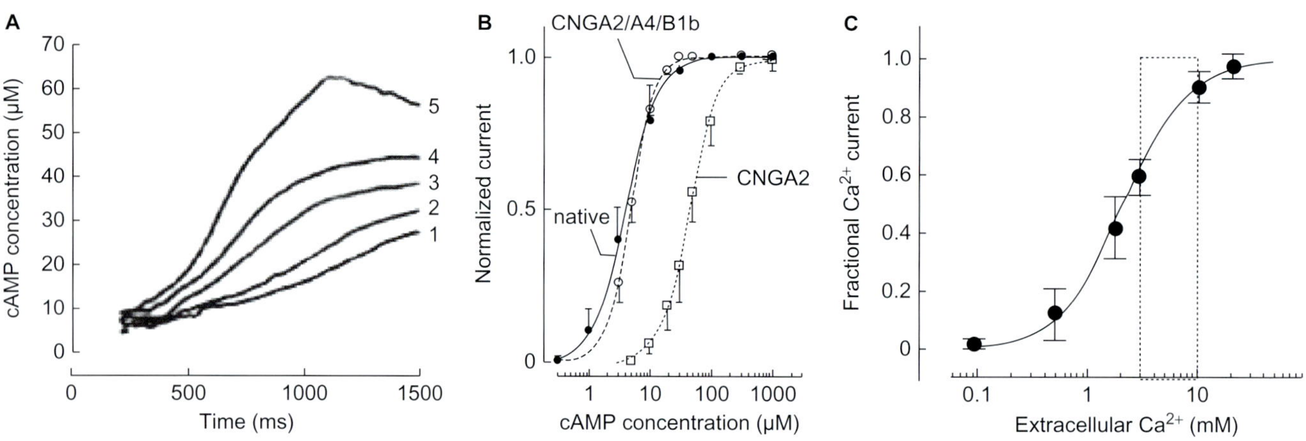

FIGURE 20.4 Generation of the odor-induced Ca^{2+} signal in OSN cilia. **A.** Increase of ciliary cAMP concentration during stimulation with various odor concentrations (highest concentration: trace 5) in a newt OSN. The cAMP concentration was calculated from the measured cAMP-induced current. The rate of cAMP synthesis depends on the stimulus intensity and ranges from 10 to 100 μM/s. Altered from Takeuchi and Kurahashi (2005). With permission of *The Journal of Neuroscience*. **B.** The cAMP-gated cation channels have two auxiliary subunits (CNGA4 and CNGB1b) which render the native channels sensitive to low concentrations of cAMP. Channels formed solely from the principal subunit CNGA2 require >10 μM cAMP to open. The dose–response relations of native channels (solid line) and of heterologously expressed channels with all three subunits (dotted line) show channel activation at 1–10 μM cAMP. **C.** cAMP-gated channels conduct a mix of monovalent cations and Ca^{2+}. The Ca^{2+} fraction depends on the extracellular Ca^{2+} concentration which, *in vivo*, is 3–10 mM (box). This figure illustrates that the cAMP-dependent current uses Ca^{2+} as the main charge carrier and injects Ca^{2+} into the cilia. The data were obtained from rat CNGA2/A4/B1b channels expressed in HEK 293 cells.

Zheng and Zagotta, 2004). Channel gating is cooperative (Nache et al., 2005; Biskup et al., 2007) and contributes to the non-linear dependence of the receptor current on stimulus intensity (Fig. 20.3A). Open cAMP-gated channels conduct a mixture of Ca^{2+} and monovalent cations into the cilia. The Ca^{2+} fraction of this current increases with the Ca^{2+} concentration on the extracellular side of the channel (Fig. 20.4C; Dzeja

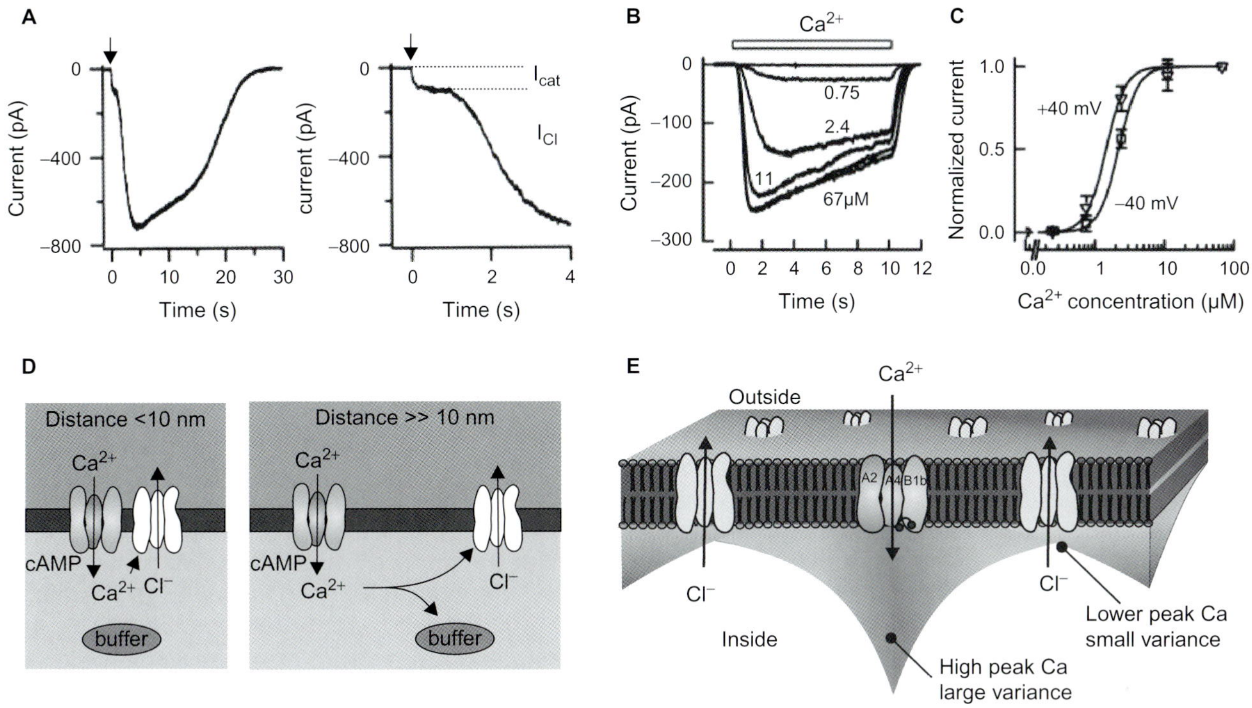

FIGURE 20.5 Coordinated activation of cation channels and chloride channels in OSNs. **A.** Photorelease of 8-Br-cAMP, a hydrolysis-resistant analog of cAMP, in mouse OSNs induces a transient current which is conducted by the two types of transduction channels. Immediately upon flash photolysis (arrow), the cAMP-gated channels open and permit Ca^{2+} entry (I_{cat}). After a certain delay (note the extended time scale in the right panel), sufficient Ca^{2+} has accumulated inside the cilia to open Ca^{2+}-dependent Cl^- channels. I_{Cl} then augments the currents about ten-fold. Altered from Boccaccio and Menini (2007). With permission of the American Physiological Society. **B.** The Ca^{2+}-activated Cl^- channels respond to micromolar concentrations of intracellular Ca^{2+}. Applying various Ca^{2+} concentrations to an apical membrane patch excised from a rat OSN induces currents which saturate at $>10\,\mu M$ Ca^{2+}. Note the slow current decline which is reversible by exposure to Ca^{2+}-free solution (not shown). **C.** Ca^{2+}-dose–response relations at the indicated voltages were fitted with a Hill coefficient of 3, indicating cooperative channel activation by Ca^{2+}. The Ca^{2+} concentration for half-maximal activation was 1.4–1.5 μM. Altered from Reisert et al. (2003). With permission of the Rockefeller University Press. **D.** Schematic illustration of the proximity assay for Ca^{2+}-permeable channels and Ca^{2+}-gated channels. Soluble Ca^{2+} buffer can only interrupt the functional coupling of the two channels if they are separated for >10 nm. The response of the channels to varying concentrations of a Ca^{2+} buffer provides quantitative information about the distance between the channels. **E.** Two-dimensional model for the distribution of cAMP-gated Ca^{2+}-permeable channels and Ca^{2+}-gated Cl^- channels in the ciliary membrane of rat OSNs. Eight Cl^- channels surround each cAMP-gated channel at a mean distance of 120 nm. The intracellular Ca^{2+} concentration peaks within a microdomain at the cAMP-gated channel, and the Ca^{2+} profiles originating from multiple cAMP-gated channels overlap at each Cl^- channel. This model provides a quantitative description of transduction speed, amplification gain and noise suppression in rat OSNs (Reisert et al., 2003).

et al., 1999). Because *in vivo* the cilia are embedded in a layer of mucus with a free Ca^{2+} concentration of 3–10 mM, the fractional Ca^{2+} current is high (box in Fig. 20.4C). Ca^{2+} influx through the cAMP-gated channels causes a Ca^{2+} signal inside the cilia, the trigger for the Ca^{2+}-dependent Cl^- current.

Ca^{2+}-activated Cl^- channels in OSN cilia need >1 μM free Ca^{2+} to be activated (Kleene, 1993). This Ca^{2+} level builds up rapidly, as several studies have shown by monitoring the time course of Cl^- channel activation (Lowe and Gold, 1993; Reisert and Matthews, 1998; Boccacchio et al., 2007). Activating the cAMP-gated channels initially gives rise to a small cation current (I_{cat}) carried by both Na^+ and Ca^{2+} (Fig. 20.5A). Within 10–100 ms, the inflowing Ca^{2+} triggers a large Cl^- current (I_{Cl}) which usually exceeds I_{cat} more than ten-fold. The high Ca^{2+} permeability of cAMP-gated channels, combined with the small volume of OSNs cilia (diameter: 0.2 μm; volume: ~0.2 fL; Lindemann, 2001), ensures the rapid rise of the ciliary Ca^{2+} concentration. Chloride-based signal amplification thus operates on a time scale similar to that of the highly efficient phototransduction cascade, and OSNs can respond to odors as fast as photoreceptors can respond to light. Importantly, the Ca^{2+}-induced Cl^- current shows only weak inactivation. In the presence of a constant Ca^{2+} concentration, the current declines by <30% over 10 s (Fig. 20.5B; Reisert et al., 2003). Within the time frame of a sniff, Ca^{2+}-activated Cl^- channels thus stereotypically respond to the ciliary Ca^{2+} concentration, turning a Ca^{2+} rise into an increasing Cl^- current. It is presently not known if the Cl^- channels are subject to regulation on longer time scales. A detailed analysis of the Ca^{2+} sensitivity of these channels yielded similar results in amphibia and mammals. The concentration for half-maximal activation is 1–2 μM free Ca^{2+} and shows only little voltage-dependence (Fig. 20.5C). The activation is cooperative with Hill coefficients near 3, and maximal activation is reached at 10 μM Ca^{2+}. The steep stimulus dependence of the odor-induced current (Hill coefficient near 5; Fig. 20.3A) thus reflects the coordinated activation of the two transduction channels, each cooperatively gated by its ligand cAMP and Ca^{2+}, respectively.

The basis of the high gain of this system became evident from biophysical analyses of the transduction channels in frog (Kleene, 1997) and rat (Reisert et al., 2003) cilia. In rat OSNs, variance analysis of cAMP- and Ca^{2+}-dependent currents revealed that both channel types had similarly small single-channel conductances of 0.5–1.5 pS at physiological ion concentrations. However, channel densities differed considerably, with estimations of 8 cAMP-gated channels per μm^2 and 62 Cl^- channels per μm^2 of ciliary membrane. This result indicates that the high amplification gain does not result from differences in single-channel conductance but from the fact that each cAMP-gated channel can activate on average 8 Cl^- channels. To understand how the functional coupling of a single cation channel and 8 Cl^- channels is organized in the cilia, Reisert et al. (2003) examined the hypothesis that the channels are organized in a supramolecular complex, a spatial organization that would maintain the Ca^{2+} well (a cation channel) and the Ca^{2+} sensors (8 Cl^- channels) in close proximity. Such a complex would keep diffusion distances short and would optimize the efficacy of functional coupling. The decisive experiment to obtain information about the proximity of the channels is to examine its sensitivity to soluble Ca^{2+} buffers (Fig. 20.5D). If the channels are located at very short distances (<10 nm), soluble buffers cannot capture Ca^{2+} ions before they bind to a Cl^- channel. In contrast, if the channels are separated by 10 nm, buffer compounds can bind Ca^{2+} before it reaches a Cl^- channel. The sensitivity of the functional coupling to soluble buffers is, therefore, a quantitative measure of proximity between Ca^{2+} well and Ca^{2+} sensor (Neher, 1998). Using HEDTA as soluble Ca^{2+} buffer, Reisert et al. (2003) determined a mean distance of 120 nm between cAMP-gated channels and Ca^{2+}-activated Cl^- channels in the ciliary membrane of rat OSNs. This result rules out the notion of a supramolecular complex with the two transduction channels as constituents. But the analysis led to a computational model which explains the currently available data fairly well. The model is based on a regular spatial distribution of transduction channels in the ciliary membrane with a uniform distance of 120 nm between cation channels and Cl^- channels, and with an eight-fold excess of the latter (Fig. 20.5E; see Reisert et al., 2003 for quantitative details). During odor detection, Ca^{2+} rises sharply in the immediate vicinity of the cation channels, as cAMP activates them. However, because the cation channels undergo frequent transitions between open and closed states, the local Ca^{2+} concentration shows a large variance, rising when the channel opens, falling when it closes. The mean Ca^{2+} concentration decreases exponentially with distance from the cation channel. Consequently, Cl^- channels, at a mean distance of 120 nm, experience only a fraction of this Ca^{2+} microdomain. However, as the model assumes an even distribution of channels over the cilia, the individual microdomains of at least 11 cation channels overlap at the site of each Cl^- channel. This overlap is an important feature of the model because it explains the striking observation that the Cl^- currents amplify current without amplifying noise (Kleene, 1997). The variance of local Ca^{2+} concentrations, which is high directly at the cation channel, is considerably smoothed out by the spatial and temporal averaging of Ca^{2+} fluctuations at the site of the Cl^- channel, as the residual contributions of 11 cation channels add up. The Cl^- current, which faithfully reports the local Ca^{2+} concentration, shows little variance because there is little variance in the local Ca^{2+} concentration.

This model represents a concept for chloride-based signal amplification as it operates in the sensory cilia of OSNs. The model provides a comprehensive explanation for the rapidity and the high cooperativity of the odor response, the high gain of signal amplification and the low-noise property of this system. A prerequisite for future work on this model is the molecular identification of the ciliary Ca^{2+}-activated Cl^- channels. With the identification of this protein, structural work on the ciliary membrane will become possible to complement the biophysical data. A first candidate is the protein bestrophin 2, which was shown to be expressed in olfactory cilia (Pifferi et al., 2006). This protein displays Cl^- channel function when expressed in HEK 293 cells (Qu et al., 2004) and can be activated by Ca^{2+}. Various functional properties of bestrophin 2 do not match the features of native Ca^{2+}-activated Cl^- channels from OSNs (Pifferi et al., 2006) but this does not exclude a role of this protein in the formation of the native channel. While other proteins with putative Ca^{2+}-activated Cl^-

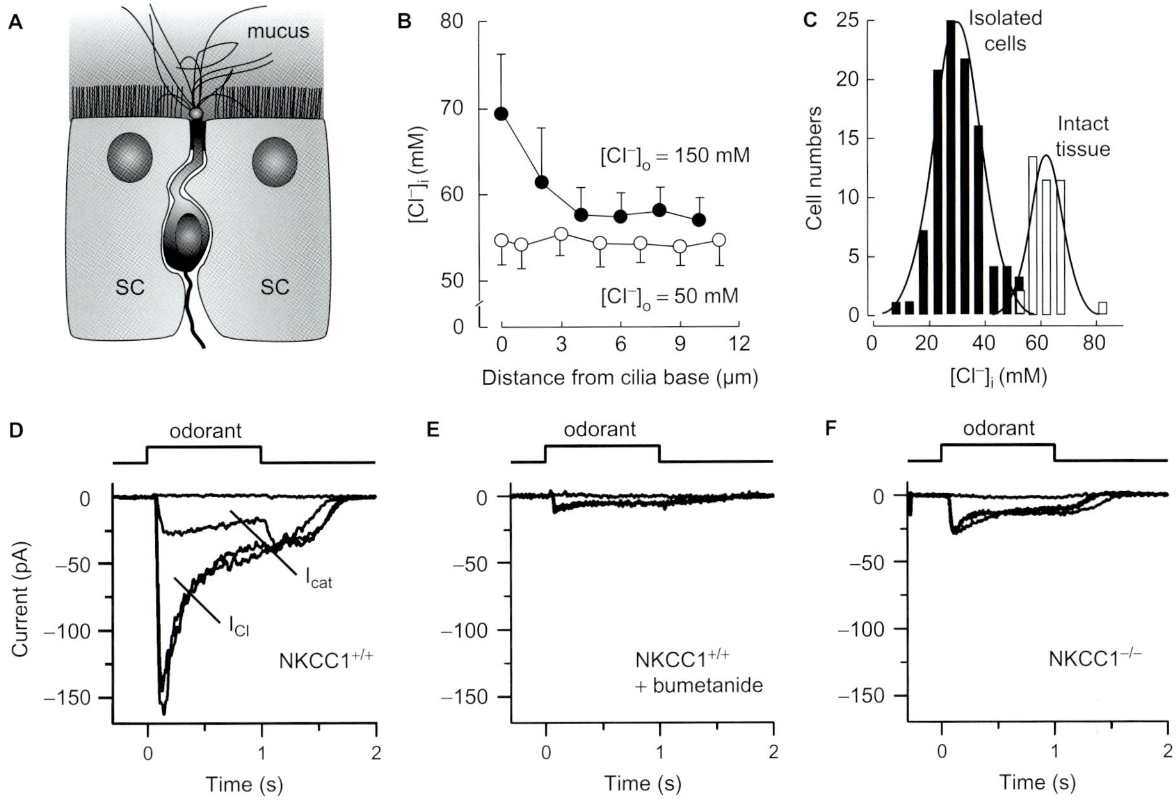

FIGURE 20.6 Chloride accumulation in OSNs. **A.** Cartoon of an OSN with its sensory cilia embedded in the mucus layer on the epithelial surface. **B.** Chloride concentrations measured at various points of the dendritic knob and distal dendrite. At 50 mM extracellular Cl^-, the intracellular Cl^- concentration is uniformly near 55 mM (open circles). Raising outside Cl^- to 150 mM (closed circles) imposes a standing Cl^- gradient with highest Cl^- levels near the cilia. **C.** Distribution of intracellular Cl^- concentrations in dendritic knobs of rat OSNs *in situ* measured by two-photon fluorescence lifetime imaging using the Cl^- indicator MQAE (Kaneko et al., 2004). Mean values are 62±6 mM in intact tissue and 30±8 mM in isolated cells. **D.** Contribution of NKCC1 to Cl^- accumulation in mouse OSNs. Suction-electrode recordings from isolated OSNs during stimulation with the odorant heptanal (100 μM). Recordings from NKCC1$^{+/+}$ mice display a large Cl^- component of the receptor current (left) which is totally suppressed by the NKCC1-blocker bumetanide (50 μM; center). An OSN from a NKCC1$^{-/-}$ knockout mouse shows no measurable Cl^- current. The absence of I_{Cl} in the single-cell recordings (but not in the EOG, see text) reflects the absence of NKCC1-mediated Cl^- accumulation in the knockout mouse. Altered from Reisert et al. (2005). With permission of Cell Press.

channel function appear weaker candidates today, members of the CLCA (Loewen and Forsyth, 2005) and the Tweety (Suzuki, 2006) protein families have not been definitely excluded from the race for the OSN transduction channel. For further discussion on molecular aspects of Ca^{2+}-activated Cl^- channels, see Chapters 1 and 13 in this volume.

III. CHLORIDE ACCUMULATION IN OLFACTORY SENSORY NEURONS

OSNs are probably the only neurons located at the body surface. With the sensory cilia, they expose a large part of their plasma membrane to the external environment. The cilia are embedded in a mucus layer (Fig. 20.6A), and the ionic distribution which is relevant for signal transduction is that between the mucus and the ciliary lumen. To function as signal amplifiers in the transduction cascade, Ca^{2+}-activated Cl^- channels have to conduct a depolarizing Cl^- efflux out of the cilia into the mucus. This can only happen if the Cl^- equilibrium potential, E_{Cl}, in the unstimulated OSN is less negative than the resting membrane potential ($V_m \sim -70$ mV). An estimate for E_{Cl} was obtained in an electron microscopic study where the *in vivo* Cl^- distribution in the olfactory epithelium was preserved in a cryosection by a rapid-freezing technique (Reuter et al., 1998). Local element contents were determined by energy-dispersive X-ray microanalysis, and the corresponding original water content was derived from electron-scattering analysis. The *in vivo* E_{Cl} was estimated to be near 0 mV, with intracellular and mucosal Cl^- concentrations of 69 ± 19 mM and 55 ± 11 mM, respectively. This result indicates that the electrical potential drives Cl^- ions out of the cilia when the channels open at negative values of V_m. To establish and maintain the difference between resting voltage and E_{Cl}, OSNs have to accumulate Cl^- against an electrochemical gradient. In fact, Cl^- accumulation in OSN needs to be particularly effective because the cilia have such a small volume (~0.2 fL in rat). Without a robust accumulation pathway, the cilia would rapidly lose all intracellular Cl^- and would cease to function (Lindemann, 2001). However, this is not the case, as Ca^{2+}-dependent Cl^- currents can persist over several seconds (Reisert and Matthews, 1998; Boccaccio and Menini, 2007). Cl^--imaging experiments suggested that the mechanism which replenishes intracellular Cl^- is located in the apical part of OSNs (Kaneko et al., 2004). Local Cl^- concentrations were measured at various distances from the base of the cilia down to the distal dendritic segment. In this apical region of the OSN, a standing Cl^- gradient was established when the olfactory epithelium was kept in a Ringer solution containing 150 mM Cl^- (Fig. 20.6B). The direction of that gradient points to the cilia or the dendritic knob as the predominant site of Cl^- uptake. The apical intracellular Cl^- concentration was 54±4 mM in rat and 37±7 mM in mouse OSNs, consistent with the X-ray analysis by Reuter et al. (1998). Apical Cl^- accumulation is quite uniform in OSNs as long as they are in the epithelium. When cells are isolated, the mean Cl^- concentration decreases to about 30 mM and show higher variability (Fig. 20.6C). Thus, cell isolation compromises Cl^- accumulation in OSNs.

Active Cl^- accumulation is often accomplished by Cl^- transporters such as the Na^+-K^+-$2Cl^-$ cotransporter NKCC1, the Na^+-Cl^- cotransporter NCC or the Cl^-/HCO_3^- exchanger (Payne et al., 2003; Gamba, 2005; Alper, 2006). The individual contributions of these (or other) Cl^- uptake mechanisms to Cl^- accumulation in OSNs are not yet fully understood. Cl^- imaging showed that apical Cl^- uptake is, in part, Na^+-dependent and can be suppressed by bumetanide, an inhibitor of NKCC1 (Kaneko et al., 2004). Moreover, in isolated OSNs of NKCC1$^{-/-}$ knockout mice, the large Cl^- component of the receptor current was almost completely suppressed, an effect that could be mimicked by applying bumetanide to NKCC1$^{+/+}$ wildtype mice (Fig. 20.6D–F; Reisert et al., 2005). These results clearly indicate a contribution of NKCC1 to Cl^- accumulation. But other Cl^- transports must also be involved, as NKCC1$^{-/-}$ mice still generate odor-induced field potentials (electro olfactogram, EOG) which represent the combined receptor currents of many OSNs in a patch of olfactory epithelium. The EOG can almost completely be blocked by niflumic acid, an inhibitor of the OSN Cl^- channels (Kleene, 1993), indicating its origin in the Cl^--based receptor current (Nickell et al., 2007). EOGs can be measured in NKCC1$^{-/-}$ mice, and the animals display normal olfactory sensitivity in operant behavioral assays (Nickell et al., 2006; Smith et al., 2008). A pharmacological investigation of the EOG revealed that NKCC1 contributes roughly 50% to Cl^- accumulation, while the remaining 50% has yet to be identified (Nickell et al., 2007). The data, so far, support the hypothesis that OSNs employ various Cl^- uptake mechanisms to maintain an elevated intracellular Cl^- level as requisite for efficient chloride-based signal amplification.

IV. BEYOND THE NOSE

Chloride-based signal amplification is not a phenomenon specific for OSNs. It is a uniquely useful way of translating Ca^{2+} signals into changes in transmembrane voltage, because the amplification gain can be adjusted in each individual cell by setting the intracellular Cl^- concentration at an appropriate level. Such flexibility cannot be achieved with cation-based mechanisms, because the distribution of Na^+, K^+ and Ca^{2+} concentrations is determined by multiple physiological constraints. As two examples of chloride-based signal amplification outside the nose, I briefly describe in the following paragraph its well-documented role in vascular smooth-muscle contraction, as well as its hypothetical function in the generation of peripheral nociceptive signals.

The vascular smooth muscle of the arterial system adjusts blood pressure through narrowing or widening arteries under the control of various metabotropic systems. These include the α1 adrenergic pathway which promotes vasoconstriction via phospholipase C and the β2 adrenergic system which promotes vasodilation via adenylyl cyclase. Vasoconstriction is associated with the generation of a Ca^{2+} signal in the smooth muscle cell. Through activation of myosin light chain kinase, this Ca^{2+} signal triggers the phosphorylation of the regulatory light chain of myosin II which leads to contraction (Eddinger and Meer, 2007). The Ca^{2+} signal is initiated by IP_3-mediated Ca^{2+} release from intracellular stores, but the largest contribution comes from L-type Ca^{2+} channels which open upon depolarization of the plasma membrane. Ca^{2+}-activated Cl^- channels are expressed in smooth muscle cells and open when the intracellular Ca^{2+} concentration rises (Criddle et al., 1997). Both Ca^{2+} release and Ca^{2+} influx induce Cl^- currents (Fig. 20.7A). As in OSNs, the physiological role of these channels is linked to the

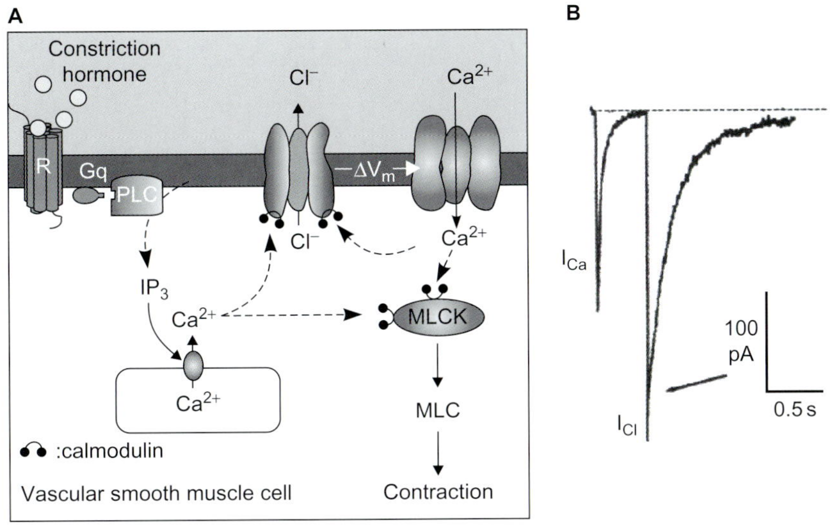

FIGURE 20.7 Chloride-based signal amplification in vascular smooth muscle cells. **A.** A vasoconstriction hormone activates phospholipase C (PLC) through a metabotropic hormone receptor (R) and a GTP-binding protein (Gq). This leads to Ca^{2+} release from Ca^{2+} stores sensitive to the second messenger inositol 1,4,5-trisphosphate (IP3) and the consequent opening of Ca^{2+}-activated Cl^- channels. Cl^- efflux causes depolarization of the plasma membrane (ΔVm) which promotes opening of voltage-gated Ca^{2+} channels. The resulting rise of the intracellular Ca^{2+} concentration both boosts the activity of the Cl^- channels and activates myosin light chain kinase (MLCK) which, in turn, phosphorylates myosin light chains (MLC) and, thereby, triggers contraction. Ca^{2+}-activated Cl^- currents can, thus, amplify depolarization and enhance Ca^{2+} influx through Ca^{2+} channels. **B.** Current recording from a portal vein myocyte illustrating how Ca^{2+} influx through a voltage-gated Ca^{2+} channel (I_{Ca}, elicited by a depolarizing pulse) can induce a large inward current carried by Cl^- (I_{Cl}). Altered from Greenwood et al. (2001). With permission of Blackwell Publishing.

specific mode of chloride homeostasis. The intracellular Cl^-concentration in vascular smooth muscle cell is in the range of 30–50 mM, the result of Cl^- accumulation by NKCC1, a Cl^-/HCO_3^- exchanger, and a third mechanism which is not yet identified at the molecular level (Chipperfield and Harper, 2000; Davis et al., 2000). With an E_{Cl} of -25 to -40 mV, opening of Cl^- channels causes depolarization from the resting voltage and is sufficient to activate voltage-gated Ca^{2+} channels, or to boost Ca^{2+} channel activation by increasing or prolonging depolarization. Extensive examinations have revealed that the Ca^{2+}-dependent Cl^- currents play an important role in the regulation of vasoconstriction, as they augment the excitatory signal and promote muscle contraction (Leblanc et al., 2005; see Chapter 13). Ca^{2+}-activated Cl^- channels in vascular smooth muscle and OSNs may be closely related at the structural level. It is possible that they are encoded by the same gene family. But this question is still unsolved. Members of the two candidate protein families CLCA (Loewen and Forsyth, 2005) and bestrophin (Hartzell et al., 2005; Leblanc et al., 2005) are being studied, but there is not yet any conclusive indication for a role of these channels in smooth muscle contraction (see Chapter 13 for further discussion). While the molecular identity of the channels remains enigmatic, important progress in the exploration of regulatory mechanisms that target Ca^{2+}-activated Cl^- channels came from experiments with smooth muscle cells (Fig. 20.7B). It appears that the channels are subject to feedback regulation through the Ca^{2+}-calmodulin-dependent protein kinase CaMKII. The kinase seems to phosphorylate the channels (or some regulatory protein) thereby causing a current reduction (Greenwood et al., 2001). The phosphatase calcineurin reverses this inhibitory effect (Ledoux et al., 2003). Most interestingly, such work on channel regulation yields insights into details of the gating machinery (e.g. Angermann et al., 2006) which may be critical for the molecular identification of the channel protein. Like in OSN research, that goal is paramount also in vascular smooth muscle research because knowledge of the channel genes will open the way to decisive experimentation with RNAi, knockout mice and protein biochemistry.

The peripheral endings of nociceptive neurons (nociceptors) detect stimuli which are intense enough to damage tissue (Woolf and Ma, 2007). These stimuli gate transduction channels, each of which is specialized for an individual stimulus quality. Thus, heat, cold, acid and mechanical stress directly operate transduction channels without any metabotropic amplification. The transduction channels are cation selective and conduct Ca^{2+} into the sensory ending (Oh, 2006). It has become apparent in recent years that depolarizing Cl^- currents contribute to the generation of sensory signals in the nociceptive terminals, as discussed in detail

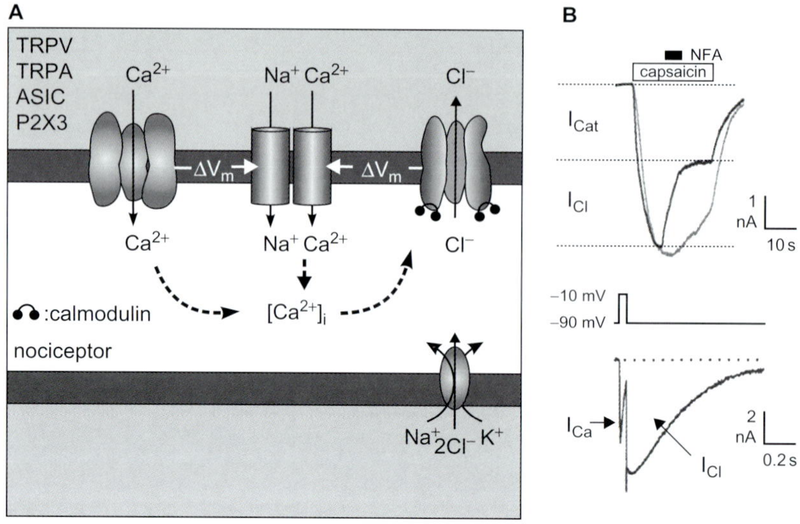

FIGURE 20.8 Chloride-based signal amplification in nociceptive endings. **A.** Noxious stimuli activate Ca^{2+}-permeable transduction channels (TRPV, TRPA: transient-receptor-potential type channels; ASIC: acid-sensitive ion channel family; P2X3: purinergic ATP-sensitive channels). The ensuing depolarization (ΔV_m) promotes opening of voltage-gated Na^+ and Ca^{2+} channels and, hence, electrical excitation of the nociceptor. Ca^{2+} influx also opens Cl^- channels which conduct a depolarizing Cl^- efflux and, thereby, amplify the excitatory receptor potential. A Na^+, K^+, $2Cl^-$ cotransporter supplies intracellular Cl^- for this amplification mechanism. **B.** Current recordings from isolated somatosensory neurons illustrate the relation between Ca^{2+} influx and Ca^{2+}-dependent Cl^- currents. The upper recording shows capsaicin-induced currents. Roughly half of the current amplitude is carried by cations (I_{cat}; TRPV1-mediated current); the rest is Cl^- current which is sensitive to niflumic acid (NFA, 100 μM). The lower recording was obtained from a chick DRG neuron (Kenyon and Goff, 1998). A 20 ms depolarizing pulse opens voltage-gated Ca^{2+} channels and generates Ca^{2+} influx (I_{Ca}). This Ca^{2+} signal activates Cl^- channels which greatly amplify the depolarizing current. With permission of Harcourt Brace.

in Chapter 22. In particular, it was shown that excitatory Cl^- currents induce pain behavior in mice (Ault and Hildebrand, 1994; Granados-Soto et al., 2005), that Cl^- uptake into sensory endings is necessary for normal pain behavior (Willis et al., 2004; Laird et al., 2004), and that somatosensory neurons express Ca^{2+}-activated Cl^- channels (Kenyon and Goff, 1998; Lee et al., 2005; see Chapter 13). Together, these and other publications (e.g. Price et al., 2005) support the hypothesis of chloride-based signal amplification in nociception (Fig. 20.8A). The depolarization caused by opening of transduction channels is accompanied by the generation of a Ca^{2+} signal in the sensory endings (Gover et al., 2003, 2007). This Ca^{2+} signal opens Cl^- channels which augment the depolarization by conducting Cl^- efflux. The high intracellular Cl^- concentration needed to drive Cl^- efflux is supported by NKCC1 (Alvarez-Leefmans et al., 1988, 2001; Sung et al., 2000; Granados-Soto et al., 2005; Rocha-González et al., 2008) but, like in OSNs and vascular myocytes, NKCC1 may not be the only Cl^- uptake pathway. Moreover, Ca^{2+} influx through T-type voltage-gated Ca^{2+} channels in the sensory endings (Jevtovic-Todorovic and Todorovic, 2006) promotes the activation of Cl^- channels which, in turn, boost the Ca^{2+} channels through depolarizing the membrane. Presently available evidence thus supports the notion that Ca^{2+}-induced Cl^- currents amplify the depolarizing receptor potential in nociceptive endings. Unfortunately, functional studies on the relation between transduction channels, voltage-gated Ca^{2+} channels and Ca^{2+}-activated Cl^- channels (Fig. 20.8B) are restricted to the somata of nociceptors isolated from dorsal root ganglia. The sensory endings themselves are not easily accessible to optical or electrophysiological investigations because they are small and embedded in tissue. However, new approaches are being developed to gain insight into the transduction processes in sensory fibers (e.g. Gover et al., 2007), and the molecular dissection of nocicepetive transduction pathways is well on its way (Oh, 2006). Details about Cl^- accumulation, Ca^{2+}-activated Cl^- channels and peripheral signal amplification will hopefully become available soon.

V. CONCLUSION

Chloride-based signal amplification is best understood in OSNs and in vascular smooth muscle cells. Both cell types employ Ca^{2+}-activated Cl^- channels to boost membrane depolarization under the control of intracellular Ca^{2+} signals. A similar amplification mechanism appears to operate in peripheral nociceptive endings of the pain system. While the genes that encode the Ca^{2+}-activated Cl^- channels are not yet identified, patch-clamp studies have revealed their basic functional properties. The critical role of these channels in sensory function and in the control of blood pressure provides the incentive for intense efforts in the search for their molecular identity.

A most promising development is the recent discovery that the TMEM16 family of membrane proteins forms calcium-activated chloride channels (Caputo et al., 2008; Schroeder et al., 2008; Yang et al., 2008), and that one isoform, TMEM16B, displays all the functional hallmarks of the olfactory channel (Pifferi et al., 2009). This finding may well be a breakthrough for the molecular identification of chloride-based signal amplification.

References

Alper, S.L. (2006). Molecular physiology of SLC4 anion exchangers. *Exp. Physiol.* **91**, 153–161.

Alvarez-Leefmans, F.J., Gamino, S.M., Giraldez, F., and Nogueron, I. (1988). Intracellular chloride regulation in amphibian dorsal root ganglion neurones studied with ion-selective microelectrodes. *J. Physiol.* **406**, 225–246.

Alvarez-Leefmans, F.J., León-Olea, M., Mendoza-Sotelo, J., Alvarez, F.J., Antón, B., and Garduno, R. (2001). Immunolocalization of the Na(+)-K(+)-2Cl(−) cotransporter in peripheral nervous tissue of vertebrates. *Neurosci.* **104**, 569–582.

Angermann, J.E., Sanguineti, A.R., Kenyon, J.L., Leblanc, N., and Greenwood, I.A. (2006). Mechanism of the inhibition of Ca^{2+}-activated Cl^- currents by phosphorylation in pulmonary arterial smooth muscle cells. *J. Gen. Physiol.* **128**, 73–87.

Ault, B. and Hildebrand, L.M. (1994). $GABA_A$ receptor-mediated excitation of nociceptive afferents in the rat isolated spinal cord-tail preparation. *Neuropharmacol.* **33**, 109–114.

Bhandawat, V., Reisert, J., and Yau, K.W. (2005). Elementary response of olfactory receptor neurons to odorants. *Science* **308**, 1931–1934.

Biskup, C., Kusch, J., Schulz, E., Nache, V., Schwede, F., Lehmann, F., Hagen, V., and Benndorf, K. (2007). Relating ligand binding to activation gating in CNGA2 channels. *Nature* **446**, 440–443.

Boccaccio, A. and Menini, A. (2007). Temporal development of cyclic nucleotide-gated and Ca^{2+}-activated Cl^- currents in isolated mouse olfactory sensory neurons. *J. Neurophysiol.* **98**, 153–160.

Bönigk, W., Bradley, J., Sesti, F., Müller, F., Boekhoff, I., Ronnett, G.V., Kaupp, U.B., and Frings, S. (1999). The native rat olfactory cyclic nucleotide-gated channel is composed of three distinct subunits. *J. Neuroscience.* **19**, 5332–5347.

Bradley, J., Bönigk, W., Yau, K.-W., and Frings, S. (2004). Calmodulin permanently associates with rat olfactory CNG channels under native conditions. *Nature Neurosci.* **7**, 705–710.

Caputo, A., Caci, E., Ferrera, L., Pedemonte, N., Barsanti, C., Sondo, E., Pfeffer, U., Ravazzolo, R., Zegarra-Moran, O., and Galietta, L.J.V. (2008). TMEM16A, a membrane protein associated with calcium-dependent chloride channel activity. *Science.* **322**, 590–594.

Chipperfield, A.R. and Harper, A.A. (2000). Chloride in smooth muscle. *Prog. Biophys. Mol. Biol.* **74**, 175–221.

Criddle, D.N., de Moura, R.S., Greenwood, I.A., and Large, W.A. (1997). Inhibitory action of niflumic acid on noradrenaline- and 5-hydroxytryptamine-induced pressor responses in the isolated mesenteric vascular bed of the rat. *Br. J. Pharmacol.* **120**, 813–818.

Davis, J.P., Chien, P.F., Chipperfield, A.R., Gordon, A., and Harper, A.A. (2000). The three mechanisms of intracellular chloride accumulation in vascular smooth muscle of human umbilical and placental arteries. *Pflügers Arch. – Eur. J. Physiol.* **441**, 150–154.

Dzeja, C., Hagen, V., Kaupp, U.B., and Frings, S. (1999). Ca^{2+} permeation in cyclic nucleotide-gated channels. *EMBO J.* **18**, 131–144.

Eddinger, T.J. and Meer, D.P. (2007). Myosin II isoforms in smooth muscle: heterogeneity and function. *Am. J. Physiol. Cell. Physiol.* **293**, C493–C508.

Elsaesser, R. and Paysan, J. (2007). The sense of smell, its signalling pathways, and the dichotomy of cilia and microvilli in olfactory sensory cells. *BMC Neurosci.* **8** (Suppl. 3), S1.

Firestein, S., Picco, C., and Menini, A. (1993). The relation between stimulus and response in olfactory receptor cells of the tiger salamander. *J. Physiol. Lond.* **468**, 1–10.

Friedrich, R.W. (2006). Mechanisms of odor discrimination: neurophysiological and behavioral approaches. *Trends Neurosci.* **29**, 40–47.

Gamba, G. (2005). Molecular physiology and pathophysiology of electroneutral cation-chloride cotransporters. *Physiol. Rev.* **85**, 423–493.

Gover, T.D., Kao, J.P.Y., and Weinreich, D. (2003). Calcium signaling in single peripheral sensory nerve terminals. *J. Neurosci.* **23**, 4793–4797.

Gover, T.D., Moreira, T.H.V., Kao, J.P.Y., and Weinreich, D. (2007). Calcium regulation in individual peripheral sensory nerve terminals of the rat. *J. Physiol. Lond.* **578**, 481–490.

Granados-Soto, V., Arguelles, C.F., and Alvarez-Leefmans, F.J. (2005). Peripheral and central antinociceptive action of Na^+-K^+-$2Cl^-$ cotransporter blockers on formalin-induced nociception in rats. *Pain* **114**, 231–238.

Greenwood, I.A., Ledoux, J., and Leblanc, N. (2001). Differential regulation of Ca(2+)-activated Cl(−) currents in rabbit arterial and portal vein smooth muscle cells by Ca(2+)-calmodulin-dependent kinase. *J. Physiol. (Lond.)* **534**, 395–408.

Hartzell, C., Putzier, I., and Arreola, J. (2005). Calcium-activated chloride channels. *Annu. Rev. Physiol.* **67**, 719–758.

Jevtovic-Todorovic, V. and Todorovic, S.M. (2006). The role of peripheral T-type calcium channels in pain transmission. *Cell Calcium* **40**, 197–203.

Kaneko, H., Putzier, I., Frings, S., Kaupp, U.B., and Gensch, T. (2004). Chloride accumulation in mammalian olfactory sensory neurons. *J. Neurosci.* **24**, 7931–7938.

Kellenberger, S. and Schild, D. (2002). Epithelial sodium channel/degenerin family of ion channels: a variety of functions for a shared structure. *Physiol. Rev.* **82**, 735–767.

Kenyon, J.L. and Goff, H.R. (1998). Temperature-dependence of Ca^{2+} current, Ca^{2+}-activated Cl^- current and Ca^{2+} transients in sensory neurons. *Cell Calcium* **24**, 35–48.

Kleene, S.J. (1993). Origin of the chloride current in olfactory transduction. *Neuron* **11**, 123–132.

Kleene, S.J. (1997). High-gain, low-noise amplification in olfactory transduction. *Biophys. J.* **73**, 1010–1017.

Kleene, S.J. and Gesteland, R.C. (1981). Dissociation of frog olfactory epithelium with N-ethylmaleimide. *Brain Res.* **229**, 536–540.

Kleene, S.J. and Gesteland, R.C. (1991). Calcium-activated chloride conductance in frog olfactory cilia. *J. Neurosci.* **11**, 3624–3629.

Kurahashi, T. and Yau, K.W. (1993). Co-existence of cationic and chloride components in odorant-induced current of vertebrate olfactory receptor cells. *Nature* **363**, 71–74.

Laird, J.M.A., García-Nicas, E., Delpire, E.J., and Cervero, F. (2004). Presynaptic inhibition and spinal pain processing in mice: a possible role for the NKCC1 cation-chloride co-transporter in hyperalgesia. *Neurosci. Lett.* **361**, 200–203.

Leblanc, N., Ledoux, J., Saleh, S., Sanguinetti, A., Angermann, J., O'Driscoll, K., Brotton, F., Perrino, B.A., and Greenwood, I.A. (2005). Regulation of calcium-activated chloride channels in smooth muscle cells: a complex picture is emerging. *Can. J. Physiol. Pharmacol.* **83**, 541–556.

Ledoux, J., Greenwood, I.A., Villeneuve, L.R., and Leblanc, N. (2003). Modulation of Ca^{2+}-dependent Cl^- channels by calcineurin in rabbit coronary arterial myocytes. *J. Physiol.* **552**, 701–714.

Lee, M.-G., MacGlashan, D.W., and Undem, B.J. (2005). Role of chloride channels in bradykinin-induced guinea pig airway vagal C-fibre activation. *J. Physiol.* **566**, 205–212.

Leinders-Zufall, T., Rand, M.N., Shepherd, G.M., Greer, C.A., and Zufall, F. (1997). Calcium entry through cyclic nucleotide-gated channels in individual cilia of olfactory receptor cells: spatiotemporal dynamics. *J. Neurosci.* **17**, 4136–4148.

Lindemann, B. (2001). Predicted profiles of ion concentrations in olfactory cilia in the steady state. *Biophys. J.* **80**, 1712–1721.

Loewen, M.E. and Forsyth, G.W. (2005). Structure and function of CLCA proteins. *Physiol. Rev.* **85**, 1061–1092.

Lowe, G. and Gold, G.H. (1993). Nonlinear amplification by calcium-dependent chloride channels in olfactory receptor cells. *Nature* **366**, 283–286.

Mombaerts, P. (2006). Axonal wiring in the mouse olfactory system. *Annu. Rev. Cell. Dev. Biol.* **22**, 713–737.

Mori, K., Takahashi, Y.K., Igarashi, K.M., and Yamaguchi, M. (2006). Maps of odorant molecular features in the mammalian olfactory bulb. *Physiol. Rev.* **86**, 409–433.

Nache, V., Schulz, E., Zimmer, T., Kusch, J., Biskup, C., Koopmann, R., Hagen, V., and Benndorf, K. (2005). Activation of olfactory-type cyclic nucleotide-gated channels is highly cooperative. *J. Physiol.* **569**, 91–102.

Neher, E. (1998). Vesicle pools and Ca^{2+} microdomains: new tools for understanding their roles in neurotransmitter release. *Neuron* **20**, 2100–2113.

Nickell, W.T., Kleene, N.K., Gesteland, R.C., and Kleene, S.J. (2006). Neuronal chloride accumulation in olfactory epithelium of mice lacking NKCC1. *J. Neurophysiol.* **95**, 2003–2006.

Nickell, W.T., Kleene, N.K., and Kleene, S.J. (2007). Mechanisms of neuronal chloride accumulation in intact mouse olfactory epithelium. *J. Physiol.* **583**, 1005–1020.

Oh, U. (2006). The nociceptive membrane. *Curr. Top. Membr.* **57** Academic Press, Amsterdam.

Payne, J.A., Rivera, C., Voipio, J., and Kaila, K. (2003). Cation-chloride co-transporters in neuronal communication, development and trauma. *Trends Neurosci.* **26**, 199–206.

Pifferi, S., Pascarella, G., Boccaccio, A., Mazzatenta, A., Gustincich, S., Menini, A., and Zucchelli, S. (2006). Bestrophin-2 is a candidate calcium-activated chloride channel involved in olfactory transduction. *Proc. Nat. Acad. Sci. USA* **103**, 12929–12934.

Pifferi, S., Dibattista, M., and Menini, A. (2009). TMEM16B induces chloride currents activated by calcium in mammalian cells. *Plügers Arch. – Eur. J. Physiol.* (in press).

Price, T.J., Cervero, F., and de Koninck, Y. (2005). Role of cation-chloride-cotransporters (CCC) in pain and hyperalgesia. *Curr. Top. Med. Chem.* **5**, 547–555.

Qu, Z., Fischmeister, R., and Hartzell, C. (2004). Mouse bestrophin-2 is a bona fide Cl^- channel: identification of a residue important in anion binding and conduction. *J. Gen. Physiol.* **123**, 323–325.

Reisert, J. and Matthews, H.R. (1998). Na^+-dependent Ca^{2+} extrusion governs response recovery in frog olfactory receptor cells. *J. Gen. Physiol.* **112**, 529–535.

Reisert, J., Bauer, P.J., Yau, K.W., and Frings, S. (2003). The Ca-activated Cl channel and its control in rat olfactory receptor neurons. *J. Gen. Physiol.* **122**, 349–363.

Reisert, J., Lai, J., Yau, K.W., and Bradley, J. (2005). Mechanism of excitatory Cl^- response in mouse olfactory receptor neurons. *Neuron* **45**, 553–561.

Reuter, D., Zierold, K., Schröder, W., and Frings, S. (1998). A depolarizing chloride current contributes to chemo-electrical transduction in olfactory sensory neurons *in situ*. *J. Neurosci.* **18**, 6623–6630.

Rocha-González, H.I., Mao, S., and Alvarez-Leefmans, F.J. (2008). Na^+,K^+,$2Cl^-$ cotransport and intracellular chloride regulation in rat primary sensory neurons: thermodynamic and kinetic aspects. *J. Neurophysiol.* **100**, 169–184.

Sato, K. and Suzuki, N. (2000). The contribution of a Ca(2+)-activated Cl^- conductance to amino-acid-induced inward current responses of ciliated olfactory neurons of the rainbow trout. *J. Exp. Biol.* **203**, 253–262.

Schaefer, A.T. and Margrie, T.W. (2007). Spatiotemporal representations in the olfactory system. *Trends Neurosci.* **30**, 92–100.

Schroeder, B.C., Cheng, T., Jan, Y.N., and Jan, L.Y. (2008). Expression cloning of TMEM16A as a calcium-activated chloride channel subunit. *Cell.* **134**, 1019–1029.

Sicard, G. and Holley, A. (1984). Receptor cell responses to odorants: similarities and differences among odorants. *Brain Res.* **292**, 282–296.

Smith, D.W., Thach, S., Marshall, E.L., Mendoza, M.G., and Kleene, S.J. (2008). Mice lacking NKCC1 have normal olfactory sensitivity. *Physiol. Behav.* **93**, 44–49.

Sung, K.W., Kirby, M., McDonald, M.P., Lovinger, D.M., and Delpire, E. (2000). Abnormal GABAA receptor-mediated currents in dorsal root ganglion neurons isolated from Na-K-2Cl cotransporter null mice. *J. Neurosci.* **20**, 7531–7538.

Suzuki, N. (2006). The Drosophila tweety family: molecular candidates for large-conductance Ca^{2+}-activated Cl^- channels. *Exp. Physiol.* **91**, 141–147.

Takeuchi, H. and Kurahashi, T. (2005). Mechanisms of signal amplification in the olfactory sensory cilia. *J. Neurosci.* **25**, 11084–11091.

Venkatachalam, K. and Montell, C. (2007). TRP channels. *Ann. Rev. Biochem.* **76**, 387–417.

Verhagen, J.V., Wesson, D.W., Netoff, T.I., White, J.A., and Wachowiak, M. (2007). Sniffing controls an adaptive filter of sensory input to the olfactory bulb. *Nature Neurosci.* **10**, 631–639.

Vosshall, L.B. and Stocker, R.F. (2007). Molecular architecture of smell and taste in Drosophila. *Ann. Rev. Neurosci.* **30**, 505–533.

Willis, E.F., Clough, G.F., and Church, M.K. (2004). Investigation into the mechanisms by which nedocromil, furosemide and bumetanide inhibit the histamin-induced itch and flare response in human skin *in vivo*. *Clin. Exp. Allergy* **34**, 450–455.

Woolf, C.J. and Ma, Q. (2007). Nociceptors – noxious stimulus detectors. *Neuron* **55**, 353–364.

Yang, Y.D., Cho, H., Koo, J.Y., Tak, M.H., Cho, Y., Shim, W.-S., Park, S.P., Lee, J., Lee, B., Kim, B.-M., Raouf, R., Shin, Y.K., and Oh, U. (2008). TMEM16A confers receptor-activated calcium-dependent chloride conductance. *Nature* **455**, 1210–1215.

Zheng, J. and Zagotta, W.N. (2004). Stoichiometry and assembly of olfactory cyclic nucleotide-gated channels. *Neuron* **43**, 411–421.

CHAPTER

21

Cochlear and Vestibular Function and Dysfunction

Daniel C. Marcus and Philine Wangemann

OUTLINE

I. INTRODUCTION

Sound, acceleration and gravity are mechanical stimuli that are detected by the cochlea and the vestibular labyrinth of the inner ear. The inner ear transduces the mechanical stimuli into electrical signals that are ultimately encoded into neuronal action potentials. The encoded information is transmitted by afferent neurons to central processing regions in the brain. The sensory process is guarded by intricate feedback mechanisms that include systems of efferent innervation that terminate on the afferent nerves and on the sensory cells in the inner ear. The mechanosensory transduction in the cochlea and in the vestibular labyrinth depends on large electrochemical gradients and an unusual fluid composition in the luminal compartment of the inner ear. This chapter focuses on cellular ion transport mechanisms in the inner ear, their dependence on Cl^- and their physiological role in generating the major electrochemical gradients on which sensory transduction depends.

II. SENSORY TRANSDUCTION

The inner ear consists of the coiled cochlea and the vestibular labyrinth. Three semicircular canals that originate from the utricle, the saccule and the endolymphatic duct and sac make up the vestibular labyrinth (Fig. 21.1). The luminal fluid in the inner ear is called endolymph. Endolymph is an unusual extracellular fluid in that it contains in most parts of the inner ear a high K^+ and a low Na^+ and Ca^{2+} concentration (Table 21.1). Endolymph is enclosed by a heterogeneous epithelium, which separates endolymph from perilymph. Perilymph is a more conventional extracellular fluid that contains Na^+, K^+ and Ca^{2+} concentrations similar to other extracellular fluids, such as plasma and cerebrospinal fluid (Table 21.1). The utricle, the ampullae of the three semicircular canals and the saccule contain patches of sensory cells that detect rotational and linear acceleration including gravity. Sensory cells are also placed along the length of the cochlea for the detection of sound. Mechanical stimulation of the inner ear sensory organs is derived from the appropriate stimuli, which is sound for the cochlea and acceleration including gravity for the vestibular labyrinth.

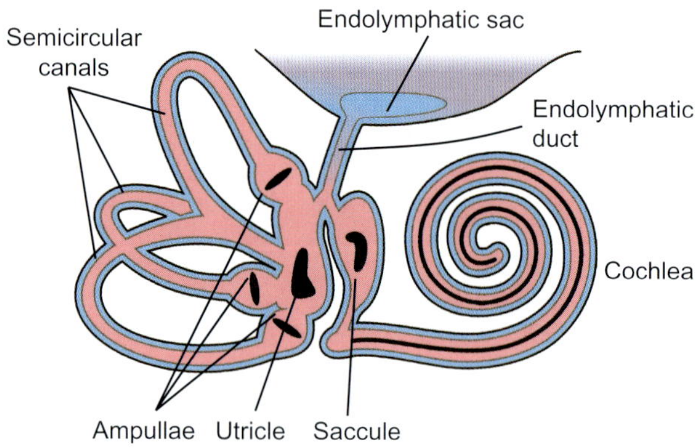

FIGURE 21.1 Diagram of inner ear organs, including the cochlea, vestibular labyrinth (utricle, saccule, ampullae of the semicircular canals) and endolymphatic sac.

A. Cochlear Transduction

Sound waves are pressure variations consisting of compressions and rarefactions of air molecules. Sound enters the outer ear, sets the tympanic membrane in motion and is conducted by middle ear bones to the oval window, from where it enters the fluid–filled cochlea (Fig. 21.2, top and middle panels).

The mechanical properties of the cochlea ensure the generation of a standing wave that displaces the basilar membrane within the cochlea (Fig. 21.2, bottom panel). The envelope of this standing wave peaks at a place along the cochlea that depends on the frequency of the sound. High frequencies peak at the base of the cochlea and low frequencies at the apex. The cochlea thus functions as a mechanical frequency analyzer. Sensory cells are resting on the basilar membrane in a structure called the organ of Corti that stretches along the entire length of the cochlea (Fig. 21.3). Mechanical stimulation of sensory cells located at the base of the cochlea leads to the sensation of high frequency sound, and stimulation of sensory cells located in the apex of the cochlea leads to the sensation of low frequency sound. This tonotopic organization of the cochlea is at least in part maintained throughout the central processing of auditory stimuli.

The cochlea consists of three open fluid compartments, scala vestibuli, scala media and scala tympani (Fig. 21.3). Scala media is lined by about 12 different types of epithelial cells that include the sensory cells in the organ of Corti. The organ of Corti consists of inner hair cells, which are the sensory cells, and of outer hair cells, which amplify the mechanical stimulus (see below). In addition, the organ of Corti contains a number of supporting cells, rests on the basilar membrane and is covered by the tectorial membrane (Fig. 21.3). The basilar and tectorial membranes provide mechanical input to the hair cells.

TABLE 21.1 Fluid composition of perilymph, cochlear and vestibular endolymph, as well as endolymph of the endolymphatic sac, cerebrospinal fluid and plasma (Wangemann, 2008)

	Cochlear perilymph	Cochlear endolymph	Utricular endolymph	Sac endolymph	Cerebrospinal fluid	Plasma
Na^+ (mM)	148	1.3	9	129	149	145
K^+ (mM)	4.2	157	149	10	3.1	5.0
Cl^- (mM)	119	132	–	124	129	106
HCO_3^- (mM)	21	31	–	–	19	18
Ca^{2+} (mM)	1.3	0.023	0.28	–	–	2.6
Protein (mg/dl)	178	38	–	–	24	4238
pH	7.3	7.5	7.5	6.9	7.3	7.3

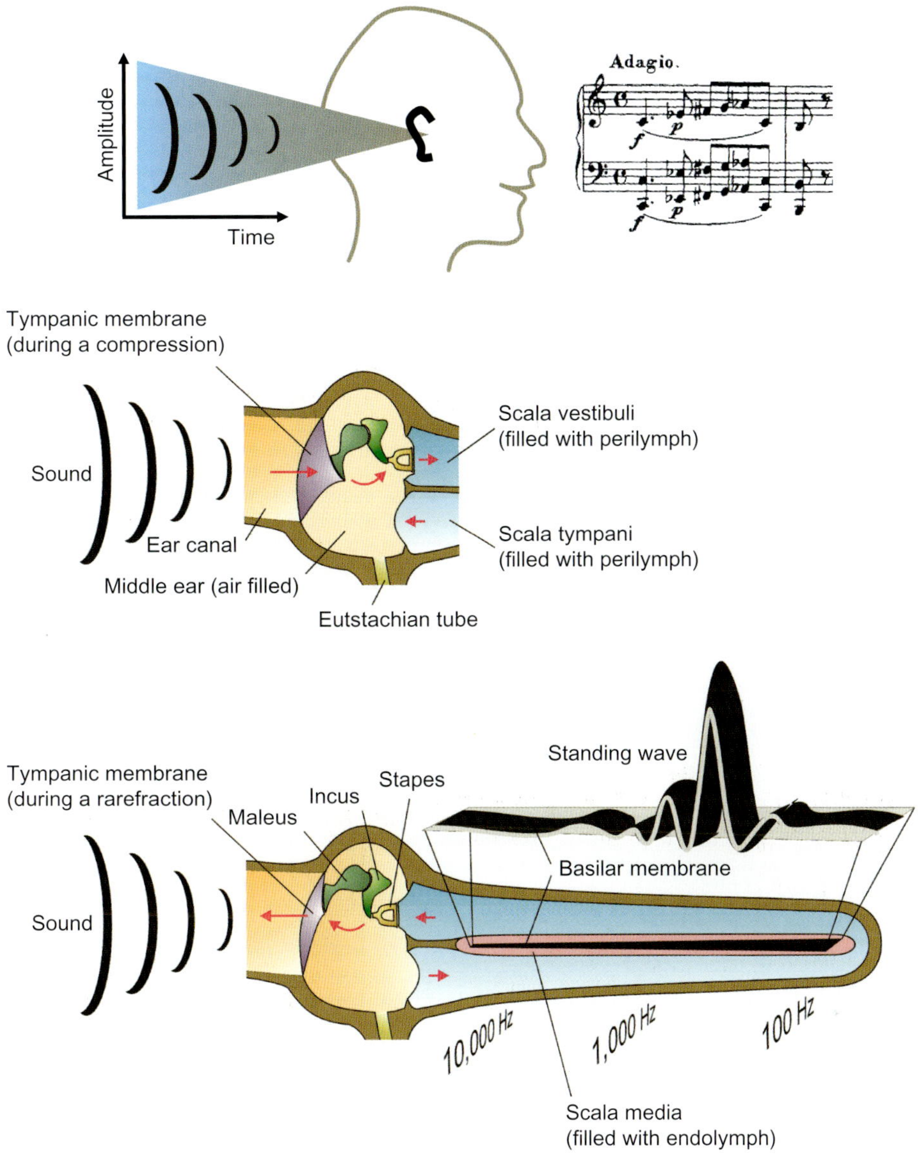

FIGURE 21.2 Diagrams of the physical pathway and effects of sounds from the environment to the cochlea. (Upper panel) Sound waves enter the outer ear and (middle panel) are transmitted by the tympanic membrane to the middle ear ossicles. The sound waves are then transmitted to scala vestibuli of the inner ear through the oval window. Acoustic energy enters the cochlea (lower panel; diagrammed uncoiled) where the frequency content of the sound waves is analyzed by the creation of standing waves along the basilar membrane and excites receptors and associated neurons at locations that code for the frequencies comprising the sound.

Sound-induced vibrations of the basilar membrane cause bending of the stereocilia on the apical membrane of the hair cells, which modulates the ionic currents through the transduction channels (Fig. 21.4). The modulated current through the transduction channel at the apical side of the hair cells leads to receptor potentials, which are changes in the membrane voltage. Stereocilia in the apical membrane of hair cells are oriented in a pattern resembling organ pipes. Bending of the stereocilia toward the longer stereocilia opens the transduction channel whereas bending toward the shorter stereocilia closes the channel.

The main consequence of receptor potentials in inner hair cells is the modulation of the release of the neurotransmitter glutamate and stimulation of type I afferent dendrites (Fig. 21.4). Type I afferent dendrites transmit the primary acoustic input to the auditory brainstem. The activity of afferent dendrites is modulated by efferent fibers that terminate on the afferents near the base of the inner hair cells. The main consequence of receptor potentials in inner hair cells differs drastically from that in outer hair cells.

The main consequences of receptor potentials in outer hair cells are piezo-electric length changes of

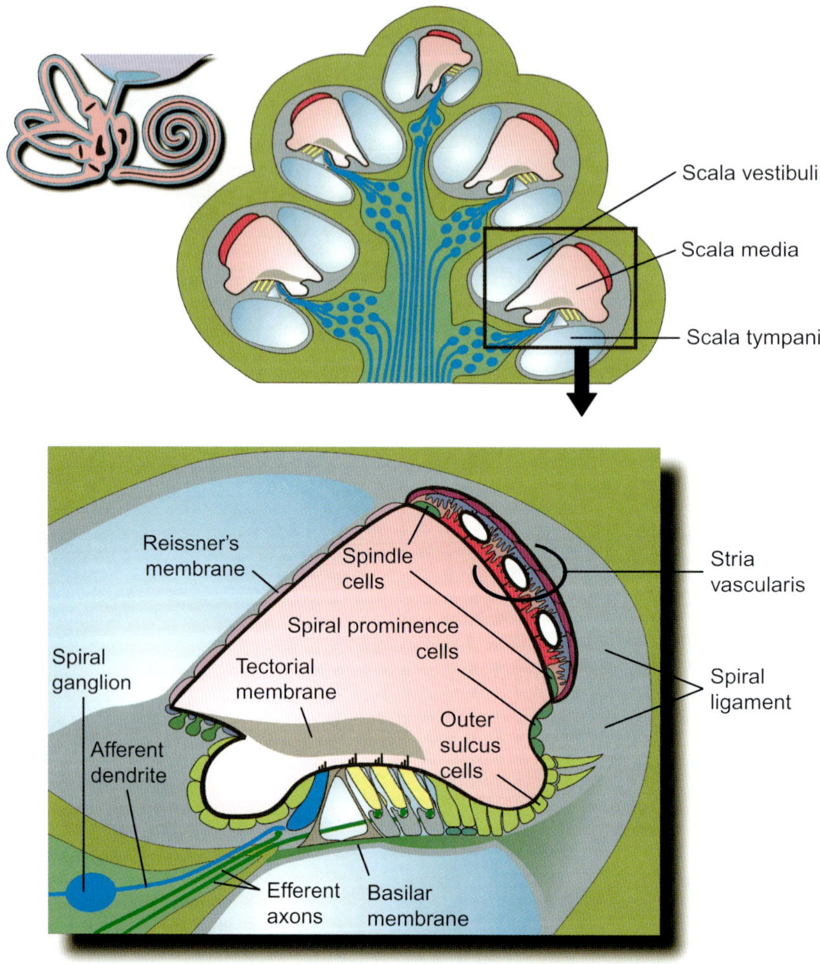

FIGURE 21.3 Diagram of a cross-section of the coiled cochlea. The cochlear lumen (pink) is filled with endolymph, a high-K/low-Na^+/low-Ca^{2+} extracellular fluid, which is produced and maintained by the epithelial cells bounding the cochlear lumen and which sustains the acoustic transduction process.

the outer hair cell that lead to an amplification of the sound-induced vibrations of the basilar membrane (Fig. 21.4). Outer hair cells express in their lateral wall the voltage-sensitive protein prestin (SLC26A5). The high density of prestin molecules causes a shortening of the cell during membrane-potential depolarizations and a lengthening during membrane-potential hyperpolarizations. The resulting amplification of basilar membrane vibrations is necessary for the high sensitivity and the sharpness of frequency discrimination of the mammalian cochlea.

B. Vestibular Transduction

The vestibular labyrinth houses multiple sensory organs for the detection of rotational and linear acceleration including gravity. The three semicircular canals in association with their ampullae, which house the sensory hair cells (Fig. 21.1), detect rotational acceleration and provide directional information by virtue of the orientation of the three canals in x–y–z orthogonal planes. Stereocilia of hair cells protrude into gelatinous cupulae that span the lumen of the ampullae (Fig. 21.5). Rotational head movements against the luminal fluid resting in the semicircular canals lead to a displacement of the stereocilia. The utricle and saccule house sensory hair cells to detect linear acceleration including gravity. Stereocilia of these hair cells protrude into an extracellular matrix that is fortified by $CaCO_3$ crystals that comprise an inertial mass during linear acceleration (Fig. 21.5). The spatial orientation of the hair cells imparts directionality to the neural signal emanating from its basal synapse.

Acceleration-induced bending of the stereocilia in the apical membrane of the vestibular hair cells modulates the ionic currents through transduction channels that are thought to function analogous to their equivalents in the cochlea. The modulated currents through

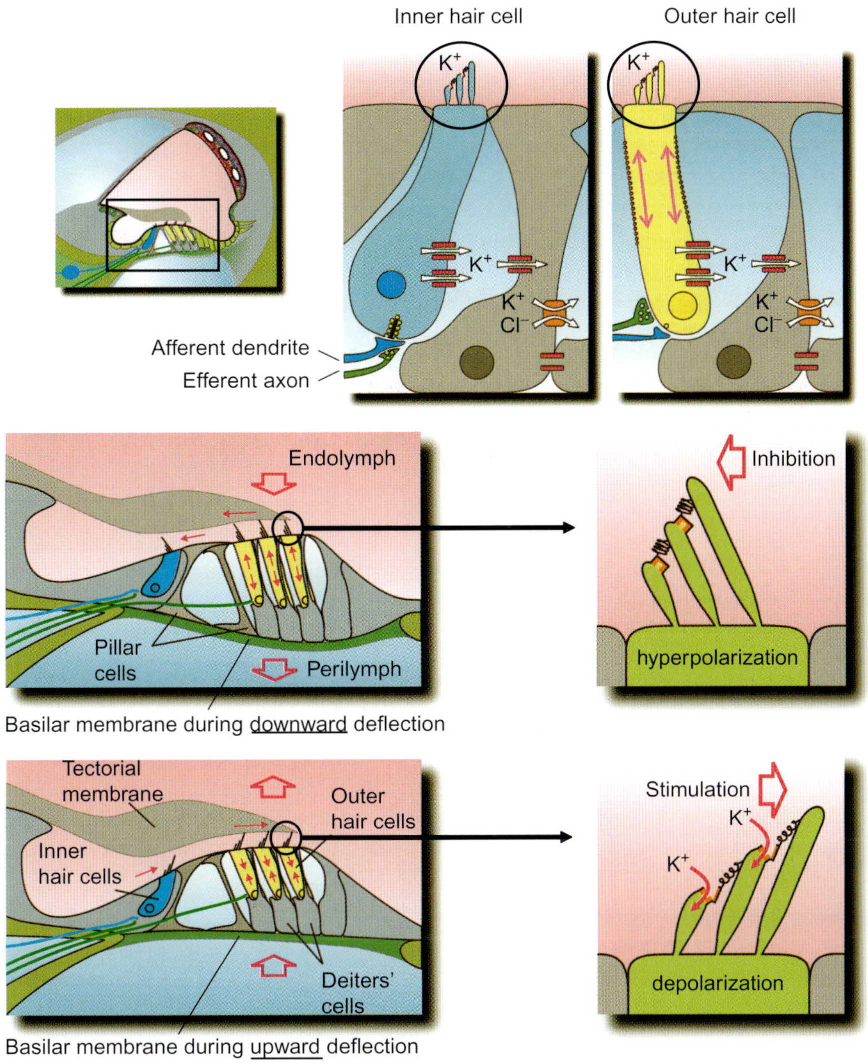

FIGURE 21.4 Diagram of the cochlear sensory structure, the organ of Corti and its inner and outer hair cells. (Top panel) The location of the organ of Corti (rectangle; left) and partial detail of an inner hair cell with supporting cell (middle) and an outer hair cell with supporting Deiters' cell (right) with salient transporters involved in transduction and putative K^+ buffering (see text). (Middle panel) Downward deflection of basilar membrane (left) in response to sound is shown with movement of the outer hair cell stereocilia (right), leading to afferent inhibition. (Bottom panel) Upward deflection of basilar membrane (left) in response to sound is shown with movement of the outer hair cell stereocilia (right), leading to afferent stimulation. Arrows indicate movements of tectorial membrane, subtectorial fluid, basilar membrane (open arrow), K^+ flux through transduction channels at the tips of stereocilia and length changes in outer hair cells.

the transduction channels lead to receptor potentials which modulate the activity of the afferent dendrites innervating vestibular hair cells (Fig. 21.6). There are two types of hair cells in the vestibular system. Type I vestibular hair cells are bottle shaped and contacted by a calyx-shaped afferent nerve terminal and type II vestibular hair cells are more cylindrical and contacted by several simple bouton-shaped afferent nerve terminals. The activity of afferent dendrites is modulated by efferent fibers that terminate on the afferent dendrites or directly on the hair cells.

III. IONIC BASIS OF SENSORY TRANSDUCTION IN THE COCHLEA AND THE VESTIBULAR LABYRINTH

The current through the transduction channel is mainly carried by K^+ since the channel is cation selective and because K^+ is the most abundant cation in endolymph (Table 21.1). A small fraction of the current through the transduction channel is carried by Ca^{2+}. This component of the current is required for

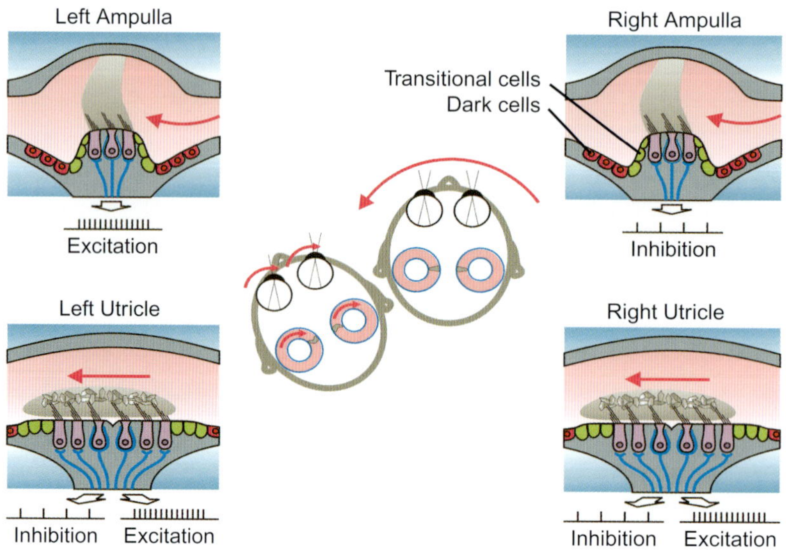

FIGURE 21.5 Vestibular transduction of rotational and linear acceleration. Rotation of the head (center panel) is detected by the ampullae of the semicircular canals (upper left and right panels) and the signals are coupled centrally to control of the eye muscles (vestibulo-optical reflex; VOR). Linear acceleration (lower left and right panels) excites or inhibits neurons projecting to the brain.

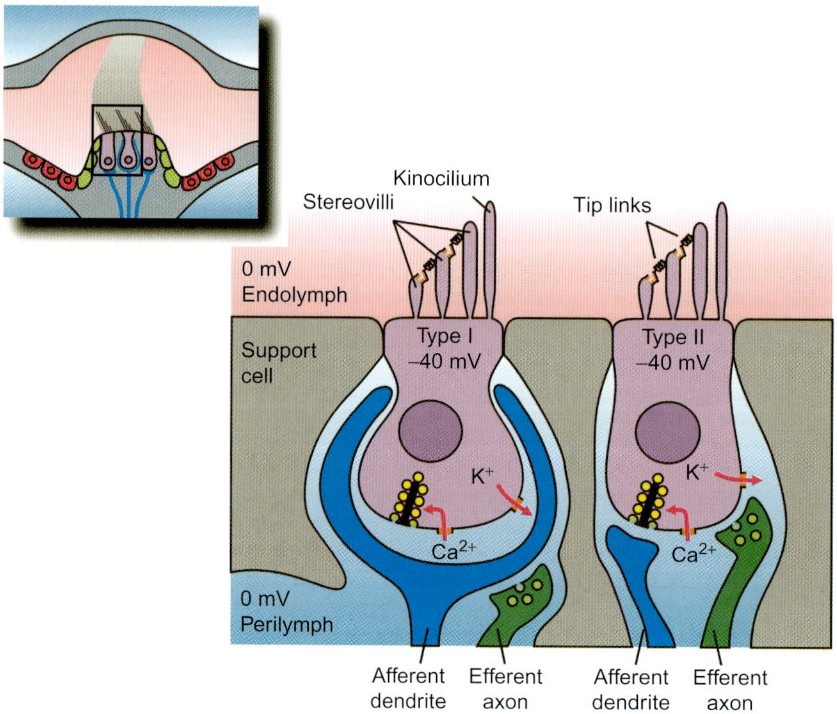

FIGURE 21.6 Vestibular hair cells and innervation. Type I and Type II hair cells are shown with their salient K^+ and Ca^{2+} ion channels and neural connections.

the control of Ca^{2+}-dependent adaptation processes in the stereocilia (Hudspeth et al., 2000).

Sensory transduction depends on the electrochemical driving force that drives K^+ across the transduction channel into the hair cells. This driving force is larger in the cochlea than in the vestibular system due to the presence of a large lumen-positive transepithelial voltage, the endocochlear potential, in the cochlea. The magnitude of the endocochlear potential is 80–100 mV. This voltage contributes to the exquisite sensitivity of the cochlea. A transepithelial voltage of equivalent magnitude is lacking in the vestibular system. K^+ enters the hair cells via the transduction channel and leaves the hair cells via K^+ channels in the basolateral membrane. Since K^+ is the main cation in the cytosol, the relative change of intracellular ionic concentrations during sensory transduction is minimal.

K^+ is secreted into endolymph at great energetic expense (see below). The expense is justified by the advantage of utilizing K^+ to carry the transduction current rather than Na^+, which is the most abundant cation in most other extracellular fluids (Table 21.1). If the transduction current were to be carried by Na^+, Na^+ loading of the cytosol would require Na^+/K^+-ATPase to remove the Na^+ across the basolateral cell membrane, which would require the support of highly active metabolic machinery. High metabolic activity would require proximity to vasculature and mechanical noise associated with vascular blood flow incompatible with sensitive stimulus detection. A solution to this problem is to utilize K^+ as charge carrier for sensory transduction. Both influx of K^+ into and efflux of K^+ from the hair cells occur down electrochemical gradients and the location of sensory transduction in the organ of Corti is separated from the location of K^+ secretion in stria vascularis, an epithelial structure located in the lateral wall of the cochlea (Fig. 21.7).

The use of K^+ as the main charge carrier has a potential problem for sensitive neural transmission since dendritic nerve endings maintain their resting membrane potential with K^+ channels. Elevations of K^+ in the small extracellular spaces between the basolateral membrane of the hair cells and the synaptic terminals can depolarize neuronal membranes and interfere with neuronal transmission. Elevations of K^+ near cochlear and vestibular hair cells have been observed during sound or mechanical stimulation (Johnstone et al., 1989; Valli et al., 1990). The magnitude of these elevations may be limited by K^+ buffering. Mechanisms for K^+ buffering (see below) may include simple diffusion into open fluid spaces such as the scala tympani in the cochlea and uptake of K^+ into neighboring supporting cells and subsequent discharge at one or more sites remote from the hair cells and nerve terminals (Spicer and Schulte, 1996).

A. Cl^- Dependence of Sensory Transduction

Sensory transduction in the cochlea and the vestibular labyrinth depends on Cl^- on all levels including amplification of basilar membrane vibrations and the generation of the endocochlear potential in the cochlea as well as cochlear and vestibular neurotransmission and ion homeostasis necessary for providing K^+ as the main charge carrier of sensory transduction.

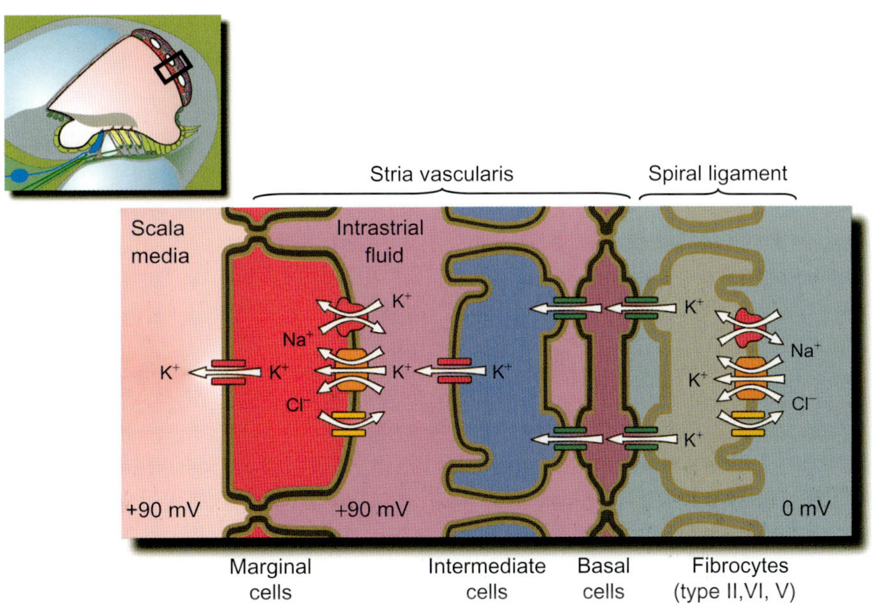

FIGURE 21.7 Cell transport model for K^+ secretion and generation of the endocochlear potential. Stria vascularis (upper panel, rectangle) secretes K^+ that is first taken up by fibrocytes in the spiral ligament of the lateral wall and subsequently passed through a syncytium of basal cells and intermediate cells to the intrastrial space between the marginal and intermediate cells. K^+ is taken up from the intrastrial space by the marginal cells and secreted into endolymph of scala media. The transporters responsible for K^+ secretion by the marginal cells are identical to those employed by vestibular dark cells. The endocochlear potential is generated by K^+ channels in the intermediate cells (see text).

B. Cl^- Dependence of Cochlear Amplification

The outer hair cell motility that is mediated by prestin is dependent on the intracellular anion concentration, although the biophysical mechanism of this dependence is currently a matter of discussion (Ashmore 2008; He et al., 2006; Muallem and Ashmore, 2006; Oliver et al., 2001). The putative intracellular anion binding site is normally thought to be occupied by anions such as Cl^- or HCO_3^- (Santos-Sacchi et al., 2006). The mechanism by which Cl^- or HCO_3^- support prestin function has been investigated by several groups. It has been proposed that prestin transports a bound anion from the intracellular compartment *partially* across the cell membrane, without releasing the anion to the extracellular side. The partial translocation would result in the structural change of prestin to the expanded state (He et al., 2006; Oliver et al., 2001).

Alternatively, it has been proposed that prestin transports a bound anion from the intracellular compartment *fully* across the cell membrane, thereby functioning as an anion transporter (Muallem and Ashmore, 2006). Two variations of the proposed transport scheme are prestin acting as a uniporter or as an anion exchanger. The uniporter model transports Cl^- (or HCO_3^-) as the sole charge carrier across the cell membrane. By contrast, the anion exchanger has the particular properties that Cl^- (or HCO_3^-) is transported from the cytosolic side to the extracellular side along with the transmembrane movement of positively charged amino acids in the prestin molecule, such as during a conformational change. Thus, two positive intrinsic charges accompany each Cl^- (or HCO_3^-) during movement from inside to outside the cell. The return cycle of the exchanger binds two anion charges (two monovalent or one divalent anion) that "neutralize" the intrinsic charge. Kinetic modeling of the anion exchanger conformed best to many reported experimental observations (Muallem and Ashmore, 2006).

Interestingly, there is evidence for a competition between Cl^- and salicylate (aspirin) for the anion binding site on the prestin molecule (Santos-Sacchi et al., 2006). Prestin bound to salicylate becomes unresponsive to receptor potentials and this may be the primary cause for aspirin-induced hearing loss and tinnitus (Kakehata and Santos-Sacchi, 1996; Mongan et al., 1973; Myers and Bernstein, 1965). Even though salicylate acutely interferes with the outer hair cell motor, chronic exposure of outer hair cells leads to an up-regulation of mRNA and protein expression of prestin (Yu et al., 2008). This up-regulation may maintain outer hair cell function and hearing in people chronically using aspirin but may also account for increased tinnitus from the resulting imbalance of inner and outer hair cell function (Yu et al., 2008).

Cl^- supports cochlear amplification not only by being essential for prestin function but also by supporting the coordination of outer hair cells, which is necessary for the high sensitivity and the sharpness of frequency discrimination of the mammalian cochlea. Contractions and elongations of outer hair cells are modulated via a local neuronal network consisting of GABAergic efferents (Thiers et al., 2008). This GABAergic system includes the ionotropic $GABA_A$ and $GABA_C$ receptors that are ligand-gated channels permeable to Cl^-. The importance of the GABAergic innervation in the cochlea is underscored by the observation of hearing loss in mice that lack $GABA_A$ receptor subunits (Maison et al., 2006).

C. Cl^- Dependence of K^+ Cycling and K^+ Buffering

Sensory transduction depends on a high concentration of K^+ in endolymph and, at least in the cochlea, on the presence of a large transepithelial potential called the endocochlear potential (see below). K^+ that serves as charge carrier for sensory transduction and exits the hair cells at the basolateral membrane is recycled into endolymph via remotely located epithelia, stria vascularis in the cochlea and dark cells in the vestibular labyrinth. Radiotracer and perfusion studies have shown that K^+ transport in the cochlea occurs in a nearly closed cycle (Konishi et al., 1978; Marcus et al., 1981; Sterkers et al., 1982; Wada et al., 1979).

K^+ exiting the hair cells is moved away from the neuronal elements in the basal region of the hair cells. Simple diffusion into the open fluid spaces of scala tympani is most likely the most important mechanism, which is supported by current and K^+ flux measurements (Johnstone et al., 1989; Zidanic and Brownell, 1990). In addition, supporting cells in the vicinity of hair cells may provide a cellular buffer mechanism that depends on inward-rectifying KCNJ10 K^+ channels as K^+ uptake mechanisms and less rectifying channels or the KCl cotransporters KCC3 (SLC12A6) and KCC4 (SLC12A7) as K^+ exit mechanisms (Fig. 21.4) (Boettger et al., 2002, 2003; Hibino et al., 1997; Nenov et al., 1998; Ruttiger et al., 2004). A similar cellular mechanism for K^+ buffering has been proposed in Müller glial cells of the retina (Kofuji et al., 2002; Reichenbach et al., 1992). Whether KCC3 and KCC4 act as K^+ release mechanisms or as uptake mechanisms, as proposed earlier, awaits experimental determination.

Locally buffered or not, K^+ reaches the cochlear lateral wall either by diffusion across the open perilymph spaces or by diffusion from cell to cell via gap junctions starting from the supporting cells next to the hair cells and ending with root cells in the spiral ligament of the cochlea (Spicer and Schulte, 1996). Regardless of the pathway of K^+ from the basolateral side of the hair cells to the spiral ligament, K^+ is taken up by specialized fibrocytes, fibrocytes types II, IV and V, in the spiral ligament. Although functional data from fibrocytes are lacking, K^+ uptake into fibrocytes is hypothesized to function like the well-studied uptake mechanism in the basolateral membrane of strial marginal cells and vestibular dark cells. Accordingly, K^+ is taken up via the Na^+/K^+-ATPase and the Na^+-K^+-$2Cl^-$ cotransporter NKCC1 (SLC12A2) with Cl^- being recycled across the membrane via the Cl^- channels CLCNKA and CLCNKB and Na^+ via the Na^+/K^+-ATPase (Crouch et al., 1997; Maehara et al., 2003; Mizuta et al., 1997; Qu et al., 2006; Schulte and Adams, 1989). K^+ is then shuttled via gap junctions from fibrocyte to fibrocyte and from there into stria vascularis (Spicer and Schulte, 1996). These gap junctions consist of the connexins GJB2 (Cx26) and GJB6 (Cx30) (Kikuchi et al., 1995; Lautermann et al., 1998; Xia et al., 1999).

D. Cl^- Dependence of K^+ Secretion and the Endocochlear Potential

Marginal cells of stria vascularis and vestibular dark cells are homologous epithelial cells in the cochlea and the vestibular labyrinth that secrete K^+ into endolymph. K^+ is taken up by strial marginal cells from the intrastrial space and by vestibular dark cells from perilymph (Fig. 21.7). The uptake mechanism consists of the Na^+-K^+-$2Cl^-$ cotransporter NKCC1 in conjunction with the Na^+/K^+-ATPase ATPA1/ATPB1 (Marcus and Marcus, 1987; Wangemann et al., 1995a). Cl^- that is brought into the cells by the Na^+-K^+-$2Cl^-$ cotransporter is recycled across the basolateral membrane through Cl^- channels (Wangemann, 1995; Wangemann and Marcus, 1992). Cl^- channels include ClC-K pore-forming α-subunits and barttin β-subunits (CLCNKA/BSND and CLCNKB/BSND) (Ando and Takeuchi, 2000; Estevez et al., 2001; Qu et al., 2006; Sage and Marcus, 2001). Uptake of K^+ across the basolateral membrane is followed by secretion of K^+ via K^+ channels in the apical membrane (Marcus and Shen, 1994; Shen et al., 1997). The K^+ channel consists of KCNQ1 pore-forming α-subunits and KCNE1 β-subunits (Barhanin et al., 1996; Sanguinetti et al., 1996). The importance of these transporters is underscored by the observation of deafness in mice that lack KCNQ1 (Casimiro et al., 2001; Lee et al., 2000), KCNE1 (Letts et al., 2000; Vetter et al., 1996), NKCC1 (Delpire et al., 1999; Dixon et al., 1999; Flagella et al., 1999), BSND (Rickheit et al., 2008) and in humans that bear mutations of BSND, CLCNKA and CLCNKB, KCNE1 or KCNQ1 or receive high doses of the furosemide-type loop diuretics that block NKCC1.

K^+ secretion by strial marginal cells is intimately associated with the generation of the endocochlear potential, although the transepithelial potential across the marginal cell layer itself is no more than a few mV. Marginal cells are part of stria vascularis, which is a two-layered epithelial structure in the lateral wall of the cochlea (Fig. 21.7). The endocochlear potential is generated across the basal cell barrier of stria vascularis. The basal cell barrier consists of the basal cells themselves, strial intermediate cells and fibrocytes of the spiral ligament. Intermediate cells and fibrocytes are electrically linked to the basal cells via gap junctions (Kikuchi et al., 1994, 1995; Takeuchi and Ando, 1998). The endocochlear potential originates from a K^+ equilibrium potential that is generated by the K^+ channel KCNJ10 located in strial intermediate cells in conjunction with the high K^+ concentration inside intermediate cells and the low K^+ concentration in the intrastrial spaces between intermediate cells and marginal cells (Marcus et al., 2002). The high K^+ concentration in intermediate cells is ensured by K^+ uptake via the Na^+-K^+-$2Cl^-$ cotransporter NKCC1 and the Na^+/K^+-ATPase ATPA1/ATPB1 in spiral ligament fibrocytes and the low K^+ concentration in the intrastrial spaces is ensured by K^+ uptake by strial marginal cells.

The outer membrane of strial basal cells that faces spiral ligament is thought to be depolarized, thereby "grounding" the cytosol of the basal cells and the connected intermediate cells. Consequently, the voltage that is generated across the KCNJ10 expressing membranes of the intermediate cells is observed with little change as the transepithelial voltage across the basal cell barrier. Further, since the marginal cell layer contributes little to the trans-strial voltage, the voltage generated across the basal cell barrier can be detected in the intrastrial space and with little change in scala media (Nin et al., 2008; Salt et al., 1987). The importance of KCNJ10 channel for the generation of the endocochlear potential and for hearing is underscored by the observation that mice lacking the KCNJ10 also lack the endocochlear potential and are deaf (Marcus et al., 2002).

E. Cl^- Dependence of HCO_3^- Secretion and pH Regulation

The secretion of HCO_3^- into endolymph leads to an endolymphatic pH that is more alkaline than the pH of perilymph (Table 21.1). HCO_3^- is secreted into endolymph by the Cl^-/HCO_3^- exchanger pendrin (SLC26A4) that is localized in the apical membrane of strial spindle cells, spiral prominence and outer sulcus epithelial cells in the cochlea and transitional cells in the vestibular labyrinth (Everett et al., 1999; Wangemann et al., 2004) (Figs 21.3 and 21.5). The importance of pendrin is underscored by the fact that mice lacking pendrin are deaf and have vestibular dysfunction (Everett et al., 2007; Nakaya et al., 2007; Wangemann et al., 2007). Deafness in mice lacking pendrin is the consequence of an enlarged endolymph volume that requires higher rates of K^+ secretion leading to oxidative stress in stria vascularis and consequently to the loss of the K^+ channel KCNJ10 and the loss of the endocochlear potential (Singh and Wangemann, 2007; Wangemann et al., 2004; Wangemann et al., 2007). Further contributors to deafness in these mice are the more acidic pH of endolymph that enhances K^+ secretion by marginal cells (Wangemann et al., 1995b) and inhibits Ca^{2+} absorption mechanisms. Inhibition of Ca^{2+} absorption leads to high endolymphatic Ca^{2+} concentrations that are likely to initiate the demise of the cochlear hair cells (Everett et al., 2001; Nakaya et al., 2007; Wangemann et al., 2007). Mutations of SLC26A4 are the second most prevalent genetic cause of hearing loss in humans and the most prevalent cause of hearing losses with a post-natal onset that provide a window of opportunity for a clinical intervention (Hilgert et al., 2008).

Important systems in the inner ear that are sensitive to (luminal) extracellular pH include the tectorial membrane, the K^+ secretory mechanism that depends on the K^+ channel KCNQ1/KCNE1 and the Ca^{2+} absorptive transport system that depends on the epithelial Ca^{2+} channels TRPV5 and TRPV6. The tectorial membrane is involved in transferring mechanical stimuli to the hair cells and its size and shape depend on the surrounding pH (Kronester-Frei, 1979; Weiss and Freeman, 1997). The K^+ channel KCNQ1/KCNE1, unlike many other K^+ channels, is stimulated by extracellular acidification (Heitzmann et al., 2007) and the Ca^{2+} channels TRPV5 and TRPV6 are blocked by acidification (Nakaya et al., 2007; Yeh et al., 2003).

In addition to pendrin, there are several other acid and base transporters expressed in cochlear and vestibular epithelia including H^+-ATPase, Na^+/H^+ exchangers and monocarboxylate transporters. H^+-ATPase consists of many subunits. Subunits of H^+-ATPase are expressed in interdental cells of the spiral limbus, marginal cells in the cochlea and in a population of epithelial cells of the endolymphatic sac. Functional evidence for H^+-ATPase activity has been obtained so far only for the endolymphatic sac (Couloigner et al., 2000). Whether at least a minimum complement of subunits that are necessary to support H^+-ATPase activity is expressed in the cochlea is currently unclear. Functional evidence for H^+-ATPase activity in the cochlea awaits experimental demonstration. The role of H^+-ATPase in cochlear function is currently unclear since mice lacking ATP6V1B1 have normal hearing (Dou et al., 2003). In contrast, human patients bearing mutations of either ATP6V1B1 or ATP6V0A4 suffer from progressive hearing loss (Karet et al., 1999; Stover et al., 2002). Unlike mice, human patients suffer from metabolic acidosis and from Ca^{2+} loss from bone in addition to hearing loss. It is conceivable that hearing loss is an epiphenomenon related to bone loss.

In addition to transepithelial acid or base transport, some epithelial cells bordering endolymph may generate an H^+ flux into endolymph that is fed by a metabolic source of acid. Vigorous aerobic metabolism is required to support K^+ secretion in stria vascularis (Konishi et al., 1978; Thalmann et al., 1977). Strial marginal cells express several acid extruding mechanisms in their apical membranes including the Na^+/H^+ exchanger (NHE3) and a monocarboxylate transporter (MCT1 or MCT2) (Bond et al., 1998; Shimozono et al., 1997, 1998). Whether these transporters generate a measurable H^+ flux awaits experimental determination.

F. Cl^- Secretion

Transepithelial Cl^- secretion has so far only been observed in the semicircular canal duct epithelium of the vestibular labyrinth (Milhaud et al., 2002). Cellular mechanisms leading to Cl^- secretion are only partially understood. The physiological significance of this transport likely lies in providing an anion pathway parallel to K^+ secretion. Support for the concept of parallel transport comes from the finding that K^+ secretion and Cl^- secretion are both under the control of adrenergic receptors. Norepinephrine stimulates K^+ secretion via β_1-adrenergic receptors and Cl^- secretion via β_2-adrenergic receptors (Milhaud et al., 2002; Wangemann et al., 2000).

G. Cl^- Independent Absorption of Na^+ and Ca^{2+}

There are several essential transepithelial transport systems in the inner ear for which there is no known dependence on Cl^-. Endolymphatic Na^+ is maintained at its low levels by active absorption which takes place in outer sulcus cells and Reissner's membrane epithelial cells in the cochlea and in transitional cells and semicircular canal epithelial cells in the vestibular labyrinth. Na^+ is taken up from endolymph via Na^+-permeable channels and transported across the basolateral membrane via the Na^+/K^+-ATPase with K^+ recycling through K^+-permeable channels. Outer sulcus cells of the cochlea and transitional cells of the vestibular labyrinth employ an amiloride-insensitive non-selective cation channel for the uptake of Na^+ across the apical cell membrane (Lee et al., 2001; Marcus and Chiba, 1999). Reissner's membrane epithelial cells in the cochlea and semicircular canal duct cells in the vestibular labyrinth employ an amiloride-sensitive channel (Lee and Marcus, 2003; Pondugula et al., 2004, 2006).

The endolymphatic Ca^{2+} is regulated by a balance between secretory and absorptive transport processes. Few sites of Ca^{2+} transport have so far been identified and supported by functional evidence. These include Ca^{2+} secretion via the plasma membrane Ca^{2+}-ATPase PMCA2 in stereocilia of cochlear and vestibular hair cells (Wood et al., 2004; Yamoah et al., 1998) and Ca^{2+} absorption via the Ca^{2+} channels TRPV5 and TRPV6 in semicircular canal duct cells (Nakaya et al., 2007).

Acknowledgements

This work was supported by NIH grants R01-DC000212 (DCM) and R01-DC001098 (PW).

References

Ando, M. and Takeuchi, S. (2000). mRNA encoding "ClC-K1, a kidney Cl^--channel" is expressed in marginal cells of the stria vascularis of rat cochlea: its possible contribution to Cl^- currents. *Neurosci. Lett.* **284**, 171–174.

Ashmore, J. (2008). Cochlear outer hair cell motility. *Physiol. Rev.* **88**, 173–210.

Barhanin, J., Lesage, F., Guillemare, E., Fink, M., Lazdunski, M., and Romey, G. (1996). K_vLQT1 and IsK (minK) proteins associate to form the I_{Ks} cardiac potassium current. *Nature* **384**, 78–80.

Boettger, T., Hubner, C.A., Maier, H., Rust, M.B., Beck, F.X., and Jentsch, T.J. (2002). Deafness and renal tubular acidosis in mice lacking the K-Cl co-transporter Kcc4. *Nature* **416**, 874–878.

Boettger, T., Rust, M.B., Maier, H., Seidenbecher, T., Schweizer, M., Keating, D.J., Faulhaber, J., Ehmke, H., Pfeffer, C., Scheel, O., Lemcke, B., Horst, J., Leuwer, R., Pape, H.C., Volkl, H., Hubner, C.A., and Jentsch, T.J. (2003). Loss of K-Cl co-transporter KCC3 causes deafness, neurodegeneration and reduced seizure threshold. *EMBO J.* **22**, 5422–5434.

Bond, B.R., Ng, L.L., and Schulte, B.A. (1998). Identification of mRNA transcripts and immunohistochemical localization of Na/H exchanger isoforms in gerbil inner ear. *Hearing Res.* **123**, 1–9.

Casimiro, M.C., Knollmann, B.C., Ebert, S.N., Vary, J.C., Jr., Greene, A.E., Franz, M.R., Grinberg, A., Huang, S.P., and Pfeifer, K. (2001). Targeted disruption of the Kcnq1 gene produces a mouse model of Jervell and Lange-Nielsen Syndrome. *Proc. Natl. Acad. Sci. USA* **98**, 2526–2531.

Couloigner, V., Teixeira, M., Hulin, P., Sterkers, O., Bichara, M., Escoubet, B., Planelles, G., and Ferrary, E. (2000). Effect of locally applied drugs on the pH of luminal fluid in the endolymphatic sac of guinea pig. *Am. J. Physiol. Regul. Integr. Comp. Physiol.* **279**, R1695–R1700.

Crouch, J.J., Sakaguchi, N., Lytle, C., and Schulte, B.A. (1997). Immunohistochemical localization of the Na-K-Cl co-transporter (NKCC1) in the gerbil inner ear. *J. Histochem. Cytochem.* **45**, 773–778.

Delpire, E., Lu, J., England, R., Dull, C., and Thorne, T. (1999). Deafness and imbalance associated with inactivation of the secretory Na-K-2Cl co-transporter. *Nature Gen.* **22**, 192–195.

Dixon, M.J., Gazzard, J., Chaudhry, S.S., Sampson, N., Schulte, B.A., and Steel, K.P. (1999). Mutation of the Na-K-Cl co-transporter gene Slc12a2 results in deafness in mice. *Hum. Mol. Genet.* **8**, 1579–1584.

Dou, H., Finberg, K., Cardell, E.L., Lifton, R., and Choo, D. (2003). Mice lacking the B1 subunit of H^+-ATPase have normal hearing. *Hearing Res.* **180**, 76–84.

Estevez, R., Boettger, T., Stein, V., Birkenhager, R., Otto, E., Hildebrandt, F., and Jentsch, T.J. (2001). Barttin is a Cl^- channel beta-subunit crucial for renal Cl^- reabsorption and inner ear K^+ secretion. *Nature* **414**, 558–561.

Everett, L.A., Belyantseva, I.A., Noben-Trauth, K., Cantos, R., Chen, A., Thakkar, S.I., Hoogstraten-Miller, S.L., Kachar, B., Wu, D.K., and Green, E.D. (2001). Targeted disruption of mouse Pds provides insight about the inner-ear defects encountered in Pendred syndrome. *Hum. Mol. Genet.* **10**, 153–161.

Everett, L.A., Morsli, H., Wu, D.K., and Green, E.D. (1999). Expression pattern of the mouse ortholog of the Pendred's syndrome gene (Pds) suggests a key role for pendrin in the inner ear. *Proc. Nat.Acad. Sci. USA* **96**, 9727–9732.

Flagella, M., Clarke, L.L., Miller, M.L., Erway, L.C., Giannella, R.A., Andringa, A., Gawenis, L.R., Kramer, J., Duffy, J.J., Doetschman, T., Lorenz, J.N., Yamoah, E.N., Cardell, E.L., and Shull, G.E. (1999). Mice lacking the basolateral Na-K-2Cl cotransporter have impaired epithelial chloride secretion and are profoundly deaf. *J. Biol. Chem.* **274**, 26946–26955.

He, D.Z., Zheng, J., Kalinec, F., Kakehata, S., and Santos-Sacchi, J. (2006). Tuning in to the amazing outer hair cell: membrane wizardry with a twist and shout. *J. Membr. Biol.* **209**, 119–134.

Heitzmann, D., Koren, V., Wagner, M., Sterner, C., Reichold, M., Tegtmeier, I., Volk, T., and Warth, R. (2007). KCNE beta subunits determine pH sensitivity of KCNQ1 potassium channels. *Cell Physiol. Biochem.* **19**, 21–32.

Hibino, H., Horio, Y., Inanobe, A., Doi, K., Ito, M., Yamada, M., Gotow, T., Uchiyama, Y., Kawamura, M., Kubo, T., and Kurachi, Y. (1997). An ATP-dependent inwardly rectifying potassium channel, K_{AB}-2 (Kir4.1), in cochlear stria vascularis of inner ear: its specific subcellular localization and correlation with the formation of endocochlear potential. *J. Neurosci.* **17**, 4711–4721.

Hilgert, N., Smith, R.J., and Van Camp, G. (2008). Forty-six genes causing nonsyndromic hearing impairment: which ones should be analyzed in DNA diagnostics? *Mutat. Res.*

Hudspeth, A.J., Choe, Y., Mehta, A.D., and Martin, P. (2000). Putting ion channels to work: mechanoelectrical transduction, adaptation, and amplification by hair cells. *Proc. Natl. Acad. Sci. USA* **97**, 11765–11772.

Johnstone, B.M., Patuzzi, R., Syka, J., and Sykova, E. (1989). Stimulus-related potassium changes in the organ of Corti of guinea-pig. *J. Physiol.* **408**, 77–92.

Kakehata, S. and Santos-Sacchi, J. (1996). Effects of salicylate and lanthanides on outer hair cell motility and associated gating charge. *J. Neurosci.* **16**, 4881–4889.

Karet, F.E., Finberg, K.E., Nelson, R.D., Nayir, A., Mocan, H., Sanjad, S.A., Rodriguez-Soriano, J., Santos, F., Cremers, C.W., Di Pietro, A., Hoffbrand, B.I., Winiarski, J., Bakkaloglu, A., Ozen, S., Dusunsel, R., Goodyer, P., Hulton, S.A., Wu, D.K., Skvorak, A.B., Morton, C.C., Cunningham, M.J., Jha, V., and Lifton, R.P. (1999). Mutations in the gene encoding B1 subunit of H^+-ATPase cause renal tubular acidosis with sensorineural deafness. *Nature Gen.* **21**, 84–90.

Kikuchi, T., Adams, J.C., Paul, D.L., and Kimura, R.S. (1994). Gap junction systems in the rat vestibular labyrinth: immunohistochemical and ultrastructural analysis. *Acta Otolaryngol. (Stockh.)* **114**, 520–528.

Kikuchi, T., Kimura, R.S., Paul, D.L., and Adams, J.C. (1995). Gap junctions in the rat cochlea: immunohistochemical and ultrastructural analysis. *Anat. Embryol. (Berl.)* **191**, 101–118.

Kofuji, P., Biedermann, B., Siddharthan, V., Raap, M., Iandiev, I., Milenkovic, I., Thomzig, A., Veh, R.W., Bringmann, A., and Reichenbach, A. (2002). Kir potassium channel subunit expression in retinal glial cells: implications for spatial potassium buffering. *Glia* **39**, 292–303.

Konishi, T., Hamrick, P.E., and Walsh, P.J. (1978). Ion transport in guinea pig cochlea. I. Potassium and sodium transport. *Acta Otolaryngol.* **86**, 22–34.

Kronester-Frei, A. (1979). The effect of changes in endolymphatic ion concentrations on the tectorial membrane. *Hearing Res.* **1**, 81–94.

Lautermann, J., ten Cate, W.J., Altenhoff, P., Grummer, R., Traub, O., Frank, H., Jahnke, K., and Winterhager, E. (1998). Expression of the gap-junction connexins 26 and 30 in the rat cochlea. *Cell Tissue Res.* **294**, 415–420.

Lee, J.H., Chiba, T., and Marcus, D.C. (2001). P2 × 2 receptor mediates stimulation of parasensory cation absorption by cochlear outer sulcus cells and vestibular transitional cells. *J. Neurosci.* **21**, 9168–9174.

Lee, J.H. and Marcus, D.C. (2003). Endolymphatic sodium homeostasis by Reissner's membrane. *Neuroscience* **119**, 3–8.

Lee, M.P., Ravenel, J.D., Hu, R.J., Lustig, L.R., Tomaselli, G., Berger, R.D., Brandenburg, S.A., Litzi, T.J., Bunton, T.E., Limb, C., Francis, H., Gorelikow, M., Gu, H., Washington, K., Argani, P., Goldenring, J.R., Coffey, R.J., and Feinberg, A.P. (2000). Targeted disruption of the Kvlqt1 gene causes deafness and gastric hyperplasia in mice. *J. Clin. Invest.* **106**, 1447–1455.

Letts, V.A., Valenzuela, A., Dunbar, C., Zheng, Q.Y., Johnson, K.R., and Frankel, W.N. (2000). A new spontaneous mouse mutation in the Kcne1 gene. *Mamm.Genome* **11**, 831–835.

Maehara, H., Okamura, H.O., Kobayashi, K., Uchida, S., Sasaki, S., and Kitamura, K. (2003). Expression of CLC-KB gene promoter in the mouse cochlea. *Neuroreport* **14**, 1571–1573.

Maison, S.F., Rosahl, T.W., Homanics, G.E., and Liberman, M.C. (2006). Functional role of GABAergic innervation of the cochlea: phenotypic analysis of mice lacking GABA(A) receptor subunits alpha 1, alpha 2, alpha 5, alpha 6, beta 2, beta 3, or delta. *J. Neurosci.* **26**, 10315–10326.

Marcus, D.C. and Chiba, T. (1999). K^+ and Na^+ absorption by outer sulcus epithelial cells. *Hearing Res.* **134**, 48–56.

Marcus, D.C., Marcus, N.Y., and Thalmann, R. (1981). Changes in cation contents of stria vascularis with ouabain and potassium-free perfusion. *Hearing Res.* **4**, 149–160.

Marcus, D.C. and Shen, Z. (1994). Slowly activating, voltage-dependent K^+ conductance is apical pathway for K^+ secretion in vestibular dark cells. *Am. J. Physiol.* **267**, C857–C864.

Marcus, D.C., Wu, T., Wangemann, P., and Kofuji, P. (2002). KCNJ10 (Kir4.1) potassium channel knockout abolishes endocochlear potential. *Am. J. Physiol.: Cell Physiol.* **282**, C403–C407.

Marcus, N.Y. and Marcus, D.C. (1987). Potassium secretion by nonsensory region of gerbil utricle in vitro. *Am. J. Physiol.* **253**, F613–F621.

Milhaud, P.G., Pondugula, S.R., Lee, J.H., Herzog, M., Lehouelleur, J., Wangemann, P., Sans, A., and Marcus, D.C. (2002). Chloride secretion by semicircular canal duct epithelium is stimulated via β2-adrenergic receptors. *Am. J. Physiol.: Cell Physiol.* **283**, C1752–C1760.

Mizuta, K., Adachi, M., and Iwasa, K.H. (1997). Ultrastructural localization of the Na-K-Cl cotransporter in the lateral wall of the rabbit cochlear duct. *Hearing Res.* **106**, 154–162.

Mongan, E., Kelly, P., Nies, K., Porter, W.W., and Paulus, H.E. (1973). Tinnitus as an indication of therapeutic serum salicylate levels. *Jama* **226**, 142–145.

Muallem, D. and Ashmore, J. (2006). An anion antiporter model of prestin, the outer hair cell motor protein. *Biophys. J.* **90**, 4035–4045.

Myers, E.N. and Bernstein, J.M. (1965). Salicylate ototoxicity; a clinical and experimental study. *Arch. Otolaryngol.* **82**, 483–493.

Nakaya, K., Harbidge, D.G., Wangemann, P., Schultz, B.D., Green, E., Wall, S.M., and Marcus, D.C. (2007). Lack of pendrin $HCO3^-$ transport elevates vestibular endolymphatic $[Ca^{2+}]$ by inhibition of acid-sensitive TRPV5 and TRPV6 channels. *Am. J. Physiol. Renal Physiol.* **292**, F1314–F1321.

Nenov, A.P., Chen, C., and Bobbin, R.P. (1998). Outward rectifying potassium currents are the dominant voltage activated currents present in Deiters' cells. *Hearing Res.* **123**, 168–182.

Nin, F., Hibino, H., Doi, K., Suzuki, T., Hisa, Y., and Kurachi, Y. (2008). The endocochlear potential depends on two K^+ diffusion potentials and an electrical barrier in the stria vascularis of the inner ear. *Proc. Natl. Acad. Sci. USA* **105**, 1751–1756.

Oliver, D., He, D.Z., Klocker, N., Ludwig, J., Schulte, U., Waldegger, S., Ruppersberg, J.P., Dallos, P., and Fakler, B. (2001). Intracellular anions as the voltage sensor of prestin, the outer hair cell motor protein. *Science* **292**, 2340–2343.

Pondugula, S.R., Raveendran, N.N., Ergonul, Z., Deng, Y., Chen, J., Sanneman, J.D., Palmer, L.G., and Marcus, D.C. (2006). Glucocorticoid regulation of genes in the amiloride-sensitive sodium transport pathway by semicircular canal duct epithelium of neonatal rat. *Physiol. Gen.* **24**, 114–123.

Pondugula, S.R., Sanneman, J.D., Wangemann, P., Milhaud, P.G., and Marcus, D.C. (2004). Glucocorticoids stimulate cation absorption by semicircular canal duct epithelium via epithelial sodium channel. *Am. J. Physiol. Renal Physiol.* **286**, F1127–F1135.

Qu, C., Liang, F., Hu, W., Shen, Z., Spicer, S.S., and Schulte, B.A. (2006). Expression of CLC-K chloride channels in the rat cochlea. *Hearing Res.* **213**, 79–87.

Reichenbach, A., Henke, A., Eberhardt, W., Reichelt, W., and Dettmer, D. (1992). K+ ion regulation in retina. *Can. J. Physiol. Pharmacol.* **70** (Suppl.), S239–S247.

Rickheit, G., Maier, H., Strenzke, N., Andreescu, C.E., De Zeeuw, C.I., Muenscher, A., Zdebik, A.A., and Jentsch, T.J. (2008). Endocochlear potential depends on Cl^- channels: mechanism underlying deafness in Bartter syndrome IV. *EMBO J.*

Ruttiger, L., Sausbier, M., Zimmermann, U., Winter, H., Braig, C., Engel, J., Knirsch, M., Arntz, C., Langer, P., Hirt, B., Muller, M., Kopschall, I., Pfister, M., Munkner, S., Rohbock, K., Pfaff, I., Rusch, A., Ruth, P., and Knipper, M. (2004). Deletion of the Ca^{2+}-activated potassium (BK) alpha-subunit but not the BKbeta1-subunit leads to progressive hearing loss. *Proc. Natl. Acad. Sci. USA* **101**, 12922–12927.

Sage, C.L. and Marcus, D.C. (2001). Immunolocalization of ClC-K chloride channel in strial marginal cells and vestibular dark cells. *Hearing Res.* **160**, 1–9.

Salt, A.N., Melichar, I., and Thalmann, R. (1987). Mechanisms of endocochlear potential generation by stria vascularis. *Laryngoscope* **97**, 984–991.

Sanguinetti, M.C., Curran, M.E., Zou, A., Shen, J., Spector, P.S., Atkinson, D.L., and Keating, M.T. (1996). Coassembly of K_vLQT1 and minK (IsK) proteins to form cardiac I_{Ks} potassium channel. *Nature* **384**, 80–83.

Santos-Sacchi, J., Song, L., Zheng, J., and Nuttall, A.L. (2006). Control of mammalian cochlear amplification by chloride anions. *J. Neurosci.* **26**, 3992–3998.

Schulte, B.A. and Adams, J.C. (1989). Distribution of immunoreactive Na^+,K^+-ATPase in gerbil cochlea. *J. Histochem. Cytochem.* **37**, 127–134.

Shen, Z., Marcus, D.C., Sunose, H., Chiba, T., and Wangemann, P. (1997). I_{sK} channel in strial marginal cells: voltage-dependence, ion-selectivity, inhibition by 293B and sensitivity to clofilium. *Aud. Neurosci.* **3**, 215–230.

Shimozono, M., Liu, J., Scofield, M.A., and Wangemann, P. (1998). Vestibular dark cells contain an H^+/monocarboxylate$^-$ cotransporter in their apical and basolateral membrane. *J. Membr. Biol.* **163**, 37–46.

Shimozono, M., Scofield, M.A., and Wangemann, P. (1997). Functional evidence for a monocarboxylate transporter (MCT) in strial marginal cells and molecular evidence for MCT1 and MCT2 in stria vascularis. *Hearing Res.* **114**, 213–222.

Singh, R. and Wangemann, P. (2007). Free radical stress mediated loss of Kcnj10 protein expression in stria vascularis contributes to deafness in Pendred syndrome mouse model. *Am. J. Physiol. Renal. Physiol.*

Spicer, S.S. and Schulte, B.A. (1996). The fine structure of spiral ligament cells relates to ion return to the stria and varies with place-frequency. *Hearing Res.* **100**, 80–100.

Sterkers, O., Saumon, G., Tran Ba Huy, P., and Amiel, C. (1982). K, Cl, and H_2O entry in endolymph, perilymph, and cerebrospinal fluid of the rat. *Am. J. Physiol.* **243**, F173–F180.

Stover, E.H., Borthwick, K.J., Bavalia, C., Eady, N., Fritz, D.M., Rungroj, N., Giersch, A.B., Morton, C.C., Axon, P.R., Akil, I., Al-Sabban, E.A., Baguley, D.M., Bianca, S., Bakkaloglu, A., Bircan, Z., Chauveau, D., Clermont, M.J., Guala, A., Hulton, S.A., Kroes, H., Li, V.G., Mir, S., Mocan, H., Nayir, A., Ozen, S., Rodriguez-Soriano, J., Sanjad, S.A., Tasic, V., Taylor, C.M., Topaloglu, R., Smith, A.N., and Karet, F.E. (2002). Novel ATP6V1B1 and ATP6V0A4 mutations in autosomal recessive distal renal tubular acidosis with new evidence for hearing loss. *J. Med.Genet.* **39**, 796–803.

Takeuchi, S. and Ando, M. (1998). Dye-coupling of melanocytes with endothelial cells and pericytes in the cochlea of gerbils. *Cell Tissue Res.* **293**, 271–275.

Thalmann, R., Thalmann, I., Ise, I., and Paloheimo, S. (1977). Noxious effects upon cochlear metabolism. *Laryngoscope* **87**, 699–721.

Thiers, F.A., Nadol, J.B., Jr., and Liberman, M.C. (2008). Reciprocal synapses between outer hair cells and their afferent terminals: evidence for a local neural network in the mammalian cochlea. *J. Assoc. Res. Otolaryngol.*

Valli, P., Zucca, G., and Botta, L. (1990). Perilymphatic potassium changes and potassium homeostasis in isolated semicircular canals of the frog. *J. Physiol.* **430**, 585–594.

Vetter, D.E., Mann, J.R., Wangemann, P., Liu, J., McLaughlin, K.J., Lesage, F., Marcus, D.C., Lazdunski, M., Heinemann, S.F., and Barhanin, J. (1996). Inner ear defects induced by null mutation of the isk gene. *Neuron* **17**, 1251–1264.

Wada, J., Kambayashi, J., Marcus, D.C., and Thalmann, R. (1979). Vascular perfusion of the cochlea: effect of potassium-free and rubidium-substituted media. *Arch. Otorhinolaryngol.* **225**, 79–81.

Wangemann, P. (1995). Comparison of ion transport mechanisms between vestibular dark cells and strial marginal cells. *Hearing Res.* **90**, 149–157.

Wangemann, P. (2008). Cochlear homeostasis and homeostatic disorders. In *Handbook of Auditory Research. Auditory Trauma and Protection* (Schacht, J., Popper, A.N., and Fay, R.R., eds), **31**, pp. 49–100. Springer.

Wangemann, P., Itza, E.M., Albrecht, B., Wu, T., Jabba, S.V., Maganti, R.J., Lee, J.H., Everett, L.A., Wall, S.M., Royaux, I.E., Green, E.D., and Marcus, D.C. (2004). Loss of KCNJ10 protein expression abolishes endocochlear potential and causes deafness in Pendred syndrome mouse model. *BMC Med.* **2**, 30.

Wangemann, P., Liu, J., and Marcus, D.C. (1995a). Ion transport mechanisms responsible for K^+ secretion and the transepithelial voltage across marginal cells of stria vascularis in vitro. *Hearing Res.* **84**, 19–29.

Wangemann, P., Liu, J., and Shiga, N. (1995b). The pH-sensitivity of transepithelial K^+ transport in vestibular dark cells. *J. Membr. Biol.* **147**, 255–262.

Wangemann, P., Liu, J., Shimozono, M., Schimanski, S., and Scofield, M.A. (2000). K^+ secretion in strial marginal cells is stimulated via β1-adrenergic receptors but not via β2-adrenergic or vasopressin receptors. *J. Membr. Biol.* **175**, 191–202.

Wangemann, P. and Marcus, D.C. (1992). The membrane potential of vestibular dark cells is controlled by a large Cl^- conductance. *Hearing Res.* **62**, 149–156.

Wangemann, P., Nakaya, K., Wu, T., Maganti, R.J., Itza, E.M., Sanneman, J.D., Harbidge, D.G., Billings, S., and Marcus, D.C. (2007). Loss of cochlear HCO_3^- secretion causes deafness via endolymphatic acidification and inhibition of Ca^{2+} reabsorption in a Pendred syndrome mouse model. *Am. J. Physiol. Renal Physiol.* **292**, F1345–F1353.

Weiss, T.F. and Freeman, D.M. (1997). Equilibrium behavior of an isotropic polyelectrolyte gel model of the tectorial membrane: effect of pH. *Hearing Res.* **111**, 55–64.

Wood, J.D., Muchinsky, S.J., Filoteo, A.G., Penniston, J.T., and Tempel, B.L. (2004). Low endolymph calcium concentrations in deafwaddler2J mice suggest that PMCA2 contributes to endolymph calcium maintenance. *J. Assoc. Res. Otolaryngol.* **5**, 99–110.

Xia, A., Kikuchi, T., Hozawa, K., Katori, Y., and Takasaka, T. (1999). Expression of connexin 26 and Na,K-ATPase in the developing mouse cochlear lateral wall: functional implications. *Brain Res.* **846**, 106–111.

Yamoah, E.N., Lumpkin, E.A., Dumont, R.A., Smith, P.J., Hudspeth, A.J., and Gillespie, P.G. (1998). Plasma membrane Ca^{2+}-ATPase extrudes Ca^{2+} from hair cell stereocilia. *J. Neurosci.* **18**, 610–624.

Yeh, B.I., Sun, T.J., Lee, J.Z., Chen, H.H., and Huang, C.L. (2003). Mechanism and molecular determinant for regulation of rabbit transient receptor potential type 5 (TRPV5) channel by extracellular pH. *J. Biol. Chem.* **278**, 51044–51052.

Yu, N., Zhu, M.L., Johnson, B., Liu, Y.P., Jones, R.O., and Zhao, H.B. (2008). Prestin up-regulation in chronic salicylate (aspirin) administration: an implication of functional dependence of prestin expression. *Cell Mol. Life Sci.* **65**, 2407–2418.

Zidanic, M. and Brownell, W.E. (1990). Fine structure of the intracochlear potential field. I. The silent current. *Biophys. J.* **57**, 1253–1268.

CHAPTER

22

Chloride Transporters in Presynaptic Inhibition, Pain and Neurogenic Inflammation

Francisco Javier Alvarez-Leefmans

OUTLINE

I. INTRODUCTION

Primary sensory neurons (PSNs) are the cells that convey virtually all somatic and visceral information to the spinal cord and brainstem. The peripheral processes of PSNs collect information from somatic (skin, muscles and joints) and visceral sensory receptors. Their central processes conduct these signals into the spinal cord and brainstem where their central terminals make synapses with higher order neurons. Within the spinal cord and brainstem these central terminals are subject to significant GABAergic presynaptic control, a crucial mechanism in gating the flow of nociceptive and other sensory information.

Unlike mature central neurons, adult PSNs are depolarized by GABA. This GABA-mediated depolarization is due to Cl^- efflux through anion channels coupled to $GABA_A$ receptors. This is possible because, unlike most mature central nervous system (CNS) neurons, PSNs maintain their intracellular Cl^- concentration ($[Cl^-]_i$) above electrochemical equilibrium. This results in an outwardly directed Cl^- electrical gradient across their plasma membranes. It is hypothesized that this Cl^- gradient, the focus of the first part of this chapter, makes possible the depolarization of the intraspinal terminals of PSNs by GABA released from dorsal horn interneurons. These interneurons make axo-axonic synapses with the central terminals of primary afferents. GABA-mediated depolarization of these central terminals, known as primary afferent depolarization (PAD), is a key factor in producing presynaptic inhibition.

The prevailing notion is that PAD inactivates Na^+ channels, thereby reducing the amplitude of the action potential as it invades the primary afferent terminals resulting in decreased transmitter release from these terminals. The amplitude of PAD affects the strength of presynaptic inhibition, and both phenomena are determined by the magnitude of the driving force for Cl^-, defined as the difference ($E_{Cl} - E_m$), where E_{Cl} is the Cl^- equilibrium potential and E_m the resting membrane potential. Further, there is evidence suggesting that following peripheral nerve or tissue injury, E_{Cl} shifts toward more positive (depolarizing) values, enhancing PAD and turning it into an excitatory response. When PAD reaches firing threshold of PSNs, the resulting action potentials, known as *dorsal root reflexes* (DRRs), cause pain and neurogenic inflammation. In sum, a key factor regulating PAD amplitude, presynaptic inhibition and DRRs is the $[Cl^-]_i$ in primary afferents. Clearly, understanding the mechanisms that regulate $[Cl^-]_i$ in PSNs is a fundamental problem in sensory physiology.

Work originated in our laboratory first established that $[Cl^-]_i$ in PSNs is maintained at levels above the predicted electrochemical equilibrium due to functional expression of a Na^+-K^+-$2Cl^-$ cotransporter mechanism (NKCC) that actively accumulates Cl^-. Subsequent molecular studies identified this cotransporter protein as NKCC1, one of the two known NKCC isoforms. Studies in rats and NKCC1-null mice ($NKCC1^{-/-}$) show that NKCC1 is involved in nociception, pain and neurogenic inflammation. Specifically, behavioral studies show that pain threshold increases significantly in $NKCC1^{-/-}$ mice, and pharmacological blockers of NKCCs have central and peripheral antinociceptive actions in rats. Further, it has been suggested that peripheral inflammatory mediators shift the Cl^- equilibrium potential (E_{Cl}) of PSNs to more depolarizing values, increasing PAD amplitude and generating DRRs. These DRRs cause hyperalgesia and neurogenic inflammation, both of which can be blocked with pharmacological inhibitors of NKCCs. The mechanisms underlying the depolarizing shift in E_{Cl} are not yet understood. Phosphorylation of NKCC1 and increase surface expression have been suggested as possible mechanisms mediating this putative enhancement of Cl^- uptake in PSNs.

NKCC is not only expressed in the cell bodies of primary sensory neurons located in dorsal root ganglia (DRG), but also in the membrane of sensory myelinated axons, and in various parts of the Schwann cells. Its distinctive presence and distribution in the paranodal region of Schwann cells suggest that it could also be involved in K^+ uptake from the periaxonal space, particularly at the paranodal region of myelinated axons, which is a zone prone to extracellular K^+ accumulation. The second part of this chapter discusses the possible involvement of NKCC in regulating the extracellular ionic environment and the excitability of myelinated axons.

A. The Discovery of Presynaptic Inhibition: A Kuhnian Paradigm Shift in Concepts and Ideas about the Structure and Function of Synapses

In a plenary lecture given at a workshop on "Integrative Physiology" held in Montpellier in 2004 on the occasion of the retirement of his disciple Denis Noble, physiologist Otto Hutter (1924–) gave an insightful account of the development of the ideas underlying chemical synaptic transmission, which began to emerge just over a century ago (Hutter, 2006). Hutter vividly remembered attending the series of lectures on "Chemical Transmission in the Nervous System" that Nobel Laureate Sir Henry

Dale (1875–1968) gave between 1943 and 1946, at the Royal Institution. Dale was a very influential figure of his time; he shaped the concept of what later was called the "Dale's Principle", based on the notion that a neuron is a metabolic unit and operates at all its synapses by the same chemical transmission mechanism. The concept was later modified by stating that a neuron releases one and the same transmitter from all its branches (Hokfelt, 2009).

"In retrospect", Hutter said, "Dale's message was a seductively simple one. In the peripheral nervous system, nerve-endings were either cholinergic or adrenergic. Receptors were postsynaptic, as John Langley (1907) had shown (Langley, 1907). Autonomic ganglia were relay stations, the rate of transmitter release being set centrally. Regarding the central nervous system, Dale pointed to the axon reflex. And he suggested that if only the vasodilator substance released by sensory nerve endings in the skin could be identified, it would prove to be also the transmitter at the central terminals of the sensory nerve (Dale, 1935). Dale based this suggestion on the metabolic unity of the neurone, and it became known as Dale's Principle (Eccles, 1957)."

In essence, the tenets of neurobiology during the first six decades of the 20th century can be summarized as follows. (1) A neuron released the same transmitter from all its branches. (2) Neurotransmitter receptors were postsynaptic. (3) Axo-axonic synapses were not part of the list of interneuronal contacts that Ramón y Cajal (1852–1934) accepted in his posthumously published monograph "Neuron Theory or Reticular Theory?" (Ramón y Cajal, 1954). (4) Synaptic inhibition in the vertebrate nervous system was only postsynaptic and hyperpolarizing (Coombs et al., 1955) (see also Chapter 5). (5) Cl^- was passively distributed across the plasma membrane of neurons, depolarizations were mediated by cation inward currents, and hyperpolarizations by K^+ outward currents.

But, as Hutter pointed out, the world was changing. "'Modernism' was being challenged by 'post-modernism', a reaction against principles and practices of established modernism. The key features of modernism – 'a search of order out of chaos, for system, for structure, for coherence, for predictability, for universally applicable grand narratives' – were being replaced with those of the post-modern era in which we now supposedly live – 'pluralism, fragmentation, ambivalence, unpredictability'. In post-modernism the emphasis is on the local, not the universal, on endless variety: dogmas and grand narratives are fodder for deconstruction, everything goes! Now curiously enough – and without any obvious causal connection to the sociological scene – the physiological scene, it seems to me, also underwent a post-modernist transformation, starting in the 1960s."

Hutter thinks that one can identify the advent of the new epoch in our conception of the structure and function of the nervous system in the discovery of presynaptic inhibition in the vertebrate spinal cord. Presynaptic inhibition was discovered by Karl Frank (1916–1993) and Michelangelo Fuortes (1917–1977) working at the National Institute of Neurological Diseases and Blindnes, at NIH in Bethesda, MD. Their findings were presented at the 78th annual meeting of the American Physiological Society in Chicago on April 18, 1957, and were published as an abstract in the Federation Proceedings (Frank and Fuortes, 1957). This may be one of the most cited abstracts in the neuroscience literature. The phenomenon they described was confirmed, and its functional significance was considerably expanded by Sir John Eccles (1903–1997) and his collaborators. Starting in the early 1960s, they produced a series of epoch-making papers that laid out our current concepts on this phenomenon (Eccles et al., 1961, 1962).

The way in which the facts developed supports Hutter's points of view; Langley's receptor concept had to be modified and extended to "presynaptic receptors" (Draguhn et al., 2008; Kullmann et al., 2005; Levy, 1977; Trigo et al., 2008). The finding of synapses upon axon terminals of primary afferents, that is, axo-axonic synapses, challenged Cajal's accepted views (Gray, 1962, 1963). GABA, the inhibitory transmitter that hyperpolarized neurons, now depolarized primary afferent terminals (Alvarez-Leefmans, 1990; Levy, 1977; Nicoll and Alger, 1979; Rudomin and Schmidt, 1999; Willis, 2006). Since GABA depolarizations were mediated by Cl^- this implied that $[Cl^-]_i$ in primary afferents was higher than electrochemical equilibrium (De Groat et al., 1972; Nishi et al., 1974), in sharp contrast to the prevailing notion that Cl^- was passively distributed (see Chapter 5 "passive distribution dogma"). This non-passive distribution of Cl^- implied the existence of a "Cl^- pump" (discussed in Chapter 5), as first suggested in the case of vertebrate primary afferent neurons by Nishi and coworkers (Gallagher et al., 1978; Nishi et al., 1974). Later work done in our lab directly measuring $[Cl^-]_i$ and E_m was the first to demonstrate that intracellular Cl^- was higher than electrochemical equilibrium, and that this non-equilibrium distribution was generated and maintained not by an ATP-driven pump, but by a bumetanide-sensitive Na^+-K^+-$2Cl^-$ cotransporter (Alvarez-Leefmans et al., 1988). This set the functional basis for Eccles' proposal that PAD *causes* presynaptic inhibition (Eccles, 1964). The precise mechanism by which PAD reduces transmitter release is still not clear and, as discussed below, this field is currently in a state of effervescence. It is likely that there is more than one mechanism of Cl^- accumulation. Further, up to now there is only indirect evidence that the mechanisms that operate in the cell body are the same as those operating at the terminals. The way in which this history has unfolded is the subject of two excellent reviews (Burke, 2006; Willis, 2006).

B. Axo-axonic Synapses and the Enigma of PAD Interneurons

To account for presynaptic inhibition, Eccles postulated the existence of presynaptic contacts on synaptic terminals, that is, axo-axonic synapses (Eccles, 1961). At the time Eccles first proposed them, there was no evidence for the postulated axo-axonal contacts, but he was undeterred by this fact. The story goes that in a seminar at Oxford University in 1961, Eccles confidently predicted that axo-axonic synapses would soon be found "because the anatomists are good boys and always find what they are told to look for" (Cowan and Kandel, 2001). The discovery of synapses upon axon terminals in the spinal cord by the cytologist George Gray (1924–1999), using electron microscopy (EM) provided the necessary morphological substrate for presynaptic inhibition (Gray, 1962).

Before the electron microscope era, the existence of axo-axonic synapses in the vertebrate spinal cord was first postulated by the Italian neuroanatomist Nello Beccari (1883–1957), in the case of the Mauthner's cell fiber collaterals of teleosts (Beccari, 1920). These axo-axonic contacts were later confirmed by Bodian, also using light microscopy (Bodian, 1952, 1972). Cajal knew about Beccari's finding from reading the well-known histology book written by Giuseppe Levi (1872–1965) and published in 1927 (Levi, 1927). However, when it came to comment on this issue in his final work (Ramón y Cajal, 1954), Cajal very cautiously stated "I have not been able to verify *dendro-dendritic* and *axo-axonal* connections as described by Levi in fish…" "I do not deny their possibility; I only declare that in thirty five years of patient work with impregnation methods I have not been convinced of their existence. Nevertheless, I believe that the case probably is due to deceiving appearances to which the best investigators are at times victims" (Fig. 22.1).

The existence of axo-axonal synapses on primary afferents in the spinal cord is now well documented (Fyffe and Light, 1984; Pierce and Mendell, 1993). Axo-axonal synapses have been found between presynaptic axon terminals coming from GABAergic interneurons, and terminals of group Ia muscle spindle afferents (Fig. 22.2), group Ib afferents from Golgi tendon organs, group II muscle spindle afferents and cutaneous afferents (for extensive and critical reviews see Alvarez, 1998; Rudomin and Schmidt, 1999; Willis, 2006). There is a great deal of electrophysiological and pharmacological evidence supporting the existence of these GABAergic interneurons (Rudomin, 2009). However, despite strenuous efforts, their morphological identity has not been ascertained. The axon terminals of these GABAergic interneurons are known as P boutons. There is abundant evidence that PAD is produced by GABA released from the P boutons (Rudomin and Schmidt, 1999). More recently, using a combined approach of immunocytochemistry with confocal immunofluorescence and EM, its has been possible to demonstrate that P boutons can be distinguished from other GABAergic terminals in the ventral horn of rat and mice spinal cord by their high levels GAD65, one of the isoforms of glutamic acid decarboxylase (GAD), the enzyme that synthesizes GABA. By carrying out retrograde labeling from lamina IX in mice that express green fluorescent protein under the control of the GAD65 promoter, it has been possible to demonstrate that the cells of origin of the P boutons are located in the medial part of laminae V and VI. Further, it was shown that identified Ia afferent fibers labeled by ionophoretic injection of neurobiotin received contacts from GAD65 terminals (Hughes et al., 2005). Altogether these data constitute compelling evidence supporting the notion that P boutons represent the major output of these GABAergic interneurons within the ventral horn, which is consistent with the view that presynaptic inhibition of proprioceptive afferents is mediated by specific populations of GABAergic interneurons. These results are in full agreement with earlier observations showing that PAD of proprioceptive terminals can be elicited using local intraspinal electrical microestimulation of laminae V and VI (Jankowska et al., 1981).

Although there is abundant evidence for PAD and presynaptic inhibition via GABAergic axo-axonic synapses in primary afferent fibers that innervate muscle stretch receptors or cutaneous mechanoreceptors, it has been more difficult to demonstrate a similar mechanism in unmyelinated C fibers and in the small thinly myelinated Aδ fibers that carry nociceptive information. Nevertheless, PAD of fine afferents, including C fibers, has been obtained by excitability testing, by recording DRRs from populations of Aδ fibers and from individual Aδ and C fibers (for references see Willis, 2006). Further, GABA depolarizes the cell bodies of DRG neurons belonging to Aδ and C fibers (Desarmenien et al., 1984; Labrakakis et al., 2003). In sum, PAD results from a GABA-mediated

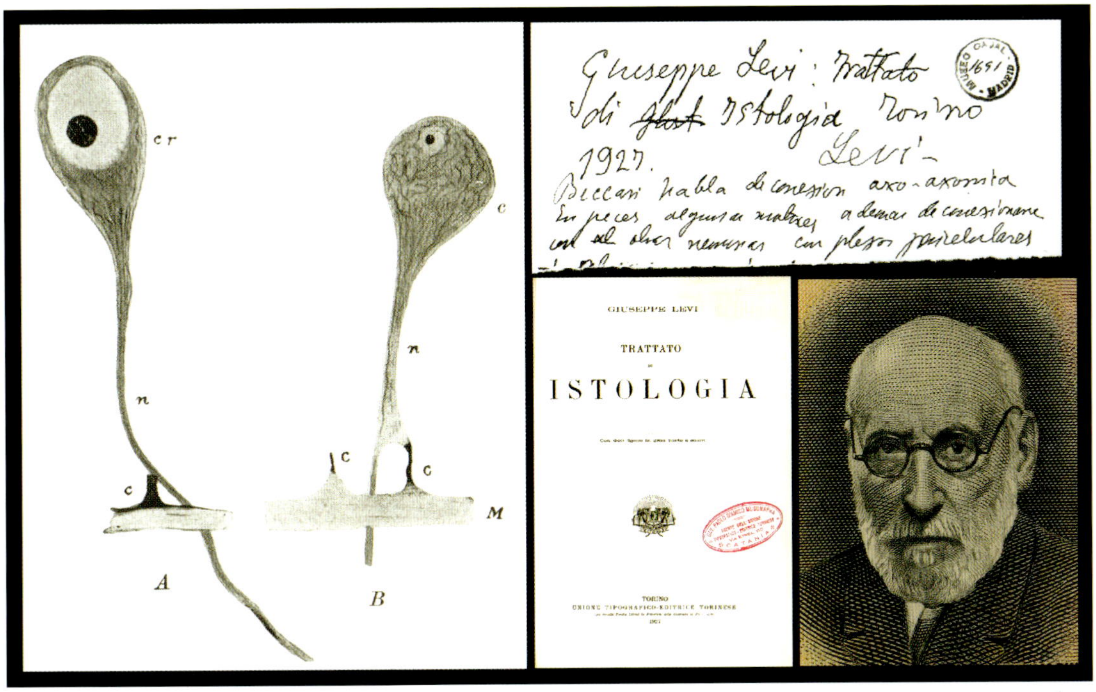

FIGURE 22.1 First histological claims for axo-axonic synapses in vertebrate spinal cord. **Left panel**: drawings from Beccari (1920), reproduced in Levi (1927), showing axo-axonic contacts (c) between the axon (n) of some spinal motoneurons (cr) and the collaterals of the Mauthner cell axon (M) in salmon larvae (*Salmo iredeus*). **Upper right panel**: fragment of a piece of paper with handwriting by Cajal (ca. 1933), where he made a note of the reference to Levi's book on histology, "Giuseppe Levi: Trattato di Istologia, Torino 1927," and below wrote that Beccari talked about axo-axonic contacts between some spinal neurons in fishes... ("Beccari habla de conexion axo-axonica. En peces algunas motrices ademas de conexionarse con otras neuronas con plexos pericelulares..."). Cajal wrote this note when he was preparing his book *Neuron Theory or Reticular Theory?*, first published in Spanish in 1933. The original of this note is kept at the Cajal Museum in Madrid. (Reproduced with permission from Herederos de Santiago Ramón y Cajal.) **Bottom left panel**: title page from Levi's book from which Cajal obtained the information about the presumed axo-axonic contacts. **Bottom right panel**: Ramón y Cajal as he appeared printed on the "50 pesetas" notes that circulated in Spain in 1935 during the Spanish Republic, one year after his death (courtesy of María Ángeles Ramón y Cajal).

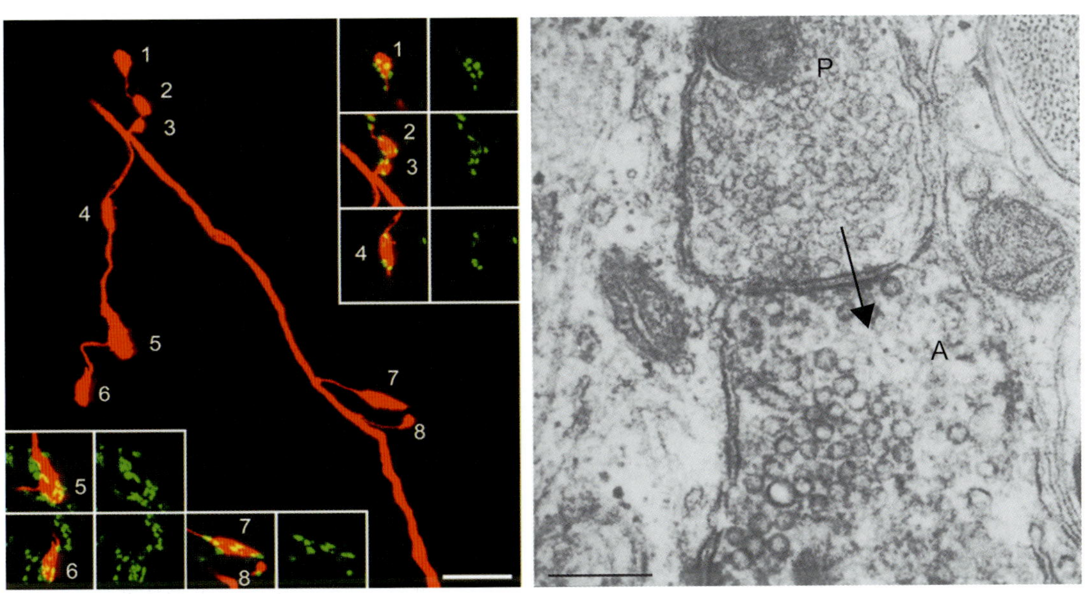

FIGURE 22.2 "P" boutons in mammalian spinal cord. **Left**: Association of GABAergic terminals from interneurons located in deep medial dorsal horn (laminae V–VI), with a neurobiotin-injected Ia afferent fiber in rat spinal cord. The main image is a projection of confocal scans through part of a single Ia collateral with eight varicosities in lamina IX (neurobiotin appears red). The remaining pairs of images show the association of GAD65 immunostaining (green) with each varicosity. All of the Ia varicosities are contacted by at least one GAD65-immunoreactive bouton. The main image is a projection of 37 optical sections at 0.5 μm z-spacing, whereas those in the insets are projected from 5, 13, 6, 5, 5 and 9 optical sections, respectively. Calibration bar: 10 μm. (Courtesy of Professor Andrew Todd. Reproduced with permission from Hughes et al., 2005.) **Right**: Electron micrograph of an HRP-labeled Ia synaptic bouton in the lumbosacral spinal cord of the cat. The "P"-type bouton (P) is presynaptic to the labeled bouton A. (Courtesy of Professor Robert Fyffe. Modified from Fyffe and Light, 1984.)

conductance increase to Cl^- in or near the central terminals of PSNs. GABA is released from spinal interneurons, often called "PAD interneurons" that make axo-axonic or dendro-axonic synapses with the central terminals of primary afferents (Hughes et al., 2005; Rudomin and Schmidt, 1999; Willis, 2006).

II. THE CELLULAR AND MOLECULAR BASIS OF PRESYNAPTIC INHIBITION AND PAD

A. Adult Primary Sensory Neurons are Depolarized by GABA over their Entire Surface Membrane due to an Outward Cl^- Gradient

The realization that GABA acting via $GABA_A$ receptors coupled to anion channels depolarizes some neurons and hyperpolarizes others has challenged the classic view that considered this amino acid to be a purely inhibitory neurotransmitter (Bowery and Smart, 2006; Owens and Kriegstein, 2002). Further, *the GABA-induced depolarizations can be excitatory or inhibitory.* GABA depolarizes and excites CNS neurons during embryonic and early postnatal development (Owens and Kriegstein, 2002). Such depolarizations are thought to be crucial for developmental processes like neurogenesis, neuronal migration, neurite outgrowth, synapse formation and maintenance of early network activity (Ben-Ari et al., 2007; Owens and Kriegstein, 2002; Represa and Ben-Ari, 2005; Wang and Kriegstein, 2009). As development proceeds GABA becomes hyperpolarizing and inhibitory in mature central neurons. Both depolarizing and hyperpolarizing GABA responses are generated by Cl^- currents through anion channels coupled to $GABA_A$-receptors. The polarity and magnitude of these currents are determined primarily by $[Cl^-]_i$ which is maintained either above or below electrochemical equilibrium by Cl^- transporters expressed in the neuronal plasma membrane (Alvarez-Leefmans, 1990, 2001; Blaesse et al., 2009; Payne et al., 2003).

The hyperpolarizing shift in the reversal potential of $GABA_A$ currents (E_{GABA}) during postnatal development is due to a relatively abrupt downward shift in $[Cl^-]_i$, as demonstrated by direct $[Cl^-]_i$ measurements in cortical and retinal neurons at different postnatal ages (Berglund et al., 2006; Zhang et al., 2006b). Because of the abruptness of this remarkable phenomenon, it is often referred to as "the developmental Cl^- switch". This developmental polarity change in GABAergic transmission in cortical neurons is correlated with changes in the temporal and spatial expression of Cl^- transporters that actively accumulate or extrude Cl^- (Ben-Ari et al., 2007; Kakazu et al., 1999; Owens and Kriegstein, 2002; Stein et al., 2004; Yamada et al., 2004; Zhu et al., 2005).

Immature neurons predominantly express isoform 1 of the Na^+-K^+-$2Cl^-$ cotransporter (NKCC1), a transport protein that mediates active Cl^- uptake across the membrane, maintaining the $[Cl^-]_i$ above electrochemical equilibrium (Achilles et al., 2007). There is evidence that some immature neurons also express a Na^+-independent anion exchanger that can also actively accumulate Cl^- (Becker et al., 2003; Gonzalez-Islas et al., 2009). The current paradigm holds that with the appearance of KCC2 at later development stages (~P7–P10), $[Cl^-]_i$ is lowered. KCC2 is one of the four isoforms of the K^+-Cl^- cotransporter proteins that actively extrude Cl^- (see Chapter 17). KCC2 maintains the equilibrium potential for Cl^- (E_{Cl}) at values more negative than the transmembrane potential (E_m), generating an *inward* Cl^- gradient that makes possible hyperpolarizing inhibition by GABA or glycine in the CNS, as discussed in detail in Chapter 17 in this volume.

Primary sensory neurons are a notable exception to the above scheme; unlike CNS neurons they are depolarized by GABA throughout adulthood (Alvarez-Leefmans et al., 1998; Davidoff and Hackman, 1985; Levy, 1977). As in immature cortical neurons, GABA depolarizations in PSNs are primarily due to Cl^- efflux through $GABA_A$ receptor-channels. GABA depolarizes the cell bodies (De Groat et al., 1972; Desarmenien et al., 1984; Gallagher et al., 1978; Nishi et al., 1974) as well as the peripheral (Ault and Hildebrand, 1994; Bhisitkul et al., 1987; Carlton et al., 1999) and central processes of PSNs (Curtis and Lodge, 1982; Labrakakis et al., 2003; Rudomin and Schmidt, 1999; Willis, 1999). Further, GABA depolarizes non-myelinated sensory axons in peripheral nerve (Brown and Marsh, 1978). Myelinated sensory axons in dorsal roots are also depolarized by GABA or the $GABA_A$ agonist muscimol, and the depolarization is blocked with bicuculline, indicating that they are mediated by $GABA_A$ receptors (Bhisitkul et al., 1990, 1987; Morris et al., 1983). Interestingly mammalian ventral root axons and their intraspinal and peripheral terminals are not depolarized by GABA (Bhisitkul et al., 1987; Curtis and Gynther, 1986; Smart, 1980). The latter indicates selectivity in the action of GABA and is consistent with the fact that adult motoneurons, whose axons form the ventral roots, maintain E_{Cl} at more negative values than the E_m. This is why GABA and glycine hyperpolarize spinal motoneurons (Curtis et al., 1968). The GABA hyperpolarization in motoneurons is sensitive to bicuculline and thus is mediated by $GABA_A$ receptors (Krnjevic et al., 1977). However, GABA does not produce any changes in E_m of motor axons. In contrast, the GABA-induced depolarizations of rat dorsal root fibers is present in neonatal rats but is most pronounced in adult animals (Kocsis and Sakatani, 1995). In sum GABA depolarizes the cell bodies as well as the peripheral and central processes of primary afferent neurons via $GABA_A$-receptor-channels. This suggests that an outward Cl^- gradient is maintained throughout the plasma membrane of these neurons.

Why does GABA depolarize primary afferent axons, the gate of entry of virtually all somatic and visceral information, but not ventral root motor axons? One possibility is that motor axons do not express $GABA_A$ receptors. Pharmacological evidence supports this view. Alternatively, it could be that motor axons do have $GABA_A$ receptors but Cl^- is at equilibrium ($E_{Cl} = E_m$) and therefore it is not possible to observe any changes in E_m (see Chapter 5). This appears to be unlikely, at least in the motor nerve terminals of frog motoneurons, since GABA does not produce any detectable effects on E_m when altering the transmembrane Cl^- gradient (Smart, 1980). These findings suggest that there is a fundamental difference between primary afferents fibers, their terminals and their somata and those of motoneurons, with respect to embryological origin, intracellular ion content, pharmacological properties and possibly also in the mechanism of transmitter release (Curtis and Gynther, 1986).

B. Functional Significance of the Outward Cl^- Gradient in the Central Terminals of Primary Afferent Neurons: PAD and Presynaptic Inhibition

This outward Cl^- gradient is functionally highly significant, as it makes possible the GABA-mediated depolarization of the central terminals of primary afferent neurons (PAD). As we have seen, PAD of the central terminals of PSNs is produced by GABA released from P boutons at axo-axonic synapses, where the postsynaptic element is a terminal from a primary afferent fiber (Fig. 22.3). It has also been proposed that PAD or tonic PAD may also be produced via paracrine transmission where GABA would act on presynaptic extrajunctional receptors (Alvarez, 1998; Kullmann et al., 2005). Whatever the mechanism, there is evidence that *PAD is a major cause of presynaptic inhibition, and not a simple epiphenomenon associated with it* (French et al., 2006; Lamotte D'Incamps et al., 1998; Lev-Tov et al., 1983, 1988). GABAergic

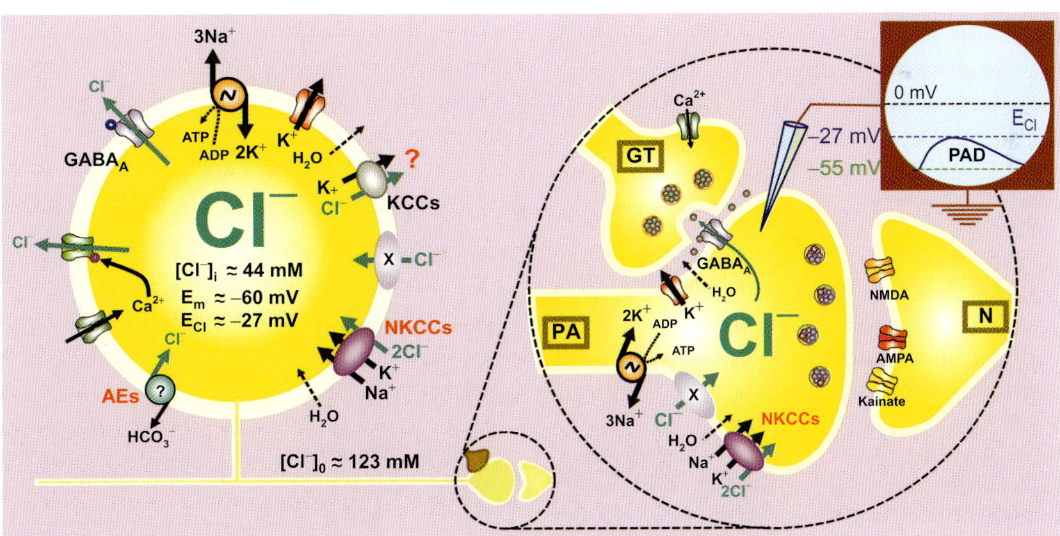

FIGURE 22.3 Schematic diagram depicting the known and hypothetical transporters and channels involved in intracellular Cl^- regulation in primary sensory neurons, and mode of generation of PAD and presynaptic inhibition. The left circle represents a DRG neuron showing measured values of intracellular Cl^- concentration, $[Cl^-]_i$; membrane potential (E_m) and Cl^- equilibrium potential, E_{Cl}. The Na^+/K^+ ATPase maintains the Na^+ and K^+ gradients across the membrane. AEs, anion exchangers; NKCCs, Na^+-K^+-$2Cl^-$ cotransporters and KCCs, are K^+-Cl^- cotransporters 1, 3 or 4 whose contribution, if any, to intracellular Cl^- regulation in these neurons under euvolemic conditions is not clear. KCC2 is not expressed in DRG neurons. X represents other yet to be identified active Cl^- uptake transporters. The right diagram encircled by a dashed line represents a magnified intraspinal primary afferent terminal from which an intracellular micropipette connected to an oscilloscope records E_m. PA, primary afferent; GT, GABAergic terminal from an interneuron; N, postsynaptic neuron with glutamate receptor channels. Note that PAD (blue trace in the oscilloscope screen) blocks the release of neurotransmitter from the PA terminal.

presynaptic control is a crucial mechanism in the gating of afferent flow of information of various sensory modalities including proprioception and pain (Cervero et al., 2003; Labrakakis et al., 2003; Pierrot-Deseilligny and Burkey, 2005; Willis, 2006, 1999).

The mechanism by which PAD decreases neurotransmitter release from the boutons of primary afferents is not clear. This is mainly because of technical limitations; it has not been possible to study the terminals directly either by electrophysiological or by optical methods. The prevailing idea, based on modeling (Cattaert et al., 2001; French et al., 2006; Graham and Redman, 1994; Lamotte D'Incamps et al., 1998; Segev, 1990) and studies in presynaptic terminals of invertebrates (Cattaert and El Manira, 1999; Cattaert et al., 1999, 2001) and vertebrates (Lamotte D'Incamps et al., 1998; Zhang and Jackson, 1993), is that GABA decreases or blocks action potential invasion into the primary afferent terminals thereby reducing or blocking Ca^{2+} influx into the terminals with a consequent decrease in the release of neurotransmitter. There are basically two hypotheses that are not mutually exclusive to explain how GABA achieves this effect via $GABA_A$ presynaptic receptors coupled to anion channels. The first hypothesis states that the increase in membrane conductance to Cl^- produced by activation of $GABA_A$ receptors decreases the input resistance of the terminal arborization thereby decreasing the amplitude of the invading action potential by a shunting mechanism (e.g. Cattaert and El Manira, 1999), as discussed in Chapter 5. However, computer simulations with realistic parameters suggest that relatively large synaptic conductances are required to induce a significant reduction in amplitude of the action potential (Graham and Redman, 1994; Segev, 1990). The second hypothesis, which has received more support, holds that PAD causes a decreased in spike height primarily by Na^+ channel inactivation produced by the depolarization (French et al., 2006; Lamotte D'Incamps et al., 1998). In both hypotheses the action potential invading the primary afferent terminals is either decreased in amplitude or blocked. Further, as suggested by Graham and Redman (1994), PAD might also have a more direct effect on transmitter release by inactivating the Ca^{2+} current at the terminals. The end result of all these possibilities is reduction or block of Ca^{2+} influx decreasing the amount of transmitter released from the primary afferent terminals and consequently reducing monosynaptic transmission to second order spinal neurons (Lamotte D'Incamps et al., 1998; Lev-Tov et al., 1983, 1988; Pierrot-Deseilligny and Burkey, 2005; Rudomin and Schmidt, 1999).

C. The GABA-mediated PAD is Determined by the Magnitude of the Outward Cl^- Gradient across the Plasma Membrane of PSNs

The above-mentioned observations set the stage for the following hypotheses: (1) PSNs maintain E_{Cl} at values less negative than E_m generating an outward driving force for Cl^- which makes possible $GABA_A$-mediated PAD and presynaptic inhibition. Thus, PSNs are endowed with active Cl^- uptake mechanisms that keep $[Cl^-]_i$ at levels above the predicted electrochemical equilibrium; (2) since GABA application depolarizes all parts of PSNs, the outward Cl^- gradient must be maintained throughout their entire cell surface, including the cell body and the peripheral and central processes, and hence the Cl^- uptake mechanisms must be expressed in the entire cell surface; (3) PAD amplitude and presynaptic inhibition are influenced by $[Cl^-]_i$; (4) regulation of Cl^- transport mechanisms that maintain the Cl^- gradient determines PAD amplitude and the strength of presynaptic inhibition. Changes in Cl^- regulatory mechanism increase $[Cl^-]_i$ shifting E_{Cl} to more depolarized values and transforming PAD into an excitatory phenomenon. When PAD amplitude reaches the firing threshold of the terminal it produces dorsal root reflexes (DRRs), pain and neurogenic inflammation, as discussed below; (5) the outward Cl^- gradient contributes to the generator potential in nociceptive peripheral terminals, as discussed below. Altogether, these hypotheses and their implications highlight the significance of understanding the mechanisms that regulate intracellular Cl^- in PSNs, and in particular those that generate and maintain the outward Cl^- gradient.

The level of intracellular Cl^-, and hence the magnitude of the outward Cl^- gradient, is the result of a functional balance between Cl^- channels and carriers present in the plasma membrane. The outward Cl^- gradient reflects the fact that $[Cl^-]_i$ is maintained above electrochemical equilibrium in PSNs. Hence, the magnitude of the Cl^- gradient and, logically, the amplitude of PAD and the strength of presynaptic inhibition must be determined by the level of intracellular Cl^- in PSNs, specifically in their intraspinal terminals. The level of intracellular Cl^- results from a functional balance between Cl^- channels (passive leak pathways) and various carriers present in the plasma membrane of PSNs (Fig. 22.3).

There are various transporters that could potentially accumulate Cl^- in PSNs. From a functional perspective, these active Cl^- loaders can be divided into two groups: those that are driven by the Na^+ gradient (Na^+dependent) and those that are Na^+independent. The Na^+-driven Cl^- cotransporters are members of the

SLC12A gene family of electroneutral cation-chloride-coupled cotransporters (CCCs) that comprises the thiazide-sensitive Na^+-Cl^- cotransporter NCC (SLC12A3) and two isoforms of the Na^+-K^+-$2Cl^-$ cotransporters (NKCCs), as discussed in Chapters 11 and 16 in this volume. Our published and unpublished pharmacological tests show that a thiazide-sensitive NCC does not play a role in Cl^- accumulation in DRG neurons (Rocha-González et al., 2008), and therefore it will not be considered further. We have shown that a Na^+-K^+-$2Cl^-$ cotransport mechanism is expressed in primary sensory neurons, as will be discussed below.

The Na^+ – independent Cl^- loaders belong to the SLC4A gene family (Romero et al., 2004) that encodes for the Anion-Exchangers (AEs) 1–4, and to some members of the SLC26A family (Mount and Romero, 2004), as discussed in Chapter 4. We have evidence showing that some DRG neurons express an AE, as will be reported in a forthcoming publication.

Regarding the possible contribution of a Cl^- leak conductance that would impact the $[Cl^-]_i$, the ubiquitous ClC-2 anion channel that stabilizes E_m in many neurons is not functionally expressed in DRG cells (Staley et al., 1996). In amphibian DRG cells the resting Cl^- conductance is negligible (Alvarez-Leefmans, 1990; Deschenes et al., 1976). In spite of its importance, little is known about so-called "background" Cl^- channels in PSNs (Strupp and Grafe, 1991). It is conceivable, however, that the intraspinal terminals of primary sensory neurons have a relatively high resting Cl^- conductance due to putative tonic signaling mediated by extrasynaptic $GABA_A$ receptors (Kullmann et al., 2005). A similar situation may occur in the cell bodies where GABA appears to be released from satellite cells (Hayasaki et al., 2006). But for now, this is just mere speculation. GABA-gated or Ca^{2+}-activated anion channels would be expected to transiently increase the membrane permeability to Cl^- and may also determine the $[Cl^-]_i$ in PSNs. The carriers and channels that may be involved in intracellular Cl^- regulation in PSNs are illustrated in Fig. 22.3.

D. Primary Sensory Neurons are Endowed with a Na^+-K^+-$2Cl^-$ Cotransporter that Generates and Maintains the Outward Cl^- Gradient

Embryonic and adult PSNs of various species, including humans, are depolarized by GABA (Alvarez-Leefmans, 1990; Alvarez-Leefmans et al., 1998; De Groat, 1972; De Groat et al., 1972; Deschenes et al., 1976; Dunlap, 1984; Eccles et al., 1963; Gallagher et al., 1978, 1983; Levy, 1977; Nishi et al., 1974; Rohrbough and Spitzer, 1996; Sung et al., 2000; Valeyev et al., 1999). These $GABA_A$-mediated depolarizations must be an evolutionary meaningful phenomenon for sensory signaling since they are preserved from lower vertebrates to humans.

Primary sensory neurons were the first vertebrate neurons in which, more than 35 years ago, GABA was found to have a depolarizing action (De Groat, 1972; De Groat et al., 1972; Eccles et al., 1963; Nishi et al., 1974; reviewed in Alvarez-Leefmans, 1990; Levy, 1977; Rudomin and Schmidt, 1999; Willis, 2006). GABA depolarizes the central terminals of primary afferents not only when applied exogenously, but also through GABAergic axo-axonic and probably dendro-axonic synapses that produce PAD (Alvarez, 1998; Alvarez-Leefmans et al., 1998; Rudomin and Schmidt, 1999; Todd and Ribeiro-da-Silva, 2005; Willis and Coggeshall, 2004), as discussed above.

Earlier workers realized that for this depolarization to ensue, $[Cl^-]_i$ in PSNs had to be higher than predicted for electrochemical equilibrium, and that maintenance of this gradient required an active Cl^- uptake mechanism. At that time (mid 1970s) the concept of secondary active transport (Crane, 1977) was not as widely held as that of "pumps", i.e. primary active transport mechanisms *directly* energized by ATP hydrolysis (see Chaper 5 in this volume). Accordingly, Nishi, Feltz, Gallagher and others proposed that PSNs were endowed with an "inward directed chloride pump" (Deschenes et al., 1976; Gallagher et al., 1978, 1983; Nishi et al., 1974; reviewed in Alvarez-Leefmans et al., 1998).

Work originated in our laboratory on amphibian dorsal root ganglia first established that the depolarizing action of GABA is possible because $[Cl^-]_i$ in PSNs is higher than predicted for electrochemical equilibrium, due to functional expression of a Na^+-K^+-$2Cl^-$ cotransport mechanism (NKCC) that actively accumulates Cl^- (Alvarez-Leefmans et al., 1986, 1988). Subsequent studies using molecular methods identified it as NKCC1 (Kanaka et al., 2001; Plotkin et al., 1997b; Sung et al., 2000). We measured $[Cl^-]_i$ and membrane potential (E_m) in single DRG neurons using double-barreled ion-selective microelectrodes, and determined E_{Cl} from directly measured intra- and extracellular Cl^- activities. E_{Cl} was -33 ± 1 mV, i.e. ~25 mV positive to the resting E_m (-58 ± 1). DRG cells that were depleted of intracellular Cl^- and then exposed to Ringer's solution containing physiological $[Cl^-]$ actively accumulated Cl^- without changes in E_m, i.e. the transmembrane Cl^- fluxes were electrically silent (see Chapter 5 and Fig. 5.4 in this volume). The net inward Cl^- fluxes were coupled to Na^+ and K^+, and were blocked by bumetanide. We concluded that the inward active Cl^- transport was not a "chloride pump" but an electroneutral Na^+-K^+-$2Cl^-$ cotransporter mechanism similar to that described a few years earlier in tumor cell lines (Geck and Heinz, 1986; Geck et al., 1980) and in squid axons (Russell, 1983, 2000). We also showed that both net Cl^- influx and efflux were mediated by NKCC consistent with the characteristic reversibility of active transport, i.e. the transporter can work in forward and backward modes and the net direction of cotransport is determined by the sum of the chemical potential gradients of the transported ions (Lytle et al., 1998; Russell, 2000).

The expression of NKCC protein by Western blot analyses and immunocytochemistry was confirmed in DRG cells of frog, cat, rat and mouse (Alvarez-Leefmans et al., 2001; Plotkin et al., 1997a; Sung et al., 2000), and was identified as NKCC1 in both mouse (Sung et al., 2000) and rat (Plotkin et al., 1997a). Another important advance supporting a critical role of NKCC1 in maintaining $[Cl^-]_i$ above electrochemical equilibrium in PSNs were the studies in DRG neurons isolated from NKCC1 knockout mice (Sung et al., 2000). E_{GABA} measured with gramicidin-perforated patches was more negative in DRG cells from NKCC1$^{-/-}$ than in those from wild type, suggesting that NKCC1$^{-/-}$ cells do not actively accumulate Cl^-. Altogether these studies confirmed that NKCC1 is a transporter responsible for active accumulation of Cl^- in PSNs.

However, it is not known if NKCC1 is the only transport mechanism that determines $[Cl^-]_i$ in these cells. Functional studies of Cl^- regulation by direct measurement of changes in $[Cl^-]_i$ in rat DRG neurons confirmed that $[Cl^-]_i$ is maintained above electrochemical equilibrium in these mammalian cells. However, only 65% of the Na^+-dependent active Cl^- uptake was inhibited by bumetanide, suggesting that DRG neurons are also endowed with Na^+-dependent Cl^- transport mechanisms other than NKCC1 (denoted as transporter "X" in Figs 22.3, 22.6 and 22.10). In the same study (Rocha-González et al., 2008) we ruled out that one of the additional mechanisms was a thiazide-sensitive NCC (Gamba, 2005); neither chlorothiazide (0.1–1 mM) nor metolazone (10 μM) affected the Na^+-dependent Cl^- uptake. However, pharmacological sensitivity of transport molecules to so-called specific blockers varies widely. In this context it is worth remembering the case of the ouabain-sensitivity of the Na^+/K^+ ATPase. The rodent α1 isoform is *1,000-fold* less sensitive to glycoside inhibition than other isoforms. Based on this glycoside sensitivity it would have been wrongly concluded that the Na^+/K^+ ATPase does not exist in rodents! It is the presence of the charged residues, R111 and D122, in the rat α isoform that accounts for the decreased ouabain sensitivity observed for the rat (Price and Lingrel, 1988). Most species have uncharged residues in this position (e.g. sheep α1 has Q111 and N122), and these species are sensitive to cardiac glycosides. Considering that the current state of our knowledge of transporters of the SLC12A family lags about 20 years behind the studies of the Na^+/K^+ ATPase, it would not be surprising if we find proteins of the SLC12A family with mutations that make them more or less sensitive to thiazides for the case of NCC, or to bumetanide for the case of NKCCs.

E. Intracellular $[Cl^-]$ is Maintained above Electrochemical Equilibrium in Most DRG Neurons Irrespective of their Phenotype

Primary sensory neurons (PSNs) exhibit great phenotypic diversity (e.g. size, type of ion channels, physiological features such as electrical properties and sensory modality, enzymes and peptides content, surface molecules, etc.). This diversity reflects their functional specialization as conduits for various sensory modalities (e.g. temperature, proprioception, pain); there is a tight correlation between a primary sensory neuron's phenotype and the sensory receptor it innervates (Lawson, 2005; Willis and Coggeshall, 2004). The cell bodies of PSNs are located in dorsal root ganglia (DRGs) and in cranial root ganglia. DRG neurons may be nociceptive or non-nociceptive; may have unmyelinated (C) fibers or myelinated (A) fibers; and may respond to mechanical, thermal and/or chemical stimuli.

Physiological studies have shown that nociceptive DRG cells are associated with unmyelinated C fibers and small, thinly myelinated Aδ fibers. The majority of C fibers (over 90% in human cutaneous nerves) are called polymodal nociceptors because they respond to more than one sensory modality, that is, they respond to noxious mechanical, thermal or chemical stimuli. Many Aδ fibers respond to intense mechanical stimuli, or are mechanothermal (heat-sensitive) nociceptors, while the remainder are associated with low threshold mechanoreceptors or cold receptors. The larger myelinated fibers Aα/β (Groups I and II) innervate non-nociceptive, low-threshold mechanoreceptors (Julius and Basbaum, 2001).

When classified by their size (soma cross-sectional area, CSA) DRG neurons yield bimodal distributions representing two overlapping populations of cells (Lawson, 2005): small and large. Nociceptive neurons fall within the small size population. Small size DRG neurons can be divided into two subpopulations that have been designated as "peptidergic" and "non-peptidergic" (Julius and Basbaum, 2001; Kashiba et al., 2001; Snider and McMahon, 1998). The peptidergic group constitutively express neuropeptides such as calcitonin gene-related peptide (CGRP) and substance P, and do not bind isolectin-B4 (IB4). In contrast, the non-peptidergic subpopulation do not express neuropeptides or do so at low levels, and is characteristically labeled with IB4 (Alvarez and Fyffe, 2000; Silverman and Kruger, 1990; Snider and McMahon, 1998). IB4 positive ($IB4^+$) primary afferents project to inner lamina II (LIIi) of the dorsal horn, whereas peptidergic, IB4 negative ($IB4^-$) afferents project to lamina I and outer lamina II (LIIo) of the dorsal horn (Silverman and Kruger, 1990). These differences between the $IB4^+$ and $IB4^-$ nociceptive afferents suggest that they play somewhat different roles in neurogenic inflammation and the dorsal horn circuitry underlying PAD, as discussed below.

Lawson has further classified the DRG cells into three groups according to their CSA, functional properties (sensory properties and conduction velocity), IB4 binding (Fang et al., 2005) and immunoreactivity to a number of biochemical markers (Lawson, 2005). In brief, neurons are divided into small (CSA up to $400\,\mu m^2$); medium ($>400\,\mu m^2$ and up to $800\,\mu m^2$) and large ($>800\,\mu m^2$). Although the boundaries between each group overlap ($\pm 100\,\mu m^2$) this is a convenient classification that parallels the classification derived from physiological studies. One of the main conclusions from Lawson's group (Fang et al., 2006) is that all strongly $IB4^+$ cells are C-nociceptive DRG neurons having CSA of up to $400\,\mu m^2$; although they spread to $\sim 600\,\mu m^2$, the size range of Aδ nociceptors. In fact, another study provided evidence that $IB4^+$ neurons are mostly C-nociceptive but some innervate high threshold Aδ nociceptors (Gerke and Plenderleith, 2001). Thus, $IB4^+$ cells are non-peptidergic C- and Aδ-nociceptive neurons. $IB4^-$ small and medium size are peptidergic nociceptors whereas large (mostly $>800\,\mu m^2$) $IB4^-$ cells innervate low threshold cutaneous mechanoreceptors, Golgi tendon organs and muscle spindles.

We have used fluorescence imaging microscopy to measure $[Cl^-]_i$ and cell water volume (CWV) in real time, in neurons dissociated from rat DRG cells (P0–P21). $[Cl^-]_i$ and CWV were measured with the fluorescent indicators MQAE (Kaneko et al., 2001; Koncz and Daugirdas, 1994; Maglova et al., 1998) and calcein (Alvarez-Leefmans et al., 1995; Crowe et al., 1995), respectively (Fig. 22.4). We found that irrespective of their size, postnatal age or IB4 binding capacity, $[Cl^-]_i$ in DRG neurons is maintained above electrochemical equilibrium primarily through an NKCC mechanism. Basal $[Cl^-]_i$ was $44.2 \pm 1.2\,mM$ (mean ± SE; SD 9.9 mM, $n = 75$ neurons) and E_{Cl} was $-27.0 \pm 0.7\,mV$ (SD 6 mV). This $[Cl^-]_i$ is three to four times higher than predicted for a passive distribution which for cells having an E_m between −60 and −55 mV (Hong et al., 2004; Petruska et al., 2000; Yu and Kocsis, 2005; Zheng et al., 2007) would have been ~11.9 and ~14.5 mM, respectively, as calculated from the equation: $[Cl^-]_i^{eq} = [Cl^-]_o \exp(E_m F/RT)$, where $[Cl^-]_i^{eq}$ is the $[Cl^-]_i$ at electrochemical equilibrium, $[Cl^-]_o$ is the extracellular $[Cl^-]$ (123 mM), and E_m, R, T, and F have their usual meaning (see Chapter 5). Hence, in rat DRG neurons, E_{Cl} was 28 to 33 mV positive to E_m, in agreement with the values found in amphibian neurons (Alvarez-Leefmans et al., 1988).

We studied the effect of changes in $[Cl^-]_o$ on $[Cl^-]_i$ and CWV. On isosmotic removal of external Cl^- (replaced with gluconate), cells lost Cl^- and shrank. On returning to control solution, cells reaccumulated Cl^- and recovered CWV. Cl^- reaccumulation and CWV recovery in Cl^--depleted cells was studied by restoring control $[Cl^-]_o$ first in the absence of external Na^+ (by isosmotic replacement of Na^+ with glucamine) for 15 to 30 min, followed by the control isosmotic solution (containing physiological $[Na^+]_o$ and $[Cl^-]_o$). We found that Cl^- reaccumulation had two components: Na^+-dependent (SDC) and Na^+-independent (SIC), as shown in Fig. 22.4A. The SIC stabilized at $[Cl^-]_i = 13.2 \pm 1.2\,mM$ ($n = 27$), and remained stable throughout the duration of exposure to the Na^+-free ($0Na^+$) solution. The $[Cl^-]_i$ at which the SIC stabilized corresponds to an E_{Cl} of $60.5 \pm 3\,mV$, which is close to or equal to E_m, suggesting that it was mediated through a passive Cl^- leak conductance (i.e. a channel). In contrast, the SDC was due to *active* Cl^- uptake. Bumetanide blocked 65% of the SDC ($IC_{50} = 5.7\,\mu M$) suggesting that most of the active Cl^- influx was mediated by NKCC (Russell, 2000). CWV recovery had also SIC and SDC. However, the SIC was negligible or absent and virtually all CWV recovery occurred through Na^+-dependent bumetanide-sensitive mechanisms (Fig. 22.4).

Rat DRG neurons of different soma area and IB4 binding capacity have the same features described in our previous study (Rocha-González et al., 2008), namely $[Cl^-]_i$ is above electrochemical equilibrium irrespective of phenotype and Cl^- accumulation has SDC and SIC (Fig. 22.5). The fact that $[Cl^-]_i$ is above equilibrium in virtually *all* DRG neurons is in agreement with the results reported by the Frings' lab (Gilbert et al., 2007), and with the immunolabeling data. Our values for $[Cl^-]_i$ are, in general, slightly

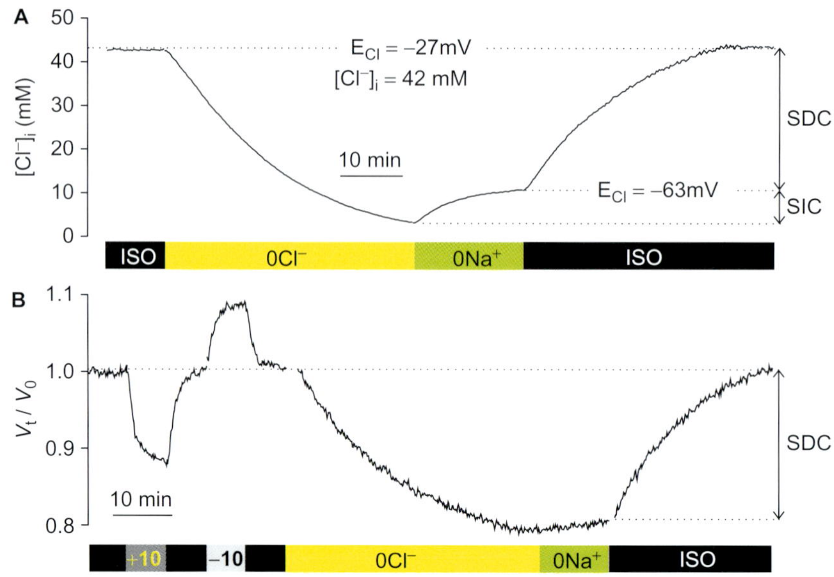

FIGURE 22.4 Changes in $[Cl^-]_i$ and cell water volume (CWV) measured with fluorescent indicators dyes MQAE (A) and calcein (B) in dissociated rat DRG neurons. **A.** External Na^+ and Cl^- requirement for intracellular Cl^- accumulation following Cl^- depletion. To measure changes in $[Cl^-]_i$ the neuron was loaded with the fluorescent dye MQAE (N-(ethoxycarbonylmethyl)-6-methoxyquinolinium bromide). Basal $[Cl^-]_i$ was 42 mM ($E_{Cl} = -27$ mV). Note the changes in $[Cl^-]_i$ following isosmotic removal and readdition of external Cl^-, in the absence and in the presence of external Na^+. On removal of external Cl^- ($0Cl^-$), the cell (491 μm^2) was depleted of Cl^-. On restoring external Cl^- (123 mM) in the absence of external Na^+ ($0Na^+$) there was a small increase in $[Cl^-]_i$ denoted SIC (sodium independent component) that reached steady-state at predicted electrochemical equilibrium ($[Cl^-]_i = 10.7$ mM, $E_{Cl} = -63$ mV). On exposure to the isosmotic control solution (ISO), $[Cl^-]_i$ recovered to initial values. This sodium-dependent component of active Cl^- accumulation is denoted as SDC. **B.** Changes in CWV on removal and readdition of external Cl^-, in the absence and in the presence of external Na^+, as in A. Osmotic calibration pulses ($\pm 10\%$ anisosmotic) preceded exposure to $0Cl^-$. On exposure to $0Cl^-$, the neuron (470 μm^2) shrank by 20.6% of its initial CWV (V_t/V_0). On reaching steady-state shrinkage, the cell was exposed to $0Na^+$. CWV recovery in $0Na^+$ (SIC) was negligible (0.06%/min). On returning to ISO control, the cell fully recovered its volume (SDC). Modified from Rocha-González, Mao and Alvarez-Leefmans (2008).

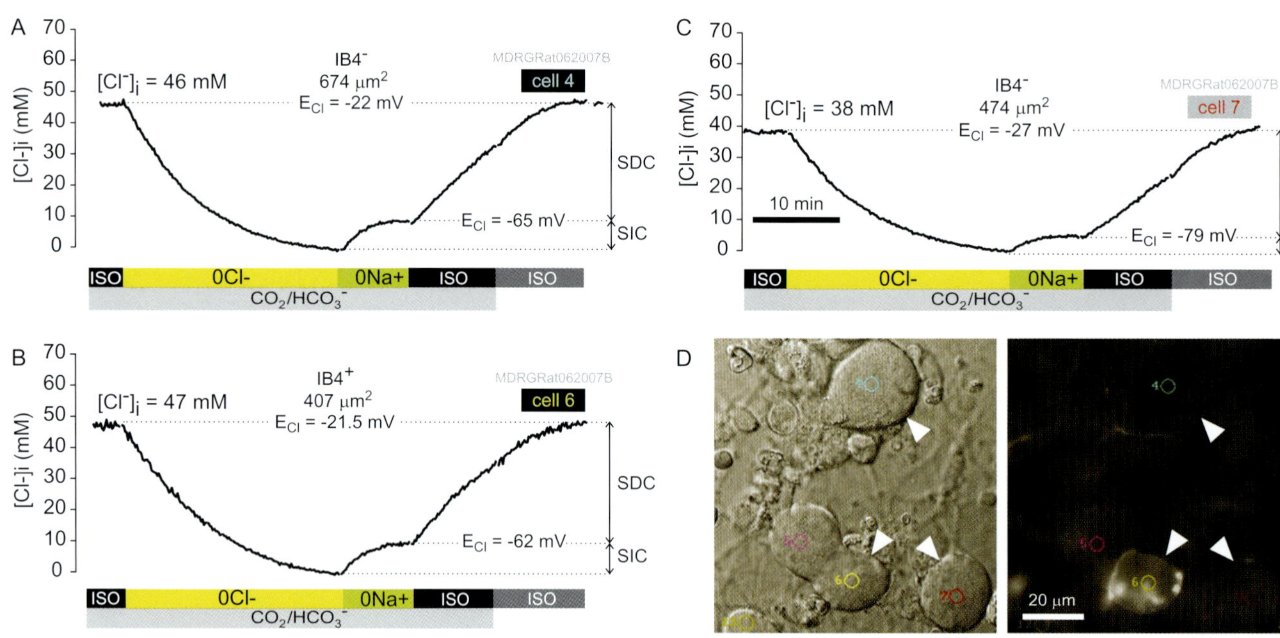

FIGURE 22.5 Intracellular $[Cl^-]_i$ in acutely dissociated DRG neurons of identified phenotype. Cells were loaded with MQAE. **A – C.** Indicated on top of each trace are: IB4-labeling capacity (+ or −), the number in μm^2 is the cross-sectional area for each cell and E_{Cl} the Cl^- equilibrium potential. SDC and SIC are the sodium-dependent and -independent components of Cl^- accumulation. The nomenclature for solution's composition is the same as in Fig. 22.4. Cells were maintained in CO_2/HCO_3^- buffered solutions during the times indicated. **D.** Left, differential interference contrast (DIC) image of the cells. Arrowheads indicate the cells from which the recordings in A – C were obtained. The small colored circles on each cell are the digital pinhole areas from which MQAE fluorescence was measured. The cell number is indicated on the left of each circle. Right, shows IB4 fluorescent labeling (FITC) of the cells in the left panel. Only cell 6 (yellow pinhole) was $IB4^+$. The composition of the ISO solution was (in mM): 125 Na^+, 108 Cl^-, 24.8 HCO_3^-, 5% CO_2, 3 K^+, 1 Ca^{2+}, 1.25 Mg^{2+}. In the $0Cl^-$ solution, external Cl^- was isosmotically replaced with gluconate. In the $0Na^+$ solution external Na^+ was isosmotically replaced with the chloride salt of N-methyl-d-glucamonium. (Unpublished data from Mao and Alvarez-Leefmans.)

lower than those reported by them, most likely due to methodological differences in calibration procedure. Nevertheless, the basic findings are essentially the same. Our results are also in agreement with the only systematic functional study that has looked at differences in $[Cl^-]_i$ (deduced from E_{GABA}) between PSNs of different phenotype (Desarmenien et al., 1984). This study was done in intact DRG from rats, using single-electrode voltage-clamp measurements of E_{GABA} in Aδ, Aβ and C DRG neurons identified on the basis of their diameter and conduction velocity. Although all DRG cells were depolarized by GABA, E_{GABA} varied considerably between cells. Whether this variability in E_{GABA} reflects different $[Cl^-]_i$ and/or Cl^- regulatory mechanisms between cells of different phenotype or differences in $GABA_A$-receptor-channel biophysical and molecular properties is not known. Direct measurement of $[Cl^-]_i$ in cells with identified phenotype (Gilbert et al., 2007; Rocha-González et al., 2008) has revealed variability in $[Cl^-]_i$ in agreement with the early inferential measurements of Desarmenien and collaborators (1984).

F. NKCC1 Protein is Expressed in Virtually all DRG Cells Irrespective of their Size

Work done by us (Alvarez-Leefmans et al., 2001) and by others (Plotkin et al., 1997a; Sung et al., 2000) using immunohistochemistry and Western blot analyses (Plotkin et al., 1997a), knockout models (Sung et al., 2000), PCR (Galan and Cervero, 2005; Price et al., 2006) and *in situ* hybridization (Kanaka et al., 2001; Toyoda et al., 2005) confirmed the presence of NKCC in mammalian DRG and identified it as NKCC1. However, as discussed above, PSNs exhibit great phenotypic diversity reflecting their functional specialization in signal transmission of various sensory modalities. The question arises as to whether NKCC1 is expressed in *all* DRG cells or is confined to specialized neuronal subtypes. Previous work (Plotkin et al., 1997a; Sung et al., 2000) showed that the NKCC1 protein is present in *all* DRG cells irrespective of their size. Our study was in agreement with this conclusion. However, the monoclonal antibody we used (T4) is not isoform specific (Lytle et al., 1995). We were aware of this shortcoming as stated in our paper (Alvarez-Leefmans et al., 2001), and this is why we referred to T4 immunoreactivity as NKCC and not NKCC1. We have re-examined the expression pattern of NKCC1 in rat and mouse DRG cells using confocal immunofluorescence and an isoform-specific antibody. Rat and mouse DRGs probed with a well-characterized affinity-purified anti-NKCC1 polyclonal antibody (Plotkin et al., 1997a) raised against amino acids 938–1011 of the carboxy terminus of NKCC1 (Kaplan et al., 1996) revealed widespread NKCC1 immunoreactivity in virtually *all* sensory neurons in a given ganglion. We have confirmed the specificity of the affinity purified NKCC1 antibody by using the most stringent control, i.e. the lack of immunoreactivity in DRG from $NKCC1^{-/-}$ mouse as compared to wild-type animals. Double immunolabeling of NKCC1 and NeuN (an antibody used as a pan-neuronal marker) confirmed that *all* DRG neurons express NKCC1, an observation that is consistent with the overwhelming evidence that: (i) GABA depolarizes all PSNs; (ii) virtually all neurons dissociated from rat DRG cells have higher than equilibrium $[Cl^-]_i$ irrespective of their size, postnatal age or IB4-labeling capacity and (iii) the mechanism of Cl^- accumulation has ion requirement and pharmacology consistent with NKCC. The results from this study will appear in a forthcoming publication from our lab.

In spite of the overwhelming evidence showing that NKCC1 protein is expressed in all DRG neurons, a recent study has challenged this view by claiming that NKCC1 mRNA is detectable only in "small and medium diameter sensory neurons" and that the NKCC1 protein does not reside in the cell bodies of PSNs (Price et al., 2009, 2006). The authors speculate that "NKCC1 protein is rapidly trafficked to central and peripheral terminals". However, neither tissue fixation protocols nor antibody validation were presented. We have found, as discussed below, that NKCCs' immunoreactivity is strongly influenced by the level of fixation, which could explain their failure to detect NKCC1.

G. Identification of NKCC1 by Immunocytochemistry: The Issue of Reliable Antibodies

In a recent review on cation-chloride cotransporter function in neurons, Blaesse and coworkers concluded, albeit without supporting evidence, that there are "no reliable immunohistochemical techniques" to study NKCC1 (Blaesse et al., 2009). These authors claimed that the monoclonal antibody T4 is "a useful tool in immunoblots" but gives conflicting results in immunohistochemistry. They praise a polyclonal antibody that "detects the phosphorylated NKCC1, thereby providing data on both the presence and functional state of NKCC1." They were referring to the phospho-antibody R5, a polyclonal antibody raised in rabbit against a 16-amino acid peptide, Tyrh208-Argh223-Lys (YYLR**T***FGHN**T***MDAVPR**K**) in which phosphothreonines Thr212 and Thr217 (in human NKCC1) were incorporated directly during synthesis (Flemmer et al., 2002). This latter antibody has proven an invaluable tool to study the role of phosphorylation on cotransporter activity, but it is not – and was never claimed to be – isoform specific. In fact, R5 has been used to detect not only phosphorylated NKCC1 (Flemmer et al., 2002) but phophorylated NKCC2 (Gimenez and Forbush, 2003) and NCC (Pacheco-Alvarez et al., 2006). Similar concerns about reliability of anti-NKCC1 antibodies were raised in another recent review (Price et al., 2009).Thus a re-evaluation of these issues is warranted in the present chapter.

There are several commercial and non-commercial antibodies raised against NKCC proteins, but not all of them have been validated with rigorous criteria. The key issue is to make sure that the antibody specifically reveals the antigen of interest and not a binding artifact. Criteria for validating antibody specificity have been discussed in several publications (Rhodes and Trimmer, 2006, 2008; Saper, 2005; Saper and Sawchenko, 2003). The gold standard is to stain the tissue with and without the antigen of interest. This can be achieved by staining tissues of wild-type mouse versus a knockout animal. The test is not free of problems because mouse monoclonal antibodies have to be tested against mouse tissues and this can result in severe non-specific staining in the knockout, leading to wrong conclusions. There are, however, ways to circumvent this problem (Brown et al., 2004; Hierck et al., 1994). Less stringent validation tests include immunolabeling of tissues in which the presence of the antigen of interest has been ascertained by independent methods (positive controls), and adsorption controls in which an excess amount of antigen is reacted with the antibody, thereby rendering it incapable of binding to the target antigen in the tissue.

Besides validation of specificity, a well-known problem worth considering is the post-fixation protocol; excessive post-fixation can alter the epitope of interest. This problem is particularly evident with NKCC1, as will be reported in a forthcoming paper from our group. Finally, another no less important problem is the epitope accessibility due to multimeric interactions, posttranslational modifications of protein antigens or orientation of the antigen. In these cases, it is necessary to "unmask" the epitopes for which there are several methods. The detail discussion of these problems is beyond the scope of the present chapter but the reader should consult extensive specialized monographs dealing with these issues (Harlow and Lane, 1988, 1999; Hockfield et al., 1993).

1. Monoclonal Antibody T4

The most widely used antibody to study NKCC1 has been T4, a mouse monoclonal antibody that was generated against a fusion protein fragment that recognizes the last 310 amino acids of the C-terminus of hNKCC1, that is, from M902 to S1212 (Lytle et al., 1995). However, T4 recognizes both NKCC1 and NKCC2 isoforms (Alvarez-Leefmans et al., 2001; Lytle et al., 1995; Marty et al., 2002; Munoz et al., 2007; Yan et al., 2003) and potentially all their splice variants. Interestingly, the percent of identity of these 310 amino acid residues in the C-terminus of human NKCC1 with respect to hNKCC2 and hNCC are: NKCC1 vs. NKCC1 = 100%; NKCC1 vs. NKCC2 = 52.7%; and NKCC1 vs. NCC = 40.3%. The identities between human, rat and mice orthologs is >98%, which may explain the cross-reactivity between species of T4. This antibody has given reliable results both in Western blot analyses and in immunocytochemistry. However, because it is not isoform specific, it is always assumed that in non-renal tissues the protein detected is NKCC1 based on the tenet that NKCC2 is not expressed outside kidney. This is why, when using T4, it is less committal to use the acronym NKCC without specifying the isoform. Lack of T4 immunostaining of tissues from NKCC1 KO animals or from *in vivo* knockdown of NKCC1 strongly suggests that the patterns of immunoreactivity that have been reported by many groups are mostly, if not exclusively, due to NKCC1 (see below). However, because of the problems discussed above (i.e. mouse monoclonal antibodies tested against mouse tissues), background immunoreactivity in KO mouse tissues is an issue that must be considered before any definitive disqualifying statements can be made.

Nevertheless, validation tests of specificity of NKCC1 detection by T4, comparing cells or tissues from wild-type ($NKCC1^{+/+}$) and KO ($NKCC1^{-/-}$) animals, have been done in cultured astrocytes (Su et al., 2002); in the cell body and axons of mouse hippocampal pyramidal cells (Chen et al., 2005) and in adult mouse retinal neurons (Li et al., 2008). In these studies both immunocytochemistry and Western blot analyses reveal NKCC1-specific staining using T4. It should be noted that T4 recognizes denatured NKCC proteins, and thus best results are obtained using 1% SDS (Alvarez-Leefmans et al., 2001; Li et al., 2008). This could explain why T4 detected NKCC in Western blots in various ferret tissues (cerebellum, hippocampus, retina, cerebral cortex) but gave "nonspecific immunostaining in the fixed ferret retina" (Zhang et al., 2006a); in this study the authors did not report the use of denaturing agents in their immunostaining procedure. In a more recent paper, the same group (see Chapter 19) obtained conflicting results with T4 in mouse (Zhang et al., 2007). Other authors, using short hairpin RNA (shRNA) against different regions of mouse NKCC1 demonstrated knock-down expression of NKCC1 in both Western blot analyses of HEK cells and immunostaining of newly generated granule cells using T4 (Ge et al., 2006). Similar results have recently been obtained in electroporated cortical neurons in which NKCC1-shRNA reduced expression of NKCC1 as demonstrated with T4 (Wang and Kriegstein, 2008). Finally, we have successfully used T4 with postembedding immunolabeling with colloidal gold particles, as shown in Fig. 22.12. Thus this antibody may be used to determine subcellular distribution pattern of NKCCs. T4 antibody is commercially available from the Developmental Studies Hybridoma Bank, The University of Iowa, Iowa City, IA.

2. *Polyclonal NKCC1 Antibodies*

There are several commercial and non-commercial polyclonal antibodies raised against NKCC1. The best characterized and validated of these polyclonal antibodies so far is the one developed by the group of Delpire (Kaplan et al., 1996). This antibody was raised against amino acids 938–1011 of the carboxyl terminus of NKCC1. It has been tested against KO DRG neurons (Sung et al., 2000), choroid plexus epithelial cells (Piechotta et al., 2002) which are known to exhibit high expression levels of NKCC1, and in mouse retinal Muller cells (Zhang et al., 2007). Using this antibody, NKCC1 expression was found in rat DRG neurons of all sizes (see above section II.F), and in the cell bodies and dendrites of cortical neurons (Plotkin et al., 1997a).

Another series of non-commercial polyclonal antibodies was described by Turner's group at NIH (Kurihara et al., 1999). One of these antibodies, α-wNT, was raised against a fusion protein corresponding to amino acids 3–202 of the amino-terminal domain of rat NKCC1. This antibody has been used for immunolocalization experiments of NKCC1 in rat parotid acinar cells. The antibody detected NKCC1 in the basolateral membrane of mouse parotid acinar cells, and its specificity was tested using the same tissue from NKCC1$^{-/-}$ animals (Evans et al., 2000). Further, using the same antibodies, immunoperoxidase labeling revealed abundant expression of NKCC1 in the basolateral domain of secretory coils of rat, mouse and human sweat glands. In contrast, NKCC1 labeling was absent from rat, mouse and human epidermis. Immunoelectron microscopy demonstrated abundant NKCC1 labeling of the basolateral plasma membrane of mouse sweat glands, with no labeling of the apical plasma membranes or intracellular structures (Nejsum et al., 2005).

Other non-commercial polyclonal NKCC1 antibodies have been developed by the group of Lytle (McDaniel et al., 2005). These antibodies were raised in rabbit, against the C-terminal domain of NKCC1 (TEFS-2) or the amino-terminal (NT) domain of hNKCC1. The NT antibody has been tested in cultured oligodendrocytes and gives the same pattern of immunoreactivity as that obtained with T4 (Wang et al., 2003). Both NT and TEFS antibodies have been tested in DRG cells by the group of Cervero (Price et al., 2006). They confirmed the previously described expression of NKCC1 in satellite cells (Alvarez-Leefmans et al., 2001). However, Price and collaborators were unable to find NKCC1 expression in DRG neuronal cell bodies using either NT or TEFS-2 polyclonal antibodies. This is in sharp contrast to what has been observed by many other groups using different antibodies (Alvarez-Leefmans et al., 2001; Funk et al., 2008; Plotkin et al., 1997a; Sung et al., 2000). The reason for the discrepancy is likely to be methodological, for instance, Price and coworkers did not report the use of SDS in their protocol. As mentioned above for the case of T4 monoclonal antibody, and confirmed by the group of Matthews, SDS treatment is necessary to be able to reveal the expression of NKCC1 using either T4 or TEFS-2 (Del Castillo et al., 2005). These authors studied the expression of NKCC1 in the basolateral membrane of cultured T84 human intestinal epithelial cells, known to be a rich source of NKCC1 (Lytle et al., 1995). They confirmed that NKCC1 is expressed in the basolateral membrane of these cells. Interestingly, TEFS-2 and T4 revealed the same pattern of immunoreactivity. On the other hand, up to now, polyclonal antibodies NT and TEFS-2 have not yet been tested in KO tissues and therefore cannot yet be considered validated antibodies. Finally, there are a series of NKCC1 antibodies that are commercially available but they have not been validated and therefore are not considered here.

H. Functional Interactions between NKCC1 and other Putative Cl^- Carriers and Channels in Primary Sensory Neurons

Is NKCC1 the only transport mechanism determining Cl^- in PSNs? This is a fundamental and unresolved issue. It is not known if active Cl^- accumulation in PSNs is effected only by NKCC1 or also by other transporters, and whether this accumulation is offset by coexistent active Cl^- extrusion mechanisms or passive leaks, and if so, to what degree. The SLC12A family of CCCs also includes four genes (SLC12A4-7) encoding for the electroneutral K^+-Cl^- cotransporters (KCC1–KCC4). Under physiological conditions, KCCs function as active Cl^- extruders. Only under non-physiological high levels of extracellular K^+ can KCCs reverse their normal operation and accumulate Cl^- (DeFazio et al., 2000; Kakazu et al., 2000). Therefore KCCs cannot be considered as playing a role in active Cl^- *accumulation* in PSNs under physiological conditions. However, KCCs may influence the levels of intracellular Cl^- by counteracting Cl^- uptake mechanisms. This is crucial inasmuch as functional interaction between Cl^- extruders and loaders (Gillen and Forbush, 1999; Lytle and McManus, 2002) coexpressed in a cell ultimately determines $[Cl^-]_i$, as discussed in Chapter 5. Equally important in this context is the fact that cytoplasmic Cl^- is a negative feedback regulator of NKCC1 activity in many cell types (Breitwieser et al., 1990, 1996; Flatman, 2002; Gamba, 2005; Gillen and Forbush, 1999; Haas et al., 1995; Lytle and Forbush, 1996; Russell, 2000) including DRG neurons, as discussed below.

The consequences of functional interaction between KCCs and NKCCs in mammalian cells were demonstrated in an epithelial cell-line (HEK293) transfected with rabbit KCC1. Overexpression of KCC1 led to a decrease in $[Cl^-]_i$ that resulted in activation of endogenous NKCC (Gillen and Forbush, 1999). A steep relation between cell Cl^- and NKCC activity was found over the physiological ranges supporting a primary role for $[Cl^-]_i$ in activation of NKCC. Functional interactions between NKCCs and putative KCCs in PSNs could affect $[Cl^-]_i$ and the activity levels of NKCCs. For instance, decreased activity of a KCC coexpressed with NKCC could result in a positive shift in E_{Cl} and hence in PAD amplitude. KCC2, also known as "the neuronal-specific isoform of K^+-Cl^- cotransporters", plays a key role in adult central neurons by extruding Cl^- and keeping $[Cl^-]_i$ below electrochemical equilibrium thereby making possible hyperpolarizing inhibition produced by GABA and glycine (Payne, 1997; Payne et al., 2003; Rivera et al., 1999), as discussed in detail in Chapter 17. Whether or not PSNs express KCC2 is controversial. Three reports claim that DRG cells express it (Funk et al., 2008; Gilbert et al., 2007; Lu et al., 1999) but others have failed to find either the protein or the transcript (Coull et al., 2003; Geng et al., 2009; Kanaka et al., 2001; Rivera et al., 1999; Toyoda et al., 2005). Another KCC isoform, KCC4, is expressed in trigeminal ganglia and the nuclei of other cranial nerves, but its function is unknown and its expression decreases with postnatal development (Karadsheh et al., 2004). KCC1, the housekeeping isoform of the KCCs (Adragna et al., 2004; Ochiai et al., 2004; Su et al., 1999; Zhang et al., 2005), is expressed in brain tissue (Su et al., 1999), in trigeminal ganglion neurons (Toyoda et al., 2005) and DRG (Gilbert et al., 2007), but its functional impact on isosmotic (euvolemic) Cl^- regulation is unknown. KCC3, another KCC isoform with splice variants expressed in brain, has been implicated in a peripheral neuropathy but is minimally expressed in DRG cells (Howard et al., 2002; Pearson et al., 2001). When KCC1, KCC3 or KCC4 are expressed in *Xenopus* oocytes, unlike KCC2, they show minimal K^+-Cl^- cotransport activity under isotonic conditions, yet are activated by cell swelling (Mercado et al., 2000; Mount et al., 1999), suggesting that they are involved in cell volume control rather than Cl^- homeostasis.

As discussed in section II.E, direct measurement of $[Cl^-]_i$ in rat DRG neurons showed that when the cells are equilibrated in isosmotic Na^+-free solution ($0Na^+$), there is a Na^+ independent accumulation of Cl^-, denoted as SIC (sodium independent component). Examples of SIC can be observed in Figs 22.4A and 22.5. We found (Rocha-González et al., 2008) that this relatively small SIC of the net Cl^- influx reached steady-state when $[Cl^-]_i$ was ~13 mM, which corresponds to an E_{Cl} of ~ -60 mV. That is, the SIC reached equilibrium at a value close to E_m, suggesting that it may reflect a passive Cl^--leak conductance. As already discussed above, this is intriguing because the resting Cl^- conductance of DRG cells is known to be negligibly small (Alvarez-Leefmans, 1990; Deschenes et al., 1976; Nishi et al., 1974). Moreover, DRG cells lack the ubiquitous ClC-2 inward rectifying Cl^- channels that stabilize E_m in other neurons (Staley et al., 1996). However, DRG cells are endowed with Ca^{2+}-activated Cl^- channels (Frings et al., 2000; Hartzell et al., 2005) as discussed in Chapter 13. Given that the SIC becomes evident upon removal of external Na^+, and that this maneuver leads to an increase in $[Ca^{2+}]_i$ due to reversal of Na^+/Ca^{2+} exchange (Verdru et al., 1997), we proposed that a rise in $[Ca^{2+}]_i$ activates a Cl^- conductance that otherwise would be quiescent. We have verified that in $0Na^+$ there is a rise in $[Ca^{2+}]_i$ during the SIC of Cl^- influx. However, the nature and properties of this putative Cl^- conductance in DRG cells are still to be determined.

Another feature of the SIC relevant in the context of whether DRG neurons lack an active Cl^- extrusion mechanism is that $[Cl^-]_i$ stabilized at a value that corresponded to an E_{Cl} close to that expected for E_m, and remained stable over the time course of exposure to the Na^+-free solution (up to 30 min). This suggests that DRG cells are not endowed with mechanisms of active Cl^- extrusion in isosmotic medium; if such mechanisms were functionally expressed, in the absence of external Na^+ the $[Cl^-]_i$ should decrease with time, and it does not. Consistent with this interpretation are the majority of reports showing that DRG cells lack the protein and the transcript of KCC2 (Coull et al., 2003; Kanaka et al., 2001; Rivera et al., 1999; Toyoda et al., 2005). As already mentioned, other K^+-Cl^- cotransporter isoforms (KCC1, KCC3 and KCC4) have been detected in PSNs (Boettger et al., 2003; Gilbert et al., 2007; Geng et al., 2009; Toyoda et al., 2005), but their function in these cells is unknown. KCC1, 3 and 4 do not seem to be active in isosmotic media when expressed in *Xenopus* oocytes, but they are activated by hyposmotic swelling and hence it is thought that their primary function is cell volume control rather than Cl^- homeostasis, although this is controversial (Gamba, 2005). Further, CWV measurements in DRG neurons reveal that volume changes are negligible during the SIC of the Cl^- influx (Fig. 22.4B). We suggested that the SIC of the Cl^- influx does not occur through KCC reversal, otherwise this should be reflected in osmotic cell swelling upon restoring external Cl^- in the absence of external Na^+. Moreover, the volume measurements suggest that Cl^- entering in the absence of external Na^+ must be exchanged for another ion and/or is entering via channels in such a way that there is no net influx of osmotically active particles.

Of course, a limitation of all these studies is that they are done in the cell bodies of primary sensory neurons. Whether the peripheral and central terminals behave in similar ways is purely inferential at this stage. An uneven distribution of Cl^- channels and transporters along primary sensory neurons could generate intracellular Cl^- gradients as those observed in other neurons (Chapters 7 and 19 in this volume). Nevertheless, for more than 40 years, DRG neurons have been successfully used as a model of what happens in the central and peripheral terminals, while finer methods are developed and/or applied to study directly the peripheral and central endings.

III. POSSIBLE ROLE OF NKCC1 IN PAIN, HYPERALGESIA AND NEUROGENIC INFLAMMATION

A. NKCC1 has been Proposed to be Involved in the Generation and Maintenance of Hyperalgesic States and Neurogenic Inflammation

Tissue injury and inflammation cause pain, hyperalgesia (enhanced pain sensitivity in response to stimuli that are normally painful) and allodynia (pain sensation evoked by stimuli that normally do not provoke pain). Hyperalgesia occurs not only at the site of injury (*primary hyperalgesia*) but also in surrounding uninjured areas (*secondary hyperalgesia*). Allodynia produced by mechanical stimulation, known as touch-evoked pain or tactile allodynia, can also be evoked at the injury site or at sites adjacent to or remote from the injury (Cervero and Laird, 1996; Treede et al., 1992). Primary hyperalgesia is due to sensitization of peripheral nociceptors (Meyer and Campbell, 1981; Raja et al., 1984), whereas secondary hyperalgesia is due to central sensitization that involves alterations in sensory processing in the spinal cord (Cervero et al., 2003). Tactile allodynia is caused by central sensitization (Ji et al., 2003; Sandkuhler, 2000; Woolf and Salter, 2000).

Acute peripheral inflammation and other forms of tissue injury producing persistent activation of nociceptors can lead to excessive PAD that reaches the firing threshold of nociceptive afferents triggering dorsal root reflexes (DRRs) (Lin et al., 1999, 2000; Rees et al., 1996, 1994, 1995; Sluka et al., 1995; Valencia-de Ita et al., 2006). This means that PAD, normally an inhibitory process, can be transformed into an excitatory one if the depolarization is large enough to trigger spikes in nociceptive afferents (Fig. 22.6). Moreover, substantial evidence supports the notion that DRRs conducted centrifugally can produce neurogenic inflammation and hyperalgesia (Lin et al., 1999, 2000; Rees et al., 1996, 1994, 1995; Sluka et al., 1995; Willis, 2006, 1999). The IB4-, peptidergic cells may play a key role in this process since substance P and CGRP are released at their peripheral endings causing vasodilation, plasma extravasation and edema, and thus adding their actions to those of other inflammatory mediators. Following our suggestion (Alvarez-Leefmans et al., 1998) that NKCC1 could modulate PAD amplitude, Willis (Willis, 1999) and later Cervero and coworkers (Cervero et al., 2003; Price et al., 2005) proposed that the enhancement of GABA-mediated PAD leading to DRRs could result from a positive shift in E_{Cl} due to an increase in $[Cl^-]_i$ brought about by up-regulation of NKCC1 activity following phosphorylation of this protein in nociceptive PSNs. These hypotheses have received strong support from a number of recent observations (Funk et al., 2008; Galan and Cervero, 2005; Pieraut et al., 2007; Pitcher et al., 2007; Price et al., 2009; Rocha-González et al., 2008; Valencia-de Ita et al., 2006).

Cervero and Laird (Cervero and Laird, 1996; Cervero et al., 2003) proposed a model for tactile allodynia involving presynaptic interactions similar to those of the "Gate Control Theory" (Melzack and Wall, 1965). The model (Fig. 22.7) is based on evidence that activation of large myelinated mechanoreceptive

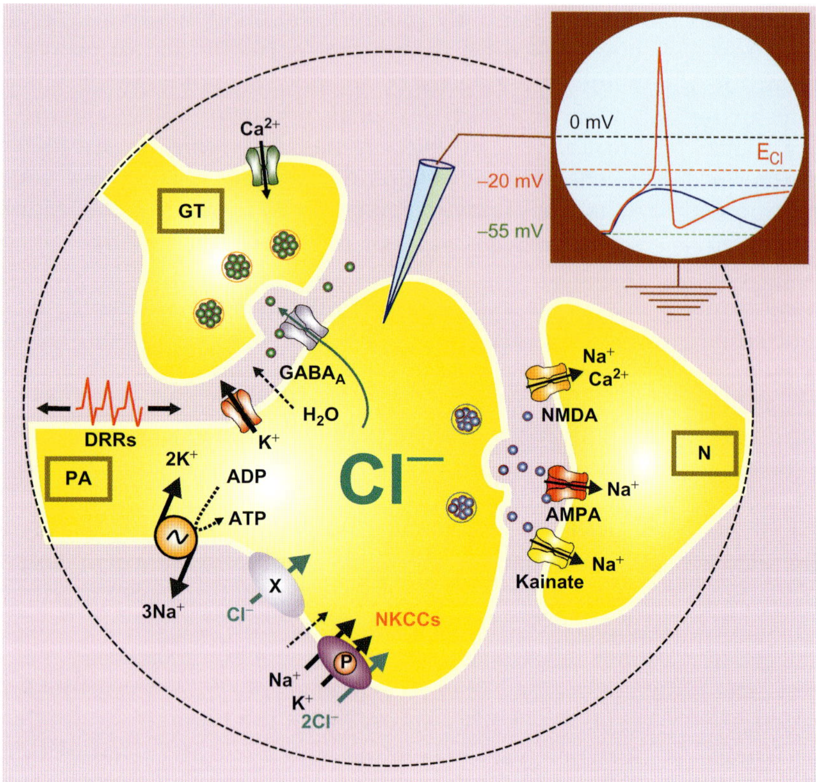

FIGURE 22.6 Schematic diagram illustrating the hypothesis of how activation of NKCCs and other yet to be identified active Cl^- uptake transport systems (X) increase $[Cl^-]_i$ in nociceptive presynaptic terminals (PA), shifting E_{Cl} toward depolarizing threshold values. Instead of PAD, GABA released from interneurons (GT) produces action potentials in the primary afferent (PA) terminals, as illustrated in the intracellular recording shown in the schematic representation of an oscilloscope screen. This results in dorsal root reflexes (DRRs) being conducted towards the periphery (antidromically) and towards the presynaptic terminal (orthodromically). The latter produces release of neurotransmitter from the presynaptic terminal, that activate the post-synaptic spinal neuron (N). Other symbols and nomenclature are the same as in the legend of Fig. 22.3.

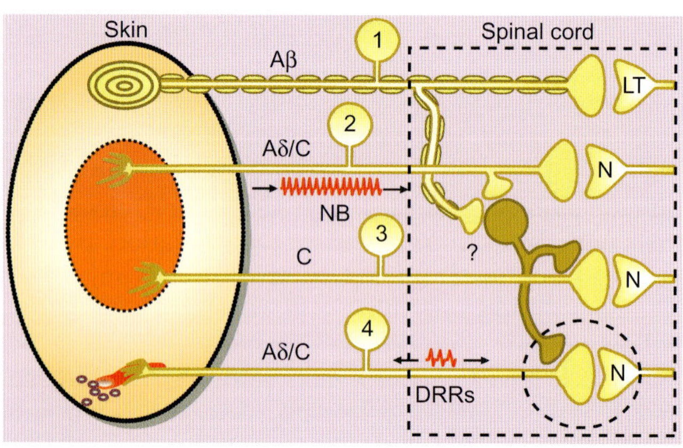

FIGURE 22.7 Hypothetical mechanisms of touch-evoked pain, hyperalgesia and neurogenic inflammation involving presynaptic interactions between primary sensory neurons. **Left oval**: skin patch with site of injury (orange) and surrounding uninjured area (peach). **Dashed rectangle**: dorsal horn with second order nociceptive neurons (N) and low threshold neurons (LT). Nociceptive barrage (NB) conducted through nociceptive primary afferent (cell 2). Dorsal root reflexes (DRRs). **Dashed circle** is magnified in Fig. 22.6. Under normal conditions, stimulation of low-threshold mechanoreceptors connected to Aβ afferents (cell 1) evokes PAD in the terminals of nociceptive afferents (Ao- and C fibers, cells 3 and 4) via GABAergic interneurons (mustard cell), reducing the effectiveness of nociceptive transmission. Following tissue injury and inflammation, the PAD evoked by tactile stimuli may become sufficiently large to evoke DRRs in nociceptive afferents. The DRRs conducted centripetally can excite neurons in the dorsal horn that are normally driven by nociceptors (N) and evoke mechanical allodynia. DRRs conducted centrifugally produce neurogenic inflammation (due to release of vasogenic mediators (cell 4) and hyperalgesia.

afferents can produce PAD in C fibers (Calvillo, 1978; Calvillo et al., 1982; Carstens et al., 1979; Fitzgerald and Woolf, 1981; Hentall and Fields, 1979; Lin et al., 1999, 2000). Under normal circumstances, stimulation of low threshold mechanoreceptors connected to Aβ afferents evokes PAD in the terminals of nociceptive afferents (Aδ and C fibers) via GABAergic interneurons, reducing the effectiveness of nociceptive transmission. A similar mechanism is often invoked to explain the common observation that rubbing a skin region close to an injury site reduces pain perception. Following tissue injury and inflammation, the PAD evoked by tactile stimuli may become sufficiently large to evoke DRRs in nociceptive afferents. The DRRs conducted centripetally may excite neurons in the dorsal horn that are normally driven by nociceptors and evoke mechanical allodynia (Cervero and Laird, 1996; Cervero et al., 2003; Pitcher et al., 2007; Price et al., 2005). Clearly, the switch from touch to pain upon activation of low-threshold mechanoreceptors is a remarkable sensory phenomenon. Study of its underlying mechanisms has generated considerable interest, largely due to its relevance in pain management.

Several studies provide evidence that NKCC1 is involved in nociceptive processing and in the generation and maintenance of hyperalgesic states and neurogenic inflammation. The evidence can be summarized as follows. (1) $NKCC1^{-/-}$ mice exhibit increased pain threshold to noxious heat and reduced touch-evoked pain (allodynia) following intradermal capsaicin injection (Laird et al., 2004; Sung et al., 2000). (2) In the formalin model of tissue injury-induced pain (Allen and Yaksh, 2004; Dubuisson and Dennis, 1977), we found that local peripheral or spinal intrathecal administration of NKCC blockers (bumetanide, piretanide and furosemide) has a significant antinociceptive effect (Granados-Soto et al., 2005), an observation that has been recently verified by other groups (Cramer et al., 2008; Pitcher et al., 2007). These observations support the notion that NKCC is implicated in nociceptive processing both intraspinally and peripherally. (3) In human skin, bumetanide and furosemide inhibit itch and flare responses to histamine (Willis and Coggeshall, 2004). NKCC inhibitors also block cough produced by irritants of the tracheal mucosa where NKCC1 is expressed in mechanosensitive vagal sensory terminals (Mazzone and McGovern, 2006, 2008). (4) Noxious visceral stimulation with intracolonic capsaicin in mice induces an increase in the membrane fraction of NKCC1 and its phosphorylation in lumbosacral spinal cord (Galan and Cervero, 2005). The changes are rapid (start ten minutes after intracolonic capsaicin injection) and transient, occurring only in the region of the spinal cord that receives afferent input from the colon. It was proposed that NKCC1 phosphorylation could be the cause of enhancement of GABA-mediated PAD that leads to DRRs in nociceptive afferents (Galan and Cervero, 2005). (5) Behavioral and electrophysiological studies show that the DRRs conducted centrifugally reach the terminals of fine sensory nerve fibers (Aδ and C), some of which release neuromediators (e.g. Substance P and CGRP) involved in the development of acute neurogenic inflammation (Chen et al., 2004; Lin et al., 1999, 2000; Rees et al., 1996, 1994, 1995; Sluka et al., 1995; Valencia-de Ita et al., 2006; Willis, 2006, 1999). (6) Spinal bumetanide reduces DRR activity, mechanical allodynia and hyperalgesia produced by intradermal injection of capsaicin (Valencia-de Ita et al., 2006).

1. *How does NKCC1 Up-regulation Transform PAD into DRRs?*

There are several possibilities. The nociceptive barrage occurring upon tissue injury leads to an enhanced activation of GABAergic pathways in the dorsal horn. This persistent activation of $GABA_A$ receptors in the central terminals of nociceptive primary afferents produces a transient decrease in intra-terminal $[Cl^-]$, leading to phosporylation of NKCCs and thereby increasing transport activity that in turn results in a rebound increase in intra-terminal $[Cl^-]$. An increase in $[Cl^-]_i$ beyond basal levels results in a depolarizing shift in E_{GABA} to values reaching spike threshold, thus explaining DRRs (Fig. 22.6). This increase in $[Cl^-]_i$ beyond basal levels could be achieved by changing the set point of NKCC1, making it less sensitive to inhibition by intracellular Cl^-. It could also be achieved by an increase in surface expression of NKCC1 via trafficking from vesicular compartments. Another possibility is trafficking of a second hypothetical active Cl^- uptake mechanism (denoted "X" in Fig. 22.6) that has a different Cl^- set point than NKCC1. None of these processes are mutually exclusive.

In many cell types there is a tight relationship between NKCC activity, cell water volume (CWV) and $[Cl^-]_i$ (Hoffmann et al., 2007; Russell, 2000). With the exception of the work of Russell and collaborators in squid axons, this relationship has been studied only in non-neuronal cell populations, and no direct measurements of $[Cl^-]_i$ and CWV had ever been made at the single cell level until our recently published study (Rocha-González et al., 2008). Moreover, as discussed in Chapter 2 in this volume, Russell and collaborators showed in dialyzed squid axons that intracellular Cl^- is a negative feedback regulator of NKCC activity (Breitwieser et al., 1990, 1996). This was later confirmed in some non-neuronal mammalian cells (reviewed in Kahle et al., 2006; Russell, 2000). This mechanism would be crucial in sensory neurons not only for keeping CWV and $[Cl^-]_i$ constant, but also because changes in the set point of a negative feedback mechanism like this could underlie the depolarizing shift in E_{Cl} that is proposed to occur in PSNs following somatic injury and inflammation of skin or nerves, as discussed above. To study how intracellular Cl^- influences and is influenced by NKCC activity the relation between $[Cl^-]_i$, $[Cl^-]_o$ and CWV in rat DRG neurons was determined in experiments like those illustrated in Fig. 22.8.

It was found that in rat DRG neurons the rate of active Cl^- uptake (which is proportional to JCl^-_{in}, the net Cl^- influx) falls with increasing $[Cl^-]_i$ and becomes negligible when $[Cl^-]_i$ reaches its physiological basal level (Figs 22.8A and 22.9A). This occurs in the face of a net free energy for NKCC that strongly favors net Cl^- uptake ($\Delta G = -2.8\,kJ \cdot mol^{-1}$) under physiological conditions (see section II.E, Chapter 5 in this volume). The observed saturation kinetics of active Cl^- uptake with respect to $[Cl^-]_i$ suggests a negative feedback system in which intracellular Cl^- regulates its own influx thereby keeping $[Cl^-]_i$ constant, above electrochemical equilibrium but below the value that would attain if NKCC reached thermodynamic equilibrium (Fig. 22.9B). Thus, kinetic constraints inactivate the cotransporter, even in the presence of a substantial thermodynamic driving force. The relevance of these observations is that this

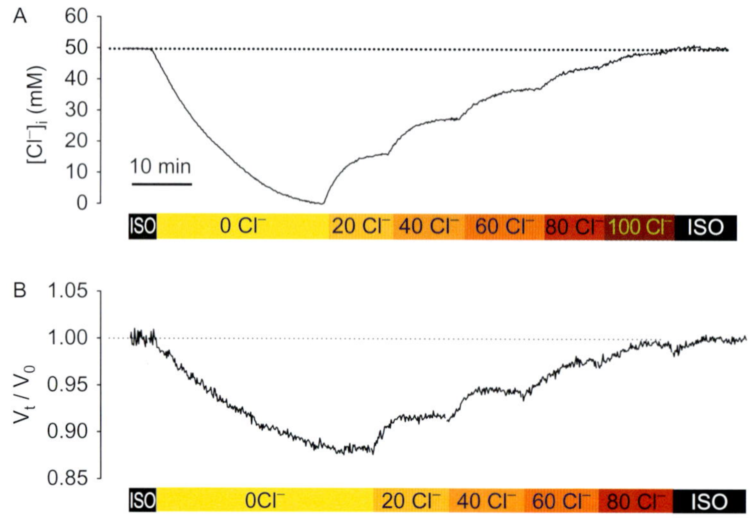

FIGURE 22.8 Effects of changes in $[Cl^-]_o$ on $[Cl^-]_i$ and relative cell water volume in DRG neurons. Cells were exposed to $0Cl^-$ until they were depleted of intracellular Cl^-, and then $[Cl^-]_o$ was increased in steps of ~20 mM keeping the osmolality constant between solutions. **A.** Changes in $[Cl^-]_i$ (initial $[Cl^-]_i = 50$ mM; $E_{Cl} = -23$ mV. **B.** Changes in relative cell water volume (V_t/V_0). Intracellular Cl^- depletion during exposure to $0Cl^-$ (A) occurs in parallel with a decrease in V_t/V_0 (B). Both isosmotic cell shrinkage and intracellular Cl^- depletion in 0Cl solution are reversed in stepwise fashion upon exposure to isosmotic solutions having equal increments in $[Cl^-]_o$, suggesting a tight coupling between intracellular Cl^- accumulation and CWV recovery. Both CWV and $[Cl^-]_i$ recovered when $[Cl^-]_o$ was ~100 mM. Increasing $[Cl^-]_o$ beyond 100 mM did not result in further changes in CWV or $[Cl^-]_i$ indicating that the mechanisms involved in these processes reached saturation. Bottom bars in A and B indicate the periods of exposure to each solution as well as their $[Cl^-]$ in mM. Superfusion with each of these solutions lasted until a new steady state for $[Cl^-]_i$ and CWV was reached. Data in A came from a neuron loaded with MQAE and in B from a neuron loaded with calcein. Modified from Rocha-González et al. (2008).

negative feedback mechanism maintains $[Cl^-]_i$ and CWV at a constant set point under basal conditions. We propose that this set point, representing a kinetic brake, can change under certain conditions (e.g. phosphorylation and/or trafficking) thus endowing the system with the possibility of shifting E_{Cl} (Fig. 22.9). The consequences of this shift in E_{Cl} has been mentioned above: it could turn GABA from depolarizing to excitatory triggering DRRs.

B. NKCC1 in Peripheral Nociceptive Endings may Contribute to the Nociceptive Generator Potential

To be detected, all noxious stimuli must generate action potentials in the afferent nerve fiber. As in other receptors, all-or-none spikes must arise from a depolarizing generator potential. Sensory neurons are endowed with Ca^{2+}-activated Cl^- channels which upon activation by a rise in intracellular Ca^{2+} produce membrane depolarization due to Cl^- efflux (Frings et al., 2000; Hartzell et al., 2005; Kenyon and Scott, 2002; Mayer, 1985; Scott et al., 1995), as discussed in great detail in Chapters 13 and 20 in this volume. We proposed that, as in olfactory neurons (Reuter et al., 1998), this putative Ca^{2+}-activated-Cl^--mediated depolarization could be an important element of the generator potential in nociceptive terminals (Granados-Soto et al., 2005). This hypothesis requires the presence of an outward Cl^- gradient in the peripheral endings of sensory neurons. Several observations support this hypothesis which is illustrated in Figure 22.10. (1) GABA or $GABA_A$ agonists depolarize peripheral sensory axons (Bhisitkul et al., 1987) including nociceptive afferents and their terminals (Ault and Hildebrand, 1994; Carlton et al., 1999), suggesting that the outward Cl^- gradient is maintained not only in the central terminals and the soma but also in the peripheral processes. (2) A number of algogens, some of which are components of the so-called "inflammatory soup" produce depolarization and spike activity in nociceptors. A common feature of many inflammatory mediators such as serotonin, prostaglandin E_2, bradykinin, histamine, protons or ATP, is that they increase intracellular $[Ca^{2+}]$ (Burnstock, 2000; Dunn et al., 2001; Gover et al., 2003; Linhart et al., 2003; Nicolson et al., 2002; Reeh and Kress, 2001). A rise in Ca^{2+} within the small volume of sensory terminals would be sufficient to activate Cl^- channels producing a depolarizing inward current mediated by Cl^- efflux (Currie et al., 1995). (3) We have shown that local peripheral pharmacological inhibition of NKCCs with loop diuretics in the hindpaw of rats produces anti-nociception in the formalin test (Granados-Soto et al., 2005). The peripheral anti-nociceptive effect of loop diuretics could be due to dissipation of the Cl^- gradient in nociceptive terminals due to NKCC inhibition. This in turn would result in abolition of Ca^{2+}-activated Cl^--mediated depolarizations, thereby blocking transmission of information at nociceptive terminals expressing Ca^{2+}-activated Cl^- channels. (4) Local furosemide and bumetanide block histamine-induced itch and flare in human skin (Willis et al., 2004). Loop diuretics also block cough produced by irritants of the tracheal mucosa where NKCC1 is expressed in mechanosensitive vagal sensory terminals (Mazzone and McGovern, 2006).

IV. Na^+-K^+-Cl^- COTRANSPORTER EXPRESSION IN PERIPHERAL AXONS AND SCHWANN CELLS

The pioneering studies of Russell in squid giant axons demonstrated the expression of an NKCC mechanism in these invertebrate unmyelinated fibers (Russell, 2000). The unique stoichiometry of the cephalopode isoform $2Na^+$:$1K^+$:$3Cl^-$ suggests differences with respect to vertebrate isoforms which have a stoichiometry of $1Na^+$:$1K^+$:$2Cl^-$ (see Chapter 5 in this volume). Unfortunately, the squid protein has not been cloned and thus it is not possible to have clues on what makes it different at the molecular level from the vertebrate isoforms. Cloning of the squid NKCC protein would be important to further understanding the evolution of the cation-coupled-chloride cotransport proteins family in the animal kingdom. Moreover, determining the amino acid sequence of squid NKCC would shed light on the molecular determinants of different stoichiometries. The function of NKCC in squid axons appears to be related to cell volume regulation. Russell proposed that "NKCC is an essential component in a feedback mechanism designed to maintain cell volume at some constant set-point". The evidence presented in the previous section (Figs 22.8 and 22.9) in mammalian DRG neurons is fully consistent with this view.

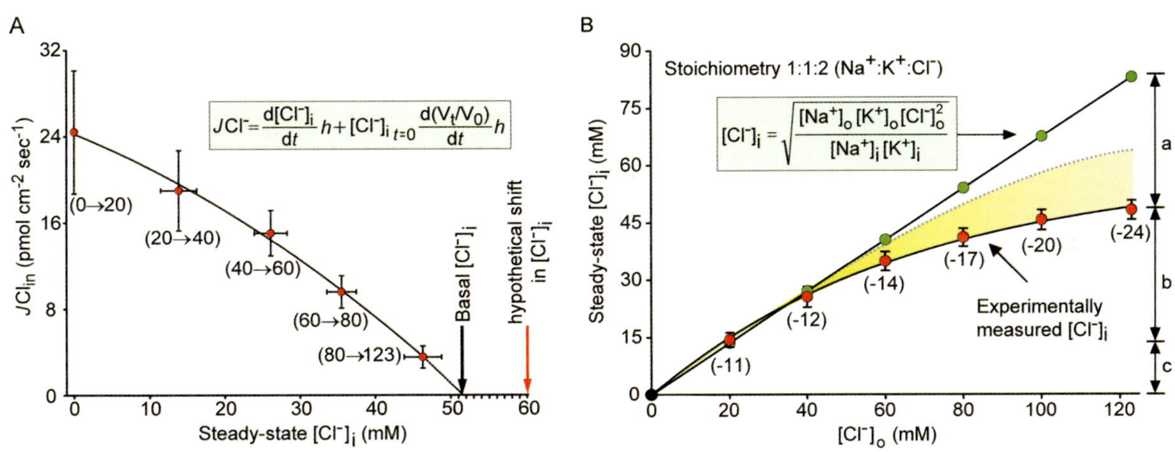

FIGURE 22.9 Intracellular $[Cl^-]$ regulates Cl^- influx through putative negative feedback. **A.** Net Cl^- influx (JCl_{in}) as a function of steady-state $[Cl^-]_i$ for ~20 mM increments in $[Cl^-]_o$. The experimental protocol followed to obtain the data is shown in Fig. 22.8. JCl_{in} was calculated using the boxed equation where: $d[Cl^-]_i/dt$ and $d(V_t/V_o)/dt$ are the initial rates of change in $[Cl^-]_i$ and relative cell water volume, respectively, h is the cell volume-to-surface-ratio (cm) assuming spherical shape. Numbers in parenthesis indicate the increments in $[Cl^-]_o$ in mM (the first number is the initial $[Cl^-]_o$ and the second is the final $[Cl^-]_o$ for each step change in $[Cl^-]_o$). Note that $J\,Cl_{in}$ decreases as $[Cl^-]_i$ increases. Bars on each point are SE. JCl_{in} measured when the external solution was changed from 0 to 20 mM Cl^- (0→20) is the sum of SDC and SIC Cl^- influx. Other points reflect Na^+-dependent "uphill" JCl_{in}. Black arrow indicates the extrapolated $[Cl^-]_i$ (~50 ± 5 mM) at which JCl_{in} becomes negligible. This value is close to the measured basal $[Cl^-]_i$. Red arrow: possible shit in $[Cl^-]_i$ resulting from a change in set point due to a change (a decrease) in Cl^- sensitivity of NKCC or associated regulatory proteins (e.g. kinases). **B.** Steady-state $[Cl^-]_i$ as a function of $[Cl^-]_o$. Green circles denote $[Cl^-]_i$ (calculated from the boxed equation) when NKCC attains thermodynamic equilibrium. NKCC stoichiometry was assumed to be 1:1:2. $[Na^+]_i$ = 10 mM and $[K^+]_i$ = 135 both assumed to be kept constant by the Na^+/K^+ pump. Red circles are actual steady-state $[Cl^-]_i$ measured for various $[Cl^-]_o$ (n = 11). Values in parenthesis correspond to E_{Cl} (in mV). Bracketed arrows: **a**, theoretical range in which $[Cl^-]_i$ could increase by NKCC upregulation. It is the difference between the maximal theoretical $[Cl^-]_i$ if NKCC attained thermodynamic equilibrium (83 mM) and the value measured (48.3±2.5 mM) at physiological $[Cl^-]_o$ (123 mM). This range represents the hypothetical "window" in which $[Cl^-]_i$ could be shifted by changes in the kinetic brake (set point); **b**, range of $[Cl^-]_i$ above electrochemical equilibrium; **c**, range of $[Cl^-]_i$ below electrochemical equilibrium. The dotted gray line and the shadowed area represent a hypothetical shift in set point and therefore in E_{Cl}.

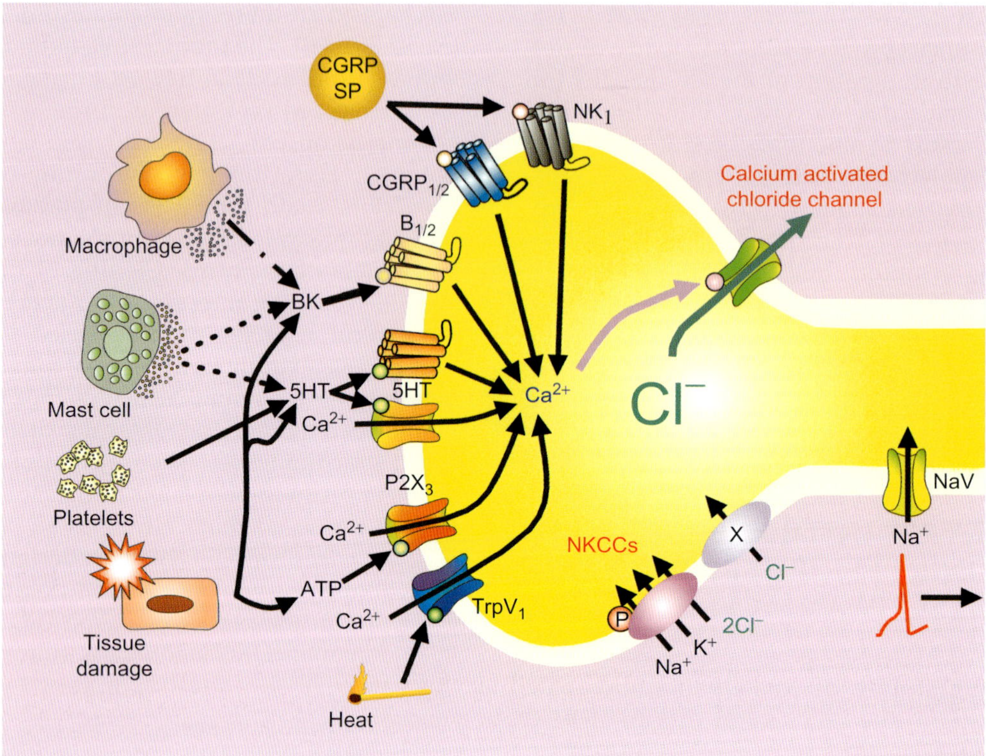

FIGURE 22.10 Schematic representation of the hypothetical processes occurring within a primary afferent nociceptor that leads to spike initiation. We propose that NKCC and other yet to be identified active Cl^- uptake transport systems (X), maintain an outward Cl^- gradient across the nociceptive terminal plasma membrane. This Cl^- gradient plays a key role in producing the nociceptor generator potential, a local depolarization that reaches threshold and initiate an all or none action potential. Nociceptors have receptors for various inflammatory mediators. A common feature of many inflammatory mediators such as serotonin (5HT), bradykinin (BK), ATP or activation of TRPV1 receptor channels is that all of them ultimately increase the concentration of intracellular Ca^{2+} in the nociceptive terminal. The increase in Ca^{2+} triggers a depolarizing Cl^- efflux via Ca^{2+}-activated Cl^- channels. When the depolarization reaches threshold, an action potential is generated.

Work originated in our laboratory provided the first description of NKCC in vertebrate myelinated axons and their associated Schwann cells (Alvarez-Leefmans et al., 2001). We used a combined approach including confocal immunofluorescence, three-dimensional reconstructions and electron microscopy. To label NKCC we used T4 monoclonal antibody (Lytle et al., 1995), and to label Schwann cells we used an antibody against the S-100 protein (Pelc et al., 1986). NKCC immunoreactivity was found prominently at the nodes of Ranvier (Fig. 22.11A–C). Immunoreactivity was particularly intense at the nodal and paranodal regions immediately adjacent to the nodal gap (Fig. 22.11A and C). This particular area corresponds to the so-called myelin sheath attachment segment, according to the nomenclature of Berthold and Rydmark (Berthold and Rydmark, 1983). NKCC immunoreactivity was also systematically observed in the outermost membrane region of the paranodal Schwann cell (arrowheads in Fig. 22.11A). In longitudinal sections of nerve fibers, the NKCC immunoreactivity at the nodal–paranodal area appeared as two bands in the nodal gap, perpendicular to the longitudinal axis of the fiber (arrow in Fig. 22.11A). Upon 40° rotation of three-dimensional confocal reconstructions, the immunoreactivity at the nodal–paranodal region revealed that the bands were actually donut-shaped structures surrounding the axon (Fig. 22.11B, arrow pointing to the node of Ranvier). These donut-shaped regions correspond to the well-known collars of Schwann cell cytoplasm which are known to be packed with Schwann cell microvilli (Fig. 22.11D), reminiscent of the brush border membranes of kidney absorptive epithelial cells (Berthold and Rydmark, 1983).

Double labeling experiments with S-100 and T4 confirmed that the NKCC immunoreactivity seen at the nodal–paranodal interface was indeed located at the Schwann cell (Fig. 22.11C). Immunoreactivity was clearly observed in the nodal region, in the area corresponding to the collar-shaped structure surrounding the axon, which is formed by Schwann cell cytoplasmic pockets and microvilli. In fibers cut longitudinally as the one shown in Fig. 22.11C, S-100 immunolabeling revealed that the collar seen in the 3D reconstructions

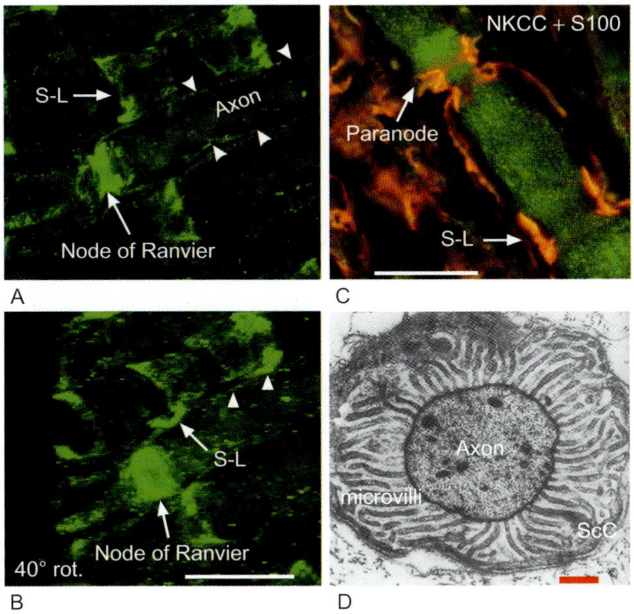

FIGURE 22.11 NKCC expression at the nodes of Ranvier in cat myelinated sensory fibers. **A.** Two-dimensional view of a bundle of myelinated axons. The image was reconstructed from 15 optical sections taken at steps of 1 μm on the *z*-axis. Intense NKCC immunoreactivity is observed in the paranodal region of the nodes of Ranvier, in the incisures of Schmidt-Lanterman (S-L) and in the outer Schwann cell layer (arrowheads). Immunoreactivity was also found in axons (Ax). **B.** 40° rotation, showing three-dimensional views of immunoreactive regions. In the 40° rotation, the immunoreactivity in the paranodal region appears as a donut-shaped structure surrounding the axon at the node of Ranvier. These donut-shaped structures correspond to the corona of Schwann cell microvilli observed in electron micrographs (D). Immunoreactive Schmidt-Lanterman incisures appear as funnel-shaped structures. **C.** Double immunolabeling of NKCC (FITC, green) and S-100 immunolabeling of the Schwann cell (colocalization appears in orange). **D.** Electron micrograph of a cross-section through a nodal axon segment showing a corona of Schwann cell microvilli. Scale bars in B and C are 20 μm. Scale bar in B applies to A. Scale bar in D = 0.5 μm. Figures A–C are modified from Alvarez-Leefmans et al. (2001) and reproduced with permission. Figure D was modified from Berthold and Rydmark (1983).

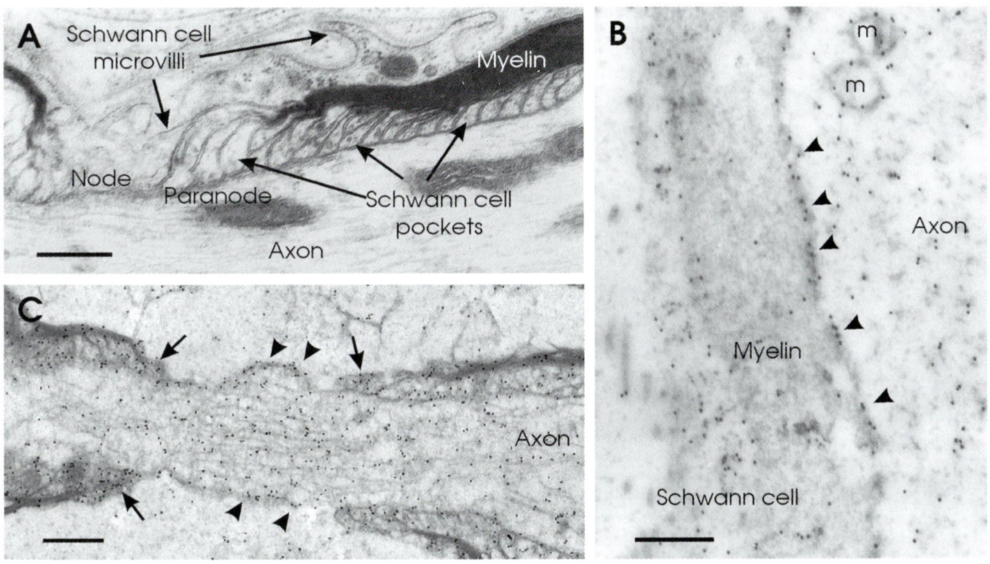

FIGURE 22.12 Ultrastructural distribution of NKCC immunoreactivity in rat myelinated axons. **A.** Ultrastructural features of the node of Ranvier in a DRG axon. At the node, the axon is loosely covered by microvilli-like cytoplasmic protrusions of the Schwann cell. In the paranodal region, pockets of Schwann cell cytoplasm that arise from the opening of the major dense line surround the axon. Material from Epon-Araldite embedding. **B.** Myelinated axon in which NKCC-immunoreactive sites appear labeled with 10 nm colloidal gold particles. Intense immunoreactivity is found along the axonal plasma membrane (arrowheads), and to a lesser extent in the Schwann cell and in the axoplasm. Membrane-bound organelles like mitochondria (m) are also surrounded by NKCC immunoreactivity. Unicryl embedding, non-osmicated tissue, T4 mAb 1:1000. **C.** NKCC immunoreactivity in the node of Ranvier is strongly associated with the paranodal regions (arrows), particularly on the membrane of the Schwann cell paranodal pockets. In the axon, immunoreactivity is found along the axonal plasma membrane, including the nodal axolemma (arrowheads), and throughout the axoplasm. Axoplasmic immunoreactivity appears somewhat concentrated in the nodal region. Unicryl embedding, osmicated tissue, T4 mAb 1:1000. Scale bars 0.5 mm (A, B), 1 mm (C). Reproduced with permission from Alvarez-Leefmans et al. (2001).

(Fig. 22.11B) appeared as a spiny bracelet structure (Fig. 22.11C, Paranode) similar to that described in classical studies (Landon and Hall, 1976; Nageotte, 1922). Double-labeling studies also confirmed that the myelin sheath was immunonegative but the axon was NKCC immunoreactive (Fig. 22.11A). Electron microscopy revealed that the latter immunoreactivity was located in the axoplasm and in the axolemma (Fig. 22.12B). As the axon passes from the end of the paranodal bulb through the collars formed by the

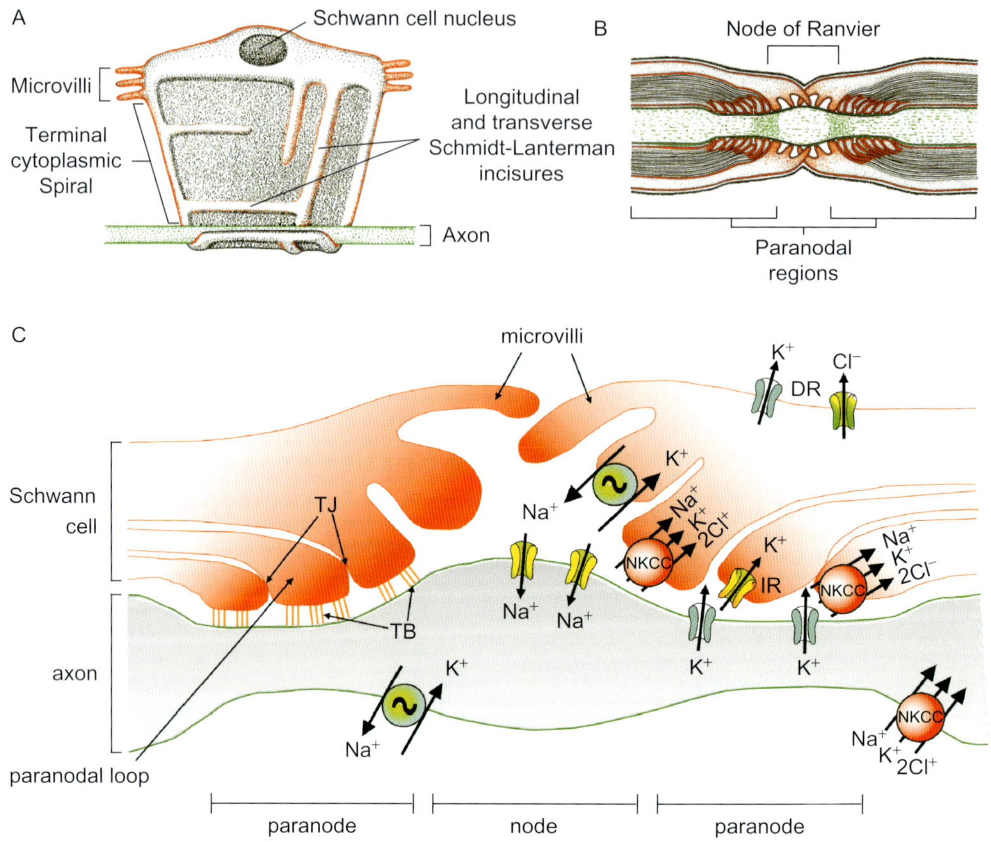

FIGURE 22.13 Diagrams showing NKCC location in various elements of a myelinated sensory fiber and the possible functional significance of NKCC in axons and Schwann cells. **A.** Unrolling of the Schwann cell sheath from an axon. The white areas indicate the presence of Schwann cell cytoplasm. The areas with black points indicate compact myelin. The unrolled sheath shows the transverse and longitudinal Schmidt-Lanterman incisures. Note the nucleus of the Schwann cell in the outer cytoplasmic belt, the microvilli and the terminal cytoplasmic spiral. NKCC immunoreactivity is indicated in red for the Schwann cell and in green for the axon. **B.** When rolled back around the axon, the terminal cytoplasmic spiral runs around the paranodal attachment segment many times, creating Schwann cell pockets or paranodal loops that, together with microvilli, overhang the paranodal and nodal axon, respectively. When transversally cut, this nodal–paranodal bracelet forms collars surrounding the axon, like the ones shown in Fig. 22.11B and D. **C.** Functional significance of NKCC in axons and Schwann cells. We propose that NKCC located in the Schwann cell may be involved in periaxonal K^+ homeostasis. External K^+ ion accumulation occurs in the paranodal areas of the Schwann cell, in particular during repetitive firing. During each action potential, Na^+ enters the axon through voltage-gated channels located in the nodal area, and K^+ exits the axon via voltage-gated channels located in the nodal–paranodal interface. The preferential location of NKCC in the paranodal areas of the Schwann cell suggests that it could be involved in the uptake of K^+ released from the axon and then in siphoning it to areas far away from the periaxonal space, thereby preventing periaxonal K^+ accumulation and controlling axonal excitability. Uptake of K^+ by the Schwann cell can also occur through inward rectifier K^+ channels (IR). The release of K^+ taken up by the Schwann cell may occur through delayed rectifier channels (DR) located in the outer membrane, opposite to the periaxonal space. "Tight junctions" (TJ) restrict access to the extracellular spaces within the myelin, and "transverse junctions" (TB) maintain a narrow periaxonal space in the paranodal region. NKCC in axons may be involved in cell volume control and intracellular Cl^- regulation and, in conjunction with the Na^+/K^+ pump, in the reuptake of K^+ released during nerve activity. A and B are reproduced with permission from Alvarez-Leefmans et al. (2001).

Schwann cell, its diameter is abruptly reduced to about one-third of its internodal value (Fig. 22.11C, green). NKCC immunoreactivity in axons appeared with a punctate pattern and was located both in the axoplasm and in the axolemma region.

Treating the tissue with the reducing agent β-mercapto-ethanol unmasked NKCC immunoreactivity in the incisures of Schmidt-Lanterman (S-L in Fig. 22.11B–C). The latter are inclusions of Schwann cytoplasm within the myelin which pursue a spiral course across the sheath, forming cylindrico-conical segments at the internodes, and so connecting the external and internal layers of Schwann cell cytoplasm (Ghabriel and Allt, 1981; Landon and Hall, 1976), as illustrated in the unrolled Schwann cell in Fig. 22.13A. In 3D rotations, the funnel shape of the NKCC-immunoreactive incisures of Schmidt-Lanterman could be more clearly appreciated (Fig. 22.11B). The longitudinally sectioned

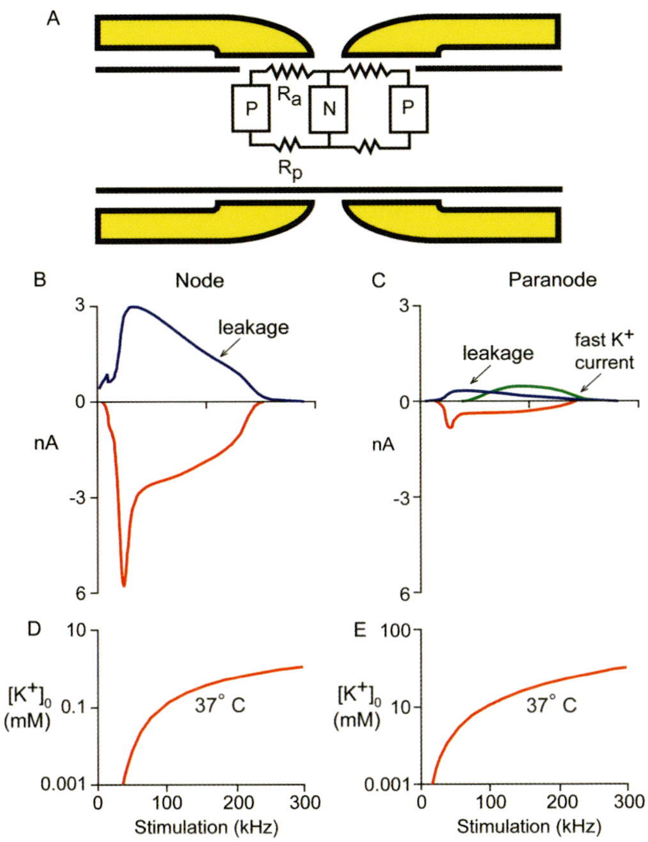

FIGURE 22.14 Computer simulation of K^+ accumulation in the nodal and paranodal periaxonal compartments of a myelinated fiber during non-propagating action potentials, assuming that the only K^+ clearance mechanism is diffusion. **A.** Equivalent circuit used in the simulation. Values used in the model are described in Chiu (1991). **B.** and **C.** Ionic currents during a single non-propagating action potential in the nodal region (B) and in the paranodal region (C). **D.** and **E.** Activity-dependent accumulation of K^+ in the nodel gap (D) and in the periaxonal space between the paranodal axon and the paranodal myelin (E). Note the different scales for $[K^+]_o$ between D and E. If there were no K^+ clearance mechanisms other than difussion, the periaxonal K^+ accumulation would be two orders of magnitude higher in the paranode (10 mM), depolarizing the node and compromising action potential conduction. (Modified from Chiu, 1995; reproduced with permission from Oxford University Press.)

Schmidt-Lanterman incisures can also be appreciated in the S-100 and T4 double immunolabeling (Fig. 22.11C).

The distribution of immunoreactivity in rat myelinated axons was studied ultrastructurally and was found to be very similar to that observed using confocal microscopy in cat myelinated axons (Fig. 22.12). Membrane immunostaining was detected in Schwann cell paranodal pockets and microvilli, and along the axonal membrane at the nodal and paranodal regions (Fig. 22.12B–C). NKCC was found in both the axoplasm and in the membrane region of axons, as well as in three well-defined regions of the Schwann cell: the external surface along the internodal segment (paranode), the nodal–paranodal region (myelin sheath attachment segment) and the incisures of Schmidt-Lanterman. These findings are summarized in Fig. 22.13A and B. The location of NKCC in these structures suggests that it may be playing a role in periaxonal K^+ homeostasis. Frankenhaeuser and Hodgkin first suggested that K^+ released by axons during activity could be transiently accumulated in the periaxonal space between Schwann cells and axons (Frankenhaeuser and Hodgkin, 1956). Since extracellular K^+ has a profound influence on excitability, efficient K^+ clearance is critical for nerve function.

It has been suggested that periaxonal K^+ is partly regulated by the Schwann cells. Two mechanisms have been proposed: inwardly rectifying K^+ channels (Chiu, 1991, 1995) and the Na^+/K^+ pump (Ransom et al., 2000), both of which are present in the plasma membrane of Schwann cells (Fig. 22.13C). We proposed that NKCC exhibits the kinetic features required for an efficient extracellular K^+ buffer, as discussed also in Chapter 5. The immunolocalization of NKCC suggests that NKCC may be another major mechanism involved in periaxonal K^+ buffering in peripheral myelinated axons (Fig. 22.13C).

Modeling activity-dependent K^+ accumulation in various compartments of mammalian myelinated fibers, Chiu showed that the largest extracellular K^+ accumulation is expected to occur at the adjacent 2–4 μm length of periaxonal space at the nodal–paranodal interface (Chiu, 1991), the region in which NKCC is highly concentrated (Figs 22.11B–D and 22.14). Indeed, this is the region where Schwann cell microvilli prominently express NKCC. These microvilli represent a dramatic increase in the ratio of surface area to volume seen in typical absorptive epithelia (Fig. 22.11D). The total membrane area of the microvilli is about ten to 15 times that of the nodal area (Berthold and Rydmark, 1983). Activity-dependent paranodal K^+ accumulation of the magnitude predicted by the model (Fig. 22.14E) may cause sufficient depolarization to compromise repetitive conduction of impulses. In the paranodal region, diffusion alone is not sufficient to avoid K^+ accumulation, as is the case for the nodal region. The features of the paranodal Schwann cell are well designed to act as a powerful periaxonal K^+ buffering system. Experiments to test this hypothesis, using electrophysiological methods, are much needed.

Finally, the presence of NKCC in the axonal membrane deserves further comment. As already mentioned, at the time in which our work was published (2001), NKCC had only been reported in unmyelinated squid axons, as already mentioned (Russell, 2000). The presence of NKCC in mammalian peripheral axons and terminals (see below) suggests that the protein is transported from the cell bodies to distal regions of sensory neurons. Further, the fact that GABA depolarizes sensory axons, including myelinated or unmyelinated (see above, section II.A), suggests that the NKCC protein is also functional as an active Cl^- uptake system, maintaining an outward Cl^- gradient over the entire length of axon. NKCC in peripheral sensory axons may be involved in reuptake of K^+ following nerve activity, as well as in cell volume control. The functional meaning of depolarizing $GABA_A$ extrajunctional receptors in sensory axons is still unknown. GABA in the periphery might play a role in the control of excitability and action potential propagation (Verdier et al., 2003). Interestingly, expression of NKCC1 in vagal sensory endings innervating the airways has been demonstrated (Mazzone and McGovern, 2006). Depolarzing Cl^- currents determine nerve ending activation underlying cough reflexes. In central neurons, GABA mediates a depolarizing response at the axon initial segments of cortical pyramidal cells (Szabadics et al., 2006). As in peripheral axons, the outward Cl^- gradient is caused by NKCC1 (Khirug et al., 2008). The functional meaning of these interactions in cortical neurons is also intriguing, and remains to be elucidated.

V. CONCLUSIONS

The evidence discussed supports the notion that $[Cl^-]_i$ in primary afferent neurons plays important roles in sensory signaling. Active Cl^- uptake transport systems like NKCC1 maintain $[Cl^-]_i$ above electrochemical equilibrium, setting up an outward Cl^- gradient along the entire neuronal surface. This Cl^- gradient makes possible $GABA_A$-mediated PAD and presynaptic inhibition in the central terminals of primary afferents. Since GABA application depolarizes all parts of PSNs, the outward Cl^- gradient must be maintained throughout their entire cell surface, including the cell body and the peripheral and central processes; hence, active Cl^- uptake mechanisms must be expressed in the entire cell surface. Changes in Cl^- regulatory mechanism may increase $[Cl^-]_i$ thereby shifting E_{Cl} to more depolarized values and transforming PAD into an excitatory phenomenon. When PAD amplitude reaches the firing threshold of the terminal it produces dorsal root reflexes (DRRs), pain and neurogenic inflammation. Further, there is evidence supporting the idea that the outward Cl^- gradient may contribute to the generator potential in nociceptive peripheral terminals, making possible action potential initiation in nociceptors. Due to technical difficulties, most studies on Cl^- transporters in sensory neurons have been done in the cell bodies. Although the cell bodies have been a useful model of what occurs in the intraspinal and peripheral terminals, what happens in these other less accessible parts of the cells remains largely inferential.

Acknowledgements

The work done by the author has been supported by The National Institute of Neurological Disorders and Stroke Grant NS29227 and by Wright State University Boonshoft School of Medicine Research Challenge and Seed Grant Programs. The author thanks Professor Francisco Alvarez for his invaluable long-term collaboration and Professor Robert Fyffe for enriching discussions of some aspects of the work summarized in this chapter. Special thanks to laboratory team members Drs Shihong Mao and Mauricio Di Fulvio for their participation in some of the work discussed.

References

Achilles, K., Okabe, A., Ikeda, M., Shimizu-Okabe, C., Yamada, J., Fukuda, A., Luhmann, H.J., and Kilb, W. (2007). Kinetic properties of Cl^- uptake mediated by Na^+-dependent K^+-$2Cl^-$ cotransport in immature rat neocortical neurons. *J. Neurosci.* **27**, 8616–8627.

Adragna, N.C., Di Fulvio, M., and Lauf, P.K. (2004). Regulation of K-Cl cotransport: from function to genes. *J. Membr. Biol.* **201**, 109–137.

Allen, J.W. and Yaksh, T.L. (2004). Tissue injury models of persistent nociception in rats. *Methods Mol. Med.* **99**, 25–34.

Alvarez, F.J. (1998). Anatomical basis for presynaptic inhibition of primary sensory neurons. In *Presynaptic Inhibition and Neural Control* (Rudomin, P., Romo, R., and Mendell, L.M., eds), pp. 13–49. Oxford University Press, New York.

Alvarez, F.J. and Fyffe, R.E. (2000). Nociceptors for the 21st century. *Curr. Rev. Pain* **4**, 451–458.

Alvarez-Leefmans, F.J. (1990). Intracellular Cl^- regulation and synaptic inhibition in vertebrate and invertebrate neurons. In *Chloride Channels and Carriers in Nerve, Muscle, and Glial Cells* (Alvarez-Leefmans, F.J. and Russell, J.M., eds), pp. 109–158. Plenum Press, New York.

Alvarez-Leefmans, F.J. (2001). Intracellular chloride regulation. In *Cell Physiology Sourcebook: A Molecular Approach* (Sperelakis, N., ed.), pp. 301–318. Academic Press, San Diego.

Alvarez-Leefmans, F.J., Altamirano, J., and Crowe, W.E. (1995). Use of ion-selective microelectrodes and fluorescent probes to measure cell volume. *Methods Neurosci.* **27**, 361–391.

Alvarez-Leefmans, F.J., Gamiño, S.M., and Giraldez, F. (1986). Direct demonstration that chloride ions are not passively distributed across the membrane of dorsal root ganglion cells of the frog: preliminary studies on the nature of the chloride pump. *Biophys. J.* **49**, 413a.

Alvarez-Leefmans, F.J., Gamino, S.M., Giraldez, F., and Nogueron, I. (1988). Intracellular chloride regulation in amphibian dorsal root ganglion neurones studied with ion-selective microelectrodes. *J. Physiol.* **406**, 225–246.

Alvarez-Leefmans, F.J., Leon-Olea, M., Mendoza-Sotelo, J., Alvarez, F.J., Anton, B., and Garduno, R. (2001). Immunolocalization of the Na^+-K^+-$2Cl^-$ cotransporter in peripheral nervous tissue of vertebrates. *Neuroscience* **104**, 569–582.

Alvarez-Leefmans, F.J., Nani, A., and Márquez, S. (1998). Chloride transport, osmotic balance, and presynaptic inhibition. In *Presynaptic Inhibition and Neural Control* (Rudomin, P., Romo, R., and Mendell, L.M., eds), pp. 50–79. Oxford University Press, New York.

Ault, B. and Hildebrand, L.M. (1994). GABAA receptor-mediated excitation of nociceptive afferents in the rat isolated spinal cord-tail preparation. *Neuropharmacology* **33**, 109–114.

Beccari, N. (1920). Peculiari modalita nelle connessione di alcuni neuroni del sistema nervoso centrale dei Pesci. Ulteriori ricerche sulle collaterali delle fibre delle Mauthner. *Arch. Ital. Anat. e Embriol.* **17**, 239–283.

Becker, M., Nothwang, H.G., and Friauf, E. (2003). Differential expression pattern of chloride transporters NCC, NKCC2, KCC1, KCC3, KCC4, and AE3 in the developing rat auditory brainstem. *Cell Tissue Res.* **312**, 155–165.

Ben-Ari, Y., Gaiarsa, J.L., Tyzio, R., and Khazipov, R. (2007). GABA: a pioneer transmitter that excites immature neurons and generates primitive oscillations. *Physiol. Rev.* **87**, 1215–1284.

Berglund, K., Schleich, W., Krieger, P., Loo, L.S., Wang, D., Cant, N.B., Feng, G., Augustine, G.J., and Kuner, T. (2006). Imaging synaptic inhibition in transgenic mice expressing the chloride indicator, Clomeleon. *Brain Cell Biol.* **35**, 207–228.

Berthold, C.H. and Rydmark, M. (1983). Electron microscopic serial section analysis of nodes of Ranvier in lumbosacral spinal roots of the cat: ultrastructural organization of nodal compartments in fibres of different sizes. *J. Neurocytol.* **12**, 475–505.

Bhisitkul, R.B., Kocsis, J.D., Gordon, T.R., and Waxman, S.G. (1990). Trophic influence of the distal nerve segment on GABAA receptor expression in axotomized adult sensory neurons. *Exp. Neurol.* **109**, 273–278.

Bhisitkul, R.B., Villa, J.E., and Kocsis, J.D. (1987). Axonal GABA receptors are selectively present on normal and regenerated sensory fibers in rat peripheral nerve. *Exp. Brain Res.* **66**, 659–663.

Blaesse, P., Airaksinen, M.S., Rivera, C., and Kaila, K. (2009). Cation-chloride cotransporters and neuronal function. *Neuron* **61**, 820–838.

Bodian, D. (1952). Introductory survey of neurons. *Cold Spring Harb. Symp. Quant. Biol.* **17**, 1–13.

Bodian, D. (1972). Synaptic diversity and characterization by electron microscopy. In *Structure and Function of Synapses* (Pappas, G.D. and Purpura, D.P., eds), pp. 45–65. Raven Press, New York.

Boettger, T., Rust, M.B., Maier, H., Seidenbecher, T., Schweizer, M., Keating, D.J., Faulhaber, J., Ehmke, H., Pfeffer, C., Scheel, O., Lemcke, B., Horst, J., Leuwer, R., Pape, H.C., Volkl, H., Hubner, C.A., and Jentsch, T.J. (2003). Loss of K-Cl co-transporter KCC3 causes deafness, neurodegeneration and reduced seizure threshold. *Embo J.* **22**, 5422–5434.

Bowery, N.G. and Smart, T.G. (2006). GABA and glycine as neurotransmitters: a brief history. *Br. J. Pharmacol.* **147** (Suppl. 1), S109–S119.

Breitwieser, G.E., Altamirano, A.A., and Russell, J.M. (1990). Osmotic stimulation of Na^+-K^+-$2Cl^-$ cotransport in squid giant axon is $[Cl^-]_i$ dependent. *Am. J. Physiol.* **258**, C749–C753.

Breitwieser, G.E., Altamirano, A.A., and Russell, J.M. (1996). Elevated $[Cl^-]_i$, and $[Na^+]_i$ inhibit Na^+ -K^+-$2Cl^-$ cotransport by different mechanisms in squid giant axons. *J. Gen. Physiol.* **107**, 261–270.

Brown, D.A. and Marsh, S. (1978). Axonal GABA-receptors in mammalian peripheral nerve trunks. *Brain Res.* **156**, 187–191.

Brown, J.K., Pemberton, A.D., Wright, S.H., and Miller, H.R. (2004). Primary antibody-Fab fragment complexes: a flexible alternative to traditional direct and indirect immunolabeling techniques. *J. Histochem. Cytochem.* **52**, 1219–1230.

Burke, R.E. (2006). John Eccles' pioneering role in understanding central synaptic transmission. *Prog. Neuro. Biol.* **78**, 173–188.

Burnstock, G. (2000). P2X receptors in sensory neurones. *Br. J. Anaesth.* **84**, 476–488.

Calvillo, O. (1978). Primary afferent depolarization of C fibres in the spinal cord of the cat. *Can. J. Physiol. Pharmacol.* **56**, 154–157.

Calvillo, O., Madrid, J., and Rudomin, P. (1982). Presynaptic depolarization of unmyelinated primary afferent fibers in the spinal cord of the cat. *Neuroscience* **7**, 1389–1409.

Carlton, S.M., Zhou, S., and Coggeshall, R.E. (1999). Peripheral GABA(A) receptors: evidence for peripheral primary afferent depolarization. *Neuroscience* **93**, 713–722.

Carstens, E., Tulloch, I., Zieglgansberger, W., and Zimmermann, M. (1979). Presynaptic excitability changes induced by morphine in single cutaneous afferent C- and A-fibers. *Pflugers Arch.* **379**, 143–147.

Cattaert, D. and El Manira, A. (1999). Shunting versus inactivation: analysis of presynaptic inhibitory mechanisms in primary afferents of the crayfish. *J. Neurosci.* **19**, 6079–6089.

Cattaert, D., El Manira, A., and Bevengut, M. (1999). Presynaptic inhibition and antidromic discharges in crayfish primary afferents. *J. Physiol. Paris* **93**, 349–358.

Cattaert, D., Libersat, F., and El Manira, A.A. (2001). Presynaptic inhibition and antidromic spikes in primary afferents of the crayfish: a computational and experimental analysis. *J. Neurosci.* **21**, 1007–1021.

Cervero, F. and Laird, J.M. (1996). Mechanisms of touch-evoked pain (allodynia): a new model. *Pain* **68**, 13–23.

Cervero, F., Laird, J.M., and Garcia-Nicas, E. (2003). Secondary hyperalgesia and presynaptic inhibition: an update. *Eur. J. Pain* **7**, 345–351.

Chen, H., Luo, J., Kintner, D.B., Shull, G.E., and Sun, D. (2005). Na^+-dependent chloride transporter (NKCC1)-null mice exhibit less gray and white matter damage after focal cerebral ischemia. *J. Cereb. Blood Flow Metab.* **25**, 54–66.

Chen, J.H., Weng, H.R., and Dougherty, P.M. (2004). Sensitization of dorsal root reflexes in vitro and hyperalgesia in neonatal rats produced by capsaicin. *Neuroscience* **126**, 743–751.

Chiu, S.Y. (1991). Functions and distribution of voltage-gated sodium and potassium channels in mammalian Schwann cells. *Glia* **4**, 541–558.

Chiu, S.Y. (1995). Schwann cell functions in saltatory conduction. In *Neuroglia* (Kettenmann, H. and Ransom, B.R., eds), pp. 777–792. Oxford University Press, New York.

Coombs, J.S., Eccles, J.C., and Fatt, P. (1955). The specific ionic conductances and the ionic movements across the motoneuronal membrane that produce the inhibitory post-synaptic potential. *J. Physiol.* **130**, 326–374.

Coull, J.A., Boudreau, D., Bachand, K., Prescott, S.A., Nault, F., Sik, A., De Koninck, P., and De Koninck, Y. (2003). Trans-synaptic shift in anion gradient in spinal lamina I neurons as a mechanism of neuropathic pain. *Nature* **424**, 938–942.

Cowan, W.M. and Kandel, E.R. (2001). A brief history of synapses and synaptic transmission. In *Synapses* (Cowan, W.M., Südhof, T.C., Stevens, C.F. and H.H.M. Institute, eds), Johns Hopkins University Press, Baltimore

Cramer, S.W., Baggott, C., Cain, J., Tilghman, J., Allcock, B., Miranpuri, G., Rajpal, S., Sun, D., and Resnick, D. (2008). The role of cation-dependent chloride transporters in neuropathic pain following spinal cord injury. *Mol. Pain* **4**, 36.

Crane, R.K. (1977). The gradient hypothesis and other models of carrier-mediated active transport. *Rev. Physiol. Biochem. Pharmacol.* **78**, 99–159.

Crowe, W.E., Altamirano, J., Huerto, L., and Alvarez-Leefmans, F.J. (1995). Volume changes in single N1E-115 neuroblastoma cells measured with a fluorescent probe. *Neuroscience* **69**, 283–296.

Currie, K.P., Wootton, J.F., and Scott, R.H. (1995). Activation of Ca^{2+}-dependent Cl^- currents in cultured rat sensory neurones by flash photolysis of DM-nitrophen. *J. Physiol.* **482** (Pt 2), 291–307.

Curtis, D.R. and Gynther, B.D. (1986). On the probable absence of GABA receptors on the terminations of motor axon collaterals in the cat spinal cord. *Exp. Brain Res.* **64**, 114–118.

Curtis, D.R., Hosli, L., Johnston, G.A., and Johnston, I.H. (1968). The hyperpolarization of spinal motoneurones by glycine and related amino acids. *Exp. Brain Res.* **5**, 235–258.

Curtis, D.R. and Lodge, D. (1982). The depolarization of feline ventral horn group Ia spinal afferent terminations by GABA. *Exp. Brain Res.* **46**, 215–233.

Dale, H.H. (1935). Pharmacology and nerve-endings. *Proc. R. Soc. Med.* **28**, 319–332.

Davidoff, R.A. and Hackman, J.C. (1985). GABA: presynaptic actions. In *Neurotransmitter Actions in the Vertebrate Nervous System* (Rogawski, M.A. and Barker, J.L., eds), pp. 3–32. Plenum Press, New York.

De Groat, W.C. (1972). GABA-depolarization of a sensory ganglion: antagonism by picrotoxin and bicuculline. *Brain Res.* **38**, 429–432.

De Groat, W.C., Lalley, P.M., and Saum, W.R. (1972). Depolarization of dorsal root ganglia in the cat by GABA and related amino acids: antagonism by picrotoxin and bicuculline. *Brain Res.* **44**, 273–277.

DeFazio, R.A., Keros, S., Quick, M.W., and Hablitz, J.J. (2000). Potassium-coupled chloride cotransport controls intracellular chloride in rat neocortical pyramidal neurons. *J. Neurosci.* **20**, 8069–8076.

Del Castillo, I.C., Fedor-Chaiken, M., Song, J.C., Starlinger, V., Yoo, J., Matlin, K.S., and Matthews, J.B. (2005). Dynamic regulation of Na^+-K^+-$2Cl^-$ cotransporter surface expression by PKC-ε in Cl^--secretory epithelia. *Am. J. Physiol. Cell Physiol.* **289**, C1332–C1342.

Desarmenien, M., Santangelo, F., Loeffler, J.P., and Feltz, P. (1984). Comparative study of GABA-mediated depolarizations of lumbar A delta and C primary afferent neurones of the rat. *Exp. Brain Res.* **54**, 521–528.

Deschenes, M., Feltz, P., and Lamour, Y. (1976). A model for an estimate in vivo of the ionic basis of presynaptic inhibition: an intracellular analysis of the GABA-induced depolarization in rat dorsal root ganglia. *Brain Res.* **118**, 486–493.

Draguhn, A., Axmacher, N., and Kolbaev, S. (2008). Presynaptic ionotropic GABA receptors. *Results Probl. Cell Differ.* **44**, 69–85.

Dubuisson, D. and Dennis, S.G. (1977). The formalin test: a quantitative study of the analgesic effects of morphine, meperidine, and brain stem stimulation in rats and cats. *Pain* **4**, 161–174.

Dunlap, K. (1984). Functional and pharmacological differences between two types of GABA receptor on embryonic chick sensory neurons. *Neurosci. Lett.* **47**, 265–270.

Dunn, P.M., Zhong, Y., and Burnstock, G. (2001). P2X receptors in peripheral neurons. *Prog. NeuroBiol.* **65**, 107–134.

Eccles, J.C. (1957). *The Physiology of Nerve Cells*. Johns Hopkins Press, Baltimore.

Eccles, J.C. (1961). The mechanism of synaptic transmission. *Ergeb. Physiol.* **51**, 299–430.

Eccles, J.C. (1964). *The Physiology of Synapses*. Springer, Berlin.

Eccles, J.C., Eccles, R.M., and Magni, F. (1961). Central inhibitory action attributable to presynaptic depolarization produced by muscle afferent volleys. *J. Physiol.* **159**, 147–166.

Eccles, J.C., Schmidt, R., and Willis, W.D. (1963). Pharmacological studies on presynaptic inhibition. *J. Physiol.* **168**, 500–530.

Eccles, J.C., Schmidt, R.F., and Willis, W.D. (1962). Presynaptic inhibition of the spinal monosynaptic reflex pathway. *J. Physiol.* **161**, 282–297.

Evans, R.L., Park, K., Turner, R.J., Watson, G.E., Nguyen, H.V., Dennett, M.R., Hand, A.R., Flagella, M., Shull, G.E., and Melvin, J.E. (2000). Severe impairment of salivation in $Na^+/K^+/2Cl^-$ cotransporter (NKCC1)-deficient mice. *J. Biol. Chem.* **275**, 26720–26726.

Fang, X., Djouhri, L., McMullan, S., Berry, C., Waxman, S.G., Okuse, K., and Lawson, S.N. (2006). Intense isolectin-B4 binding in rat dorsal root ganglion neurons distinguishes C-fiber nociceptors with broad action potentials and high Nav1.9 expression. *J. Neurosci.* **26**, 7281–7292.

Fang, X., McMullan, S., Lawson, S.N., and Djouhri, L. (2005). Electrophysiological differences between nociceptive and non-nociceptive dorsal root ganglion neurones in the rat in vivo. *J. Physiol.* **565**, 927–943.

Fitzgerald, M. and Woolf, C.J. (1981). Effects of cutaneous nerve and intraspinal conditioning of C-fibre afferent terminal excitability in decerebrate spinal rats. *J. Physiol.* **318**, 25–39.

Flatman, P.W. (2002). Regulation of Na-K-2Cl cotransport by phosphorylation and protein–protein interactions. *Biochim. Biophys. Acta* **1566**, 140–151.

Flemmer, A.W., Gimenez, I., Dowd, B.F., Darman, R.B., and Forbush, B. (2002). Activation of the Na-K-Cl cotransporter NKCC1 detected with a phospho-specific antibody. *J. Biol. Chem.* **277**, 37551–37558.

Frank, K. and Fuortes, M.G.F. (1957). Presynaptic and postsynaptic inhibition of monosynaptic reflexes. *Fed. Proc.* **16**, 39–40.

Frankenhaeuser, B. and Hodgkin, A.L. (1956). The after-effects of impulses in the giant nerve fibres of Loligo. *J. Physiol.* **131**, 341–376.

French, A.S., Panek, I., and Torkkeli, P.H. (2006). Shunting versus inactivation: simulation of GABAergic inhibition in spider mechanoreceptors suggests that either is sufficient. *Neurosci. Res.* **55**, 189–196.

Frings, S., Reuter, D., and Kleene, S.J. (2000). Neuronal $Ca2^+$-activated Cl^- channels –homing in on an elusive channel species. *Prog. NeuroBiol.* **60**, 247–289.

Funk, K., Woitecki, A., Franjic-Wurtz, C., Gensch, T., Mohrlen, F., and Frings, S. (2008). Modulation of chloride homeostasis by inflammatory mediators in dorsal root ganglion neurons. *Mol. Pain* **4**, 32.

Fyffe, R.E. and Light, A.R. (1984). The ultrastructure of group Ia afferent fiber synapses in the lumbosacral spinal cord of the cat. *Brain Res.* **300**, 201–209.

Galan, A. and Cervero, F. (2005). Painful stimuli induce in vivo phosphorylation and membrane mobilization of mouse spinal cord NKCC1 co-transporter. *Neuroscience* **133**, 245–252.

Gallagher, J.P., Higashi, H., and Nishi, S. (1978). Characterization and ionic basis of GABA-induced depolarizations recorded in vitro from cat primary afferent neurones. *J. Physiol.* **275**, 263–282.

Gallagher, J.P., Nakamura, J., and Shinnick-Gallagher, P. (1983). The effects of temperature, pH and Cl-pump inhibitors on GABA responses recorded from cat dorsal root ganglia. *Brain Res.* **267**, 249–259.

Gamba, G. (2005). Molecular physiology and pathophysiology of electroneutral cation-chloride cotransporters. *Physiol. Rev.* **85**, 423–493.

Ge, S., Goh, E.L., Sailor, K.A., Kitabatake, Y., Ming, G.L., and Song, H. (2006). GABA regulates synaptic integration of newly generated neurons in the adult brain. *Nature* **439**, 589–593.

Geck, P. and Heinz, E. (1986). The Na-K-2Cl cotransport system. *J. Membr. Biol.* **91**, 97–105.

Geck, P., Pietrzyk, C., Burckhardt, B.C., Pfeiffer, B., and Heinz, E. (1980). Electrically silent cotransport on Na^+, K^+ and Cl^- in Ehrlich cells. *Biochim. Biophys. Acta* **600**, 432–447.

Geng, Y., Hoke, A., and Delpire, E. (2009). The Ste20 Kinases Ste20-related Proline-Alanine-rich Kinase and Oxidative-stress

Response 1 Regulate NKCC1 Function in Sensory Neurons. *J Biol Chem* **284**, 14020–14028.

Gerke, M.B. and Plenderleith, M.B. (2001). Binding sites for the plant lectin Bandeiraea simplicifolia I-isolectin B(4) are expressed by nociceptive primary sensory neurones. *Brain Res.* **911**, 101–104.

Ghabriel, M.N. and Allt, G. (1981). Incisures of Schmidt-Lanterman. *Prog. NeuroBiol.* **17**, 25–58.

Gilbert, D., Franjic-Wurtz, C., Funk, K., Gensch, T., Frings, S., and Mohrlen, F. (2007). Differential maturation of chloride homeostasis in primary afferent neurons of the somatosensory system. *Int. J. Dev. Neurosci.* **25**, 479–489.

Gillen, C.M. and Forbush, B., 3rd (1999). Functional interaction of the K-Cl cotransporter (KCC1) with the Na-K-Cl cotransporter in HEK-293 cells. *Am. J. Physiol.* **276**, C328–C336.

Gimenez, I. and Forbush, B. (2003). Short-term stimulation of the renal Na-K-Cl cotransporter (NKCC2) by vasopressin involves phosphorylation and membrane translocation of the protein. *J. Biol. Chem.* **278**, 26946–26951.

Gonzalez-Islas, C., Chub, N., and Wenner, P. (2009). NKCC1 and AE3 appear to accumulate chloride in embryonic motoneurons. *J. NeuroPhysiol.* **101**, 507–518.

Gover, T.D., Kao, J.P., and Weinreich, D. (2003). Calcium signaling in single peripheral sensory nerve terminals. *J. Neurosci.* **23**, 4793–4797.

Graham, B. and Redman, S. (1994). A simulation of action potentials in synaptic boutons during presynaptic inhibition. *J. NeuroPhysiol.* **71**, 538–549.

Granados-Soto, V., Arguelles, C.F., and Alvarez-Leefmans, F.J. (2005). Peripheral and central antinociceptive action of Na^+-K^+-$2Cl^-$ cotransporter blockers on formalin-induced nociception in rats. *Pain* **114**, 231–238.

Gray, E.G. (1962). A morphological basis for pre-synaptic inhibition? *Nature* **193**, 82–83.

Gray, E.G. (1963). Electron microscopy of presynaptic organelles of the spinal cord. *J. Anat.* **97**, 101–106.

Haas, M., McBrayer, D., and Lytle, C. (1995). [Cl−]i-dependent phosphorylation of the Na-K-Cl cotransport protein of dog tracheal epithelial cells. *J. Biol. Chem.* **270**, 28955–28961.

Harlow, E. and Lane, D. (1988). *Antibodies: A Laboratory Manual.* Cold Spring Harbor Laboratory, Cold Spring Harbor, NY.

Harlow, E. and Lane, D. (1999). *Using Antibodies: A Laboratory Manual.* Cold Spring Harbor Laboratory Press, Cold Spring Harbor, NY.

Hartzell, C., Putzier, I., and Arreola, J. (2005). Calcium-activated chloride channels. *Annu. Rev. Physiol.* **67**, 719–758.

Hayasaki, H., Sohma, Y., Kanbara, K., Maemura, K., Kubota, T., and Watanabe, M. (2006). A local GABAergic system within rat trigeminal ganglion cells. *Eur. J. Neurosci.* **23**, 745–757.

Hentall, I.D. and Fields, H.L. (1979). Segmental and descending influences on intraspinal thresholds of single C-fibers. *J. NeuroPhysiol.* **42**, 1527–1537.

Hierck, B.P., Iperen, L.V., Gittenberger-De Groot, A.C., and Poelmann, R.E. (1994). Modified indirect immunodetection allows study of murine tissue with mouse monoclonal antibodies. *J. Histochem. Cytochem.* **42**, 1499–1502.

Hockfield, S., Carlson, S., Evans, C., Levitt, P., Pintar, J., and Silberstein, L. (1993). *Selected Methods for Antibody and Nucleic Acid Probes.* Cold Spring Harbor Laboratory Press, Plainview, NY.

Hoffmann, E.K., Schettino, T., and Marshall, W.S. (2007). The role of volume-sensitive ion transport systems in regulation of epithelial transport. *Comp. Biochem. Physiol. A Mol. Integr. Physiol.* **148**, 29–43.

Hokfelt, T. (2009). Coexistence of neuromessenger molecules – a perspective. In *Co-Existence and Co-Release of Classical Neurotransmitters* (Gutierrez, R. ed.), pp. 1–13. Springer, New York.

Hong, S., Morrow, T.J., Paulson, P.E., Isom, L.L., and Wiley, J.W. (2004). Early painful diabetic neuropathy is associated with differential changes in tetrodotoxin-sensitive and -resistant sodium channels in dorsal root ganglion neurons in the rat. *J. Biol. Chem.* **279**, 29341–29350.

Howard, H.C., Mount, D.B., Rochefort, D., Byun, N., Dupre, N., Lu, J., Fan, X., Song, L., Riviere, J.B., Prevost, C., Horst, J., Simonati, A., Lemcke, B., Welch, R., England, R., Zhan, F.Q., Mercado, A., Siesser, W.B., George, A.L., Jr., McDonald, M.P., Bouchard, J.P., Mathieu, J., Delpire, E., and Rouleau, G.A. (2002). The K-Cl cotransporter KCC3 is mutant in a severe peripheral neuropathy associated with agenesis of the corpus callosum. *Nat. Genet.* **32**, 384–392.

Hughes, D.I., Mackie, M., Nagy, G.G., Riddell, J.S., Maxwell, D.J., Szabo, G., Erdelyi, F., Veress, G., Szucs, P., Antal, M., and Todd, A.J. (2005). P boutons in lamina IX of the rodent spinal cord express high levels of glutamic acid decarboxylase-65 and originate from cells in deep medial dorsal horn. *Proc. Natl. Acad. Sci. USA* **102**, 9038–9043.

Hutter, O.F. (2006). Some deconstructed dogmas. *Prog. Biophys. Mol. Biol.* **90**, 5–12.

Jankowska, E., McCrea, D., Rudomin, P., and Sykova, E. (1981). Observations on neuronal pathways subserving primary afferent depolarization. *J. NeuroPhysiol.* **46**, 506–516.

Ji, R.R., Kohno, T., Moore, K.A., and Woolf, C.J. (2003). Central sensitization and LTP: do pain and memory share similar mechanisms?. *Trends Neurosci.* **26**, 696–705.

Julius, D. and Basbaum, A.I. (2001). Molecular mechanisms of nociception. *Nature* **413**, 203–210.

Kahle, K.T., Rinehart, J., Ring, A., Gimenez, I., Gamba, G., Hebert, S.C., and Lifton, R.P. (2006). WNK protein kinases modulate cellular Cl^- flux by altering the phosphorylation state of the Na-K-Cl and K-Cl cotransporters. *Physiology (Bethesda)* **21**, 326–335.

Kakazu, Y., Akaike, N., Komiyama, S., and Nabekura, J. (1999). Regulation of intracellular chloride by cotransporters in developing lateral superior olive neurons. *J. Neurosci.* **19**, 2843–2851.

Kakazu, Y., Uchida, S., Nakagawa, T., Akaike, N., and Nabekura, J. (2000). Reversibility and cation selectivity of the K^+-Cl^- cotransport in rat central neurons. *J. NeuroPhysiol.* **84**, 281–288.

Kanaka, C., Ohno, K., Okabe, A., Kuriyama, K., Itoh, T., Fukuda, A., and Sato, K. (2001). The differential expression patterns of messenger RNAs encoding K-Cl cotransporters (KCC1, 2) and Na-K-2Cl cotransporter (NKCC1) in the rat nervous system. *Neuroscience* **104**, 933–946.

Kaneko, H., Nakamura, T., and Lindemann, B. (2001). Noninvasive measurement of chloride concentration in rat olfactory receptor cells with use of a fluorescent dye. *Am. J. Physiol. Cell Physiol.* **280**, C1387–C1393.

Kaplan, M.R., Plotkin, M.D., Brown, D., Hebert, S.C., and Delpire, E. (1996). Expression of the mouse Na-K-2Cl cotransporter, mBSC2, in the terminal inner medullary collecting duct, the glomerular and extraglomerular mesangium, and the glomerular afferent arteriole. *J. Clin. Invest.* **98**, 723–730.

Karadsheh, M.F., Byun, N., Mount, D.B., and Delpire, E. (2004). Localization of the KCC4 potassium-chloride cotransporter in the nervous system. *Neuroscience* **123**, 381–391.

Kashiba, H., Uchida, Y., and Senba, E. (2001). Difference in binding by isolectin B4 to trkA and c-ret mRNA-expressing neurons in rat sensory ganglia. *Brain Res. Mol. Brain Res.* **95**, 18–26.

Kenyon, J. and Scott, R. (2002). Calcium-activated chloride channels as Ca^{2+} sensors with particular reference to the modulation of neuronal excitability. *Curr. Topics Membr.* **53**, 131–162.

Khirug, S., Yamada, J., Afzalov, R., Voipio, J., Khiroug, L., and Kaila, K. (2008). GABAergic depolarization of the axon initial segment in cortical principal neurons is caused by the Na-K-2Cl cotransporter NKCC1. *J. Neurosci.* **28**, 4635–4639.

Kocsis, J.D. and Sakatani, K. (1995). Modulation of axonal excitability by neurotransmitter receptors. In *The Axon: Structure,*

Function, and Pathophysiology (Waxman, S.G., Kocsis, J.D., and Stys, P.K., eds), pp. 281–295. Oxford University Press, New York.

Koncz, C. and Daugirdas, J.T. (1994). Use of MQAE for measurement of intracellular $[Cl^-]$ in cultured aortic smooth muscle cells. *Am. J. Physiol.* **267**, H2114–H2123.

Krnjevic, K., Puil, E., and Werman, R. (1977). Bicuculline, benzyl penicillin, and inhibitory amino acids in the spinal cord of the cat. *Can. J. Physiol. Pharmacol.* **55**, 670–680.

Kullmann, D.M., Ruiz, A., Rusakov, D.M., Scott, R., Semyanov, A., and Walker, M.C. (2005). Presynaptic, extrasynaptic and axonal GABAA receptors in the CNS: where and why? *Prog. Biophys. Mol. Biol.* **87**, 33–46.

Kurihara, K., Moore-Hoon, M.L., Saitoh, M., and Turner, R.J. (1999). Characterization of a phosphorylation event resulting in upregulation of the salivary Na^+-K^+-$2Cl^-$ cotransporter. *Am. J. Physiol.* **277**, C1184–C1193.

Labrakakis, C., Tong, C.K., Weissman, T., Torsney, C., and MacDermott, A.B. (2003). Localization and function of ATP and GABAA receptors expressed by nociceptors and other postnatal sensory neurons in rat. *J. Physiol.* **549**, 131–142.

Laird, J.M., Garcia-Nicas, E., Delpire, E.J., and Cervero, F. (2004). Presynaptic inhibition and spinal pain processing in mice: a possible role of the NKCC1 cation-chloride co-transporter in hyperalgesia. *Neurosci. Lett.* **361**, 200–203.

Lamotte D'Incamps, B., Meunier, C., Monnet, M.L., Jami, L., and Zytnicki, D. (1998). Reduction of presynaptic action potentials by PAD: model and experimental study. *J. Comput. Neurosci.* **5**, 141–156.

Landon, D.N. and Hall, S. (1976). The myelinated nerve fibre. In *The Peripheral Nerve* (Landon, D.N., ed.), pp. 1–105. Chapman & Hall, London.

Langley, J.N. (1907). On the contraction of muscle chiefly in relation to the presence of "receptive" substances: Part I. *J. Physiol.* **36**, 347–384.

Lawson, S. (2005). The peripheral sensory nervous system: dorsal root ganglion neurons Chap. 8. In *Peripheral Neuropathy* (Dyck, P.J. and Thomas, P.K., eds), pp. 163–202. W.B. Saunders; Elsevier, Philadelphia, PA.

Levi, G. (1927). *Trattato de Istologia.* Unione Tipografico-Editrice Torinese, Torino.

Lev-Tov, A., Fleshman, J.W., and Burke, R.E. (1983). Primary afferent depolarization and presynaptic inhibition of monosynaptic group Ia EPSPs during posttetanic potentiation. *J. NeuroPhysiol.* **50**, 413–427.

Lev-Tov, A., Meyers, D.E., and Burke, R.E. (1988). Modification of primary afferent depolarization in cat group Ia afferents following high frequency intra-axonal tetanization of individual afferents. *Brain Res.* **438**, 328–330.

Levy, R.A. (1977). The role of GABA in primary afferent depolarization. *Prog. NeuroBiol.* **9**, 211–267.

Li, B., McKernan, K., and Shen, W. (2008). Spatial and temporal distribution patterns of Na-K-2Cl cotransporter in adult and developing mouse retinas. *Vis. Neurosci.* **25**, 109–123.

Lin, Q., Wu, J., and Willis, W.D. (1999). Dorsal root reflexes and cutaneous neurogenic inflammation after intradermal injection of capsaicin in rats. *J. NeuroPhysiol.* **82**, 2602–2611.

Lin, Q., Zou, X., and Willis, W.D. (2000). Adelta and C primary afferents convey dorsal root reflexes after intradermal injection of capsaicin in rats. *J. NeuroPhysiol.* **84**, 2695–2698.

Linhart, O., Obreja, O., and Kress, M. (2003). The inflammatory mediators serotonin, prostaglandin E2 and bradykinin evoke calcium influx in rat sensory neurons. *Neuroscience* **118**, 69–74.

Lu, J., Karadsheh, M., and Delpire, E. (1999). Developmental regulation of the neuronal-specific isoform of K-Cl cotransporter KCC2 in postnatal rat brains. *J. NeuroBiol.* **39**, 558–568.

Lytle, C. and Forbush, B., 3rd (1996). Regulatory phosphorylation of the secretory Na-K-Cl cotransporter: modulation by cytoplasmic Cl. *Am. J. Physiol.* **270**, C437–C448.

Lytle, C. and McManus, T. (2002). Coordinate modulation of Na-K-2Cl cotransport and K-Cl cotransport by cell volume and chloride. *Am. J. Physiol. Cell Physiol.* **283**, C1422–C1431.

Lytle, C., McManus, T.J., and Haas, M. (1998). A model of Na-K-2Cl cotransport based on ordered ion binding and glide symmetry. *Am. J. Physiol.* **274**, C299–C309.

Lytle, C., Xu, J.C., Biemesderfer, D., and Forbush, B., 3rd (1995). Distribution and diversity of Na-K-Cl cotransport proteins: a study with monoclonal antibodies. *Am. J. Physiol.* **269**, C1496–C1505.

Maglova, L.M., Crowe, W.E., Altamirano, A.A., and Russell, J. M. (1998). Human cytomegalovirus infection stimulates Cl^-/ HCO−3 exchanger activity in human fibroblasts. *Am. J. Physiol.* **275**, C515–C526.

Marty, S., Wehrle, R., Alvarez-Leefmans, F.J., Gasnier, B., and Sotelo, C. (2002). Postnatal maturation of Na^+, K^+, $2Cl^-$ cotransporter expression and inhibitory synaptogenesis in the rat hippocampus: an immunocytochemical analysis. *Eur. J. Neurosci.* **15**, 233–245.

Mayer, M.L. (1985). A calcium-activated chloride current generates the after-depolarization of rat sensory neurones in culture. *J. Physiol.* **364**, 217–239.

Mazzone, S.B. and McGovern, A.E. (2006). Na^+-K^+-$2Cl^-$ cotransporters and Cl^- channels regulate citric acid cough in guinea pigs. *J. Appl. Physiol.* **101**, 635–643.

Mazzone, S.B. and McGovern, A.E. (2008). Immunohistochemical characterization of nodose cough receptor neurons projecting to the trachea of guinea pigs. *Cough* **4**, 9.

McDaniel, N., Pace, A.J., Spiegel, S., Engelhardt, R., Koller, B.H., Seidler, U., and Lytle, C. (2005). Role of Na-K-2Cl cotransporter-1 in gastric secretion of nonacidic fluid and pepsinogen. *Am. J. Physiol. Gastrointest. Liver Physiol.* **289**, G550–G560.

Melzack, R. and Wall, P.D. (1965). Pain mechanisms: a new theory. *Science* **150**, 971–979.

Mercado, A., Song, L., Vazquez, N., Mount, D.B., and Gamba, G. (2000). Functional comparison of the K + -Cl− cotransporters KCC1 and KCC4. *J. Biol. Chem.* **275**, 30326–30334.

Meyer, R.A. and Campbell, J.N. (1981). Myelinated nociceptive afferents account for the hyperalgesia that follows a burn to the hand. *Science* **213**, 1527–1529.

Morris, M.E., Di Costanzo, G.A., Fox, S., and Werman, R. (1983). Depolarizing action of GABA (gamma-aminobutyric acid) on myelinated fibers of peripheral nerves. *Brain Res.* **278**, 117–126.

Mount, D.B., Mercado, A., Song, L., Xu, J., George, A.L., Jr., Delpire, E., and Gamba, G. (1999). Cloning and characterization of KCC3 and KCC4, new members of the cation-chloride cotransporter gene family. *J. Biol. Chem.* **274**, 16355–16362.

Mount, D.B. and Romero, M.F. (2004). The SLC26 gene family of multifunctional anion exchangers. *Pflugers Arch.* **447**, 710–721.

Munoz, A., Mendez, P., DeFelipe, J., and Alvarez-Leefmans, F.J. (2007). Cation-chloride cotransporters and GABA-ergic innervation in the human epileptic hippocampus. *Epilepsia* **48**, 663–673.

Nageotte, J. (1922). *L'Organisation de la Matiere.* Librairie Felix Alcan, Paris.

Nejsum, L.N., Praetorius, J., and Nielsen, S. (2005). NKCC1 and NHE1 are abundantly expressed in the basolateral plasma membrane of secretory coil cells in rat, mouse, and human sweat glands. *Am. J. Physiol. Cell Physiol.* **289**, C333–C340.

Nicoll, R.A. and Alger, B.E. (1979). Presynaptic inhibition: transmitter and ionic mechanisms. *Int. Rev. NeuroBiol.* **21**, 217–258.

Nicolson, T.A., Bevan, S., and Richards, C.D. (2002). Characterisation of the calcium responses to histamine in capsaicin-sensitive and capsaicin-insensitive sensory neurones. *Neuroscience* **110**, 329–338.

Nishi, S., Minota, S., and Karczmar, A.G. (1974). Primary afferent neurones: the ionic mechanism of GABA-mediated depolarization. *Neuropharmacology* **13**, 215–219.

Ochiai, H., Higa, K., and Fujise, H. (2004). Molecular identification of K-Cl cotransporter in dog erythroid progenitor cells. *J. Biochem. (Tokyo)* **135**, 365–374.

Owens, D.F. and Kriegstein, A.R. (2002). Is there more to GABA than synaptic inhibition? *Nat. Rev. Neurosci.* **3**, 715–727.

Pacheco-Alvarez, D., Cristobal, P.S., Meade, P., Moreno, E., Vazquez, N., Munoz, E., Diaz, A., Juarez, M.E., Gimenez, I., and Gamba, G. (2006). The Na^+:Cl^- cotransporter is activated and phosphorylated at the amino-terminal domain upon intracellular chloride depletion. *J. Biol. Chem.* **281**, 28755–28763.

Payne, J.A. (1997). Functional characterization of the neuronal-specific K-Cl cotransporter: implications for [K+]o regulation. *Am. J. Physiol.* **273**, C1516–C1525.

Payne, J.A., Rivera, C., Voipio, J., and Kaila, K. (2003). Cation-chloride co-transporters in neuronal communication, development and trauma. *Trends Neurosci.* **26**, 199–206.

Pearson, M.M., Lu, J., Mount, D.B., and Delpire, E. (2001). Localization of the K^+-Cl^- cotransporter, KCC3, in the central and peripheral nervous systems: expression in the choroid plexus, large neurons and white matter tracts. *Neuroscience* **103**, 481–491.

Pelc, S., Gompel, C., and Simonet, M.L. (1986). S-100 protein expression in satellite and Schwann cells in neuroblastoma. An immunohistochemical and ultrastructural study. *Virchows Arch. B Cell Pathol. Incl. Mol. Pathol.* **51**, 487–495.

Petruska, J.C., Napaporn, J., Johnson, R.D., Gu, J.G., and Cooper, B.Y. (2000). Subclassified acutely dissociated cells of rat DRG: histochemistry and patterns of capsaicin-, proton-, and ATP-activated currents. *J. NeuroPhysiol.* **84**, 2365–2379.

Piechotta, K., Lu, J., and Delpire, E. (2002). Cation chloride cotransporters interact with the stress-related kinases Ste20-related proline-alanine-rich kinase (SPAK) and oxidative stress response 1 (OSR1). *J. Biol. Chem.* **277**, 50812–50819.

Pieraut, S., Laurent-Matha, V., Sar, C., Hubert, T., Mechaly, I., Hilaire, C., Mersel, M., Delpire, E., Valmier, J., and Scamps, F. (2007). NKCC1 phosphorylation stimulates neurite growth of injured adult sensory neurons. *J. Neurosci.* **27**, 6751–6759.

Pierce, J.P. and Mendell, L.M. (1993). Quantitative ultrastructure of Ia boutons in the ventral horn: scaling and positional relationships. *J. Neurosci.* **13**, 4748–4763.

Pierrot-Deseilligny, E. and Burkey, D. (2005). *The Circuitry of the Human Spinal Cord. Its Role in Motor Control and Movement Disorders*. Cambridge University Press, New York.

Pitcher, M.H., Price, T.J., Entrena, J.M., and Cervero, F. (2007). Spinal NKCC1 blockade inhibits TRPV1-dependent referred allodynia. *Mol. Pain* **3**, 17.

Plotkin, M.D., Kaplan, M.R., Peterson, L.N., Gullans, S.R., Hebert, S.C., and Delpire, E. (1997a). Expression of the Na^+-K^+-$2Cl^-$ cotransporter BSC2 in the nervous system. *Am. J. Physiol.* **272**, C173–C183.

Plotkin, M.D., Snyder, E.Y., Hebert, S.C., and Delpire, E. (1997b). Expression of the Na-K-2Cl cotransporter is developmentally regulated in postnatal rat brains: a possible mechanism underlying GABA's excitatory role in immature brain. *J. NeuroBiol.* **33**, 781–795.

Price, E.M. and Lingrel, J.B. (1988). Structure-function relationships in the Na,K-ATPase alpha subunit: site-directed mutagenesis of glutamine-111 to arginine and asparagine-122 to aspartic acid generates a ouabain-resistant enzyme. *Biochemistry* **27**, 8400–8408.

Price, T.J., Cervero, F., and de Koninck, Y. (2005). Role of cation-chloride-cotransporters (CCC) in pain and hyperalgesia. *Curr. Top. Med. Chem.* **5**, 547–555.

Price, T.J., Cervero, F., Gold, M.S., Hammond, D.L., and Prescott, S.A. (2009). Chloride regulation in the pain pathway. *Brain Res. Rev.* **60**, 149–170.

Price, T.J., Hargreaves, K.M., and Cervero, F. (2006). Protein expression and mRNA cellular distribution of the NKCC1 cotransporter in the dorsal root and trigeminal ganglia of the rat. *Brain Res.* **1112**, 146–158.

Raja, S.N., Campbell, J.N., and Meyer, R.A. (1984). Evidence for different mechanisms of primary and secondary hyperalgesia following heat injury to the glabrous skin. *Brain* **107** (Pt 4), 1179–1188.

Ramón y Cajal, S. (1954). *Neuron Theory or Reticular Theory?* Consejo Superior de Investigaciones Científicas, Madrid.

Ransom, C.B., Ransom, B.R., and Sontheimer, H. (2000). Activity-dependent extracellular K^+ accumulation in rat optic nerve: the role of glial and axonal Na^+ pumps. *J. Physiol.* **522** (Pt 3), 427–442.

Reeh, P.W. and Kress, M. (2001). Molecular physiology of proton transduction in nociceptors. *Curr. Opin. Pharmacol.* **1**, 45–51.

Rees, H., Sluka, K.A., Lu, Y., Westlund, K.N., and Willis, W.D. (1996). Dorsal root reflexes in articular afferents occur bilaterally in a chronic model of arthritis in rats. *J. NeuroPhysiol.* **76**, 4190–4193.

Rees, H., Sluka, K.A., Westlund, K.N., and Willis, W.D. (1994). Do dorsal root reflexes augment peripheral inflammation? *Neuroreport* **5**, 821–824.

Rees, H., Sluka, K.A., Westlund, K.N., and Willis, W.D. (1995). The role of glutamate and GABA receptors in the generation of dorsal root reflexes by acute arthritis in the anaesthetized rat. *J. Physiol.* **484** (Pt 2), 437–445.

Represa, A. and Ben-Ari, Y. (2005). Trophic actions of GABA on neuronal development. *Trends Neurosci.* **28**, 278–283.

Reuter, D., Zierold, K., Schroder, W.H., and Frings, S. (1998). A depolarizing chloride current contributes to chemoelectrical transduction in olfactory sensory neurons in situ. *J. Neurosci.* **18**, 6623–6630.

Rhodes, K.J. and Trimmer, J.S. (2006). Antibodies as valuable neuroscience research tools versus reagents of mass distraction. *J. Neurosci.* **26**, 8017–8020.

Rhodes, K.J. and Trimmer, J.S. (2008). Antibody-based validation of CNS ion channel drug targets. *J. Gen. Physiol.* **131**, 407–413.

Rivera, C., Voipio, J., Payne, J.A., Ruusuvuori, E., Lahtinen, H., Lamsa, K., Pirvola, U., Saarma, M., and Kaila, K. (1999). The K^+/Cl^- co-transporter KCC2 renders GABA hyperpolarizing during neuronal maturation. *Nature* **397**, 251–255.

Rocha-González, H.I., Mao, S., and Alvarez-Leefmans, F.J. (2008). Na^+, K^+, $2Cl^-$ cotransport and intracellular chloride regulation in rat primary sensory neurons: thermodynamic and kinetic aspects. *J. NeuroPhysiol.* **100**, 169–184.

Rohrbough, J. and Spitzer, N.C. (1996). Regulation of intracellular Cl^- levels by Na^+-dependent Cl^- cotransport distinguishes depolarizing from hyperpolarizing $GABA_A$ receptor-mediated responses in spinal neurons. *J. Neurosci.* **16**, 82–91.

Romero, M.F., Fulton, C.M., and Boron, W.F. (2004). The SLC4 family of HCO3− transporters. *Pflugers Arch.* **447**, 495–509.

Rudomin, P. (2009). In search of lost presynaptic inhibition. *Exp. Brain Res.* **196**, 139–151.

Rudomin, P. and Schmidt, R.F. (1999). Presynaptic inhibition in the vertebrate spinal cord revisited. *Exp. Brain Res.* **129**, 1–37.

Russell, J.M. (1983). Cation-coupled chloride influx in squid axon, Role of potassium and stoichiometry of the transport process. *J. Gen. Physiol.* **81**, 909–925.

Russell, J.M. (2000). Sodium-potassium-chloride cotransport. *Physiol. Rev.* **80**, 211–276.

Sandkuhler, J. (2000). Learning and memory in pain pathways. *Pain* **88**, 113–118.

Saper, C.B. (2005). An open letter to our readers on the use of antibodies. *J. Comp. Neurol.* **465**, 161–163.

Saper, C.B. and Sawchenko, P.E. (2003). Magic peptides, magic antibodies: guidelines for appropriate controls for immunohistochemistry. *J. Comp. Neurol.* **465**, 161–163.

Scott, R.H., Sutton, K.G., Griffin, A., Stapleton, S.R., and Currie, K.P. (1995). Aspects of calcium-activated chloride currents: a neuronal perspective. *Pharmacol. Ther.* **66**, 535–565.

Segev, I. (1990). Computer study of presynaptic inhibition controlling the spread of action potentials into axonal terminals. *J. NeuroPhysiol.* **63**, 987–998.

Silverman, J.D. and Kruger, L. (1990). Selective neuronal glycoconjugate expression in sensory and autonomic ganglia: relation of lectin reactivity to peptide and enzyme markers. *J. Neurocytol.* **19**, 789–801.

Sluka, K.A., Rees, H., Westlund, K.N., and Willis, W.D. (1995). Fiber types contributing to dorsal root reflexes induced by joint inflammation in cats and monkeys. *J. NeuroPhysiol.* **74**, 981–989.

Smart, T.G. (1980). Do motor-nerve terminals have gamma-aminobutyric acid receptors? *Br. J. Pharmacol.* **71**, 279–286.

Snider, W.D. and McMahon, S.B. (1998). Tackling pain at the source: new ideas about nociceptors. *Neuron* **20**, 629–632.

Staley, K., Smith, R., Schaack, J., Wilcox, C., and Jentsch, T.J. (1996). Alteration of GABAA receptor function following gene transfer of the CLC-2 chloride channel. *Neuron* **17**, 543–551.

Stein, V., Hermans-Borgmeyer, I., Jentsch, T.J., and Hubner, C.A. (2004). Expression of the KCl cotransporter KCC2 parallels neuronal maturation and the emergence of low intracellular chloride. *J. Comp. Neurol.* **468**, 57–64.

Strupp, M. and Grafe, P. (1991). A chloride channel in rat and human axons. *Neurosci. Lett.* **133**, 237–240.

Su, G., Kintner, D.B., and Sun, D. (2002). Contribution of Na^+-K^+-Cl^- cotransporter to high $[K^+]_o$-induced swelling and EAA release in astrocytes. *Am. J. Physiol. Cell Physiol.* **282**, C1136–C1146.

Su, W., Shmukler, B.E., Chernova, M.N., Stuart-Tilley, A.K., de Franceschi, L., Brugnara, C., and Alper, S.L. (1999). Mouse K-Cl cotransporter KCC1: cloning, mapping, pathological expression, and functional regulation. *Am. J. Physiol.* **277**, C899–C912.

Sung, K.W., Kirby, M., McDonald, M.P., Lovinger, D.M., and Delpire, E. (2000). Abnormal $GABA_A$ receptor-mediated currents in dorsal root ganglion neurons isolated from Na-K-2Cl cotransporter null mice. *J. Neurosci.* **20**, 7531–7538.

Szabadics, J., Varga, C., Molnar, G., Olah, S., Barzo, P., and Tamas, G. (2006). Excitatory effect of GABAergic axo-axonic cells in cortical microcircuits. *Science* **311**, 233–235.

Todd, A. and Ribeiro-da-Silva, A. (2005). Molecular architecture of the dorsal horn, Chap. 3. In *The Neurobiology of Pain* (Hunt, S. and Koltzenburg, M., eds), pp. 65–94. Oxford University Press.

Toyoda, H., Yamada, J., Ueno, S., Okabe, A., Kato, H., Sato, K., Hashimoto, K., and Fukuda, A. (2005). Differential functional expression of cation-Cl^- cotransporter mRNAs (KCC1, KCC2, and NKCC1) in rat trigeminal nervous system. *Brain Res. Mol. Brain Res.* **133**, 12–18.

Treede, R.D., Meyer, R.A., Raja, S.N., and Campbell, J.N. (1992). Peripheral and central mechanisms of cutaneous hyperalgesia. *Prog. NeuroBiol.* **38**, 397–421.

Trigo, F.F., Marty, A., and Stell, B.M. (2008). Axonal $GABA_A$ receptors. *Eur. J. Neurosci.* **28**, 841–848.

Valencia-de Ita, S., Lawand, N.B., Lin, Q., Castaneda-Hernandez, G., and Willis, W.D. (2006). Role of the Na^+-K^+-$2Cl^-$ cotransporter in the development of capsaicin-induced neurogenic inflammation. *J. NeuroPhysiol.* **95**, 3553–3561.

Valeyev, A.Y., Hackman, J.C., Holohean, A.M., Wood, P.M., Katz, J.L., and Davidoff, R.A. (1999). GABA-induced Cl^- current in cultured embryonic human dorsal root ganglion neurons. *J. NeuroPhysiol.* **82**, 1–9.

Verdier, D., Lund, J.P., and Kolta, A. (2003). GABAergic control of action potential propagation along axonal branches of mammalian sensory neurons. *J. Neurosci.* **23**, 2002–2007.

Verdru, P., De Greef, C., Mertens, L., Carmeliet, E., and Callewaert, G. (1997). Na^+-Ca^{2+} exchange in rat dorsal root ganglion neurons. *J. NeuroPhysiol.* **77**, 484–490.

Wang, D.D. and Kriegstein, A.R. (2008). GABA regulates excitatory synapse formation in the neocortex via NMDA receptor activation. *J. Neurosci.* **28**, 5547–5558.

Wang, D.D. and Kriegstein, A.R. (2009). Defining the role of GABA in cortical development. *J. Physiol.* **587**, 1873-1879.

Wang, H., Yan, Y., Kintner, D.B., Lytle, C., and Sun, D. (2003). GABA-mediated trophic effect on oligodendrocytes requires Na-K-2Cl cotransport activity. *J. NeuroPhysiol.* **90**, 1257–1265.

Willis, E.F., Clough, G.F., and Church, M.K. (2004). Investigation into the mechanisms by which nedocromil sodium, frusemide and bumetanide inhibit the histamine-induced itch and flare response in human skin in vivo. *Clin. Exp. Allergy* **34**, 450–455.

Willis, W. and Coggeshall, R. (2004). *Sensory Mechanisms of the Spinal Cord. Primary Afferent Neurons and the Spinal Dorsal Horn*, 3rd edition. Kluwer Academic/Plenum Publishers, New York.

Willis, W.D. (2006). John Eccles' studies of spinal cord presynaptic inhibition. *Prog. NeuroBiol.* **78**, 189–214.

Willis, W.D., Jr. (1999). Dorsal root potentials and dorsal root reflexes: a double-edged sword. *Exp. Brain Res.* **124**, 395–421.

Woolf, C.J. and Salter, M.W. (2000). Neuronal plasticity: increasing the gain in pain. *Science* **288**, 1765–1769.

Yamada, J., Okabe, A., Toyoda, H., Kilb, W., Luhmann, H.J., and Fukuda, A. (2004). Cl^- uptake promoting depolarizing GABA actions in immature rat neocortical neurones is mediated by NKCC1. *J. Physiol.* **557**, 829–841.

Yan, Y., Dempsey, R.J., Flemmer, A., Forbush, B., and Sun, D. (2003). Inhibition of Na^+-K^+-$2Cl^-$ cotransporter during focal cerebral ischemia decreases edema and neuronal damage. *Brain Res.* **961**, 22–31.

Yu, K. and Kocsis, J.D. (2005). Schwann cell engraftment into injured peripheral nerve prevents changes in action potential properties. *J. NeuroPhysiol.* **94**, 1519–1527.

Zhang, J.J., Misri, S., Adragna, N.C., Gagnon, K.B., Fyffe, R.E., and Lauf, P.K. (2005). Cloning and expression of sheep renal K-CI cotransporter-1. *Cell Physiol. Biochem.* **16**, 87–98.

Zhang, L.L., Delpire, E., and Vardi, N. (2007). NKCC1 does not accumulate chloride in developing retinal neurons. *J. NeuroPhysiol.* **98**, 266–277.

Zhang, L.L., Fina, M.E., and Vardi, N. (2006a). Regulation of KCC2 and NKCC during development: membrane insertion and differences between cell types. *J. Comp. Neurol.* **499**, 132–143.

Zhang, L.L., Pathak, H.R., Coulter, D.A., Freed, M.A., and Vardi, N. (2006b). Shift of intracellular chloride concentration in ganglion and amacrine cells of developing mouse retina. *J. NeuroPhysiol.* **95**, 2404–2416.

Zhang, S.J. and Jackson, M.B. (1993). GABA-activated chloride channels in secretory nerve endings. *Science* **259**, 531–534.

Zheng, J.H., Walters, E.T., and Song, X.J. (2007). Dissociation of dorsal root ganglion neurons induces hyperexcitability that is maintained by increased responsiveness to cAMP and cGMP. *J. NeuroPhysiol.* **97**, 15–25.

Zhu, L., Lovinger, D., and Delpire, E. (2005). Cortical neurons lacking KCC2 expression show impaired regulation of intracellular chloride. *J. NeuroPhysiol.* **93**, 1557–1568.

CHAPTER

23

Modulation of Chloride Homeostasis by Microglia

Yves De Koninck

I. INTRODUCTION

A large body of evidence indicates that injury either to the peripheral or central nervous system (CNS) causes a number of changes in the CNS, which lead to abnormal information processing and hyperexcitability. A decrease in synaptic inhibition, or "disinhibition", appears to be an important substrate of several pathophysiological conditions of the CNS, such as traumatic injuries, epilepsy and chronic pain (Prince, 1999; Toth et al., 1997; Woolf and Salter, 2000).

This disinhibition results in a general increased excitability of individual neurons and networks of neurons; it has also the potential to allow inputs to be relayed through pathways that are normally maintained silent by inhibition, destabilizing neuronal circuits.

For decades, a neuron-centric view has predominated to explain changes in excitability underlying CNS pathologies, but recent work has uncovered neuroimmune interactions as important contributing mechanisms. Interactions between the immune and nervous systems occur at multiple levels, where different types of immune/glial cells and immune-derived substances are implicated at different stages of the pathogenesis of a wide array of syndromes, including spinal cord injury, multiple sclerosis, neuropathic pain, epilepsy and brain ischemia as well as neurodegenerative diseases such as Alzheimer's and Parkinson's disease, and amyotrophic lateral sclerosis (Hanisch and Kettenmann, 2007; Marchand et al., 2005; Nguyen et al., 2002; Rock et al., 2008; Scholz et al., 2008). Importantly, many of these neuroimmune interactions occur within the CNS parenchyma. Further, not only central, but also peripheral nerve injuries trigger important central neuroinflammatory responses (Marchand et al., 2005; Nguyen et al., 2002; Rock et al., 2008; Scholz et al., 2008). Yet, while immune cells of the CNS, notably microglia, are recognized as active participants in the control of neuronal function, the mechanisms by which microglia signal to neurons to alter their excitability had remained elusive until recently.

In this chapter we show that modulation of neuronal Cl^- homeostasis serves as a mechanistic link between central neuroimmune response and neuronal excitability: via the release of BDNF acting on neuronal TrkB, microglia cause a depolarizing shift in reversal potential for $GABA_A$ and glycine receptor-mediated currents effectively impairing, and in some case even inverting, inhibition. This results in an imbalance of inputs towards excitation, raising neuronal network excitability. Thus, by regulating Cl^- homeostasis in their neighboring neurons, microglial cells can effectively control neuronal network excitability. The findings outlined in this chapter also highlight the fact that Cl^- homeostasis does not only change during development, or as a consequence of direct injury to neurons, but that it can be regulated by signaling mechanisms in adult tissue.

This microglia-to-neuron communication pathway was unveiled at the level of the spinal cord in an experimental model of chronic pain. Thus, to describe the cascade of neuron-to-microglia-to-neuron signaling mechanisms that eventually lead to altered Cl^- homeostasis, we will focus mainly on spinal mechanism underlying neuropathic pain, but these findings are likely to be relevant to other syndromes that involve neuroimmune interactions in other CNS areas including epilepsy, multiple sclerosis, spinal cord injury and brain ischemia.

II. ALTERED CHLORIDE HOMEOSTASIS IN THE SPINAL DORSAL HORN AS A SUBSTRATE OF NEUROPATHIC PAIN

A. Blocking Spinal Inhibition Causes Pain Hypersensitivity

There is ample evidence indicating that decreasing inhibition (disinhibition) causes pain hypersensitivity. The largest body of evidence comes from work in the spinal cord where blocking inhibition mediated by GABA and glycine replicates symptoms of neuropathic and inflammatory pain (Sherman et al., 1997; Sherman and Loomis, 1994; Sivilotti and Woolf, 1994; Sorkin and Puig, 1996; Yaksh, 1989). The network of local inhibitory neurons within the dorsal horn represses a large amount of established excitatory connections and suppression of this inhibitory control unmasks a profound network of polysynaptic excitatory input to relay neurons. This results in a general increased excitability of the network and has the potential for allowing inputs to be relayed through pathways that do not convey these inputs in normal conditions (e.g. normally nociceptive specific pathways). For example, after blockade of $GABA_A$ and glycine receptors, spinal lamina I and lamina II neurons with normally little or no low threshold input can be seen to receive considerable inputs from low threshold afferents (comparable in magnitude to that from high threshold afferents) via unmasked polysynaptic links (Baba et al., 2003; Torsney and MacDermott, 2006). Nociceptive specific lamina I neurons show responses to innocuous touch after blockage of intrinsic $GABA_A$/glycine inhibition (Keller et al., 2007) and nociceptive specific thalamic neurons display responses to innocuous input following blockade of glycine receptors at the lumbar spinal level indicating subliminal low threshold input to normally nociceptive specific spinal output pathways (Sherman et al., 1997). Because there are normally no monosynaptic inputs to lamina I from low threshold afferents, these findings confirm that the network is organized in such a way that $GABA_A$/glycine inhibitory neurons repress existing excitatory connections allowing a crosstalk between low and high threshold pathways. The same principles can be transposed to other brain centers where disinhibition can directly raise the excitability of individual cells, but also alter the flow of information through networks by unmasking connections. However, the fact that blocking inhibition causes pain hypersensitivity does not necessarily imply that disinhibition occurs in pathological conditions. Hence, several recent studies have been aimed at identifying substrates of impaired inhibition in experimental models of chronic pain.

B. Pre- vs. Postsynaptic Mechanisms

Furthermore, blocking inhibition using the approaches highlighted above may affect both pre- and postsynaptic inhibition at the spinal level, i.e. inhibition of primary afferents vs. that of neurons intrinsic to the dorsal horn. Primary afferent depolarization is an important mechanism of inhibitory control of incoming signals into the spinal cord. Plasticity of Cl^- homeostasis in primary afferents and its impact in nociception are discussed in Chapter 22, in particular in the context of acute tissue injury, somatic pain and neurogenic inflammation. In the present chapter, we will focus on disinhibition of dorsal horn neurons, principally under neuropathic pain conditions.

C. Conventional Substrates of Dorsal Horn Disinhibition

Several studies have shown that peripheral nerve injury causes loss of inhibition impinging on spinal dorsal horn neurons. A potential underlying mechanism is the degeneration of inhibitory interneurons (Ibuki et al., 1997; Moore et al., 2002; Scholz et al., 2005). However, it has been argued that pain hypersensitivity after nerve injury may occur without any apparent loss of interneurons (Polgar et al., 2004, 2005). Altered expression of GABA (or its synthesizing enzyme GAD) and/or $GABA_A$ receptors is another potential mechanism, but findings appear to differ depending on chronic pain models and animal strains (Castro-Lopes et al., 1993, 1994, 1995; Moore et al., 2002; Somers and Clemente, 2002). Thus, while functional disinhibition appears to be a substrate of pain hypersensitivity, the substrate has remained elusive when considering conventional measures of inhibition such as transmitter levels (or its synthesizing enzymes), receptor levels or neuronal counts.

D. Impaired Chloride Extrusion as a Novel Mechanism of Dorsal Horn Disinhibition

In parallel to the above-mentioned studies, we have identified another mechanism to account for dorsal horn disinhibition in neuropathic pain conditions. This mechanism involves a decrease in the expression of the K^+-Cl^- cotransporter KCC2 in spinal dorsal horn neurons following peripheral nerve injury (Coull et al., 2003). Because primary afferents do not express KCC2 in normal conditions, as discussed in Chapter 22, the loss of KCC2 pointed to a mechanism specifically affecting postsynaptic inhibition onto dorsal horn neurons (Coull et al., 2003). In functional terms, this resulted in a shift in E_{GABA} to more depolarized values, effectively eliminating the hyperpolarizing action of GABA and glycine (Fig. 23.1). In most cases, the shift in E_{GABA} was sufficient to invert GABA/glycine-mediated hyperpolarization to depolarization and, in a subset of cells, it effectively converted

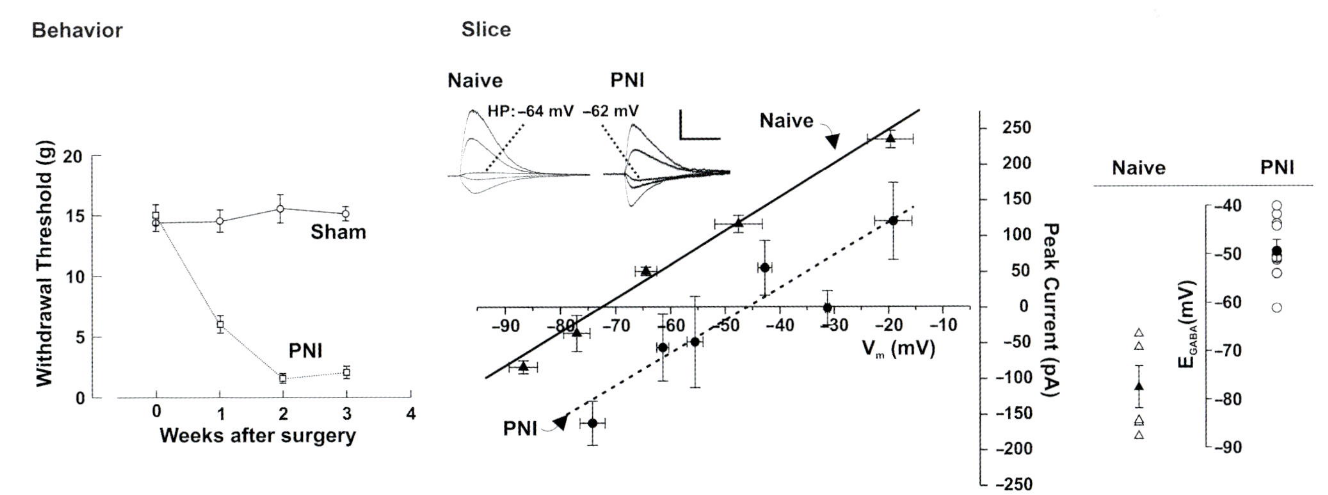

FIGURE 23.1 Peripheral nerve injury causes a decrease in paw withdrawal threshold and a concurrent depolarizing shift in E_{GABA} in superficial dorsal horn neurons. The graph on the left shows that peripheral nerve injury (PNI), but not sham surgery (Sham), caused a significant reduction in the nociceptive withdrawal threshold to mechanical stimulation of the hindpaw in rats. The middle graph shows I/V relations of the mean peak current in response to GABA, at different holding potentials (V_m) (inset scale: $x = 1.0$s; $y = 0.6$nA). The records were obtained using perforated patch (gramicidin) clamp recordings. The intersection with the V_m axis indicates E_{GABA}. Nerve injury caused a depolarizing shift in E_{GABA} of more than 20mV. Horizontal standard error bars represent interneuron differences in holding potentials. The graph on the right shows the distribution of E_{GABA} measured across the population (solid symbol = mean $E_{GABA} \pm$ SEM). Modified from (Coull et al., 2003); reproduced with permission.

inhibition into net excitation mediated by these neurotransmitters (Coull et al., 2003). Local blockade or knockdown of spinal KCC2 in intact rats markedly reduced nociceptive threshold, confirming that disruption of chloride homeostasis in superficial dorsal horn neurons was sufficient to replicate the symptoms of neuropathic pain (Coull et al., 2003). This was consistent with an increase in excitability of spinal relay neurons (Prescott et al., 2006). We also found that blocking Cl^- transport unmasked innocuous input to normally nociceptive specific spinal lamina I projection neurons, indicating a substrate for tactile allodynia (pain sensation in response to a normally innocuous stimulus) characteristic of neuropathic pain (Keller et al., 2007) (Fig. 23.2). Because nociceptive relay neurons in spinal lamina I do not receive direct

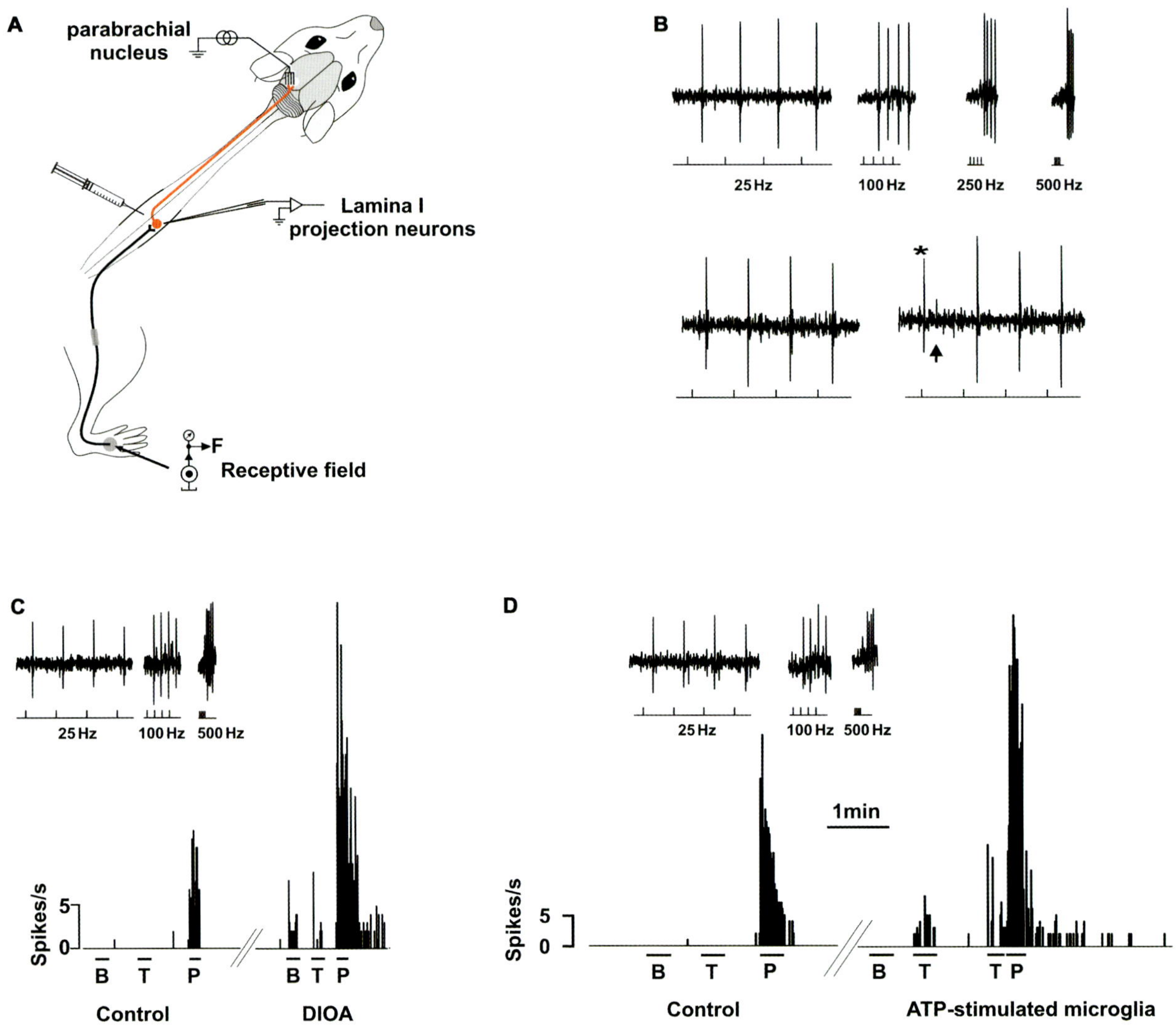

FIGURE 23.2 Blockade of cation-chloride cotransporter and local application of ATP-stimulated microglia unmasks low threshold sensory input to normally selectively high threshold projection neurons. **A.** Schematic representation of the experimental setting to perform single unit extracellular recording from single antidromically identified lamina I projection neurons. Cells were recorded from control animals and from animals that received a chronic constriction injury of the sciatic nerve. **B.** Confirmation of recording from a lamina I projection neuron. *Top*: Extracellular single unit recordings from a lamina I neuron showing one-for-one following of a train of antidromic stimuli (lower traces mark the stimulus) delivered from an electrode positioned in the lateral parabrachial nucleus. *Bottom*: Collision of the first of four antidromic action potentials (25 Hz) with an orthodromic action potential (*) occurring within the critical interval. The arrow points to the position where the first antidromic action potential would have occurred in absence of the orthodromic action potential (as in the trace on the left). **C.** In a nociceptive specific lamina I projection neuron that responded only to pinch (P) stimulation in control conditions, blockade of cation-chloride cotransporters with local spinal administration of DIOA unmasked innocuous input (brush B and touch T). The unmasking of low threshold is interpreted as the result of blocking KCC2 action because even if DIOA also blocks NKCC1 as suggested by other authors (Gillen and Forbush, 1999), the latter effect would work against exaggerated depolarization in primary afferents and thus prevent rather than produce crosstalk between them as proposed for inflammatory insult (Pitcher et al., 2007). See discussion in Keller et al. (2007). **D.** Local spinal administration of ATP-stimulated microglia also unmasked innocuous input to a normally nociceptive specific lamina I projection neuron. Modified from Keller et al. (2007); reproduced with permission.

input from low threshold afferents, the latter finding indicated that the disruption of inhibition by the impaired Cl^- transport in the superficial dorsal horn effectively unmasked indirect (likely polysynaptic) connections between innocuous inputs and these nociceptive relay neurons, similar to what has been shown with local spinal blockade of $GABA_A$ and/or glycine receptor mediated inhibition (Torsney and MacDermott, 2006) (Fig. 23.3).

E. Loss of KCC2 as a Result of Intercellular Signaling Mechanisms

Importantly, the loss of KCC2 expression observed after peripheral nerve injury occurred in neurons intrinsic to the dorsal horn and therefore not the directly injured neurons (Coull et al., 2003) (Fig. 23.1). This indicated that, in contrast to other traumatic injuries (Nabekura et al., 2002; Van den Pol et al., 1996), altered

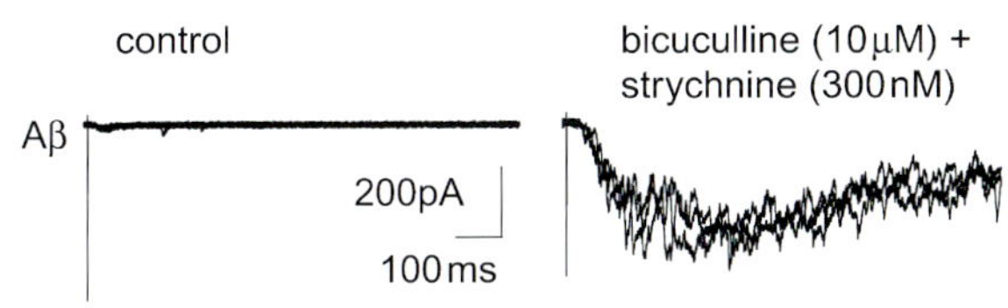

FIGURE 23.3 Blocking $GABA_A$ and glycine receptor-mediated transmission unmasks polysynaptic coupling between low threshold afferents and spinal lamina I neurons. Whole cell patch clamp recording from lamina I neurons in a spinal cord slice with dorsal roots attached. Stimulation of the dorsal root at Aβ strength elicited no response in the cells. After bath application of bicuculline and strychnine a large polysynaptic response was recorded in response to a stimulus at the same Aβ strength. Modified from Torsney and MacDermott (2006); reproduced with permission.

Cl^- homeostasis in spinal neurons after peripheral nerve injury involved a signaling mechanism between the injured afferents and dorsal horn neurons. The concept that Cl^- homeostasis is modulated by intercellular signaling introduces a novel perspective to the role of the Cl^- transport mechanism: a substrate for activity-regulated plasticity of CNS network excitability.

III. BDNF, A SIGNALING MOLECULE CONTROLLING KCC2 EXPRESSION

Modulating ion gradients as a means to control the strength of synaptic transmission provides a novel perspective on synaptic plasticity. Indeed, changes in ion gradients, especially Cl^- gradients, have been well documented throughout development (see Chapters 7 and 19) but little consideration had been given until recently to the possibility that ion gradients could be actively modulated in adult tissue, as discussed in Chapter 22.

Initial demonstration of a potential intercellular signaling event modulating Cl^- homeostasis in adult tissue came from the observation that application of brain-derived neurotrophic factor (BDNF) onto dissociated hippocampal cell cultures caused a selective shift in reversal potential of inhibitory synaptic currents (IPSCs) in a subpopulation of neurons (Wardle and Poo, 2003). Subsequently, it was shown that exposing hippocampal slice cultures to BDNF caused down-regulation of both KCC2 message and protein on a rapid time scale (within an hour; Fig. 23.4) with a corresponding deficit in Cl^- extrusion capacity

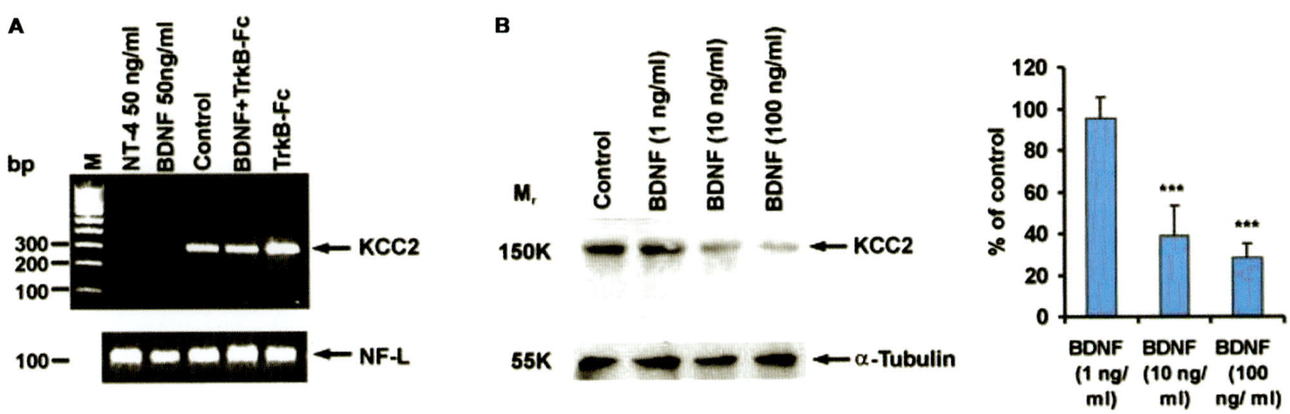

FIGURE 23.4 Exogenous BDNF and NT-4 decrease KCC2 mRNA and protein expression in organotypic hippocampal slices. **A.** Representative RT-PCR experiment (out of four similar ones), showing the down-regulatory effect of BDNF and NT-4 on KCC2 mRNA expression. The effects of BDNF and NT-4 were inhibited by 200 ng/ml of TrkB-Fc. The amplified light neurofilament (NF-L) fragment indicates an equal amount of mRNA in the reactions. Note the increase in KCC2 mRNA brought about by exposure to TrkB-Fc only. **B.** Western blot analysis showing the dose-dependent effect of BDNF on KCC2 protein expression. *Left*: A representative Western blot. *Right*: The average normalized optical densities displayed as a percentage of control (right, mean ± SD, $n = 5$). Modified from Rivera et al. (2002); reproduced with permission.

(Rivera et al., 2002). Kindling-induced epileptogenesis caused a concurrent up-regulation of BDNF and down-regulation of KCC2 with similar temporal profiles. Consistent with these observations, we found that BDNF application onto adult spinal cord slices caused a depolarizing shift in E_{GABA}, decreasing hyperpolarizing inhibition mediated by $GABA_A$ and glycine receptors (Fig. 23.5). In a subset of neurons, BDNF application even caused GABAergic currents to switch to net excitatory events (Coull et al., 2005). We then tested whether endogenous BDNF was responsible for the shift in E_{GABA} after nerve injury. For this, we either removed (or mopped up) endogenous BDNF using a TrkB-Fc fusion protein (Mannion et al., 1999; Thompson et al., 1999) or blocked its binding to TrkB receptors using a function blocking anti-TrkB antibody (Balkowiec and Katz, 2000) in spinal slices taken from rats with peripheral nerve injury. We found that acute blockade of BDNF-TrkB signaling reversed the depolarizing shift in E_{GABA} (Coull et al., 2005) (Fig. 23.5).

These findings indicated: (1) that endogenously released BDNF was causing the depolarizing shift in E_{GABA} and (2) that after peripheral nerve injury, E_{GABA} is maintained at a depolarized level in dorsal horn neurons by tonically released BDNF at the spinal level. Thus, in the absence of TrkB receptor activity, Cl^- is maintained at normal levels by regulatory mechanisms. Maintenance of intracellular $[Cl^-]$ at a given level is thus an ongoing process that can be readily modulated to control the strength and polarity of $GABA_A$/glycine-receptor mediated transmission. Importantly, for therapeutic considerations in the nerve injury model we studied, because Cl^- homeostasis in the CNS is altered by a tonically released modulator, impaired inhibition (disinhibition) can be reversed by drug treatments.

IV. MECHANISM OF BDNF-TRKB MEDIATED ALTERED CHLORIDE HOMEOSTASIS

Two intracellular second messenger pathways by which BDNF-TrkB signaling controls KCC2 expression have been identified; the Shc/FRS-2 (src homology 2 domain-containing transforming protein/fibroblast growth factor receptor substrate 2) pathway, and the phospholipase C-γ (PLCγ)-cyclic adenosine monophosphate response element-binding protein (CREB)

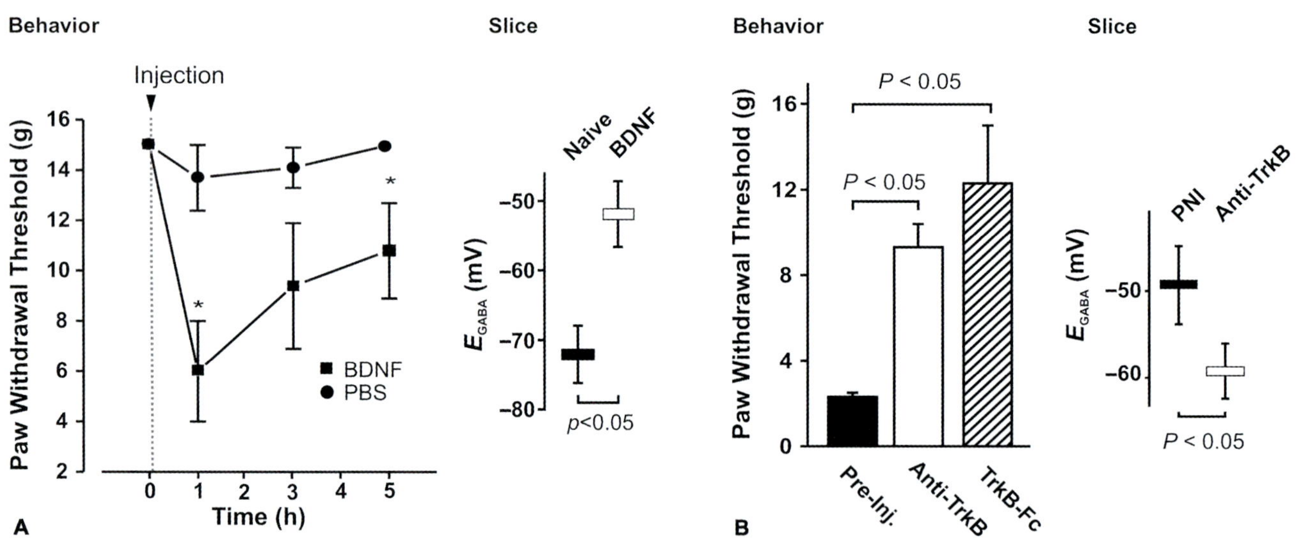

FIGURE 23.5 Increased concentrations of BDNF in the dorsal horn cause pain hypersensitivity and a depolarizing shift in E_{GABA}. Blocking BDNF-TrkB signaling reverses both of these effects in rats with nerve injury. **A.** *Left graph*: Intrathecal delivery of recombinant human BDNF (20 µg) to the lumbar dorsal horn of intact rats led to a significant and transient decrease in paw withdrawal threshold within 1 hour of injection, compared to saline control (PBS), which did not elicited significant changes. *Right graph*: Depolarizing shift in mean E_{GABA} measured in lamina I neurons in spinal cord slices treated with BDNF (50 ng/ml; for >90 min) vs. slices in control ACSF (Naïve). **B.** *Left histograms*: Intrathecal administration of either anti-TrkB (a function-blocking receptor antibody) or TrkB-Fc (a BDNF-neutralizing fusion protein) to the lumbar dorsal horn of rats that displayed robust allodynia in response to peripheral nerve injury (PNI) caused a significant increase in paw withdrawal threshold. *Right graph*: pooled results from gramicidin patch-clamp recordings from lamina I neurons in slices taken from PNI rats in control and after anti-TrkB antibody treatment. Slices we superfused for 2 hours with either control ACSF (control) or anti-TrkB (1 µg/ml). E_{GABA} was shifted to depolarizing potentials in slices taken from PNI rats, whereas it was hyperpolarized from rest in slices perfused with anti-TrkB. Modified from Coull et al. (2005); reproduced with permission.

pathway (Rivera et al., 2004) (Fig. 23.6). Interestingly, activation of both Shc and PLCγ cascades by TrkB causes down-regulation of KCC2, whereas activation of the Shc pathway in the absence of PLCγ activation leads to up-regulation of KCC2 (Rivera et al., 2004) (Fig. 23.6). This finding is particularly interesting because there is evidence supporting the notion that BDNF plays an important role in the ontogeny of intracellular Cl^- regulation by promoting the up-regulation of KCC2 during the early postnatal period (Aguado et al., 2003). The bidirectional action of BDNF on KCC2 expression, depending on whether both Shc and PLCγ are involved, provides a potential mechanism for the divergent action of BDNF in early developmental stages compared to in adulthood, which is reminiscent of the opposite effects of BDNF on $GABA_A$-receptor trafficking in immature and adult brain tissue (Mizoguchi et al., 2003). Thus, while BDNF signaling in early developmental stage appears to promote maturation of inhibition and thus a decrease in the excitatory action of GABA and glycine, in adult tissue it causes weakening of inhibition and sometimes even its inversion into excitation. In other words, if BDNF secretion is involved in a repair

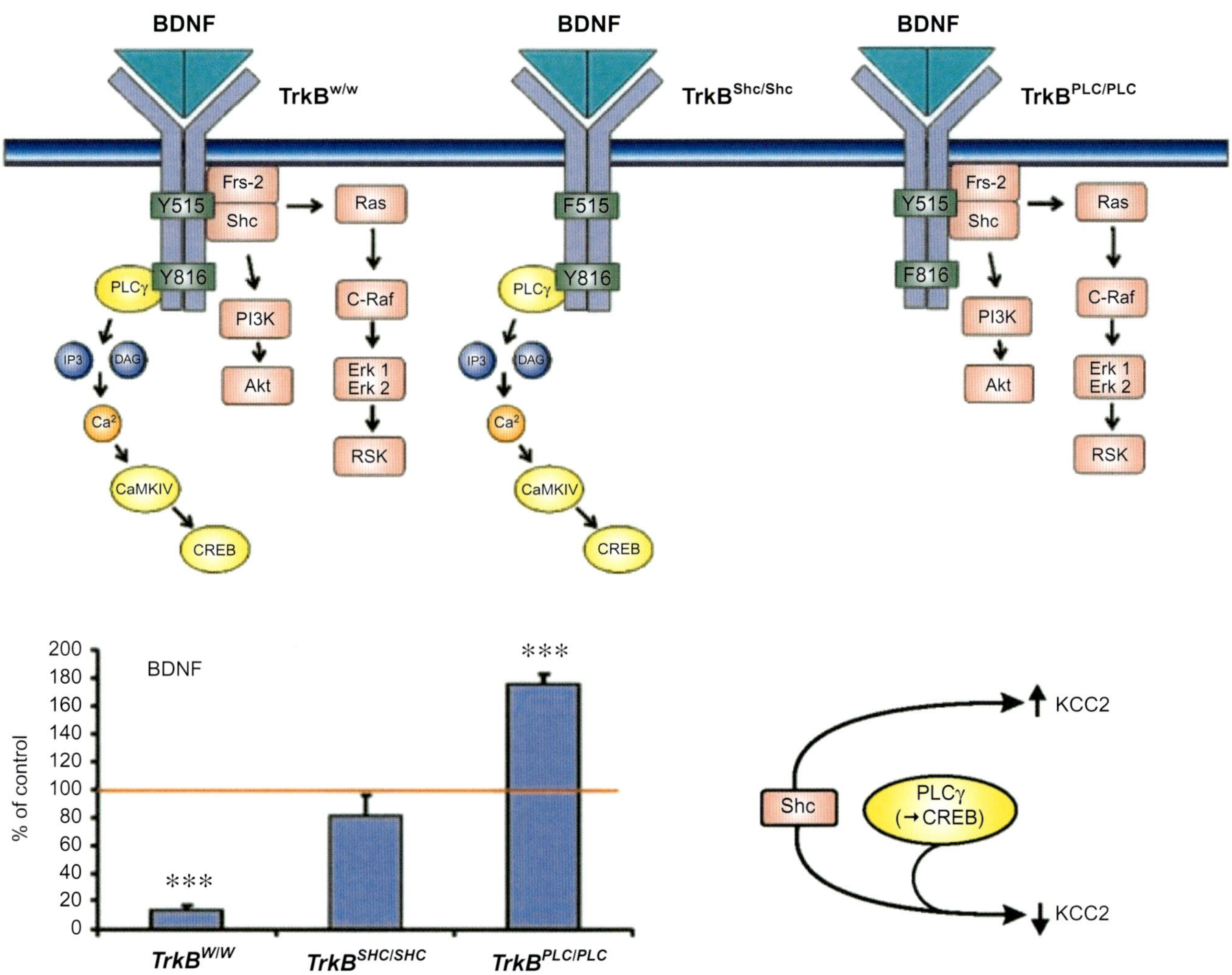

FIGURE 23.6 Both PLCγ and Shc docking sites of TrkB are required for BDNF-TrkB-mediated KCC2 down-regulation. In contrast the Shc site acting in isolation mediates KCC2 up-regulation. The schematic diagram at the top represents mutant TrkB receptors with a point mutation at either tyrosine 515 (Shc site) in the juxtamembrane region (Minichiello et al., 1998) or at tyrosine 816 (PLCγ site) in the C-terminal region. In the $trk^{BPLC/PLC}$ and $trk^{BSHC/SHC}$ mutant receptors, tyrosine (Y816) and tyrosine (Y515) were replaced by phenylalanine (F816) and phenylalanine (F515), respectively. The bar graph in the lower left shows a quantification of the Western blot band intensities from acute hippocampal slices after exposure to BDNF for 3–4 hours (means ± SEM; *t* test; ***$p < 0.001$; **$p < 0.01$). The schematic diagram on the right summarizes the interpretation of the data: activation of both Shc- and PLCγ-coupled intracellular cascades is required for trkB-mediated down-regulation of KCC2, whereas activation of the Shc pathway in the absence of PLCγ activation leads to up-regulation of KCC2. Modified from Rivera et al. (2004); reproduced with permission.

response following injury, a side effect of the action of the neurotrophin in adult tissue is to weaken the Cl^- extrusion capacity of the cell, yielding an imbalance of synaptic inputs towards enhanced excitation; opposite to what BDNF does during development. Other intracellular signaling pathways associated with TrkB activation, such as the PKA pathway (Qiu et al., 2002) may be involved in regulating KCC2 activity and perhaps other cotransporters (Gamba, 2005) but this remains to be studied more directly (see Chapter 18).

V. MICROGLIA CONTROL NEUROTRANSMITTER-MEDIATED NEURONAL EXCITABILITY BY ALTERING ANION HOMEOSTASIS

A. Active Sensors and Effectors in Normal and Pathologic Conditions

Within the CNS, microglia and astrocytes represent two highly reactive intraparenchymal cell populations. Microglial cells are resident macrophages of the CNS. In normal conditions they fulfill a constitutive "surveillance" function (Hanisch and Kettenmann, 2007). They acquire a different reactive profile – often termed activation – early in response to injury, infections, ischemia, brain tumors or neurodegeneration. Activated microglia change from a ramified to an ameboid morphology, proliferate, increase expression of cell surface markers/receptors, migrate to areas of damage, become phagocytotic and produce/release pro-inflammatory substances or cell-signaling mediators (Coull et al., 2005; Echeverry et al., 2008; Hanisch and Kettenmann, 2007; Haynes et al., 2006; Koizumi et al., 2007). Astrocytes also respond to different types of insult and this activation is characterized by morphological changes, increased production of intermediate filaments such as glial fibrillary acidic protein (GFAP), increased production of signaling molecules and alterations in their homeostatic activity (Maragakis and Rothstein, 2006).

B. Multiple Sources and Multiple Phenotypes

Importantly, both microglial and astrocytic activations are multidimensional. There are many different activation states, with various components expressed with different time courses and intensities that are dependent on the stimulus that triggers the activation mechanisms (Hanisch and Kettenmann, 2007). Recent results also revealed that nerve injury-induced spinal microglial response can include not only activation of pre-existing resident microglia, but also the generation of new microglia from proliferation (Echeverry et al., 2008) and from recruitment of peripheral monocytes (Zhang et al., 2007). Both resident and bone marrow-derived microglia may be involved in the generation of neuronal hyperexcitability in the CNS. The ability of blood-borne monocytes to populate the CNS parenchyma and differentiate into microglia has been observed in adult animals, especially in certain special pathological conditions, although the exact conditions enabling CNS infiltration remain a subject of debate (Soulet and Rivest, 2008).

C. ATP-stimulated Microglia Alter Neuronal E_{GABA}

While activated microglial cells are known to release a number of pro-inflammatory and signaling molecules, the direct mechanisms by which microglia alter neuronal excitability in different pathological conditions has remained largely elusive. Peripheral nerve injury induces spinal microglial/astrocytic activation in several models of pain hypersensitivity (Colburn et al., 1999; Fu et al., 1999; Tsuda et al., 2003; Zhuang et al., 2005). Following nerve injury, spinal activated microglia express *de novo* the $P2X_4$ receptor, and blocking $P2X_4$ receptors reverse the tactile allodynia associated with the injury (Tsuda et al., 2003). The finding by Tsuda et al. (2003) that it is possible to administer locally, onto the spinal cord, ATP-stimulated microglia to cause tactile allodynia provided an ideal framework to test whether microglia may exert their action via altering neuronal E_{GABA} (Coull et al., 2005). Indeed, we found that local spinal administration of ATP-stimulated microglia, but not unstimulated microglia, in control animals, caused a significant shift in E_{GABA} of superficial spinal dorsal horn neurons to more depolarized values (Fig. 23.7). This shift in E_{GABA} effectively unmasked low threshold input to normally nociceptive specific lamina I projection neurons as did nerve injury, pharmacological blockade of cation chloride cotransport, or blockade of $GABA_A$ receptors (Fig. 23.2). The fact that microglia caused a shift in E_{GABA} similar to that seen with BDNF administration (see above and Fig. 23.5) suggested that these cells may act on the mechanisms regulating the Cl^- gradient by releasing this neurotrophin. Assays in microglial cell cultures indeed confirmed that ATP stimulation promotes the secretion of BDNF from microglia (Fig. 23.8). More direct evidence that microglia signal directly to neurons via BDNF came from the finding that treatment of microglia with siRNA against BNDF prior to ATP

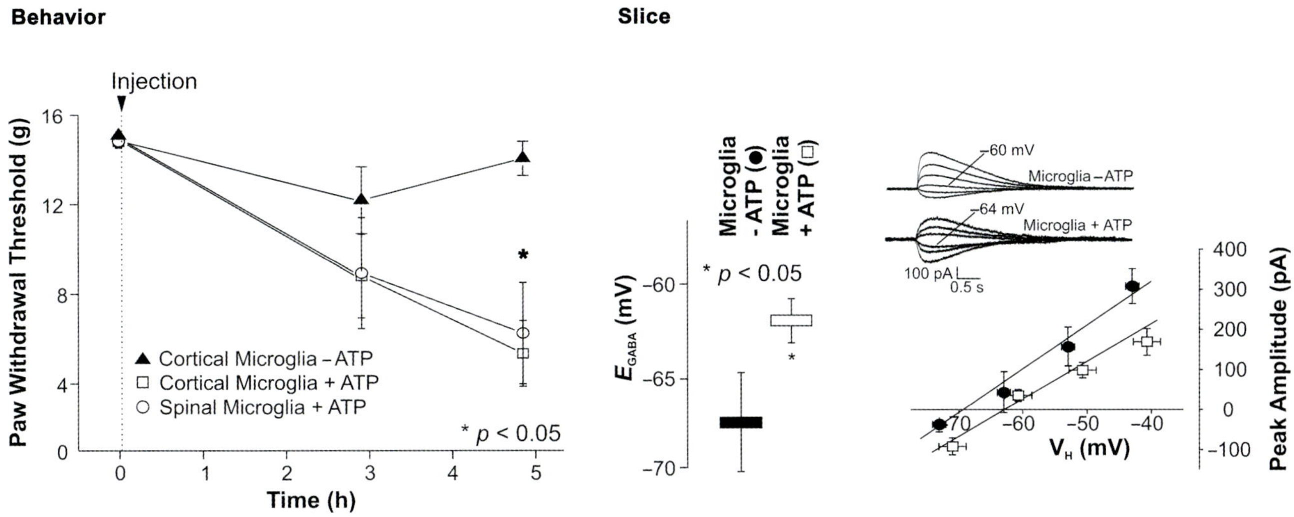

FIGURE 23.7 Spinal delivery of ATP-stimulated microglia caused tactile allodynia and a depolarizing shift in E_{GABA} in spinal lamina I neurons. *Left* (Behavior): Intrathecal injections of ATP-stimulated microglia (of cortical or spinal origin), but not resting microglia, caused a significant decrease in the paw withdrawal threshold in rats (values are mean ± SEM). *Right* (Slice): Comparison of the mean E_{GABA} recorded in lamina I neurons from resting microglia- and ATP-stimulated microglia-injected rats (there were no significant changes in resting membrane potential of the cells). The I/V relation on the right is a plot of the mean peak current evoked by GABA at various holding transmembrane potentials (V_H). *Inset*: Representative GABA-induced currents at different holding potentials. Recordings were made on lamina I neurons in spinal cord slices taken from rats treated with ATP-stimulated or resting microglia. The horizontal standard error bars represent interneuron differences in holding potentials. Modified from Coull et al. (2005); reproduced with permission.

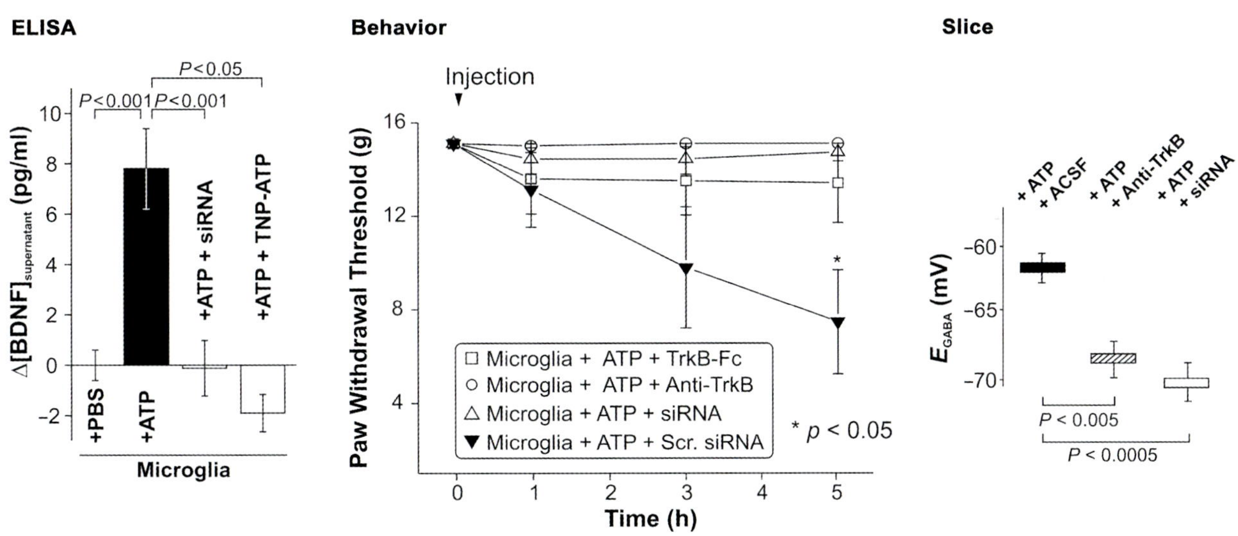

FIGURE 23.8 Microglia-derived BDNF triggers both allodynia and a depolarizing shift in E_{GABA} in spinal lamina I neurons. *Left panel*: ELISA-based measurement of BDNF protein in the supernatant of cultured microglia 5 hours after treatment with phosphate buffered saline vehicle (PBS), ATP, ATP + TNP-ATP (10 μM) or ATP after pre-treatment with BDNF siRNA. *Middle panel* (Behavior): Local spinal delivery of ATP-stimulated microglia neither incubated with anti-TrkB or TrkB-Fc, nor lipofected with BDNF interfering RNA (siRNA) caused a significant change in paw withdrawal threshold. Lipofection of ATP-stimulated microglia with a scrambled version of the interfering RNA (Scr. siRNA) did, however, cause a significant decrease in withdrawal threshold 5 hours after injection. *Right panel* (Slice): The mean E_{GABA} measured from lamina I neurons taken from rats that had received local spinal delivery of either ATP-stimulated microglia mixed with anti-TrkB or ATP-stimulated microglia lipofected with BDNF siRNA was significantly more negative than that measured from LI neurons from rats that were injected with ATP-stimulated microglia. Modified from Coull et al. (2005); reproduced with permission.

stimulation prevented their behavioral effect and their effect on the neuronal Cl^- gradient (Fig. 23.8). Thus neurotrophin BDNF appears to be the direct mediator of microglia–neuron signaling and the agent responsible for the alteration of neuronal Cl^- homeostasis. This suggests that the mechanism by which microglia

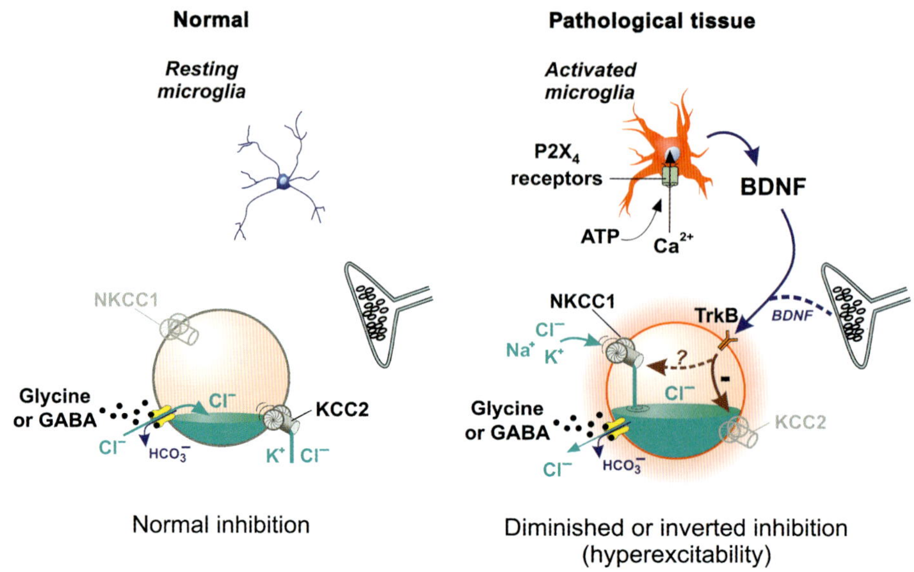

FIGURE 23.9 Summary of the cascade of events underlying modulation of intracellular Cl^- homeostasis by endogenous agents. *Left panel*: Under normal conditions, intracellular $[Cl^-]$ is low such that E_{Cl} is more negative than E_m. When GABA or glycine opens $GABA_A$/glycine anion-permeable channels a net outward current results, causing hyperpolarizing inhibition. This outward current is carried primarily by Cl^- influx. *Right panel*: Under pathological conditions, microglia in the CNS become activated (McMahon et al., 2005) and begin to express purinergic $P2X_4$ receptors (Tsuda et al., 2003). Stimulation of $P2X_4$ receptors causes microglia to secrete brain-derived neurotrophic factor (BDNF) (Coull et al., 2005). BDNF, in turn, acts on TrkB receptors in CNS neurons down-regulating KCC2, which results in decreased Cl^- extrusion capacity of the neurons (Rivera et al., 2002, 2004). In some conditions, enhanced BDNF release originates from neighboring nerve terminals (Zhang et al., 2008; Zhao et al., 2006). Intracellular Cl^- accumulates until E_{Cl} reaches E_m, causing a collapse in the transmembrane Cl^- gradient and a decrease in hyperpolarizing inhibition. In fact, in most cases, the net current is inverted, and GABA (as well as glycine) becomes depolarizing and excitatory (Coull et al., 2003, 2005). This suggests that Cl^- is actively accumulated, presumably by unbalanced NKCC1 activity and/or that HCO_3^- permeability through $GABA_A$ channels, albeit small, brings E_{GABA} more positive than E_m because $E_{HCO_3^-}$ is much more positive than E_m. Blockade of microglia-BDNF-TrkB signaling restores normal intracellular Cl^- levels (Coull et al., 2005). Up-regulation of NKCC1 may occur and act synergistically with down-regulation of KCC2 (Gamba, 2005; Kahle et al., 2005, 2006). Modified from De Koninck (2007); reproduced with permission.

can control how neurons integrate synaptic inputs is BDNF release (Fig. 23.9).

D. Ongoing Microglial Stimulation After Nerve Injury

Acute administration of the P2X receptor antagonist TNP-ATP (at doses specific for $P2X_4$ receptors) to spinal cord slices taken from rats that had received nerve injury produced a shift in E_{GABA} towards more *hyperpolarized* values, thus effectively reversing the collapse in anion gradient that characterized neuropathic pain (Coull et al., 2005). This indicates that the altered anion homeostasis is maintained by continuous signaling from microglia in the tissue, consistent with the observation of tonic spinal BDNF secretion in chronic neuropathic animals (see above and Fig. 23.5).

E. $P2X_4$ to BDNF Coupling

Enhanced spinal release of BDNF after peripheral nerve injury appears to be specifically linked to $P2X_4$ receptor activation by ATP. Indeed, in $P2X_4$ null mice, nerve injury failed to cause tactile allodynia and application of ATP onto microglial cells failed to cause BDNF release (Fig. 23.10) (Ulmann et al., 2008). Stimulation of $P2X_4$ causes a p38-MAPK-dependent increase in expression of BDNF and release by microglia in culture (Trang et al., 2008).

F. Microglia, Not Sensory Nerves, are the Source of BDNF in Neuropathic Pain

While microglial cells release BDNF, this does not necessarily mean that they represent the sole, or even the main source of BDNF causing KCC2 down-regulation. In the spinal cord, for example, an important source of BDNF can be the terminals of peptidergic afferents (Heppenstall and Lewin, 2001; Mannion et al., 1999; Thompson et al., 1999). The results of an interesting recent study serve as a test for whether BDNF from sensory afferents represent a significant player in causing spinal hyperexcitability. Selective deletion of BDNF from small diameter afferents effectively prevented

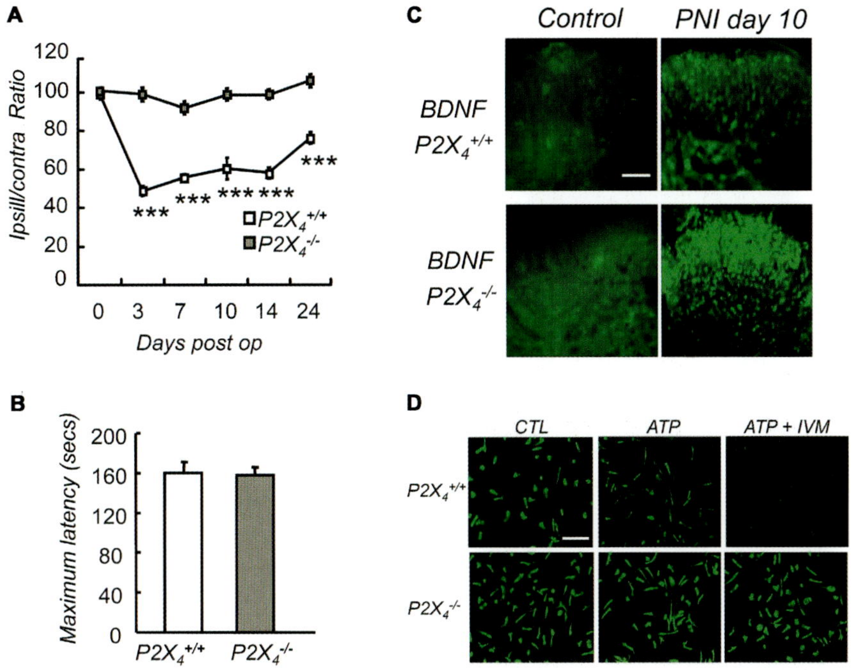

FIGURE 23.10 Lack of mechanical hyperalgesia and altered BDNF signaling in $P2X_4$-deficient mice after peripheral nerve injury. **A.** Sciatic nerve ligature caused mechanical hypersensitivity (hyperalgesia) in wildtype mice ($P2X_4^{+/+}$), but not in $P2X_4^{-/-}$ mice. The graph shows the time course of mechanical hypersensitivity caused by sciatic nerve ligature expressed as ipsi/contralateral ratios (n = 15 mice, two-way ANOVA, $***p < 0.005$). **B.** Measurements of sensorimotor coordination using the rotarod test showed no difference between $P2X_4^{+/+}$ and $P2X_4^{-/-}$ mice ruling out a potential motor deficit in the $P2X_4^{-/-}$ mice. **C.** In sham-operated mice, BDNF immunoreactivity was not detected in either $P2X_4^{+/+}$ or $P2X_4^{-/-}$ animals in the dorsal horn of the spinal cord (left panels). Ten days post-PNI (PNI day 10), BDNF immunoreactivity was slightly increased in wildtype mice, whereas a strong immunoreactivity was observed in the $P2X_4^{-/-}$ mice. Scale bar 100 μm. **D.** ATP-evoked release of BDNF from wildtype and $P2X_4^{-/-}$ microglia primary cultures. Intracellular BDNF content of microglia was analyzed by immunostaining. Stimulation of a culture with ATP (100 μM) alone or co-applied with ivermectin (IVM, a positive allosteric modulator of $P2X_4R$; 3 μM) induced a strong decrease of BDNF immunoreactivity in a primary microglial culture from wildtype mice, but not from cultures from $P2X_4^{-/-}$ mice. Scale bar 100 μm. Modified from Ulmann et al. (2008); reproduced with permission.

tactile allodynia due to peripheral inflammation (Zhao et al., 2006). This is consistent with the earlier findings that enhanced BDNF release occurs from small diameter sensory afferents after peripheral inflammation (Heppenstall and Lewin, 2001; Mannion et al., 1999; Thompson et al., 1999). However, this deletion of BDNF from peptidergic afferents had no effect on nerve injury-induced tactile allodynia, consistent with the idea that BDNF originates from another source (e.g. microglia) after nerve injury (Coull et al., 2005).

G. BDNF from other Sources Affect KCC2 Expression

As mentioned above, while in neuropathic pain conditions, microglia appear to be the source of enhanced BDNF release causing disruption of Cl^- regulation in dorsal horn neurons, following peripheral inflammation, the source of enhanced BDNF release appears to be the small diameter sensory afferents (Heppenstall and Lewin, 2001; Mannion et al., 1999; Thompson et al., 1999; Zhao et al., 2006). This raised the question of whether enhanced BDNF release from different sources results in similar effects on neuronal Cl^- homeostasis.

A recent study showed down-regulation of KCC2 in the spinal cord after peripheral inflammation (Zhang et al., 2008), consistent with enhanced BDNF release centrally, in the spinal cord, from peptidergic sensory fibers (Heppenstall and Lewin, 2001; Mannion et al., 1999; Thompson et al., 1999; Zhao et al., 2006). KCC2 down-regulation was prevented by pretreatment with K252a. While K252a may affect several intracellular signaling pathways, this result is consistent with the hypothesis that BDNF-TrkB signaling is responsible for down-regulation of KCC2. Interestingly, the oligomeric form of this transporter appeared to be more specifically down-regulated after inflammation (Zhang et al., 2008). This is interesting in view of the suggestion that the oligomeric form of KCC2 may be the active form (Blaesse et al., 2006).

In summary, regardless of the source of BDNF, neuronal or glial, activation of TrkB receptors by this neurotrophin appears to cause disruption of Cl^- homeostasis. Thus, cation-coupled Cl^- cotransporters may represent the common pathway by which enhanced network excitability occurs in CNS pathologies involving increased BDNF expression (Rivera et al., 2002, 2004).

VI. WHAT ACTIVATES MICROGLIA?

While blockade/knockdown or knockout of $P2X_4$ receptors prevents the development of nerve injury-induced tactile allodynia, neither intervention affects microglial activation (Tsuda et al., 2003; Ulmann et al., 2008). Thus, other mechanisms, upstream of ATP signaling, must trigger microglial activation after injury.

A. Microglial Activation/Migration is Spatially Defined and Results from Intercellular Signaling

We and others have found that microglial activation and migration occurs in a spatially delimited area within the spinal cord parenchyma following peripheral nerve injury: essentially the spinal territory occupied by the central terminals of injured afferents (Beggs and Salter, 2007; Zhang et al., 2007; Zhang and De Koninck, 2006) (Fig. 23.11). Yet, unlike with nerve injury, only a weak microglial reaction occurs within the spinal gray matter following rhizotomy even though a strong activation occurs in the spinal dorsal funiculus (Scholz et al., 2008). This suggests that although degeneration of central terminals is sufficient to elicit microglial activation, it does not account for the intraparenchymal response within the dorsal horn after peripheral nerve injury. Thus, in the latter case, microglial activation and chemotaxis must be the results of a local signaling mechanism between injured nerve terminals and microglia. Several potential candidates for intercellular signaling cascades have been recently identified, as outlined below and summarized in Fig. 23.12.

B. Fractalkine to CX3CL1 Signaling

In the spinal cord dorsal horn, the chemokine fractalkine is expressed in neurons and primary sensory

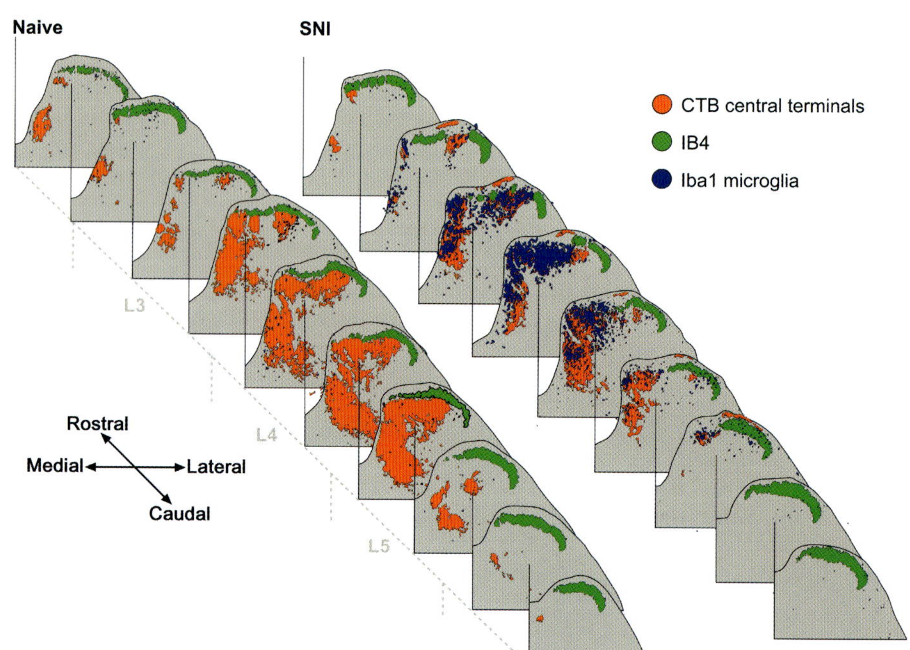

FIGURE 23.11 Microglial activation and migration in the spinal dorsal horn parenchyma occurs selectively in the CNS territory occupied by the central terminals of injured sensory fibers. Somatotopic nature of the microglial response to peripheral nerve injury (SNI; spared nerve injury model; Decosterd and Woolf, 2000). Regions of activated microglia (detected by iba1 immunolabeling) are indicated in blue. The B subunit of cholera toxin (CTB; indicated in red) was injected into the transected nerve branches (tibial and common peroneal) of the SNI animals to label the terminal field of myelinated primary afferents in the spinal cord. CTB is a transganglionic tracer that binds specifically to the GM1 ganglioside on the surface of myelinated, but not unmyelinated, primary afferent fibers. IB4 labeling (indicated in green) was used to identify the terminal field of non-peptidergic C-afferent fibers. Note how the pattern of microglial activation follows, to a large extent, the territory where primary afferent labeling is diminished after nerve injury. Modified from Beggs and Salter (2007); reproduced with permission.

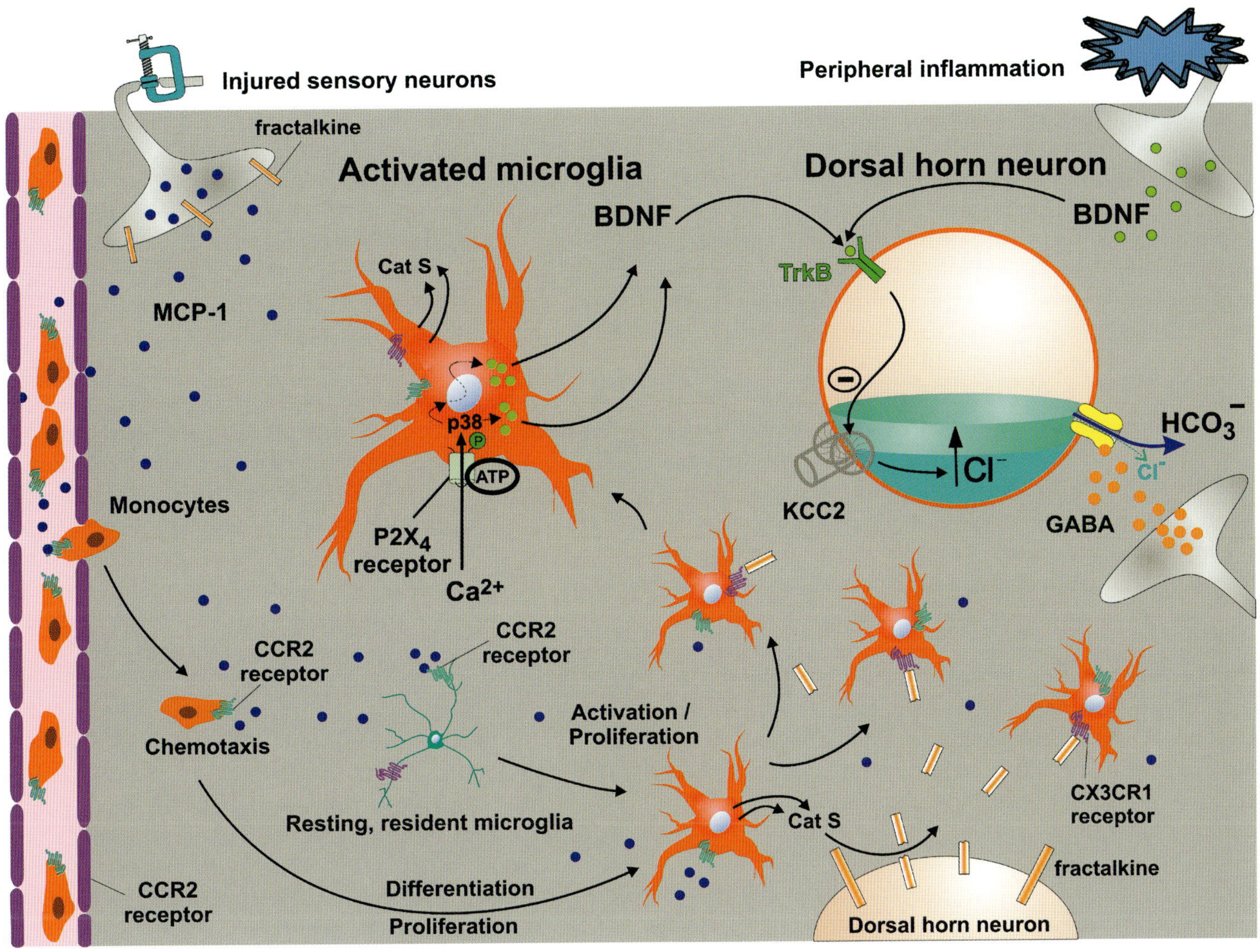

FIGURE 23.12 Summary of CNS neuroimmune interactions involved in activation of microglia and release of BDNF. Pain hypersensitivity is caused by a neuron-to-microglia-to-neuron signaling cascade that results in increased neuronal network excitability in the spinal dorsal horn. See the text for a detailed description of the evidence implicating each of the signaling events depicted here. Briefly, following nerve injury, damaged sensory neurons up-regulate the matrix metalloprotease MMP-9 (Kawasaki et al., 2008) and the chemokine MCP1 (also named CCL2; Zhang and De Koninck, 2006). MMP-9 cleaves the cytokine IL-1β to activate microglia while MCP-1 cause both microglial activation and chemotaxis of circulating monocytes via activation of the CCR2 receptor (Abbadie et al., 2003; Zhang et al., 2007). Stimulated monocytes infiltrate the spinal cord parenchyma and differentiate into activated microglia (Zhang et al., 2007). Activated microglia proliferate and release cathepsin S (CatS), which cleaves fractalkine tethered to the extracellular surface of neurons (Clark et al., 2007). Cleaved fractalkine acts on CX3CR1 receptors on microglia to amplify microglial activation and up-regulate p38 MAP kinase (Ji and Suter, 2007; Zhuang et al., 2007). Activated microglia express *de novo* the P2X4 receptors (Tsuda et al., 2003). Selective stimulation of P2X4 receptors by endogenous ATP causes the release of brain-derived neurotrophic factor (BDNF) in a p38 MAP kinase-dependent fashion (Trang et al., 2008; Ulmann et al., 2008). In turn, BDNF acts on neuronal tyrosine kinase B (TrkB) receptors to down-regulate the K^+-Cl^- cotransporter KCC2 in dorsal horn neurons (Coull et al., 2005; Rivera et al., 2002, 2004). After peripheral inflammation, enhanced BDNF release originates from sensory nerve terminals (Zhang et al., 2008; Zhao et al., 2006). Thus, regardless of the source of BDNF, the final common pathway appears to be a loss of KCC2 expression, causing a decrease in Cl^- extrusion capacity and consequent accumulation of intracellular Cl^- to electrochemical equilibrium levels. This effectively abolishes Cl^- gradients across neuronal membrane and thus impairs the efficacy of inhibitory $GABA_A^-$ and glycine receptor-mediated inhibitory transmission. In some cases Cl^- is accumulated above equilibrium probably by NKCC1. This, together with the outward flux of HCO3− may invert the action of GABA and glycine into net depolarizing excitation (Cordero-Erausquin et al., 2005; Coull et al., 2005). This event leads to a loss of spinal inhibition (disinhibition) and thus increases the excitability of dorsal horn neuronal networks (Keller et al., 2007; Prescott et al., 2006). Modified from Zhang and De Koninck (2008); reproduced with permission.

terminals, whereas its receptor, CX3CR1, is expressed primarily on microglia (Verge et al., 2004). This chemokine is unique in that it is tethered to the extracellular surface of neurons and can be cleaved to form a diffusible signal (Chapman et al., 2000). Lysosomal cysteine protease cathepsin S (CatS) expressed on activated spinal microglia, has been shown to be responsible for the cleavage of neuronal transmembrane fractalkine into active soluble fractalkine (Clark et al., 2007). After nerve injury, CatS-expressing activated microglia in spinal cord dorsal horn innervated by damaged fibers release CatS, which then liberates soluble fractalkine from primary afferent terminals and surrounding spinal neurons. The released fractalkine feeds back onto microglia cells via the CX3CR1 receptors activating the p38 MAPK pathway. Both exogenous/endogenous fractalkine and CatS have pronociceptive effects and inhibition of CatS enzymatic activity and antagonizing CX3CR1 receptor with a neutralizing antibody reversed pain hypersensitivity in animals with nerve injury (Barclay et al., 2007; Milligan et al., 2004). CatS-induced hyperalgesia is lost in CX3CR1 null mice (Clark et al., 2007). Thus, while neurons are the source of fractalkine signaling to CX3CR1 on microglia, the initiating signal appears to be microglial CatS. Thus, activated microglia might be signaling to other microglia via the cleavage of fractalkine on neurons. The concurrent increase in CatS expression by microglia and fractalkine by neurons could serve as a mechanism of amplification and coincidence detection. Because it is activated microglia that appears to release CatS, the latter is likely not the initiating mechanism causing the activation of microglia. Another upstream signaling event must occur that triggers microglial activation, which is then likely amplified and perhaps maintained by CatS-fractalkine-CX3CR1 signaling.

C. Matrix Metalloproteases (MMPs)

A recent study demonstrated that after nerve injury, matrix metalloproteinase (MMP)-9 expression is induced in dorsal root ganglion (DRG) sensory neurons within hours following injury, and returned to baseline three days after injury, whereas MMP-2 is induced in DRG satellite cells and in spinal astrocytes at later times (from days to weeks after injury) (Kawasaki et al., 2008). Although the cellular distribution and the time course are different, both MMPs require a common molecular player, the cytokine IL-1β. MMP-9 induces neuropathic pain through IL-1β cleavage and microglial activation at early times, whereas MMP-2 maintains neuropathic pain through IL-1β cleavage and astrocyte activation at later times (Kawasaki et al., 2008). MMP-9 may thus be a good candidate as initial trigger of microglial activation.

D. MCP-1 to CCR2 Signaling

The monocyte chemoattractant protein-1 (MCP-1) is a member of the CC chemokine family (also termed CCL2) that specifically attracts and activates monocytes to sites of inflammation (Leonard et al., 1991). Absent in normal CNS, MCP-1 is up-regulated in DRG sensory neurons following chronic constriction of the sciatic nerve (Tanaka et al., 2004; Zhang and De Koninck, 2006). The MCP-1 expressed in DRG neurons is transported to the spinal dorsal horn (Zhang and De Koninck, 2006) where it is released in response to electrical stimulation of sensory nerves (Thacker et al., 2008). CCR2, the receptor of MCP-1, is up-regulated in spinal microglia after peripheral nerve injury. Further, transgenic mice lacking CCR2 do not develop tactile allodynia after nerve injury (Abbadie et al., 2003).

Both temporally and spatially, the MCP-1 induction in terminals of damaged sensory neurons closely correlates with the surrounding spinal microglial cells, which are subsequently activated (Zhang and De Koninck, 2006). This nerve injury-induced spinal microglial activation is completely abolished in mice lacking CCR2 (Zhang et al., 2007). Local spinal injection of exogenous MCP1 induces microglial activation and this activation is lost in CCR2 knockout mice (Zhang et al., 2007). Altogether, these results implicate MCP-1 as a trigger for spinal microglial activation after peripheral nerve injury.

In addition to activation of microglia residing in the spinal cord, blood-borne monocytes/macrophages have the ability to infiltrate the spinal cord, proliferate and differentiate into activated microglia (Zhang et al., 2007). MCP-1/CCR2 signaling is also involved in the neuron-to-monocytes/macrophages crosstalk from the CNS to the periphery since neutralization of MCP-1 at the spinal level prevented monocyte/macrophage infiltration after nerve injury (Zhang et al., 2007). Thus, both activation of resident microglia and monocyte/macrophage infiltration appear to be due to a signaling mechanism within the spinal cord. Other evidence implicates MCP-1/CCR2 in the recruitment of monocytes/macrophages and activated lymphocytes into the CNS in a variety of inflammatory, infective and traumatic conditions (Huang et al., 2001; Kelder et al., 1998; Rancan et al., 2001). CCR^{2+} monocytes were identified as direct circulating precursors of microglia responsible for CNS infiltration (Mildner et al., 2007). MCP-1 may also facilitate chemotaxis from the periphery to the CNS by altering the expression of tight junction-associated proteins in brain and spinal cord microvascular endothelial cells, thereby increasing blood–brain/spinal cord barrier permeability (Song and Pachter, 2004; Stamatovic et al., 2003).

MCP-1 is not only a necessary mediator for spinal microglial activation, its action is necessary for the development of mechanical allodynia. Indeed, mice lacking the CCR2 (CCR2 KO) do not develop pain hypersensitivity following nerve injury (Zhang et al., 2007). By selectively deleting CNS and peripheral CCR2, respectively, we found that either source of microglia – resident or blood derived – was sufficient to cause pain hypersensitivity. Thus, infiltration of monocytes/macrophages into the CNS under pathological and/or traumatic conditions may cause a disruption of Cl^- homeostasis and alter neuronal excitation mediated by ligand-gated anion channels.

E. Does Microglial Activation Always Cause a Disruption of Chloride Homeostasis?

As mentioned above, microglial activation is highly diverse and dynamic (Hanisch and Kettenmann, 2007) and expression of $P2X_4$ receptors and/or BDNF secretion only occurs in specific conditions. Several other phenotypes have been observed in pathologies (Avignone et al., 2008; Hanisch and Kettenmann, 2007; Pocock and Kettenmann, 2007). It thus remains to be determined if microglia is a source of enhanced BDNF in conditions other than nerve injury induced pain.

Similarly, it remains to be determined whether BDNF-TrkB signaling is the only intercellular signaling pathway involved in disruption of neuronal Cl^- homeostasis. Another possibility to consider is whether BDNF-TrkB signaling affects the activity of other cotransporters such as NKCC1. In several cases, opposite regulation of NKCC1 and KCC2 by intracellular signaling pathways has been observed (De Koninck, 2007), as discussed in Chapter 18. The same may occur in response to activation of TrkB receptors, although a recent study reported no change in NKCC1 expression in the dorsal horn after peripheral inflammation which causes enhanced spinal BDNF release (Zhang et al., 2008).

VII. THERAPEUTIC IMPLICATIONS

The results of the studies outlined in this chapter open interesting avenues for therapeutic interventions. Drugs aimed at restoring normal Cl^- homeostasis may present advantages over current therapeutic strategies because they do not affect neuronal excitability directly, but rather modulate the actions of endogenous inhibitory neurotransmitters. Restoring endogenous inhibition rather than imposing inhibition (actively depressing excitability) may also yield more specific therapeutic treatments with less detrimental side effects (De Koninck, 2007).

The finding that normal expression of KCC2 is restored upon blocking BDNF-TrkB or ATP-$P2X_4$ signaling *after* the pathological changes have developed in the case of nerve injury-induced pain hypersensitivity (Fig. 23.5) is encouraging because it indicates that there may be targets to restore cation-Cl^- cotransporter function (Coull et al., 2005). Targeting BDNF-TrkB signaling directly may not be attractive for therapeutics because this neurotrophin is involved in several functions throughout the CNS. On the other hand, specifically preventing activated microglia from secreting BDNF – by targeting $P2X_4$ receptors, p38-MAPK kinase signaling or by preventing BDNF transcription selectively in microglia – may represent a more promising strategy.

The identification of the complex cascade of events that lead to microglial activation also offers a wide array of novel targets for therapeutics. But, because activation of microglial cells is part of the normal immune reaction, preventing them from playing their role in the central inflammatory response to injury may be highly detrimental. Thus, the discovery of specific signaling mechanisms in which microglia participate under certain conditions of activation, and of selected downstream effector mechanisms such as Cl^- homeostasis, provides potential avenues to selectively control certain side effects of the central inflammatory response without affecting other aspects of neuroimmune function.

Acknowledgements

The author acknowledges support from the Canadian Institutes of Health (CIHR), the Natural Sciences and Engineering Research Council of Canada (NSERC), Neuroscience Canada Foundation, the Ontario Neurotrauma Foundation and the Krembil Foundation. The author is also a *Chercheur National* of the Fonds de la Recherche en Santé du Québec (FRSQ).

References

Abbadie, C., Lindia, J.A., Cumiskey, A.M., Peterson, L.B., Mudgett, J.S., Bayne, E.K., DeMartino, J.A., MacIntyre, D.E., and Forrest, M.J. (2003). Impaired neuropathic pain responses in mice lacking the chemokine receptor CCR2. *Proc. Natl. Acad. Sci. USA* **100**, 7947–7952.

Aguado, F., Carmona, M.A., Pozas, E., Aguilo, A., Martinez-Guijarro, F.J., Alcantara, S., Borrell, V., Yuste, R., Ibanez, C.F., and Soriano, E. (2003). BDNF regulates spontaneous correlated activity at early developmental stages by increasing synaptogenesis and expression of the K^+/Cl^- co-transporter KCC2. *Development* **130**, 1267–1280.

Avignone, E., Ulmann, L., Levavasseur, F., Rassendren, F., and Audinat, E. (2008). Status epilepticus induces a particular microglial activation state characterized by enhanced purinergic signaling. *J. Neurosci.* **28**, 9133–9144.

Baba, H., Ji, R.R., Kohno, T., Moore, K.A., Ataka, T., Wakai, A., Okamoto, M., and Woolf, C.J. (2003). Removal of GABAergic inhibition facilitates polysynaptic A fiber-mediated excitatory transmission to the superficial spinal dorsal horn. *Mol. Cell Neurosci* **24**, 818–830.

Balkowiec, A. and Katz, D.M. (2000). Activity-dependent release of endogenous brain-derived neurotrophic factor from primary sensory neurons detected by ELISA in situ. *J. Neurosci.* **20**, 7417–7423.

Barclay, J., Clark, A.K., Ganju, P., Gentry, C., Patel, S., Wotherspoon, G., Buxton, F., Song, C., Ullah, J., Winter, J., Fox, A., Bevan, S., and Malcangio, M. (2007). Role of the cysteine protease cathepsin S in neuropathic hyperalgesia. *Pain* **130**, 225–234.

Beggs, S. and Salter, M.W. (2007). Stereological and somatotopic analysis of the spinal microglial response to peripheral nerve injury. *Brain Behav. Immun.* **21**, 624–633.

Blaesse, P., Guillemin, I., Schindler, J., Schweizer, M., Delpire, E., Khiroug, L., Friauf, E., and Nothwang, H.G. (2006). Oligomerization of KCC2 correlates with development of inhibitory neurotransmission. *J. Neurosci.* **26**, 10407–10419.

Castro-Lopes, J.M., Malcangio, M., Pan, B.H., and Bowery, N.G. (1995). Complex changes of $GABA_A$ and $GABA_B$ receptor binding in the spinal cord dorsal horn following peripheral inflammation or neurectomy. *Brain Res.* **679**, 289–297.

Castro-Lopes, J.M., Tavares, I., and Coimbra, A. (1993). GABA decreases in the spinal cord dorsal horn after peripheral neurectomy. *Brain Res.* **620**, 287–291.

Castro-Lopes, J.M., Tölle, T.R., Pan, B., and Zieglgänsberger, W. (1994). Expression of GAD mRNA in spinal cord neurons of normal and monoarthritic rats. *Brain Res. Mol. Brain Res.* **26**, 169–176.

Chapman, G.A., Moores, K., Harrison, D., Campbell, C.A., Stewart, B.R., and Strijbos, P.J. (2000). Fractalkine cleavage from neuronal membranes represents an acute event in the inflammatory response to excitotoxic brain damage. *J. Neurosci.* **20**, RC87.

Clark, A.K., Yip, P.K., Grist, J., Gentry, C., Staniland, A.A., Marchand, F., Dehvari, M., Wotherspoon, G., Winter, J., Ullah, J., Bevan, S., and Malcangio, M. (2007). Inhibition of spinal microglial cathepsin S for the reversal of neuropathic pain. *Proc. Natl. Acad. Sci. USA* **104**, 10655–10660.

Colburn, R.W., Rickman, A.J., and DeLeo, J.A. (1999). The effect of site and type of nerve injury on spinal glial activation and neuropathic pain behavior. *Exp. Neurol.* **157**, 289–304.

Cordero-Erausquin, M., Coull, J.A., Boudreau, D., Rolland, M., and De Koninck, Y. (2005). Differential maturation of GABA action and anion reversal potential in spinal lamina I neurons; impact of chloride extrusion capacity. *J. Neurosci.* **25**, 9613–9623.

Coull, J.A., Beggs, S., Boudreau, D., Boivin, D., Tsuda, M., Inoue, K., Gravel, C., Salter, M.W., and De Koninck, Y. (2005). BDNF from microglia causes the shift in neuronal anion gradient underlying neuropathic pain. *Nature* **438**, 1017–1021.

Coull, J.A., Boudreau, D., Bachand, K., Prescott, S.A., Nault, F., Sik, A., De Koninck, P., and De Koninck, Y. (2003). Trans-synaptic shift in anion gradient in spinal lamina I neurons as a mechanism of neuropathic pain. *Nature* **424**, 938–942.

De Koninck, Y. (2007). Altered chloride homeostasis in neurological disorders: a new target. *Curr. Opin. Pharmacol.* **7**, 93–99.

Decosterd, I. and Woolf, C.J. (2000). Spared nerve injury: an animal model of persistent peripheral neuropathic pain. *Pain* **87**, 149–158.

Echeverry, S., Shi, X.Q., and Zhang, J. (2008). Characterization of cell proliferation in rat spinal cord following peripheral nerve injury and the relationship with neuropathic pain. *Pain* **135**, 37–47.

Fu, K.Y., Light, A.R., Matsushima, G.K., and Maixner, W. (1999). Microglial reactions after subcutaneous formalin injection into the rat hind paw. *Brain Res.* **825**, 59–67.

Gamba, G. (2005). Molecular physiology and pathophysiology of electroneutral cation-chloride cotransporters. *Physiol Rev.* **85**, 423–493.

Gillen, C.M. and Forbush, B., III (1999). Functional interaction of the K-Cl cotransporter (KCC1) with the Na-K-Cl cotransporter in HEK-293 cells. *Am. J. Physiol* **276**, C328–C336.

Hanisch, U.K. and Kettenmann, H. (2007). Microglia: active sensor and versatile effector cells in the normal and pathologic brain. *Nat. Neurosci.* **10**, 1387–1394.

Haynes, S.E., Hollopeter, G., Yang, G., Kurpius, D., Dailey, M.E., Gan, W.B., and Julius, D. (2006). The P2Y12 receptor regulates microglial activation by extracellular nucleotides. *Nat. Neurosci.* **9**, 1512–1519.

Heppenstall, P.A. and Lewin, G.R. (2001). BDNF but not NT-4 is required for normal flexion reflex plasticity and function. *Proc. Natl. Acad. Sci. USA* **98**, 8107–8112.

Huang, D.R., Wang, J., Kivisakk, P., Rollins, B.J., and Ransohoff, R.M. (2001). Absence of monocyte chemoattractant protein 1 in mice leads to decreased local macrophage recruitment and antigen-specific T helper cell type 1 immune response in experimental autoimmune encephalomyelitis. *J. Exp. Med.* **193**, 713–726.

Ibuki, T., Hama, A.T., Wang, X.T., Pappas, G.D., and Sagen, J. (1997). Loss of GABA-immunoreactivity in the spinal dorsal horn of rats with peripheral nerve injury and promotion of recovery by adrenal medullary grafts. *Neuroscience* **76**, 845–858.

Ji, R.R. and Suter, M.R. (2007). p38 MAPK, microglial signaling, and neuropathic pain. *Mol. Pain* **3**, 33.

Kahle, K.T., Rinehart, J., de Los, H.P., Louvi, A., Meade, P., Vazquez, N., Hebert, S.C., Gamba, G., Gimenez, I., and Lifton, R.P. (2005). WNK3 modulates transport of Cl^- in and out of cells: implications for control of cell volume and neuronal excitability. *Proc. Natl. Acad. Sci. USA* **102**, 16783–16788.

Kawasaki, Y., Xu, Z.Z., Wang, X., Park, J.Y., Zhuang, Z.Y., Tan, P. H., Gao, Y.J., Roy, K., Corfas, G., Lo, E.H., and Ji, R.R. (2008). Distinct roles of matrix metalloproteases in the early- and late-phase development of neuropathic pain. *Nat. Med.* **14**, 331–336.

Kelder, W., McArthur, J.C., Nance-Sproson, T., McClernon, D., and Griffin, D.E. (1998). Beta-chemokines MCP-1 and RANTES are selectively increased in cerebrospinal fluid of patients with human immunodeficiency virus-associated dementia. *Ann. Neurol.* **44**, 831–835.

Keller, A.F., Beggs, S., Salter, M.W., and De Koninck, Y. (2007). Transformation of the output of spinal lamina I neurons after nerve injury and microglia stimulation underlying neuropathic pain. *Mol. Pain* **3**, 27.

Koizumi, S., Shigemoto-Mogami, Y., Nasu-Tada, K., Shinozaki, Y., Ohsawa, K., Tsuda, M., Joshi, B.V., Jacobson, K.A., Kohsaka, S., and Inoue, K. (2007). UDP acting at P2Y6 receptors is a mediator of microglial phagocytosis. *Nature* **446**, 1091–1095.

Leonard, E.J., Skeel, A., and Yoshimura, T. (1991). Biological aspects of monocyte chemoattractant protein-1 (MCP-1). *Adv. Exp. Med. Biol.* **305**, 57–64.

Mannion, R.J., Costigan, M., Decosterd, I., Amaya, F., Ma, Q.P., Holstege, J.C., Ji, R.R., Acheson, A., Lindsay, R.M., Wilkinson, G.A., and Woolf, C.J. (1999). Neurotrophins: peripherally and centrally acting modulators of tactile stimulus-induced inflammatory pain hypersensitivity. *Proc. Natl. Acad. Sci. USA* **96**, 9385–9390.

Maragakis, N.J. and Rothstein, J.D. (2006). Mechanisms of disease: astrocytes in neurodegenerative disease. *Nat. Clin. Pract. Neurol.* **2**, 679–689.

Marchand, F., Perretti, M., and McMahon, S.B. (2005). Role of the immune system in chronic pain. *Nat. Rev. Neurosci.* **6**, 521–532.

McMahon, S.B., Cafferty, W.B., and Marchand, F. (2005). Immune and glial cell factors as pain mediators and modulators. *Exp. Neurol.* **192**, 444–462.

Mercado, A., Broumand, V., Zandi-Nejad, K., Enck, A.H., and Mount, D.B. (2006). A C-terminal domain in KCC2 confers constitutive K^+-Cl^- cotransport. *J. Biol. Chem.* **281**, 1016–1026.

Mildner, A., Schmidt, H., Nitsche, M., Merkler, D., Hanisch, U.K., Mack, M., Heikenwalder, M., Bruck, W., Priller, J., and Prinz, M. (2007). Microglia in the adult brain arise from Ly-6ChiCCR2+ monocytes only under defined host conditions. *Nat. Neurosci.* **10**, 1544–1553.

Milligan, E.D., Zapata, V., Chacur, M., Schoeniger, D., Biedenkapp, J., O'Connor, K.A., Verge, G.M., Chapman, G., Green, P., Foster, A.C., Naeve, G.S., Maier, S.F., and Watkins, L.R. (2004). Evidence that exogenous and endogenous fractalkine can induce spinal nociceptive facilitation in rats. *Eur. J. Neurosci.* **20**, 2294–2302.

Minichiello, L., Casagranda, F., Tatche, R.S., Stucky, C.L., Postigo, A., Lewin, G.R., Davies, A.M., and Klein, R. (1998). Point mutation in trkB causes loss of NT4-dependent neurons without major effects on diverse BDNF responses. *Neuron* **21**, 335–345.

Mizoguchi, Y., Ishibashi, H., and Nabekura, J. (2003). The action of BDNF on GABA(A) currents changes from potentiating to suppressing during maturation of rat hippocampal CA1 pyramidal neurons. *J. Physiol* **548**, 703–709.

Moore, K.A., Kohno, T., Karchewski, L.A., Scholz, J., Baba, H., and Woolf, C.J. (2002). Partial peripheral nerve injury promotes a selective loss of GABAergic inhibition in the superficial dorsal horn of the spinal cord. *J. Neurosci.* **22**, 6724–6731.

Nabekura, J., Ueno, T., Okabe, A., Furuta, A., Iwaki, T., Shimizu-Okabe, C., Fukuda, A., and Akaike, N. (2002). Reduction of KCC2 expression and $GABA_A$ receptor-mediated excitation after in vivo axonal injury. *J. Neurosci.* **22**, 4412–4417.

Nguyen, M.D., Julien, J.P., and Rivest, S. (2002). Innate immunity: the missing link in neuroprotection and neurodegeneration? *Nat. Rev. Neurosci* **3**, 216–227.

Pitcher, M.H., Price, T.J., Entrena, J.M., and Cervero, F. (2007). Spinal NKCC1 blockade inhibits TRPV1-dependent referred allodynia. *Mol. Pain* **3**, 17.

Pocock, J.M. and Kettenmann, H. (2007). Neurotransmitter receptors on microglia. *Trends Neurosci.* **30**, 527–535.

Polgar, E., Gray, S., Riddell, J.S., and Todd, A.J. (2004). Lack of evidence for significant neuronal loss in laminae I–III of the spinal dorsal horn of the rat in the chronic constriction injury model. *Pain* **111**, 144–150.

Polgar, E., Hughes, D.I., Arham, A.Z., and Todd, A.J. (2005). Loss of neurons from laminas I–III of the spinal dorsal horn is not required for development of tactile allodynia in the spared nerve injury model of neuropathic pain. *J. Neurosci.* **25**, 6658–6666.

Prescott, S.A., Sejnowski, T.J., and De Koninck, Y. (2006). Reduction of anion reversal potential subverts the inhibitory control of firing rate in spinal lamina I neurons: a biophysical basis for neuropathic pain. *Mol. Pain* **2**, 32.

Prince, D.A. (1999). Epileptogenic neurons and circuits. *Adv. Neurol.* **79**, 665–684.

Qiu, J., Cai, D., and Filbin, M.T. (2002). A role for cAMP in regeneration during development and after injury. *Prog. Brain Res.* **137**, 381–387.

Rancan, M., Otto, V.I., Hans, V.H., Gerlach, I., Jork, R., Trentz, O., Kossmann, T., and Morganti-Kossmann, M.C. (2001). Upregulation of ICAM-1 and MCP-1 but not of MIP-2 and sensorimotor deficit in response to traumatic axonal injury in rats. *J. Neurosci. Res.* **63**, 438–446.

Rivera, C., Li, H., Thomas-Crusells, J., Lahtinen, H., Viitanen, T., Nanobashvili, A., Kokaia, Z., Airaksinen, M.S., Voipio, J., Kaila, K., and Saarma, M. (2002). BDNF-induced TrkB activation down-regulates the K^+-Cl^- cotransporter KCC2 and impairs neuronal Cl^- extrusion. *J. Cell Biol.* **159**, 747–752.

Rivera, C., Voipio, J., Thomas-Crusells, J., Li, H., Emri, Z., Sipila, S., Payne, J.A., Minichiello, L., Saarma, M., and Kaila, K. (2004). Mechanism of activity-dependent downregulation of the neuron-specific K-Cl cotransporter KCC2. *J. Neurosci.* **24**, 4683–4691.

Rock, R.B., Gekker, G., Aravalli, R.N., Hu, S., Sheng, W.S., and Peterson, P.K. (2008). Potentiation of HIV-1 expression in microglial cells by nicotine: involvement of transforming growth factor-beta1. *J. Neuroimmune. Pharmacol.* **3**, 143–149.

Scholz, J., Abele, A., Marian, C., Haussler, A., Herbert, T.A., Woolf, C.J., and Tegeder, I. (2008). Low-dose methotrexate reduces peripheral nerve injury-evoked spinal microglial activation and neuropathic pain behavior in rats. *Pain.* **6**.

Scholz, J., Broom, D.C., Youn, D.H., Mills, C.D., Kohno, T., Suter, M.R., Moore, K.A., Decosterd, I., Coggeshall, R.E., and Woolf, C.J. (2005). Blocking caspase activity prevents transsynaptic neuronal apoptosis and the loss of inhibition in lamina II of the dorsal horn after peripheral nerve injury. *J. Neurosci.* **25**, 7317–7323.

Sherman, S.E. and Loomis, C.W. (1994). Morphine insensitive allodynia is produced by intrathecal strychnine in the lightly anesthetized rat. *Pain* **56**, 17–29.

Sherman, S.E., Luo, L., and Dostrovsky, J.O. (1997). Spinal strychnine alters response properties of nociceptive-specific neurons in rat medial thalamus. *J. Neurophysiol.* **78**, 628–637.

Sivilotti, L. and Woolf, C.J. (1994). The contribution of $GABA_A$ and glycine receptors to central sensitization: disinhibition and touch-evoked allodynia in the spinal cord. *J. Neurophysiol.* **72**, 169–179.

Somers, D.L. and Clemente, F.R. (2002). Dorsal horn synaptosomal content of aspartate, glutamate, glycine and GABA are differentially altered following chronic constriction injury to the rat sciatic nerve. *Neurosci Lett.* **323**, 171–174.

Song, L. and Pachter, J.S. (2004). Monocyte chemoattractant protein-1 alters expression of tight junction-associated proteins in brain microvascular endothelial cells. *Microvasc. Res.* **67**, 78–89.

Sorkin, L.S. and Puig, S. (1996). Neuronal model of tactile allodynia produced by spinal strychnine: effects of excitatory amino acid receptor antagonists and a μ-opiate receptor agonist. *Pain* **68**, 283–292.

Soulet, D. and Rivest, S. (2008). Bone-marrow-derived microglia: myth or reality? *Curr. Opin. Pharmacol.* **6**.

Stamatovic, S.M., Keep, R.F., Kunkel, S.L., and Andjelkovic, A.V. (2003). Potential role of MCP-1 in endothelial cell tight junction "opening": signaling via Rho and Rho kinase. *J. Cell Sci.* **116**, 4615–4628.

Tanaka, T., Minami, M., Nakagawa, T., and Satoh, M. (2004). Enhanced production of monocyte chemoattractant protein-1 in the dorsal root ganglia in a rat model of neuropathic pain: possible involvement in the development of neuropathic pain. *Neurosci. Res.* **48**, 463–469.

Thacker, M.A., Clark, A.K., Bishop, T., Grist, J., Yip, P.K., Moon, L.D., Thompson, S.W., Marchand, F., and McMahon, S.B. (2008). CCL2 is a key mediator of microglia activation in neuropathic pain states. *Eur. J. Pain.* **6**.

Thompson, S.W., Bennett, D.L., Kerr, B.J., Bradbury, E.J., and McMahon, S.B. (1999). Brain-derived neurotrophic factor is an endogenous modulator of nociceptive responses in the spinal cord. *Proc. Natl. Acad. Sci. USA* **96**, 7714–7718.

Torsney, C. and MacDermott, A.B. (2006). Disinhibition opens the gate to pathological pain signaling in superficial neurokinin 1 receptor-expressing neurons in rat spinal cord. *J. Neurosci.* **26**, 1833–1843.

Toth, Z., Hollrigel, G.S., Gorcs, T., and Soltesz, I. (1997). Instantaneous perturbation of dentate interneuronal networks by a pressure wave-transient delivered to the neocortex. *J. Neurosci.* **17**, 8106–8117.

Trang, T., Beggs, S., and Salter, M.W. (2008). P2X4 receptor activation leads to release and synthesis of brain-derived neurotrophic factor in cultured microglia. 12th World Congress on Pain.

Tsuda, M., Shigemoto-Mogami, Y., Koizumi, S., Mizokoshi, A., Kohsaka, S., Salter, M.W., and Inoue, K. (2003). P2X4 receptors induced in spinal microglia gate tactile allodynia after nerve injury. *Nature* **424**, 778–783.

Ulmann, L., Hatcher, J.P., Hughes, J.P., Chaumont, S., Green, P.J., Conquet, F., Buell, G.N., Reeve, A.J., Chessell, I.P., and Rassendren, F. (2008). Up-regulation of P2X4 receptors in spinal microglia following peripheral nerve injury mediates BDNF release and neuropathic pain. *J. Neurosci.* **28**, 11263–11268.

Van den Pol, A.N., Obrietan, K., and Chen, G. (1996). Excitatory actions of GABA after neuronal trauma. *J. Neurosci.* **16**, 4283–4292.

Verge, G.M., Milligan, E.D., Maier, S.F., Watkins, L.R., Naeve, G.S., and Foster, A.C. (2004). Fractalkine (CX3CL1) and fractalkine receptor (CX3CR1) distribution in spinal cord and dorsal root ganglia under basal and neuropathic pain conditions. *Eur. J. Neurosci.* **20**, 1150–1160.

Wardle, R.A. and Poo, M.M. (2003). Brain-derived neurotrophic factor modulation of GABAergic synapses by postsynaptic regulation of chloride transport. *J. Neurosci.* **23**, 8722–8732.

Woolf, C.J. and Salter, M.W. (2000). Neuronal plasticity: increasing the gain in pain. *Science* **288**, 1765–1769.

Yaksh, T.L. (1989). Behavioral and autonomic correlates of the tactile evoked allodynia produced by spinal glycine inhibition: effects of modulatory receptor systems and excitatory amino acid antagonists. *Pain* **37**, 111–123.

Zhang, J. and De Koninck, Y. (2006). Spatial and temporal relationship between monocyte chemoattractant protein-1 expression and spinal glial activation following peripheral nerve injury. *J. Neurochem.* **97**, 772–783.

Zhang, J. and De Koninck, Y. (2008). Central neuroglial interactions in the pathophysiology of neuropathic pain. In *Functional Chronic Pain Syndromes: Presentation and Pathophysiology.* (Mayer, E.A. and Bushnell, M.C., eds). IASP Press, Seattle.

Zhang, J., Shi, X.Q., Echeverry, S., Mogil, J.S., De Koninck, Y., and Rivest, S. (2007). Expression of CCR2 in both resident and bone marrow-derived microglia plays a critical role in neuropathic pain. *J. Neurosci.* **27**, 12396–12406.

Zhang, W., Liu, L.Y., and Xu, T.L. (2008). Reduced potassium-chloride co-transporter expression in spinal cord dorsal horn neurons contributes to inflammatory pain hypersensitivity in rats. *Neuroscience* **152**, 502–510.

Zhao, J., Seereeram, A., Nassar, M.A., Levato, A., Pezet, S., Hathaway, G., Morenilla-Palao, C., Stirling, C., Fitzgerald, M., McMahon, S.B., Rios, M., and Wood, J.N. (2006). Nociceptor-derived brain-derived neurotrophic factor regulates acute and inflammatory but not neuropathic pain. *Mol. Cell Neurosci.* **31**, 539–548.

Zhuang, Z.Y., Gerner, P., Woolf, C.J., and Ji, R.R. (2005). ERK is sequentially activated in neurons, microglia, and astrocytes by spinal nerve ligation and contributes to mechanical allodynia in this neuropathic pain model. *Pain* **114**, 149–159.

Zhuang, Z.Y., Kawasaki, Y., Tan, P.H., Wen, Y.R., Huang, J., and Ji, R.R. (2007). Role of the CX3CR1/p38 MAPK pathway in spinal microglia for the development of neuropathic pain following nerve injury-induced cleavage of fractalkine. *Brain Behav. Immun.* **21**, 642–651.

CHAPTER

24

Cation-chloride Cotransporters as Pharmacological Targets in the Treatment of Epilepsy

Kristopher T. Kahle and Kevin J. Staley

OUTLINE

I. INTRODUCTION

Seizures occurring in the neonatal period, during intractable temporal lobe epilepsy, or after hypoxic-ischemic injury all have poor treatment responses to classic anti-epileptic medications that stimulate $GABA_A$ receptors. While these seizure syndromes at first glance appear to have very little in common, they have recently been linked to potentially treatable abnormalities in neuronal Cl^- transport.

In the mammalian central nervous system (CNS), the intracellular Cl^- concentration ($[Cl^-]_i$) in neurons determines the polarity and magnitude, and hence the strength, of GABAergic neurotransmission. $[Cl^-]_i$ is primarily determined by the functional activity of *SLC12A* cation-chloride cotransporters (CCCs), including the Na^+-K^+-$2Cl^-$ cotransporter NKCC1, which mediates active Cl^- uptake) and the K^+-Cl^- cotransporter KCC2, which mediates active Cl^- extrusion. Mature CNS neurons of adults maintain $[Cl^-]_i$ at values lower than electrochemical equilibrium due to relatively low NKCC1 expression but robust KCC2 expression, such that GABA-induced allosteric activation of the $GABA_A$-gated anion channels results in Cl^-

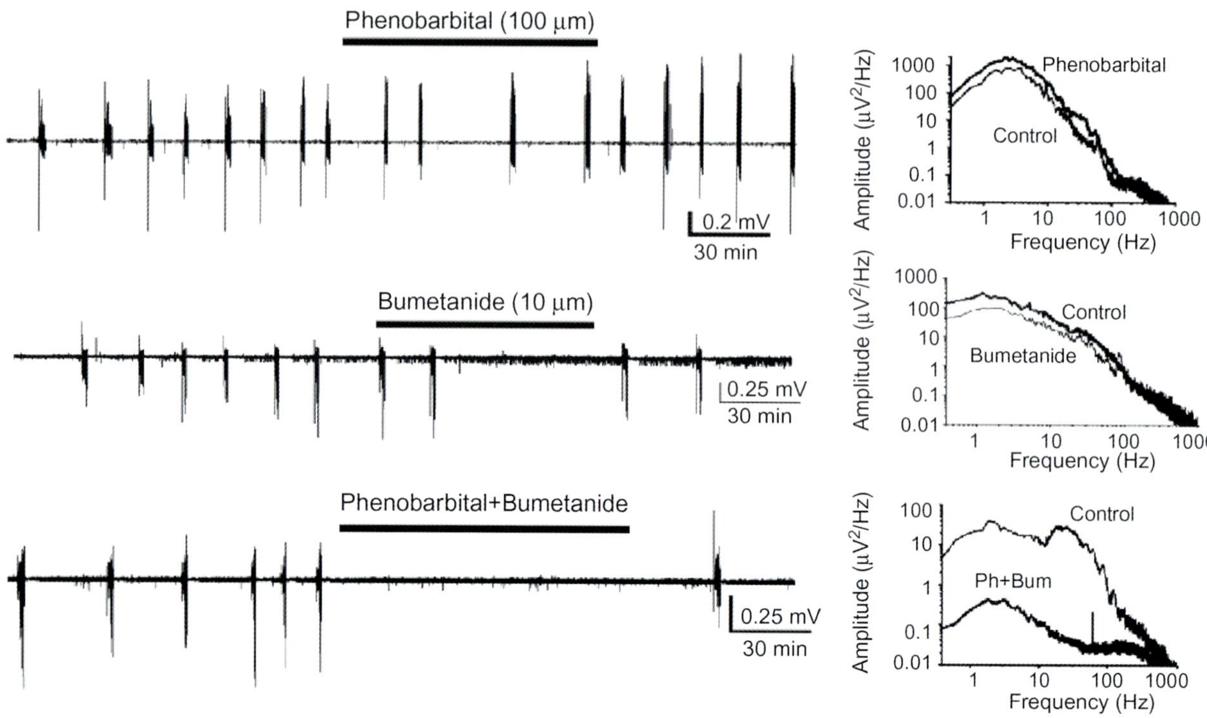

FIGURE 24.1 Synergistic effects of the NKCC1 inhibitor bumetanide and the GABAergic anticonvulsant phenobarbital on seizures induced by low $Mg^{2+}{}_o$ in the intact hippocampal preparation from postnatal day 5 rat pups. *Top panel*: Seizures induced by low $Mg^{2+}{}_o$ perfusate are modestly reduced by 100 μM phenobarbital. *Left*: Extracellular recording of epileptiform activity before during and after perfusion of phenobarbital for the time period indicated by the horizontal bar. *Right*: Power spectrum of extracellular recording before and during perfusion of phenobarbital. *Middle panel*: A similar experiment in which NKCC1 was blocked by perfusing 10 μM bumetanide for the indicated time period. *Bottom panel*: Phenobarbital and bumetanide demonstrate synergistic anticonvulsant effects when perfused together. Reprinted with permission from Dzhala et al. (2008).

influx that results in membrane hyperpolarization. In many adults with seizures, barbiturates and benzodiazepines are able to decrease seizure activity by increasing the duration and/or frequency of the opening of $GABA_A$ receptor-anion channels.

In contrast to mature CNS neurons of healthy adults, normal immature neurons of neonates, and neurons of adult patients suffering from intractable temporal lobe epilepsy or hypoxic-ischemic brain injury, appear to have $[Cl^-]_i$ that is higher than electrochemical equilibrium due to high functional NKCC1 expression and minimal KCC2 expression, such that activation of $GABA_A$ receptors elicits an efflux of Cl^- that results in membrane depolarization. This depolarizing action of GABA can become excitatory thus contributing to ictogenesis in these seizure syndromes, and has a tendency to render barbiturates and benzodiazepines ineffective, because these drugs (in contrast to their action on normal mature neurons, which harbor lower than passive $[Cl^-]_i$) favor electrodiffusional efflux of Cl^- down their electrochemical gradient, thereby depolarizing and potentially exciting neurons.

Recent data have shown that the NKCCs' inhibitor bumetanide, a drug already approved by the Food and Drug Administration as a diuretic, shows efficacy against seizures in the immature rodent brain, and thus may be clinically useful in the treatment of human neonatal seizures. Recent studies have also shown that bumetanide is able to quell seizures in models of temporal lobe epilepsy due to mesial temporal sclerosis and hypoxic-ischemic encephalopathy. This chapter will discuss the recent work that has begun to elucidate the role of the CCCs in fostering excitatory GABAergic neurotransmission in the immature and adult epileptic brain, and how the pharmacologic inhibition of certain cotransporters might hold promise for the treatment of seizures whose pathogenesis is attributable to derangements in Cl^- transport.

II. THE MAGNITUDE AND POLARITY OF GABAergic NEUROTRANSMISSION IS DETERMINED BY THE INTRANEURONAL CONCENTRATION OF Cl^-

In the adult cortex, GABA is the main *inhibitory* neurotransmitter, where its activity is essential for maintaining appropriate electrical activity by balancing inputs that are triggered by excitatory neurotransmitters like glutamate. In contrast, in the neonatal cortex, GABA is depolarizing and can become *excitatory*, playing important roles in neuronal development and the activity-dependent wiring of circuits by triggering the depolarizing inputs that underlie large-scale spontaneous electrical activity (Ben-Ari, 2001, 2002).

In neurons, GABA binding to $GABA_A$ receptors triggers conformational changes in these receptors that result in the opening of anion-permeable channels, thus allowing passive, electrodiffusional influx or efflux of Cl^- depending on the sign and magnitude of the driving force for Cl^-, which is the difference between E_m, the resting membrane potential, and E_{Cl}, the equilibrium potential for Cl^-. When $[Cl^-]_i$ is higher than electrochemical equilibrium, E_{Cl} is positive relative to E_m, and the opening of anion channels by $GABA_A$-receptor activation results in Cl^- efflux, which consequently depolarizes the neuron. When $[Cl^-]_i$ is lower than electrochemical equilibrium, E_{Cl} is negative relative to the E_m, and activation of the $GABA_A$ receptor results in Cl^- influx, leading to neuronal membrane hyperpolarization. When GABA depolarizes neurons, the depolarization increases the probability of firing action potentials, whereas hyperpolarization is associated with a decreased chance of it firing an action potential. There are, however, exceptions to these considerations such as shunting inhibition mediated by depolarizing conductances (Staley and Mody, 1992; Jean-Xavier et al., 2007), voltage-dependent inactivation of Na^+ conductances (see Chapter 22) and action potentials triggered by rebound or H currents after hyperpolarizing conductances (Andersen et al., 1964), so it is sometimes preferable to assay the net effect of endogenously-released GABA on the rate of action potentials in the neural network rather than extrapolate from the effect of GABA on membrane potential of single cells (Dzhala and Staley, 2003).

III. A DEVELOPMENTAL SWITCH IN CATION-CHLORIDE COTRANSPORTER EXPRESSION RENDERS GABA HYPERPOLARIZING DURING NEURONAL MATURATION

A developmental switch in neuronal driving force for Cl^- parallels the maturation of GABAergic signaling from excitatory to inhibitory (Ben-Ari, 2001, 2002). In immature CNS neurons GABA plays an important role in neuronal growth, development, and the activity-dependent wiring of circuits by triggering the excitatory postsynaptic responses that underlie large-scale spontaneous electrical activity in the neonatal central and peripheral nervous system. In contrast to its action in the adult, as mentioned above during this early time period GABA behaves as an excitatory neurotransmitter due to the higher than passive $[Cl^-]_i$ characteristic of immature neurons. This depolarization generates action potentials that directly activate voltage-dependent calcium channels, and indirectly activate N-methyl d-aspartate (NMDA) receptors by removing the voltage-dependent block by magnesium ions in the receptor pore. The subsequent GABA-induced elevations of intracellular calcium promote neuronal survival and differentiation, and are important for the genesis and maintenance of synaptic connections (Ge et al., 2006; Wang and Kriegstein, 2008). However, during later stages of development, excitatory GABAergic inputs become inhibitory. A shift in the relative predominance of neuronal NKCC1 to KCC2 activity is now recognized as the mechanism that underlies the excitatory-to-inhibitory developmental transition of GABAergic signaling (Rivera et al., 1999).

Active Cl^- uptake in neurons is mediated largely via NKCC1. In embryonic and early postnatal life, neurons show robust expression of NKCC1 but minimal expression of KCC2 (Plotkin et al., 1997; Clayton et al., 1998; Wang et al., 2002). Removal of excitatory GABAergic signaling during development by sustained pharmacological inhibition or genetic disruption of NKCC1 results in a lack of morphological maturation of cortical neurons, fewer mature spinal neurons, reduced dendritic arborization, disrupted motor activity, fewer motor neurons and interneurons, a reduction in the elaboration of axonal tracts and smaller brains and spinal cords (Ge et al., 2006; Cancedda et al., 2007; Reynolds et al., 2008).

Active Cl^- extrusion in CNS neurons is mediated largely by KCC2. In many species, this cotransporter is expressed at very low levels at birth (Stein et al., 2004). In rat hippocampal and neocortical pyramidal neurons, a negative shift in the GABA reversal potential (E_{GABA}) is paralleled by a robust increase in KCC2 expression near the end of the second postnatal week (Rivera et al., 1999). Dzhala et al. demonstrated that KCC2 expression in the human neocortex begins to increase at 40 weeks after conception (Dzhala et al., 2005). This increase in KCC2 expression counteracts NKCC1 activity to the point that E_{Cl} becomes negative relative to E_m, thereby rendering GABAergic signals hyperpolarizing. *Kcc2*$^{-/-}$ mice have seizures and die at birth, in part owing to malfunction of the respiratory center (Hubner et al., 2001), as discussed in detail in Chapter 17.

A major question surrounding the developmental shift in polarity of GABAergic responses is what controls the transformation of increased KCC2 activity relative to NKCC1 activity. GABA itself, by acting as a self-limiting trophic factor, might promote this switch by triggering specific intracellular cascades that up-regulate KCC2 gene expression (Ganguly et al., 2001). However, the increase in KCC2 and the negative shift in the GABA reversal potential can take place in the presence of $GABA_A$ receptor antagonists, suggesting that some factor other than GABA is necessary for the transition to take place. Recent work suggests this other factor might be cholinergic signaling. Spontaneous cholinergic activity, by triggering Ca^{2+}-signaling cascades that up-regulate KCC2 downstream of the acetylcholine nicotinic receptor, has been shown to be important for the GABAergic excitation-to-inhibition transition (Liu et al., 2006). Synergistic with the newly established inhibitory GABAergic signaling, cholinergic signaling then triggers later events of neuronal development. Factors acting on tyrosine kinase receptors such as BDNF have also been implicated in the up-regulation of KCC2 during the excitatory-to-inhibitory transition of GABAergic signaling (Aguado et al., 2003).

IV. THE ROLE OF NKCC1 IN NEONATAL SEIZURES

Neonatal seizures, or epileptic episodes suffered by infants in the first 28 days of life, occur in 1 to 2% of patients in neonatal intensive care units, and are the most common manifestation of an acute neurologic disorder in newborn infants (Volpe, 2001). Most commonly caused by hypoxic-ischemic encephalopathy, hemorrhage, or cerebral infarction, the presence of neonatal seizures often portends severe neurological dysfunction later in life, with high rates of adult epilepsy and long-term cognitive and motor deficits in survivors. In animal models, neonatal seizures have been shown to be injurious to the development of the brain, inducing synaptic reorganization, altering synaptic plasticity, and priming cortical neurons to increased damage from seizures sustained later in life (Rennie and Boylan, 2003; Ben-Ari and Holmes, 2006). Thus, the prompt diagnosis and successful treatment of neonatal seizures is important for improving long-term neurologic outcome.

While seizure activity in adults is usually clinically obvious and the EEG reflects coordinated seizure activity, diagnosing seizures in neonates is difficult because seizures are often behaviorally subtle (Silverstein and Jensen, 2007). Because many electrographic neonatal seizures are not clinically evident (Murray et al., 2008), EEG is the current gold standard for diagnosis, with a discharge duration of 10 seconds (versus 3 seconds in older age groups) required to diagnose an electrographic seizure. However, because EEGs are not immediately available in many neonatal intensive care units, the initial diagnosis and treatment of seizures is often based on clinical assessment or single-channel amplitude-integrated EEG (Shellhaas et al., 2007). Standard EEGs are usually performed *after* the administration of anti-epileptic drugs (Silverstein and Jensen, 2007). Unfortunately, it is an all-too-common occurrence for electrographic seizures to persist in encephalopathic neonates despite anti-convulsant drug levels being "therapeutic" (Scher, 2003).

Conventional anti-epileptic drugs have limited utility in treating neonatal seizures (Booth and Evans, 2004). Barbiturates and benzodiazepines, $GABA_A$ receptor agonists that are efficacious for treating adult seizures, are currently among the first-line drugs for neonatal seizures; however, they are often ineffective, and have been shown to actually *potentiate* seizure activity in the immature brain. Phenytoin has been used with similarly limited success. Barbiturates and benzodiazepines have also been known to produce a phenomenon termed *electroclinical dissociation* or *uncoupling* in neonates, whereby the overt clinical manifestations of seizures (e.g. convulsions) are inhibited, but EEG-documented cortical seizure activity is either unaffected or exacerbated (Scher et al., 2003; Murray et al., 2008). This insidious effect of the GABA agonists has the potential for great harm, as it provides physicians with a false sense of security that seizures are under control, while cortical seizure activity – and its associated detrimental effects – rages on. Because the developmental up-regulation of KCC2 expression lags the up-regulation expression of KCC2 in subcortical structures (Stein et al., 2004), one mechanism for electroclinical dissociation is the differential effect of GABA in the cortex (where it is excitatory at birth) vs. brainstem and spinal cord (where, by extrapolation from rodent studies, GABA may be inhibitory at birth). Under these conditions, seizures would be frequently manifested only in the cortex, and administration of GABAergic drugs would only increase the disparity between EEG manifestations of cortical seizure activity and clinical seizures, as is currently widely observed. Because seizure activity is thought to exacerbate pre-existing neuronal injury in experimental (Wirrell et al., 2001) and human (Miller et al., 2002) neonatal seizures, an open question is whether effective treatment of seizures in injured human neonates will improve their poor neurodevelopmental outcome (Tekgul et al., 2006; Pisani et al., 2007).

There have been few prospective studies or randomized, controlled trials of the anti-epileptic drugs that are currently used to treat neonatal seizures. To date, the only randomized trial of anti-epileptic drugs for the treatment of neonatal seizures compared the current first-line drug phenobarbital to phenytoin in the treatment of electrographic seizures (Painter et al., 1999). In this study, the majority of neonates had asphyxia, infarction or hemorrhage as the etiology of their seizures. Complete control of electrographic seizures was achieved with either drug in only ~25% of neonates whose seizure frequency was increasing. Seizure control was achieved in another 15% of newborns when both agents were used concurrently. Preliminary studies of anti-epileptic drugs other than phenobarbital and phenytoin have shown only modest efficacy in smaller cohorts of neonates, though sufficiently powered randomized trials are needed to conclusively demonstrate whether any of these drugs are truly effective. The lack of evidence-based treatment recommendations, coupled with the paucity of data regarding the underlying pathophysiology and natural history of neonatal seizures, has made their current management far from optimal.

Increasing synaptic excitation and/or decreasing synaptic inhibition can cause neurons to become hyperactive. Because neurons harbor a multiplicity of connections with other neurons, the electrical firing of even a small population of hyperexcitable neurons, when synchronized, can progressively entrain larger neural networks until seizures ensue. In neonates, while GABA-mediated excitation plays a role in neuronal development, it also renders the developing brain particularly susceptible to seizures. In the adult cortex, excitatory glutaminergic signaling is balanced by inhibitory GABAergic signaling. However, in the neonate, the "additional" depolarization due to $GABA_A$ receptor activation likely adds to the excitation already initiated by glutamate neurotransmission to tip the balance of excitation/inhibition toward excessive excitation and a propensity to seizure activity. GABA-mediated excitation has been shown to support epileptogenesis in the developing hippocampus (Dzhala and Staley, 2003), and also decrease the seizure threshold of neonates (Khalilov et al., 2003; Khazipov et al., 2004). The excitatory nature of GABA signaling in immature neurons also explains why GABA agonists like barbiturates and benzodiazepines are often ineffective in reducing neonatal seizures, and might even exacerbate electrographic seizure activity. Clearly, new anti-epileptic treatment strategies are needed for neonatal seizures.

Because the higher than passive $[Cl^-]_i$ of immature neurons is due to robust activity of NKCC1, this cotransporter is currently being explored as a target for novel anticonvulsant strategies for neonatal seizures. In theory, inhibition of NKCC1, by reducing $[Cl^-]_i$, could reduce the GABA-mediated excitation of immature neonatal neurons, or even possibly convert the GABA response to inhibitory.

Dzhala and collaborators were the first to test this hypothesis in a recent study (Dzhala et al., 2005). Previous groups had established that NKCC1 expression in rats is highest in cortical neurons during the first postnatal week, begins to decrease at P14, and then drops to the low levels that are found in adults (Plotkin et al., 1997; Wang et al., 2002). Conversely, the expression of KCC2 is minimal at birth in rat cortical neurons is low during the first postnatal week and at P14 attains an expression level that is comparable with adults (Lu et al., 1999). In rats, the higher expression levels of NKCC1 earlier in development correlate with an increased $[Cl^-]_i$ (as judged from measurements of E_{GABA}) and with the presence of excitatory GABAergic signaling (Owens et al., 1996; Yamada et al., 2004). Dzhala and co-workers demonstrated that a similar expression pattern is present in the human cortex, with high NKCC1 and low KCC2 expression during the neonatal time period and before the end of the first year of life – a timescale not indifferent from the rat, when adjusted for the greater length of time required for the development of the human cortex relative to the rodent brain (Dzhala et al., 2005). These data supported the hypothesis that, similar to the rat, GABA is depolarizing and excitatory in immature human cortical neurons, and neonates may be susceptible to seizures due to the excitatory effects of GABA during development. Because NKCC1 is known to establish the elevated levels of $[Cl^-]_i$ that underlie excitatory GABAergic signaling in immature neurons, these data predicted that bumetanide, an inhibitor of NKCC1, might be an effective treatment for neonatal seizures.

This hypothesis was tested by Staley and colleagues in a series of *in vitro* and *in vivo* physiological studies (Dzhala et al., 2005). They found that NKCC1 blockade by bumetanide inhibited cortical seizure activity in neonatal rats both *in vitro* and *in vivo*, and this inhibition was observed at doses that have already been extensively tested in human neonates in diuresis studies (Sullivan et al., 1996; Lopez-Samblas et al., 1997). Specifically, they demonstrated that pharmacologic inhibition of NKCC1 by bumetanide (1) produced a negative shift in E_{GABA}, suggesting a decrease in $[Cl^-]_i$, (2) inhibited GABA-dependent synchronous excitatory activity in the immature hippocampus, (3) suppressed interictal and ictal-like activity in immature hippocampal slices *in vitro* and (4) attenuated kainate-induced seizure activity *in vivo* in neonatal rats. These anti-convulsant effects of bumetanide were shown to be specific, as they were (1) achieved at low doses that selectively block NKCC1, (2) did not affect epileptiform activity in brain slices from *NKCC1*$^{-/-}$ mice (indicating that inhibition of NKCC1 is the mechanism by which bumetanide exerts its GABA-dependent anticonvulsant effects) and (3) did not depress epileptiform activity in mature neurons (where the expression of NKCC1 is less than 10% of that in neonatal tissue; Dzhala et al., 2005).

Given these promising findings, it seemed reasonable to combine bumetanide, which blocks the excitatory effect of GABA in immature neurons (most likely by decreasing $[Cl^-]_i$) with phenobarbital, and a GABA agonist, which opens $GABA_A$ receptor-associated chloride channels. Theoretically, such an increase in GABA-mediated conductance in neurons already targeted by bumetanide would serve to increase shunting inhibition and maximize the anticonvulsant power of the GABA system.

The efficacy of bumetanide, in combination with the GABA-enhancing anticonvulsant phenobarbital, was recently tested for the treatment of recurrent tonic-clonic epileptiform activity in the intact immature hippocampus *in vitro* (Dzhala et al., 2008). In this study, a low-magnesium model of neonatal seizures in the intact immature hippocampal formation *in vitro* was employed. Such a model has the benefits of not altering the energy gradient for NKCC1 and preserving longitudinal intrahippocampal connections. Recurrent seizures were induced in the intact hippocampal preparation by a continuous 5-hour exposure to low-magnesium solution, and the anti-convulsant efficacy of phenobarbital, bumetanide, and the combination of these drugs was then studied. While phenobarbital failed to abolish or depress recurrent seizures in 70% of immature hippocampi, phenobarbital *in combination with bumetanide* abolished seizures in 70% of immature hippocampi, and significantly reduced the frequency, duration and power of seizures in the remaining 30% of immature hippocampi. Taken together, these *in vitro* and *in vivo* studies suggested that bumetanide, alone or in combination with other drugs such as phenobarbital, might be useful in the treatment of neonatal seizures in humans (Dzhala et al., 2005, 2008).

V. ROLE OF NKCC1 AND KCC2 IN TEMPORAL LOBE EPILEPSY AND SEIZURES FOLLOWING HYPOXIC-ISCHEMIC INJURY

Because neurons function in interconnected circuits, increasing neuronal excitation (via increased glutaminergic signaling) or decreasing neuronal inhibition (via decreased inhibitory GABAergic signaling or increased excitatory GABAergic signaling) can cause the circuit to become hyperactive. Due to the myriad connections among neurons, when the electrical firing of even a small population of hyperexcitable neurons becomes synchronized, progressively larger networks become entrained so that seizures ensue. Recent studies have implicated excitatory GABAergic signaling in the genesis of temporal lobe epilepsy and seizures that occur after ischemic-hypoxemic insult. A feature common to both of these disorders is that neurons adopt transmembrane chloride gradients that phenotypically resemble those of immature neurons, rendering GABA activity excitatory instead of inhibitory. This shift results from an pathologic increase in the relative abundance and/or activity of NKCC1 to KCC2, which elevates $[Cl^-]$i above its equilibrium potential.

A. Temporal Lobe Epilepsy

Medically intractable temporal lobe epilepsy (TLE), a common clinical entity, often necessitates the removal of epileptic foci, with the risks of morbidity accompanying a major neurosurgical procedure. Mesial temporal sclerosis (MTS) is a common histological and radiographic finding in the brains of patients with TLE and is characterized by cell loss and gliosis in the CA1 area (i.e. Sommer's sector) of the hippocampal formation with an apparently intact subiculum. Despite years of research, the mechanism underlying epileptigenesis in TLE associated with medial sclerosis is still unknown. Novel remedies are needed for TLE because current pharmacotherapeutics are frequently ineffective or wrought with side effects.

Cohen et al. first suggested that excitatory GABAergic signaling due to perturbed neuronal chloride transport might play a role in the epileptigenesis of TLE (Cohen et al., 2002). In temporal lobe slices from TLE patients with MTS, Cohen et al. demonstrated the presence of excitatory interictal depolarizations in subicular GABAergic neurons. This is pathological, because GABA is normally an inhibitory transmitter in the subiculum and other regions of the adult hippocampus. These subicular GABA-triggered excitatory impulses were shown to serve as a nidus for the spread of other interictal depolarizations in adjacent regions of the hippocampus during seizure activity. Subsequent immunolabeling studies in the human temporal lobe of patients with TLE with MTS showed that in the subicular regions where interictal discharges were recorded by Cohen et al. (2002), there was decreased expression of KCC2 in neurons that expressed NKCC1 and GABAergic hyperinnervation (Muñoz et al., 2007). The authors suggested that in TLE patients with hippocampal sclerosis, the lack of KCC2 in NKCC1-expressing cells may contribute to the depolarizing responses induced by GABAergic signaling through $GABA_A$ receptors in neurons within the subiculum and its transitional region with CA1. Studies *in vitro* and *in vivo* in animals and humans have demonstrated that bumetanide and furosemide are effective at reducing seizure activity (Palma et al., 2006; Huberfeld et al., 2007; Hochman et al., 1995; Hochman and Schwartzkroin, 2000). In patients undergoing temporal lobectomy for TLE associated with MTS, intravenous administration of furosemide sharply reduces interictal spike frequency; and a large epidemiological study has shown that diuretic use is associated with a decreased risk of developing epilepsy (Hesdorffer et al., 2001; Haglund and Hochman, 2005). These data have prompted investigators to explore the specific role of NKCC1 and the K-Cl cotransporters in this disease more closely.

The contribution of altered NKCC1 and KCC2 activity to epileptigenesis has been explored in model systems and in adult humans. Knockout of KCC2 in *Drosophila melanogaster* results in flies having an increased susceptibility for seizures due to excitatory GABAergic signaling (Hekmat-Scafe et al., 2006). $KCC2^{-/-}$ mice exhibit generalized seizures (and die shortly after birth) due to hyperexcitatory GABA responses in the CA1 region of the hippocampus, and exhibit gliosis and neuronal loss in the hippocampus and temporal lobe (Woo et al., 2002). Several groups have demonstrated increased NKCC1 and/or decreased KCC2 expression in the subiculum, compared with the hippocampus proper or the temporal lobe neocortex, in patients with TLE associated with MTS versus normal individuals (Palma et al., 2006; Huberfeld et al., 2007; Muñoz et al., 2007; Sen et al., 2007). Subicular GABA reversal potentials have been shown to be more depolarized compared with the other hippocampal regions in TLE patients with MTS, and bumetanide shifts these depolarizing reversal potentials to more hyperpolarizing values (Palma et al., 2006; Muñoz et al., 2007). A significant number of subicular neurons in hippocampal slices from TLE patients harbor excitatory GABAergic potentials and elicit interictal depolarizations due to a lack of KCC2 expression; bumetanide administration produces a hyperpolarizing shift in E_{GABA} and suppresses epileptic activity in these neurons (Huberfeld et al., 2007). E_{GABA} is shifted toward positive values with respect to E_m immediately after a severe episode of experimental status epilepticus (Khalilov et al., 2003), suggesting that alterations in Cl^- transport might underlie its pathogenesis (Young et al., 1990). Increased NKCC1 and/or altered KCC2 expression has been noted in medically intractable epilepsies associated with malformations of cortical development, including focal cortical dysplasia, hemimegalencephaly and ganglioglioma (Aronica et al., 2007; Munakata et al., 2007; Shimizu-Okabe et al., 2007).

Together, these data suggest that activation of NKCC1-mediated Cl^- uptake and/or decreased KCC2-mediated Cl^- extrusion with the attending change in the neuronal GABA reversal potential drives epileptogenesis and seizure intractability in epilepsy associated with MTS and/or developmental malformations.

B. Hypoxic-Ischemic Seizures

Similar to neonatal seizures and various types of TLE, seizures due to hypoxic-ischemic encephalopathy also respond poorly to anticonvulsants that activate $GABA_A$ receptors (Young et al., 1990). In these patients, seizures do not occur during the acute hypoxic episode, but rather during the period of cerebral reperfusion. In rat models of cerebral hypoxic-ischemic injury (such as oxygen-glucose deprivation (OGD)), re-oxygenation causes a prolonged rise in the intracellular concentration of chloride in hippocampal neurons that renders GABAergic neurons hyperexcitable (Inglefield and Schwartz-Bloom, 1998; Galeffi et al., 2004). This rise in intracellular chloride was associated with an increase in the expression of the phosphorylated, active form of NKCC1, and was completely prevented by administration of low doses of bumetanide (Pond et al., 2006). Bumetanide also prevents OGD-induced neuronal damage and/or cell swelling, and glutamate receptor activation and high extracellular levels of potassium – factors known to contribute to the pathogenesis of ischemic damage and neuronal excitotoxicity – induce the serine-threonine phosphorylation (and activation) of NKCC1 (Schomberg et al., 2001; Su et al., 2002a, 2002b; Beck et al., 2003; Yan et al. 2003; Pond et al., 2004; Chen and Sun, 2005). These data suggest that a pathologic increase in NKCC1 activity due to an increase in cotransporter serine-threonine phosphorylation contributes to the ischemia-induced accumulation of intracellular chloride that triggers the neuronal hyperexcitability, excitotoxicity and injury to neurons following ischemia. These concepts are discussed in depth in Chapter 25 of this volume.

C. Mechanism of Altered CCC Expression in Adult Seizure Syndromes

Whether the changes in NKCC1 and/or KCC2 activity that promote GABAergic hyperexcitability in seizures are the primary events underlying disease pathogenesis, or are merely one part of a complex secondary cascade of events that propagates seizure activity, is unknown. Moreover, the mechanism behind the changes in CCC function and/or expression in these seizure types is unclear. However, it is known that neuronal trauma (van den Pol et al., 1996; Nabekura et al., 2002) as well as excitotoxicity and ischemia lead to a de-differentiation of the neuronal phenotype in regard to ion transport properties and the composition of $GABA_A$ subunits (Cohen et al., 2003; Payne et al., 2003). This "immature" molecular phenotype (including high NKCC1 expression, low KCC2 expression, elevated levels of intracellular Cl^-, and depolarizing GABAergic responses), while potentially advantageous for neuronal repair, might predispose neurons to seizure activity. A critical factor that might explain the shift of adult neurons to an immature phenotype following seizure activity/excitotoxicity (or other stresses like ischemia or trauma) appears to be the brain-derived neurotrophic factor (BDNF) and its tyrosnine-kinase receptor TrkB.

Seizures themselves might produce long-term depolarizing shifts in E_{GABA} by triggering activity-dependent down-regulation of KCC2 (and/or up-regulation of NKCC1) via BDNF signaling through TrkB. Sustained epileptic activity leads to an increase in hippocampal expression of BDNF and the TrkB receptor (Binder et al., 2001; Rivera et al., 2002; Wake et al., 2007). Following *in vivo* seizure kindling, KCC2 shows a rapid, dramatic decrease in expression in the same parts of the epileptic hippocampus that exhibit robust BNDF-TrkB up-regulation. This seizure-induced down-regulation of KCC2 requires the activation of the TrkB receptor and the two major TrkB-mediated signaling cascades, the PLC and Shc activated pathways. Down-regulation of the tyrosine phosphorylation and activity of KCC2 are also seen after seizures, and precedes the BDNF decreases in KCC2 protein or mRNA expression. Whether seizures trigger activity-dependent up-regulation of NKCC1 is currently unknown.

Why do adult neurons stressed by seizures adopt the neonatal features of NKCC1 up-regulation and/or KCC2 down-regulation and the associated depolarizing shifts in GABAergic currents? One hypothesis is that this phenotype is an attempt to provide repair mechanisms and enable the targeting of neurons during damage-related rewiring. In the chronically epileptic human hippocampus, the depolarizing and excitatory GABA responses that characterize subicular interictal activity resemble the early large-scale spontaneous discharges that have been implicated in the formation of neuronal networks. While excitatory GABAergic neurotransmission during development leads to the activation of voltage-gated calcium channels and subsequent increases in intracellular calcium that play a role in neuronal growth, maturation and repair, such excitatory GABA neurotransmission might maladaptively contribute to seizure activity and neuronal damage from excitotoxicity.

VI. CATION-CHLORIDE COTRANSPORTERS AS THERAPEUTIC TARGETS FOR SEIZURES

The CCCs are targets of some of the most common drugs in medicine – the "loop" and "thiazide" diuretics – and are mutated in several inherited human diseases (see Hebert et al., 2004 and Chapters 2, 3 and 16 in this volume). Both bumetanide and furosemide, well-known "loop" diuretics, are capable of inhibiting the cation-chloride cotransporters *in vitro* and *in vivo*. Bumetanide has an approximately 500-fold greater affinity for NKCC1 (inhibition constant [K_i] of approximately 0.1 μM) than for KCC2 (K_i of approximately 25–50 μM). Furosemide inhibits NKCC1 and KCC2 with equal potency (K_i of approximately 25–50 μM). Therefore, at low doses (2–10 μM), bumetanide is a relatively specific inhibitor of NKCC1. The accumulation of bumetanide in the central nervous after systemic administration has not been directly measured, but the drug's high lipid:water partition coefficient and its documented anticonvulsant effects *in vivo* suggest that the drug is able to cross the blood–brain barrier (Dzhala et al., 2005, 2008).

At low concentrations (2–10 μM), bumetanide has well-established pharmacokinetic and pharmacodynamic properties in adult humans with few side effects (Staley, 2006). Because the expression patterns of NKCC1 during development are similar in the human and rat cortex (Dzhala et al., 2005), bumetanide might be useful for the treatment of seizures in human neonates. Bumetanide has been extensively used in both healthy and critically ill human term and pre-term infants to treat fluid volume overload due to cardiac and/or pulmonary disease with few side effects other than minor electrolyte imbalances, so extrapolation from these studies might help guide the design of any potential pilot studies or clinical trials (Sullivan et al., 1996; Lopez-Samblas et al., 1997).

However, it will be important to investigate whether the pharmacokinetics of bumetanide are altered by any of the underlying diseases which are responsible for triggering seizures in neonates, because many term newborns with refractory seizures have hypoxic-ischemic encephalopathy from perinatal asphyxia, which is often accompanied by multiorgan dysfunction (including hepatic and renal failure); such organ dysfunction can dramatically affect drug metabolism (Clancy, 2000; McBride et al., 2000; Tekgul et al., 2006).

Perhaps the best new treatment strategy for neonatal seizures, and one that could be easily tested, is a combination regimen that would include bumetanide with a barbiturate like phenobarbital. In the neonatal rat brain, phenobarbital has been rendered a more effective anticonvulsant by co-administering it with bumetanide, which in theory reverses the Cl^- gradient in immature neurons to a level such that $GABA_A$ potentiation by phenobarbital results in synaptic inhibition (Dzhala et al., 2008). If such a trial were to take place, neonates with persistent seizures despite an initial loading dose of phenobarbital (the current standard of care) could be offered bumetanide along with the second dose of phenobarbital. Continuous EEG monitoring of patients could then be used to determine whether bumetanide reduces seizures compared with controls (i.e. those neonates treated with phenobarbital alone). Pilot studies are now under way evaluating the efficacy of bumetanide administered with phenobarbital for the treatment of neonatal seizures.

The combination of bumetanide and a barbiturate should obviate the need to use the high doses of barbiturate or benzodiazepine that have been associated with significant side effects such as apoptotic neurodegeneration in the developing brain, and late cognitive/behavioral impairment (Farwell et al., 1990; Reinisch et al., 1995; Bittigau et al., 2002, 2003). Moreover, because of its longstanding safe use in newborns as a diuretic, the low doses of bumetanide that are required to inhibit NKCC1 are not anticipated to produce short- or long-term side effects. However, caution must be exercised as work proceeds, and studies should be done to determine any potential side effects of inhibiting NKCC1 in the neonatal nervous system, since GABA-mediated excitation is important for neuronal development (Ben-Ari, 2002). To date, bumetanide-mediated inhibition of NKCC1 in the brain, for periods of time that would far exceed the duration that would be used for the treatment of neonatal seizures, have been shown to have few developmental side effects (Ge et al., 2006; Cancedda et al., 2007). These modest but measurable side effects must be weighed against the well-known detrimental effects of persistent seizures in the immature brain on cortical development, and the absence of knowledge regarding the developmental effects of NKCC1 inhibition in the setting of such seizures (Holmes and Ben Ari, 1998).

A human trial of bumetanide is also taking place in adults with medically intractable and surgically unresectable TLE associated with mesial temporal sclerosis or different cortical malformations (Maa et al., 2007). Bumetanide might also be useful for seizures that occur secondary to hypoxic-ischemic encephalopathy. In these conditions, coupling bumetanide with an agent that inhibits its action in the periphery to limit systemic side effects (e.g. renal electrolyte or fluid disturbances) might optimize desired pharmacotherapeutic effects.

Currently, there are no drugs that are specific inhibitors of K-Cl cotransport, nor are there drugs that differentiate between multiple K-Cl cotransporter isoforms. Furosemide and DIOA (R(+)-[(2-n-butyl-6, 7-dichloro-2-cyclopentyl-2,3-dihydro-1-oxo-1H-inden-5-yl)oxy] acetic acid) have been used to inhibit K-Cl cotransporters experimentally *in vitro*, but are more potent blockers of NKCC1 than of KCC2 and also inhibit carbonic anhydrases (Payne et al., 2003). What may be more clinically efficacious than K-Cl inhibitors are K-Cl cotransporter agonists. A specific KCC2 agonist might, in combination with NKCC1 inhibitors (like bumetanide), synergistically augment GABA-mediated inhibition for the treatment of seizures or other conditions in which neuronal hyperexcitability is due to derangement in cellular Cl^- gradients. Augmenting KCC2 activity might be challenging, as it involves designing drugs that increase the functional expression of this protein. However, because KCC2 is repressed by BDNF, an indirect method of increasing KCC2 expression might be to inhibit BDNF or its receptor on neurons (TrkB).

Finally, the endogenous regulators of NKCC1 and the K-Cl cotransporters should be explored for their roles in the pathogenesis of diseases which exhibit altered CCC activity. BDNF, released upon seizure-like activity, causes down-regulation of KCC2 which results in decreased Cl^- extrusion capacity of neurons that propagates further seizure episodes (Rivera et al., 2004). Because inhibition of BDNF action via blockage of the TrkB receptor reverses depolarizing shifts in E_{GABA} (Cordero-Erausquin et al., 2005), inhibitors of TrkB, by increasing the endogenous expression of KCC2, might help treat different seizure syndromes. Additionally, targeting the volume/chloride-sensitive regulatory kinases of the CCCs, the WNKs or SPAK/OSR1 might have utility. The role of these kinases in the mammalian nervous system is only beginning to be explored; however, *in vitro* experiments in mammalian cells and *in vivo* experiments in lower organisms suggest that these serine-threonine kinases – initially characterized in the kidney – might prove relevant for CNS physiology (Kahle et al., 2005). Moreover, drugs that target the CCCs in the nervous system might have specific advantages over current drugs for seizures, because they would restore endogenous GABAergic inhibition rather than actively depressing excitability by imposing exogenous inhibition (e.g. by raising GABA levels or triggering excessive stimulation of the $GABA_A$ receptor). This might yield more specific treatments with fewer detrimental side effects (De Koninck, 2007).

References

Adragna, N.C., Di Fulvio, M., and Lauf, P.K. (2004). Regulation of K-Cl cotransport: from function to genes. *J. Membr. Biol.* **201**, 109–137.

Aguado, F., Aguado, F., Carmona, M.A., Pozas, E., Aguiló, A., Martínez-Guijarro, F.J., Alcantara, S., Borrell, V., Yuste, R., Ibañez, C.F., and Soriano, E. (2003). BDNF regulates spontaneous correlated activity at early developmental stages by increasing synaptogenesis and expression of the K^+/Cl^- co-transporter KCC2. *Development.* **130**, 1267–1280.

Andersen, P., Brooks, C.M., Eccles, J.C., and Sears, T.A. (1964). The ventro-basal nucleus of the thalamus: potential fields, synaptic transmission and excitability of both presynaptic and postsynaptic components. *J. Physiol.* **174**, 348–369.

Aronica, E., Boer, K., Redeker, S., Spliet, W.G., van Rijen, P.C., Troost, D., and Gorter, J.A. (2007). Differential expression patterns of chloride transporters, Na^+-K^+-$2Cl^-$-cotransporter and K^+-Cl^--cotransporter, in epilepsy-associated malformations of cortical development. *Neuroscience* **145**, 185–196.

Beck, J., Lenart, B., Kintner, D.B., and Sun, D. (2003). Na-K-Cl cotransporter contributes to glutamate-mediated excitotoxicity. *J. Neurosci.* **23**, 5061–5068.

Ben-Ari, Y. (2001). Developing networks play a similar melody. *Trends Neurosci.* **24**, 353–360.

Ben-Ari, Y. (2002). Excitatory actions of GABA during development: the nature of the nurture. *Nat. Rev. Neurosci.* **3**, 728–739.

Ben-Ari, Y. and Holmes, G.L. (2006). Effects of seizures on developmental processes in the immature brain. *Lancet Neurol.* **5**, 1055–1063.

Binder, D.K., Croll, S.D., Gall, C.M., and Scharfman, H.E. (2001). BDNF and epilepsy: too much of a good thing? *Trends Neurosci.* **24**, 47–53.

Bittigau, P., Sifringer, M., Genz, K., Reith, E., Pospischil, D., Govindarajalu, S., Dzietko, M., Pesditschek, S., Mai, I., Dikranian, K., Olney, J.W., and Ikonomidou, C. (2002). Antiepileptic drugs and apoptotic neurodegeneration in the developing brain. *Proc. Natl. Acad. Sci. USA* **99**, 15089–15094.

Bittigau, P., Sifringer, M., and Ikonomidou, C. (2003). Antiepileptic drugs and apoptosis in the developing brain. *Ann. NY Acad. Sci.* **993**, 103–114.

Booth, D. and Evans, D.J. (2004). Anticonvulsants for neonates with seizures. *Cochrane Database Syst. Rev.* 2004; CD004218.

Cancedda, L., Fiumelli, H., Chen, K., and Poo, M.M. (2007). Excitatory GABA action is essential for morphological maturation of cortical neurons in vivo. *J. Neurosci.* **27**, 5224–5235.

Chen, H. and Sun, D. (2005). The role of Na-K-Cl co-transporter in cerebral ischemia. *Neurol. Res.* **27**, 280–286.

Clancy, R.R. (2006). Summary proceedings from the neurology group on neonatal seizures. *Pediatrics* **117**, 23–27.

Clayton, G.H., Owens, G.C., Wolff, J.S., and Smith, R.L. (1998). Ontogeny of cation-Cl^- cotransporter expression in rat neocortex. *Brain Res. Dev. Brain Res.* **109**, 281–292.

Cohen, I., Navarro, V., Clemenceau, S., Baulac, M., and Miles, R. (2002). On the origin of interictal activity in human temporal lobe epilepsy in vitro. *Science* **298**, 1418–1421.

Cohen, I., Navarro, V., Le Duigou, C., and Miles, R. (2003). Mesial temporal lobe epilepsy: a pathological replay of developmental mechanisms? *Biol. Cell* **95**, 329–333.

Cordero-Erausquin, M., Coull, J.A., Boudreau, D., Rolland, M., and De Koninck, Y. (2005). Differential maturation of GABA action and anion reversal potential in spinal lamina I neurons: impact of chloride extrusion capacity. *J. Neurosci.* **25**, 9613–9623.

De Koninck, Y. (2007). Altered chloride homeostasis in neurological disorders: a new target. *Curr. Opinion in Pharmacol.* **7**, 93–99.

Delpire, E. and Gagnon, K.B. (2008). SPAK and OSR1: STE20 kinases involved in the regulation of ion homoeostasis and volume control in mammalian cells. *Biochem. J.* **409**, 321–331.

Dzhala, V.I. and Staley, K.J. (2003). Excitatory actions of endogenously released GABA contribute to initiation of ictal epileptiform activity in the developing hippocampus. *J. Neurosc.* **23**, 1840–1846.

Dzhala, V.I. et al. (2005). NKCC1 transporter facilitates seizures in the developing brain. *Nature Med.* **11**, 1205–1213.

Dzhala, V.I., Brumback, A.C., and Staley, K.J. (2008). Bumetanide enhances phenobarbital efficacy in a neonatal seizure model. *Ann. Neurol.* **63**, 222–235.

Farwell, J.R., Lee, Y.J., Hirtz, D.G. et al. (1990). Phenobarbital for febrile seizures – effects on intelligence and on seizure recurrence. *N. Engl. J. Med.* **322**, 364–369.

Flatman, P.W. (2002). Regulation of Na-K-2Cl cotransport by phosphorylation and protein-protein interactions. *Biochim. Biophys. Acta* **1566**, 140–151.

Galeffi, F., Sah, R., Pond, B.B., George, A., and Schwartz-Bloom, R.D. (2004). Changes in intracellular chloride after oxygen-glucose deprivation of the adult hippocampal slice: effect of diazepam. *J. Neurosci.* **24**, 4478–4488.

Gamba, G. (2005). Molecular physiology and pathophysiology of electroneutral cation-chloride cotransporters. *Physiol. Rev.* **85**, 423–493.

Ganguly, K., Schinder, A.F., Wong, S.T., and Poo, M. (2001). GABA itself promotes the developmental switch of neuronal GABAergic responses from excitation to inhibition. *Cell* **105**, 521–532.

Ge, S., Goh, E.L., Sailor, K.A., Kitabatake, Y., Ming, G.L., and Song, H. (2006). GABA regulates synaptic integration of newly generated neurons in the adult brain. *Nature* **439**, 589–593.

Haglund, M.M. and Hochman, D.W. (2005). Furosemide and mannitol suppression of epileptic activity in the human brain. *J. Neurophysiol.* **94**, 907–918.

Hebert, S.C., Mount, D.B., and Gamba, G. (2004). Molecular physiology of cation-coupled Cl^- cotransport: the SLC12 family. *Pflugers Arch.* **447**, 580–593.

Hekmat-Scafe, D.S., Lundy, M.Y., Ranga, R., and Tanouye, M.A. (2006). Mutations in the K^+/Cl^- cotransporter gene kazachoc (kcc) increase seizure susceptibility in Drosophila. *J. Neurosci.* **26**, 8943–8954.

Hesdorffer, D.C., Stables, J.P., Hauser, W.A., Annegers, J.F., and Cascino, G. (2001). Are certain diuretics also anticonvulsants? *Ann. Neurol.* **50**, 458–462.

Hochman, D.W., Baraban, S.C., Owens, J.W., and Schwartzkroin, P.A. (1995). Dissociation of synchronization and excitability in furosemide blockade of epileptiform activity. *Science* **270**, 99–102.

Hochman, D.W. and Schwartzkroin, P.A. (2000). Chloride-cotransport blockade desynchronizes neuronal discharge in the "epileptic" hippocampal slice. *J. Neurophysiol.* **8**, 406–417.

Holmes, G.L. and Ben Ari, Y. (1998). Seizures in the developing brain: perhaps not so benign after all. *Neuron* **21**, 1231–1234.

Huang, E.J. and Reichardt, L.F. (2003). Trk receptors: roles in neuronal signal transduction. *Ann. Rev. Biochem.* **72**, 609–642.

Huberfeld, G., Wittner, L., Clemenceau, S., Baulac, M., Kaila, K., Miles, R., and Rivera, C. (2007). Perturbed chloride homeostasis and GABAergic signaling in human temporal lobe epilepsy. *J. Neurosci.* **27**, 9866–9873.

Hubner, C.A., Stein, V., Hermans-Borgmeyer, I., Meyer, T., Ballanyi, K., and Jentsch, T.J. (2001). Disruption of KCC2 reveals an essential role of K-Cl cotransport already in early synaptic inhibition. *Neuron* **30**, 515–524.

Inglefield, J.R. and Schwartz-Bloom, R.D. (1998). Activation of excitatory amino acid receptors in the rat hippocampal slice increases intracellular Cl^- and cell volume. *J. Neurochem.* **71**, 1396–1404.

Jean-Xavier, C., Mentis, G.Z., O'Donovan, M.J., Cattaert, D., and Vinay, L. (2007). Dual personality of GABA/glycine-mediated depolarizations in immature spinal cord. *Proc. Natl. Acad. Sci. USA* **104**, 11477–11482.

Kahle, K.T., Rinehart, J., de Los Heros, P., Louvi, A., Meade, P., Vazquez, N., Hebert, S.C., Gamba, G., Gimenez, I., and Lifton, R.P. (2005). WNK3 modulates transport of Cl− in and out of cells: implications for control of cell volume and neuronal excitability. *Proc. Natl. Acad. Sci. USA* **102**, 16783–16788.

Kahle, K.T., Rinehart, J., Ring, A., Gimenez, I., Gamba, G., Hebert, S.C., and Lifton, R.P. (2006). WNK protein kinases modulate cellular Cl^- flux by altering the phosphorylation state of the Na-K-Cl and K-Cl cotransporters. *Physiology (Bethesda)* **21**, 326–335.

Kahle, K.T., Ring, A.M., and Lifton, R.P. (2008). Molecular physiology of the WNK kinases. *Annu. Rev. Physiol.* **70**, 329–355.

Karadsheh, M.F., Byun, N., Mount, D.B., and Delpire, E. (2004). Localization of the KCC4 potassium-chloride cotransporter in the nervous system. *Neuroscience* **123**, 381–391.

Khalilov, I., Holmes, G.L., and Ben-Ari, Y. (2003). In vitro formation of a secondary epileptogenic mirror focus by interhippocampal propagation of seizures. *Nat. Neurosci.* **6**, 1079–1085.

Khazipov, R., Khalilov, I., Tyzio, R., Morozova, E., Ben-Ari, Y., and Holmes, G.L. (2004). Developmental changes in GABAergic actions and seizure susceptibility in the rat hippocampus. *Eur. J. Neurosci.* **19**, 590–600.

Liu, Z., Neff, R.A., and Berg, D.K. (2006). Sequential interplay of nicotinic and GABAergic signaling guides neuronal development. *Science* **314**, 1610–1613.

Lopez-Samblas, A.M., Adams, J.A., Goldberg, R.N., and Modi, M.W. (1997). The pharmacokinetics of bumetanide in the newborn infant. *Biol. Neonate* **72**, 265–272.

Lu, J., Karadsheh, M., and Delpire, E. (1999). Developmental regulation of the neuronal-specific isoform of K-Cl cotransporter KCC2 in postnatal rat brains. *J. Neurobiol.* **39**, 558–568.

Maa, E., Bainbridge, J., Spitz, M.C., and Staley, K.J. (2007). Oral bumetanide add-on therapy in refractory temporal lobe epilepsy. *Epilepsia* **48** (Suppl. 6) [abstract # 3.222].

McBride, M.C., Laroia, N., and Guillet, R. (2000). Electrographic seizures in neonates correlate with poor neurodevelopmental outcome. *Neurology* **55**, 506–513.

Mercado, A., Broumand, V., Zandi-Nejad, K., Enck, A.H., and Mount, D.B. (2006). A C-terminal domain in KCC2 confers constitutive K^+-Cl^- cotransport. *J. Biol. Chem.* **281**, 1016–1026.

Miller, S.P., Weiss, J., Barnwell, A., and Ferriero, D.M. (2002). Seizure-associated brain injury in term newborns with perinatal asphyxia. *Neurology* **58**, 542–548.

Munakata, M., Watanabe, M., Otsuki, T., Nakama, H., Arima, K., Itoh, M., Nabekura, J., Iinuma, K., and Tsuchiya, S. (2007). Altered distribution of KCC2 in cortical dysplasia in patients with intractable epilepsy. *Epilepsia* **48**, 837–844.

Muñoz, A., Mendez, P., DeFelipe, J., and Alvarez-Leefmans, F.J. (2007). Cation-chloride cotransporters and GABA-ergic innervation in the human epileptic hippocampus. *Epilepsia* **48**, 663–673.

Murray, D.M., Boylan, G.B., Ali, I., Ryan, C.A., Murphy, B.P., and Connolly, S. (2008). Defining the gap between electrographic seizure burden, clinical expression and staff recognition of neonatal seizures. *Arch. Dis. Child Fetal Neonatal Ed.* **93**, F187–F191.

Nabekura, J., Ueno, T., Okabe, A., Furuta, A., Iwaki, T., Shimizu-Okabe, C., Fukuda, A., and Akaike, N. (2002). Reduction of KCC2 expression and GABAA receptor-mediated excitation after in vivo axonal injury. *J. Neurosci.* **22**, 4412–4417.

Owens, D.F., Boyce, L.H., Davis, M.B., and Kriegstein, A.R. (1996). Excitatory GABA responses in embryonic and neonatal cortical slices demonstrated by gramicidin perforated-patch recordings and calcium imaging. *J. Neurosci.* **16**, 6414–6423.

Painter, M.J., Scher, M.S., Stein, A.D., Armatti, S., Wang, Z., Gardner, J.C., Paneth, N., Minnigh, B., and Alvin, J. (1999). Phenobarbital compared with phenytoin for the treatment of neonatal seizures. *N. Engl. J. Med.* **341**, 485–489.

Palma, E., Amici, M., Sobrero, F., Spinelli, G., Di Angelantonio, S., Ragozzino, D., Mascia, A., Scoppetta, C., Esposito, V., Miledi, R., and Eusebi, F. (2006). Anomalous levels of Cl− transporters in the hippocampal subiculum from temporal lobe epilepsy patients make GABA excitatory. *Proc. Natl. Acad. Sci. USA* **103**, 8465–8468.

Pathak, H.R., Weissinger, F., Terunuma, M., Carlson, G.C., Hsu, F.C., Moss, S.J., and Coulter, D.A. (2007). Disrupted dentate granule cell chloride regulation enhances synaptic excitability during development of temporal lobe epilepsy. *J. Neurosci.* **27**, 14012–14022.

Payne, J.A., Stevenson, T.J., and Donaldson, L.F. (1996). Molecular characterization of a putative K-Cl cotransporter in rat brain. A neuronal-specific isoform. *J. Biol. Chem.* **271**, 16245–16252.

Payne, J.A., Rivera, C., Voipio, J., and Kaila, K. (2003). Cation-chloride co-transporters in neuronal communication, development and trauma. *Trends Neurosci.* **26**, 199–206.

Pearson, M.M., Lu, J., Mount, D.B., and Delpire, E. (2001). Localization of the K^+-Cl^- cotransporter, KCC3, in the central and peripheral nervous systems: expression in the choroid plexus, large neurons and white matter tracts. *Neuroscience* **103**, 481–491.

Pisani, F., Cerminara, C., Fusco, C., and Sisti, L. (2007). Neonatal status epilepticus vs recurrent neonatal seizures: clinical findings and outcome. *Neurology* **69**, 2177–2185.

Plotkin, M.D., Kaplan, M.R., Peterson, L.N., Gullans, S.R., Hebert, S.C., and Delpire, E. (1997). Expression of the Na^+-K^+-$2Cl^-$ cotransporter BSC2 in the nervous system. *Am. J. Physiol.* **272**, C173–C183.

Pond, B.B., Galeffi, F., Ahrens, R., and Schwartz-Bloom, R.D. (2004). Chloride transport inhibitors influence recovery from oxygen-glucose deprivation-induced cellular injury in adult hippocampus. *Neuropharmacology* **47**, 222–253.

Pond, B.B., Berglund, K., Kuner, T., Feng, G., Augustine, G.J., and Schwartz-Bloom, R.D. (2006). The chloride transporter Na^+-K^+-Cl^- cotransporter isoform-1 contributes to intracellular chloride increases after in vitro ischemia. *J. Neurosci.* **26**, 1396–1406.

Reinisch, J.M., Sanders, S.A., Mortensen, E.L., and Rubin, D.B. (1995). In utero exposure to phenobarbital and intelligence deficits in adult men. *Jama* **274**, 1518–1525.

Rennie, J.M. and Boylan, G.B. (2003). Neonatal seizures and their treatment. *Curr. Opin. Neurol.* **16**, 177–181.

Reynolds, A., Brustein, E., Liao, M., Mercado, A., Babilonia, E., Mount, D.B., and Drapeau, P. (2008). Neurogenic role of the depolarizing chloride gradient revealed by global overexpression of KCC2 from the onset of development. *J. Neurosci.* **28**, 1588–1597.

Rivera, C., Voipio, J., Payne, J.A., Ruusuvuori, E., Lahtinen, H., Lamsa, K., Pirvola, U., Saarma, M., and Kaila, K. (1999). The K^+/Cl^- co-transporter KCC2 renders GABA hyperpolarizing during neuronal maturation. *Nature* **397**, 251–255.

Rivera, C., Voipio, J., Thomas-Crusells, J., Li, H., Emri, Z., Sipila, S., Payne, J.A., Minichiello, L., Saarma, M., and Kaila, K. (2004). Mechanism of activity-dependent downregulation of the neuron-specific K-Cl cotransporter KCC2. *J. Neurosci.* **24**, 4683–4691.

Rivera, C., Li, H., Thomas-Crusells, J., Lahtinen, H., Viitanen, T., Nanobashvili, A., Kokaia, Z., Airaksinen, M.S., Voipio, J., Kaila, K., and Saarma, M. (2002). BDNF-induced TrkB activation down-regulates the K^+-Cl^- cotransporter KCC2 and impairs neuronal Cl− extrusion. *J. Cell Biol.* **159**, 747–752.

Scher, M.S., Alvin, J., Gaus, L., Minnigh, B., and Painter, M.J. (2003). Uncoupling of EEG-clinical neonatal seizures after antiepileptic drug use. *Pediatr. Neuro.* **28**, 277–280.

Schomberg, S.L., Su, G., Haworth, R.A., and Sun, D. (2001). Stimulation of Na-K-2Cl cotransporter in neurons by activation of non-NMDA ionotropic receptor and group-I mGluRs. *J. Neurophysiol.* **85**, 2563–2575.

Sen, A., Martinian, L., Nikolic, M., Walker, M.C., Thom, M., and Sisodiya, S.M. (2007). Increased NKCC1 expression in refractory human epilepsy. *Epilepsy Res.* **74**, 220–227.

Shellhaas, R.A., Soaita, A.I., and Clancy, R.R. (2007). Sensitivity of amplitude-integrated electroencephalography for neonatal seizure detection. *Pediatrics* **120**, 770–777.

Shimizu-Okabe, C., Okabe, A., Kilb, W., Sato, K., Luhmann, H.J., and Fukuda, A. (2007). Changes in the expression of cation-Cl− cotransporters, NKCC1 and KCC2, during cortical malformation induced by neonatal freeze-lesion. *Neurosci. Res.* **59**, 288–295.

Silverstein, F.S. and Jensen, F.E. (2007). Neonatal seizures. *Ann. Neurol.* **62**, 112–120.

Song, L., Mercado, A., Vazquez, N., Xie, Q., Desai, R., George, A.L., Jr., Gamba, G., and Mount, D.B. (2002). Molecular, functional, and genomic characterization of human KCC2, the neuronal K-Cl cotransporter. *Brain Res. Mol. Brain Res.* **103**, 91–105.

Staley, K.J. and Mody, I. (1992). Shunting of excitatory input to dentate gyrus granule cells by a depolarizing GABAA receptor-mediated postsynaptic conductance. *J. Neurophysiol.* **68**, 197–212.

Staley, K.J. (2006). Wrong-way chloride transport: is it a treatable cause of some intractable seizures? *Epilepsy Curr.* **6**, 124–127.

Stein, V., Hermans-Borgmeyer, I., Jentsch, T.J., and Hubner, C.A. (2004). Expression of the KCl cotransporter KCC2 parallels neuronal maturation and the emergence of low intracellular chloride. *J. Comp. Neurol.* **468**, 57–64.

Su, G., Kintner, D.B., Flagella, M., Shull, G.E., and Sun, D. (2002a). Astrocytes from Na^+-K^+-Cl^- cotransporter-null mice exhibit absence of swelling and decrease in EAA release. *Am. J. Physiol. Cell Physiol.* **282**, C1147–C1160.

Su, G., Kintner, D.B., and Sun, D. (2002b). Contribution of Na^+-K^+-Cl^- cotransporter to high-$[K^+]_o$-induced swelling and EAA release in astrocytes. *Am. J. Physiol. Cell Physiol.* **282**, C1136–C1146.

Sullivan, J.E., Witte, M.K., Yamashita, T.S., Myers, C.M., and Blumer, J.L. (1996). Pharmacokinetics of bumetanide in critically ill infants. *Clin. Pharmacol. Ther.* **60**, 405–413.

Tekgul, H., Gauvreau, K., Soul, J., Murphy, L., Robertson, R., Stewart, J., Volpe, J., Bourgeois, B., and du Plessis, A.J. (2006). The current etiologic profile and neurodevelopmental outcome of seizures in term newborn infants. *Pediatrics* **117**, 1270–1280.

van den Pol, A.N., Obrietan, K., and Chen, G. (1996). Excitatory actions of GABA after neuronal trauma. *J. Neurosci.* **16**, 4283–4292.

Volpe, J.J. (2001). Neonatal seizures. In *Neurology of the Newborn*, pp. 178–214. W.B. Saunders, Philadelphia.

Wake, H., Watanabe, M., Moorhouse, A.J., Kanematsu, T., Horibe, S., Matsukawa, N., Asai, K., Ojika, K., Hirata, M., and Nabekura, J. (2007). Early changes in KCC2 phosphorylation in response to neuronal stress result in functional downregulation. *J. Neurosci.* **27**, 1642–1650.

Wang, C., Shimizu-Okabe, C., Watanabe, K., Okabe, A., Matsuzaki, H., Ogawa, T., Mori, N., Fukuda, A., and Sato, K. (2002). Developmental changes in KCC1, KCC2, and NKCC1 mRNA expressions in the rat brain. *Brain Res. Dev. Brain Res.* **139**, 59–66.

Wang, D.D. and Kriegstein, A.R. (2008). GABA regulates excitatory synapse formation in the neocortex via NMDA receptor activation. *J. Neurosci.* **28**, 5547–5558.

Wirrell, E.C., Armstrong, E.A., Osman, L.D., and Yager, J.Y. (2001). Prolonged seizures exacerbate perinatal hypoxic-ischemic brain damage. *Pediatr. Res.* **50**, 445–454.

Woo, N.S., Lu, J., England, R., McClellan, R., Dufour, S., Mount, D.B., Deutch, A.Y., Lovinger, D.M., and Delpire, E. (2002). Hyperexcitability and epilepsy associated with disruption of the mouse neuronal-specific K-Cl cotransporter gene. *Hippocampus* **12**, 258–268.

Yamada, J., Okabe, A., Toyoda, H., Kilb, W., Luhmann, H.J., and Fukuda, A. (2004). Cl− uptake promoting depolarizing GABA actions in immature rat neocortical neurones is mediated by NKCC1. *J. Physiol.* **557**, 829–841.

Yan, Y., Dempsey, R.J., Flemmer, A., Forbush, B., and Sun, D. (2003). Inhibition of Na^+-K^+-Cl^- cotransporter during focal cerebral ischemia decreases edema and neuronal damage. *Brain Res.* **961**, 22–31.

Young, G.B., Gilbert, J.J., and Zochodne, D.W. (1990). The significance of myoclonic status epilepticus in postanoxic coma. *Neurology* **40**, 1843–1848.

CHAPTER

25

The Role of Cation-Chloride Transporters in Brain Ischemia

Dandan Sun, Douglas B. Kintner and Brooks B. Pond

OUTLINE

I. INTRODUCTION

Stroke is the third leading cause of death and disability in the USA. Presently, the only FDA-approved treatment for ischemic stroke is the clot-busting drug tissue plasminogen activator (tPA), which must be administered within a three-hour window from the onset of symptoms to be effective. Only 3 to 5% of stroke patients are candidates for this treatment (Qureshi, 1996). The lack of effective treatments for stroke demands a better understanding of the cellular and molecular mechanisms underlying cell death following ischemia.

Ischemic stroke is the inevitable consequence of transient or permanent reduction of regional cerebral blood flow to brain tissue. Because of the high energy demand of brain tissue, a reduction of blood flow below ~50% will cause serious perturbations in tissue metabolism (Hossmann, 1994). The resulting tissue damage will depend on the severity and length of the brain perfusion deficit. The brain damage in the infarct core is considered to be irreversible in focal stroke. Within minutes, there is a complete collapse of energy production, dissipation of ion gradients, cessation of macromolecular synthesis and loss of cell structure (Lo et al., 2003). On the other hand, the penumbra, which surrounds the infarct core, retains residual blood flow and can remain viable for hours or even days (Mongin, 2007). The penumbra tissue is the target for therapeutic intervention.

Loss of ion homeostasis plays a central role in pathogenesis of ischemic cell damage. Ischemia-induced perturbation of ion homeostasis leads to intracellular accumulation of Na^+ and Ca^{2+} and subsequent activation of proteases, phospholipases and formation of oxygen and nitrogen-free radicals (Lipton, 1999; Siesjo, 1992). This cascade of signal transduction events results in long-term functional and structural changes in membrane and cytoskeletal integrity and eventual cell death (Lipton, 1999; Siesjo, 1992). Secondary active ion transport proteins are important in maintaining steady-state intracellular ionic concentrations. These include the Na^+-K^+-$2Cl^-$ cotransporter isoform 1 (NKCC1), K^+-Cl^- cotransporters (KCCs) and the Na^+/Ca^{2+} exchanger (NCX). During ischemia, cells lose K^+ and gain Ca^{2+}, Na^+, Cl^-, along with H_2O (Siesjo, 1992). This may result from a halt in the ATP synthesis and extensive Na^+-K^+-ATPase pump failure after anoxia. Both ion conductances and ion transporters could also affect ionic homeostasis during ischemia. Considerable research effort has been centered on the roles of passive fluxes via cation and anion conductances in dissipation of the ion concentration gradients (Lipton, 1999; Mongin, 2007). This chapter will focus on the recent studies on the role of cation-chloride cotransporters in ischemia-induced dissipation of ionic homeostasis. Results from both *in vitro* and *in vivo* experimental studies suggest that these ion transport proteins are potential targets to reduce or prevent ischemia-mediated loss of ionic homeostasis.

II. Cl^- TRANSPORTERS IN DISSOCIATED BRAIN CELLS FOLLOWING *IN VITRO* ISCHEMIA

A. Cl^- Transporters in Neuronal Ionic Dysregulation, Swelling and Death

Cl^- movement has been shown to be a central component of the acute excitotoxic response in neurons (Nicklas et al., 1987; Olney et al., 1986; Rothman, 1985). Acute excitotoxicity is thought to be mediated by excessive depolarization of the postsynaptic membrane. Depolarization results in an osmotic imbalance which is countered by an influx of Cl^-, Na^+ and water and leads to cell lysis. Removal or reduction of Cl^- from the extracellular medium during excitatory amino acid (EAA) exposure completely eliminates the acute excitotoxic response in hippocampal (Rothman, 1985) and retinal neurons (Nicklas et al., 1987; Olney et al., 1986).

NKCC1 is important in maintenance of intracellular Cl^- ($[Cl^-]_i$) and contributes to GABA-mediated depolarization in adult sensory neurons (Alvarez-Leefmans et al., 1988; Alvarez-Leefmans, 2001; Sung et al., 2000) and immature CNS neurons (Misgeld et al., 1986; Sun and Murali, 1998). Thus, NKCC1 may affect neuronal excitability through regulation of $[Cl^-]_i$ (Jang et al., 2001; Sung et al., 2000). In addition, activation of the ionotropic glutamate NMDA receptor, the AMPA receptor and the metabotropic glutamate receptor (group-I) significantly stimulates NKCC1 activity in SH-SY5Y neuroblastoma cells (Sun and Murali, 1998) (Fig. 25.1). Furthermore, glutamate treatment of rat cortical neurons for 24 h results in ~50% cell death (Fig. 25.2). This is accompanied by an increase in $[Na^+]_i$ and in ^{36}Cl content (Beck et al., 2003). Interestingly, inhibition of NKCC1 activity with its potent inhibitor bumetanide abolishes glutamate-mediated neurotoxicity and significantly attenuates oxygen and glucose deprivation (OGD)-induced neuronal death (Beck et al., 2003). Furthermore, NMDA-mediated increases in ^{36}Cl

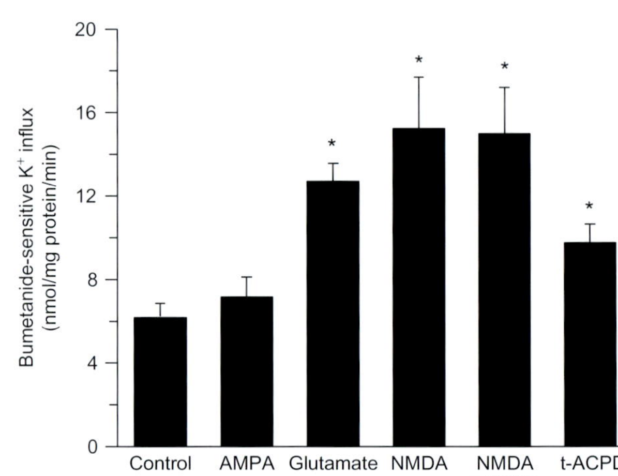

FIGURE 25.1 Stimulation of NKCC1 by activation of both ionotropic and metabotropic glutamate receptors. SH-SY5Y cells were preincubated in AMPA, glutamate, NMDA, trans-ACPD or NMDA plus trans-ACPD for 10 min. Bumetanide-sensitive ^{86}Rb influx rates were determined over the next 3 min. Data are means ± SE, * $p < 0.05$ vs. control. Adapted from Sun and Murali (1998).

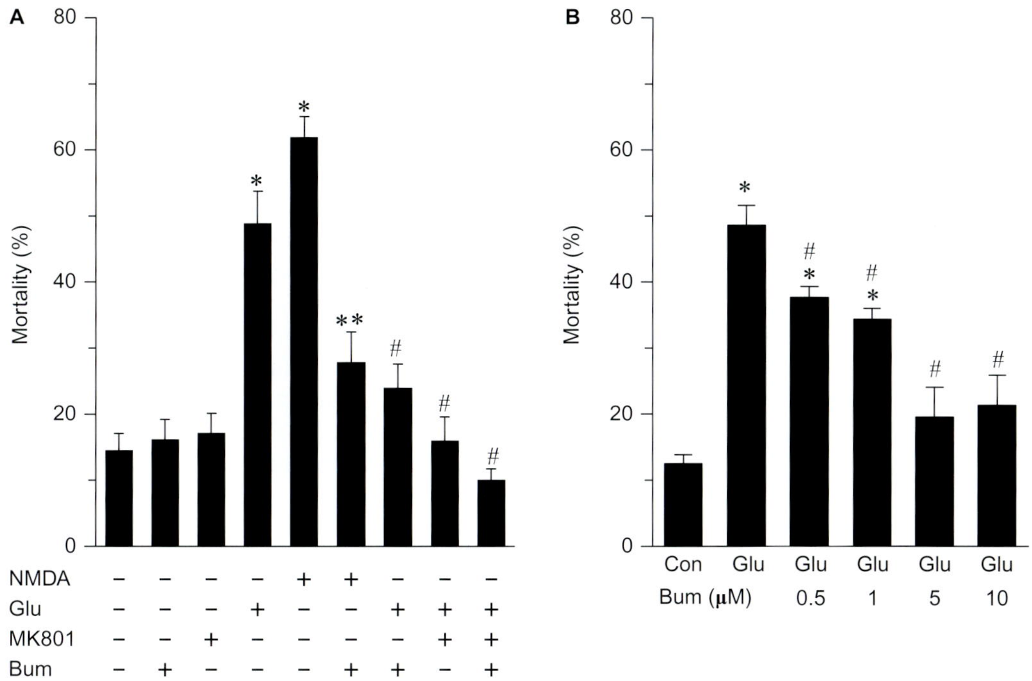

FIGURE 25.2 Inhibition of NKCC1 decreases glutamate-mediated neurotoxicity. Cell mortality in cultured neurons was assessed by propidium iodide and calcein staining after 24 h treatment with glutamate or NMDA. **A.** Data are means ± SE. * $p < 0.05$ vs. control, ** $p < 0.05$ vs. NMDA, # $p < 0.05$ vs. glutamate. **B.** Bumetanide was added 30 min prior to the glutamate treatment and cells were assessed at 24 h post-treatment. * $p < 0.05$ vs. control, # $p < 0.05$ vs. glutamate. Adapted from Beck et al. (2003).

content and intracellular Na^+ overload are reduced in the presence of bumetanide (Beck et al., 2003). Bumetanide also blocks NMDA-induced neuronal swelling (Beck et al., 2003). Thus, NKCC1 may contribute to Na^+ and Cl^- overload during glutamate-mediated acute excitotoxicity. This view is further supported by the similar levels of neuroprotection exhibited in neurons from NKCC1 null ($NKCC1^{-/-}$) mice (Chen et al., 2005). No additional protective effects were found when 10 μM bumetanide is administrated in $NKCC1^{-/-}$ neurons (Chen et al., 2005), indicating that neuroprotection exhibited in $NKCC1^{-/-}$ neurons indeed result from loss of NKCC1 function. In addition, in neurons with injured axons, Cl^- accumulation is attributable to NKCC1-mediated Cl^- influx and a decrease in Cl^- extrusion by KCC2 (Nabekura et al., 2002). Cl^- accumulation causes a switch in GABA action from inhibitory to excitatory and leads to excitotoxicity (Nabekura et al., 2002). Blockage of Na^+ and Cl^- entry by removal of extracellular Na^+ and Cl^- abolishes excitotoxic dendritic injury and NMDA-mediated neurodegeneration in retinal ganglion cells (Chen et al., 1998; Hasbani et al., 1998). Down-regulation of KCC2 and a subsequent decrease in GABA-mediated inhibition may also contribute to excitotoxicity in injured neurons (Toyoda et al., 2003). These results imply that stimulation of NKCC1 activity and down-regulation of KCC2 are important in ischemic neuronal damage.

B. Cl^- Transporters in Ionic Dysregulation, Swelling, Excitatory Amino Acid Release and Damage in Astrocytes

Glial cells outnumber neurons by ten times and occupy up to 50% of the human brain volume (Tower and Young, 1973). For many years, glial cells were thought to function only as passive support cells, e.g. bringing nutrients to and removing wastes from the neurons. In contrast, neurons were credited with carrying out the critical functions of information processing, plasticity, learning and memory in the central nervous system (CNS) (Floyd and Lyeth, 2007). However, this view has undergone a dramatic transformation with the knowledge that glial cells play critical roles in the glutamate–glutamine shuttle, lactic acid supply for neuronal energy metabolism, regulation of synaptic and perisynaptic glutamate levels, regulation of ion concentrations (H^+, K^+ and Ca^{2+}) in the intracellular and extracellular spaces, and the coupling of neuronal activity and cerebral blood flow

(Chen and Swanson, 2003; Schurr, 2006). Therefore, alterations of astrocyte function after ischemia could have a significant impact on cerebral brain damage.

Early loss of astrocytes is a principal feature of hyperglycemic and complete ischemia conditions (Pulsinelli et al., 1982). Rapid loss of astrocytes also occurs in animal models of normoglycemic ischemia (Lukaszevicz et al., 2002; Pluta, 1987) and in optical nerve white matter under ischemic conditions (Stys, 2004). Despite the importance of astrocytes in neurodegeneration, the mechanisms underlying ischemic astrocyte damage are poorly understood. The following research findings illustrate the role of cation-chloride cotransporters in ischemia-induced astrocyte damage.

Both astrocytes and the choroid plexus epithelia have been suggested to play an important role in K^+ uptake in the CNS. Colocalization of NKCC1 and glial fibrillary acidic protein (astrocyte marker protein) was detected in astrocytes of cortex, corpus callosum, hippocampus and cerebellum of rat (Yan et al., 2001). Expression of NKCC1 was also observed in the perivascular astrocytes of cortical cortex, cerebral white matter and hippocampus (Yan et al., 2001). One of the major functions of astrocytes in the CNS is to buffer the extracellular K^+ concentration by transporting the K^+ released by active neurons (Paulson and Newman, 1987). Astrocytes take up K^+, which can then leave astrocytes at remote locations where extracellular K^+ concentrations have not risen, or K^+ can efflux at the endfeet of the astrocyte where it is subsequently transported by blood vessels (Paulson and Newman, 1987). NKCC1 in the processes of the perivascular astrocytes is colocalized with aquaporin 4, the latter is abundantly expressed in the astrocyte endfeet (Nielsen et al., 1997). Therefore, these data suggest that NKCC1 may play a role in transport of K^+ from brain to blood to restore the extracellular K^+ concentration after neuron firing cessation.

In addition to K^+ uptake, NKCC1 may also function in the regulation of intracellular Cl^- in astrocytes. In acutely isolated hippocampal astrocytes from mature Sprague-Dawley rat, intracellular Cl^- level is higher than it is predicted by passive distribution and the Cl^- equilibrium potential (E_{Cl}) is far more positive than the membrane potential (Fraser et al., 1995). Application of GABA in the CA3 region of the neuron-free rat hippocampal slice induced depolarization of astrocytes (MacVicar et al., 1989). A similar $GABA_A$ receptor-induced depolarization has also been observed in acutely isolated astrocytes from adult Sprague-Dawley rat or cultured mature hippocampal astrocytes from neonatal Wistar rat (Bekar and Walz, 1999; Fraser et al., 1995). These findings suggest that NKCC1 may play a role in maintaining high intracellular Cl^- levels in astrocytes.

1. *High $[K^+]_o$-mediated Astrocyte Swelling*

Astrocytes undergo rapid swelling in a number of acute pathological states, such as ischemia and traumatic brain injury (Kimelberg, 2005). Unresolved astrocyte swelling will have detrimental effects such as reduction of extracellular space, a decrease of normal cerebral blood flow, and an accumulation of excitatory amino acids such as glutamate (Maxwell et al., 1994; Rutledge and Kimelberg, 1996). In addition, swollen astrocytes would have a diminished capacity to perform their normal homeostatic functions. Considerable work has been focused on the role of cation-chloride cotransporters in maintenance and regulation of astrocyte volume.

As previously mentioned, NKCC1 is found in astrocytes (Yan et al., 2001), and the cotransporter in these cells could function in clearing excessive extracellular potassium ($[K^+]_o$) as mentioned above (Walz, 2000). During ischemia, one significant pathophysiological change in the CNS is an elevation of $[K^+]_o$. A few minutes of anoxia/ischemia raises $[K^+]_o$ to $\sim$60 mM (far above ceiling value) (Siesjo, 1992). NKCC1 in astrocytes may play a role in K^+ uptake under high $[K^+]_o$ conditions. In cultured astrocytes, 75 mM $[K^+]_o$ causes NKCC1-mediated K^+ influx to be stimulated by $\sim$80% (Su et al., 2002a). The high $[K^+]_o$-induced activation of NKCC1 is completely abolished by either removal of extracellular Ca^{2+} or by blocking the L-type voltage dependent Ca^{2+} channels with nifedipine (Su et al., 2000). Additionally, intracellular ^{36}Cl accumulation increases significantly in response to 75 mM $[K^+]_o$ and is abolished by inhibition of NKCC1 activity either by bumetanide or genetic ablation (Su et al., 2002a, b). These data suggest that the cotransporter activity is stimulated under high $[K^+]_o$ via Ca^{2+}-mediated signal transduction pathways.

Seventy-five mM $[K^+]_o$ also triggers cell swelling in $NKCC1^{+/+}$ astrocytes by 20% (Su et al., 2002a). This high $[K^+]_o$-mediated swelling is abolished by inhibition of NKCC1 activity, either with bumetanide or genetic ablation (Su et al., 2002a, b). In addition, high $[K^+]_o$-induced astrocyte swelling is also observed in the rat optic nerve model (MacVicar et al., 2002). Furosemide and bumetanide reversibly suppress the high $[K^+]_o$-induced astrocyte swelling in enucleated nerves (MacVicar et al., 2002). Additionally, astrocytes in optic nerves exhibited a Ca^{2+}-independent swelling during ischemia that was blocked by bumetanide (Thomas et al., 2004). Taken together, these studies suggest that NKCC1 activation leads to astrocyte swelling.

2. High $[K^+]_o$-mediated Excitatory Amino Acid Release from Astrocytes

Glutamate is the principal excitatory neurotransmitter in the mammalian CNS. Elevation of glutamate in the extracellular space and excessive activation of NMDA receptors cause cell death (Olney, 1969). The glutamate-induced excitotoxicity plays a major role in ischemic brain damage (Arundine and Tymianski, 2004).

Astrocytes maintain a large transmembrane glutamate gradient, with $[Glu]_i$ concentrations of 2–10 mM, whereas $[Glu]_o$ is approximately 1 μM (Erecinska and Silver, 1990). This gradient results from the high density of Na^+-dependent glutamate transporters in astrocytes, which regulate synaptic transmission by controlling glutamate diffusion and concentration in the extracellular space (Ni et al., 2007; Oliet et al., 2001). Therefore, reduced glutamate uptake or release of sequestered glutamate from astrocytes may contribute to excitotoxicity under pathophysiological conditions (Kimelberg, 2005). For example, one consequence of high $[K^+]_o$-induced astrocyte swelling will be the release of glutamate via activation of volume-sensitive organic anion channels (VSOACs) (Abdullaev et al., 2006; Su et al., 2002a). VSOACs are ubiquitously expressed chloride channels which are permeable to a variety of small organic anions, including the amino acids taurine, glutamate and aspartate (Strange et al., 1996). In response to swelling, VSOAC-mediated efflux of amino acids can function as a regulatory volume decrease mechanism in astrocytes (Rutledge et al., 1998). Release of $[^{14}C]$-D-aspartate from astrocytes, which serves as a surrogate for glutamate, is negligible under control conditions (Rutledge et al., 1998; Su et al., 2002b). However, high $[K^+]_o$ triggers a significant release of $[^{14}C]$-D-aspartate from astrocytes. High $[K^+]_o$-induced release of $[^{14}C]$-D-aspartate is abolished by the anion channel blocker DIDS, suggesting an involvement of VSOACs (Su et al., 2002b). Volume-sensitive anion channels are discussed in greater detail in Chapter 15. When NKCC1 is inhibited by bumetanide or genetic ablation, this high $[K^+]_o$-induced aspartate release is reduced by 30–50% (Fig. 25.3) (Su et al., 2002a, b). On the other hand, stimulation of NKCC1 under high $[K^+]_o$ conditions leads to increases in intracellular Na^+ which could slow or even reverse Na^+-dependent glutamate transporters and increase glutamate release from astrocytes. Taken together, astrocyte NKCC1 stimulation will contribute to excitotoxic injury by increasing swelling-induced glutamate release or decreasing glutamate clearance from the extracellular space.

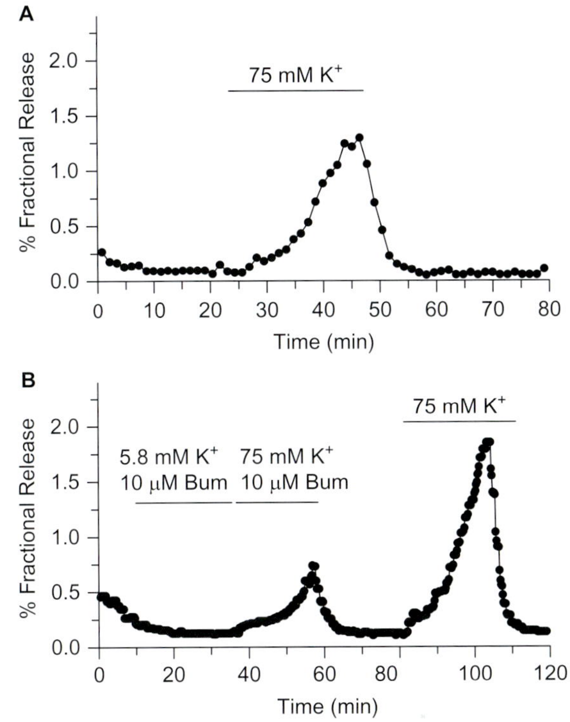

FIGURE 25.3 Effect of bumetanide on the high-$[K^+]_o$-mediated release of preloaded D-$[^{14}C]$aspartate from astrocytes. **A.** Cells loaded with D-$[^{14}C]$aspartate were perfused with 75 mM $[K^+]_o$ for 20 min. **B.** Astrocytes were exposed to bumetanide for 20 min and then exposed to 75 mM $[K^+]_o$ + bumetanide. Cells were then washed with normal buffer and re-exposed to 75 mM $[K^+]_o$. Adapted from Su et al. (2002a).

3. Na^+ and Cl^- Dysregulation in Astrocytes following Oxygen and Glucose Deprivation

Loss of the trans-plasma membrane Na^+ gradient is one of the key elements in promoting cellular damage in astrocytes during ischemia (Longuemare et al., 1999). An increase in intracellular Na^+ is found in rat spinal cord astrocytes (Rose et al., 1998), rat cortical astrocytes (Longuemare et al., 1999) and mouse cortical astrocytes (Silver et al., 1997) when these cells are exposed to glucose deprivation, NaN_3-mediated chemical hypoxia or simulated ischemia. While a decrease in Na^+/K^+-ATPase activity during ischemia will lead to reduced Na^+ extrusion, the activation of several Na^+ entry pathways, including NKCC1, also contributes to ischemia-induced loss of the Na^+ homeostasis. OGD in astrocytes results in both an increase in NKCC1 phosphorylation and ^{86}Rb influx (Kintner et al., 2004). This stimulation of NKCC1 activity is accompanied by an accumulation of ^{36}Cl and 3.5-fold increase in $[Na^+]_i$ in astrocytes (Lenart et al., 2004). Either bumetanide or genetic ablation of NKCC1 significantly reduces the OGD/reoxygenation-mediated

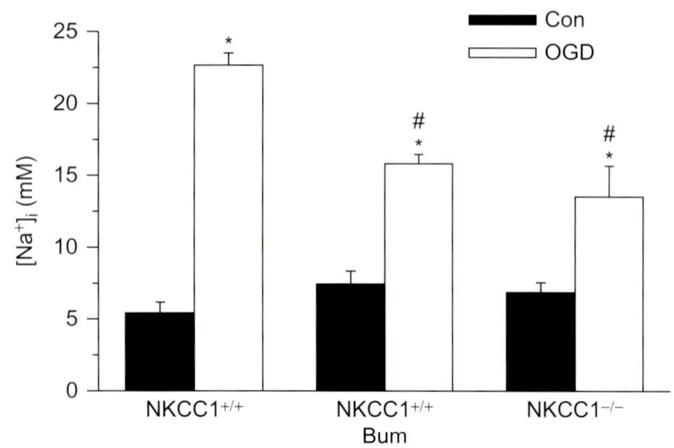

FIGURE 25.4 NKCC1 contributes to Na^+ accumulation in neurons after OGD/reoxygenation. Either NKCC1$^{+/+}$ or NKCC1$^{-/-}$ neurons were exposed to 1 h OGD and assessed for $[Na^+]_i$ at 1 h reoxygenation. In bumetanide studies, bumetanide was present during entire OGD/reoxygenation. Data are means ± SD. * $p < 0.05$ vs. control; # $p < 0.05$ vs. NKCC1$^{+/+}$ OGD. Adapted from Beck et al. (2003).

increase in $[Na^+]_i$ and $[Cl^-]_i$ (Fig. 25.4) (Kintner et al., 2007a).

Given that the extracellular ionic and oxygen composition change significantly during hypoxia/ischemia with a lower concentration of Na^+, Cl^-, Ca^{2+} and a higher concentration of K^+ as well as H^+ (Bondarenko and Chesler, 2001), it is important to examine whether NKCC1 still functions in this altered extracellular ionic environment. Thus, an alternative *in vitro* hypoxic model is to perfuse cells with hypoxic, acidic, ion shifted-Ringers (HAIR) that mimics the extracellular ionic and oxygen composition *in vivo* (Bondarenko and Chesler, 2001). During 5 min of HAIR treatment, $[Na^+]_i$ decreases in astrocytes due to the reduced Na^+ concentration in the HAIR buffer (Fig. 25.5). Interestingly, $[Na^+]_i$ decreases even further in NKCC1$^{-/-}$ astrocytes during HAIR treatment, which suggests that during HAIR treatment NKCC1 is still inwardly directed (Kintner et al., 2007a). When the astrocytes are returned to normoxic buffer, $[Na^+]_i$ raises rapidly over the first 5 min and plateaus to a value ~4-fold of control and is sustained over the following 40 min (Fig. 25.5). Bumetanide or genetic ablation

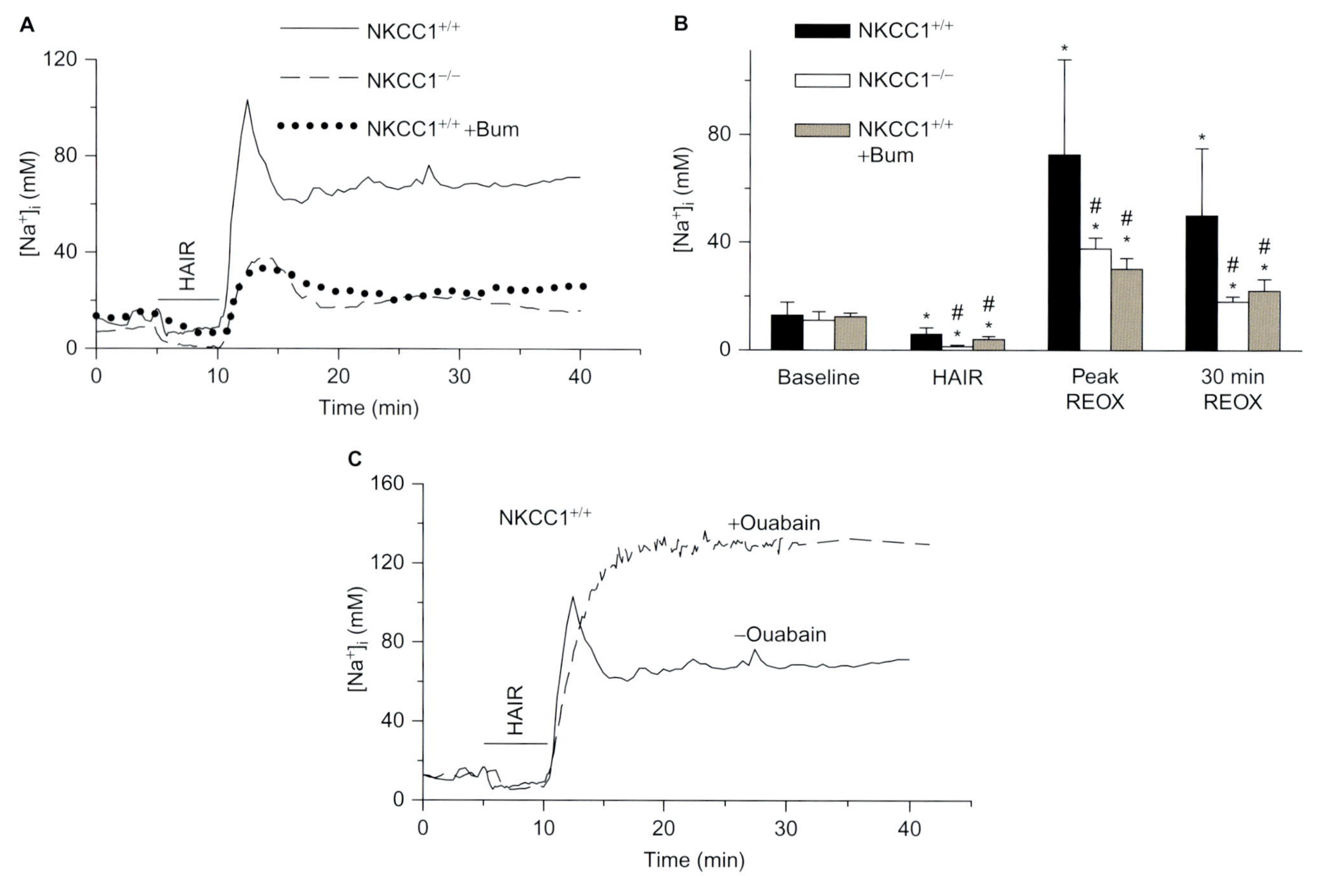

FIGURE 25.5 Changes in $[Na^+]_i$ in astrocytes following HAIR/reoxygenation. **A.** $[Na^+]_i$ was monitored in single NKCC1$^{+/+}$, NKCC1$^{-/-}$ or NKCC1$^{+/+}$ astrocytes treated with bumetanide during 5 min HAIR and 30 min reoxygenation. **B.** Summary of data for A. Data are means ± SE. * $p < 0.05$ vs. baseline, # $p < 0.05$ vs. NKCC1$^{+/+}$ HAIR/reoxygenation. **C.** $[Na^+]_i$ was monitored in single NKCC1$^{+/+}$ astrocytes following HAIR/reoxygenation with or without 1 mM ouabain. Adapted from Kintner et al. (2007).

of NKCC1 reduces this hypoxia-mediated rise in $[Na^+]_i$ by ~65% (Kintner et al., 2007a). The HAIR-mediated increases in $[Na^+]_i$ are even doubled when Na^+/K^+-ATPase activity is blocked with ouabain (Kintner et al., 2007a). This suggests that the NKCC1 mediates an Na^+ overload that cannot be offset by Na^+/K^+-ATPase activity.

4. *Ca^{2+} Dysregulation in Astrocytes following Oxygen and Glucose Deprivation*

NKCC1-mediated increases in $[Na^+]_i$ following *in vitro* ischemia have been linked to Ca^{2+} loading and subsequent mitochondrial damage. As a consequence of increases in $[Na^+]_i$, NCX-mediated Ca^{2+} extrusion is reduced or NCX may function in a reverse mode, which leads to increases in $[Ca^+]_i$ (Fig. 25.6). Thus, NKCC1 and the reverse-mode operation of NCX (NCX_{rev}) contribute to intracellular Na^+ and Ca^{2+} overload in astrocytes following *in vitro* ischemia (Kintner et al., 2007a; Lenart et al., 2004). In addition, Na^+ and Ca^{2+} overload in ischemic astrocytes leads to a dissipation of the mitochondrial membrane potential and release of mitochondrial cytochrome C, which can be attenuated by bumetanide or with genetic ablation of NKCC1 (Kintner et al., 2007a). These findings suggest that the concerted activities of multiple ion transport proteins are important in the perturbations of Na^+ and Ca^{2+} homeostasis and in astrocyte damage in response to hypoxia/ischemia.

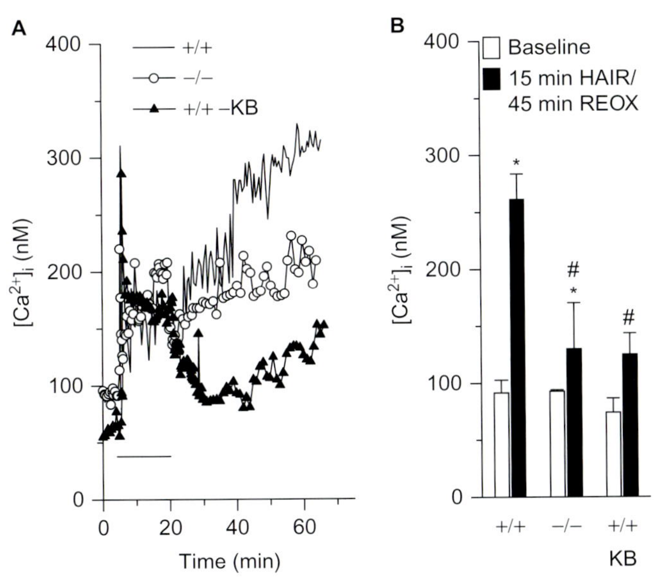

FIGURE 25.6 Increase in $[Ca^{2+}]_i$ following HAIR/reoxygenation depends on NKCC1 and NCX activity. **A.** $[Ca^{2+}]_i$ was monitored in NKCC1$^{+/+}$ or NKCC1$^{-/-}$ astrocytes during 15 min HAIR and 45 min reoxygenation. In studies involving KB-R7943, the drug was present 30 min before HAIR and during reoxygenation. **B.** Summary data of changes in $[Ca^{2+}]_i$ at 45 min reoxygenation. Data are means ± SE. * $p < 0.05$ vs. control, # $p < 0.05$ vs. NKCC1$^{+/+}$ HAIR/reoxygenation. Adapted from Kintner et al. (2007).

C. Cl^- Transporters in Excitotoxicity in Oligodendrocytes

The primary role of oligodendrocytes in the CNS is the production of myelin, which forms an insulating sheath around the axons of neurons and allows for efficient electrical transmission (Waxman and Ritchie, 1993). Loss of oligodendrocyte function is involved in a number of diseases, such as multiple sclerosis, amylotrophic lateral sclerosis, Huntington's disease, Alzheimer's disease and Parkinson's disease. Likewise, early ischemic damage to white matter involves oligodendrocyte and axon damage (Valeriani et al., 2000).

NKCC1 plays a role in oligodendrocyte damage following activation of non-NMDA glutamate receptors. AMPA/kainate receptor-mediated excitotoxicity results in spinal cord white matter injury in a process that is linked to oligodendrocyte damage and axonal demyelination (Park et al., 2004; Tekkok and Goldberg, 2001). Indeed, the vulnerability of immature oligodendrocytes to ischemic injury is a major component of the brain injury associated with cerebral palsy (Wilke et al., 2004). Oligodendrocytes are thought to express primarily non-NMDA-type ionotropic glutamate receptors, although recent studies suggest that NMDA-type glutamate receptors may predominate in processes while AMPA/kainate receptors predominate on the soma (Salter and Fern, 2005).

Exposing cultured oligodendrocytes to AMPA plus the AMPA desensitization blocker cyclothiazide (CTZ) leads to a transient rise in $[Ca^{2+}]_i$, which is followed by a sustained intracellular $[Ca^{2+}]_i$ overload, NKCC1 phosphorylation and a NKCC1-mediated Na^+ influx (Chen et al., 2007). In the presence of a specific AMPA receptor inhibitor 6-cyano-7-nitroquinoxaline-2,3-dione (CNQX), AMPA/CTZ fails to elicit any changes in $[Ca^{2+}]_i$. The AMPA/CTZ-induced sustained $[Ca^{2+}]_i$ rise leads to mitochondrial Ca^{2+} accumulation, release of cytochrome C from mitochondria and cell death (Chen et al., 2007). The AMPA/CTZ-elicited $[Ca^{2+}]_i$ increase, mitochondrial damage and cell death are significantly reduced by inhibiting NKCC1 or NCX_{rev}. These data suggest that in cultured oligodendrocytes, activation of AMPA receptors leads to NKCC1 phosphorylation that enhances NKCC1-mediated Na^+ influx. The latter triggers NCX_{rev}-mediated $[Ca^{2+}]_i$ overload and compromises mitochondrial function and cellular viability (Chen et al., 2007). Such a link between Na^+ overload and activation of NCX_{rev} has been observed under other conditions (Stys, 2005).

Indeed, NCX_{rev} causes Ca^{2+} influx after NMDA and non-NMDA receptor activation in depolarized and glucose-deprived neurons (Czyz and Kiedrowski, 2002; Hoyt et al., 1998). Furthermore, increase in intracellular Na^+ and the subsequent induction of NCX_{rev} have also been found in mechanical strain injury of astrocytes (Floyd et al., 2005) and in axonal damage (Stys, 2005).

Theoretical thermodynamic analysis of NCX_{rev} in oligodendrocytes predicts that the flux through NCX is near zero under control conditions and a resting potential of −90 mV (Fig. 25.7). During AMPA treatment, $[Ca^{2+}]_i$ rises to 192 nmol/L and 224 nmol/L at 1 min and 30 min, respectively. Although this increase in $[Ca^{2+}]_i$ may favor the forward-mode operation of NCX, the significant concurrent increase in $[Na^+]_i$ strongly favors an inwardly directed Ca^{2+} current via NCX_{rev} (Fig. 25.7). The simulation predicts that if plasma membrane potential remains at −90 mV, NCX will function in the reverse mode when $[Na^+]_i$ increases to ~25 mmol/L with $[Ca^{2+}]_i$ of 192–224 nmol/L. In fact, $[Na^+]_i$ is elevated to ~134 mmol/L at 1 min and 46 mmol/L at 30 min of AMPA/CTZ exposure, which favors NCX_{rev} (Chen et al., 2007).

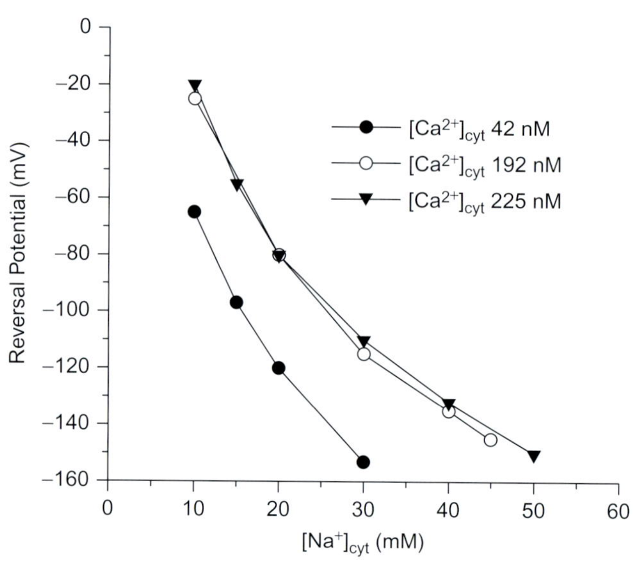

FIGURE 25.7 Thermodynamics of NCX activity after AMPA/CTZ exposure. It is assumed that NCX is an electrogenic mechanism that transports Na^+ and Ca^{2+} with a stoichiometry of 3:1. The LabHEART simulation generated values of NCX current at different voltages, using known and assumed values for intracellular and extracellular Na^+ and Ca^{2+}. The reversal potential under each ionic condition was the voltage when NCX current equals zero. The reversal potential was then plotted against the $[Na^+]_i$ used to generate the current/voltage curve. $[Ca^{2+}]_i$ was set at 42, 192 and 225 mM for control, 1 min or 30 min after AMPA/CTZ application, respectively. Adapted from Chen et al. (2007).

III. Cl^- TRANSPORTERS IN BRAIN SLICES FOLLOWING *IN VITRO* ISCHEMIA

A. *In vitro* Brain Slice Model for Global Cerebral Ischemia

The brain slice is a widely used model for investigating mechanisms underlying anoxia- or ischemia-induced neuronal damage because it maintains the *in vivo* anatomical microenvironment. Acute brain slices and organotypic slice culture are two common methods. Acute brain slices are isolated in an artificial cerebral spinal fluid (aCSF) or Ringer solution equilibrated with O_2/CO_2. Ischemic episodes are induced in the acute brain slice by submerging them in a glucose-free aCSF equilibrated with N_2/CO_2. Although the acute brain slice can be maintained only for a short time, slices can be prepared routinely from mature adult tissue, where the Cl^- transport systems and the negative shift in the E_{Cl} have been established (Kaila, 1994). Hence, they have been particularly useful in gaining insight into the role of Cl^- transporters in cerebral ischemia. On the other hand, organotypic brain slices can be maintained for longer periods of time in culture, but they are typically prepared from postnatal (P0–P7) rats and mice. In order to model cerebral ischemia, the organotypic brain slice can be submerged in a glucose-free culture medium that is infused with N_2/CO_2 or placed in an anoxic culture chamber (Noraberg et al., 2005). In either slice model, this insult is termed *in vitro* ischemia or OGD. Brain slice models exhibit the same selective vulnerability as seen *in vivo*; for instance, OGD results in significant damage to the CA1 area of the hippocampus in the slice (Lipton, 1999; Noraberg et al., 2005).

B. Disruption of Cl^- Homeostasis is Associated with *in vitro* Ischemia

Ionic disturbances during and following *in vitro* ischemia are particularly important initiators of neuronal damage that lead to irreversible cell injury and activation of cell death pathways. Specifically, there is substantial evidence indicating that the increases in intracellular Na^+ and Ca^{2+} are important initiators of ischemia-induced cell death. Although the role of elevated intracellular Cl^- in ischemia has not been studied as thoroughly as that of other ions, Cl^- or Cl^- transporters has been demonstrated as a major contributor to ischemia-induced neuronal injury (see Table 25.1).

TABLE 25.1 Studies investigating the role of Cl^- transporters in the oxygen and glucose-deprived brain slice

Animal	Insult	Damage outcome measured	Cl^- measurement method	Cl^- transport inhibitors	Citation
ACUTE HIPPOCAMPAL SLICE					
Rat (P14–P19)	OGD	CA1 field responses, cell death, cell swelling	MEQ	Picrotoxin	Inglefield, and Schwartz-Bloom (1998)
Adult rat	OGD	CA1 field responses, interstitial voltage	Electron probe X-ray microanalysis	N/A	Taylor et al. (1999)
Adult rat	Hypoxia	Spreading depression-like depolarization	N/A	Furosemide, DIDS, DNDS	Muller (2000)
Adult rat	OGD	N/A	Electron probe X-ray microanalysis	N/A	Lopachin et al. (2001)
Rat (P10–P15)	OGD	N/A	MEQ	Furosemide, bumetanide	Yamada et al. (2001)
Adult rat	OGD	ATP recovery; CA1 morphology	N/A	Furosemide, bumetanide, DIOA, ethacrynic acid	Pond et al. (2004)
Adult rat	OGD	CA1 field responses, Cl^- changes	MEQ	Bicuculline, picrotoxin	Galeffi et al. (2004)
Adult rat	Hypoxia/hypoglycemia	CA1 field responses	N/A	Bumetanide, DIOA	Busse et al. (2005)
Adult rat	OGD	Edema formation	N/A	Furosemide, bumetanide, bicuculline, DIDS, niflumic acid	Kumar et al. (2006)
Adult mouse	OGD	Cl^- changes, CA1 morphology	Clomeleon	Furosemide, bumetanide, picrotoxin	Pond et al. (2006)
ORGANOTYPIC SLICE CULTURE					
Rat (P10)	Hypoxia/ hypoglycemia	Cell death	N/A	Bumetanide, DIOA	Busse et al. (2005)
Mouse (P6–P8)	Hypoxia/ hypoglycemia or OGD	Cell death	N/A	DIDS	Yao et al. (2007)
Mouse (P6–P8)	Hypoxia	Cell death, LDH	N/A	DIDS	Xue et al. (2008)

MEQ, 6-methoxy-N-ethylquinolinium iodide; **DIDS**, 4,4′-diisothiocyano-2,2′-stilbene disulfonic acid; **DNDS**, 4,4-dinitro stilbene-2,2-disulfonate; **DIOA**, dihydroindenyloxyalkanoic acid; **LDH**, lactate dehydrogenase activity (measure of cell injury)

Several studies have indicated that OGD in the brain slice induces an accumulation of $[Cl^-]_i$ in area CA1 hippocampal neurons (Galeffi et al., 2004; Inglefield and Schwartz-Bloom, 1998; Lopachin et al., 2001; Pond et al., 2006; Taylor et al., 1999). In acute brain slice models, Taylor and colleagues found that $[Cl^-]_i$ started to increase during OGD and accumulated throughout reoxygenation, reaching levels twice as high as those found immediately after OGD (Taylor et al., 1999). In another study, Cl^- content in the cytoplasm, mitochondria and nuclei was increased by approximately 50% after 10 min OGD and by approximately 300% at 30 min reoxygenation following 10 min OGD (Lopachin et al., 2001). However, these initial studies using electron probe X-ray microanalysis did not allow continuous monitoring of changes in $[Cl^-]_i$

(Table 25.1). Recently, imaging of $[Cl^-]_i$ has been developed using the Cl^- sensitive fluorescent dyes 6-methoxy-N-ethylquinolinium chloride (MEQ) and N-(ethoxycarbonylmethyl)-6-methoxyquinolinium bromide (MQAE); or the genetically-encoded Cl^- indicator, Clomeleon, described in Chapter 7 (Berglund et al., 2004; Kuner and Augustine, 2000). Continuous imaging of slices reveals that increases in $[Cl^-]_i$ are biphasic (Galeffi et al., 2004; Pond et al., 2004). $[Cl^-]_i$ rises during OGD and recovers partially upon onset of reoxygenation; however, this is followed by a second and more prolonged rise in $[Cl^-]_i$ within the next hour (Galeffi et al., 2004; Pond et al., 2004). Interestingly, this sustained rise in $[Cl^-]_i$ following OGD can directly induce neuronal damage. In one study, CA1 hippocampal neurons were dialyzed with various $[Cl^-]_o$ using the ionophores nigericin and tributyltin, and the morphological changes were examined. The degree of CA1 neuronal damage was significantly increased with the elevated $[Cl^-]_i$ (Pond et al., 2006). In addition, if the reoxygenation buffer following OGD is maintained with low $[Cl^-]_o$, the secondary rise in $[Cl^-]_i$ is completely blocked and neuronal damage is attenuated (Newell et al., 1995; Pond et al., 2006). It has also been shown that preventing the rise in $[Cl^-]_i$ is neuroprotective following oxidative stress (Wake et al., 2007). Together, these findings indicate that the OGD-induced rise in $[Cl^-]_i$ is both necessary and sufficient to produce neuronal damage in ischemic brain slices.

Disruptions of Cl^- homeostasis can contribute to ischemic damage via multiple mechanisms. First, synaptic inhibition by the inhibitory neurotransmitters GABA and glycine depends on the transmembrane Cl^- gradient, which maintains an E_{Cl} that is negative relative to the membrane potential. If the rise in $[Cl^-]_i$ induces a shift in the E_{Cl} such that the E_{Cl} becomes more positive relative to the membrane potential, GABA and glycine receptor activation will cause Cl^- efflux and membrane depolarization. This may cause GABA and glycine to be less inhibitory or even excitatory, depending on the membrane potential and the E_{Cl} (Kaila, 1994; Schwartz-Bloom and Sah, 2001; Thompson and Gahwiler, 1989). Therefore, the loss of $GABA_A$-mediated inhibition due to the rise in $[Cl^-]_i$ may increase neuronal excitotoxicity after *in vitro* as well as *in vivo* ischemic episodes. Specifically, the OGD-induced $[Cl^-]_i$ accumulation may reduce $GABA_A$ responses in area CA1 pyramidal neurons during reoxygenation (Galeffi et al., 2004; Inglefield and Schwartz-Bloom, 1998). Similarly, in hippocampal slices, anoxia suppresses GABA-mediated inhibitory postsynaptic currents (IPSCs) as a result of a positive shift in E_{Cl} (Katchman et al., 1994). These *in vitro* studies are supported by studies using *in vivo* models of cerebral ischemia. In rats subjected to focal cerebral ischemia, there is a depolarizing shift in the reversal potential for $GABA_A$-mediated inhibitory postsynaptic potentials (IPSPs) in neocortex, resulting in a reduction of GABA-mediated inhibition (Mittmann et al., 1998). In addition to the effects on inhibitory neurotransmission, elevated $[Cl^-]_i$ may also cause damage by ultrastructural changes to the smooth endoplasmic reticulum (Rosenberg and Lucas, 1996) or affect pH regulation via coupling to function of the Cl^-/HCO_3^- exchange (Schlue and Deitmer, 1988).

C. Role of Cl^- Transporters in Disruptions of Cl^- Homeostasis following *in vitro* Ischemia

Given the apparent significance of the OGD-mediated rise in $[Cl^-]_i$, considerable effort has focused on the underlying mechanisms. The transmembrane Cl^- gradient is maintained by Cl^- transport mechanisms, including Cl^- efflux via KCC2 (Sung et al., 2000), the Na^+-dependent Cl^-/HCO_3^- exchanger (Schlue and Deitmer, 1988), the Cl^- ATPase (Schwartz-Bloom and Sah, 2001) and the voltage-dependent Cl^- channel, ClC2 (Muller, 2000). Therefore, OGD-induced Cl^- increases could be due to reduced function or reversal of the Cl^- efflux transporters. In turn, the activation of Cl^- influx mechanisms may also contribute to elevations in intracellular Cl^-. Chloride influx may occur through the Cl^-/HCO_3^- exchanger, the $GABA_A$ receptor, the glycine receptor and NKCC1 (Schwartz-Bloom and Sah, 2001). Therefore, OGD-induced Cl^- increases could result from altered function of Cl^- efflux and/or influx pathways.

Recent studies indicate that the rise in $[Cl^-]_i$ during OGD is likely to be due to a number of different Cl^- transport mechanisms (Kumar et al., 2006; Pond et al., 2006; Slemmer et al., 2004;). OGD induces a severe membrane depolarization in neurons, leading to an increase in membrane permeability (Hansen, 1985; Tanaka et al., 1997). While Cl^- fluxes play a role in triggering this depolarization, inhibition of Cl^- transport is unable to prevent its onset (Muller, 2000). As the membrane potential approaches zero, an inwardly directed electrochemical potential is imposed on Cl^-, driving Cl^- into the cell through any available Cl^- channel. Therefore, it is likely that multiple Cl^- transporters and channels are responsible for Cl^- entry during this time.

In contrast, the rise in $[Cl^-]_i$ during reoxygenation has been attributed to specific Cl^- transporters. This rise in $[Cl^-]_i$ within 1 h reoxygenation can be completely prevented by the loop diuretics, bumetanide and furosemide, while the GABA-gated Cl^- channel blocker picrotoxin has no effect (Pond et al., 2006). Since bumetanide, at concentrations selective for NKCC1, completely prevented the secondary rise in $[Cl^-]_i$, it is likely that NKCC1 activity is primarily responsible for these changes. In fact, increased phosphorylation of NKCC1 has been demonstrated after OGD, consistent with the timing of changes of $[Cl^-]_i$ (Pond et al., 2006). Phosphorylation of NKCC1 increases the transporter's activity (Darman and Forbush, 2002) and could lead to the accumulation of intracellular Cl^- following OGD.

Other Cl^- transport mechanisms may also contribute to the OGD-mediated secondary rise in $[Cl^-]_i$. It is known that OGD causes significant increases in extracellular K^+, which could drive Cl^- influx via NKCC1 and/or reverse KCC2 to bring Cl^- into neurons (DeFazio et al., 2000; Payne, 1997; Payne et al., 2003). In the case of KCC2, it has been demonstrated that OGD induces a substantial decrease in KCC2 protein levels in rat hippocampal slices (Galeffi et al., 2004). A recent study indicates that oxidative stress causes the dephosphorylation of tyrosine residues on KCC2; interestingly, this dephosphorylation corresponded to a loss of functional KCC2 activity and preceded decreases in KCC2 expression (Wake et al., 2007). Since KCC2 serves as one of the primary Cl^- extrusion mechanisms in neurons, the decrease in its activity and/or expression may compromise the ability of neurons to restore the Cl^- gradient after OGD.

D. Role of Cl^- Transporters in Neuronal Damage following *in vitro* Ischemia

There have been several studies investigating the role of Cl^- transporters in neuronal damage associated with *in vitro* ischemia. Due to the fact that aberrant activity of NKCC1 and KCC2 may contribute to $[Cl^-]_i$ increases following OGD, studies have focused on these transporters. The NKCC inhibitors bumetanide and furosemide promote the biochemical (recovery of ATP levels) and morphological recovery of hippocampal slices exposed to OGD (Pond et al., 2004). In addition, bumetanide also reduces neuronal damage in organotypic hippocampal slice cultures exposed to hypoxia and hypoglycemia (Busse et al., 2005). Prevention of KCC2 loss following OGD or oxidative stress and the subsequent maintenance of normal intracellular Cl^- levels has been shown to be neuroprotective (Galeffi et al., 2004; Wake et al., 2007).

In addition to NKCC1 and KCC2, other Cl^- transport mechanisms have been investigated for their role in OGD-induced neuronal damage. Several studies indicate the importance of the Cl^-/HCO_3^- exchanger and the Na^+-dependent Cl^-/HCO_3^- exchanger in OGD-induced neuronal damage. These transport mechanisms are important regulators of both intracellular Cl^- as well as pH (Schlue and Deitmer, 1988). An inhibitor of the Cl^-/HCO_3^- and the Na^+-dependent Cl^-/HCO_3^- exchangers, DIDS, has been shown to be neuroprotective in organotypic slices (Xue et al., 2007; Yao et al., 2007;). However, DIDS does not specifically inhibit these exchange mechanisms alone, making interpretation of these results difficult (Yao et al., 2007). Another Cl^- transport mechanism that has been investigated is the GABA-gated Cl^- channel. However, the GABA-gated Cl^- channel does not appear to be involved in the accumulation of $[Cl^-]_i$ during reoxygenation (Galeffi et al., 2004; Pond et al., 2006). Thus, the GABA-gated Cl^- channel inhibitor picrotoxin is not neuroprotective after OGD (Pond et al., 2004).

In summary, *in vitro* ischemia in the brain slice results in a biphasic increase in $[Cl^-]_i$ during OGD and reoxygenation. The severe membrane depolarization that occurs during OGD probably imposes a strong inwardly directed electrochemical potential on Cl^-, driving Cl influx via Cl^- channels. However, the rise in $[Cl^-]_i$ during the reoxygenation period is most likely due to the aberrant activity of NKCC1 and KCC2. NKCC1 activity is increased following OGD, as evidenced by an increase in phosphorylation, and blocking NKCC1 with bumetanide prevents the rise in $[Cl^-]_i$. Furthermore, a loss of functional KCC2 appears to occur following OGD, which would prevent neurons from restoring the Cl^- gradient after OGD. These data further support the notion that Cl^- transporters could be novel targets for preventing neuronal damage after ischemia. The case of bumetanide is exceedingly interesting as the loop diuretic in combination with the $GABA_A$ receptor potentiator phenobarbital also prevents the development of neonatal seizures (see Chapter 24).

IV. Cl^- TRANSPORTERS FOLLOWING *IN VIVO* CEREBRAL ISCHEMIA

While *in vitro* studies are useful for demonstrating proof of principle for a role of NKCC1 in cerebral ischemia, it is important to extend these studies by using *in vivo* animal models. The most widely used model of focal cerebral ischemia is the endovascular suture occlusion of the middle cerebral artery (MCAO) in rodents. In this model, a filament is advanced through the external carotid artery into the internal carotid artery to block blood flow to the middle cerebral artery. After a period of ischemia, the filament can be withdrawn, and blood flow restored and the animal allowed to recover.

A. Changes in NKCC1 Protein Expression

Expression of NKCC1 detected with T4 antibody immunoreactive signals is observed in cell bodies and dendrites of pyramidal neurons in cerebral cortex of sham rats. Two hours of focal cerebral ischemia and 24h reperfusion lead to increased NKCC1 immunostaining in neurons scattered in the ischemic cortex and striatum (Yan et al., 2001). Immunoblotting revealed that expression of NKCC1 protein was increased following 2h focal ischemia in cerebral cortex and striatum. A sustained up-regulation of NKCC1 in cortex was detected at 4, 8, 12 and 24h of reperfusion (Yan et al., 2001). An increase in the phosphorylated NKCC1 (NKCC1-p) was also found at 4 and 8h of reperfusion in the ipsilateral cortex (Yan et al., 2001). Because the change in phosphorylation state of NKCC1 is a major regulatory mechanism that stimulates NKCC1 activity, these findings suggest that elevated NKCC1 expression and activity may be involved in cerebral ischemic damage. Up-regulation of NKCC1 is found in hippocampus in gerbils after 5min global ischemia and 24h reperfusion (Kang et al., 2002). However, in the same model there is also a transient down-regulation of NKCC1 protein expression at 12h reperfusion (Kang et al., 2002), which may be a compensatory reaction to protect neurons from excitotoxicity.

B. Bumetanide-mediated Neuroprotection

Two hours of MCAO and 24h reperfusion result in an average infarction volume of 238.3 mm^3 in spontaneous hypertensive rats (Yan et al., 2001). However, when 100 μM bumetanide is continuously infused via microdialysis into the cortex during MCAO and reperfusion, the resulting infarct was reduced by ~25% (Fig. 25.8). Edema induced by MCAO was also reduced by 79% in the bumetanide-treated brains (Yan et al., 2001). This indicates that NKCC1 contributes to cerebral ischemic cell damage and edema *in vivo*. However, no protection is found if bumetanide is only administered during 24h reoxygenation (Yan et al., 2003, 2001), implying that activation of NKCC1 may play a role in the early stage of ischemic damage. In rats subjected to permanent MCAO, intravenous administration of bumetanide (7.6–30.4 mg/kg) immediately before occlusion attenuates edema formation as determined by magnetic resonance imaging

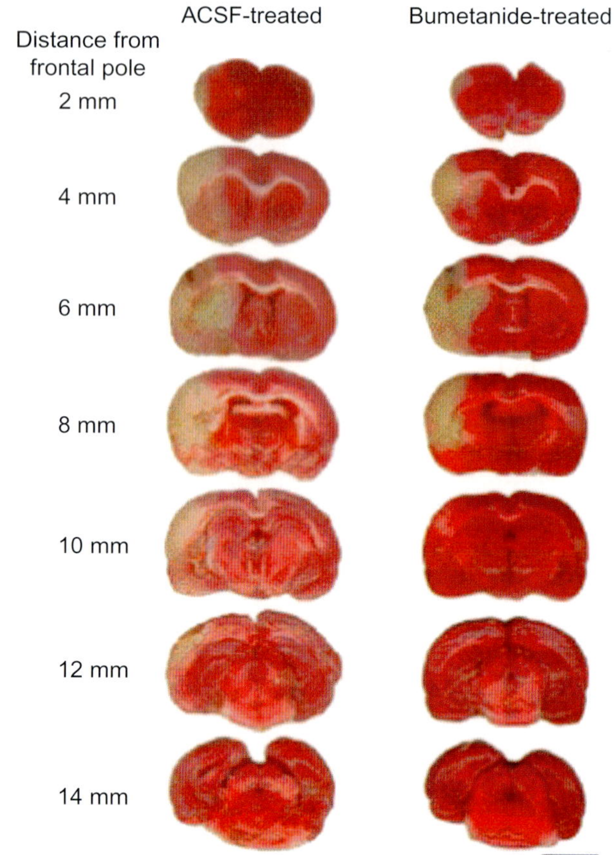

FIGURE 25.8 Pre-ischemic bumetanide treatment reduces infarct volume. Spontaneous hypertensive rats were treated with either artificial-CSF (ACSF) or bumetanide 1h prior to 2h MCAO. Representative TTC-stained coronal sections are shown at different levels posterior to the frontal pole from either ACSF-treated or bumetanide-treated brains. Scale bar 5.5 mm. Adapted from Yan et al. (2003).

(O'Donnell et al., 2004). This further suggests a role for NKCC1 in the edema formation during cerebral ischemia.

C. Less Ischemic Damage in NKCC1-null Mice

In addition to pharmacological blockage of NKCC1, genetic ablation of NKCC1 can be used to further establish that NKCC1 contributes to ischemic damage. Two hours MCAO and 10 or 24h reperfusion caused infarction (~85 mm^3) in NKCC1 wild-type ($NKCC1^{+/+}$) mice (Fig. 25.9). Infarction volume in $NKCC1^{-/-}$ mice was reduced by ~30% to 46% (Chen et al., 2005). Heterozygous mutant ($NKCC1^{+/-}$) mice also showed ~28% reduction in infarction (Chen et al., 2005). Brain edema was significantly increased after 2h MCAO and 24h reperfusion in $NKCC1^{+/+}$ mice. In contrast, there was ~50% less edema formation in either $NKCC1^{+/-}$ or $NKCC1^{-/-}$ mice (Chen et al.,

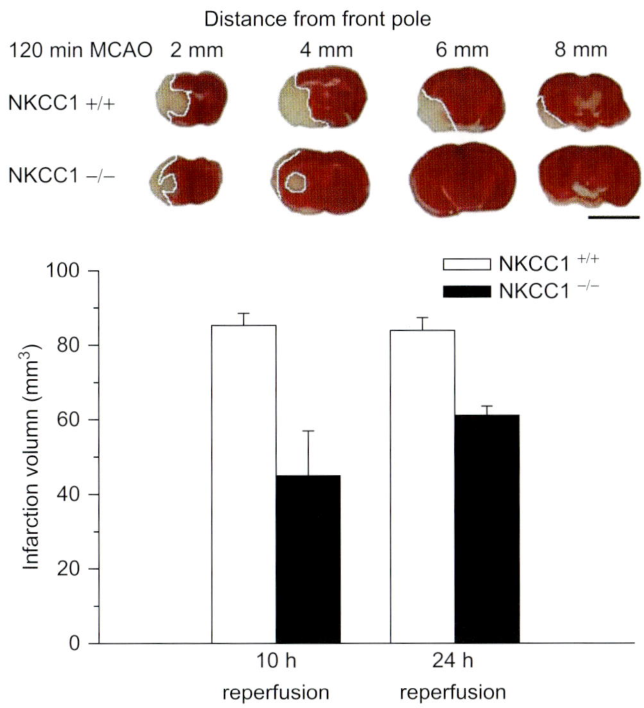

FIGURE 25.9 Brain infarction is decreased in NKCC1$^{-/-}$ mice. **A.** Coronal sections of brains from NKCC1$^{+/+-}$ or NKCC1$^{-/-}$ mice after 2h MCAO and 10 or 24h reperfusion were stained with TCC and infarct volume was calculated. **B.** Summarized infarct volume. Data are means±SD, $p < 0.05$ vs. NKCC1$^{+/+}$. Scale bar 5mm. Adapted from Chen et al. (2005).

2005). These results are consistent with the neuroprotective effects in rats mediated by pharmacological inhibition of NKCC1 with bumetanide.

White matter ion homeostasis, especially Na^+ and K^+, is crucial to normal axonal functions and dissipation of ionic homeostasis will lead to irreversible injury. Amyloid precursor protein (APP) is transported by fast anterograde axon transport along microtubules and has recently been shown as a sensitive marker of axonal disruption in white matter during brain ischemia (Yam et al., 2000). Disruption of this transport as a consequence of cytoskeleton derangement results in APP accumulation, indicating the dysfunction or discontinuity of the axons (Dewar et al., 1999; Koo et al., 1990). A 2 to 4h focal ischemia causes cytoskeletal breakdown and disturbance of fast axonal transport in myelinated fiber tracts (Dewar et al., 1999). NKCC1$^{+/+}$ brains exhibit significant accumulation of APP in the ipsilateral internal capsule after 2h focal ischemia or 10h reperfusion. In contrast, APP accumulation is reduced by ~55% in NKCC1$^{-/-}$ brains after 2h focal ischemia and 10h reperfusion (Chen et al., 2005). White matter degeneration is often assumed to be secondary to neuronal damage. However, no changes in myelin basic protein expression were observed after 2h MCAO and 24h reperfusion (Chen et al., 2005). This implies that APP accumulation occurs early and is likely a result of primary axonal ischemic damage. Taken together, these results suggest that NKCC1 plays a role not only in gray matter but also in white matter ischemic damage.

The *in vitro* studies in astrocytes or in oligodendrocytes show that NKCC1 activation leads to Na^+ overload, which triggers NCX_{rev} after ischemia and excitotoxicity (Chen et al., 2007; Kintner et al., 2007a; Lenart et al., 2004). Although these findings support that NKCC1 and NCX_{rev} may collectively contribute to ischemic cell damage, it has not yet been tested in *in vivo* ischemic models until recently. Luo et al. (Luo et al., 2007) examined whether reduction of NKCC1 and NCX1 could provide neuroprotection after *in vitro* or *in vivo* ischemia. Both neurons and astrocytes cultured from double heterozygous (NKCC1$^{+/-}$/NCX1$^{+/-}$) mice showed considerably reduced expression and function of NKCC1 and NCX1 (Luo et al., 2007). In addition, as compared to NCX1$^{+/+}$ mice, infarct volume at either 24 or 72h reperfusion following 30min MCAO is reduced by ~50% in NKCC1$^{+/-}$/NCX1$^{+/-}$ mice (Luo et al., 2007).

V. CONCLUSIONS

Loss of ion homeostasis plays a central role in pathogenesis of ischemic cell damage. Ischemic-induced perturbation of ion homeostasis leads to intracellular accumulation of Na^+ and Ca^{2+} and subsequent activation of proteases, phospholipases and formation of oxygen- and nitrogen-free radicals (Lipton, 1999; Siesjo, 1992). This cascade of signal transduction events results in long-term functional and structural changes in membrane and cytoskeletal integrity and eventual cell death. Secondary active ion transport proteins (NKCC1, KCC2 and NCX) are important in maintaining steady-state intracellular ionic concentrations (Fig. 25.10) and (Kintner et al., 2007b). During ischemia, cells lose K^+ and gain Ca^{2+}, Na^+, Cl^-, along with H_2O. This may result from a halt in ATP synthesis and extensive pump failure after anoxia. Considerable research effort has been centered on roles of passive fluxes via cation and anion conductances in dissipation of the ion concentration gradients. Results from recent *in vitro* and *in vivo* experimental studies demonstrate the role of cation-chloride transporters in ischemia-induced dissipation of ionic homeostasis. Therefore, these ion transport proteins could be a target for the future development of therapeutic treatment of ischemic cerebral damage.

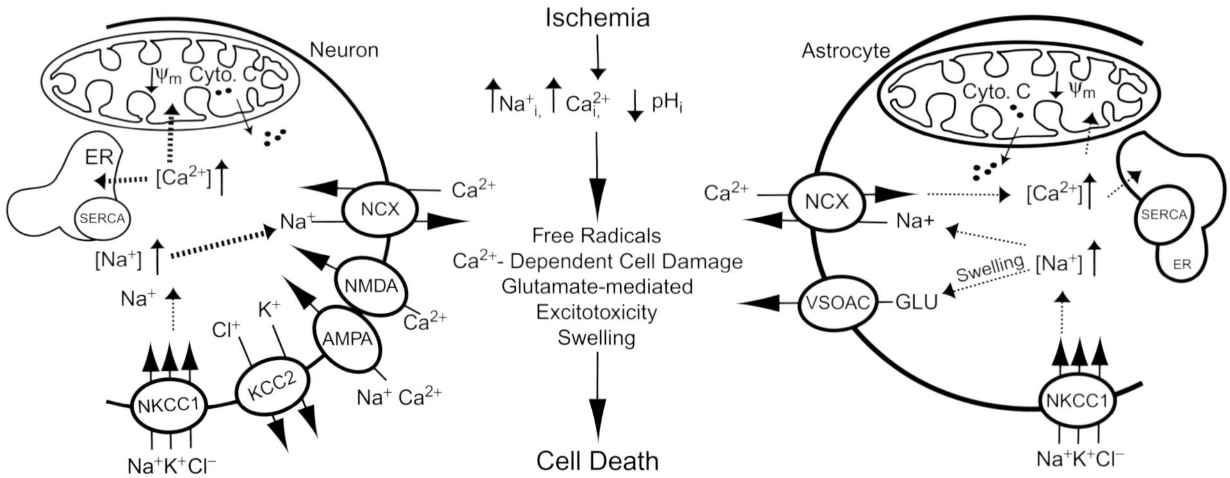

FIGURE 25.10 Schematic describing a model of ischemia-induced perturbation of ionic homeostasis. Ischemia leads to intracellular accumulation of Ca^{2+} and Na^{+} and subsequent activation of proteases, phospholipases and formation of oxygen- and nitrogen-free radicals. This cascade of signal transduction events results in long-term functional and structural changes in membrane and cytoskeletal integrity and eventual cell death. Secondary active ion transport proteins, such as Na^{+}-dependent chloride transporter (NKCC1), K^{+}-Cl^{-} transporter (KCC2) and Na^{+}/Ca^{2+} exchanger (NCX), are important in maintaining steady-state intracellular ionic concentrations. Altered functions of these ion transporters may contribute to cell damage. **ER**, endoplasmic reticulum; **AMPA**, alpha-amino-3-hydroxy-5-methyl-4-isoxazolepropionic acid; **NMDA**, N-methyl-D-aspartic acid; **VSOAC**, volume-sensitive organic osmolyte-anion channel; **GLU**, glutamate; **SERCA**, Sarco/Endoplasmic Reticulum Ca^{2+}-ATPase; **Cyto. C**, cytochrome C; Ψ_m, mitochondrial membrane potential.

Acknowledgement

This work was supported in part by NIH grants (RO1NS48216, NS038118) and AHA EIA 0540154 (Dandan Sun).

References

Abdullaev, I.F., Rudkouskaya, A., Schools, G.P., Kimelberg, H.K., and Mongin, A.A. (2006). Pharmacological comparison of swelling-activated excitatory amino acid release and Cl^{-} currents in cultured rat astrocytes. *J. Physiol.* **572**, 677–689.

Alvarez-Leefmans, F.J., Gamiño, S.M., Giraldez, F., and Nogueron, I. (1988). Intracellular chloride regulation in amphibian dorsal root ganglion neurones studied with ion-selective microelectrodes. *J. Physiol.* **406**, 225–246.

Alvarez-Leefmans, F.J. (2001). *Intracellular chloride regulation*. In *Cell Physiology Sourcebook: A Molecular Approach*. (N. Sperelakis, Ed.), pp. 301–318. Academic Press, San Diego.

Arundine, M. and Tymianski, M. (2004). Molecular mechanisms of glutamate-dependent neurodegeneration in ischemia and traumatic brain injury. *Cell Mol. Life Sci.* **61**, 657–668.

Beck, J., Lenart, B., Kintner, D.B., and Sun, D. (2003). Na-K-Cl cotransporter contributes to glutamate-mediated excitotoxicity. *J. Neurosci.* **23**, 5061–5068.

Bekar, L.K. and Walz, W. (1999). Evidence for chloride ions as intracellular messenger substances in astrocytes. *J. Neurophysiol.* **82**, 248–254.

Berglund, K., Dunbar, R.L., Lee, P., Feng, G., and Augustine, G.J. (2004). *Imaging synaptic inhibition with Clomeleon, a genetically encoded chloride indicator*. In: Imaging in neuroscience and development : a laboratory manual, edited by Rafael Yuste, Arthur Konnerth. Cold Spring Harbor Laboratory Press, Cold Spring Harbor, NY. p 84.

Bondarenko, A. and Chesler, M. (2001). Calcium dependence of rapid astrocyte death induced by transient hypoxia, acidosis, and extracellular ion shifts. *Glia* **34**, 143–149.

Bondarenko, A. and Chesler, M. (2001). Rapid astrocyte death induced by transient hypoxia, acidosis, and extracellular ion shifts. *Glia* **34**, 134–142.

Busse, S., Breder, J., Dinkel, K., Reymann, K.G., and Schroder, U.H. (2005). Inhibitors of cation-chloride-cotransporters affect hypoxic/hypoglycemic injury in hippocampal slices. *Brain Res.* **1046**, 116–121.

Chen, H., Kintner, D.B., Jones, M., Matsuda, T., Baba, A., Kiedrowski, L., and Sun, D. (2007). AMPA-mediated excitotoxicity in oligodendrocytes: role for Na^{+}-K^{+}-Cl^{-} co-transport and reversal of Na^{+}/Ca^{2+} exchanger. *J. Neurochem.* **102**, 1783–1795.

Chen, H., Luo, J., Kintner, D.B., Shull, G.E., and Sun, D. (2005). Na^{+}-dependent chloride transporter (NKCC1)-null mice exhibit less gray and white matter damage after focal cerebral ischemia. *J. Cereb. Blood Flow Metab.* **25**, 54–66.

Chen, Q., Olney, J.W., Lukasiewicz, P.D., Almli, T., and Romano, C. (1998). Ca^{2+}-independent excitotoxic neurodegeneration in isolated retina, an intact neural net: a role for Cl^{-} and inhibitory transmitters. *Mol. Pharmacol.* **53**, 564–572.

Chen, Y. and Swanson, R.A. (2003). Astrocytes and brain injury. *J. Cereb. Blood Flow Metab.* **23**, 137–149.

Czyz, A. and Kiedrowski, L. (2002). In depolarized and glucose-deprived neurons, Na^{+} influx reverses plasmalemmal K^{+}-dependent and K^{+}-independent Na^{+}/Ca^{2+} exchangers and contributes to NMDA excitotoxicity. *J. Neurochem.* **83**, 1321–1328.

Dewar, D., Yam, P., and McCulloch, J. (1999). Drug development for stroke: importance of protecting cerebral white matter. *Eur. J. Pharmacol.* **375**, 41–50.

Erecinska, M. and Silver, I.A. (1990). Metabolism and role of glutamate in mammalian brain. *Prog. Neurobiol.* **35**, 245–296.

Floyd, C.L., Gorin, F.A., and Lyeth, B.G. (2005). Mechanical strain injury increases intracellular sodium and reverses Na^{+}/Ca^{2+} exchange in cortical astrocytes. *Glia* **51**, 35–46.

Floyd, C.L. and Lyeth, B.G. (2007). Astroglia: important mediators of traumatic brain injury. *Prog. Brain Res.* **161**, 61–79.

Fraser, D.D., Duffy, S., Angelides, K.J., Perez-Velazquez, J.L., Kettenmann, H., and MacVicar, B.A. (1995). $GABA_A$/benzodiazepine receptors in acutely isolated hippocampal astrocytes. *J. Neurosci.* **15**, 2720–2732.

Galeffi, F., Sah, R., Pond, B.B., George, A., and Schwartz-Bloom, R. D. (2004). Changes in intracellular chloride after oxygen-glucose deprivation of the adult hippocampal slice: effect of diazepam. *J. Neurosci.* **24**, 4478–4488.

Hasbani, M.J., Hyrc, K.L., Faddis, B.T., Romano, C., and Goldberg, M.P. (1998). Distinct roles for sodium, chloride, and calcium in excitotoxic dendritic injury and recovery. *Exp. Neurol.* **154**, 241–258.

Hossmann, K.A. (1994). Viability thresholds and the penumbra of focal ischemia. *Ann. Neurol.* **36**, 557–565.

Hoyt, K.R., Arden, S.R., Aizenman, E., and Reynolds, I.J. (1998). Reverse Na^+/Ca^{2+} exchange contributes to glutamate-induced intracellular Ca^{2+} concentration increases in cultured rat forebrain neurons. *Mol. Pharmacol.* **53**, 742–749.

Inglefield, J.R. and Schwartz-Bloom, R.D. (1998). Activation of excitatory amino acid receptors in the rat hippocampal slice increases intracellular Cl^- and cell volume. *J. Neurochem.* **71**, 1396–1404.

Jang, I.S., Jeong, H.J., and Akaike, N. (2001). Contribution of the Na-K-Cl cotransporter on $GABA_A$ receptor-mediated presynaptic depolarization in excitatory nerve terminals. *J. Neurosci.* **21**, 5962–5972.

Kaila, K. (1994). Ionic basis of $GABA_A$ receptor channel function in the nervous system. *Prog. Neurobiol.* **42**, 489–537.

Kang, T.C., An, S.J., Park, S.K., Hwang, I.K., Yoon, D.K., Shin, H.S., and Won, M.H. (2002). Changes in Na^+-K^+-Cl^- cotransporter immunoreactivity in the gerbil hippocampus following transient ischemia. *Neurosci. Res.* **44**, 249–254.

Katchman, A.N., Vicini, S., and Hershkowitz, N. (1994). Mechanism of early anoxia-induced suppression of the $GABA_A$-mediated inhibitory postsynaptic current. *J. Neurophysiol.* **71**, 1128–1138.

Kimelberg, H.K. (2005). Astrocytic swelling in cerebral ischemia as a possible cause of injury and target for therapy. *Glia* **50**, 389–397.

Kintner, D.B., Luo, J., Gerdts, J., Ballard, A.J., Shull, G.E., and Sun, D. (2007a). Role of Na^+-K^+-Cl^- cotransport and Na^+/Ca^{2+} exchange in mitochondrial dysfunction in astrocytes following in vitro ischemia. *Am. J. Physiol. Cell Physiol.* **292**, C1113–C1122.

Kintner, D.B., Su, G., Lenart, B., Ballard, A.J., Meyer, J.W., Ng, L.L., Shull, G.E., and Sun, D. (2004). Increased tolerance to oxygen and glucose deprivation in astrocytes from Na^+/H^+ exchanger isoform 1 null mice. *Am. J. Physiol. Cell Physiol.* **287**, C12–C21.

Kintner, D.B., Wang, Y., and Sun, D. (2007b). Role of membrane ion transport proteins in cerebral ischemic damage. *Front. Biosci.* **12**, 762–770.

Koo, E.H., Sisodia, S.S., Archer, D.R., Martin, L.J., Weidemann, A., Beyreuther, K., Fischer, P., Masters, C.L., and Price, D.L. (1990). Precursor of amyloid protein in Alzheimer disease undergoes fast anterograde axonal transport. *Proc. Natl. Acad. Sci. USA* **87**, 1561–1565.

Kumar, V., Naik, R.S., Hillert, M., and Klein, J. (2006). Effects of chloride flux modulators in an in vitro model of brain edema formation. *Brain Res.* **1122**, 222–229.

Kuner, T. and Augustine, G.J. (2000). A genetically encoded ratiometric indicator for chloride: capturing chloride transients in cultured hippocampal neurons. *Neuron* **27**, 447–459.

Lenart, B., Kintner, D.B., Shull, G.E., and Sun, D. (2004). Na-K-Cl cotransporter-mediated intracellular Na^+ accumulation affects Ca^{2+} signaling in astrocytes in an in vitro ischemic model. *J. Neurosci.* **24**, 9585–9597.

Lipton, P. (1999). Ischemic cell death in brain neurons. *Physiol. Rev.* **79**, 1431–1568.

Lo, E.H., Dalkara, T., and Moskowitz, M.A. (2003). Mechanisms, challenges and opportunities in stroke. *Nat. Rev. Neurosci.* **4**, 399–415.

Longuemare, M.C., Rose, C.R., Farrell, K., Ransom, B.R., Waxman, S.G., and Swanson, R.A. (1999). K^+-induced reversal of astrocyte glutamate uptake is limited by compensatory changes in intracellular Na^+. *Neuroscience* **93**, 285–292.

Lopachin, R.M., Gaughan, C.L., Lehning, E.J., Weber, M.L., and Taylor, C.P. (2001). Effects of ion channel blockade on the distribution of Na, K, Ca and other elements in oxygen-glucose deprived CA1 hippocampal neurons. *Neuroscience* **103**, 971–983.

Lukaszevicz, A.C., Sampaio, N., Guegan, C., Benchoua, A., Couriaud, C., Chevalier, E., Sola, B., Lacombe, P., and Onteniente, B. (2002). High sensitivity of protoplasmic cortical astroglia to focal ischemia. *J. Cereb. Blood Flow Metab.* **22**, 289–298.

Luo, J., Wang, Y., Chen, H., Kintner, D.B., Cramer, S.W., Gerdts, J.K., Chen, X., Shull, G.E., Philipson, K.D., and Sun, D. (2007). A concerted role of Na^+-K^+-$2Cl^-$ cotransporter and Na^+/Ca^{2+} exchanger in ischemic damage. *J. Cereb. Blood Flow Metab.* **28**, 737–746.

MacVicar, B.A., Feighan, D., Brown, A., and Ransom, B. (2002). Intrinsic optical signals in the rat optic nerve: role for K^+ uptake via NKCC1 and swelling of astrocytes. *Glia* **37**, 114–123.

MacVicar, B.A., Tse, F.W., Crichton, S.A., and Kettenmann, H. (1989). GABA-activated Cl^- channels in astrocytes of hippocampal slices. *J. Neurosci.* **9**, 3577–3583.

Maxwell, W.L., Bullock, R., Landholt, H., and Fujisawa, H. (1994). Massive astrocytic swelling in response to extracellular glutamate – a possible mechanism for post-traumatic brain swelling? *Acta Neurochir. Suppl (Wien.)* **60**, 465–467.

Misgeld, U., Deisz, R.A., Dodt, H.U., and Lux, H.D. (1986). The role of chloride transport in postsynaptic inhibition of hippocampal neurons. *Science* **232**, 1413–1415.

Mittmann, T., Qu, M., Zilles, K., and Luhmann, H.J. (1998). Long-term cellular dysfunction after focal cerebral ischemia: in vitro analyses. *Neuroscience* **85**, 15–27.

Mongin, A.A. (2007). Disruption of ionic and cell volume homeostasis in cerebral ischemia: the perfect storm. *Pathophysiology* **14**, 183–193.

Muller, M. (2000). Effects of chloride transport inhibition and chloride substitution on neuron function and on hypoxic spreading-depression-like depolarization in rat hippocampal slices. *Neuroscience* **97**, 33–45.

Nabekura, J., Ueno, T., Okabe, A., Furuta, A., Iwaki, T., Shimizu-Okabe, C., Fukuda, A., and Akaike, N. (2002). Reduction of KCC2 expression and $GABA_A$ receptor-mediated excitation after in vivo axonal injury. *J. Neurosci.* **22**, 4412–4417.

Newell, D.W., Barth, A., Papermaster, V., and Malouf, A.T. (1995). Glutamate and non-glutamate receptor mediated toxicity caused by oxygen and glucose deprivation in organotypic hippocampal cultures. *J. Neurosci.* **15**, 7702–7711.

Ni, Y., Malarkey, E.B., and Parpura, V. (2007). Vesicular release of glutamate mediates bidirectional signaling between astrocytes and neurons. *J. Neurochem.* **103**, 1273–1284.

Nicklas, W.J., Zeevalk, G., and Hyndman, A. (1987). Interactions between neurons and glia in glutamate/glutamine compartmentation. *Biochem. Soc. Trans.* **15**, 208–210.

Nielsen, S., Nagelhus, E.A., Amiry-Moghaddam, M., Bourque, C., Agre, P., and Ottersen, O.P. (1997). Specialized membrane domains for water transport in glial cells: high-resolution immunogold cytochemistry of aquaporin-4 in rat brain. *J. Neurosci.* **17**, 171–180.

Noraberg, J., Poulsen, F.R., Blaabjerg, M., Kristensen, B.W., Bonde, C., Montero, M., Meyer, M., Gramsbergen, J.B., and Zimmer, J. (2005). Organotypic hippocampal slice cultures for studies of brain damage, neuroprotection and neurorepair. *Curr. Drug Targets. CNS Neurol. Disord.* **4**, 435–452.

O'Donnell, M.E., Tran, L., Lam, T.I., Liu, X.B., and Anderson, S.E. (2004). Bumetanide inhibition of the blood–brain barrier Na-K-Cl cotransporter reduces edema formation in the rat middle cerebral artery occlusion model of stroke. *J. Cereb. Blood Flow Metab.* **24**, 1046–1056.

Oliet, S.H., Piet, R., and Poulain, D.A. (2001). Control of glutamate clearance and synaptic efficacy by glial coverage of neurons. *Science* **292**, 923–926.

Olney, J.W. (1969). Brain lesions, obesity, and other disturbances in mice treated with monosodium glutamate. *Science* **164**, 719–721.

Olney, J.W., Price, M.T., Samson, L., and Labruyere, J. (1986). The role of specific ions in glutamate neurotoxicity. *Neurosci. Lett.* **65**, 65–71.

Park, E., Velumian, A.A., and Fehlings, M.G. (2004). The role of excitotoxicity in secondary mechanisms of spinal cord injury: a review with an emphasis on the implications for white matter degeneration. *J. Neurotrauma* **21**, 754–774.

Paulson, O.B. and Newman, E.A. (1987). Does the release of potassium from astrocyte endfeet regulate cerebral blood flow? *Science* **237**, 896–898.

Pluta, R. (1987). Resuscitation of the rabbit brain after acute complete ischemia lasting up to one hour: pathophysiological and pathomorphological observations. *Resuscitation* **15**, 267–287.

Pond, B.B., Berglund, K., Kuner, T., Feng, G., Augustine, G.J., and Schwartz-Bloom, R.D. (2006). The chloride transporter Na^+-K^+-Cl^- cotransporter isoform-1 contributes to intracellular chloride increases after in vitro ischemia. *J. Neurosci.* **26**, 1396–1406.

Pond, B.B., Galeffi, F., Ahrens, R., and Schwartz-Bloom, R.D. (2004). Chloride transport inhibitors influence recovery from oxygen-glucose deprivation-induced cellular injury in adult hippocampus. *Neuropharmacology* **47**, 253–262.

Pulsinelli, W.A., Brierley, J.B., and Plum, F. (1982). Temporal profile of neuronal damage in a model of transient forebrain ischemia. *Ann. Neurol.* **11**, 491–498.

Qureshi, N. (1996). Tissue plasminogen activator for acute ischemic stroke. *N. Engl. J. Med.* **334**, 1406.

Rose, C.R., Waxman, S.G., and Ransom, B.R. (1998). Effects of glucose deprivation, chemical hypoxia, and simulated ischemia on Na^+ homeostasis in rat spinal cord astrocytes. *J. Neurosci.* **18**, 3554–3562.

Rosenberg, L.J. and Lucas, J.H. (1996). Reduction of NaCl increases survival of mammalian spinal neurons subjected to dendrite transection injury. *Brain Res.* **734**, 349–353.

Rothman, S.M. (1985). The neurotoxicity of excitatory amino acids is produced by passive chloride influx. *J. Neurosci.* **5**, 1483–1489.

Rutledge, E.M., Aschner, M., and Kimelberg, H.K. (1998). Pharmacological characterization of swelling-induced D-[^{3}H]aspartate release from primary astrocyte cultures. *Am. J. Physiol.* **274**, 1–20.

Rutledge, E.M. and Kimelberg, H.K. (1996). Release of [^{3}H]-D-aspartate from primary astrocyte cultures in response to raised external potassium. *J. Neurosci.* **16**, 7803–7811.

Salter, M.G. and Fern, R. (2005). NMDA receptors are expressed in developing oligodendrocyte processes and mediate injury. *Nature* **438**, 1167–1171.

Schlue, W.R. and Deitmer, J.W. (1988). Ionic mechanisms of intracellular pH regulation in the nervous system. *Ciba Found. Symp.* **139**, 47–69.

Schurr, A. (2006). Lactate: the ultimate cerebral oxidative energy substrate? *J. Cereb. Blood Flow Metab.* **26**, 142–152.

Schwartz-Bloom, R.D. and Sah, R. (2001). Gamma-aminobutyric acid(A) neurotransmission and cerebral ischemia. *J. Neurochem.* **77**, 353–371.

Siesjo, B.K. (1992). Pathophysiology and treatment of focal cerebral ischemia. Part I: Pathophysiology. *J. Neurosurg.* **77**, 169–184.

Siesjo, B.K. (1992). Pathophysiology and treatment of focal cerebral ischemia. Part II: Mechanisms of damage and treatment. *J. Neurosurg.* **77**, 337–354.

Silver, I.A., Deas, J., and Erecinska, M. (1997). Ion homeostasis in brain cells: differences in intracellular ion responses to energy limitation between cultured neurons and glial cells. *Neuroscience* **78**, 589–601.

Strange, K., Emma, F., and Jackson, P.S. (1996). Cellular and molecular physiology of volume-sensitive anion channels. *Am. J. Physiol.* **270**, C711–C730.

Stys, P.K. (2004). White matter injury mechanisms. *Curr. Mol. Med.* **4**, 113–130.

Stys, P.K. (2005). General mechanisms of axonal damage and its prevention. *J. Neurol. Sci.* **233**, 3–13.

Su, G., Haworth, R.A., Dempsey, R.J., and Sun, D. (2000). Regulation of Na^+-K^+-Cl^- cotransporter in primary astrocytes by dibutyryl cAMP and high $[K^+]_o$. *Am. J. Physiol. Cell Physiol.* **279**, C1710–C1721.

Su, G., Kintner, D.B., Flagella, M., Shull, G.E., and Sun, D. (2002b). Astrocytes from Na^+-K^+-Cl^- cotransporter-null mice exhibit absence of swelling and decrease in EAA release. *Am. J. Physiol. Cell Physiol.* **282**, C1147–C1160.

Su, G., Kintner, D.B., and Sun, D. (2002a). Contribution of Na^+-K^+-Cl^- cotransporter to high-[K(+)](o)- induced swelling and EAA release in astrocytes. *Am. J. Physiol. Cell Physiol.* **282**, C1136–C1146.

Sun, D. and Murali, S.G. (1998). Stimulation of Na^+-K^+-$2Cl^-$ cotransporter in neuronal cells by excitatory neurotransmitter glutamate. *Am. J. Physiol. Cell Physiol.* **275**, C772–C779.

Sung, K.W., Kirby, M., McDonald, M.P., Lovinger, D.M., and Delpire, E. (2000). Abnormal $GABA_A$ receptor-mediated currents in dorsal root ganglion neurons isolated from Na-K-2Cl cotransporter null mice. *J. Neurosci.* **20**, 7531–7538.

Taylor, C.P., Weber, M.L., Gaughan, C.L., Lehning, E.J., and Lopachin, R.M. (1999). Oxygen/glucose deprivation in hippocampal slices: altered intraneuronal elemental composition predicts structural and functional damage. *J. Neurosci.* **19**, 619–629.

Tekkok, S.B. and Goldberg, M.P. (2001). Ampa/kainate receptor activation mediates hypoxic oligodendrocyte death and axonal injury in cerebral white matter. *J. Neurosci.* **21**, 4237–4248.

Thomas, R., Salter, M.G., Wilke, S., Husen, A., Allcock, N., Nivison, M., Nnoli, A.N., and Fern, R. (2004). Acute ischemic injury of astrocytes is mediated by Na-K-Cl cotransport and not Ca^{2+} influx at a key point in white matter development. *J. Neuropathol. Exp. Neurol.* **63**, 856–871.

Thompson, S.M. and Gahwiler, B.H. (1989). Activity-dependent disinhibition. II. Effects of extracellular potassium, furosemide, and membrane potential on E_{Cl^-} in hippocampal CA3 neurons. *J. Neurophysiol.* **61**, 512–523.

Tower, D.B. and Young, O.M. (1973). The activities of butyrylcholinesterase and carbonic anhydrase, the rate of anaerobic glycolysis, and the question of a constant density of glial cells in cerebral cortices of various mammalian species from mouse to whale. *J. Neurochem.* **20**, 269–278.

Toyoda, H., Ohno, K., Yamada, J., Ikeda, M., Okabe, A., Sato, K., Hashimoto, K., and Fukuda, A. (2003). Induction of NMDA and

GABAA receptor-mediated Ca^{2+} oscillations with KCC2 mRNA downregulation in injured facial motoneurons. *J. Neurophysiol.* **89**, 1353–1362.

Valeriani, V., Dewar, D., and McCulloch, J. (2000). Quantitative assessment of ischemic pathology in axons, oligodendrocytes, and neurons: attenuation of damage after transient ischemia. *J. Cereb. Blood Flow Metab.* **20**, 765–771.

Wake, H., Watanabe, M., Moorhouse, A.J., Kanematsu, T., Horibe, S., Matsukawa, N., Asai, K., Ojika, K., Hirata, M., and Nabekura, J. (2007). Early changes in KCC2 phosphorylation in response to neuronal stress result in functional downregulation. *J. Neurosci.* **27**, 1642–1650.

Walz, W. (2000). Role of astrocytes in the clearance of excess extracellular potassium. *Neurochem. Int.* **36**, 291–300.

Waxman, S.G. and Ritchie, J.M. (1993). Molecular dissection of the myelinated axon. *Ann. Neurol.* **33**, 121–136.

Wilke, S., Thomas, R., Allcock, N., and Fern, R. (2004). Mechanism of acute ischemic injury of oligodendroglia in early myelinating white matter: the importance of astrocyte injury and glutamate release. *J. Neuropathol. Exp. Neurol.* **63**, 872–881.

Xue, J., Zhou, D., Yao, H., and Haddad, G.G. (2008). Role of transporters and ion channels in neuronal injury under hypoxia. *Am. J. Physiol. Regul. Integr. Comp. Physiol.* **294**, R451–R457.

Yam, P.S., Dunn, L.T., Graham, D.I., Dewar, D., and McCulloch, J. (2000). NMDA receptor blockade fails to alter axonal injury in focal cerebral ischemia. *J. Cereb. Blood Flow Metab.* **20**, 772–779.

Yamada, Y., Fukuda, A., Tanaka, M., Shimano, Y., Nishino, H., Muramatsu, K., Togari, H., and Wada, Y. (2001). Optical imaging reveals cation-Cl^- cotransporter-mediated transient rapid decrease in intracellular Cl^- concentration induced by oxygen-glucose deprivation in rat neocortical slices. *Neurosci. Res.* **39**, 269–280.

Yan, Y., Dempsey, R.J., Flemmer, A., Forbush, B., and Sun, D. (2003). Inhibition of Na^+-K^+-Cl^- cotransporter during focal cerebral ischemia decreases edema and neuronal damage. *Brain Res.* **961**, 22–31.

Yan, Y.P., Dempsey, R.J., and Sun, D. (2001). Expression of Na^+-K^+-Cl^- cotransporter in rat brain during development and its localization in mature astrocytes. *Brain Res.* **911**, 43–55.

Yan, Y.P., Dempsey, R.J., and Sun, D. (2001). Na^+-K^+-Cl^- cotransporter in rat focal cerebral ischemia. *J. Cereb. Blood Flow Metab.* **21**, 711–721.

Yao, H., Sun, X., Gu, X., Wang, J., and Haddad, G.G. (2007). Cell death in an ischemic infarct rim model. *J. Neurochem.* **103**, 1644–1653.

CHAPTER

26

Chloride Transport in Glioma Growth and Cell Invasion

Harald Sontheimer

OUTLINE

I. INTRODUCTION

Brain tumors fall into two principal categories, primary and secondary. Primary tumors are often called gliomas and originate in the brain. Secondary or metastatic brain tumors are peripheral cancers that invade the brain. Together they account for well over 100,000 new cancer cases diagnosed each year in the USA, of which approximately 40,000 are primary tumors (according to data from the Central Brain Tumor Registry of the United States, CBTRUS). In addition to their dissimilar origin, primary and secondary brain tumors differ in many aspects of their etiology and biology. For example, metastatic cancers are easily distinguishable from normal brain tissue as they represent the new growth of cancerous tissue with the properties of the organ it originated from. Hence, they present as liver or lung cells growing within brain and tumors typically grow as confined solid masses. This is not the case for primary brain tumors which often lack clear boundaries between normal and malignant brain tissue. A representative example is illustrated in Fig. 26.1A, which shows a cerebral glioma with characteristic diffuse margins. An important difference between these two cancer types relates to how the tumors spread and form metastasis. Metastatic brain tumors disseminate hematogenously throughout the body and enter the brain through the vasculature. By comparison, primary brain tumors rarely metastasize into the periphery but instead spread within the brain often reaching distant sites such as the contralateral brain hemispheres or the spinal cord, as illustrated

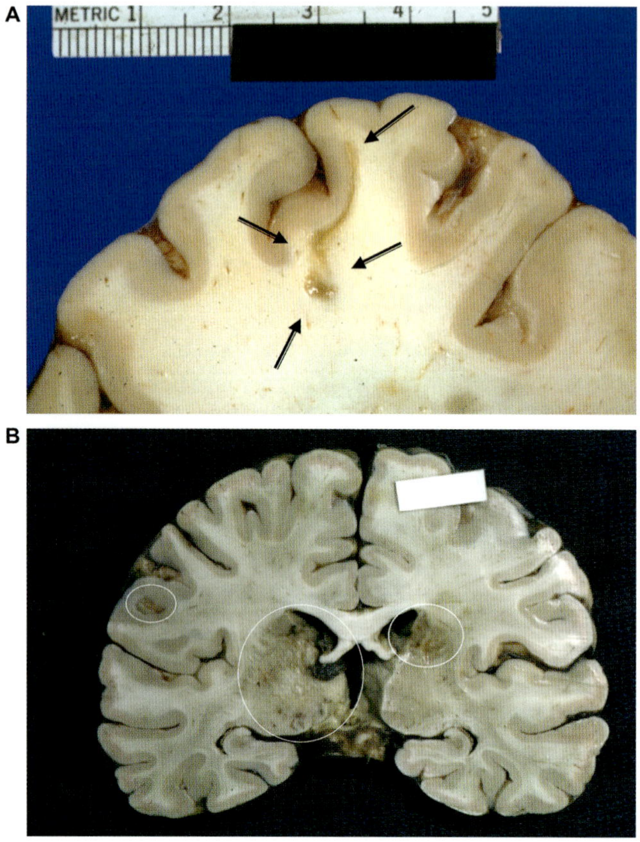

FIGURE 26.1 Primary glioma at autopsy. **A.** Poorly defined margins are characteristic of cerebral gliomas, like the one shown in this example (arrows). **B.** Although gliomas rarely metastasize outside the brain, they often present secondary tumors in other parts of the brain, often distant from the site of the primary tumor. These secondary tumors are highlighted in B by white ovals. Copyrighted images: University of Alabama at Birmingham, Department of Pathology.

in the example shown in Fig. 26.1B. These cancers spread by active cell migration without extravasating into the vasculature. This is reminiscent of neuronal and glial cell migration during brain development or stem cells and microglial cells in the adult brain suggesting that some of the underlying mechanisms of migration may be shared between these cells.

II. GLIOMAS AND THEIR LINEAGE

Although primary brain tumors can originate from a number of growth competent cells in the brain or spinal cord, the majority of them appear to derive from glial cells or their precursors. Reflecting this presumed relationship these tumors are collectively called gliomas. These include a diverse group of cancers that may not always be of defined lineage. Among the distinguishing features are immuno-positivity for certain glial associated antigens (Kleihues et al., 1995), for example to glial fibrillary acidic protein (GFAP), myelin associated glycoprotein (MAG), myelin basic protein (MBP), S100 beta or vimentin. GFAP and/or S100 positive cells are frequently termed astrocytomas, MBP- or MAG-positive cells; Oligodendrogliomas and cells that stain for both sets of markers are mixed gliomas. While these names imply a known and well-defined lineage relationship of these tumors with normal glial cell type or their progenitor cells, such a relationship has not actually been demonstrated and the cell types of origin remain controversial. In studies addressing this question, investigators have transfected glial progenitor cells with known mutations in oncogenes and tumor suppressor genes and have been able to induce a malignant transformation that yielded tumor growth in mice, suggesting that committed glial progenitor cells may indeed be the most likely cell type of origin (Dai et al., 2001).

Gliomas exhibit many of the characteristic features of systemic cancers which include mutations in the tumor suppressor genes P16 and P53, and amplification and overexpression of certain oncogenic growth factor receptors including EGF-R or PDGF-R (Von Deimling et al., 1995). As with other cancers, angiogenesis or the induction of new blood vessels in response to the release of vascular endothelial growth factor is common (Plate and Risau, 1995). Furthermore, the release of matrix degrading enzymes that facilitate the remodeling of the tumor associated extracellular space is common and facilitates cell invasion (Giese et al., 1994).

A glioma diagnosis is almost always fatal as current treatment options are limited (Butowski et al., 2006). By the time a tumor is detectable, it has frequently seeded tumor cells throughout the nervous system, and upon surgery these cells can quickly give rise to recurrent malignancies. The diffuse pattern of cellular invasion illustrated in Fig. 26.1 not only makes complete surgical resection impossible, but also limits focal treatments such as exogenous beam irradiation as cells remote from the tumor will escape the radiation. Upon recurrence, many gliomas become even more malignant. Recurrence is believed to result from cells that have invaded surrounding brain areas. Surprisingly, little is known about the underlying mechanisms. For example, pathways of cell migration are poorly understood as are molecules involved in chemotaxis and path finding. These aspects of glioma biology are promising areas for future research as they may yield more effective therapeutic tools. An important aspect of tumor biology that has been well studied in recent years and which will be discussed in greater detail in this chapter pertains to biophysical and biomechanical adaptations that support the migration and invasion of gliomas into normal brain tissue. Some of these findings may pertain to other migratory cells in the brain and even to other cancers.

III. GLIOMA MIGRATION AND INVASION

As illustrated in Fig. 26.1, the boundaries between a primary glioma and normal brain tissue are often difficult to delineate at a macroscopic level. At a microscopic level, thousands of glioma cells will have diffusely invaded the surrounding areas of brain tissue, and, over time, they will have spread to very distant sites. Wherever possible, invading glioma cells appear to take advantage of other structures in the brain to migrate. For example, they frequently migrate along nerve fiber bundles or, as illustrated in Fig. 26.2A, along blood vessels. Whether the spaces along these structures are more favorable for migration, or whether there are other guidance cues, or a more slippery extracellular matrix, is not entirely clear. Without question, however, the narrowness of the extracellular space provides a significant impediment to cell migration. At the electron microscopic level, invading cells appear elongated, wedge shaped, and with an overall shrunken appearance (Fig. 26.2B). This has led to the hypothesis that glioma cells may dynamically adjust their cell volume as they invade. As illustrated in cartoon form in Fig. 26.2C, and further discussed below, recent findings support this hypothesis and suggest an important role for Cl^- channels and transporters in this context.

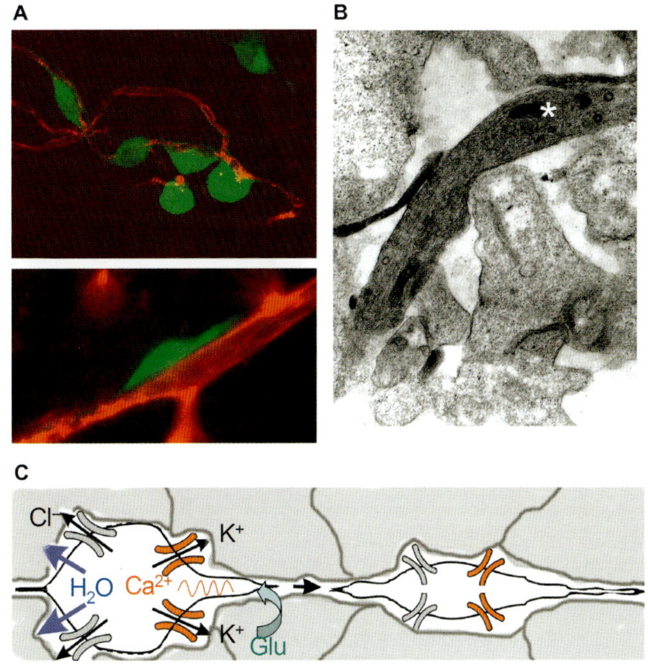

FIGURE 26.2 Invading glioma cells *in situ*. **A.** Confocal images of invading D54MG cells stably expressing EGFP (green). The invading glioma cells adhere to blood vessels (red). Cells often exhibit an elongated wedge-shaped appearance as shown in the lower panel. **B.** The elongated shape is quite apparent in this electron micrograph that captured an invading glioma cell (*, recognized because of the abundance of ribosomes and other organelles that incorporate lead citrate and give a darker appearance) extending between normal brain cells, likely astrocytes (based on their large nuclei and abundance of electron dense, glycogen deposits throughout the cytoplasm). **C.** Cell shrinkage requires water efflux which is driven by the concomitant efflux of Cl^- and K^+ through their respective ion channels. Glutamate is shown as a possible motogenic stimulus acting via AMPA receptors that raise intracellular $[Ca^{2+}]$ which may in turn activate BK^+ channels. (Panel B is reproduced with permission from Soroceanu et al., 1999.)

IV. Cl^- TRANSPORT AND CELL VOLUME REGULATION IN GLIOMA CELLS

As extensively discussed in Chapter 15 in this book, all eukaryotic cells have developed powerful mechanisms to maintain a constant cell volume even when extracellular osmotic conditions change. Glioma cells are not an exception; when exposed to a 50% hyposmotic challenge they regulate their volume back to baseline within just a few minutes. As illustrated in Fig. 26.3A, this regulatory volume decrease (RVD) is inhibited by drugs known to block Cl^- channels including NPPB, Cd^{2+} and DIDS with almost complete inhibition by NPPB and Cd^{2+} (Ernest et al., 2005). The remaining volume regulatory response is inhibited by drugs that block the K^+-Cl^- cotransporters (Ernest et al., 2005). Furthermore, volume regulation is supported when Cl^- is replaced by halide ions such as I^- or Br^- with known permeability to Cl^- channels, but not when gluconate substitutes for Cl^- (Ernest, 2007). These studies suggest that RVD in glioma cells utilizes Cl^- as osmolyte, which is released from the cell through Cl^- channels.

An important question is whether Cl^- may similarly act as an osmolyte during cell volume changes associated with cell invasion, a process that is very different from cell volume changes elicited by osmotic challenges. Under hyposmotic conditions, a gradient for Cl^- efflux is favored by the dilution of extracellular ions with water, whereas under isosmotic conditions, this is not the case, unless the cell has a sufficiently high $[Cl^-]_i$. Hence, the hypothesized cell shrinkage of invading cells requires that intracellular Cl^- be accumulated so that an outward directed

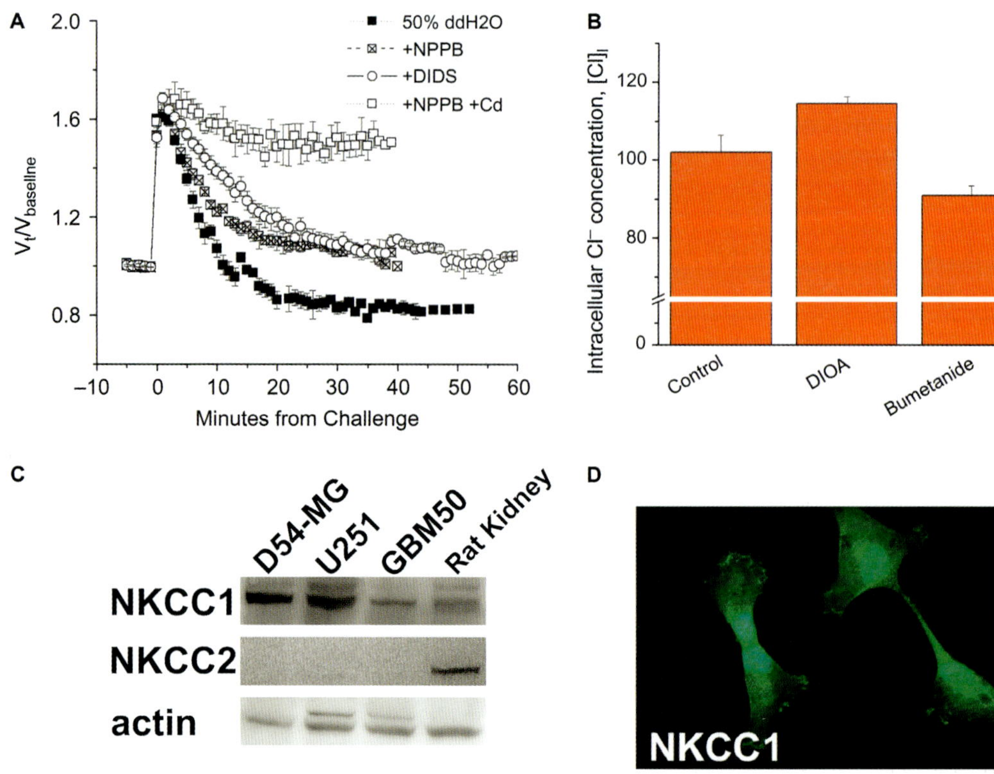

FIGURE 26.3 Volume regulation in glioma cells. **A.** On exposure to a 50% hypotonic challenge, glioma cells swell and regulate their volume back to baseline or even below the baseline level. This regulatory volume decrease is partially inhibited by NPPB or DIDS and is almost completely inhibited by NPPB and Cd^{2+}. **B.** Glioma cells maintain an elevated intracellular $[Cl^-]$ which is accumulated via the bumetanide-sensitive Na^+-K^+-Cl^- cotransporter, NKCC1. Pharmacological inhibition of the cotransporter by 20 μM bumetanide causes a decrease in $[Cl^-]_i$. Exposure to 40 μM DIOA causes an increase in $[Cl^-]_i$ above control levels, presumably by inhibition of KCC mediated Cl^- efflux. **C.** Western blot of lysates from two glioma cell-lines D54-MG and U251, and from samples obtained from one patient (GBM50), show prominent expression of NKCC1 but absence of NKCC2. Rat kidney lysates were used as control for NKCC2. **D.** Immunostaining also shows prominent membrane associated labeling for NKCC1 in representative U251 glioma cells. Antibodies directed against NKCC1 and NKCC2 were from Alpha Diagnostics, and were used at a dilution of 1:500. (Panels A and B are reproduced with permission from Ernest et al. (2005), and panel C is reproduced with permission from Ernest and Sontheimer (2007).)

electrochemical gradient for Cl^- is maintained. Using the Cl^- sensitive fluorescence indicator SPQ, intracellular $[Cl^-]$ was measured in glioma cells and found to be around 100 mM (Ernest, 2007), a value far greater than that typically observed in mature central neurons (7–10 mM), glial cells (30–40 mM) or mature primary afferent neurons (~45 mM) (see Chapters 7, 19 and 22 in this volume). These findings were recently confirmed by patch-clamp studies at the single cell level in which the reversal potential of Cl^- currents was used as an indirect indicator of $[Cl^-]_i$. Since glioma cells in culture lack ligand-gated Cl^- channels, such as the $GABA_A$ receptor-channel widely expressed in neurons, recombinant ligand-gated non-inactivating Cl^- channels (GABA-rho) were introduced into glial cells, and stable cell-lines expressing GABA-gated Cl^- channels were created. Gramicidin-perforated patch recordings allowed determination of the reversal potential of the GABA-induced currents (E_{GABA}). $[Cl^-]_i$ was estimated from E_{GABA} (Habela et al., 2009). These studies indicated an intracellular $[Cl^-]$ of 105 mM in glioma cells, a value close to that determined using SPQ. Hence, glioma cells maintain a steep outward directed gradient for Cl^-. In most cells, Cl^- is actively accumulated via the Na^+-K^+-$2Cl^-$ cotransporter (NKCC1), which is a widely expressed Cl^- importer (Chapters 2, 16 and 19 in this volume). Western blot and immunostaining analyses of several glioma cell-lines, including those obtained from acute patient biopsies, demonstrated prominent expression of NKCC1 and absence of NKCC2 (see Fig. 26.3C and 26.3D as well as Ernest and Sontheimer, 2007). Gliomas also express KCC1 and KCC3 (Ernest et al., 2005). Consistent with NKCC1 being principally responsible for the accumulation of intracellular Cl^- above electrochemical equilibrium in gliomas, pharmacological inhibition of the cotransporter with bumetanide causes a significant drop in intracellular

Cl^- (Fig. 26.3B). As we will be discussing below, high intracellular $[Cl^-]$ is possibly required for immature neurons and glioma cells to migrate. High Cl^- might facilitate water extrusion and cell shrinkage, processes necessary for migrating cells to navigate through confined extracellular spaces.

V. CHANGES IN CELL VOLUME OF INVADING CELL REQUIRE Cl^- EFFLUX VIA ClC CHANNELS

As illustrated in Fig. 26.2C migratory glioma cells appear to undergo profound changes in cell volume as they invade surrounding tissues. We hypothesize that these spontaneously occurring changes in cell volume are driven by efflux of Cl^- and obligatory water. The notion that a favorable outward Cl^- gradient is established by NKCC1 was experimentally tested in a recent study (Habela et al., 2009). Glioma cells were stably transfected with GABA-rho channels, and subjected to volume measurements using a Coulter Counter. Application of GABA induced opening of Cl^- channels which caused progressive cell volume decrease. This cell shrinkage was not observed in untransfected cells, or in the absence of GABA. This suggests that opening of Cl^- channels causes efflux of Cl^- associated with obligatory water loss, and cell shrinkage. These experiments also suggest that Cl^- efflux is sufficient to induce a volume decrease in glioma cells. Interestingly, these studies were made by inserting a ligand-gated Cl^- channel which could be activated on demand, but glioma cells express a significant resting Cl^- conductance (Ransom et al., 2001). Indeed, when recorded using perforated patch-clamp technique to avoid disturbing cytosolic Cl^-, glioma cells exhibit a resting Cl^- conductance sensitive to NPPB and DIDS. These currents are outwardly rectifying, show time-dependent inactivation at positive potentials and exhibit the following anion permeability sequence: $I^- > Br^- > Cl^-$. However, although the currents could be potentiated by cell swelling, this was not required for current activation. Because Cl^- channel inhibitors still lack specificity, attributing the inhibitory effect of NPPB and DIDS to a specific ion channel is not yet possible. As a first step towards the molecular identification of the Cl^- channels expressed in glioma cells, Western blots using lysates obtained from gliomas isolated from patients were probed with antibodies directed against epitopes of cloned Cl^- channels. These studies demonstrated the presence of ClC-2, ClC-3 and ClC-5 proteins in all gliomas examined (Olsen et al., 2003). Further, immunolabeling studies showed that ClC-3 staining was predominant in invading processes of isolated glioma cells. In an effort to further identify the Cl^- channels functioning in gliomas, cells were treated with antisense oligonucleotides to known members of the ClC Cl^- channels' super family. These studies showed prominent expression of currents attributable to ClC-2 and ClC-3, respectively (Olsen et al., 2003). ClC-2 currents, known to be sensitive to Cd^{2+}, inwardly rectifying, and potentiated by a negative prepulse to $-120\,mV$, were selectively eliminated in glioma cells treated with ClC-2 antisense oligonucleotides (Fig. 26.4A). As expected, these currents were unaltered by ClC-3 antisense oligonucleotides.

ClC-3 channels giving rise to outwardly rectifying currents that show time-dependent inactivation and are sensitive to NPPB were greatly reduced in glioma cells treated with ClC-3 antisense oligonucleotides (Fig. 26.4B). These data are consistent with both ClC-2 and ClC-3 channels being functionally expressed in gliomas. However, functional expression of ClC-3 channels in the plasma membrane is controversial, as discussed in detail in Chapter 12 of this volume. ClC-3 knockout mice primarily show CNS pathology associated with loss of synaptic vesicles in hippocampal neurons (Stobrawa et al., 2001). Thus, whether ClC-3 protein is able to generate functional channels in the plasma membrane has been questioned. Immunogold electron microscopy of human gliomas, however, shows immunoreactivity associated with both plasma membrane and intracellular vesicles (Fig. 26.4C). Further, ClC-3 in cultured glioma cells colocalizes to the β-subunit of cholera-toxin, which binds to lipid raft domains, arguing for membrane localization of the Cl^- channel (see merged signals in Fig. 26.4D).

Figure 26.5A shows that currents with the biophysical signature of ClC-3 are inhibited in a dose-dependent fashion by chlorotoxin (Cltx), a peptide isolated from the venom of the scorpion *Leiurus quinquestriatus* (DeBin et al., 1993). This toxin might inhibit Cl^- channels with some specificity (McFerrin and Sontheimer, 2005). Importantly, as shown in Fig. 26.5B–C, when cells were challenged to cross a transwell barrier that mimics the spatial constraints of the extracellular space in the brain, cell migration across the barrier was inhibited when the Cl^- conductance was blocked with NPPB (Ransom et al., 2001), Cd^{2+} or Cltx (Soroceanu et al., 1999). Of all these drugs, NPPB and Cltx were the most effective inhibitors of cell migration in the transwell assays. Cltx in both its native and recombinant form inhibited transwell migration in a dose-dependent fashion (Deshane et al., 2003). Further, a fluorescently labeled Cltx showed binding to the cell surface of human malignant glioma cells in patient biopsies. Based on these data we proposed ClC-3 as a

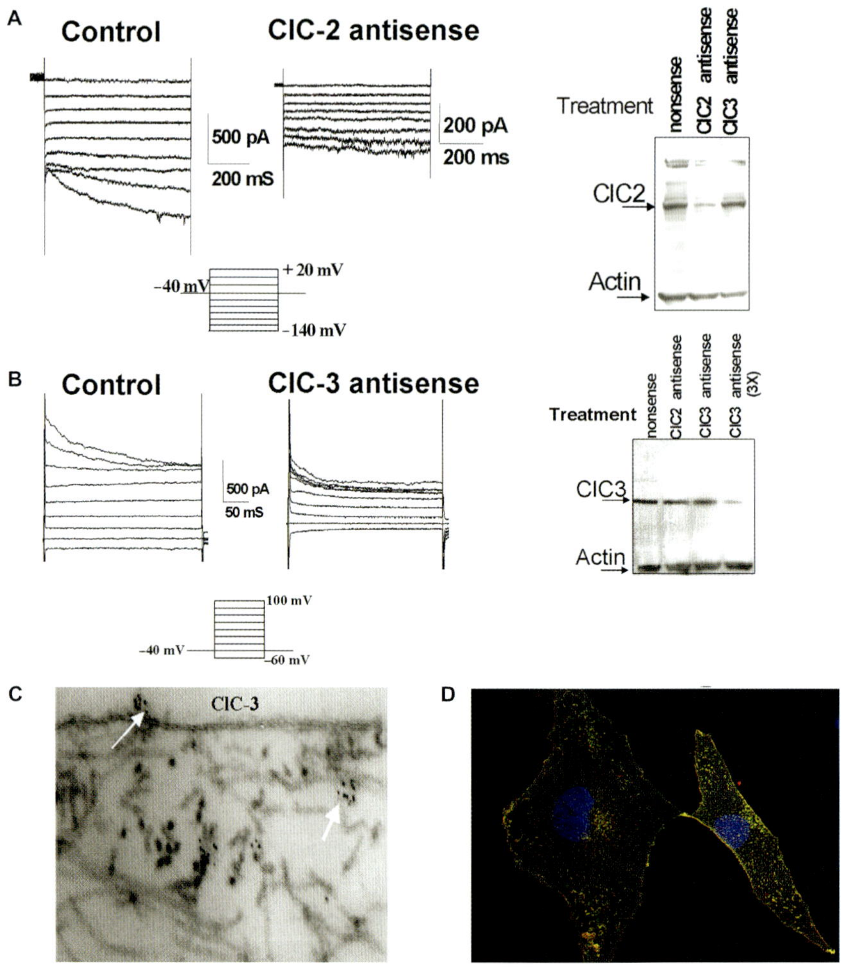

FIGURE 26.4 Glioma cells express functional ClC-2 and ClC-3 channels. Using specific antisense oligonucleotides to ClC-2 (**A**) and ClC-3 (**B**), it was possible to identify currents attributable to these channels, respectively. The Western blots illustrate effective reduction in corresponding protein expression following antisense treatment, demonstrating specificity of the effects observed in the membrane currents. **C.** Immuno-gold EM with antibodies to ClC-3 show clusters of channels at the cell surface (thin white arrow) as well as in intracellular vesicles (thick white arrow). **D.** Merged image of triple immunolabeling of cultured D54-MG (human glioma cell line). ClC-3 antibody (labeled with alexa 546) shows that this protein colocalizes with lipid rafts which are identified by immunolabeling of the beta subunit of cholera-toxin (fluorescein-conjugated cholera-toxin β subunit). Nuclei were counterstained blue with DAPI. (Panels A and B were modified from Olsen et al. (2003); C is an unpublished image; and D is reproduced with permission from McFerrin and Sontheimer (2006).)

prime candidate for mediating the Cl^- fluxes required to accomplish the cell shrinkage needed for glioma cell migration. The data further suggest that ClC-3 may be a biological target of Cltx and that the latter might be a potent inhibitor of glioma cell migration.

VI. MECHANISM OF CHLOROTOXIN ACTION ON GLIOMA INVASION

The finding that Cltx inhibited Cl^- channels (Fig. 26.5A) and was effective in preventing cell invasion in transwell assays (Fig. 26.5B) prompted further studies on its mechanism of action; Cltx has a potential clinical use as an anti-invasive drug. While biophysical studies suggested that Cltx inhibits Cl^- channels in glioma cells, it took up to 15 minutes to achieve its maximal effect. This long delay questioned a direct action on the channels; channel-specific toxins typically act almost instantaneously. Using a His-tagged recombinant Cltx, Deshane and collaborators were able to isolate a protein complex and analyzed it by mass-spectroscopy (Deshane et al., 2003). These studies showed that matrix-metalloproteinase-2 (MMP-2), a 72 kD protein that is highly expressed on the surface of invading glioma cells, could be the primary binding site for Cltx. However, the isolated protein complex

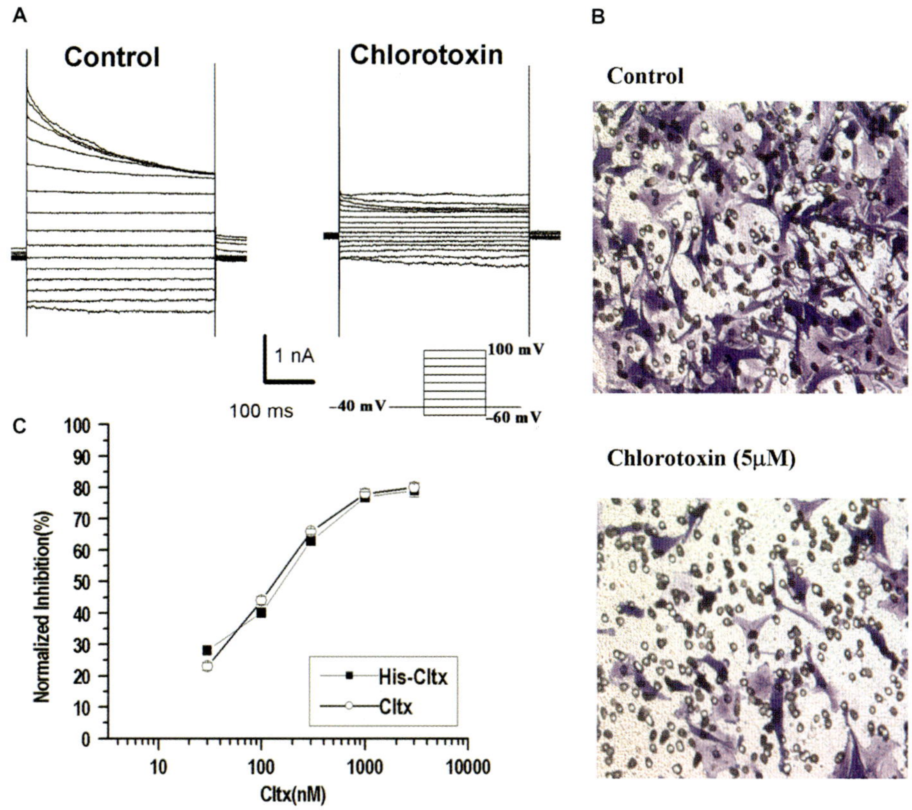

FIGURE 26.5 Inhibition of Cl^- channels with chlorotoxin retards glioma cell migration. **A.** Representative currents in response to voltage steps, recorded before (control) and 15 minutes after application of 1 μM chlorotoxin. Outwardly rectifying, inactivating Cl^- currents were recorded by whole-cell patch-clamp in D54-MG glioma cells using 20 mV voltage steps ranging from −120 to 160 mV. **B.** To show that a chlorotoxin-sensitive Cl^- conductance is required for migration across a spatial barrier, D54-MG glioma cells were plated on the upper surface of a Transwell insert with 8 μm pores and allowed to migrate for 4 hours towards vitronectin coated on the bottom of the filter insert (top left). Under control conditions, most cells migrated successfully, indicated by crystal violet staining of cells (control). In the presence of 5 μM chlorotoxin only a few cells migrated through the filter. **C.** Chlorotoxin inhibits glioma cell migration. Dose–response curve of D54-MG cells treated with His-Cltx or commercial Cltx peptide (Alomone) and analyzed by matrigel invasion assay at 24 h post-treatment. Half maximal inhibition (IC50) for Cltx was 184 nM. Percent inhibition was calculated as the decrease in the number of migrated cells normalized to control. (Panel A reproduced with permission from McFerrin and Sontheimer (2006); panels B and C are reproduced with permission from Deshane et al. (2003).)

also contained ClC-3 channels and several regulators of MMP-2. To further investigate how Cltx may have decreased Cl^- channel function, cell surface expression was examined by biotinylation. These studies showed that upon application of Cltx, membrane associated ClC-3 channels gradually disappeared and were almost undetectable at the surface after 30 minutes (McFerrin and Sontheimer, 2005). Further, it was observed that the plasma membrane channels ended up in intracellular caveolar vesicles. In the presence of filipin, a sterol-binding drug that prevents the formation of caveolar vesicles, Cltx lost its effect on ClC-3 channel internalization. This suggested that binding of the toxin induces the internalization of ClC-3 channels together with Cltx into caveolar raft vesicles. These findings explain the intracellular trapping of Cltx observed in other studies, including those in humans, as discussed below.

VII. CLINICAL USE OF CHLOROTOXIN

In light of the specific binding of Cltx to cultured glioma cells, it was logical to explore the biological activity of this molecule in animal models of malignant glioma. Using a radiolabeled peptide we demonstrated its specific binding to human gliomas xenografted into the cerebrum of immunocompromised mice (Soroceanu et al., 1998). These studies were followed by screening of human tissues searching for specific binding of Cltx (Lyons et al., 2002). These studies, which examined over 100 samples, revealed binding of Cltx to gliomas of all malignancy grades, as well as to tumors that share an embryological relationship with them. The latter includes primarily cancers originating from neuroectodermally derived tissues such as melanoma or small lung cell carcinomas.

In contrast, non-malignant tissues were universally negative. In 2002, a synthetically manufactured Cltx was approved by the US Food and Drug Administration (FDA) to be examined in 18 patients in a phase I study. Like in the previous preclinical studies, this trial used a radiolabeled form of the peptide which was introduced through an intrathecal catheter. The radiolabel could then be detected by whole-body gamma camera scans (Fig. 26.6A), or with greater resolution, by SPEC imaging (Fig. 26.6B). In this study, fluid samples were collected to determine the pharmacokinetics of the molecule. The data from this clinical study were published in 2006 (Mamelak et al., 2006). Sample images like those shown in Fig. 26.6 were published in 2005 (Hockaday et al., 2005). The safety and localization data gathered in the phase I trial justified further use of Cltx in 59 patients, in a phase II clinical study which concluded recently. Preliminary data released from this trial showed a significant increase in mean survival, following administration of three or six doses of Cltx. Importantly, imaging studies such as those illustrated in Fig. 26.6 suggest that Cltx is retained at the tumor for 5–8 days. This observation is in good agreement with the internalization of Cltx together with ClC-3 and MMP-2 into caveolar vesicles. The therapeutic efficacy of Cltx is therefore, in all likelihood, due to (1) the internalization of ClC-3 channels and decreased cell migration and (2) the trapping of the radiolabel toxin which could have its own effect on cellular DNA.

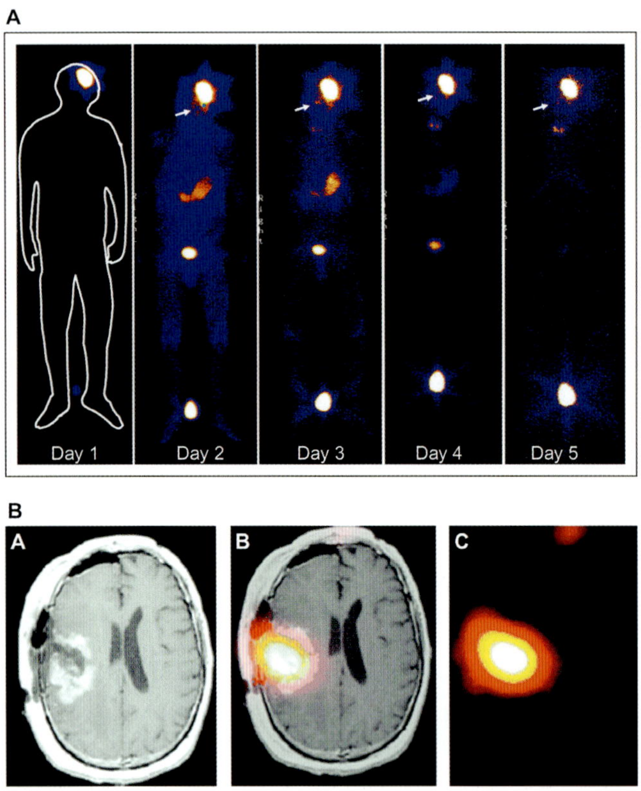

FIGURE 26.6 The Cl^- channel inhibitor chlorotoxin localizes to gliomas *in vivo*. **A.** A single dose of ^{131}I-chlorotoxin given to a patient in a phase I clinical study shows tumor-specific localization in whole-body scans performed over a 5 day period (modified from Shen et al., 2005). **B.** Overlay of MRI and SPECT images showing tumor-specific retention of chlorotoxin, 8 days after administration of the drug. Axial view of T1-Wc (left), coregistered (middle), and SPECT (right). (Reproduced with permission from Hockaday et al., 2005.)

VIII. CELL VOLUME CHANGES ASSOCIATED WITH CELL PROLIFERATION

In addition to being highly invasive, primary brain tumors also exhibit relentless growth, with mitotic indices suggesting that over 30% of high-grade gliomas are in the active process of cell division. As cells divide, they give rise to two daughter cells of approximately half the volume of the parent cell. Yet, within just a few hours, cell size and volume are restored in both daughter cells. Surprisingly, little is known about cell volume changes occurring in dividing cells in general (see Chapter 27 in this volume). In a recent study, we imaged complete cycles of cell division using three-dimensional time-lapsed video microscopy following individual cells from birth through to the next cell division giving rise to new daughter cells (Fig. 26.7A, B). In this study, cell volume was obtained from 200 to 400 serial sections at each time point, allowing relatively accurate cell volume measurements for the entire cell cycle (Fig. 26.7C). We demonstrated a reduction in cell volume prior to the M-phase of the cell cycle (Fig. 26.7D), a phenomenon which we termed "pre-mitotic volume condensation" (Habela and Sontheimer, 2007). Regardless of the cell volume that a cell maintains during interphase, it condenses to a volume of approximately 6000-fL prior to entering into M-phase, approximately 6 h before giving rise to two daughter cells of approximately 3000-fL volume (Fig. 26.7D). The condensed cells have already synthesized the cell membrane of the two daughter cells, as this is readily visible by the thickened membrane (Fig. 26.8A). This finding was entirely unexpected, as the common assumption has been that cells grow in size continuously until division occurs. A contraction of the cytoplasmic volume was not expected. Furthermore, the fact that the cell

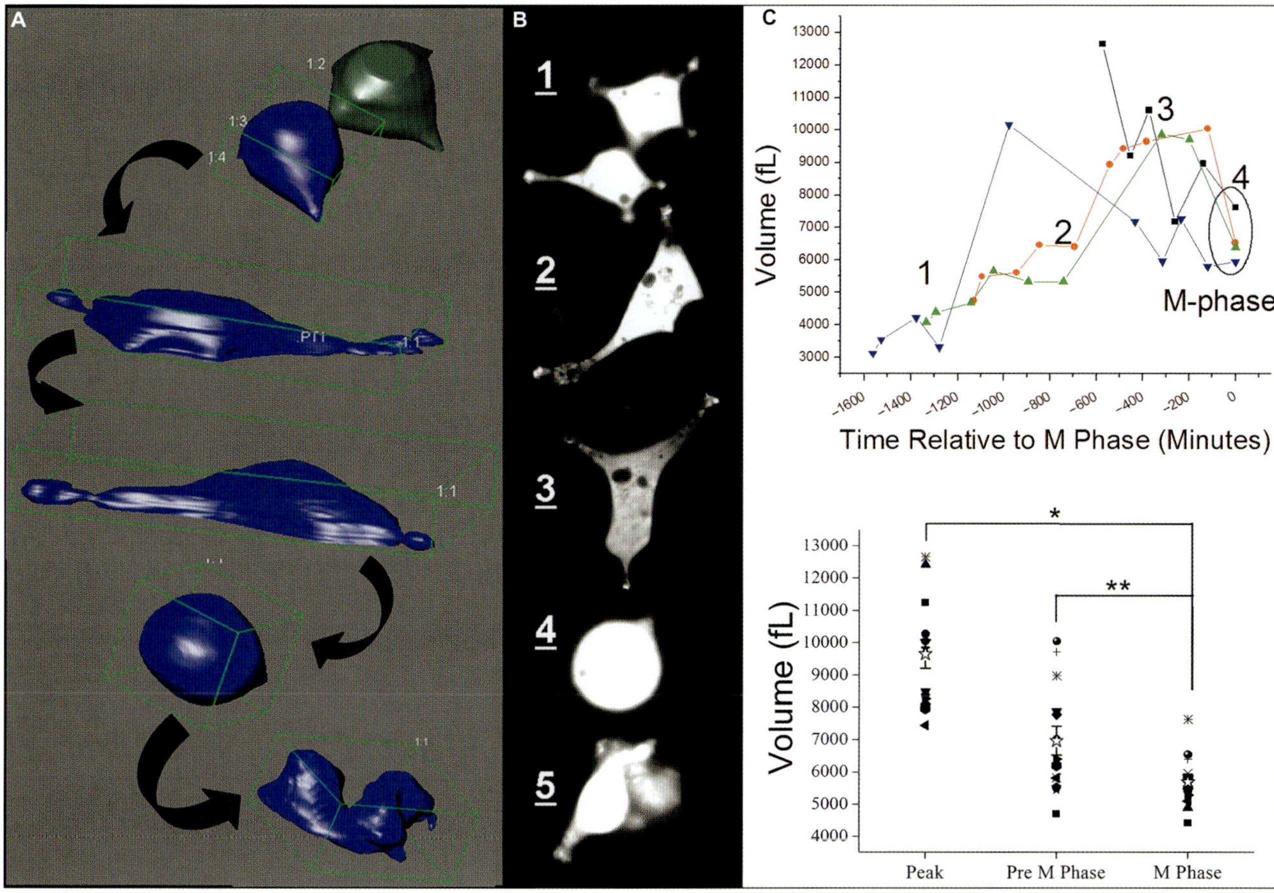

FIGURE 26.7 3D time-lapse imaging of glioma cells division allows accurate determination of cell volume throughout the cell cycle process. **A.** 3D projections created from image z-stacks computed from 200 optical sections such as those shown in B, and rendered in 3D using ImagePro. This program also computed volumes in fL for each 3D rendered cell. **B.** Sections from the z-stack used to generate the corresponding projections shown in A. Images are in chronological order from 1 to 5. **C.** Volume measurements (in fL) at specific time points relative to division are shown for four cells including the cell in A (green triangle symbols). Time points 1 through 4 correspond to projections 1–4 in B. For each cell, M-phase was set at $t = 0$ minutes. Note the convergence in volumes immediately before M-phase, where volumes are tightly clustered around 6000 fL. **D.** Cells assume a volume of ~6000 fL as they reach M-phase, regardless of their volume during interphase ($n = 14$ cells). (Reproduced with permission from Habela and Sontheimer, 2007.)

membrane thickens as the cell volume condenses suggests that, at this stage, cells have membrane folds ready to be unfolded once a cell division and separation of two daughter cells has occurred. Upon division, to achieve normal volume, each daughter cell only needs to reaccumulate water through the uptake of Na^+ and Cl^-, presumably via NKCC1. The volume changes that may occur through the cell cycle are illustrated in Fig. 26.8B. Importantly, studies that directly compared intracellular $[Cl^-]$ in M-phase cells versus the bipolar interphase cells showed a 40% reduction in $[Cl^-]_i$ in the condensed M-phase cells, suggesting that Cl^- efflux is mechanistically linked to the cell volume reduction (Habela et al., 2009). Closer examination also showed that cytoplasmic condensation is accompanied by condensation of nuclear chromatin and indeed, the two processes appear to occur in close synchrony (Habela and Sontheimer, 2007). The initial condensation of the cytoplasm and hence the chromosomal condensation are mediated by the efflux of Cl^- through the same ClC-3 channels that are involved in cell volume decreases associated with invading cells since shRNA knock-down of ClC-3 impaired cell condensation (Habela et al., 2008). While pharmacological studies have long suggested a role for Cl^- channels in cell division, these studies are the first to ascribe a mechanistic role to these channels in cell division; they mediate cytoplasmic condensation through water loss, a necessary step for cells to enter the M-phase.

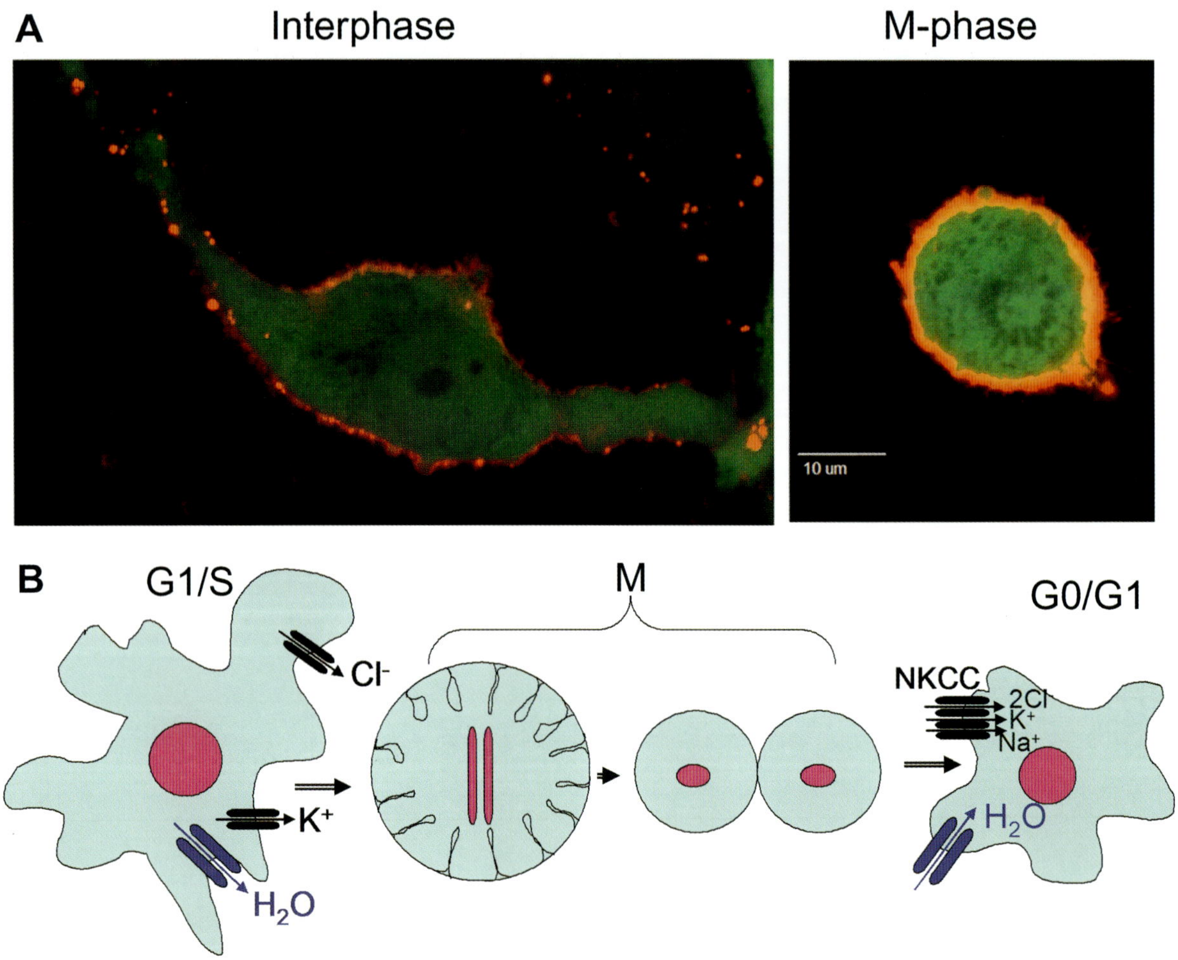

FIGURE 26.8 A. Membrane thickening of M-phase cells following volume condensation are displayed by labeling cells expressing cytosolic eGFP (green) with membrane bound DiI (red). **A.** In interphase, the cell membrane associated DiI is a thin membrane layer surrounding the cytoplasm. In M-phase, this area is thickened suggesting a ruffled membrane. **B.** Changes in cell volume during the cell cycle illustrated in cartoon form. As cells progress through G1/S, they increase their plasma membrane area and overall cell volume. As they progress to the M-phase they condense their cytoplasmic volume but maintain their cell membrane area which becomes thickened and folded. Cell division into two daughter cells divides membrane and cytoplasm equally between daughter cells. The acquisition of new membrane is accompanied by uptake of Na^+, Cl^- and water through NKCC1, which establishes the normal cell size/volume. (Reproduced with permission from Habela and Sontheimer, 2007.)

IX. CONCLUSIONS

Taken together, the data discussed in this chapter suggest that Cl^- channels and transporters cooperate to support dynamic changes in cell volume that govern cell proliferation and cell migration/invasion. The outward electrochemical gradient for Cl^- is established by the Na^+-K^+-$2Cl^-$ cotransporter, NKCC1, and this gradient permits rapid Cl^- efflux through Cl^- channels and obligatory water movement across the plasma membrane, ultimately leading to cell volume reduction. It is conceivable that similar mechanisms operate in other migratory cells, e.g. developing neurons, stem cells, and other cell types which migrate during development prior to settling down, maturing and forming tissues. Furthermore, a similar role for Cl^- channels in the control of cell growth and proliferation may widen the therapeutic potential for Cl^- channel blockers as anti-cancer reagents.

Acknowledgements

The author is grateful for the continued support by grants from the National Institutes of Health RO1 NS-31234, RO1 NS-52634, NS-36692 and P50-CA97247.

References

Butowski, N.A., Sneed, P.K., and Chang, S.M. (2006). Diagnosis and treatment of recurrent high-grade astrocytoma. *J. Clin. Oncol.* **24**, 1273–1280.

Dai, C., Celestino, J.C., Okada, Y., Louis, D.N., Fuller, G.N., and Holland, E.C. (2001). PDGF autocrine stimulation dedifferentiates cultured astrocytes and induces oligodendrogliomas and oligoastrocytomas from neural progenitors and astrocytes in vivo. *Genes Dev.* **15**, 1913–1925.

DeBin, J.A., Maggio, J.E., and Strichartz, G.R. (1993). Purification and characterization of chlorotoxin, a chloride channel ligand from the venom of the scorpion. *Am. J. Physiol.* **264**, C361–C369.

Deshane, J., Garner, C.C., and Sontheimer, H. (2003). Chlorotoxin inhibits glioma cell invasion via matrix metalloproteinase-2. *J. Biol. Chem.* **278**, 4135–4144.

Ernest, N.J. and Sontheimer, H. (2007). Extracellular glutamine is a critical modulator for regulatory volume increase in human glioma cells. *Brain Res.* **1144**, 231–238.

Ernest, N.J., Weaver, A.K., Van Duyn, L.B., and Sontheimer, H.W. (2005). Relative contribution of chloride channels and transporters to regulatory volume decrease in human glioma cells. *Am. J. Physiol. Cell Physiol.* **288**, C1451–C1460.

Giese, A., Rief, M.D., Loo, M.A., and Berens, M.E. (1994). Determinants of human astrocytoma migration. *Cancer Res.* **54**, 3897–3904.

Habela, C.W., Ernest, N.J., Swindall, A.F., and Sontheimer, H. (2009). Chloride accumulation drives volume dynamics underlying cell proliferation and migration. *J. Neurophysiol.* **101**, 750–757.

Habela, C.W., Olsen, M.L., and Sontheimer, H. (2008). ClC3 is a critical regulator of the cell cycle in normal and malignant glial cells. *J. Neurosci.* **28**, 9205–9217.

Habela, C.W. and Sontheimer, H. (2007). Cytoplasmic volume condensation is an integral part of mitosis. *Cell Cycle* **6**, 1613–1620.

Hockaday, D.C., Shen, S., Fiveash, J., Raubitschek, A., Colcher, D., Liu, A., Alvarez, V., and Mamelak, A.N. (2005). Imaging glioma extent with 131I-TM-601. *J. Nucl. Med.* **46**, 580–586.

Kleihues, P., Soylemezoglu, F., Schaueble, B., Scheithauer, B.W., and Burger, P.C. (1995). Histopathology, classification and grading of gliomas. *Glia* **15**, 211–221.

Lyons, S.A., O'Neal, J., and Sontheimer, H. (2002). Chlorotoxin, a scorpion-derived peptide, specifically binds to gliomas and tumors of neuroectodermal origin. *Glia* **39**, 162–173.

Mamelak, A.N., Rosenfeld, S., Bucholz, R., Raubitschek, A., Nabors, L.B., Fiveash, J.B., Shen, S., Khazaeli, M.B., Colcher, D., Liu, A., Osman, M., Guthrie, B., Schade-Bijur, S., Hablitz, D.M., Alvarez, V.L., and Gonda, M.A. (2006). Phase I single-dose study of intracavitary-administered iodine-131-TM-601 in adults with recurrent high-grade glioma. *J. Clin. Oncol.* **24**, 3644–3650.

McFerrin, M.B. and Sontheimer, H. (2006). A role for ion channels in glioma cell invasion. *Neuron Glia Biol.* **2**, 39–49.

Olsen, M.L., Schade, S., Lyons, S.A., Amarillo, M.D., and Sontheimer, H. (2003). Expresssion of voltage-gated chloride channels in human glioma cells. *J. Neurosci.* **23**, 5572–5582.

Plate, K.H. and Risau, W. (1995). Angiogenesis in malignant gliomas. *Glia* **15**, 339–347.

Ransom, C.B., O'Neal, J.T., and Sontheimer, H. (2001). Volume-activated chloride currents contribute to the resting conductance and invasive migration of human glioma cells. *J. Neurosci.* **21**, 7674–7683.

Shen, S., Khazaeli, M.B., Gillespie, G.Y., and Alvarez, V.L. (2005). Radiation dosimetry of 131I-chlorotoxin for targeted radiotherapy in glioma-bearing mice. *J. Neurooncol.* **71**, 113–119.

Soroceanu, L., Gillespie, Y., Khazaeli, M.B., and Sontheimer, H. (1998). Use of chlorotoxin for targeting of primary brain tumors. *Cancer Res.* **58**, 4871–4879.

Soroceanu, L., Manning, T.J., Jr, and Sontheimer, H. (1999). Modulation of glioma cell migration and invasion using Cl^- and K^+ ion channel blockers. *J. Neurosci.* **19**, 5942–5954.

Stobrawa, S.M., Breiderhoff, T., Takamori, S., Engel, D., Schweizer, M., Zdebik, A.A., Bosl, M.R., Ruether, K., Jahn, H., Draguhn, A., Jahn, R., and Jentsch, T.J. (2001). Disruption of ClC-3, a chloride channel expressed on synaptic vesicles, leads to a loss of the hippocampus. *Neuron* **29**, 185–196.

Von Deimling, A., Louis, D.N., and Wiestler, O.D. (1995). Molecular pathways in the formation of gliomas. *Glia* **15**, 328–338.

CHAPTER

27

The Sodium-Potassium-Chloride Cotransporter, Human Cytomegalovirus and the Cell Cycle

John M. Russell

OUTLINE

I. INTRODUCTION

Viruses, being incomplete life forms, must usurp the genetic and biochemical machinery of their host cell in order to replicate (e.g. Sanchez and Spector, 2008). Thus, the study of virally mediated perturbations of host cell functions has the potential of yielding insights, even revealing unexpected functions of known proteins. The interface between normal cell biology and the subversion/inhibition/stimulation of carriers by pathogens has been the sporadic focus of research for some time. Most of these reports have been on the Na^+/H^+ exchanger (NHE), the Na^+-K^+-Cl^- cotransporter (NKCC) and the sodium pump (Na^+-K^+-ATPase; e.g. Benos et al., 1994; del Castillo et al., 1991; Voss et al., 1996; O'Brien et al., 1996, Fons et al., 2000; Patton et al., 2000; Schaefer et al., 1984). In addition to these ion transporters, our group has identified significant effects of human cytomegalovirus (HCMV) infection on the Cl^-/HCO_3^- exchanger (CBE), as well as on an as yet poorly understood acid extrusion mechanism. These reports from a variety of cell types and caused by several different viruses suggest that carriers may be important targets of viral pathogens. This is most likely due to the fact that viruses require host cells whose basic life processes are more or less intact if they are to serve as suitable replicative factories for the pathogen. However, the virus must "customize" the cellular functions to meet its specific needs as well as to defeat cellular defense mechanisms such as antigen recognition and apoptosis.

II. HCMV BASICS

HCMV infection is widespread, affecting 50–90% of the adult population (Becker et al., 1993; Britt and Alford, 1996; Staras et al., 2006). Although usually benign, HCMV has the potential to be deadly under certain conditions. The most obvious morphological characteristic of infection with this pathogen is cell swelling, termed cytomegaly, a 2–3-fold enlargement of the host cell. It is this cell enlargement feature that led to our original interest in the possible role of NKCC in the HCMV infection cycle. The virus is spread by contact with body fluids such as blood, saliva, semen, tears, breast milk and vaginal secretions (Staras et al., 2006).

III. HCMV PATHOLOGY

HCMV, a beta-herpes virus, is an opportunistic virus like other members of the Herpes family. Following the primary infection, it remains latent, hidden in cells of the salivary glands, kidneys, bone cells as well as in blood cells such as lymphocytes and macrophages (Becker et al., 1993). Thus, infection with HCMV is endemic. In most people, the primary infection goes virtually unnoticed and the latent infection is asymptomatic. However, under the appropriate conditions, HCMV infection can cause pneumonitis, encephalitis, bronchiolitis, retinitis, hepatitis and gastroenteritis (Alford and Britt, 1990; Ettinger et al., 1993; Bando et al., 1995). Most serious health concerns will arise from one of two situations. (1) A compromised immune system permits this usually benign virus to rapidly "bloom" into serious infections. For this reason, the AIDS epidemic greatly stimulated interest in HCMV. Recent advances in antiviral therapy have reduced the death rate among AIDS sufferers caused by HCMV infection to well below the 25% seen in the late 1980s. Nevertheless, HCMV still strongly impacts the care and progress of AIDS. Similarly, organ transplant patients with pharmacologically induced suppression of their immune systems are plagued with recurrent HCMV overgrowth occurrences. (2) Perinatal infection is a serious concern because HCMV crosses the placental barrier, and can infect the fetus. In this way, HCMV causes a significant number of congenital neurological abnormalities such as mental retardation, sensorineural hearing loss and even death (Alford and Britt, 1990). In the USA, maternal HCMV infection is *the* leading viral cause of congenital mental retardation with ~1 in 750 live births in the USA experiencing permanent neurological damage as a result of congenital HCMV infection (Ho, 1991; Becker et al., 1993; Center for Disease Control website: http://www.cdc.gov/cmv/facts.htm). *In vitro* results demonstrate that HCMV can infect, and replicate, within human fetal neuroepithelial precursor cells (McCarthy et al., 2000). This supports the view that the fetal human nervous system is vulnerable to HCMV. In addition, there is evidence implicating this virus in re-stenosis of coronary arteries following angioplasty (Speir et al., 1994; Zhou et al., 1996).

IV. BIOLOGY OF THE VIRUS

Studies on permissive cultured human host cells (human fibroblasts are the preferred cell culture model) show that HCMV (a double-stranded, 230 kb DNA genome, surrounded by a protein capsid, which is surrounded by a protein tegument which in turn is surrounded by a lipid envelope) enters host cells and rapidly sheds its external protein coat. Formed virions can no longer be seen ("viral eclipse") at that point. Viral gene expression occurs in three phases following exposure (post-exposure; PE) of the host cell to the virus: an immediate early (IE) phase (0–2 h PE) expresses genes involved in escape from host-cell immune surveillance, and for transcription of the early phase genes (2–12 h PE). The early-phase genes encode the viral proteins necessary for viral DNA synthesis. Finally, there are the late phase genes (12–24 h PE) which encode the structural proteins necessary for synthesizing the virion. In a productive infection, that is, one resulting in viral replication, viral DNA replication begins about 24–30 h after the initial viral exposure (Bresnahan et al., 1996) and newly synthesized virions begin to appear in the host cell cytoplasm about 50h PE. Maximal virus yields occur 97h PE and later (Albrecht et al., 1989). Concurrent with the viral replication events outlined above, infection leads to stereotypical changes of host cell size (Fig. 27.1). Early in the infection (6–16h PE), the host cell usually "rounds up" and may actually become smaller (Albrecht et al., 1989). However, beginning ~24h PE, the host cell begins to enlarge (Albrecht et al., 1989). The enlargement (cytomegaly) progresses through at least 72–96h PE. By 72h PE, the host cell volume may be increased 2–3-fold compared to mock-infected cells (mock-infection exposes cells to cellular debris from uninfected cells to control for effects of the inevitable cell debris accompanying viral production).

V. HCMV PARTIALLY ACTIVATES HOST CELL CYCLE

HCMV infection is accompanied by a variety of effects in the host cell that are analogous to activation of the host cell cycle by mitogens (e.g. Albrecht et al., 1989; Boldogh et al., 1991; Sanchez and Spector, 2008). Such a mitogen-like action is a general feature of herpes viruses (reviewed in de Beeck and Caillet-Fauquet, 1997). Host cells infected with HCMV while in either G_0- or G_1-phases of the cell cycle halt their progress in the cell cycle at the G_1/S transition and do not progress to host cell division. Rather, viral replication events begin to occur. On the other hand, cells infected while in S-phase continue to progress through the cell cycle, divide, then halt at the G_1/S transition. Thereafter, the viral replication processes commence (Salvant et al., 1998). Thus, the virus "synchronizes" productively infected cells such that all are in late G_1-phase or early S-phase (e.g. reviewed in Fortunato et al., 2000).

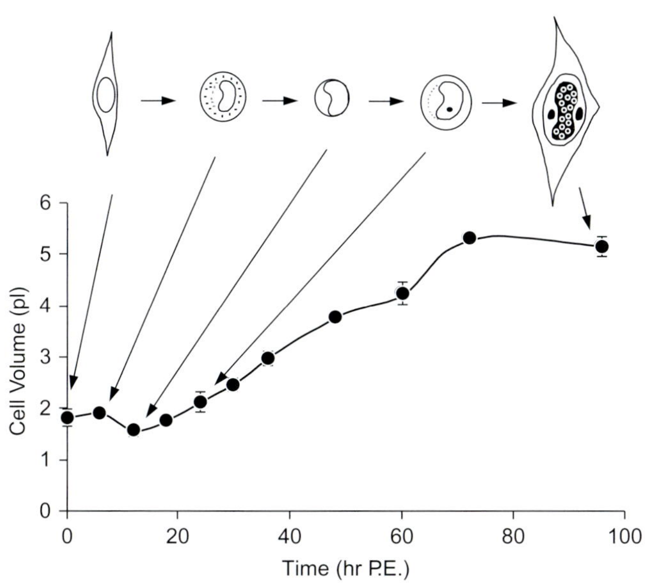

FIGURE 27.1 HCMV-infected human fibroblasts undergo a period of progressive enlargement between ~24–72h PE, resulting in a 2–3-fold increase in cell volume and a major change in morphology. Preceding this period of enlargement, there is transient cell shrinkage and cell rounding ~12h PE, which are likely due to initial cell dysfunctions invoked by cellular defenses upon viral entry. (Adapted from Albrecht et al., 1989.)

This effect of the virus is particularly interesting when one considers that *in vivo* HCMV tends to infect epithelial and endothelial cells (Weller, 1971) which are terminally differentiated cells which ordinarily have their macromolecular synthetic machinery strongly repressed. It appears that this DNA virus partially up-regulates the cell division machinery in order to facilitate synthesis of its own DNA (Salvant et al., 1998). Thus, the virus' strategy is to stimulate the host cell to enter the cell cycle sufficiently to permit viral DNA synthesis, but not to permit host cell DNA synthesis. It is therefore highly likely that viral DNA synthesis depends strongly on host cell functions that optimize the cell for DNA synthesis (Sinclair et al., 2000). Interestingly, some cells when exposed to virus during the G_1-phase are able to complete the cell cycle upon viral infection and these cells will not support viral replication; they undergo what is called an abortive infection (Albrecht et al., 1989). Clearly, the ability of the virus to halt the host cell cycle machinery in a kind a G_1/S-phase is critical for viral replication.

The basis of HCMV's ability to partially activate the cell cycle has been the object of considerable work. To summarize, the virus induces the transcription of FOS, JUN and MYC within minutes of infection to an extent comparable with stimulation by serum (Boldogh et al., 1991). It alters the expression levels of several key regulatory proteins including cyclins A, B, D and E, and the tumor-suppressor proteins p53 and Rb (reviewed in Fortunato et al., 2000). Although the virus stimulates significant increases in the levels of p53, its targets, p21 and MDM2, are not activated, probably as a result of p53 being bound by a viral protein (Fortunato et al., 2000). Nevertheless, the virally bound p53 is still able to respond to cellular DNA damage to repress the cell cycle (Bonin and McDougall, 1997). Expression of cyclin D_1 protein is inhibited by the virus (Bresnahan et al., 1996), whereas cyclin E expression is stimulated by an immediate early gene product (IE86) as is the protein kinase activity of cyclin E/Cdk2 (Bresnahan et al., 1998; Sinclair et al., 2000). Cyclin E levels are significantly elevated by 17–24h PE (Bresnahan et al., 1997; Sinclair et al., 2000). Furthermore, pharmacological inhibition of cyclin E-activated Cdk4 prevented HCMV replication (Bresnahan et al., 1997).

Since viral DNA synthesis begins 24–30h PE, it is likely that the HCMV-"activated" cell has traversed its pseudo-G_1-phase by this time. Figure 27.2 summarizes some of these effects and compares them to the normal cell cycle events.

The normal cell cycle progress during the G_1-to-S-phase involves first the rapid, but relatively short-lived, up-regulation of FOS, JUN, MYC. Then, cyclin D_1 is up-regulated and its gene levels remain high throughout the G_1-phase. At about the time of the G_1 checkpoint, cyclin E levels begin to increase, peaking as the cell moves into the S-phase of the cycle (e.g. Roussel, 1998). During late G_1-phase, there is a gradual increase in cell size of 30–40%. HCMV apparently does all of these *except* activating cyclin D_1 and causing the late G_1 cell size increase (cell size may actually transiently decrease during the "mitogenic activation" period; see Fig. 27.1).

It has been demonstrated that cells transfected with the IE86 (immediate early protein 86Kd) message could no longer proliferate, but tended to accumulate in the S-phase (Bresnahan et al., 1998; Murphy et al., 2000). This ability to halt the cell cycle is apparently independent of Rb, p53 or p21CIP (Weibusch and Hagemeier, 1999). Another viral protein has also been shown to have a role in the cell cycle effect of the virus. This protein, UL69, is part of the tegument of the virus particle and therefore is present in the host cell from the moment of viral penetration. Because this protein is part of the viral particle itself, UV-irradiated virus has the ability to block the host cell cycle, and does so in the G_1-phase (Lu and Shenk, 1999; Hayashi et al., 2000). Using both transfection and transduction techniques on two different cell lines, Lu and Shenk (1999) demonstrated that UL69 expression in serum-stimulated human cells largely prevented cell proliferation and the cells tended to be halted in the G_1-phase. In an elegant study in which they constructed mutant HCMVs that either lacked the UL69 message or lacked both the UL69 message and the UL69 tegument protein, Hayashi et al. (2000) showed that HCMV without either the message or the tegument UL69 could not halt the host cell cycle and consequently replicated very slowly. The mutant containing only tegument UL69 (with no message for UL69) was able to halt the host cell cycle in G_1. Thus, at least two viral proteins have been identified as being capable of inducing cell cycle halt in the G_1/S transition region of the cell cycle.

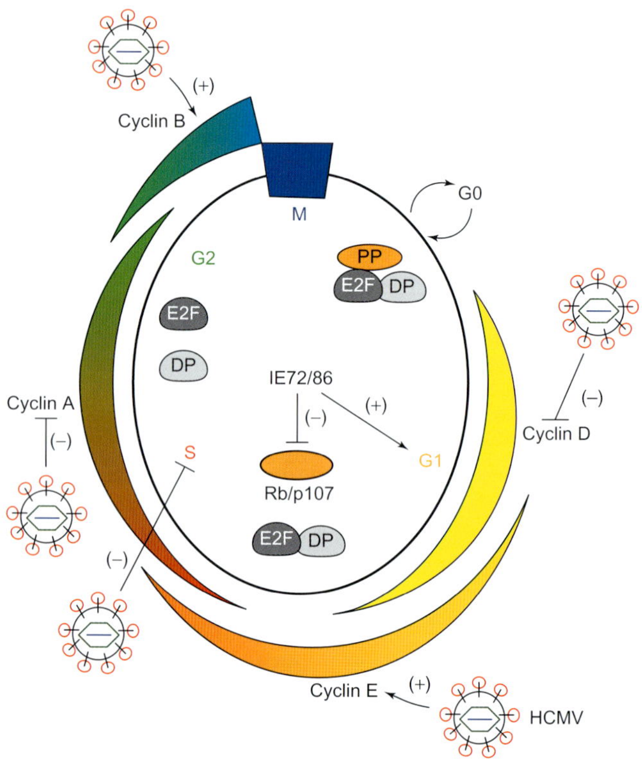

FIGURE 27.2 Summary of HCMV effects on the normal factors that regulate progression through the cell cycle. By activating cyclins E and B and inhibiting cyclins D and A, the virus facilitates viral DNA replication while inhibiting host cell DNA replication. HCMV stimulates the host cell through the G_1-phase by activating cyclin E and inhibiting the repressive pocket protein (PP) component of the E2F complex. Cell cycle progress is halted at that point by inhibiting progression into S-phase likely due to the inhibition of cyclin A. The viral tegument protein UL69 plus immediate early proteins, IE1-72 (IE72) and IE2-86 (IE86) are partially responsible for activating G_1. (Reproduced from Fortunato et al., 2000, with permission.)

In summary, HCMV, as part of its replication strategy, induces a pseudo-G_1-phase in the host cell. This period occurs before viral replication begins and before cytomegaly starts to develop. This phase is characterized by most of the same changes observed in uninfected cells stimulated by a mitogen to enter the cell cycle and traversing the G_1-phase of that cycle. Three effects are conspicuously missing, however. First, cyclin D_1 activity is not increased. Second, the cell volume does not increase by 30–40% (see below section VIII.A) rather the cell volume decreases ~25%. Third, unlike with true mitogen stimulation, NKCC activity is not stimulated, but rather is inhibited and its protein expression in the plasmalemma all but vanished (see below section VIII). In addition to the down-regulation of cyclin D_1, we suggest that these latter two differences represent possible important changes signaling permission to enter the S-phase.

VI. HCMV AND OTHER HOST CELL ION TRANSPORTERS

Our lab has examined the effects of HCMV infection on the function and protein levels of several plasma membrane ion transporters in addition to the NKCC, including the Na^+/K^+ ATPase, two isoforms of the Na^+/H^+ exchanger (NHE1 and NHE2) and the Cl^-/HCO_3^- exchanger.

A. Effects on the Sodium Pump

Using ouabain-sensitive ^{86}Rb uptake as a measure of sodium pump activity, Albrecht and collaborators demonstrated two phases of response by the pump to HCMV infection (Nokta et al., 1988). First, there was a ~50% inhibition in the period 12–25h PE (the period of development of the pseudo-G_1-phase). Thereafter, paralleling the development of cytomegaly, the pump exhibited progressive stimulation. The early period of reduced sodium pump activity also corresponds to the period during which the HCMV-infected cell actually becomes smaller. We followed up the preceding study with studies of ouabain binding and Western blotting to determine whether the functional changes were paralleled by changes in the number of pump sites or amount of sodium pump protein in order to obtain an estimate of changes (if any) in the number of sodium pump sites (Altamirano et al., 1994; Maglova et al., 1998a). Both approaches revealed no change in the number of pump sites or protein during the early period of infection during which the sodium pump function was decreased. However, both techniques revealed an increase in sodium pump sites (^3H-ouabain-binding sites) and sodium pump protein (detected by the α-5 antibody to the α-subunit of the sodium pump) during the later period of increased sodium pump activity. When both the ouabain-binding and Western-blot data were corrected for the effects of changes in cell numbers and protein/cell occasioned by HCMV infection, we found that they both showed an increase in sodium pump sites of about 4–7-fold, which is in the expected range for the increase of surface area for the cytomegalic cells.

B. Effects on the Na^+/H^+ Exchanger

We have reported that HCMV infection has several effects on the Na^+/H^+ exchanger function (Crowe et al., 1997). At 72h post-exposure (PE), we demonstrated that HCMV infection caused the pH_i vs. acid extrusion rate to shift about 0.15pH units to the alkaline side. This has the operational effect of increasing the proton extrusion rate at normal values of pH_i by a factor of about two-fold. In addition to being stimulated by a decrease of pH_i, the Na^+/H^+ exchanger is often stimulated by cell shrinkage (e.g. Grinstein et al., 1989; Wakabayashi et al., 1992). We reported that mock-infected cells have only a small response to cell shrinkage but that 72h after HCMV infection, the response is greatly increased (Crowe et al., 1997). In unpublished studies, we have observed that this increase in volume sensitivity is observed as early as ~18h PE. We measured the $K_{0.5}$ and V_{max} of the Na^+/H^+ exchanger as a function of external Na^+ concentration and found that HCMV infection increased the V_{max} about 80% while it reduced the affinity for external Na^+ from 25mM to 79mM. Thus, at 72h PE, the virus clearly stimulates Na^+/H^+ exchanger transport activity, while reducing its apparent affinity for Na^+ and increasing its V_{max}.

These kinetic effects may be related to our observations that suggest HCMV causes the human fibroblast cell line to express isoform 2 of the Na^+/H^+ exchanger (NHE2). Briefly, it is known that growth factors stimulate the NHE1 isoform by changing the sensitivity to intracellular pH, whereas the NHE2 isoform responds by an increase of its V_{max} (Wakabayashi et al., 1992; Nath et al., 1996). We have identified both kinds of effects (see above) in the HCMV-infected cells. Both NHE1 and NHE2 are reported to respond to cell shrinkage by stimulation (e.g. Kapus et al., 1994; Wakabayashi et al., 1992). Using RT-PCR primers for both NHE1 and NHE2 we found that HCMV-infected MRC-5 human fibroblasts (which ordinarily express only the NHE1 isoform; Siczkowski and Ng, 1995) produced cDNA bands for both NHE1 and NHE2 whereas mock-infected cells only provided bands for NHE1. Semi-quantitative Western blots of mock- and HCMV-infected cells revealed that the level of NHE1 protein expression was reduced ~50% by HCMV infection, whereas NHE2 protein levels were nearly doubled (Maglova et al., 2001). Thus, our evidence suggests that HCMV infection causes a change in the ratio of these two NHE isoforms and we know from a functional point of view that total Na^+/H^+ exchanger's acid extrusion ability is increased 72h PE (increased V_{max}).

C. Effects on the Cl^-/HCO_3^- Exchanger

We have reported that HCMV infection stimulates a Cl^-/HCO_3^- exchanger (CBE) activity in MRC-5 human fibroblasts (Maglova et al., 1998a). Mock or uninfected cells have very little CBE activity as judged from effects of replacing external Cl^- and treatment with H_2-DIDS on pH_i or by measuring changes in intracellular Cl^- concentration in the presence or absence of CO_2/HCO_3^-. Figure 27.3 shows that pH_i sensitivity of the CBE activity was changed by HCMV

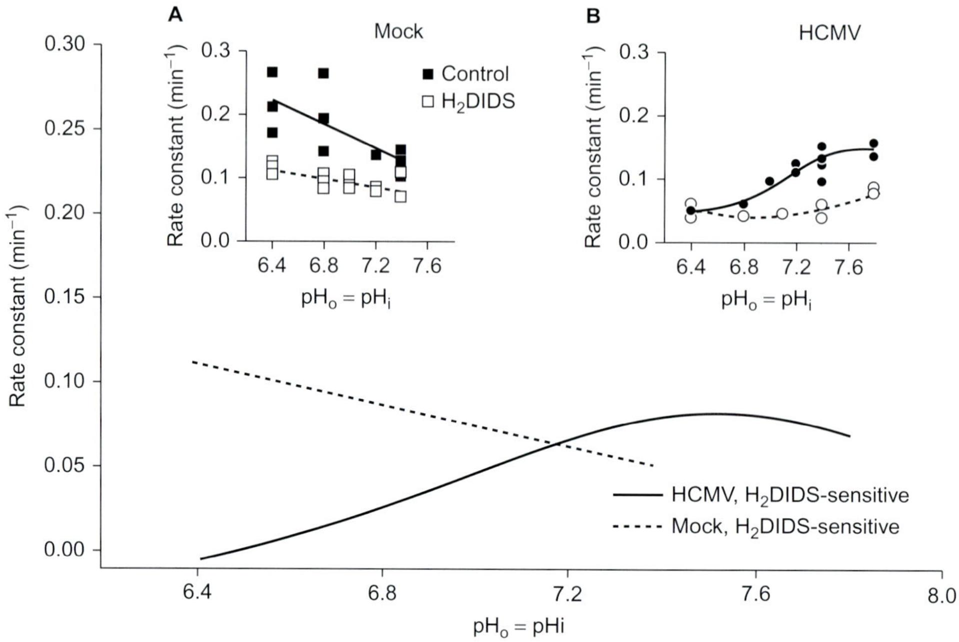

FIGURE 27.3 Intracellular pH (pH_i) dependence of net Cl^- efflux mediated by Cl^-/NO_3^- exchange. A high-K^+-nigericin technique was used to "clamp" pH_i to various values between 6.4 and 7.8. The external Cl^- was completely replaced with NO_3^- and the rate of loss of cellular Cl^- was used to determine an exponential efflux rate constant in the absence and presence of 100 μM H_2DIDS. **A.** Results of individual experiments on mock-infected cells. Lines (solid and dashed) are linear best fits to rate constant vs. pH_i data. **B.** Results of individual experiments on HCMV-infected cells (72 h after exposure). *Bottom*: pH_i dependence of H_2DIDS-sensitive rate constant of net Cl^- efflux. Lines were determined as difference between lines in A and B for net Cl^- efflux under control and H_2DIDS conditions. pH_o, extracellular pH. (Redrawn from Maglova et al., 1998a.)

infection from being slightly inhibited by alkaline pH_i (a characteristic of the AE1 isoform; Jiang et al., 1997) to being stimulated (a characteristic of the AE2 isoform; Jiang et al., 1997). In addition, we have reported that exposure of HCMV-infected cells to mild hyperosmotic conditions in the presence of CO_2/HCO_3^- results in pH_i alkalinization which is a characteristic of the AE2 isoform as well (Jiang et al., 1997). This result suggests that the dominant anion exchanger isoform in the host cell was changed from AE1 to AE2 by the HCMV infection. Importantly, in the presence of CO_2/HCO_3^- the $[Cl^-]_i$ was higher than in its absence, suggesting that the CBE was mediating net Cl^- uptake by the HCMV-infected cells.

TABLE 27.1 Effects of HCMV infection (72 h PE) on intracellular water space and ion concentrations using radioisotope equilibrium

	Mock-infected cells	HCMV-infected cells
Water space (pL/cell) (7)*	2.8±0.6#	7.2 ± 1.3
$[K^+]_i$ (mM) (7)	149.8 ± 10.4	113.9 ± 9.3
τ_K (min) (7)	46.1 ± 8.8	130.3 ± 15.3
$[Na^+]_i$ (mM) (5)	29.8 ± 7.5	19.7 ± 5.4
τ_{Na} (min) (5)	<20	<20
$[Cl^-]_i$ (mM) (10)	42.6 ± 9.8	64.9 ± 9.9
τ_{Cl} (min) (10)	<20	<20

values are mean values ± SEM
*numbers in parentheses indicate number of separate experiments

VII. HCMV EFFECTS ON NKCC

A. HCMV Infection Increases Cell Volume and $[Cl^-]_i$

The data in Table 27.1 are taken from HCMV-infected MRC-5 fibroblasts 72 h PE (at a time when cytomegaly is well established) that were bathed in a CO_2/HCO_3^--free medium. As expected, the cell volume (measured here as the [^{14}C]3-O-methyl-D-glucose distribution space) increased. The data also show the significant effects of HCMV infection on the intracellular concentrations of the three major inorganic ions. The intracellular concentrations of both K^+ (measured using ^{86}Rb uptake) and Na^+ (measured using ^{22}Na uptake) are slightly reduced whereas the $[Cl^-]_i$ (measured using ^{36}Cl uptake) has increased ~50%. As noted above, when CO_2 and HCO_3^- are present, the $[Cl^-]_i$ is even higher (~90 mM) as a result of the Cl^-/HCO_3^- exchanger activity. We have also measured $[Cl^-]_i$ using MQAE, a fluorescent dye quenched by Cl^- (Maglova et al., 1998a, b) and obtained very similar values for $[Cl^-]_i$ in the absence of CO_2/HCO_3^-. High (above normal) $[Cl^-]_i$ values have been reported to inhibit cellular protein synthesis (Weber et al., 1977) and HCMV is known to inhibit host cell protein synthesis (Stinski, 1977). It is also of interest to note that HCMV infection had no detectable effect on the rate at which ^{22}Na or ^{36}Cl reached isotopic equilibrium whereas in the case of ^{86}Rb, the time constant for reaching isotopic equilibrium was greatly increased indicating that K^+ transmembrane pathways were very significantly blocked.

B. HCMV Inhibits NKCC Ion Transport Activity

In the light of the results shown in Table 27.1, we were surprised to find that HCMV infection abolishes NKCC function. We determined this by measuring $[Cl^-]_i$ using the Cl^- sensitive fluorescent dye, MQAE. We first depleted the cells of Cl^- by bathing them in a Cl^--free medium, then measured the rate of net Cl^- uptake that occurred upon reintroduction of Cl^--containing medium. This protocol was carried out in the absence or presence of 10 μM bumetanide. Figure 27.4 shows that by 24 h PE, bumetanide-sensitive net Cl^- uptake is abolished by HCMV infection. As this time point before cytomegaly begins to develop, it is clear that NKCC can play no role in the virally mediated cell swelling. HCMV infection has been reported to block host cell protein production (Stinski, 1977), so we examined the effects of infection on NKCC protein levels as determined with semi-quantitative Western blots. We discovered that by 24 h PE, host cell plasmalemmal NKCC protein was reduced by ~85% (Fig. 27.5B), but that total cellular NKCC protein amounts were virtually unchanged (Fig. 27.5A). The plasmalemma results shown in this figure were obtained using biotinylation to tag the membrane NKCC. Essentially, identical results were obtained using cellular fractionation techniques to obtain membrane-enriched fractions.

C. HCMV Effects on Cellular NKCC Protein Distribution

Using biotinylation and semi-quantitative Western blotting, we determined that ~10–20% of total cellular NKCC protein resides in the plasmalemma of mock-infected MRC-5 fibroblasts. Using immunostaining, we determined that the remainder of the cellular NKCC protein is rather homogeneously distributed throughout the cytoplasm visualized as very small punctuate areas (Maglova et al., 2004). HCMV infection changed this pattern in two significant ways. The first way has already been described: NKCC protein nearly disappears from the plasmalemma. The second change is that the uniform punctuate distribution is progressively altered to an increasingly aggregated pattern so that by ~50 h PE, NKCC protein was almost entirely found in a single, large perinuclear structure (Maglova et al., 2004). This unanticipated finding was confirmed using three different anti-NKCC antibodies. These structures can be disintegrated by treatment with 3 μM nocodazole (which prevents the

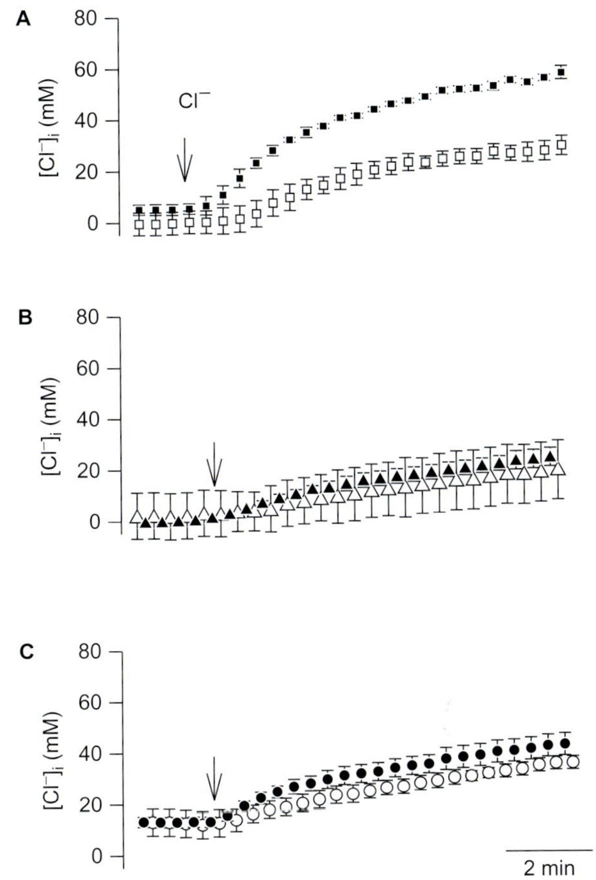

FIGURE 27.4 Effect of HCMV infection on net Cl^- uptake by mock- and HCMV-infected human fibroblasts. The cells were Cl^- depleted then re-exposed to Cl^--containing medium in the presence or absence of bumetanide. The increase of $[Cl^-]_i$ upon re-exposure to Cl^--containing medium (as determined using the Cl^--sensitive fluorescent dye, MQAE) was followed as a function of time. Filled circles = control; open circles = cells treated with 10 μM bumetanide. **A.** Mock-infected cells. **B.** 24 h PE HCMV-infected cells. **C.** 72 h PE HCMV-infected cells. (Redrawn from Maglova et al., 1998b.)

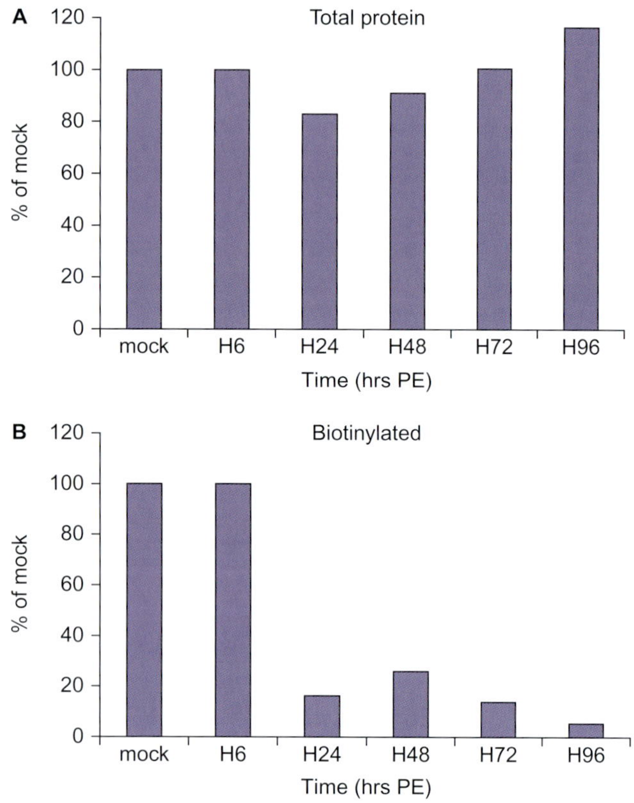

FIGURE 27.5 Time course of NKCC protein levels following HCMV infection. **A.** The levels of total cellular NKCC protein in HCMV-infected human fibroblasts at different times following infection (hours PE) were similar to that found in mock-infected cells, whereas **B.** the plasma membrane levels (biotinylated fraction) in HCMV-infected cells, decreased rapidly by 24h PE, compared to mock-infected cells. Similar results were obtained using cellular fractionation techniques to obtain plasmalemmal-enriched membrane fractions (adapted from Maglova et al., 2004).

polymerization of microtubules) and we can find no evidence of involvement by vimentin (Maglova et al., unpublished observations).

Perinuclear accumulations of membrane and/or cytosolic proteins have been reported in a variety of cells. One type, termed "aggresomes", has been associated with so-called "conformational diseases" (e.g. Garcia-Mata et al., 2002). Aggresomes have been described as perinuclear structures that form under conditions favoring protein production over degradation (Wigley et al., 1999; Rajan et al., 2001). An aggresome may, in some cases, serve as a repository of aggregated proteins preparatory to being degraded as they are closely associated with proteasomes, lysosomes and chaperones (e.g. Johnston et al., 1998; Garcia-Mata et al., 1999, 2002; Junn et al., 2002). Alternatively, or in addition, they may represent a pathologic response as there are diseases characterized by the development of aggresome-like structures, e.g. systemic amyloidosis, most neurodegenerative diseases including possibly retinitis pigmentosa (Kopito, 2000; Illing et al., 2002).

Alternatively, the perinuclear accumulation of NKCC protein may be associated with the perinuclear accumulation of viral proteins that is typical of infection by large DNA viruses. It has been suggested that these structures play a role in viral assembly (Heath et al., 2002). In the particular case of HCMV infections, a perinuclear structure containing several HCMV tegument and envelope proteins has been described (e.g. Sanchez et al., 2000a, b). These accumulations of viral proteins share some common features with aggresomes, but also have some very different properties. As with aggresomes, the accumulations of HCMV protein have a characteristic perinuclear location and form around the microtubule organizing center. However, unlike aggresomes, they are not surrounded by vimentin, and they are disrupted by nocodazole (Sanchez et al., 2000a, b). As noted above, we have evidence that the NKCC-containing perinuclear structure is disrupted by nocodazole and is not associated with vimentin. Furthermore, we have reported (Maglova et al., 2003) that the NKCC protein in the perinuclear structure is colocalized with several HCMV tegument proteins including pp65. We also have evidence (unpublished immunostaining) that NHE1 protein is found in the perinuclear structure. The possibility that host cell NKCC protein shares a structure with HCMV tegument and envelope proteins raises interesting questions as to why host cell ion transport proteins would be located in "viral factories"; i.e. is NKCC protein found in the protein coats of formed virions?

D. Summary of HCMV Effects on NKCC and Other Ion Transporters

Our results have demonstrated that HCMV infection has significant effects on all four ion transport processes we have investigated. Following infection the activities of the sodium pump, the NHE and the CBE are all either maintained or increased whereas the activity of NKCC is profoundly reduced. It is clear from the foregoing results that NKCC cannot be involved with the development of cytomegaly despite the fact that it is functionally well designed to do so. Our data on cellular ion content indicate that ~2/3 of the osmolytes needed to account for the cell swelling caused by HCMV infection are accounted for by the increased cellular contents of Na^+, K^+ and Cl^-. The identity of the remaining additional cellular osmolytes is presently unknown. Our results are consistent with the NHE together with the CBE acting to provide a net import mechanism for NaCl. In support of this conclusion is an early report by Fons et al. (1991) showing that treatment with amiloride (an inhibitor of the NHE) prevented the development of cytomegaly as well as viral replication.

It is equally clear from our results that the virus does not indiscriminately block the protein synthesis of all these host cell membrane proteins as might be suspected given that HCMV is reported to significantly block host cell protein synthesis (Stinski, 1977). We therefore draw the tentative conclusion that the HCMV effects on NKCC function and the location of the NKCC protein are targeted effects by the virus that presumably favor the replication of the virus. The obvious question is: Why might this be the case? We suggest it is related to the host cell cycle.

VIII. NKCC AND THE NORMAL CELL CYCLE: POSSIBLE LINK TO EFFECT OF HCMV

Considerable evidence suggests that NKCC plays a role in the normal cell cycle. What follows is a brief summary of the findings that support such a role.

A. Mitogenic Stimulation Increases NKCC Transport Activity Before DNA Synthesis Begins

Resting animal cells stimulated to enter the cell cycle exhibit an almost immediate, but relatively short-lived, increase in their uptake of K^+ (and Na^+) across the plasma membrane (reviewed in Russell, 2000). Two studies followed the time course of ^{86}Rb uptake (used as a radioactive tracer for K^+ uptake) from the time of the mitogen application until DNA synthesis begins. Both studies (Tupper et al., 1977 and Stiernberg et al., 1983) reported that 4–5 hours after application of a mitogen, the initial increase of ^{86}Rb uptake had declined to pre-mitogenic stimulation levels. However, beginning 6–8 hours after mitogen application a second gradually developing increase in ^{86}Rb uptake occurred that continued over the next several hours. DNA synthesis then began 1–3 hours after the peak of the second increase of ^{86}Rb uptake. The early stimulation of ^{86}Rb uptake is at least partially mediated by the NKCC (e.g. Panet et al., 1994, 2000; Berman et al., 1995; Bussolati et al., 1996). We have confirmed this two-stage stimulation of ^{86}Rb uptake by serum-starved MRC-5 human fibroblasts when exposed to fetal calf serum and found that bumetanide treatment (10 μM) will largely prevent both phases of stimulation of the uptake (Maglova et al., 2001). It is possible that this second stage of NKCC activity is necessary for entry into the S-phase of the cell cycle. It may be related to the 20–40% increase of cell size (Killander and Zetterberg, 1965; Tupper et al., 1977; Takahashi et al., 1993; Pusch et al., 1997; Soucek et al., 1997) that occurs during the G_1-phase of the cell cycle. Cell volume is known to be one of the important variables monitored at the normal G_1 checkpoint (e.g. Alberts et al., 2003).

B. Overexpression of the NKCC Promotes Cell Proliferation

Further evidence in support of an important role for NKCC function in the cell cycle comes from the use of mouse Balb/c 3T3 cells stably transfected with NKCC1 cDNA (Panet at al., 2000). The transfected clones had a ten-fold higher bumetanide-sensitive ^{86}Rb uptake rate than control cells. Importantly, NKCC-transfected cells had several characteristics of immortalized, transformed cells. Among these were enhanced DNA synthesis (8–10-fold higher than in the untransfected cells) and proliferation of confluent cells. Finally, it is of interest that others have reported up-regulation of NKCC gene expression, protein and transport activity in proliferating cells (e.g. Raat et al., 1996).

We described earlier the effects of the HCMV to partially activate the host cell cycle (see section V) to move the host cell to the G1/S-phase and no further. This strategy apparently activates the necessary cellular DNA-replicating machinery without permitting the host cell to actually begin its own DNA replication. Thus, a reasonable working hypothesis is that part of the HCMV effect to partially stimulate the host cell cycle may be to inactivate the NKCC.

As discussed earlier in section V, UL69 is capable of halting the host cell cycle at the G_1/S stage of the cell cycle (Lu and Shenk, 1999; Hayashi et al., 2000). UL69 is part of the tegument of the virus particle and therefore is present in the host cell from the moment of viral penetration. We have asked whether the UV-irradiated virus will also functionally inactivate the host cell NKCC. Preliminary results have shown that exposure of MRC-5 fibroblasts to UV-irradiated HCMV viral particles reduces the amount of plasmalemmal NKCC protein (determined with semi-quantitative Western blotting) by ~30–40%. Although this effect is less than that seen when the fibroblasts are treated with intact viral particles (~85%), it suggests that an intact viral genome may not be absolutely required for the down-regulation of plasmalemmal NKCC protein.

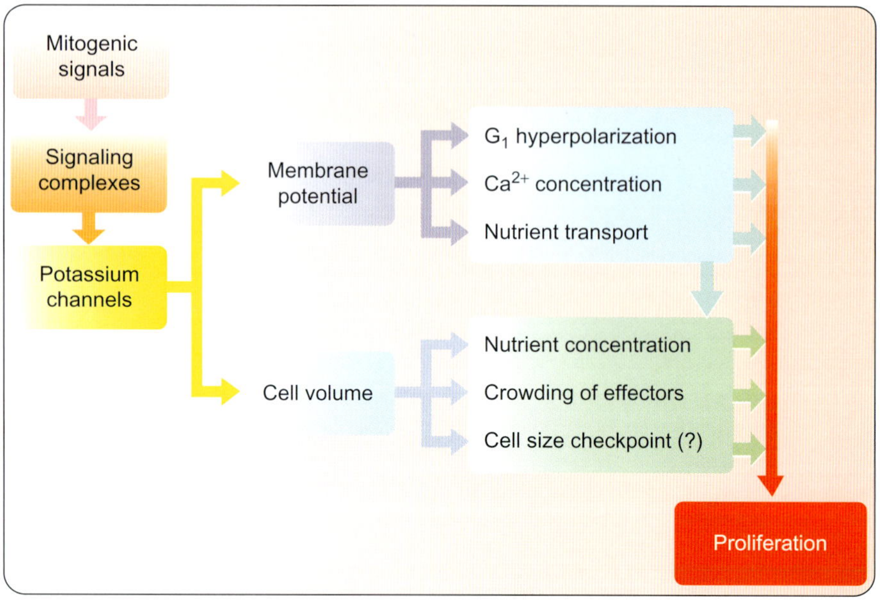

FIGURE 27.6 Possible explanations for role of K^+ channels in the normal cell cycle. In common with NKCC cell volume mechanism, this scheme includes a role for K^+ channels in cell volume changes prior to the G_1 cell cycle checkpoint. (Adapted from Pardo, 2004.)

The effect of HCMV to profoundly decrease NKCC plasma membrane levels and its ability to cause perinuclear NKCC-rich structures may both have important neurological consequences considering the critical roles that the cotransporter is believed to play in nervous system development. First, the NKCC1 isoform is prominently expressed in the inner ear and NKCC1 knockout mice have severe hearing and balance problems (Delpire et al., 1999). Furthermore, it is known that HCMV infection during human pregnancy is now the leading cause of virally mediated neurological birth defects and prominent among these defects are loss of hearing and problems with balance. Thus, HCMV infection may result in neurodevelopmental problems that are, at least partially, caused by HCMV effects on NKCC function and/or protein expression. Second, there are numerous reports linking HCMV to dementias, including those secondary to HIV infection (e.g. Goplen et al., 2001) and those termed vascular dementia (e.g. Lin et al., 2002a, b). This is particularly interesting in view of our preliminary observation of a large NKCC-protein-rich perinuclear accumulation and the well-known observation that a large number of human neurodegenerative diseases are characterized by cytoplasmic inclusion bodies in nerve and glial cells (e.g. Alves-Rodrigues et al., 1998; Tran and Miller, 1999).

IX. WHY IS NKCC "TARGETED" BY HCMV?

We have presented evidence that links the activity of the NKCC to the normal development of the cell cycle and shown that the HCMV, which acts as a mitogen, but halts the host cell at the G_1/S-phase, functionally disables the NKCC and causes the NKCC protein to be aggregated into a perinuclear structure. The question to be addressed is how might these phenomena be related? At least three different mechanisms are consistent with available findings. Whether one, two or all three are involved cannot be resolved at this time.

First, NKCC may be involved in the normal 30–45% increase in cell volume observed when normal mammalian cells undergo the G_1-phase of the cell cycle. By inactivating the NKCC function, the virus prevents this cell volume increase, an increase that may be necessary for the cell to pass the G_1 checkpoint thereby preventing host cell DNA synthesis. Because treatment with bumetanide acts mainly to slow, but not prevent, cell division, it seems unlikely that this explanation can be the complete one, although it may contribute.

A second possibility is that by functionally inactivating NKCC, K^+ uptake is reduced thereby compromising the activity of K^+ channels (Fig. 27.6). The literature contains numerous examples of cell cycle blockade by a variety of K^+ channel blockers (e.g. see reviews by Wonderlin and Strobl, 1996 and Pardo, 2004). We have shown that HCMV infection has the following two effects that may relate to reduced K^+ channel activity: (1) the $[K^+]_i$ is reduced by infection. The G_1-phase is normally characterized by a transient depolarization (Wonderlin and Strobl, 1996), an effect that would be reduced if $[K^+]_i$ is reduced and (2) that the time to reach isotopic equilibrium is increased by nearly three-fold. This latter effect suggests K^+ channels may be inactivated by the virus. Either effect would be expected to inhibit progress of the cell cycle from G_1 to S.

The third possibility follows from the work of Panet et al. (2002, 2006). They have reported that NKCC, activated by a variety of mitogens, plays a role in the activation of Mitogen Activated Protein Kinase (MAPK or MEK). Inhibition of NKCC by bumetanide inhibits the phosphorylation of these kinases. As these kinases are integral to the normal cell cycle progression, it follows that by inhibiting the NKCC, HCMV can block the cell cycle.

Although much work remains to be done to fully understand the role of the NKCC in HCMV infection, it is clear that this membrane ion transporter is targeted by the virus and, until proven otherwise, this targeting strongly suggests that the NKCC plays a crucial role in normal cell physiology that must be inhibited in order for HCMV proliferation to proceed. The best current evidence suggests that this role is in support of the progression of the cell cycle from G_1 to S.

Acknowledgements

The author gratefully acknowledges the many contributions of his colleagues Dr. William E. Crowe and Dr. Lilia M. Maglova to the work discussed herein. The work from the author's laboratory was supported by NIH grants NS11946 and NS040905.

References

Alberts, B., Bray, D., Hopkin, K., Johnson, A., Lewis, J., Raff, M., Roberts, K., and Walter, P. (2003). Chapter 18, figure 18-1. In Essential Cell Biology, 2nd edition, p. 612, Garland Press, NY.

Albrecht, T., Boldogh, I., Fons, M., Lee, C.H., AbuBakar, S., Russell, J.M., and Au, W.W. (1989). Cell-activation responses to cytomegalovirus infection: relationship to the phasing of CMV replication and to the induction of cellular damage. *Subcell. Biochem.* **15**, 157–202.

Alford, C.A. and Britt, W.J. (1990). *Cytomegalovirus*. Raven Press, Ltd, New York, NY.

Altamirano, A.A., Fons, M.P., Russell, J.M., Cragoe, E.J., Jr., and Albrecht, T. (1994). Human cytomegalovirus infection increases the number of ouabain-binding sites in human fibroblasts. *Virology* **199**, 151–159.

Alves-Rodrigues, A., Gregori, L., and Figueiredo-Pereira, M.E. (1998). Ubiquitin, cellular inclusions and their role in neurodegeneration. *Trends Neurosci.* **21**, 516–520.

Bando, K., Paradis, I., Similo, S., Konishi, H., Komatsu, K., Zullo, T., Yousem, S., Close, J., Zeevi, A., and Duquesnoy, R. (1995). Obliterative brochiolitis after lung and heart-lung transplantation An analysis of risk factors and management. *J. Thorac. Cardiovasc. Surg.* **110**, 4–13.

Becker, Y., Darai, G., and Huang, E.-S. (1993). *Molecular aspects of human cytomegalovirus diseases, Frontiers of Virology* 2. Springer-Verlag.

Benos, D.J.H., Hahn, B.H., Bubien, J.K., Ghosh, S.K., Mashburn, N.A., Chaikin, M.A., Shaw, G.M., and Benveniste, E.N. (1994). Envelope glycoprotein gp120 of human immunodeficiency virus type 1 alters ion transport in astrocytes: implications for AIDS dementia complex. *Proc. Nat. Acad. Sci. USA* **91**, 494–498.

Berman, E., Sharon, I., and Atlan, H. (1995). An early transient increase of intracellular Na^+ may be one of the first components of the mitogenic signal Direct detection by ^{23}Na-NMR spectroscopy in quiescent 3T3 mouse fibroblasts stimulated by growth factors. *Biochim. Biophys. Acta* **1239**, 177–185.

Bonin, L. and McDougall, J.K. (1997). Human cytomegalovirus IE2 86-kilodalton protein binds p53 but does not abrogate G_1 checkpoint function. *J. Virol.* **71**, 5861–5870.

Bresnahan, W.A., Boldogh, I., Thompson, E.A., and Albrecht, T. (1996). Human cytomegalovirus inhibits cellular DNA synthesis and arrests productively infected cells in late G_1. *Virology* **224**, 150–160.

Bresnahan, W.A., Boldogh, I., Chi, P., Thompson, E.A., and Albrecht, T. (1997). Inhibition of cellular Cdk2 activity blocks human cytomegalovirus replication. *Virology* **231**, 239–247.

Bresnahan, W.A., Albrecht, T., and Thompson, W.A. (1998). The cyclin E promoter is activated by human cytomegalovirus 86-kDa immediate early protein. *J. Biol. Chem.* **273**, 22075–22082.

Britt, W. and Alford, C. (1996). Cytomegalovirus. In *Field's Virology* (Fields, B.N. et al., eds), pp. 2493–2523. Lippincott Williams & Wilkins.

Bussolati, O., Uggeri, J., Belleti, S., Dall'Asta, V., and Gazzola, G.C. (1996). The stimulation of Na,K,Cl cotransport and of system A for neutral amino acid transport is a mechanism for cell volume increase during the cell cycle. *FASEB J.* **10**, 920–926.

Crowe, W.E., Altamirano, A.A., and Russell, J.M. (1997). Human cytomegalovirus infection enhances osmotic stimulation of Na^+/H^+ exchange in human fibroblasts. *Am. J. Physiol.* **273**, C1739–C1748.

de Beeck, A.O. and Caillet-Fauquet, P. (1997). Viruses and the cell cycle. *Prog. Cell Cycle Res.* **3**, 1–19.

del Castillo, J.R., Ludert, J.E., Sanchez, A., Ruiz, M.-C., Michelangeli, F., and Liprandi, F. (1991). Rotavirus infection alters Na^+ and K^+ homeostasis in MA-104 cells. *J. Gen. Virol.* **72**, 541–547.

Delpire, E., Lu, J., England, R., Dull, C., and Thorne, T. (1999). Deafness and imbalance associated with inactivation of the secretory Na-K-2Cl co-transporter. *Nat. Genet.* **22**, 192–195.

Ettinger, N., Bailey, T., Trulock, E., Storch, G., Anderson, D., Raab, S., Spitznagel, E., Dresler, C., and Cooper, J. (1993). Cytomegalovirus infection and pneumonitis Impact after isolated lung transplantation. *Am. Rec. Respir. Dis.* **147**, 1017–1023.

Fons, M., Nokta, M., Cerruti-Sola, S., and Albrecht, T. (2000). Amiloride inhibition of human cytomegalovirus replication. *Proc. Soc. Exp. Biol. Med.* **196**, 89–96.

Fortunato, E.A., McElroy, A.K., Sanchez, V., and Spector, D.H. (2000). Exploitation of cellular signaling and regulatory pathways by human cytomegalovirus. *Trends Microbiol.* **8**, 111–119.

Garcia-Mata, R., Bebo, Z., Sorscher, E.J., and Sztul, E.J. (1999). Characteristics and formation of aggresome formation by a cytosolic GFP-chimera. *J. Cell Biol.* **146**, 1239–1254.

Garcia-Mata, R., Gao, Y.S., and Sztul, E. (2002). Hassles with taking out the garbage: aggravating aggresomes. *Traffic* **3**, 388–396.

Goplen, A.K., Liestol, K., Dunlop, O., Bruun, J.N., and Maehlen, J. (2001). Dementia in AIDS patients in Oslo: the role of HIV encephalitis and CMV encephalitis. *Scand. J. Infect. Dis.* **33**, 755–758.

Grinstein, S., Rotin, D., and Mason, M.J. (1989). Na^+/H^+ exchange and growth factor-induced cytosolic pH changes Role in cellular proliferation. *Biochim. Biophys. Acta* **988**, 73–97.

Hayashi, M.L., Blankenship, C., and Shenk, T. (2000). Human cytomegalovirus UL69 protein is required for efficient accumulation of infected cells in the G_1 phase of the cell cycle. *Proc. Nat. Acad. Sci. USA* **97**, 2692–2696.

Heath, C.M., Windsor, M., and Wileman, T. (2002). Aggresomes resemble sites specialized for virus assembly. *J. Cell Biol.* **153**, 449–455.

Ho, M. (1991). *Cytomegalovirus: Biology and Infection*. Plenum Publishing Corp, NY.

Illing, M.E., Rajan, R.S., Bence, N.F., and Kopito, R.R. (2002). A rhodopsin mutant linked to autosomal dominant retinitis pigmentosa is prone to aggregate and interacts with the ubiquitin system. *J. Biol. Chem.* **277**, 34150–34160.

Jiang, L., Chenova, M.N., and Alprer, S.L. (1997). Secondary regulatory volume increase conferred on *Xenopus* oocytes by expression of AE2 anion exchanger. *Am. J. Physiol. Cell* **272**, C191–C202.

Johnston, J.A., Ward, C.L., and Kopito, R.R. (1998). Aggresomes: a cellular response to misfolded proteins. *J. Cell Biol.* **143**, 1883–1898.

Junn, E., Lee, S.S., Suhr, U.T., and Mouradian, M.M. (2002). Parkin accumulation in aggresomes due to proteasome impairment. *J. Biol. Chem.* **277**, 47870–47877.

Kapus, A., Grinstein, S., Wasan, S., Kandasamy, R., and Orlovski, J. (1994). Functional characterization of three isoforms of the Na^+/H^+ exchanger stably expressed in Chinese hamster ovary cells ATP dependence, osmotic sensitivity, and its role in cell proliferation. *J. Biol. Chem.* **269**, 23544–23552.

Killander, D. and Zetterberg, A. (1965). A quantitative cytochemical investigation of the relationship between cell mass and initiation of DNA synthesis in mouse fibroblasts in vitro. *Exp. Cell Res.* **40**, 12–20.

Kopito, R.R. (2000). Aggresomes, inclusion bodies and protein aggregation. *Trends Cell. Biol.* **10**, 524–530.

Lin, W.R., Wozniak, M.A., Cooper, R.J., Wilcock, G.K., and Itzhaki, R.F. (2002a). Herpesviruses in brain and Alzheimers disease. *J. Pathol.* **197**, 395–402.

Lin, W.R., Wozniak, M.A., Wilcock, G.K., and Itzhaki, R.F. (2002b). Cytomegalovirus is present in a very high proportion of brains from vascular dementia patients. *Neurobiol. Dis.* **9**, 82–87.

Lu, M. and Shenk, T. (1999). Human cytomegalovirus UL69 protein induces cells to accumulate in G_1 phase of the cell cycle. *J. Virol.* **73**, 676–683.

Maglova, L.M., Crowe, W.E., Altamirano, A.A., and Russell, J.M. (1998a). Human cytomegalovirus infection stimulates Cl^-/HCO_3^- exchanger activity in human fibroblasts. *Am. J. Physiol.* **275**, C515–C526.

Maglova, L.M., Crowe, W.E., Smith, P.R., Altamirano, A.A., and Russell, J.M. (1998b). Na-K-Cl cotransport in human fibroblasts is inhibited by cytomegalovirus infection. *Am. J. Physiol. Cell Physiol.* **275**, C1330–C1341.

Maglova, L.M., Crowe, W.E., and Russell, J.M. (2001). UV-irradiated human cytomegalovirus (HCMV) reduces Na,K,Cl co-transporter NKCC) activity in MRC-5 cells. *Mol. Biol. Cell* **12**, 1521.

Maglova, L.M., Crowe, W.E., and Russell, J.M. (2003). Na-K-Cl-cotransporter protein is redistributed after cytomegalovirus infection (Abstract). *FASEB J.* **17**, A904.

Maglova, L., Crowe, W.E., and Russell, J.M. (2004). Perinuclear localization of Na-K-Cl, cotransporter protein after human cytomegalovirus infection. *Am. J. Physiol. Cell Physiol.* **286**, C1324–C1334.

Maglova, L., Crowe, W.E., Coolican, S., and Russell, J.M. (2001). Reciprocal changes of expression of NHE1 and NHE2 isoforms caused by human cytomegalovirus infection in human MRC–5 fibroblasts. *Biophys. J.* **80**, 182a.

McCarthy, M., Auger, D., and Whittemore, S.R. (2000). Human cytomegalovirus causes productive infection and neuronal injury in differentiating fetal human central nervous system neuroepithelial precursor cells. *J. Human Virol.* **3**, 215–228.

Murphy, E.A., Streblow, D.N., Nelson, J.A., and Stinski, M.F. (2000). The human cytomegalovirus IE86 protein can block cell cycle progression after inducing transition into the S phase of permissive cells. *J. Virol.* **74** (15), 7108–7118.

Nath, S.K., Huang, C.Y, Levine, S.A., Yun, C.H.C., Montrose, M.H., and Donowitz, M. (1996). Hyperosmolarity inhibits the Na^+/H^+ exchanger isoforms NHE2 and NHE3: an effect opposite to that on NHE1. *Am. J. Physiol.* **270**, G431–G441.

Nokta, M., Fons, M.P., Eaton, D.C., and Albrecht, T. (1988). Cytomegalovirus: sodium entry and development of cytomegaly in human fibroblasts. *Virology* **164**, 411–419.

O'Brien, W.J., Palmer, M.L., Guy, J., and Taylor, J.L. (1996). Endothelial barrier function and Na^+/K^+-ATPase pump density in herpetic stromal disease. *Invest. Opthalmol. Vis. Sci.* **37**, 29–36.

Panet, R., Markus, M., and Atlan, H. (1994). Bumetanide and furosemide inhibited vascular endothelial cell proliferation. *J. Cell. Physiol.* **158**, 121–127.

Panet, R., Markus, M., and Atlan, H. (2000). Overexpression of the $Na^+/K^+/Cl^-$ cotransporter gene induces cell proliferation and phenotypic transformation in mouse fibroblasts. *J. Cell. Physiol.* **182**, 109–118.

Panet, R., Eliash, M., and Atlan, H. (2002). $Na^+/K^+/Cl^-$ cotransporter activates mitogen-activated protein kinase in fibroblasts and lymphocytes. *J. Cell. Physiol.* **190**, 227–237.

Panet, R., Eliash, M., and Atlan, H. (2006). $Na^+/K^+/Cl^-$ cotransporter activates MAP-kinase cascade downstream to protein kinase C, and upstream to MEK. *J. Cell. Physiol.* **206**, 578–585.

Pardo, L.A. (2004). Voltage-gated potassium channels in cell proliferation. *Physiology* **19**, 285–292.

Patton, H.K., Zhou, Z.-H., Bubien, J.K., Benveniste, E.N., and Benos, D.J. (2000). Gp120-induced alterations of human astrocyte function: Na^+/H^+ exchange K^+ conductance, and glutamate flux. *Am. J. Physiol. Cell.* **279**, C700–C708.

Pusch, O., Bernachek, G., Eilers, M., and Hengstschläger, M. (1997). Activation of c-Myc uncouples DNA replication from activation of G1 cyclin-dependent kinases. *Oncogene* **15**, 649–656.

Raat, N.J.H., Delpire, E., van Os, C.H., and Bindels, R.J.M. (1996). Culturing induced expression of basolateral Na^+-K^+-$2Cl^-$ cotransporter BSC2 in proximal tubule, aortic epithelium and vascular smooth muscle. *Pflügers Arch.* **431**, 458–460.

Rajan, R.S., Illing, M.E., Bence, N.F., and Kopito, R.R. (2001). Specificity in intracellular protein aggregation and inclusion body formation. *Proc. Nat. Acad. Sci USA* **98**, 13060–13065.

Roussel, M.F. (1998). Key effectors of signal transduction and G1 progression. *Adv. Cancer Res.* **74**, 1–24.

Russell, J.M. (2000). The sodium-potassium chloride cotransporter. *Physiol. Rev.* **80**, 211–276.

Salvant, B.S., Fortunato, E.A., and Spector, D.H. (1998). Cell cycle dysregulation by effects on cyclin transcription. *J. Virol.* **72**, 3729–3741.

Sanchez, V., Sztul, E., and Britt, W.J. (2000a). Human cytomegalovirus pp28 (UL99) localizes to a cytoplasmic compartment, which overlaps the endoplasmic reticulum-Golgi-intermediate compartment. *J. Virol.* **74**, 3842–3851.

Sanchez, V., Greis, K.D., Sztul, E., and Britt, W.J. (2000b). Accumulation of virion tegument and envelope proteins in a stable cytoplasmic compartment during human cytomegalovirus replication: characterization of a potential site of virus assembly. *J. Virol.* **74**, 975–986.

Sanchez, V. and Spector, D.H. (2008). Subversion of cell cycle regulatory pathways. *Curr. Top. Microbiol. Immunol.* **325**, 243–262.

Schaefer, A., Geck, P., Zibirre, R., Kühne, J., and Koch, G. (1984). Alterations of ^{86}Rb fluxes in poliovirus-infected HeLa cells and their dependence on virus replication. *Virology* **136**, 457–461.

Siczkowski, M. and Ng, L.L. (1995). Culture density and the activity, abundance and phosphorylation of the Na^+/H^+ exchanger isoform 1 in human fibroblasts. *Biochem. Biophys. Res. Comm.* **209**, 191–197.

Sinclair, J., Baillie, J., Bryant, L., and Caswell, R. (2000). Human cytomegalovirus mediates cell cycle progression through G_1 into early S phase in terminally differentiated cells. *J. Gen. Virol.* **81**, 1553–1565.

Soucek, T., Pusch, O., Hengstschläger-Ottnad, E., Adams, P.D., and Hengstschläger, M. (1997). Deregulated expression of E2F-1 induces cyclin A- and E-associated kinase activities independently from cell cycle position. *Oncogene* **14**, 2251–2257 Review.

Speir, E., Modali, R., Huang, E., Leon, M.B., Shawl, F., Finkel, T., and Epstein, S.E. (1994). Potential role of human cytomegalovirus and p53 interaction in coronary restenosis. *Science* **265**, 391–394.

Staras, S.A., Dollard, S.C., Radford, K.W., Flanders, W.D., Pass, R.F., and Cannon, M.J. (2006). Seroprevalence of cytomegalovirus infection in the United States, 1988–1994. *Clin. Infect. Dis.* **43**, 1143–1151.

Stiernberg, J., LaBelle, E.F., and Carney, D.H. (1983). Demonstration of a late amiloride-sensitive event as a necessary step in initiation of DNA synthesis by thrombin. *J. Cell. Physiol.* **117**, 272–281.

Stinski, M. (1977). Synthesis of proteins and glycoproteins in cells infected with human cytomegalovirus. *J. Virol.* **23**, 751–767.

Takahashi, A., Yamaguchi, H., and Miyamoto, H. (1993). Change in K^+ current of HeLa cells with progression of the cell cycle studied by patch clamp technique. *Am. J. Physiol. Cell Physiol.* **265**, C328–C336.

Tran, P.B. and Miller, R.J. (1999). Aggregates in neurodegenerative disease: crowds and power? *Trends Neurosci.* **22**, 194–197.

Tupper, J.T., Zorgniotti, F., and Mills, B. (1977). Potassium transport and content during G_1 S phase following serum stimulation of 3T3 cells. *J. Cell. Physiol.* **91**, 429–440.

Voss, T.G., Gatti, P.L., Fermin, C.D., and Garry, R.F. (1996). Reduction of human immunodeficiency virus production and cytopathic effects by inhibitors of the $Na^+/K^+/2Cl^-$ cotransporter. *Virology* **219**, 291–294.

Wakabayashi, S., Fafournoux, P., Sardet, C., and Pouyssegur, J. (1992). The Na^+/H^+ antiporter domain mediates growth factor signals and controls "H^+-sensing". *Proc. Nat. Acad. Sci. USA* **89**, 2424–2428.

Weber, L.A., Hickey, E.D., Maroney, P.A., and Baglioni, C. (1977). Inhibition of protein synthesis by Cl^-. *J. Biol. Chem.* **252**, 4007–4010.

Weibusch, L. and Hagemeier, C. (1999). Human cytomegalovirus 86-kilodalton IE2 protein blocks cell cycle progression in G_1. *J. Virol.* **73**, 9274–9283.

Weller, T.H. (1971). The cytomegaloviruses: ubiquitous agents with protean clinical manifestations. *N. Eng. J. Med.* **285**, 203–214.

Wigley, W.C., Fabunmi, R.F., Lee, M.G., Marino, C.R., Muallem, S., DeMartino, G.N., and Thomas, P.J. (1999). Dynamic association of proteasomal machinery with the centrosome. *J. Cell Biol.* **145**, 481–490.

Wonderlin, W.F. and Strobl, J.S. (1996). Potassium channels, proliferation and G_1 progression. *J. Membr. Biol.* **154**, 91–107.

Zhou, Y.F., Leon, M.B., Waclawiw, M.A., Popma, J.J., Yu, Z.X., Finkel, T., and Epstein, S.E. (1996). Association between prior cytomegalovirus infection and the risk of restenosis after coronary atherectomy. *N. Engl. J. Med.* **335**, 624–630.

PART V

CATION-CHLORIDE COTRANSPORT IN CHOROID PLEXUS AND BLOOD–BRAIN BARRIER

CHAPTER

28

Chloride Transporters as Water Pumps: Elements in a New Model of Epithelial Water Transport

Nanna MacAulay, Steffen Hamann and Thomas Zeuthen

OUTLINE

I. INTRODUCTION

It is increasingly clear that cotransporters and uniporters, in addition to their transport of minerals and organic molecules, also transport water. In epithelial cells and endothelial cells of the blood–brain barrier, cation-Cl^- cotransporters play an important role in the transport of ions. Typically, water or fluid follows the movement of ions, and the question arises about the nature of the mechanism underlying water movement. We believe that the transport of water is mediated by these ion transporters, and that proteins

TABLE 28.1 Passive water permeability per transport protein

Transporter	L_p per transporter [10^{-14} cm^3 s^{-1}]	References
AQP1	1.4–6	Yang and Verkman (1997); Zampighi et al. (1995); Zeidel et al. (1992)
AQP0	0.015–0.25	Yang and Verkman (1997); Zampighi et al. (1995)
rSGLT1	1.4	Loo et al. (1999)
NKCC1	4[a]	Hamann et al. (2005)
NKCC2	0	Hamann et al. (2005)
EAAT1	0.2	MacAulay et al. (2002a)
KCC4	0.01[a]	Zeuthen (1994); Zeuthen (1991b)
hGAT1	0.7	Loo et al. (1999)
NaDC1	1.5	Meinild et al. (2000)
MCT1	0.3	Hamann et al. (2003); Zeuthen et al. (1996)
GLUT2	0.09	Zeuthen et al. (2007)
GLUT1(human)	0.2[b]	Zeuthen unpublished
UT-B	7.5	Yang and Verkman (2002)

[a]These values were estimated as given in the text below.
[b]Calculated from the glucose uptake, the glucose permeability P_s ($1.5\,10^{-6}$ cm s^{-1}), and the water permeability L_p ($6.5\,10^{-5}$ cm s^{-1}) for human GLUT1 expressed in *Xenopus* oocytes at room temperature. The turnover rate was 151 s^{-1} at room temperature and 1116 s^{-1} at 37°C (Simpson et al., 2007). The measurements were performed in analogy to the study of GLUT2 (Zeuthen et al., 2007).
[c]Calculated from the Gibbs equation and the near zero reflection coefficient for NaCl (Hamann et al., 2005).

TABLE 28.2 Number of water molecules cotransported per transport cycle: the coupling ratio CR.

Protein	Process	CR	References
KCC4	KCl cotransport	500	Zeuthen (1994); Zeuthen (1991a)
NKCC1	Na^+-K^+-Cl cotransport	590[c]	Hamann et al. (2005)
NKCC2	Na^+-K^+-Cl cotransport	0	Hamann et al. (2005)
hSGLT1	Na^+-glucose cotransport	230	Meinild et al. (1998); Zeuthen et al. (2006)
rSGLT1	Na^+-glucose cotransport	380	Meinild et al. (1998); Zeuthen et al. (2006)
hGAT1	Na^+-Cl^--GABA cotransport	330	MacAulay et al. (2002b)
EAAT1	Na^+-glutamate cotransport	425	MacAulay et al. (2001)
NaDC-1	Na^+-dicarboxylate cotransport	176	Meinild et al. (2000)
MCT1	H^+-lactate cotransport	500	Hamann et al. (2003); Zeuthen et al. (1996)
GLUT2	Glucose transport	40–110	Zeuthen et al. (2007)

directly couple water movement to the transport of substrate. The combination of water, ion and organic nutrient transport has also been found for a variety of other, Cl^- independent, cotransporters; see Tables 28.1 and 28.2, and some recent reviews (Loo et al., 2002; Zeuthen and MacAulay, 2002).

Water and substrate transport in epithelial and endothelial cell layers present some fundamental problems. In the lumen of the small intestine, for example, the presence of high concentrations of glucose and other nutrients, of up to 200 mOsm hyperosmolar to the epithelial cells just after a meal, constitutes a potential threat for osmotic back-flux and dehydration. Despite these adverse osmotic gradients, water is being reabsorbed (Pappenheimer, 1998; Zeuthen et al., 2007). As another example, the endothelial cell layer of the brain capillaries have to be relatively impermeable to water in order to avoid brain edema, yet the water permeability of the glucose transporters (GLUT1) and of NKCC1 conveys a significant water permeability. On top of these epithelial problems, there is also the problem of the cells themselves: how they regulate their own volume (water content), intracellular osmolarity, and ion concentrations in the face of the large transepithelial concentration gradients and fluxes.

In this chapter, we argue that water transport through KCC and NKCC is central for the coupling of substrate and water transport across epithelia and endothelia. If, for example, influxes change some intracellular parameters, such as the osmolarity or ion concentrations, these changes must be sensed by the cells, and efflux rates adjusted in order to normalize these internal parameters. If the input mechanisms are distributed to one side of the cell and the output mechanisms are balanced to the other, transcellular transport will be in steady state and the integrity of the cell maintained. Accordingly, volume regulation plays a key role in epithelial and endothelial water transport. The importance of the KCC (K^+-Cl^- cotransporter) and the NKCC1 (Na^+-K^+-$2Cl^-$ cotransporter) lies in their ability to mediate uphill water transport. In addition, they are well known to be involved in volume regulation in non-polarized cells. The nature of the primary osmosensing signal(s) is still under debate, while some of the components of the regulatory pathways, i.e. the kinases that regulate KCC and NKCC, have been described (Delpire and Gagnon, 2008; Kahle et al., 2008; Kahle et al., 2006).

II. WATER TRANSPORT BY COTRANSPORTERS

It is generally accepted that some cotransporters carry water in the process of solute translocation, although the precise molecular mechanism is unknown. They carry water in two distinct ways: (1) in analogy to aquaporins, they act as simple channels in which water is driven by an external osmotic driving force and (2) they also function as water pumps in which a water flux is energized by the movement of substrate. Thus, water transport via cotransporters is bimodal, with a passive component and a secondary active component (Fig. 28.1). These two components are discussed in the next sections.

A. The Passive Component

As seen in Table 28.1, the passive water permeability per copy of cotransporter or uniporter can be high, even compared with that of aquaporin water channels. Some transporters have water permeabilities that are comparable to AQP1, e.g. NKCC1 and the urea uniporter UT-B. The least water permeable transporters have permeabilities comparable to AQP0. This is the case for the glucose uniporters (GLUT), and the K^+-Cl^- cotransporter, KCC4.

These values show that the cotransporters, when their number is high, can contribute a significant fraction of the passive water permeability of a given membrane. This would explain the experimental finding that humans lacking AQP1 (Colton null individuals) are without a significant clinical phenotype; the lack of AQP1 only becomes noticeable under abnormal stress of water deprivation (King et al., 2004). Likewise, the mouse knockout of AQP4 in the astroglial endfeet seems to be important only under conditions of brain edema (Amiry-Moghaddam et al., 2004; Manley et al., 2000). The effects of aquaporin removal have recently been studied in detail in the nematode worm *Caenorhabditis elegans* (Huang et al., 2007). This organism expresses four aquaporin molecules, and knockout or deletion of all or combinations of one, two or three of these water transport proteins did not produce any noticeable phenotype in regard to water homeostasis. This points toward the role of other membrane proteins as water transporters and supports the finding that the prime function of several aquaporins is not water transport, but rather that of other substrates such as urea, glycerol and ammonia (for reviews, see Hill et al., 2004; King et al., 2004; Litman et al., 2008).

It should be emphasized that the permeabilities listed in Table 28.1 are defined operationally from the water flux initiated by a given osmotic gradient. Some transporters (e.g. SGLT1; Loo et al., 1999) have a water permeability with a low activation energy, which indicates that they behave as conventional water channels, whereas others have a high activation energy, indicating that movements of ions or conformational changes in the protein are associated with the osmotically induced water transport (e.g. KCC and NKCC1; Hamann et al., 2005; Zeuthen, 1994).

Is it possible to estimate the contribution of the lipid component, the aquaporins, and solute cotransporters to the passive water permeability of a given membrane? A useful insight can be obtained from a comparison between the mammalian small intestine and kidney proximal tubule. These two epithelia have roughly the same rate of transepithelial Na^+ and water transport per surface area. The major difference is the passive water permeability (L_p), with the kidney epithelium being around ten times more water permeable than the intestine epithelium, due to the presence of AQP1 (Table 28.3). Both epithelia also

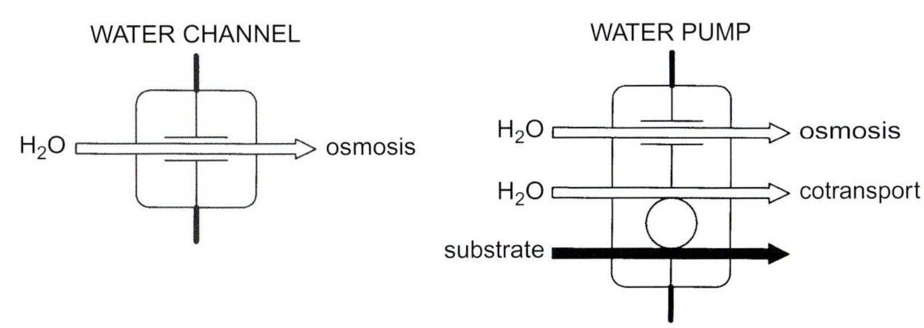

FIGURE 28.1 Molecular water channels and water pumps. Water transport across cell membranes takes place through several routes and mechanisms. Water can cross the membrane via water channels such as aquaporins, driven by a transmembrane difference in water chemical potential. Water can also be transported through membrane proteins such as cotransporters. In these proteins, water is cotransported together with the other substrates, i.e. the transporters function as molecular water pumps in which the free energy contained in the substrate gradient can be transferred to the uphill transport of water. Some cotransporters also have significant passive water permeability, i.e. they also function as water channels.

TABLE 28.3 Passive water permeability of epithelial and endothelial cell layers

Cell layer	Passive water permeability [10^{-2} cm sec^{-1}]
Proximal tubule (rat, apparent area)	20
Small intestine (rat jejunum)	2.5
Gall bladder (rabbit)	0.5
Choroid plexus (frog)	≤0.5
BBB (man)	0.1
Frog skin	0.1

For references see House (1974); Zeuthen (1996)

have Na^+-coupled substrate transporters, in the apical (entry) membrane, e.g. SGLT1 for glucose entry. In the intestinal epithelium, roughly two-thirds of the water influx across the apical membrane takes place via cotransporters, and the remainder one-third via the lipid bilayers. In the kidney epithelium, roughly one-third of the water influx takes place via the cotransporters while two-thirds enter via the AQP1 (Zeuthen et al., 2001). It would appear that there is only a modest gain in inserting a large number of aquaporins, but the real advantage may lie elsewhere. Indeed, with the given data, it is possible to estimate the osmolarity difference between the intracellular milieu, relative to the apical bathing solution to drive the water movement. In the small intestine, it was 7 mOsm, while the kidney proximal tubule cells required only 0.5 mOsm to drive the same influx of water. Clearly, as it will cost more energy to maintain a high intracellular osmolarity; it could be concluded that the aquaporins do not increase the rate of water transport, but they make it cheaper.

B. The Active Component

In this section, we consider the transport of substrate leading to cotransport of water (i.e. secondary active water transport). It has been intensely debated whether the coupling takes place by a mechanism inside the protein or whether it is a consequence of conventional unstirred layers effects in the cytoplasm. There is now agreement that the latter alternative is untenable. Indeed, when the substrate is transported into a cell it diffuses away from the cotransporter, and the build-up at the inside (or *trans* side) of the membrane will depend upon the diffusion coefficient. But as the diffusion coefficient is relatively high in the cytoplasm, the substrate concentration at the inside of the membrane remains fairly low and insufficient to induce significant passive osmotic water transport. This question has been investigated in detail for the Na^+-coupled glucose transporter (SGLT1) and the glucose uniport (GLUT2) expressed in the *Xenopus laevis* oocytes. The diffusion coefficient for Na^+ and glucose in the cytoplasm is between one-half to one-fifth of the free solution diffusion coefficient in agreement with what is found in other cell types (Charron et al., 2006; Lapointe, 2007; Zeuthen et al., 2002; Zeuthen and Zeuthen, 2007). But this is still much too high to support significant conventional unstirred layer effects, which would require the diffusion coefficients to be three orders of magnitude lower than in free solutions. This can be ascertained by correlating water and Na^+ fluxes at a high resolution (20 pL and 1 s; Zeuthen et al., 2006). Consequently, transport of water by a mechanism intrinsic to the cotransporter is the best explanation for the observed coupling between water and substrate (Naftalin, 2008; Zeuthen et al., 2006, 2007; Zeuthen and Zeuthen, 2007). Similar arguments for the lack of unstirred layer effects have been addressed in the studies of KCC and NKCC1 (Hamann et al., 2005; Zeuthen, 1994).

The molecular nature of the water cotransport component is supported by a number of experimental facts. (i) For a given cotransporter, the coupling ratio (CR) between the number of water molecules transported per substrate molecule is constant under a variety of experimental conditions, irrespective of whether the transport is driven by electrical or concentration gradients. (ii) The CR is specific for a given isoform, for example the human SGLT1 transports 230 water molecules, while the rabbit SGLT1 transports 380 per glucose molecule. (iii) This phenomenon has been observed using a number of independent methods, ranging from fluorescence, ion-selective microelectrodes and two microelectrode voltage-clamp combined with high-resolution volume measurements (references in Table 28.2). (iv) Secondary active water transport has been found in transporters in their native tissue, as well as in transporters heterologously expressed in mammalian cell-lines and *Xenopus laevis* oocytes.

The numbers in Table 28.2 compare well to the amounts of water being shifted between aqueous enzymes and their surrounding bulk water. Glucose binding in hexokinase as well as in α-amylase is associated with the exchange of about 100 water molecules with the bulk solution (Quin et al., 1995; Rand and Fuller, 1992; Steitz et al., 1981); O_2 binding in hemoglobin is associated with the exchange of 60–75 water molecules (Colombo et al., 1992). Moreover, certain voltage-gated anion channels shift an impressive 660 and 1320 water molecules when opening or closing the permeation pathway (Zimmerberg and Parsegian,

1986). Thus it would appear that the transporters listed in Table 28.2 can be viewed as membrane-bound enzymes that take up water and substrate from one side of the membrane and deliver them to the other.

The mechanism of substrate-coupled water transport must be sought in the transport protein itself. In the present context of Cl^--dependent cotransporters, we will focus on KCC4 (K^+-Cl^- cotransporter), NKCC1 (Na^+-K^+-Cl^- cotransporter), and the GABA transporter GAT1. The water transporting properties of other cotransporters have been reviewed elsewhere (Zeuthen and MacAulay, 2002). As evidence for cotransport of water, we will address at the following specific questions:

- Is there a fixed stoichiometry between the water and the substrate transport?
- Can energy be transferred between the substrate and water fluxes, i.e. can substrate transport lead to uphill water transport?
- Do the fluxes exhibit saturation?
- Is there any link between the value and properties of the passive water permeability and the ability for active water transport?

C. The K^+-Cl^- Cotransporter, KCC

The choroid plexus epithelium is a convenient place to study KCC. The epithelium is similar to other leaky epithelia but it transports salt and water in the opposite direction, i.e. from the blood and into the cerebrospinal fluid. Since epithelial KCC transporters are primarily situated in the exit membrane, i.e. the apical membrane facing the cerebrospinal fluid, the choroid plexus epithelium has the experimental advantage that KCC in the exit membrane can be exposed abruptly to changes in the composition of the bathing solution. The intracellular driving forces for KCC can be followed at a high time resolution by means of ion-selective, double-barreled microelectrodes that record the K^+ and Cl^- activities. The cell volume is monitored likewise: The K^+ electrodes are very sensitive to choline ions or tetra-methylammonium ions; by having a low concentration of these ions inside the cell, the ion concentration acts as a volume marker. Most data have been obtained in the choroid plexus epithelium from the salamander *Necturus maculosus* (Zeuthen, 1991a&b, 1994). In terms of location, this KCC would be equivalent to the KCC4 isoform from mammals (Karadsheh et al., 2004); SLC12A7 in the SLC12 family (Gamba, 2005). At physiological osmolarity, half of the capacity for water transport across the exit membrane is contributed by KCC, the other most likely by AQP1 (Zeuthen, 1994).

The findings for KCC can be summarized as follows (Zeuthen, 1991a&b; 1994; Zeuthen, 1995). (i) Water transport by KCC is abolished when Cl^- is removed from the bathing solutions. (ii) Compared to the other cotransporters listed in Table 28.1, the passive water permeability per KCC protein is rather low, of the order of $10^{-16}\,cm^3 s^{-1}$. The number is estimated from the maximal rate of transport of the Cl^--induced K^+ transport and assuming a turnover number around $100\,s^{-1}$; the latter seems reasonable, considering the high activation energy of $27\,kcal\,mol^{-1}$ (Ellory and Hall, 1988). (iii) The flow of water, K^+ and Cl^- via the KCC are strictly coupled in a ratio of 500 water molecules per K^+ and Cl^- ion. This applies irrespective of whether the flux was induced by perturbations in the K^+, the Cl^- or the osmotic gradient. (iv) Cotransport of K^+ and Cl^- can drive water transport uphill. As illustrated in Fig. 28.2, a KCl-driven water

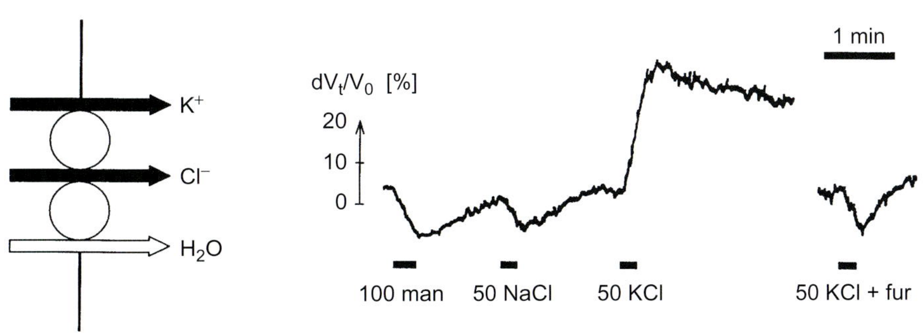

FIGURE 28.2 KCC is a molecular water pump. In K^+-Cl^- cotransport, the flux of one K^+ ion and one Cl^- ion is coupled to the flux of water molecules in a ratio of 1:1:500. The trace demonstrates the coupling between a downhill flux of KCl and an uphill transport of water. The addition of 100 mM of mannitol or 50 mM of NaCl (=100 mOsm) to the external bathing solution caused cell shrinkage. Addition of 50 mM of KCl, however, caused cell swelling, despite the fact that the external osmolarity was now higher than the intracellular osmolarity by 100 mOsm. The intracellular concentrations of K^+ and Cl^- only changed by a few mM during the exposure to KCl. When KCC was blocked by furosemide (fur) the cell shrank in response to the addition of 50 mM of KCl. Data are from the choroid plexus epithelium of *Necturus maculosus* (Zeuthen, 1994).

flux can proceed into the cell despite the extracellular osmolarity being about 100 mOsm higher than that of the intracellular compartment. In other experiments, it was found that osmotic gradients as high as 200 mOsm could be matched by KCl-induced influxes of water (Zeuthen, 1994). (v) The water fluxes showed simple first order kinetics: The water fluxes induced by KCl were half-saturated at KCl concentrations of 25 mmol l^{-1} which is close to the K_m determined for the KCC transporters (Mercado et al., 2000). The osmotically induced fluxes of water, K^+ and Cl^- saturated at osmotic gradients of about 200 mOsm.

The properties listed above show that the KCC is a key player in transepithelial water transport. With 500 water molecules transported per K^+ ion and 2 K^+ transported for each 3 Na^+ by the Na^+/K^+-ATPase, it follows that these two transporters in unison, given similar turnover rates, would produce a secretion with about 333 water molecules per Na^+ ion. This is two-thirds of the water required to explain the observed isosmotic transport which requires about 500 water molecules per Na^+ ion. Importantly, the KCC would also explain the ability of leaky epithelia to perform uphill transport of water. Certain epithelia, such as small intestine and gallbladder, can absorb water against osmotic gradients of up to 200 mOsm, in good agreement with the properties of the KCC. An epithelial model with the KCC localized to the exit membrane is shown in Figs 28.7 and 28.8; for reviews see Zeuthen (2002, 1996).

D. The Na^+-K^+-$2Cl^-$ Cotransporters, NKCC1 and NKCC2

The NKCC mediates electroneutral cotransport of Na^+, K^+ and two Cl^- ions. The cotransporter has been found in two isoforms, NKCC1 and NKCC2; SLC12A2 and SLC12A1 in the SLC12 family (Gamba, 2005). NKCC1 is present in most mammalian cells, while NKCC2 is found predominantly in the thick ascending limb of Henle in the kidney, for reviews see Delpire and Mount (2002), Gamba (2005), Haas and Forbush (2000) and Russell (2000). NKCC1 is a key protein in secretory epithelial cells, where it couples the movements of Cl^- and K^+ to the inwardly directed Na^+ gradient set up by the Na^+/K^+-ATPase, Figs 28.7 and 28.8. The water transport properties of NKCC1 have been studied in the ciliary body of the mammalian eye (Hamann et al., 2005; Hamann et al., unpublished) which participate in the transport water from the blood into the aqueous humor. The ciliary body consists of two cell layers: the non-pigmented and the pigmented epithelium; the latter contains NKCC1 in its basolateral membrane (Dunn et al., 2001). When grown as a single layer of cells on a glass plate, the basolateral membrane of the pigmented epithelium faces upwards and thereby allows the transport properties of NKCC1 to be investigated during abrupt changes in the composition of the bathing solution. At physiological osmolarity, half the osmotic water permeability of the basolateral membrane of the pigmented epithelium can be attributed to NKCC1.

The data for NKCC1 can be summarized as follows (Hamann et al., 2005; Hamann et al., unpublished). (i) The capacity for water transport by NKCC1 is inhibited by bumetanide. (ii) The water transport by NKCC1 has a high Arrhenius activation energy of 21 kcal mol^{-1}, which is more than four times that of an aqueous pore. The activation energy for water transport is close to that of the ion transport process of 18 kcal mol^{-1} (Ellory and Hall, 1988). The high value of the activation energy strongly suggests that the transport of water involves conformational changes of the protein. (iii) The unit water permeability is relatively high about $4 10^{-14} cm^3 s^{-1}$ and of the same order as that of AQP1 (Table 28.1). This estimate is based upon the number of transporters required to maintain maximal transport rates of ions at a turnover rate of 100 s^{-1}. The latter estimate seems reasonable given the high activation energy of 18 kcal mol^{-1}. (iv) The reflection coefficient for NaCl and KCl is nearly zero. This means that when the osmolarity of the solution is raised by addition of NaCl or KCl there is little net water transport as compared to the case when the flow is induced by the addition of mannitol. One interpretation is that the increased salt concentration results in an inward movement of water coupled to the influx of salt, which in turn compensate for the osmotic efflux. In other words, the increased influx of ions gives rise to an influx of water. (V) In NKCC1, ion fluxes are tightly coupled to water fluxes which can be deduced from the capacity of uphill transport of water (Hamann et al., unpublished). In the experiment illustrated in Fig. 28.3, epithelial tissues were exposed to Cl^--free solutions, which induced rapid cell shrinkage. When Cl^- ions were returned to the bathing solutions, there was a rapid influx of water (i.e. re-swelling of the cells) which could proceed uphill against osmotic gradients of more than 50 mOsm imposed by mannitol.

NKCC2 behaves entirely differently than NKCC1. When studied in cell cultures from the thick ascending loop of Henle (TALH), it showed ion fluxes similar to that of NKCC1, but it showed no capacity for water transport (Hamann et al., 2005).

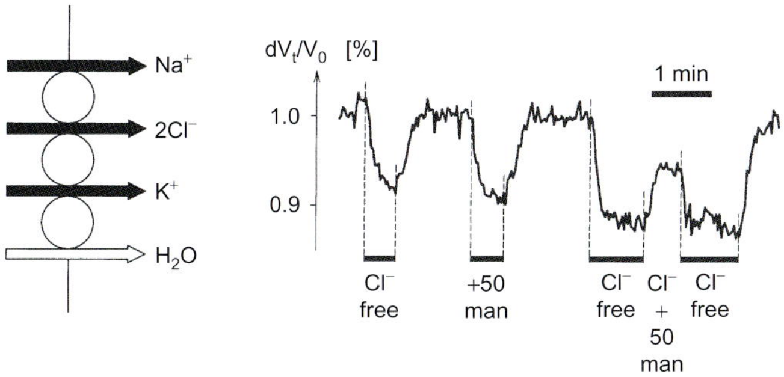

FIGURE 28.3 NKCC1 is a molecular water pump. Human pigmented ciliary epithelial cells where exposed to abrupt changes in bathing solution composition. The resulting volume changes can be attributed to NKCC1 located to the apical membrane, as described in Hamann et al. (2005). In the first experiment (Cl^- free), isosmotic substitution of Cl^- by the inert anion gluconate caused rapid cell shrinkage. When Cl^- was returned, the cells returned to their original volume. In the second experiment, addition of 50 mOsm of mannitol (+50 man) to the bathing solution produced a rapid cell shrinkage. In the last experiment, the cells were first shrunk by removal of Cl^- (Cl^- free). Then Cl^- was returned, simultaneously with the addition of 50 mOsm of mannitol (Cl^- + 50 man). The question posed by this experiment is whether the Cl^--dependent influx of water can overcome the osmotic efflux generated by the mannitol. The experiment shows clearly the Cl^--dependent influx can proceed inwards against the osmotic gradient of 50 mOsm imposed by the mannitol (Hamann et al., unpublished).

E. The Na^+-Cl^--coupled GABA Transporter, GAT1

In order to maintain synaptic transmission, the neurotransmitter GABA is removed from the synaptic cleft by means of Na^+-Cl^--coupled re-uptake. Four GABA cotransporter subtypes have been described, GAT-1, GAT-2, GAT-3 and BGT-1 (Borden, 1996). The transport of GABA is electrogenic, coupled to the movement of two Na^+ and one Cl^- ion for each GABA (Kavanaugh et al., 1992; Radian and Kanner, 1983). The transporter belongs to the SLC6 family of solute carriers (Hediger et al., 2004).

The water transport properties of GAT-1 were studied while expressed in *Xenopus* oocytes. The volume of the oocyte was followed by a CCD camera and the rate of water movements could be recorded at high volume (10 pL) and temporal (1 s) resolutions (Zeuthen et al., 2006). The measurements were combined with a two-electrode voltage clamp in order to correlate water movements and transport currents. When the bathing solution was made hyperosmolar, the rate of shrinkage defined the passive water permeability. The contribution of the GAT-1 to the water permeability of the oocyte was around $3 10^{-6}$ cm sec^{-1}(osm l^{-1})$^{-1}$ equivalent to $1.6 10^{-4}$ cm s^{-1} (Loo et al., 1999; MacAulay et al., 2002b). This was obtained at an expression level of GAT-1 that gave currents of around 350 nA at clamp potentials of −50 mV. Both the water permeability and the clamp current were abolished by the inhibitor SKF89976A. In addition to water, GAT-1 was permeable to urea (Loo et al., 1999). This has also been reported for other cotransporters: the Na^+-coupled glutamate cotransporter (MacAulay et al., 2002a), the Na^+-coupled glucose cotransporter (SGLT1) and the iodide cotransporter (NIS) (Leung et al., 2000). This supports the notion of an aqueous pathway through the transporter.

In addition to its passive water permeability, GAT-1 also cotransports water along with the GABA as a secondary active transport (Loo et al., 1996; MacAulay et al., 2002b). In analogy with what has been observed for a wide range of cotransporters (Table 28.2), the volume of the GAT-1 expressing oocyte increased linearly with time in the presence of GABA under voltage clamp conditions. The volume changes correlated closely with the integrated GABA-induced currents, which equals the charge that has entered the oocyte (Fig. 28.4). The correspondence between the volume change and the charge demonstrates that a fixed number of water molecules enters per translocated charge or, synonymously, per turnover of the transporter. The calculated coupling ratio amounts to 330 water molecules per turnover. If Li^+ replaced Na^+ in the bathing solution, both cotransport of GABA as well as of water were abolished; instead the cotransporter acted as a passive electrical leak in which the current is carried by Li^+ (MacAulay et al., 2002b). Such leak currents did not give rise to cotransport of water. This shows that the cotransport of water depends on Na^+-mediated cotransport of GABA by a mechanism intrinsic to the cotransporter and not by the accumulation of ions in an unstirred layer in the cytoplasm of the oocyte. If transport into the cytoplasm had given rise to unstirred layers, such effects should also be observed with the Li^+ currents.

F. What is the Molecular Mechanism of Water Cotransport?

The ability of cotransport proteins for passive transport of water (and urea) is good evidence for an aqueous pathway through the proteins (Guan and Kaback, 2006). The capacity of the pathway and thereby the passive permeability may vary among various conformational states of the cotransporter (MacAulay et al., 2002a), but in all cases, the fluxes are driven by externally applied gradients.

While the concept of a passive permeability is relatively simple, the more difficult question remains: Which

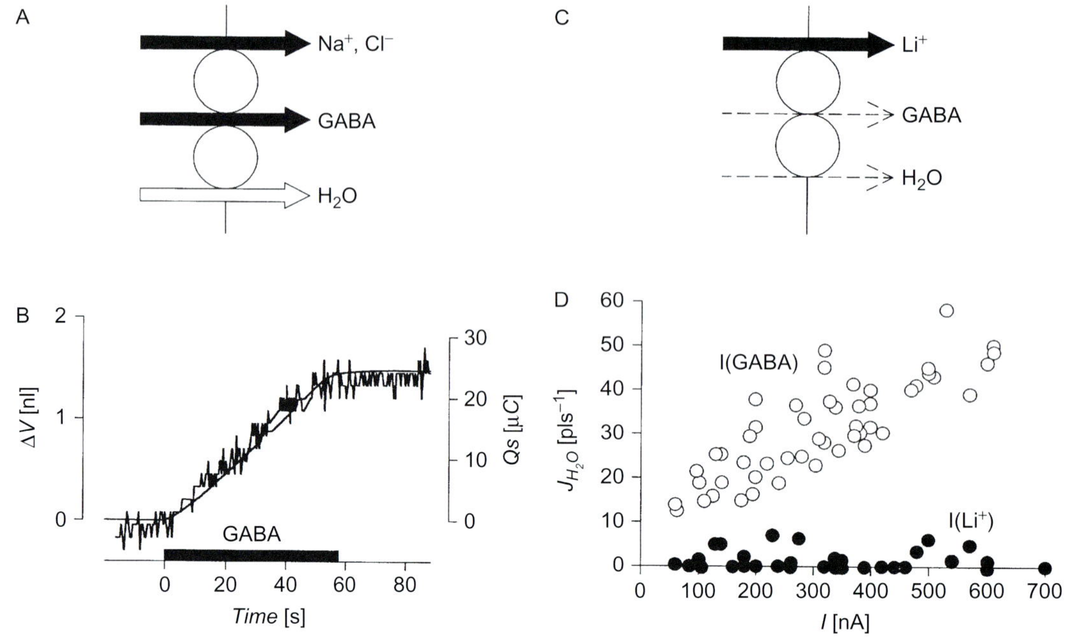

FIGURE 28.4 Cotransport of water by the Na^+-coupled GABA transporter GAT1. **A.** In the presence of Na^+ and Cl^- the cotransport of GABA is associated by a flux of water in a strict stoichiometric relationship. **B.** A GAT-1-expressing oocyte was clamped to −50 mV and 100 μM GABA was isotonically added to the test solution (black bar). Accordingly, there was no osmotic driving force across the membrane. The jagged line in the figure represents the volume of the oocyte and the straight line represents the total amount of charges translocated by the GABA transport. **C.** If Na^+ is replaced by Li^+, the transporter carries a large Li^+ leak current but GABA and H_2O are not cotransported. In (**D**) GAT-1-expressing oocytes were clamped to varying membrane potentials (from −30 to −140 mV). The leak current ($I_{Li}+$) obtained with 100 mM Li^+ ($I_{Li} - I_{Ch}$) or the GABA current I_{GABA} obtained with 100 μM GABA ($I_{Na+GABA} - I_{Na}$) gave currents in the range 50 to 700 nA, n = 6–7. The accompanying water flux (J_{H2O}) is plotted versus this current for the leak current (black circles) and the GABA-induced current (white circles).

structures and conformational changes in cotransport proteins could explain cotransport of water in which energy from the downhill movements of ions and other non-aqueous substrates are coupled to the uphill movements of water? Two different mechanisms have been suggested, the mobile barrier model (Zeuthen, 2000, 1994; Zeuthen and MacAulay, 2002; Zeuthen and Stein, 1994) and the osmotic coupling mechanism (Naftalin, 2008; Zeuthen, 2000, 1991a; Zeuthen et al., 2007; Zeuthen and Stein, 1994). The two mechanisms are set up in schematic form in Fig. 28.5A and B. In a final model, the two mechanisms may operate simultaneously.

The first mechanism is based upon the alternating access model (Mitchell, 1957) and can be divided into five steps, see Fig. 28.5A: (i) when the substrate is applied to the *cis* side of the transporter, it binds inside an aqueous cavity, (ii) conformational changes are elicited which occlude the substrates together with a number of water molecules, (iii) the cavity opens to the *trans* side and the substrates exits, (iv) in the absence of substrates, the transporters attain a closed state with a small or no aqueous cavity and as a consequence, a number of water molecules equivalent to the closure will follow the substrates, and (v) the protein returns to its initial state.

The second mechanism (Fig. 28.5B) is analogous to the three-compartment model for transepithelial water transport (Curran and Macintosh, 1962; Naftalin, 2008; Zeuthen and Stein, 1994). In the first version of the model, the coupling took place in a cellular compartment; in its molecular version, the coupling space is in the transport molecule, associated with the aqueous cavity in which the substrate binds. This space becomes hyperosmolar when the substrate leaves the binding site and becomes thermodynamically free inside the cavity. This situation remains until the substrate exits into the bulk solution on the *trans* side. As long as the substrate remains in the cavity, the resulting hyperosmolarity will pull water from the *cis* compartment, across the protein and into the *trans* compartment. This mode of operation can be formulated thermodynamically as follows. The barrier that separates the *cis* compartment from the cavity constitutes a selectivity filter for the substrate and functions as a semipermeable membrane defined by a given osmotic

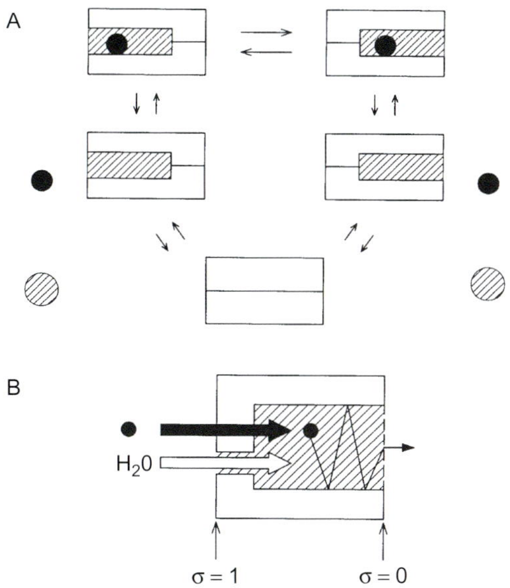

FIGURE 28.5 Molecular mechanism for water transport through cotransport proteins. **A.** *The cotransport mechanism.* Here the mobile barrier model (Mitchell, 1957) is adapted for cotransport of non-aqueous substrates (black spheres) and water (hatched). Hydration of the access cavity allows entry of the substrate from the *cis* side and binding inside the cavity. Subsequently, a conformational change in the protein shifts the opening of the cavity from the *cis* side to the *trans* side, which allows the substrate to leave towards the *trans* side. The transition from being open towards the *trans* side to being open to the *cis* side takes place via a closed conformation in which the aqueous cavity in the protein is small. **B.** *The osmotic coupling mechanism.* The substrate (black spheres) is transported from the *cis* compartment into a static aqueous cavity in the protein which is permanently open to the *trans* side. The cavity is limited to the *cis* side by a semi-permeable membrane (reflection coefficient $\sigma = 1$). The *cis* barrier has passive water permeability and water is transported by osmosis from the *cis* side into the cavity. The *trans* side of the protein is characterized by a low reflection coefficient ($\sigma = 0$) and a certain permeability to the substrate. This model is based upon the two barrier model of Curran and Macintosh (1962), see text.

water permeability (L_p) and a reflection coefficient (σ) close to one. As a result, water enters the cavity from the *cis* side by osmosis as long as the substrate occupies the cavity. The other barrier of the protein is located at the *trans* side of the protein. This barrier can also be described by two parameters: a permeability that defines the delayed exit of the substrate from the cavity and a reflection coefficient (σ). Since the barrier discriminates only little between substrate and water, the reflection coefficient is close to zero. Accordingly, water is being expelled from the cavity of the protein by hydrostatic forces generated within the protein. It should be emphasized that the physical problems associated with the exit of a substrate from an aqueous cavity outlined in Fig. 28.5B are also encountered in the mobile barrier model when the substrate leaves the protein prior to closure (Fig. 28.5A). Accordingly, a third model could be obtained by combining the models of Figs 28.5A and 28.5B.

The wide aqueous cavities required by the models above are in fact present in membrane proteins such as cotransporters, as revealed by recent high-resolution structures (for reviews see DeFelice, 2004; Guan and Kaback, 2006). The oxalate/formamate antiporter (OxlT) (Hirai et al., 2002), the bacterial H^+-lactose symporter LacY (Abramson et al., 2003), the bacterial inorganic phosphate (P_i)/glycerol-3-phosphate antiporter (GlpT) (Huang et al., 2003), the bacterial Na^+-glutamate cotransporter Glt_{Ph} (Yernool et al., 2004) and the leucine transporter LeuT (Singh et al., 2007; Yamashita et al., 2005) all have large hydrophilic cavities which contain the substrate binding site and have linear dimensions of up to 50 Å. Recently, structural analysis of the Ca^{2+} pump (Olesen et al., 2007) has also revealed aqueous cavities through which Ca^{2+} ions exit into the outside solutions. These are precisely the features required by the models for coupling between water and substrates described above.

Conformational changes in proteins involve movements of complete domains relative to each other. This will change the amount of water held by the protein in two ways. First, cavities that contain bulk water may change their size. Second, the surface area of the protein may change and loosely attached surface water will be exchanged with the surroundings (Parsegian, 2002; Rand, 2002). Such effects have been studied in experiments where enzymatic activity was monitored as a function of the chemical potential for water in the external medium. The concept has been applied to a number of enzymes, both membrane-bound and water soluble, and has demonstrated a wide range of numbers for exchangeable water molecules, from ten to more than a thousand, see Table 28.2 and associated text. From simple physical consideration, such values are not inconceivable. The volume of 500 molecules of bulk water is about 15,300 A^3. Given a molecular weight of the KCC of about 125 kD (Gamba, 2005; Russell, 2000), this water would constitute less than 10% of the protein volume, assuming a density of the protein of about 1. Surface water is more compact than bulk water as one water molecule occupies an area of 10 A^2 (Colombo et al., 1992). Given this value, 500 water molecules would occupy a surface of 5000 A^2, equivalent to the surface of a 100 A long, 15 A wide tube. These estimates from membrane-bound transporters are not incompatible with those obtained from aqueous enzymes, α-amylase, hexokinase and hemoglobin, described above. Studies of α-amylase show that between 5 and 10% of the protein volume is occupied by water, equivalent to 300–500 water molecules;

about 90 of the water molecules are exchanged with bulk water during conformational changes.

In conclusion, the concept of cotransporters as molecular water pumps is supported by our present knowledge of the structure of cotransport proteins. To escape the proteins, substrates have to move through aqueous cavities, which will give rise to local hyper-osmolarities and intra-molecular osmosis. In addition, these cavities may open and close during transport. The two mechanisms outlined in Fig. 28.5 could, combined or in isolation, contribute to the coupling between substrate and water transport observed in cotransporters.

III. A MOLECULAR MODEL OF EPITHELIAL WATER TRANSPORT BASED UPON COTRANSPORTERS AND THEIR REGULATION

Cells strive to maintain a constant intracellular milieu; osmolarity, pH, cell volume, etc. within narrow limits. Under steady-state conditions, any influx of ions and water has to be matched precisely by identical effluxes. In non-polarized cells, such as erythrocytes or muscle cells, the influx and efflux mechanisms are distributed evenly over the membrane and there is no vectorial or trans-cellular transport (Fig. 28.6A). In polarized or epithelial cells, the situation is different. Some influx mechanisms are located at the one membrane and some efflux mechanisms to the other membrane. (Please note that ion and fluid movement can go in either direction, depending on the type of epithelium: secretive versus absorptive.) The asymmetry gives rise to the transepithelial transport. In short: net influxes across one membrane have to be matched exactly by net effluxes across the other. The critical question is how are the fluxes made to match?

In the following example, we consider a simplified epithelial cell and a simplified transport situation (Fig. 28.6B). The influx of NaCl and water at the entry membrane are assumed to be uncontrolled, i.e. they only follow the prevailing electrochemical gradients. For most leaky epithelia this assumption can be justified. Reductions of external Na^+ leave the net flux of Na^+ mostly unaffected, due to the relatively low value of K_m of the entry mechanism of about 10 mM (Baerentsen et al., 1983). The effluxes, however, are maintained by Na^+/K^+ATPase, KCC and NKCC1. The ATPase maintains the Na^+ and K^+ gradients that energize the transports. In the present context, it is not assumed that transport be regulated other than from responding to the given Na^+ and K^+ gradients. The activity of KCC and NKCC1, however, are assumed to be controlled by signals originating from sensors of some vital intracellular parameters, as, for example, the osmolarity. The simple epithelial model given

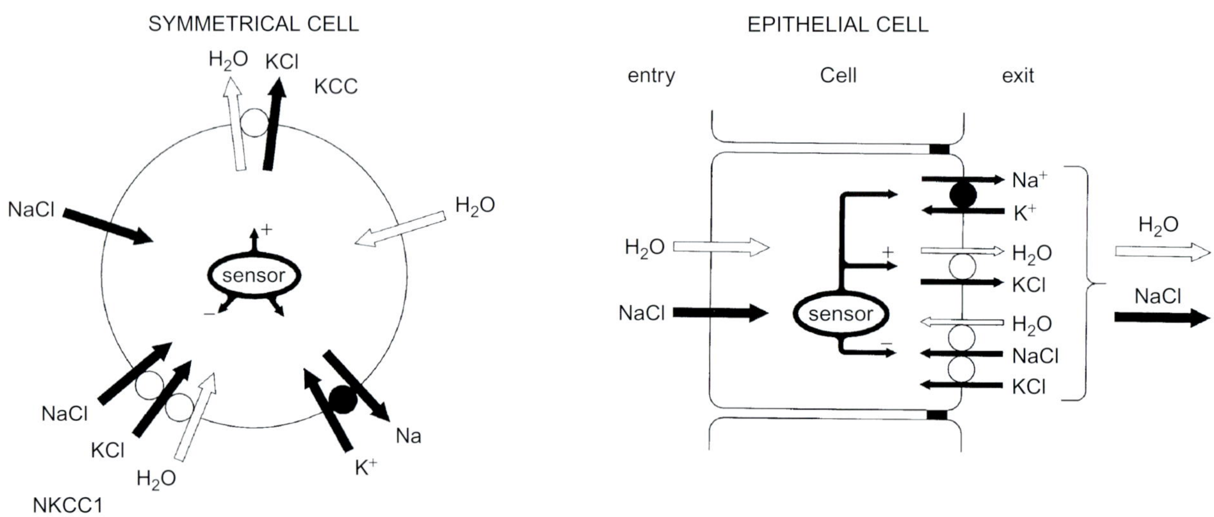

FIGURE 28.6 Transepithelial transport is unilateral volume regulation. To maintain intracellular parameters constant (i.e. osmolarity and Na^+ concentration) cells have to maintain effluxes that match the influxes. **A.** In symmetrical or unpolarized cells the membrane transporters are distributed evenly over the membrane. **B.** In polarized or epithelial cells all (or a fraction of) the influx mechanisms are localized to the entry membrane. It follows from the principle of mass balance that matching efflux mechanisms are localized at the exit membrane. This leads to transepithelial transport. In the illustrated example there is influx of water and NaCl. Effluxes are maintained by a combination of the Na^+/K^+-ATPase, the K^+-Cl^- cotransporter, and the Na^+-K^+-$2Cl^-$ cotransporter. The model requires that a set of intracellular parameters are monitored and that the appropriate signals are sent to the membrane transporters. In the present example, only transporters at the exit membrane are regulated, i.e. by changing their level of phosphorylation (see text).

in Fig. 28.6B is a good representation of the choroid plexus epithelium and other leaky epithelia, the main function of which is to transport water and NaCl.

A. Regulation of KCC and NKCC

Changes in cell volume can be detrimental to cells and they therefore have fine-tuned volume-regulation mechanisms that are based, at least in part, on the concerted activities of NKCC and KCC. When cells shrink upon exposure to hypertonic extracellular fluid, the activity of NKCC increases while the activity of KCC decreases (Adragna et al., 2004; Gamba, 2005; Russell, 2000) and transport of Na^+, K^+ and Cl^- into the cell is augmented (Fig. 28.6A). As described above, water is transported within the NKCC1 which, together with the gradual intracellular accumulation of osmolytes (Na^+, K^+ and Cl^-), increases the rate of regulatory volume increase. Conversely, when cells swell upon exposure to hypotonic media, the activity of KCC increases, while the activity of NKCC decreases (Adragna et al., 2004; Russell, 2000) and K^+ and Cl^- are exported out of the cell, together with around 500 molecules of water per turnover of the protein, as described above. Thereby regulatory volume decrease is initiated and the cell volume reverts to its original size.

Somehow the cells manage to sense the change in osmolarity and/or the intracellular Cl^- concentration and convey a signal to KCC and NKCC1 to increase or decrease their activity, respectively. It has been shown that upon phosphorylation, the activity of NKCC increases, whereas the activity of KCC decreases (Adragna et al., 2004; Flatman, 2002; Russell, 2000). KCC and NKCC have been shown to be regulated by a variety of kinases and phosphatases but in the present context, we will focus on a strong candidate for this concerted volume regulatory pattern: the WNK kinase family (With No K = lysine), reviewed by (Delpire and Gagnon, 2008; Kahle et al., 2008, 2006). WNK3 has been shown to colocalize with NKCC1 and the two isoforms KCC1 and KCC2 in Cl^--transporting epithelia and does indeed increase the transport activity of NKCC1 and decrease the transport activity of KCC1-2 expressed in *Xenopus* oocytes (Kahle et al., 2005). Moreover, the osmolarity-dependent change in activity of NKCC1 and KCC1/2 is bypassed by WNK3, which altogether point to WNK3 as being part of the osmosensing mechanism by which the cell maintains its volume during exposure to osmotic stress (Kahle et al., 2006, 2005), possibly via an indirect mechanism involving phosphorylation of a phosphatase (de Los Heros et al., 2006). Other members of the WNK family also play a role in the osmosensing response in a signaling pathway involving SPAK (Ste20SPS1-related, proline alanine-rich kinase) and OSR1 (oxidative stress-responsive kinase-1) (Delpire and Gagnon, 2008; Kahle et al., 2006): WNK1/WNK4 phosphorylate SPAK/OSR1, which in turn phosphorylate NKCC and KCC and thereby lead to an up-regulation of NKCC and a down-regulation of KCC (Gagnon et al., 2006; Garzon-Muvdi et al., 2007). WNK1 has been shown to be autophosphorylated and activated by hypertonicity (Lenertz et al., 2005; Zagorska et al., 2007), and the SPAK/OSR1-dependent phosphorylation of NKCC1 can be initiated by hyperosmolar stress (Vitari et al., 2006), which taken together points to WNK1 as a possible osmosensor candidate as well as an effector. It has been shown that SPAK translocates from the cytoplasm to the cytoskeleton upon a hyperosmotic challenge (Tsutsumi et al., 2000) and as SPAK has been shown to physically interact with NKCC1 and WNK4 (Gagnon et al., 2006), this stress-dependent translocation may be one of the osmosensing signals required to initiate regulatory volume changes. Altogether, the WNK family and the SPAK/OSR-1 kinases may serve as osmosensors as well as the kinases initiating the regulatory volume changes by directly or indirectly changing the level of phosphorylation on NKCC and KCC1-2. This topic is covered in more details in Chapter 11. Other candidates for osmosensors are stretch-activated ion channels (Voisin and Bourque, 2002), volume-activated ion channels (Grunnet et al., 2002) and aquaporins (Hill et al., 2004; Wells, 1998).

B. The Direction of Water Transport in Leaky Epithelia and Transport by KCC and NKCC Coincide

The oppositely directed regulation of KCC and NKCC1 by phosphorylation provides a simple explanation as to how absorption in a given epithelial cell can be changed into secretion and vice versa. If a phosphorylation is induced (by cell shrinkage, for example) NKCC would be up-regulated, and KCC down-regulated and vice versa (Fig. 28.7).

Epithelia working under normal physiological conditions can be classified according to whether they have KCC or NKCC1 as the dominating Cl^--dependent cotransporter colocalized with the Na^+/K^+-ATPase, Fig. 28.8 (Russell, 2000; Zeuthen, 1996). KCC has been located to the exit membrane in a number of water transporting (leaky) epithelia; gall bladder, kidney proximal tubule, small intestine and choroid plexus (Adragna et al., 2004; Karadsheh et al., 2004; Zeuthen, 1994). The transporter is energized, in part, by the K^+ gradient built up by the Na^+/K^+-ATPase. Importantly,

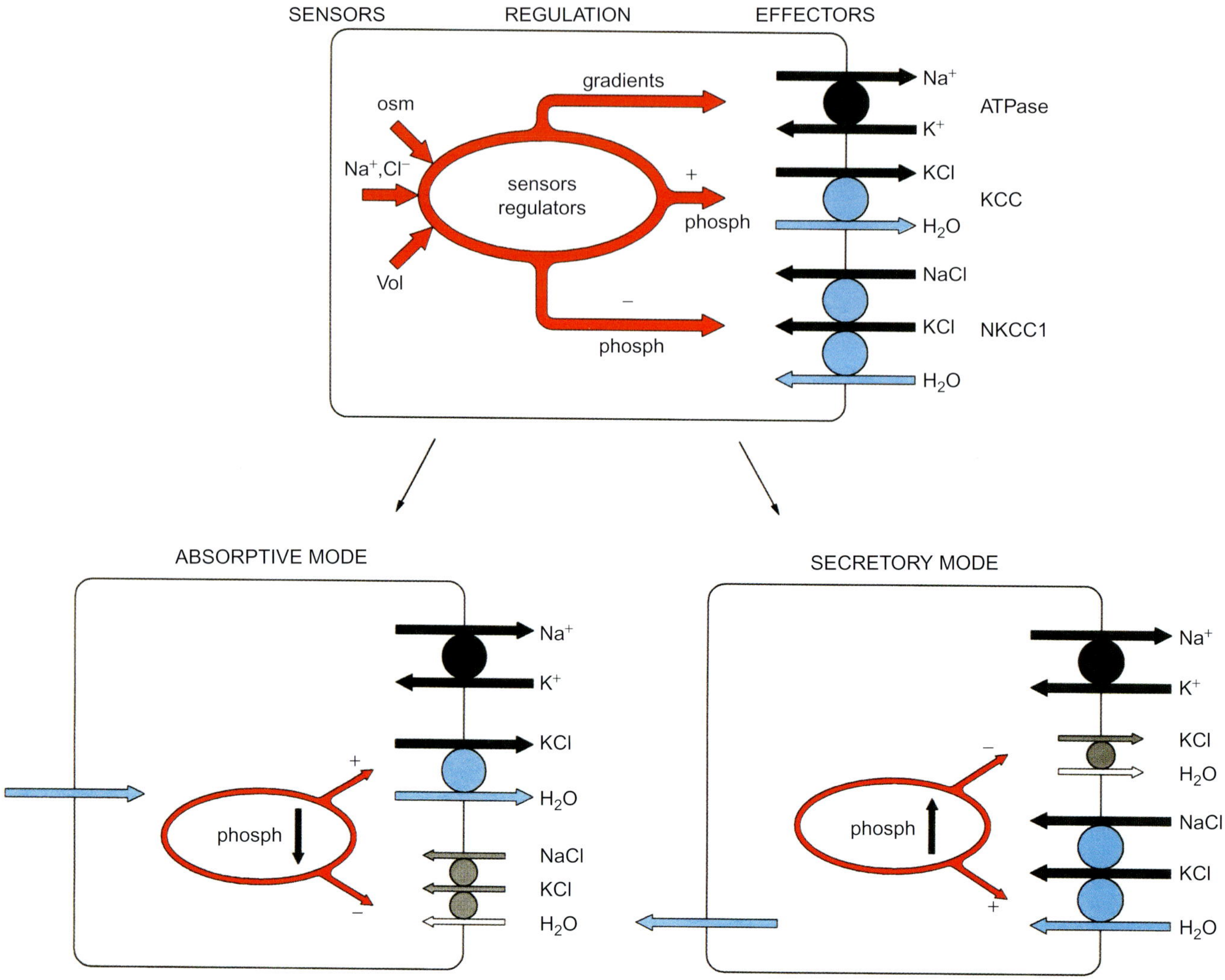

FIGURE 28.7 Regulation of KCC and NKCC in epithelial cells. The KCC and the NKCC1 are oppositely regulated by phosphorylation: NKCC1 is up-regulated while KCC is down-regulated. In our simplified model the Na^+/K^+-ATPase is not directly regulated, but its transport rate is determined by the gradients of Na^+ and K^+ concentrations. The nature of the sensors of epithelial cell parameters is debated; sensors of osmolarity, Na^+ activity, Cl^- activity and cell volume are indicated. The general regulatory model (top) can be adjusted by dephosphorylation into an absorptive mode (lower left) or by phosphorylation into a secretory mode (lower right).

there is always a KCC in the exit membrane, the transport direction of which is the same as the water transport of the whole epithelium (Zeuthen, 1996). In situations where additional transport functions are required, there is evidence that NKCC1 is present or can be induced in the exit membrane: choroid plexus (Johanson et al., 1990; Karadsheh et al., 2004), small intestine (MacLeod and Hamilton, 1990) and kidney proximal tubule (Aronson, 1989). NKCC1 is located primarily to the entry membrane of a number of secretory epithelia, such as the glandular acini and the upper airways (Silva et al., 1977). It has also been found in the absorptive retinal pigment epithelium (Fig. 28.8). The transporter is energized, at least in part, by the Na^+ gradient built up by the Na^+/K^+-ATPase. In analogy with the KCC, the direction of the ion and water transport by the NKCC is always identical to that of the whole epithelium.

Thus, among the water transporting epithelia, the direction of water transport and that of ion and water transport by KCC and NKCC coincide. In contrast, the direction of active ion transport by the Na^+/K^+-ATPase as well as the ultrastructure of the epithelium give no clue as to the direction of water transport. It is tempting to conclude that KCC and NKCC are central for the coupling between ion and water transport in these epithelia. But how well do the simple models of Figs 28.6, 28.7 and 28.8 explain some important physiological and well-established transport properties?

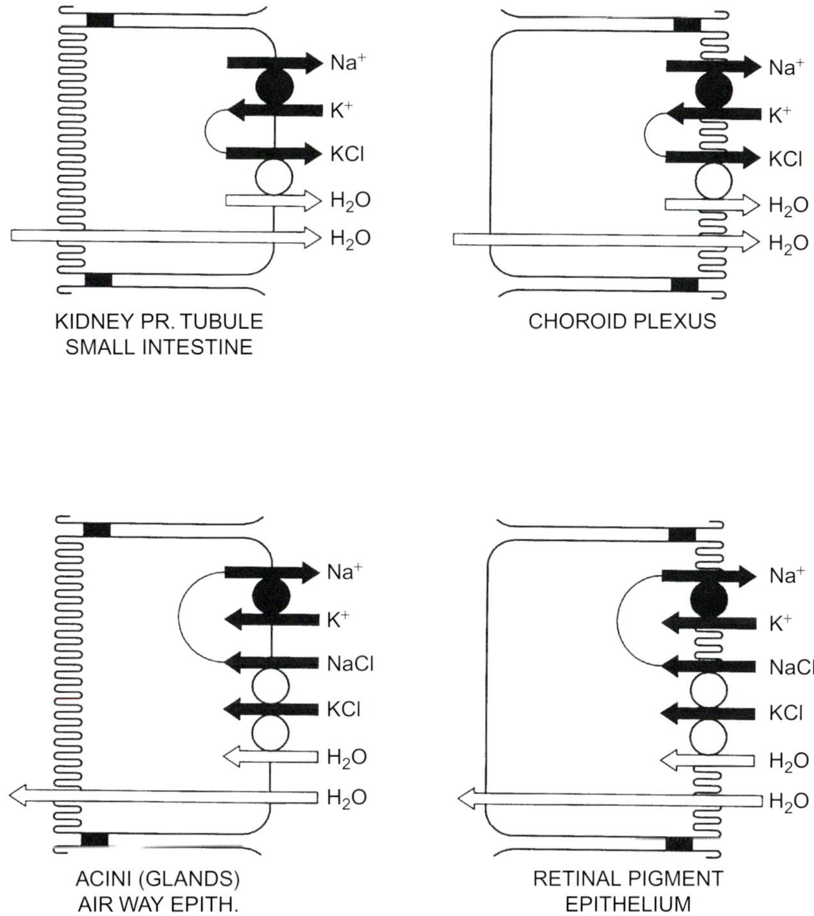

FIGURE 28.8 Transepithelial water transport and cotransport of ions and water by the KCC and the NKCC take place in the same direction. Water transporting epithelia can be grouped according to their Cl^--dependent cotransporters (KCC or NKCC), their ultrastructure (microvilli and junctions), the direction of water transport and the localization of the Na^+/K^+-ATPase (filled circle). In absorptive leaky epithelia, such as gall bladder, small intestine, kidney proximal tubule and choroid plexus, the direction of water transport by KCC (KCl-H_2O cotransport) and of transepithelial water transport coincide. In other types of epithelia, primarily secretory, such as airway epithelia, acini from glands, and retinal pigment epithelium, the direction of water transport by NKCC (Na^+-K^+-$2Cl^-$-H_2O cotransport) and the transepithelial water transport coincide. There is no correlation between the direction of transepithelial water transport and the ultrastructure (orientation of the microvilli and of the lateral spaces, and the location of the junctional complex, *Zonulae occludence*), or the direction of transport by the Na^+/K^+ ATPase.

C. KCC Plays a Key Role for the Coupling of Salt and Water Transport in Leaky Epithelia

A model for water transport in leaky epithelia should explain two groups of well-established findings: isotonic transport and the ability for uphill water transport (Zeuthen, 2002). In humans, the most prominent example of isotonic transport is seen in the proximal tubule of the kidney, which reabsorbs about 135 liters of plasma-like ultra-filtrate each day. The most prominent example of uphill transport is in the small intestine just after a meal, where the epithelium is able to reabsorb water against an osmotic gradient of 200 mOsm, largely made up by an excess of glucose (Pappenheimer, 1998; Zeuthen, 2008).

An epithelium transports isotonically when it transports a solution with the same tonicity (or osmolarity) as that of the bathing solution. In the simplest situation, the epithelium is bathed on its entry side in the test solution (osmolarity π_{entry}, Fig. 28.9), while the solution on the exit side consists of the transported solution itself (osmolarity π_{exit}). For isotonic transport the two osmolarities equal each other ($\pi_{entry} = \pi_{exit}$). This deceptively simple situation in fact poses at least three very difficult questions: How can the flux of water proceed without any transepithelial osmotic gradient? What are the mechanisms that ensure that the salt flux and the water flux proceed in a ratio equal to that in the bathing solution? If the cell is hyperosmolar to the solution at the entry side, how does water come out of the cell?

The problems of explaining isotonic transport by exclusively osmotic mechanisms are illustrated in Fig. 28.9, where the *experimentally* determined osmotic profile through the epithelium is shown in Fig. 28.9 A: the osmolarity of the cellular compartment (π_{cell}) is higher than both the entry compartment (π_{entry}) and the exit compartment (π_{exit}), while the entry bath and the exit bath have the same osmolarity ($\pi_{entry} = \pi_{exit}$). It is generally accepted that the cell is hyperosmolar relative to the entry compartment (Zeuthen, 1982, 1983) and that the entry of water, at least in principle, could be explained by simple osmosis. (Please note that a significant fraction of the water may enter in cotransport with organic substrates (Table 28.2), but that is not the issue here.) Thus, water can get easily in by osmosis, but how does it get out? The osmotic driving force for water exit points to the wrong direction.

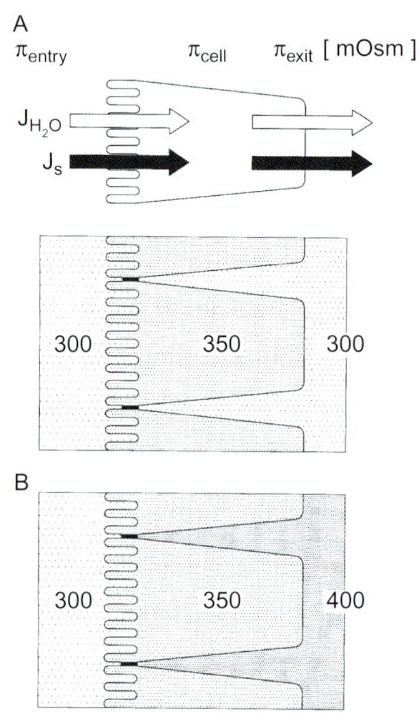

FIGURE 28.9 Osmotic models for water transport across leaky epithelia are inadequate. In leaky epithelia such as the choroid plexus, small intestine and proximal tubule, the flux of water (J_{H2O}) follows the flux of salt (J_S). **A.** Experiments show that during isotonic transport the osmolarity of the solution at the exit side (the transported solution, π_{exit}) equals that of the solution at the entry side (π_{entry}), and that the osmolarity of the cell (π_{cell}) is higher than both. The numbers (in mOsm) are based on data from the mammalian small intestine (Zeuthen, 2002). Clearly, J_{H2O} can enter the cell by osmosis, but how does water get out? The problem is to explain how water moves uphill, from the cell of osmolarity π_{cell} into the exit compartment of osmolarity π_{exit}; the osmotic gradient has the wrong sign. **B.** If the transepithelial water transport is assumed to take place by osmosis across both the entry and the exit membrane, the osmotic profile through the cellular pathway must increase in two steps. As a consequence, the osmolarity of the transported solution (π_{exit}) needs to be around 400 mOsm and the transport is not isotonic in conflict with the experimental data, see text.

In a steady state, the entry of water would have to be matched by the exit. If the exit of water also took place through osmosis, it would require that the osmolarity of the solution in the exit compartment be larger than that of the cell and therefore even larger than that of the solution in the entry compartment ($\pi_{entry} < \pi_{cell} < \pi_{exit}$, Fig. 28.9B). But these two stepwise increases in osmolarity would be incompatible with isotonic transport, i.e. with the osmolarity of the exit bath being equal to that of the entry bath ($\pi_{exit} = \pi_{entry}$). The solution in the exit compartment would always be hyperosmolar relative to the solution bathing the entry side.

Two hypotheses have been advanced to rescue the idea of water transport by purely osmotic mechanisms, and both are clearly insufficient. In the first hypothesis, the osmotic permeabilities of the two membranes are assumed to be so high that the hyperosmolarity of the transported solution would in effect be very small. This scheme is untenable for two reasons. First, passive water permeabilities of epithelial cell membranes are not, in general, very high. Second, most leaky epithelia are able to transport water uphill, from the entry to the exit solution, against the chemical potential gradient of water (see below); in case of highly water permeable membranes the passive back-flux of water would be prohibitively large. In the second hypothesis, it is assumed that the osmolarity in lateral intercellular spaces or in some subepithelial compartment is higher than that of the cell. This would ensure the transport of water from the cell into the exit solution. Despite significant experimental efforts, no such hyperosmolar compartments have yet been found. On the contrary, detailed studies with ion-selective microelectrodes in intact lateral intercellular spaces have shown that the osmolarity in the lateral intercellular spaces is isosmotic with the solution bathing the entry membrane as well as the exit membrane (Ikonomov et al., 1985; Zeuthen, 1983). This agrees well with the fact that lateral intercellular spaces are too short and wide to restrict solute diffusion, and that subepithelial structures are not tight enough to impede significantly the movements of salt away from the epithelium (Hill, 1975), for reviews see (Hill, 2008; Zeuthen, 2002).

The central problem is therefore to explain how water moves, uphill, from the cell into the exit solution against the direction of the water chemical potential difference. The inclusion of a KCC which mediates K^+-Cl^--H_2O cotransport at the exit membrane solves this problem directly. There is sufficient energy contained in the outwardly directed gradients of K^+ and Cl^- to mediate an efflux of water against osmotic gradients as high as 200 to 300 mOsm (Zeuthen, 1994,

1991a, b). KCC may also contribute a major portion of the total transepithelial water transport. First, the tonicity of the solution transported by KCC is close to that of the bathing solutions. Physiological saline for amphibians contains about 110 mM of NaCl. The ratio between KCl and water transport in the amphibian KCC is fixed at 500 water molecules per pair of K^+ and Cl^-, equivalent to 110 mM (Table 28.2). Second, in case the K^+ flux through KCC is balanced precisely by influx of K^+ through the Na^+/K^+ ATPase, it can be estimated that the water transported by KCC is about two-thirds of the transepithelial water flux.

The ability for uphill water transport by KCC gives a direct explanation to the perplexing property of uphill water transport. It was discovered more than a hundred years ago that epithelia could transport water uphill, against the direction of the water chemical potential gradient (Ludwig, 1861; Reid, 1892, 1901). Only when sufficient amounts of an impermeable osmolyte were added to the entry solution was the transepithelial water transport arrested. But the amount to be added could be surprisingly high, with values ranging from 20 mOsm in the kidney proximal tubule (Green et al., 1991) to 200 mOsm in the small intestine (Pappenheimer, 1998), for a review see Zeuthen (2002). The proximal tubule has high passive water permeability due to its high number of aquaporins and the osmotic back-flux induced by the added osmolyte becomes appreciable at low concentrations. In contrast, the intestine has relatively low passive water permeability and it takes much higher luminal hyperosmolarities to nullify the net transport of water. The ability of KCC to transport water against osmotic gradients of up to 200 to 300 mOsm would give a direct explanation of the ability of the whole epithelium for uphill transport of water. For quantitative considerations see Zeuthen (2002, 1996).

The concept of KCC being central to the coupling between salt and water transport in leaky epithelia suggests several experiments in which the role of the KCC and its regulation could be tested. It is, for example, well established that the ability for isotonic transport remains when epithelia are adapted to transport in dilute solutions: As illustrated in Fig. 28.9A the osmolarity of the solution secreted at the exit side equals that of the bathing solution on the entry side. In fact, the situation is qualitatively the same as encountered under physiological conditions, but the important point is that the rate of water transport goes up in proportion to the degree of dilution, while the rate of NaCl transport remains constant (Diamond, 1964; Hill and Hill, 1978; Whitlock and Wheeler, 1964; Whittembury and Hill, 1982; Zeuthen, 1982, 1983), see Fig. 28.10. Interestingly, leaky epithelia tolerate reversibly major dilutions. The dilution likely leads to cell swelling which causes an up-regulation of KCC (Adragna et al., 2004).

IV. ROLE OF COTRANSPORTERS AND UNIPORTERS IN A TIGHT EPITHELIUM, THE BLOOD–BRAIN BARRIER

Epithelia are divided into leaky and tight epithelia on the basis of the permeability of their junctions (*Zonulae occludence*). Leaky epithelia are responsible for large rates of transport of plasma-like solutions and can support paracellular ion fluxes in order to maintain electroneutrality. In contrast, tight epithelia have to maintain transepithelial concentration differences. This is carried out by well-controlled transporters of specific ions and substrates located to the cell membranes, whereas the junctions are relatively impermeable. But how is the water transported across a tight epithelium controlled? The ion and substrate transporters have an inherent capacity for water transport (Tables 28.1 and 28.2) so there will always be a given water permeability associated with the ion and substrate transport. This situation will be reviewed using as example the tight endothelium of the brain.

A. The Water Permeability of the Endothelial Wall can be Explained by GLUT1 and NKCC1

The volume of the brain and its surrounding cerebrospinal fluid (CSF) is under strict homeostatic control. An increased influx of water from the blood under pathological conditions leads to an increased intercranial pressure and reduced cerebral blood flow, and this may become fatal. The brain-CSF system is separated from the rest of the body by the blood–brain barrier (BBB), the major component of which is the endothelial wall of the brain capillaries. The BBB serves two, in certain respects conflicting, purposes: it protects the brain from changes in plasma composition and it provides pathways for nutrients and exchange of respiratory gases.

To protect the brain from volume changes, the endothelial wall is rather water impermeable. When compared to other cell layers (Table 28.3), it is seen that the water permeability of the BBB is among the lowest when compared to other cell layers, and similar to that of the frog skin.

However, the BBB cannot be completely impermeable, as it has to provide the pathway for nutrients

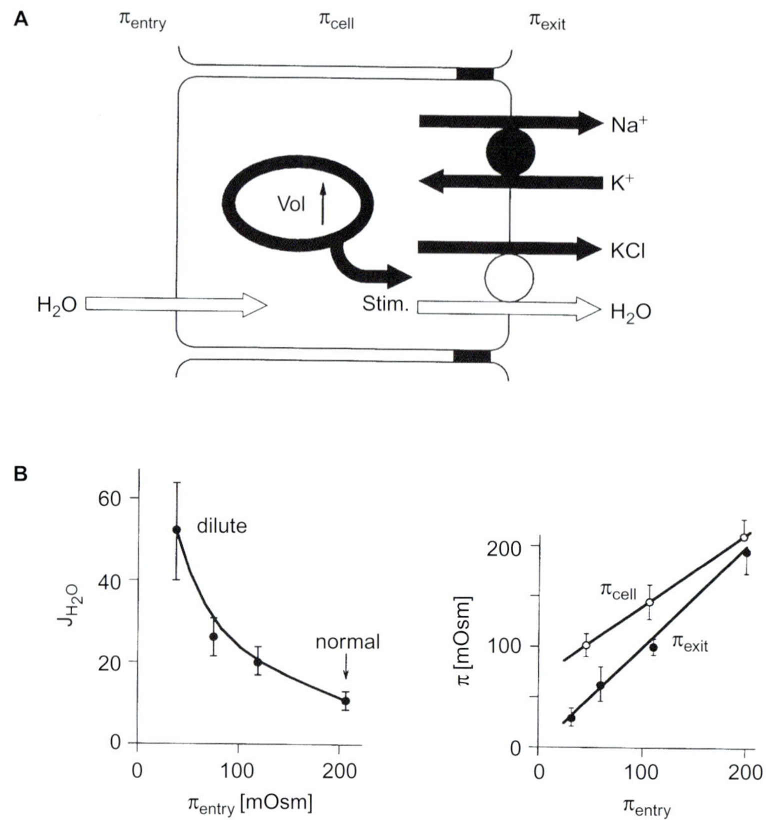

FIGURE 28.10 Isotonic transport in dilute solutions. An example of up-regulated volume transport in leaky epithelia. When an epithelium is adapted to transport in dilute solutions the cells will tend to swell and KCC will be up-regulated. The prediction of increased volume transport is supported by experimental data. **A.** The rate of water transport (J_{H2O} in units of $\mu L cm^{-2} h^{-1}$) is larger in epithelia adapted to dilute solutions, while the rate of NaCl transport is unaffected (not shown). **B.** Simultaneous measurements of osmolarity show that the osmolarity of epithelial cell (π_{cell}) is larger than that of the mucosal bathing solution (osmolarity π_{entry}) as well as that of the secreted solution (osmolarity π_{exit}). The transport remained isosmotic under all conditions: π_{exit} equaled π_{entry}. The osmolarity of the solution in the lateral intercellular spaces was the same as that in the secreted solution. Data from amphibian gall bladders (Zeuthen, 1983, 1982, 1981).

and gases. As a consequence, the area of the endothelial wall that supplies a given volume of the brain has to be large. The surface-to-volume ratio for the brain is about 100 cm^2 of endothelial membrane per cm^3 of brain (Paulson et al., 1977). Given this high ratio, water transport across the endothelial cell layer becomes significant. For example, a gradient of 1 mOsm (equivalent to 0.3% of the human plasma osmolarity) across the endothelial wall would lead to a 16% change in brain volume per day.

Which proteins are responsible for the water permeability of the BBB? Several of the water-permeable transporters (or other isoforms thereof) listed in Tables 28.1 and 28.2 are present in the endothelial wall (Fig. 28.11): NKCC1 (O'Donnell et al., 2004), GLUT1, MCT1, GAT-2 (Takanaga et al., 2001), EAAT1-3 (O'Kane et al., 1999), as well as the large amino acid transporter LAT1 (Matsuo et al., 2000). The water permeability of hLAT1/4F2hc-expressing oocytes was 40% larger than that of the non-injected oocytes, while the ^{14}C-leucine uptake was seven-fold larger (MacAulay, unpublished results). It can be estimated that the water permeability of these transporters is sufficient to explain the water permeability of the BBB. Take glucose transport by GLUT1 as an example. At first sight, the water permeability per protein seems small (Table 28.1), but as the number of proteins needed to supply the brain with glucose is large, the total water permeability becomes significant. In other words, it is the ratio of water permeability to glucose permeability that is important. As a first step, the transport capacity can be inferred from the rate of glucose consumption. From an oxygen consumption of $2.6 10^{-8}$ mol O_2/gram*s, it follows that $2.6 10^{-8}$ mol glucose is consumed per gram*s by aerobic metabolism. The capillary surface is 100 cm^2 per gram of brain tissue (Paulson et al., 1977), so the glucose transport across the brain capillaries has to supply $1.3 10^{-10}$ mol glucose/cm^2 * s. If we assume a plasma glucose of 10 mM (equivalent to 5 mM in whole blood) and a

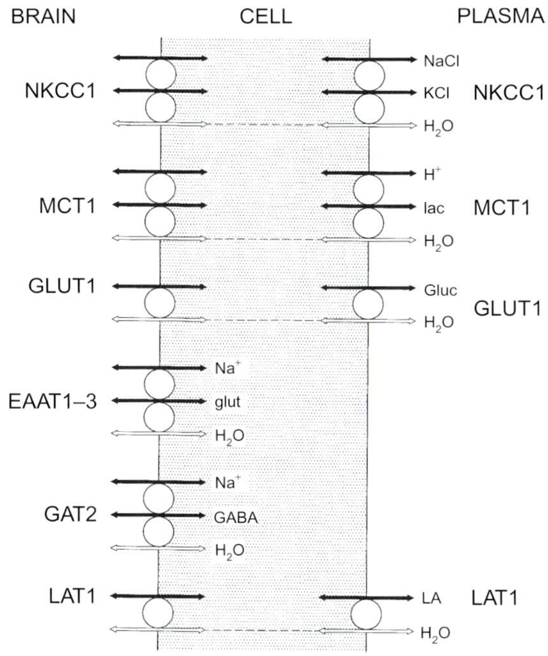

FIGURE 28.11 Transport proteins with capacity for water transport in the endothelium of the blood–brain barrier. Some transporters, such as NKCC1, the glucose uniport GLUT1, the lactate transporter MCT1 and the large amino acid transporter LAT1, are found in both the luminal membrane facing the blood plasma and in the membrane facing the brain (for NKCC1: 80% in the luminal membrane and 20% in the abluminal membrane (O'Donnell et al., 2004)). Others pathways have so far only been localized to one of the membrane aspects, i.e. the glutamate transporters EAAT1-EAAT3, the GABA transporter GAT2.

glucose concentration of 5 mM in the brain interstitium (Simpson et al., 2007), the gradient across the endothelial wall amounts to about 5 mM. Accordingly, the P_{GLUC} can be calculated as $0.52*10^{-4}$ cm s^{-1}. The second step is to estimate the water permeability (L_p) associated with this P_{GLUC} by means of the ratio of L_p/P_{GLUC} which for GLUT1 is about 40:1 at room temperature (see Table 1, Zeuthen unpublished). At body temperature (37°C) the ratio would be closer to 6:1, considering the high temperature coefficient of glucose transport (Simpson et al., 2007). These values were obtained from expression studies in *Xenopus laevis* oocytes in a manner analogous to the study of GLUT2 for which L_p/P_{GLUC} was 10:1 at room temperature (Zeuthen et al., 2007). Accordingly, the L_p contributed by the glucose transporter can be estimated at $0.3\,10^{-3}$ cm s^{-1} for 37°C (or $2.1\,10^{-3}$ cm s^{-1} at room temperature) which is in good agreement with the overall L_p measured in humans *in vivo* of $1.04\,10^{-3}$ cm s^{-1} (Paulson et al., 1977).

It is not possible to give an estimate of the relative contributions of the other transporters listed in Fig. 28.11 to the overall passive water permeability of the BBB. It may be inferred, however, that the contribution of the NKCC1 is significant, as the water permeability of NKCC1 is comparable to that of AQP1 (Table 28.1), and with this transport protein being present in the endothelium (O'Donnell et al., 2004). Thus, sizeable water fluxes are expected to take place through this protein.

B. The Role of NKCC1 during Brain Edema Formation

Brain edema is a potentially life-threatening condition that is due to an increase in brain water content and in intra-cranial pressure. Brain edema may accompany diverse conditions, such as brain tumors, brain hemorrhage, severe head trauma, meningitis, stroke (ischemia), liver failure, etc. The path of water entry initially depends on the cause of the edema, as brain edemas can be divided into (at least) two major kinds; vasogenic edema and cytotoxic edema (Go, 1997). Vasogenic edema is characterized by the breakdown of the blood–brain barrier, such as during the course of meningitis and tumor formation. This perforation of the BBB would allow water, ions, plasma, etc. to diffuse freely into the brain and exert its damaging effects. Cytotoxic brain edema arises typically following the lack of oxygen supply to a part of the brain. This leads to a reduction of the ATP levels and thereby reduces the function of the Na^+/K^+-ATPase leading to a general rundown of the Na^+ and K^+ gradients, a depolarization of the neurons and glia cells, and increased intracellular osmolarity. The water entering into the brain parenchyma via the BBB under this pathological condition may, at least in part, go through NKCC1 in the brain endothelia. In support of this notion is the lack of (Nagelhus et al., 1999; Nielsen et al., 1997) or limited (Amiry-Moghaddam et al., 2004) expression of aquaporin 4 (AQP4) in the endothelial cells that comprise the BBB. AQP4 is, however, highly expressed in the endfeet of the glia cells (Nielsen et al., 1997), where they confer a large water permeability to the glial membrane facing the endothelial cells. It was recently shown that NKCC1 is implicated in the formation of cytotoxic brain edema during middle cerebral artery occlusion (MCAO) in rats (O'Donnell et al., 2004): Intravenous administration of bumetanide (a selective NKCC inhibitor) during the MCAO reduced the edema formation by around 50%, pointing to NKCC1 as an important player in the water permeability of the blood–brain barrier.

C. Vasopressin and its Role in Brain Edema

During conditions that lead to brain edema, increased levels of the anti-diuretic peptide hormone vasopressin have been detected in the blood stream (Barreca et al., 2001; Miura et al., 1989; Tenedieva et al., 1994). In experimental studies, this hormone worsens the outcome of the brain edema, whether it is given intravenously or directly into the ventricles (Chung et al., 2003; Rosenberg et al., 1990; Vajda et al., 2001). This is probably due to the fact that activation of the vasopressin receptor in the BBB increases the overall permeability of the barrier to water as well as to Na^+ and K^+ (Nagao, 1998). The effect of vasopressin has been assigned to a G-protein-coupled receptor, the vasopressin receptor, type 1a (V1aR) (Rosenberg et al., 1990), which upon activation initiates a downstream activation of phospholipase C, release of calcium and activation of PKC (Lolait et al., 1995). Antagonists of this receptor have in several studies been shown to improve the outcome of inflicted brain edema (Kagawa et al., 1996; Nagao et al., 1994; Rosenberg et al., 1992; Vakili et al., 2005). A V1aR-induced increase in water flow has also been detected in cortical slices by optical measurements (Niermann et al., 2001).

What could be the molecular targets that would lead to an increase in water transport (and Na^+ and K^+ transport) upon activation of the vasopressin receptor? We propose that NKCC1 is the major candidate for the vasopressin-induced increased water and ion transport. As mentioned above, the cotransporter contributes to the edema formation, it does transport water, and it is expressed in endothelial cells. Phosphorylation of NKCC1 upon activation of V1aR increases the activity of the transporter in endothelial cells (O'Donnell et al., 2005) and cultured astrocytes (Katay et al., 1998), which would then lead to an increase in water transport across the membrane (Hamann et al., 2005, and unpublished, see above).

V. CONCLUDING REMARKS

The rate of water transport varies markedly among different cell layers: high rates are seen in the epithelia of the small intestine and of the proximal tubule of the kidney; whereas in contrast, the endothelial cell layer of the blood–brain barrier has to minimize water fluxes. Disturbances of these functions lead to life-threatening conditions: whole body dehydration or an increase in intra-cranial pressure. Despite the vital importance of understanding water movement, there is no general model for epithelial and endothelial water transport.

We presented here a new model of epithelial water transport based upon unilateral cellular volume regulation. The key building blocks are the Cl^--coupled cotransporters KCC (in absorptive epithelia) and NKCC1 (in secretory epithelia). The coupling between salt and water transport takes place within these proteins. As they are located to one side of the cell only, transepithelial transport will ensue. In order to match influx with efflux, the rate of transport by KCC and NKCC are tightly regulated by a variety of kinases. This regulation is key to achieving cellular homeostasis, i.e. control of volume, osmolarity and ion activities. For simplicity, other transporters, such as ion and water channels, are not included in the model.

The ability for water transport by KCC and NKCC is shared by several other cotransport mechanisms, as well as by some uniporters. The transport is bimodal: secondary active water transport along with the substrate, and passive water transport upon an osmotic challenge. The coupling between water and substrate takes place through a mechanism intrinsic to the protein, probably associated with the substrate.

The blood–brain barrier (BBB) is a tight epithelium that contains virtually no aquaporins. We propose that the transport proteins, which are located in the endothelial cells that comprise the BBB, are responsible for the majority of the water transport between blood and brain, with NKCC1 and GLUT1 being the principal contributors.

Acknowledgements

Thanks for support from: Danish Medical Research Council, Løvens Foundation, Einer Willumsens Legat, The Velux Foundation, The Merck Foundation, Kurt Bønnelykke and hustru's Foundation, Kleinsmed S.H.A. Schrøder's Foundation, Michaelsens Foundation, The Lundbeck Foundation. Svend Christofferse is thanked for artwork.

References

Abramson, J., Smirnova, I., Kasho, V., Verner, G., Kaback, H.R., and Iwata, S. (2003). Structure and mechanism of the lactose permease of *Escheria coli*. *Science* **301**, 610–615.

Adragna, N.C., Di Fulvio, M., and Lauf, P.K. (2004). Regulation of K-Cl cotransport: from function to genes. *J. Membrane Biol.* **201**, 109–137.

Amiry-Moghaddam, M., Xue, R., Haug, F.M., Neely, J.D., Bhardwaj, A., Agre, P., Adams, M.E., Froehner, S.C., Mori, S., and Ottersen, O.P. (2004). Alpha-syntrophin deletion removes the perivascular but not endothelial pool of aquaporin-4 at the blood–brain barrier and delays the development of brain edema in an experimental model of acute hyponatremia. *FASEB J.* **18**, 542–544.

Aronson, P.S. (1989). The renal proximal tubule: a model for diversity of anion exchangers and stilbene-sensitive anion transporters. *Annu. Rev. Physiol.* **51**, 419–441.

Baerentsen, H., Giraldez, F., and Zeuthen, T. (1983). Influx mechanisms for Na^+ and Cl^- across the brush border membrane of leaky epithelia: a model and microelectrode study. *J. Membr. Biol.* **75**, 205–218.

Barreca, T., Gandolfo, C., Corsini, G., Del, S.M., Cataldi, A., Rolandi, E., and Franceschini, R. (2001). Evaluation of the secretory pattern of plasma arginine vasopressin in stroke patients. *Cerebrovasc. Dis.* **11**, 113–118.

Borden, L.A. (1996). GABA transporter heterogeneity: pharmacology and cellular localization. *Neurochem. Int.* **29**, 335–356.

Charron, F.M., Blanchard, M.G., and Lapointe, J.-Y. (2006). Intracellular hypertonicity is responsible for water flux associated with Na^+/glucose cotransport. *Biophys. J.* **90**, 3546–3554.

Chung, C., Vaquero, J., Gottstein, J., and Blei, A.T. (2003). Vasopressin accelerates experimental ammonia-induced brain edema in rats after portacaval anastomosis. *J. Hepatol.* **39**, 193–199.

Colombo, M.F., Rau, D.C., and Parsegian, V.A. (1992). Protein solvation in allosteric regulation: a water effect on hemoglobin. *Science* **256**, 655–659.

Curran, P.F. and Macintosh, J.R. (1962). A model system for biological water transport. *Nature* **193**, 347–348.

de Los Heros, P., Kahle, K.T., Rinehart, J., Bobadilla, N.A., Vazquez, N., San, C.P., Mount, D.B., Lifton, R.P., Hebert, S.C., and Gamba, G. (2006). WNK3 bypasses the tonicity requirement for K-Cl cotransporter activation via a phosphatase-dependent pathway. *Proc. Natl. Acad. Sci. USA* **103**, 1976–1981.

DeFelice, L.J. (2004). Transporter structure and mechanism. *Trends Neurosci.* **27**, 352–359.

Delpire, E. and Gagnon, K.B. (2008). SPAK and OSR1: STE20 kinases involved in the regulation of ion homoeostasis and volume control in mammalian cells. *Biochem. J.* **409**, 321–331.

Delpire, E. and Mount, D.B. (2002). Human and murine phenotypes associated with defects in cation-chloride cotransport. *Annu. Rev. Physiol.* **64**, 803–843.

Diamond, J.M. (1964). Transport of salt and water in rabbit and guinea pig gall bladder. *J. Gen. Physiol.* **48**, 1–14.

Dunn, J.J., Lytle, C., and Crook, R.B. (2001). Immunolocalization of the Na-K-Cl cotransporter in bovine ciliary epithelium. *Invest. Ophthalmol. Vis. Sci.* **42**, 343–353.

Ellory, J.C. and Hall, A.C. (1988). Human red cell volume regulation in hypotonic media. *Comp. Biochem. Physiol.* **90A**, 533–537.

Flatman, P.W. (2002). Regulation of Na-K-2Cl cotransport by phosphorylation and protein-protein interactions. *Biochim. Biophys. Acta* **1566**, 140–151.

Gagnon, K.B., England, R., and Delpire, E. (2006). Volume sensitivity of cation-Cl^- cotransporters is modulated by the interaction of two kinases: Ste20-related proline-alanine-rich kinase and WNK4. *Am. J. Physiol. Cell Physiol.* **290**, C134–C142.

Gamba, G. (2005). Molecular physiology and pathophysiology of electroneutral cation-chloride cotransporters. *Physiol. Rev.* **85**, 423–493.

Garzon-Muvdi, T., Pacheco-Alvarez, D., Gagnon, K.B., Vazquez, N., Ponce-Coria, J., Moreno, E., Delpire, E., and Gamba, G. (2007). WNK4 kinase is a negative regulator of K^+-Cl^- cotransporters. *Am. J. Physiol. Renal Physiol.* **292**, F1197–F1207.

Go, K.G. (1997). The normal and pathological physiology of brain water. *Adv. Tech. Stand. Neurosurg.* **23**, 47–142.

Green, R., Giebisch, G., Unwin, R., and Weinstein, A.M. (1991). Coupled water transport by rat proximal tubule. *Am. J. Physiol.* **261**, F1046–F1054.

Grunnet, M., MacAulay, N., Jorgensen, N.K., Jensen, S., Olesen, S.P., and Klaerke, D.A. (2002). Regulation of cloned, Ca^{2+}-activated K^+ channels by cell volume changes. *Pflugers Arch.* **444**, 167–177.

Guan, L. and Kaback, H.R. (2006). Lessons from lactose permease. *Annu. Rev. Biophys. Biomol. Struct.* **35**, 67–91.

Haas, M. and Forbush, B., III (2000). The Na-K-Cl cotransporter of secretory epithelia. *Annu. Rev. Physiol.* **62**, 515–534.

Hamann, S., Herrea-Perez, J.J., Bundgaard, M., Alvarez-Leefmans, F.J., and Zeuthen, T. (2005). Water transport in NKCC1. *J. Physiol.* **568**, 123–135.

Hamann, S., Kiilgaard, J.F., la Cour, M., Prause, J.U., and Zeuthen, T. (2003). Cotransport of H^+, lactate and H_2O in porcine retinal pigment epithelial cells. *Exp. Eye Res.* **76**, 1–12.

Hediger, M.A., Romero, M.F., Peng, J.B., Rolfs, A., Takanaga, H., and Bruford, E.A. (2004). The ABCs of solute carriers: physiological, pathological and therapeutic implications of human membrane transport proteins. *Introduction. Pflugers Arch.* **447**, 465–468.

Hill, A.E. (1975). Solute-solvent coupling in epithelia: a critical examination of the standing-gradient osmotic flow theory. *Proc. R. Soc. Lond. B* **190**, 99–114.

Hill, A.E. (2008). Fluid transport: a guide for the perplexed. *J. Membr. Biol.* **223**, 1–11.

Hill, A.E., Shachar-Hill, B., and Shachar-Hill, Y. (2004). What are aquaporins for? *J. Membr. Biol.* **197**, 1–32.

Hill, B.S. and Hill, A.E. (1978). Fluid transfer by necturus gall bladder epithelium as a function of osmolarity. *Proc. R. Soc. Lond. B* **200**, 151–162.

Hirai, T., Heymann, J.A.W., Shi, D., Sarker, R., Maloney, P.C., and Subramaniam, S. (2002). Three-dimensional structure of a bacterial oxalate transporter. *Nature Struct. Biol.* **9**, 597–600.

House, C.R. (1974). *Water Transport in Cells and Tissues*. Edward Arnold, London.

Huang, C.G., Lamitima, T., Agre, P., and Strange, K. (2007). Functional analysis of the aquaporin gene family in *Caenorhabditis elegans*. *Am. J. Physiol.* **292**, C1867–C1873.

Huang, Y., Lemieux, M.J., Song, J., Auer, M., and Wang, D.-N. (2003). Structure and mechanism of the glycerol-3-phosphate transporter from *Escherichia coli*. *Science*, **5633**, 616–620.

Ikonomov, O., Simon, M.X., and Frömter, E. (1985). Electrophysiological studies on lateral intercellular spaces of necturus gallbladder epithelium. *Pflugers Arch.* **403**, 301–307.

Johanson, C.E., Sweeney, S.M., Parmelee, J.T., and Epstein, M.H. (1990). Cotransport of sodium and chloride by the adult mammalian choroid plexus. *Am. J. Physiol.* **258**, C211–C216.

Kagawa, M., Nagao, S., and Bemana, I. (1996). Arginine vasopressin receptor antagonists for treatment of vasogenic brain edema: an experimental study. *J. Neurotrauma* **13**, 273–279.

Kahle, K.T., Rinehart, J., de Los Heros, H.P., Louvi, A., Meade, P., Vazquez, N., Hebert, S.C., Gamba, G., Gimenez, I., and Lifton, R. P. (2005). WNK3 modulates transport of Cl^- in and out of cells: implications for control of cell volume and neuronal excitability. *Proc. Natl. Acad. Sci. USA* **102**, 16783–16788.

Kahle, K.T., Rinehart, J., Ring, A., Gimenez, I., Gamba, G., Hebert, S.C., and Lifton, R.P. (2006). WNK protein kinases modulate cellular Cl^- flux by altering the phosphorylation state of the Na-K-Cl and K-Cl cotransporters. *Physiology (Bethesda)* **21**, 326–335.

Kahle, K.T., Ring, A.M., and Lifton, R.P. (2008). Molecular physiology of the WNK kinases. *Annu. Rev. Physiol.* **70**, 329–355.

Karadsheh, M.F., Byun, N., Mount, D.B., and Delpire, E. (2004). Localization of the KCC4 potassium-chloride cotransporter in the nervous system. *Neuroscience* **123**, 381–391.

Katay, L., Latzkovits, L., Fonagy, A., Janka, Z., and Lajtha, A. (1998). Effects of arginine vasopressin and atriopeptin on chloride uptake in cultured astroglia. *Neurochem. Res.* **23**, 831–836.

Kavanaugh, M.P., Arriza, J.L., North, R.A., and Amara, S.G. (1992). Electrogenic uptake of gamma-aminobutyric acid by a cloned transporter expressed in Xenopus oocytes. *J. Biol. Chem.* **267**, 22007–22009.

King, L.S., Kozono, D., and Agre, P. (2004). From structure to disease: the evolving tale of aquaporin biology. *Nat. Rev. Mol. Cell Biol.* **5**, 687–698.

Lapointe, J.Y. (2007). Response to Zeuthen and Zeuthen's comment to the editor: enough local hypertonicity is enough. *Biophys. J.* **93**, 1417–1419.

Lenertz, L.Y., Lee, B.H., Min, X., Xu, B.E., Wedin, K., Earnest, S., Goldsmith, E.J., and Cobb, M.H. (2005). Properties of WNK1 and implications for other family members. *J. Biol. Chem.* **280**, 26653–26658.

Leung, D.W., Loo, D.F., Hirayama, B.A., Zeuthen, T., and Wright, E.M. (2000). Urea transport by cotransporters. *J. Physiol.* **528.2**, 251–257.

Litman, T., Søgaard, R., and Zeuthen, T. (2008). Ammonia and urea permeability of mammalian aquaporins. In *Handbook of Experimental Pharmacology* (Beitz, E., ed.). Springer.

Lolait, S.J., O'Carroll, A.M., and Brownstein, M.J. (1995). Molecular biology of vasopressin receptors. *Ann. NY Acad. Sci.* **771**, 273–292.

Loo, D.D.F., Wright, E.M., and Zeuthen, T. (2002). Water pumps. *J. Physiol.* **542.1**, 53–60.

Loo, D.D.F., Zeuthen, T., Chandy, G., and Wright, E.M. (1996). Cotransport of water by the Na^+/glucose cotransporter. *Proc. Natl. Acad. Sci. USA* **93**, 13367–13370.

Loo, D.F., Hirayama, B.A., Meinild, A.-K., Chandy, G., Zeuthen, T., and Wright, E. (1999). Passive water and ion transport by cotransporters. *J. Physiol.*, **518**, 195–202.

Ludwig, C. (1861). *Lehrbuch der Physiologie des Menschen*. C.F. Wintersche Verlagshandlung, Leipzig und Heidelberg.

MacAulay, N., Gether, U., Klaerke, D.A., and Zeuthen, T. (2001). Water transport by the Na^+-coupled glutamate cotransporter. *J. Physiol.* **530**, 367–378.

MacAulay, N., Gether, U., Klaerke, D.A., and Zeuthen, T. (2002a). Passive water and urea permeability of a human Na^+-glutamate cotransporter expressed in *Xenopus* oocytes. *J. Physiol.* **530**, 367–378.

MacAulay, N., Zeuthen, T., and Gether, U. (2002b). Conformational basis for the Li^+-induced leak current in the rat gamma-aminobutyric acid (GABA) transporter-1. *J. Physiol.* **544**, 447–458.

MacLeod, R.J. and Hamilton, J.R. (1990). Regulatory volume increase in mammalian jejunal villus cells is due to bumetanide-sensitive NaKCl2. *Am. J. Physiol.* **258**, G665–G674.

Manley, G.T., Fujimura, M., Ma, T., Noshita, N., Filiz, F., Bollen, A.W., Chan, P., and Verkman, A.S. (2000). Aquaporin-4 deletion in mice reduces brain edema after acute water intoxication and ischemic stroke. *Nat. Med.* **6**, 159–163.

Matsuo, H., Tsukada, S., Nakata, T., Chairoungdua, A., Kim, D.K., Cha, S.H., Inatomi, J., Yorifuji, H., Fukuda, J., Endou, H., and Kanai, Y. (2000). Expression of a system L neutral amino acid transporter at the blood–brain barrier. *Neuroreport* **11**, 3507–3511.

Meinild, A.-K., Klaerke, D.A., Loo, D.D.F., Wright, E.M., and Zeuthen, T. (1998). The human Na^+/glucose cotransporter is a molecular water pump. *J. Physiol.* **508.1**, 15–21.

Meinild, A.-K., Loo, D.F.F., Pajor, A., Zeuthen, T., and Wright, E.M. (2000). Water transport by the renal Na^+/dicarboxylate cotransporter. *Am. J. Physiol.* **278**, F777–F783.

Mercado, A., Song, L., Vázquez, N., Mount, D.B., and Gamba, G. (2000). Functional comparison of the K^+-Cl^- cotransporters KCC1 and KCC4. *J. Biol. Chem.* **275**, 30326–30334.

Mitchell, P. (1957). A general theory of membrane transport from studies of bacteria. *Nature* **180**, 134–136.

Miura, M., Takagi, S., Matsukado, Y., and Ushio, Y. (1989). Influence of vasopressin level on osmotic pressure and sodium concentration in plasma and cerebrospinal fluid in patients with intracranial lesions. *Neurol. Med. Chir. (Tokyo)* **29**, 806–810.

Naftalin, R.J. (2008). Osmotic water transport with glucose in GLUT2 and SGLT. *Biophys. J.* **94**, 3912–3923.

Nagao, S. (1998). [Vasopressin and blood–brain barrier]. *No To Shinkei* **50**, 809–815.

Nagao, S., Kagawa, M., Bemana, I., Kuniyoshi, T., Ogawa, T., Honma, Y., and Kuyama, H. (1994). Treatment of vasogenic brain edema with arginine vasopressin receptor antagonist – an experimental study. *Acta Neurochir. Suppl. (Wien.)* **60**, 502–504.

Nagelhus, E.A., Horio, Y., Inanobe, A., Fujita, A., Haug, F.M., Nielsen, S., Kurachi, Y., and Ottersen, O.P. (1999). Immunogold evidence suggests that coupling of K^+ siphoning and water transport in rat retinal Muller cells is mediated by a coenrichment of Kir4.1 and AQP4 in specific membrane domains. *Glia* **26**, 47–54.

Nielsen, S., Nagelhus, E.A., miry-Moghaddam, M., Bourque, C., Agre, P., and Ottersen, O.P. (1997). Specialized membrane domains for water transport in glial cells: high-resolution immunogold cytochemistry of aquaporin-4 in rat brain. *J. Neurosci.* **17**, 171–180.

Niermann, H., miry-Moghaddam, M., Holthoff, K., Witte, O.W., and Ottersen, O.P. (2001). A novel role of vasopressin in the brain: modulation of activity-dependent water flux in the neocortex. *J. Neurosci.* **21**, 3045–3051.

O'Donnell, M.E., Duong, V., Suvatne, J., Foroutan, S., and Johnson, D.M. (2005). Arginine vasopressin stimulation of cerebral microvascular endothelial cell Na-K-Cl cotransporter activity is V1 receptor and [Ca] dependent. *Am. J. Physiol. Cell Physiol.* **289**, C283–C292.

O'Donnell, M.E., Tran, L., Lam, T.I., Liu, X.B., and Anderson, S.E. (2004). Bumetanide inhibition of the blood–brain barrier Na-K-Cl cotransporter reduces edema formation in the rat middle cerebral artery occlusion model of stroke. *J. Cereb. Blood Flow Metab.* **24**, 1046–1056.

O'Kane, R.L., Martinez-Lopez, I., DeJoseph, M.R., Vina, J.R., and Hawkins, R.A. (1999). Na^+-dependent glutamate transporters (EAAT1, EAAT2, and EAAT3) of the blood–brain barrier. A mechanism for glutamate removal. *J. Biol. Chem.* **274**, 31891–31895.

Olesen, C., Picard, M., Winther, A.-M.L., Gyrup, C., Morth, J.P., Oxvig, C., Møller, J.V., and Nissen, P. (2007). The structural basis of calcium transport by the calcium pump. *Nature* **450**, 1036–1039.

Pappenheimer, J.R. (1998). Scaling of dimensions of small intestines in non-ruminant eutherian mammals and its significance for absorptive mechanisms. *Comp. Biochem. Physiol.* **121**, 45–58.

Parsegian, A. (2002). Protein-water interactions. *Int. Rev. Cytol.* **215**, 1–30.

Paulson, O.B., Hertz, M.M., Bolwig, T.G., and Lassen, N.A. (1977). Filtration and diffusion of water across the blood–brain barrier in man. *Microvascular Res.* **13**, 124.

Quin, M., Haser, R., and Payan, F. (1995). Carbohydrate binding sites in a pancreatic a-amylase-substrate complex, derived from X-ray structure analysis at 2.1A resolution. *Protein Sci.* **4**, 747–755.

Radian, R. and Kanner, B.I. (1983). Stoichiometry of sodium- and chloride-coupled gamma-aminobutyric acid transport by synaptic plasma membrane vesicles isolated from rat brain. *Biochemistry* **22**, 1236–1241.

Rand, R.P. (2002). The lipid-water interface: revelations by osmotic stress. *Int. Rev. Cytol.* **215**, 33–46.

Rand, R.P. and Fuller, N.L. (1992). Water as an inhibiting ligand in yeast hexokinase. *Biophys. J.* **61**, A345.

Reid, E.W. (1892). Report on experiments upon "absorbtion without osmosis". *BMJ* **1**, 323–326.

Reid, E.W. (1901). Transport of fluid by certain epithelia. *J. Physiol.* **26**, 436–444.

Rosenberg, G.A., Estrada, E., and Kyner, W.T. (1990). Vasopressin-induced brain edema is mediated by the V1 receptor. *Adv. Neurol.* **52**, 149–154.

Rosenberg, G.A., Scremin, O., Estrada, E., and Kyner, W.T. (1992). Arginine vasopressin V1-antagonist and atrial natriuretic peptide reduce hemorrhagic brain edema in rats. *Stroke* **23**, 1767–1773.

Russell, J.M. (2000). Sodium-potassium-chloride cotransport. *Physiol. Rev.* **80**, 211–276.

Silva, P., Stoff, J., Field, M., Fine, L., Forrest, J.N., and Epstein, F.H. (1977). Mechanism of active chloride secretion by shark rectal gland: role of Na-K-ATPase in chloride transport. *Am. J. Physiol.* **233**, F298–F306.

Simpson, I.A., Carruthers, A., and Vannucci, S.J. (2007). Supply and demand in cerebral energy metabolism: the role of nutrient transporters. *J. Cereb. Blood Flow Metab.* **27**, 1766–1791.

Singh, S.K., Yamashita, A., and Gouaux, E. (2007). Antidepressant binding site in a bacterial homologue of neurotransmitter transporters. *Nature* **448**, 952–956.

Steitz, T.A., Shoham, M., and Bennett, W.S., Jr. (1981). Structural dynamics of yeast hexokinase during catalysis. *Phil. Trans. R. Soc. Lond.* **293**, 43–52.

Takanaga, H., Ohtsuki, S., Hosoya, K., and Terasaki, T. (2001). GAT2/BGT-1 as a system responsible for the transport of gamma-aminobutyric acid at the mouse blood-brain barrier. *J. Cereb. Blood Flow Metab.* **21**, 1232–1239.

Tenedieva, V.D., Lyamin, P.V., and Nepomnyaschi, V.P. (1994). The plasma and CSF vasopressin levels in brain tumors with brain edema. *Acta Neurochir. Suppl. (Wien.)* **60**, 387–389.

Tsutsumi, T., Ushiro, H., Kosaka, T., Kayahara, T., and Nakano, K. (2000). Proline- and alanine-rich Ste20-related kinase associates with F-actin and translocates from the cytosol to cytoskeleton upon cellular stresses. *J. Biol. Chem.* **275**, 9157–9162.

Vajda, Z., Pedersen, M., Doczi, T., Sulyok, E., Stodkilde-Jorgensen, H., Frokiaer, J., and Nielsen, S. (2001). Effects of centrally administered arginine vasopressin and atrial natriuretic peptide on the development of brain edema in hyponatremic rats. *Neurosurgery* **49**, 697–704.

Vakili, A., Kataoka, H., and Plesnila, N. (2005). Role of arginine vasopressin V1 and V2 receptors for brain damage after transient focal cerebral ischemia. *J. Cereb. Blood Flow Metab.* **25**, 1012–1019.

Vitari, A.C., Thastrup, J., Rafiqi, F.H., Deak, M., Morrice, N.A., Karlsson, H.K., and Alessi, D.R. (2006). Functional interactions of the SPAK/OSR1 kinases with their upstream activator WNK1 and downstream substrate NKCC1. *Biochem. J.* **397**, 223–231.

Voisin, D.V. and Bourque, C.W. (2002). Integration of sodium and osmosensory signals in vasopressin neurons. *Trends Neurosci.* **25**, 199–205.

Wells, T. (1998). Vesicular osmometers, vasopressin secretion and aquaporin-4: a new mechanism for osmoreseption? *Mol. Cell. Endocrinol.* **136**, 103–107.

Whitlock, R.T. and Wheeler, H.O. (1964). Coupled transport of solute and water across rabbit gallbladder epithelium. *J. Clin. Invest* **43**, 2249–2265.

Whittembury, G. and Hill, B.S. (1982). Fluid reabsorption by Necturus proximal tubule perfused with solutions of normal and reduced osmolarity. *Proc. R. Soc.* **215**, 411–431.

Yamashita, A., Singh, S.K., Kawate, T., Jin, Y., and Gouaux, E. (2005). Crystal structure of a bacterial homologue of Na^+/Cl^--dependent neurotransmitter transporters. *Naturen* **437**, 215–223.

Yang, B. and Verkman, A.S. (1997). Water and glycerol permeabilities of aquaporin 1-5 and MIP determined quantitatively by expression of epitope-tagged constructs in *Xenopus* oocytes. *J. Biol. Chem.* **272**, 14146–16140.

Yang, B. and Verkman, A.S. (2002). Analysis of double knockout mice lacking aquaporin-1 and urea transporter UT-B. *J. Biol. Biochem.* **277**, 36782–36786.

Yernool, D., Boudker, O., Jin, Y., and Gouaux, E. (2004). Structure of a glutamate transporter homologue from *Pyrococcus horikosh ii. Nature* **431**, 811–818.

Zagorska, A., Pozo-Guisado, E., Boudeau, J., Vitari, A.C., Rafiqi, F.H., Thastrup, J., Deak, M., Campbell, D.G., Morrice, N.A., Prescott, A.R., and Alessi, D.R. (2007). Regulation of activity and localization of the WNK1 protein kinase by hyperosmotic stress. *J. Cell Biol.* **176**, 89–100.

Zampighi, G.A., Kreman, M., Boorer, K.J., Loo, D.D.F., Bezanilla, F., Chandy, G., Hall, J.E., and Wright, E.M. (1995). A method for determining the unitary functional capacity of cloned channels and transporters expressed in *Xenopus laevis* oocytes. *J. Membr. Biol.* **148**, 65–78.

Zeidel, M.L., Ambudkar, S.V., Smith, B.L., and Agre, P. (1992). Reconstitution of functional water channels in liposomes containing purified red cell CHIP28 protein. *Biochemistry* **31**, 7436–7440.

Zeuthen, T. (2000). Molecular water pumps. *Rev. Physiol. Biochem. Pharmacol.* **141**, 97–151.

Zeuthen, T., Belhage, B., and Zeuthen, E. (2006). Water transport by Na^+-coupled cotransporters of glucose (SGLT1) and of iodide (NIS). The dependence of substrate size studied at high resolution. *J. Physiol.* **570.3**, 485–499.

Zeuthen, T. (1996). *Molecular Mechanisms of Water Transport*. Springer, Berlin, R.G. Landes Company, Texas.

Zeuthen, T. (1995). Molecular mechanisms for passive and active transport of water. *Int. Rev. Cytol.* **160**, 99–161.

Zeuthen, T. (1994). Cotransport of K^+, Cl^- and H_2O by membrane proteins from choroid plexus epithelium of *Necturus maculosus*. *J. Physiol.* **478**, 203–219.

Zeuthen, T. (1991a). Secondary active transport of water across ventricular cell membrane of choroid plexus epithelium of necturus maculosus. *J. Physiol.* **444**, 153–173.

Zeuthen, T. (1991b). Water permeability of ventricular cell membrane in choroid plexus epithelium from *Necturus maculosus*. *J. Physiol.* **444**, 133–151.

Zeuthen, T. (1981). Isotonic transport and intracellular osmolarity in the necturus gall-bladder epithelium. In Ussing, H.H., Bindslev, N., Lassen, N.A., and Sten-Knudsen, O., eds, pp. 313–331. Munksgaard, Copenhagen.

Zeuthen, T. (2002). General models for water transport across leaky epithelia. *Int. Rev. Cytol.* **215**, 285–317.

Zeuthen, T. (1982). Relations between intracellular ion activities and extracellular osmolarity in necturus gallbladder epithelium. *J. Membr. Biol.* **66**, 109–121.

Zeuthen, T. (1983). Ion activities in the lateral intercellular spaces of gallbladder epithelium transporting at low external osmolarities. *J. Membr. Biol.* **76**, 113–122.

Zeuthen, T. (2008). Molecular water pumps – or how water can move uphill across epithelia. *Physiology News* **68**, 3–5.

Zeuthen, T., Hamann, S., and la Cour, M. (1996). Cotransport of H^+, lactate and H_2O by membrane proteins in retinal pigment epithelium of bullfrog. *J. Physiol.* **497.1**, 3–17.

Zeuthen, T. and MacAulay, N. (2002). Cotransporters as molecular water pumps. *Int. Rev. Cytol.* **215**, 259–284.

Zeuthen, T., Meinild, A.-K., Loo, D.D.F., Wright, E.M., and Klaerke, D.A. (2001). Isotonic transport by the Na^+-glucose cotransporter SGLT1. *J. Physiol.* **531.3**, 631–644.

Zeuthen, T. and Stein, W.D. (1994). Co-transport of salt and water in membrane proteins: membrane proteins as osmotic engines. *J. Membr. Biol.* **137**, 179–195.

Zeuthen, T. and Zeuthen, E. (2007). The mechanism of water transport in Na^+-coupled glucose transporters expressed in *Xenopus* oocytes. *Biophys. J.* **93**, 1413–1416.

Zeuthen, T., Zeuthen, E., and Klaerke, D.A. (2002). Mobility of ions, sugar, and water in the cytoplasm of *Xenopus* oocytes expressing Na^+-coupled sugar transporters (SGLT1). *J. Physiol.* **542.1**, 71–87.

Zeuthen, T., Zeuthen, E., and MacAulay, N. (2007). Water transport by GLUT2 expressed in *Xenopus laevis* oocytes. *J. Physiol.* **549.2**, 345–361.

Zimmerberg, J. and Parsegian, V.A. (1986). Polymer inaccessible volume changes during opening and closing of a voltage-dependent ionic channel. *Nature* **323**, 36.

CHAPTER

29

Ion Transport in Choroid Plexus

Peter D. Brown, Sarah L. Davies and Ian D. Millar

OUTLINE

I. INTRODUCTION TO THE CEREBROSPINAL FLUID AND THE CHOROID PLEXUSES

The cerebrospinal fluid (CSF) fills the ventricles of the brain, the spinal canal and the subarachnoid space (Fig. 29.1A). In humans, the CSF has a total volume of about 140 ml representing 40% of the extracellular fluid in the central nervous system. The CSF is separated from the majority of neuronal tissue in the brain by the ependyma (which lines the ventricles and canals), and the pia (which covers the external surface of the brain). The composition of the CSF does, however, influence neuronal activity, notably in the central chemoreceptors of the medulla oblongata which

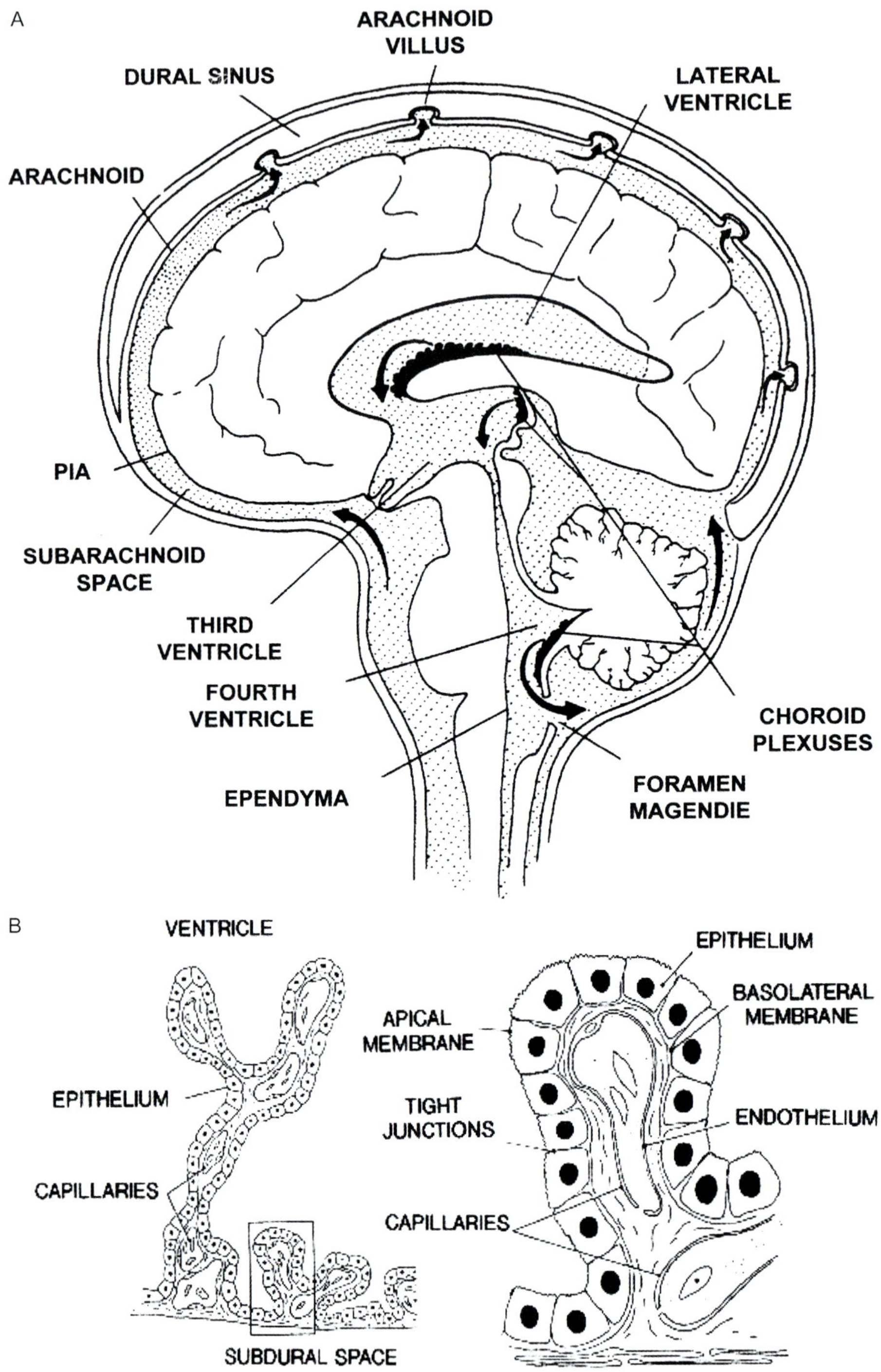

FIGURE 29.1 The locations and structure of the choroid plexuses in the human brain. **A.** The choroid plexuses are shown as the solid black structures, and the location of the cerebrospinal fluid they secrete is shown as the shaded areas. **B.** Branched structure of the choroid plexus with villi projecting into the ventricle of the brain. Each plexus consists of a network of capillaries covered by a single layer of cuboidal epithelial cells.

modulate ventilation by responding to changes in CSF pH. In addition to helping maintain a stable environment for neuronal activity, the CSF has a number of other important functions: (i) it provides mechanical support for the brain, reducing its effective weight by more than 60% (Segal, 1993); (ii) it acts as an excretory pathway for the brain by providing a "sink" into which products of metabolism or synaptic activity are diluted and subsequently removed (Segal, 1993); (iii) it is an important route by which nutrients can reach the central nervous system (Johanson, 1999); (iv) it also has a putative role as a route for the movement of hormones and transmitters between different areas of the brain (Johanson, 1999). Most of these functions require that the composition of the CSF is carefully regulated.

The composition of CSF differs significantly from that of plasma (Table 29.1). A major difference is that of the protein concentration, which is greatly reduced in CSF, i.e. ~0.4% of that in plasma. CSF is not just an ultrafiltrate of plasma; the concentrations of K^+ and amino acids are lower in CSF, compared to plasma levels, and $[Cl^-]$ is 10% to 20% higher in CSF than in plasma. The CSF is constantly produced, and in humans the total volume is replaced about four times each day. Thus, approximately 600 ml of CSF is produced every 24 hours (Wright, 1978). Most of the CSF is produced by the four choroid plexuses (one in each ventricle of the brain; see Fig. 29.1A. The process by which the choroid plexuses secrete CSF is carefully coordinated so that concentrations of certain ions remain essentially constant even when their plasma concentrations are varied experimentally, e.g. K^+ (Husted and Reed, 1976; Keep et al., 1987), HCO_3^- (Husted and Reed, 1977) and Ca^{2+} (Murphy et al., 1986).

II. AN OVERVIEW OF CSF SECRETION BY THE CHOROID PLEXUSES

The choroid plexuses have a branched morphology with numerous villi projecting into the ventricles of the brain. Each villus is composed of a single layer of epithelial cells overlying a core of connective tissue

TABLE 29.1 Solute concentrations (mM) in human plasma and CSF

Solute	Plasma	CSF
Na^+	152	141
K^+	4.4	2.9
Ca^{2+}	0.2 (ionized)	1.1 (ionized)
Mg^{2+}	0.6 (ionized)	1.1 (ionized)
Cl^-	102	113–120
HCO_3^-	24†	22
Ph	7.4	7.4
Proteins	7 g/dl	0.03 g/dl

Brain extracellular fluid resembles CSF, but both differ from plasma. Plasma is 92% water, and CSF is 99% water. Data sources: Irani (2009); Boron and Boulpaep (2009); Katzman and Pappius (1973). †Value in mixed venous blood. In arterial blood is ~22 mM.

and blood capillaries (Fig. 29.1B). These capillaries, unlike those in the majority of the cerebral circulation, are fenestrated so they offer little resistance to the movement of small molecules, ions and water (Segal, 1993). A barrier between the blood and brain, however, is formed by the epithelial cells which are linked by tight junctions (Fig. 29.1B). The epithelial cells of the choroid plexuses therefore form what is known as the blood–CSF barrier. Note that this barrier is very different from the blood–brain barrier (BBB), which is formed by the brain microvessel endothelial cells. Both the fine structure of the BBB as well as the ion and fluid transport across this barrier are discussed in Chapter 30.

Fluid secretion across epithelial barriers such as the choroid plexuses has been found to be dependent on the unidirectional transport of ions. This creates an osmotic gradient which drives the movement of water. Unidirectional transport of ions (either secretion or absorption) is achieved due to the polarized expression of membrane transport proteins in the apical and basolateral membranes of the epithelial cells. In the majority of epithelia Na^+ and Cl^- are the main ions secreted, but in the choroid plexus HCO_3^- also makes an important contribution (Fig. 29.2A). Furthermore, choroid plexus epithelial cells are also thought to mediate the net absorption of K^+ from the CSF to the blood (Fig. 29.2A). The transport of K^+ is not only important in maintaining the $[K^+]$ in the CSF (Zeuthen and Wright, 1981), but is also linked to the process of CSF secretion.

To understand the process of secretion by any epithelial tissue, it is essential to have knowledge of the expression and membrane localization of ion channels and transporters. In the last two decades excellent progress has been made in the identification of proteins involved in Na^+ and K^+ transport by the choroid plexus. The last five years have seen an explosion in the amount of data on HCO_3^- transporter expression. These data are discussed in the sections III to V of this chapter. The chapter will then move on to discuss the potential roles of cation-chloride cotransporters in the choroid plexus (sections VI to VIII).

III. CATION CHANNELS AND CATION TRANSPORTERS EXPRESSION IN THE CHOROID PLEXUS

Many of the early studies of ion transport by the choroid plexus focused on the roles of Na^+ and K^+. These experiments determined the rates at which radioactive isotopes penetrated the CSF, or how the

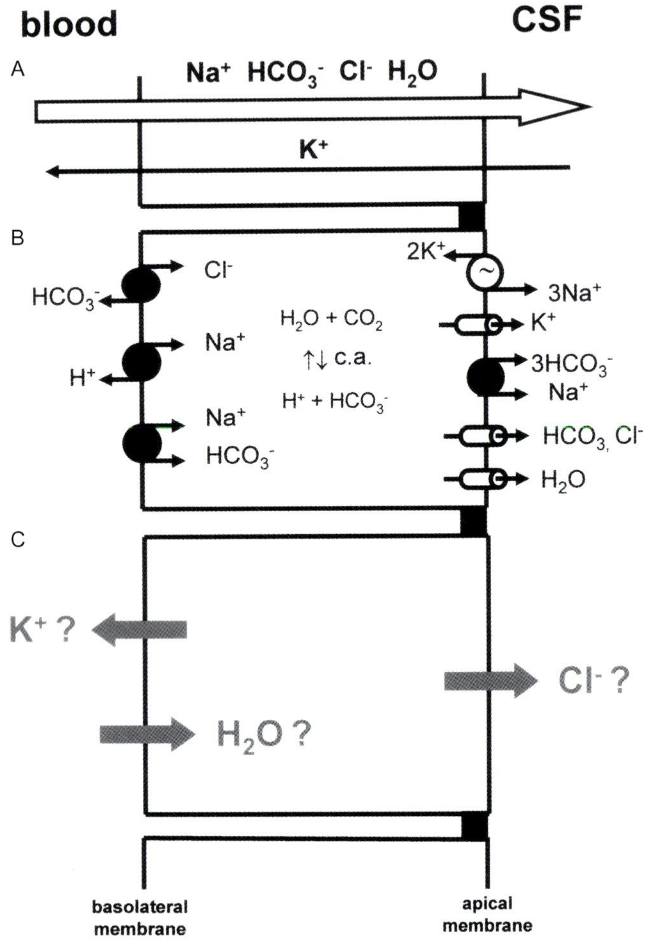

FIGURE 29.2 Ion transport by the choroid plexus. **A.** Major fluxes of ions across the choroid plexus epithelium. **B.** Ion transporters expressed in the choroid plexus and thought to be involved in Na^+, HCO_3^-, H_2O secretion and K^+ absorption. Data on cation-chloride cotransporter expression have been omitted for clarity, and c.a. = carbonic anhydrase. **C.** Remaining questions about ion and H_2O transport by the choroid plexus epithelium.

rate of CSF secretion was affected by various transport inhibitors. The data obtained have more recently been refined and elaborated upon by the use of molecular localization methods.

A. Na^+/K^+ ATPase

Activity of the Na^+/K^+ ATPase or "sodium pump" is known to be closely associated with the secretion of CSF, because inhibitors such as ouabain reduce the rate of fluid secretion and the movement of Na^+ into the CSF (Davson and Segal, 1970; Wright, 1978; Pollay et al., 1985). Various studies have determined the expression of the α_1, β_1 and β_2 subunits in rat choroid plexus (Watts et al., 1991; Zlokovic et al., 1993; Klarr et al., 1997). Ouabain binding studies (Quinton et al., 1973) and immunocytochemical studies (Masuzawa et al., 1984b; Ernst et al., 1986) revealed that the α_1 subunit, and presumably the two β subunits, are expressed at the apical membrane of rat and mouse choroid plexus (Fig. 29.2B). Most recently, Praetorius and Nielsen (2006) have determined that Na^+/K^+ ATPase is also located in the apical membrane of human choroid plexus.

B. K^+ Channels

On the basis of measurements of ion fluxes, membrane potential and intracellular ion activities with microelectrodes, Zeuthen and Wright (1981) first suggested that K^+ channels are expressed in both the apical and basolateral membranes of amphibian choroid plexus. Using whole-cell patch-clamp methods, we have identified two K^+ conductances in rat choroid plexus: a time-independent, inward-rectifying conductance (Kir) observed at hyperpolarizing membrane potentials, and a time-dependent outward-rectifying conductance (Kv) observed at depolarizing membrane potentials (Kotera and Brown, 1994). The inward-rectifying conductance is now known to be carried by Kir7.1 channels (Döring et al., 1998), which are expressed in the apical membrane of the choroid plexus (Nakamura et al., 1999). The Kv conductance is thought to be mainly carried by Kv1.1 and Kv1.3 channels, which are also expressed in the apical membrane (Speake et al., 2004). To date, K^+ channels have not been identified in the basolateral membrane of the mammalian choroid plexus.

C. Na^+/H^+ Exchange

The expression of an Na^+/H^+ exchanger (NHE) in the basolateral membrane could provide a potential route for Na^+ influx during CSF secretion, as well as helping to regulate intracellular pH. Indeed, several functional studies have indicated that an NHE is expressed in mammalian choroid plexus (Murphy and Johanson, 1989; Murphy and Johanson, 1990; Mayer and Sanders-Bush, 1993). Data on the molecular identity of the NHE, however, are equivocal. NHE1 expression was indicated in the pig lateral ventricle choroid plexus by the kinetics of amiloride binding, and by RT-PCR methods (Kalaria et al., 1998). However, immunocytochemical methods have failed to detect NHE1 protein expression in rat choroid plexus (Alper et al., 1994). Thus, an NHE or NHE-like protein is expressed in the choroid plexus, but the molecular identity of this protein remains to be determined.

IV. HCO_3^- TRANSPORTERS AND CSF SECRETION

It has been recognized for many years that the process of CSF secretion is dependent on bicarbonate (HCO_3^-). This was initially indicated by the observation that CSF secretion was reduced by carbonic anhydrase inhibitors (Maren and Broder, 1970). Subsequent experiments by Saito and Wright determined that secretion is dependent on the presence of HCO_3^- at the basolateral membrane of the epithelium (Saito and Wright, 1983, 1984). The last few years have seen significant progress in the molecular identification of a whole family of HCO_3^- transport proteins: the SLC4 family (see Romero et al., 2004 and Chapter 4 in this volume). A number of these SLC4 proteins, in addition to two HCO_3^- permeable anion channels and several isoforms of carbonic anhydrase, have now been identified in choroid plexus (Fig. 29.2B).

A. Carbonic Anhydrases

The carbonic anhydrases are a group of enzymes which catalyze the production of HCO_3^- and H^+ from H_2O and CO_2. Although not directly involved in ion transport, it is thought that carbonic anhydrases have an important role in CSF secretion. The main evidence for this is that acetazolamide, a carbonic anhydrase inhibitor, reduces CSF secretion in rats by as much as 50% (Maren and Broder, 1970; Vogh et al., 1987). Histochemical methods were first used to demonstrate carbonic anhydrase expression in the cytoplasm and membrane of the rat choroid plexus (Masuzawa and Sato, 1983). Immunocytochemical methods have subsequently identified CAII in the cytoplasm of rat choroid plexus epithelial cells (Masuzawa et al., 1984a), and CAIII in the cytoplasm of rat and human choroid plexus (Nogradi et al., 1993). As for the membrane bound isoform CAIV, functional studies suggest that it contributes little to CSF secretion in the rat (Maren et al., 1997).

B. Basolateral Membrane HCO_3^- Transporters

Lindsey et al. (1990) demonstrated expression of the AE2 isoform of the Cl^-/HCO_3^- exchanger in the choroid plexus epithelium. In their immunolocalization studies, they showed that the AE2 protein is expressed exclusively in the basolateral membrane of mouse choroid plexus cells, an observation which has since been confirmed in rat choroid plexus (Alper et al., 1994; Wu et al., 1998).

In 1997, the first Na^+-HCO_3^- cotransporter (NBC) was identified at the molecular level (Romero et al., 1997). Since then, a number of NBCs have been identified in the choroid plexus epithelium. Praetorius et al. (2004) reported mRNA expression of an electroneutral Na^+-HCO_3^- cotransporter (NBCn-1 or SLC4A7) in rat and mouse choroid plexus. The NBCn-1 protein was then visualized by immunocytochemical methods in the basolateral membrane of rat and mouse choroid plexus (Praetorius et al., 2004). NCBE (SLC4A10) is also expressed in choroid plexus. The protein was originally described as an Na-dependent Cl^-/HCO_3^- exchanger (Wang et al., 2000), but it is now thought to be a second electroneutral Na^+-HCO_3^- cotransporter (NBCn2; Parker et al., 2008). RT-PCR and Western blot analyses indicate that NBCn2 is indeed expressed in rat and mouse choroid plexus (Praetorius et al., 2004). Immunocytochemical localization studies showed that the protein is expressed in the basolateral membrane (Praetorius et al., 2004). Finally, SLC4A8, the putative Na^+-dependent Cl^-/HCO_3^- exchanger (NDCBE), is also expressed in the basolateral membrane of choroid plexus epithelial cells from fetal but not adult rats (Chen et al., 2008).

C. Apical Membrane HCO_3^- Transporters

Bicarbonate transporters are also found in the apical membrane of choroid plexus epithelial cells. RT-PCR and Western blot analyses have determined that the electrogenic Na^+-HCO_3^- cotransporter NBCe2 (SLC4A5), but not NBCe1 (SLC4A4), is expressed in rodent choroid plexus (Praetorius et al., 2004; Bouzinova et al., 2005). The NBCe2 protein intriguingly appears to be expressed in the apical membrane of both rat and mouse choroid plexus, where it could potentially contribute to HCO_3^- efflux; that is if it transports $3HCO_3^-:1Na^+$, as described by Sassani et al. (2002). However, NBCe2 is also known to mediate HCO_3^- influx with a 2:1 stoichiometry in some cells (Virkki et al., 2002). Using whole-cell patch-clamp methods to measure the small currents generated by the NBCe2, we have determined that the NBCe2 in mouse choroid plexus has a 3:1 stoichiometry. Thus, the cotransporter is likely to contribute to HCO_3^- efflux into the cerebrospinal fluid in mice and possibly rats (Millar and Brown, 2008). A final twist to this story, however, is that NBCe2 does not appear to be expressed in human choroid plexus (Damkier et al., 2007). NaBC1, another member of the SLC4 family (SLC4A11), is instead expressed in apical membrane of human choroid plexus (Damkier et al., 2007), but the function of NaBC1 is as yet unknown (Romero et al., 2004; Pushkin and Kurz, 2006; Damkier et al., 2007).

D. Anion Channels

Anion channels located at the apical membrane play a key role in many secretory epithelia (Jentsch et al., 2002). Indeed, Saito and Wright (1984) predicted that HCO_3^- permeable channels have a central role in CSF secretion by amphibian choroid plexus. This observation was supported by data from single channel recording methods which identified anion channels with conductances of about 20 pS in the apical membrane of both amphibian (Christensen et al., 1989) and mammalian choroid plexus (Garner and Brown, 1992). The properties of these anion channels in mammalian choroid plexus have been further studied using whole-cell recording methods. Two anion conductances have been characterized in these experiments: an inward-rectifying conductance and a volume-activated anion conductance.

The *inward-rectifying anion conductance* was first identified in rat choroid plexus (Kibble et al., 1996), then subsequently in mouse (Kibble et al., 1997) and porcine tissue (Kajita et al., 2000). When first identified, it was thought to be the largest conductance in the choroid plexus, and furthermore it was reported to have a uniquely high permeability to HCO_3^- (P_{HCO_3}:P_{Cl} = 1.5; Kibble et al., 1996). Given these properties, and the HCO_3^- permeable channel hypothesis of Saito and Wright (1984), it was suggested that these channels may have an important role in CSF secretion (Speake et al., 2001; Brown et al., 2004). Recent data, however, have challenged this conclusion, as: (i) the presence of NBCe2 currents in the whole-cell conductance (Millar and Brown, 2008) means that the high P_{HCO3} may represent a significant overestimate since the experiments were performed in the presence of Na^+ (Kibble et al., 1996); (ii) the properties of the inward-rectifying conductance are distinct from those of the apical membrane channels identified in single channel studies; and (iii) cation substitution experiments indicate that a significant component of the inward-rectifying conductance is carried by non-selective cation channels (Millar and Brown, 2008). In conclusion the properties and possibly the membrane localization of the channel may not be consistent with a role in CSF secretion. Furthermore, the contribution of the channels to the whole-cell conductance is much smaller than originally anticipated. Thus, it now appears likely that these channels make less of a contribution to CSF secretion than had previously been believed.

The second anion conductance identified by whole-cell methods in rat and mouse choroid plexus is carried by *volume-activated anion channels* (Kibble et al., 1996, 1997). The properties of this volume-activated conductance have not been studied in great detail, but some properties (e.g. P_{HCO_3}:P_{Cl} = 0.6; I.D. Millar and P.D. Brown; unpublished data) suggest it may be carried by the single channels identified in the apical membrane (P_{HCO3}:P_{Cl} = 0.5; C. Garner and P.D. Brown; unpublished data). The volume-activated channels, however, make only a very minor contribution to the whole-cell conductance at normal cell volumes (Kibble et al., 1996, 1997). Thus, although they may be located in the apical membrane, it seems unlikely that they make a significant contribution to the process of CSF secretion.

V. A MODEL FOR CSF SECRETION: THE RELATIONSHIP BETWEEN Na^+, HCO_3^- AND Cl^- TRANSPORT IN THE CHOROID PLEXUS

The expression of the proteins involved in ion transport by the choroid plexus are summarized in Fig. 29.2B. This section of the review discusses the interactions between these different proteins to bring about CSF secretion.

A. Cation Transport at the Apical Membrane

As stated above, the Na^+/K^+ ATPase has a central role in the secretory process; it is the main route for Na^+ efflux across the apical membrane, and also generates the ion gradients which energize the activities of other transporters. Thus, the Na^+ gradient drives the accumulation of Cl^- and HCO_3^- influx across the basolateral membrane. The Na^+/K^+ ATPase is also the most efficient mechanism for K^+ absorption from the CSF. In fact it is so efficient that, to prevent the removal of all K^+ from the CSF, most of the K^+ pumped into the cell at the apical membrane simply leaks back out of the cell across the same membrane. This is achieved by the diffusion of K^+ through the inward-rectifying and delayed-rectifying K^+ channels expressed in the apical membrane. The combined actions of the K^+ channels and Na^+/K^+ ATPase not only maintains the concentration of K^+ in the CSF, but also regulates intracellular K^+ activity and generates an intracellular negative membrane potential. This, as we shall see later, is essential in driving anion efflux across the apical membrane. In amphibian choroid plexus, membrane potential values of around −45 mV have been measured (Zeuthen and Wright, 1981; Saito and Wright, 1984), whereas in mammals, values of −60 mV and −35 mV have been reported for rabbit and rat, respectively (Welch and Sadler, 1965; S.L. Greenwood and P.D. Brown, unpublished data).

B. Cation and Anion Transport at the Basolateral Membrane

A combination of NBCn-1 (SLC4A7) and NBCn-2 (SLC4A10) Na^+-HCO_3^- cotransporters mediate Na^+ and HCO_3^- entry through the basolateral membrane. These secondary active transport mechanisms for HCO_3^-, in conjunction with the production of HCO_3^- by the action of several carbonic anhydrases, result in the accumulation of HCO_3^- within the cell. The importance of the NBCn-2 to CSF secretion has been highlighted recently by experiments showing that the volume of the brain ventricles is significantly reduced (presumably due to a decrease in CSF secretion) in mice in which the Slc4a10 gene has been disrupted (Jacobs et al., 2008). The net effect of Na^+-HCO_3^- cotransport activity is to elevate the intracellular concentration of HCO_3^-, which can also be demonstrated by the very rapid intracellular alkalinization that ensues when choroid plexus cells are exposed to HCO_3^--containing solutions (Bouzinova et al., 2005). This, in turn, is thought to promote the activity of AE2 (SLC4A2), which mediates electroneutral Cl^- accumulation in exchange for HCO_3^-. This hypothesis is supported by the observation that DIDS (an inhibitor of both AEs and NBCs) reduces Cl^- uptake into the choroids plexus and reduces CSF production (Deng and Johanson, 1989). Furthermore, patch-clamp studies of rat choroid plexus show a DIDS-sensitive shift in the reversal potential for Cl^- channels in cell-attached patches, when cells are bathed in HCO_3^- solutions, which is consistent with an increase in the intracellular Cl^- activity (Brown and Garner, 1993). Estimates based on these data suggest that the intracellular $[Cl^-]$ in rat choroid plexus increases from 30 mM in HCO_3^--free Hepes-buffered solutions to 80 mM in the presence of HCO_3^- (Brown and Garner, 1993). This latter value is not too dissimilar to the 67 mM measured by $^{36}Cl^-$ distribution in rat tissue (Smith and Johanson, 1985). In summary, the combined actions of the anion transporters in the basolateral membrane appear to mediate the HCO_3^--dependent accumulation of Na^+ and Cl^-.

C. Anion Transport at the Apical Membrane

The mechanism of anion transport at the apical membrane is still not fully defined. As suggested above, the accumulation of HCO_3^- and Cl^- by the choroid plexus cells, coupled to the negative membrane potential, means that there are very significant electrochemical gradients set to drive the efflux of both anions. The gradient for HCO_3^- is certainly sufficient to drive efflux via the NBCe2, as discussed by Millar and Brown (2008). Cl^- and HCO_3^- efflux via the anion channels expressed in choroid plexus may also contribute to CSF secretion, but it is now debatable whether the conductance of these channels is sufficient to account for all of Cl^- efflux into the CSF (see section IV).

VI. REMAINING QUESTIONS ABOUT THE MECHANISM OF CSF SECRETION: POTENTIAL ROLES FOR CATION-CHLORIDE COTRANSPORTERS

Figure 29.2B summarizes our current understanding about ion transport in the choroid plexus. By contrast, Fig. 29.2C identifies three important questions about the mechanism of the CSF secretion which remained to be answered. These questions are:

A. What is the Route for Cl^- Efflux at the Apical Membrane?

In previous models of CSF secretion we have proposed that the inward-rectifying anion conductance could account for the efflux of both Cl^- and HCO_3^- from the cells into the CSF (Brown and Garner, 1993; Speake et al., 2001). Recent data, however, indicate that the inward-rectifying conductance may have a much less significant role in CSF secretion (Millar and Brown, 2008). Thus, other transport proteins may contribute to anion efflux, e.g. HCO_3^- efflux may be mediated by NBCe2. Alternative routes for Cl^- efflux, however, remain to be identified.

B. How does K^+ Traverse the Basolateral Membrane of the Choroid Plexus Epithelium?

Zeuthen and Wright (1981) reported that approximately 90% of the K^+ pumped into bullfrog choroid plexus cells by the Na^+/K^+ ATPase leaks back out of the cell via K^+ channels in the apical membrane. The remaining 10% of K^+ was proposed to exit the cell via K^+ channels in the basolateral membrane. This minor fraction of K^+ efflux at the basolateral is important because it helps explain the net transport of K^+ across the choroid plexus epithelium from the CSF to the blood. A similar model is hypothesized for the mammalian choroid plexus, where Kir7.1, Kv1.3 and Kv1.6 K^+ channels, which are the major contributors to the whole-cell K^+ conductance, are all expressed in the apical membrane. K^+ channel proteins have not been identified in the basolateral membrane. Thus, an as yet uncharacterized K^+ channel or some other K^+ transporter must be expressed in the basolateral membrane.

C. By what Mechanism is Water Transported across the Basolateral Membrane of the Choroid Plexus?

The transport of water across epithelia, including that of the choroid plexus, is generally considered to occur as the result of osmosis in response to gradients created by ion transport. To facilitate the transcellular movements of water by osmosis most epithelia express aquaporin water channels in both their apical and basolateral membranes. Consistent with this hypothesis, the apical membrane of the choroid plexus epithelium contains high concentrations of aquaporin 1 (AQP1; Nielsen et al., 1993; Speake et al., 2003). Indeed, CSF secretion is reduced in AQP1 knockout mice (Oshio et al., 2005). However, aquaporin expression in the basolateral membrane is much less well defined. Small amounts of AQP1 are expressed in the basolateral membrane (Praetorius, 2007), but whether so few channels, compared to the apical membrane, are able to account for all the transport of water in CSF secretion is unknown. Other aquaporins are expressed in choroid plexus cells, i.e. AQP4 and AQP11, but neither of these proteins is located in the basolateral membrane (Speake et al., 2003; Praetorius, 2007). Thus, it is possible that other water-transporting proteins may be expressed in the basolateral membrane thereby contributing to the influx of water.

The answers to each of these questions remain to be determined experimentally, but here we hypothesize that some members of the SLC12 family of cation-chloride cotransporters could make a contribution to each of the above processes. For instance, a KCC could contribute to Cl^- efflux at the apical membrane and/or K^+ efflux at the basolateral membrane. Furthermore, as discussed in Chapter 28 of this volume and in Zeuthen and Hamann (2008), members of the cation-chloride cotransporter family (SLC12) may also transport water. The final two sections of this chapter review the available data on cation-chloride cotransporter expression in the choroid plexuses and discuss the evidence for their function in the choroid plexus and CSF secretion.

VII. THE EXPRESSION OF CATION-CHLORIDE COTRANSPORTERS (SLC12 FAMILY) IN CHOROID PLEXUS

Cation-chloride cotransporters have a number of potential roles in CSF secretion. However, the expression of these transporters in the choroid plexus and their role in CSF secretion has been controversial for many years (Javaheri, 1991). It is now known that several members of the cation-chloride cotransporter family are expressed in both the apical and basolateral membrane of mammalian choroid plexus (see Table 29.2). The data supporting the expression of cation-chloride cotransporters in choroid plexus are briefly reviewed below.

A. The Na^+-K^+-$2Cl^-$ Cotransporters

NKCC1 (SLC12A2) – Expression of NKCC1 has been demonstrated in the rat choroid plexus by *in situ* hybridization and Western blot analysis (Plotkin et al., 1997). Immunocytochemical studies found that the NKCC1 protein is expressed in the apical membrane of rat (Plotkin et al., 1997; Wu et al., 1998) and mouse (Piechotta et al., 2002) choroid plexus cells.

NKCC2 (SLC12A1) – NKCC2 is thought to be expressed exclusively in the kidney, so expression has not been specifically examined in the adult choroid plexus. However, *in situ* hybridization studies of NKCC2 expression during fetal development clearly show that NKCC2 is not expressed in the choroid plexus (Igarashi et al., 1995). This is in marked contrast to NKCC1, which is strongly expressed in the fetal choroid plexus (Li et al., 2002; Vanden Heuvel et al., 2006).

B. The Na^+-Cl^- Cotransporter

NCC (SLC12A3) – The thiazide-sensitive Na^+-Cl^- cotransporter (NCC) is expressed mainly in the distal convoluted tubule of the kidney, but is also found in some non-renal tissues, e.g. placenta, brain and small intestine (Chang et al., 1996). We have examined the expression NCC in rat choroid plexus by RT-PCR. Figure 29.3 shows that a 292 bp product is detected by PCR with mRNA isolated from rat kidney, but not with mRNA from rat choroid plexus.

C. The K^+-Cl^- Cotransporters

KCC1 (SLC12A4) – *In situ* hybridization studies have shown that mRNA encoding KCC1 is expressed in the choroid plexus of rats and mice (Kanaka et al., 2001; Li et al., 2002; Le Rouzic et al., 2006). To date protein expression has not been investigated so that it

TABLE 29.2 Molecular localization of cation-chloride cotransporters in rat and mouse choroid plexus

HUGO designated name	Common name	Species	mRNA, protein or both	Apical or basolateral	References
SLC12A1	NKCC2	M (f)	Not expressed	–	Igarashi et al. (1995)
		R		n.d.	–
SLC12A2	NKCC1	M	Protein	Apical	Piechotta et al. (2002)
		R	Both	Apical	Plotkin et al. (1997)
SLC12A3	NCC	M	n.d.	–	–
		R	Not expressed	–	Fig. 29.3
SLC12A4	KCC1	M (f)	mRNA	n.d.	Li et al. (2002)
		R	mRNA	n.d.	Kanaka et al. (2001); Li et al. (2002)
SLC12A5	KCC2	M (f)	Not expressed	–	Li et al. (2002)
		R	Not expressed	–	Le Rouzic et al. (2006)
SLC12A6	KCC3	M	Both	Basolateral	Pearson et al. (2001)
		R	mRNA	–	Le Rouzic et al. (2006)
SLC12A7	KCC4	M	Both	Apical	Li et al. (2002); Karadsheh et al. (2004)
		R	mRNA	–	Le Rouzic et al. (2006)

Data are for adult tissue except where stated (f = fetal). mRNA was detected by in situ hybridization or northern analysis except for NCC which was determined by RT-PCR. Protein expression was by Western analysis and immunocytochemistry. Apical localization of NKCC1 was first determined by functional methods (Keep et al., 1994). NKCC1 expression has recently been determined in the apical membrane of human choroid plexus (Praetorius and Nielsen, 2006). n.d. = not determined.

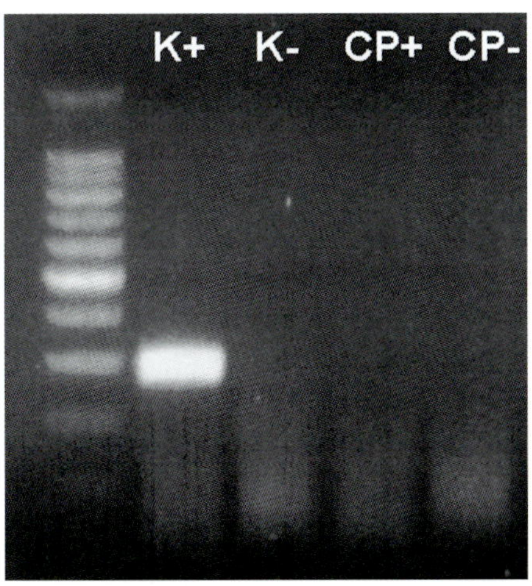

FIGURE 29.3 RT-PCR for NCC in rat kidney and choroid plexus. Primers based on the rat sequence (accession number NM_019345). The sense primer was CACCTTGTACCTACGCAC C and antisense TGCATCGGATCATCACTCCC. The primers yielded a product of 292bp from mRNA extracted from kidney but not choroid plexus.

is not yet known whether the transporter is expressed in the apical or basolateral membrane.

KCC2 (SLCA5) – KCC2 is expressed only in CNS neurons (see Chapters 5 and 22) and there is no evidence for the expression of KCC2 in rat choroid plexus epithelium, either by *in situ* hybridization (Kanaka et al., 2001) or by RT-PCR (Davies and Brown, unpublished data).

KCC3 (SLC12A6) – KCC3 expression in mouse choroid plexus has been determined by Northern blot, Western blot and immunofluorescence analyses. The protein is immunolocalized to the basolateral membrane (Pearson et al., 2001).

KCC4 (SLC12A7) – The expression of mRNA for KCC4 has been determined in rat choroid plexus by *in situ* hybridization (Li et al., 2002; Le Rouzic et al., 2006), and the KCC4 protein localized to the apical membrane of mouse choroid plexus by immunocytochemistry (Karadsheh et al., 2004).

VIII. FUNCTIONS OF CATION-CHLORIDE COTRANSPORTERS IN CHOROID PLEXUS EPITHELIAL CELLS

The previous section describes how molecular localization methods have resolved some of the confusion created by earlier studies of cation-chloride cotransporter expression in the choroid plexus. The actual functions of the four cotransporters identified in the choroid plexus have, however, proved more difficult to resolve. This is mainly because there is still a lack of specific inhibitors with which to differentiate between the cation-chloride cotransporters. The use in future experiments of molecular methodologies such as siRNA to transiently reduce protein expression, or gene knockout mice may help resolve this problem. To date, however, no impairments to central nervous system function, which could be attributed to major disturbances in CSF production, have been reported in mice lacking: NKCC1 (Delpire et al., 1999; Flagella et al., 1999; Sung et al., 2000); KCC1 (Rust et al., 2007); KCC3 (Howard et al., 2002; Boettger et al., 2003) or KCC4 (Boettger et al., 2002).

A. NKCC1 Activity in the Choroid Plexus

NKCC1 is known to play a major role in ion secretion in many epithelia (Russell, 2000). In these secretory epithelia the NKCC1 is located in the basolateral membrane where it mediates Cl^- influx driven by the inwardly directed Na^+ gradient. This "secondary active transport" results in the accumulation of Cl^- above electrochemical equilibrium, so that there is a driving force for Cl^- efflux via anion channels in the apical membrane. For many years investigators anticipated that Na^+-coupled Cl^- cotransporters (either NKCC1 or NCC) would have similar roles in the basolateral membrane of the choroid plexus (for review see Javaheri, 1991). The finding that NKCC1 is expressed in the apical membrane of the choroid plexus therefore required a complete reappraisal of the mechanism of CSF secretion.

In virtually all cells where it has been studied, NKCC1 mediates ion influx because of the dominance of the inward Na^+ and Cl^- chemical gradients against the outward K^+ chemical gradient. The suggestion that NKCC1 could mediate ion influx across the apical membrane of the choroid plexus, however, appears counter-intuitive, because both Na^+ and Cl^- will be transported in the "wrong direction" in terms of explaining Na^+ and Cl^- transport into the CSF. In addition, the transport of ions into the choroid plexus cells would create an osmotic gradient for H_2O absorption rather than secretion. There are, however, some experimental data to support the hypothesis that NKCC1 mediates ion influx. For instance, Plotkin et al. (1997), who first demonstrated the apical localization of NKCC1, also found that $^{86}Rb^+$ (a radioactive tracer for K^+) influx into rat choroid plexus cells in primary culture is inhibited by about 90% using the loop diuretic bumetanide at a concentration of 50 μM. A subsequent study by the same group found that 100 μM bumetanide causes a 10% decrease in cell volume, suggesting that ion influx via the cotransporter is required to maintain normal cell volume (Wu et al., 1998). These data are also supported by earlier studies which found that in isolated choroid plexus tissue 100 μM bumetanide inhibits $^{86}Rb^+$ influx by 25% and causes a 25% decrease in tissue water (an indirect measure of cell volume; Bairamian et al., 1991). One criticism of all these experiments, however, is that bumetanide was used at a high concentration, e.g. 100 μM bumetanide is known to have inhibitory effects on other transporters which are expressed in the choroid plexus, e.g. KCC and AE (Culliford et al., 2003). To examine the possibility that 100 μM bumetanide may be acting non-specifically we have repeated some of these experiments using 10 μM bumetanide (i.e. a concentration which completely inhibits NKCC1 activity in most cells; Javaheri, 1991; Russell, 2000).

Figure 29.4 shows data in which isolated choroid plexus cells were superfused with HCO_3^--buffered, isotonic solutions for 12 minutes. For the 10 minute period indicated by the bar either 10 μM or 100 μM bumetanide was included in the superfusate. The data indicate that 10 μM bumetanide had little effect on cell volume, whereas 100 μM bumetanide caused a 10 to 20% decrease in cell volume similar to that observed by Wu et al. (1998). In other experiments, we have measured volume changes in cells superfused with hypertonic solutions. In control conditions this maneuver caused an immediate decrease in cell volume

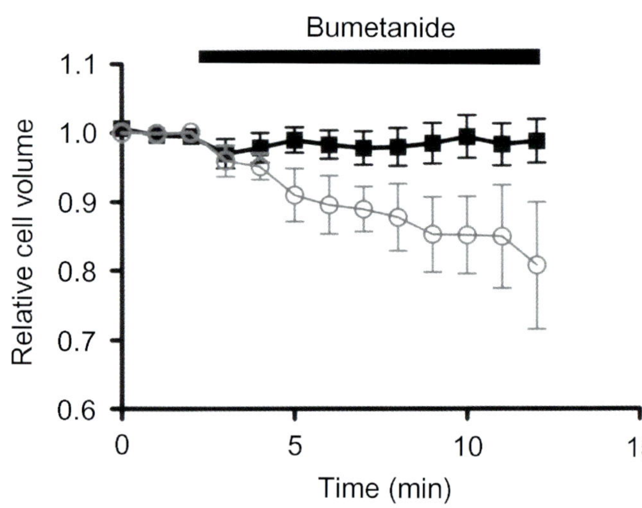

FIGURE 29.4 100 μM bumetanide but not 10 μM bumetanide causes a decrease in choroid plexus cell volume. Experiments were performed on cells isolated by dispase digestion and cell volume was measured by video imaging. The cells were superfused with an HCO_3^--buffered ringer (120 mM NaCl; 25 mM Na HCO_3; 5 mM KCl; 1 mM $CaCl_2$; 1 mM $MgCl_2$; 15 mM mannitol and 5 mM glucose, pH 7.4). Data a mean±s.e.m. of 7 (10 μM bumetanide) and 3 (100 μM) experiments.

followed by a slower recovery of volume (a regulatory volume increase) due to the transport of ions and water into the cells. Neither 10 μM nor 100 μM bumetanide had a significant effect on the regulatory volume increase; further suggesting that NKCC1 does not contribute to volume maintenance in choroid plexus cells. At present, we do not know why 100 μM bumetanide causes a decrease in cell volume, but preliminary data indicate that the cell volume decrease is only observed in CO_2/HCO_3^--buffered solutions suggesting the involvement of an HCO_3^--dependent process. Future experiments using a range of different transport inhibitors will examine this possibility in more detail.

If NKCC1 does not contribute to ion influx, as suggested above, does it mediate ion efflux? Keep et al. (1994) first suggested this possibility based mainly on calculations made using measured concentrations of intracellular ions. Such calculations, however, are very dependent on the accuracy of the ion concentrations measurements, and values in the literature vary substantially for all three of the transported ions. Experimental evidence for the hypothesis shows that relatively modest concentrations of bumetanide (10 or 30 μM) cause a 75% inhibition of K^+ efflux from rat choroid plexus (Keep et al., 1994). However, there are few other data to support this hypothesis. Indeed in our experiments, 10 μM bumetanide did not cause cell swelling when added to isotonic solutions which might be expected in NKCC1 mediating ion efflux. Thus, in conclusion the currently available data suggest that NKCC1 does not make a major contribution to ion efflux or influx at the apical membrane of the choroid plexus, and that NKCC1 is unlikely to be an additional route for Cl^- efflux in the process of CSF secretion.

An alternative explanation for the role of NKCC1 in the choroid plexus is that NKCC1 makes a vital contribution to K^+ transport and hence the regulation of $[K^+]$ in the CSF. Husted and Reed (1976) demonstrated the capacity of the choroid plexus to regulate K^+ transport by showing that $[K^+]$ in the CSF is carefully maintained in animals experimentally made hypokalemic and hyperkalemic. The most effective methods of transporting K^+ into cells is the Na^+/K^+ ATPase, and it has generally been thought to mediate K^+ influx in the choroid plexus (Zeuthen and Wright, 1981; Speake et al., 2004). The Na^+/K^+ ATPase, however, has quite a high affinity for K^+ (K_m of approximately 1 mM; Crambert et al., 2000), so is probably operating close to a maximum rate at normal K^+ concentrations (3.0 mM; Table 29.1). By contrast, NKCC1 has a higher K_m for K^+ of ~1.6–2.5 mM (Payne et al., 1995), and therefore it has the potential to mediate increased K^+ absorption if CSF $[K^+]$ were to increase above this value. Furthermore, if the driving forces for Na^+, K^+ and Cl^- transport are at equilibrium (as suggested by our experimental data), then even a small increase in CSF $[K^+]$ will alter the driving force to favor ion influx. An additional advantage of NKCC1 mediating K^+ influx is that K^+ transport will not be coupled to Na^+ secretion as is the case for Na^+/K^+ ATPase.

The hypothesis that NKCC1 mediates K^+ influx is supported by data showing that a large increase in extracellular K^+ from 3 to 25 mM causes a 40% increase in choroid plexus cell volume (Wu et al., 1998). The fact that this volume increase is inhibited by bumetanide (albeit at 100 μM; Wu et al., 1998) suggests the involvement of NKCC1 in ion influx, rather than increased Na^+/K^+ ATPase activity or decreased efflux via K^+ channels. These observations certainly merit further investigation preferably using confluent monolayers of choroid plexus cells in which solution changes and drug additions can be made to either the basolateral or apical membrane. Such model systems have been developed over the last decade (Haselbach et al., 2001), but as yet have not been used in systematic studies of inorganic ion transport across the choroid plexus epithelium.

B. KCC Function in the Choroid Plexus

KCC transporters have several potential roles in CSF secretion, e.g. Cl^- (and K^+) efflux at the apical membrane; K^+ efflux at the basolateral membrane and even the transport of H_2O at the basolateral membrane. However, just as with NKCC1, there are few definitive data on KCC function in the choroid plexus. This is primarily because of the lack of specific blockers for KCC. Furosemide has frequently been used in the past as an inhibitor of KCC, but it also affects the activities of NKCC1, AE and carbonic anhydrase (Vogh and Langham, 1981; Culliford et al., 2003). Dihydroindenyloxy alkanoic acid (DIOA) has been suggested to be a selective inhibitor of KCC (Garay et al., 1988), but it also has effects on a number of anion transport processes, including AE and NKCC1 (Culliford et al., 2003).

In vivo experiments have found that furosemide increases the K^+ content of rat choroid plexus epithelial cells (suggesting inhibition of K^+ efflux), while also reducing the rate of Cl^- transport from the blood to the CSF (Johanson et al., 1992). Similar data were also generated in microdialysis experiments, where furosemide produced a 40% reduction of Cl^- transport into the CSF (Johanson et al., 1994). These data are indicative of KCC having a role in ion efflux at the apical membrane. However, the carbonic anhydrase inhibitor acetazolamide also causes a 55% reduction of Cl^- transport (Johanson and Preston, 1994). It is therefore difficult to draw any firm conclusions from the furosemide data because it inhibits carbonic anhydrase in addition to KCC (Javaheri, 1991).

Contradictory evidence of KCC function has also been obtained from *in vitro* experiments. A key finding for KCC activity was that the efflux of $^{86}Rb^+$ from rat choroid plexus was inhibited by more than 50% when Cl^- in the artificial CSF bathing the tissue was replaced by either NO_3^- or SCN^- (Johanson and Preston, 1994). This observation suggests that the presence of Cl^- is obligatory for K^+ influx. Johanson and Preston (1994) also observed that Rb^+ efflux increases by about 10% in the presence of 0.5 mM N-ethylmaleimide (NEM), a thiol-reactive reagent known to activate KCC (Lauf et al., 1984). In this study, however, the NEM-induced increment of K^+ efflux was also observed in the absence of Cl^- (NO_3^- substitution) so that the effect of NEM may be on a protein other than KCC (Johanson and Preston, 1994). Furthermore, Keep et al. (1994) were unable to identify DIOA-sensitive components to K^+ influx or efflux in rat choroid plexus epithelial cells. Therefore, it is not clear how much K^+ efflux from the choroid plexus is mediated by KCC.

Better functional evidence for coupled efflux of K^+ and Cl^- through the apical membrane comes from experiments using ion-sensitive microelectrodes to study the choroid plexus isolated from the amphibian *Necturus maculosus* (Zeuthen, 1994). Coupled transport of K^+ and Cl was observed as interdependent, parallel changes in ion concentrations. These changes were inhibited by 0.1 mM furosemide. Furthermore this study provided some of the first evidence that K^+ and Cl^- transport may also be coupled to the movement of H_2O across the membrane (Zeuthen, 1994). These data are discussed in more detail by Thomas Zeuthen and collaborators in Chapter 28 in this volume. Whether the transport of K^+, Cl^- and H_2O is mediated by an amphibian isoform of KCC1 or KCC4 remains to be determined. If Zeuthen and his colleagues are correct these transporters may be contributing to ion and water transport at the apical membrane of the mammalian choroid plexus. The KCC3 isoform in the basolateral membrane of the mammalian choroid plexus, however, is unlikely to contribute to H_2O influx at this membrane, because the prevailing gradients strongly favor the efflux of K^+ and Cl^-.

IX. CONCLUSIONS

Much progress has been made over the last two decades in identifying ion transport protein in the mammalian choroid plexus, and in determining whether these proteins are expressed in the apical or basolateral membrane of the epithelium. However, the function of many of these transporters, particularly those of the cation-chloride cotransporter family, remains to be determined. Progress has been impaired by the lack of specific transport inhibitors and important questions remain about: the mechanisms of Cl^- and K^+ efflux at the apical and basolateral membranes, respectively, and the influx of H_2O at the basolateral membrane. The use of tools such as siRNA, gene knockout animals and monolayers of choroid plexus cells in primary culture should help provide answers in the near future.

References

Alper, S., Stuart-Tilley, A., Simmons, C., Brown, D., and Drenckhahn, D. (1994). The fodrin-ankyrin cytoskeleton of choroid plexus preferentially colocalizes with apical Na^+K^+-ATPase rather than with basolateral anion exchanger AE2. *J. Clin. Invest* **93**, 1430–1438.

Bairamian, D., Johanson, C.E., Parmelee, J.T., and Epstein, M.H. (1991). Potassium cotransport with sodium and chloride in the choroid plexus. *J. Neurochem.* **56**, 1623–1629.

Boettger, T., Hübner, C.A., Maier, H., Rust, B.M., Beck, F.X., and Jentsch, T.J. (2002). Deafness and renal tubular acidosis in mice lacking the K-Cl co-transporter KCC4. *Nature* **416**, 874–878.

Boettger, T., Rust, M.B., Maier, H., Seidenbecher, T., Schweizer, M., Keating, D.J., Faulhaber, J., Ehmke, H., Pfeffer, C., Scheel, O., Lemcke, B., Horst, J., Leuwer, R., Pape, H.C., Völkl, H., Hübner, C.A., and Jentsch, T.J. (2003). Loss of K-Cl co-transporter KCC3 causes deafness, neurodegeneration and reduced seizure threshold. *EMBO J.* **22**, 5422–5434.

Boron, W.F. and Boulpaep, E.L. (2009). *Medical Physiology*, Second Edition. Elsevier Inc, Saunders.

Bouzinova, E.V., Praetorius, J., Virkki, L.V., Nielsen, S., Boron, W.F., and Aalkjaer, C. (2005). Na^+-dependent HCO_3^- uptake into the rat choroid plexus epithelium is partially DIDS sensitive. *Am. J. Physiol.* **289**, C1448–C1456.

Brown, P.D., Davies, S.L., Speake, T., and Millar, I.D. (2004). Molecular mechanisms of cerebrospinal fluid production. *Neuroscience* **129**, 957–970.

Brown, P.D. and Garner, C. (1993). Cerebrospinal fluid secretion: the transport of fluid and electrolytes by the choroid plexus. In *The Subcommissural Organ* (Oksche, A., Rodriguez, E., and Fernandez-Llebrez, P., eds), pp. 233–242. Springer International, Heidelberg.

Chang, H., Tashiro, K., Hirai, M., Ikeda, K., Kurokawa, K., and Fujita, T. (1996). Identification of a cDNA encoding a thiazide-sensitive sodium-chloride cotransporter from the human and its mRNA expression in various tissues. *Biochem. Biophys. Res. Comm.* **223**, 324–328.

Chen, L.M., Kelly, M.L., Parker, M.D., Bouyer, P., Gill, H.S., Felie, J. M., Davis, B.A., and Boron, W.F. (2008). Expression and localization of Na-driven Cl-HCO_3 exchanger (SLC4A8) in rodent CNS. *Neuroscience* **153**, 162–174.

Christensen, O., Simon, M., and Randlev, T. (1989). Anion channels in a leaky epithelium A patch-clamp study of choroid plexus. *Pflugers Arch.* **415**, 37–46.

Crambert, G., Hasler, U., Beggah, A.T., Yu, C., Modyanov, N.N., Horisberger, J.D., Lelièvre, L., and Geering, K. (2000). Transport and pharmacological properties of nine different human Na K-ATPase isozymes. *J. Biol. Chem.* **275**, 1786–1976.

Culliford, S., Ellory, C., Lang, H.J., Englert, H., Staines, H., and Wilkins, R. (2003). Specificity of classical and putative Cl^- transport inhibitors on membrane transport pathways in human erythrocytes. *Cell Physiol. Biochem.* **13**, 181–188.

Damkier, H.H., Nielsen, S., and Praetorius, J. (2007). Molecular expression of SLC4-derived Na^+-dependent anion transporters in selected human tissues. *Am. J. Physiol.* **293**, R2136–R2146.

Davson, H. and Segal, M.B. (1970). The effects of some inhibitors and accelerators of sodium transport on the turnover of ^{22}Na in the cerebrospinal fluid and the brain. *J. Physiol.* **209**, 139–153.

Delpire, E., Lu, J., England, R., Dull, C., and Thorne, T. (1999). Deafness and imbalance associated with inactivation of the secretory Na-K-2Cl co-transporter. *Nat. Genet.* **22**, 192–195.

Deng, Q.S. and Johanson, C.E. (1989). Stilbenes inhibit exchange of chloride between blood, choroid plexus and the cerebrospinal fluid. *Brain Res.* **510**, 183–187.

Döring, F., Derst, C., Wischmeyer, E., Karschin, C., Schneggenburger, R., Daut, J., and Karschin, A. (1998). The epithelial inward rectifier channel Kir 7.1 displays unusual K^+ permeation properties. *J. Neurosci.* **18**, 8625–8636.

Ernst, S.A., Palacios, J.R., and Siegel, G.J. (1986). Immunocytochemical localization of Na^+, K^+-ATPase catalytic polypeptide in mouse choroid plexus. *J. Histochem. Cytochem.* **34**, 189–195.

Flagella, M., Clarke, L.L., Miller, M.L., Erway, L.C., Giannella, R.A., Andringa, A., Gawenis, L.R., Kramer, , Duffy, J.J., Doetschman, T., Lorenz, J.N., Yamoah, E.N., Cardell, E.L., and Shull, G.E. (1999). Mice lacking the basolateral Na-K-2Cl cotransporter have impaired epithelial chloride secretion and are profoundly deaf. *J. Biol. Chem.* **274**, 26946–26955.

Garay, R.P., Nazaret, C., Hannaert, P.A., and Cragoe, E.J. (1988). Demonstration of a K^+,Cl^--cotransport system in human red cells by its sensitivity to (dihydroindenyl)oxyalkanoic acids: regulation of cell swelling and distinction from the bumetanide-sensitive Na^+,K^+,Cl^--cotransport system. *Mol. Pharmacol.* **33**, 696–701.

Garner, C. and Brown, P.D. (1992). Two types of chloride channel in the apical membrane of rat choroid plexus epithelial cells. *Brain Res.* **591**, 137–145.

Haselbach, M., Wegener, J., Decker, S., Engelbertz, C., and Galla, H. J. (2001). Porcine choroid plexus epithelial cells in culture: regulation of barrier properties and transport processes. *Microsc. Res. Tech.* **52**, 137–152.

Howard, H.C., Mount, D.B., Rochefort, D., Byun, N., Dupré, N., Lu, J., Fan, X., Song, L., Rivière, J.-B., Prévost, C., Welch, R., England, R., Zhan, F.Q., Mercado, A., Siesser, W.B., George, A.L., Horst, J., Simonati, A., McDonald, M.P., Bouchard, J.-P., Mathieu, J., Delpire, E., and Rouleau, G.A. (2002). Mutations in the K-Cl cotransporter KCC3 cause a severe peripheral neuropathy associated with agenesis of the corpus callosum. *Nat. Genet.* **32**, 384–392.

Husted, R.F. and Reed, D.J. (1976). Regulation of cerebrospinal fluid potassium by the cat choroid plexus. *J. Physiol.* **259**, 213–221.

Husted, R.F. and Reed, D.J. (1977). Regulation of cerebrospinal fluid bicarbonate by the cat choroid plexus. *J. Physiol.* **267**, 411–428.

Igarashi, P., Vanden Heuvel, G.B., Payne, J.A., and Forbush, B. (1995). Cloning, embryonic expression, and alternative splicing of a murine kidney-specific Na-K-Cl cotransporter. *Am. J. Physiol.* **269**, F405–F418.

Irani, D.N. (2009). Properties and composition of normal cerebrospinal fluid Chapter 10. In *Cerebrospinal Fluid in Clinical Practice* (Irani, D.N., ed.), pp. 69–89. Saunders, Elsevier Inc.

Jacobs, S., Ruusuvuori, E., Sipilä, S.T., Haapanen, A., Damkier, H.H., Kurth, I., Hentschke, M., Schweizer, M., Rudhard, Y., Laatikainen, L.M., Tyynelä, J., Praetorius, J., Voipio, J., and Hübner, C.A. (2008). Mice with targeted Slc4a10 gene disruption have small brain ventricles and show reduced neuronal excitability. *Proc. Natl. Acad. Sci. USA* **105**, 311–316.

Javaheri, S. (1991). Role of NaCl cotransport in cerebrospinal fluid production: effects of loop diuretics. *J. Appl. Physiol.* **71**, 795–800.

Jentsch, T.J., Stein, V., Weinreich, F., and Zdebik, A.A. (2002). Molecular structure and physiological function of chloride channels. *Physiol. Rev.* **82**, 503–568.

Johanson, C., Palm, D., Dyas, M., and Knuckey, N. (1994). Microdialysis analysis of effects of loop diuretics and acetazolamide on chloride transport from blood to CSF. *Brain Res.* **641**, 121–126.

Johanson, C. and Preston, J. (1994). Potassium efflux from infant and adult rat choroid plexuses: effects of CSF anion substitution N-ethylmaleimide and Cl transport inhibitors. *Neurosci. Lett.* **169**, 207–211.

Johanson, C.E. (1999). Choroid plexus. In *Encyclopedia of Neuroscience* (Adelman, G. and Smith, B.H., eds), pp. 384–387. Elsevier Science, New York.

Johanson, C.E., Murphy, V.A., and Dyas, M. (1992). Ethacrynic acid and furosemide alter Cl,K, and Na distribution between blood, choroid plexus, CSF, and brain. *Neurochem. Res.* **17**, 1079–1085.

Kajita, H., Omori, K., and Matsuda, H. (2000). The chloride channel ClC-2 contributes to the inwardly rectifying Cl^- conductance in cultured porcine choroid plexus epithelial cells. *J. Physiol.* **523**, 313–324.

Kalaria, R.N., Premkumar, D.R., Lin, C.W., Kroon, S.N., Bae, J. Y., Sayre, L.M., and LaManna, J.C. (1998). Identification and expression of the Na^+/H^+ exchanger in mammalian cerebrovascular and choroidal tissues: characterization by amiloride-sensitive [^{3}H]MIA binding and RT-PCR analysis. *Brain Res. Mol. Brain Res.* **58**, 178–187.

Kanaka, C., Ohno, K., Okabe, A., Kuriyama, K., Itoh, T., Fukuda, A., and Sato, K. (2001). The differential expression patterns of messenger RNAs encoding K-Cl cotransporters (KCC1,2) and Na-K-2Cl cotransporter (NKCC1) in the rat nervous system. *Neuroscience* **104**, 933–946.

Karadsheh, M.F., Byun, N., Mount, D.B., and Delpire, E. (2004). Localization of the KCC4 potassium-chloride cotransporter in the nervous system. *Neuroscience* **123**, 381–391.

Kattzman, R. and Pappius, H.M. (1973). *Brain Electrolytes and Fluid Metabolism*. The Williams and Wilkins Co, Baltimore.

Keep, R., Cawkwell, R., and Jones, H. (1987). Choroid plexus structure and function in young rats on a high-potassium diet. *Brain Res.* **413**, 45–52.

Keep, R.F., Xiang, J., and Betz, A.L. (1994). Potassium cotransport at the rat choroid plexus. *Am. J. Physiol.* **267**, C1616–C1622.

Kibble, J.D., Garner, C., Kajita, H., Colledge, W.H., Evans, M.J., Radcliff, R., and Brown, P.D. (1997). Whole-cell Cl^- conductances in mouse choroid plexus epithelial cells do not require CFTR expression. *Am. J. Physiol.* **272**, C1899–C1907.

Kibble, J.D., Tresize, A.O., and Brown, P.D. (1996). Properties of the cAMP-activated Cl^- conductance in choroid plexus epithelial cells isolated from the rat. *J. Physiol.* **496**, 69–80.

Klarr, S., Ulanski, L.J., Stummer, W., Xiang, J., Betz, A.L., and Keep, R.F. (1997). The effects of hypo- and hyperkalemia on choroid plexus potassium transport. *Brain Res.* **758**, 39–44.

Kotera, T. and Brown, P.D. (1994). Two types of potassium current in rat choroid plexus epithelial cells. *Pflügers Archiv.* **237**.

Lauf, P.K., Adragna, N.C., and Garay, R.P. (1984). Activation by N-ethylmaleimide of a latent K^+-Cl^- flux in human red blood cells. *Am. J. Physiol.* **246**, C385–C390.

Le Rouzic, P., Ivanov, T.R., Stanley, P.J., Baudoin, F.M., Chan, F., Pinteaux, E., Brown, P.D., and Luckman, S.M. (2006). KCC3 and KCC4 expression in rat adult forebrain. *Brain Res.* **1110**, 39–45.

Li, H., Tornberg, J., Kaila, K., Airaksinen, M.S., and Rivera, C. (2002). Patterns of cation-chloride cotransporter expression during embryonic rodent CNS development. *Eur. J. Neurosci.* **16**, 2358–2370.

Lindsey, A.E., Schneider, K., Simmons, D.M., Baron, R., Lee, B.S., and Kopito, R.R. (1990). Functional expression and subcellular localization of an anion exchanger cloned from choroid plexus. *Proc. Natl. Acad. Sci. USA* **87**, 5278–5282.

Maren, T.H. and Broder, L.E. (1970). The role of carbonic anhydrase in anion secretion into cerebrospinal fluid. *J. Pharmacol. Exp. Ther.* **172**, 197–202.

Maren, T.H., Conroy, C.W., Wynns, G.C., and Godman, D.R. (1997). Renal and cerebrospinal fluid formation pharmacology of a high molecular weight carbonic anhydrase inhibitor. *J. Pharmacol. Exp. Ther.* **280**, 98–104.

Masuzawa, T., Hasegawa, T., Nakahara, N., Iida, K., and Sato, F. (1984a). Localization of carbonic anhydrase in rat choroid plexus epithelial cells. *Ann. NY Acad. Sci.* **429**, 405–407.

Masuzawa, T., Ohta, T., Kawamura, M., Nakahara, N., and Sato, F. (1984b). Immunohistochemical localization of Na^+, K^+-ATPase in the choroid plexus. *Brain Res.* **302**, 357–362.

Masuzawa, T. and Sato, F. (1983). The enzyme histochemistry of the choroid plexus. *Brain Res.* **106**, 55–99.

Mayer, S.E. and Sanders-Bush, E. (1993). Sodium-dependent antiporters in choroid plexus epithelial cultures from rabbit. *J. Neurochem.* **60**, 1308–1316.

Millar, I.D. and Brown, P.D. (2008). NBCe2 exhibits a 3 HCO_3^-:1 Na^+ stoichiometry in mouse choroid plexus epithelial cells. *Biochem. Biophys. Res. Comm.* **373**, 550–554.

Murphy, V., Smith, Q., and Rapoport, S. (1986). Homeostasis of brain and cerebrospinal fluid calcium concentrations during chronic hypo- and hypercalcemia. *J. Neurochem.* **47**, 1735–1741.

Murphy, V.A. and Johanson, C.E. (1989). Alteration of sodium transport by the choroid plexus with amiloride. *Biochim. Biophys. Acta* **979**, 187–192.

Murphy, V.A. and Johanson, C.E. (1990). Na^+-H^+ exchange in choroid plexus and CSF in acute metabolic acidosis or alkalosis. *Am. J. Physiol.* **258**, F1528–F1537.

Nakamura, N., Suzuki, Y., Sakuta, H., Ookata, K., Kawahara, K., and Hirose, S. (1999). Inwardly rectifying K^+ channel Kir7.1 is highly expressed in thyroid follicular cells, intestinal epithelial cells and choroid plexus epithelial cells: implication for a functional coupling with Na^+,K^+-ATPase. *Biochem. J.* **342**, 329–336.

Nielsen, S., Smith, B.L., Christensen, E.I., and Agre, P. (1993). Distribution of the aquaporin CHIP in secretory and resorptive epithelia and capillary endothelia. *Proc. Natl. Acad. Sci. USA* **90**, 7275–7279.

Nogradi, A., Kelly, C., and Carter, N.D. (1993). Localization of acetazolamide-resistant carbonic anhydrase III in human and rat choroid plexus by immunocytochemistry and in situ hybridization. *Neurosci. Lett.* **151**, 162–165.

Oshio, K., Watanabe, H., Song, Y., Verkman, A.S., and Manley, G. T. (2005). Reduced cerebrospinal fluid production and intracranial pressure in mice lacking choroid plexus water channel Aquaporin-1. *FASEB J.* **19**, 76–78.

Parker, M.D., Musa-Aziz, R., Rojas, J.D., Choi, I., Daly, C.M., and Boron, W.F. (2008). Characterization of human SLC4A10 as an electroneutral Na/HCO_3 cotransporter (NBCn2) with Cl^- self-exchange activity. *J. Biol. Chem.* **283**, 12777–12788.

Payne, J.A., Xu, J.C., Haas, M., Lytle, C.Y., Ward, D., and Forbush, B. (1995). Primary structure, functional expression, and chromosomal localization of the bumetanide-sensitive Na-K-Cl cotransporter in human colon. *J. Biol. Chem.* **270**, 17977–17985.

Pearson, M., Lu, J., Mount, D.B., and Delpire, E. (2001). Localization of the K^+-Cl^- cotransporter, KCC3, in the central and peripheral nervous systems: expression in the choroid plexus, large neurons and white matter tracts. *Neuroscience* **103**, 481–491.

Piechotta, K., Lu, J., and Delpire, E. (2002). Cation chloride cotransporters interact with the stress-related kinases Ste20-related proline-alanine-rich kinase (SPAK) and oxidative stress response 1 (OSR1). *J. Biol. Chem.* **277**, 50812–50819.

Plotkin, M.D., Kaplan, M.R., Peterson, L.N., Gullans, S.R., Hebert, S.C., and Delpire, E. (1997). Expression of the Na^+-K^+-$2Cl^-$ cotransporter BSC2 in the nervous system. *Am. J. Physiol.* **272**, C173–C183.

Pollay, M., Hisey, B., Reynolds, E., Tomkins, P., Stevens, A., and Smith, R. (1985). Choroid plexus Na^+/K^+-activated adenosine triphophatase and cerebrospinal fluid secretion. *Neurosurgery* **17**, 768–772.

Praetorius, J. (2007). Water and solute secretion by the choroid plexus. *Pflugers Arch.* **454**, 1–18.

Praetorius, J., Nejsum, L.N., and Nielsen, S. (2004). A SCL4A10 gene product maps selectively to the basolateral membrane of choroid plexus epithelial cells. *Am. J. Physiol.* **286**, C601–C610.

Praetorius, J. and Nielsen, S. (2006). Distribution of sodium transporters and aquaporin-1 in the human choroid plexus. *Am. J. Physiol. Cell Physiol.* **291**, C59–C67.

Pushkin, A. and Kurtz, I. (2006). SLC4 base (HCO_3^-, CO_3^{2-}) transporters: classification, function, structure, genetic diseases, and knockout models. *Am J. Physiol. Renal Physiol.* **290**, F580–F599.

Quinton, P.M., Wright, E.M., and Tormey, J.M. (1973). Localization of sodium pumps in the choroid plexus epithelium. *J. Cell Biol.* **58**, 724–730.

Romero, M.F., Fulton, C.M., and Boron, W.F. (2004). The SLC4 family of HCO_3^- transporters. *Pflugers Arch.* **447**, 495–509.

Romero, M.F., Hediger, M.A., Boulpaep, E.L., and Boron, W.F. (1997). Expression cloning and characterization of a renal electrogenic Na^+/HCO_3^- cotransporter. *Nature* **387**, 409–413.

Russell, J.M. (2000). Sodium-potassium-chloride cotransport. *Physiol. Rev.* **80**, 211–276.

Rust, M.B., Alper, S.L., Rudhard, Y., Shmukler, B.E., Vicente, R., Brugnara, C., Trudel, M., Jentsch, T.J., and Hübner, C.A. (2007). Disruption of erythroid K-Cl cotransporters alters erythrocyte volume and partially rescues erythrocyte dehydration in SAD mice. *J. Clin. Invest* **117**, 1708–1717.

Saito, Y. and Wright, E.M. (1983). Bicarbonate transport across the frog choroid plexus and its control by cyclic nucleotides. *J. Physiol.* **336**, 635–648.

Saito, Y. and Wright, E.M. (1984). Regulation of bicarbonate transport across the brush border membrane of the bull-frog choroid plexus. *J. Physiol.* **350**, 327–342.

Sassani, P., Pushkin, A., Gross, E., Gomer, A., Abuladze, N., Dukkipati, R., Carpenito, G., and Kurtz, I. (2002). Functional characterization of NBC4: a new electrogenic sodium-bicarbonate cotransporter. *Am. J. Physiol.* **282**, C408–C416.

Segal, M.B. (1993). Extracellular and cerebrospinal fluid. *J. Inherit. Metabol. Dis.* **16**, 617–638.

Smith, Q. and Johanson, C. (1985). Active transport of chloride by lateral ventricle choroid plexus of the rat. *Am. J. Physiol.* **249**, F470–F477.

Speake, T., Freeman, L.J., and Brown, P.D. (2003). Expression of aquaporin 1 and aquaporin 4 water channels in rat choroid plexus. *Biochim. Biophys. Acta* **1609**, 80–86.

Speake, T., Kibble, J.D., and Brown, P.D. (2004). Kv1.1 and Kv1.3 channels contribute to the delayed-rectifying K^+ conductance in rat choroid plexus epithelial cells. *Am. J. Physiol.* **286**, C620–C622.

Speake, T., Whitwell, C., Kajita, H., and Brown, P.D. (2001). Mechanism of CSF secretion by the choroid plexus. *Micros. Res. Tech.* **52**, 49–59.

Sung, K.W., Kirby, M., McDonald, M.P., Lovinger, D.M., and Delpire, E. (2000). Abnormal $GABA_A$ receptor-mediated currents in dorsal root ganglion neurons isolated from Na-K-2Cl cotransporter null mice. *J. Neurosci.* **20**, 7531–7538.

Vanden Heuvel, G.B., Payne, J.A., Igarashi, P., and Forbush, B. (2006). Expression of the basolateral Na-K-Cl cotransporter during mouse nephrogenesis and embryonic development. *Gene Expr. Patterns* **6**, 1000–1006.

Virkki, L.V., Wilson, D.A., Vaughan-Jones, R.D., and Boron, W.F. (2002). Functional characterization of human NBC4 as an electrogenic Na^+-HCO_3^- cotransporter (NBCe2). *Am. J. Physiol.* **282**, C1278–C1289.

Vogh, B., Godman, D., and Maren, T. (1987). The effect of $AlCl_3$ and other acids on cerebrospinal fluid production: a correction. *J. Pharmacol. Exp. Ther.* **243**, 35–39.

Vogh, B.P. and Langham, M.R. (1981). The effect of furosemide and bumetanide on cerebrospinal fluid formation. *Brain Res.* **221**, 171–183.

Wang, C.Z., Yano, H., Nagashima, K., and Seino, S. (2000). The Na^+-driven Cl^-/HCO_3^- exchanger. Cloning, tissue distribution, and functional characterization. *J. Biol. Chem.* **275**, 35486–35490.

Watts, A.G., Sanchez-Watts, G., Emanuel, J.R., and Levenson, R. (1991). Cell-specific expression of mRNAs encoding Na^+, K^+-ATPase alpha- and beta-subunit isoforms within the rat central nervous system. *Proc, Natl, Acad, Sci, USA* **88**, 7425–7429.

Welch, K. and Sadler, K. (1965). Electrical potentials of choroid plexus of the rabbit. *J. Neurosurg.* **22**, 344–351.

Wright, E.M. (1978). Transport processes in the formation of the cerebrospinal fluid. *Rev. Physiol. Pharmacol.* **83**, 1–34.

Wu, Q., Delpire, E., Hebert, S.C., and Strange, K. (1998). Functional demonstration of Na^+-K^+-$2Cl^-$ cotransporter activity in isolated, polarized choroid plexus cells. *Am. J. Physiol* **275**, C1565–C1572.

Zeuthen, T. (1994). Cotransport of K^+, Cl^- and H_2O by membrane proteins from choroid plexus epithelium of *Necturus maculosus*. *J. Physiol.* **478**, 203–219.

Zeuthen, T. and Wright, E.M. (1981). Epithelial potassium transport: tracer and electrophysiological studies in choroid plexus. *J. Membrane Biol.* **60**, 105–128.

Zlokovic, B.V., Makic, J.B., Wang, L., McComb, J.G., and McDonough, A. (1993). Differential expression of Na,K-ATPase α and β subunit isoforms at the blood–brain barrier and the choroid plexus. *J. Biol. Chem.* **268**, 8019–8025.

CHAPTER

30

Ion and Water Transport across the Blood–Brain Barrier

Martha E. O'Donnell

OUTLINE

I. INTRODUCTION

This chapter will provide an overview of blood–brain barrier (BBB) structure and function with an emphasis on Cl^- transporters and channels and their role in health and disease. The field of blood–brain barrier research has grown somewhat more slowly than that of neuroscience. However, we now know that the BBB not only plays a central role in maintaining normal brain function but that it also participates in a range of neurological diseases. This, together with the fact that BBB endothelial cells are more readily accessible than parenchymal cells of the brain for targeted therapies, makes it imperative that we work to understand cellular processes underlying BBB function. Among these processes, BBB regulation of water and electrolyte movement into and out of the brain is of paramount importance. Thus, while many studies have focused on the role of BBB transporters in nutrient and drug movement across the barrier as well as the role of the barrier in inflammation, this chapter will focus solely on BBB ion transporters and channels. For other aspects of BBB function, the reader is referred to the many excellent reviews that are available.

A. Structure of the Blood–Brain Barrier

The blood–brain barrier is comprised of brain microvessel endothelial cells. It is a highly specialized endothelium that provides an important anatomic and functional barrier between the blood and the brain (Betz, 1986; Bradbury, 1984; Goldstein and Betz, 1986). The BBB has unique features that set it apart from endothelial cells of other vascular beds. This includes the presence of complex tight junctions that create a barrier with very limited paracellular solute flux and a high electrical resistance between blood and brain on the order of 2000 ohm · cm^2, making this barrier comparable to tight epithelia such as those found in the distal kidney and colon. Like other tight epithelial barriers, BBB endothelial cells have a polarized distribution of membrane transporter and channel proteins between the apical and basolateral plasma membrane. The barrier also has very few pinocytotic vesicles, such that there is little vesicular traffic that could move solutes between blood and brain. This means that the vast majority of solutes must move across the BBB via a transendothelial rather than a paracellular route and thus it is BBB transporters that determine what moves between blood and brain.

While microvessel endothelial cells alone comprise the BBB, the abluminal surface of these cells is ensheathed by foot processes of perivascular astrocytes with only the endothelial basal lamina separating the two cell types (Fig. 30.1). The exception to this is the occasional presence of pericytes found within the basal lamina of the BBB (Goldstein and Betz, 1986). Numerous studies have demonstrated that astrocytes and BBB endothelial cells have a close functional as well as anatomic association. For example, astrocytes secrete factors that induce and maintain the BBB phenotype (Beck et al., 1984, 1990, 1986; Betz et al., 1980; Betz and Goldstein, 1986; Janzer, 1993; Janzer and Raff, 1987). The factors released have not been well characterized although IL-6 appears to be among astrocyte factors that induce BBB properties (Sun et al., 1997). BBB phenotype characteristics induced by astrocyte factors include an increase in the complexity of tight junctions and thus an increase in electrical resistance across the barrier. Astrocytes, as well as astrocyte conditioned medium, also increase the expression of several membrane proteins that appear to be important in BBB function. This will be considered in more detail below in the discussion of BBB functions (Section I.B).

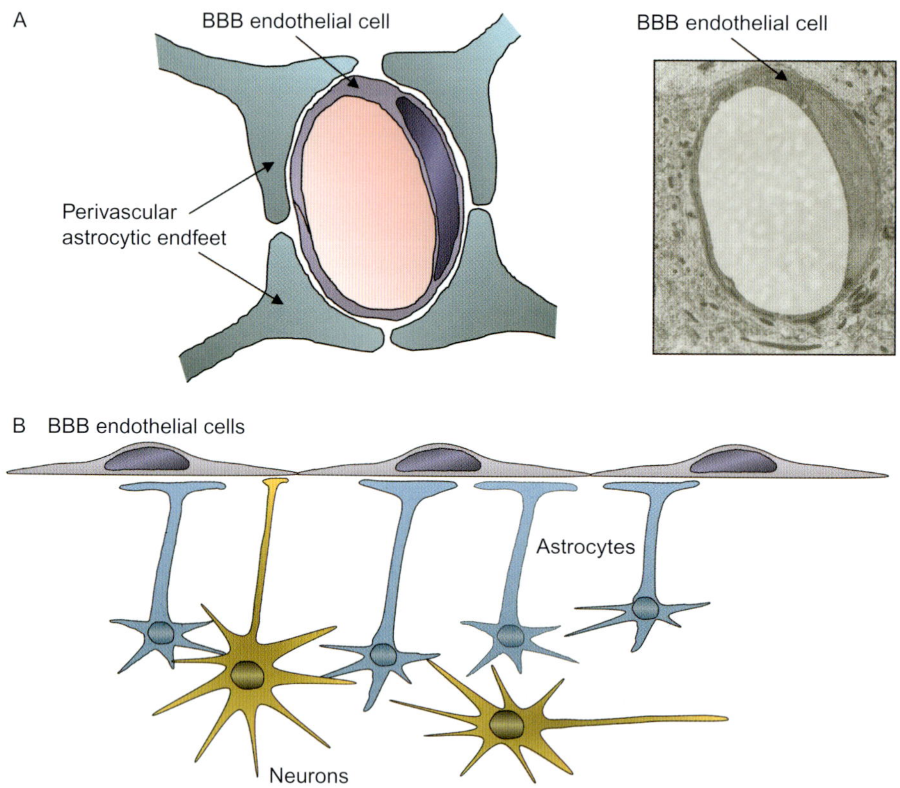

FIGURE 30.1 The blood–brain barrier with abluminal surface ensheathed by perivascular astrocytes. **A.** The blood–brain barrier (BBB) is comprised solely of brain microvascular endothelial cells. However, BBB endothelial cells have a close anatomic and functional relationship with perivascular astrocytes. The electron micrograph in the right panel shows a BBB endothelial cell. Note the lumen of the microvessel in the center and the dense cellular brain parenchyma surrounding the microvessel. The abluminal surface of the BBB endothelial cells is covered by endfeet of perivascular astrocytes, illustrated by the cartoon in the left panel. **B.** Blood–brain barrier endothelial cells operate within a "neural environment". Together with astrocytes and neurons, BBB endothelial cells form the neurovascular unit, defined as the astrocytes and neurons associated with each BBB endothelial cell or, conversely, the astrocytes and BBB endothelial cells associated with each neuron. Note that some neuronal processes terminate directly on BBB endothelial cells.

BBB endothelial cells also have important functional associations with neurons. Indeed, the observation that BBB endothelial cells operate in a neural/astrocyte environment has led to the term "neurovascular unit" which can be thought of as the neurons, astrocytes and pericytes associated with a BBB endothelial cell or, conversely, the BBB endothelial cells, pericytes and astrocytes associated with each neuron (Stroke Progress Review Group, 2002).

B. Functions of the Blood–Brain Barrier

The BBB has many important functions that are vital for maintenance of normal brain activity. This includes transport of nutrients (e.g. glucose and amino acids) into the brain and removal of waste products from the brain. The BBB is also well known for its robust ability to transport drugs in and out of the brain. The export of drugs from the brain by BBB drug transporters is perhaps most widely recognized because it presents a barrier to many different drug therapies developed to treat a variety of brain pathologies, such as epilepsy, tumors and Alzheimer's disease (Dallas et al., 2006; Kusuhara and Sugiyama, 2005; Löscher and Patschka, 2005; Miller et al., 2008; Neuwelt et al., 2008). As a barrier that separates the brain compartment from the blood compartment, the BBB also protects neurons and glial cells of the brain from toxins and from acute fluctuations in plasma ion concentrations. With respect to the latter, the BBB, together with the choroid plexus epithelium, serves the vitally important function of regulating both composition and volume of brain interstitial fluid (Chapters 28 and 29). This is critical for two reasons. First, the electrically excitable neurons are quite sensitive to changes in ionic composition of the extracellular fluid such that, for example, elevation of extracellular [K^+] can depolarize the cells, causing action potentials to fire inappropriately. Second, the brain sits within a rigid cranium and thus there is very little room for expansion of tissue volume without causing significant brain damage (Simard et al., 2007).

BBB endothelial cells are structurally and functionally polarized, with a number of enzymes and transport systems asymmetrically distributed between apical and basolateral surfaces of the cells. These include, for example, the Na^+/K^+ ATPase (primarily abluminal), the Na^+-K^+-Cl^- cotransporter (primarily luminal), gamma glutamyl transpeptidase (primarily luminal), amino acid transporters (either luminal or abluminal, depending on the transporters) and P-glycoprotein (Betz, 1986; Betz and Goldstein, 1986; Bradbury, 1984; Gaillard et al., 2000; O'Donnell et al., 2004; Sánchez del Pino et al., 1995a, b) (luminal or abluminal, depending on the transporter), to name a few. Consequently, the BBB, like many epithelia but unlike other endothelial barriers, is organized in a manner that promotes vectorial, i.e. transendothelial, solute and water flux. This asymmetry is the basis for the ability of BBB endothelial cells to tightly regulate the movement of nutrients, electrolytes and other solutes between blood and brain (Betz, 1986; Betz and Goldstein, 1986; Bradbury, 1984). The remainder of this chapter will focus on ion transporters of the BBB. A number of excellent reviews are available that cover BBB transport of other solutes, including nutrient and drug transporters (Vannucci et al., 1997; Löscher and Patschka, 2005; Leybaert, 2005; Hawkins et al., 2006; Dallas et al., 2006; Leybaert et al., 2007; Pardridge, 2007; Banks, 2008; Simpson et al., 2008; Miller et al., 2008).

II. ION TRANSPORT OF THE BLOOD–BRAIN BARRIER

A. Overview: Water and Electrolyte Transport across the Blood–Brain Barrier

The ion transporters of the BBB that regulate brain interstitial fluid volume and composition have yet to be fully elucidated. At the simplest level, we know that BBB endothelial cells possess the ability to transport Na^+, Cl^-, K^+ and water between blood and brain (Betz et al., 1994), as well as to regulate intracellular pH and volume of the cells. With respect to transendothelial transport, the BBB constantly secretes Na^+ and Cl^- from blood into brain such that in the healthy, normoxic brain, the BBB produces up to 30% of the brain interstitial fluid, with choroid plexus epithelial cells producing the remainder (Bradbury, 1984; Cserr et al., 1989; Keep, 1993). In the case of cerebral ischemia, however, this BBB secretion of brain interstitial fluid is greatly increased, contributing to the pathology of ischemic stroke, as will be discussed in Section III below. It should be noted that BBB-mediated secretion and absorption occur in apical to basolateral and basolateral to apical directions, respectively, since these processes refer to direction relative to blood. The BBB also functions in regulating brain extracellular [K^+], transporting K^+ between brain and blood as needed (Betz et al., 1994) for maintaining an appropriately low brain interstitial fluid [K^+] to support neuronal function. Thus, the BBB maintains extracellular [K^+] at about 3 mM in the brain while plasma [K^+] tends to range between 3 and 6 mM (Keep, 1993; Stummer et al., 1994, 1995; Keep et al., 1995a, b, 1999).

B. Chloride Transporters and Channels of the Blood–Brain Barrier

Despite the long held understanding that the BBB secretes brain interstitial fluid (primarily Na^+ and Cl^-) and absorbs K^+, we don't yet fully know the complement of transporters and channels involved in this phenomenon. Indeed, much remains to be discovered about ion transporters and channels that must be present in BBB endothelium to support not only vectorial transport but also intracellular pH and volume regulation of these cells. With respect to Na^+ secretion and K^+ absorption across the BBB, early studies demonstrated an abundance of an abluminally located Na^+/K^+ pump (Betz et al., 1980, 1994; Betz and Goldstein, 1986). As a primary active transporter that pumps Na^+ out of the cells and K^+ into the cells, the Na^+/K^+ pump not only maintains appropriate transmembrane gradients for Na^+ and K^+ but also provides a means of driving both Na^+ secretion and K^+ absorption. Early studies also provided evidence for a luminal Na^+- and K^+-selective, amiloride-sensitive channel (Vigne et al., 1989) and thus, it has been hypothesized that the pump and the cation channel may work together to transport Na^+ from blood into brain and/or transport K^+ from brain into blood (Fig. 30.2). In recent years a more complete understanding of BBB transporters and channels has begun to emerge. It is not possible to provide a comprehensive review of all BBB ion transporters and channels known to date and how they may function in health and disease. Thus, I will focus here on the transporters and channels thought to participate in Na^+ secretion (and K^+ absorption) across the BBB, as well as pH and volume regulation.

1. Secretion of Na^+, Cl^- and Water

a. Na^+-K^+-Cl^- Cotransport

It has been recognized for a number of years that because the BBB secretes Na^+ and Cl^- from blood into brain, the BBB must have pathways for both Na^+ and Cl^- entry at the luminal membrane. In studies examining Na^+ uptake into brain of intact rats, Betz and co-workers demonstrated previously that the Na^+-K^+-Cl^- cotransport inhibitor furosemide inhibits a Cl^--dependent Na^+ uptake into the brain, which led them to hypothesize that a luminal Na^+-Cl^- cotransporter participates in secreting Na^+, Cl^- and water across the BBB (Betz, 1983a, b). In other studies, isolated rat cerebral microvessels were shown to have a furosemide-inhibitable and Na^+- and Cl^--dependent cellular K^+ influx, indicating the presence of Na^+-K^+-Cl^- cotransport (Lin, 1985, 1988). Using the potent and more selective Na^+-K^+-Cl^- cotransport inhibitor bumetanide, several studies have subsequently provided good evidence for the presence of Na^+-K^+-Cl^- cotransport in cerebral microvascular endothelial cells (CMEC) of several species. Our early studies of BBB endothelial cells showed that cultured bovine CMEC exhibit Na^+-K^+-Cl^- cotransporter activity, evaluated as bumetanide-sensitive, Na^+- and Cl^--dependent K^+ influx (O'Donnell, 1993; O'Donnell et al., 1995a; Sun et al., 1995) and that cotransporter activity is also present in freshly isolated rat cerebral microvessels (O'Donnell et al., 2005). Bumetanide-sensitive, Na^+- and Cl^--dependent K^+ influxes were also observed in cultured rat CMEC by Spatz and co-workers (Kawai et al., 1996a,b, 1997) and Vigne and co-workers (Vigne et al., 1994). We and others have also shown that Na^+-K^+-Cl^- cotransporter activity is also observed

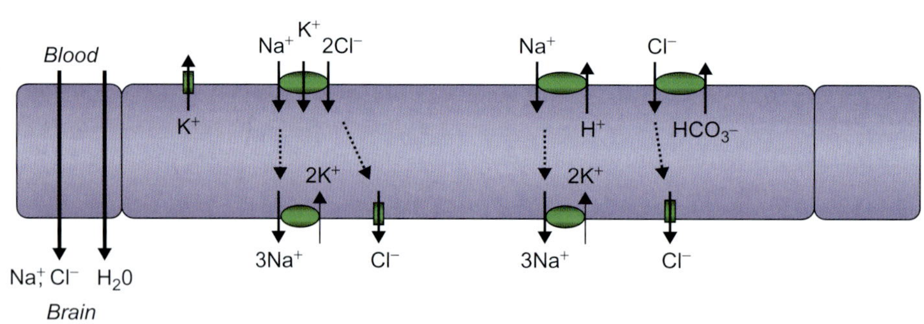

FIGURE 30.2 Hypothesized pathways for BBB secretion of Na^+ and Cl^- into the brain and absorption of K^+ into blood. Evidence suggests that a luminal Na^+-K^+-Cl^- cotransporter works in conjunction with an abluminal Na^+/K^+ pump and a Cl^- efflux pathway (such as a Cl^- channel) to mediate vectorial transport of Na^+ and Cl^- with osmotically obliged water from blood into brain. At the same time, K^+ can be absorbed from brain into blood as it enters via an Na^+/K^+ pump and exits via a luminal K^+ channel. A luminal Na^+/H^+ exchanger is also hypothesized to participate in BBB secretion of Na^+ by working with an abluminal Na^+/K^+ pump to secrete Na^+ across the BBB. In this case, Cl^- accompanying Na^+ is hypothesized to enter the cell at the luminal membrane via Cl^-/HCO_3^- exchanger and/or the Na^+-K^+-Cl^- cotransporter.

in cultured human CMEC (O'Donnell et al., 2005; Spatz et al., 1997). In studies conducted to provide a molecular characterization of the endothelial Na^+-K^+-Cl^- cotransporter (NKCC), Northern blot analyses revealed the presence of NKCC1 RNA in bovine CMEC (Yerby et al., 1997). By Western blot analyses, abundant NKCC protein has also been demonstrated in both cultured CMEC and in freshly isolated rat brain microvascular endothelial cells (O'Donnell et al., 1995b; Sun et al., 1995; Yerby et al., 1997).

Evidence that the Na^+-K^+-Cl^- cotransporter may be of importance to BBB function is suggested by the observation that coculture of CMEC with astrocytes or exposure of CMEC to astrocyte-conditioned medium increases the expression of Na^+-K^+-Cl^- cotransporter protein in the CMEC (Sun et al., 1995). In these early studies using CMEC grown on permeable tissue culture filter inserts, we found that astrocyte-conditioned medium and coculture with astrocytes also increased the amount of specific [^{3}H]-bumetanide binding in the CMEC, an index of Na^+-K^+-Cl^- cotransporter protein abundance (Sun et al., 1995). In these studies, approximately 90% of specific [^{3}H]-bumetanide binding and 90% of bumetanide-sensitive K^+ influx occurred at the apical versus basolateral membrane of the CMEC monolayers, suggesting that the cotransporter may serve as a luminal Na^+ and Cl^- uptake pathway in the BBB (Sun et al., 1995). We have also found immunoelectron-microscopy evidence for the presence of Na^+-K^+-Cl^- cotransporter protein in the luminal BBB membrane *in situ* (O'Donnell et al., 2004; Brillault et al., 2007). This will be discussed in more detail in Section II. Numerous previous studies have demonstrated that the Na^+-K^+-Cl^- cotransporter plays a major role in salt and water transport across a variety of epithelia (Chipperfield, 1996; Haas, 1994; O'Grady et al., 1987; Russell, 2000). Together with the above findings regarding Na^+-K^+-Cl^- cotransporter protein expression and activity in BBB endothelial cells, this suggests that NKCC functions by transporting Na^+, Cl^- and water across BBB endothelial cells as well, as shown in Fig. 30.2. This will be explored further in the context of BBB NaCl secretion in stroke (Section III).

b. Cl^- Channels

Secretion of Na^+, Cl^- and water across the BBB requires a Cl^- exit pathway at the abluminal membrane of the cells. Previous studies of salt absorbing epithelia have shown that an apical Na^+-K^+-Cl^- cotransporter functionally coupled to a basolateral Na^+/K^+ pump and a basolateral Cl^- channel mediates Na^+ and Cl^- absorption (O'Grady et al., 1987; Paulais and Teulon, 1990; Hebert et al., 1996; Russell, 2000; Fischer et al., 2007). While we know that the BBB Na^+/K^+ pump can move Na^+ across the abluminal membrane into the brain, less information is available about how Cl^- exits these cells. However, von Weikersthal and collaborators have described a volume-sensitive Cl^- current in BBB endothelial cells (von Weikersthal et al., 1999). In those studies, Cl^- currents of rat brain CMECs were evaluated by whole-cell voltage-clamp methods. Cell swelling induced by exposure to hypotonic media caused a large outwardly rectifying Cl^- current that could be blocked by the Cl^- channel blocker NPPB and also by DIDS. This swelling-activated current was found to be dependent on intracellular ATP but independent of intracellular Ca^{2+}. When the rat CMECs were subjected to an RT-PCR analysis of possible volume-sensitive Cl^- channels, it was found that the cells expressed the transcripts for ClC-2, ClC-3 and ClC-5 (Chapters 12 and 15). Whether this volume-sensitive Cl^- channel does, in fact, function as the BBB abluminal exit pathway for Cl^-, working in conjunction with the luminal Na^+-K^+-Cl^- cotransporter and abluminal Na^+/K^+ pump to mediate secretion of Na^+ and Cl^- from blood into brain, will require further studies. In keeping with this, previous studies have suggested that some Cl^- channels of the CLC family are sensitive to small changes in volume (Strange et al., 1996), but the molecular identity of volume-activated Cl^- channels is a highly controversial issue, as discussed in detail in Chapters 15. Further, ClC-3 and ClC-5 are now known to be Cl^-/H^+ exchangers rather than channels, and they reside in endolysosomal compartments, as discussed in Chapter 12 in this volume. Thus, if even a small increase in BBB cell volume occurs as the Na^+-K^+-Cl^- cotransporter is stimulated to increase influx of Na^+, K^+ and Cl^-, it is possible that an abluminal Cl^- channel, whatever its molecular identity turns out to be, would be activated allowing Cl^- efflux at the abluminal membrane. It should be noted that maintaining a negative cell interior is also essential for promoting Cl^- efflux via Cl^- channels. In this regard, we know that the cells possess an amiloride-sensitive Na^+- and K^+-permeable channel (Vigne et al., 1989), inwardly rectifying K^+ channels (Hoyer et al., 1991), and ATP-sensitive K^+ channels (Janigro et al., 1993). The relative contribution of these channels to ion transport across the BBB has yet to be determined.

c. Na^+/H^+ Exchange, Cl^-/HCO_3^- Exchangers and Na^+-HCO_3^- Cotransport

It is unlikely that the Na^+-K^+-Cl^- cotransporter is the only luminal pathway by which Na^+ and Cl^- enter the BBB endothelial cell for NaCl secretion. We and others have found evidence of a Na^+/H^+ exchanger in BBB endothelial cells (Hsu et al., 1996; Kalaria et al., 1998; Kawai et al., 1997; Lam and O'Donnell, 2008; Sipos et al., 2005; Spatz et al., 1997; Vigne et al., 1991) (to be discussed in more detail below in Section III). The Na^+/H^+ exchanger is known to function in vectorial transport of Na^+ across epithelia, and also in intracellular pH and volume regulation in a variety of cell types (Boron, 2004; Chesler, 2003; Pedersen et al., 2006). With respect to vectorial transport of Na^+, Cl^- and water, in salt absorbing epithelia an apical Na^+/H^+ exchanger can functionally couple with apical Cl^-/HCO_3^- exchange to bring Na^+ and Cl^- into the cell at the apical membrane (Friedman and Andreoli, 1982). Evidence has also been provided for the presence of Cl^-/HCO_3^- exchange in BBB endothelial cells, although its cellular location has not been established (Taylor et al., 2006). Nevertheless, the observations that both Na^+/H^+ exchange and Cl^-/HCO_3^- exchange are present in the BBB raise the possibility that these two exchangers also participate in BBB secretion of Na^+ and Cl^- into the brain (as shown in Fig. 30.2).

There are additional possibilities for entry of Na^+ and Cl^- at the BBB luminal membrane. These include a Na^+-driven Cl^-/HCO_3^- exchanger (Na^+-dependent Cl^-/HCO_3^- exchanger) and a Na^+/HCO_3^- cotransporter (Taylor et al., 2006). In a recent study by Taylor et al., RT-PCR analysis of rat brain CMEC revealed the abundant presence of transcripts for the AE2 isoform of Cl^-/HCO_3^- exchange and the NBCn1 and NBCn2 isoforms of electroneutral Na^+/HCO_3^- cotransporters. The AE1 and AE3 isoforms of Cl^-/HCO_3^- exchange were also detected but at lower levels, as were the electrogenic Na^+-HCO_3^- cotransporter NBCe2, and the Na^+-dependent Cl^-/HCO_3^- exchangers NDCBE and NCBE (Taylor et al., 2006). In this study, rat CMEC were also evaluated to determine which of these transporter proteins might be active in the cells. At basal intracellular pH levels (~ pH 7.1) the rat CMEC appeared to have Na^+/HCO_3^- cotransporter activity as well as Na^+-independent Cl^-/HCO_3^- exchange activity, and possibly also Na^+-dependent Cl^-/HCO_3^- exchange activity. While these studies examined the HCO_3^- transporters from the perspective of pH regulation, the transporters have been found to participate in HCO_3^- secretion or absorption in polarized epithelial cells. Thus it is possible that one or more of these HCO_3^- transporters may participate in vectorial transport of HCO_3^- across the BBB or, in the case of Cl^-/HCO_3^- exchangers, also vectorial Cl^- transport. For example, in pancreatic duct, an apical Na^+-HCO_3^- cotransporter is thought to be functionally coupled to a basolateral Cl^-/HCO_3^- exchanger to mediate HCO_3^- secretion into the duct lumen (Romero, 2005; Romero et al., 2004). It has been suggested that such an arrangement of these transporters in the luminal and abluminal BBB membranes could similarly participate in secretion of HCO_3^- from blood into brain across the BBB (Taylor et al., 2006). This is because the BBB determines the composition of brain interstitial fluid which contains HCO_3^- as well as Na^+, Cl^- and K^+, and thus there must be mechanisms for BBB secretion of HCO_3^- into the brain (Taylor et al., 2006). Clarifying whether Na^+-dependent Cl^-/HCO_3^- exchange and/or Na^+/HCO_3^- cotransport may participate in secretion of Na^+, Cl^- and possibly HCO_3^- by the BBB clearly requires further investigation, not the least of which will be to determine the plasma membrane distribution of these transporters in BBB endothelial cells.

2. *Regulation of Intracellular pH*

A number of studies have now provided evidence that a Na^+/H^+ exchanger participates in regulation of intracellular pH in BBB endothelial cells (Vigne et al., 1991; Schmid et al., 1992; Hsu et al., 1996; Lam et al., 2006; Taylor et al., 2006; Hom et al., 2007; Nicola et al., 2007; Lam and O'Donnell, 2008). This will be examined further in Section III. In addition, Taylor and co-workers have found that both Na^+-dependent and Na^+-independent Cl^-/HCO_3^- exchangers as well as Na^+-HCO_3^- cotransport appear to contribute to pH regulation of BBB endothelial cells (Nicola et al., 2007; Taylor et al., 2006). A more complete understanding of how these HCO_3^- transporters may contribute to BBB intracellular pH regulation in health and disease awaits future studies.

3. *Regulation of Cell Volume*

Previous studies have shown that BBB endothelial cells have a robust ability to regulate their cell water volume via both regulatory volume increase (RVI) and regulatory volume decrease (RVD) mechanisms (Kempski et al., 1985; DePetris et al., 2001). The Na^+-K^+-Cl^- cotransporter is well known for its ability to regulate cell volume in a variety of cell types (Chipperfield, 1996; O'Donnell, 1993; O'Grady et al., 1987; Russell, 2000), as also discussed in Chapters 2, 22 and 27. In previous studies we showed that Na^+-K^+-Cl^- cotransporter participates in cell volume regulation of aortic endothelial cells, mediating RVI following hyperosmotic media-induced cell shrinkage (O'Donnell, 1993). In those studies we also found that stimulation of the cotransporter by vasoactive agents, including vasopressin, angiotensin II and bradykinin, can drive increases in cell volume. With respect to BBB, we and others have shown that CMEC Na^+-K^+-Cl^- cotransporter activity is stimulated by hyperosmotic media (Vigne et al., 1994; O'Donnell et al., 1995a), suggesting that it may be involved in RVI (see Chapter 2). Na^+/H^+ exchange also participates in cell volume regulation and, when coupled to Cl^-/HCO_3^-, has been shown to mediate increases in volume, especially the RVI that follows hypertonic medium-induced cell shrinkage (Kregenow, 1981; Eveloff and Warnock, 1987; Sarkadi and Parker, 1991; McManus et al., 1995; Chesler, 2003; Pedersen et al., 2006). Whether the Na^+/H^+ exchanger and/or Cl^-/HCO_3^- exchange contribute to regulation of BBB endothelial cell volume remains to be determined.

III. DISEASES OF THE BLOOD–BRAIN BARRIER

A. Overview

While BBB ion transporters are important for BBB function in the normal, healthy brain, they can also participate in pathophysiological processes occurring during central nervous system injury and disease. Numerous studies in recent years have provided strong evidence that CNS disorders and pathological states involve BBB dysfunction (Grant et al., 1998). These include, for example, epilepsy, brain tumors, diabetes, multiple sclerosis, Alzheimer's disease, traumatic head injury, and both ischemic and hemorrhagic stroke. Because it is simply not possible to cover the many brain pathological states that very likely involve altered function of BBB ion transporters, I will focus on ischemic stroke. This is pertinent because we know that BBB regulation of brain interstitial fluid volume and composition is altered during ischemic stroke, leading to cerebral edema formation. We also know that edema is a major contributing factor to the morbidity and mortality of stroke (Kato et al., 1987; Klatzo, 1994; Bronner et al., 1995; Kimelberg, 1995). The role of Cl^- transporters in the pathophysiology of ischemic brain damage is also discussed in Chapter 25 in this volume.

B. Stroke

A number of events unfold during ischemic stroke that together lead to brain damage. Some of these are initiated within minutes and others develop over hours or even days. Some of the earliest events include rapid development of hypoxia, as well as release of arginine vasopressin (AVP) from extrahypothalamic neuronal processes that terminate on cerebral microvessels (Dóczi et al., 1984; Jójárt et al., 1984; Sorensen et al., 1985; Landgraf, 1992; Dóczi, 1993; Baringa, 1996). Another event that is initiated very quickly is development of cerebral edema. In the early hours of ischemic stroke, brain edema forms by a process involving net uptake of Na^+ and water across the barrier from blood into brain (Gotoh et al., 1985; Hatashita and Hoff, 1990; Schielke et al., 1991; Menzies et al., 1993; Kimelberg, 1995; Betz, 1996). This early ischemia-induced brain edema formation occurs in the presence of an intact BBB, with breakdown of the barrier occurring approximately 4–6 hours after the onset of ischemia (Gotoh et al., 1985; Hatashita and Hoff, 1990; Menzies et al., 1993). Also, it has been found that the net uptake of cations and water into the brain occurring during these early hours accounts for the majority of edema formation (Betz, 1996) and further that ischemia triggers an increased secretion of Na^+, Cl^- and water across the BBB from blood into brain (Kimelberg, 1999; Menzies et al., 1993; Schielke et al., 1991). At the same time, astrocytes are stimulated by ischemia to take up Na^+, Cl^- and water from the brain interstitium and rapidly swell (Bourke et al., 1980; Gotoh et al., 1985; Iadecola, 1999; Kempski et al., 1991; Kimelberg, 1995, 1999). The total amount of Na and water in the brain increases during ischemia (Menzies et al., 1993; Schielke et al., 1991). This means that astrocytes don't simply swell by taking up ions and water from the extracellular space, i.e. that astrocyte swelling cannot by itself account for the observed brain edema. Rather, the astrocytes take up ions and water coming from the blood into the brain as the result of increased BBB secretion. In this regard, the increased BBB secretion of Na^+, Cl^- and water facilitates astrocyte swelling. This close functional relationship between BBB and astrocytes with respect to water and electrolyte movements during ischemic stroke is underscored by the observation that aquaporin 4 water channels are remarkably concentrated at perivascular endfeet that appose the abluminal BBB surface (Amiry-Moghaddam et al., 2003). As edema continues to form during stroke, the progression of swelling reduces flow in additional blood vessels, expanding the ischemic zone. Understanding the cellular mechanisms underlying ischemia-induced edema formation is essential for identifying therapies that can reduce or abolish edema.

1. *A Role for BBB Na^+-K^+-Cl^- Cotransport in Ischemia-induced Cerebral Edema*

The apparent predominantly luminal membrane location of the BBB Na^+-K^+-Cl^- cotransporter, together with evidence that it may be stimulated by hypoxia and AVP, makes it a good candidate for participating in ischemia-induced hypersecretion of Na^+, Cl^- and water across the BBB during stroke. Thus, we have hypothesized that a luminal Na^+-K^+-Cl^- cotransporter is stimulated by ischemic conditions, including hypoxia and AVP, and that increased cotransporter-mediated Na^+ and Cl^- influx leads to hypersecretion of the ions as they exit the cell via the abluminal Na^+/K^+ pump and an abluminal Cl^- channel (as shown in Fig. 30.3). It should be noted here that the Na^+/K^+ pump is highly sensitive to an increase in intracellular $[Na^+]$ such that as the cotransporter brings more Na^+ into the cell, the pump is stimulated to transport the increased Na^+ out at the abluminal membrane. Also, because Cl^- flux through the Cl^- channel is determined by the electrochemical gradient for Cl^-, an increase in intracellular $[Cl^-]$ will increase Cl^- efflux via Cl^- channels. The studies discussed in the following sections address the possibility that the BBB Na^+-K^+-Cl^- cotransporter participates in edema formation during ischemic stroke.

a. Plasma Membrane Distribution of the BBB Na^+-K^+-Cl^- Cotransporter *in situ*

If the Na^+-K^+-Cl^- cotransporter participates in ischemia-induced edema formation by functioning as the luminal entry pathway for BBB secretion of Na^+ and Cl^-, it should be present in the luminal membrane of BBB endothelial cells *in situ*. We have tested this using perfusion-fixed rat brains and immunoelectron microscopy with specific primary antibodies that recognize the Na^+-K^+-Cl^- cotransporter protein and secondary antibodies conjugated to gold particles. These studies revealed that in healthy, normoxic rat brain, the Na^+-K^+-Cl^- is indeed present in BBB *in situ* and further that it is distributed 80% in the luminal membrane and 20% in the abluminal membrane (O'Donnell et al., 2004). This is consistent with the cotransporter participating in the low rate of Na^+, Cl^- and water secretion across the BBB that occurs under normoxic conditions. To determine whether the cotransporter is present in luminal membranes under ischemic conditions as well, we subjected rats to middle cerebral artery occlusion (MCAO), a well-established rat model of stroke (Schielke et al., 1991; Zea Longa et al., 1989; Betz et al., 1995; O'Donnell et al., 2004). Here, we perfusion fixed the brains after 90 min of permanent MCAO and searched for the presence of Na^+-K^+-Cl^- protein in BBB endothelial cells *in situ*, both in ipsilateral (occluded) ischemic brain and in the contralateral normoxic brain (Fig. 30.4). We found that for both the normoxic and ischemic brain regions, the distribution of Na^+-K^+-Cl^- protein between luminal and abluminal BBB membranes is approximately 80% and 20%, respectively (Brillault et al., 2007).

b. Reduction of Ischemia-induced Edema and Infarct by Inhibition of the BBB Na^+-K^+-Cl^- Cotransporter

If the luminal BBB Na^+-K^+-Cl^- cotransporter is a significant contributor to ischemia-induced brain edema formation, then intravenous (i.v.) administration of bumetanide to inhibit the luminal cotransporter is predicted to reduce or abolish edema formation following MCAO in the rat. One way to do this is to administer bumetanide i.v., subject rats to MCAO and then evaluate brain water at various times after the start of MCAO. In this case, each rat only provides one time point. These studies have been

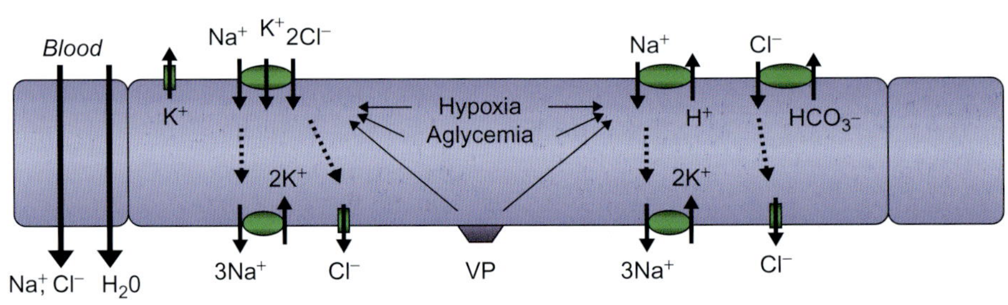

FIGURE 30.3 Blood–brain barrier "hypersecretion" of Na^+ and Cl^- during ischemic stroke: hypothesized roles of Na^+-K^+-Cl^- cotransport and Na^+/H^+ exchange. Previous studies provide evidence that a luminal BBB Na^+-K^+-Cl^- cotransporter is stimulated by ischemic factors, including hypoxia, aglycemia and arginine vasopressin (AVP), to participate in increased secretion of Na^+, Cl^- and water into the brain during the early hours of stroke. By this model, Na^+ and Cl^- entering at the luminal membrane are predicted to exit the BBB at the abluminal membrane by the Na^+/K^+ pump and a Cl^- channel. Evidence suggests that ischemic factors also increase secretion of Na^+ into the brain by stimulating activity of a luminal Na^+/H^+ exchanger. Cl^- secretion is hypothesized to accompany Na^+/H^+ exchange-mediated Na^+ secretion, with Cl^- entering at the luminal membrane (e.g. by Cl^-/HCO_3^- exchange or even the Na^+-K^+-Cl^- cotransporter) and exiting via the abluminal Cl^- efflux pathway (such as a Cl^- channel).

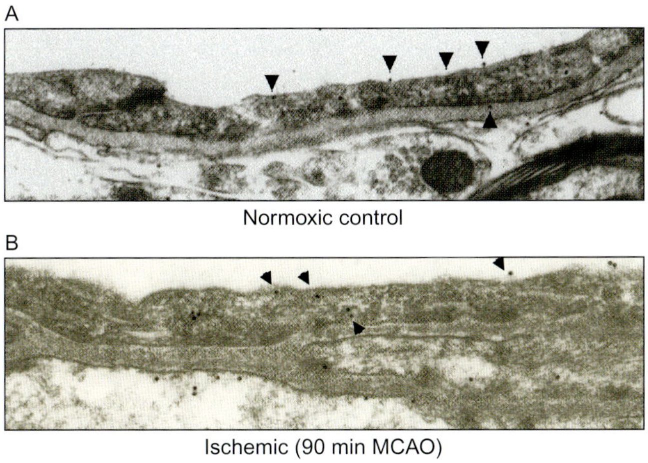

FIGURE 30.4 Immuno-electron microscopic localization of BBB Na^+-K^+-Cl^- cotransport protein *in situ*. Perfusion-fixed rat brains were labeled with primary antibody that recognizes Na^+-K^+-Cl^- cotransporter protein (T4 at a 1:7500 dilution), then gold particle-conjugated secondary antibody. Figures shown are representative micrographs. Vessel lumens are at top of each image with astrocyte and neuronal elements below the basal lamina underlying the endothelium. Locations of gold particles in luminal and abluminal membranes are indicated by arrowheads. **A.** Immuno-electron micrograph from normoxic cortex. **B.** Immuno-electron micrograph from ischemic cortex of a rat subjected to 90 min MCAO (then brain immediately perfusion-fixed). This image is from cortical region of interest 3 shown in Fig. 30.5A. (Brillault et al., 2007).

done with good success by various laboratories to test effects of different drugs on brain water (Betz et al., 1994; Ennis et al., 1996; Menzies et al., 1993; O'Donnell et al., 2004; Schielke et al., 1991). Another approach is to use magnetic resonance diffusion-weighted imaging (MR DWI) to evaluate brain edema in living rats over a time course following MCAO (O'Donnell et al., 2004). This allows monitoring of brain edema in real time and has the advantage that data can be obtained both before and after treatment for each rat. Using MR DWI with MCAO, one can also visualize the location of the edema as hyperintensity in the spin echo image. In the image shown in Fig. 30.5A, taken from a rat 90 min after the start of left MCAO, a hyperintense signal can be seen in the left hemisphere, indicating the presence of cytotoxic edema (O'Donnell et al., 2004). Apparent diffusion coefficient (ADC) values can be determined for specific regions of interest (ROI) as desired. It has been shown that by this method, a drop in ADC values provides a good index of edema formation (Hasegawa et al., 1994; Knight et al., 1994; Liu et al., 2001; Moseley et al., 1990). Comparing ADC values for ROIs in ipsilateral ischemic brain with those of contralateral normoxic brain provides an internal control for each rat. Figure 30.5B is a plot of ADC ratios for left (ischemic)/right (normoxic) brain for ROI 1 over a time course following induction of MCAO. Note that Sham-operated rats, undergoing MCAO surgery but never occluded, show L/R ADC ratios of 1, as expected. For rats subjected to permanent MCAO, the L/R ADC ratios drop, indicating the presence of edema. However, in rats administered i.v. bumetanide (in this case 30 mg/kg) immediately before the start of MCAO (to inhibit the BBB Na^+-K^+-Cl^- cotransporter), the fall in ADC ratios is attenuated significantly, indicating reduction of edema formation. Note also that this reduction is sustained throughout the 3-hour imaging period, and thus blocking the cotransporter does not simply slow edema formation but causes a quick onset, sustained inhibition of edema formation (O'Donnell et al., 2004). These findings are in good agreement with gravimetric analyses of brain water in rats subjected to 3 hours of permanent MCAO and treated with bumetanide or vehicle (O'Donnell et al., 2004), i.e. brain water was significantly increased following MCAO in rats treated with vehicle but not those treated with bumetanide.

We have also found that administering bumetanide reduces brain infarct volume, assessed as TTC (2,3,5-triphenyltetrazolium chloride) staining of brain slices after 3 hours of permanent MCAO (Figure 30.5C). In this figure, note the large TTC-defined lesion in the left hemisphere. The total infarct volume in these studies was 40% of the ipsilateral hemisphere in vehicle-treated rats and 18% to 25% of the ipsilateral hemisphere in bumetanide-treated rats (for doses ranging from 7.6 to 30 mg/kg i.v.). Identifying luminal BBB proteins that participate in CNS pathophysiology, given the fact that such targets appear to be readily accessible to i.v. administration of drugs, even in the case of reduced cerebral blood flow in ischemic stroke, suggests that this is a promising therapeutic avenue worthy of pursuit (see Chapter 25 in this volume).

c. Stimulation of BBB Na^+-K^+-Cl^- Cotransport Activity by Ischemic Factors

If the BBB Na^+-K^+-Cl^- cotransporter participates in ischemia-induced "hypersecretion" of Na^+, Cl^- and water from blood into brain, then it is predicted that one or more factors that are present during cerebral ischemia will increase activity of the cotransporter. There is now good evidence that this is in fact the case, as will be discussed in the following sections.

i. AVP

Several studies have provided compelling evidence that a central release of AVP occurs during cerebral ischemia and also that AVP containing extrahypothalamic neuronal processes terminate on brain microvessels (Baringa, 1996; Dóczi, 1993; Dóczi et al., 1984;

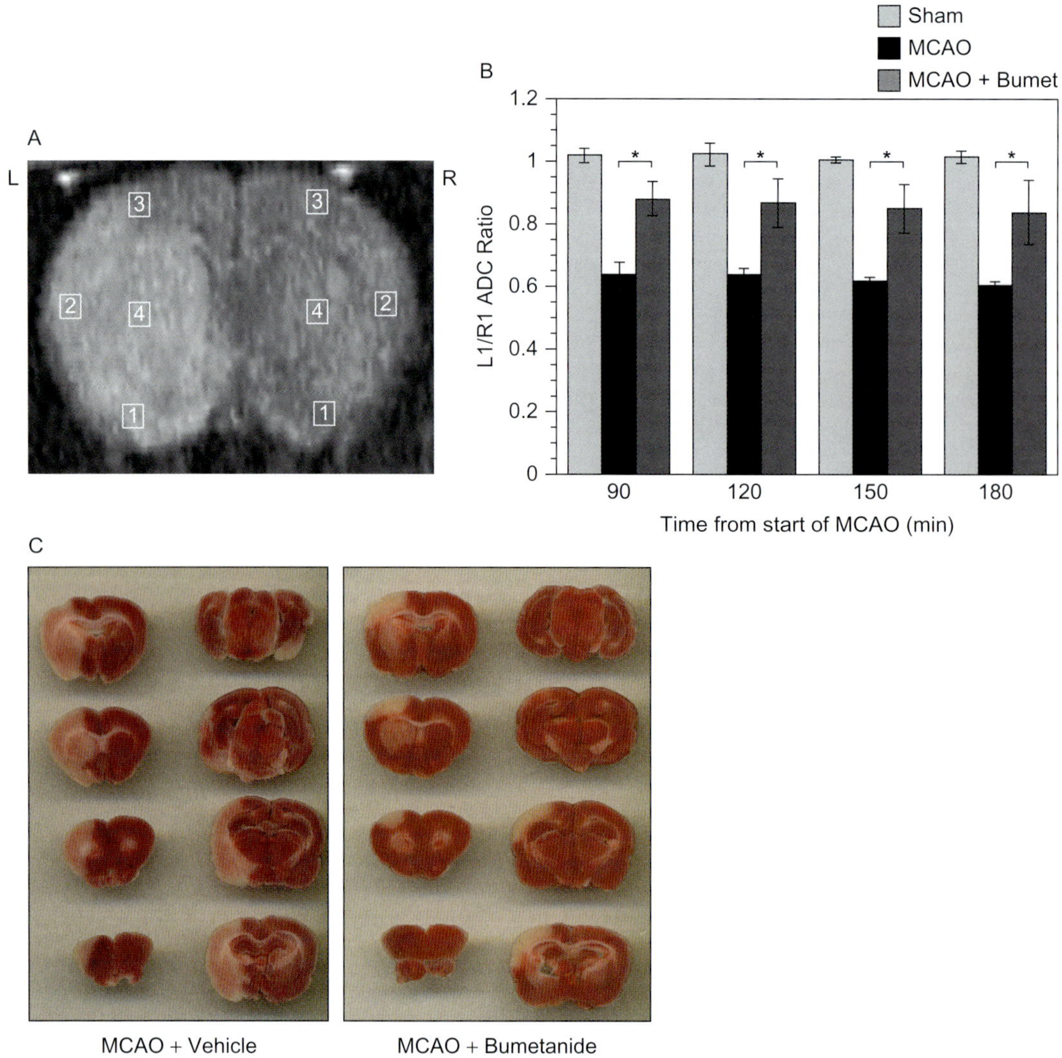

FIGURE 30.5 Magnetic resonance imaging assessment of edema in focal ischemia. **A.** Magnetic resonance diffusion-weighted spin echo image (2 mm slice) of a rat brain acquired 90 min after initiation of middle cerebral artery occlusion (MCAO). A relative hyperintensity (edema) is seen on the left side (L) of the image. Apparent diffusion coefficient (ADC) values were determined for the regions of interest (ROI) indicated by the boxes (L1–L4 and R1–R4) and ratios of left/right ADC values calculated for each ROI. **B.** Bumetanide attenuation of MCAO-induced cerebral edema. Rats were subjected to MCAO or Sham surgery, with bumetanide (30 mg/kg) or vehicle administered i.v. 20 min prior to start of occlusion. Values shown are ratios of left (occluded) to right (non-occluded control) ADC values for cortical ROI 1. ADC ratios of MCAO animals fall below 1.0, indicating the presence of edema. Animals given bumetanide (Bumet) then subjected to MCAO exhibit less edema formation, as indicated by the attenuated fall in ADC ratios. ADC values for Sham animals were not affected by bumetanide (not shown). Data are means±SD of four to nine experiments for the three conditions. For all four ROI evaluated, ADC ratios for MCAO plus bumetanide were found to be significantly different from ADC ratios for MCAO at all times (* $P < 0.0001$). **C.** Effects of bumetanide on infarct size assessed with TTC (2,3,5-triphenyltetrazolium chloride) staining. At the conclusion of magnetic resonance diffusion-weighted imaging (MR DWI) experiments, TTC staining of brain slices was performed as described in O'Donnell et al. (2004). Images are representative brain slices from rats treated i.v. for 20 min with vehicle (left panel) or 30 mg/kg bumetanide (right panel) followed by 180 min of permanent MCAO.

Jójárt et al., 1984; Landgraf, 1992; Sorensen et al., 1985) and that AVP receptors are present in brain microvessels (Kretzschmar and Ermisch, 1989; Pearlmutter et al., 1988). An AVP receptor antagonist has been shown to attenuate ischemia-induced brain infarct in rats (Shuaib et al., 2002). Further, Brattleboro rats, which are genetically deficient in AVP, show reduced edema formation in MCAO and administering AVP to those rats restores MCAO-induced edema formation (Dickinson and Betz, 1992). In studies of bovine CMEC, we have found that AVP does indeed stimulate Na^+-K^+-Cl^- cotransporter activity of the cells, evaluated as bumetanide-sensitive K^+ influx. This stimulation occurs with an EC_{50} of about 1 nM and is seen with 5 min of exposure to AVP (Fig. 30.6). Na^+-K^+-Cl^- cotransporter activity of human CMEC and freshly isolated bovine cerebral microvessels is also stimulated by AVP (O'Donnell et al., 2005). Prolonged exposures to AVP of 4 to 36 h also results in increased Na^+-K^+-Cl^- cotransporter activity (O'Donnell et al., 2005). The AVP effect on the CMEC cotransporter appears to be mediated via a V1-type AVP receptor because V1 but not V2 agonists also stimulate cotransporter activity in the cells and V1 but not V2 receptor antagonists inhibit AVP stimulation of cotransporter activity in the cells (O'Donnell et al., 2005). V1 AVP receptors are known to be linked to activation of phospholipase C (PLC) and elevation of intracellular $[Ca^{2+}]$. Consistent with this, our studies showed that the AVP stimulation of CMEC cotransporter activity is attenuated by PLC inhibitors and by treatments that prevent elevation of intracellular $[Ca^{2+}]$.

ii. Hypoxia

An important factor to consider in ischemia-induced hypersecretion of ions by the BBB is hypoxia. Due to the low O_2 store in the brain, hypoxia develops rapidly during an ischemic episode. In focal ischemia, as occurs with MCAO, the O_2 level can vary from zero in the core area to 20–50% of normal in the adjacent,

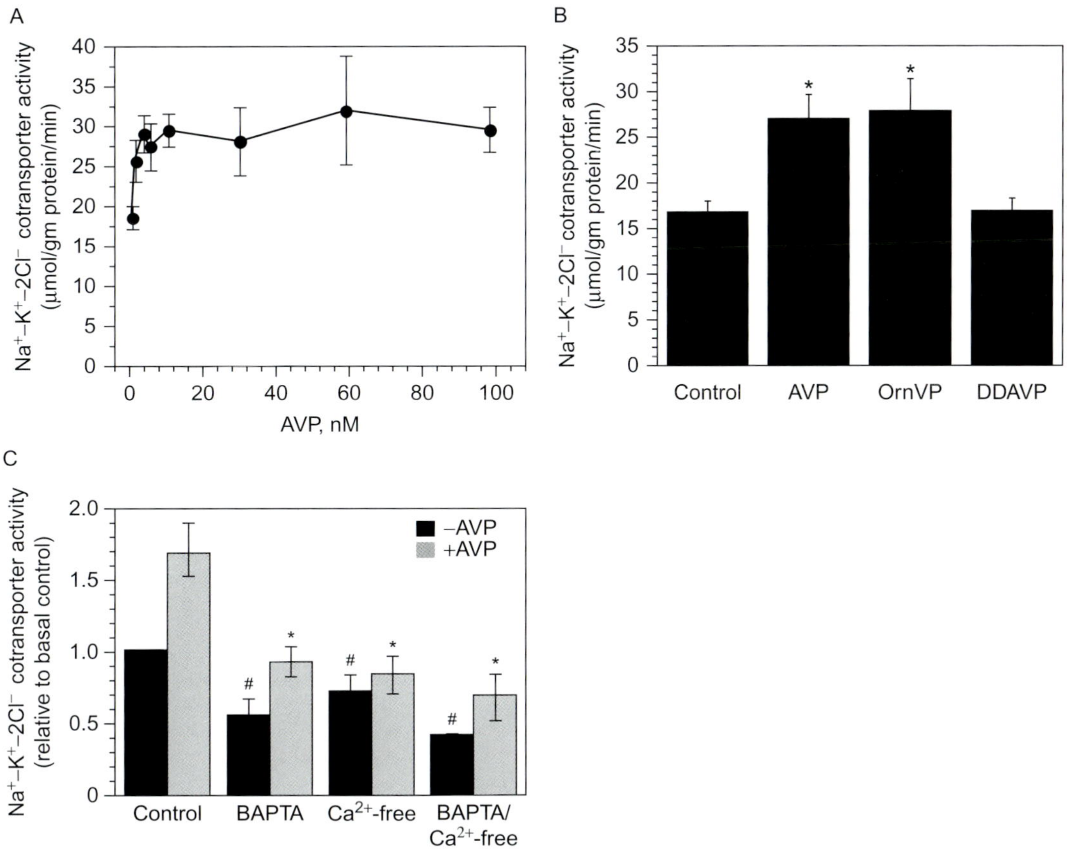

FIGURE 30.6 Arginine vasopressin stimulation of Na^+-K^+-Cl^- cotransport activity of cerebral microvascular endothelial cells. **A.** Dose-dependent stimulation by arginine vasopressin (AVP) of cerebral microvascular endothelial cells (CMEC). Cells were pretreated for 5 minutes with bumetanide (0 or 10 μM) ± AVP (1 to 100 nM), then assayed in the same media containing ^{86}Rb for 5 minutes. Cotransport activity was assessed as described previously (O'Donnell et al., 2005). Data represent means ± SEM of 3–7 experiments. **B.** Effects of V1 and V2 agonists on Na^+-K^+-Cl^- cotransport activity of CMEC. Cells were pretreated for 5 minutes in media containing AVP (0 or 10 nM), the V1 agonist Ornithine VP (Orn VP; 0 or 10 nM), or the V2 agonist DDAVP (deamino-Cys1,D-Arg8)-vasopressin (0 or 10 nM), followed by a 5 minute incubation in media with or without bumetanide (10 μM) ± AVP, Orn VP and/or DDAVP (all at 10 nM). Na^+-K^+-Cl^- cotransporter activity was then assayed for 5 minutes in the same media. Values shown represent means ± SEM of 13, 13, 7 and 8 experiments for control, AVP, Orn VP and DDAVP, respectively. * Significantly different from control (by ANOVA). **C.** Ca^{2+}-dependence of AVP effects on Na^+-K^+-Cl^- cotransport activity of CMEC. Cells were exposed to the membrane-permeable form of the intracellular Ca^{2+} buffer BAPTA-AM (5 μM) at 37 °C for 20 min, then incubated for 5 min in BAPTA-free and/or Ca-free medium, followed by incubation for another 5 min in the same medium also containing AVP (100 nM) ± bumetanide (10 μM), and finally assayed for 5 min in the same medium. Values shown are means ± SEM of four experiments for BAPTA, Ca^{2+}-free, and BAPTA/Ca^{2+}-free conditions and eight experiments for control. #, Significantly different from control without AVP. *, Significantly different from control with AVP (AVP with no inhibitor). In addition, control AVP-stimulated cotransporter activity is significantly different than AVP-stimulated cotransporter activity in the presence of BAPTA, Ca^{2+}-free medium or BATPA/ Ca^{2+}-free medium (the difference between the dark and light bar in each case). All comparisons were done by Student's *t* tests.

penumbral regions (Baringa, 1996). In determining whether the BBB Na^+-K^+-Cl^- cotransporter is stimulated by hypoxia to participate in brain edema formation, one must thus ask whether hypoxia can stimulate cotransporter activity of the cells, and if so, then at what levels and over what time course this occurs.

Early studies examining the effects of hypoxia on BBB endothelial cells assessed the effects of treating the cells with oligomycin, considered to be a form of chemical hypoxia because it acts on mitochondria, reducing ATP levels. It was found that in rat CMEC exposed to oligomycin (1 μg/ml) for 5 to 60 min, Na^+-K^+-Cl^- cotransporter activity, evaluated as bumetanide-sensitive K^+ influx, is significantly increased (Kawai et al., 1996a, b). Our studies have revealed that oligomycin (1 μg/ml) also significantly stimulates Na^+-K^+-Cl^- cotransporter activity of bovine CMEC (a 1.8-fold increase) (Foroutan et al., 2005). Rat CMEC have also been tested for the effect of a 24 h exposure to hypoxia (O_2 level of <2%) and it was found that, like oligomycin, true hypoxia also increases CMEC cotransporter activity (Kawai et al., 1996a). We have further evaluated the effects of hypoxia on CMEC Na^+-K^+-Cl^- cotransporter activity, testing both the sensitivity of the cotransporter to different levels of hypoxia and the time course of hypoxia effects. For this, we used a hypoxia chamber in which temperature, O_2 levels and CO_2 levels are precisely controlled and we varied O_2 levels (in the chamber atmosphere and assay media) from the control level of 19% down to 0.5% O_2 (inclusion of 5% CO_2 in the hypoxic chamber reduces O_2 below the atmospheric O_2 level of 20%). In those studies, we found that exposing the CMEC to 7.5%, 3%, 2% or 0.5% O_2 for 30 min significantly increased Na^+-K^+-Cl^- cotransporter activity (Fig. 30.7A). When cells were

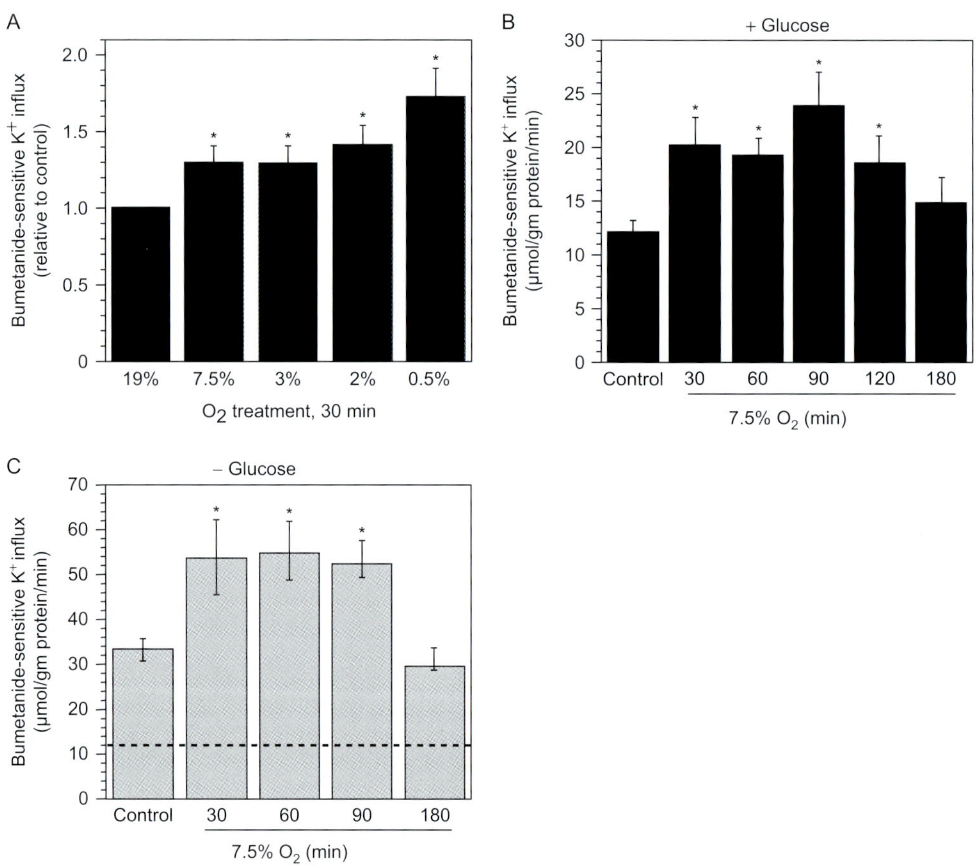

FIGURE 30.7 **A.** Effects of exposure to varying degrees of hypoxia on Na^+-K^+-Cl^- cotransporter activity of cerebral microvascular endothelial cells. CMEC monolayers were subjected to varying O_2 levels in glucose-containing media at 37°C in a hypoxia chamber, as described by us previously (Foroutan et al., 2005). Na^+-K^+-Cl^- cotransporter activity was measured as ouabain-insensitive, bumetanide-sensitive K^+ influx. **B. and C.** Effects of exposing CMEC to 7.5% O_2 for varying times in glucose-containing (B) and glucose-free (C) media, on Na^+-K^+-Cl^- cotransporter activity. CMEC were exposed to 19% O_2 (control) for 30 min or to 7.5% O_2 for varying times in media containing or lacking glucose, then Na^+-K^+-Cl^- cotransporter activity assessed as in A. Data are means±SEM of at least five experiments; * Significantly different from control by ANOVA. Note the dashed line in C depicting the control level of Na^+-K^+-Cl^- cotransporter activity in the presence of glucose (control values in B).

exposed to 7.5% O_2 for up to 180 min cotransporter activity was significantly elevated above the normoxic control through 120 min and then activity began to fall toward normoxic levels (Fig. 30.7B). This is a level of hypoxia as might be found in the penumbral region of an ischemic area (Kato et al., 1987; Baringa, 1996). At 2% O_2, a more severe hypoxia as can be found in the ischemic core, cotransporter activity was increased above control at 30 min, then fell toward control levels at 120 min. This suggests that hypoxia stimulation of the cotransporter is more prolonged in the ischemic penumbra than the core, consistent with the observation that increases in water and Na content are greater in the penumbra than the core (Kato et al., 1987). We and others have shown previously that the Na^+-K^+-Cl^- cotransporter is phosphorylated when activated, both in CMEC (Sun and O'Donnell, 1996) and in a variety of other cell types (O'Donnell et al., 1995b; Muzyamba et al., 1999; Flemmer et al., 2002; Giménez and Forbush, 2003; Flatman, 2005). Using an antibody that recognizes phosphorylated NKCC (p-NKCC) (Flemmer et al., 2002) and Western blot analyses, we observed significant increases in the abundance of p-NKCC but not total NKCC following 30 and 90 min treatment of CMEC with 7.5% O_2 (1.6- and 1.8-fold, respectively) (Foroutan et al., 2005) (Fig. 30.8). We also found that AVP (10 min, 100 nM) significantly increased the abundance of p-NKCC in the cells (Foroutan et al., 2005). It is noteworthy that previous studies evaluating the effects of oligomycin and 24 h exposures to severe hypoxia (<2%) found that for both treatments, cellular ATP levels fell significantly (~65% and 50% reductions, respectively). However, when we evaluated CMEC ATP levels following treatment with more moderate hypoxia or shorter exposures to severe hypoxia, we found ATP levels were not significantly reduced below control. In these studies, cell ATP did not fall after 30 min to 4 h exposures to either 7.5% or 2% O_2. Only after a 24 h exposure to 2% O_2 was ATP significantly reduced (by 40%). This suggests that the signaling pathway by which hypoxia stimulates CMEC Na^+-K^+-Cl^- cotransporter activity does not involve a drop in cell ATP levels (Foroutan et al., 2005).

An important question to address in testing the hypothesis that a luminal Na^+-K^+-Cl^- cotransporter functionally couples with the abluminal Na^+/K^+ pump to "hypersecrete" Na^+, Cl^- and water during cerebral ischemia is whether, in fact, the pump is still active in BBB endothelial cells under conditions of ischemia. Consistent with our observation that cell ATP levels do not fall more than 50% even after 24 h of severe hypoxia, in our studies we have found that none of the hypoxia treatments we examined reduced activity of the Na/K pump. In evaluating Na^+-K^+-Cl^- cotransporter activity we measured the effects

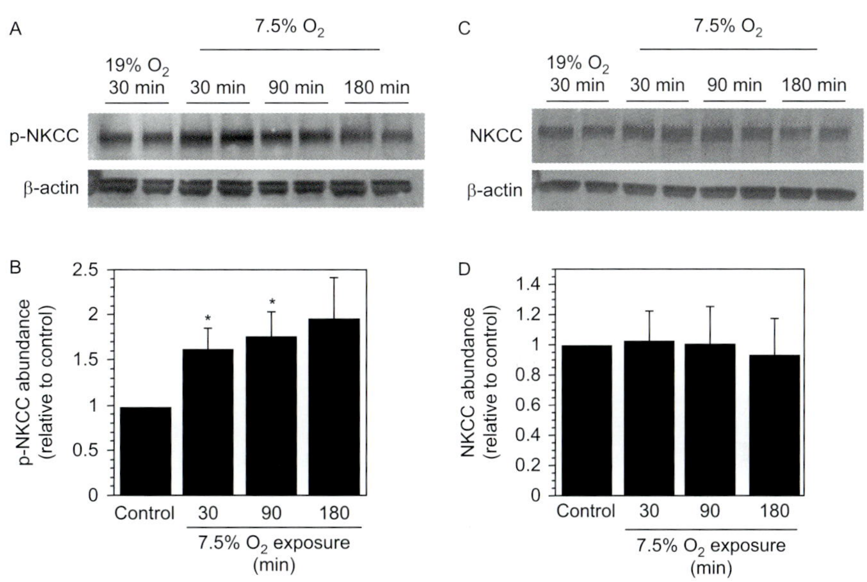

FIGURE 30.8 Increased expression of phosphorylated Na^+-K^+-Cl^- in cerebral microvascular endothelial cells following exposure to hypoxia. Western blot analyses were conducted to evaluate total Na^+-K^+-Cl^- (NKCC) and phosphorylated Na^+-K^+-Cl^- (p-NKCC), using T4 and R5 antibodies, respectively (42). **A. and C.** Representative Western blots of p-NKCC abundance and total NKCC abundance, respectively, in CMEC exposed to 19% O_2 for 30 minutes or to 7.5% O_2 for 30, 90 or 180 minutes. **B. and D.** Quantitation of p-NKCC abundance and total NKCC abundance, respectively, by densitometric analysis of Western blots following exposures to hypoxia. Note that 30 and 90 min exposures to 7.5% O_2 significantly increased p-NKCC without altering total NKCC abundance.

of both bumetanide and ouabain on K^+ influx in the CMEC. Thus, we evaluated cotransporter activity in the absence of ouabain as well as in the presence of ouabain, and we also evaluated Na^+/K^+ pump activity in the absence and in the presence of bumetanide. We found that ouabain-sensitive, bumetanide-insensitive pump activity (i.e. measured in the presence of bumetanide) was not altered by hypoxia treatments. However, if we measured Na^+/K^+ pump activity in the *presence* of Na^+-K^+-Cl^- cotransporter activity (i.e. in the absence of bumetanide), hypoxia actually increased Na^+/K^+ pump activity (Foroutan et al., 2005). This would be expected to occur if increased Na^+-K^+-Cl^- cotransporter activity delivers more Na^+ to the cell, thereby stimulating the Na^+/K^+ pump; as mentioned, the latter is highly sensitive to elevation of intracellular $[Na^+]$. This finding also supports the hypothesis that the cotransporter and the pump are functionally coupled during hypoxia.

iii. Aglycemia

Ischemic factors include not only AVP and hypoxia but also aglycemia. To determine whether aglycemia, independent of other ischemic factors, may stimulate BBB Na^+-K^+-Cl^- cotransporter activity, we evaluated CMEC bumetanide-sensitive K^+ influx in glucose-free media. We found that exposure to glucose-free, normoxic media increased cotransporter activity by 2.6-fold and further that this effect of aglycemia was additive with the effects of hypoxia. Thus, aglycemia caused a ~2.6-fold elevation of cotransporter activity on top of the stimulation observed with 7.5% O_2 exposures of 30 to 90 min (Fig. 30.7C). Note that the dashed line in the graph of Fig. 30.7C represents the normoxic control cotransporter activity level that we observed in the presence of glucose, as shown in the graph on the left (Fig. 30.7B). Thus, exposing CMEC to 7.5% O_2 in glucose-free media for 30 min caused a 4.2-fold increase in cotransporter activity.

Studies by Abbruscato and co-workers have examined the possibility that the fraction of Na^+-K^+-Cl^- cotransporter present in the BBB abluminal membrane is important for K^+ absorption during cerebral ischemia. In this regard, they have hypothesized that K^+ enters the BBB at the abluminal surface not only through the Na^+/K^+ pump but also through the Na^+-K^+-Cl^- cotransporter and then exits at the luminal membrane via a K^+ channel, allowing the BBB to offset the ischemia-induced elevation of brain extracellular $[K^+]$ by transporting K^+ from brain to blood even if the Na^+/K^+ pump starts to fail in prolonged severe hypoxia (Abbruscato et al., 2004; Yang et al., 2006). The results of these studies provided evidence that abluminal Na^+-K^+-Cl^- cotransporter activity is increased following 6 h of hypoxia (<3% O_2)/aglycemia by not after 24 h of hypoxia/aglycemia (Abbruscato et al., 2004).

iv. Endothelin

The peptide endothelin-1 (ET-1) is also increased during ischemia (Barone et al., 1994; Bian et al., 1994; Kawai et al., 1995; Lampl et al., 1997; Spatz et al., 1995; Suzuki et al., 1990). BBB endothelial cells both synthesize ET-1 and exhibit ET_A receptors (Bacic et al., 1992; Stanimirovic et al., 1994; Vigne et al., 1991; Yoshimoto et al., 1990). ET-1 has been hypothesized to modulate BBB functions during ischemia (Kawai et al., 1996b, 1995). Further, ET_A receptor antagonists have been shown to reduce brain edema injury following both transient and permanent MCAO in rats (Barone et al., 2000; Matsuo et al., 2001). Thus, it is possible that, like AVP, ET-1 is another peptide released during ischemia that may stimulate the BBB Na^+-K^+-Cl^- cotransporter to participate in ischemia-induced edema formation. To test this hypothesis, Kawai, Spatz and co-workers have evaluated the effects of endothelin on bumetanide-sensitive K^+ influx of rat on the human CMEC. They found that the cotransporter in CMEC of both rat and human is readily stimulated by ET-1 at low nanomolar concentrations (Kawai et al., 1995, 1996b, 1997; Spatz et al., 1997, 1998). Stimulation of rat CMEC Na^+-K^+-Cl^- cotransporter activity by ET-1 has also been reported by Vigne and co-workers (Vigne et al., 1994, 1995).

d. Estradiol Reduction of Ischemia Effects on the BBB Na^+-K^+-Cl^- Cotransporter

Estradiol provides neuroprotection in stroke, attenuating ischemia-induced edema and brain damage (Dubal et al., 1998; Zhang et al., 1998; Tuong et al., 1998; Rusa et al., 1999; Linford et al., 2000; Hurn and Macrae, 2000; Lee and McKewen, 2001; Dhandapani and Brann, 2002). We have found that treating ovariectomized (OVX) rats with estradiol (17β-estradiol, either 7 days or 30 min) reduces edema and infarct induced by 210 min of permanent MCAO, assessed as MR DWI-determined ADC values and by TTC staining (O'Donnell et al., 2006). This raises the possibility that since bumetanide inhibition of the BBB Na^+-K^+-Cl^- cotransporter reduces edema and infarct in permanent MCAO, estradiol may act, at least in part, by reducing ischemia stimulation of the cotransporter. Thus, we evaluated estradiol effects on bovine CMEC Na^+-K^+-Cl^- cotransporter activity. Here, we found that estradiol (1–100 nM) significantly reduced cotransporter activity whether the treatment time was 5 min or 3 h. Further, estradiol abolished stimulation of the cotransporter by AVP or by oligomycin (O'Donnell et al., 2006). Further, Western blot analysis revealed that estradiol also reduced abundance of the CMEC Na^+-K^+-Cl^- cotransporter protein, both following just one hour of estradiol treatment and after 7 days of treatment.

e. Hypoxia Effects on BBB Endothelial Cell Volume

The above observations collectively support the hypothesis that ischemic factors stimulate the BBB cotransporter to participate in edema formation through hypersecretion of Na^+, Cl^- and water from blood into brain, where the astrocytes swell as they take up the increased ions and water presented to them. However, it is important to consider whether the cotranporter could be contributing to edema in another manner, e.g. by causing the BBB endothelial cells themselves to swell. This is possible given that the Na^+-K^+-Cl^- cotransporter is known to be involved in regulation of intracellular volume for a variety of cell types, including endothelial cells, and some factors that stimulate cotransporter activity have been shown to drive increases in intracellular volume (O'Donnell, 1993). Thus, one must ask whether, and under what conditions, ischemic factor stimulation of BBB cotransporter activity could lead to swelling of the BBB endothelial cells. Such an event could, e.g., compress the microvessel lumen, further reducing blood flow and/or it could participate in breakdown of the BBB, which would contribute to edema formation (vasogenic edema in the latter case). We have addressed this question, using bovine CMEC and radioisotopic equilibration of [^{3}H]-3-0-methyl-D-glucose, to assess intracellular water space along with [^{14}C]-sucrose to correct for trapped extracellular water space. When the cells were exposed to hypoxia, significant increases in CMEC volume did not occur before 3h (7.5% or 3% O_2) or 4h (1% O_2) of treatment (Fig. 30.9). Cells exposed to 1% O_2 didn't swell before 4h and the swelling observed at 5h was less than that of cells exposed to 7.5% O_2. This is not entirely surprising, given the observation that more cerebral swelling occurs in peripheral regions of an ischemic zone (Kato et al., 1987). We also assessed net Na^+ and K^+ uptake by the CMEC during 7.5% O_2 treatments and found that cell Na content did not increase until 3h of exposure (Brillault et al., 2007). This is consistent with the Na^+-K^+-Cl^- cotransporter participating in vectorial transport of ions across the BBB during the early hours of ischemia, with cell swelling occurring more gradually. Treating CMEC with 7.5% O_2

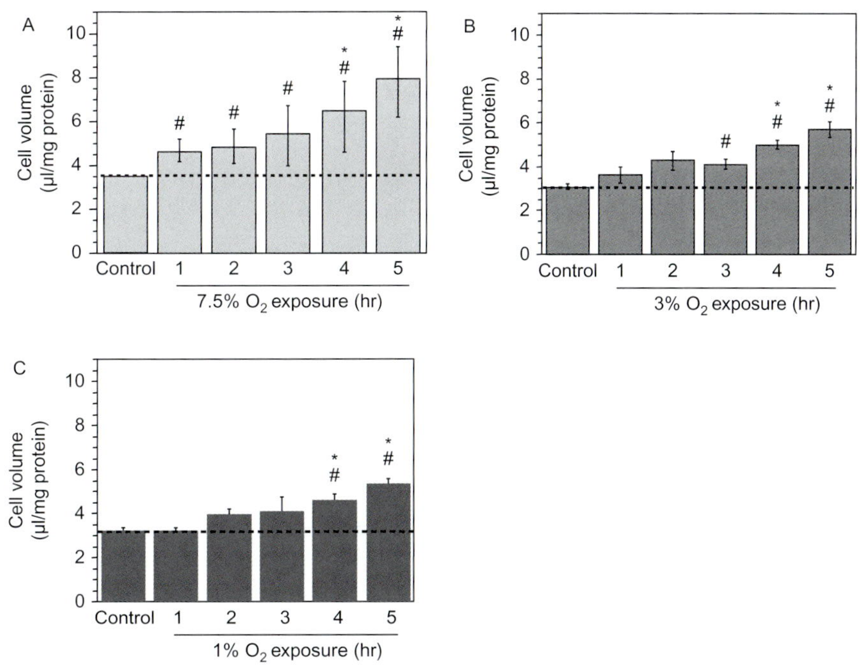

FIGURE 30.9 Effect of hypoxia on cerebral microvascular endothelial cell volume. CMEC monolayers were subjected to 19% O_2 (control) or to 7.5%, 3% or 1% O_2 (A, B and C, respectively) for 1 to 5h in glucose-containing DMEM-HEPES medium at 37°C using a hypoxia chamber. Cell volume was subsequently assessed using radioisotopic equilibration of [^{3}H]-3-o-methyl-D-glucose and [^{14}C]-sucrose for total and extracellular trapped space, respectively. Data are means±SEM of three experiments. * Significantly different from control by ANOVA, $p < 0.05$. # Significantly different from control by Student's paired t test, $p < 0.05$.

in the presence of bumetanide reduced cell swelling observed after 5 h, suggesting that the cotransporter in some manner contributes to CMEC swelling. Overall, these observations are consistent with cotransporter-mediated hypersecretion that starts within minutes of ischemia onset and also a cotransporter-mediated cell swelling with a slower onset and increasing over time, contributing to more widespread edema.

2. *A Role for Other Na^+ and Cl^- Transporters and Channels in Ischemia-induced Cerebral Edema?*

a. Na^+/H^+ Exchange

Previous studies have also suggested a role for a luminal Na^+/H^+ exchanger in ischemia-stimulated BBB secretion of Na^+, Cl^- and water. Ennis and co-workers found that i.v. administration of dimethylamiloride (DMA), an Na^+/H^+ exchange inhibitor, reduced ^{22}Na uptake by normal (non-ischemic) rat brain, evaluated by *in situ* perfusion methods (Ennis et al., 1996). Further, Betz and co-workers showed that DMA also attenuated ischemia-induced increases in brain water and Na following 4 h of permanent MCAO in rats (Betz et al., 1995). Several studies have provided evidence for the presence of Na^+/H^+ exchanger (NHE) RNA and protein in brain microvascular endothelial cells. Multiple isoforms of NHE have been described in various cell types, with NHE1–5 found in plasma membrane and NHE6–9 in organelle membranes (Chesler, 2003; Nakamura et al., 2005). mRNA for NHE1 has been demonstrated for rat CMEC (Kalaria et al., 1998) and mRNA for NHE2, NHE3 and NHE4 isoforms has also been reported to be present in the cells (Sipos et al., 2005). It has also been shown that basal intracellular pH (pH_i) of piglet CMEC is decreased by amiloride (Hsu et al., 1996) and that pH_i recovery from an acid load is inhibited by amiloride or ethylisopropylamiloride (EIPA) in piglet and rat CMEC (Sipos et al., 2005), suggesting that a Na^+/H^+ exchanger participates in pH regulation of BBB endothelial cells. In recent studies evaluating bovine and rat CMEC for the presence of NHE isoforms, Western blot analysis revealed that both NHE1 and NHE2 isoforms are present in bovine CMEC as well as in freshly isolated rat cerebral microvessels (Lam and O'Donnell, 2008). If either of these NHE isoforms participates in ischemia-induced edema formation by providing a luminal membrane BBB entry pathway for Na^+, then the NHE proteins should be present in the luminal membrane. Using perfusion-fixed rat brain with immuno-electron microscopy and specific antibodies for NHE1 and NHE2, we have found that both isoforms are present in the BBB endothelial cells *in situ*, with approximately 80% of both NHE1 and NHE2 in the luminal membrane compared to the abluminal membrane (Lam and O'Donnell, 2008). We have also evaluated Na^+/H^+ exchange activity in bovine CMEC, using the NH_4^+ prepulse method with the pH-sensitive dye BCECF to assess Na^+-dependent, Na^+/H^+ exchange inhibitor-sensitive H^+ flux (Boyarsky et al., 1988; McLean et al., 2000). For these experiments, we used HOE642 (cariporide), which is a potent, highly selective inhibitor of NHE1 that also inhibits NHE2 activity (at IC_{50} values ~150–200-fold higher than for NHE1) (Masereel et al., 2003; Scholz et al., 1995). We observed an Na^+-dependent, cariporide-sensitive H^+ flux in CMEC with an IC_{50} value of ~7 μM, consistent with the presence of both NHE1 and NHE2 activity in the cells. Exposing the CMEC to 30 min of either hypoxia (2%), aglycemia or AVP (1 to 100 nM) significantly increased Na^+/H^+ exchanger activity (Fig. 30.10). Other studies have shown that in CMEC an EIPA-sensitive ^{22}Na influx is stimulated by endothelin, another factor present during cerebral ischemia (Kawai et al., 1997; Spatz et al., 1997; Vigne et al., 1991). Thus, it appears that the CMEC Na^+/H^+ exchanger, like the Na^+-K^+-Cl^- cotransporter is stimulated by ischemic factors, suggesting that the exchanger may also participate in the BBB response to ischemia. Whether the increased Na^+-K^+-Cl^- exchanger activity is due to stimulation of NHE1, NHE2 or both, necessitates further investigation.

If a luminal BBB Na^+/H^+ exchanger participates in ischemia-induced edema formation, then inhibition of the exchanger by intravenous administration of cariporide is predicted to reduce edema and infarct in rats subjected to MCAO. Indeed, in experiments evaluating ADC values during permanent MCAO, rats treated with cariporide immediately before initiation of MCAO had significantly attenuated reductions in edema formation, assessed as the fall in ipsilateral/contralateral ADC ratios (Tran et al., 2004). They also exhibited significantly reduced infarct volumes following 180 min of MCAO, assessed as TTC-lesion volume.

While these studies collectively support the hypothesis that a luminal BBB Na^+/H^+ exchanger is stimulated during cerebral ischemia to participate in increased secretion of Na^+ and water into the brain, it is also possible that the exchanger may also, or even instead, contribute to ischemia-induced BBB endothelial cell swelling. In cell volume studies we found that when CMEC are exposed to 7.5% O_2 for 5 h in the presence of cariporide, CMEC swelling was significantly reduced. Further, when the cells were treated with both bumetanide (10 μM) and cariporide (10 μM)

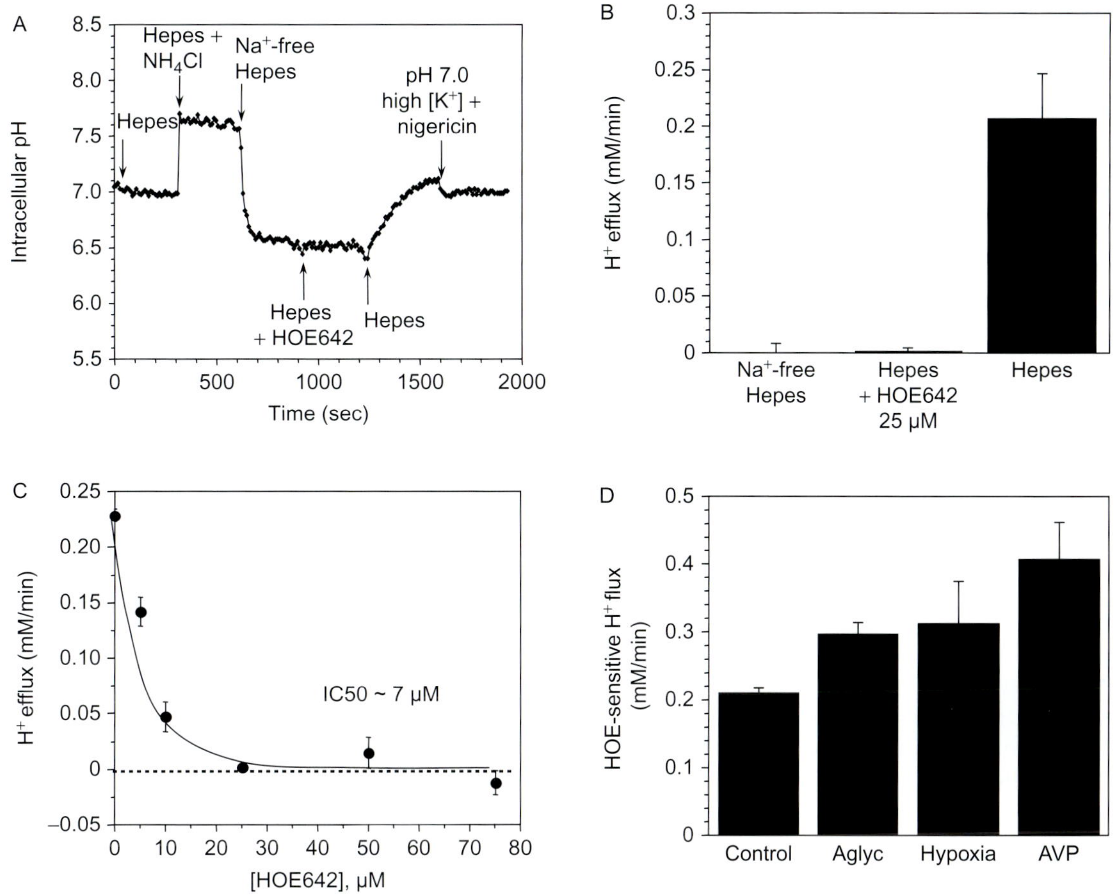

FIGURE 30.10 Na^+/H^+ exchange activity in cerebral microvascular endothelial cells. **A.** Representative NH_4Cl prepulse experiment showing the presence of an Na^+-dependent and HOE642 (25 μM)-sensitive pH recovery from intracellular acidification following removal of the NH_4Cl. Note that all Hepes-buffered media contain Na^+ except for the one labeled "Na^+-free Hepes". The trace shown represents the average value of 30 cells in the field of view. **B.** H^+ net flux values calculated from prepulse experiments. Data shown are means±SEM of three experiments. **C.** Dose response of HOE642 inhibition of Na^+-dependent H^+ efflux in CMEC ($n = 3$). The IC_{50} value is higher than predicted for NHE1 alone, consistent with the possibility that NHE2 also contributes to the observed Na^+/H^+ exchange activity in the cells. **D.** CMEC were exposed for 30 min to hypoxic media (~2% O_2), or to normoxic media that were either glucose-free (Aglyc) or contained AVP (100 nM), then Na^+/H^+ exchange activity assessed as in A. Values are means±SEM of three experiments each.

during 5 h of moderate hypoxia (7.5% O_2), the increase in cell volume was nearly abolished (Brillault et al., 2007). Clarification of the processes by which the Na^+-K^+-Cl^- cotransporter and Na^+/H^+ exchanger contribute to hypoxia-induced swelling of CMEC will require further study.

Davis and co-workers have also demonstrated the presence of Na^+/H^+ exchange protein in BBB endothelial cells and provided evidence that the exchanger may play a role in the increase in ischemia-induced brain damage in hypertensive animals (Hom et al., 2007). In those studies, brain microvessel abundance of NHE1 protein was increased in spontaneously hypertensive rats (SHR) following development of hypertension but not in age-matched normotensive control rats (WKY). Further, intracerebroventricular administration of the Na^+/H^+ exchange inhibitor DMA significantly reduced TTC-defined infarct following 4 h of permanent MCAO in the hypertensive SHR animals (Hom et al., 2007). This is consistent with the observation that the severity of brain damage in ischemic stroke is generally more pronounced in hypertensive than normotensive individuals (Barone et al., 1992; Kannel et al., 1970; Li et al., 2005; Liu et al., 2005; Sacco et al., 1991; Stokes et al., 1989).

b. Cl^- Channels and Cl^-/HCO_3^- Exchangers

The pathways that mediate increased secretion of Cl^- across the BBB during ischemic stroke are not well defined and, as discussed above for NaCl secretion occurring during normoxic conditions (Section II), the available evidence suggests that an abluminal volume-activated Cl^- channel may be the abluminal Cl^- efflux pathway (von Weikersthal et al., 1999). Thus, increased Na^+ and Cl^- influx at the luminal membrane, mediated by Na^+-K^+-Cl^- cotransporter and perhaps also Na^+/H^+ exchange, is predicted to stimulate increased Na^+ and Cl^- efflux at the abluminal membrane via Na^+/K^+ pump and Cl^- channel. To clarify whether this Cl^- channel may function in this capacity during ischemic stroke will require other studies, including determination of whether the volume-activated Cl^- channel is indeed present in the abluminal membrane under ischemic conditions. Given that estradiol appears to attenuate ischemia-induced edema formation and also to block ischemic factor stimulation of the BBB Na^+-K^+-Cl^- cotransporter, it is noteworthy that the rat CMEC volume-activated Cl^- current was also found to be blocked by the estrogen antagonist tamoxifen (von Weikersthal et al., 1999). This suggests that the edema reducing effects of estradiol may include inhibition of an abluminal volume-sensitive Cl^- channel.

Recent studies providing evidence for the presence of Na^+-independent and Na^+-dependent Cl^-/HCO_3^- exchangers as well as Na^+-HCO_3^- cotransporters in BBB endothelial cells suggest additional pathways for Cl^- and also HCO_3^- secretion (Nicola et al., 2007; Taylor et al., 2006). Thus, as discussed in Section II above for ion secretion by normoxic BBB, it is possible that a polarized distribution of these Cl^- and HCO_3^- transporters in the luminal and abluminal membranes provides for vectorial transport of the anions across the BBB. It should also be recognized that an abluminal Na^+-dependent Cl^-/HCO_3^- exchanger could also participate in Cl^- efflux into the brain. While the BBB transporters discussed here are thought to participate in increased secretion of Na^+, Cl^- and probably also HCO_3^- into the brain during ischemia, they also very likely function in regulation of intracellular pH during ischemic challenge. In this regard, Nicola and collaborators have shown that in response to an increased acid load in the CMEC, as can occur during hypoxia, the cells can regulate their pH by Na^+/H^+ exchange-mediated H^+ efflux and also by Na^+-HCO_3^- cotransporter-mediated HCO_3^- influx, with both mechanisms working to correct mild acid loads and Na^+/H^+ exchange primarily handling large acid loads (Nicola et al., 2007). Clarifying the roles of Na^+-dependent and Na^+-independent Cl^-/HCO_3^- exchange, as well as Na^+-HCO_3^- cotransport, at the BBB clearly requires further investigation, not the least of which will be to determine the plasma membrane distribution (i.e. luminal, abluminal or both) of these transporters in BBB endothelial cells.

IV. CONCLUSIONS

BBB ion transporters play a vital role in brain electrolyte and water distribution in both health and disease. We now have evidence that a luminal BBB Na^+-K^+-Cl^- cotransport participates in the increased secretion of Na^+, Cl^- and water that contributes to edema formation during ischemia and probably also participates in the normal lower rate of NaCl secretion into the healthy, normoxic brain. Mounting evidence also suggests a BBB Na^+/H^+ exchanger serves as a luminal Na^+ entry pathway for secretion of Na^+ into the brain, under both normoxic and ischemic conditions. Na^+ entering the BBB endothelial cell via the exchanger may be accompanied by Cl^- entering the cell via Cl^-/HCO_3^- exchange as well as Na^+-K^+-Cl^- cotransport. The recently discovered presence of Na^+-dependent Cl^-/HCO_3^- exchange and Na^+-HCO_3^- cotransport in BBB endothelial cells suggests the possibility that these transporters may also participate in NaCl secretion across the BBB. Candidates for abluminal efflux of Cl^- that accompanies Na^+/K^+ pump-mediated efflux of Na^+ during NaCl secretion include a volume-activated Cl^- channel and possibly also Na^+-dependent Cl/HCO_3 exchange. However, it is not yet known whether these are present in the abluminal BBB membrane. In addition to vectorial transport, other important ion transporter and channel BBB functions include regulation of intracellular pH and/or volume. The Na^+-K^+-Cl^- cotransporter very likely participates in BBB volume regulation, as does the Na^+/H^+ exchanger and Cl^-/HCO_3^- exchanger. Further, there is evidence that the Na^+/H^+ exchanger, Na^+-HCO_3^- cotransporter and both Na^+-dependent and Na^+-independent Cl^-/HCO_3^- exchangers all contribute to regulation of intracellular pH of BBB endothelial cells. Additional studies are needed to clarify the roles of these transporters and channels in BBB function. Given the importance of the BBB in brain water and electrolyte distribution in both health and disease, understanding how these transporters and channels are distributed in the cells and how they respond to changing conditions in the neuron microenvironment is essential for developing new therapies to treat ischemic stroke and other brain pathologies.

References

Abbruscato, T.J., Lopez, S.P., Roder, K., and Paulson, J.R. (2004). Regulation of blood–brain barrier Na,K,2Cl-cotransporter through phosphorylation during in vitro stroke conditions and nicotine exposure. *J. Pharmacol. Exper. Therapeut.* **310**, 459–468.

Amiry-Moghaddam, M., Otsuka, T., Hurn, P.D., Traystman, R.J., Haug, F.-M., Froehner, S.C., Adams, M.E., Neely, J.D., Agre, P.,

Ottersen, O.P., and Bhardwaj, A. (2003). An α-syntrophin-dependent pool of AQ4 in astroglial endfeet confers bidirectional water flow between blood and brain. *Proc. Natl. Acad. Sci.* **100**, 2106–2111.

Bacic, F., Uematsu, S., McCarron, R.M., and Spatz, M. (1992). Secretion of immunoreactive endothelin-1 by capillary and microvascular endothelium of human brain. *Neurochem. Res.* **17**, 699–702.

Banks, W.A. (2008). The blood–brain barrier: connecting the gut and the brain. *Regulatory Peptides* **149**.

Baringa, M. (1996). Finding new drugs to treat stroke. *Science* **272**, 646–666.

Barone, F.C., Globus, M.Y.-T., Price, W.J., White, R.F., Storer, B.L., Feuerstein, G.Z., Busto, R., and Ohlstein, E.H. (1994). Endothelin levels increase in rat focal and global ischemia. *J. Cereb. Blood Flow Metab.* **14**, 337–342.

Barone, F.C., Ohlstein, E.H., Hunter, A.J., Campbell, C.A., Hadingham, S.H., Parsons, A.A., Yang, Y., and Shohami, E. (2000). Selective antagonism of endothelin-A-receptors improves outcome in both head trauma and focal stroke in rat. *J. Cardio. Pharmacol.* **36**, S357–S361.

Barone, F.C., Price, W.J., White, R.F., Willette, R.N., and Feuerstein, G.Z. (1992). Genetic hypertension and increased susceptibility to berebral ischemia. *Neurosci. Biobehav. Rev.* **16**, 219–233.

Beck, D., Vinters, J.V., Hart, M.N., and Cancilla, P.A. (1984). Glial cells influence polarity of the blood–brain barrier. *J. Neuropathol. Exper. Neurol.* **43**, 219–224.

Beck, D.W., Dunbar, E.C., and Jr., M. M. P. (1990). Glioma cell influence on cerebral endothelial cell Na^+-K^+-ATPase. *Adv. Neurol.*, **52**, 415–420.

Beck, D.W., Roberts, R.L., and Olson, J.J. (1986). Glial cells influence membrane-associated enzyme activity at the blood–brain barrier. *Brain Res.* **381**, 131–187.

Betz, A.L. (1983a). Sodium transport from blood to brain: inhibition by furosemide and amiloride. *J. Neurochem.* **41**, 1158–1164.

Betz, A.L. (1983b). Sodium transport in capillaries isolated from rat brain. *J. Neurochem.* **41**, 1150–1157.

Betz, A.L. (1986). Transport of ions across the blood–brain barrier. *Fed. Proc.* **45**, 2050–2054.

Betz, A.L. (1996). Alterations in cerebral endothelial cell function in ischemia. *Adv. Neurol.* **71**, 301–313.

Betz, A.L., Ennis, S.R., Ren, X., Schielke, G.P., and Keep, R.F. (1995). *Blood–Brain Barrier Sodium Transport and Brain Edema Formation.* Plenum Press, New York.

Betz, A.L., Firth, J.A., and Goldstein, G.W. (1980). Polarity of the blood–brain barrier: distribution of enzymes between the luminal and antiluminal membranes of the brain capillary endothelial cells. *Brain Res.* **192**, 17–28.

Betz, A.L. and Goldstein, G.W. (1986). Specialized properties and solute transport in brain capillaries. *Annu. Rev. Physiol.* **48**, 250–421.

Betz, A.L., Keep, R.F., Beer, M.E., and Ren, X. (1994). Blood–brain barrier permeability and brain concentration of sodium, potassium, and chloride during focal ischemia. *J. Cereb. Blood Flow Metab.* **14**, 29–37.

Bian, L.-G., Zhang, T.-X., Zhao, W.-G., Shen, J.-K., and Yang, G.-Y. (1994). Increased endothelin-1 in the rabbit model of middle cerebral artery occlusion. *Neurosci. Lett.* **174**, 47–50.

Boron, W.F. (2004). Regulation of intracellular pH. *Adv. Physiol. Edu.* **28**, 160–179.

Bourke, R.S., Kimelberg, H.K., Nelson, L.R., Barron, K.D., Auen, E.L., Popp, A.J., and Waldman, J.B. (1980). Biology of glial swelling in experimental brain edema. *Adv. Neurol.: Brain Edema* **28**, 99–109.

Boyarsky, G., Ganz, M.B., Sterzel, R.B., and Boron, W.F. (1988). pH regulation in single glomerula mesangial cells. I. Acid extrusion in the absence and presence of HCO_3^-. *Am. J. Physiol. Cell Physiol.* **255**, C844–C856.

Bradbury, M.W.B. (1984). The structure and function of the blood–brain barrier. *Fed. Proc.* **43**, 186–190.

Brillault, J., Lam, T.I., Rutkowsky, J.M., Foroutan, S., and O'Donnell, M.E. (2007). Hypoxia effects on cell volume and ion uptake of cerebral microvascular endothelial cells. *Am. J. Physiol. Cell Physiol.* **294**, C88–C96.

Bronner, L.L., Kanter, D.S., and Manson, J.E. (1995). Primary prevention of stroke. *New Eng. J. Med.* **333**, 1392–1400.

Chesler, M. (2003). Regulation and modulation of pH in the brain. *Physiol. Rev.* **83**, 1183–1221.

Chipperfield, A.R. (1996). The (Na^+-K^+-Cl^-) co-transport system. *Clin. Sci.* **71**, 465–476.

Cserr, H.F., DePasquale, M., Patlak, C.S., and Pullen, R.G.L. (1989). Convection of cerebral interstitial fluid and its role in brain volume regulation. *Ann. NY Acad. Sci.* **481**, 123–134.

Dallas, S., Miller, D.S., and Bendayan, R. (2006). Multidrug resistance-associated proteins: expression and function in the central nervous system. *Pharm. Rev.* **58**, 140–161.

DePetris, L., Luchetti, A., and Emma, F. (2001). Cell volume regulation and transport mechanisms across the blood–brain barrier: implications for the management of hypernatraemic states. *Eur. J. Pediat.* **160**, 71–77.

Dhandapani, K.M. and brann, D.W. (2002). Protective effects of estrogen and selective estrogen receptor modulators in the brain. *Biol. Repro.* **67**, 1379–1385.

Dickinson, L.D. and Betz, A.L. (1992). Attenuated development of ischemic brain edema in vasopressin-deficient rats. *J. Cereb. Blood Flow Metab.* **12**, 681–690.

Dóczi, T. (1993). Volume regulation of the brain tissue – a survey. *Acta Neurochirurgica* **121**, 1–8.

Dóczi, T., László, F.A., Szerdahelyi, P., and Joó, F. (1984). The role of vasopressin in brain oedema formation: further evidence obtained from Brattleboro diabetes insipidus rat with subarachnoid haemorrhage. *Neurosurgery* **14**, 436–440.

Dubal, D.B., Kashon, M.L., Pettigrew, L.C., Ren, J.M., Finkelstein, S.P., Rau, S.W., and Wise, P.M. (1998). Estradiol protects agains ischemic injury. *J. Cereb. Blood Flow Metab.* **18**, 1253–1258.

Ennis, S.R., Ren, X., and Betz, A.L. (1996). Mechanisms of sodium transport at the blood–brain barrier studies with in situ perfusion of rat brain. *J. Neurochem.* **66**, 756–763.

Eveloff, J.L. and Warnock, D.G. (1987). Activation of ion transport systems during cell volume regulation. *Am. J. Physiol.* **252**, F1–F10.

Fischer, H., Illik, B., Finkbeiner, W.E., and Widdicombe, J.H. (2007). Basolateral Cl channels in primary airway cultures. *Am. J. Physiol. Lung Cell. Mol. Physiol.* **292**, L1432–L1443.

Flatman, P.W. (2005). Activation of ferret erythrocyte Na^+-K^+-$2Cl^-$ cotransport by deoxygenation. *J. Physiol.* **563.2**, 421–431.

Flemmer, A.W., Giménez, I., Dowd, B.F.X., Darman, R.B., and Forbush, B. (2002). Activation of the Na-K-Cl cotransporter NKCC1 detected with a phospho-specific antibody. *J. Biol. Chem.* **277**, 37551–37558.

Foroutan, S., Brillault, J., Forbush, B., and O'Donnell, M.E. (2005). Moderate to severe ischemic conditions increase activity and phosphorylation of the cerebral microvascular endothelial cell Na-K-Cl cotransporter. *Am. J. Physiol. Cell Physiol.* **289**, C1492–C1501.

Friedman, P.A. and Andreoli, T.E. (1982). CO2-stimulated NaCl absorption in the mouse renal cortical thick ascending limb of Henle. *J. Gen. Physiol.* **80**, 683–711.

Gaillard, P.J., Sandt, I.C.J.v.d., Voorwinden, L.H., Vu, D., Nielsen, J.L., Boer, A.G. d., and Breimer, D.D. (2000). Astrocytes increase the functional expression of P-glycoprotein in an in vitro model of the blood–brain barrier. *Pharm. Res.*, **17**, 1198–1205.

Giménez, I. and Forbush, B. (2003). Short-term stimulation of the renal Na-K-Cl cotransporter (NKCC2) by vasopressin involves phosphorylation and membrane translocation of the protein. *J. Biol. Chem.* **278**, 26946–26951.

Goldstein, G.W. and Betz, A.L. (1986). The blood–brain barrier. *Scientific American* **255**, 74–83.

Gotoh, O., Asano, T., Koide, T., and Takakura, K. (1985). Ischemic brain edema following occlusion of the middle cerebral artery in the rat. I: The time course of the brain water, sodium and potassium contents and blood–brain barrier permeability to 125I-albumin. *Stroke* **16**, 101–109.

Grant, G.A., Abbott, N.J., and Janigro, D. (1998). Understanding the physiology of the blood–brain barrier: in vitro models. *News Physiol. Sci.* **13**, 287–293.

Haas, M. (1994). The Na-K-Cl cotransporters. *Am. J. Physiol.* **267**, C869–C885.

Hasegawa, Y., Fisher, M., Latour, L.L., Dardzinski, B.J., and Sotak, C.H. (1994). MRI diffusion mapping of reversible and irreversible ischemic injury in focal brain ischemia. *Neurology* **44**, 1484–1490.

Hatashita, S. and Hoff, J. (1990). Brain edema and cerebrovascular permeability during cerebral ischemia in rats. *Stroke* **21**, 582–588.

Hawkins, R.A., O'Kane, R.L., Simpson, I.A., and Vina, R. (2006). Structure of the blood–brain barrier and its role in the transport of amino acids. *J. Nutri* **136**, 218A–226S.

Hebert, S.C., Gamba, G., and Kaplan, M. (1996). The electroneutral Na^+-K^+-Cl^- cotransport family. *Kidney Int.* **49**, 1638–1641.

Hom, S., Fleegal, M.A., Egleton, R.D., Campos, C.R., Hawkins, B.T., and Davis, T.P. (2007). Comparative changes in the blood–brain barrier and cerebral infarction of SHR and WKY rats. *Am. J. Physiol. Reg. Integr. Comp. Physiol.* **292**, R1881–R1892.

Hoyer, J., Popp, R., Meyer, J., Galla, H.-J., and Gögelein, J. (1991). Angiotensin II, vasopressin and GTP[γ-S] inhibit inward-rectifying K^+ channels in porcine cerebral capillary endothelial cells. *J. Membr. Biol.* **123**, 55–62.

Hsu, P., Haffner, J., Albuquerque, M.L., and Leffler, C.W. (1996). pHi in piglet cerebral microvascular endothelial cells: recovery from an acid load. *Proc. Soc. Experi. Biol. Med.* **212**, 256–262.

Hurn, P.D. and Macrae, I.M. (2000). Estrogen as a neuroprotectant in stroke. *J. Cereb. Blood Flow Metab.* **20**, 631–652.

Iadecola, C. (1999). Mechanisms of cerebral ischemic damage. In *Cerebral Ischemia: Molecular and Cellular Pathophysiology* (Walz, W., ed.), pp. 3–34. Humana Press, Totowa, NJ.

Janigro, D., West, B.A., Gordon, E.L., and Winn, H.R. (1993). ATP-sensitive K^+ channels in rat aorta and brain microvascular endothelial cells. *Am. J. Physiol. Cell Physiol.* **265**, C812–C821.

Janzer, R.C. (1993). The blood–brain barrier: Cellular basis. *J. Inher. Metab. Dis.* **16**, 639–647.

Janzer, R.C. and Raff, M.C. (1987). Astrocytes induce blood–brain barrier properties in endothelial cells. *Nature* **325**, 253–256.

Jójárt, I., Joó, F., Siklós, L., and László, F.A. (1984). Immunoelectron histochemical evidence for innervation of brain microvessels by vasopressin-immunoreactive neurons in the rat. *Neurosci. Lett.* **51**, 259–264.

Kalaria, R.N., Premkumar, D.R., Lin, C.W., Kroon, S.N., Bae, J.Y., Sayre, L.M., and LaManna, J.C. (1998). Identification and expression of the Na^+/H^+ exchanger in mammalian cerebrovascular and choroidal tissues: characterization by amiloride-sensitive [3 H]MIA binding and RT-PCR analysis. *Brain Res. Mol. Brain Res.* **58**, 178–187.

Kannel, W.B., Wolf, P.A., Verter, J., and McNamara, P.M. (1970). Epidemiologic assessment of the role of blood pressure in stroke The Framingham study. *J. Am. Med. Assoc.* **214**, 301–310.

Kato, H., Kogure, K., Sakamoto, N., and Watanabe, T. (1987). Greater disturbance of water and ion homeostasis in the periphery of experimental focal cerebral ischemia. *Exper. Neurol.* **96**, 118–126.

Kawai, N., McCarron, R.M., and Spatz, M. (1996a). Effect of hypoxia on Na^+-K^+-Cl^- cotransport in cultured brain capillary endothelial cells of the rat. *J. Neurochem.* **66**, 2572–2579.

Kawai, N., McCarron, R.M., and Spatz, M. (1996b). Na^+-K^+-Cl^- cotransport system in brain capillary endothelial cells: response to endothelin and hypoxia. *Neurochem. Res.* **21**, 1259–1266.

Kawai, N., McCarron, R.M., and Spatz, M. (1997). The effect of endothelins on ion transport systems in cultured rat brain capillary endothelial cells. *Acta Neurochir.* **70**, 138–140.

Kawai, N., Yamamoto, T., Yamamoto, H., McCarron, R.M., and Spatz, M. (1995). Endothelin 1 stimulates Na^+,K^+-ATPase and Na^+-K^+-Cl^- cotransport through ET_A receptors and protein kinase C-dependent pathway in cerebral capillary endothelium. *J. Neurochem.* **65**, 1588–1596.

Keep, R.F. (1993). Potassium transport at the blood–brain and blood–CSF barriers. In *Frontiers in Cerebral Vascular Biology: Transport and Its Regulation* (Drewes, L.R. and Betz, A.L., eds), pp. 43–54. Plenum Press, New York.

Keep, R.F., Ennis, S.R., Beer, M.E., and Betz, A.L. (1995a). Developmental changes in blood–brain barrier potassium permeability in the rat: relation to brain growth. *J. Physiol.* **488**, 439–448.

Keep, R.F., II, Ulanski, L.J., Ziang, J., Ennis, S.R., and Betz, A.L. (1999). Blood–brain barrier mechanisms involved in brain calcium and potassium homeostasis. *Brain Res.* **815**, 200–205.

Keep, R.F., Stummer, W., Klarr, S.A., Furspan, P.B., and Betz, A.L. (1995b). Mechanisms and control of blood–brain potassium transport. In *New Concepts of a Blood–Brain Barrier* (Greenwood, J., Begley, D.J., and Segal, M.B., eds), pp. 169–174. Plenum, NY.

Kempski, O., Rosen, S.V., Weight, H., Staub, F., Peters, J., and Baethmann, A. (1991). Glial ion transport and volume control. In *Glial–Neuronal Interaction* (Abbott, N.J., ed.), pp. 306–317. The New York Academy of Sciences, New York, NY.

Kempski, O., Spatz, M., Valet, G., and Baethmann, A. (1985). Cell volume regulation of cerebrovascular endothelium in vitro. *J. Cell. Physiol.* **123**, 51–54.

Kimelberg, H.K. (1995). Current concepts of brain edema Review of laboratory investigations. *J. Neurosurg.* **83**, 1051–1059.

Kimelberg, H.K. (1999). Cell swelling in cerebral ischemia. In *Cerebral Ischemia: Molecular and Cellular Pathophysiology* (Walz, W., ed.), pp. 5–8. Humana Press, Totowa, NJ.

Klatzo, I. (1994). Evolution of brain edema concepts. *Acta Neurochir. [Suppl.]* **60**, 3–6.

Knight, R.A., Dereski, M.O., Helpern, J.A., Ordidge, R.J., and Chopp, M. (1994). Magnetic resonance imaging assessment of evolving focal cerebral edema Comparison with histopathology in rats. *Stroke* **25**, 1252–1262.

Kregenow, F.M. (1981). Osmoregulatory salt transporting mechanisms: control of cell volume in anisotonic media. *Annu. Rev. Physiol.* **43**, 493–505.

Kretzschmar, R. and Ermisch, A. (1989). Arginine-vasopressin binding to isolated hippocampal microvessels of rats with different endogenous concentrations of the neuropeptide. *Exper. Clin. Endocrinol.* **94**, 151–156.

Kusuhara, H. and Sugiyama, Y. (2005). Active efflux across the blood–brain barrier: role of the solute carrier family. *J. Am. Soc. Exper. Neurotherapeut.* **2**, 73–85.

Lam, T.I. and O'Donnell, M.E. (2008). BBB Na/H exchange: plasma membrane distribution of NHE1 and NHE2 isoforms and stimulation by arginine vaspopressin. *FASEB J.* **22**, A734.3.

Lam, T.I., Wise, P.M., and O'Donnell, M.E. (2009). Cerebral Microvascular Endothelial Cell Na/H Exchange: Evidence for the Presence of NHE1 and NHE2 Isoforms and Regulation by Arginine Vasopressin. *Am J Physiol Cell Physiol.* (in press).

Lampl, Y., Fleminger, G., Gilad, R., Galron, R., Sarova-Pinhas, E., and Sokilovsky, M. (1997). Endothelin in cerebrospinal fluid and plasma of patients in the early stage of ischemic stroke. *Stroke* **28**, 1951–1955.

Landgraf, R. (1992). Central release of vasopressin: stimuli, dynamics, consequences. *Prog. Brain Res.* **91**, 29–39.

Lee, S.J. and McKewen, B.S. (2001). Neurotrophic and neuroprotective actions of estrogens and their therapeutic implications. *Annu. Rev. Pharmacol. Toxicol.* **41**, 569–591.

Leybaert, L. (2005). Neurobarrier coupling in the brain: a partner of neurovascular and neurometabolic coupling? *J. Cereb. Blood Flow Metab.* **25**, 2–16.

Leybaert, L., DeBock, M., VanMoorhem, M., Decrock, E., and DeVuyst, E. (2007). Neurobarrier coupling in the brain: adjusting glucose entry with demand. *J. Neurosci. Res.* **85**, 213–220.

Li, C., Engstrom, G., Hedblad, B., Berglund, G., and Janzon, L. (2005). Blood pressure control and risk of stroke: a population-based prospective cohort study. *Stroke* **36**, 725–730.

Lin, J.D. (1985). Potassium transport is isolated cerebral microvessels from the rat. *Jap. J. Physiol.* **35**, 817–830.

Lin, J.D. (1988). Effect of osmolarity on potassium transport in isolated cerebral microvessels. *Life Sci.* **43**, 325–333.

Linford, N., Wade, C., and Dorsa, D. (2000). The rapid effects of estrogen are implicated in estrogen-mediated neuroprotection. *J. Neurocytol.* **29**, 367–374.

Liu, K.-F., Li, F., Tatlisumak, T., Garcia, J.H., Sotak, C.H., Fisher, M., and Fenstermacher, J.D. (2001). Regional variations in the apparent diffusion coefficient and the intracellular distribution of water in rat brain during acute focal ischemia. *Stroke* **32**, 1897–1905.

Liu, X.F., vanMelle, G., and Bogousslavsky, J. (2005). Analysis of risk factors in 3901 patients with stroke. *Chin. Med. Sci. J.* **20**, 35–39.

Löscher, W. and Patschka, H. (2005). Blood–brain barrier active efflux transporters: ATP-binding cassette gene family. *J. Am. Soc. Experi. Neurotherapeut.* **2**, 86–98.

Löscher, W. and Potschka, H. (2005). Drug resistance in brain diseases and the role of drug efflux transporter. *Nature Rev. Neurosci.* **6**, 591–602.

Masereel, B., Pochet, L., and Laeckmann, D. (2003). An overview of inhibitors of Na^+/H^+ exchanger. *Eur. J. Med. Chem.* **38**, 547–554.

Matsuo, Y., Mihara, S., Ninomiya, M., and Fujimoto, M. (2001). Protective effect of endothelin type A receptor antagonist on brain edema and injury after transient middle cerebral artery occlusion in rats. *Stroke* **32**, 2143–2148.

McLean, L.A., Roscoe, J., Jørgensen, N.K., Gorin, F.A., and Cala, P.M. (2000). Malignant gliomas display altered pH regulation by NHE1 compared with nontransformed astrocytes. *Am. J. Physiol. Cell Physiol.* **278**, C676–C688.

McManus, M.L., Churchwell, K.B., and Strange, K. (1995). Regulation of cell volume in health and disease. *New Eng. J. Med.* **333**, 1260–1266.

Menzies, S.A., Betz, A.L., and Hoff, J.T. (1993). Contributions of ions and albumin to the formation and resolution of ischemic brain edema. *J. Neurosurg.* **78**, 257–266.

Miller, D.S., Bauer, B., and Hartz, A.M.S. (2008). Modulation of P-glycoprotein at the blood–brain barrier: opportunities to improve central nervous system pharmacotherapy. *Pharmacol. Rev.* **60**, 196–209.

Moseley, M.E., Cohen, Y., Mintorovitch, J., Chileuitt, L., Shimizu, H., Kucharczyk, J., Wendland, M.F., and Weinstein, P.R. (1990). Early detection of regional cerebral ischemia in cats: comparison of diffusion- and T2-weighted MRI and spectroscopy. *Magn. Reson. Med.* **14**, 330–346.

Muzyamba, M.C., Cossins, A.R., and Gibson, J.S. (1999). Regulation of Na-K-2Cl cotransport in turkey red cells: the role of oxygen tension and protein phosphorylation. *J. Physiol.* **517.2**, 421–429.

Nakamura, N., Tanaka, S., Teko, Y., Mitsui, K., and Kanazawa, H. (2005). Four Na^+/H^+ exchanger isoforms are distributed to golgi and post golgi compartments and are involved in organelle pH regulation. *J. Biol. Chemi.* **280**, 1561–1572.

Neuwelt, E., Abbott, N.J., Abrey, L., Banks, W.A., Blakley, B., Davis, T., Engelhardt, B., Grammas, P., Nedergaard, M., Nutt, J., Pardridge, W., Rosenberg, G.A., Smith, Q., and Drewes, L.R. (2008). Strategies to advance translational research into brain barriers. *Lancet Neurol.* **7**, 84–96.

Nicola, P.A., Taylor, C.J., Wang, S., Barrand, M.A., and Hladky, S.B. (2007). Transport activities involved in intracellular pH recovery following acid and alkali challenges in rat brain microvascular endothelial cells. *Pflugers Arch. – Eur. J. Physiol.* **456**, 801–812.

O'Donnell, M.E. (1993). Role of Na-K-Cl cotransport in vascular endothelial cell volume regulation. *Am. J. Physiol.* **264**, C1316–C1326.

O'Donnell, M.E., Duong, V., Suvatne, S., Foroutan, S., and Johnson, D.M. (2005). Arginine vasopressin stimulation of cerebral microvascular endothelial cell Na-K-Cl cotransport activity is V1 receptor- and [Ca]-dependent. *Am. J. Physiol. Cell Physiol.* **289**, C283–C292.

O'Donnell, M.E., Lam, T.I., Tran, L.Q., Foroutan, S., and Anderson, S.E. (2006). Estradiol reduces activity of the blood–brain barrier Na-K-Cl cotransporter and decreases edema formation in permanent middle cerebral artery occlusion. *J. Cereb. Blood Flow Metab.* **26**, 1234–1249.

O'Donnell, M.E., Martinez, A., and Sun, D. (1995a). Cerebral microvascular endothelial cell Na-K-Cl cotransport: regulation by astrocyte-conditioned medium. *Am. J. Physiol.* **268**, C747–C754.

O'Donnell, M.E., Martinez, A., and Sun, D. (1995b). Endothelial Na-K-Cl cotransport regulation by tonicity and hormones: phosphorylation of cotransport protein. *Am. J. Physiol.* **269**, C1513–C1523.

O'Donnell, M.E., Tran, L., Lam, T., Liu, X.B., and Anderson, S.E. (2004). Bumetanide inhibition of the blood–brain barrier Na-K-Cl cotransporter reduces edema formation in the rat middle cerebral artery occlusion model of stroke. *J. Cereb. Blood Flow Metab.* **24**, 1046–1056.

O'Grady, S.M., Palfrey, H.C., and Field, M. (1987). Characteristics and functions of Na-K-Cl cotransport in epithelial tissues. *Am. J. Physiol.* **253**, C177–C192.

Pardridge, W.M. (2007). Drug targeting to the brain. *Pharmacol. Res.* **24**, 1733–1744.

Paulais, M. and Teulon, J. (1990). cAMP-activated chloride channel in the basolateral membrane of the thick ascending limb of the mouse kidney. *J. Membr. Biol.* **113**, 253–260.

Pearlmutter, A.F., Szkrybalo, M., Kim, Y., and Harik, S.I. (1988). Arginine vasopressin receptors in pig cerebral microvessels, cerebral cortex and hippocampus. *Neurosci. Lett.* **87**, 121–126.

Pedersen, S.F., O'Donnell, M.E., Anderson, S.E., and Cala, P.M. (2006). Physiology and pathophysiology of Na^+/H^+ exchange and Na^+-K^+-$2Cl^-$ cotransporter in the heart, brain and blood. *Am. J. Physiol. Reg. Integr. Comp. Physiol.* **291**, R1–R25.

Report of the Stroke Progress Review Group (2002). National Institutes of Neurological Disorders and Stroke 1–116. Available at http://www.ninds.nih.gov/about_ninds/04_2002_Stroke_PRG_Report.htm.

Romero, M.F. (2005). Molecular pathophysiology of SLC4 bicarbonate transporters. *Curr. Opin. Nephrol. Hyperten.* **14**, 495–501.

Romero, M.F., Fulton, C.M., and Boron, W.F. (2004). The SLC4 family of HCO_3^- transporters. *Pflugers Arch. – Eur. J. Physiol.* **447**, 495–509.

Rusa, R., Alkayed, N.J., Crain, B.J., Traystman, R.J., Kimes, A.S., London, E.D., Klaus, J.A., and Hurn, P.D. (1999). β-estradiol reduces stroke injury in estrogen-deficient animals. *Stroke* **30**, 1665–1670.

Russell, J.M. (2000). Sodium-potassium-chloride cotransport. *Physiologic. Rev.* **80**, 211–276.

Sacco, S.E., Whisnant, J.P., Broderick, J.P., Phillips, S.J., and O'Fallon, W.M. (1991). Epidemioloical characteristics of lacunar infarcts in a population. *Stroke* **22**, 1236–1241.

Sánchez del Pino, M.M., Hawkins, R.A., and Peterson, D.R. (1995a). Biochemical discrimination between luminal and abluminal enzyme and transport activities of the blood–brain barrier. *J. Biol. Chem.* **270**, 14907–14912.

Sánchez del Pino, M.M., Peterson, D.R., and Hawkins, R.A. (1995b). Neutral amino acid transport characterization of isolated luminal and abluminal membranes of the blood–brain barrier. *J. Biol. Chem.* **270**, 14913–14918.

Sarkadi, B. and Parker, J.C. (1991). Activation of ion transport pathways by changes in cell volume. *Biochim. Biophys. Acta* **1071**, 407–427.

Schielke, G.P., Moises, H.C., and Betz, A.L. (1991). Blood to brain sodium transport and interstitial fluid potassium concentration during focal ischemia in the rat. *J. Cereb. Blood Flow Metab.* **11**, 466–471.

Schmid, A., Scholz, W., Lang, H.J., and Popp, R. (1992). Na^+/H^+ exchange in porcine cerebral capillary endothelial cells is inhibited by a benzoylguanidine derivative. *Biochem. Biophys. Res. Commun.* **184**, 112–117.

Scholz, W., Albus, U., Counillon, L., Gögelein, H., Lang, H.-J., Linz, W., Weichert, A., and Schölkens, B.A. (1995). Protective effects of HOE642, a selective sodium-hydrogen exchange subtype 1 inhibitor, on cardiac ischaemia and reperfusion. *Cardiovasc. Res.* **29**, 260–268.

Shuaib, A., Wang, C.X., Yang, T., and Noor, R. (2002). Effects of nonapeptide V1 vasopressin receptor antagonist SR-49059 on infraction volume and recovery of function in a focal embolic stroke model. *Stroke* **33**, 3033–3038.

Simard, J.M., Kent, T.A., Chen, M., Tarasov, K.V., and Gerzanich, V. (2007). Brain oedema in focal ischemia: molecular pathophysiology and theoretical implications. *Lancet Neurol.* **6**, 258–268.

Simpson, I.A., Carruthers, A., and Vannucci, S.J. (2008). Supply and demand in cerebral energy metabolism: the role of nutrient transporter. *J. Cereb. Blood Flow Metab.* **27**, 1766–1791.

Sipos, H., Törocsik, B., Tretter, L., and Adam-Vizi, V. (2005). Impaired regulation of pH homeostasis by oxidative stress in rat brain capillary endothelial cells. *Cell. Mol. Neurobiol.* **25**, 141–151.

Sorensen, P.S., Gjerris, A., and Hammer, M. (1985). Cerebrospinal fluid vasopressin in neurological and psychiatric disorders. *J. Neurol. Neurosurg. Psychiat.* **48**, 50–57.

Spatz, M., Kawai, N., Bembry, J., Lenz, F., and McCarron, R.M. (1998). Human brain capillary endothelium: modulation of K+ efflux and K^+, Ca^{2+} uptake by endothlin. *Neurochem. Res.* **23**, 1125–1132.

Spatz, M., Merkel, K.N., Bembry, J., and McCarron, R.M. (1997). Functional properties of cultured endothelial cells derived from large microvessels of human brain. *Am. J. Physiol.* **272**, C231–C239.

Spatz, M., Stanimirovic, D., Strasser, A., and McCarron, R.M. (1995). Nitro-L-arginine augments the endothelin-1 content of cerebrospinal fluid induced by cerebral ischemia. *Brain Res.* **684**, 99–102.

Stanimirovic, D.B., Yamamoto, T., Uematsu, S., and Spatz, M. (1994). Endothelin-1 receptor binding and cellular signal transduction in cultured human brain endothelial cells. *J. Neurochem.* **62**, 592–601.

Stokes, J., Kannel, W.B., Wolf, P.A., D'Agostino, R.B., and Cupples, L.A. (1989). Blood pressure as a risk factor for cardiovascular disease. The Framingham study – 30 years of follow-up. *Hypertension* **13**, 113–118.

Strange, K., Emma, F., and Jackson, P.S. (1996). Cellular and molecular physiology of volume-sensitive anion channels. *Am. J. Physiol. Cell Physiol.* **270**, C711–C730.

Stummer, W., Betz, A.L., and Keep, R.F. (1995). Mechanisms of brain ion homeostasis during acute and chronic variations of plasma potassium. *J. Cereb. Blood Flow Metab.* **15**, 336–344.

Stummer, W., Keep, R.F., and Betz, A.L. (1994). Rubidium entry into brain and cerebrospinal fluid during acute and chronic alterations in plasma potassium. *Am. J. Physiol.* **266**, H2239–H2246.

Sun, D., Lytle, C., and O'Donnell, M.E. (1995). Astroglial cell-induced expression of Na-K-Cl cotransporter in brain microvascular endothelial cells. *Am. J. Physiol.* **269**, C1506–C1512.

Sun, D., Lytle, C., and O'Donnell, M.E. (1997). IL-6 secreted by astroglial cells regulates Na-K-Cl cotransport in brain microvessel endothelial cells. *Am. J. Physiol.* **272**, C1829–C1835.

Sun, D. and O'Donnell, M.E. (1996). Astroglial-mediated phosphorylation of Na-K-Cl cotransporter in brain microvessel endothelial cells. *Am. J. Physiol.* **271**, C620–C627.

Suzuki, H., Sato, S., Suzuki, Y., Takekoshi, K., Ishihara, N., and Shimoda, S. (1990). Increased endothelin concentration in CSF from patients with subarachnoid hemorrhage. *Acta Neurolog. Scand.* **81**, 553–554.

Taylor, C.J., Nicola, P.A., Wang, S., Barrand, M.A., and Hladky, S.B. (2006). Transporters involved in regulation of intracellular pH (pHi) in primary cultured rat brain endothelial cells. *J. Physiol.* **576**, 769–785.

Tran, L.Q., Anderson, S.E., and O'Donnell, M.E. (2004). HOE-642 and bumetanide reduce edema formation and infarct following permanent rat middle cerebral artery occlusion. *FASEB J.* **18**, A1069.

Tuong, T.J.K., Traystman, R.J., and Hurn, P.D. (1998). Estrogen-mediated neuroprotection after experimental stroke in male rats. *Stroke* **29**, 1666–1670.

Vannucci, S.J., Maher, F., and Simpson, I.A. (1997). Glucose transporter proteins in brain: delivery of glucose to neurons and glia. *Glia* **21**, 2–21.

Vigne, P., Breittmayer, J.P., and Frelin, C. (1995). Sensitization by calyculin A of brain capillary endothelial cells to endothelin-1. *Brit. J. Pharmacol.* **114**, 1014–1016.

Vigne, P., Champigny, G., Marsault, R., Barbry, P., Frelin, C., and Lazdunski, M. (1989). A new type of amiloride-sensitive cationic channel in endothelial cells of brain microvessels. *J. Biol. Chem.* **264**, 7663–7668.

Vigne, P., Farre, A.L., and Frelin, C. (1994). Na^+-K^+-Cl^- cotransporter of brain capillary endothelial cells. Properties and regulation by endothelins, hyperosmolar solutions, calyculin A, and interleukin-1. *J. Biol. Chem.* **269**, 19925–19930.

Vigne, P., Ladoux, A., and Frelin, C. (1991). Endothelins activate Na^+/H^+ exchange in brain capillary endothelial cells via a high affinity endothelin-3 receptor that is not coupled to phospholipase C. *J. Biol. Chem.* **266**, 5925–5928.

von Weikersthal, S.F., Barrand, M.A., and Hladky, S.B. (1999). Functional and molecular characterization of a volume-sensitive chloride current in rat brain endothelial cells. *J. Physiol.* **516**, 75–84.

Yang, T., Roder, K.E., Bhat, G.J., Thekkumkara, T.J., and Abbruscato, T.J. (2006). Protein kinase C family members as a target for regulation of blood–brain barrier Na,K,2Cl-cotransporter during in vitro stroke conditions and nicotine exposure. *Pharmaceut. Res.* **23**, 291–302.

Yerby, T.R., Vibat, C.R.T., Sun, D., Payne, J.A., and O'Donnell, M.E. (1997). Molecular characterization of the Na-K-Cl cotransporter of bovine aortic endothelial cells. *Am. J. Physiol.* **273**, C188–C197.

Yoshimoto, S., Ishizaki, Y., Kurihara, H., Sasaki, T., Yoshizumi, M., Yanagisawa, M., Yazaki, , Masaki, T., Takakura, K., and Murota, S. (1990). Cerebral microvessel endothelium is producing endothelin. *Brain Res.* **508**, 283–285.

Zea Longa, E., Weinstein, P.R., Carlson, S., and Cummins, R. (1989). Reversible middle cerebral artery occlusion without craniectomy in rats. *Stroke* **20**, 84–91.

Zhang, Y.-Q., Shi, J., Rajakumar, G., Day, A.L., and Simpkins, J.W. (1998). Effects of gender and estradiol treatment on focal brain ischemia. *Brain Res.* **784**, 321–324.

Index

Q

R